radio
handbook

twenty-second edition

William I. Orr, W6SAI

Howard W. Sams & Co., Inc.
4300 WEST 62ND ST. INDIANAPOLIS, INDIANA 46268 USA

International Standard Book Number: 0-672-21874-7
Library of Congress Catalog Card Number: 40-33904

Edited by: *Welborn Associates*
Illustrated by: *D. B. Clemons*

Printed in the United States of America.

Preface to the 22nd Edition

The editor and staff of *Radio Handbook* are pleased to present the twenty-second edition of this popular work. At the time the previous edition was published, communication services including the Amateur Radio Service were awaiting the completion of the World Administrative Radio Conference (WARC), held in Geneva, Switzerland, under the auspices of the International Telecommunications Union.

Happily, the Amateur Radio Service emerged from the Conference greatly strengthened, with little loss of spectrum space and with three new amateur hf assignments at 10, 18, and 24 MHz. In addition, the 160-meter assignment was expanded in many areas of the world. Vhf and uhf assignments, the region of tomorrow's great amateur expansion, remained relatively unscathed.

Today, amateur radio stands upon a new frontier, looking forward to major technical advances in the coming decades: spread-spectrum modulation, digital communications, and speech synthesis, to name a few. *Radio Handbook* has helped bring amateur and commercial radio communication from where they *were* to where they *are*, and the editor, staff and publisher look forward to the next decade of development with anticipation.

The great technical advances made in the past have been reflected in the pages of *Radio Handbook*. And, in the future, this publication will remain in the forefront of communication progress.

William I. Orr, W6SAI

ACKNOWLEDGMENTS

The editor wishes to thank the following individuals and companies for their assistance and contributions in compiling this 22nd edition of the *Radio Handbook*:

Bob Artigo, KN6J
William Ayer, K6VQ
Bob Bounds, K3EG
A. J. F. Clement, W6KPC
Stuart Cowan, W2LX
Coleman Ellsworth, W6OXP
Jim Garland, W8ZR
Bill Hillard, K6OPZ
George Jacobs, W3ASK
John Keith, WB5DJE
Bob Magnani, K6QXY
Ed Marriner, W6XM
Jack McCullough, W6CHE
Bobby McDonald, WB7CLV

Peter Morley, K6SRG
Henry Olson, W6GXN
B. A. Ontiveros, W6FFF
Dick Rasor, W6EDE
Rodney Reynolds, VK3AAR
George Smith, W4AEO
Bob Sutherland, W6PO
T. H. Tenney, Jr., W1NLB
Bob Welborn, W9PBW
Amidon Associates
EIMAC, division of Varian
J. W. Miller, division of Bell Industries
Motorola Semiconductor
Sunair Electronics

CONTENTS

GLOSSARY OF TERMS

Symbol	Notation
A	Amperes (ac, rms, or dc)
$\hat{A}$	Amplifier voltage gain
Å	Angstrom unit
a	Amperes (peak)
ac	Alternating current
a-m	Amplitude modulation
C	Capacitance
c.f.m.	Cubic feet per minute
C_{gg}	Capacitance grid to ground
C_{gk}, C_{gp}, etc.	Tube capacitance between indicated electrodes
C_{in}	Input capacitance
C_k	Capacitance between cathode and ground
cm	Centimeter
C_N	Neutralizing capacitance
C_{out}	Output capacitance
C_{pg2}	Capacitance, plate to screen
c-w	Continuous wave
dB	Decibel
dc	Direct current
E	Voltage (ac, rms, or dc)
e	Peak voltage
E_b	Average plate voltage
e_b	Instantaneous plate voltage
$e_{b\ max}$	Peak plate voltage
$e_{b\ min}$	Minimum instantaneous plate voltage referenced to ground
e_{cmp}	Maximum positive grid voltage
E_{co}	Cutoff-bias voltage
E_{c1}	Average grid #1 voltage
E_{c2}	Average grid #2 voltage
E_{c3}	Average grid #3 voltage
e_{c1}	Instantaneous grid #1 voltage
e_{c2}	Instantaneous grid #2 voltage
e_{c3}	Instantaneous grid #3 voltage
E_f	Filament voltage
e_g	Rms value of exciting voltage
e_p	Instantaneous plate voltage (ac) referenced to E_b
$e_{p\ max}$	Peak ac plate voltage referenced to E_b
E_{sig}	Applied signal voltage (dc)
e_{sig}	Applied signal voltage (ac)
e_k	Instantaneous cathode voltage
$e_{k\ max}$	Peak cathode voltage
emf	Electromotive force
F	Farad, magnetomotive force
f	Frequency (in Hertz)
fil	Filament
G	Giga (10^9)
g, g_1, g_2, etc.	Grid (number to identify, starting from cathode)
$g_{2,4}$	Grids having common pin connection
GHz	Gigahertz (10^9 cycles per second)
G_m or S_m	Transconductance (grid-plate)
H	Henry
Hz	Hertz
i	Peak current
I	Current (ac, rms or dc)
I_b	Average dc plate current
$I_{b\ max}$	Peak signal dc plate current
i_b	Instantaneous plate current
$i_{b\ max}$	Peak plate current
I_{bo}	Idling plate current
I_c	Average dc grid current current
i_p	Instantaneous ac plate current referred to I_b
$i_{p\ max}$	Peak ac plate current referred to I_b
i_1 etc.	Fundamental component of r-f plate current
$i_{1\ max}$	Peak fundamental component of r-f plate current
I_1	Single tone dc plate current
I_2 etc.	Two-tone, etc., dc plate current
I_{c1}, $c2$, etc.	Average grid #1, #2, etc. current
I_f	Filament current
i_{g1} i_{g2} etc.	Instantantous grid current
$i_{g1\ max}$, etc.	Peak grid current
I_k	Average cathode current
i_k	Instantaneous cathode current
$i_{k\ max}$	Peak cathode current
K	Cathode, dielectric constant
k	Kilo(10^3), coefficient of coupling
kHz	Kilohertz
kV	Peak kilovolts
kVac	Ac kilovolts
kVdc	Dc kilovolts
kW	Kilowatts
λ	Wavelength

Symbol	Notation	Symbol	Notation
L	Inductance	R_k	Resistance in series with the cathode
M	Mutual inductance	R_L	Load resistance
M	Mega (10^6)	rms	Root mean square
m	Meter	R_p	Resistance in series with plate
m	One thousandth		
mm	Millimeter	r_p	Dynamic internal plate resistance
mA	Milliamperes		
Meg or meg	Megohm	S_c or G_c	Conversion transconductance
mH	Millihenry	S_m or G_m	Transconductance
MHz	Megahertz	SSB	Single sideband
m.m.f.	Magnetomotive force	SWR	Standing-wave ratio
Mu or μ	Amplification factor, micro	T	Temperature (°C)
mV	Millivolts	t	Time (seconds)
MW	Megawatts	θ	Conduction angle
mW	Milliwatts	μ	Micro (10^{-6}) or amplification factor
NF	Noise figure		
N_p	Efficiency	μ	Amplification Factor
Ω	Ohms	μA	Microampere
p	Pico (10^{-12})	μmho	Micromho
P_d	Average drive power	μF	Microfarad
p_d	Peak drive power	μH	Microhenry
P_{ft}	Average feedthrough power	μs	Microsecond
p_{ft}	Peak feedthrough power	μV	Microvolt
pF	Picofarad	μ_2	Grid-screen amplification factor
PEP	Peak envelope power		
P_{g1}, P_{g2}, etc.	Power dissipation of respective grids	V	Volt(s), (ac, rms, or dc)
P_i	Power input (average)	v	Peak volts
p_i	Peak power input	Vac	Ac volts
P_o	Power output (average)	Vdc	Dc volts
p_o	Peak power output	VSWR	Voltage standing-wave ratio
P_p	Plate dissipation		
Q	Figure of merit	W	Watts
Q_L	Loaded Q	Z	Impedance
R	Resistance	Z_g	Grid impedance
r	Reflector	Z_i	Input impedance
r-f	Radio frequency	Z_k	Cathode impedance
R_g	Resistance in series with the grid	Z_L	Load impedance
		Z_o	Output impedance
r_g	Dynamic internal grid resistance	Z_p	Impedance in plate circuit
		Z_s	Screen bypass impedance

CHAPTER ONE

Introduction to Amateur Radio Communication

The field of *radio* is a division of the much larger field of electronics. Radio itself is such a broad study that it is still further broken down into a number of smaller fields of which only short-wave or high-frequency radio is covered in this book. Specifically the field of communication on frequencies from 1.8 to 1296 MHz is taken as the subject matter for this work.

The largest group of persons interested in the subject of high-frequency communication is the more than 800,000 radio amateurs located in nearly all countries of the world. Strictly speaking, a *radio amateur* is anyone noncommercially interested in radio, but the term is ordinarily applied only to those hobbyists possessing transmitting equipment and a license to operate from their government.

It was for the radio amateur, and particularly for the serious and more advanced amateur, that most of the equipment described in this book was developed. The design principles behind the equipment for high-frequency and vhf radio communication are of course the same whether the equipment is to be used for commercial, military, or amateur purposes. The principal differences lie in construction practices, and in the tolerances and safety factors placed on components.

With the increasing complexity of high-frequency and vhf communication, resulting primarily from increased utilization of the available spectrum, it becomes necessary to delve more deeply into the basic principles underlying radio communication, both from the standpoint of equipment design and operation and from the standpoint of signal propagation. Thus, it will be found that this edition of the RADIO HANDBOOK has been devoted in greater proportion to the teaching of the principles of equipment design and signal propagation. Also included are expanded and revised sections covering solid state devices and the principles of operation of modern equipment. The mathematics chapter, in addition, has been revised in the light of the modern pocket electronic calculator. All of these factors, of course, are reflected in the changing picture of amateur radio today.

1-1   Amateur Radio

Amateur radio is a fascinating scientific hobby with many facets. At the same time it is a public service as well as a recognized Radio Service and, as such, is assigned specific bands of frequencies by the *International Telecommunications Union*, to which body the United States of America is a signatory power.

From a few thousand amateurs at the end of World War I, amateur radio has grown into a world wide institution of communicators and experimenters joined in the common interest of communication by means of radio. So strong is the fascination offered by this hobby that many executives, engineers and military and commercial electronic experts, as well as students and citizens not otherwise engaged in the field of electronics are united by the common bond of amateur radio.

Radio amateurs have rendered much public service, especially in the United States, through furnishing emergency communications to and from the outside world in cases where a natural disaster has isolated an area by severing all normal means of communication. Amateurs have innumerable records of service and heroism on such occasions. The amateur's fine record of performance with the "wireless" equipment of World War I was surpassed by his outstanding service in World War II.

The induction of thousands of radio amateurs in the Armed Forces during 1940-1945 and the explosion of electronic technology during that period created an expansion of amateur radio, the direct result of which is that many of those amateurs are now the leaders of our modern electronics industry. It is through the continuing expansion of amateur radio in the future that many of tomorrow's engineers, technicians and electronic executives will come.

The Amateur Radio Service has been proven to be a national and international resource of great benefit to all nations and to mankind. In addition, of equal importance is the effect of the service as a stimulus to economic growth and scientific knowledge. Radio amateurs continue to play a significant role in the development of the state of the radio art and are continuing to make major contributions both to basic radio theory and to practical applications thereof.

In recent years radio amateurs have contributed to the state of the art in numerous ways including the discovery in 1934 of reflection and refraction of vhf signals in the lower atmosphere, the development and adaptation of SSB techniques for widespread usage, the achievement of random "moonbounce" communication between amateurs and the development of the OSCAR series of satellites and the relatively inexpensive equipment and technique for communicating through the satellites.

Continuing into the closing quarter of the Twentieth Century, the status of amateur radio in the communities of the world emphasize to the beginning radio amateur that his hobby is the gateway to a career in the expanding field of electronics, if he wishes it, and that amateur radio is indeed an impressive introduction to one of the most exciting fields of endeavor in this century.

1-2   International Regulations

The domestic regulatory pattern of the United States agrees with the international agreements established by the International Telecommunications Union and to which the United States is a signatory power. The frequency bands reserved for the Amateur Radio Service are included in the ITU frequency allocations table, as one of the services to which frequencies are made available. In the lower-frequency amateur bands, the international allocations provide for joint use of the bands by several services in addition to the amateur service in various areas of the world.

Article I of the ITU Radio Regulations defines the amateur service as: *"A service of self-training, intercommunication, and technical investigations carried on by amateurs, that is, by duly authorized persons interested in radio technique solely with a personal aim and without a pecuniary interest."* Within this concept, the U.S. radio regulations governing radio amateur licensing and regulation are formulated.

By reciprocal treaty, the United States now has a number of agreements with other countries permitting amateurs of one country to operate in the other. One the other hand, by international agreement, notification to the ITU may forbid international communications with radio amateurs of certain countries.

The World Administrative Radio Conference (WARC-79) In December, 1979 the World Administrative Radio Conference of the International Telecommunications Union made important changes in the international Table of Frequency Allocations for users of the radio spectrum. Many of these changes affected amateur radio.

Certain small segments of the radio-frequency spectrum between 1800 kHz and 250 GHz are reserved for operation of amateur radio stations. These segments are in general agreement throughout the world, although certain portions of different amateur assignments may be shared with other services or used for other purposes in various geographic regions. For purposes of definition, a chart of the regions, as defined in the Frequency Allocation Chart is reproduced in Chart 1.

Region 1 includes the area limited on the east by line A, on the west by line B, excluding any portion of the territory of Iran which lies between these limits. It also includes that part of the Territory of Turkey and the Union of Soviet Socialist Republics lying outside of these limits, the territory of the Mongolian People's Republic and the area to the north of the U.S.S.R. which lies between lines A and C.

Region 2 includes the area limited on the east by line B and on the west by line C.

Region 3 includes the area limited on the east by line C and on the west by line A, except the territories of the Mongolian People's Republic, Turkey, the territory of the U.S.S.R. and the area to the north of the U.S.S.R. It also includes that part of the territory of Iran lying outside of those limits.

As can be seen from the map, the Americas including Greenland and Hawaii fall in Region 2.

Footnotes to the allocations chart modify it in many countries, particularly for the fixed and mobile services in the uhf region and many amateur bands in the hf, vhf, and uhf region. A summary of the amateur bands follows.

1-3   The Amateur Bands

The designated bands in the allocations table are indicated as *primary, permitted,* and *secondary* with regard to the assignments. Primary and permitted services have equal rights, except that, in the preparation of frequency plans, the primary service shall have prior choice of frequencies. The secondary service shall not cause harmful interference to the primary or permitted services and cannot claim protection from harmful interference from stations in these services.

**160 Meters
(1800 kHz–2000 kHz)** In Region 1 the 160-meter amateur assignment is complex. The primary assignment is limited to 1810–1850 kHz. Footnotes provide an alternative allocation of 1810–1830 kHz to the fixed and mobile services in a number of countries. Other countries may allocate the amateur assignment in the bands of 1715–1800 kHz or 1850-2000 kHz on a nonin-

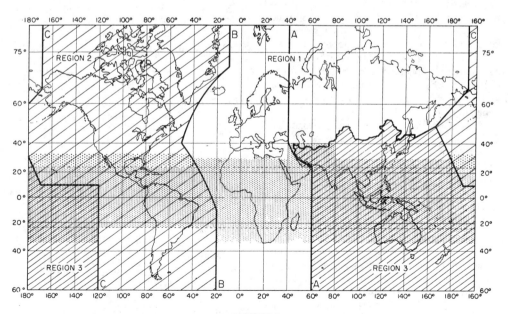

CHART 1

ITU WORLD FREQUENCY ALLOCATIONS

terference basis to fixed and mobile services of other countries. Final settlement of the assignments is subject to the condition that existing stations in other services presently in this band may receive replacement assignments.

In Region 2, the 160-meter band covers 1800–2000 kHz. The portion 1800–1850 kHz is a primary assignment, subject to the limitations imposed by the Loran A chain of stations, which will cease operation at the end of 1982. The 1850–2000 kHz portion of the band is shared with fixed and mobile services including radiolocation.

In Region 3, the 160-meter band covers 1800–2000 kHz and is shared with fixed and mobile services including radionavigation, subject to the requirements of Loran A.

80 Meters
(3500 kHz–4000 kHz) In Region 1, the 80-meter band is restricted to 3500–3800 kHz and in some countries is shared with fixed and mobile, and radiodetermination systems having a radiated mean power of less than 50 watts.

In Region 2 the band 3500–3750 kHz is an exclusive amateur assignment, with the proviso that 3500–3750 kHz is also allocated to the fixed and mobile services in some Central and South American countries. The band 3750–4000 kHz is allocated to the amateur and fixed and mobile services and also to the radiolocation service in some South American countries. In addition, in Canada and Greenland the band 3950–4000 kHz is also allocated to the broadcast service on a primary basis.

In Region 3, the band encompasses 3500–3900 kHz on a shared basis with the fixed and mobile services.

40 Meters
(7000 kHz–7300 kHz) In all three Regions the band 7000–7100 kHz is an exclusive amateur assignment, with the exception of certain African countries where the band 7000–7050 kHz is allocated to the fixed service on a primary basis. In Regions 1 and 3, the band 7100–7300 kHz is allocated to the broadcasting service, whereas in Region 2 it is allocated to the amateur service providing this use does not impose constraints

on the broadcasting service intended for use within Regions 1 and 3.

30 Meters
(10,100–10,150 kHz) This is a new band assigned at WARC-79 to the Amateur Radio Service. Occupancy of the band will be permitted when the services presently in this range can be moved elsewhere. Primary assignment is to the fixed service, with secondary service assigned to the Amateur Service in all regions.

20 Meters
(14,000–14,350 kHz) In all Regions the band 14,000–14,250 is assigned to the Amateur Service on a primary basis. The band 14,250–14,350 kHz is also assigned in a similar fashion, with the exception that certain countries in Regions 1 and 3 allocate the band to the fixed service on a primary basis.

17 Meters
(18,068–18,168 kHz) This new, narrow amateur band has been assigned on a primary basis to the Amateur Service for future occupancy, subject to the completion of satisfactory transfer of all assignments to stations in the fixed service operating in this band. In the U.S.S.R. this band is also allocated to the fixed service on a primary basis.

15 Meters
(21,000–21,450 kHz) In all regions this band is assigned to the Amateur Service on a primary basis.

12 Meters
(24,890–24,990 kHz) This new, narrow amateur band has been assigned on a primary basis to the Amateur Service for future occupancy, subject to the completion of satisfactory transfer of all assignments to stations in the fixed service operating in this band. In the U.S.S.R. this band is also allocated to the fixed and land mobile services on a primary basis.

10 Meters
(28–29.7 MHz) In all regions this band is assigned to the Amateur Service on a primary basis.

6 Meters (50–54 MHz)   In Region 1, this band is allocated to the broadcast service, with the exception of certain African countries where the band is allocated to the Amateur Service on a primary basis. In Regions 2 and 3 the band is allocated to the Amateur Service with certain restrictions imposed by various countries for broadcasting purposes.

2 Meters (144–148 MHz)   In all regions, the band 144–146 MHz is allocated to the Amateur Service on a primary basis, with an additional allocation in certain countries to the fixed and mobile services. In Region 2, the band 146–148 MHz is assigned to the Amateur Service on a primary basis. In Region 3, the band 146–148 MHz is assigned to the Amateur, fixed and mobile services, with allocation in certain countries to the fixed and mobile services on a primary basis.

1¼ Meters (220–225 MHz)   In Region 2, the allocation is 220–225 MHz with the band 216–225 allocated to the radiolocation service on a primary basis until January, 1990. No amateur operation is permitted in this band in Regions 1 and 3.

420–450 MHz   The amateur assignment in this band is complex. In the United States, the band allocated by footnote to the Allocations Table is 420–450 MHz on a shared basis and in Canada 430–450 MHz. In other countries the band is divided between amateur, radiolocation, fixed, and mobile services on a splinter basis. In New Zealand an additional allocation of 610–620 MHz is assigned to the Amateur Service on a secondary basis.

902–928 MHz   In Region 2, this band is assigned to the Amateur Service on a secondary basis, shared with radiolocation and fixed and mobile services. The assignment also applies to industrial, scientific, and medical applications. As of writing, the band has not yet been authorized for amateur use in any countries. The band is assigned to other services in Regions 1 and 3.

1240–1300 MHz   This band has a primary assignment of radiolocation and radio navigation with a secondary assignment to the Amateur Radio service. In many countries the primary service is fixed and mobile or radionavigation, with Amateur as a secondary service.

2300–2450 MHz   Primary service in this band is fixed and mobile, plus radiolocation. The band 2400–2500 MHz is also designated for industrial, scientific and medical services.

3300–3500 MHz   The band 3300–3400 MHz is assigned to radiolocation on a primary basis, with Amateur as a secondary service in Regions 2 and 3. The band is footnoted by many countries for different categories of service and is not available to amateurs in all countries. The band 3400–3500 MHz is assigned to the Amateur Service on a secondary basis in Region 2.

Additional Amateur Service Allocations in the uhf-shf Spectrum   The following bands are available on a secondary allocation to the Amateur Service. Exact assignment varies from country to country.

5650–5925 MHz (Fixed-satellite and Radiolocation primary)

10.0–10.5 GHz (Fixed, mobile, and Radiolocation primary)

24.0–24.25 GHz (Amateur Service primary in 24.0–24.05 GHz)

47.0–47.2 GHz (Amateur Service primary)

75.5–81.0 GHz (Amateur Service primary in 75.5–76 GHz)

119.98–120.02 GHz (Amateur Service secondary)

142.0–149 GHz (Amateur Service primary in 142–144 GHz)

241–250 GHz (Amateur Service primary in 248–250 GHz)

A graph of the high-frequency bands for the Amateur Service in the United States is given in Chart 2. The 17- and 12-meter allocations shown in Chart 2 may become effective in the mid 1980s. The 30-meter allocation becomes effective January 1, 1982,

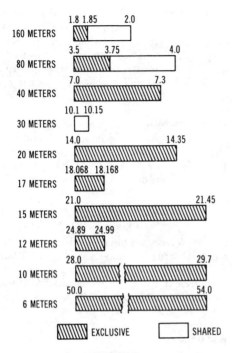

CHART 2

**AMATEUR RADIO HF FREQUENCY
ALLOCATIONS IN THE
UNITED STATES**

subject to adoption of the necessary changes
in FCC regulations.

1-4   Characteristics
of the Amateur Bands

The high-frequency amateur bands are
characterized by ionospheric propagation.
Groundwave propagation, important on 160
meters and to a lesser extent on 80 meters,
grows progressively less important on the
higher frequency bands.

The 160-meter band is least affected by
the 11-year solar sunspot cycle. The *maximum usable frequency* (MUF) even during
years of decreased sunspot activity rarely
drops below 4 MHz, therefore this band is
not subject to the violent fluctuations found
on the higher frequency bands. Long distance contacts on this band are limited by
ionospheric absorption which is quite high

in daylight hours. At night the absorption
is often low enough to permit distant contacts during the winter season.

The 80-meter band has low ionospheric
absorption during the years of minimum
sunspot activity and long distance contacts
are possible during the winter night hours.
Daytime operation is limited, in general, to
contacts of 500 miles or less. During high
portions of the sunspot cycle, increased
ionospheric absorption will tend to degrade
the long distance possibilities of this band.

The 40-meter band is high enough in frequency to be severely affected by the 11-year
sunspot cycle. During years of minimum
activity, the MUF may drop below 7 MHz
and the band becomes erratic, dropping completely out during the dark hours. As the
MUF rises, the skip distance increases, especially during the winter months. During
periods of high MUF, daylight skip distance
is quite long and the band is open for long-
distance communication during the dark
hours.

The new 30-meter band is expected to resemble the 40-meter band in many respects.

The 20-meter band is high enough in frequency to be severely affected at the bottom
of the solar cycle, yet still provides good
daylight long-distance communication during daylight hours. During the summer
months, the band is active until the late
evening hours, but during the winter months
the band is only good during the daylight
hours.

As the sunspot count rises and the MUF
increases, the 20-meter band opens for longer
hours during the winter. Maximum skip distance increases and "long-path" signals (180
degrees opposite the Great Circle path) are
useful for communication.

The band is susceptible to "fadeout"
caused by solar disturbances, and all except
local signals may completely drop out for
periods of a few hours to a day or so.

The new 17-meter band is expected to resemble the 20-meter band in many respects.

The 15-meter band is useful during low
portions of the sunspot cycle, particularly
during the late fall and early spring months.
North-south communication paths will remain open even though east-west paths may
be closed. As the sunspot count and MUF
rise, the band may remain open 24 hours a

day, especially in equatorial areas of the world. As in the case of 20 meters, long-path openings to remote areas of the world will be useful during years having high sunspot numbers. The new 12-meter band is expected to resemble the 15-meter band in many respects.

The 10-meter band supports excellent worldwide communication during periods of high MUF. The combination of long skip and low ionospheric absorption make reliable long-distance communication with low-power equipment possible. The great width of the band (1700 kHz) provides room for a large number of stations. The long skip (1500 miles) prevents nearby amateurs from hearing each other, thus reducing the interference level. During the summer months, sporadic-E (short-skip) signals up to about 1200 miles will be heard. Extremely long daylight skip distance is common on this band and during periods of high MUF the band will support intercontinental communication well into the evening hours.

The 6-meter band is considered a local band except at the peak period of a high sunspot cycle when long distance communication is possible. Sporadic-E propagation also provides beyond-horizon contacts. The proximity of the band to television channel 2 often causes interference problems to amateurs and viewers. Interest in this band wanes during periods of lesser solar activity as contacts are normally restricted to ground-wave communication.

The vhf bands are the least affected by the vagaries of the sunspot cycle and the ionosphere. Their predominant use is for reliable communication over short distances. Much long-distance, weak-signal operation takes place in the three lowest vhf bands. Experimental *moonbounce* (earth-moon-earth) transmissions, meteor scatter, sporadic-E, aurora reflection, and other exotic modes of communication take place in these bands. Satellite communication through the use of OSCAR satellites also takes place. Vhf fm repeaters are popular for short distance communication. The higher vhf bands are useful for wideband tv transmission, spread-spectrum communication, and other interesting modes of communication.

The shf bands are largely unexplored by radio amateurs because of the past unavail-ability of equipment, but more and more experimenters are investigating these frequencies as radio amateurs forge ahead into the microwave regions.

1-5 Amateur Stations and Operator Licenses

Every radio transmitting station in the United States (with the exception of certain low-power communication devices) must have a license from the Federal Government before being operated; some classes of stations must have a permit from the government even before being constructed. And every operator of a licensed transmitting station must have an operator's license before operating a transmitter. There are no exceptions. Similar laws apply in practically every major country.

Classes of Amateur Operator Licenses The Amateur Radio Service in the United States is in the process of going through a major change in the license structure. At the time of publication of this Handbook, there exist six classes of amateur operator licenses authorized by the Federal Communications Commission. These classes differ in many important respects, so each will be discussed briefly.

Novice Class—The Novice Class license is available to any U.S. Citizen or national who has not previously held an amateur license of any class issued by an agency of the U.S. Government, military or civilian. The license is valid for a period of five years and is renewable.

The examination may be taken only by mail, under the direct supervision of an amateur holding a General Class license or higher, or a commercial radiotelegraph licensee. The examination consists of a code test at a speed of 5 words per minute, plus a written examination on the rules and regulations essential to beginners operation, including sufficient elementary radio theory for the understanding of these rules.

Technician Class—The Technician Class exists for the purpose of encouraging a greater interest in experimentation and development of the higher frequencies among experimenters and would-be radio amateurs. This Class of license is available to any U.S. Citizen or national. The examination is similar to that given for the General Class license, except that the code test in sending and receiving is at a speed of 5 words per minute.

General Class—The General Class license is the standard radio amateur license and is available to any U.S. Citizen or national. The license is valid for a period of five years and is renewable on proper application. Applicants for the General Class license must take the examination before an FCC representative (with certain exceptions) discussed under the Conditional Class license). Code speed for the General Class license is 13 words per minute.

Conditional Class—The Conditional Class license is equivalent to the General Class license in the privileges accorded by its use. This license is issued to an applicant who: (1) lives more than 175 miles airline distance from the nearest point at which the FCC conducts examinations twice yearly, or oftener; (2) is unable to appear for examination because of physical disability to travel; (3) is unable to appear for examination because of military service; (4) is temporarily resident outside the United States, its territories, or possessions for a year or more. The Conditional Class license may be taken only by mail and is renewable.

Advanced Class—The Advanced Class license is available to any U.S. citizen or national. The license is valid for a period of five years and is renewable on proper application.

Amateur Extra Class—The Amateur Extra Class license is the highest-grade amateur license issued by the FCC and the recipient, on request, may receive a special diploma-type certificate from the District FCC Engineer-in-Charge. The license is valid for a period of five years and is renewable.

Each license class provides certain operating privileges. The Rules and Regulations governing the Amateur Service in the United States are in a state of flux and the license requirements and privileges for all classes are subject to change.

A comprehensive coverage of United States licensing procedure for radio amateurs and applicable rules and regulations may be found in *The Radio Amateur's License Manual*, published by the American Radio Relay League, Newington, Conn. 06111.

The Amateur Station License The station license authorizes the radio apparatus of the radio amateur for a particular address and designates the official call sign to be used. The license is a portion of the combined station-operator license normally issued to the radio amateur. Authorization is included for portable or mobile operation within the continental limits of the United States, its territories or possessions, on any amateur frequency authorized to the class of license granted the operator. The station license must be modified on a permanent change in address. The station license is customarily renewed with the operator license.

Reciprocal Licensing (USA) Under Public Law 92-81 resident aliens who have filed a *Declaration of Intention to Become A Citizen* may apply for amateur station and operator licenses. Other special rules apply to resident aliens. Visiting amateurs from certain countries may operate their own stations using their own calls upon receiving permission from the FCC.

1-6 Starting Your Study

When you start to prepare yourself for the amateur examination you will find that the circuit diagrams, tube and transistor characteristic curves, and formulas appear confusing and difficult to understand. But after a few study sessions one becomes sufficiently familiar with the notation of the diagrams and the basic concepts of theory and operation so that the acquisition of further knowledge becomes easier and even fascinating.

Since it takes a considerable time to become proficient in sending and receiving code, it is a good idea to intersperse technical study sessions with periods of code practice. Many short code-practice sessions benefit one more than a small number of longer sessions. Alternating between one study and the other keeps the student from getting "stale" since each type of study serves as a sort of respite from the other.

When you have practiced the code long enough you will be able to follow the gist of the slower-sending stations. Many stations send very slowly when working other stations at great distances. Stations repeat their calls many times when calling other stations before contact is established, and one need not have achieved much code proficiency to make out their calls and thus determine their location.

The Code The applicant for any class of amateur operator license must be able to receive the Continental Code (sometimes called the International Morse Code). The speed required for the receiving test may be either 5, 13, or 20 words per minute, depending on the class of license, assuming an average of five characters to the word in each case. The receiving test runs for five minutes, and one minute of errorless reception must be accomplished within the five-minute interval.

Approximately 30% of amateur applicants fail to pass the test. It should be expected that nervousness and excitement will, at least to some degree, temporarily lower the applicant's code ability. The best insurance against this is to master the code at a little greater than the required speed under ordinary conditions. Then if you slow down a little due to nervousness during a test, the result will not prove fatal.

Memorizing the Code There is no shortcut to code proficiency. To memorize the alphabet entails but a few evenings of diligent application, but considerable time is required to build up speed. The exact time required depends on the individual's ability and the regularity of practice.

While the speed of learning will naturally vary greatly with different individuals, about 70 hours of practice (no practice period to be over 30 minutes) will usually suffice to bring a speed of about 13 wpm; 16 wpm requires about 120 hours; 20 wpm, 175 hours.

Since code reading requires that individual letters be recognized instantly, any memorizing scheme which depends on orderly sequence, such as learning all *"dah"* letters and all *"dit"* letters in separate groups, is to be discouraged. Before beginning with a code practice set it is necessary to memorize the whole alphabet perfectly. A good plan is to study only two or three letters a day and to drill with those letters until they become part of your consciousness. Mentally translate each day's letters into their sound equivalent wherever they are seen, on signs, in papers, indoors and outdoors. Tackle two additional letters in the code chart each day, at the same time reviewing the characters already learned.

Avoid memorizing by routine. Be able to sound out any letter immediately without so much as hesitating to think about the letters preceding or following the one in question. Know C, for example, apart from the sequence ABC. Skip about among all the characters learned, and before very long sufficient letters will have been acquired to enable you to spell out simple words to yourself in *"dit dahs."* This is interesting exercise, and for that reason it is good to memorize all the vowels first and the most common consonants next.

Actual code practice should start only when the entire alphabet, the numerals, period, comma, and question mark have been memorized so thoroughly that any one can be sounded without the slightest hesitation. Do not bother with other punctuation or miscellaneous signals until later.

Sound — Not Sight Each letter and figure *must* be memorized by its *sound* rather than its appearance. Code is a system of sound communication, the same as is the spoken word. The letter A, for example, is one short and one long sound in combination sounding like *dit dah,* and it must be remembered as such, and not as "dot dash."

Practice Time, patience, and regularity are required to learn the code properly. Do not expect to accomplish it within a few days.

Don't practice too long at one stretch; it does more harm than good. Thirty minutes at a time should be the limit.

Lack of regularity in practice is the most common cause of lack of progress.

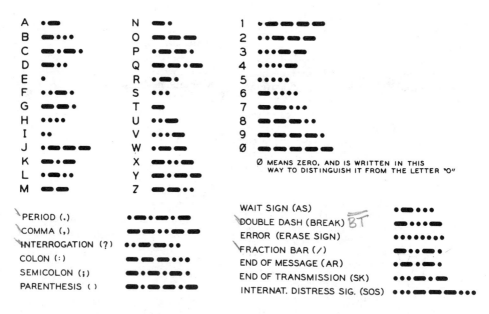

Ø MEANS ZERO, AND IS WRITTEN IN THIS
WAY TO DISTINGUISH IT FROM THE LETTER "O"

PERIOD (.)
COMMA (,)
INTERROGATION (?)
COLON (:)
SEMICOLON (;)
PARENTHESIS ( )

WAIT SIGN (AS)
DOUBLE DASH (BREAK) *BT*
ERROR (ERASE SIGN)
FRACTION BAR (/)
END OF MESSAGE (AR)
END OF TRANSMISSION (SK)
INTERNAT. DISTRESS SIG. (SOS)

Figure 1

The Continental (or International Morse) Code is used for substantially all nonautomatic radio
communication. DO NOT memorize from the printed page; code is a language of SOUND, and
must not be learned visually; learn by listening as explained in the text.

Irregular practice is very little better than no practice at all. Write down what you have heard; then forget it; *do not look back.* If your mind dwells even for an instant on a signal about which you have doubt, you will miss the next few characters while your attention is diverted.

While various automatic code machines, cassette tapes, etc., will give you practice, by far the best practice is to obtain a study companion who is also interested in learning the code. When you have both memorized the alphabet you can start sending to each other. Practice with a key and oscillator or key and buzzer generally proves superior to all automatic equipment. Two such sets, operated between two rooms are fine—or between your house and his will be just that much better. Avoid talking to your partner while practicing. If you must ask him a question, do it in code. It makes more interesting practice than confining yourself to random practice material.

When two co-learners have memorized the code and are ready to start sending to each other for practice, it is a good idea to enlist the aid of an experienced operator for the first practice session or two so that they will get an idea of how properly formed characters sound.

During the first practice period the speed should be such that substantially solid copy can be made without strain. Never mind if this is only two or three words per minute. In the next period the speed should be increased slightly to a point where nearly all of the characters can be caught only through conscious effort. When the student becomes proficient at this new speed, another slight

Figure 2

These code characters are used in languages
other than English. They may occasionally be
encountered so it is well to know them.

increase may be made, progressing in this manner until a speed of about 16 words per minute is attained if the object is to pass the amateur 13-word per minute code test. The margin of 3 wpm is recommended to overcome a possible excitement factor at examination time. Then when you take the test you don't have to worry about the "jitters" or an "off day."

Speed should not be increased to a new level until the student finally makes solid copy with ease for at least a five-minute period at the old level. How frequently increases of speed can be made depends on individual ability and the amount of practice. Each increase is apt to prove disconcerting, but remember "you are never learning when you are comfortable."

A number of amateurs are sending code practice on the air on schedule once or twice each week; excellent practice can be obtained after you have bought or constructed your receiver by taking advantage of these sessions.

If you live in a medium-size or large city, the chances are that there is an amateur-radio club in your vicinity which offers free code-practice lessons periodically.

Skill When you listen to someone speaking you do not consciously think how his words are spelled. This is also true when you read. In code you must train your ears to read code just as your eyes were trained in school to read printed matter. With enough practice you acquire skill, and from skill, speed. In other words, it becomes a *habit*, something which can be done without conscious effort. Conscious effort is fatal to speed; we can't think rapidly enough; a speed of 25 words a minute, which is a common one in commercial operations, means 125 characters per minute or more than two per second, which leaves no time for conscious thinking.

Perfect Formation of Characters When transmitting on the code practice set to your partner, concentrate on the *quality* of your sending, *not* on your speed. Your partner will appreciate it and he could not copy you if you speeded up anyhow.

If you want to get a reputation as having an excellent "fist" on the air, just re-

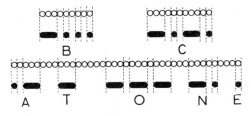

Figure 3

Diagram illustrating relative lengths of dashes and spaces referring to the duration of a dot. A dash is exactly equal in duration to three dots; spaces between parts of a letter equal one dot; those between letters, three dots; space between words, five dots. Note that a slight increase between two parts of a letter will make it sound like two letters.

member that speed alone won't do the trick. Proper execution of your letters and spacing will make much more of an impression. Fortunately, as you get so that you can send evenly and accurately, your sending speed will automatically increase. Remember to try to see how *evenly* you can *send,* and how *fast* you can *receive.* Concentrate on making signals properly with your key. Perfect formation of characters is paramount to everything else. Make every signal right no matter if you have to practice it hundreds or thousands of times. Never allow yourself to vary the slightest from perfect formation once you have learned it.

If possible, get a good operator to listen to your sending for a short time, asking him to criticize even the slightest imperfections.

Timing It is of the utmost importance to maintain uniform spacing in characters and combinations of characters. Lack of uniformity at this point probably causes beginners more trouble than any other single factor. Every dot, every dash, and every space must be correctly timed. In other words, accurate timing is absolutely essential to intelligibility, and timing of the spaces between the dots and dashes is just as important as the lengths of the dots and dashes themselves.

The characters are timed with the dot as a "yardstick." A standard dash is three times as long as a dot. The spacing between parts of the same letter is equal to one dot, the space between letters is equal to three dots, and that between words equal to five dots.

The rule for spacing between letters and words is not strictly observed when sending slower than about 10 words per minute for the benefit of someone learning the code and desiring receiving practice. When sending at, say, 5 wpm., the individual letters should be made the same as if the sending rate were about 10 wpm., except that the spacing between letters and words is greatly exaggerated. The reason for this is obvious. The letter *L*, for instance, will then sound exactly the same at 10 wpm. as at 5 wpm., and when the speed is increased above 5 wpm. the student will not have to become familiar with what may seem to him like a new sound, although it is in reality only a faster combination of dots and dashes. At the greater speed he will merely have to learn the identification of the *same* sound without taking as long to do so.

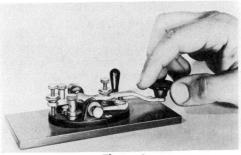

Figure 4

**PROPER POSITION OF THE FINGERS
FOR OPERATING A TELEGRAPH KEY**

The fingers hold the knob and act as a cushion. The hand rests lightly on the key. The muscles of the forearm provide the power, the wrist acting as the fulcrum. The power should not come from the fingers, but rather from the forearm muscles.

Be particularly careful of letters like *B*. Many beginners seem to have a tendency to leave a longer space after the dash than that which they place between succeeding dots, thus making it sound like *TS*. Similarly, make sure that you do not leave a longer space after the first dot in the letter *C* than you do between other parts of the same letter: otherwise it will sound like *NN*.

Sending vs. Receiving Once you have memorized the code thoroughly you should concentrate on increasing your *receiving* speed. True, if you have to practice with another newcomer who is learning the code with you, you will both have to do some sending. But don't attempt to practice *sending* just for the sake of increasing your sending *speed*.

When transmitting code to your partner so that he can practice, concentrate on the *quality* of your sending, not on your speed.

Because it is comparatively easy to learn to send rapidly, especially when no particular care is given to the quality of sending, many operators who have just received their licenses get on the air and send mediocre (or worse) code at 20 wpm when they can barely receive good code at 13. Most old-timers remember their own period of initiation and are only too glad to be patient and considerate if you tell them that you are a newcomer. But the surest way to incur their scorn is to try to impress them with your "lightning speed," and then to request them to send more slowly when they come back at you at the same speed.

Stress your copying ability; never stress your sending ability. It should be obvious that if you try to send faster than you can receive, your ear will not recognize any mistakes which your hand may make.

Using the Key Figure 4 shows the proper position of the hand, fingers and wrist when manipulating a telegraph or radio key. The forearm should rest naturally on the desk. It is preferable that the key be placed far enough back from the edge of the table (about 18 inches) that the elbow can rest on the table. Otherwise, pressure of the table edge on the arm will tend to hinder the circulation of the blood and weaken the ulnar nerve at a point where it is close to the surface, which in turn will tend to increase fatigue considerably.

The knob of the key is grasped lightly with the thumb along the edge; the index and third fingers rest on the top towards the front or far edge. The hand moves with a free up and down motion, the wrist acting as a fulcrum. The power must come entirely from the arm muscles. The third and index fingers will bend slightly during the sending but not because of deliberate effort to manipulate the finger muscles. Keep your finger muscles just tight enough to act as a cushion for the arm motion and let the slight movement of the fingers take care of itself. The key's spring is adjusted to the individual wrist and should be neither too stiff nor too loose. Use a moderately stiff tension at first and gradually lighten it as you become more proficient. The separation between the contacts must be the proper amount for the desired speed, being somewhat under 1/16 inch for slow speeds and slightly closer together (about 1/32 inch) for faster speeds. Avoid extremes in either direction.

Do not allow the muscles of arm, wrist or fingers to become tense. Send with a full, free arm movement. Avoid like the plague any finger motion other than the slight cushioning effect mentioned above.

Stick to the regular handkey for learning code. No other key is satisfactory for this purpose. Not until you have thoroughly mastered both sending and receiving at the maximum speed in which you are interested should you tackle any form of automatic or semiautomatic key such as the *Vibroplex* ("bug") or an electronic key.

Difficulties Should you experience difficulty in increasing your code speed after you have once memorized the characters, there is no reason to become discouraged. It is more difficult for some people to learn code than for others, but there is no justification for the contention sometimes made that "some people just can't learn the code." It is not a matter of intelligence, so don't feel ashamed if you seem to experience a little more than the usual difficulty in learning code. Your reaction time may be a little slower or your coordination not so good. If this is the case, remember *you can still learn the code.* You may never learn to send and receive at 40 wpm, but you can learn sufficient speed for all noncommercial purposes (and even for most commercial purposes) if you have patience, and refuse to be discouraged by the fact that others seem to pick it up more rapidly.

When the sending operator is sending just a bit too fast for you (the best speed for

practice), you will occasionally miss a signal or a small group of them. When you do, leave a blank space; do not spend time futilely trying to recall it; dismiss it, and center attention on the next letter; otherwise you'll miss more. Do not ask the sender any questions until the transmission is finished.

To prevent guessing and get equal practice on the less common letters, depart occasionally from plain language material and use a jumble of letters in which the usually less commonly used letters predominate.

As mentioned before, many students put a greater space after the dash in the letter B, than between other parts of the same letter so it sounds like TS. C, F, Q, V, X, Y, and Z often give similar trouble. Make a list of words or arbitrary combinations in which these letters predominate and practice them, both sending and receiving until they no longer give you trouble. Stop everything else and stick to them. So long as these characters give you trouble you are not ready for anything else.

Follow the same procedure with letters which you may tend to confuse such as F

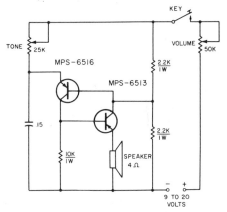

Figure 5

Two inexpensive "hobby"-type transistors and a 9-volt battery, plus a handful of parts make up a code-practice oscillator. Volume and tone are controlled by the potentiometers. Low-impedance earphones may be substituted for the speaker, if desired.

and L, which are often confused by beginners. Keep at it until you *always* get them right without having to stop *even an instant* to think about it.

If you do not instantly recognize the sound of any character, you have not learned it; go back and practice your alphabet further. You should never have to omit writing down every signal you hear except when the transmission is too fast for you.

Write down what you hear, not what you think it should be. It is surprising how often the word which you guess will be wrong.

Copying Behind All good operators copy several words behind, that is, while one word is being received, they are writing down or typing, say the fourth or fifth previous word. At first this is very difficult, but after sufficient practice it will be found actually to be easier than copying close up. It also results in more accurate copy and enables the receiving operator to capitalize and punctuate copy as he goes along. It is not recommended that the beginner attempt to do this until he can send and receive accurately and with ease at a speed of at least 12 words a minute.

It requires a considerable amount of training to disassociate the action of the subconscious mind from the direction of the conscious mind. It may help some in obtaining this training to write down two columns of short words. Spell the first word in the first column out loud while writing down the first word in the second column. At first this will be a bit awkward, but you will rapidly gain facility with practice. Do the same with all the words, and then reverse columns.

Next try speaking aloud the words in the one column while writing those in the other column; then reverse columns.

After the foregoing can be done easily, try sending with your key the words in one column while spelling those in the other. It won't be easy at first, but it is well worth keeping after if you intend to develop any real code proficiency. Do *not* attempt to catch up. There is a natural tendency to close up the gap, and you must train yourself to overcome this.

Next have your code companion send you a word either from a list or from straight text; do not write it down yet. Now have him send the next word; *after* receiving this second word, write down the first word.

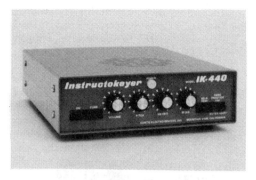

Figure 6

INSTRUCTOKEYER TEACHES CODE

This solid-state keyer is ideal for teaching large code classes. It provides random groups of Morse letters, numbers, punctuation and word spaces at random, in a sequence which never exactly repeats itself. Code speed is adjustable from 4 to 50 w p m. (Photo courtesy Curtis Electro Devices, Inc., Box 4090, Mountain View, CA 94040.)

After receiving the third word, write the second word; and so on. Never mind how slowly you must go, even if it is only two or three words per minute. *Stay behind.*

It will probably take quite a number of practice sessions before you can do this with any facility. After it is relatively easy, then try staying two words behind; keep this up until it is easy. Then try three words, four words, and five words. The more you practice keeping received material in mind, the easier it will be to stay behind. It will be found easier at first to copy material with which one is fairly familiar, then gradually switch to less familiar material.

Learning Aids A variety of learning aids are available to help the would-be amateur learn the code. Tape cassettes are available from several sources that contain both code practice and theory for the Novice examination. Other cassettes are available that contain code practice at speeds up to 21 words per minute. Long-playing code records (33⅓ r.p.m.) are also sold by several concerns that specialize in training aids.

The Maxim Memorial Station, W1AW, operated by amateurs at the American Radio Relay League (plus other amateur stations in the United States) transmits practice messages in Morse Code on various amateur bands. Transmission speeds vary from 5 to 20 words per minute. Copying "live" code off the air is a very effective means of increasing receiving speed.

Table 1. Class D CB Frequencies

MHz	Channel	MHz	Channel
26.965	1	27.215	21
26.975	2	27.225	22
26.985	3	27.255	23
27.005	4	27.235	24
27.015	5	27.245	25
27.025	6	27.265	26
27.035	7	27.275	27
27.055	8	27.285	28
27.065	9	27.295	29
27.075	10	27.305	30
27.085	11	27.315	31
27.105	12	27.325	32
27.115	13	27.335	33
27.125	14	27.345	34
27.135	15	27.355	35
27.155	16	27.365	36
27.165	17	27.375	37
27.175	18	27.385	38
27.185	19	27.395	39
27.205	20	27.405	40

Once you can copy about 10 wpm you can also get receiving practice by listening to slow-sending stations on your receiver. Many amateur stations send slowly particularly when working far distant stations. When receiving conditions are particularly poor many commercial stations also send slowly, sometimes repeating every word. Until you can copy around 10 wpm your receiver isn't much use, and either another operator or a cassette or records is necessary for getting receiving practice after you have once memorized the code.

As a good key may be considered an investment it is wise to make a well-made key your first purchase. Regardless of what type code-practice set you use, you will need a key, and later on you will need one to key your transmitter. If you get a good key to begin with, you won't have to buy another one later.

The key should be rugged and have fairly heavy contacts. Not only will the key stand up better, but such a key will contribute to the "heavy" type of sending so desirable for radio work. Morse (telegraph) operators use

a "light" style of sending and can send somewhat faster when using this light touch. But, in radio work static and interference are often present; and a slightly heavier dot is desirable. If you use a husky key, you will find yourself automatically sending in this manner.

An example of the audio-oscillator type of code-practice set is illustrated in Figure 5. Two inexpensive "hobby"-type transistors are used and the unit is powered by a 9-volt transistor radio battery. Low-impedance (4–8 ohms) earphones may be substituted for the speaker, if desired. The oscillator may be built up on a phenolic circuit board.

A new training aid for large code classes in the *Instructokeyer* (Figure 6), a solid state device which sends random groups of Morse letters, numbers, punctuation and word spaces in an ever-changing sequence which never exactly repeats. Code speed is adjustable from 4 to 50 wpm. Code groups are of varying lengths but average five characters per group. A rear panel switch selects alphabet only or full alphanumeric code groups. The *Instructokeyer* provides an infinite variety of code groups allowing unlimited practice for higher proficiency.

The Personal Radio Service In 1977 the Federal Communications Commission expanded the Citizens Band in the United States to include 40 channels extending between 26.965 MHz and 27.410 MHz (Table 1).

In addition, the FCC established a band between 49.82 MHz and 49.90 MHz for low-power communication devices, such as the popular 100 mW "walkie-talkies." Specific channels of 49.830, 49.845, 49.860, 49.875 and 49.890 MHz are assigned for this service. Either amplitude or frequency modulation can be used as long as the emissions are confined within a 20-kHz channel centered at the carrier frequency.

Addresses of FCC District Offices
Listed below are the addresses and telephone numbers of the FCC district offices. This list also includes offices in Puerto Rico and the District of Columbia (Washington, D.C.).

Anchorage District Office, Engineer in Charge, Federal Communications Commission, 1011 E. Tudor Rd., Room 240, P.O. Box 2955, Anchorage, Alaska 99510 (907) 276-7453, (907) 278-5233 [1]

Atlanta District Office, Engineer in Charge, Federal Communications Commission, Room 440, Massell Building, 1365 Peachtree Street, NE, Atlanta, Georgia 30309 (404) 881-3084/5, (404) 881-7381 [1]

Baltimore District Office, Engineer in Charge, Federal Communications Commission, 1017 Federal Building, 31 Hopkins Plaza, Baltimore, Maryland 21201 (301) 962-2728/9, (301) 982-2727 [1]

Beaumont Office, Engineer in Charge, Federal Communications Commission, Jack Brooks Federal Building, Rm. 323, 300 Willow Street, Beaumont, Texas 77701 (713) 838-0271

Boston District Office, Engineer in Charge, Federal Communications Commission, 1800 Customhouse, 165 State Street, Boston, Massachusetts 02109 (617) 223-8809 (617) 223-0889, (617) 223-6607/8 [1]

Buffalo District Office, Engineer in Charge, Federal Communications Commission, 1307 Federal Building, 111 West Huron Street, Buffalo, New York 14202 (716) 846-4511/2 (716) 856-8250 [1]

Chicago District Office, Engineer in Charge Federal Communications Commission, 230 S. Dearborn St., Room 3935, Chicago, Illinois 60804 (312) 353-0195/6, (312) 353-0197 [1]

Cincinnati Office, Engineer in Charge, Federal Communications Commission, 3620 Winton Road, Cincinnati, Ohio 45231 (513) 521-1790, (513) 521-1716 [1]

Dallas District Office, Engineer in Charge, Federal Communications Commission, Earle Cabell Federal Building, U.S. Courthouse, Room 13E7, 1100 Commerce Street, Dallas, Texas 75242 (214) 787-0781, (214) 767-0764 [1]

Denver District Office, Engineer in Charge Federal Communications Commission, The Executive Tower, Room 2925, 1405 Curtis Street, Denver, Colorado 80202 (303) 837-5137/8, (303) 837-4053 [1]

Detroit District Office, Engineer in Charge Federal Communications Commission, 1064 Federal Building, 231 W. Lafayette Street, Detroit, Michigan 48226 (313) 226-8078/9, (313) 228-8077 [1]

Honolulu District Office, Engineer in Charge Federal Communications Commission, Prince Kuhio Federal Bldg. 300 Ala Moana Blvd., Room 7304, P.O. Box 50223 Honolulu, Hawaii 98850 (808) 546-5640

Houston District Office, Engineer in Charge Federal Communications Commission, New Federal Office Building, 515 Rusk Ave., Room 5636 Houston, Texas 77002 (713) 226-3624 (713) 226-4306 [1]

Kansas City District Office, Engineer in Charge, Federal Communications Commission, Brywood Office Tower, Room 320, 8800 East 63rd Street, Kansas City, Missouri 64133 (816) 926-5111 (816) 356-4050

Long Beach District Office, Engineer in Charge, Federal Communications Commission, 3711 Long Beach Blvd., Room 501, Long Beach, California 90807 (213) 426- 4451, (213) 426-7888 [1], (213) 426-7963 [1]

Miami District Office, Engineer in Charge, Federal Communications Commission, 51 S.W. First Ave., Room 919, Miami, Florida 33130 (305) 350-5542, (306) 350-5541 [1]

New Orleans District Office, Engineer in Charge, Federal Communications Commission, 1007 F. Edward Hebert Federal Bldg., 800 South Street, New Orleans, Louisiana 70130 (504) 589-2095/6, (504) 589-2094 [1]

New York District Office, Engineer in Charge, Federal Communications Commission, 201 Varick Street, New York, New York 10014 (212) 620-3437/8 (212) 620-3435 [1], (212) 620-3436 [1]

Norfolk District Office, Engineer in Charge Federal Communications Commission, Military Circle, 870 N. Military Highway Norfolk, Virginia 23502 (804) 441-8472, (804) 461-4000 [1]

Philadelphia District Office, Engineer in Charge, Federal Communications Commission, 11425 James A. Byrne Federal Courthouse, 801 Market Street, Philadelphia, Pennsylvania 19108 (215) 597-4411/2, (215) 597-4410 [1]

Pittsburgh Office, Engineer in Charge, Federal Communications Commission, 3755 William Penn Highway, Monroeville, Pennsylvania 15146 (412) 823-3380, (412) 823-3553 [1]

Portland District Office, Engineer in Charge, Federal Communications Commission, 1782 Federal Building, 1220 S.W. Third Avenue, Portland, Oregon 97204 (503) 221-4114, (503) 221-3097 [1]

St. Paul District Office, Engineer in Charge, Federal Communications Commission, 691 Federal Bldg. & U.S. Courthouse, 316 North Robert Street, St. Paul, Minnesota 55101 (612) 725-7810, (612) 725-7819 [1]

San Diego Office, Engineer in Charge, Federal Communications Commission, 7840 El Cajon Blvd., Room 406, La Mesa, California 92041 (714) 293-6478, (714) 293-5460 [1]

San Francisco District Office, Engineer in Charge,, Federal Communications Commission, 323-A Customhouse, 555 Battery Street, San Francisco, California 94111 (415) 556-7701/2, (415) 556-7700 [1]

San Juan District Office, Engineer in Charge, Federal Communications Commission, San Juan Field Office, 747 Federal Building, Hato Ray, Puerto Rico, 00918 (809) 753-4008, (809) 753-4567

Savannah Office, Engineer in Charge, Federal Communications Commission, 238 Post Office Building and Courthouse, P.O. Box 8004, (125 Bull Street), Savannah, Georgia 31412 (912) 232-4321

Seattle District Office, Engineer in Charge, Federal Communications Commission, 3256 Federal Building, 915 Second Avenue, Seattle, Washington 98174 (208) 442-7853/4, (206) 442-7610 [1]

Tampa Office, Engineer in Charge, Federal Communications Commission, ADP Building, Room 601, 1211 N. Westshore Blvd., Tampa, Florida 33607 (813) 228-2872, (813) 228-2805 [1]

Washington District Office, Engineer in Charge, 6525 Belcrest Road, Room 901-B, P.O. Box 1789, Hyattsville, Maryland 20788 (301) 436-7591, (301) 436-7590 [1]

[1] Recorded information.

CHAPTER TWO

Direct-Current Circuits

All naturally occurring matter (excluding artificially produced radioactive substances) is made up of 92 fundamental constituents called *elements*. These elements can exist either in the free state such as iron, oxygen, carbon, copper, tungsten, and aluminum, or in chemical unions commonly called *compounds*. The smallest unit which still retains all the original characteristics of an element is the *atom*.

Combinations of atoms, or subdivisions of compounds, result in another fundamental unit, the *molecule*. The molecule is the smallest unit of any compound. All reactive elements when in the gaseous state also exist in the molecular form, made up of two or more atoms. The nonreactive gaseous elements helium, neon, argon, krypton, xenon, and radon are the only gaseous elements that ever exist in a stable monatomic state at ordinary temperatures.

2-1   The Atom

An atom is an extremely small unit of matter—there are literally billions of them making up so small a piece of material as a speck of dust. To understand the basic theory of electricity and hence of radio, we must go further and divide the atom into its main components, a positively charged *nucleus* and a cloud of negatively charged particles that surround the nucleus. These particles, swirling around the nucleus in elliptical orbits at an incredible rate of speed, are called *orbital electrons*.

The Nucleus   The *nucleus* of the atom has been split open by applying high energy, primarily with accelerators. The nucleus is composed of *protons* and *neutron* existing in an "atmosphere" of *mesons*, of which there are many types. This basic knowledge led to the release of nuclear energy through fission and fusion processes.

Despite the knowledge that the nucleus is complex, the picture of the planetary atom is still valid, as more than one concept is required to explain matter in its various states.

The various particles and states of matter are developments of nuclear force which, taken with gravitational, electromagnetic, and interaction forces are responsible for the order, shape, and change in the visible world we see and the invisible world beyond our senses.

Orbital Electrons   It is on the behavior of the electrons when freed from the atom, that depends the study of electricity and radio, as well as allied sciences.

The atoms of different elements differ in respect to the charge on the positive nucleus and in the number of electrons revolving around this charge. They range all the way from hydrogen, having a net charge of one on the nucleus and one orbital electron, to uranium with a net charge of 92 on the nucleus and 92 orbital electrons. The number of orbital electrons is called the *atomic number* of the element.

The electron may be considered as a minute negatively charged particle, having a mass of 9×10^{-28} gram, and a charge of -1.59×10^{-19} coulomb. Electrons are always identical, regardless of the source from which they are obtained.

Action of the Electrons From the foregoing it must not be thought that the electrons revolve in a haphazard manner around the nucleus. Rather, the electrons in an element having a large atomic number are grouped into rings having a definite number of electrons. The only atoms in which these rings are completely filled are those of the inert gases mentioned before; all other elements have one or more uncompleted rings of electrons. If the uncompleted ring is nearly empty, the element is *metallic* in character, being most metallic when there is only one electron in the outer ring. If the incomplete ring lacks only one or two electrons, the element is usually *nonmetallic*. Elements with a ring about half completed will exhibit both nonmetallic and metallic characteristics; carbon, silicon, germanium, and arsenic are examples. Such elements are called *semiconductors*.

In metallic elements these outer ring electrons are rather loosely held. Consequently, there is a continuous helter-skelter movement of these electrons and a continual shifting from one atom to another. The electrons which move about in a substance are called *free electrons*, and it is the ability of these electrons to drift from atom to atom which makes possible the *electric current*.

Conductors, Semiconductors, and Insulators If the free electrons are numerous and loosely held, the element is a good *conductor*. On the other hand, if there are few free electrons (as is the case when the electrons in an outer ring are tightly held), the element is a poor conductor. If there are virtually no free electrons, the element is a good *insulator*.

Materials having few free electrons are classed as semiconductors and exhibit conductivity approximately midway between that of good conductors and good insulators.

2-2 Fundamental Electrical Units and Relationships

Basic Electrical Dimensions, Units, and Symbols Electrical dimensions, units, and qualities are expressed as letters, combinations of letters, and other characters that may be used in place of the proper names for these characteristics. In addition, various prefixes are added to the symbols to indicate multiples or submultiples of units (Table 1).

The international system of fundamental units which covers mechanics, electricity, and magnetism is designated the *Rational MKS (meter-kilogram-second) System*.

In this system, length is measured in *meters*, mass in *kilograms*, and time in *seconds*. A summary of important dimensions is given in Table 2.

The MKS System is a subsystem of the International System of Units (1960). To unite the mechanical system with electricity and magnetism, the *coulomb* is taken as a fourth fundamental unit.

Fundamental and Secondary Units Electrical measurements expressed in the MKS System are traceable to the *National Bureau of Standards* in the United States. Aside from the meter, kilogram, and second, the major electrical unit is the coulomb (Q), a unit of charge (6.28×10^{18} electron charges). The coulomb is defined as an *ampere-second*, or that steady

TABLE 1.

PREFIXES TO ELECTRICAL DIMENSIONS

MULTIPLE	PREFIX	SYMBOL
10^{12}	tera	T
10^{9}	giga	G
10^{6}	mega	M
10^{3}	kilo	k
10^{2}	hecto	h
10	deka	da
10^{-1}	deci	d
10^{-2}	centi	c
10^{-3}	milli	m
10^{-6}	micro	μ
10^{-9}	nano	n
10^{-12}	pico	p
10^{-15}	femto	f
10^{-18}	atto	a

TABLE 2
FUNDAMENTAL DIMENSIONS

DIMENSION	EQUIVALENT
Meter	3.281 feet—one foot = 0.3048 meter
Kilometer	1000 meters = 0.6214 statute miles
Centimeter	10^{-2} meter = 0.3937 inch
Meter	10^{10} angstrom units (A)
Kilogram	1000 grams = 2.205 pounds
Gram	3.527×10^{-2} ounces
Coulomb	1 ampere flowing for 1 second

current flowing through a solution of silver nitrate, which will deposit silver at the rate of 1.118×10^{-6} kilograms per second.

Secondary, or *derived units*, are based on the above listed fundamental units. The rate of current flow is the *ampere* (I), whose dimensions are in coulombs per second. The unit of energy or work is the *joule* (J) whose dimensions are volts × coulombs. The unit of power is the *watt* (W), whose dimensions are joules per second. The electrical pressure that moves a coulomb of charge past a measuring point is the *volt* (E or V), whose dimensions are joules per coulomb.

The unit of opposition to current flow is the *ohm* (R), whose dimensions are volts per ampere. Two units express charge storage in a circuit. The first is the *farad* (F), a unit of capacitance whose dimensions are coulombs per volt. The second is the *henry*

(H), a unit of inductance whose dimensions are volts per ampere-second. These and other electrical units are summarized in Table 3. Other complex quantities may be built up from these units.

The Electrostatic Force An electrified particle is specified by its mass at rest and by the magnitude and sign of the electric charge. In addition to the electron charge mentioned earlier, the charge of a proton is equal, but of opposite sign.

Associated with each electric charge is a *force field* which tends to impart motion to other charges in the field. The field surrounding a particle is represented by lines of force that originate at the center of the charge and radiate outward in all directions. The force of attraction or repulsion between two electric charges is proportional to the product of the charge magnitudes and inversely proportional to the distance between them and to the characteristic of the medium, described as the *permittivity*:

$$F = \frac{Q_1 \times Q_2}{4\,\pi\,\varepsilon\,l^2}$$

where,

F equals the force in Newtons,
Q equals the numerical value of charge in Coulombs,
π equals 3.14,

TABLE 3. ELECTRICAL UNITS

CHARACTERISTIC	SYMBOL	UNIT	DESCRIPTION
Charge	Q or q	coulomb	6.28×10^{18} electric charges
Voltage	E or e V or v	Volt	potential difference (joules per coulomb)
Current	I or i	Ampere	electrons in motion (coulombs per second)
Resistance	R or r	Ohm	electrical resistance (volts per ampere)
Conductance	G or g	mho	reciprocal of resistance
Energy	J	Joule	quantity of work (volts x coulombs)
Power	W	Watt	unit of power (joules per second)
Storage	F	Farad	unit of charge storage (coulombs per volt)
Storage	H	Henry	unit of inductance (volts per ampere-second)

ε equals permittivity in Farads per meter,
l equals the distance between charges in meters.

In the case of charges in a vacuum, the permittivity is 8.85×10^{-32}. Permittivity is also termed *dielectric constant*.

A representation of a two-dimensional electric field about isolated electric charges is shown in figure 1.

The electric potential difference between two points in an electrostatic field is equal to the work done in transferring a unit of positive charge from one point to the other. The *potential* or voltage of a point may be expressed as the ratio of the energy of transfer to the charge. Maximum work is done on a charge that moves along the lines of electric force, whereas no work is done on a charge that moves perpendicular to lines of electric force. The potential difference between two points in the MKS system has the dimension of joules per coulomb, and is termed the volt.

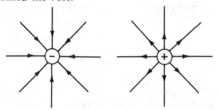

Figure 1

THE ELECTRIC FIELD

Line of force about electric charges. A negative sign indicates an excess of negative charges and a positive sign indicates a deficiency of negative charges.

Electromotive Force: The free electrons in a
Potential Difference conductor move constantly about and change their position in a haphazard manner. To produce a drift of electrons, or *electric current*, along a wire it is necessary that there be a difference in "pressure" or *potential* between the two ends of the wire. This *potential difference* can be produced by connecting a source of *electrical potential* to the ends of the wire.

As will be explained later, there is an excess of electrons at the negative terminal of a battery and a deficiency of electrons at the positive terminal, due to chemical action.

When the battery is connected to the wire, the deficient atoms at the positive terminal attract free electrons from the wire in order for the positive terminal to become neutral. The attracting of electrons continues through the wire, and finally the excess electrons at the negative terminal of the battery are attracted by the positively charged atoms at the end of the wire. Other sources of electrical potential (in addition to a battery) are: an electrical generator (dynamo), a thermocouple, an electrostatic generator, a photoelectric cell, and a crystal or piezoelectric generator.

Thus it is seen that a potential difference is the result of a difference in the number of electrons between the two (or more) points in question. The force or pressure due to a potential difference is termed the *electromotive* force, usually abbreviated *e.m.f.* It is expressed in volts.

It should be noted that for there to be a potential difference between two bodies or points it is not necessary that one have a positive charge and the other a negative charge. If two bodies each have a negative charge, but one more negative than the other, the one with the lesser negative charge will act as though it were positively charged *with respect to the other body*. It is the *algebraic* potential difference that determines the force with which electrons are attracted or repulsed, the potential of the earth being taken as the zero reference point.

The Electric   The *electric current* through a
Current   conductor is the time rate at which negative charges (electrons) flow through it. The flow may be induced by the application of an electromotive force. This flow, or drift, is in addition to the irregular movements of the electrons. However, it must not be thought that each free electron travels from one end of the circuit to the other. On the contrary, each free electron travels only a short distance before colliding with an atom; this collision generally knocks off one or more electrons from the atom, which in turn move a short distance and collide with other atoms, knocking off other electrons. Thus, in the general drift of electrons along a wire carrying an electric current, each electron travels only a short distance and the excess of electrons at one

end and the deficiency at the other are balanced by the source of the e.m.f. When this source is removed the state of normalcy returns; there is still the rapid interchange of free electrons between atoms, but there is no general trend or "net movement" in either one direction or the other—in other words, no current flows.

In electronics, the terms "electron flow" and "current" are synonymous and the current flow in a conductor is the electron drift from the negative terminal of the source voltage, through the conductor to the positive terminal of the source.

The number of free electrons in a conductor is a function of temperature so that the electrical properties of conductors are a function of temperature. In general, the resistivity to the flow of current in a conductor increases with temperature.

Conductors include those materials that have a large number of free electrons. Most metals (those elements which have only one or two electrons in their outer ring) are good conductors. Silver, copper, and aluminum, in that order, are the best of the common metals used as conductors at normal temperatures, having the greatest conductivity, or lowest resistance, to the flow of a electric current (Table 4).

TABLE 4. TABLE OF RESISTIVITY

Material	Resistivity in Ohms per Circular Mil-Foot	Temp. Coeff. of resistance per °C. at 20° C.
Aluminum	17	0.0049
Brass	45	0.003 to 0.007
Cadmium	46	0.0038
Chromium	16	0.00
Copper	10.4	0.0039
Iron	59	0.006
Silver	9.8	0.004
Zinc	36	0.0035
Nichrome	650	0.0002
Constantan	295	0.00001
Manganin	290	0.00001
Monel	255	0.0019

Resistance *Resistance* is that property of an electrical circuit which determines for a given current the rate at which electric energy is converted into heat or radiant energy. Generally speaking, resistance is an opposition to current flow in a material, and is one of its physical properties.

The unit of resistance is the *ohm* (Ω). Every substance has a *specific resistance*; usu-

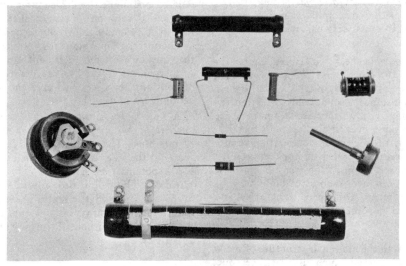

Figure 2
TYPICAL RESISTORS

Shown above are various types of resistors used in electronic circuits. The larger units are power resistors. On the left is a variable power resistor. Three precision-type resistors are shown in the center with two small composition resistors beneath them. At the right is a composition-type potentiometer, used for audio circuitry.

ally expressed as *ohms per mil-foot*, which is determined by the material's molecular structure and temperature. A mil-foot is a piece of material one circular mil in area and one foot long. Another measure of resistivity frequently used is expressed in the units *microhms per centimeter cube*. The resistance of a uniform length of a given substance is directly proportional to its length and specific resistance, and inversely proportional to its cross-sectional area. A wire with a certain resistance for a given length will have twice as much resistance if the length of the wire is doubled. For a given length, doubling the cross-sectional area of the wire will *halve* the resistance, while doubling the *diameter* will reduce the resistance to *one fourth*. This is true since the cross-sectional area of a wire varies as the square of the diameter. The relationship between the resistance and the linear dimensions of a conductor may be expressed by the following equation:

$$R = \frac{rl}{A}$$

where,

R equals resistance in ohms,
r equals resistivity in *ohms per mil-foot*,
l equals length of conductor in feet,
A equals cross-sectional area in circular mils.

For convenience, two larger units the *kilohm* (10^3 ohms) and the *megohm* (10^6 ohms) are often used.

The resistance also depends on temperature, rising with an increase in temperature for most substances (including most metals), due to increased electron acceleration resulting in a greater number of impacts between electrons and atoms. However, in the case of some substances such as carbon and glass the temperature coefficient is negative and the resistance decreases as the temperature increases.

Insulators In the molecular structure of many materials such as glass, ceramic, and mica all electrons are tightly held within their orbits and there are comparatively few free electrons. This type of material will conduct an electric current only with great difficulty and is termed an *insulator*. An insulator is said to have high electrical resistance.

An insulator is classified as a material having a resistivity of greater than 10^9 ohms per centimeter. A semiconductor is classified as having a resistivity from 10^{-4} ohms to 10^9 ohms per centimeter. A conductor is classified as having a resistivity of less than 10^{-4} ohms per centimeter.

An important property of an insulator is its power loss, or the ratio of the energy dissipated in the insulator to the energy stored. This ratio is termed the *dissipation factor*. The better insulators have a very low dissipation factor over a very wide frequency range. R-f measurement of the dissipation factor provides a guide as to the excellence of an insulator; some materials that are a satisfactory insulator for low frequencies may have a high dissipation factor at higher frequencies and are relatively worthless as an insulator.

Secondary Electrical Units These units are the *volt*, the *ampere*, and the *ohm*. They were mentioned in the preceding paragraphs, but were not completely defined in terms of fixed, known quantities.

The fundamental unit of *current*, or *rate of flow* of electricity is the ampere. A current of one ampere will deposit silver from a specified solution of silver nitrate at a rate of 1.118 milligrams per second.

The international standard for the ohm is the resistance offered by a uniform column of mercury at 0° C., 14.4521 grams in mass, of constant cross-sectional area and 106.300 centimeters in length.

A volt is *the e.m.f. that will produce a current of one ampere through a resistance of one ohm.* The standard of electromotive force is the Weston cell which at 20° C. has a potential of 1.0183 volts across its terminals. This cell is used only for reference purposes in a bridge circuit, since only an infinitesimal amount of current may be drawn from it without disturbing its characteristics.

Ohm's Law The relationship between the electromotive force (voltage), the flow of current (amperes), and the resistance which impedes the flow of current (ohms), is very clearly expressed in a simple but highly valuable law known as *Ohm's*

Law. This law states that *the current in amperes is equal to the voltage in volts divided by the resistance in ohms.* Expressed as an equation:

$$I = \frac{E}{R}$$

If the voltage (*E*) and resistance (*R*) are known, the current (*I*) can be readily found. If the voltage and current are known, and the resistance is unknown, the resistance (*R*) is equal to $\frac{E}{I}$. When the voltage is the unknown quantity, it can be found by multiplying $I \times R$. These three equations are all secured from the original by simple transposition. The expressions are here repeated for quick reference:

$$I = \frac{E}{R} \quad R = \frac{E}{I} \quad E = IR$$

where,

I is the current in amperes,
R is the resistance in ohms,
E is the electromotive force in volts.

Taken in a broader sense, Ohm's Law expresses a *ratio* of voltage to current when the circuit resistance is known. This concept is important in transmission-line studies and antenna work.

Conductance Instead of speaking of the resistance of a circuit, the *conductance* may be referred to as a measure of the ease of current flow. Conductance is the reciprocal of resistance $\frac{1}{R}$ and is measured in *mhos* (ohms spelled backwards) and is designated by the letter *G*.

The relation between resistance and conductance is:

$$G = \frac{1}{R}, R = \frac{1}{G} \text{ or } I = EG$$

In electronics work, a small unit of conductance, which is equal to one-millionth of a mho, frequently is used. It is called a *micromho.*

Application of Ohm's Law All electrical circuits fall into one of three classes: *series circuits, parallel circuits,* and *series-parallel circuits.* A series circuit is one in which the current flows in a single continuous path and is of the same value at every point in the circuit (figure 3). In a

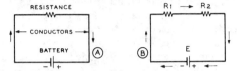

Figure 3

SIMPLE SERIES CIRCUITS

At (A) the battery is in series with a single resistor. At (B) the battery is in series with two resistors, the resistors themselves being in series. The arrows indicate the direction of electron flow.

parallel circuit there are two or more current paths between two points in the circuit, as shown in figure 4. Here the current divides at A, part going through R_1 and part through R_2, and combines at B to return

Figure 4

SIMPLE PARALLEL CIRCUIT

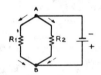

The two resistors R_1 and R_2 are said to be in parallel since the flow of current is offered two parallel paths. An electron leaving point A will pass either through R_1 or R_2, but not through both, to reach the positive terminal of the battery. If a large number of electrons are considered, the greater number will pass through whichever of the two resistors has the lower resistance.

to the battery. Figure 5 shows a series-parallel circuit. There are two paths between

Figure 5

SERIES-PARALLEL CIRCUIT

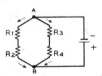

In this type of circuit the resistors are arranged in series groups, and these groups are then placed in parallel.

points A and B as in the parallel circuit, and in addition there are two resistances in series in each branch of the parallel combination. Two other examples of series-parallel arrangements appear in figure 6. The way in which the current splits to flow through the parallel branches is shown by the arrows.

In every circuit, each of the parts has some resistance: the batteries or generator, the connecting conductors, and the apparatus itself. Thus, if each part has some resistance, no matter how little, and a current is flowing through it, there will be a voltage drop across it. In other words, there will be a potential difference between the two ends of the circuit element in question. This drop in voltage is equal to the product of the current and the resistance hence it is called the *IR drop*.

Internal   The source of voltage has an *in-*
Resistance   *ternal* resistance, and when connected into a circuit so that current flows, there will be an *IR* drop in the source just as in every other part of the circuit. Thus, if the terminal voltage of the source could be measured in a way that would cause no current to flow, it would be found to be more than the voltage measured when a current flows by the amount of the *IR* drop in the source. The voltage measured with no current flowing is termed the *no load* voltage; that measured with current flowing is the *load* voltage. It is apparent that a voltage source having a low internal resistance is most desirable.

Resistances The current flowing in a series
in Series   circuit is equal to the voltage impressed divided by the *total* resistance across which the voltage is impressed. Since the same current flows through every part of the circuit, it is only necessary to add all the individual resistances to obtain the total resistance. Expressed as a formula:

$$R_{\text{Total}} = R_1 + R_2 + R_3 + \ldots + R_N$$

Of course, if the resistances happened to be all the same value, the total resistance would be the resistance of one multiplied by the number of resistors in the circuit.

Resistances Consider two resistors, one of
in Parallel 100 ohms and one of 10 ohms, connected in parallel as in figure 4, with a potential of 10 volts applied across each resistor, so the current through each can be easily calculated.

$$I = \frac{E}{R}$$

$E = 10$ volts
$R_1 = 100$ ohms $I_1 = \dfrac{10}{100} = 0.1$ ampere

$E = 10$ volts
$R_2 = 10$ ohms $I_2 = \dfrac{10}{10} = 1.0$ ampere

Total current $= I_1 + I_2 = 1.1$ ampere

Until it divides at A, the entire current of 1.1 amperes is flowing through the conductor from the battery to A, and again from B through the conductor to the battery. Since this is more current than flows through the smaller resistor it is evident that the resistance of the parallel combination must be less than 10 ohms, the resistance of the smaller resistor. This value can be found by applying Ohm's law:

$$R_T = \frac{E}{I}$$

$E = 10$ volts
$I = 1.1$ amperes $R_T = \dfrac{10}{1.1} = 9.09$ ohms

The resistance of the parallel combination is 9.09 ohms.

The following is a simple formula for finding the effective resistance of two resistors connected in parallel.

$$R_T = \frac{R_1 \times R_2}{R_1 + R_2}$$

where,
R_T equals unknown resistance,
R_1 equals resistance of the first resistor.
R_2 equals resistance of the second resistor.

If the effective value required is known, and it is desired to connect one unknown resistor in parallel with one of known value, a transposition of the above formula will simplify the problem of obtaining the unknown value:

$$R_2 = \frac{R_1 \times R_T}{R_1 - R_T}$$

where,
R_T equals effective value required,
R_1 equals the known resistor,
R_2 equals value of the unknown resistance necessary to give R_T when in parallel with R_1.

The resultant value of placing a number of unlike resistors in parallel is equal to the reciprocal of the sum of the reciprocals of the various resistors. This can be expressed as:

$$R_T = \cfrac{1}{\cfrac{1}{R_1} + \cfrac{1}{R_2} + \cfrac{1}{R_3} + \ldots \cfrac{1}{R_n}}$$

The effective value of placing any number of unlike resistors in parallel can be determined from the above formula. However, it is commonly used only when there are three or more resistors under consideration, since the simplified formula given before is more convenient when only two resistors are being used.

From the above, it also follows that when two or more resistors of the same value are placed in parallel, the effective resistance of the paralleled resistors is equal to the value of one of the resistors divided by the number of resistors in parallel.

The effective value of resistance of two or more resistors connected in parallel is *always* less than the value of the lowest resistance in the combination. It is well to bear this simple rule in mind, as it will assist greatly in approximating the value of paralleled resistors.

Resistors in Series-Parallel To find the total resistance of several resistors connected in series-parallel, it is usually easiest to apply either the formula for series resistors or the parallel resistor formula first, in order to reduce the original arrangement to a simpler one. For instance, in figure 5 the series resistors should be added in each branch, then there will be but two resistors in parallel to be calculated. In figure 6A the paralleled resistors should be reduced to the equivalent series value, and then the series resistance value can be added.

Resistance in series-parallel can be solved by combining the series and parallel formulas into one similar to the following (refer to figure 6B):

$$R_T = \cfrac{1}{\cfrac{1}{R_1 + R_2} + \cfrac{1}{R^3 + R^4}}$$

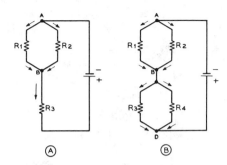

Figure 6

OTHER COMMON SERIES-PARALLEL CIRCUITS

Voltage Dividers A *voltage divider* is a series of resistors across a source of voltage from which various lesser values of voltage may be obtained by connections to various points along the resistors.

A voltage divider serves a most useful purpose in electronic equipment because it offers a simple means of obtaining voltages of different values from a common power supply source. It may also be used to obtain very low voltages of the order of .01 to .001 volt with a high degree of accuracy, even though a means of measuring such voltages is lacking, since with a given current the voltage across a resistor in a voltage divider is proportional to the resistance value. If the source voltage is accurately known, and the resistance can be measured, the voltage at any point along a resistor string is known, provided no current is drawn from the tap-on point unless this current is taken into consideration.

Voltage Divider Calculations Proper design of a voltage divider for any type of electronic equipment is a relatively simple matter. The first consideration is the amount of "bleeder" current to be drawn. In addition, it is also necessary that the desired voltage and the exact current at each tap on the voltage divider be known.

Figure 7 illustrates the flow of current in a simple voltage-divider and load circuit. The light arrows indicate the flow of bleeder current, while the heavy arrows indicate the flow of the load current. The design of a combined bleeder resistor and voltage di-

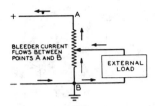

Figure 7

SIMPLE VOLTAGE-DIVIDER CIRCUIT

The arrows indicate the manner in which the current flow divides between the voltage divider itself and the external load circuit.

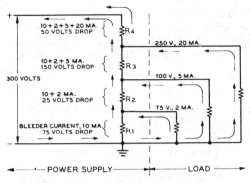

Figure 8

MORE COMPLEX VOLTAGE DIVIDER ILLUSTRATING KIRCHHOFF'S LAW

The method for computing the values of the resistors is discussed in the text.

vider, such as is commonly used in radio equipment, is illustrated in the following example:

A power supply delivers 300 volts and is conservatively rated to supply all needed current and still allow a bleeder current of 10 milliamperes. The following voltages are wanted: 75 volts at 2 milliamperes, 100 volts at 5 milliamperes, and 250 volts at 20 milliamperes. The required voltage drop across R_1 is 75 volts, across R_2 25 volts, across R_3 150 volts, and across R_4 it is 50 volts. These values are shown in the diagram of figure 8. The respective current values are also indicated. Apply Ohm's law:

$$R_1 = \frac{E}{I} = \frac{75}{.01} = 7500 \text{ ohms}$$

$$R_2 = \frac{E}{I} = \frac{25}{.012} = 2083 \text{ ohms}$$

$$R_3 = \frac{E}{I} = \frac{150}{.017} = 8823 \text{ ohms}$$

$$R_4 = \frac{E}{I} = \frac{50}{.037} = 1351 \text{ ohms}$$

$$R_{\text{Total}} = 7500 + 2083 + 8823 + 1351 = 19,757 \text{ ohms}$$

A 20,000 ohm resistor with three adjustable taps may be used, the wattage being equal to that maximum value required by any single resistor in the string. If four separate resistors are chosen, their "rounded" values would be: R_1, 7500 ohms; R_2, 2000 ohms; R_3, 8800 ohms and R_4, 1400 ohms. The power dissipated in each resistor is approximately 0.15 watt, 0.3 watt, 2.6 watts, and 1.9 watts, respectively, as discussed in a following section.

Kirchhoff's Laws   Ohm's law is all that is necessary to calculate the values in simple circuits, such as the preceding examples; but in more complex problems, involving several loops, or more than one voltage in the same closed circuit, the use of *Kirchhoff's laws* will greatly simplify the calculations. These laws are merely rules for applying Ohm's law.

Kirchhoff's first law is concerned with net current to a point in a circuit and states that:

At any point in a circuit the current flowing toward the point is equal to the current flowing away from the point.

Stated in another way: if currents flowing to the point are considered positive, and those flowing from the point are considered negative, the sum of all currents flowing toward and away from the point — taking signs into account — is equal to zero. Such a sum is known as an *algebraic sum;* such that the law can be stated thus: *The algebraic sum of all currents entering and leaving a point is zero.*

Figure 9 illustrates this first law. If the effective resistance of the network of resistors is 5 ohms, then 4 amperes flow toward point A, and 2 amperes flow away through the two 5-ohm resistors in series. The remaining 2 amperes flow away through

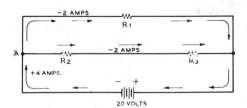

Figure 9
ILLUSTRATING KIRCHOFF'S FIRST LAW

The current flowing toward point "A" is equal to the current flowing away from point "A."

the 10-ohm resistor. Thus, there are 4 amperes flowing to point A and 4 amperes flowing away from the point. If R_T is the effective resistance of the network (5 ohms), $R_1 = 10$ ohms, $R_2 = 5$ ohms, $R_3 = 5$ ohms, and $E = 20$ volts, the following equation can be set up:

$$\frac{E}{R_T} - \frac{E}{R_1} - \frac{E}{R_2 + R_3} = 0$$

$$\frac{20}{5} - \frac{20}{10} - \frac{20}{5+5} = 0$$

$$4 - 2 - 2 = 0$$

Kirchhoff's second law is concerned with net voltage drop around a closed loop in a circuit and states that:

In any closed path or loop in a circuit the sum of the IR drops must equal the sum of the applied e.m.f.'s.

The second law also may be conveniently stated in terms of an algebraic sum as: *The algebraic sum of all voltage drops around a closed path or loop in a circuit is zero.* The applied e.m.f.'s (voltages) are considered positive, while *IR* drops taken in the direction of current flow (including the internal drop of the sources of voltage) are considered negative.

Figure 10 shows an example of the application of Kirchhoff's laws to a comparatively simple circuit consisting of three resistors and two batteries. First, an arbitrary direction of current flow in each closed loop of the circuit is assumed, drawing an arrow to indicate the assumed direction of current flow. Then the sum of all *IR* drops plus battery drops around each loop are equated to

zero. One equation for each unknown to be determined is required. Then the equations are solved for the unknown currents in the general manner indicated in figure 10. If the answer comes out positive, the direction of current flow originally assumed was correct. If the answer comes out negative, the current flow is in the opposite direction to the arrow which was drawn originally. This is illustrated in the example of figure 10, where the direction of flow of I_1 is opposite to the direction assumed in the sketch.

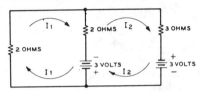

1. SET VOLTAGE DROPS AROUND EACH LOOP EQUAL TO ZERO.

    $I_1 2_{(OHMS)} + 2(I_1 - I_2) + 3 = 0$ (FIRST LOOP)

    $-6 + 2(I_2 - I_1) + 3I_2 = 0$ (SECOND LOOP)

2. SIMPLIFY

    $2I_1 + 2I_1 - 2I_2 + 3 = 0 \qquad 2I_2 - 2I_1 + 3I_2 - 6 = 0$

    $\dfrac{4I_1 + 3}{2} = I_2 \qquad\qquad 5I_2 - 2I_1 - 6 = 0$

    $\qquad\qquad\qquad\qquad \dfrac{2I_1 + 6}{5} = I_2$

3. EQUATE

    $\dfrac{4I_1 + 3}{2} = \dfrac{2I_1 + 6}{5}$

4. SIMPLIFY

    $20I_1 + 15 = 4I_1 + 12$

    $I_1 = -\dfrac{3}{16}$ AMPERE

5. RE-SUBSTITUTE

    $I_2 = \dfrac{-\frac{12}{16} + 3}{2} = \dfrac{2\frac{1}{4}}{2} = 1\frac{1}{8}$ AMPERE

Figure 10
ILLUSTRATING KIRCHOFF'S SECOND LAW
The voltage drop around any closed loop in a network is equal to zero.

Power in Resistive Circuits In order to cause electrons to flow through a conductor, constituting a current flow, it is necessary to apply an electromotive force (voltage) across the circuit. Less power is expended in creating a small current flow through a given resistance than in creating a large one; so it is necessary to have a unit of power as a reference.

The unit of electrical power is the *watt*, which is the rate of energy consumption when an e.m.f. of 1 volt forces a current of 1 ampere through a circuit. The power in a resistive circuit is equal to the product of the voltage applied across, and the current flowing in, a given circuit. Hence: P (watts) $= E$ (volts) $\times I$ (amperes).

Since it is often convenient to express power in terms of the resistance of the circuit and the current flowing through it, a substitution of IR for E ($E = IR$) in the above formula gives: $P = IR \times I$ or $P = I^2R$. In terms of voltage and resistance, $P = E^2/R$. Here, $I = E/R$ and when this is substituted for I the original formula becomes $P = E \times E/R$, or $P = E^2/R$. To repeat these three expressions:

$$P = EI, P = I^2R, \text{ and } P = E^2/R$$

where,

P equals power in watts,
E equals electromotive force in volts,
I equals current in amperes.

To apply the above equations to a typical problem: The voltage drop across a resistor in a power amplifier stage is 50 volts; the current flowing through the resistor is 150 milliamperes. The number of watts the resistor will be required to dissipate is found from the formula: $P = EI$, or $50 \times .150 = 7.5$ watts (.150 ampere is equal to 150 milliamperes). From the foregoing it is seen that a 7.5-watt resistor will safely carry the required current, yet a 10- or 20-watt resistor would ordinarily be used to provide a safety factor.

Efficiency and Energy   The *efficiency* of any device is the ratio of the usable power output to the power input. For electrical devices, the equation is:

$$E = \frac{P_o}{P_{in}}$$

where,

E equals efficiency expressed as a decimal,
P_o equals power output in watts,
P_{in} equals power input in watts.

Electrical energy is work and may be expressed in *watt-hours* (Wh):

$$J = PT$$

where,

J equals energy in watt-hours,
P equals power in watts,
T equals time in hours.

In industry, energy is usually measured in *kilowatt-hours* (kWh).

Power, Energy and Work   It is important to remember that power (expressed in watts, horsepower, etc.), represents the *rate* of energy consumption or the *rate* of doing work. But when we pay

Figure 11

MATCHING OF RESISTANCES

To deliver the greatest amount of power to the load, the load resistance R_L, should be equal to the internal resistance of the battery R_I.

our electric bill to the power company we have purchased a specific *amount* of *energy* or *work* expressed in the common units of *kilowatt-hours*. Thus *rate* of energy consumption (watts or kilowatts) multiplied by *time* (seconds, minutes, or hours) gives us total energy or work. Other units of energy are the watt-second, Btu, calorie, erg, and joule.

Heating Effect   Heat is generated when a source of voltage causes a current to flow through a resistor (or, for that matter, through any conductor). As explained earlier, this is due to the fact that heat is given off when free electrons collide with the atoms of the material. More heat is generated in high-resistance materials than in those of low resistance, since the free electrons must strike the atoms harder to knock off other electrons. As the heating effect is a function of the current flowing and the resistance of the circuit, the power expended in heat is given by the formula:

$$P = I^2R.$$

Lethal Electric Currents   While the examples given in the preceding pages have been concerned with relatively low voltages, certain electronic equipments contain extremely high voltages which are a deadly hazard. The human body is very sensitive to electric currents and appreciation of the dangerous effects of electric shock is necessary to maintain eternal vigilance in matters pertaining to electrical safety.

Alternating current, in particular, is especially dangerous, since a current of only a few milliamperes flowing through the body will cause muscular contraction, resulting in the inability of the victim to release his grasp on a live conductor. The maximum current at which a person is still capable of releasing a conductor by using muscles affected by the current is termed the *let-go current*. Currents only slightly in excess of this value may "freeze" the victim to the circuit with lethal effects. The average let-go current, found by experiment at the University of California in carefully controlled tests, was approximately 16 milliamperes for men and 10.5 milliamperes for women. Safe let-go currents for men and women are considered to be 9 and 6 milliamperes, respectively.

A severe electrical shock can produce ventricular fibrillation, or heart spasm, in a human which can bring death within minutes. Resuscitation techniques must be applied immediately if the victim is to be saved.

The accepted treatment consists of prompt rescue and immediate and continuous application of artificial respiration, preferably the mouth-to-mouth method.

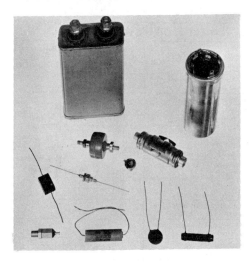

Figure 12

TYPICAL FIXED CAPACITORS

The two large units are high-value filter capacitors. Shown beneath these are various types of bypass capacitors for r-f and audio applications.

If the rescuer has been trained, artificial respiration should be combined with closed-chest cardiac massage, and resuscitation continued all the time the victim is being transported to the hospital. In addition to cardiac arrest, high currents may produce fatal damage to the central nervous system and deep burns.

Experimenters working with solid-state circuits often develop sloppy work habits, adjusting and modifying the equipment while the primary power is left running. This is a dangerous habit, because if the individual turns to work with vacuum-tube circuits or high-voltage power supplies, he may inadvertently expose himself to lethal high-voltage circuits. For safety's sake, electronic equipment of any size or power should never be worked on, or tested, unless the power is removed. If tests are to be run under operating conditions, the experimenter should be well clear of the equipment before the power is turned on.

2-3   Electrostatics and Capacitors

Electrical energy can be stored in an *electrostatic field*. A device capable of storing energy in such a field is called a *capacitor* (in earlier usage the term *condenser* was frequently used but the IEEE standards call for the use of capacitor instead of condenser) and is said to have a certain *capacitance*. The *energy* stored in an electrostatic field is expressed in *joules* (watt-seconds) and is equal to $CE^2/2$, where C is the capacitance in *farads* (a unit of capacitance to be discussed) and E is the potential in volts. The *charge* (Q) is equal to CE, the charge being expressed in *coulombs*.

Capacitance and Capacitors   Two conducting areas, or plates, separated from each other by a thin layer of insulating material (called a *dielectric*, in this case) form a *capacitor*. When a source of dc potential is momentarily applied across these plates, they may be said to become charged. If the same two plates are then joined together momentarily by means of a switch, the capacitor will *discharge*.

When the potential was first applied, electrons immediately flowed from one plate to the other through the battery or such

Figure 13

At top left are three variable air capacitors intended for hf/vhf use. At the right is a small variable vacuum capacitor intended for high-voltage service. Across the bottom are (left to right): two sub-miniature variable split-stator capacitors, a precision "plunger" capacitor, a compression mica capacitor, and a variable ceramic trimming capacitor.

source of dc potential as was applied to the capacitor plates. However, the circuit from plate to plate in the capacitor was *incomplete* (the two plates being separated by an insulator) and thus the electron flow ceased, meanwhile establishing a shortage of electrons on one plate and a surplus of electrons on the other.

When a deficiency of electrons exists at one end of a conductor, there is always a tendency for the electrons to move about in such a manner as to re-establish a state of balance. In the case of the capacitor herein discussed, the surplus quantity of electrons on one of the capacitor plates cannot move to the other plate because the circuit has been broken; that is, the battery or dc potential was removed. This leaves the capacitor in a *charged* condition; the capacitor plate with the electron *deficiency* is *positively* charged, the other plate being *negative*.

In this condition, a considerable stress exists in the insulating material (dielectric) which separates the two capacitor plates, due to the mutual attraction of two unlike potentials on the plates. This stress is known

TABLE 5. DIELECTRIC MATERIALS			
Material	Dielectric Constant 10 MHz	Power Factor 10 MHz	Softening Point Fahrenheit
Aniline-Formaldehyde Resin	3.4	0.004	260°
Barium Titianate	1200	1.0	—
Castor Oil	4.67		
Cellulose Acetate	3.7	0.04	180°
Glass, Window	6-8	Poor	2000°
Glass, Pyrex	4.5	0.02	
Kel-F Fluorothene	2.5	0.6	—
Methyl-Methacrylate-Lucite	2.6	0.007	160°
Mica	5.4	0.0003	
Mycalex Mykroy	7.0	0.002	650°
Phenol-Formaldehyde, Low-Loss Yellow	5.0	0.015	270°
Phenol-Formaldehyde Black Bakelite	5.5	0.03	350°
Porcelain	7.0	0.005	2800°
Polyethylene	2.25	0.0003	220°
Polystyrene	2.55	0.0002	175°
Quartz, Fused	4.2	0.0002	2600°
Rubber Hard-Ebonite	2.8	0.007	150°
Steatite	6.1	0.003	2700°
Sulfur	3.8	0.003	236°
Teflon	2.1	.0006	—
Titanium Dioxide	100-175	0.0006	2700°
Transformer Oil	2.2	0.003	
Urea-Formaldehyde	5.0	0.05	260°
Vinyl Resins	4.0	0.02	200°
Wood, Maple	4.4	Poor	

known as *electrostatic* energy, as contrasted with *electromagnetic* energy in the case of an inductor. This charge can also be called *potential energy* because it is capable of performing work when the charge is released through an external circuit. The charge is proportional to the voltage but the energy is proportional to the voltage squared, as shown in the following example.

The charge represents a definite amount of electricity, or a given number of electrons. The potential energy possessed by these electrons depends not only on their number, but also on their potential, or voltage. Thus, a 1-μF capacitor charged to 1000 volts possesses twice as much *potential energy* as does a 2-μF capacitor charged to 500 volts, though the charge (expressed in coulombs: $Q = CE$) is the same in either case.

The Unit of Capac- If the external circuit of
itance: The Farad the two capacitor plates is completed by joining the terminals together with a piece of wire, the electrons will rush immediately from one plate to the other through the external circuit and establish a state of equilibrium. This latter phenomenon explains the *discharge* of a capacitor. The amount of stored energy in a charged capacitor is dependent on the charging potential, as well as a factor which takes into account the *size* of the plates, *dielectric thickness, nature* of the dielectric, and the *number* of plates. This factor, which is determined by the foregoing, is called the *capacitance* of a capacitor and is expressed in *farads*.

The farad has the dimensions of one coulomb of electricity added to a capacitor by an applied voltage of one volt. Since this unit is too large for practical use in electronics, a smaller unit, the *microfarad* (10^{-6} farad) abbreviated μF, is used. A smaller unit, the *picofarad* (10^{-12} farad) abbreviated pF, is also used in the communication industry.

Dielectric Although any substance which has
Materials the characteristics of a good insulator may be used as a dielectric material, commercially manufactured capacitors make use of dielectric materials which have been selected because their char-

acteristics are particularly suited to the job at hand. Air is a very good dielectric material, but an air-spaced capacitor contains a large volume per unit of capacitance, as the dielectric constant of air is only slightly greater than one.

Certain materials such as lucite and other plastics dissipate considerable energy when used as capacitor dielectrics. This energy loss is expressed in terms of the *power factor*, or dissipation factor, of the capacitor which represents the portion of the input volt-amperes lost in the dielectric.

Better materials such as mylar, polystyrene, mica, ceramic, and titanium dioxide are especially well suited for dielectric material, and capacitors made of these materials are discussed at length in chapter 17.

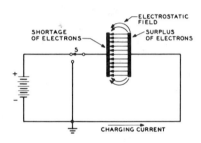

Figure 14

SIMPLE CAPACITOR

Illustrating the imaginary lines of force representing the paths along which the repelling force of the electrons would act on a free electron located between the two capacitor plates.

Dielectric The capacitance of a capacitor is
Constant determined by the thickness and nature of the dielectric material between plates. Certain materials offer a greater capacitance than others, depending on their physical makeup and chemical constitution. This property is expressed by a constant K, called the *dielectric constant*. ($K = 1$ for air.)

Dielectric If the charge becomes too great
Breakdown for a given thickness of a certain dielectric, the capacitor will break down, i.e., the dielectric will puncture. It is for this reason that capacitors are rated in the manner of the amount of voltage they will safely withstand as well

as the capacitance in microfarads. This rating is commonly expressed as the *dc working voltage (dcwv)*.

The breakdown voltage of a dielectric at 50 Hz is substantially the same as for dc conditions, however, as the frequency is raised a lowering of the breakdown voltage below the dc value occurs. Typically, at 3.5 MHz the breakdown voltage in air for a given gap is about 80 percent of the dc value and at 14 MHz it is about 75 percent of the dc value. In the vhf region, at small gap lengths, the breakdown voltage resembles that for the high frequency region until a critical potential is reached when, for a further increase in gap length, there is a decrease in breakdown voltage. This is thought to be a function of oscillations of electrons in the gap.

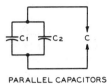
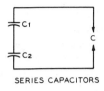

PARALLEL CAPACITORS   SERIES CAPACITORS

Figure 15

CAPACITORS IN SERIES AND PARALLEL

where,

C equals capacitance in picofarads,
K equals dielectric constant of spacing material,
A equals area of dielectric in square inches,
t equals thickness of dielectric in inches.

This formula indicates that the capacitance is *directly* proportional to the area of the plates and *inversely* proportional to the thickness of the dielectric (spacing between the plates). This simply means that when the area of the plate is doubled, the spacing between plates remaining constant, the capacitance will be doubled. Also, if the area of the plates remains constant, and the plate spacing is doubled the capacitance will be reduced to half.

The above equation also shows that capacitance is directly proportional to the dielectric constant of the spacing material. An air-spaced capacitor that has a capacitance of 100 pF in air would have a capacitance of 467 pF when immersed in castor oil, because the dielectric constant of castor oil is 4.67 times as great as the dielectric constant of air.

Where the area of the plate is definitely set, when it is desired to know the spacing needed to secure a required capacitance,

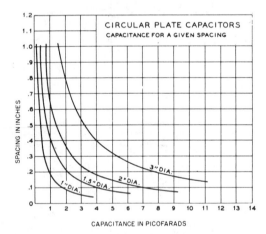

CIRCULAR PLATE CAPACITORS
CAPACITANCE FOR A GIVEN SPACING

SPACING IN INCHES

CAPACITANCE IN PICOFARADS

CHART 1

Through the use of this chart it is possible to determine the required plate diameter (with the necessary spacing established by peak voltage considerations) for a circular-plate capacitor. The capacitance given is between adjacent faces of the two plates.

$$t = \frac{A \times 0.2248 \times K}{C}$$

where all units are expressed just as in the preceding formula. This formula is not confined to capacitors having only square or rectangular plates, but also applies when the plates are circular in shape. The only change will be the calculation of the *area* of such circular plates; this area can be computed by squaring the *radius* of the plate, then multiplying by π (3.14).

Calculation of Capacitance The capacitance of two parallel plates may be determined with good accuracy by the following formula:

$$C = 0.2248 \times K \times \frac{A}{t}$$

The capacitance of a multiplate capacitor can be calculated by taking the capacitance of one section and multiplying this by the number of dielectric spaces. In such cases, however, the formula gives no consideration to the effects of edge capacitance; so the capacitance as calculated will not be entirely accurate. These additional capacitances will be but a small part of the effective total capacitance, particularly when the plates are reasonably large and thin, and the final result will, therefore, be within practical limits of accuracy.

Capacitors in Parallel and in Series Equations for calculating capacitances of capacitors in *parallel* connections are the same as those for resistors in *series*.

$$C_T = C_1 + C_2 + \ldots + C_n$$

Capacitors in *series* connection are calculated in the same manner as are resistors in *parallel* connection.

The formulas are repeated: (1) For two or more capacitors of *unequal* capacitance in series:

$$C_T = \frac{1}{\dfrac{1}{C_1} + \dfrac{1}{C_2} + \dfrac{1}{C_3}}$$

or,

$$\frac{1}{C_T} = \frac{1}{C_1} + \frac{1}{C_2} + \frac{1}{C_3}$$

(2) *Two* capacitors of *unequal* capacitance in series:

$$C_T = \frac{C_1 \times C_2}{C_1 + C_2}$$

(3) Three capacitors of *equal* capacitance in series:

$$C_T = \frac{C_1}{3}$$

where,

C_1 is the common capacitance.

(4) Three or more capacitors of *equal* capacitance in series.

$$C_T = \frac{\text{Value of common capacitance}}{\text{Number of capacitors in series}}$$

(5) Six capacitors in series-parallel:

$$C_T = \frac{1}{\dfrac{1}{C_1} + \dfrac{1}{C_2}} + \frac{1}{\dfrac{1}{C_3} + \dfrac{1}{C_4}} + \frac{1}{\dfrac{1}{C_5} + \dfrac{1}{C_6}}$$

Capacitors in AC and DC Circuits When a capacitor is connected into a direct-current circuit, it will block the dc, or stop the flow of current. Beyond the initial movement of electrons during the period when the capacitor is being charged, there will be no flow of current because the circuit is effectively broken by the dielectric of the capacitor.

Strictly speaking, a very small current may actually flow because the dielectric of the capacitor may not be a perfect insulator. This minute current flow is the leakage current previously referred to and is dependent on the internal dc resistance of the capacitor. This leakage current is usually quite noticeable in most types of electrolytic capacitors.

When an alternating current is applied to a capacitor, the capacitor will charge and discharge a certain number of times per second in accordance with the frequency of the alternating voltage. The electron flow in the charge and discharge of a capacitor when an ac potential is applied constitutes an alternating current, in effect. It is for this reason that a capacitor will pass an alternating current yet offer practically infinite opposition to a direct current. These two properties are repeatedly in evidence in electronic circuits.

Voltage Rating of Capacitors in Series Any good, modern-dielectric capacitor has such a high internal resistance that the exact resistance will vary considerably from capacitor to capacitor even though they are made by the same manufacturer and are of the same rating. Thus, when 1000 volts dc are connected across two 1-μF 500-volt capacitors in series, the chances are that the voltage will divide unevenly; one capacitor will receive more than 500 volts and the other less than 500 volts.

By connecting a half-megohm, 2-watt composition resistor across each capacitor, the voltage will be equalized (figure 16).

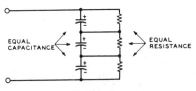

Figure 16

**SHOWING THE USE OF VOLTAGE
EQUALIZING RESISTORS ACROSS
CAPACITORS CONNECTED IN SERIES**

Energy Storage   Energy stored in a capacitor is:

$$J = \frac{CE^2}{2}$$

where,

J is the unit of energy in Joules,
C is the capacitance in farads,
E is the average charge in volts.

The energy stored in a large capacitance, high voltage capacitor is formidable, the charge varying with the square of the voltage. Experimenters are cautioned to stay clear of large, high voltage capacitors and not touch them until the charge has been dissipated by shorting the terminals.

Capacitors in   When two capacitors are con-
Series on AC   nected in series, *alternating*
                    voltage pays no heed to the relatively high internal resistance of each capacitor, but divides across the capacitors in inverse proportion to the *capacitance*. Because, in addition to the dc voltage across a capacitor in a filter or audio amplifier circuit there is usually an ac or audio-frequency voltage component, it is inadvisable to series-connect capacitors of unequal capacitance even if dividers are provided to keep the dc voltages within the ratings of the individual capacitors.

For instance, if a 500-volt 1-μF capacitor is used in series with a 4-μF 500-volt capacitor across a 250-volt ac supply, the 1-μF capacitor will have 200 ac volts across it and the 4-μF capacitor only 50 volts. An equalizing divider, to do any good in this case, would have to be of very low resistance because of the comparatively low impedance of the capacitors to alternating

current. Such a divider would draw excessive current and be impracticable.

The safest rule to follow is to use only capacitors of the same capacitance and voltage rating and to install matched high-resistance proportioning resistors across the various capacitors to equalize the dc voltage drop across each capacitor. This holds regardless of how many capacitors are series-connected.

Electrolytic   *Electrolytic capacitors* use a very
Capacitors   thin film of oxide as the dielectric, and are polarized; that is, they have a positive and a negative terminal which must be properly connected in a circuit; otherwise, the oxide will break down and the capacitor will overheat. The unit then will no longer be of service. When electrolytic capacitors are connected in series, the positive terminal is always connected to the positive lead of the power supply; the negative terminal of the capacitor connects to the *positive* terminal of the *next* capacitor in the series combination.

2-4   Magnetism and Electromagnetism

The common bar or horseshoe magnet is familiar to most people. The magnetic field which surrounds it causes the magnet to attract other magnetic materials, such as iron nails or tacks. Exactly the same kind of magnetic field is set up around any conductor carrying a current, but the field exists only while the current is flowing.

Magnetic Fields   Before a potential, or voltage, is applied to a conductor there is no external field, because there is no general movement of the electrons in one direction. However, the electrons do progressively move along the conductor when an e.m.f. is applied, the direction of motion depending on the polarity of the e.m.f. Since each electron has an electric field about it, the flow of electrons causes these fields to build up into a resultant external field which acts in a plane at right angles to the direction in which the current is flowing. This field is known as the *mag-*

netic field. The magnetic field is composed of *magnetic lines of force,* as illustrated in figure 17.

The direction of this magnetic field depends entirely on the direction of electron drift, or current flow, in the conductor. When the flow is toward the observer, the field about the conductor is clockwise; when the flow is away from the observer, the field is counterclockwise. This is easily remembered if the left hand is clenched, with the thumb outstretched and pointing in the direction of electron flow. The fingers then indicate the direction of the magnetic field around the conductor.

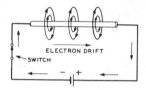

Figure 17

LEFT-HAND RULE

Showing the direction of the magnetic lines of force produced around a conductor carrying an electric current.

Each electron adds its field to the total external magnetic field, so that the greater the number of electrons moving along the conductor, the stronger will be the resulting field. The strength of the field, thus, is directly proportional to the current flowing in the conductor.

One of the fundamental laws of magnetism is that *like poles repel one another and unlike poles attract one another.* This is true of current-carrying conductors as well as of permanent magnets. Thus, if two conductors are placed side by side and the current in each is flowing in the same direction, the magnetic fields will also be in the same direction and will combine to form a larger and stronger field. If the current flow in adjacent conductors is in opposite directions, the magnetic fields oppose each other and tend to cancel.

The magnetic field around a conductor may be considerably increased in strength by winding the wire into a coil. The field around each wire then combines with those of the adjacent turns to form a total field

through the coil which is concentrated along the axis of the coil and behaves externally in a way similar to the field of a bar magnet.

If the left hand is held so that the thumb is outstretched and parallel to the axis of a coil, with the fingers curled to indicate the direction of electron flow around the turns of the coil, the thumb then points in the direction of the north pole of the magnetic field.

The Magnetic Circuit In the magnetic circuit, the units which correspond to current, voltage, and resistance in the electrical circuit are *flux, magnetomotive force,* and *reluctance.*

Flux; Flux Density As a current is made up of a drift of electrons, so is a magnetic field made up of lines of force, and the total number of lines of force in a given magnetic circuit is termed the *flux.* The flux depends on the material, cross section, and length of the magnetic circuit, and it varies directly as the current flowing in the circuit. The unit of flux is the *weber,* or *maxwell* and the symbol is ϕ (phi).

Flux density is the number of lines of force per unit area. It is expressed in *gauss* if the unit of area is the square centimeter (1 gauss = 1 line of force per square centimeter), or in *lines per square inch.* The symbol for flux density is B if it is expressed in gauss, or B if expressed in lines per sq. in.

Magnetomotive Force The force which produces a flux in a magnetic circuit is called *magnetomotive force.* It is abbreviated m.m.f. and is designated by the letter F. The unit of magnetomotive force is the *gilbert,* which is equivalent to $1.26 \times NI$, where N is the number of turns and I is the current flowing in the circuit in amperes.

The m.m.f. necessary to produce a given flux density is stated in gilberts per centimeter (oersteds) (H), or in ampere-turns per inch (H).

Reluctance Magnetic reluctance corresponds to electrical resistance, and is the property of a material that opposes the creation of a magnetic flux in the material. It is expressed in *rels*, and the symbol is the letter R. A material has a reluctance of 1 rel when an m.m.f. of 1 ampere-turn (NI) generates a flux of 1 line of force in it. Combinations of reluctances are treated the same as resistances in finding the total effective reluctance. The *specific reluctance* of any substance is its reluctance per unit volume.

Except for iron and its alloys, most common materials have a specific reluctance very nearly the same as that of a vacuum, which, for all practical purposes, may be considered the same as the specific reluctance of air.

The relations between flux, magnetomotive force, and reluctance are exactly the same as the relations between current, voltage, and resistance in the electrical circuit. These can be stated as follows:

$$\phi = \frac{F}{R} \quad R = \frac{F}{\phi} \quad F = \phi R$$

where,

ϕ equals flux,

F equals m.m.f.,

R equals reluctance.

Permeability *Permeability* expresses the ease with which a magnetic field may be set up in a material as compared with the effort required in the case of air. Iron, for example, has a permeability of around 2000 times that of air, which means that a given amount of magnetizing effort produced in an iron core by a current flowing through a coil of wire will produce 2000 times the *flux* density that the same magnetizing effect would produce in air. It may be expressed by the ratio B/H or B/H. In other words,

$$\mu = \frac{B}{H} \quad \text{or} \quad \mu = \frac{B}{H}$$

where μ is the permeability, B is the flux density in gausses, B is the flux density in lines per square inch, H is the m.m.f. in gilberts per centimeter (*oersteds*), and H is the m.m.f. in *ampere-turns* per inch.

It can be seen from the foregoing that permeability is inversely proportional to the specific reluctance of a material.

Saturation Permeability is similar to *electric conductivity*. This is, however, one important difference: the permeability of magnetic materials is not independent of the magnetic current (flux) flowing through it, although electrical conductivity is substantially independent of the electric current in a wire. When the flux density of a magnetic conductor has been increased to the *saturation point*, a further increase in the magnetizing force will not produce a corresponding increase in flux density.

B-H Curve To simplify magnetic circuit calculations, a magnetization curve may be drawn for a given unit of material. Such a curve is termed a B-H curve, and may be determined by experiment. When the current in an iron-core coil is first applied, the relation between the winding current and the core flux is shown at A-B in figure 18. If the current is then reduced to zero, reversed, brought back again to zero and reversed to the original direction, the flux passes through a typical hysteresis loop as shown.

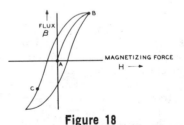

Figure 18

TYPICAL HYSTERESIS LOOP
(B-H CURVE = A-B)

Showing relationship between the current in the winding of an iron-core inductor and the core flux. A direct current flowing through the inductance brings the magnetic state of the core to some point on the hysteresis loop, such as C.

The magnetism remaining in a material after the magnetizing force is removed is called *residual magnetism*. *Retentivity* is the property which causes a magnetic material to have residual magnetism after having been magnetized.

Hystereisis is the characteristic of a magnetic system which causes a loss of power due to the fact that a reverse magnetizing force must be applied to reduce the residual magnetism to zero. This reverse force is termed *coercive force*. Hysteresis loss is apparent in transformers and chokes by the heating of the core.

Inductance   *Inductance* (L) is the property of an electrical circuit whereby changes in current flowing in the circuit produce changes in the magnetic field such that a counter-e.m.f. is set up in that circuit or in neighboring ones. If the counter-e.m.f. is set up in the original circuit, it is called *self-inductance* and if it it is set up in neighboring circuit it is called *mutual inductance*.

The unit of inductance is the *henry* (H) and is defined as that value of inductance in which an induced e.m.f. of one volt is produced when the inducing current is varied at the rate of one ampere per second. The henry is commonly subdivided into several smaller units, the *millihenry* (10^{-3} henry) abbreviated *mH*, the microhenry (10^{-6} henry) abbreviated μH and the *nanohenry* (10^{-9} henry), abbreviated *nH*.

The storage of energy in a magnetic field is expressed in *joules* and is equal to $LI^2/2$ and the dimensions are in watt-seconds.

Mutual Inductance   When one coil is near another, a varying current in one will produce a varying magnetic field which cuts the turns of the other coil, inducing a current in it. This induced current is also varying, and will therefore induce another current in the first coil. This reaction between two coupled circuits is called *mutual inductance*, and can be calculated and expressed in henrys. The symbol for mutual inductance is M. Two circuits thus joined are said to be *inductively coupled*.

The magnitude of the mutual inductance depends on the shape and size of the two circuits, their positions and distances apart, and the permeability of the medium. The extent to which two inductors are coupled is expressed by a relation known as *coefficient of coupling* (k). This is the ratio of the mutual inductance actually present to the maximum possible value.

Thus, when k is 1, the coils have the maximum degree of mutual induction.

The mutual inductance of two coils can be formulated in terms of the individual inductances and the coefficient of coupling:

$$M = k \sqrt{L_1 \times L_2}$$

For example, the mutual inductance of two coils, each with an inductance of 10 henrys and a coupling coefficient of 0.8 is:

$$M = 0.8 \sqrt{10 \times 10} = 0.8 \times 10 = 8$$

The formula for mutual inductance is $L = L_1 + L_2 + 2M$ when the coils are poled so that their fields add. When they are poled so that their fields buck, then $L = L_1 + L_2 - 2M$ (figure 19).

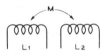

Figure 19

MUTUAL INDUCTANCE

The quantity M represents the mutual inductence between the two coils L_1 and L_2.

Inductors in Parallel   Inductors in parallel are combined exactly as are resistors in parallel, provided that they are far enough apart so that the mutual inductance is entirely negligible.

Inductors in Series   Inductors in series are additive, just as are resistors in series, again provided that no mutual inductance exists. In this case, the total inductance L is:

$$L = L_1 + L_2 + \ldots , \text{etc.}$$

Where mutual inductance does exist:

$$L = L_1 + L_2 + 2M$$

where,

M is the mutual inductance.

This latter expression assumes that the coils are connected in such a way that all flux linkages are in the same direction, i.e., additive. If this is not the case and the mutual linkages *subtract* from the self-linkages, the following formula holds:

$$L = L_1 + L_2 - 2M$$

where,

M is the mutual inductance.

Core Material　Ordinary magnetic cores cannot be used for radio frequencies because the *eddy current and hysteresis losses* in the core material become enormous as the frequency is increased. The principal use for conventional magnetic cores is in the audio-frequency range below approximately 15,000 Hertz, whereas at very low frequencies (50 to 60 Hertz) their use is mandatory if an appreciable value of inductance is desired.

An air-core inductor of only 1 henry inductance would be quite large in size, yet values as high as 500 henrys are commonly available in small iron-core chokes. The inductance of a coil with a magnetic core will vary with the amount of current (both ac and dc) which passes through the coil. For this reason, iron-core chokes that are used in power supplies have a certain inductance rating at a *predetermined value of direct current.*

The permeability of air does not change with flux density; so the inductance of iron-core coils often is made less dependent on flux density by making part of the magnetic path air, instead of utilizing a closed loop of iron. This incorporation of an *air gap* is necessary in many applications of iron-core coils, particularly where the coil carries a considerable dc component. Because the permeability of air is so much lower than that of iron, the air gap need *comprise* only a small fraction of the magnetic circuit in order to provide a substantial proportion of the total reluctance.

Inductors at　Inductors of all forms are
Radio Frequencies　used at frequencies up into the microwave region. Air, iron, ferrite and brass are common core materials and the coils may either be the solenoid type, or toroidal. The design and use of these coils is covered in chapter 17 of this handbook.

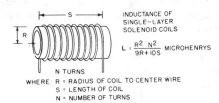

WHERE: R = RADIUS OF COIL TO CENTER WIRE
S = LENGTH OF COIL
N = NUMBER OF TURNS

Figure 20

FORMULA FOR CALCULATING INDUCTANCE

Through the use of the equation and the sketch shown above the inductance of single-layer solenoid coils can be calculated with an accuracy of about one percent for the types of coils normally used in the hf and vhf range.

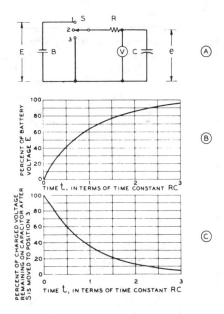

Figure 21

TIME CONSTANT OF AN RC CIRCUIT

Shown at (A) is the circuit upon which is based the curves of (B) and (C). (B) shows the rate at which capacitor C will charge from the instant at which switch S is placed in position 1. (C) shows the discharge curve of capacitor C from the instant at which switch S is placed in position 3.

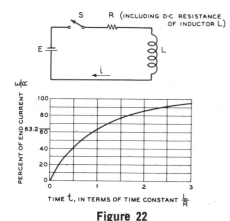

Figure 22

TIME CONSTANT OF AN RL CIRCUIT

Note that the time constant for the increase in current through an R L circuit is identical to the rate of increase in voltage across the capacitor in an R C circuit.

2-5 RC and RL Transients

A voltage divider may be constructed as shown in figure 21. Kirchhoff's and Ohm's Laws hold for such a divider. This circuit is known as an *RC circuit*.

Time Constant- RC and RL Circuits When switch S in figure 21 is placed in position 1, a voltmeter across capacitor C will indicate the manner in which the capacitor will become charged through the resistor R from battery B. If relatively large values are used fo R and C, and if a high-impedance voltmeter which draws negligible current is used to measure the voltage (e), the rate of charge of the capacitor may actually be plotted with the aid of a timer.

Voltage Gradient It will be found that the voltage (e) will begin to rise rapidly from zero the instant the switch is closed. Then, as the capacitor begins to charge, the rate of change of voltage across the capacitor will be found to decrease, the charging taking place more and more slowly as capacitor voltage e approaches battery voltage E. Actually, it will be found that in any given interval a constant percentage of the remaining difference between e and E will be delivered to the capacitor as an increase in voltage. A voltage which changes in this manner is said to increase *logarithmically*, or follows an *exponential* curve.

Time Constant A mathematical analysis of the charging of a capacitor in this manner would show that the relation-

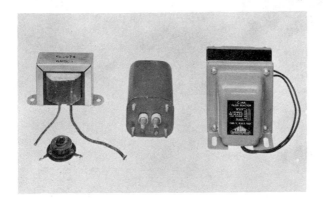

Figure 23

TYPICAL IRON-CORE INDUCTANCES

At the right is an upright mounting filter choke intended for use in low-powered transmitters and audio equipment. At the center is a hermetically sealed inductance for use under poor environmental conditions. To the left is an inexpensive receiving-type choke, with a small iron-core r-f choke directly in front of it.

ship between battery voltage E and the voltage across the capacitor (e) could be expressed in the following manner:

$$e = E \left(1 - \varepsilon^{-t/RC}\right)$$

where e, E, R, and C have the values discussed above, $\varepsilon = 2.718$ (the base of Naperian or natural logarithms), and t represents the time which has elapsed since the closing of the switch. With t expressed in seconds, R and C may be expressed in farads and ohms, or R and C may be expressed in microfarads and megohms. The product RC is called the *time constant* of the circuit, and is expressed in seconds. As an example, if R is one megohm and C is one microfarad, the time constant RC will be equal to the product of the two, or one second.

When the elapsed time (t) is equal to the time constant of the RC network under consideration, the exponent of ε becomes -1. Now ε^{-1} is equal to $1/\varepsilon$, or $1/2.718$,

which is 0.368. The quantity $(1 - 0.368)$ then is equal to 0.632. Expressed as percentage, the above means that the voltage across the capacitor will have increased to 63.2 percent of the battery voltage in an interval equal to the time constant or RC product of the circuit. Then, during the next period equal to the time constant of the RC combination, the voltage across the capacitor will have risen to 63.2 per cent of the remaining difference in voltage, or 86.5 per cent of the applied voltage (E).

RL Circuit In the case of a series combination of a resistor and an inductor, as shown in figure 22, the current through the combination follows a very similar law to that given above for the voltage appearing across the capacitor in an RC series circuit. The equation for the current through the combination is:

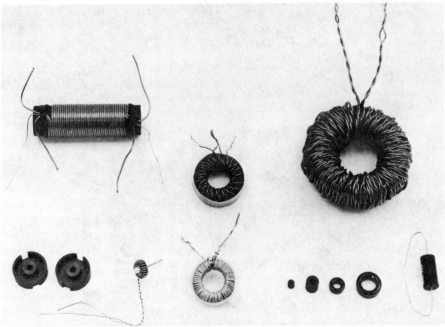

Figure 24
TRIFILAR, TOROIDAL, AND CUP-CORE INDUCTORS

At top left is a trifilar (three-winding) filament choke wound on a ferrite rod. To the right are two toroid inductors with bifilar windings on ferrite cores. At the lower left is a ferrite cup-core assembly, with two miniature ferrite toroid inductors at the center. To the lower right are typical miniature ferrite toroid cores and an encapsulated ferrite-core r-f choke.

$$i = \frac{E}{R} \ (1 - \varepsilon^{-t/RL})$$

where i represents the current at any instant through the series circuit, E represents the applied voltage, and R represents the total resistance of the resistor and the dc resistance of the inductor in series. Thus the time constant of the RL circuit is L/R, with R expressed in ohms and L expressed in henrys.

Voltage Decay When the switch in figure 21 is moved to position 3 after the capacitor has been charged, the capacitor voltage will drop in the manner shown in figure 21-C. In this case the voltage across the capacitor will decrease to 36.8 percent of the initial voltage (will make 63.2 per cent of the total drop) in a period of time equal to the time constant of the RC circuit.

CHAPTER THREE

Alternating Current, Impedance, and Resonant Circuits

The study of electromagnetic waves and radio transmission begins with the observation of electrons in motion, which constitutes an electric current. Of paramount importance is a type of current whose direction of flow reverses periodically. The reversal may take place at a low rate, or it may take place millions of times a second, in the case of communication frequencies. This type of current is termed *alternating current (ac)*.

3-1 Alternating Current

An alternating current is one whose amplitude of current flow periodically rises from zero to a maximum in one direction, decreases to zero, changes its direction, rises to maximum in the opposite direction, and decreases to zero again. This complete process, starting from zero, passing through two maximums in opposite directions, and returning to zero again, is called a *cycle*. The number of times per second that a current passes through the complete cycle is called the *frequency (f)* of the current. One and one-quarter cycles of an alternating current wave are illustrated diagrammatically in figure 1.

Frequency Spectrum   At present the usable frequency range for alternating electrical currents extends over the *electromagnetic spectrum* from about 15 cycles per second to perhaps 30,000,-000,000 cycles per second. It is cumbersome to use a frequency designation in c.p.s. for

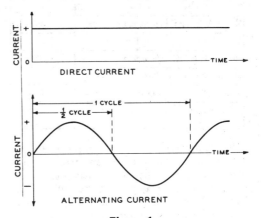

Figure 1

**ALTERNATING CURRENT
AND DIRECT CURRENT**

Graphical comparison between unidirectional (direct) current and alternating current as plotted against time.

enormously high frequencies, so three common units which are multiples of one cycle per second are established and are universally used by engineers.

The unit of frequency measurement is the *Hertz* (Hz) and is one cycle per second. The standard metric prefixes of *kilo* (10^3), *mega* (10^6) and *giga* (10^9) are used with the basic unit.

The frequencies between 15 and 30,000 Hz are termed *audio frequencies* (a-f) since a portion of this range is audible to the human ear. Frequencies in the vicinity of 60 Hz are also called *power frequencies* since they are commonly used to distribute electric power to the consumer.

The frequencies falling between 3 kHz and 30 GHz are termed *radio frequencies* (r-f) since they are commonly used in radio communication and the allied arts. The radio spectrum is divided into eight frequency bands, each one of which is ten times as high in frequency as the one just below it in the spectrum. The present spectrum, with classifications, is given in Table 1.

TABLE 1.

FREQUENCY CLASSIFICATION

FREQUENCY	CLASSIFICATION	DESIGNATION
3 to 30 kHz	Very-low frequency	VLF
30 to 300 kHz	Low frequency	LF
300 to 3000 kHz	Medium frequency	MF
3 to 30 MHz	High frequency	HF
30 to 300 MHz	Very-high frequency	VHF
300 to 3000 MHz	Ultrahigh frequency	UHF
3 to 30 GHz	Superhigh frequency	SHF
30 to 300 GHz	Extremely high frequency	EHF

For industrial and military purposes the vhf and uhf portions of the spectrum are divided into sub-bands which have no international standing at present, but which are commonly used in the United States. These bands in round numbers are listed in Table 2.

TABLE 2.

RADAR FREQUENCY BANDS

DESIGNATION	FREQUENCY (GHz)
L-band	1.0 - 2.0
S-band	2.0 - 4.0
C-band	4.0 - 8.0
X-band	8.0 - 12.0
K_u-band	12.0 - 18.0
K-band	18.0 - 27.0
K_a-band	27.0 - 40.0
Millimeter	40.0 - 300.0

Generation of Alternating Current Faraday discovered that if a conductor which forms part of a closed circuit is moved through a magnetic field so as to cut across the lines of force, a current will flow in the conductor. He also discovered that, when a conductor in a second closed circuit is brought near the first conductor and the current in the first one is varied, a current will flow in the second conductor. This effect is known as *induction*, and the currents so generated are *induced currents*. In the latter case it is the lines of force which are moving and cutting the second conductor, due to the varying current strength in the first conductor.

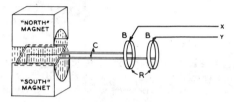

Figure 2

THE ALTERNATOR

Semi-schematic representation of the simplest form of the alternator.

A current is induced in a conductor if there is a relative motion between the conductor and a magnetic field, its direction of flow depending on the direction of the relative motion between the conductor and the field, and its strength depends on the intensity of the field, the rate of cutting lines of force, and the number of turns in the conductor.

The Alternator A machine that generates an alternating current is called an *alternator* or *ac generator* (figure 2). It consists of two magnets, the opposite poles of which face each other and which have a common radius. Between the two poles a magnetic field exists and a rotating conductor is suspended in the field in such a way that induced voltage may be taken off by means of collector rings (R) and brushes (B) to an external circuit (X-Y).

When driven from an external source, the direction of current flow in the rotating conductor is continuously changing from positive to negative and back again at a rate determined by the speed of rotation and the construction of the alternator.

The current does not increase directly as the angle of rotation, but rather as the *sine* of the angle and the current is said to vary *sinusoidally* (figure 3).

The rate at which the complete cycle of current reversal takes place is called the *frequency,* expressed in cycles per second, or *Hertz* (abbreviated *Hz*). The magnitude of the cycle is called the *amplitude,* which is measured at the peak of the cycle.

Another term used with reference to an alternating current is the *period* of the wave, which is defined as the reciprocal of the frequency:

$$Period = \frac{1}{Frequency}$$

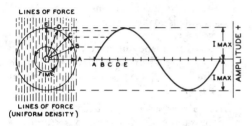

LINES OF FORCE

LINES OF FORCE
(UNIFORM DENSITY)

Figure 3
OUTPUT OF THE ALTERNATOR

Graph showing sine-wave output current of the alternator of figure 2.

The arrow rotating to the left in figure 3 represents a conductor rotating in a constant magnetic field of uniform density. The arrow also can be taken as a *vector* representing the strength of the magnetic field. This means that the length of the arrow is determined by the strength of the field (number of lines of force), which is constant. If the arrow is rotating at a constant rate (that is, with constant *angular velocity*), then the voltage developed across the conductor will be proportional to the rate at which it is cutting lines of force, which rate is proportional to the vertical distance between the tip of the arrow and the horizontal base line.

If EO is taken as unity, or a voltage of 1, then the voltage (vertical distance from tip of arrow to the horizontal base line) at point C for instance may be determined simply by referring to a table of sines and looking up the sine of the angle which the arrow makes with the horizontal.

When the arrow has traveled from point A to point E, it has traveled 90 degrees or one quarter cycle. The other three quadrants are not shown because their complementary or mirror relationship to the first quadrant is obvious.

It is important to note that time units are represented by *degrees* or quadrants. The fact that AB, BC, CD, and DE are equal chords (forming equal quadrants) simply means that the arrow (conductor or vector) is traveling at a constant speed, because these points on the radius represent the passage of equal units of time. A sine wave plotted against time is shown in figure 4.

The frequency of the generated voltage is proportional to the speed of rotation of the alternator, and to the number of magnetic poles in the field. Alternators may be built to produce radio frequencies up to 100 kHz, and some such machines are still used for stand-by low-frequency communication. By means of multiple windings, three-phase output may be obtained from large industrial alternators.

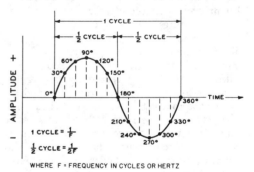

WHERE F = FREQUENCY IN CYCLES OR HERTZ

Figure 4
THE SINE WAVE

Illustrating one cycle of a sine wave. One complete cycle of alternation is broken up into 360 degrees. Then one-half cycle is 180 degrees, one-quarter cycle is 90 degrees, and so on down to the smallest division of the wave. A cosine wave has a shape identical to a sine wave but is shifted 90 degrees in phase—in other words the wave begins at full amplitude, the 90-degree point comes at zero amplitude, the 180-degree point comes at full amplitude in the opposite direction of current flow, etc.

Radian Notation The value of **an ac wave** varies **continuously, as** shown in figure 1. It is important **to know** the amplitude of the wave in terms **of the**

peak ·amplitude at any instant in the cycle. It is convenient mathematically to divide the cycle either into *electrical degrees* (360° represents one cycle) or into *radians*. A radian is an arc of a circle equal to the radius of the circle, there being 2π radians per cycle (figure 5).

.Both radian notation and electrical-degree notation are used in discussions of alternating-current circuits. However, trigonometric tables are much more readily available in terms of degrees than radians, so the following simple conversions are useful.

$$2\pi \text{ radians } = 1 \text{ cycle} = 360°$$

$$\pi \text{ radians } = 1/2 \text{ cycle} = 180°$$

$$1 \text{ radian } = \frac{1}{2\pi}\text{cycle} = 57.3°$$

When the conductor in the simple alternator of figure 2 has made one complete revolution it has generated one cycle and has rotated through 2π radians. The expression $2\pi f$ then represents the number of radians in one cycle multiplied by the number of cycles per second (the frequency) of the alternating voltage or current.

In technical literature the expression $2\pi f$ is often replaced by ω, (*omega*). Velocity multiplied by time gives the distance travelled, so $2\pi ft$ (or ωt) represents the angular distance through which the rotating conductor or the rotating vector has travelled since the reference time $t = 0$. In the case of a sine wave the reference time $t = 0$ represents the instant when the voltage or the current, whichever is under discussion, also is equal to zero.

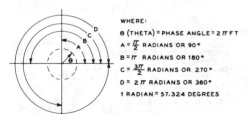

WHERE:

θ (THETA) = PHASE ANGLE = $2\pi FT$

$A = \frac{\pi}{2}$ RADIANS OR 90°

$B = \pi$ RADIANS OR 180°

$C = \frac{3\pi}{2}$ RADIANS OR 270°

$D = 2\pi$ RADIANS OR 360°

1 RADIAN = 57.324 DEGREES

Figure 5

ILLUSTRATING RADIAN NOTATION

The radian is a unit of phase angle, equal to 57.324 degrees. It is commonly used in mathematical relationships involving phase angles since such relationships are simplified when radian notation is used.

Instantaneous Value of Voltage or Current The instantaneous voltage or current is proportional to the sine of the angle through which the rotating vector has travelled since reference time $t = 0$. Thus, when the peak value of the ac wave amplitude (either voltage or current amplitude) is known, and the angle through which the rotating vector has travelled is established, the amplitude of the wave at this instant can be determined through use of the following expression:

$$e = E_{\text{max}} \sin 2\pi ft$$

where,

    e equals the instantaneous voltage,
    E_{max} equals maximum peak value of voltage,
    f equals frequency in hertz,
    t equals period of time which has elapsed since $t = 0$ (expressed as a fraction of one second).

It is often easier to visualize the process of determining the instantaneous amplitude by ignoring the frequency and considering only one cycle of the ac wave. In this case, for a sine wave, the expression becomes:

$$e = E_{\text{max}} \sin \theta$$

where θ represents the angle through which the vector has rotated since time (and amplitude) were zero. As examples:

$$\text{when } \theta = 60°$$

$$\sin \theta = 0.866$$

$$\text{and } e = 0.866 \, E_{\text{max}}$$

· · · · · · · · · · ·

$$\text{when } \theta = 1 \text{ radian}$$

$$\sin \theta = 0.8415$$

$$\text{and } e = 0.8415 \, E_{\text{max}}$$

Effective Value of an Alternating Current The instantaneous value of an alternating current or voltage varies continuously throughout the cycle, so some value of an ac wave must be chosen to establish a relationship between the effectiveness of an ac and a dc voltage or current. The heating value of an alternating current has been chosen to establish the reference between the *effective* values of ac and dc. Thus *an alternating current will have an effective value of 1 ampere when it produces the same heat in a resistor as does 1 ampere of direct current.*

The effective value is derived by taking the instantaneous values of current over a cycle of alternating current, squaring these values, taking an average of the squares, and then taking the square root of the average. By this procedure, the effective value becomes known as the *root mean square*, or rms, value. This is the value that is read on ac voltmeters and ac ammeters. The rms value is 70.7 percent of the peak or maximum instantaneous value (for sine waves only) and is expressed as follows:

$$E_{eff} \text{ or } E_{rms} = 0.707 \times E_{max}, \text{ or}$$

$$I_{eff} \text{ or } I_{rms} = 0.707 \times I_{max}$$

The following relations are extremely useful in radio and power work:

$$E_{rms} = 0.707 \times E_{max}, \text{ and}$$

$$E_{max} = 1.414 \times E_{rms}$$

Rectified Alternating Current or Pulsating Direct Current If an alternating current is passed through a rectifier, it emerges in the form of a current of *varying amplitude* which flows in *one* direction only. Such a current is known as *rectified ac or pulsating dc*. A typical wave form of a pulsating direct current as would be obtained from the output of a full-wave rectifier is shown in figure 6.

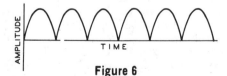

Figure 6

FULL-WAVE RECTIFIED SINE WAVE

Waveform obtained at the output of a full-wave rectifier being fed with a sine wave and having 100 percent rectification efficiency. Each pulse has the same shape as one-half cycle of a sine wave. This type of current is known as pulsating direct current.

Measuring instruments designed for dc operation will not read the peak or instantaneous maximum value of the pulsating dc output from the rectifier; they will read only the *average value*. This can be explained by assuming that it could be possible to cut off some of the peaks of the waves, using the cutoff portions to fill in the spaces that are open, thereby obtaining an *average* dc value. A milliammeter and voltmeter connected to the adjoining circuit, or across the output of the rectifier, will read this average value. It is related to *peak* value by the following expression:

$$E_{avg} = 0.636 \times E_{max}$$

It is thus seen that the average value is 63.6 percent of the peak value.

Relationship Between Peak, RMS, or Effective, and Average Values To summarize the three most significant values of an ac sine wave: the peak value is equal to 1.41 times the rms or effective, and the rms value is equal to 0.707 times the peak value; the average value of a full-wave rectified ac wave is 0.636 times the peak value, and the average value of a rectified wave is equal to 0.9 times the rms value.

rms	$= 0.707 \times$ peak
average	$= 0.636 \times$ peak
. . . .	
average	$= 0.9 \quad \times$ rms
rms	$= 1.11 \quad \times$ average
. . . .	
peak	$= 1.414 \times$ rms
. . . .	
peak	$= 1.57 \quad \times$ average

Applying Ohm's Law to Alternating Current Ohm's law applies equally to direct or alternating current, *provided* the circuits under consideration are purely resistive, that is, circuits which have neither inductance nor capacitance. When capacitive or inductive reactance is introduced in the circuit, Ohm's law still applies, but additional considerations are involved; these will be discussed in a later paragraph.

3-2   Reactive Circuits

As was stated in Chapter Two, when a changing current flows through an inductor a back- or counterelectromotive force is developed, opposing any change in the initial current. This property of an inductor causes it to offer opposition or *impedance* to a change in current. The measure of impedance offered by an inductor to an alternating current of a given frequency is known as its *inductive reactance*. This is expressed as X_L and is shown in figure 7.

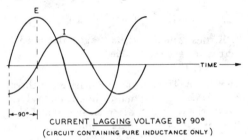

CURRENT <u>LAGGING</u> VOLTAGE BY 90°
(CIRCUIT CONTAINING PURE INDUCTANCE ONLY)

Figure 7

LAGGING PHASE ANGLE

Showing the manner in which the current lags the voltage in an ac circuit containing pure inductance only. The lag is equal to one quarter cycle or 90 degrees.

$$X_L = 2\pi f L$$

where,

X_L equals inductive reactance expressed in ohms,
π equals 3.14,
f equals frequency in Hertz,
L equals inductance in henrys.

Inductive Reactance It is often necessary to
at Radio Frequencies compute inductive reactance at radio frequencies. The same formula may be used, but to make it less cumbersome the inductance is expressed in *millihenrys* and the frequency in *kilohertz*. For higher frequencies and smaller values of inductance, frequency is expressed in *megahertz* and inductance in *microhenrys*. The basic equation need not be changed, since the multiplying factors for inductance and frequency appear in numerator and denominator, and are cancelled out. However, it is not possible in the same equation to express L in millihenrys and f in Hertz without conversion factors.

Capacitive   Inductive reactance is the mea-
Reactance   sure of the ability of an inductor
                to offer impedance to the flow
of an alternating current. Capacitors have a similar property although in this case the opposition is to any change in the voltage across the capacitor. This property is called *capacitive reactance* and is expressed as follows:

$$X_C = \frac{1}{2\pi f C}$$

where,

X_C equals capacitive reactance in ohms,
π equals 3.14,
f equals frequency in Hertz,
C equals capacitance in farads.

Capacitive Re-   Here again, as in the case
actance at   of inductive reactance,
Radio Frequencies   the units of capacitance
                    and frequency can be
converted into smaller units for practical problems encountered in radio work. The equation may be written:

$$X_C = \frac{10^6}{2\pi f C}$$

where,

f equals frequency in megahertz,
C equals capacitance in picofarads.

Phase   When an alternating current flows
            through a purely resistive circuit, it
will be found that the current will go through maximum and minimum in perfect step with the voltage. In this case the current is said to be in step, or *in phase* with the voltage. For this reason, Ohm's law will apply equally well for *ac* or *dc* where pure resistances are concerned, provided that the same values of the wave (either peak or rms) for both voltage and current are used in the calculations.

However, in calculations involving alternating currents the voltage and current are not necessarily in phase. The current through the circuit may lag behind the voltage, in which case the current is said to have *lagging* phase. Lagging phase is caused by inductive reactance. If the current reaches its maximum value ahead of the voltage (figure 8) the current is said to have a *leading* phase. A leading phase angle is caused by capacitive reactance.

In an electrical circuit containing reactance only, the current will either lead or lag the voltage by 90°. If the circuit contains inductive reactance only, the current will lag the voltage by 90°. If only capaci-

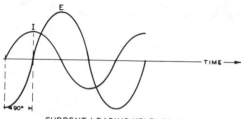

CURRENT <u>LEADING</u> VOLTAGE BY 90°
(CIRCUIT CONTAINING PURE CAPACITANCE ONLY)

Figure 8

LEADING PHASE ANGLE

Showing the manner in which the current leads the voltage in an ac circuit containing pure capacitance only. The lead is equal to one-quarter cycle or 90 degrees.

tive reactance is in the circuit, the current will lead the voltage by 90°.

Reactances in Combination Inductive and capacitive reactance have exactly opposite effects on the phase relation between current and voltage in a circuit and when they are used in combination their effects tend to neutralize each other. The combined effect of a capacitive and an inductive reactance is often called the *net reactance* of a circuit. The net reactance (X) is found by subtracting the capacitive reactance from the inductive reactance ($X = X_L - X_C$).

The result of such a combination of pure reactances may be either positive, in which case the positive reactance is greater so that the net reactance is inductive, or it may be negative in which case the capacitive reactance is greater so that the net reactance is capacitive. The net reactance may also be zero in which case the circuit is said to be *resonant*. The condition of resonance will be discussed in a later section. Note that inductive reactance is always taken as being positive while capacitive reactance is always taken as being negative.

TABLE 3. Quantities, Units, and Symbols

Symbol	Quantity	Unit	Abbreviation
f	Frequency	hertz	Hz
λ	Wavelength	meter	M
X_L	Inductive Reactance	ohm	Ω
X_C	Capacitive Reactance	ohm	Ω
Q	Figure of merit	$\dfrac{\text{reactance}}{\text{resistance}}$	—
z	Impedance	ohm	Ω

e =	instantaneous value of voltage
E_{max} =	peak value of voltage
i =	instantaneous value of current
I_{max} =	peak value of current
θ =	phase angle, expressed in **degrees**
E_{eff} **or** E_{rms} =	effective or rms value of voltage
I_{eff} **or** I_{rms} =	effective or rms value of current
j =	vector operator (90° rotation)

Impedance; Circuits Containing Reactance and Resistance Pure reactances introduce a phase angle of 90° between voltage and current; pure resistance introduces no phase shift between voltage and current. Therefore it is not correct to add a reactance and a resistance directly. When a reactance and a resistance are used in combination, the resulting phase angle of current flow with respect to the impressed voltage lies somewhere between plus or minus 90° and 0° depending on the relative magnitudes of the reactance and the resistance.

The term *impedance* is a general term which can be applied to any electrical entity which impedes the flow of current. The term may be used to designate a resistance, a pure reactance, or a complex combination of both reactance and resistance. The designation for impedance is Z. An impedance must be defined in such a manner that both its magnitude and its phase angle are established. The designation may be accomplished in either of two ways—one of which is convertible into the other by simple mathematical operations.

The j Operator The first method of designating an impedance is actually to specify both the resistive and the reactive component in the form $R + jX$. In this form R represents the resistive component in ohms and X represents the reactive component. The j merely means that the X component is reactive and thus cannot be added directly to the R component. Plus jX means that the reactance is positive or inductive, while if minus jX were given it would mean that the reactive component was negative or capacitive.

Figure 9 illustrates a vector $(+A)$ lying along the positive X-axis of the usual X-Y coordinate system. If this vector is multiplied by the quantity (-1), it becomes $(-A)$ and its position now lies along the X-axis in the negative direction. The *operator* (-1) has caused the vector to rotate through an angle of 180 degrees. Since (-1) is equal to $(\sqrt{-1} \times \sqrt{-1})$, the same result may be obtained by operating on the vector with the operator $(\sqrt{-1} \times \sqrt{-1})$. However if the vector is op-

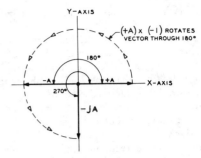

Figure 9

Operation on the vector (+A) by the quantity (−1) causes vector to rotate through 180 degrees.

erated on but once by the operator $(\sqrt{-1})$, it is caused to rotate only 90 degrees (figure 10). Thus the operator $(\sqrt{-1})$ rotates a vector by 90 degrees. For convenience, this operator is called the *j operator*. In like fashion, the operator $(-j)$ rotates the vector of figure 9 through an angle of 270 degrees, so that the resulting vector $(-jA)$ falls on the $(-Y)$ axis of the coordinate system.

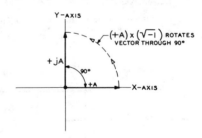

Figure 10

Operation on the vector (+A) by the quantity √−1, or j, causes vector to rotate through 90 degrees.

Polar Notation The second method of representing an impedance is to specify its absolute magnitude and the phase angle of current with respect to voltage, in the form $Z \angle \theta$. Figure 11 shows graphically the relationship between the two common ways of representing an impedance.

The construction of figure 11 is called an *impedance diagram*. Through the use of such a diagram we can add graphically a resistance and a reactance to obtain a value for the resulting impedance in the scalar form. With zero at the origin, resistances

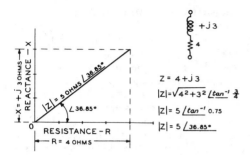

$$Z = 4 + j3$$
$$|Z| = \sqrt{4^2 + 3^2} \; \underline{/tan^{-1} \tfrac{3}{4}}$$
$$|Z| = 5 \; \underline{/tan^{-1} \, 0.75}$$
$$|Z| = 5 \; \underline{/36.85°}$$

Figure 11

THE IMPEDANCE TRIANGLE

Showing the graphical construction of a tri-angle for obtaining the net (scalar) impedance resulting from the connection of a resistance and a reactance in series. Shown also alongside is the alternative mathematical procedure for obtainting the values associated with the tri-angle.

are plotted to the right, positive values of reactance (inductive) in the upward direc-tion, and negative values of reactance (ca-pacitive) in the downward direction.

Note that the resistance and reactance are drawn as the two sides of a right triangle, with the hypotenuse representing the result-ing impedance. It is possible to determine mathematically the value of a resultant im-pedance through the familiar right-triangle relationship—*the square of the hypotenuse is equal to the sum of the squares of the other two sides:*

$$Z^2 = R^2 + X^2$$

or,

$$|Z| = \sqrt{R^2 + X^2}$$

Note also that the angle θ included between R and Z can be determined from any of the following trigonometric relationships:

$$\sin \theta = \frac{X}{|Z|}$$

$$\cos \theta = \frac{R}{|Z|}$$

$$\tan \theta = \frac{X}{R}$$

One common problem is that of determining the *scalar magnitude* of the impedance, $|Z|$, and the phase angle θ, when resistance and reactance are known; hence, of converting from the $Z = R + jX$ to the $|Z| \angle \theta$ form.

In this case two of the expressions just given can be used:

$$|Z| = \sqrt{R^2 + X^2}$$

$$\tan \theta = \frac{X}{R}, \; (\text{or } \theta = \tan^{-1}\frac{X}{R} \;)$$

The inverse problem, that of converting from the $|Z| \angle \theta$ to the $R + jX$ form is done with the following relationships, both of which are obtainable by simple division from the trigonometric expressions just given for determining the angle θ:

$$R = |Z| \cos \theta$$

$$jX = |Z| j \sin \theta$$

By simple addition these two expressions may be combined to give the relationship between the two most common methods of indicating an impedance:

$$R + jX = |Z| \, (\cos \theta + j \sin \theta)$$

In the case of impedance, resistance, or re-actance, the unit of measurement is the ohm; thus, the ohm may be thought of as a unit of *opposition to current flow*, with-out reference to the relative phase angle be-tween the applied voltage and the current which flows.

Further, since both capacitive and in-ductive reactance are functions of fre-quency, impedance will vary with fre-quency. Figure 12 shows the manner in which $|Z|$ will vary with frequency in an RL series circuit and in an RC series circuit.

Series RLC Circuits In a series circuit con-taining R, L, and C, the impedance is determined as discussed before except that the reactive component in the expressions defines the net reactance—that is, the difference between X_L and X_C. ($X_L - X_C$) may be substituted for X in the equations:

$$|Z| = \sqrt{R^2 + (X_L - X_C)^2}$$

$$\theta = \tan^{-1} \frac{(X_L - X_C)}{R}$$

A series RLC circuit thus may present an impedance which is capacitively reactive if the net reactance is capacitive, inductively

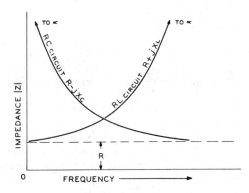

Figure 12

IMPEDANCE—FREQUENCY GRAPH FOR RL AND RC CIRCUITS

The impedance of an RC circuit approaches infinity as the frequency approaches zero (dc), while the impedance of a series RL circuit approaches infinity as the frequency approaches infiinity. The impedance of an RC circuit approaches the impedance of the series resistor as the frequency approaches infinity, while the impedance of a series RL circuit approaches the resistance as the frequency approaches zero.

reactive if the net reactance is inductive, or resistive if the capacitive and inductive reactances are equal.

Addition of Complex Quantities The addition of complex quantities (for example, impedances in series) is quite simple if the quantities are in the rectangular form. If they are in the polar form they only can be added graphically, unless they are converted to the rectangular form by the relationships previously given. As an example of the addition of complex quantities in the rectangular form, the equation for the addition impedance is:

$$(R_1 + jX_1) + (R_2 + jX_2) = (R_1 + R_2) + j(X_1 + X_2)$$

For example if we wish to add the impedances $(10 + j50)$ and $(20 - j30)$ we obtain:

$$(10 + j50) + (20 - j30)$$
$$= (10 + 20) + j[50 + (-30)]$$
$$= 30 + j(50 - 30)$$
$$= 30 + j20$$

Multiplication and Division of Complex Quantities It is often necessary in solving certain types of circuits to multiply or divide two complex quantities. It is a much simpler mathematical operation to multiply or divide complex quantities if they are expressed in the polar form. If the quantities are given in the rectangular form they should be converted to the polar form before multiplication or division is begun. Then the multiplication is accomplished by *multiplying* the $|Z|$ terms together and *adding* algebraically the $\angle \theta$ terms, as:

$$(|Z_1| \angle \theta_1) (|Z_2| \angle \theta_2) = |Z_1| |Z_2| (\angle \theta_1 + \angle \theta_2)$$

For example, suppose that the two impedances $|20| \angle 43°$ and $|32| \angle -23°$ are to be multiplied. Then:

$$(|20| \angle 43°) (|32| \angle -23°) = |20 \cdot 32| (\angle 43° + \angle -23°)$$
$$= 640 \angle 20°$$

Division is accomplished by *dividing* the denominator into the numerator, and *subtracting* the angle of the denominator from that of the numerator, as:

$$\frac{|Z_1| \angle \theta_1}{|Z_2| \angle \theta_2} = \frac{|Z_1|}{|Z_2|} (\angle \theta_1 - \angle \theta_2)$$

For example, suppose that an impedance of $|50| \angle 67°$ is to be divided by an impedance of $|10| \angle 45°$. Then:

$$\frac{|50| \angle 67°}{|10| \angle 45°} = \frac{|50|}{|10|} (\angle 67° - \angle 45°) = |5| (\angle 22°)$$

Ohm's Law for Complex Quantities The simple form of Ohm's law used for dc circuits may be stated in a more general form for application to ac circuits involving either complex quantities or simple resistive elements. The form is:

$$I = \frac{E}{Z}$$

in which, in the general case, I, E, and Z are complex (vector) quantities. In the simple case where the impedance is a pure resistance with an ac voltage applied, the equation simplifies to the familiar $I = E/R$. In any case the applied voltage may be expressed either as peak, rms, or average; the resulting current always will be in the same type used to define the voltage.

In the more general case vector algebra must be used to solve the equation. And, since either division or multiplication is involved, the complex quantities should be expressed in the polar form. As an example, take the case of the series circuit shown in figure 13 with 100 volts applied. The im-

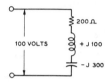

Figure 13

SERIES RLC CIRCUIT

pedance of the series circuit can best be obtained first in the rectangular form, as:

$$200 + j(100 - 300) = 200 - j200$$

Now, to obtain the current we must convert this impedance to the polar form.

$$|Z| = \sqrt{200^2 + (-200)^2}$$

$$= \sqrt{40,000 + 40,000}$$

$$= \sqrt{80,000}$$

$$= 282\ \Omega$$

$$\theta = \tan^{-1}\frac{X}{R} = \tan^{-1}\frac{-200}{200} = \tan^{-1}(-1)$$

$$= -45°$$

Therefore, $Z = 282 \angle -45°$

Note that in a series circuit the resulting impedance takes the sign of the largest reactance in the series combination.

Where a slide rule is being used to make the computations, the impedance may be found without any addition or subtraction operations by finding the angle θ first, and then using the trigonometric equation below for obtaining the impedance:

$$\theta = \tan^{-1}\frac{X}{R} = \tan^{-1}\frac{-200}{200} = \tan^{-1}(-1)$$

$$= -45°.$$

Then, Z equals $\dfrac{R}{\cos \theta}$

and $\cos -45° = 0.707$

$$|Z| = \frac{200}{0.707} = 282\ \text{ohms}$$

Since the applied voltage will be the reference for the currents and voltages within the circuit, it may be defined as having a zero phase angle: $E = 100 \angle 0°$. Then:

$$I = \frac{100 \angle 0°}{282 \angle -45°} = 0.354 \angle 0° - (-45°)$$

$$= 0.354 \angle 45°\ \text{amperes}$$

This same current must flow through all three elements of the circuit, since they are in series and the current through one must already have passed through the other two. The voltage drop across the resistor (whose phase angle of course is 0°) is:

$$E = IR$$

$$E = (0.354 \angle 45°)(200 \angle 0°)$$

$$= 70.8 \angle 45°\ \text{volts}$$

The voltage drop across the inductive reactance is:

$$E = IX_L$$

$$E = (0.354 \angle 45°)(100 \angle 90°)$$

$$= 35.4 \angle 135°\ \text{volts}$$

Similarly, the voltage drop across the capacitive reactance is:

$$E = IX_C$$

$$E = (0.354 \angle 45°)(300 \angle -90°)$$

$$= 106.2 \angle -45°$$

Note that the voltage drop across the capacitive reactance is greater than the supply voltage. This condition often occurs in a series RLC circuit, and is explained by the fact that the drop across the capacitive reactance is cancelled to a lesser or greater extent by the drop across the inductive reactance.

It is often desirable in a problem such as the above to check the validity of the answer by adding vectorially the voltage drops across the components of the series circuit to make sure that they add up to the supply voltage or (to use the terminology of Kirchhoff's Second Law) to make sure that the voltage drops across all elements of the circuit, including the source taken as negative, is equal to zero.

In the general case of the addition of a number of voltage vectors in series it is best to resolve the voltages into their in-phase and out-of-phase components with respect to the supply voltage. Then these components may be added directly:

$$E_R = 70.8 \angle 45°$$
$$= 70.8 (\cos 45° + j \sin 45°)$$
$$= 70.8 (0.707 + j0.707)$$
$$= 50 + j50$$

. . . .

$$E_L = 35.4 \angle 135°$$
$$= 35.4 (\cos 135° + j \sin 135°)$$
$$= 35.4 (-0.707 + j0.707)$$
$$= -25 + j25$$

. . . .

$$E_C = 106.2 \angle 45°$$
$$= 106.2 (\cos -45° + j \sin -45°)$$
$$= 106.2 (0.707 - j0.707)$$
$$= 75 - j75$$

. . . .

$$E_R + E_L + E_C = (50 + j50)$$
$$+ (-25 + j25) + (75 - j75)$$

$$= (50 - 25 + 75) +$$
$$j(50 + 25 - 75)$$
$$E_R + E_L + E_C = 100 + j0$$
$$= 100 \angle 0°,$$

which is equal to the supply voltage.

Checking by Construction on the Complex Plane It is frequently desirable to check computations involving complex quantities by constructing vectors representing the quantities on the complex plane. Figure 14 shows such a construction for the quantities of the problem just completed. Note that the answer to the problem may be checked by constructing a parallelogram with the voltage drop

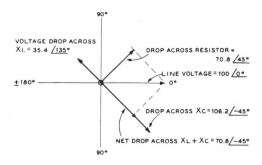

Figure 14

Graphical construction of the voltage drops associated with the series RLC circuit of figure 13.

across the resistor as one side and the net voltage drop across the capacitor plus the inductor (these may be added algebraically as they are 180° out of phase) as the adjacent side. The vector sum of these two voltages, which is represented by the diagonal of the parallelogram, is equal to the supply voltage of 100 volts at zero phase angle.

Resistance and Reactance in Parallel In a series circuit, such as just discussed, the current through all the elements which go to make up the series circuit is the same. But the voltage drops across each of the components are, in general, different from one another. Conversely, in a parallel RLC or RX circuit the voltage is, obviously, the same across each of the elements. But the currents through each of the elements are usually different.

There are many ways of solving a problem involving paralleled resistance and reactance; several of these ways will be described. In general, it may be said that the impedance of a number of elements in parallel is solved using the same relations as are used for solving resistors in parallel, except that complex quantities are employed. The basic relation is:

$$\frac{1}{Z_{\text{Total}}} = \frac{1}{Z_1} + \frac{1}{Z_2} + \frac{1}{Z_3} + \ldots .,$$

or when only two impedances are involved:

$$Z_{\text{Total}} = \frac{Z_1 Z_2}{Z_1 + Z_2}$$

As an example, using the two-impedance relation, take the simple case, illustrated in figure 15, of a resistance of 6 ohms in parallel with a capacitive reactance of 4 ohms. To simplify the first step in the computation it is best to put the impedances in the polar form for the numerator, since multiplication is involved, and in the rectangular form for the addition in the denominator.

$$Z_{Total} = \frac{(6 \angle 0°)\ (4 \angle -90°)}{6 - j4}$$

$$= \frac{24 \angle -90°}{6 - j4}$$

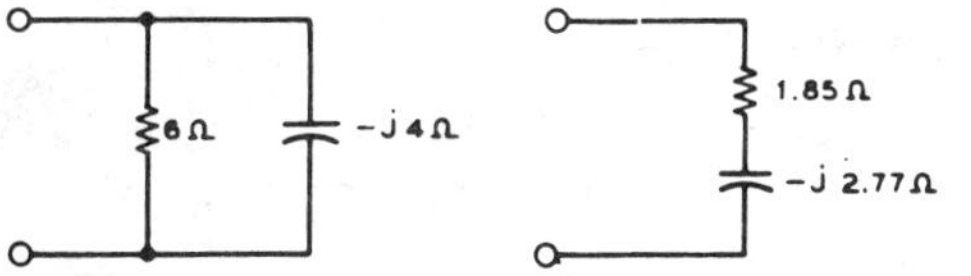

PARALLEL
CIRCUIT

EQUIVALENT SERIES
CIRCUIT

Figure 15

THE EQUIVALENT SERIES CIRCUIT

Showing a parallel RC circuit and the equivalent series RC circuit which represents the same net impedance as the parallel circuit.

Then the denominator is changed to the polar form for the division operation:

$$\theta = \tan^{-1} \frac{-4}{6} = \tan^{-1} -0.667 = -33.7°$$

$$|Z| = \frac{6}{\cos -33.7°} = \frac{6}{0.832} = 7.21 \text{ ohms}$$

$$6 - j4 = 7.21 \angle -33.7°$$

Then:

$$Z_{Total} = \frac{24 \angle -90°}{7.21 \angle -33.7°} = 3.33 \angle -56.3°$$

$$= 3.33 (\cos -56.3° + j \sin -56.3°)$$

$$= 3.33 [0.5548 + j (-0.832)]$$

$$= 1.85 - j 2.77$$

Equivalent Series Circuit Through the series of operations in the previous paragraph a circuit composed of two impedances in parallel has been converted into an *equivalent series circuit*

composed of impedances in series. An equivalent series circuit is one which, as far as the terminals are concerned, acts identically to the original parallel circuit; the current through the circuit and the power dissipation of the resistive elements are the same for a given voltage at the specified frequency.

The mathematical conversion from series to parallel equivalent and vice-versa is important in antenna and circuit studies, as certain test equipment makes one form of measurement and others make the opposite form. This conversion exercise may be required to compare the two types of data.

It is possible to check the equivalent series circuit of figure 15 with respect to the original circuit by assuming that one volt ac (at the frequency where the capacitive reactance in the parallel circuit is 4 ohms) is applied to the terminals of both the series and parallel circuits.

In the parallel circuit the current through the resistor will be $\frac{1}{6}$ ampere (0.166 amp) while the current through the capacitor will be $j\ \frac{1}{4}$ ampere ($+ j$ 0.25 amp). The total current will be the sum of these two currents, or $0.166 + j$ 0.25 amp. Adding these vectorially, as follows:

$$|I| = \sqrt{0.166^2 + 0.25^2} = \sqrt{0.09}$$
$$= 0.3 \text{ amp.}$$

The dissipation in the resistor will be $1^2/6$ = 0.166 watts.

In the case of the equivalent series circuit the current will be:

$$|I| = \frac{E}{|Z|} = \frac{1}{3.33} = 0.3 \text{ amp}$$

And the dissipation in the resistor will be:

$$W = I^2R = 0.3^2 \times 1.85$$

$$= 0.09 \times 1.85$$

$$= 0.166 \text{ watts}$$

Thus the equivalent series circuit checks exactly with the original parallel circuit.

Parallel RLC Circuits In solving a more complicated circuit made up of more than two impedances in parallel it is possible to use either of two methods of solution. These methods are called the *ad-*

mittance method and the *assumed-voltage* method. However, the two methods are equivalent since both use the sum-of-reciprocals equation:

$$\frac{1}{Z_{Total}} = \frac{1}{Z_1} + \frac{1}{Z_2} + \frac{1}{Z_3} \ldots$$

In the admittance method we use the relation $Y = 1/Z$, where $Y = G + jB$; Y is called the *admittance*, defined above, G is the *conductance* or R/Z^2 and B is the *susceptance* or $-X/Z^2$. Then $Y_{total} = 1/Z_{total} = Y_1 + Y_2 + Y_3 \ldots$ In the assumed-voltage method we multiply both sides of the equation above by E, the assumed voltage, and add the currents, as:

$$\frac{E}{Z_{Total}} = \frac{E}{Z_1} + \frac{E}{Z_2} + \frac{E}{Z_3} \ldots = I_{z_1} + I_{z_2} + I_{z_3} \ldots$$

Then the impedance of the parallel combination may be determined from the relation:

$$Z_{Total} = E/I_{Z\,Total}$$

AC Voltage Dividers   Voltage dividers for use with alternating current are quite similar to dc voltage dividers. However, since capacitors and inductors as well as resistors oppose the flow of ac current, voltage dividers for alternating voltages may take any of the configurations shown in figure 16.

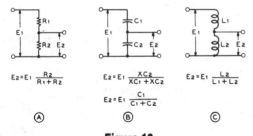

$$E_2 = E_1 \frac{R_2}{R_1 + R_2}$$

$$E_2 = E_1 \frac{X_{C2}}{X_{C1} + X_{C2}}$$

$$E_2 = E_1 \frac{C_1}{C_1 + C_2}$$

$$E_2 = E_1 \frac{L_2}{L_1 + L_2}$$

(A)   (B)   (C)

Figure 16

SIMPLE AC VOLTAGE DIVIDERS

Since the impedances within each divider are of the same type, the output voltage is in phase with the input voltage. By using combinations of different types of impedances, the phase angle of the output may be shifted in relation to the input phase angle

at the same time the amplitude is reduced. Several dividers of this type are shown in figure 17. Note that the ratio of output voltage is equal to the ratio of the output impedance to the total divider impedance. This relationship is true only if negligible current is drawn by a load on the output terminals.

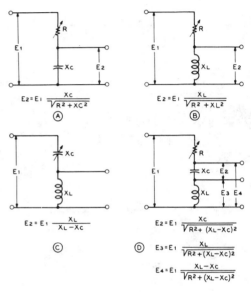

$$E_2 = E_1 \frac{X_C}{\sqrt{R^2 + X_C^2}}$$

(A)

$$E_2 = E_1 \frac{X_L}{\sqrt{R^2 + X_L^2}}$$

(B)

$$E_2 = E_1 \frac{X_L}{X_L - X_C}$$

(C)

$$E_3 = E_1 \frac{X_L}{\sqrt{R^2 + (X_L - X_C)^2}}$$

$$E_4 = E_1 \frac{X_L - X_C}{\sqrt{R^2 + (X_L - X_C)^2}}$$

(D)

Figure 17

COMPLEX AC VOLTAGE DIVIDERS

Reactive Power   In a reactive circuit composed of R, L, and C energy is stored in the electric field of the capacitor and in the magnetic field of the inductor. This stored energy is responsible for the fact that voltages across reactances in series can often be larger than the voltage applied to them, as illustrated in figure 14.

The power stored in a reactance is equal to I^2X, where X is the reactance value in ohms. This stored power is not lost, as is the power dissipated in resistance, but is power that is transferred back and forth between the circuit and the field of the reactance. The general formula for reactive power is:

$$P = E I \cos \theta$$

Where θ is as defined in polar notation (phase angle) and $\cos \theta$ is called the *power factor*.

3-3   Resonant Circuits

A series circuit such as shown in figure 18 is said to be in *resonance* when the applied frequency is such that the capacitive reactance is exactly balanced by the inductive reactance. At this frequency the two reactances will cancel, and the impedance of the circuit will be at a minimum so that maximum current will flow. The net impedance of a series circuit at resonance is equal to the resistance which remains in the circuit after the reactances have been cancelled.

Resonant Frequency   Some resistance is always present in a circuit because it is possessed in some degree by both the inductor and the capacitor. If the fre-

Figure 18

SERIES-RESONANT CIRCUIT

quency of the alternator E is varied from nearly zero to some high frequency, there will be one particular frequency at which the inductive reactance and capacitive reactance will be equal. This is known as the *resonant frequency*, and in a series circuit it is the frequency at which the circuit current will be a maximum. Such series-resonant circuits are chiefly used when it is desirable to allow a certain frequency to pass through the circuit (low impedance to this frequency), while at the same time the circuit is made to offer considerable opposition to currents of other frequencies.

If the values of inductance and capacitance both are fixed, there will be only one resonant frequency.

If either the inductance or the capacitance are made variable, the circuit may then be changed or *tuned*, so that a number of combinations of inductance and capacitance can resonate at the same frequency. This can be

more easily understood when one considers that inductive reactance and capacitive reactance change in opposite directions as the frequency is varied. For example, if the frequency were to remain constant and the values of inductance and capacitance were then changed, the following combinations would have equal *reactance*:

L	X_L	C	X_C
.265	100	26.5	100
2.65	1000	2.65	1000
26.5	10,000	.265	10,000
265.00	100,000	.0265	100,000
2,650.00	1,000,000	.00265	1,000,000

Frequency is constant at 60 Hz.

L is expressed in henrys.

C is expressed in microfarads (10^{-6} farad).

Frequency   From the formula for reso-
of Resonance   nance ($2\pi fL = 1/2\pi fC$) the resonant frequency is determined by use of the following equation:

$$f = \frac{1}{2\pi \sqrt{LC}}$$

where,

f equals frequency in hertz,
L equals inductance in henrys,
C equals capacitance in farads.

It is more convenient to express *L* and *C* in smaller units for radio communication work, as shown in the following equation:

$$f = \frac{10^6}{2\pi \sqrt{LC}}$$

where,

f equals frequency in kilohertz (kHz),
L equals inductance in microhenries (μH),
C equals capacitance in picofarads (pF).

Impedance of Series   The impedance across
Resonant Circuits   the terminals of a series-resonant circuit (figure 18) is:

$$Z = \sqrt{r^2 + (X_L - X_C)^2}$$

where,

Z equals impedance in ohms,
r equals resistance in ohms,
X_C equals capacitive reactance in ohms,
X_L equals inductive reactance in ohms.

From this equation, it can be seen that the impedance is equal to the vector sum of the circuit resistance and the *difference* be-

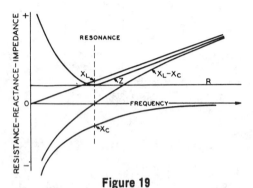

Figure 19

**IMPEDANCE OF A
SERIES-RESONANT CIRCUIT**

Showing the variation in reactance of the separate elements and in the net impedance of a series resonant circuit (such as figure 18) with changing frequency. The vertical line is drawn at the point of resonance ($X_L - X_C = 0$) in the series circuit.

tween the two reactances. Since at the resonant frequency X_L equals X_C, the difference between them (figure 19) is zero, so that at resonance the impedance is simply equal to the resistance of the circuit; therefore, because the resistance of most normal radio-frequency circuits is of a very low order, the impedance is also low.

At frequencies higher and lower than the resonant frequency, the difference between the reactances will be a definite quantity and will add with the resistance to make the impedance higher and higher as the circuit is tuned off the resonant frequency.

If X_C should be greater than X_L, then the term ($X_L - X_C$) will give a negative number. However, when the difference is squared the product is always positive. This means that the smaller reactance is subtracted from the larger, regardless of whether it be capacitive or inductive, and the difference is squared.

Current and Voltage in Series-Resonant Circuits Formulas for calculating currents and voltages in a series-resonant circuit are similar to those of Ohm's Law.

$$I = \frac{E}{Z} \quad E = IZ$$

The complete equations are:

$$I = \frac{E}{\sqrt{r^2 + (X_L - X_C)^2}}$$

$$E = I \sqrt{r^2 + (X_L - X_C)^2}$$

Inspection of the above formulas will show the following to apply to series-resonant circuits: When the impedance is low, the current will be high; conversely, when the impedance is high, the current will be low.

Since the impedance is very low at the resonant frequency, it follows that the current will be a maximum at this point. If a graph is plotted of the current versus the frequency either side of resonance, the resultant curve is known as a *resonance curve*. Such a curve is shown in figure 20, the

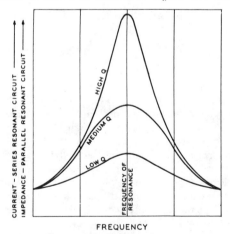

FREQUENCY

Figure 20

RESONANCE CURVE

Showing the increase in impedance at resonance for a parallel-resonant circuit, and similarly, the increase in current at resonance for a series-resonant circuit. The sharpness of resonance is determined by the Q of the circuit, as illustrated by a comparison between the three curves.

frequency being plotted against *current* in the series-resonant circuit.

Several factors will have an effect on the shape of this resonance curve, of which resistance and *L-to-C* ratio are the important considerations. The lower curves in figure 20 show the effect of adding increasing values of resistance to the circuit. It will be seen that the peaks become less and less prominent as the resistance is increased; thus, it can be said that the *selectivity* of the circuit is thereby *decreased*. Selectivity in this case can be defined as the ability of a circuit to discriminate against frequencies adjacent to (both above and below) the resonant frequency.

Voltage Across Coil and Capacitor in Series Circuit Because the ac or r-f voltage across a coil and capacitor is proportional to the reactance (for a given current), the actual voltages across the coil and across the capacitor may be many times greater than the *terminal* voltage of the circuit. At resonance, the voltage across the coil (or the capacitor) is Q times the applied voltage. Since the Q (or *merit factor*) of a series circuit can be in the neighborhood of 100 or more, the voltage across the capacitor, for example, may be high enough to cause flashover, even though the applied voltage is of a value considerably below that at which the capacitor is rated.

Circuit Q — Sharpness of Resonance An extremely important property of a capacitor or an inductor is its *factor-of-merit*, more generally called its Q. It is this factor, Q, which primarily determines the sharpness of resonance of a tuned circuit. This factor can be expressed as the ratio of the reactance to the resistance, as follows:

$$Q = \frac{2\pi fL}{R}$$

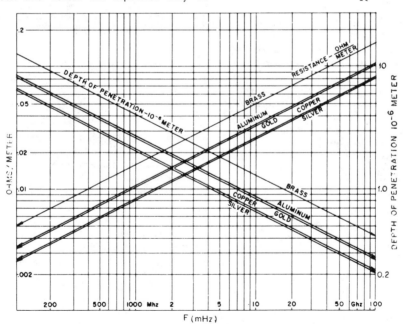

Figure 21

SKIN EFFECT, OR DEPTH OF PENETRATION IN METAL

The resistance and r-f loss in a conductor increase with the square root of frequency because the layer in which current flows decreases in thickness as the frequency increases.

where,

R equals total resistance.

Skin Effect   The actual resistance in a wire or an inductor can be far greater than the dc value when the coil is used in a radio-frequency circuit; this is because the current does not travel through the entire cross section of the conductor, but has a tendency to travel closer and closer to the surface of the wire as the frequency is increased. This is known as the *skin effect*.

In the hf region, skin effect limits the depth of electron flow in a conductor to a few thousandths of an inch. The resistance and r-f losses in a conductor increase with the square root of the frequency and become of increasing importance above 100 MHz (figure 21).

Variation of Q   Examination of the equation
with Frequency   for determining Q might seem to imply that even though the resistance of an inductor increases with frequency, the inductive reactance does likewise, so that the Q might be a constant. Actually, however, it works out in practice that the Q of an inductor will reach a relatively broad maximum at some particular frequency. Thus, coils normally are designed in such a manner that the peak in their curve of Q-versus-frequency will occur at the normal operating frequency of the coil in the circuit for which it is designed.

The Q of a capacitor ordinarily is much higher than that of the best coil. Therefore, it usually is the merit of the coil that limits the overall Q of the circuit.

At audio frequencies the core losses in an iron-core inductor greatly reduce the Q from the value that would be obtained simply by dividing the reactance by the resistance. Obviously the core losses also represent circuit resistance, just as though the loss occurred in the wire itself.

Parallel   In radio circuits, parallel reso-
Resonance   nance (more correctly termed *antiresonance*) is more frequently encountered than series resonance; in fact, it is the basic foundation of receiver and

transmitter circuit operation. A circuit is shown in figure 22.

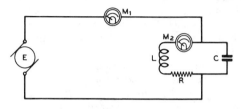

Figure 22

PARALLEL-RESONANT CIRCUIT

The inductance L and capacitance C comprise the reactive elements of the parallel-resonant (antiresonant) tank circuit, and the resistance R indicates the sum of the r-f resistance of the coil and capacitor, plus the resistance coupled into the circuit from the external load. In most cases the tuning capacitor has much lower r-f resistance than the coil and can therefore be ignored in comparison with the coil resistance and the coupled-in resistance. The instrument M_1 indicates the "line current" which keeps the circuit in a state of oscillation—this current is the same as the fundamental component of the plate current of a class-C amplifier which might be feeding the tank circuit. The instrument M_2 indicates the "tank current" which is equal to the line current multiplied by the operating Q of the tank circuit.

The "Tank"   In this circuit, as contrasted
Circuit   with a circuit for series resonance, L (inductance) and C (capacitance) are connected in *parallel*, yet the *combination* can be considered to be in series with the remainder of the circuit. This combination of L and C, in conjunction with R, the resistance which is principally included in L, is sometimes called a *tank circuit* because it effectively functions as a storage tank when incorporated in electronic circuits.

Contrasted with series resonance, there are two kinds of current which must be considered in a parallel-resonant circuit: (1) the *line current*, as read on the indicating meter M_1, (2) the *circulating current* which flows within the parallel LCR portion of the circuit.

At the resonant frequency, the line current (as read on the meter M_1) will drop to a very low value although the circulating current in the LC circuit may be quite large. The parallel-resonant circuit acts in a distinctly opposite manner to that of a series-resonant circuit, in which the current is at a maximum and the impedance is minimum at resonance. It is for this reason that in a parallel-resonant circuit the principal consideration is one of impedance rather than current. It is also significant that the *impedance* curve for *parallel* circuits is very nearly identical to that of the *current* curve for *series* resonance. The impedance at resonance is expressed as:

$$Z = \frac{(2\pi fL)^2}{R}$$

where,

Z equals impedance in ohms,
L equals inductance in henrys,
f equals frequency in hertz,
R equals resistance in ohms.

Or, impedance can be expressed as a function of Q as:

$$Z = 2\pi fLQ$$

showing that the impedance of a circuit is directly proportional to its effective Q at resonance.

The curves illustrated in figure 20 can be applied to parallel resonance. Reference to the curve will show that the effect of adding resistance to the circuit will result in both a broadening out and lowering of the peak of the curve. Since the voltage of the circuit is directly proportional to the impedance, and since it is this voltage that is applied to a detector or amplifier circuit, the impedance curve must have a sharp peak in order for the circuit to be *selective*. If the curve is broadtopped in shape, both the desired signal and the interfering signals at close proximity to resonance will give nearly equal voltages, and the circuit will then be *nonselective*; that is, it will tune broadly.

Effect of L/C Ratio in Parallel Circuits In order that the highest possible voltage can be developed across a parallel-resonant circuit, the impedance of this circuit must be very high. The impedance will be greater with conventional coils of limited Q when the ratio of inductance to capacitance is great, that is, when L is large as compared with C. When the resistance of the circuit is very low, X_L will equal X_C at

maximum impedance. There are innumerable ratios of L and C that will have *equal* reactance, at a given resonant frequency, exactly as in the case in a series-resonant circuit.

In practice, where a certain value of inductance is tuned by a variable capacitance over a fairly wide range in frequency, the L/C ratio will be small at the lowest-frequency end and large at the high-frequency end. The circuit, therefore, will have unequal gain and selectivity at the two ends of the band of frequencies which is being tuned. Increasing the Q of the circuit (lowering the resistance) will obviously increase *both* the selectivity and gain.

Circulating Tank Current at Resonance The Q of a circuit has a definite bearing on the circulating tank current at resonance. This tank current is very nearly the value of the line current multiplied by the effective circuit Q. For example: an r-f line current of 0.050 ampere, with a circuit Q of 100, will give a circulating tank current of approximately 5 amperes. From this it can be seen that both the inductor and the connecting wires in a circuit with a high Q must be of very low resistance, particularly in the case of high-power transmitters, if heat losses are to be held to a minimum.

Because the voltage across the tank at resonance is determined by the Q, it is possible to develop very high peak voltages across a high-Q tank with but little line current.

3-4 Coupled Circuits

Two circuits are said to be *coupled* if they have a common impedance through which the current flowing in one circuit affects the current in the second circuit. The common impedance may be resistive, capacitive, or inductive or a combination of these. The coefficient of coupling (k) was briefly discussed in Chapter Two. Coupled circuits are useful as unwanted frequencies may easily be rejected and power efficiently transferred from a primary circuit to a secondary circuit by

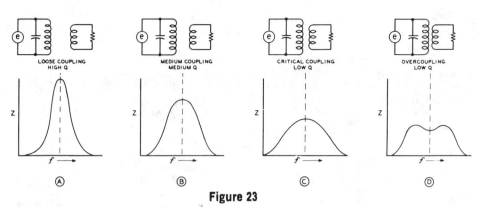

Figure 23

EFFECT OF COUPLING ON CIRCUIT IMPEDANCE AND Q

means of the proper degree and type of coupling.

In the case of simple inductive coupling of a parallel-resonant circuit (figure 23), such as an antenna coupler, the impedance and effective Q of the parallel-tuned circuit is decreased as the coupling becomes tighter. This effect is the same as if an actual resistance were added in series with the parallel-tuned circuit. The resistance thus coupled into the parallel circuit can be considered as being *reflected* from the output circuit to the input, or driver, circuit.

The behavior of *coupled circuits* depends largely on the amount of coupling, as shown in figure 23. The coupled current in the secondary circuit is small, varying with frequency, being maximum at the resonant frequency of the circuit. As the coupling is increased between the two circuits, the secondary resonance curve becomes broader and the resonant amplitude increases, until the reflected resistance is equal to the primary resistance. This point is called the *critical coupling point*. With greater coupling, the secondary resonance curve becomes broader and develops double resonance humps, which become more pronounced and farther apart in frequency as the coupling between the two circuits is increased.

In the case where both primary and secondary circuits are tuned, the configuration can be given wideband characteristics by shunting the primary and secondary resonant circuits by resistances as shown in figure 24 in order to achieve low values of circuit Q and by adjusting the coefficient of coupling.

If the primary and secondary circuits are identical, with equal shunt resistances, so that Q valves are equal and the coupling is critical, the half-power bandwidth is:

$$\text{Half-power bandwidth} = \sqrt{2} \text{ half-power bandwidth of a single-tuned stage}$$

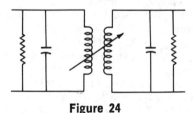

Figure 24

WIDEBAND COUPLED CIRCUIT

Bandwidths determined by shunting resistances and coefficient of coupling.

Compared with a single-tuned stage having the same half-power bandwidth the double-tuned circuit has a response that is flatter near the resonant frequency and has steeper sides, as shown in figure 25.

The bandwidth is defined as that frequency band over which the power response does not drop to less than one-half, or — 3 dB, of the power response at resonance. This corresponds to the voltage response dropping to 70.7 percent of its maximum value. This is commonly referred to as the "3-dB bandwidth" figure.

Wideband tuned amplifiers of this type are useful for passing signals modulated by short pulses or by television video signals and are characterized by the ability to repro-

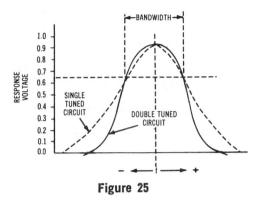

Figure 25

IMPROVED RESPONSE OF DOUBLE-TUNED CIRCUIT

Bandwidth is measured between 0.707 voltage response points.

duce abrupt changes in amplitude of the passing signal.

The coupled circuit can act as an impedance matching device, depending on the coupling between the circuits and the degree of secondary loading of the circuit.

Instead of magnetic coupling, two resonant circuits may be coupled through a common circuit element as shown in figure 26. The degree of coupling is a function of the common element, which may be adjusted to provide a *bandpass* circuit, suitable for wideband applications.

A form of inductive coupling is *link coupling* where two circuits are coupled by means of small linking coils (figure 27). The degree of coupling is adjusted by alter-

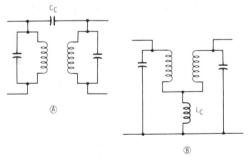

Figure 26

COUPLING THROUGH COMMON CIRCUIT ELEMENT

A—Capacitive coupling through C_C
B—Inductive coupling through L_C

ing the position of the coils with respect to the resonant circuits. Additional data on link-coupled circuits is given in chapter 11.

Impedance and Resonance in Antenna Systems The preceding discussion has been limited to the study of lumped circuits; that is, circuits containing discrete elements of resistance, inductance, and capacitance arranged in series or parallel configuration. An antenna, on the other hand, has distributed quantities of resistance, inductance, and capacitance throughout the length of the radiator. For the sake of study and computation, the distributed values are commonly considered to be lumped into discrete components and electrically equivalent circuits for a given antenna can be expressed and manipulated in terms of the equivalent lumped constants.

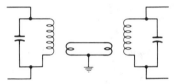

Figure 27

LINK COUPLING

Coupling is adjusted by altering position of link coils

For example, a dipole operating near the first resonant frequency bears an electrical resemblance to a series lumped circuit. Below resonance, the antenna may be defined in terms of a series RC circuit, at resonance in terms of a series-resonant circuit, and above resonance in terms of a series RL circuit. The plot of figure 19, in fact, may be compared to the characteristics of a dipole at near resonance. Transmission lines, in addition, may be expressed in terms of lumped constants for convenience, and some of the more important electrical characteristics of antennas and transmission lines are discussed in later chapters of this handbook.

3-5 Nonsinusoidal Waves and Transients

Pure sine waves, discussed previously, are basic wave shapes. Waves of many different and complex shapes are used in electronics, particularly square waves, sawtooth waves, and peaked waves.

Wave Composition Any periodic wave (one that repeats itself in definite time intervals) is composed of sine waves of different frequencies and amplitudes, added together. The sine wave which has the same frequency as the complex, periodic wave is called the *fundamental*. The frequencies higher than the fundamental are called *harmonics,* and are always a whole number of times higher than the fundamental. For example, the frequency twice as high as the fundamental is called the *second harmonic.*

The Square Wave Figure 28 compares a square wave with a sine wave (A) of the same frequency. If another sine wave (B) of smaller amplitude, but three times the frequency of A, called the third harmonic, is added to A, the resultant

wave (C) more nearly approaches the desired square wave.

This resultant curve (figure 29) is added to a fifth-harmonic curve (D), and the sides of the resulting curve (E) are steeper than before. This new curve is shown in figure 30 after a 7th-harmonic component has been added to it, making the sides of the composite wave even steeper. Addition of more higher odd harmonics will bring the resultant wave nearer and nearer to the desired square-wave shape. The square wave will be achieved if an infinite number of odd harmonics are added to the original sine wave.

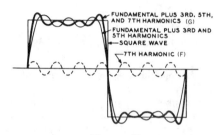

Figure 30

RESULTANT WAVE, COMPOSED OF FUNDAMENTAL, THIRD, FIFTH, AND SEVENTH HARMONICS

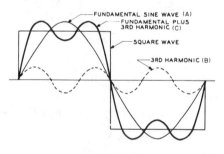

Figure 28

**COMPOSITE WAVE—
FUNDAMENTAL
PLUS THIRD HARMONIC**

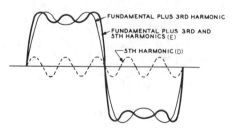

Figure 29

**THIRD-HARMONIC WAVE PLUS
FIFTH HARMONIC**

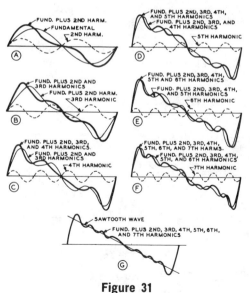

Figure 31

**COMPOSITION OF A SAWTOOTH
WAVE**

Irregular Waveforms In the same fashion, a *sawtooth wave* is made up of different sine waves (figure 31). The addition of all harmonics, odd and even, produces the sawtooth waveform.

Figure 32 shows the composition of a *peaked wave*. Note how the addititon of each successive harmonic makes the peak of the resultant higher, and the sides steeper.

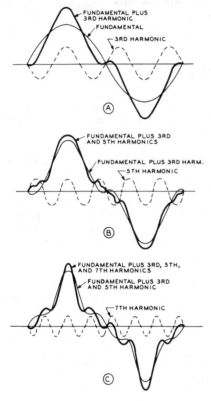

Figure 32

COMPOSITION OF A PEAKED WAVE

The three preceding examples show how a complex periodic wave is composed of a fundamental wave and different harmonics. The shape of the resultant wave depends on the harmonics that are added, their relative amplitudes, and relative phase relationships. In general, the steeper the sides of the waveform, the more harmonics it contains.

AC Transient Circuits If an ac voltage is substituted for the dc input voltage in the *RC* transient circuits discussed in Chapter 2, the same principles may be applied in the analysis of the transient behavior. An *RC* coupling circuit is designed to have a long time constant with respect to the lowest frequency it must pass. Such a circuit is shown in figure 33. If a nonsinusoidal voltage is to be passed unchanged through the coupling circuit, the time constant must be long with respect to the period of the lowest frequency contained in the voltage wave.

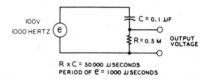

Figure 33

RC COUPLING CIRCUIT WITH LONG TIME CONSTANT

RC Differentiator and Integrator An *RC* voltage divider that is designed to distort the input waveform is known as a *differentiator* or *integrator*, depending on the locations of the output taps. The output from a differentiator is taken across the resistance, while the output from an integrator is taken across the capacitor. Such circuits will change the shape of any

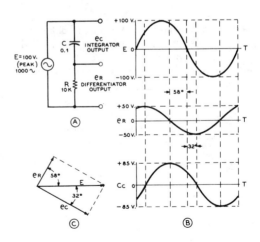

Figure 34

RC DIFFERENTIATOR AND INTEGRATOR ACTION ON A SINE WAVE

complex ac waveform that is impressed on them. This distortion is a function of the value of the time constant of the circuit as compared to the period of the waveform. Neither a differentiator nor an integrator can change the shape of a pure sine wave, they will merely shift the phase of the wave (figure 34). The differentiator output is a sine wave leading the input wave, and the integrator output is a sine wave which lags the input wave. The sum of the two outputs at any instant equals the instantaneous input voltage.

Square-Wave Input   If a square-wave voltage is impressed on the circuit of figure 35, a square-wave voltage output may be obtained across the integrating capacitor if the time constant of the circuit allows the capacitor to become fully charged. In this particular case, the capacitor never fully charges, and as a result the output of the integrator has a smaller amplitude than the input. The differentiator output has a maximum value greater than the input amplitude, since the voltage left on the capacitor from the previous half wave will add to the input voltage. Such a circuit, when used as a differentiator, is often called a *peaker*. Peaks of twice the input amplitude may be produced.

Sawtooth-Wave Input   If a back-to-back sawtooth voltage is applied to an *RC* circuit having a time constant one-sixth the period of the input voltage,

the result is shown in figure 36. The capacitor voltage will closely follow the input voltage, if the time constant is short, and the integrator output closely resembles the input. The amplitude is slightly reduced and there is a slight phase lag. Since the voltage across the capacitor is increasing at a constant rate, the charging and discharging current is constant. The output voltage of the differentiator, therefore, is constant during each half of the sawtooth input.

Miscellaneous   Various voltage waveforms
Inputs   other than those represented here may be applied to short-time-constant *RC* circuits for the purpose of producing across the resistor an output voltage with an amplitude *proportional to the rate of change* of the input signal. The shorter the *RC* time constant is made with respect to the period of the input wave, the more nearly the voltage across the capacitor conforms to the input voltage. Thus, the differentiator output becomes of particular importance in very short-time-constant *RC* circuits. Differentiator outputs for various types of input waves are shown in figure 37.

Square-Wave Test   The application of a
for Audio Equipment square-wave input signal to audio equipment, and the observation of the reproduced output signal on an oscilloscope will provide a quick and accurate check of the overall operation of audio equipment.

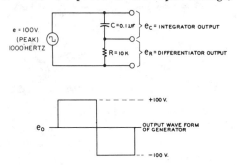

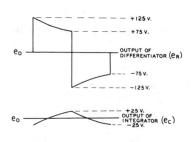

Figure 35

**RC DIFFERENTIATOR AND INTEGRATOR ACTION ON
A SQUARE WAVE**

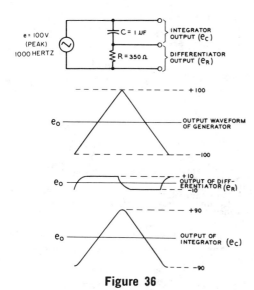

Figure 36

**RC DIFFERENTIATOR AND
INTEGRATOR ACTION ON
A SAWTOOTH WAVE**

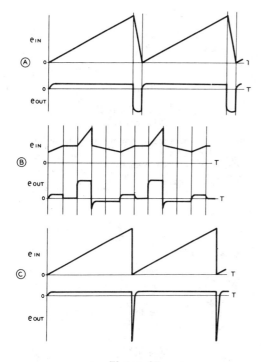

Figure 37

Differentiator outputs of short-time-constant RC
circuits for various input voltage wave-shapes.
The output voltage is proportional to the rate
of change of the input voltage.

Low-frequency and high-frequency response, as well as transient response can be examined easily.

If the amplifier is deficient in low-frequency response, the flat top of the square wave will be canted, as in figure 38. If the high-frequency response is inferior, the rise time of the output wave will be retarded (figure 39).

An amplifier with a limited high- and low-frequency response will turn the square wave into the approximation of a sawtooth wave (figure 40).

3-6   Transformers

When two coils are placed in such inductive relation to each other that the lines of force from one cut across the turns of the other inducing a current, the combination can be called a *transformer*. The name is derived from the fact that energy is transformed from one winding to another. The inductance in which the original flux is produced is called the *primary;* the inductance which *receives* the induced current is called the *secondary*. In a radio-receiver power transformer, for example, the coil through which the 120-volt ac passes is the *primary,* and the coil from which a higher or lower voltage than the ac line potential is obtained is the *secondary*.

Transformers can have either air or magnetic cores, depending on the frequencies at which they are to be operated. The reader should thoroughly impress on his mind the fact that current can be transferred from one circuit to another *only* if the primary current is changing or alternating. From this it can be seen that a power transformer cannot possibly function as such when the primary is supplied with nonpulsating dc.

A power transformer usually has a magnetic core which consists of laminations of iron, built up into a square or rectangular form, with a center opening or window. The secondary windings may be several in number, each perhaps delivering a different voltage. The secondary voltages will be proportional to the turns ratio and the primary voltage.

Transformers are used in alternating-current circuits to transfer power at one volt-

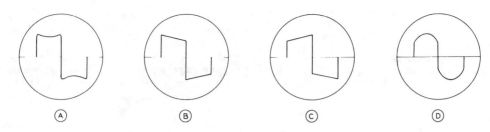

Figure 38

Amplifier deficient in low-frequency response will distort square wave applied to the input circuit, as shown. A 60-Hz square wave may be used.
A: Drop in gain at low frequencies
B: Leading phase shift at low frequencies
C: Lagging phase shift at low frequencies
D: Accentuated low-frequency gain

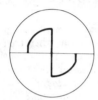

Figure 39

Output waveshape of amplifier having deficiency in high-frequency response. Tested with 10-kHz square wave.

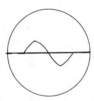

Figure 40

Output waveshape of amplifier having limited low-frequency and high-frequency response. Tested with 1 kHz square wave.

age and impedance to another circuit at another voltage and impedance. There are three main classifications of transformers: those made for use in power-frequency circuits, those made for audio-frequency applications, and those made for radio frequencies.

The Transformation Ratio   In a perfect transformer all the magnetic flux lines produced by the primary winding link crosses every turn of the secondary winding (figure 41). For such a transformer, the ratio of the primary and secondary voltages is the same as the ratio of the number of turns in the two windings:

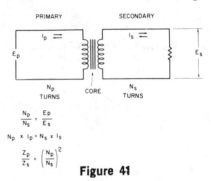

$$\frac{N_p}{N_s} = \frac{E_p}{E_s}$$

$$N_p \times I_p = N_s \times I_s$$

$$\frac{Z_p}{Z_s} = \left(\frac{N_p}{N_s}\right)^2$$

Figure 41

THE LOW-FREQUENCY TRANSFORMER

Power is transformed from the primary to the secondary winding by means of the varying magnetic field. The voltage induced in the secondary for a given primary voltage is proportional to the ratio of secondary to primary turns. The impedance transformation is proportional to the square of the primary to secondary turns ratio.

$$\frac{N_\mathrm{P}}{N_\mathrm{S}} = \frac{E_\mathrm{P}}{E_\mathrm{S}}$$

where,

N_P equals number of turns in the primary,
N_S equals number of turns in the secondary,
E_P equals voltage across the primary,
E_S equals voltage across the secondary.

In practice, the transformation ratio of a transformer is somewhat less than the turns ratio, since unity coupling does not exist between the primary and secondary windings.

Ampere Turns (NI) The current that flows in the secondary winding as a result of the induced voltage must produce a flux which exactly equals the primary flux. The magnetizing force of a coil is expressed as the product of the number of turns in the coil times the current flowing in it:

$$N_P \times I_P = N_S \times I_S, \text{ or } \frac{N_P}{N_S} = \frac{I_S}{I_P}$$

where,

I_P equals primary current,
I_S equals secondary current.

It can be seen from this expression that when the voltage is stepped up, the current is stepped down, and vice versa.

Leakage Reactance Since unity coupling does not exist in a practical transformer, part of the flux passing from the primary circuit to the secondary circuit follows a magnetic circuit acted on by the primary only. The same is true of the secondary flux. These leakage fluxes cause *leakage reactance* in the transformer, and tend to cause the transformer to have poor voltage regulation. To reduce such leakage reactance, the primary and secondary windings should be in close proximity to each other. The more expensive transformers have interleaved windings to reduce inherent leakage reactance.

Impedance Transformation In the ideal transformer, the impedance of the secondary load is reflected back into the primary winding in the following relationship:

$$Z_P = N^2 Z_S, \text{ or } N = \sqrt{Z_P / Z_S}$$

where,

Z_P equals reflected primary impedance,
N equals turns ratio of transformer,
Z_S equals impedance of secondary load.

Thus any specific load connected to the secondary terminals of the transformer will be transformed to a different specific value appearing across the primary terminals of the transformer. By the proper choice of turns ratio, any reasonable value of secondary load impedance may be "reflected" into the primary winding of the transformer to produce the desired transformer primary impedance. The phase angle of the primary "reflected" impedance will be the same as the phase angle of the load impedance. A capacitive secondary load will be presented to the transformer source as a capacitance, a resistive load will present a resistive "reflection" to the primary source. Thus the primary source "sees" a transformer load entirely dependent on the secondary load impedance and the turns ratio of the transformer (figure 41).

The Auto-transformer The type of transformer in figure 42, when wound with heavy wire over an iron core, is a common device in primary power circuits for the purpose of increasing or decreasing the line voltage. In effect, it is merely a continuous winding with taps taken at various points along the winding, the input voltage being applied to the bottom and also to one tap on the winding. If the output is taken from this same tap, the voltage ratio will be 1 to 1; i.e., the input voltage will be the same as the output voltage. On the other hand, if the output tap is moved down toward the common terminal, there will be a stepdown in the turns ratio with a consequent stepdown in voltage. The initial setting of the middle input tap is chosen so that the number of turns will have sufficient

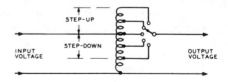

Figure 42

THE AUTOTRANSFORMER

Schematic diagram of an autotransformer showing the method of connecting it to the line and to the load. When only a small amount of step up or step down is required, the autotransformer may be much smaller physically than would be a transformer with a separate secondary winding. Continuously variable autotransformers (Variac and Powerstat) are widely used commercially.

reactance to hold the no-load primary current to a reasonable value.

Leakage Reactance In a practical transformer, *leakage reactance* is caused by the small amount of magnetic flux which is not common to both windings. The more closely the transformer windings are interleaved, the less leakage reactance there is, up to the limits imposed by the layout and design of an individual transformer.

Current flowing through the leakage reactance causes a voltage drop in the transformer which degrades the voltage regulation of the unit and results in a lower secondary voltage under load than would be indicated by the turns ratio of the transformer. In some transformer designs, the leakage reactance may be controlled by a small amount of dc power flowing in a special winding permitting adjustment of the secondary voltage.

3-7   The Toroid Coil

The *toroid coil* is doughnut shaped, with the winding covering the entire core. Air-core toroid coils were used in some of the broadcast receivers of the mid-thirties. Modern toroid coils have a ferrite or powdered-iron core to provide high permeability and good Q-versus-frequency characteristics. In addition, the toroid configuration provides self-shielding as the magnetic field is contained within the coil. This permits greater component density and virtually eliminates the need of external shielding around the coil.

Ferromagnetic core material is useful from the low audio frequency range up through several hundred megahertz and it has a large permeability range. Power levels up to 50 kW have been achieved in transformers and inductors using this core material.

The limitations of ferromagnetic material are that it is temperature sensitive and expands and contracts with temperature changes. Since the distributed capacitance of a winding changes rapidly with temperature, it must be kept to a minimum to avoid changing the inductance. In addition, core permeability changes with temperature.

Iron-Powder Toroidal Cores *Iron-powder toroidal cores* are available in numerous sizes ranging from 0.05" to more than 5" in outside diameter. There are two basic material groups: carbonyl irons and hydrogen-reduced irons. The carbonyl irons are used for their temperature stability. Their permeability (μ) range is from under 3μ to 35μ and they are suitable for use over the frequency range of 50 kHz to 200 MHz. They provide a combination of good stability and high Q.

The hydrogen-reduced iron cores have a permeability range of 35μ to 90μ. Somewhat lower Q can be expected from this group of cores and they are mainly used for low-frequency work, such as audio and power-line filters.

Ferrite-Powder Toroidal Cores Ferrite toroidal cores are available in sizes ranging from 0.1" to more than 5" in outside diameter, with a permeability range from 20μ to more than $10,000\mu$. They are suitable for use in a variety of rf circuit applications providing a combination of high inductance values with minimum inductor size.

There are two basic material groups: nickle zinc and manganese zinc. The nickle-zinc ferrite core has a permeability range of 20μ to 800μ. This core material is useful over the frequency range of 500 kHz to 100 MHz and provides high Q and moderate stability. It is commonly employed in low-power, high-inductive resonant circuits.

The manganese-zinc core has a permeability range of 800μ to 5000μ. It is suitable over the frequency range of 1 kHz to 1 MHz, providing high Q and moderate saturation flux density. Cores of this type are widely used for switched-mode power conversion at frequencies from 20 kHz to 100 kHz.

Choice of Core Material In rf power circuits the iron-powder material has distinct advantages over the ferrite equivalent. The iron provides a higher flux density for a given cross-sectional area than the ferrite and less change in permeability with respect to temperatures. Ferrite, on the other hand, provides a permeability factor as high as 5000, whereas the upper permeability for iron powder is about 75.

For either core material, each toroid with a given size, shape and permeability has an *inductance index* (A_L) which is used to determine the inductance of the winding:

$$A_L = \frac{L \times 10^4}{N^2}$$

where,

A_L equals microhenries per turn,
L equals the measured inductance in microhenries
N equals the number of turns.

The index is determined by placing a few turns of wire on the core and measuring the inductance. Once the index is determined, it may be used to calculate the number of turns required for a desired inductance value. In most cases, the index is specified by the manufacturer.

Toroid Coil Inductance   The inductance of a toroid coil is a function of the core material and the number of turns in the winding. The inductance index (A_L) is given for various cores and the resulting coil inductance (or coil turns) is:

$$N = 100 \sqrt{\frac{L(\mu H)}{A_L}}$$

where,

N equals number of turns required,
L equals coil inductance in microhenries,
A_L equals microhenries/per turn

A wire chart is used to determine if the required number of turns will fit on a chosen core size.

When A_L is given in *millihenries per turn*, the coil inductance (or coil turns) is:

$$N = 1000 \sqrt{\frac{L_{(mH)}}{A_L}}$$

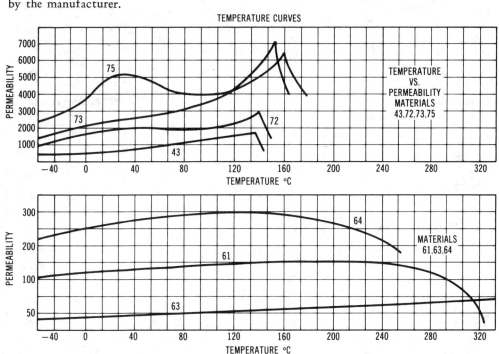

Courtesy of Amidon Associates, Inc.

Figure 43

TEMPERATURE VERSUS PERMEABILITY OF FERRITE TOROIDAL CORES

Table 4. Ferrite Toroidal Cores

MATERIAL	68	67	61	64	33	43	77	73	75
μ	20	40	125	250	800	850	1800	2500	5000
RANGE (MHz)	80-180	10-80	.2-15	.2-4	.001-1.0	1.0-50	.001-1	5-50	.001-1

FERRITE MATERIAL #63 — PERMEABILITY 40

Core number	OD	ID	Hgt	A_e	l_e	V_e	A_s	A_w	A_L
FT- 23 -63	.230	.120	.060	.0213	1.34	.0287	.81	.073	7.9
FT- 37 -63	.375	.187	.125	.0761	2.15	.1630	2.49	.177	17.7
FT- 50 -63	.500	.281	.188	.1330	3.02	.4010	4.71	.400	22.0
FT- 50A-63	.500	.312	.250	.1516	3.68	.5589	6.02	.522	24.0
FT- 50B-63	.500	.312	.500	.3030	3.18	.9640	9.74	.493	48.0
FT- 82 -63	.825	.516	.250	.2458	5.25	1.2900	10.97	1.368	22.4
FT-114 -63	1.142	.750	.295	.3750	7.42	2.7900	18.84	2.830	25.4

FERRITE MATERIAL #61 — PERMEABILITY 125

Core number	OD	ID	Hgt	A_e	l_e	V_e	A_s	A_w	A_L
FF- 23 -61	.230	.120	.060	.0213	1.34	.0287	.81	.073	24.8
FT- 37 -61	.375	.187	.125	.0761	2.75	.1630	2.49	.177	55.3
FT- 50 -61	.500	.281	.188	.1330	3.02	.4010	4.71	.400	68.0
FT- 50A-61	.500	.312	.250	.1516	3.68	.5589	6.02	.522	75.0
FT- 50B-61	.500	.312	.500	.3030	3.18	.9640	9.74	.493	150.0
FT- 82 -61	.825	.516	.250	.2458	5.25	1.2900	10.97	1.368	73.3
FT-114 -61	1.142	.750	.295	.3750	7.42	2.7900	18.84	2.830	79.3
FT-114A-61	1.142	.610	.320	.4026	6.27	2.527	16.78	1.880	101.0

FERRITE MATERIAL #43 — PERMEABILITY 850

Core number	OD	ID	Hgt	A_e	l_e	V_e	A_s	A_w	A_L
FT- 23 -43	.230	.120	.060	.0213	1.34	.0287	.81	.073	188.0
FT- 37 -43	.375	.187	.125	.0761	2.75	.1630	2.49	.177	420.0
FT- 50 -43	.500	.281	.188	.1330	3.02	.4010	4.71	.400	523.0
FT- 50A-43	.500	.312	.250	.1516	3.68	.5589	6.02	.522	570.0
FT- 50B-43	.500	.312	.500	.3030	3.18	.9640	9.74	.493	1140.0
FT- 82 -43	.825	.516	.250	.2458	5.25	1.2900	10.97	1.368	557.0
FT-114 -43	1.142	.750	.295	.3750	7.42	2.7900	18.84	2.830	603.0

FERRITE MATERIAL #72 — PERMEABILITY 2000

Core number	OD	ID	Hgt	A_e	l_e	V_e	A_s	A_w	A_L
FT- 23 -72	.230	.120	.060	.0213	1.34	.0287	.81	.073	396.0
FT- 37 -72	.375	.187	.125	.0761	2.15	.1630	2.49	.177	884.0
FT- 50 -72	.500	.281	.188	.1330	3.02	.4010	4.71	.400	1100.0
FT- 50A-72	.500	.312	.250	.1516	3.68	.5589	6.02	.522	1200.0
FT- 50B-72	.500	.312	.500	.3030	3.18	.9640	9.74	.493	2400.0
FT- 82 -72	.825	.516	.250	.2458	5.25	1.2900	10.97	1.368	1172.0
FT-114 -72	1.142	.750	.295	.3750	7.42	2.7900	18.84	2.830	1268.0
FT-114A-72	1.142	.610	.320	.4026	6.27	2.5270	16.78	1.880	1610.0

FERRITE MATERIAL #75 — PERMEABILITY 5000

Core number	OD	ID	Hgt	A_e	l_e	V_e	A_s	A_w	A_L
FT- 23 -75	.230	.120	.060	.0213	1.34	.0287	.81	.073	990.0
FT- 37 -75	.375	.187	.125	.0761	2.15	.1630	2.49	.177	2210.0
FT- 50 -75	.500	.281	.188	.1330	3.02	.4010	4.71	.400	2750.0
FT- 50A-75	.500	.312	.250	.1516	3.68	.5589	6.02	.522	2990.0
FT- 50B-75	.500	.312	.500	.3030	3.18	.9640	9.74	.493	5990.0
FT- 82 -75	.825	.516	.250	.2458	5.25	1.2900	10.97	1.368	2930.0
FT-114 -75	1.142	.750	.295	.3750	7.42	2.7900	18.84	2.830	3170.0

OD = Outer diameter (inches)
ID = Inner diameter (inches)
Hgt = Height (inches)
A_e = Effective cross sectional area (cm)^2
l_e = Effective magnetic path length (cm)

V_e = Effective magnetic volume (cm)^3
A_s = Surface area for cooling (cm)^2
A_w = Total window area (cm)^2
A_L = Inductance (mH per 1000 turns)

Key to FERRITE TOROIDAL CORE part numbers

FT — Ferrite Toroid
50 — Outer diameter
61 — Material

A_L values ±20%

$Turns = 1000 \sqrt{\dfrac{L(mH)}{A_L}}$

Practical Toroidal Cores A range of ferrite cores is shown in Table 4. This information is provided by *Amidon Associates*, 12033 Otsego St., North Hollywood, CA 91607. Representation of permeability versus temperature is shown in figure 43. A similar range of iron-powder cores is shown in Table 5.

The power capability of an iron-powder toroid core is related to the ability of the core to dissipate heat and the upper limit of saturation of the core. Generally speaking, a core which is used in a low impedance, broadband circuit (such as found in a medium power, solid-state h-f amplifier) will have about ten times the power handling capability of the same core used in a high impedance resonant circuit. For example, a 2-inch outer diameter iron-powder toroid having a permeability of 10 can safely handle about 700 watts in a low impedance circuit but only about 70 watts in a high impedance, tuned tank circuit.

Power capability of an iron-powder core may be increased by increasing the core area. This may be done by stacking several cores, wrapping the core assembly with glass tape before placing the winding on the core stack. It is also possible to use *Teflon*-insulated wire for the winding to prevent flashover at high r-f voltage. Formvar coated magnet wire may be used for power levels less than 500 watts in low impedance circuits.

R-F Transformers Conventional transformer theory is applicable to devices working in the vhf spectrum but the physical design of the transformer is considerably different from audio or power-type units. Different core materials are used for the high-frequency designs and the design of the windings is critical if broadband characteristics are desired.

To achieve maximum bandwidth for a given transformer, the core material and the inductance of the winding must be specified. For the hf range (2–30 MHz) a core permeability of 950 is often used. A high value of permeability is chosen so that a minimum number of turns are used to reduce winding resistance. The winding should have an impedance of about four times the load impedance of the transformer at the lowest frequency of operation. This is readily achieved with a high permeability core.

A different form of hf transformer is the *transmission line transformer*. In this design a multiwire transmission line is wound on the core. The impedance transformation ratio depends on the characteristic impedance of the line, the core material and the reactance of the winding at the lowest operating frequency. The number of turns, and the ratio of the turns, unlike conventional transformers, are not a factor in the design.

Broadband transformers are useful in low-impedance circuits where external reactances have a minor effect on transformer performance. Their use in high-impedance circuits is impaired by such external influences.

3-8 Wave Filters

There are many applications where it is desirable to discriminate between frequency bands, passing one band and rejecting the other, or to pass all frequencies. Circuits which accomplish this are termed *wave filters*, or simply *filters*. Filters differ from simple resonant circuits in that they provide attenuation over the *stopband*. Attenuation may be made as large as necessary if a sufficient number of filter sections of proper design are employed.

Filter Operation A filter acts by virtue of its property of offering very high impedance to the undesired frequencies, while offering but little impedance to the

TABLE 5. IRON-POWDER TOROIDAL CORES

Physical dimensions

Core Size	Outer Diam. (in.)	Inner Diam. (in.)	Height (in.)	Cross Sect. Area cm²	Mean Length cm	Core Size	Outer Diam. (in.)	Inner Diam. (in.)	Height (in.)	Cross Sect. Area cm²	Mean Length cm
T-200	2.000	1.250	.550	1.330	12.97	T- 50	.500	.303	.190	.121	3.20
T-184	1.840	.950	.710	2.040	11.12	T- 44	.440	.229	.159	.107	2.67
T-157	1.570	.950	.570	1.140	10.05	T- 37	.375	.205	.128	.070	2.32
T-130	1.300	.780	.437	.930	8.29	T- 30	.307	.151	.128	.065	1.83
T-106	1.060	.560	.437	.706	6.47	T- 25	.255	.120	.096	.042	1.50
T- 94	.942	.560	.312	.385	6.00	T- 20	.200	.088	.067	.034	1.15
T- 80	.795	.495	.250	.242	5.15	T- 16	.160	.078	.060	.016	0.75
T- 68	.690	.370	.190	.196	4.24	T- 12	.125	.062	.050	.010	0.74

IRON POWDER TOROIDAL CORES A_L VALUES (μh/100 turns)

Core Size	41-Mix Green μ=75 1-10 kHz	3-Mix Grey μ=35 .05-.5 MHz	15-Mix Rd & Wh μ=25 .1-2 MHz	1-Mix Blue μ=20 .5-5 MHz	2-Mix Red μ=10 1-30 MHz	6-Mix Yellow μ=8 2-50 MHz	10-Mix Black μ=6 10-100 MHz	12-Mix Gn & Wh μ=3 20-250 MHz	0-Mix Tan μ=1 50-300 MHz
T-200 —	755	360	NA	NA	120	105	NA	NA	NA
T-184 —	1640	720	NA	NA	240	195	NA	NA	NA
T-157 —	970	420	NA	NA	140	115	NA	NA	NA
T-130 —	785	330	215	200	110	96	NA	NA	15.0
T-106 —	900	405	330	280	135	116	NA	NA	19.2
T- 94 —	590	248	NA	160	84	70	58	32	10.6
T- 80 —	450	180	170	115	55	45	34	22	8.5
T- 68 —	420	195	180	115	57	47	32	21	7.5
T- 50 —	320	175	135	100	50	40	31	18	6.4
T- 44 —	229	180	160	105	57	42	33	NA	6.5
T- 37 —	308	110	90	80	42	30	25	15	4.9
T- 30 —	375	110	93	85	43	36	25	16	6.0
T- 25 —	225	100	85	70	34	27	19	13	4.5
T- 20 —	175	90	65	52	27	22	16	10	3.5
T- 16 —	130	61	NA	44	22	19	13	8	3.0
T- 12 —	112	60	NA	48	24	19	12	7.5	3.0

NA—Not available in that size.
Add MIX number to CORE SIZE in space provided (—) for complete part number.

Turns = 100 √ desired L (μH) ÷ A_L Value (above)

NUMBER of TURNS vs. WIRE SIZE and CORE SIZE

Approximate maximum number of turns—single layer wound—tnameled wire

Wire Size	T-200	T-130	T-106	T-94	T-80	T-68	T-50	T-37	T-25	T-12
10	33	20	12	12	10	6	4	1		
12	43	25	16	16	14	9	6	3		
14	54	32	21	21	18	13	8	5	1	
16	69	41	28	28	24	17	13	7	2	
18	88	53	37	37	32	23	18	10	4	1
20	111	67	47	47	41	29	23	14	6	1
22	140	86	60	60	53	38	30	19	9	2
24	177	109	77	77	67	49	39	25	13	4
26	223	137	97	97	85	63	50	33	17	7
28	281	173	123	123	108	80	64	42	23	9
30	355	217	154	154	136	101	81	54	29	13
32	439	272	194	194	171	127	103	68	38	17
34	557	346	247	247	218	162	132	88	49	23
36	683	424	304	304	268	199	162	108	62	30
38	875	544	389	389	344	256	208	140	80	39
40	1103	687	492	492	434	324	264	178	102	51

Actual number of turns may vary slightly according to tightness of wind.

IRON - POWDER MATERIAL vs. FREQUENCY RANGE

For best 'Q', select larger cores from the lower portion of the material frequency range, and smaller cores from the upper portion of the frequency range.

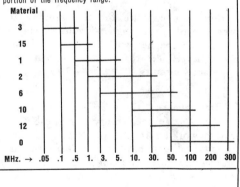

Material: 3, 15, 1, 2, 6, 10, 12, 0

MHz. → .05 .1 .5 1. 3. 5. 10. 30. 50. 100 200 300

desired frequencies. This will also apply to dc with a superimposed ac component, as dc can be considered as an alternating current of zero frequency so far as filter discussion goes.

Figure 44 illustrates the important characteristics of an electric filter. The filter *passband* is defined as the frequency region to the points at which the response is at-

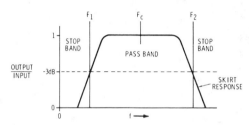

Figure 44

FREQUENCY RESPONSE OF REPRESENTATIVE FILTER

The cutoff frequencies (F₁ and F₂) of the filter are at the —3 dB points on the curve, which are 0.707 of the maximum voltage or 0.5 of the maximum power. Filters are designated as low-pass, high-pass or bandpass. The filter illustrated is a bandpass filter.

tenuated 3 dB. The points are termed the *cutoff frequencies* of the filter.

Basic Filters Early work done for the telephone companies standardized filter designs around the *constant-k* and *m-derived* filter families. The constant-*k* filter is one in which the input and output impedances are so related that their arithmetical product is a constant (k^2). The *m-derived* filter is one in which the series or shunt element is resonated with a reactance of the opposite sign. If the complementary reactance is added to the series arm of the filter, the device is said to be *shunt-derived*; if added to the shunt arm, it is said to be *series-derived*.

The basic filters are made up of elementary filter sections (*L-sections*) which consist of a series element (Z_A) and a parallel element (Z_B) as shown in figure 45. A number of L-sections can be combined into a basic filter section, called a *T network*, or a π *network*. Both the *T* and π networks may be divided in half to form half sections.

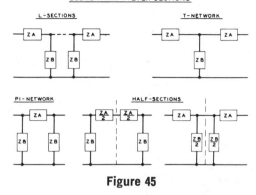

Figure 45

Complex filters may be made from these basic filter sections.

Each impedance of the *m*-derived section is related to a corresponding impedance in the constant-*k* section by some factor which is a function of the constant *m*. In turn, *m* is a function of the ratio between the cutoff frequency and the frequency of infinite attenuation, and will have some value between zero and one. As the value of *m* approaches zero, the sharpness of cutoff increases, but the less will be the attenuation at several times cutoff frequency. A value of 0.6 may be used for *m* in most applications. The "notch" frequency is determined by the resonant frequency of the tuned filter element. The amount of attenuation obtained at the "notch" when a derived section is used is determined by the effective Q of the resonant arm (figure 46).

Filter Assembly Constant-*k* sections and *m*-derived sections may be cascaded to obtain the combined characteristics of sharp cutoff and good remote frequency attenuation. Such a filter is known as a *composite* filter. The amount of attenuation will depend on the number of filter sections used, and the shape of the transmission curve depends on the type of filter sections used. All filters have some *insertion loss*. This attenuation is usually uniform to all frequencies within the passband. The insertion loss varies with the type of filter, the Q of the components, and the type of termination employed.

The basic data for classic pi-section filters is shown in figure 47.

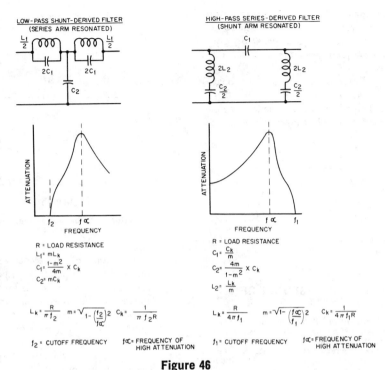

Figure 46

TYPICAL LOW-PASS AND HIGH-PASS FILTERS ILLUSTRATING SHUNT AND SERIES DERIVATIONS

3-8 Modern Filter Design

The constant-k and m-derived filters of traditional image-parameter design have been surpassed by newer techniques and designs based upon Butterworth and Chebyshev polynomials. Optimized filter configurations for sharp-cutoff filters (often using less components than the more traditional design filter) can be derived from filter tables based upon the new designs. This technique is well suited to computer programming which generates a file of precalculated and cataloged designs normalized to a cutoff frequency of one Hz, or one radian per second (1/6.28 Hz or 0.16 Hz) and terminations of one ohm. The catalog may be readily adapted to a specific use by scaling the normalized parameters to the cutoff frequency and terminating resistance desired. To scale frequency, all L and C

values are divided by the new frequency and to scale impedance, all R and L values are multiplied, and all C values divided, by the new impedance level. The filter response remains the same after scaling as before.

The *Butterworth* filter has a smooth response and does not exhibit any *passband ripple*. Its stopband, or cutoff, contains no point of infinite rejection except at infinite frequency. The steepness of the cutoff response depends on the number of poles in the filter.

The *Chebyshev* filter exhibits a steeper cutoff slope than a Butterworth filter of the same number of poles, but has a known amount of passband ripple. The *ellipticfunction* filter has a steeper cutoff slope than the Butterworth and exhibits infinite rejection frequencies in the stopband (figure 48).

Detailed information on filter design may be found in *A Handbook on Electrical Fil-*

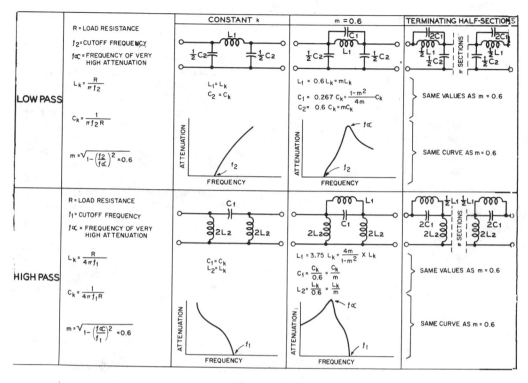

Figure 47

PI-SECTION FILTER DESIGN

Through the use of the curves and equations which accompany the diagrams in the illustration above it is possible to determine the correct values of inductance and capacitance for the usual types of pi-section filters.

ters—*Theory and Practice*, White Electromagnetics, Rockville, MD; *Reference Data for Radio Engineers* (sixth edition), Howard W. Sams & Co., Inc.; and *Approximation Methods for Electronic Filter Design*, Daniels, McGraw-Hill Book Co. These books may be available at the engineering library of any large university.

Computer-Designed Filters   Designing a filter is time consuming and requires specialized knowledge, and the designs frequently yield circuits with nonstandard components. The chart of figure 49 is based on selections from computer-calculated filter designs. They will work at frequencies from 1 kHz to 100 MHz, and use standard capacitor values.

Thirty-six designs (18 low-pass and 18 high-pass) of five-element circuits were chosen for tabulation, and they were normalized for 50-ohm terminations and a 0.1- to 1-MHz frequency range. To select a filter, simply choose a frequency nearest the desired 3-dB cutoff frequency (f_{co}). Read the L and C component values from the table, and assemble the components in accordance with the appropriate diagram. Although the filter tabulation covers directly only a 0.1- to 1-MHz frequency range and 50-ohm terminations, filter parameters for other cutoff frequencies and termination impedances can easily be determined by a simple scaling operation.

Termination of input and output with equal impedances makes possible equal values for the inductors ($L_2 = L_4$) and capacitors ($C_1 = C_5$). This simplifies component selection. Also a π configuration for the low-pass filter, and T for the high-pass, minimizes the number of inductors.

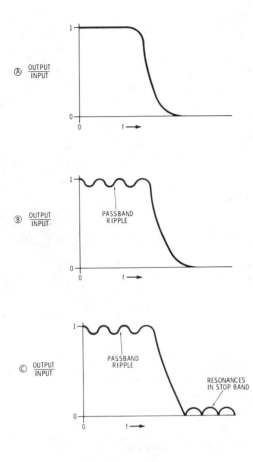

Figure 48

PASSBAND OF MODERN FILTERS

A—Butterworth filter
B—Chebyshev filter
C—Elliptic-function filter

The tabulated filter cutoff frequencies (f_{co} in megahertz at -3 dB) have been selected to provide values to within about 15% of any value in the 0.1- to 1-MHz range. The designs are keyed to indicate three levels of standard capacitor use. For example, those with the symbol "0" have all capacitors of the more common standard sizes. Where the choice of cutoff frequency is flexible, selection of designs with a greater number of the more common standard capacitance values makes component procurement easier. Inductor values are non-

standard, but this should present no problem, since inductors are often hand-wound or available with a slug adjustment.

Filter attenuation slope, VSWR (voltage standing-wave ratio), and passband ripple are interrelated. In the first octave after cutoff, the tabulated designs have a minimum and maximum attenuation slope that lies between 30 and 40 dB/octave. The minimum and maximum values of VSWR and passband ripple are 1.00 to 1.29 and zero to 0.079 dB, respectively. The attenuation slope increases as the filter VSWR and passband ripple increase. Beyond 3 f_{co} the attenuation slope becomes 30 dB/octave and is independent of the VSWR. Because the VSWR and passband ripple of these designs are low, they should prove adequate for most ordinary filter requirements. Attenuation curves plotted for the filters are normalized in terms of f/f_{co} for low-pass filters or f_{co}/f for high-pass.

For termination resistances other than 50 ohms and cutoff frequencies outside the 0.1- to 1-MHz range, use the scaling equations shown with the tabulations. However, to retain the new capacitor values in standard sizes, the resistance or frequency multipliers, F or R, must each be an integral power of 10. For example, if a 500-ohm, 2-kHz low-pass filter is required, the resistance and frequency multipliers are $R = 10$ and $F = 10^{-2}$. The tabulated 0.20-MHz low-pass filter design would be selected. The corresponding capacitances and inductances—.01 μF, .033 μF, and 65.5 μH—then become 0.1 μF, 0.33 μF, and 65.5 mH, respectively.

To match a 500-ohm filter to a 600-ohm line, two minimum-loss, 500/600-ohm L-pads can be installed, one at each end of the filter. For instance, each pad could consist of a series-connected, 240-ohm resistor and a shunt-connected, 1200-ohm resistor. The insertion loss of these two pads is approximately 7.5 dB.

Though capacitors and inductors with tolerances of 5 or 10% can be used, the actual cutoff frequency obtained will vary accordingly from the tabulated f_{co} values.

(The preceding section material and illustration are reprinted with permission from *Electronic Design*, Hayden Publishing Co., Inc. Rochelle Park, NJ 07662. The material was compiled by E. E. Wetherhold.)

Filter Chart

Scaling Equations

For cutoff frequencies outside the 0.1 to 1 MHz range and termination other than 50 Ω, use the following scaling equations:

$$L' = L \left(\frac{R}{F} \right), \quad C' = \frac{C}{(R \cdot F)}$$

L' & C' = New Component Values
L & C = Tabulated Values

Multiplier $R = \dfrac{R'}{50}$

Where R' is a new termination resistance chosen to make R an integral power of ten.

Multiplier $F = \dfrac{f'_{co}}{f_{co}}$

Where f'_{co} is a new cutoff frequency and f_{co} is a tabulated cutoff frequency, both chosen to make F an integral power of ten.

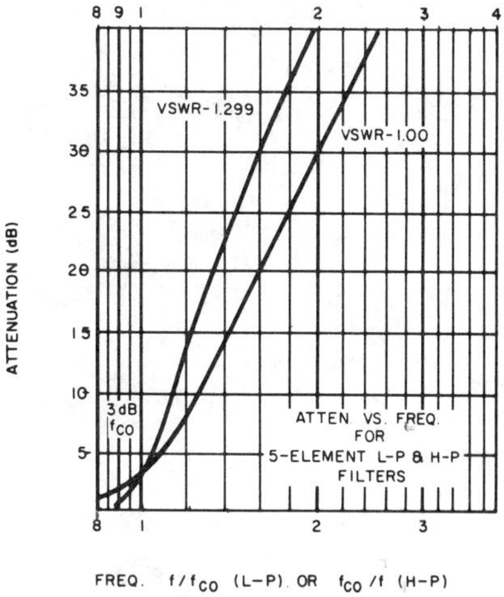

Low-pass Filters

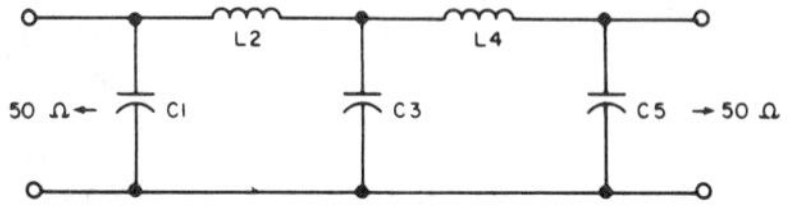

High-pass Filters

*	f_{co} 3dB	VSWR	$C_{1,5}$	C_3	$L_{2,4}$
Key	(MHz)		μF		μH
Δ	0.10	1.299	0.039	0.068	125.0
x	0.11	1.020	0.022	0.056	119.0
o	0.14	1.083	0.022	0.047	98.5
x	0.17	1.260	0.022	0.039	73.7
o	0.19	1.062	0.015	0.033	70.7
o	0.20	1.000	0.010	0.033	65.5
x	0.24	1.010	0.010	0.027	56.8
o	0.29	1.000	0.0068	0.022	44.5
x	0.35	1.010	0.0068	0.018	38.6
o	0.42	1.000	0.0047	0.015	30.8
Δ	0.47	1.273	0.0082	0.015	27.0
x	0.53	1.020	0.0047	0.012	25.3
x	0.57	1.273	0.0068	0.012	22.4
o	0.64	1.083	0.0047	0.010	21.0
x	0.71	1.151	0.0047	0.0091	18.5
x	0.76	1.020	0.0033	0.0082	17.8
x	0.85	1.051	0.0033	0.0075	16.0
o	0.95	1.105	0.0033	0.0068	14.1

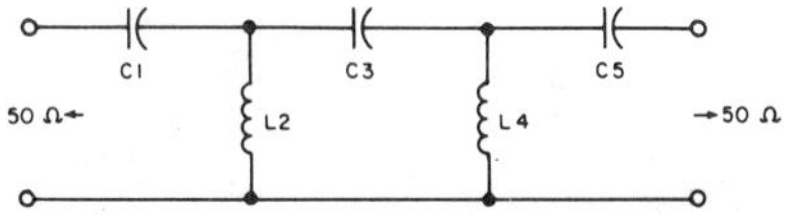

*	f_{co} 3dB	VSWR	$C_{1,5}$	C_3	$L_{2,4}$
Key	(MHz)		μF		μH
o	0.10	1.073	0.033	0.015	45.0
x	0.13	1.210	0.022	0.012	38.3
Δ	0.14	1.286	0.018	0.010	34.7
o	0.16	1.000	0.033	0.010	31.5
x	0.18	1.235	0.015	0.0082	27.1
x	0.20	1.151	0.015	0.0075	23.8
o	0.23	1.000	0.022	0.0068	21.0
x	0.26	1.030	0.015	0.0062	18.1
x	0.30	1.151	0.010	0.0050	15.9
o	0.34	1.000	0.015	0.0047	14.3
x	0.41	1.020	0.010	0.0039	11.6
o	0.48	1.105	0.0068	0.0033	10.0
x	0.53	1.051	0.0068	0.0030	8.77
x	0.60	1.020	0.0068	0.0027	7.88
x	0.65	1.139	0.0047	0.0024	7.32
o	0.72	1.083	0.0047	0.0022	6.56
x	0.85	1.210	0.0033	0.0018	5.75
x	0.96	1.116	0.0033	0.0016	4.93

*Key

o — C_1, C_3, and C_5 are common standard values.

x — C_1 & C_5 are common standard values; C_3 is a less-common standard value.

Δ — C_1 & C_5 are less-common standard values; C_3 is a common standard value.

Figure 49

CHAPTER FOUR

Semiconductor Devices

Part I—Diodes and Bipolar Devices

One of the earliest detection devices used in radio was the galena crystal, a crude example of a *semiconductor*. More modern examples of semiconductors are the selenium and silicon rectifiers, the germanium diode, and numerous varieties of the transistor and integrated circuit. All of these devices offer the interesting property of greater resistance to the flow of electrical current in one direction than in the opposite direction. Typical conduction curves for some semiconductors are shown in figure 1. The *transistor,* a three-terminal device, moreover, offers current amplification and may be used for a wide variety of control functions including amplification, oscillation, and frequency conversion.

Semiconductors have important advantages over other types of electron devices. They are very small, light and require no filament voltage. In addition, they consume very little power, are rugged, and can be made impervious to many harsh environmental conditions.

Transistors are capable of usable amplification into the microwave region and provide hundreds of watts of power capacity at frequencies through the vhf range.

Common transistors are current-operated devices whereas vacuum tubes are voltage-operated devices so that direct comparisons between the two may prove to be misleading, however economic competition exists between the two devices and the inexpensive and compact transistor has taken over most of the functions previously reserved for the more expensive vacuum tube.

4-1 Atomic Structure of Germanium and Silicon

Since the mechanism of conduction of a semiconductor is different from that of a vacuum tube, it is well to briefly review the atomic structure of various materials used in the manufacture of solid-state devices.

It was stated in an earlier chapter that the electrons in an element having a large atomic number are conveniently pictured as being grouped into rings, each ring having a definite number of electrons. Atoms in which these rings are completely filled are termed *inert gases,* of which helium and argon are examples. All other elements have one or more incomplete rings of electrons. If the incomplete ring is loosely bound, the electrons may be easily removed, the element is called *metallic,* and is a conductor of electric current. Copper and iron are examples of conductors. If the incomplete ring is tightly bound, with only a few electrons missing, the element is called *nonmetallic,* and is an insulator (nonconductor) to electric current. A group of elements, of which germanium, gallium, and silicon are examples, fall between these two sharply defined groups and exhibit both metallic and nonmetallic characteristics. Pure germanium or silicon may be considered to be a good insulator. The addition of certain impurities in carefully controlled amounts to the pure element will alter the conductivity of the material. In addition, the choice of the impurity can change the direction of conduc-

4.1

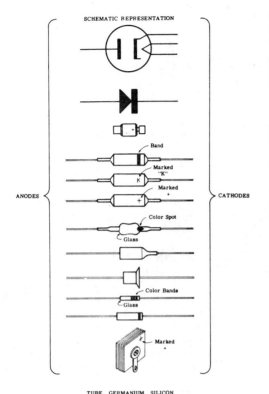

SCHEMATIC REPRESENTATION

ANODES

CATHODES

Band

Marked
"K"

Marked
+

Color Spot
Glass

Color Bands
Glass

Marked
+

TUBE, GERMANIUM, SILICON
AND SELENIUM DIODES

MILLIAMPERES

SILICON
JUNCTION

SCHOTTKY
BARRIER
DIODE

1N34A

VOLTS

Figure 1

DIODE CHARACTERISTICS
AND CODING

The semiconductor diode offers greater resistance to the flow of current in one direction than in the opposite direction. Note expansion of negative current and positive voltage scales. Diode coding is shown above, with notations usually placed on cathode (positive) end of unit.

tivity through the element, some impurities increasing conductivity to positive potentials and others increasing conductivity to negative potentials. Early transistors were

mainly made of germanium but most modern transistors possessing power capability are made of silicon. Experimental transistors are being made of gallium arsenide which combines some of the desirable features of both germanium and silicon.

Both germanium and silicon may be "grown" in a diamond lattice crystal configuration, the atoms being held together by bonds involving a shared pair of electrons (figure 2). Electrical conduction within the crystal takes place when a bond is broken, or when the lattice structure is altered to obtain an excess electron by the addition of an impurity. When the impurity is added, it may have more or less loosely held electrons than the original atom, thus allowing an electron to become available for conduction, or creating a vacancy, or *hole*, in the shared electron bond. The presence of a hole encourages the flow of electrons and may be considered to have a positive charge, since it represents the absence of an electron. The hole behaves, then, as if it were an electron, but it does not exist outside the crystal.

4-2   Mechanism of Conduction

There exist in semiconductors both negatively charged electrons and absence of electrons in the lattice (holes), which behave as though they had a positive electrical charge equal in magnitude to the negative charge on the electron. These electrons and holes drift in an electrical field with a velocity which is proportional to the field itself:

$$V_{dh} = \mu_h E$$

where,
  V_{dh} equals drift velocity of hole,
  E equals magnitude of electric field,
  μ_h equals mobility of hole.

In an electric field the holes will drift in a direction opposite to that of the electron and with about one-half the velocity, since the hole mobility is about one-half the electron mobility. A sample of a semiconductor,

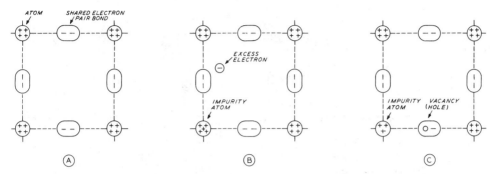

Figure 2

SEMICONDUCTOR CRYSTAL LATTICE

Silicon and germanium lattice configuration made up of atoms held by bonds involving a shared pair of electrons. Conduction takes place when bond is altered to provide excess electron (B) or to create electron vacancy or conducting "hole" (C).

such as germanium or silicon, which is both chemically pure and mechanically perfect will contain in it approximately equal numbers of holes and electrons and is called an *intrinsic* semiconductor. The intrinsic resistivity of the semiconductor depends strongly on the temperature, being about 50 ohm cm for germanium at room temperature. The intrinsic resistivity of silicon is about 65,000 ohm cm at the same temperature.

If, in the growing of the semiconductor crystal, a small amount of an impurity, such as phosphorus is included in the crystal, each atom of the impurity contributes one free electron. This electron is available for conduction. The crystal is said to be *doped* and has become electron-conducting in nature and is called N *(negative)-type* silicon. The impurities which contribute electrons are called *donors*. N-type silicon has better conductivity than pure silicon in one direction, and a continuous stream of electrons will flow through the crystal in this direction as long as an external potential of the correct polarity is applied across the crystal.

Other impurities, such as boron add one hole to the semiconducting crystal by accepting one electron for each atom of impurity, thus creating additional holes in the semiconducting crystal. The material is now said to be hole-conducting, or P *(positive)-type* silicon. The impurities which create holes are called *acceptors*. P-type silicon has better conductivity than pure silicon in one direction. This direction is opposite to that of the N-type material. Either the N-type or

the P-type silicon is called *extrinsic* conducting type. The doped materials have lower resistivities than the pure materials, and doped semiconductor material in the resistivity range of .01 to 10 ohm cm is normally used in the production of transistors.

The electrons and holes are called *carriers*; the electrons are termed majority carriers, and the holes are called minority carriers.

4-3 The PN Junction

The semiconductor diode is a *PN junction*, or *junction diode* having the general electrical characteristic of figure 1 and the electrical configuration of figure 3. The anode of the junction diode is always positive type

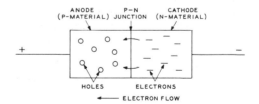

Figure 3

PN JUNCTION DIODE

P-type and N-type materials form junction diode. Current flows when P anode is positive with respect to the N cathode (forward bias). Electrons and holes are termed carriers, with holes behaving as though they have a positive charge.

(P) material while the cathode is always negative-type (N) material. Current flow occurs when the P-anode is positive with respect to the N-cathode. This state is termed *forward bias*. Blocking occurs when the P-anode is negative with respect to the N-cathode. This is termed *reverse* bias. When no external voltage is applied to the PN junction, the energy barrier created at the junction prevents diffusion of carriers across the junction. Application of a positive potential to the P-anode effectively reduces the energy barrier, and application of a negative potential increases the energy barrier, limiting current flow through the junction.

In the forward-bias region shown in figure 1, current rises rapidly as the voltage is increased, whereas in the reverse-bias region current is much lower. The junction, in other words is a high-resistance element in the reverse-bias direction and a low-resistance element in the forward-bias direction.

Junction diodes are rated in terms of average and peak-inverse voltage in a given environment, much in the same manner as thermionic rectifiers. Unlike the latter, however, a small *leakage current* will flow in the reverse-biased junction diode because of a few hole-electron pairs thermally generated in the junction. As the applied inverse voltage is increased, a potential will be reached at which the leakage current rises abruptly at an *avalanche voltage* point. An increase in inverse voltage above this value can result in the flow of a large reverse current and the possible destruction of the diode.

Maximum permissible forward current in the junction diode is limited by the voltage drop across the diode and the heat-dissipation capability of the diode structure. Power diodes are often attached to the chassis of the equipment by means of a *heat-sink* to remove excess heat from the small junction. Silicon diode rectifiers exhibit a forward voltage drop of 0.4 to 0.8 volts, depending on the junction temperature and the impurity concentration of the junction. The forward voltage drop is not constant, increasing directly as the forward current increases. Internal power loss in the diode increases as the square of the current and thus increases rapidly at high current and temperature levels.

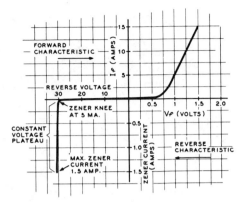

Figure 4

ZENER-DIODE CHARACTERISTIC CURVE

Between zener-knee and point of maximum current, the zener voltage is essentially constant at 30 volts. Units are available with zener voltages from approximately 4 to 200.

After a period of conduction, a silicon rectifier requires a finite time interval to elapse before it may return to the reverse-bias condition. This *reverse recovery time* imposes an upper limit on the frequency at which a silicon rectifier may be used. Operation at a frequency above this limit results in overheating of the junction and possible destruction of the diode because of the power loss during the period of recovery.

The Zener Diode The *zener diode* (reference diode) is a PN junction that can be used as a constant-voltage reference, or as a control element. It is a silicon element operated in the reverse-bias avalanche breakdown region (figure 4). The break from nonconductance to conductance is very sharp and at applied voltages greater than the breakdown point, the voltage drop across the diode junction becomes essentially constant for a relatively wide range of currents. This is the *zener control region*. Zener diodes are available in ratings to 50 watts, with zener voltages ranging from approximately 4 volts to 200 volts.

Thermal dissipation is obtained by mounting the zener diode to a heat sink composed of a large area of metal having free access to ambient air.

The zener diode has no ignition potential as does a gas regulator tube, thus eliminating the problems of relaxation oscillation and high firing potential, two ailments of the gas tube. Furthermore, the zener regulator or combinations can be obtained for almost any voltage or power range, while the gas tubes are limited to specific voltages and restricted current ranges.

Actually, only the zener diode having a voltage rating below approximately 6.8 volts is really operating in the zener region. A higher voltage zener diode displays its constant voltage characteristic by virtue of the *avalanche effect*, which has a very sharp knee (figure 4). A diode for a voltage below 6.8 operates in the true zener region and is characterized by a relatively soft knee.

Avalanche and zener modes of breakdown have quite different temperature characteristics and breakdown diodes that regulate in the 5.6- to 6.2-volt region often combine some of each mechanism of breakdown and have a voltage versus temperature characteristic which is nearly flat. Many of the very stable *reference diodes* are rated at 6.2 volts. Since the avalanche diode (breakdown voltage higher than 6.8 volts) displays a positive voltage-temperature slope, it is possible to temperature-compensate it with one or more series forward-biased silicon diodes (D_1) as shown in figure 5. The 1N935 series

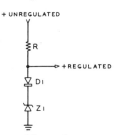

+ UNREGULATED

R

+REGULATED

D1

Z1

Figure 5

TEMPERATURE-COMPENSATED ZENER DIODE

(9 volt) is apparently of this sort, since the voltage is not 6.2 or some integer multiple thereof.

Several manufacturers have been successful in extending the avalanche mode of breakdown into the low-voltage region normally considered the domain of zener breakdown. By using such a *low-voltage avalanche* (LVA) diode instead of a zener, a sharp knee may be obtained at breakdown voltages below 6.8 volts.

National Semiconductor also has a series of 1.8 to 5.6 volt regulator diodes that display sharp knees compared to zener equivalents. These "diodes" are actually very small IC chips with a number of transistors on them. Only two leads are brought out of the package for use as a diode. The LM-103-1.8 through LM-103-5.6 comprise the diode family of 13 devices. A more complex IC is available as a 1.22 volt reference diode, the LM-113.

Silicon epitaxial transistors may also be used as zener diodes, if the current requirement is not too large. Most small, modern, silicon signal transistors have a V_{BEO} (back emitter-base breakdown voltage) between 3 and 5 volts. If the base and emitter leads are used as a zener diode, the breakdown will occur at a volt or so in excess of the V_{BEO} rating. Figure 6 shows NPN and PNP tran-

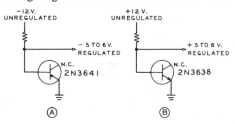

−12 V. UNREGULATED   +12 V. UNREGULATED

− 5 TO 6 V. REGULATED   + 5 TO 6 V. REGULATED

N.C. 2N3641   N.C. 2N3638

Ⓐ   Ⓑ

Figure 6

SMALL-SIGNAL SILICON TRANSISTOR USED AS ZENER DIODE

sistors used in this fashion. For safety, no more than one quarter the rated power dissipation of the transistor should be used when the device is operated this way.

All types of zener diodes are a potential source of noise, although some types are worse than others. If circuit noise is critical, the zener diode should be bypassed with a low-inductance capacitor. This noise can be evident at any frequency, and in the worst cases it may be necessary to use LC decoupling circuits between the diode and highly sensitive r-f circuits.

Junction Capacitance   The PN junction possesses capacitance as the result of opposite electric charges existing on the sides of the barrier. Junction capacitance may change with applied voltage, as shown in figure 7.

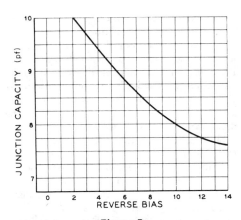

Figure 7

**JUNCTION CAPACITANCE VARIATION
WITH RESPECT TO REVERSE VOLTAGE**

A voltage-variable capacitor (*varactor* or *varicap*) is generally made of a silicon junction having a special impurity concentration to enhance the capacitance variation and to minimize series resistance losses.

The varicap and the varactor are fundamentally the same type of device, the former used in tuning resonant circuits electrically and the latter used in parametric amplifiers and frequency multipliers. Both devices have been designed to give a high-Q capacitance vs. voltage relationship at radio frequencies.

The circuit of figure 8A shows a varicap used to electrically tune a resonant circuit. This form of tuning is restricted to circuits which have a very small r-f voltage across them, such as in receiver r-f amplifier stages. Any appreciable ac voltage (compared to the dc control voltage across the device)

will swing its capacitance at the r-f rate, causing circuit nonlinearity and possible crossmodulation of incoming signals. This nonlinearity may be overcome by using two varicap devices as shown in figure 8B. In this case, the ac component increases the capacitance of one varicap while decreasing that of the other. This tuning method may be used in circuits having relatively high r-f voltages without the danger of nonlinearity.

The Varactor The varactor frequency multiplier (also called the *parametric multiplier*) is a useful vhf/uhf multiplier which requires no dc input power. The input power consists only of the fundamental-frequency signal to be multiplied and typically 50% to 70% of that r-f power is recovered at the output of the multiplier unit. Since the efficiency of a varactor multiplier drops as the square of the multiple (n), such devices are not usually used for values of n greater than five.

Examples of varactor multipliers are shown in figure 9. There are usually a number of *idlers* (series-resonant circuits) in a varactor multiplier. In general, there will be n-2 idlers. These idlers are high-Q selective short circuits which reflect undesired harmonics back into the nonlinear capacitance diode.

An interesting development in multiplier diodes is the *step-recovery* diode. Like the varactor, this device is a frequency multiplier requiring no dc input. The important difference between the step-recovery diode and the varactor is that the former is deliberately driven into forward conduction by the fundamental drive voltage. In addi-

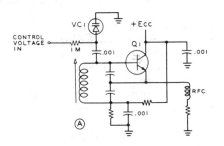
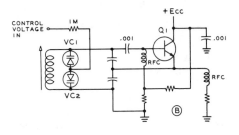

Figure 8

VOLTAGE VARIABLE CAPACITORS

A—Single varicap used to tune resonant circuit
B—Back-to-back varicaps provide increased tuning range with improved linearity

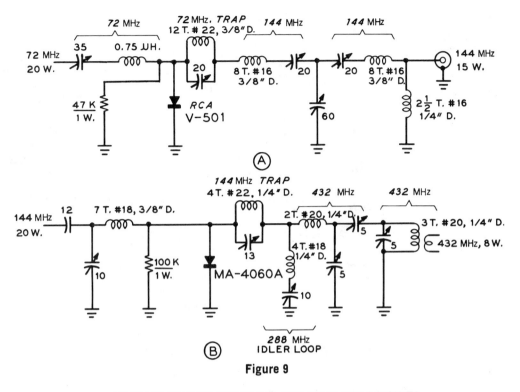

Figure 9

BASIC VARACTOR DOUBLING AND TRIPLING CIRCUITS
If "step-recovery" diode is used, idler loop may be omitted.

tion, the step-recovery diode multiplier requires no idler circuits and has an output efficiency that falls off only as $1/n$. A "times-ten" frequency multiplier could then approach 10% efficiency, as compared to a varactor multiplier whose efficiency would be in the neighborhood of 1%. A typical step-recovery multiplier is shown in figure 10. Diode multipliers are capable of providing output powers of over 25 watts at 1 GHz and several watts at 5 GHz. Experimental devices have been used for frequency multiplication at frequencies over 20 GHz, with power capabilities in the milliwatt region.

Point-Contact A rectifying junction can be
Diodes made of a metal "whisker"
touching a very small semiconductor die. When properly assembled, the die injects electrons into the metal. The contact area exhibits extremely low capacitance and *point-contact* diodes are widely

used as uhf mixers, having noise figures as low as 6 dB at 3 GHz. The 1N21-1N26 series and the 1N82 are typical versions of point-contact silicon diodes for mixer use. The germanium point-contact diode, as exemplified by the 1N34 and 1N270, has been most used as an r-f detector at vhf and lower frequencies. The germanium point-contact

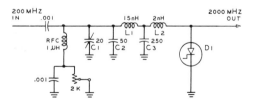

Figure 10

STEP-RECOVERY FREQUENCY MULTIPLIER

Step-recovery diode is used as multiplier. No idler circuits are required, such as used with varactor.

diode is still quite useful as a detector, but is being replaced in more modern designs by the silicon *Schottky-barrier (hot-carrier)* diode. The Schottky-barrier diode is similar to the silicon point-contact diode, with the metal-to-silicon interface made by metal deposition on silicon. This device behaves like a silicon point-contact diode, having a lower forward voltage drop than an equivalent silicon unit, good high-frequency response, and a lower noise figure.

Other Diode Devices   *Impatt, Trapatt, and Gunn* diodes are used to produce r-f directly from dc when used in microwave cavities. The PIN diode is useful as an attenuator or switch at radio frequencies. This is a PN junction with a layer of undoped (intrinsic) silicon between

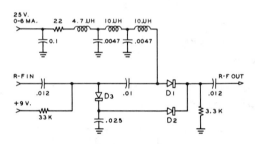

Figure 11

PIN DIODE USED AS R-F ATTENUATOR OR SWITCH

Diode D₁ appears resistive to frequencies whose period is shorter than "carrier" lifetime. Control voltage varies r-f attenuation of diode.

the P and N regions. Because of the neutral intrinsic layer, the charge carriers in the diode are relatively slow; that is, they have a long carrier lifetime. If this lifetime is long compared to the period of the radio frequency impressed on the device, the diode appears resistive to that frequency. Since PIN diodes appear resistive to frequencies whose period is shorter than their carrier lifetime, these diodes can be used as attenuators and switches. An example of such an electrically variable PIN diode attenuator is shown in figure 11.

4-4   Diode Power Devices

Semiconductor devices have ratings which are based on thermal considerations similar to other electronic devices. The majority of power lost in semiconductors is lost internally and within a very small volume of the device. Heat generated by these losses must flow outward to some form of *heat exchanger* in order to hold junction temperature to a reasonable degree. The largest amount of heat flows out through the case and mounting stud of the semiconductor and thence through the heat exchanger into the air. The heat exchanger (or *heat sink*) must be in intimate contact with the case or leads of the semiconductor to achieve maximum uniform contact and maximum heat transfer. The matching surfaces are often lubricated with a substance having good thermal conductivity to reduce oxides or galvanic products from forming on the surfaces (*Dow-Corning Silicone Grease #200* and *Corning PC-4* are often used). The latter is silicone grease loaded with zinc oxide for improved heat transfer.

Care must be exercised in the contact between dissimilar metals when mounting semiconductor devices, otherwise electrolytic action may take place at the joint, with subsequent corrosion of one or more surfaces. Many rectifiers come with plated finishes to provide a nonactive material to be placed in contact with the heat sink.

When it is necessary to electrically insulate the case of the semiconductor from the heat sink, a thin mica or plastic washer may be placed between the device and the heat sink after lubricating the surfaces with a thermal lubricant.

Diode Rectifiers   Semiconductor power rectifiers are the most-used solid-state devices in the electronics industry. Copper-oxide disc rectifiers have been used for decades, as have selenium disc rectifiers. The germanium junction rectifier, too, has been used extensively in electronics; the representative type 1N91 is still available.

Almost all new rectifier system design today uses the *silicon junction* rectifier (figure 12). This device offers the most promising range of applications; from extreme cold

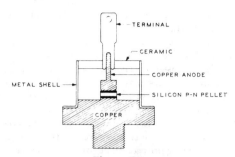

Figure 12

SILICON RECTIFIER

Silicon pellet is soldered to copper stud to provide low thermal resistance path between PN junction and heat sink. Copper anode is soldered to top of junction. Temperature of junction must be held to less than 200°C, as a result of increasing temperature on reverse current flow through junction.

to high temperature, and from a few watts of the output power to very high voltage and currents. Inherent characteristics of silicon allow junction temperatures in the order of 200°C before the material exhibits intrinsic properties. This extends the operating range of silicon devices beyond that of any other efficient semiconductor and the excellent thermal range coupled with very small size per watt of output power make silicon rectifiers applicable where other rectifiers were previously considered impractical.

Silicon Current Density The current density of a silicon rectifier is very high, and on present designs ranges from 600 to 900 amperes per square inch of effective barrier layer. The usable current

density depends on the general construction of the unit and the ability of the heat sink to conduct heat from the crystal. The small size of the crystal is illustrated by the fact that a rectifier rated at 15 dc amperes, and 150 amperes peak surge current has a total cell volume of only .00023 inch. Peak currents are extremely critical because the small mass of the cell will heat instantaneously and could reach failure temperatures within a time lapse of microseconds.

Operating Characteristics The *reverse direction* of a silicon rectifier is characterized by extremely high resistance, up to 10^9 ohms below a critical voltage point. This point of *avalanche voltage* is the region of a sharp break in the resistance curve, followed by rapidly decreasing resistance (figure 13A). In practice, the peak inverse working voltage is usually set at least 20% below the avalanche point to provide a safety factor.

A limited reverse current, usually of the order of 0.5 mA or less flows through the silicon diode duing the inverse-voltage cycle. The reverse current is relatively constant to the avalanche point, increasing rapidly as this reverse-voltage limit is passed. The maximum reverse current increases as diode temperature rises and, at the same time, the avalanche point drops, leading to a "runaway" reverse-current condition at high temperatures which can destroy the diode.

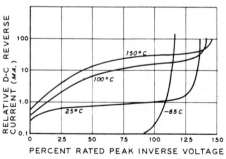

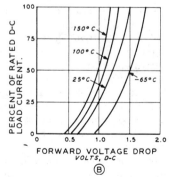

Figure 13

(A) (B)

SILICON RECTIFIER CHARACTERISTICS

A—Reverse direction of silicon rectifier is characterized by extremely high resistance up to point of avalanche voltage.
B—Threshold voltage of silicon cell is about 0.6 volt. Once device starts conducting the current increases exponentially with small increments of voltage, then nearly linearly on a very steep slope.

The forward characteristic, or resistance to the flow of forward current, determines the majority of power lost within the diode at operating temperatures. Figure 13B shows the static forward current characteristic relative to the forward voltage drop for a typical silicon diode. A small forward bias (a function of junction temperature) is required for conduction. The power loss of a typical diode rated at 0.5 ampere average forward current and operating at 100°C, for example, is about 0.6 watt during the conducting portion of the cycle. The forward voltage drop of silicon power rectifiers is carefully controlled to limit the heat dissipation in the junction.

Diode Ratings and Terms　Silicon diodes are rated in terms similar to those used for vacuum-tube rectifiers. Some of the more important terms and their definitions follow: *Peak Inverse Voltage* (PIV). The maximum reverse voltage that may be applied to a specific diode type before the avalanche breakdown point is reached.

Maximum RMS Input Voltage—The maximum rms voltage that may be applied to a specific diode type for a resistive or inductive load. The PIV across the diode may be greater than the applied rms voltage in the case of a capacitive load and the maximum rms input voltage rating must be reduced accordingly.

Maximum Average Forward Current—The maximum value of average current allowed to flow in the forward direction for a specified junction temperature. This value is specified for a resistive load.

Peak Recurrent Forward Current— The maximum repetitive instantaneous forward current permitted to flow under stated conditions. This value is usually specified for 60 Hz and a specific junction temperature.

Maximum Single-Cycle Surge Current— The maximum one-cycle surge current of a 60-Hz sine wave at a specific junction temperature. Surge currents generally occur when the diode-equipped power supply is first turned on, or when unusual voltage transients are introduced in the supply line.

Derated Forward Current—The value of direct current that may be passed through a diode for a given ambient temperature. For higher temperatures, less current is allowed through the diode.

Maximum Reverse Current—The maximum leakage current that flows when the diode is biased to the peak-inverse voltage.

Silicon diodes may be mounted on a conducting surface termed a *heat sink* that, because of its large area and heat dissipating ability, can readily dispose of heat generated in the diode junction, thereby safeguarding the diode against damage by excessive temperature.

Improved Rectifier Types　A recent silicon rectifier design has been developed having most of the advantages of silicon, but also low forward voltage drop. This device is the Schottky-barrier or hot-carrier diode in a large format for power use. For two equal volume units, the Schottky-barrier type provides a higher current rating than does the equivalent silicon unit, bought about by the lower forward voltage drop.

The Schottky-barrier device is also a very fast rectifier; operation in high-frequency inverter circuits (up to several hundred kHz) is quite practical. So far the PIV of these diodes remains quite low (less than 50 volts).

A second semiconductor rectifier which combines most of the features of the

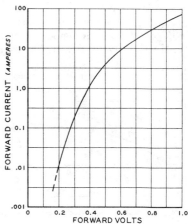

Figure 14

ION-IMPLANTED DIODE FEATURES LOW FORWARD DROP AND FAST RECOVERY TIME

Schottky-barrier and the common junction device is the *ion-implanted* diode. This diode has impurities implanted in the silicon by means of an "atom smasher." The impurity ions are fired from a particle accelerator into the silicon target wafer. The resultant silicon cystal lattice is modified in such a way as to cause the diodes made from this wafer to have a low forward drop and a fast recovery time (figure 14).

SCR Devices The *thyristor* is a generic term for that family of multilayer semiconductors that comprise *silicon controlled rectifiers* (SCR's), *Triacs, Diacs, Four Layer Diodes* and similar devices. The SCR is perhaps the most important member of the family, at least economically, and is widely used in the control of large blocks of 60-Hz power.

The SCR is a three-terminal, three-junction semiconductor, which could be thought of as a solid-state thyratron. The SCR will conduct high current in the forward direction with low voltage drop, presenting a high impedance in the reverse direction. The three terminals (figure 15) of an SCR device are *anode, cathode,* and *gate.* Without

Figure 15

THE SILICON CONTROLLED RECTIFIER

This three-terminal semiconductor is an open switch until it is triggered in the forward direction by the gate element. Conduction will continue until anode current is reduced below a critical value.

gate current the SCR is an open switch in either direction. Sufficient gate current will close the switch in the forward direction only. Forward conduction will continue even with gate current removed until anode current is reduced below a critical value. At this point the SCR again blocks open. The SCR is therefore a high-speed unidirectional switch capable of being latched on in the forward direction.

The gate signal used to trigger an SCR may be an ac wave, and the SCR may be used for dimming lights or speed control of small ac universal series-wound motors, such as those commonly used in power tools. Several power-control circuits using SCR devices and *triacs* (bidirectional triode thyristors) are shown in figure 16.

The *triac* is similar to the SCR except that when its gate is triggered on, it will conduct either polarity of applied voltage. This makes full-wave control much easier to achieve than with an SCR. An example of the triac in a full-wave power control circuit is shown in figure 16C.

The *four layer diode* is essentially an SCR without a gate electrode. As the forward voltage is increased across it, no conduction occurs until the voltage rises to the holdoff value, above which the device conducts in much the same fashion an SCR does when its holdoff voltage has been exceeded.

The *diac* is analogous to the triac with no gate electrode. It acts like a four layer diode, except that it has similar holdoff in both directions. The diac is used principally to generate trigger pulses for triac gating circuits.

The *silicon unilateral switch* (SUS) is similar to the four layer diode and the *silicon bilateral switch* (SBS) is similar to the diac. There are also a number of other variously named "trigger diodes" for use with thyristors, but they are all found to be functionally similar to the four layer diode or diac.

There exists one other thyristor of importance: it is the *silicon controlled switch* (SCS). This device has two electrodes: a *gate* to turn it on, and a second terminal called a *turn-off gate.* The SCS has, so far, only been available in low-voltage low-current versions, as exemplified by the 3N81-3N85 series.

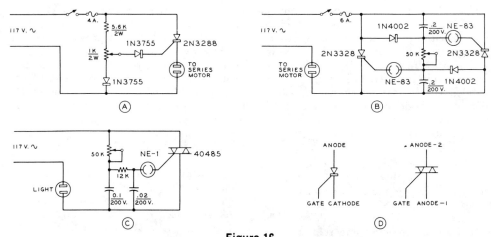

Figure 16

SCR CIRCUITS FOR MOTOR OR LIGHT CONTROL

A—Half-wave control circuit for series motor or light. B— Full-wave control circuit for series motor or light. C— Triac control light circuit. D— Symbols for SCR and Triac units.

The Unijunction Transistor   The *unijunction transistor* (UJT) was originally known as the double-base diode, and its terminal designations (emitter, base 1, base 2) still reflect that nomenclature. If a positive voltage is placed between B_2 and B_1, no conduction occurs until the emitter voltage rises to a fixed fraction of this voltage. The fixed fraction is termed η (the Greek letter *eta*) and is specified for each type of UJT. In the manner of the thyristor, when the emitter reaches η times the voltage between B_1 and B_2, the resistance between the base elements suddenly and markedly decreases. For this reason, the UJT makes a good relaxation oscillator. A simple relaxation oscillator is shown in figure 17.

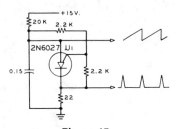

Figure 17

UNIJUNCTION TRANSISTOR SERVES AS RELAXATION OSCILLATOR

Sawtooth or spike waveforms are produced by this simple circuit using single 2N6027 PUT.

Packaged equivalents are termed *programmed unijunction transistors* (PUT).

4-5 The Bipolar Transistor

The device event in the creation of the modern semiconductor was the invention of the *transistor* in late 1947. In the last decade semiconductor devices have grown prodigiously in variety, complexity, power capability, and speed of operation. The transistor is a solid-state device having gain properties previously found only in vacuum tubes. The elements germanium and silicon are the principal materials exhibiting the proper semiconducting properties which permit their application in transistors. However, other semiconducting materials, including the compounds of Gallium and Arsenic have been used experimentally in the production of transistors.

Classes of Transistors   Thousands of type numbers of transistors exist, belonging to numerous families of construction and use. The large classes of transistors, based on manufacturing processes are:

Point Contact Transistor—The original transistor was of this class and consisted of *emitter* and *collector* electrodes touching a small block of germanium called the *base*. The base could be either N-type or P-type material and was about .05″ square. Because of the difficulty in controlling the characteristics of this fragile device, it is now considered obsolete.

Grown Junction Transistor—Crystals made by this process are grown from molten germanium or silicon in such a way as to have the closely spaced junctions imbedded in the wafer. The impurity material is changed during the growth of the crystal to produce either PNP or NPN ingots, which are then sliced into individual wafers. Junction transistors may be subdivided into *grown juncton, alloy junction,* or *drift field* types. The latter type transistor is an alloy junction device in which the impurity concentration is contained within a certain region of the base in order to enhance the high-frequency performance of the transistor.

Diffused Junction Transistor—This class of semiconductor has enhanced frequency capability and the manufacturing process has facilitated the use of silicon rather than germanium, which aids the power capability of the unit. Diffused junction transistors may be subdivided into *single diffused* (home-taxial), *double diffused, double diffused planar* and *triple diffused planar* types.

Epitaxial Transistors—These junction transistors are grown on a semiconductor wafer and photolithogaphic processes are used to define emitter and base region during growth. The units may be subdivided into *epitaxial-base, epitaxial-layer,* and *overlay* transistors. A representation of an epitaxial-layer transistor is shown in figure 18.

Field-Effect Transistors—Developed in the last decade from experiments conducted over

forty years ago, the *field-effect* (FET) *transistor* may be expected to replace many more common transistor types. This majority carrier device is discussed in a later section of this Handbook.

Manufacturing techniques, transistor end-use, and patent restrictions result in a multitude of transistors, most of which fall into the broad groups discussed previously. Transistors, moreover, may be gouped in families wherein each member of the family is a unique type, but subtle differences exist between members in the matter of end-use, gain, capacitance, mounting, case, leads, breakdown-voltage characteristics, etc. The differences are important enough to warrant individual type identification of each member. In addition, the state of the art permits transistor parameters to be economically designed to fit the various equipment, rather than designing the equipment around available transistor types. This situation results in a great many transistor types having nearly identical general characteristics. Finally, improved manufacturing techniques may "obsolete" a whole family of transistors with a newer, less-expensive family. It is recommended, therefore, that the reader refer to one of the various transistor substitution manuals for up-to-date guidance in transistor classification and substitution.

Transistor Nomenclature Semiconductors are generally divided into product groups classified as "entertainment," "industrial," and "military." The latter classifications often call for multiple testing, tighter tolerances, and quality documentation; and transistors from the same production line having less rigorous specifications often fall into the first, and least-expensive, category. Semiconductors are type numbered by several systems. The oldest standard is the JEDEC system. The first number of the identifier establishes the number of electrodes, or ports (1 = diode, 2 = triode, 3 = tetrode and 4 = heptode). The letter N stands for a semiconductor, followed by a sequential number under which the device was registered.

European manufacturers employ an identifier consisting of a type number composed of two or three letters followed by two or three numbers, the letters indicating the

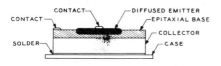

Figure 18

EPITAXIAL TRANSISTOR

Epitaxial, dual-epitaxial and overlay transistors are grown on semiconductor wafer In a lattIce structure. After fabrication, individual transistors are separated from wafer and mounted on headers. Connector wires are bonded to metalized regions and unit is sealed in an inclosure.

type of **transistor** and use and the numbers indicating the sequential number in the particular classification. Japanese transistors are usually identified by the code 2S, followed by an identifying letter and sequential number. In addition to these generally recognized codes, numerous codes adapted by individual manufacturers are also in use.

The Junction   The junction transistor is fab-
Transistor   ricated in many forms, with the planar silicon type providing the majority of units. A pictorial equivalent of a silicon planar power transistor is shown in figure 19. In this type of transistor the emitter and base junctions are often formed by a photolithographic process in selected areas of the silicon dice. Many variations of this technique and design are in use.

The transistor has three essential actions which collectively are called *transistor action*. These are: minority carrier injection,

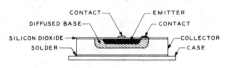

Figure 19

DIFFUSED JUNCTION TRANSISTOR

Emitter and base junctions are diffused into same side of semiconductor wafer which serves as collector. Junction heat is dissipated through solder joint between collector and package.

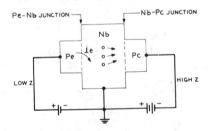

Figure 20

PICTORIAL EQUIVALENT OF PNP
JUNCTION TRANSISTOR

transport, and collection. Fig. 20 shows a simplified drawing of a PNP junction-type transistor, which can illustrate this collective action. The PNP transistor consists of a piece of N-type silicon on opposite sides of which a layer of P-type material has been grown by the fusion process. Terminals are

connected to the two P-sections and to the N-type base. The transistor may be considered as two PN junction rectifiers placed in close juxtaposition with a semiconduction crystal coupling the two rectifiers together. The left-hand terminal is biased in the forward (or conducting) direction and is called the *emitter*. The right-hand terminal is biased in the back (or reverse) direction and is called the *collector*. The operating potentials are chosen with respect to the *base terminal*, which may or may not be grounded. If an NPN transistor is used in place of the PNP, the operating potentials are reversed.

The P_e—N_b junction on the left is biased in the forward direction and holes from the P_e region are injected into the N_b region, producing therein a concentration of holes substantially greater than normally present in the material. These holes travel across the base region toward the collector, attracting neighboring electrons, finally increasing the available supply of conducting electrons in the collector loop. As a result, the collector loop possesses lower resistance whenever the emitter circuit is in operation. In junction transistors this *charge transport* is by means of diffusion wherein the charges move from a region of high concentration to a region of lower concentration at the collector. The collector, biased in the opposite direction, acts as a *sink* for these holes, and is said to collect them.

Alpha   It is known that any rectifier biased in the forward direction has a very low internal impedance, whereas one biased in the back direction has a very high internal impedance. Thus, current flows into the transistor in a low-impedance circuit, and appears at the output as current flowing in a high-impedance circuit. The ratio of a change in dc collector current to a change in emitter current is called the *current amplification*, or *alpha*:

$$\alpha = \frac{i_c}{i_e}$$

where,

    α equals current amplification,
    i_c equals change in collector current,
    i_e equals change in emitter current.

Values of alpha up to 3 or so may be obtained in commercially available point-contact transistors, and values of alpha up to about 0.999 are obtainable in junction transistors.

Beta The ratio of change in dc collector current to a change in base current (i_b) is a measure of amplification, or *beta*:

$$\beta = \frac{\alpha}{1 - \alpha} = \frac{i_c}{i_b}$$

Values of beta run to 100 or so in inexpensive junction transistors. The static dc forward current gain of a transistor in the common-emitter mode is termed the dc *beta* and may be designated β_F or h_{FE}.

Cutoff Frequencies The *alpha cutoff frequency* (f_{hfb}) of a transistor is that frequency at which the grounded base current gain has deceased to 0.7 of the gain obtainable at 1 kHz. For audio transistors the alpha cutoff frequency is about 1 MHz. For r-f and switching transistors the alpha cutoff frequency may be 50 MHz or higher. The upper frequency limit of operation of the transistor is determined by the small but finite time it takes the majority carriers to move from one electrode to the other.

The *beta cutoff frequency* (f_{hfe}) is that frequency at which the grounded-emitter current gain has decreased to 0.7 of the gain obtainable at 1 kHz. *Transconductance cutoff frequency* (f_{gm}) is that frequency at which the transconductance falls to 0.7 of that value obtainable at 1 kHz. The *maximum frequency of oscillation* (f_{max}) is that frequency at which the maximum power gain of the transistor drops to unity.

Various internal time constants and transit times limit the high-frequency response

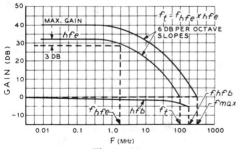

Figure 21

GAIN-BANDWIDTH CHART FOR TYPICAL HF TRANSISTOR

of the transistor and these limitations are summarized in the *gain-bandwidth product* (f_t), which is identified by the frequency at which the beta current gain drops to unity. These various cutoff frequencies and the gain-bandwith products are shown in figure 21.

The Transition Region A useful rule common to both PNP and NPN transistors is: *moving the base potential toward the collector voltage point turns the transistor on, while moving the base potential away from the collector voltage point turns the transistor off.* When fully on, the transistor is said to be *saturated.* When fully off, the transistor is said to be *cut off.* The region between these two extremes is termed the *transition region.* A transistor may be used as a switch by simply biasing the base-emitter circuit on and off. Adjusting the base-emitter bias to some point in the transition region will permit the transistor to act as a signal amplifier. For such operation, base-emitter dc bias will be about 0.3 volt for many common germanium transistors, and about 0.6 volt for silicon transistors.

Handling Transistors Used in the proper circuit under correct operating potentials the life of a transistor is practically unlimited. Unnecessary transistor failure often occurs because the user does not know how to handle the unit or understand the limitations imposed on the user by virtue of the minute size of the transistor chip. Microwave transistors, in particular, are subject to damage due to improper handling. The following simple rules will help the user avoid unnecessary transistor failures:

Know how to handle the transistor. Static discharges may damage microwave transistors or certain types of field-effect transistors because of small emitter areas in the former and the thin active layer between the channel and the gate in the latter. The transistor should always be picked up by the case and not by the leads. The FET, moreover, should be protected against static electricity by wrapping the leads with tinfoil when it is not in use, or otherwise interconnecting the leads when the unit is moved about or stored. Finally, no transistor should be inserted into or removed from a socket when power is applied to the socket pins. *Never use an ohmmeter for continuity checks.* An ohmmeter may be used at some risk to determine if certain types of transistors are open or shorted. On the low ranges, however, an ohmmeter can supply over 250 milliamperes into a low-resistance load. Many small transistors are rated at a maximum emitter current of 20 to 50 milliamperes and should be tested only in a transistor test set wherein currents and voltages are adjustable and limited. *Don't solder transistor leads unless you can do it fast.* Always use a low-wattage (20 watts or so) pencil iron and a heat sink when soldering transistors into or removing them from the circuit. Long-nose pliers grasping the lead between iron and transistor body will help to prevent transistor chip temperature from becoming excessive. Make the joint fast so that time does not permit the chip to overheat.

In-circuit precautions should also be observed. Certain transistors may be damaged by applying operating potential of reversed polarity, applying an excessive surge of transient voltage, or subjecting the equipment to excessive heat. Dissipation of heat from intermediate-size and power transistors is vital and such units should never be run without an adequate heat-sink apparatus. Finally, a danger exists when operating a transistor close to a high-powered transmitter. The input circuit of the transistorized equipment may be protected by shunting it with two small diodes back to back to limit input voltage excursions.

Transistor Symbols The electrical symbols for common three-terminal transistors are shown in figure 22. The left drawing is of a PNP transistor. The symbol for an NPN transistor is similar except that the direction of the arrow of the emitter points away from the base. This suggests that *the arrow points toward the negative terminal of the power source,* and the source potentials are reversed when going from NPN to PNP transistors, or vice-versa. As stated earlier, a useful rule-of-thumb common to both NPN and PNP transistors concerns the base-emitter bias: Moving the base toward the collector voltage turns the transistor *on,* and moving the base away from the collector voltage turns the transistor

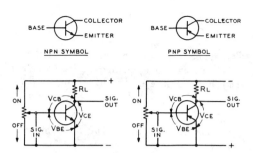

Figure 22

TRANSISTOR SYMBOLS AND BIAS

Moving the base potential toward the collector turns the transistor on. Moving the base potential away from the collector turns the transistor off. Voltage notations are: Collector-to-base voltage, V_{CB}; base-to-emitter voltage, V_{BE}; collector-to-emitter voltage, V_{CE}.

off. As shown in the illustration, capital letters are used for dc voltages. The important dc voltages existing in transistor circuitry are: *base-emitter voltage* (V_{BE}), *collector-emitter voltage* (V_{CE}), and *collector-base voltage* (V_{CB}). Signal and alternating voltages and currents are expressed by lower-case letters.

4-6 Transistor Characteristics

The transistor produces results that may be comparable to a vacuum tube, but there is a basic difference between the two devices. The vacuum tube is a voltage-controlled device whereas the transistor is a current-controlled device. A vacuum tube normally operates with its grid biased in the negative, or high-resistance, direction, and its plate biased in the positive, or low-resistance, direction. The tube conducts only by means of electrons, and has its conducting counterpart in the form of the NPN transistor, whose majority carriers are also electrons. There is no vacuum-tube equivalent of the PNP transistor, whose majority carriers are holes.

As discussed earlier, the transistor may be turned off and on by varying the bias on the base electrode in relation to the emitter potential. Adjusting the bias to some point **approximately midway between cutoff and saturation will place the transistor in the** *active* region of operation. When operated

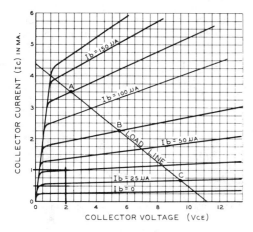

Figure 23

CHARACTERISTIC PLOT OF JUNCTION TRANSISTOR

Characteristics of junction transistor biased in active region may be expressed in terms of plot of collector voltage versus collector current. Load line and limits of operation (points A, C) are plotted, as well as operating point (B) in the manner shown in Chapter Six for vacuum-tube plots.

in this region the transistor is capable of amplification. The characteristics of a transistor biased in the active region may be expressed in terms of electrode voltages and currents as is done for vacuum tubes in Chapter Five. The plot of V_{CE} versus I_C (collector-emitter voltage versus collector current) shown in figure 23, for example, should be compared with figure 16, Chapter Five, the plot of I_b versus E_b (plate current versus plate voltage) for a pentode tube. Typical transistor graphs are discussed in this chapter, and the use of similar vacuum-tube plots is discussed in Chapter Six.

Transistor Analysis — Transistor behavior may be analyzed in terms of mathematical equations which express the relationships among currents, voltages, resistances, and reactances. These relationships are termed *hybrid parameters* and define instantaneous voltage and current values existing in the circuit under examination. The parameters permit the prediction of the behavior of the particular circuit without actually constructing the circuit.

Equivalent circuits constructed from parameter data allow formulas to be derived

for current gain, voltage gain, power gain, and other important information necessary to establish proper transistor operation. A complete discussion of hybrid parameters and transistor circuitry may be obtained in the book *Basic Theory and Application of Transistors*, technical manual TM-11-690, available from the Superintendent of Documents, U.S. Government Printing Office, Washington, D.C. 20402.

Some of the more useful parameters for transistor application are listed below:

The *resistance gain* of a transistor is expressed as the ratio of output resistance to input resistance. The input resistance of a typical transistor is low, in the neighborhood of 500 ohms, while the output resistance is relatively high, usually over 20,000 ohms. For a junction transistor, the resistance gain is usually over 50.

The *voltage gain* of a transistor is the product of *alpha* times *the resistance gain*.

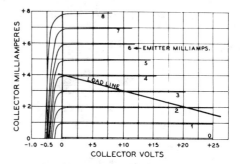

Figure 24

PLOT OF JUNCTION TRANSISTOR

Plot resembles that of a pentode tube except that emitter current, not grid voltage, defines each member of the curve family. Collector current is practically independent of collector voltage.

A junction transistor which has a value of alpha less than unity nevertheless has a resistance gain of the order of 2000 because of its extremely high output resistance, and the resulting voltage gain is about 1800 or so. For this type of transistor the *power gain* is the product of *alpha squared* times *the resistance gain* and is of the order of 400 to 500.

The output characteristics of the junction transistor are of great interest. A typical example is shown in figure 24. It is seen that the junction transistor has the characteristics of an ideal pentode vacuum tube.

The collector current is practically independent of the collector voltage. The range of linear operation extends from a minimum voltage of about 0.2 volts up to the maximum rated collector voltage. A typical load line is shown, which illustrates the very high load impedance that would be required for maximum power transfer. A common-emitter circuit is usually used, since the output impedance is not as high as when a common-base circuit is used.

Equivalent Circuit of a Transistor   As is known from network theory, the small-signal performance of any device in any network can be represented by means of an equivalent circuit. The most convenient equivalent circuit for the low-frequency small-signal performance of junc-

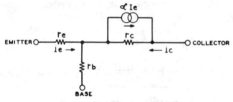

VALUES OF THE EQUIVALENT CIRCUIT

PARAMETER	JUNCTION TRANSISTOR (i_e = 1 MA., V_c = 3 V.)
r_e – EMITTER RESISTANCE	$\left(\dfrac{26}{i_e}\right)$
r_b – BASE RESISTANCE	300 Ω
r_c – COLLECTOR RESISTANCE	1 MEGOHM
α – CURRENT AMPLIFICATION	0.97

Figure 25

LOW-FREQUENCY EQUIVALENT (COMMON-BASE) CIRCUIT FOR JUNCTION TRANSISTOR

Parameter r_e is equivalent to 52/ie for silicon and 26/ie for germanium

tion transistors is shown in figure 25. r_e, r_b, and r_c are dynamic resistances which can be associated with the emitter, base, and collector regions of the transistor. The current generator aI_e, represents the transport of charge from emitter to collector.

Transistor Configurations   There are three basic transistor configurations; grounded-base connection, grounded-emitter connection, and grounded-collector connection. These correspond roughly

to grounded-grid, grounded-cathode, and grounded-plate circuits in vacuum-tube terminology (figure 26).

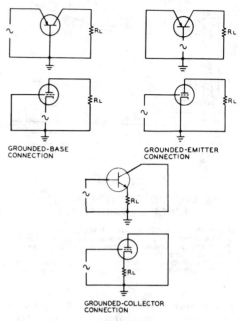

GROUNDED-BASE CONNECTION

GROUNDED-EMITTER CONNECTION

GROUNDED-COLLECTOR CONNECTION

Figure 26

COMPARISON OF BASIC VACUUM-TUBE AND TRANSISTOR CONFIGURATIONS

The grounded-base circuit has a low input impedance and high output impedance, and no phase reversal of signal occurs from input to output circuit. The grounded-emitter circuit has a higher input impedance and a lower output impedance than the grounded-base circuit, and a reversal of phase between the input and output signal occurs. This usually provides maximum voltage gain from a transistor. The grounded-collector circuit has relatively high input impedance, low output impedance, and no phase reversal of signal from input to output circuit. Power and voltage gain are both low.

Bias Stabilizaton   To establish the correct operating parameters of the transistor, a bias voltage must be established between the emitter and the base. Since transistors are temperature-sensitive devices, and since some variation in characteristics usually exists between transistors

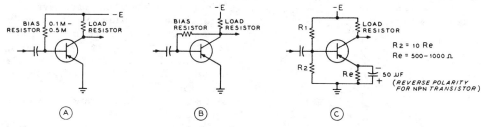

Figure 27

BIAS CONFIGURATIONS FOR TRANSISTORS

The voltage divider system of C is recommended for general transistor use. Ratio of R_1/R_2 establishes base bias, and emitter bias is provided by voltage drop across R_e. Battery polarity is reversed for NPN transistors.

of a given type, attention must be given to the bias system to overcome these difficulties. The simple *self-bias* system is shown in figure 27A. The base is simply connected to the power supply through a large resistance which supplies a fixed value of base current to the transistor. This bias system is extremely sensitive to the current-transfer ratio of the transistor, and must be adjusted for optimum results with each transistor.

When the supply voltage is fairly high and wide variations in ambient temperature do not occur, the bias system of figure 27B may be used, with the bias resistor connected from base to collector. When the collector voltage is high, the base current is increased, moving the operating point of the transistor down the load line. If the collector voltage is low, the operating point moves upward along the load line, thus providing automatic control of the base bias voltage. This circuit is sensitive to changes in ambient temperature, and may permit transistor failure when the transistor is operated near maximum dissipation ratings.

These circuits are often used in small imported transistor radios and are not recommended for general use unless the bias resistor is selected for the value of current gain of the particular transistor in use. A better bias system is shown in figure 27C, where the base bias is obtained from a voltage divider, (R_1, R_2), and an emitter resistor (R_e) is used. To prevent signal degeneration, the emitter bias resistor is bypassed with a large capacitance. A high degree of circuit stability is provided by this form of bias, providing the emitter capacitance is of

the order of 50 μF for audio-frequency applications.

Bias Circuitry Calculation   The voltage-divider bias technique illustrated in figure 27C is redrawn in generalized form in figure 28. This configuration divides the emitter resistor into two units (R_4 and R_5), one of which is bypassed. This introduction of a slight degree of feedback allows the designer more freedom to determine ac gain, while maintaining good dc stability. The assumption is made that a modern junction transistor is used having a h_{fe} of at least 40 and a low value of I_{CBO} (collector-cutoff current, emitter open). The procedure to determine bias circuitry is given in the following steps:

1. Collector current (I_c) is chosen from the data sheet.

2. Collector load resistor (R_3) is calculated so that the collector voltage is a little more than one-half the supply voltage.

3. Ac gain value (A) is chosen and emitter resistor R_4 calculated, letting $R_4 = R_3/A$.

4. Emitter resistor R_5 is calculated to raise emitter voltage (E_e) to about 10% to 15% of supply voltage:
$$R_5 = (E_e/I_e) - R_4$$

5. Total base voltage (E_b) is sum of E_e plus base-to-emitter voltage drop (about 0.7 volt for small-signal silicon devices).

6. The sum of base bias resistors R_1 and R_2 is such that one-tenth the value of the dc collector current flows through the bias circuit.

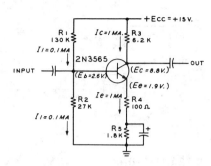

Figure 28

BIAS CIRCUITRY CALCULATION

Generalized form of voltage-divider bias technique.

7. Values of resistors R_1 and R_2 are calculated, knowing current and value of base voltage at midpoint of R_1 and R_2.
8. The ac input impedance is approximately equal to the parallel combination of R_1, R_2, and $h_{fe} \times R_4$.

To illustrate the design method, an example based on the 2N3565 is chosen. It is assumed that 1 mA of collector-emitter current flows. Collector load resistor R_3 is estimated to be 6.2K, so that the voltage drop across it is 6.2 volts, placing the collector at a potential of $15 - 6.2 - 8.8$ volts.

The data sheet of the 2N3565 shows that the range value of h_{fe} at 1 mA of collector current is 150 to 600. An ac gain value (A) of 62 may be chosen, which is well below the ultimate current gain of the device. Emitter resistor R_4 is now calculated, being equal to $R_3/A = 6200/62 = 100$ ohms. Emitter resistor R_5 is now calculated to be 1.8K, which raises the emitter voltage to 1.9 volts.

The base-emitter drop is between 0.6 to 0.7 volt for small-signal silicon devices, so this places the base at approximately 2.6 volts. Assuming no base current, the values of resistors R_1 and R_2 can now be determined as they are a simple voltage divider. The series current through R_1 and R_2 is to be one-tenth of the collector current, or 100 μA. Resistor $R_2 = 2.6V/.0001$ mA $= 26,000$ ohms and $R_1 = 15 - 2.6V/.0001$ mA $= 124,000$ ohms. These are nonstandard values of resistance so 27K and 130K are used.

Once these calculations have been completed, the approximate value of the ac input impedance may be determined. This is the parallel combination of R_1, R_2, and $h_{fe} \times R_4$. Thus, R_1 and R_2 in parallel are 22.3K and $h_{fe} \times R_4$ is 15K. Finally, 22.3K and 15K in parallel are 9K.

Actually, the ac input impedance will be higher than 9K because a minimum value of h_{fe} was used. Also, it is worth noting that the dc collector voltage is 8.8 volts. This is about half-way between $+ 15V$ and $+ 2.6V$, permitting the collector to swing ± 6 volts in response to the ac input voltage without clipping the peaks of the waveform.

This method of determining circuit parameters is quite simple and effective for RC amplifier design. With practice, the designer can juggle resistance values as calculations are made to avoid doing the design over at the end of the process.

Output Characteristic Curves Calculation of the current, voltage and power gain of a common-emitter amplifier may be accomplished by using the common-emitter output static

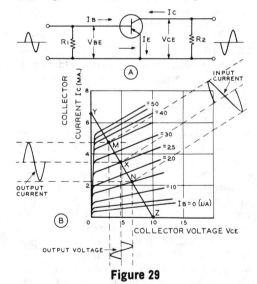

Figure 29

CHARACTERISTIC CURVES AND LOAD LINE FOR COMMON-EMITTER CIRCUIT

Calculation of current, voltage and power gain of a common-emitter transistor amplifier can be accomplished by using output characteristic curves as discussed in the text.

characteristic curves (figure 29) which plot collector current against collector voltage with the base current as a fixed value. In this example, the collector voltage supply is 10 volts, the load resistance is 1500 ohms, the input resistance is 500 ohms, the peak-to-peak input current is 20 microamperes and the *operating point* (X) is chosen at 25 microamperes of base current and 4.8 volts on the collector.

The first step is to establish a *load line* on the characteristic curves representing the voltage drop across the load resistor (R$_2$). When the collector current is zero, the total collector supply voltage (10 volts) equals the collector voltage, V$_{CE}$. Point Z (one point of the load line) then is at the 10-volt mark on the collector voltage axis (x-axis). When the collector current is zero, the total collector supply voltage (10 volts) is dropped across load resistor R$_2$. The total current (I$_c$) then is:

$$I_c = \frac{10}{1500} = 0.0066 \text{ amp} = 6.6 \text{ mA}$$

Point Y (a second point of the load line) then is at the 6.6-mA mark on the collector-current axis (y-axis). Connect points Y and Z to establish the load line. The operating point is located at point X on the load-line. Since the peak-to-peak input current is 20 microamperes, the deviation is 10 microamperes above the operating point (point M) and 10 microamperes below the operating point (point N).

The input current, output current, and output voltage waveforms may now be established by extending lines from the operating point perpendicular to the load line and to the y and x axes respectively and plotting the waveforms from each deviation point along the load-line excursions between points M and N.

Current gain (beta) in this configuration is the ratio of the change in collector current to the change in base current:

$$A_i = \frac{\Delta I_C}{\Delta I_B} = \frac{I_{C\,(max)} - I_{C\,(min)}}{I_{B\,(max)} - I_{B\,(min)}}$$

where,

A$_i$ is current gain,
I$_C$ is collector current,
I$_B$ is base current,
Δ equals a small increment.

Substituting known values in the formula:

$$\text{Current Gain } (A_i) =$$
$$\frac{4.7 - 2.1}{35 - 15} = \frac{2.6 \text{ mA}}{20 \text{ } \mu\text{A}} = 130$$

Voltage gain in this configuration is the ratio of the change in collector voltage to the change in base voltage:

$$A_v = \frac{\Delta V_{CE}}{\Delta V_{BE}} = \frac{V_{CE\,(max)} - V_{CE\,(min)}}{V_{BE\,(max)} - V_{BE\,(min)}}$$

where,

A$_v$ is voltage gain,
V$_{CE}$ is collector to emitter voltage,
V$_{BE}$ is base to emitter voltage.

(Note: The change in input voltage is the change in input current multiplied by the input impedance. In this case the input voltage is: 20 microamperes times 500 ohms, or 0.01 volt).

Therefore:

$$\text{Voltage Gain } (A_v) = \frac{6.7 - 2.7}{0.01} = 400$$

Power gain is voltage gain times current gain:

$$\text{Power gain} = 130 \times 400 = 52,000$$

Power gain in decibels is:

$$\text{Gain} = 10 \log 52,000 = 10 \times 4.7$$
$$= 47 \text{ decibels}$$

Constant-Power-Dissipation Line   Each transistor has a maximum collector power that it can safely dissipate without damage to the transistor. To ensure that the maximum collector dissipation rating is not exceeded, a *constant-power-dissipation line* (figure 30) is drawn on the characteristic curves, and the collector load resistor is selected so that its load line falls in the area bounded by the vertical and horizontal axes and the constant-power-dissipation line. The dissipation line is determined by selecting points of collector voltage and current, the products of which are equal to the maximum collector power rating of the transistor. Any load line selected so that it is tangent to the constant-power-dissipation line will ensure maximum permissible power gain of the transistor while operating within the maximum collector power-dissipation rating. This is important in the design and use of power amplifiers.

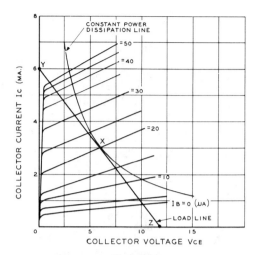

Figure 30

CONSTANT POWER-DISSIPATION LINE

Constant power-dissipation line is placed on output characteristic curves, with collector load line positioned so it falls within area bounded by vertical and horizontal axes and constant power-dissipation line. Load line tangent at (X) permits maximum power gain within maximum collector dissipation rating.

4-7 Transistor Audio Circuitry

The transistor can be connected as either a common-base, common-collector, or common-emitter stage, as discussed previously. Similar to the case for vacuum tubes, choice of transistor circuit configuration depends on the desired operating characteristics of the stage. The overall characteristics of these three circuits are summarized in figure 31. Common-emitter circuits are widely used for high gain amplification, and common-base circuits are useful for oscillator circuits and high-frequency operation, and common-collector circuits are used for various impedance transformation applications. Examples of these circuits will be given in this section.

Audio As in the case of electron-tube
Circuitry amplifiers, transistor amplifiers can be operated Class A, class AB, class B, or class C. The first three classes are used in audio circuitry. The class-A transistor amplifier is biased so that collector

current flows continuously during the complete electrical cycle, even when no drive signal is present. The class-B transistor amplifier can be biased either for collector current cutoff or for zero collector voltage. The former configuration is most often used, since collector current flows only during that half-cycle of the input signal voltage that aids the forward bias. This bias technique is used because it results in the best power efficiency. Class-B transistor amplifiers must be operated in push-pull to avoid severe signal distortion. Class-AB transistor amplifiers can be biased so that either collector current or voltage is zero for less than half a cycle of the input signal, and the above statements for class-B service also apply for the class-AB mode.

A simple small-signal voltage amplifier is shown in figure 32A. Direct-current stabilization is employed in the emitter circuit. Operating parameters for the amplifier are given in the drawing. In this case, the input impedance of the amplifier is quite low. When used with a high-impedance driving source such as a crystal microphone,

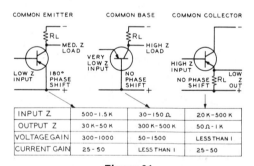

	COMMON EMITTER	COMMON BASE	COMMON COLLECTOR
INPUT Z	500–1.5 K	30–150 Ω	20 K–500 K
OUTPUT Z	30 K–50 K	300 K–500 K	50 Ω–1 K
VOLTAGE GAIN	300–1000	50–1500	LESS THAN 1
CURRENT GAIN	25–50	LESS THAN 1	25–50

Figure 31

THREE BASIC TRANSISTOR CIRCUITS

Common-emitter circuits are used for high-gain amplification, common-base circuits are useful for oscillator circuits and common-collector circuits are used for various impedance transformations.

an emitter-follower input should be employed as shown in figure 32B.

The circuit of a two-stage resistance-coupled amplifier is shown in figure 33A. The input impedance is approximately 1600 ohms. Feedback may be placed around such an amplifier from the collector of the second stage to the base of the first stage, as shown in figure 33B. A direct-coupled version of

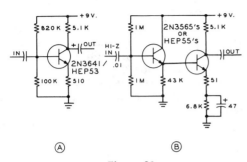

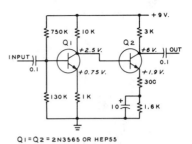

Figure 32

SMALL-SIGNAL VOLTAGE AMPLIFIERS

A—Low impedance, dc stabilized amplifier
B—Two stage amplifier features high input impedance

Figure 34

DIRECT-COUPLED TWO-STAGE AMPLIFIER

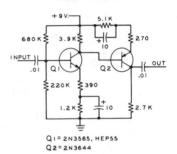

Figure 35

COMPLEMENTARY AMPLIFIER USING NPN AND PNP DEVICES

the resistance-coupled amplifier is shown in figure 34.

It is possible to employ NPN and PNP transistors in a common *complementary circuit* as shown in figure 35. There is no equivalent of this configuration in vacuum-tube technology. A variation of this interesting concept is the *complementary-symmetry* circuit of figure 36 which provides all the advantages of conventional push-pull operation plus direct coupling.

The Emitter The *emitter-follower* configura-
Follower   tion can be thought of as being
                 very much like the vacuum-tube cathode follower, since both have a high input impedance and a relatively low output impedance. The base emitter fol-

lower is shown in figure 37A. The output voltage is always 0.6 to 0.7 volt below the input (for silicon small-signal devices) and input and output impedances are approximately related by h_{fe}, the current gain of the transistor. Thus, a simple emitter follower with an emitter resistance of 500 ohms

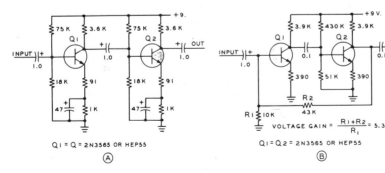

Figure 33

TWO STAGE RC AMPLIFIERS

A—Input impedance of amplifier is about 1600 ohms.
B—Feedback amplifier with feedback loop from collector of Q_2 to base of Q_1.

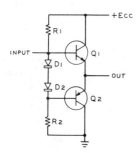

Figure 36

COMPLEMENTARY-SYMMETRY AMPLIFIER

Crossover distortion is reduced by use of diodes—D_1 and D_2. Forward voltage drop in diodes is equal to the emitter-base forward voltage drop of transistors Q_1 and Q_2.

using a transistor having an h_{fe} of 150 can have an input impedance of over 75,000 ohms. A complementary emitter follower is shown in figure 37B.

A variation of the emitter-follower design is the *Darlington pair* (figure 37C). This arrangement cascades two emitter-follower stages with dc coupling between the devices. Darlington-pair-wired dual transistors in monolithic form (for near-perfect temperature tracking) are available in both NPN and PNP pairs, even for power applications. A disadvantage of the Darlington pair emitter follower is that there are two emitter-base diode voltage drops between input and output. The high equivalent h_{fe} of the Darlington pair, however, allows for very large impedance ratios from input to output.

For power output stages another type of emitter follower is often used. A *push-pull* complementary emitter follower is shown in figure 38A. This circuit exhibits an inherent distortion in the form of a "dead zone" which exists when the input voltage is too low to turn on transistor O_1 and too high to turn on transistor Q_2. Thus, a sine wave would be distorted so as to appear as shown in figure 38B. The circuit of figure 36 corrects this problem by making the forward voltage drop in diodes D_1 and D_2 equal to the emitter-base forward voltage drop of transistors Q_1 and Q_2.

Power-Amplifier Circuits   The transistor may also be used as a class-A power amplifier as shown in figure 39.

Commercial transistors are available that will provide 50 watts of audio power when operating from a 28-volt supply. The smaller units provide power levels of a few milliwatts. The correct operating point is chosen so that the output signal can swing equally in the positive and negative directions, as shown in the collector curves of figure 39B.

The proper primary impedance of the output transformer depends on the amount of power to be delivered to the load:

$$R_P = \frac{E_c^2}{2P_o}$$

The collector current bias is:

$$I_c = \frac{2P_o}{E_c}$$

In a class-A output stage, the maximum ac power output obtainable is limited to 0.5 the allowable dissipation of the transistor. The product $I_c E_c$ determines the maximum collector dissipation, and a plot of these

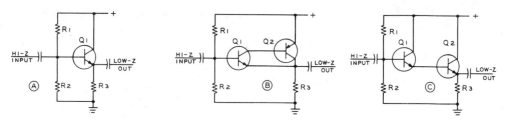

Figure 37

EMITTER-FOLLOWER CIRCUITS

A—Output voltage of emitter-follower is about 0.7 volt below input voltage
B—Complementary emitter follower
C—Darlington pair emitter follower. Q_1 and Q_2 are often on one chip

Figure 38
PUSH-PULL EMITTER-FOLLOWER OUTPUT STAGE

A—Crossover distortion exists when input voltage is too low to turn on Q₁ and too high to turn on Q₂.

B—Waveform distortion. Circuit of figure 36 corrects this problem.

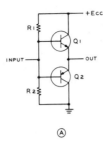

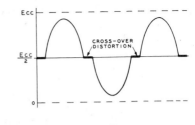

Figure 39

TYPICAL CLASS-A AUDIO AMPLIFIER

Operating point is chosen so that output signal can swing equally in a positive or negative direction without exceeding maximum collector dissipation.

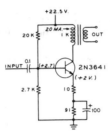

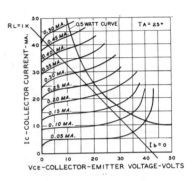

values is shown in figure 39B. The load line should always lie under the dissipation curve, and should encompass the maximum possible area between the axes of the graph for maximum output condition. In general, the load line is tangent to the dissipation curve and passes through the supply-voltage point at zero collector current. The dc operating point is thus approximately one-half the supply voltage.

The circuit of a typical push-pull class-B transistor amplifier is shown in figure 40A. Push-pull operation is desirable for transistor operation, since the even-order hamonics are largely eliminated. This permits transistors to be driven into high collector-current regions without distortion normally caused by nonlinearity of the collector. Crossover distortion is reduced to a minimum by providing a slight forward base bias in addition to the normal emitter bias. The base bias is usually less than 0.5 volt in most cases. Excessive base bias will boost the quiescent collector current and thereby lower the overall efficiency of the stage.

The operating point of the class-B amplifier is set on the $I_c = 0$ axis at the point where the collector voltage equals the supply voltage. The collector-to-collector impedance of the output transformer is:

$$R_{C-C} = \frac{2E_C^2}{P_o}$$

In the class-B circuit, the maximum ac power input is approximately equal to three times the allowable collector dissipation of

Figure 40

CLASS-B AUDIO AMPLIFIER CIRCUITRY

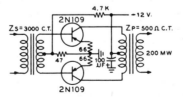

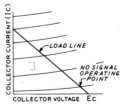

each transistor. Power transistors, such as the 2N514 have collector dissipation ratings of 80 watts and operate with class-B efficiency of about 67 percent. To achieve this level of operation the heavy-duty transistor relies on efficient heat transfer from the transistor case to the chassis, using the large thermal capacity of the chassis as a *heat sink*. An infinite heat sink may be approximated by mounting the transistor in the center of a 6″ × 6″ copper or aluminum sheet. This area may be part of a larger chassis.

The collector of most power transistors is electrically connected to the case. For applications where the collector is not grounded a thin sheet of mica may be used between the case of the transistor and the chassis.

The "Bootstrap" Circuit   The bipolar transistor in common-emitter configuration presents a low input impedance unsuitable for use with high-impedance driving sources such as a crystal microphone or a diode voltmeter probe. The *bootstrap* circuit of figure 41 provides

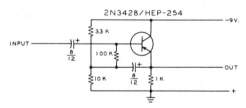

Figure 41

HIGH INPUT IMPEDANCE (BOOTSTRAP) AMPLIFIER

High input impedance provided by simple feedback circuit makes this amplifier attractive for use with crystal microphones and other high-impedance devices. Input impedance may run from 100K to 10 megohms.

a very high input impedance for these special circuits. The low-impedance base-bias network is isolated from the input circuit by the 100K resistor. The signal is fed to the base of the transistor and the output signal, taken across the emitter resistor, is also coupled to the bottom of the 100K isolating resistor via a capacitor. When a signal appears at the base, it also appears at the emitter in the same phase and almost the same amplitude. Thus, nearly identical signal voltages appear at the ends of the isolat-

ing resistor and little or no signal current flows through it. The resistor then resembles an infinitely high impedance to the signal current, thus effectively isolating the base-bias resistors. Since the isolating resistor has no effect on the bias level, the base bias remains unchanged. In practice, the signal voltage at the emitter is slightly less than at the base, thus limiting the overall effectiveness of the circuit. For example, if the emitter-follower voltage gain is 0.99, and the value of the isolating resistor is 100K, the effective resistance to the ac input signal is 100K raised to 10 megohms, an increase in value by a factor of 100 times.

4-8   R-F Circuitry

The bipolar transistor, almost from its commercial inception, proved to be operable up into the hf range. The device has been refined and improved to the point where, now, operation into the gigahertz region is feasible. External feedback circuits are often used to counteract the effects of internal transistor feedback and to provide more stable performance at high gain figures. It should be noted, however, the bipolar transistor is not like a vacuum tube or FET device and must have its base-emitter junction forward-biased to display gain. The result of this requirement is that the driving stage is driving a nonlinear diode into forward conduction by the r-f signal intended to be amplified. This indicates the bipolar device is a nonlinear amplifier, to a greater or lesser degree. If the bipolar transistor is only required to amplify one frequency at a time, and that frequency is of constant amplitude, the bipolar transistor makes a satisfactory amplifier. When an ensemble of signals of different frequencies and or amplitudes is present, the typical bipolar device will demonstrate the effect of its inherent nonlinearity in a high level of cross-modulation distortion. The fact the bipolar transistor exhibits such nonlinearity makes it useful as a frequency multiplier and mixer.

The severity of the nonlinearity of a bipolar device depends to a degree upon how it is used in a given circuit. The current gain (h_{fe}) of a transistor drops rapidly with increasing frequency (figure 21) and the tendency is to use the transmitter in a common-emitter configuration to optimize gain. This circuit configuration also unfortunately optimizes nonlinearity. The common emitter circuit may be improved by leaving a portion of the emitter resistor unbypassed as shown in figure 42. This reduces stage gain, but also reduces nonlinearity and resultant cross-modulation problems to a greater degree. The unbypassed emitter resistor also boosts the input impedance at the base of the amplifier.

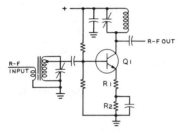

Figure 42

COMMON EMITTER R-F AMPLIFIER

Linearity is improved by leaving a portion of the emitter resistor unbypassed. Stage gain and cross modulation are both reduced.

R-F Amplifiers   A representative common-base r-f amplifier is shown in figure 43. This configuration generally has lower gain than the common-emitter circuit and is less likely to require neutralization. The linearity is better than that of the common-emitter circuit because of matching considerations. The input impedance of a common-base amplifier is in the region of 50 ohms so no voltage step up is involved in matching the transistor to the common 50-ohm antenna circuit. In the common-emitter stage the input impedance of a small hf transistor is about 500 ohms

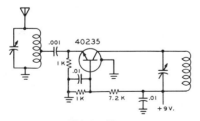

Figure 43

COMMON-BASE R-F AMPLIFIER

Linearity of this circuit is better than that of common-emitter configuration.

and a step-up impedance network must be used, causing the base voltage to be higher and aggravating the crossmodulation problem.

The relatively low gain of the common-base circuit may not be a detriment for hf operation because good receiver design calls for only enough gain to overcome mixer noise at the frequency of operation.

Mixers and Converters   As mentioned previously, the bipolar transistor is an inherently nonlinear device and, as such, can be used as an effective mixer or converter. Figure 44 shows two widely used

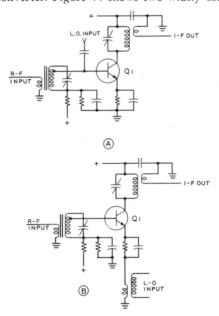

Figure 44

REPRESENTATIVE MIXER CIRCUITS

A—Base circuit injection of local oscillator.
B—Emitter injection from low-impedance source.

transistor mixer circuits. The local oscillator signal can be injected into the base cicuit in parallel with the r-f signal, or injected separately from a low-impedance source into the emitter circuit. The mixer products appear in the collector circuit and the desired one is taken from a selective output circuit.

A single transistor may be used in an *autodyne converter* circuit, as shown in figure 45. This is a common-emitter mixer with a tuned feedback circuit between

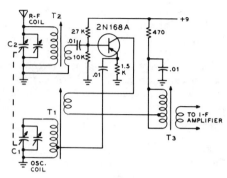

Figure 45

THE AUTODYNE CONVERTER CIRCUIT USING A 2N168A AS A MIXER

emitter and collector and is often used in inexpensive transistorized broadcast receivers. The circuit has only economy to recommend it and often requires selection of transistors to make it oscillate.

Transistor Oscillators The bipolar transistor may be used in the oscillator circuits discussed in Chapter 11 (*Generation of Radio Frequency Energy*). Because of the base-emitter diode, the oscillator is of the self-limiting type, which produces a waveform with high harmonic content. A representative NPN transistor oscillator circuit is shown in figure 46. Sufficient coup-

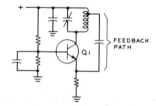

Figure 46

NPN OSCILLATOR CIRCUIT

External feedback path permits oscillation up to approximately the alpha-cutoff frequency of device.

ling between input and output circuits of the transistor via collector-base capacitance or via external circuitry will permit oscillation up to or slightly above the alpha-cutoff frequency.

Because of the relatively low impedance associated with bipolar transistors, they are best used with crystals operating in the se-

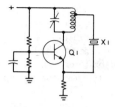

Figure 47

SERIES-MODE TRANSISTOR OSCILLATOR

Crystal is placed in feedback path and oscillates in series mode.

ries mode, as shown in figure 47. If a standard parallel-mode type crystal is used in one of these series circuits, it will oscillate at its series-resonant frequency which is slightly lower than that frequency marked on the holder.

Transistor Detectors The bipolar device can be used as an amplitude detector, very much as a diode is used since the emitter-base junction is, after all, a diode. The transistor detector offers gain, however, since current passed by the base-emitter diode is multiplied by the factor b_{fe}. The detected signal is recovered at the collector. Since germanium transistors have a lower forward conduction voltage than silicon types, they are often used in this circuit. This allows the detector to operate on a few tenths of a volt (peak) as opposed to about 0.6 volt (peak) required for a silicon transistor. The bipolar transistor can also be used as a product detector for SSB and c-w, such as shown in figure 48.

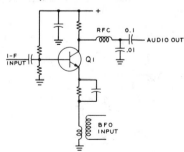

Figure 48

PRODUCT DETECTOR

Bfo is injected into the emitter circuit from a low-impedance source. Audio is recovered in the collector circuit.

The gain of a transistor amplifier stage will decrease as the emitter current is decreased. This property can be used to control the gain of an r-f or i-f amplifier strip so that weak and strong signals will produce the same audio output level. Automatic gain control voltage may be derived as described in Chapter 10 (*Radio Receiver Fundamentals*). If NPN transistors are used in the gain controlled stages, a negative agc voltage is required which reduces the fixed value of forward bias on the stage, decreasing the emitter current. If PNP transistors are used, a positive agc voltage is required.

There are also transistors especially designed for agc controlled amplifier service which are forward-biased to decrease gain. The *Fairchild SE-5003* has a gain curve as shown in figure 49 and normally operates at a collector current of 4 mA. As can be seen,

Figure 49

TYPICAL GAIN VERSUS COLLECTOR CURRENT, SE 5003

an increase in collector current will decrease the gain. This is a result of h_{fe} decreasing after it peaks at about 4 mA. All bipolar devices have a similar h_{fe} peak, but the SE-5003 is designed to peak at a low enough current so that increasing collector current beyond the h_{fe} peak value is still within the dissipation rating of the device.

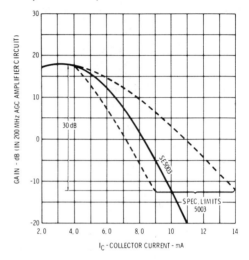

Part II—Field-Effect Devices, Integrated Circuits and Numeric Displays

4-9 Field-Effect Devices

The *junction field-effect transistor* (JFET), or unipolar transistor was explored in 1928 but it was not until 1958 that the first practical field-effect transistor was developed. This device may be most easily visualized as a bar, or channel, of semiconductor material of either N-type or P-type silicon. An ohmic contact is made to each end of the bar as shown in figure 1A, which represents an N-type field-effect transistor in its simplest form. If two P-regions are diffused into a bar of N-material (from opposite sides of the N-channel) and externally connected together electrically, a *gate* is produced. One ohmic contact is called the *source* and the other the *drain*; it matters not which if the gate diffusion is in the center of the device. If a positive voltage is applied between drain and source (figure 1B)

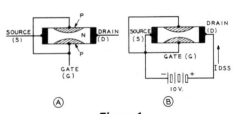

Figure 1

JUNCTION FIELD-EFFECT TRANSISTOR

A—Basic JFET is channel of N- or P-type material with contact at each end. Two P or N regions are diffused into the bar. B—If a positive voltage is applied across contacts a current flows through the gate region. Control of gate bias changes current flow from source contact to drain contact. Drain current is thus controlled by gate voltage.

and the gate is connected to the source, a current will flow. This is the most important definitive current in a field-effect device and is termed the *zero bias drain current* (I_{DSS}).

This current represents the maximum current flow with the gate-source diode at zero bias. As the gate is made more negative relative to the source, the P-region expands cutting down the size of the N-channel through which current can flow. Finally, at a negative gate potential termed the *pinch-off voltage*, conduction in the channel ceases. The region of control for negative gate voltages lies between zero and the gate-to-source cutoff voltage (V_{GS-off}). These voltages cause the gate-source junction to be *back-biased*, a condition analogous to the vacuum tube, since drain current is controlled by gate voltage. In the vacuum tube

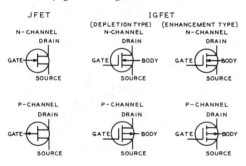

Figure 2

SYMBOLS AND NOMENCLATURE FOR FIELD-EFFECT TRANSISTORS

a potential on the grid affects the plate current, however the charge carrying the signal does not flow in the region between cathode and plate to any significant extent.

It is possible to build a P-channel JFET device that requires a negative drain voltage and is biased with positive gate voltage. Combining both N-channel and P-channel JFET's makes it possible to design complementary circuits as in the manner previously described for NPN and PNP bipolar transistors. The symbols used to depict N-channel and P-channel JFET's are shown in figure 2.

The *Insulated Gate Field-Effect Transistor* (IGFET) differs from the JFET in a number of ways. The gate element is insulated from the rest of the device and control is by means of capacitance variation. The IGFET may be visualized as in figure 3, again an N-channel device. The basic form of the device is P-type material, into which has been diffused two N-type regions

to form the source and drain. The gate is a layer of metalization laid down directly over the P-type region between source and drain, but separated from the region by a thin layer of insulating silicon dioxide (silicon nitride is also used in some types). If a positive voltage is applied to the drain, relative

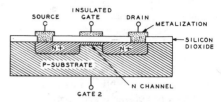

Figure 3

INSULATED-GATE FIELD-EFFECT TRANSISTOR

IGFET has insulated gate element and current control is by means of capacitance variation. Enhancement mode (positive gate control) and depletion mode (negative gate control) IGFETs are available. Gate voltage limitation is point of breakdown of oxide dielectric in the gate. Diode-protected IGFET has zener diodes on the chip to limit potential between gate and body of device.

to the source, and there is no potential difference between gate and substrate, no current will flow because the path appears as two back-to-back diodes (NP-PN). If a positive voltage is applied to the gate relative to the substrate, it will *induce* an N-region between source and drain and conduction will occur. This type of IGFET is termed an *enhancement mode* type; that is, application of forward bias to the gate enhances current flow from source to drain. (It is not possible to build an enhancement mode JFET because the gate is a diode which will conduct if forward-biased).

A *depletion mode* IGFET is built by diffusing a small N-region between the source and drain to cause conduction even if there is no voltage applied between gate and substrate. Similar to the JFET, this depletion mode IGFET must have its gate reverse-biased to reduce source-to-drain current. The depletion mode IGFET is used in the same manner as the JFET except that the gate may also be driven forward and the drain current can be increased to values even greater than the zero-bias drain current, I_{DSS}.

Gate voltage of the JFET is limited in the reverse direction by the avalanche breakdown potential of the gate-source and gate-drain circuits. In the IGFET, on the other hand, the gate voltage limitation is the point of destructive breakdown of the oxide dielectric under the gate. This breakdown must be avoided to prevent permanent damage to the oxide.

Static electricity represents the greatest threat to the gate insulation in IGFET devices. This type of charge accumulation can be avoided by wrapping the leads in tinfoil, or by otherwise connecting the leads when the devices are being transported and installed. The user of the device, moreover, may accumulate a static potential that will damage the IGFET when it is handled or installed and a grounding strap around the electrodes is recommended. Gate protection is often included within the device in the form of zener diodes on the chip between the gate and the body, forming a *diode-protected IGFET*.

FET Terminal Leads Note in figures 1 and 3 there are really four terminations associated with any FET device. In the JFET they are source, drain, and the two connections to the two P-diffusions made in the channel. In the IGFET they are source, drain, gate, and substrate. In some JFETs all four leads are brought out of the package and in others only three leads are available. In a three-lead configuration, it is considered that the two P-diffusion gate connections are tied together inside the package. In the case of the IGFET, all four leads are generally available for use; but more often than not, the substrate is externally connected to the source in the actual circuit. The advantage of the four-lead package is the ability to allow separate control ports, much like a multigrid vacuum tube.

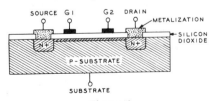

Figure 4

DUAL-GATE IGFET

Depletion type, dual-gate IGFET is intended for r-f use through the vhf range. One port is for input signal and the other for agc control.

An improved *dual-gate IGFET* of the depletion type has recently become available, intended for r-f use through the vhf range. The 3N140, 3N141, and 40673 of RCA, and the Motorola MFE-3006 and MFE-3007 are representative types. Their construction is shown in figure 4. These devices serve where dual ports are required, such as in mixers, product detectors, and agc-controlled stages, with one gate used as the signal port and the other the control port.

V-MOS The *VMOSFET* is a relatively new addition to the family of available semiconductors. It is named "V" MOS because of its physical crosssection which is in the form of a V (figure 5). This is quite different from the MOSFET cross-section shown in figure 3. Because of the vertical

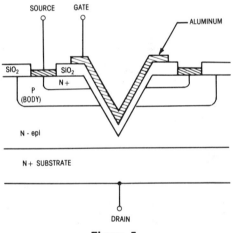

Figure 5

CROSS-SECTION OF VMOS CHANNEL

N+ material is used as substrate. The gate is positive with respect to the source resulting in an N-type channel, with electrons flowing from the source, through the channel and N-epi layer into the substrate, or drain.

penetration of the V-shaped gate, the VMOSFET is sometimes called a Vertical MOSFET.

The VMOSFET provides high-voltage, high current, and high-frequency performance which was previously unavailable in conventional designs. The first devices available were the VPM1, VPM11 and VPM12 (*Siliconix*) packaged in the TO-3 configuration. These devices are capable of currents

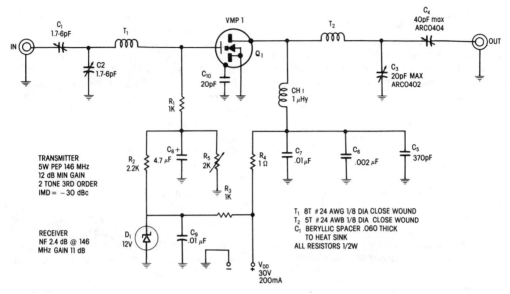

Figure 6

A 5-WATT PEP OUTPUT AMPLIFIER USING VMOSFET

This circuit may be used either as a 5-watt linear amplifier for 2 meters or as a low-noise preamplifier.

up to two amperes and potentials up to 90 volts and can produce 12 dB gain and a 2.4 dB noise figure at 146 MHz. A 5-watt PEP output linear amplifier using a VPM1 device is shown in figure 6. This amplifier has a power gain of about 12 dB.

The VMOSFET was originally developed for use as output stage devices for high fidelity audio systems to provide extremely linear characteristics that certain low-μ triode tubes (such as the 2A3) can produce with

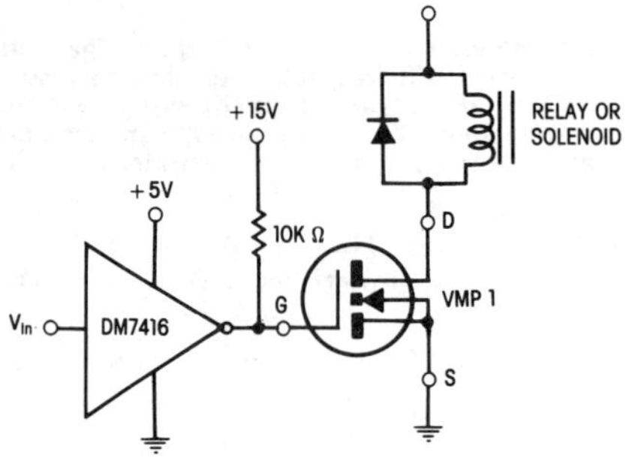

negative feedback. As such, the VMOSFET makes a very fine audio amplifier. In addition, since this device has a high impedance MOS gate, it has a perfect input impedance for interfacing MOS or PMOS logic family ICs to power load, such as a lamp or a relay. An example of this use is shown in figure 7.

The R-F Power MOSFET   Power MOSFET devices are currently available for r-f power amplifiers for service to over 175 MHz. Devices made by *Siliconix, Communications Transistor Corp.* and others provide collector dissipation ratings to 100 watts with power gain figures of about 10 dB. Typically, the power MOSFETs operate from a drain supply of 28 to 35 volts and are designed for class-AB operation. These devices are very useful as their efficiency is good and the *spectral purity* is greater than for bipolar devices of a similar power level.

The CTC BF100-35, for example, can provide over 100 watts output at 175 MHz with a drive power of about 12 watts. Power output at 28 volts is somewhat less. At the other end of the power scale, the

Figure 7

TTL LOGIC-COMPATIBLE HIGH-CURRENT RELAY DRIVER

CTC BF7-35 can provide over 7 watts output at 175 MHz, with a drive level of about 0.5 watt. Circuitry for the power MOSFET resembles that shown in figure 6.

4-10 Circuitry

JFET and depletion-mode IGFET devices are used in linear circuitry in very much the same way as are vacuum tubes, but at lower voltages. As an example, the drain characteristics of an inexpensive and popular FET (*Siliconix* E300) are shown in figure 8. The line that is labeled $V_{GS} = 0$ is the one that represents the zero-bias drain current state, or I_{DSS}. At a drain to source potential of 10 volts, I_D is 15 milliamperes and, according to the data sheet, I_D could be any value between 6 mA and 30 mA at this potential. This *spread* of I_{DSS} is fairly typical of the lower cost FETs and the curve shown is also typical, as is the value of I_D read from it.

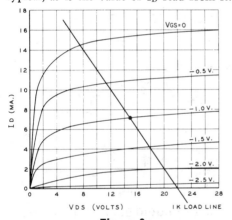

Figure 8

DRAIN CHARACTERISTICS OF E300 FET

Drain characteristic curves of FET resemble the characteristic curves of pentode vacuum tube as the current plots are nearly horizontal in slope above V_{DS} of about 6 volts. Load line is drawn on plot for gate bias of −1 volt, drain voltage of +15 volts, and drain current of 7 milliamperes.

The E300 drain characteristics look very similar in shape to the characteristics of a pentode vacuum tube; that is, at V_{ds} (drain to source potential) greater than about 6 volts, the drain current curves are nearly horizontal in slope.

A 1000-ohm load line is drawn on the characteristic plot in the same manner as one is drawn on a vacuum-tube plate characteristic curve (see Chapter 7). The load line is marked for a gate-bias voltage of −1

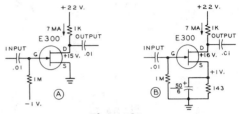

Figure 9

COMMON-SOURCE AMPLIFIERS USING E300 FET

Common-source amplifiers operating under conditions shown in figure 8. A—Separate gate bias. B—Source self-bias.

volt, a drain voltage of +15 volts, and a resting drain current of 7 milliamperes. The circuit of a *common-source* amplifier operating under these conditions is shown in figure 9.

The *common-gate* configuration shown in figure 10 may be compared in performance to the cathode-driven vacuum-tube amplifier, having a rather low value of input

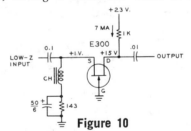

Figure 10

COMMON-GATE AMPLIFIER USING E300 FET

Input impedance of common-gate circuit is about 150 ohms. Stage gain is lower than common-source circuit.

impedance. A typical value of input impedance is approximately $1/g_{fs}$ where g_{fs} is the transconductance (similar to g_m in the vacuum tubes). The g_{fs} for the E300 device is about 6600 micromhos; so the circuit of figure 7 will have an input impedance around 150 ohms.

The FET analogy to the cathode follower is shown in figure 11. This *source follower*, shown with self-bias, has a very high input impedance and very low output impedance $(1/g_{fs})$.

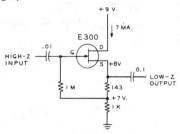

Figure 11

SOURCE-FOLLOWER AMPLIFIER USING E300 FET

Source-follower circuit has very high input impedance and low output impedance.

The FET in Specialized Circuits   The FET makes a very good r-f device because of some of its unique characteristics. In particular, the FET has a transfer characteristic that is remarkably free of third-order curvature, which ensures that intermodulation distortion and crossmodulation will be at a minimum in a properly designed circuit. A typical IGFET (depletion mode) vhf r-f amplifier is shown in figure 12.

FET devices have second order curvature in their transfer functions and operate as

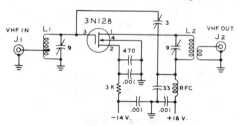

Figure 12

IGFET DEPLETION-MODE VHF AMPLIFIER

3N128 is neutralized for best circuitry stability and optimum noise figure.

good mixers having little intermodulation distortion. The use of FET devices in receivers is discussed in Chapter 10.

Aside from common usage discussed elsewhere in this handbook, the characteristics of the FET permit it to do a good job in specialized circuits. A phase-shift audio os-

cillator using the HEP 801 is shown in figure 13. This configuration employs the *tapered RC network* wherein each RC pair has the same time constant but successively higher impedance. The bridge-T and Wien bridge circuits also adapt themselves easily to the FET as shown in figures 14 and 15.

Since the FET is commonly operated in the constant-current region, it is often used as a constant-current generator with the

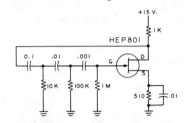

Figure 13

PHASE-SHIFT AUDIO OSCILLATOR WITH HEP 801

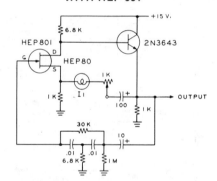

Figure 14

BRIDGE-T AUDIO OSCILLATOR USING HEP 801 AND 2N3643

I₁—Sylvania 120 MB lamp.

gate and source connected together to form a two-terminal device. A linear ramp generator using a FET in place of a transistor to charge a capacitor is shown in figure 16. A unijunction transistor is used to discharge the capacitor.

A combination FET and zener diode circuit (figure 17A) provides improved regulation since the current flow through the zener is constant. Special JFETs that serve as *constant-current diodes* are available, but the experimenter can use nearly any small

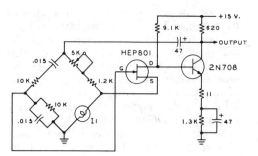

Figure 15

WIEN BRIDGE AUDIO OSCILLATOR USING HEP 801 AND 2N708

I₁—Sylvania 120 MB lamp.

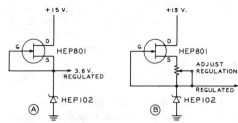

Figure 17

FET AND ZENER DIODE PROVIDE IMPROVED REGULATION

A—Constant current source. B—Variable current source.

JFET in a similar manner by connecting the gate to the source. If the FET is used with a variable resistance in the source lead, as shown in figure 17B, an adjustable but constant-current source is available.

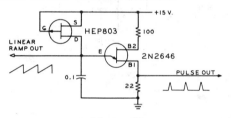

Figure 16

LINEAR RAMP GENERATOR

HEP 803 FET used as constant current source to generate linear ramp waveforms.

The enhancement-mode IGFET (P-channel) is almost exclusively used as a switch for computing or for logic circuits and the basic building block upon which one form of logic integrated circuit is based, as discussed in a later chapter. Discrete enhancement-mode IGFETs are used in *sample and hold* circuits, such as shown in figure 18. The waveform at the input is sampled only when the negative sample pulse, applied between substrate and gate, is present. The capacitor (C) is then charged to whatever value the input received during the sample pulse, and holds this value because the IGFET represents an open circuit at all other times. The voltage on the capacitor may be used to drive another FET (depletion mode) so that the input impedance of the

sensing amplifier does not discharge the capacitor to any degree during sampling times. The enhancement-mode IGFET also serves as a fast switch in chopper service or as a series switch in certain types of noise suppression devices.

As the technology of FET construction develops, JFETSs and IGFETs continue to invade new circuit areas. JFETs for 1-GHz operation are available and so are 10-watt stud-mounted types for lower-frequency power application. IGFETs are being designed for 1-GHz operation to satisfy the demands of UHF-TV reception. Some experimental FETs have been built to operate at 10 GHz. Other experimental JFETs available for low-frequency work can withstand 100 volts between source and drain.

It appears that virtually every circuit that can be realized with receiving type vacuum tubes can also be eventually duplicated with some sort of FET package and interesting variations of this efficient and inexpensive solid-state device that will apply to high-frequency communication are on the horizon.

The Fetron   A JFET called a *Fetron* has been developed that replaces a vacuum tube in a circuit directly, without requiring major modifications in the circuit. High-voltage FETS are used and the Fetron can either be a single JFET or two cascode connected JFETs in a hybrid integrated circuit. The Fetron is packaged in an oversize metal can that has the same pin configuration as the tube it replaces. The JFET characteristics can be chosen to simulate the dynamic performance of a tube. Two JFETs are required to simulate the performance of a pentode. Fetrons feature long life, low aging, and reduced power consumption as compared to an equivalent vacuum tube.

Microwave *Gallium Arsenide (GaAs) FETS*
FETs have been developed that promise superior low-noise performance for microwave applications. Typical noise figures for these devices are about 3 dB at 4 GHz, 4 dB at 8 GHz, and 5 dB at 12.5 GHz. Developmental GaAs FETs with a Schottky-barrier gate exhibit a noise figure of 3.3 dB at 10 GHz and a power gain of 9 dB. Many of these new experimental FETs have an f_{max} in excess of 30 GHz. Enhanced noise figures have been produced by cooling the FET device with liquid nitrogen to 77° K.

Recent advances in Gallium Arsenide FET technology have produced devices capable of 1 watt output at 10 GHz. These state-of-the-art GaAs FETs are very expensive, and for the present must be considered as components used in high priority systems, such as avionics.

Analog Switches The example of using an enhancement mode MOSFET as an analog switch in the sample-and-hold circuit of figure 18 shows the general concept of analog switching. A whole class of IC-packaged analog switches have been developed for this purpose, making circuit details much simpler than with the use of other devices. The control input to these new devices is usually compatible with one or more of the standard digital logic families such as TTL or CMOS. Figure 19A shows an analog gate using a DG200 (*Siliconix*) as a noise pulse gate such as employed in an i-f noise silencer. Note that TTL control is used and

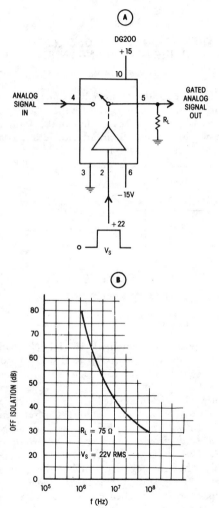

Figure 19

THE ANALOG SWITCH

A—Use of Siliconix DG200 as an analog gate. A positive pulse opens switch.
B—Isolation of DG200 vs frequency.

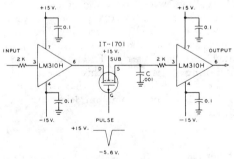

Figure 18

SAMPLE AND HOLD CIRCUIT WITH ENHANCEMENT MODE IGFET

Input waveform is sampled only when negative sample pulse applied between substrate and gate of IT-1701 IGFET is present. Capacitor C is then charged to value of input voltage and drives sensing amplifier through operational amplifier LM310H, at right. Capacitor holds charge because IGFET represents open circuit after pulse passes.

that the DG200 uses standard ± 15 volt power, as do most IC op amps and some other linear ICs.

A plot of isolation vs. frequency (figure 19B) shows that the DG200 is usable up to about 100 MHz. Also, since there are two analog gates in each DG200 package, the device can be used as a single-pole, double-throw switch for analog signals or to form a series-shunt analog gate for increased isolation, as shown in figure 20.

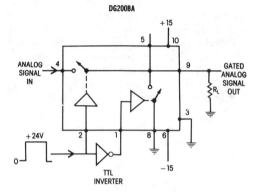

Figure 20

SERIES-SHUNT ANALOG SWITCH

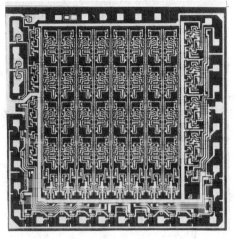

Figure 21

INTEGRATED CIRCUIT ASSEMBLY

This 36-lead integrated circuit complex is smaller than a postage stamp and includes 285 gates fabricated on a single chip. It is used for access to computer memory circuits. (Fairchild TTµL 9035).

4-11 Integrated Circuits

The *integrated circuit* (IC) comprises a family in the field of microelectronics in which small, conventional components are combined in an orderly fashion in compact, high-density assemblies (*micromodules*) as shown in figure 21. Integrated circuits may be composed of passive elements (resistors, capacitors, and interconnections), and active elements such as diodes and transistors. The IC family may be divided into *monolithic* and multichip, or *hybrid*, circuits. The former category consists of an entire circuit function constructed in a single semiconductor block. The latter consists of two or more semiconductor blocks, each containing active or passive elements interconnected to form a complete circuit and assembled in a single package.

Integrated circuits offer relief in complex systems by permitting a reduction in the number of pieces and interconnections making up the system, a reduction in overall system size, better transistor matching and potentially lower system cost.

Using very small monolithic IC's makes it possible to make thousands of circuits simultaneously. For example, several hundred *dice* (plural of die) may be produced side by side from a single silicon slice in the simultaneous processing of about a hundred slices. Each die contains a complete circuit made up of ten to one hundred or more active and inactive components.

The silicon slice is prepared by an *epitaxial* process, which is defined as "the placement of materials on a surface." Epitaxy is used to grow thin layers of silicon on the slice, the layer resistivity controlled by the addition of N-type or P-type impurities (*diffusion*) to the silicon atoms being deposited. When localized regions are diffused into the base material (*substrate*), isolated circuits are achieved. Diffusion of additional P-type or N-type regions forms transistors.

Once the die is prepared by successive diffusions, a photomasking and etching process cuts accurately sized-and-located windows in the oxide surface, setting the circuit element dimensions simultaneously on every circuit in the slice. The wafer is then coated with an insulating oxide layer which can be opened in areas to permit metalization and interconnection.

The metalization process follows next, connecting circuit elements in the substrate. Electrical *isolation barriers* (insulators) may be provided in the form of reverse-biased PN junctions, or the resistance of the substrate may be used. Dielectric insulation, making use of a formed layer around a sensitive region is also employed. Successive diffusion processes produce transistors and circuit elements of microscopic size, ready to have external leads bonded to them, and suitable for encapsulation.

Typical IC dice range in size from less than 0.02″ square up to 0.08″ × 0.2″. Many package configurations are used, the most popular being the *multipin TO-5* package, the *dual in-line package*, the *flat package*, and the inexpensive *epoxy package*.

Digital and Linear IC's   Integrated circuits may be classified in terms of their functional end-use into two families:

Digital—A family of circuits that operate effectively as "on-of" switches. These circuits are most frequently used in com-

puters to count in accord with the absence or presence of a signal.

Linear (Analog)—A family of circuits that operate on an electrical signal to change its shape, increase its amplitude, or modify it for a specific use.

The *differential amplifier* is a basic circuit configuration for ICs used in a wide variety of linear applications (figure 22). The circuit is basically a balanced amplifier in which the currents to the emitter-coupled differential pair of transistors are supplied from a constant-current source, such as a transistor. An *operational amplifier* is a high-gain direct-coupled amplifier which is designed to use feedback for control of response characteristics (figure 23). The circuit symbol for these amplifiers is a triangle with the apex pointing in the direction of operation.

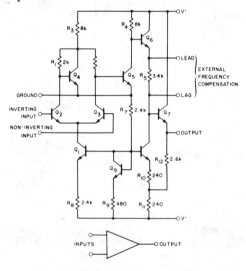

Figure 23

OPERATIONAL INTEGRATED-CIRCUIT AMPLIFIER

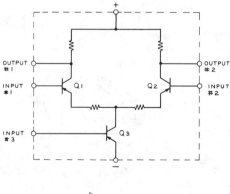

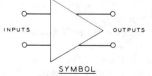

Figure 22

DIFFERENTIAL INTEGRATED-CIRCUIT AMPLIFIER

The MOSFET IC   The basic monolithic bipolar IC requires a seven-mask process; that is, seven different photographic masks (negatives) must be used in diffusion, etching, and oxidizing cycles. The necessity for all of these masks to *exactly* overlay (or register) is one very critical factor in getting the yield of an IC fabrication process up to a reasonable percentage of functional chips.

Another monolithic IC, that is more simple to fabricate, is the MOSFET type. The *MOSFET IC* is principally used in logic type functional blocks. Unlike the bipolar monolithic IC, no separate diffusion is necessary to make resistors—FETs are used as resistors as well as active devices. Since MOSFET's have capacitors inherent in them (gate to channel capacitance), the small capacitors needed are already present. So, with every device on the chip a MOSFET, only several maskings must be made. The smaller number of mask processes has the effect of increasing yields, or alternately allowing more separate elements to be put on the chip.

A simple MOS-IC circuit is shown in figure 25. This is a digital inverter, Q_1 serving as the active device and Q_2 functioning as a drain resistor. A typical MOS-IC chip has literally hundreds or thousands of circuits such as this on it, interconnected as a relatively complex circuit system block, such as a shift register.

4-12   Digital-Logic ICs

An electronic system that deals with discrete events based on digits functions on an "on-off" principle wherein the active devices in the system are either operating in one of two modes: *cutoff* (off) or *saturation* (on). Operation is based on *binary* mathematics using only the digits *zero* and *one*. In general, zero is indicated by a low signal voltage and one by a higher signal voltage. In a negative logic system the reverse is true, one being indicated by the most negative voltage.

In either case, the circuits that perform digital logic exercises may be made up of hundreds or thousands of discrete components, both active and inactive. Logic dia-

Figure 24

I-C CIRCUIT BOARD PERFORMS AS VOLTAGE REGULATOR

Complicated circuitry is reduced to printed-circuit board, eight "in-line" IC's and ten TO-5 style IC's. Transistor version would occupy many times this volume and have hundreds of discrete components. Final voltage regulator IC is at left with heat sink.

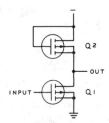

Figure 25

BASIC MOS INTEGRATED CIRCUIT

Device Q_1 serves as active device and Q_2 serves as drain resistor.

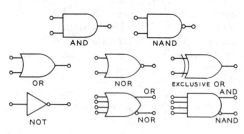

Figure 26

EXAMPLES OF SYMBOLIC LOGIC CIRCUITRY

grams show symbols based on the specific functions performed and not on the component configuration which may consist of many microscopic particles on a semiconductor chip. Typical examples of symbolic circuitry are shown in figure 26.

RTL Logic   The earliest practical IC logic form was *resistor-transistor logic* (RTL). A basic building block of RTL is the inverter or *NOT* gate (figure) 27A), whose output is the opposite or complement of the input level. The output and input levels, thus, are *not* the same. The *NOR* gate is shown in figure 27B. These gates, plus the *NAND* gate permit the designer to build up *OR* and *AND* gates, plus multivibrators and even more complicated logic functions.

The NOR gate (not OR) makes use of two or more bipolar devices. If both NOR inputs are at ground (state "0"), then the output level is at $+3.6$ volt in this example (state "1"). However, if either input A or input B is at a positive level, then the

output level drops to a voltage near ground. The logic statement expressed in binary mathematics by the NOR gate is (in Boolean algebra): $A+B=\bar{C}$, or if A or B is one, then C is *zero*. Simply, the statement says input at gate 1 or gate 2 yields a *zero* (NOR) at the output.

By adding a NOT circuit after the NOR, or OR circuit is formed (figure 27C); now if either A or B are *one*, then C is *one*. In Boolean notation: $A + B = C$.

If *one* is termed *true* and *zero* termed *false*, these terms relate the circuits to logic in the common sense of the word. An *AND* gate is shown in figure 27D.

These simple AND, OR, and NOT circuits can be used to solve complex problems, and systems may be activated by the desired combination of true and false input statements. In addition to use in logic functions, NAND, NOR, and NOT gates can be wired as astable (free-running) multivibrators, monostable (one-shot) multivibrators and Schmitt triggers. Representative examples of such functions are shown in figure 28.

DTL Logic   Some logic ICs are *diode transistor logic* (DTL) as shown in figure 29. Illustration A shows one-quarter of a quadruple-two-input NAND gate. The DTL configuration behaves differently than the RTL devices. If the two inputs of figure 29A are open ("high," or *one*), the output is "low," (or *zero*). If any input is grounded (*zero*), the output remains high. Current has to flow out of the diode inputs to place the output level at zero. This action is termed *current sinking*.

The portion of the two-input NAND gate shown in figure 29B is a member of the TTL family, all of which can be interfaced electrically with each other and with DTL as far as signal levels are concerned. It is possible to use logic ICs in linear circuits and figure 30 shows two crystal oscillators built around RTL and TTL integrated circuits.

RTL and DTL devices are inexpensive and easily used in system designs. The RTL devices require a $+3.6$-volt supply and the DTL devices require a $+5.0$-volt supply. Both these families suffer the disadvantage of

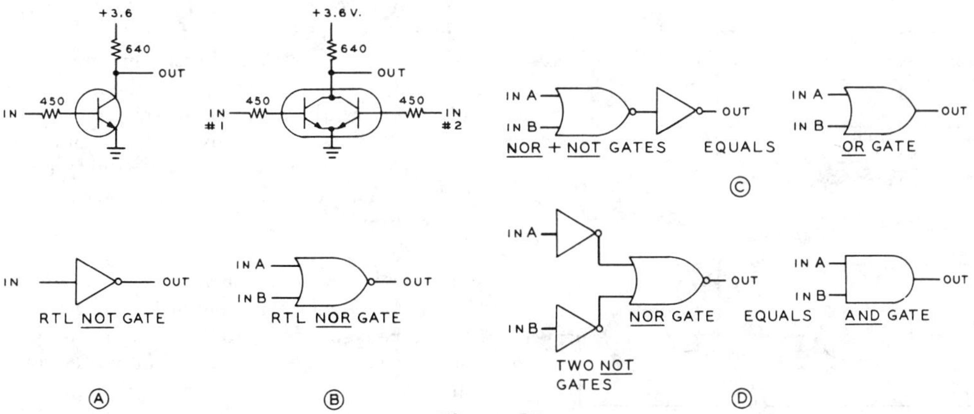

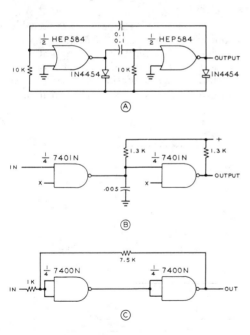

Figure 27

RTL LOGIC

A—Inverter, or NOT gate. B—Noninverting NOR gate. C—NOR plus NOT gates form OR gate. D—Two NOT gates plus NOR gate form AND gate.

low immunity to transient noise and are sensitive to r-f pickup.

Figure 28

TTL AND RTL GATES USED AS MULTIVIBRATORS AND TRIGGERS

A—Free-running multivibrator using RTL dual gate. B—Monostable multivibrator (one-shot) made from half of a TTL quad-gate. C—Schmitt trigger made from half of a TTL quad-gate.

Flip Flops and Counters A *flip flop* is a device which provides two outputs which can be driven to *zero-* and *one-*level combinations. Usually when one output is *zero*, the other is *one*. Flip-flop

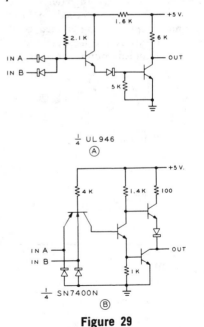

Figure 29

TTL AND DTL LOGIC GATES

A—DTL two input NAND gate using ¼ of µL 946. B—TTL two input NAND gate using ¼ of SN7400N.

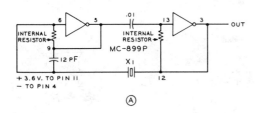

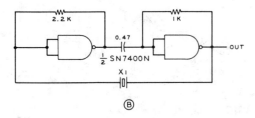

Figure 30

CRYSTAL OSCILLATORS USING RTL AND TTL INTEGRATED CIRCUITS

A—7 MHz oscillator using RTL dual buffer.
B—1 MHz oscillator using TTL gates.

devices may be interconnected to provide a *decade counter* (a divide-by-ten operation with ten input pulses required to provide one output pulse). A programmed counter can be used to divide frequencies by 2^n, 10, or any programmed number for service in

frequency counters and synthesizers. A decade divider made up of four flip flops is shown in figure 31. These flip flops are toggled or clocked devices which change state as a result of an input change.

Flip-flop devices to divide by a common integer are available on a single chip, a divide-by-ten counter such as shown being representative.

HTL Logic   Another form of DTL type logic device is designed to operate at a higher signal level for noise and transient immunity. *High Threshold Logic* (HTL) and *High Noise Immunity Logic* (HNIL) are devices often used in circuits that have relays and control power, such as those found in industrial systems. These families of ICs are generally operated from +12 to + 15 volts and special HTL HNIL devices are available to interface with the less expensive RTL, DTL, and TTL families.

ECL Logic   *Emitter-coupled logic* (ECL) is a very high speed system capable of operation as high as 1200 MHz with certain devices. A typical ECL configuration is shown in figure 32. ECL operates on the principle of nonsaturation of the internal transistors. Logic swings are reduced in am-

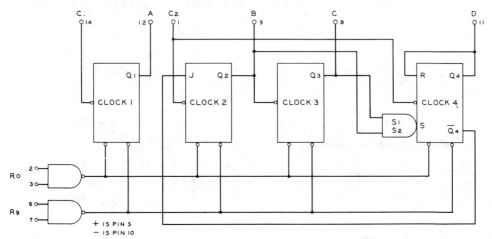

Figure 31

SN 7490N USED AS DECADE DIVIDER

Decade divider is made up of four flip-flop devices which provide zero and one level combinations. If R_O and R_9 terminals are grounded and terminals 1 and 12 jumpered, input frequency applied to terminal 14 will be divided by 10 and appear at terminal 11. Output waveform has 20% on-cycle.

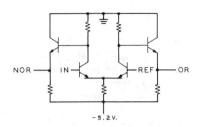

Figure 32

HIGH SPEED ECL LOGIC CIRCUIT

ECL device operates up to 350 MHz with non-saturation of internal transistors.

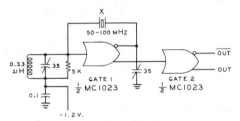

Figure 33

ECL CRYSTAL-CONTROLLED OSCILLATOR

Frequency range is 50 MHz to 100 MHz dependent on crystal and resonant circuit tuning.

plitude and the fact that the stored charge of a saturated transistor does not have to be discharged results in the speed increase. ECL is, by convention, operated from a — 5.2 volt source and the swing from *zero* to *one* in logic levels is comparatively small; *zero* being — 1.55 volt and *one* being — 0.75 volt. This is still considered to be "positive" logic because the most negative voltage level is defined as *zero*.

Representative nonlogic IC usage as a crystal-controlled oscillator and an astable multivibrator is shown in figure 33. Interface ICs are available to or from ECL and RTL, DTL, and TTL.

4-13 MOS Logic

Digital MOS devices have been recently developed that handle logic problems whose solution is impractical in other logic fami-

lies, such as problems requiring very high capacity memories. *Complementary MOS* (CMOS) will interface directly with RTL, DTL, TTL, or HTL if operated on a common power buss. Because of the low power consumption of CMOS, it is widely used for the frequency-divider IC in quartz-crystal-controlled watches.

A typical CMOS inverter is shown in figure 34. It makes use of a P-channel, N-channel pair (both enhancement-mode types). If the gates are high (*one*), then the N-channel MOSFET is on and the P-channel is off, so the output is low (*zero*). If the gates are low (*zero*), then the P-channel MOSFET is on and the N-channel is off, so the output is high (*one*). Note that in

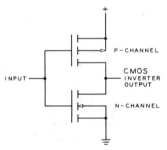

Figure 34

CMOS INVERTER

CMOS device makes use of P-channel, N-channel, enhancement-mode devices and provides low current consumption which is proportional to switching speed.

either state one device or the other is off and the inverter pair draws only a very small leakage current, with appreciable current being drawn only during the transition from *one* to *zero* and vice versa. The more transitions per second, the higher is the average current drawn, thus the power consumption of CMOS is directly proportional to the frequency at which it is switched.

As a result of the low power consumption and the simplifications of MOS-type fabrication CMOS is moving rapidly through medium scale integration (MSI), with hundreds of FETs per chip, into large scale integration (LSI), with thousands of FETs per chip—all in one package and at a relatively low cost.

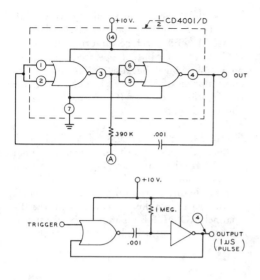

Figure 35

CMOS GATES USED AS MULTIVIBRATORS

A—Astable multivibrator using CD4001/D dual gates.
B—One-shot multivibrator using dual CMOS gates.

The CMOS devices now available allow for quite a large variety of circuitry, and like the types previously discussed, they may be used in nonlogic ways. Figure 35 shows how CMOS gates may be used as an astable multivibrator and a one-shot multivibrator.

CMOS is now available in two families; the original CD4000 series by *RCA* (second-sourced by at least six other suppliers) and the 74C00 family originated by *National Semiconductor*. The latter family has the same terminals and generally the same usage rules as the popular 7400 TTL logic family. Both CMOS families are compatible in logic levels and it remains to be seen which will become the dominant family.

**P-MOS Memory)
Logic** Conventional *P-MOS* (P-channel, enhancement mode) logic provides low cost, high capacity *shift registers* and *mem-*

ories. The shift register is a unique form of memory device which has one input and one output, plus a *clock* (timing) input. One commonly used P-MOS shift register has 256 bits of storage in it. The shift register may be compared to a piece of pipe just long enough to hold 256 marbles which are randomly colored white and black. The black marbles indicate a *one* value and the white marbles indicate a *zero* value. The sum of marbles makes up a 256-bit binary *word.* The pipe is assumed to be opaque so the sequence of marbles cannot be seen. In order to determine the binary word, it is necessary to push 256 marbles in at the input end of the pipe and observe each marble exiting from the output, noting the binary sequence of the marbles. Each marble pushed in the pipe is the equivalent of a clock pulse. In a real shift register the output is wired back to the input, 256 clock pulses are triggered, and the content of the register is read and the binary word is loaded back into the register.

The shift register form of memory represents a valid way of storing binary information but it is slow because interrogating the register takes as many clock pulses as the register is long. To speed up access to the content of a memory, it is possible to array the bits of storage in better ways.

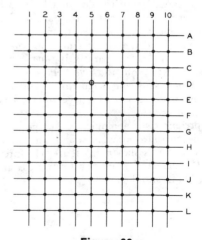

Figure 36

FERRITE-CORE MEMORY

Representation of core memory showing cores and sensing wires. Address of representative sample core is D-5. This configuration is termed a matrix.

A more efficient organization of a large memory bank is the use of a *ferrite-core memory,* such as shown in figure 36. A bit of information can be permanently stored in a core by having it magnetized or not magnetized. If the memory has a 30 × 30 *matrix,* there are 900 cores and 900 bits of storage. Any X-line and Y-line combination locates one particular core; this location is referred to as the core *address.*

If, instead of ferrite memory cores, a large number of MOS two-state circuits are arranged in a similar matrix, an *IC memory* is produced. Most small ICs, however, are pin-limited by their packaging and to bring out 60 leads from one package is a mechanical problem. The common package has 10 leads brought out for addressing purposes; five leads for the X-line, and five for the Y-line. By using all the lines in X and Y to define a location, $2^5 = 32$ X and Y coordinates are available, thus the total bit storage is thus $2^5 \times 2^5 = 1024$ bits of information.

The Random-Access Memory A *random-access* memory device (RAM) is organized in the above fashion and 32 × 32 is a common size. These memories can be written-into and read-out of, and are used for purposes where the stored information is of a changing nature, such as in signal processing systems. For this reason a RAM is often referred to as a *scratch-pad memory.*

There is a feature about MOS devices which is unique and which allows the manufacture of shift registers and RAMs that are unlike any other semiconductor memory. Since the gate of a MOSFET is a capacitor it will store a charge, making a complete two-state flip flop to store *ones* and *zeros* unnecessary if the data rate is high enough. Such a *dynamic register* will only hold data for about one millisecond. Each cell of the dynamic shift register is simpler than a cell of a static shift register so the dynamic type permits more bits on a chip and is cheaper per bit to manufacture.

The Read-Only Memory The *read-only memory* (ROM) can only be programmed once and is read in sequence. Certain ROMs, however, are made in reprogrammable versions, where the stored information can be changed. The ROM is used in a type of Morse code automatic keyer which employs a 256-bit device custom-programmed to send a short message, such as: CQ CQ DE W6SAI K. This type of program is permanently placed in the chip matrix in the manufacturing process by a photomask process. However, at least one semiconductor manufacturer makes a *programmable* ROM (pROM) that may be programmed in the field. The way in which a pROM is programmed is by subjecting the bits desired to be *zeros* to a pulse of current which burns out a fusible link of nichrome on the chip. Some manufacturers will program a pROM for the buyer to his specification for a nominal charge.

Another type of pROM has been developed that is not only programmable, but which may be erased and reprogrammed. The *avalanche-induced charge-migration* pROM is initially all *zeros.* By pulsing high current into each location where a *one* is desired, the device is programmed. This charge is apparently permanent, until a flash of ultraviolet light is directed through the quartz window atop the chip. Following the ultraviolet erasure, the pROM can be

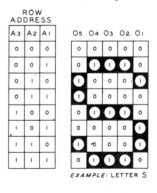

EXAMPLE: LETTER S

CHARACTER ADDRESS

	A4	A5	A6	A7	A8	A9
ASCII CHARACTER	1	1	0	0	1	0

Figure 37

TELETYPE-TO-CODE CONVERTER

Signetics 2513 ROM device produces letters and figures on screen of a cathode-ray tube from an ASCII teletype code input. ROM illustrates letter "S" readout.

programmed again. Some pROMs are available in up to 2048 bits, with 4096-bit capacity expected shortly.

Other ROM Devices There are several standard ROMS available that have factory mask programs of potential interest to the radio amateur. The *character generator* is useful for presenting letters and numerals on a cathode-ray tube such as is done in various electronic RTTY (radio teletype) terminal units. An example of such an ROM is the *Signetics 2513* which creates readable characters from an ASCII 8-level teletype code used in most time-shared computer terminals (figure 37).

Radio amateurs use the older 5-level *Baudot code* in their RTTY systems, but another ROM device can make the translation from Baudot to ASCII code. Still another

ROM is now available to generate "The quick brown fox jumps over the lazy dog 1 2 3 4 5 6 7 8 9 0."

4-14　　　Linear ICs

The *linear integrated circuit* is a device whose internal transistors operate in the amplification region rather than snapping back and forth from one state to another (such as cutoff to saturation). Some linear ICs are designed to replace nearly all the discrete components used in earlier composite equipment. Others perform unique functions heretofore unavailable.

Operational amplifiers, differential amplifiers and *diode-transistor arrays* are important members of the linear IC family.

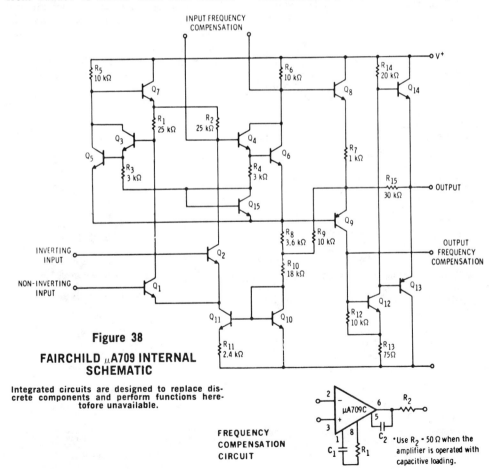

Figure 38

FAIRCHILD μA709 INTERNAL SCHEMATIC

Integrated circuits are designed to replace discrete components and perform functions heretofore unavailable.

The *Fairchild* μA700 series of linear monolith IC devices and particularly the μA709, are the most widely used linear IC types and more recent IC *operational amplifiers* (op-amps) are compatible in their pin configuration to this basic family of devices. The basic μA709 schematic is shown in figure 38, along with the equivalent op-amp symbol. Compensating networks may be required for stable operation and some of the newer op-amps have the necessary compensation built inside the package.

The Operational The perfect operational am-
Amplifier plifier is a high-gain dc coupled amplifier having two
differential inputs of infinite impedance, infinite gain, zero output impedance, and no phase shift. (Phase shift is 180° between the output and inverting input and 0° between the output and noninverting input).

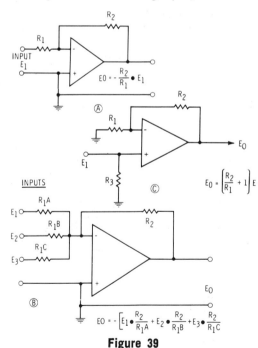

$$E_0 = -\frac{R_2}{R_1} \cdot E_1$$

Ⓐ

$$E_0 = \left(\frac{R_2}{R_1} + 1\right) E$$

Ⓒ

$$E_0 = -\left[E_1 \cdot \frac{R_2}{R_1 A} + E_2 \cdot \frac{R_2}{R_1 B} + E_3 \cdot \frac{R_2}{R_1 C}\right]$$

Ⓑ

Figure 39
OPERATIONAL AMPLIFIER (OP-AMP) SYMBOL

A—Differential amplifier in inverting mode. B—Summing amplifier. If input is applied to positive gate, output is subtractive. C—Differential amplifier using noninverting mode. R_3 is chosen to match input signal source.

Two voltages may be *added* in a differential amplifier as shown in figure 39. In illustration A, the noninverting (plus) input is grounded and the amplifier is in the *inverting* mode. The stage gain is the ratio R_2/R_1 and the input impedance is R_1. The circuit may be modified as shown in illustration C so that a noninverting gain of

$$\frac{R_2}{R_1} + 1 \text{ is obtained.}$$

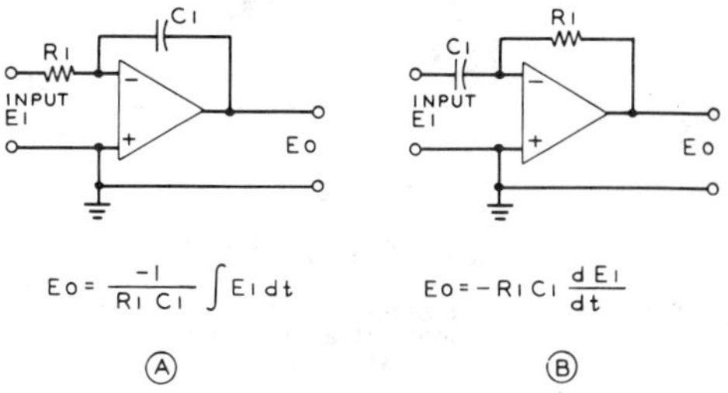

$$E_0 = \frac{-1}{R_1 C_1} \int E_1 \, dt$$

Ⓐ

$$E_0 = -R_1 C_1 \frac{dE_1}{dt}$$

Ⓑ

Figure 40
INTEGRATING AND DIFFERENTIATING AMPLIFIERS

A—Inverting integrating circuit. B—Inverting differentating circuit.

The op-amp can be connected to perform the *integral* or *differential* of the input voltage as shown in figure 40. By combining these operations in a number of coordinated op-amps an *analog computer* may be constructed. This type of machine represents the use of an electrical system as a model for a second system that is usually more difficult or more expensive to construct or measure, and that obeys the equations of the same form. The term *analog* implies similarity of relations or properties between the systems.

The Differential The *differential amplifier* is
Amplifier a dc-coupled amplifier having similar input circuits.
The amplifier responds to the difference between two input voltages or currents (figure 41). The differential amplifier may be compared to a push-pull stage fed from a constant current source.

Differential amplifiers are useful linear devices over the range from dc to the vhf spectrum and are useful as product detectors, mixers, limiters, frequency multipliers and r-f amplifiers. Various versions of the

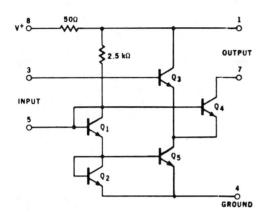

Figure 41

DIFFERENTIAL AMPLIFIER

The differential op-amp is a dual input dc coupled amplifier comparable to a push-pull stage
fed from a constant-current source.

differential amplifier are discussed in the following sections.

A widely used differential amplifier is the
r-f/i-f amplifier device used as an i-f amplifier at 10.7 MHz in f-m tuners. The Fairchild μA703, Motorola HEP-590 and the
Signetics NE-510 are typical examples of
this device. A representative amplifier-lim-
iter is shown in figure 42. These ICs can be
used for a variety of other purposes and an
a-m modulator using the HEP-590 is shown
in figure 43.

The National Semiconductor LM-373 IC
may be used for the detection of a-m, f-m,
cw, or SSB signals, as shown in figure 44.
Note that the gain of the LM-373 has been
divided into two blocks, with provisions for
insertion of an i-f bandpass filter between
the blocks.

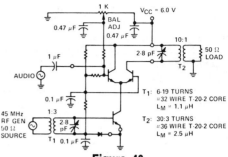

Figure 43

**HEP-590 IC USED AS
A-M MODULATOR**

Various ICs have been developed for use
as i-f/f-m detectors in TV receivers. One
unit comprises a complete 4.5-MHz TV
sound system using the quadrature method
of f-m detection similar to that employed
with the 6BN6 tube. This unit has a quadrature f-m detector, 10.7-MHz i-f, and limiter in one package (figure 45).

An IC package that is useful in signal
processing applications—especially SSB—is
shown in figure 46. The circuit is a balanced
demodulator for SSB detection.

The PLL IC   A recent development is the
phase-locked loop integrated
circuit which performs a remarkable range
of functions: selective amplifier, f-m detector, frequency multiplier, touchtone decoder,
a-m detector, frequency synthesizer, and
many more. The Signetics NE-560B shown
in figure 47 is configured as an f-m detector.
In this circuit the voltage-controlled oscillator (VCO) in the PLL locks itself into a
90° phase relationship with the incoming

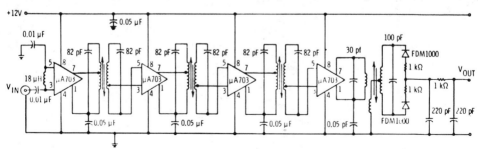

Figure 42

DIFFERENTIAL AMPLIFIERS IN R-F SERVICE

FAIRCHILD μA703 ICs used in f-m i-f amplifier and limiter.

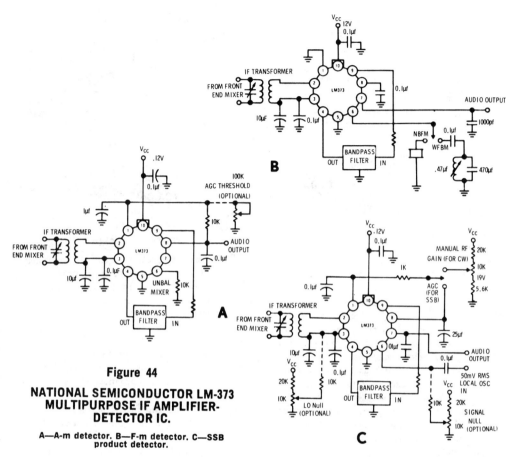

Figure 44

NATIONAL SEMICONDUCTOR LM-373 MULTIPURPOSE IF AMPLIFIER-DETECTOR IC.

A—A-m detector. B—F-m detector. C—SSB product detector.

Figure 45

SIGNETICS N5111A AS QUADRATURE F-M DETECTOR AT 10.7 MHz

carrier signal. Variations of this circuit are useful in solid-state color-TV receivers.

Diode-Transistor Arrays A category of linear ICs that is of great use comprises the *diode-transistor array* family, or *array* for short. The various types of arrays available contain a number of bipolar transistors inside the package which are more or less uncommitted to any particular configuration. Because of pin limitations there are necessarily some interconnections inside the package but there is still great flexibility to interconnect the transistors for a specific purpose. Examples of these array devices are the CA 3018, CA 3036, etc. of RCA. A voltage regulator built around the CA 3018 is shown in figure 48. Note that one of the internal transistor base-emitter junctions of the IC has been used as a breakdown diode for a voltage reference.

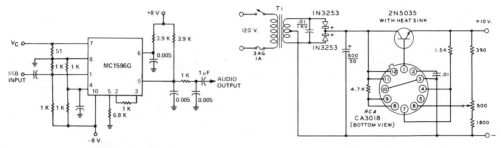

Figure 46

MC-1596G AS BALANCED MODULATOR FOR SSB DETECTION

Figure 48

RCA CA-3018 AS VOLTAGE REGULATOR

T_1—10/20/40 volts center tap. Triad F-91X. Use red and yellow leads.

This is only one of many circuits possible using an IC array.

Many other types of linear ICs exist: video amplifiers, logarithmic amplifiers, TV chroma demodulators, stereo-multiplex demodulators, squelch amplifiers, and so on. These represent special interest areas and it would be impossible to treat each category here. Looking at the large market areas wherein linear electronics is used, the experimenter will find ICs available or being designed for TV receivers, auto ignition systems, CATV distribution, a-m/f-m radios, stereo gear, and camera equipment. Doubtless many of the ICs developed for these markets will be readily usable in the radio communications field.

The area of greatest growth in recent years in linear ICs has been in power-handling capability. Most manufacturers of ICs now have IC regulators and audio output devices that will handle reasonable amounts of power without external transistors to assist them.

The *Fairchild* μA-7800 and the *National* LM-309K families of three terminal regulators are of particular interest. The LM-309K is a 5-volt regulator capable of 1 ampere of regulated output. It is only necessary to connect the common pin (the case of the TO-3 package) to ground and the unregulated input to the input pin; output is then

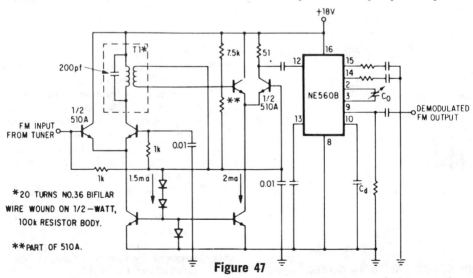

*20 TURNS NO.36 BIFILAR WIRE WOUND ON 1/2 —WATT, 100k RESISTOR BODY.

**PART OF 510A.

Figure 47

SIGNETICS NE-560B PHASE-LOCKED LOOP AS AN F-M DETECTOR

taken from the output pin (a 0.22 μF capacitor across the input is required for stability). The *Fairchild* regulators offer similar performance in a variety of positive regulated output voltages. *Motorola* offers a similar family, the MC-7800C series of devices.

Both *Fairchild* and *Motorola* also offer a 7900 series of negative, three-terminal regulators as well as *National Semiconductor*.

Raytheon has introduced the RC-4194TK in a 9-pin, TO-66 package. This device produces plus and minus regulated voltages for operating linear ICs. This regulator can pass up to 250 mA at an output voltage adjustable from zero up to 42 volts.

In the audio-amplifier area, ICs are now available at output levels up to 15 watts. The *National* LM-380 will drive an 8-ohm speaker up to 5 watts output and other devices are on the market that will drive a 4-ohm speaker up to 15 watts.

4-15   Solid-State Light Sources and Numeric Displays

A recent development is the *light-emitting diode* (LED) which promises to replace the incandescent lamp as a light source in displays—especially those subject to heavy vibration.

The first LEDs were a deep red in color and made of Gallium-Arsenide-Phosphide and produced about 30 to 100 microwatts of light power output. More recently, the green LED of Gallium-Phosphide and the amber LED of Gallium-Arsenide-Phosphide have been made available.

Small LEDs have a forward voltage drop of about 1.5 to 2.0 volts and they can be driven up to about 40 ma. The LED does not have a sudden end-of-life as does an incandescent lamp, instead the LED loses brilliance with age. Predicted life (to half brilliance) of a typical LED is 10^6 hours.

Another type of LED is the *infrared diode* which has maximum radiation at about 9000 Angstrom units (10^{-10} meters) wavelength in the near-infrared region. Because it radiates just outside the visible spectrum, the infrared produced by this Gallium-Arsenide diode is treated in the same manner as visible light, using conventional optics. The IR output of these diodes is very close to the op-

optimum sensitivity of most silicon photodiodes, light-sensitive transistors, and FETs. The IR LED can be modulated (even at megahertz rates) and serves as a transmitter in voice and data links or as an intruder alarm. A Gallium Arsenide emitter and Silicon detector may be combined in an *opti-*

Figure 49

HEWLETT-PACKARD SOLID-STATE NUMERIC INDICATORS

cally coupled isolator (opto-electronic switch) which combines the pair in an opaque, plastic package. Light then couples the input circuit of the emitter to the output circuit of the detector, with no electrical coupling between the ports. This isolator is the equivalent of a relay, with none of the mechanically fragile components.

An array of LEDs can be configured as a seven-segment display for numeric indication and integrated circuits are available that will convert the binary-coded decimal system to the seven-segment coding required for this display.

A solid-state numeric indicator is shown in figure 49. This small unit is a hybrid microcircuit consisting of a decoder-driver and an array of light-emitting diodes. The numeric indicator is enabled by a pulse and the display will follow changes on the logic inputs as long as the enable port is held at zero (low). In this mode the device is operated as a real-time display. When the enable line rises (high), the latches retain the current inputs and the display is no longer affected by changes on the logic input ports. The decimal point voltage low corresponds to point illumination.

Other Digital Displays In addition to light-emitting diodes, other forms of digital display exist. The *liquid crystal* display provides a brilliant indication that consumes very little power and can be driven at low voltage by CMOS circuitry. The display consists of a sandwich of two thin glass sheets, coated on their inner surfaces with a thin transparent conductor such as indium oxide. The conductor is etched into seven bars of the standard 7-segment display format (figure 50). Each bar, or segment, is electrically separate and can be selected by a logic driver circuit so that any numeral can be formed.

The interior of the cell is filled with a liquid crystal material whose molecular order is disturbed when an electric field is applied to the segments. The optical appearance of the crystal is thus altered to display the digits.

Because crystal displays are relatively fast, it is necessary to drive them at a frequency which is above the observable flicker rate.

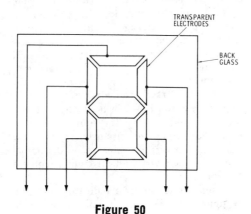

Figure 50
LIQUID-CRYSTAL DISPLAY

Typical drive frequencies are in the 30-Hz to 60-Hz range.

Gas Discharge displays provide a large digit and are composed of a glass sandwich on which the segments are silk-screened. The

unit is sealed and filled with a neon gas mixture. The display anodes and the keep-alive cathode are sequentially gated, one anode at a time to create the appropriate number or character. The display is cycled at about 80 Hz to remain flicker-free.

Large displays often make use of a matrix, or array, of special *incandescent lamps* arranged to form the desired characters when the lamps are appropriately driven. Lamps are available for this purpose in a wide range of size, color and style.

Opto-Isolators and Solid-State Relays The fact that a gallium-arsenide Infrared Emitting Diode produces an infrared output which closely matches the spectral sensitivity of a silicon transistor has led to the creation of a broad category of *opto-isolators*. This device consists of an infrared diode and silicon transistor in a single opaque package. The *Monsanto* MCT series is representative of such devices, although they are also made by others. Figure 51 illustrates the use of a simple opto-isolator. If the base of the transistor Q_1 is left uncon-

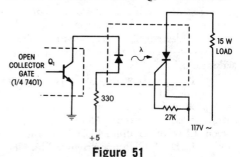

Figure 51
HALF-WAVE CONTROL OF AC CIRCUIT WITH SCR-PHOTOCOUPLER

nected, the device has only four terminals: two input and two output. The ratio of current flowing in the output to that flowing in the input is expressed as the *current transfer ratio*. This figure can be as low as thirty-five percent for simple IR diode-transistor opto-couplers and as high as five hundred percent for those using a Darlington pair as the phototransistor. A full-wave opto-isolator circuit is shown in figure 52.

The *solid-state* relay (SSR) is an extension of the diode-isolator. It is capable of alternating current control up to 50 amperes as

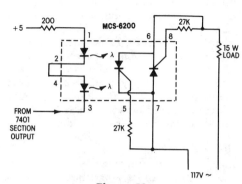

Figure 52

FULL-WAVE CONTROL OF AC CIRCUIT WITH SCR PHOTOCOUPLER

high as 440 volts. In addition most SSRs have zero-voltage turn-on and zero-current turn-off features. This aids interference problems since the SSR never turns on the ac load at the peak voltage of the cycle, nor turns it off at a current peak. Figure 53 represents a *Monsanto* solid-state relay.

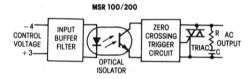

Figure 53

ELECTRICAL CIRCUIT OF SOLID-STATE RELAY

Zero voltage "on" and zero current "off" eliminates damaging high current surges, high voltage transients and avoids RFI from these sources.

4-16 The Microprocessor

A recent LSI addition to the logic IC area is the *microprocessor*. This device consists of various ICs on a chip and resembles a small-scale version of the central processor in a computer. It is thus often called a "computer on a chip." This is not literally true, since a computer comprises more than a central processor, but the microprocessor is a powerful data processing device when provided with support components, such as memories, input output elements, and signal processors.

As communication equipment becomes more complex, it is reasonable to expect to see microprocessors built in receivers, transmitters, Morse code keyers, and combinations of these as a sequence controller to make the operator-equipment interface simpler.

The Basic MPU The microprocessor unit (MPU) is a single integrated circuit chip that contains some of the processing power of a small computer. It

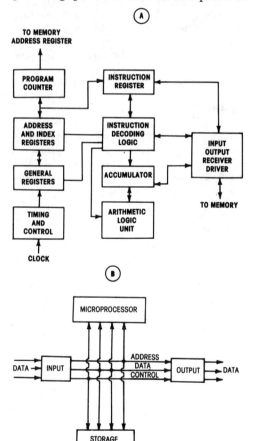

Figure 54

A—BLOCK DIAGRAM OF MICROPROCESSOR
B—BASIC MICROPROCESSOR STRUCTURE

can perform arithmetic, logic, and decision making as well as communicate with input and output devices controlled by instructions stored in the memory. The MPU can therefore replace a large number of digital integrated circuits. The flexibility of the MPU permits it to perform many tasks depending on the programming that is used.

The basic microprocessor structure is shown in figure 54. The system is interconnected through *address, data,* and *control buses.* The microprocessor develops the address and control bus signals used by the other system elements. The data bus is bidirectional and permits the microprocessor to exchange data with the other elements. The buses allow data to be exchanged or transferred between the elements of the micropocessor system in the form of ones and zeros.

The microprocessor transforms data present at its input and controls which element the data is transferred to in accordance with the program the microprocessor is executing. The input section accepts data for processing and the output section presents processed data for use. Data may be transferred to and from the input/output devices in either parallel or serial form. Series data transmission is slower but requires less cable than parallel data transmission. The storage element saves values for future use. This memory can consist of *Random Access Memories* (RAM), *Read Only Memories* (ROM), magnetic tape, floppy disc, or other digital storage devices. The RAM permits the microprocessor to read or write for it by placing the address on the address bus, then manipulating the data on the data bus. The ROM allows the microprocessor to read from the storage area (it cannot write or change the data). This permits programming material to be entered once and thereafter it may be referred to by the microprocessor. The material will not be lost when the system is powered down. This programming when entered into a ROM or PROM (Programmable Read Only Memory) is called *firmware.*

The word size of the microprocessor is usually defined as the *width* of the data bus, and the most basic classification of this device relates to its optimum *bit* handling capability. Four-bit processors are used in small control systems where speed is not required. Larger 8-, 12-, and 16-bit processors are used in more sophisticated computer systems.

CHAPTER FIVE

Electron Tubes

Part I—Principles of Operation

In the previous chapters the manner in which an electric current flows through a conductor as a result of electron drift has been discussed. This drift, which takes place between the ends of the conductor, is in addition to the normal random electron motion between the molecules of the conductor.

Devices that utilize the flow of free electrons in a vacuum are referred to as *vacuum tubes*, or, simply, *tubes or valves* (in Europe). Since the current flow in a tube takes place in an evacuated enclosure, there must be located within the enclosure a source of electrons and a collector for the emitted electrons. The source is termed the *cathode*, and the collector the *anode*, or *plate*.

Emission of electrons from a heated surface is called *thermionic* emission. This surface is called a *filament, heater,* or *cathode*.

The most efficient cathode is the oxide type which operates at an orange-red temperature (1100°K) and is used for receiving tubes and low-power transmitting tubes. The thoriated-tungsten filament is used in medium- and high-power transmitting tubes and operates at a temperature of about 2000°K.

The *heater-cathode* emitter (figure 2) provides an emitter which can be operated from alternating current yet does not introduce any hum modulation on the cathode emission. The heater may be operated at any voltage from 2.5 to 120 volts, although 6.3 and 12.6 volts are the most common values.

5-1 Tube Types

The Diode The *diode* is a two element tube made up of a cathode that emits electrons and an anode, or plate. If a source of dc voltage is placed in the external circuit between plate and cathode so that a positive potential is on the plate with respect to the cathode, electrons are drawn away from the cathode.

At moderate plate voltages the cathode current is limited by the space charge but increasing the plate voltage will increase the electron flow. The space charge is a cloud of negatively charged electrons in the vicinity of the cathode. Plate current is determined by the plate potential and is substantially independent of the electron emission of the cathode. When limited by space charge, plate current is proportional to the three-halves power of the plate voltage, the total plate current being expressed as:

$$I_b = KE_b^{3/2}$$

where,

K is a constant (perveance) determined by the geometry of the tube,

E_b is the anode voltage with respect to the cathode.

At high values of plate voltage the space charge is neutralized and all cathode electrons are attracted to the plate. The diode is now saturated and a further increase in plate voltage will cause only a small increase in plate current.

The plate current flowing in the plate-cathode area of a conducting diode represents the energy required to accelerate electrons from the potential of the cathode to that of the plate. When the accelerated electrons strike the plate, the energy associated with their velocity is released to the anode and appears as heating of the plate or anode structure.

The Triode The *triode* can be thought of as a diode with a control electrode added between cathode and plate (figure 3). The control electrode is called the *grid* and it serves as an imperfect electrostatic shield, permitting some but not all of the electrons

Figure 1

VHF AND UHF TUBE TYPES

At the left is an 8058 nuvistor tetrode, representative of the family of small vhf types useful in receivers and low-power transmitters. The second type is a 6816 planar tetrode rated at 180 watts input to 1215 MHz. The third tube from the left is a 3CX100A5 planar triode, an improved and ruggedized version of the 2C39A, and rated at 100 watts input to 2900 MHz. The fourth tube from the left is the X-843 (Eimac) planar triode designed to deliver over 100 watts at 2100 MHz. The tube is used in a grounded-grid cavity configuration. The tube to the right is a 7213 planar tetrode, rated at 2500 watts input to 1215 MHz. All of these vhf/uhf negative grid tubes make use of ceramic insulation for the lowest envelope loss at the higher frequencies and the larger ones have coaxial bases for use in resonant cavities.

to reach the plate. If the grid charge is made sufficiently negative, all electrons leaving the cathode will be repelled back to it and the plate current will be reduced to zero. The grid control voltage is called the *grid bias* and the smallest negative voltage that will cause plate current cutoff at a particular plate voltage is called the *cutoff bias.*

Plate current in a triode is the result of the net field at the cathode caused by grid and plate voltages. The ratio between the change in grid bias and change in plate voltage which will cause the same small change in plate current is called the *amplification factor* and is defined as:

$$\mu = \frac{\Delta E_b}{\Delta E_c}$$

where,

μ equals amplification factor,
E_b equals plate voltage,
E_c equals grid voltage,
Δ represents a small increment.

The amplification factor of modern triodes ranges from 3 to over 200.

The cathode current in a triode is proportional to the three-halves power of $(E_c +$

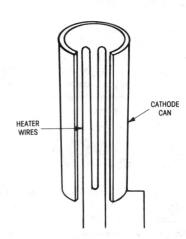

Figure 2

Cut-open view of a cathode-heater assembly. The indirectly heated cathode can is a source of electrons when heater wires reach operating temperature.

E_b/μ). The cutoff bias corresponds to $-E_b/\mu$.

Other important coefficients of the triode tube are:

Plate Resistance—The dynamic plate resistance (r_p) is the ratio of change in plate

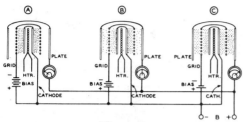

Figure 3

ACTION OF THE GRID
IN A TRIODE

(A) shows the triode tube with cutoff bias on the grid. Note that all the electrons emitted by the cathode remain inside the grid mesh. (B) shows the same tube with an intermediate value of bias on the grid. Note the medium value of plate current and the fact that there is a reserve of electrons remaining within the grid mesh. (C) shows the operation with a relatively small amount of bias which with certain tube types will allow substantially all the electrons emitted by the cathode to reach the plate.

current which a small change in plate voltage produces, and is expressed in ohms.

$$r_p = \frac{\Delta E_b}{\Delta I_b}$$

where,

I_b equals plate current

Conductance—The conductance (G_m) is the ratio of a change in plate current to a change in grid voltage which brought about the plate current change, the plate voltage being constant,

$$G_m = \frac{\Delta I_b}{\Delta E_c} = \frac{\mu}{r_p}$$

5-2 Operating Characteristics

The Load Line A *load line* is a graphical representation of the voltage on the plate of a vacuum tube and the current passing through the plate circuit of the tube for various values of plate load resistance and plate supply voltage. Figure 4 illustrates a triode tube with a resistive plate load, and a supply voltage of 300 volts. The voltage at the plate of the tube (e_b) may be expressed as:

$$e_b = E_b - \left(i_b \times R_L \right)$$

where,

E_b equals plate supply voltage,
i_b equals plate current,
R_L equals load resistance in ohms.

Assuming various values of i_b flowing in the circuit, controlled by the internal resistance of the tube (a function of the grid bias), values of plate voltage may be plotted as shown for each value of plate current (i_b). The line connecting these points is called the *load line* for the particular value of plate load resistance used. At point A on

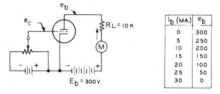

I_b (MA)	e_b
0	300
5	250
10	200
15	150
20	100
25	50
30	0

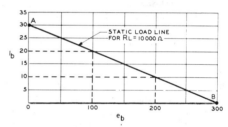

Figure 4

THE STATIC LOAD LINE FOR A
TYPICAL TRIODE TUBE WITH A
PLATE LOAD RESISTANCE OF
10,000 OHMS

the load line, the voltage across the tube is zero. This would be true for a perfect tube with zero internal voltage drop, or if the tube is short-circuited from cathode to plate. Point B on the load line corresponds to the cutoff point of the tube, where no plate current is flowing. The operating range of tube lies between these two extremes.

When the signal voltage applied to the grid has its maximum positive instantaneous value the plate current is also maximum. Reference to figure 4 shows that this maximum plate current flows through plate-load resistor R_L, producing a maximum voltage drop across it. The lower end of R_L is connected to the plate supply, and is therefore held at a constant potential of 300 volts. With maximum voltage drop across the load resistor, the upper end of R_L is at a minimum instantaneous voltage. The plate of the tube is connected to this end of R_L and is therefore at the same minimum instantaneous potential.

The Tetrode and Pentode   Additional grids can be added to a tube to create a *tetrode* (four elements) or a *pentode* (five elements). The additional grids modify the voltage and current relations within the tube in a way that is useful for many purposes. The extra grids also provide electrostatic shielding between the control grid and the plate. This shielding is important when the tube is used to amplify very high frequencies.

In the case of the tetrode, when the screen voltage is held at a constant value, it is possible to make large changes in plate voltage without appreciably affecting the plate current.

The pentode tube incorporates a *suppressor grid* which is operated at near-cathode potential. The characteristic curves of a typical pentode are shown in figure 5. The suppressor grid allows efficient operation of the tube at low values of plate voltage, increasing the lower limit to which the instantaneous plate voltage may approach under conditions of excitation.

Important coefficients of the triode and pentode tube are:

Grid-screen Mu Factor—The grid screen μ factor (μ_S) is analogous to the amplification factor in a triode, except that the screen element is substituted for the plate,

$$\mu_S = \frac{\Delta E_{C2}}{\Delta E_{C1}}$$

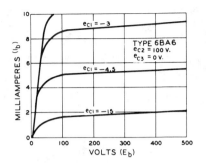

Figure 5

TYPICAL I_b VS. E_b PENTODE CHARACTERISTIC CURVES

where,

E_{C2} equals screen voltage,
E_{C1} equals grid voltage.
and I_{C2} is held constant.

Cathode current (I_k) in a tetrode or pentode is expressed by,

$$I_k = k\left(E_{C1} + \frac{E_{c2}}{\mu_s} + \frac{E_b}{\mu}\right)^{3/2}$$

Note that total cathode current is relatively independent of plate voltage in a tetrode or pentode.

Conductance—The conductance of a tetrode or pentode is equal to:

$$G_m = \frac{\Delta I_b}{\Delta E_c}$$

where,

E_{c2} and E_b are constant.

Plate resistance—The plate resistance of a tetrode or pentode is equal to:

$$r_p = \frac{\Delta E_b}{\Delta I_b}$$

where,

E_{c1} and E_{c2} are constant.

Part II—Electron Tube Amplifiers

The ideal electron tube amplifier should have an infinite input impedance, zero output impedance and a high forward gain. Thus, it takes no input power but can furnish an unlimited output power. In addition, it is unilateral, in that its input circuit is

not affected by the voltage at the output circuit. Practical amplifiers differ from this ideal in many respects.

While the advent of the transistor has limited the use of the vacuum tube in many cases, it is still widely used in special applications. The voltage handling capability of the vacuum tube satisfies the requirements for high-power circuits and for pulse generators for radar and other specialized equipment.

Knowledge of the operation of vacuum-tube circuits, however, is helpful to the experimenter and the generalized knowledge of vacuum-tube circuitry is useful in the study and application of advanced solid-state devices.

5-3 Classes and Types of Vacuum-Tube Amplifiers

Vacuum-tube amplifiers are grouped into various classes and subclasses according to the type of work they are intended to perform. The difference between the various classes is determined primarily by the angle of plate-current flow, the value of average grid bias employed, and the maximum value of the exciting signal impressed on the grid circuit.

Class-A Amplifier A *class-A amplifier* is an amplifier biased and supplied with excitation of such amplitude that plate current flows continuously (360° of the exciting voltage waveshape) and grid current does not flow at any time. Such an amplifier is normally operated in the center of the grid-voltage plate-current transfer characteristic and gives an output waveshape which is a substantial replica of the input waveshape.

Class-AB Amplifier *Class-AB* signifies an amplifier operated under such conditions of grid bias and exciting voltage that plate current flows for more than one-half the input voltage cycle but for less than the complete cycle. In other words the operating angle of plate current flow is appreciably greater than 180° but less than 360.°

Class-B Amplifier A *class-B amplifier* is biased substantially to cutoff of plate current so that plate current flows essentially over one-half the input voltage cycle. The operating angle of plate-current

flow is 180°. The class-B amplifier is excited to the extent that grid current flows.

Class-C Amplifier A *class-C amplifier* is biased to a value greater than the value required for plate-current cutoff and is excited with a signal of such amplitude that grid current flows over an appreciable period of the input-voltage waveshape. The angle of plate-current flow in a class-C amplifier is appreciably less than 180°, or in other words, plate current flows less than one-half the time. Class-C amplifiers are not capable of linear amplification as their output waveform is not a replica of the input voltage for all signal amplitudes.

Types of Amplifiers There are three general types of amplifier circuits in use. These types are classified on the basis of the *return* for the input and output circuits (figure 6). Conventional amplifiers are called *grid-driven amplifiers,* with the cathode acting as the common return for both the input and output circuits. The second type is known as a plate-return amplifier or *cathode follower* since the plate circuit is effectively at ground for the input and output signal voltages and the output voltage or power is taken between cathode and plate. The third type is called a *cathode-driven* or *grounded-grid* amplifier since the grid is effectively at ground potential for input and output signals and output is taken between grid and plate.

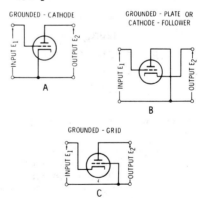

Figure 6

TYPES OF AMPLIFIERS

Bias Considerations   The average potential difference between the control grid and cathode is called the *grid bias*, or simply *bias*. The bias value is chosen to place the tube in the desired class of operation.

Bias may be obtained from an external source, such as a battery, or provided by the voltage drop across a cathode resistor, as shown in figure 7A-B. In the latter case, the cathode of the tube is placed at a positive voltage with respect to the grid. A capacitor is commonly placed across the bias resistor to provide a low impedance path to ground for the signal component of the cathode current.

A third method of providing bias is shown in figure 7C. This is termed *grid-resistor* bias. During the positive portion of the input cycle, a small amount of grid current flows from grid to cathode, charging capacitor C_c. During the negative portion of the signal cycle, the discharge path of the capacitor is through the grid resistor. Discharge time constant is quite long in comparison to the period of the input signal and only a small portion of the charge is lost. Thus the bias developed is substantially constant and the average grid potential does not follow the positive portion of the input signal.

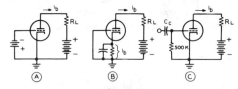

Figure 7

TYPES OF BIAS SYSTEMS

A—Fixed bias
B—Cathode bias
C—Grid-resistor bias

Distortion in Amplifiers   There are four main types of distortion that may occur in amplifiers: frequency distortion, phase distortion, amplitude distortion and intermodulation distortion.

Frequency distortion occurs when some frequency components of a signal are amplified more than others. It may occur at low frequencies if coupling capacitors are inadequate, or at high frequencies as a result of excessive circuit distributed capacities.

Phase distortion occurs when a harmonic of an input signal is shifted in time with relation to the fundamental signal. Phase shift of a sine wave has no effect on the output signal, however, when a complex wave is passed through a coupling circuit each component frequency of the wave may be shifted in phase by a different amount, so the output signal is not a reproduction of the input waveform.

Amplitude distortion occurs when a tube is operated on a nonlinear portion of its characteristic curve. In such a region, a change in grid voltage does not result in a change in plate current which is directly proportional to the grid voltage change.

Intermodulation distortion is a result of a change in stage gain with respect to signal level when the stage is driven by a complex signal having more than one frequency. This form of distortion occurs in any nonlinear device and generates spurious frequencies falling within the passband of the amplifier. The subject of intermodulation distortion is covered in Chapter 7 in more detail.

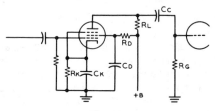

Figure 8

STANDARD CIRCUIT FOR RESISTANCE-CAPACITANCE COUPLED PENTODE AMPLIFIER STAGE

5-4   Interstage Coupling Circuits

The circuit of a *resistance-coupled* amplifier is shown in figure 8. Resistor R_L is the load across which the output signal is developed. Capacitor C_c prevents the plate voltage from being applied to the grid circuit of the next stage and its value should be large enough to offer a low reactance to the lowest frequency being amplified. The grid-resistor, R_G, should offer a very high resistance in order that the shunting effect of the resistor and coupling capacitor on the circuit is small.

The response of a *resistance-coupled* amplifier varies with frequency as shown in figure 9 for the case of a pentode tube. Amplification is constant over a midband frequency range, falling off at the lower and higher frequencies. Low-frequency falloff is caused by the reactance of the coupling capacitor and high-frequency falloff is due to shunting effect of circuit capacitances.

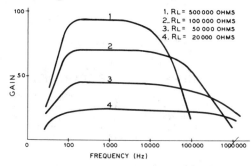

Figure 9

THE VARIATION OF STAGE GAIN WITH FREQUENCY IN AN RC-COUPLED PENTODE AMPLIFIER FOR VARIOUS VALUES OF PLATE-LOAD RESISTANCE

In the *transformer-coupled* amplifier, the load impedance in the plate circuit of the tube is the primary of a transformer, the secondary of which is connected to the grid of the succeeding stage. The transformer winding may be centertapped for push-pull service. Transformer coupling is commonly used to drive a class-B power amplifier which requires driving power combined with good driver regulation.

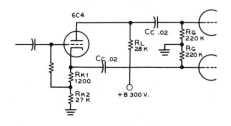

Figure 10

TYPICAL PHASE-INVERTER CIRCUIT WITH RECOMMENDED VALUES FOR CIRCUIT COMPONENTS

A *phase inverter* is a form of resistance coupling which provides voltages equal in amplitude but opposite in polarity for a push-pull stage (figure 10). There are a large number of phase-inversion circuits but the one shown is most satisfactory from the point of view of the number of components used and the accuracy of the two out-of-phase voltages. The circuit is based on the principle that a 180° phase shift occurs between the cathode and plate load circuits, (R_L and R_{K2}). Cathode bias is supplied by resistor R_{K1}.

5-5 The Feedback Amplifier

It is possible to modify the characteristics of an amplifier by feeding back a portion of the output to the input. All components, circuits, and tubes included between the point where the feedback is taken off and the point where the feedback energy is inserted are said to be included within the feedback loop. An amplifier containing a feedback loop is said to be a *feedback amplifier*. One stage or any number of stages may be included within the feedback loop. However, the difficulty of obtaining proper operation of a feedback amplifier increases with the bandwidth of the amplifier, and with the number of stages and circuit elements included within the feedback loop.

Gain and Phase Shift in Feedback Amplifiers The gain and phase shift of any amplifier are functions of frequency. For any amplifier containing a feedback loop to be completely stable, the gain of such an amplifier, as measured from the input back to the point where the feedback circuit connects to the input, must be less than unity at the frequency where the feedback voltage is in phase with the input voltage of the amplifier. If the gain is equal to or more than unity at the frequency where the feedback voltage is in phase with the input, the amplifier will oscillate. This fact imposes a limitation on the amount of feedback which may be employed in an amplifier which is to remain stable.

Types of Feedback may be either negative
Feedback or positive, and the feedback voltage may be proportional either to
output voltage or output current. The most
commonly used type of feedback with a-f or
video amplifiers is *negative feedback* proportional to output voltage. Figure 11 gives the

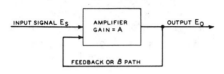

VOLTAGE AMPLIFICATION WITH FEEDBACK = $\dfrac{A}{1-A\beta}$

A = GAIN IN ABSENCE OF FEEDBACK

β = FRACTION OF OUTPUT VOLTAGE FED BACK

β IS NEGATIVE FOR NEGATIVE FEEDBACK

FEEDBACK IN DECIBELS = 20 LOG $(1-A\beta)$

= 20 LOG $\dfrac{\text{MID-FREQ. GAIN WITHOUT FEEDBACK}}{\text{MID-FREQ. GAIN WITH FEEDBACK}}$

DISTORTION WITH FEEDBACK = $\dfrac{\text{DISTORTION WITHOUT FEEDBACK}}{(1-A\beta)}$

$R_O = \dfrac{R_N}{1-A\beta\left(1+\dfrac{R_N}{R_L}\right)}$

WHERE:

R_O = OUTPUT IMPEDANCE OF AMPLIFIER WITH FEEDBACK

R_N = OUTPUT IMPEDANCE OF AMPLIFIER WITHOUT FEEDBACK

R_L = LOAD IMPEDANCE INTO WHICH AMPLIFIER OPERATES

Figure 11

**FEEDBACK AMPLIFIER
RELATIONSHIPS**

general operating conditions for feedback
amplifiers. Note that the reduction in distortion is proportional to the reduction in gain
of the amplifier, and also that the reduction
in the output impedance of the amplifier is
somewhat greater than the reduction in the
gain by an amount which is a function of
the ratio of the output impedance of the
amplifier without feedback to the load impedance. The reduction in noise and hum in
those stages included within the feedback
loop is proportional to the reduction in gain.
However, due to the reduction in gain of
the output section of the amplifier somewhat increased gain is required of the stages
preceding the stages included within the
feedback loop.

Figure 12 illustrates a very simple and effective application of negative-voltage feedback to an output pentode or tetrode
amplifier stage. The reduction in hum and

distortion may amount to 15 to 20 dB. The
reduction in the effective plate impedance
of the stage will be by a factor of 20 to
100 depending on the operating conditions.

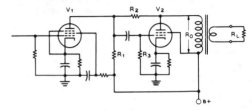

Figure 12

**SHUNT FEEDBACK CIRCUIT
FOR PENTODES OR TETRODES**

This circuit requires only the addition of one
resistor (R_2) to the normal circuit for such an
application. The plate impedance and distortion
introduced by the output stage are materially
reduced.

5-6 The Cathode-Ray Tube

The *cathode-ray tube* is a special type of
electron tube which permits the visual observation of electrical signals. It may be incorporated into an oscilloscope for use as a
test instrument or it may be the display device for radar equipment or television.

Operation of A cathode-ray tube always in-
the CRT cludes an *electron gun* for producing a stream of electrons, a
grid for controlling the intensity of the electron beam, and a *luminescent screen* for converting the impinging electron beam into
visible light. Such a tube always operates in
conjunction with either a built-in or an external means for focusing the electron stream
into a narrow beam, and a means for deflecting the electron beam in accordance with an
electrical signal.

The main electrical difference between
types of cathode-ray tubes lies in the means
employed for focusing and deflecting the
electron beam. The beam may be focused
and/or deflected either electrostatically or
magnetically, since a stream of electrons can
be acted on either by an electrostatic or a
magnetic field. In an electrostatic field the
electron beam tends to be deflected toward

the positive termination of the field (figure 13). In a magnetic field the stream tends to be deflected at right angles to the field. Further, an electron beam tends to be deflected so that it is normal (perpendicular) to the equipotential lines of an electrostatic field—and it tends to be deflected so that it is parallel to the lines of force in a magnetic field.

Large cathode-ray tubes used as *kinescopes* **in television receivers** usually are both focused and deflected magnetically. On the other hand, the medium-size CR tubes used in oscilloscopes and small television receivers usually are both focused and deflected electrostatically. Cathode-ray tubes for special applications may be focused magnetically and deflected electrostatically or vice versa.

There are advantages and disadvantages to both types of focusing and deflection. However, it may be stated that electrostatic deflection is much better than magnetic deflection when high-frequency waves are to be displayed on the screen; hence the almost universal use of this type of deflection for oscillographic work. When a tube is operated at a high value of accelerating potential so as to obtain a bright display on the face of the tube as for television or radar work, the use of magnetic deflection becomes desirable since it is relatively easier to deflect a high-velocity electron beam magnetically than electrostatically An *ion trap* is required with magnetic deflection since the heavy negative ions emitted by the cathode are not materially deflected by the magnetic field and would burn an *ion spot* in the center of the luminescent screen. With electrostatic deflection the heavy ions are deflected equally as well as the electrons in the beam so that an ion spot is not formed.

Construction of   The construction of a
Electrostatic CRT   typical electrostatic-focus, electrostatic - deflection cathode-ray tube is illustrated in the pictorial diagram of figure 13. The *indirectly heated cathode* (K) releases free electrons when heated by the enclosed filament. The cathode is surrounded by a cylinder (G) which has a small hole in its front for the passage of the electron stream. Although this element is not a wire mesh as is the usual grid, it is known by the same name because

its action is similar: it controls the electron stream when its negative potential is varied.

Next in order, is found the first *accelerating anode* (H) which resembles another disk or cylinder with a small hole in its center. This electrode is run at a high or moderately high positive voltage, to accelerate the electrons toward the far end of the tube.

The *focusing electrode* (F) is a sleeve which usually contains two small disks, each with a small hole.

After leaving the focusing electrode, the electrons pass through another *accelerating anode* (A) which is operated at a high positive potential. In some tubes this electrode is operated at a higher potential than the first accelerating electrode (H) while in other tubes both accelerating electrodes are operated at the same potential.

The electrodes which have been described up to this point constitute the *electron gun,* which produces the free electrons and focuses them into a slender, concentrated, rapidly traveling stream for projecting onto the viewing screen.

Electrostatic   To make the tube useful, means
Deflection   must be provided for deflecting the electron beam along two axes at right angles to each other. The more common tubes employ *electrostatic deflection plates,* one pair to exert a force on the beam in the vertical plane and one pair to exert a force in the horizontal plane. These plates are designated as B and C in figure 13.

Standard oscilloscope practice with small cathode-ray tubes calls for connecting one of the B plates and one of the C plates together and to the high-voltage accelerating anode. With the newer three-inch tubes and with five-inch tubes and larger, all four de-

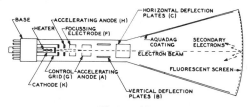

Figure 13

**TYPICAL ELECTROSTATIC
CATHODE-RAY TUBE**

flection plates are commonly used for deflection. The *positive* high voltage is grounded, instead of the negative as is common practice in amplifiers, etc., in order to permit operation of the deflecting plates at a dc potential at or near ground.

An *Aquadag* coating is applied to the inside of the envelope to attract any secondary electrons emitted by the fluorescent screen.

In the average electrostatic-deflection CR tube the spot will be fairly well centered if all four deflection plates are returned to the potential of the second anode (ground). However, for accurate centering and to permit moving the entire trace either horizontally or vertically to permit display of a particular waveform, horizontal- and vertical-centering controls usually are provided on the front of the oscilloscope.

After the spot is once centered, it is necessary only to apply a positive or negative voltage (with respect to ground) to one of the ungrounded or "free" deflector plates in order to move the spot. If the voltage is positive with respect to ground, the beam will be attracted toward that deflector plate. If it is negative, the beam and spot will be repulsed. The amount of deflection is directly proportional to the voltage (with respect to ground) that is applied to the free electrode.

With the larger-screen higher-voltage tubes it becomes necessary to place deflecting voltage on both horizontal and both vertical plates. This is done for two reasons: First, the amount of deflection voltage required by the high-voltage tubes is so great that a transmitting tube operating from a high-voltage supply would be required to attain this voltage without distortion. By using push-pull deflection with two tubes feeding the deflection plates, the necessary plate-supply voltage for the deflection amplifier is halved. Second, a certain amount of defocusing of the electron stream is always present on the extreme excursions in deflection voltage when this voltage is applied only to one deflecting plate. When the deflecting voltage is fed in push-pull to both deflecting plates in each plane, there is no defocusing because the *average* voltage acting on the electron stream is zero, even though the *net* voltage (which causes the

deflection) acting on the stream is twice that on either plate.

The fact that the beam is deflected by a magnetic field is important even in an oscilloscope which employs a tube using electrostatic deflection, because it means that precautions must be taken to protect the tube from the transformer fields and sometimes even the earth's magnetic field. This normally is done by incorporating a magnetic shield around the tube and by placing any transformers as far from the tube as possible, oriented to the position which produces minimum effect on the electron stream.

Construction of Electro- The electromagnetic
magnetic CRT cathode-ray tube allows greater definition than does the electrostatic tube. Also, electromagnetic definition has a number of advantages when a rotating radial sweep is required to give polar indications.

The production of the electron beam in an electromagnetic tube is essentially the same as in the electrostatic tube. The grid structure is similar, and controls the electron beam in an identical manner. The elements of a typical electromagnetic tube are shown in figure 14. The *focus coil* is wound on an iron core which may be moved along the neck of the tube to focus the electron beam. For final adjustment, the current flowing in the coil may be varied. A second pair of coils, the *deflection coils,* are mounted at right angles to each other around the neck of the tube. In some cases, these coils can rotate around the axis of the tube.

Two *anodes* are used for accelerating the electrons from the cathode to the screen. The second anode is a graphite coating (*Aquadag*) on the inside of the glass envelope. The function of this coating is to attract any secondary electrons emitted by

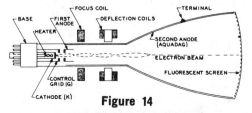

Figure 14

**TYPICAL ELECTROMAGNETIC
CATHODE-RAY TUBE**

the fluorescent screen, and also to shield the electron beam.

In some types of electromagnetic tubes, a first, or *accelerating anode* is also used in addition to the *Aquadag.*

Electromagnetic Deflection   A magnetic field will deflect an electron beam in a direction which is at right angles to both the direction of the field and the direction of motion of the beam.

In the general case, two pairs of deflection coils are used (figure 15). One pair is for horizontal deflection, and the other pair is for vertical deflection. The two coils in a pair are connected in series and are wound in such directions that the magnetic field flows from one coil, through the electron beam to the other coil. The force exerted on the beam by the field moves it to any point on the screen by the application of the proper currents to these coils.

The Trace   The human eye retains an image for about one-sixteenth second after viewing. In a CRT, the spot can be moved so quickly that a series of adjacent spots can be made to appear as a line, if the beam is swept over the path fast enough. As long as the electron beam strikes in a given place at least sixteen times a second, the spot will appear to the human eye as a source of continuous light with very little flicker.

Screen Materials— "Phosphors"   At least five types of luminescent screen materials are commonly available on the various types of CR tubes com-

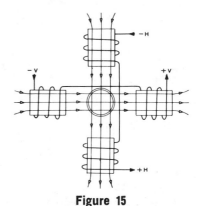

Figure 15

TWO PAIRS OF COILS ARRANGED FOR ELECTROMAGNETIC DEFLECTION IN TWO DIRECTIONS

mercially available. These screen materials are called *phosphors;* each of the five phosphors is best suited to a particular type of application. The P-1 phosphor, which has a green fluorescence with medium persistence, is almost invariably used for oscilloscope tubes for visual observation. The P-4 phosphor, with white fluorescence and medium persistence, is used on television viewing tubes (*Kinescopes*). The P-5 and P-11 phosphors, with blue fluorescence and very short persistence, are used primarily in oscilloscopes where photographic recording of the trace is to be obtained. The P-7 phosphor, which has a blue flash and a long-persistence greenish-yellow persistence, is used primarily for radar displays where retention of the image for several seconds after the initial signal display is required.

CHAPTER SIX

Special Microwave Tubes

The electron tube has been largely replaced in low-power hf and vhf communication. Aside from the lower cost and better performance of the solid-state device, the electron tube has inherent problems that limit its usefulness as an effective vhf amplifier. Among the critical tube parameters that affect vhf performance are interelectrode capacitance, lead inductance and transit time. Tubes designed to partially overcome

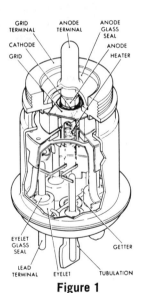

Figure 1

**CUTAWAY VIEW OF
WESTERN ELECTRIC 416-B/6280
VHF PLANAR TRIODE TUBE**

The 416-B, designed by the Bell Telephone Laboratories is intended for amplifier or frequency multiplier service in the 4000 MHz region. Employing grid wires having a diameter equal to fifteen wavelengths of light, the 416-B has a transconductance of 50,000. Spacing between grid and cathode is .0005", to reduce transit-time effects. Entire tube is gold plated.

these difficulties are expensive but can operate to over 1000 MHz (figure 1).

If all linear dimensions of an electron tube are held at a fixed ratio to each other there will be no change in electrical characteristics for a given set of voltages regardless of change in the tube dimensions. Interelectrode capacitances, lead inductance, and electron transit time, however, are in direct proportion to the magnitude of the linear dimensions.

As the frequency of operation of an electron tube is raised the lead inductance becomes so great that much of the output voltage appears within the envelope where it is unavailable and the electron transit time from filament to plate becomes an appreciable portion of the operating cycle. Plate current is no longer in phase with grid voltage and efficiency and power output drop. At the same time, the tube is more difficult to drive.

Reducing all dimensions of the tube raises the maximum operating frequency but a point is soon reached where further reduction in size becomes impractical because of mechanical and thermal problems. The upper frequency limit varies from 100 MHz for conventional tube types to about 2 GHz for specialized types such as the planar triode. Above the limiting frequency, the conventional negative-grid electron tube no longer is practicable and recourse must be taken to totally different types of devices in which electron transit time and capacitance are not limitations to operation. Three of the most important types of such microwave tube types are the *klystron*, the *magnetron*, and the *traveling-wave tube*. Variations of these basic types can produce hundreds of kilowatts of r-f power to above 100 GHz.

Figure 2

CW KLYSTRON AMPLIFIER

Varian 890H is a four-cavity vapor-cooled klystron used as a final amplifier tube in both visual and aural sections of uhf-TV transmitters. The tube covers the range of 470–566 MHz (channels 14–29). It provides a signal gain of at least 35 dB and 32 kW peak-of-sync output with less than 10 watts of r-f drive. Tube is about five feet tall and weighs 270 pounds.

6-1 The Power Klystron

The klystron is a rugged, microwave power tube in which electron transit time is used to advantage (figure 2). The klystron consists of a number of *resonant cavities* linked together by metallic sections called *drift tubes*. The drift tubes provide isolation between the cavities at the operating frequency of the tube and the output circuitry of the klystron is effectively isolated from the input circuitry, an important consideration in vhf amplifiers (figure 3).

The cathode, or *electron* gun, emits a stream of electrons which is focused into a tight beam. The beam passes through the succession of cavities and drift tubes, ultimately reaching the *collector*. The main body of the tube is usually operated at ground potential, with the cathode and associated focus electrode operated at a high negative potential. The electron beam is held on course by means of an axial magnetic field created by magnetic coils placed about the tube. The strength of the magnetic field is adjustable to permit accurate adjustment of the electron beam, which can be made to travel long distances, with less than one percent current interception by the drift tube walls.

Bunching   The electron stream leaving the cathode gun of a klystron is uniform in density, but the action of the cavities and drift tubes causes a large degree of *density modulation* to appear in the beam at the output cavity. This action, called *bunching*, is a result of the beam being exposed to the varying electric field which appears across the gaps in the cavities. Elec-

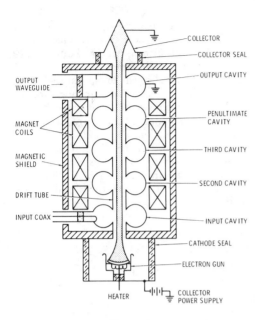

Figure 3

THE POWER KLYSTRON

Large klystrons are commonly used in uhf-TV transmitter service providing upward of 35-kW output at frequencies up to 900 MHz. The resonant cavities may be integral (as shown) or external, clamped to the drift tube which has large ceramic insulating sections covering the cavity gap.

trons passing through the gaps, when the r-f field across the gap is zero, travel in the drift regions at a velocity corresponding to the beam voltage. When the gap appears positive, the electrons are accelerated, and when the gap appears negative the electrons travel at reduced velocity. The result of this velocity modulation is that the electrons tend to bunch progressively.

The output cavity of the klystron is exposed to a series of electron bunches which are timed to arrive with a frequency equal to the resonant frequency of the cavity and a net power flow from the beam to the cavity exists. The energy is extracted from the cavity by means of a coupling loop.

The r-f power in the output cavity will be much greater than that applied in the bunching cavity. This is due to the ability of the concentrated bunches of electrons to deliver great amounts of energy to the output cavity. Since the electron beam delivers

most of its energy to the output cavity, it arrives at the collector with less total energy than it had when it passed through the input cavity. This difference in beam energy is approximately equal to the energy delivered to the output cavity.

Klystron amplifiers have been built with as many as seven cavities (that is, with five intermediate cavities). The effect of the intermediate cavities is to improve the bunching process. This results in increased amplifier gain, and to a lesser extent increased efficiency. Adding more intermediate cavities is analogous to adding more stages to an if amplifier. That is, overall amplifier gain is increased and the overall bandwidth is reduced if all the stages are tuned to the same frequency. The same effect occurs with klystron amplifier tuning and is called *synchronous tuning*. If the cavities are tuned to slightly different frequencies the gain of the klystron is reduced and the bandwidth may be appreciably increased. This is called *stagger tuning*.

Saturation The klystron is not a perfect linear amplifier at all operating levels and will saturate at strong signal levels (figure 4). Curve A shows typical performance for synchronous tuning for

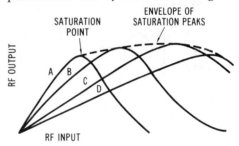

Figure 4

KLYSTRON PERFORMANCE

Plot of typical klystron amplifier performance for various tuning conditions.

maximum gain. The power output is linear with respect to the input up to about seventy percent of saturation. However, as the r-f input is increased beyond that point, the gain decreases and the tube saturates. As the r-f input is increased beyond saturation, the r-f output decreases (curves C and D).

Focusing An axial magnetic field (parallel to the axis of the klystron) is required to keep the electron beam properly formed during its travel through the r-f section. The mutual repulsion between electrons causes the beam to spread in a direction perpendicular to the axis of the tube. If this occurs, electrons will strike the drift tube and be collected there, rather than passing through the tube to the collector.

The action of the magnetic field is to exert a force on the electrons which keeps them focused into a narrow beam. Either a permanent or electromaget may be used. The beam is allowed to spread before it strikes the collector, minimizing cooling collector problems which would result from the beam remaining concentrated at the ·time of interception.

The collector is normally insulated from the r-f section of large klystron amplifiers to permit separate metering of the electrons intercepted by the drift tubes and by the collector.

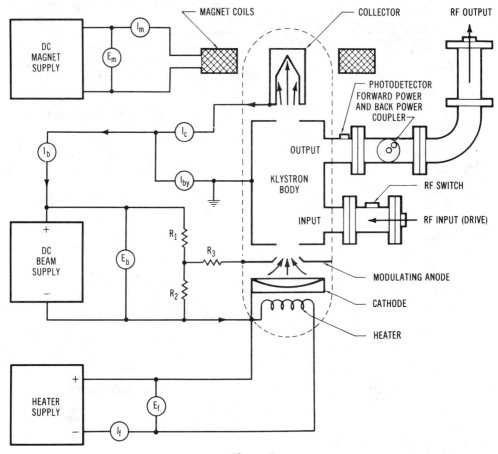

Figure 5

ASSOCIATED EQUIPMENT FOR KLYSTRON AMPLIFIER

The body of the klystron is at ground potential in order for the tube to be easily connected into the rest of the system. The dc beam supply provides acceleration voltage and current limiting resistors protect the tube in case of arc from cathode to grid. Collector is insulated from the body of the tube.

Klystron Circuitry Some of the power supplies, monitoring devices and protective devices used in a typical klystron amplifier are shown in figure 5. It is convenient to operate the tube body at ground potential eliminating danger to operating personnel. The *beam supply* provides the voltage to accelerate the electrons and form the beam and the *crowbar* system quickly discharges the beam supply in the event of an internal klystron arc or other fault condition.

Some klystrons have a grid or modulating electrode used to pulse the beam on or off or to impart intelligence to the beam. In most gridded klystron tubes the grid is never allowed to go positive with respect to the cathode.

Body, collector, and beam current are monitored separately. Body current is usually limited to one or two percent of the total beam current and excessive body current trips an overload relay which kills the entire supply system.

Figure 7

REFLEX KLYSTRON OSCILLATOR

Small klystron oscillator provides 100 mW power for use as a pump in a parametric amplifier. Klystron oscillators are available for operation up to 220 gigahertz.

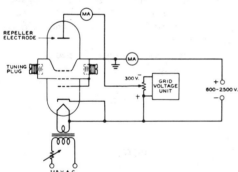

Figure 6

REFLEX KLYSTRON OSCILLATOR

A conventional reflex klystron oscillator of the type commonly used as a local oscillator in superheterodyne receivers operating above about 2000 MHz is shown above. Frequency modulation of the output frequency of the oscillator, or afc operation in a receiver, may be obtained by varying the negative voltage on the repeller electrode.

The Reflex Klystron The multicavity klystron as described in the preceding paragraphs is primarily used as a transmitting device since large amounts of power are made available in its output circuit. However, for applications where a much smaller amount of power is required — power levels in the milliwatt range — for low-power transmitters, receiver local oscillators, etc., another type of klystron having only a single cavity is more frequently used.

The theory of operation of the single-cavity klystron is essentially the same as the multicavity type with the exception that the velocity-modulated electron beam, after having left the input cavity is reflected back into the area of the cavity again by a repeller electrode as illustrated in figure 6. The potentials on the various electrodes are adjusted to a value such that proper bunching of the electron beam will take place just as a particular portion of the velocity-modulated beam re-enters the area of the resonant cavity. Since this type of klystron has only one circuit it can be used only as an oscillator and not as an amplifier. Effective modulation of the frequency of a single-cavity klystron for f-m work can be obtained by modulating the repeller electrode voltage. A representative reflex klystron is shown in figure 7.

6-2   The Magnetron

The *magnetron* is a uhf oscillator tube normally employed where very high values of peak power are required in the range of about 700 MHz to 30,000 MHz. Special magnetrons have peak power capability of

Figure 8

TUNABLE HIGH POWER MAGNETRON

The SFD 332 coaxial magnetron covers the range of 32.9 to 33.5 GHz. Minimum peak power output is 40 kW at a duty cycle of .0006. Output fitting is at the bottom and tuning adjustment is on the rear of the structure.

several megawatts at frequencies as high as 10 GHz. The normal duty cycle of such devices ranges as high as 2/10 of one percent so that the average power output is about 1000 watts (figure 8).

The magnetron is a diode having strong magnetic forces exerted on the electrons that travel from the cathode to the anode. The magnetic field is usually provided by a strong permanent magnet mounted around the magnetron so that the magnetic field is parallel with the axis of the cathode (figure 9). The magnetron is a self-contained unit, that is, it produces a microwave frequency output within its enclosure without the use of external tuned circuits.

Operation of the magnetron is based on the motion of electrons under the influence of combined electric and magnetic fields. In an electric field the force exerted by the field on an electron is proportional to field strength and the electrons tend to move from a point of negative potential toward a positive potential (figure 10). In other words, the electrons tend to move against the electric field and during electron acceleration, energy is taken from the field by the electrons.

The force exerted on an electron in a magnetic field is at right angles to both

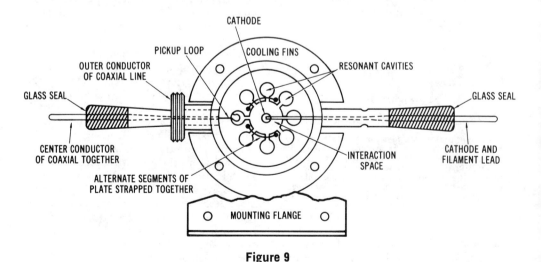

Figure 9

CUTAWAY VIEW OF A MAGNETRON

Representative magnetron is a diode having strong magnetic forces exerted on electrons traveling from cathode to anode. The magnetron has internal resonant circuitry and does not require external tuned circuits.

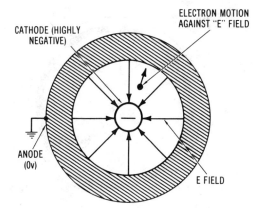

Figure 10

ELECTRON MOTION IN AN ELECTRIC FIELD (E)

The force exerted by an electric field (E) on an electron is proportional to the strength of the field. Electrons tend to move from a point of negative potential toward a positive potential, or in other words, tend to move against the E-field.

the field and the path of the electron so that the electron follows a clockwise trajectory when viewed in the direction of the field (figure 11). If the magnetic field is increased, the electron path will bend sharper. In a like manner, if the velocity of

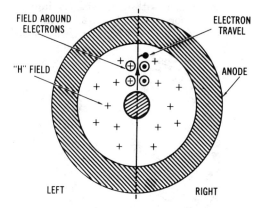

Figure 11

ELECTRON MOTION IN A MAGNETIC FIELD (H)

The force exerted on an electron in a magnetic (H) field is at right angles to both the field and the path of the electron so that the electron follows a clockwise trajectory when viewed in the direction of the field.

the electron increases, the field around it increases and its path will bend more sharply.

A basic magnetron design is shown in figure 12. The device consists of a cylindrical plate with a cathode placed coaxially inside it. The tuned circuit (not shown) is cavities physically located in the plate structure.

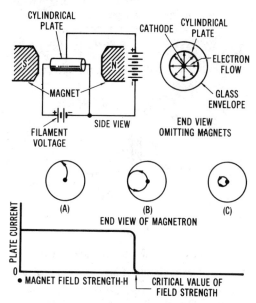

Figure 12

BASIC MAGNETRON DESIGN

The magnetron consists of a cylindrical plate with a cathode placed coaxially within it. The tuned circuits are cavities physically located in the plate (not shown). The effect of the magnetic field on a single electron is shown in the lower illustration.

Under the influence of the magnetic field, electrons leaving the filament are deflected from their normal paths and move in circular orbits within the anode cylinder (B). If the magnetic field is strong enough the electrons just miss the plate and return to the filament, the plate current dropping to zero (C). When the magnetron is adjusted to plate current cutoff, the device can produce microwave oscillations by virtue of the currents introduced electrostatically by the moving electron. This frequency is determined by the travel time of the electrons from the cathode to the plate and back again. A transfer of microwave energy to a load is made possible by connecting an external circuit between cathode and plate of the magnetron.

The Negative Resistance Magnetron   The split-anode negative resistance magnetron is a variation of the basic design which provides more power output at the higher frequencies. Its general construction is similar to the basic design except that it has a split plate (figure 13). The half-plates are operated at

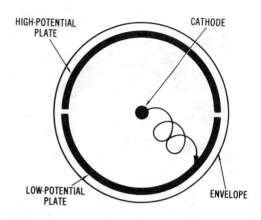

Figure 14

MOVEMENT OF ELECTRON IN A SPLIT-ANODE MAGNETRON

Electron makes a series of loops through the magnetic field until it finally falls on the low potential plate.

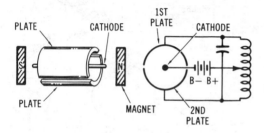

Figure 13

THE SPLIT-ANODE MAGNETRON

This design is capable of providing more output at a higher frequency than the basic magnetron. The half plates are operated at different potentials to provide an electron motion such as shown in Figure 14.

different potentials to provide an electron motion as shown in figure 14. The electron leaving the cathode and progressing toward the high-potential plate is deflected by the magnetic field at a certain radius of curvature and, after passing the split between the plates, enters the field set up by the lower-potential plate. Here the magnetic field has more effect upon the electron which is deflected at a smaller radius of curvature. The electron continues to make a series of loops through the magnetic field until it finally reaches the low potential plate. Oscillations are started by applying the proper value of magnetic field to the tube.

The Electron Resonance Magnetron   In this design the plate is constructed so as to function as a tank circuit, thus there are no external tuned cir-

cuits. The electron path is such that it exhibits a curve having a series of abrupt cusps such as shown in figure 15. In this example, an eight-segment anode is used.

This type of magnetron is the most widely used for microwave work and develops high power at good efficiency. The average power is limited by filament emission and peak power is limited by maximum voltage the device can withstand without injury. Three common types of anode blocks are shown in figure 16. The first two anode

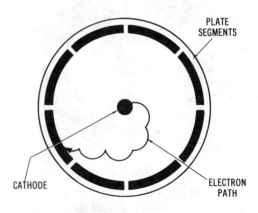

Figure 15

PATH OF SINGLE ELECTRON IN ELECTRON RESONANCE MAGNETRON

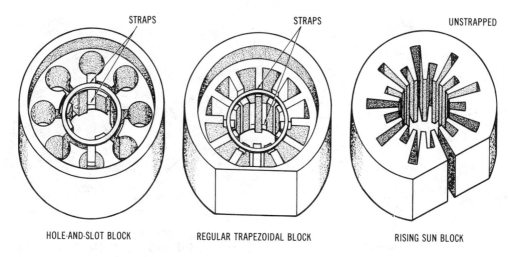

STRAPS   STRAPS   UNSTRAPPED

HOLE-AND-SLOT BLOCK   REGULAR TRAPEZOIDAL BLOCK   RISING SUN BLOCK

Figure 16

COMMON TYPES OF MAGNETRON ANODE BLOCKS

First two blocks require that alternate segments are connected, or strapped. Block on right utilized large and small trapezoid cavities that do not require strapping.

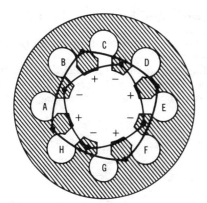

Figure 17

STRAPPING ALTERNATE SEGMENTS OF MAGNETRON

Cavities are connected in parallel due to strapping technique. Unstrapped cavities are connected in series from an electrical point of view.

sides of the slot form the plates of a capacitor while the walls of the hole act as an inductor. The hole and slot thus form a high-Q resonant circuit. The anode of a magnetron contains a number of these cavities. The cavities are operating in series as shown in figure 19 if the anode is not strapped but strapping the anode places the cavities in parallel (figure 20).

Electron flow in a multicavity magnetron is complex and the flow resembles the spokes of a wheel, the wheel rotating about the

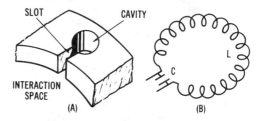

SLOT   CAVITY

INTERACTION SPACE
(A)   (B)

Figure 18

EQUIVALENT CIRCUIT OF A HOLE-AND-SLOT CAVITY

Physical appearance of resonant cavity is shown at (A) and the electrical equivalent circuit is shown at (B). The parallel sides of the slot form the capacitance while the walls of the hole act as an inductor. Hole and slot thus form a high-Q resonant circuit.

blocks operate in such a way that alternate segments must be strapped to ensure that each segment is opposite in polarity to its neighboring segment on either side as shown in figure 17. This requires an even number of cavities.

The electrical equivalent of the cavity-slot design is shown in figure 18. The parallel

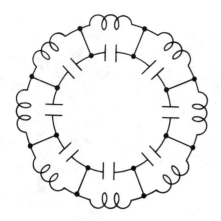

Figure 19

MAGNETRON CAVITIES CONNECTED IN SERIES

Analysis of unstrapped anode.

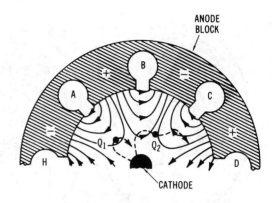

Figure 21

ELECTRON PATHS IN MAGNETRON DURING R-F OSCILLATION

Cumulative effect of many electrons forms a pattern resembling the spokes of a wheel. This pattern rotates about the cathode at an angular velocity of two segments per cycle of the r-f field.

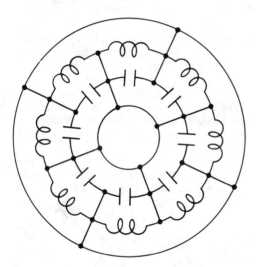

Figure 20

MAGNETRON CAVITIES CONNECTED IN PARALLEL BY STRAPPING

in the cavities that changes the surface area to volume ratio in a high current region (figure 22). The element is termed a *sprocket tuner.*

Capacitance tuning is realized by insertion of an element into the cavity slot that increases slot capacitance, decreasing the reso-

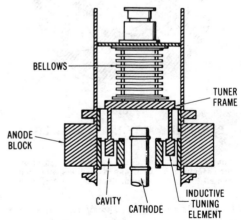

Figure 22

INDUCTIVE TUNING OF MAGNETRON

Inductive tuning element is inserted in the resonant cavity to alter surface to volume ratio in a high current region. As element is inserted, the inductance of the cavity decreases.

cathode at an angular velocity of two poles (anode segments) per cycle of the r-f field (figure 21).

The resonant frequency of a magnetron may be changed by varying the inductance or capacitance of the cavities. Inductive tuning is accomplished by inserting an element

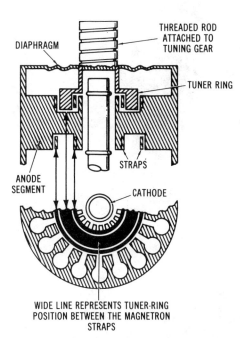

DIAPHRAGM

THREADED ROD ATTACHED TO TUNING GEAR

TUNER RING

STRAPS

ANODE SEGMENT

CATHODE

WIDE LINE REPRESENTS TUNER-RING POSITION BETWEEN THE MAGNETRON STRAPS

Figure 23

CAPACITIVE TUNING OF MAGNETRON

Metal ring inserted into double-strapped magnetron increases strap capacitance and lowers resonant frequency.

nant frequency (figure 23). The element is termed a *cookie cutter tuner*. A ten-percent frequency range may be obtained with either of the tuning methods described.

A variation of the magnetron is the *cross-field amplifier* which is a broadband, phase-stable amplifier for use in coherent radar equipment. Its high peak output and light weight make it suitable for airborne radar equipment. Another class of magnetrons includes the injected-beam, backward-wave oscillator (*Carcinatron*).

6-3 The Traveling-Wave Tube

The *traveling-wave tube* (TWT) differs from the klystron in that the r-f field is not confined to a limited region but is distributed along a wave-propagating structure. A longitudinal electron beam interacts continuously with the field of a wave traveling along this structure. In its most common form it is an amplifier, although there are related TWTs that are oscillators (figure 24).

The TWT is an amplifying device having extremely wide bandwidth and high power gain. Figure 25 is a simplified sketch of a basic helix-type TWT tube. An electron stream is produced by an electron gun, travels along the axis of the tube, and is finally collected by a suitable anode (collector). Spaced closely around the beam is a circuit, in this case a helix of tightly wound wire, capable of propagating a slow wave. The r-f energy travels along the wire at the velocity of light but, because of the helical path, the energy progresses along the length of the tube at a considerably lower velocity than is determined primarily by the pitch of the helix. In a typical low power TWT a value of about one-tenth the velocity of light is used. The velocity of the electron stream is adjusted to be approximately the same as the axial phase velocity of the wave on the helix. The result is that an interaction occurs between the electron beam and the r-f signal on the helix. The interaction is such that some electrons in the beam are slowed by the r-f field, while others are accelerated. As the velocity-modulated electrons move down through the helix they form bunches which, in turn, interact with the helix r-f wave. The result is that dc energy in the beam is given up to the helix as r-f energy, and the wave is thus amplified. Gain figures as high as 70 dB have been achieved in a TWT, with 30 dB being a common value for commercial products.

To restrain the size of the electron beam as it travels along the tube, it is necessary to provide a focusing field, either magnetic or electrostatic, strong enough to overcome the space charge effects that would otherwise cause the beam to spread. Permanent-magnet focusing structures are used in many tubes to reduce weight. The collector performs no other function than to dissipate the electrons in the form of heat as they emerge from the slow-wave structure.

While the TWT provides extremely high gain, its uniqueness is found in its broad-band capability, TWTs have been made to amplify r-f signals at frequencies over a 5-to-1 bandwidth, due principally to the non-

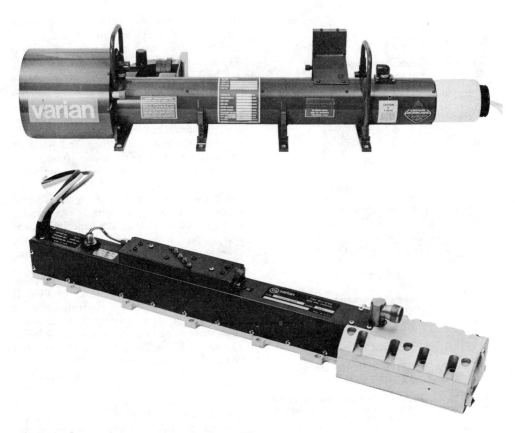

Figure 24

HIGH POWER TRAVELING-WAVE TUBE

A—The Varian VTS-5754-C1 is a cavity coupled traveling-wave tube operating in the frequency range of 3.1-3.5 gigahertz. Peak power output is 125 kW at a duty cycle of 0.06. Saturation gain is 47 dB. B—The Varian VTG-6232-F1 is a CW-mode TWT which operates over a frequency range of 2.0-4.0 gigahertz with a power output of 200 watts. It is used for commercial and electronic warfare systems.

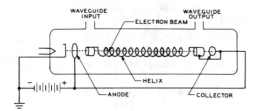

Figure 25

THE TRAVELING-WAVE TUBE

Operation of this tube is the result of interaction between the electron beam and wave traveling along the helix.

frequency-resonant construction. It is therefore not subject to the gain-bandwidth equation, which states that as the gain is increased, bandwidth is decreased.

Active electronic countermeasure (ECM) systems are designed to receive a wide band of input frequencies and retransmit range information. The TWT is ideally tailored to such a microwave system by its inherent ability to provide high gain over octave bandwidths. Additional benefits offered by the TWT are low noise capability, high power-handling capability, high efficiency and good linearity.

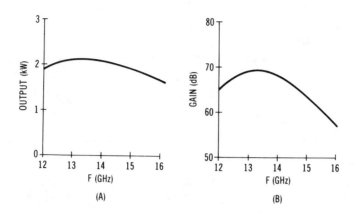

Figure 26

POWER OUTPUT AND GAIN FOR
Ku BAND TWT

**TWT type VTU-5395A1 is a conduction cooled
TWT that provides up to 2 kW peak power at a
duty cycle of .02.**

The long life characteristic of the TWT has led to its extensive use in spaceborne equipment, such as the Mariner and Pioneer missions into outer space, and commercial and military communication satellites where continuous operation is essential.

Traveling-wave tube amplifiers provide 1 to 20 watts from 1 to 18 GHz in standard octave bands. Amplifiers are available in the same frequency range with power outputs in excess of 10,000 watts. Higher-frequency devices are available in the 18–40 GHz bands and typical characteristics for a K-band amplifier are shown in figure 26.

Special families of TWT amplifiers are used in communications satellite transponders in the down-link circuit. These tubes are operated at low current densities and low temperatures to achieve a high order of reliability and long life.

CHAPTER SEVEN

Radio-Frequency Power Amplifiers

All modern radio transmitters consist of a comparatively low-level source of radio-frequency energy which is amplified in strength and mixed or multiplied in frequency to achieve the desired power level and operating frequency. Microwave transmitters may be of the self-excited oscillator type, but when it is possible to use r-f amplifiers in uhf transmitters the flexibility of their application is increased.

Radio-frequency power amplifiers are generally classified according to frequency range (hf, vhf, uhf, etc.), power level, type of tube used, and type of service (a-m, f-m, c-w, SSB). In addition, the amplifier may be classified according to mode, or dynamic operating characteristic of the tube (Class AB_1, B, or C); and according to circuitry (grid driven or cathode driven). Each mode of operation and circuit configuration has its distinct advantages and disadvantages, and no one mode or circuit is superior in all respects to any other. As a result, modern transmitting equipments employ various modes of operation, intermixed with various tubes and circuit configurations. The following portion of this chapter will be devoted to the calculation of dynamic characteristics for some of the more practical modes of tuned power amplifier operation.

7-1 Class-C R-F Power Amplifiers

It is often desired to operate the r-f power amplifier in the class-B or class-C mode since such stages can be made to give high

plate-circuit efficiency. Hence, the tube cost and cost of power to supply the stage is least for any given power output. Nevertheless, the class-C amplifier provides less *power gain* than either a class-A or class-B amplifier under similar conditions. The grid of the class-C amplifier must be driven highly positive over the small portion of the exciting signal when the instantaneous plate voltage on the tube is at its lower point, and is at a large negative potential over a major portion of the operating cycle. As a result, no plate current will flow except during the time plate voltage is very low. Comparatively large amounts of drive power are necessary to achieve this mode of operation. Class-C operational efficiency is high because no plate current flows except when the plate-to-cathode voltage drop across the tube is at its lowest value, but the price paid for stage efficiency is the large value of drive power required to achieve this mode of operation.

The gain of a class-B amplifier is higher than that of the class-C stage, and driving power is less in comparison. In addition, the class-B amplifier may be considered to be linear; that is, the output voltage is a replica of the input voltage at all signal levels up to overload. This is not true in the case of the class-C amplifier whose output waveform consist of short pulses of current, as discussed later in this chapter.

The gain of a class-A amplifier is higher than that of the class-B or class-C stage, but the efficiency is the lowest of the three modes of operation. As with the class-B stage, the class-A amplifier is considered to be linear with respect to input and output waveforms.

Relationships in Class-C Stage   The class-C amplifier is analyzed as its operation provides an all-inclusive case of the study of class-B and class-AB₁ r-f amplifiers.

The class-C amplifier is characterized by the fact that the plate current flows in pulses which, by definition, are less than one-half of the *operating cycle*. The operating cycle is that portion of the electrical cycle in which the grid is driven in a positive direction with respect to the cathode. The operating cycle is considered in terms of the plate or grid *conduction angle* (**θ**). The conduction angle is an expresion of that fraction of time (expressed in degrees of the electrical cycle) that the tube conducts plate or grid current as compared to the operating cycle of the input voltage waveform.

The theoretical efficiency of any power amplifier depends on the magnitude of the conduction angle; a tuned class-A amplifier having a large conduction angle with a maximum theoretical efficiency of 50 percent; a class-B amplifier with an angle of 180 degrees, and efficiency of 78.5 percent; and a class-C amplifier with an angle of about 160 degrees and efficiency of about 85 percent.

Figure 1 illustrates a transfer curve representing the relationships between grid and plate voltages and currents during the operating cycle of a class-C amplifier. Symbols shown in figure 1 and given in the following discussion are defined and listed in the *Glossary of Terms* included at the front of this Handbook.

The plot is of the *transfer curve* of a typical triode tube, and represents the change in plate current, (i_b) for a given amount of grid voltage (e_c). The representation is of the form of the I_b versus E_c plot for a triode shown in figure 9, chapter 5.

The *operating point*, or grid-bias level (E_c), is chosen at several times cutoff bias (E_{co}), and superimposed on the operating point is one-half cycle of the grid exciting voltage, $e_{g\ max}$. A sample point of grid voltage, e_{cx}, is shown to produce a value of instantaneous plate current, i_{bx}. All other points on the grid-voltage curve relate to

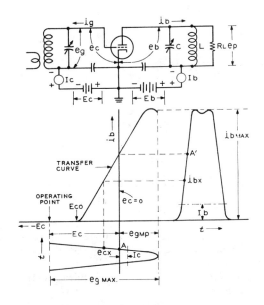

Figure 1

TRANSFER CURVE FOR OPERATING CYCLE OF CLASS-C AMPLIFIER

Typical class-C amplifier (less neutralizing circuits) is shown with various average and instantaneous voltages noted. A summary of symbols is given in the glossary of terms. The plot is of the transfer curve, representing the change in plate current for a given grid voltage. The grid signal (e_g max) is represented by a pulse of voltage along the y-axis, with the operating point determined by the amount of grid bias, E_c. As the waveform rises in amplitude, a corresponding pulse of plate current is developed across the plate load impedance, (R_L). A single point of grid voltage (A) represents a corresponding value of instantaneous plate current (A'). All other points on the grid-voltage curve relate to corresponding points on the plate-current curve.

corresponding points on the plate-current curve.

As the grid is driven considerably positive, grid current flows, causing the plate current to be "starved" at the peak of each cycle, thus the plate-current waveform pulse is slightly indented at the top. As the waveform is poor and the distortion high, class-C operation is restricted to r-f amplification where high efficiency is desirable and when the identity of the output waveform to the input waveform is relatively unimportant.

The relation between grid and plate voltages and currents is more fully detailed in the graphs of figures 2 and 3, which illustrate in detail the various voltage and current variations during one electrical cycle of the exciting signal.

Voltage at the Grid The curves of figure 2 represent the grid voltage and current variations with respect to time. The x-axis for grid voltage is E_{c1} with a secondary axis ($E_{co} = 0$) above it, the vertical distance between axes representing the fixed grid-bias voltage (E_c). At the beginning of the operating cycle ($t = 0$) the exciting voltage (e_g) is zero and increases in amplitude, until at *point A* it equals in magnitude the value of the bias voltage. At this point, the instantaneous voltage on the grid of the tube is zero with respect to the cathode, and plate current has already begun to flow (*point A* in figure 1), as the exciting signal is already greater in magnitude than the cutoff grid voltage (E_{co}). The relations are normally such that at the crest of the positive grid voltage cycle, e_{cmp} (or $e_{g\ max}$ positive), the grid is driven appreciably positive with respect to the cathode and consequently draws some grid current, i_g. The dc component of grid current, I_c, may be read on the grid meter shown in figure 1. The grid draws current only over that por-

tion of the operating cycle when it is positive with respect to the cathode (that portion of the curve above the $E_c = 0$ axis in graph A). This portion of the exciting voltage is termed the *maximum positive grid voltage* (e_{cmp}).

Voltage at the Plate The voltage at the plate of the tube responds to the changes in grid voltage as shown in figure 3. Instantaneous plate voltage (e_p), consists of the dc plate voltage (E_p) less the ac voltage drop across the plate load impedance (e_p). As the grid element becomes more positive, a greater flow of electrons reach the plate, instantaneous plate current increases, and the voltage drop across the plate load impedance (R_L) rises. The phase relations are such that the minimum instantaneous plate potential ($e_{b\ min}$) and the maximum instantaneous grid potential ($e_{g\ max}$) occur simultaneously. The corresponding instantaneous plate current (i_b) for this sequence is shown in the current plot of figure 3.

As plate current is conducted only between *points A* and *B* of the grid-voltage

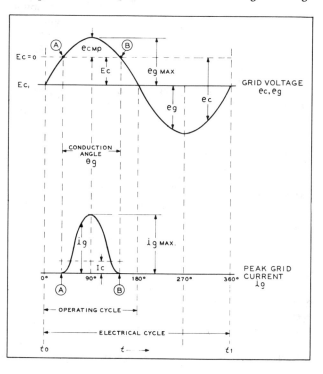

Figure 2

INSTANTANEOUS GRID VOLTAGE AND CURRENT OF A CLASS-C R-F POWER AMPLIFIER

Grid voltage and current variations with respect to time are shown. The grid is negatively biased by the amount E_c. As soon as the positive value of grid exciting voltage (e_g) exceeds E (point A) the grid starts to draw current, as it is positive with respect to the filament. Grid current flows from point A to point B of the grid voltage plot. This portion of the grid cycle is termed the conduction angle. Average value of grid current (I_c) may be read on a dc meter in series with grid return line to bias supply. For typical class-C performance, grid current flows over a portion of the operating cycle which is less than half the electrical cycle.

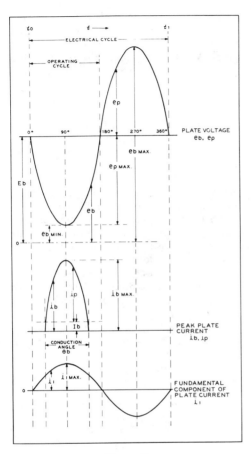

Figure 3

INSTANTANEOUS PLATE VOLTAGE AND CURRENT OF CLASS-C POWER AMPLIFIER

Instantaneous plate voltage and current responds to the changes in grid voltage shown in figure 2. As grid becomes more positive, the peak plate current rises, causing an increased voltage drop across the plate load impedance (R_L, figure 1). Maximum peak plate current flows at condition of minimum instantaneous plate voltage ($e_{b \ min}$) and maximum voltage drop across load impedance ($e_{p \ max}$). Plate-current pulse exists only over a portion of the operating cycle (conduction angle). Usable power is derived from the fundamental component of the plate current which is a sine wave developed across the resonant tank circuit. $e_{p \ max}$ equals $e_{p \ min}$.

excursion, it can be seen that the plate-current pulse exists only over a portion (θ_b) of the complete plate *operating cycle* . (The operating cycle is taken to be that half-cycle of grid voltage having a positive excursion of the drive voltage.) The opposite half of the electrical cycle is of little interest,

as the grid merely assumes a more negative condition and no flow of plate current is possible.

Peak plate current pulses, then, flow as pictured in figure 3 over the *conduction angle* of each operating cycle. *The fundamental component of plate current* (i_1) however, is a sine wave since it is developed across a resonant circuit (LC). The resonant circuit, in effect, acts as a "flywheel," holding r-f energy over the pulsed portion of the operating cycle, and releasing it during the quiescent portion of the electrical cycle.

The patterns of grid voltage and current shown in figure 2 are important in determining grid-circuit parameters, and the patterns of plate voltage and current shown in the illustrations can be used to determine plate-circuit parameters, as will be discussed later.

The various manufacturers of vacuum tubes publish data sheets listing in adequate detail various operating conditions for the tubes they manufacture. In addition, additional operating data for special conditions is often available for the asking. It is, nevertheless, often desirable to determine optimum operating conditions for a tube under a particular set of circumstances. To assist in such calculations the following paragraphs are devoted to a method of calculating various operating conditions which is moderately simple and yet sufficiently accurate for all practical purposes. It is based on wave-analysis techniques of the peak plate current of the operating cycle, adapted from Fourier analysis of a fundamental wave and its accompanying harmonics. Considerable ingenuity has been displayed in devising various graphical ways of evaluating the waveforms in r-f power amplifiers. One of these techniques, a *Tube Performance Calculator*, for class-AB₁ class-B, and Class-C service may be obtained at no cost by writing: Application Engineering Dept., Eimac Division of Varian, San Carlos, Calif. 94070.

Tank-Circuit When the plate circuit of a
Flywheel Effect class-B or class-C operated
tube is connected to a parallel-resonant circuit tuned to the same frequency as the exciting voltage for the ampli-

fier, the plate current serves to maintain this L/C circuit in a state of oscillation.

The plate current is supplied in short pulses which do not begin to resemble a sine wave, even though the grid may be excited by a sine-wave voltage. These spurts of plate current are converted into a sine wave in the plate tank circuit by virtue of the Q or *flywheel effect* of the tank.

If a tank did not have some resistance losses, it would, when given a "kick" with a single pulse, continue to oscillate indefinitely. With a moderate amount of resistance or "friction" in the circuit the tank will still have inertia, and continue to oscillate with decreasing amplitude for a time after being given a "kick." With such a circuit, almost pure sine-wave voltage will be developed across the tank circuit even though power is supplied to the tank in short pulses or spurts, so long as the spurts are evenly spaced with respect to time and have a frequency that is the same as the resonant frequency of the tank.

Another way to visualize the action of the tank is to recall that a resonant tank with moderate Q will discriminate strongly against harmonics of the resonant frequency. The distorted plate current pulse in a class-C amplifier contains not only the fundamental frequency (that of the grid excitation voltage) but also higher harmonics. As the tank offers low impedance to the harmonics and high impedance to the fundamental (being resonant to the latter), only the fundamental — a sine-wave voltage — appears across the tank circuit in substantial magnitude.

Loaded and Unloaded Q Confusion sometimes exists as to the relationship between the unloaded and the loaded Q of the tank circuit in the plate of an r-f power amplifier. In the normal case the loaded Q of the tank circuit is determined by such factors as the operating conditions of the amplifier, bandwidth of the signal to be emitted, permissible level of harmonic radiation, and such factors. The normal value of *loaded* Q for an r-f amplifier used for communications service is from perhaps 6 to 20. The *unloaded* Q of the tank circuit determines the efficiency of the output circuit and is determined by the losses in the

tank coil, its leads and switch contacts, if any, and by the losses in the tank capacitor which ordinarily are very low. The unloaded Q of a good quality large diameter tank coil in the high-frequency range may be as high as 500, and values greater than 300 are quite common.

Tank-Circuit Efficiency Since the unloaded Q of a tank circuit is determined by the minimum losses in the tank, while the loaded Q is determined by useful loading of the tank circuit from the external load in addition to the internal losses in the tank circuit, the relationship between the two Q values determines the operating efficiency of the tank circuit. Expressed in the form of an equation, the loaded efficiency of a tank circuit is:

$$\text{Tank efficiency} = \left(1 - \frac{Q_1}{Q_u}\right) \times 100$$

where,

Q_u equals unloaded Q of the tank circuit,
Q_1 equals loaded Q of the tank circuit.

As an example, if the unloaded Q of the tank circuit for a class-C r-f power amplifier is 400, and the external load is coupled to the tank circuit by an amount such that the loaded Q is 20, the tank-circuit efficiency will be: eff. $= (1 - 20/400) \times 100$, or $(1 - 0.05) \times 100$, or 95 per cent. Hence 5 percent of the power output of the class-C amplifier will be lost as heat in the tank circuit and the remaining 95 percent will be delivered to the load.

7-2   Constant-Current Curves

Although class-C operating conditions can be determined with the aid of conventional grid-voltage versus plate-current operating curves (figure 9, chapter 5), the calculation is simplified if the alternative *constant current* graph of the tube in question is used (figure 4). This representation is a graph of constant plate current on a grid-voltage versus plate-voltage plot, as previously shown in figure 10, chapter 5. The constant-current plot is helpful as the *operating line* of a

tuned power amplifier is a straight line on
a set of curves and lends itself readily to
graphic computations. Any point on the
operating line, moreover, defines the instantaneous values of plate, screen and grid current which must flow when these particular
values of plate, screen and grid voltages are
applied to the tube. Thus, by taking off the
values of the currents and plotting them
against time, it is possible to generate a
curve of instantaneous electrode currents,
such as shown in figures 1 and 2. An analysis
of the curve of instantaneous current values will derive the d c components of the
currents, which may be read on a d c ammeter. In addition, if the plate current
flows through a properly loaded resonant r-f
circuit, the amount of power delivered to

the circuit may be predicted, as well as
drive power, and harmonic components of
drive and output voltage.

A set of typical constant-current curves
for the 304-TH medium-μ triode is shown
in figure 5, with a corresponding set of
curves for the 304-TL low-μ triode shown
in figure 6. The graphs illustrate how much
more plate current can be obtained from the
low-μ tube without driving the grid into
the positive-grid region, as contrasted to the
higher-μ tube. In addition, more bias voltage is required to cut off the plate current
of the low-μ tube, as compared to the higher-μ tube for a given value of plate voltage.
With the higher value of bias, a corresponding increase in grid-voltage swing is required
to drive the tube up to the zero grid-volt-

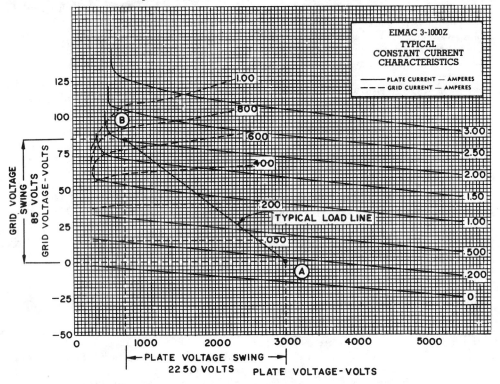

Figure 4

CONSTANT-CURRENT CHART FOR 3-1000Z HIGH-μ TRIODE

The constant-current chart is a plot of constant plate-current lines for various values of grid voltage
and plate current. At the start of operation (quiescent point A) the tube rests at a plate voltage of
3000 and zero grid voltage. At a positive grid potential of 85 volts (point B), the plate current has
increased to 2 amperes, and the plate voltage has dropped to 750, by virtue of the voltage drop
across the plate load impedance. As the grid voltage rises from zero to maximum, the operating
point passes from A to B along the load line. By examining representative samples of plate voltage
and current along the load line, typical operating characteristics may be derived for the given set
of conditions shown on the graph.

age point on the curve. Low-μ tubes thus, by definition, have lower voltage gain, and this can be seen by comparing the curves of figures 5 and 6.

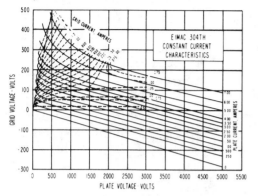

Figure 5

CONSTANT-CURRENT CHART FOR MEDIUM-μ TRIODE

Constant current plot for a 304TH triode with a μ of 20. Note that the lines of constant plate current have a greater slope than the corresponding lines of the high-μ triode (3-1000Z) and that for a given value of positive grid potential, and plate potential, the plate current of this tube is higher than that of the higher-μ tube.

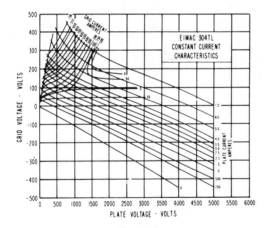

Figure 6

CONSTANT CURRENT CHART FOR LOW-μ TRIODE

Constant-current plot for a 304TL triode with a μ of 12. Note that more plate current at a given plate voltage can be obtained from the low-μ triode without driving the grid into the positive voltage region. In addition, more bias voltage is required to cut off the plate current at a given plate voltage. With this increased value of bias there is a corresponding increase in grid-voltage swing required to drive up to the zero grid-voltage point on the graph.

Low-μ (3-15) power triodes are chosen for class-A amplifiers and series-pass tubes in voltage regulators, as they operate well over a wide range of load current with low plate voltage drop. Medium-μ (15-50) triodes are generally used in r-f amplifiers and oscillators, as well as class-B audio modulators. High-μ (50-200) triodes have high power gain and are often used in cathode-driven ("grounded-grid") r-f amplifiers. If the amplification factor (μ) is sufficiently high, no external bias supply is required, and no protective circuits for loss of bias or drive are necessary. A set of constant-current curves for the 3-500Z high-μ triode is given in figure 7.

The amplification factor of a triode is a function of the physical size and location of the grid structure. The upper limit of amplification factor is controlled by grid dissipation, as high-μ grid structures require many grid wires of small diameter having relatively poorer heat-conduction qualities as compared to a low-μ structure, made up of fewer wires of greater diameter and better heat conductivity. A set of constant-current curves for the 250TH power triode with a sample load line drawn thereon is shown in figure 8.

7-3 Class-C Amplifier Calculations

In calculating and predicting the operation of a vacuum tube as a class-C radio-frequency amplifier, the considerations which determine the operating conditions are plate efficiency, power output required, maximum allowable plate and grid dissipation, maximum allowable plate voltage, and maximum allowable plate current. The values chosen for these factors will depend on the demands of a particular application of the tube.

The plate and grid currents of a class-C amplifier tube are periodic pulses, the durations of which are always less than 180 degrees. For this reason the average grid current, average plate current, power output, driving power, etc., cannot be directly calculated but must be determined by a Fourier analysis from points selected at proper intervals along the line of operation as plotted on the constant-current characteristics. This may be done either analytically or graphical-

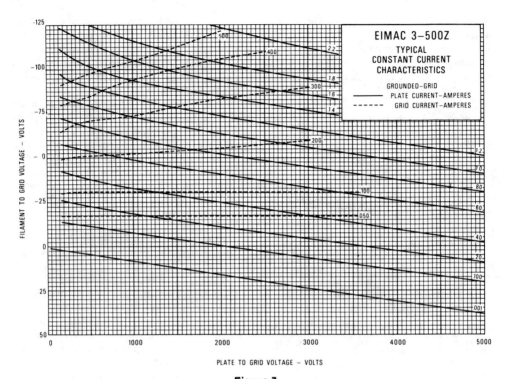

PLATE TO GRID VOLTAGE – VOLTS

Figure 7

CONSTANT-CURRENT CHART FOR HIGH-μ TRIODE

Constant current plot for a 3-500Z triode with μ of 160. The 3-500Z is considered to be "zero bias" up to a plate potential of about 3000. Resting plate current at this value of plate voltage is approximately 160 milliamperes. This plot is for grounded-grid, cathode-driven use, and grid-voltage axis is defined in terms of filament to grid voltage (negative) instead of grid-to-filament voltage (positive). Grid and screen currents are usually logged on constant-current plots, along with plate current.

ly. While the Fourier analysis has the advantage of accuracy, it also has the disadvantage of being tedious and involved.

The approximate analysis which follows has proved to be sufficiently accurate for most applications. This type of analysis also has the advantage of giving the desired information at the first trial. The system is direct in giving the desired information since the important factors, power output, plate efficiency, and plate voltage are arbitrarily selected at the beginning.

Method of Calculation The first step in the method to be described is to determine the power which must be delivered by the class-C amplifier. In making this determination it is well to remember that ordinarily from 5 to 10 percent of the power delivered by the amplifier tube or tubes will be lost in well-designed tank and coupling circuits at frequencies below 20 MHz. Above 20 MHz the tank and circuit losses are ordinarily somewhat above 10 percent.

The plate power input necessary to produce the desired output is determined by the plate efficiency: $P_i = P_o/N_p$, assuming 100-percent tank circuit efficiency.

For most applications it is desirable to operate at the highest practicable efficiency. High-efficiency operation usually requires less-expensive tubes and power supplies, and the amount of external cooling required is frequently less than for low-efficiency operation. On the other hand, high-efficiency operation usually requires more driving power and involves the use of higher plate voltages and higher peak tube voltages. The better types of triodes will ordinarily operate at a plate efficiency of 75 to 85 percent at the highest rated plate voltage, and at a plate

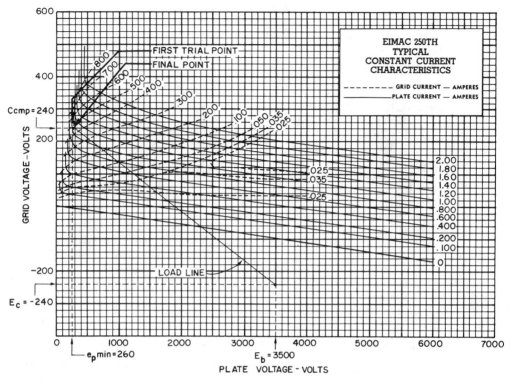

Figure 8

CONSTANT-CURRENT CHART FOR 250TH

Active portion of load line for an Eimac 250TH class-C r-f power amplifier, showing first trial point and final operating point for calculation of operating parameters at a power input of 1000 watts.

efficiency of 65 to 75 percent at intermediate values of plate voltage.

The first determining factor in selecting a tube or tubes for a particular application is the amount of plate dissipation which will be required of the stage. The total plate dissipation rating for the tube or tubes to be used in the stage must be equal to or greater than that calculated from: $P_p = P_i - P_o$.

After selecting a tube or tubes to meet the power output and plate dissipation requirements it becomes necessary to determine from the tube characteristics whether the tube selected is capable of the desired operation and, if so, to determine the driving power, grid bias, and grid dissipation.

The complete procedure necessary to determine a set of class-C amplifier operating conditions is given in the following steps:

1. Select the plate voltage, power output and efficiency.
2. Determine plate input from:
   $P_i = P_o/N_p$
3. Determine plate dissipation from:
   $P_p = (P_i - P_o) / 1.1$
   (P_p must not exceed maximum rated plate dissipation for selected tube or tubes. Tank circuit efficiency assumed to be 90%).
4. Determine average plate current (I_b) from: $I_b = P_i/E_b$.
5. Determine approximate peak plate current ($i_{b \ max}$) from:

   $i_{b \ max} = 4.9 \ I_b$ for $N_p = 0.85$
   $i_{b \ max} = 4.5 \ I_b$ for $N_p = 0.80$
   $i_{b \ max} = 4.0 \ I_b$ for $N_p = 0.75$
   $i_{b \ max} = 3.5 \ I_b$ for $N_p = 0.70$
   $i_{b \ max} = 3.1 \ I_b$ for $N_p = 0.65$

Note: A figure of $N_p = 0.75$ is often used for class-C service, and a figure of $N_p = 0.65$ is often used for class-B and class-AB service.

6. Locate the point on the constant-current chart where the constant-current plate line corresponding to the appropriate value of $i_{b\ max}$ determined in step 5 crosses the point of intersection of equal values of plate and grid voltage. (The locus of such points for all these combinations of grid and plate voltage is termed the *diode line*). Estimate the value of $e_{p\ min}$ at this point.

   In some cases, the lines of constant plate current will inflect sharply upward before reaching the diode line. If so, $e_{p\ min}$ should not be read at the diode line but at a point to the right where the plate-current line intersects a line drawn from the origin through these points of inflection.

7. Calculate $e_{b\ min}$ from:

$$e_{b\ min} = E_b - e_{p\ min}.$$

8. Calculate the ratio: $i_{1\ max} / I_b$ from:

$$\frac{i_{1\ max}}{I_b} = \frac{2\ N_p \times E_b}{e_{p\ min}}$$

   (where $i_{1\ max}$ = peak fundamental component of plate current).

9. From the ratio of $i_{1\ max} / I_b$ calculated in step 8 determine the ratio: $i_{b\ max}/I_b$ from the graph of figure 9.

10. Derive a new value for $i_{b\ max}$ from the ratio found in step 9:
    $i_{b\ max}$ = (ratio found in step 9) $\times I_b$

11. Read the values of maximum positive grid voltage, $e_{g\ max}$ and peak grid current ($i_{g\ max}$) from the chart for the values of $e_{p\ min}$ and $i_{b\ max}$ found in steps 6 and 10 respectively.

12. Calculate the cosine of one-half the angle of plate-current flow (one-half the operating cycle, $\theta_p/2$).

$$\cos\frac{\theta_p}{2} = 2.32\left(\frac{i_{1\ max}}{I_b} - 1.57\right)$$

13. Calculate the grid bias voltage (E_c) from:

$$E_c = \frac{1}{1 - \cos\dfrac{\theta_p}{2}} \times$$

$$\left[\cos\frac{\theta_p}{2}\left(\frac{e_{b\ min}}{\mu} - e_{cmp}\right) - \frac{E_b}{\mu}\right]$$

for triodes.

$$E_{c1} = \frac{1}{1 - \cos\dfrac{\theta_p}{2}} \times$$

$$\left[- e_{cmp} \times \cos\frac{\theta_p}{2} - \frac{E_{c2}}{\mu_s}\right]$$

for tetrodes, where μ_s is the grid-screen amplification factor.

14. Calculate the peak fundamental grid voltage, $e_{g\ max}$ from:
    $e_{g\ max} = e_{cmp} - (- E_c)$, using negative value of E_c.

15. Calculate the ratio $e_{g\ max}/E_c$ for the values of E_c and $e_{g\ max}$ found in steps 13 and 14.

16. Read the ratio $i_{g\ max}/I_c$ from figure 10 for the ratio $e_{g\ max} / E_c$ found in step 15.

17. Calculate the average grid current (I_c) from the ratio found in step 16 and the value of $i_g\ max$ found in step 11:

$$I_c = \frac{i_{g\ max}}{(\text{ratio found in step 16})}$$

18. Calculate approximate grid driving power from:

$$P_d = 0.9\ e_{g\ max} \times I_c$$

19. Calculate grid dissipation from:

$$P_g = P_d - (- E_c \times I_c)$$

   (P_g must not exceed the maximum rated grid dissipation for the tube or tubes selected).

Sample Calculation A typical example of class-C amplifier calculation is shown in the following example. Reference is made to figures 8, 9, and 10 in the calculation. The steps correspond to those in the previous section.

1. Desired power output—800 watts.
2. Desired plate voltage—3500 volts.
   Desired plate efficiency—80%
   ($N_p = 0.8$). $P_i = 800/0.8 = 1000$
   watts.

3. $P_p = \dfrac{1000 - 800}{1.1} = 182$ watts.

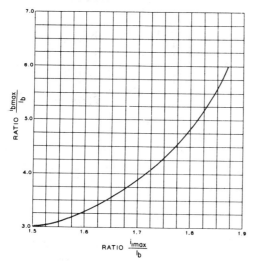

RATIO $\dfrac{i_{max}}{I_b}$

Figure 9

Relationship between the peak value of the fundamental component of the tube plate current, and average plate current; as compared to the ratio of the instantaneous peak value of tube plate current, and average plate current value.

(Use 250TH; max $P_p = 250W$;
$\mu = 37$).

4. $I_b = 1000/3500 = 0.285$ ampere (285 mA). (Maximum rated I_b for 250TH = 350 mA).
5. Approximate $i_{b\ max}$: $0.285 \times 4.5 = 1.28$ amp
6. $e_{b\ min} = 260$ volts (see figure 8, first trial point).
7. $e_{p\ min} = 3500 - 260 = 3240$ volts.
8. $i_{1\ max} / I_b = (2 \times 0.8 \times 3500) / 3240 = 1.73$.
9. $i_{b\ max} / I_b = 4.1$ (from figure 9).
10. $i_{b\ max} = 4.1 \times 0.285 = 1.17$.
11. $e_{cmp} = 240$ volts
    $i_{g\ max} = 0.43$ amp
    (Both read from final point on figure 8).
12. $\cos \dfrac{\theta_b}{2} = 2.32\ (1.73 - 1.57) = 0.37$

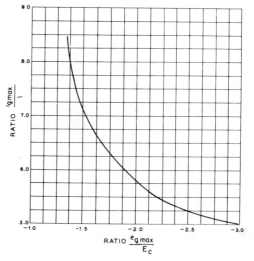

RATIO $\dfrac{e_{g\,max}}{E_c}$

Figure 10

Relationship between the ratio of the peak value of the fundamental component of the grid excitation voltage, and the average grid bias; as compared to the ratio between instantaneous peak grid current and average grid current.

$\left(\dfrac{\theta_b}{2} = 68.3° \text{ and } \theta_b = 136.6° \right)$

13. $E_c = \dfrac{1}{1 - 0.37} \times$

$$\left[0.37 \left(\dfrac{3240}{37} - 240 \right) - \dfrac{3500}{37} \right]$$

$$= -240 \text{ volts.}$$

14. $e_{g\ max} = 240 - (-240) = 480$ volts.
15. $e_{g\ max}/E_c = 480/-240 = -2$.
16. $i_{g\ max}/I_c = 5.75$ (from figure 10).
17. $I_c = 0.43/5.75 = 0.075$ amp (75mA).
18. $P_d = 0.9 \times 480 \times 0.075 = 32.5$ watts.
19. $P_g = 32.5 + (-240 \times 0.075) = 14.5$ watts (Maximum rated P_g for 250TH = 40 watts).
20. The power output of any type of r-f amplifier is equal to:

$$P_o = \dfrac{i_{1\ max} \times e_{p\ min}}{2}$$

($i_{1\ max}$ can be determined by multiplying the ratio determined in step 8 by

I_b. Thus $= 1.73 \times 0.285 = 0.495$).
$P_o = (0.495 \times 3240)/2 = 800$ watts

21. The plate load impedance of any type of r-f amplifier is equal to:

$$R_L = \frac{e_{p\ min}}{i_{1\ max}}$$

$$R_L = \frac{3240}{0.495} = 6550 \text{ ohms}$$

An alternative equation for the approximate value of R_L is:

$$R_L \cong \frac{E_b}{1.8 \times I_b}$$

$$R_L \cong \frac{3500}{1.8 \times 0.285} = 6820 \text{ ohms}$$

Q of Amplifier Tank Circuit In order to obtain proper plate tank-circuit tuning and low radiation of harmonics from an amplifier it is necessary that the plate tank circuit have the correct Q. Charts giving compromise values of Q for class-C amplifiers are given in the chapter, *Generation of R-F Energy*. However, the amount of inductance required for a special tank-circuit Q under specified operating conditions can be calculated from the following expression:

$$\omega L = \frac{R_L}{Q}$$

where,

ω equals $2\pi \times$ operating frequency,
L equals tank inductance,
R_L equals required tube load impedance,
Q equals effective tank circuit Q.

A tank circuit Q of 12 to 20 is recommended for all normal conditions. However, if a balanced push-pull amplifier is employed the tank receives two impulses per cycle and the circuit Q may be lowered somewhat from the above values.

Quick Method of Calculating Amplifier Plate Efficiency The plate-circuit efficiency of a class-B or class-C r-f amplifier is approximately equal to the product of two factors: F_1, which is equal to the ratio of $e_{p\ max}$ to E_b ($F_1 = e_{p\ max}/E_b$) and F_2, which is proportional to the one-half angle of plate current flow $\theta_b/2$.

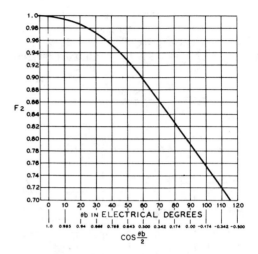

Figure 11

Relationship between factor F_2 and the half-angle of plate-current flow in an amplifier with sine-wave input and output voltage, operating at a grid-bias voltage greater than cutoff.

A graph of F_2 versus both $\theta_b/2$ and $\cos \theta_b/2$ is given in figure 11. Either $\theta_b/2$ or $\cos \theta_b/2$ may be used to determine F_2. $\cos \theta_b/2$ may be determined either from the procedure previously given for making class-C amplifier computations or it may be determined from the following expression:

$$\cos \frac{\theta_b}{2} = -\frac{\mu E_c + E_b}{\mu \times e_{g\ max} - e_{p\ max}}$$

Example of Method It is desired to know the one-half angle of plate-current flow and the plate-circuit efficiency for an 812 tube operating under the following class-C conditions which have been assumed from inspection of the data and curves given in the RCA Transmitting Tube Handbook:

1. E_b = 1100 volts
   E_c = -40 volts
   μ = 29
   $e_{g\ max}$ = 120 volts
   $e_{p\ max}$ = 1000 volts

2. $F_1 = \dfrac{e_{p\ max}}{E_b} = 0.91$

3. $\cos \dfrac{\theta_b}{2} = -\left[\dfrac{(29 \times 40) + 1100}{(29 \times 120) - 1000}\right] = \dfrac{60}{2480} = 0.025$

4. F_2 = 0.79 (by reference to figure 11)

5. N_p = $F_1 \times F_2$ = 0.91 $\times$ 0.79 = 0.72 (72 percent efficiency)

F_1 could be called the *plate-voltage-swing efficiency factor*, and F_2 can be called the *operating-angle efficiency factor* or the maximum possible efficiency of any stage running with that value of half-angle of plate current flow.

N_p is, of course, only the ratio between power output and power input. If it is desired to determine the power input, exciting power, and grid current of the stage, these can be obtained through the use of steps 7, 8, 9, and 10 of the previously given method for determining power input and output; and knowing that $i_{g\,max}$ *is* 0.095 ampere, the grid-circuit conditions can be determined through the use of steps 15, 16, 17, 18, and 19.

7-4 Class-B Radio-Frequency Power Amplifiers

Radio-frequency power amplifiers operating under class-B conditions of grid bias and excitation voltage are used in various types of applications in transmitters. The first general application is as a buffer-amplifier stage where it is desired to obtain a high value of power amplification in a particular stage without regard to linearity. A particular tube type operated with a given plate voltage will be capable of somewhat greater output for a certain amount of excitation power when operated as a class-B amplifier than when operated as a class-C amplifier.

Calculation of Operating Characteristics Calculation of the operating conditions for this type of class-B r-f amplifier can be carried out in a manner similar to that described in the previous paragraphs, except that the grid-bias voltage is set on the tube before calculation at the value: $E_c = - E_b/\mu$. Since the grid bias is set at cutoff the one-half angle of plate-current flow is 90°; hence cos $\theta_b/2$ is fixed at 0.00. The plate-circuit efficiency for a class-B r-f amplifier operated in this manner can be determined in the following manner:

$$N_p = 78.5 \times \frac{e_{p\,max}}{E_b}$$

Note: In reference to figure 3, $e_{p\,max}$ is equal in magnitude to $e_{p\,min}$ and absolute value should be used.

The "Class-B Linear" The second type of class-B r-f amplifier is the so-called *class-B linear amplifier* which is often used in transmitters for the amplification of a single-sideband signal or a conventional amplitude-modulated wave. Calculation of operating conditions may be carried out in a manner similar to that previously described with the following exceptions: The first trial operating point is chosen on the basis of the 100-percent positive modulation peak (or PEP condition) of the exciting wave. The plate-circuit and grid-peak voltages and currents can then be determined and the power input and output calculated. Then (in the case for an a-m linear) with the exciting voltage reduced to one-half for the no-modulating condition of the exciting wave, and with the same value of load resistance reflected on the tube, the a-m plate input and plate efficiency will drop to approximately one-half the values at the 100-percent positive modulation peak and the power output of the stage will drop to one-fourth the peak-modulation value. On the negative modulation peak the input, efficiency and output all drop to zero.

In general, the proper plate voltage, bias voltage, load resistance, and power output listed in the tube tables for class-B audio work will also apply to class-B linear r-f application.

Calculation of Operating Parameters for a Class-B Linear Amplifier The class-B linear amplifier parameters may be calculated from constant-current curves, as suggested, or may be derived from the E_b vs I_b curves, as outlined in this section.

Figure 12 illustrates the characteristic curves for an 813 tube. Assume the plate supply to be 2000 volts, and the screen supply to be 400 volts. To determine the operating parameters of this tube as a class-B linear SSB r-f amplifier, the following steps should be taken:

1. The grid bias is chosen so that the resting plate current will produce approximately 1/3 of the maximum plate dissipation of the tube. The maximum dissipation of the 813 is 125 watts, so the bias is set to allow one-third of this value, or 42 watts of resting dissipation. At a plate potential of 2000 volts, a plate current of 21 milliamperes will produce this figure. Referring to figure 12, a grid bias of −45 volts is approximately correct.

2. A practical class-B linear r-f amplifier runs at an efficiency of about 66% at full output (the carrier efficiency dropping to about 33% with a *modulated* exciting signal). In the case of single-sideband suppressed-carrier excitation, the linear amplifier runs at the resting or quiescent input of 42 watts with no exciting signal. The peak allowable power input to the 813 is:

$$\text{PEP input power } (p_1) =$$

$$\frac{\text{plate dissipation} \times 100}{(100 - \% \text{ plate efficiency})} =$$

$$\frac{125 \times 100}{33} = 378 \text{ watts PEP}$$

3. The maximum dc signal plate current is:

$$I_{b\,max} = \frac{p_1}{E_b} = \frac{378}{2000} = 0.189 \text{ ampere}$$

(Single-tone drive signal condition)

4. The plate-current conduction angle (θ_b) of the class-B linear amplifier is *approximately* 180°, and the peak plate-current pulses have a maximum value of about 3.14 times $I_{b\,max}$:

$$i_{b\,max} = 3.14 \times 0.189 = 0.593 \text{ amp.}$$

5. Referring to figure 12, a current of about 0.6 ampere (*Point A*) will flow at a positive grid potential of 60 volts and a minimum plate potential of 420 volts. The grid is biased at −45 volts, so a peak r-f grid voltage of 60 + 45 volts, or 105 volts, swing is required.

6. The grid driving power required for the class-B linear stage may be found by the aid of figure 13. It is one-third the product of the peak grid current times the peak grid swing.

$$P_d = \frac{0.015 \times 105}{3} = 0.525 \text{ watt}$$

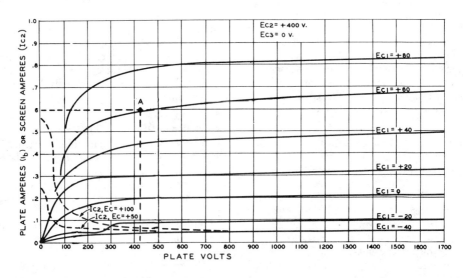

Figure 12

AVERAGE PLATE CHARACTERISTICS OF 813 TUBE

7. The single-tone (peak) power output of the 813 is:

$$P_o = .785 \, (E_b - e_{b \, min}) \times I_{b \, max}$$

$$P_o = .785 \, (2000 - 420) \times 0.189$$
$$= 235 \text{ watts PEP}$$

8. The plate load resistance is:

$$R_L \cong \frac{E_b}{1.8 \times I_b} = \frac{2000}{1.8 \times 0.188}$$

$$= 5870 \text{ ohms}$$

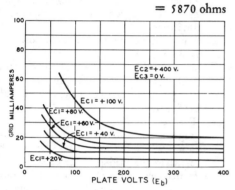

Figure 13

E_{g1} VERSUS E_b CHARACTERISTICS OF 813 TUBE

9. If a loaded plate tank circuit Q of 12 is desired, the reactance of the plate tank capacitor of a parallel tuned circuit at resonance is:

$$X_c = \frac{R_L}{Q} = \frac{5870}{12} = 490 \text{ ohms}$$

10. For an operating frequency of 4.0 MHz, the effective resonant capacitance is:

$$C = \frac{10^6}{6.28 \times 4.0 \times 490} = 81 \text{ pF}$$

11. The inductance required to resonate at 4.0 MHz with this value of capacitance is:

$$L = \frac{490}{6.28 \times 4.0} = 19.5 \text{ microhenrys}$$

Grid-Circuit Considerations 1. The maximum positive grid potential is 60 volts and the peak r-f grid voltage is 105 volts. Required peak driving power is 0.525 watt. The equivalent grid resistance of this stage is:

$$r_g = \frac{(e_{g \, max})^2}{2 \times P_d} = \frac{105^2}{2 \times 0.525}$$

$$= 10,000 \text{ ohms}$$

2. As in the case of the class-B audio amplifier the grid resistance of the linear amplifier varies from infinity to a low value when maximum grid current is drawn. To decrease the effect of this resistance excursion, a swamping resistor should be placed across the grid-tank circuit. The value of the resistor should be dropped until a shortage of driving power begins to be noticed. For this example, a resistor of 3000 ohms is used. The grid circuit load for no grid current is now 3000 ohms instead of infinity, and drops to 2300 ohms when maximum grid current is drawn.

3. A circuit Q of 15 is chosen for the grid tank. The capacitive reactance required is:

$$X_C = \frac{2300}{15} = 154 \text{ ohms}$$

4. At 4.0 MHz the effective capacitance is:

$$C = \frac{10^6}{6.28 \times 4.0 \times 154} = 259 \text{ pF}$$

5. The inductive reactance required to resonate the grid circuit at 4.0 MHz is:

$$L = \frac{154}{6.28 \times 4.0} = 6.1 \text{ microhenrys}$$

6. By substituting the loaded-grid resistance figure in the formula in the first paragraph, the peak grid driving power is now found to be approximately 2.4 watts.

Screen-Circuit Considerations By reference to the plate characteristic curve of the 813 tube, it can be seen that at a minimum plate potential of 420 volts, and a maximum plate current of 0.6 ampere, the screen current will be approximately 30 milliamperes, dropping to one or two milliamperes in the quiescent state. It is necessary to use a well-regulated screen supply to hold the screen voltage at the correct potential over this range of current excursion. The use of an electronically regulated screen supply is recommended.

7-5 Grounded-Grid and Cathode-Follower R-F Power Amplifier Circuits

The r-f power amplifier discussions of Sections 7-3 and 7-4 have been based on the assumption that a conventional grounded-cathode or cathode-return type of amplifier was in question. It is possible, however, as in the case of a-f and low-level r-f amplifiers to use circuits in which electrodes other than the cathode are returned to ground insofar as the signal potential is concerned. Both the plate-return or cathode-follower amplifier and the *grid-return* or *grounded-grid* amplifier are effective in certain circuit applications as tuned r-f power amplifiers.

Disadvantages of Grounded-Cathode Amplifiers An undesirable aspect of the operation of cathode-return r-f power amplifiers using triode tubes is that such amplifiers must be neutralized. Principles and methods of neutralizing r-f power amplifiers are discussed in the chapter *Generation of R-F Energy*. As the frequency of operation of an amplifier is increased the stage becomes more and more difficult to neutralize due to inductance in the grid and cathode leads of the tube and in the leads to the neutralizing capacitor. In other words the bandwidth of neutralization decreases as the presence of the neutralizing capacitor adds additional undesirable capacitive loading to the grid and plate tank circuits of the tube or tubes. To look at the problem in another way, an amplifier that may be perfectly neutralized at a frequency of 30 MHz may be completely out of neutralization at a frequency of 120 MHz. Therefore, if there are circuits in both the grid and plate circuits which offer appreciable impedance at this high frequency it is quite possible that the stage may develop a parasitic oscillation in the vicinity of 120 MHz.

Grounded-Grid R-F Amplifiers This condition of restricted-range neutralization of r-f power amplifiers can be greatly alleviated through the use of a *cathode-driven* or *grounded-grid* r-f stage. The grounded-grid amplifier has the following advantages:

1. The output and input capacitances of a stage are reduced to approximately one-half the value which would be obtained if the same tube or tubes were operated as a conventional neutralized amplifier.
2. The tendency toward parasitic oscillations in such a stage is greatly reduced since the shielding effect of the control grid between the filament and the plate is effective over a broad range of frequencies.
3. The feedback capacitance within the stage is the plate-to-cathode capacitance which is ordinarily very much less than the grid-to-plate capacitance. Hence neutralization is ordinarily not required in the high frequency region. If neutralization is required the neutalizing capacitors are very small in value and are cross-connected between plates and cathodes in a push-pull stage, or between the opposite end of a split plate tank and the cathode in a single-ended stage.

The disadvantages of a grounded-grid amplifier are:

1. A large amount of excitation energy is required. However, only the normal amount of energy is lost in the grid circuit of the amplifier tube; most additional energy over this amount is delivered to the load circuit as useful output.
2. The cathode of a grounded-grid amplifier stage is above r-f ground. This means that the cathode must be fed through a suitable impedance from the filament supply, or the filament transformer must be of the low capacitance type and adequately insulated for the r-f voltage which will be present.
3. A grounded-grid r-f amplifier cannot be plate modulated 100 percent unless the output of the exciting stage is modulated also. Approximately 70-per

cent modulation of the exciter stage, while the final stage is modulated 100 percent, is recommended. However the grounded-grid r-f amplifier is quite satisfactory as a class-B linear r-f amplifier for single-sideband or conventional amplitude-modulated waves or as an amplifier for a straight c-w or f-m signal.

Figure 14 shows a simplified representation of a grounded-grid zero-bias triode r-f power amplifier stage. The relationships between input and output power and the peak fundamental components of electrode voltages and currents are given below the drawing. The calculation of the complete operating conditions for a grounded-grid amplifier stage is somewhat more complex than that for a conventional amplifier because the input circuit of the tube is in series with the output circuit as far as the load is con-

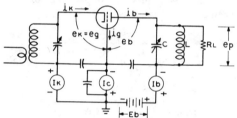

$$\text{PEP POWER TO LOAD} = \frac{(e_b \text{ MIN} + e_g \text{ MAX}) \times i_1 \text{ MAX}}{2}$$

$$\text{PEP POWER DELIVERED BY OUTPUT TUBE} = \frac{e_b \text{ MIN} \times i_1 \text{ MAX}}{2}$$

$$\text{PEP DRIVE POWER} = \frac{e_g \text{ MAX} \times i_1 \text{ MAX}}{2} + 0.9 \, (e_g \text{ MAX} \times I_c)$$

$$Z_K \cong \frac{e_g \text{ MAX}}{i_1 \text{ MAX} + 1.5 \times I_c}$$

$$R_L \cong \frac{E_b}{1.8 \times I_b}$$

Figure 14

ZERO-BIAS GROUNDED-GRID AMPLIFIER

The equations in the above figure give the relationships between the output power, drive power, feedthrough power, and input and output impedances expressed in terms of the various voltages and currents of the stage.

cerned. The primary result of this effect is, as stated before, that considerably more power is required from the driver stage. The normal power gain for a g-g stage is from 3 to 15 depending on the grid-circuit conditions chosen for the output stage. The higher

the grid bias and grid swing required on the output stage, the higher will be the requirement from the driver.

Calculation of Operating Conditions of Grounded-Grid R-F Amplifers It is most convenient to determine the operating conditions for a class-B or class-C grounded-grid r-f power amplifier in a two-step process. The first step is to determine the plate-circuit and grid-circuit operating conditions of the tube as though it were to operate as a conventional grid-driven amplifier. The second step is to then add in the additional conditions imposed on the orginal data by the fact that the stage is to operate as a grounded-grid amplifier. This step is the addition of the portion of the drive power contributed by the conversion of drive power to plate output power. This portion of the drive power is referred to as *converted drive power,* or *feedthrough power.* The latter term is misleading, as this portion of drive power does not appear in the plate load circuit of the cathode-driven stage until after it is converted to a *varying-dc* plate potential effectively in series with the main amplifier power supply. The converted drive power serves a useful function in linear amplifier service because it swamps out the undesirable effects of nonlinear grid loading and presents a reasonably constant load to the exciter.

Special constant-current curves are often used for grounded-grid operation wherein the grid drive voltage is expressed as the *cathode-to-grid voltage* and is negative in sign. It must be remembered, however, that a negative cathode voltage is equal to a positive grid voltage, and normal constant-current curves may also be employed for cathode-driven computations.

For the first step in the calculations, the procedure given in Section 7-3 is used. For this example, a 3-1000Z "zero bias" triode is chosen, operating at 3000 plate volts at 2000 watts PEP input in class-B service. Computations are as follows:

3-1000Z at 3000 volts class-B

1,2,3. $E_b = 3000$
$P_i = 2000$ watts PEP

Let $N_p = 65\%$, an average value for class-B mode

$P_o = 2000 \times 0.65 = 1300$ W PEP

$\mu = 200$

4. $I_b = \dfrac{2000}{3000} = 0.67$ amp

5. Approx. $i_{b\ max} = 3.1\ I_b$ (for $N_p = 0.65$) $= 3.1 \times 0.65 = 2.08$ amperes

6. Locate the point on the constant-current chart where the constant-current line corresponding to the appropriate value of $i_{b\ max}$ determined in step 5 inflects sharply upward. Approximate $e_{b\ min} = 500$ volts.

7. $e_{p\ min} = 3000 - 500 = 2500$ volts.

8. $\dfrac{i_{1\ max}}{I_b} = \dfrac{2 \times 0.65 \times 3000}{2500} = 1.56$

9. $\dfrac{i_{b\ max}}{I_b} = 3.13$ (from figure 9).

10. $i_{b\ max} = 3.13 \times 0.67 = 2.1$ amps.

This agrees closely with the approximation made in Step 5.

11. Read the values maximum cathode-to-filament voltage (e_k) and peak grid current ($i_{g\ max}$) from the constant-current chart for the values of $e_{b\ min}$ and $i_{b\ max}$ found in steps 6 and 10 respectively.

$$e_k = -88$$
$$i_{g\ max} = 0.8\ \text{amp}$$

12. $cos\ \dfrac{\theta_b}{2} = 2.32\ (1.56 - 1.57) = 0$

    (Conduction angle is approximately $180°$ and $cos\ 180° = 0$)

13. $E_c = 0$

14. $e_{k\ max} = -88$ volts

15-17. For zero bias class-B mode, $I_c \cong 0.25\ i_{g\ max}$. $I_c \cong 0.25 \times 0.8 = 0.2$ amp. (200 mA)

18. $p_d = 0.9 \times |88| \times 0.2 = 15.8$ watts PEP

19. $p_g = 15.8$ watts PEP

20. $i_{1\ max} =$ (Ratio of step 8) $\times I_b$
    $i_{1\ max} = 1.56 \times 0.67 = 1.06$ amp

$$P_o\ (\text{PEP}) = \dfrac{1.06 \times 2500}{2}$$
$$= 1325\ \text{watts.}$$

21. $R_L \cong \dfrac{3000}{1.8 \times 0.67} = 2500$ ohms

22. Total peak drive power,

$$p_k = \dfrac{e_k \times i_{1\ max}}{2} + p_d$$

$$p_k = \dfrac{88 \times 1.06}{2} + 15.8 \cong 61\ \text{watts PEP}$$

23. Total power output of the stage is equal to 1325 watts (contributed by 3-1000Z) plus that portion of drive power contributed by the conversion of drive power to plate output power. This is approximately equal to the first term of the equation of step 22.

$$P_o\ (\text{PEP})\ \text{total} = 1325 + 44$$
$$= 1369\ \text{watts}$$

24. Cathode driving impedance of the grounded grid stage is:

$$Z_k \cong \dfrac{e_k}{i_{1\ max} + 1.5 \times I_c}$$

$$Z_k \cong \dfrac{88}{1.06 + 0.3} = 64\ \text{ohms}$$

A summary of the typical operating parameters for the 3-1000 Z at $E_b = 3000$ are

Dc Plate Voltage	3000
Zero-Signal Plate Current (from constant-current chart)	180 mA
Max. Signal (PEP) Plate Current	670 mA
Max. Signal (PEP) Grid Current	200 mA
Max Signal (PEP) Drive Power	61 watts
Max. Signal (PEP) Power Input	2000 watts
Max. Signal (PEP) Power Output (including feedthrough power)	1369 watts
Plate Load Impedance	2500 ohms
Cathode Driving Impedance	64 ohms

Cathode Tank of G-G or C-F Power Amplifier   The cathode tank circuit for either a grounded-grid or cathode-follower r-f power amplifier may be a conventional tank circuit if the filament transformer for the stage is of the low-capacitance high-voltage type. Conventional filament transformers, however, will not operate with the high values of r-f voltage present in such a circuit. If a conventional filament transformer is to be used, the cathode tank coil may consist of two parallel heavy conductors (to carry the high filament current) bypassed at both the ground end and at the tube socket. The tuning capacitor is then placed between filament and ground. It is possible in certain cases to use two r-f chokes of special design to feed the filament current to the tubes, with a conventional tank circuit between filament and ground. Coaxial lines also may be used to serve both as cathode tank and filament feed to the tubes for vhf and uhf work.

Control-Grid Dissipation in Grounded-Grid Stages Tetrode tubes may be operated as grounded-grid (cathode-driven) amplifiers by tying the grid and screen together and operating the tube as a high-μ triode (figure 15). Combined grid and screen current, however, is a function of tube geometry and may reach destructive values under conditions of full excitation. Proper division of excitation between grid and screen should be as the ratio of the screen-to-grid amplification, which is approximately 5 for tubes such as the 4-250A, 4-400A, etc. The proper ratio of grid/screen excitation may be achieved by tapping the grid at some point on the input circuit, as shown. Grid dissipation is reduced, but the over-all level of excitation is increased about 30% over the value required for simple grounded-grid operation.

Plate-Return or Cathode-Follower R-F Power Amplifier Circuit diagram, electrode potentials and currents, and operating conditions for a cathode-follower r-f power amplifier are given in figure 16. This circuit can be used, in addition to the grounded-grid circuit just discussed, as an r-f amplifier with a triode tube and no additional neutralization circuit. However, the circuit will oscillate if the impedance from cathode to ground is allowed to become capacitive rather than

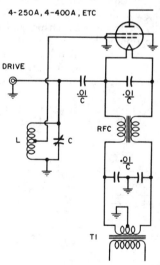

4-250A, 4-400A, ETC

DRIVE

Figure 15

TAPPED INPUT CIRCUIT REDUCES EXCESSIVE GRID DISSIPATION IN G-G CIRCUIT

C = 20 pF per meter wavelength
RFC = Dual-winding on ½-inch diameter, 3½-inch long ferrite rod. Q-1 material. (Indiana General).

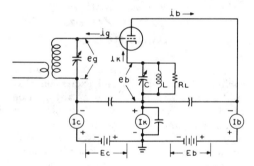

POWER OUTPUT TO LOAD $= \dfrac{e_{b\,MIN}\,(i_{1\,MAX} + 1.8\,I_c)}{2}$

POWER DELIVERED BY OUTPUT TUBE $= \dfrac{e_{b\,MIN} \times i_{1\,MAX}}{2}$

DRIVE POWER $= \dfrac{(e_{g\,MP} + e_{b\,MIN}) \times 1.8\,I_c}{2}$

$Z_G \cong \dfrac{e_{g\,MAX}}{1.8\,I_c}$

$R_L \cong \dfrac{E_b}{1.8 \times I_k}$

Figure 16

CATHODE-FOLLOWER R-F POWER AMPLIFIER

The equations show the relationship between the tube potentials and currents and the input and output power of the stage. The approximate input and output load impedances are also given.

inductive or resistive with respect to the operating frequency. The circuit is not recommended except for vhf or uhf work with coaxial lines as tuned circuits since the peak grid swing required on the r-f amplifier stage is approximately equal to the plate voltage on the amplifier tube if high-efficiency operation is desired. This means, of course, that the grid tank must be able to withstand slightly more peak voltage than the plate tank. Such a stage may not be plate modulated unless the driver stage is modulated the same percentage as the final amplifier. However, such a stage may be used as an amplifier of modulated waves (class-B linear) or as a c-w or f-m amplifier.

The design of such an amplifier stage is essentially the same as the design of a grounded-grid amplifier stage as far as the first step is concerned. Then, for the second step the operating conditions given in figure 16 are applied to the data obtained in the first step.

7-6   Class-AB₁ Radio-Frequency Power Amplifiers

Class-AB₁ r-f amplifiers operate under such conditions of bias and excitation that grid current does not flow over any portion of the input cycle. This is desirable, since distortion caused by grid-current loading is absent, and also because the stage is capable of high power gain. Stage efficiency is about 60 percent when a plate current conduction angle of 210° is chosen, as compared to 65 percent for class-B operation.

The level of static (quiescent) plate current for *lowest distortion* is quite high for class-AB₁ tetrode operation. This value is determined by the tube characteristics, and is not greatly affected by the circuit parameters or operating voltages. The maximum dc potential is therefore limited by the static dissipation of the tube, since the resting plate current figure is fixed. The static plate current of a tetrode tube varies as the 3/2 power of the screen voltage. For example, raising the screen voltage from 300 to 500 volts will double the plate current. The optimum static plate current for mini-

mum distortion is also doubled, since the shape of the E_c-I_b curve does not change.

In actual practice, somewhat lower static plate current than optimum may be employed without raising the distortion appreciably, and values of static plate current of 0.6 to 0.8 of optimum may be safely used, depending on the amount of nonlinearity that can be tolerated.

As with the class-B linear stage, the minimum plate voltage swing ($e_{b\ min}$) of the class-AB₁ amplifier must be kept above the dc screen potential to prevent operation in the nonlinear portion of the characteristic curve. A *low value* of screen voltage allows greater r-f plate voltage swing, resulting in improvement in plate efficiency of the tube. A balance between plate dissipation, plate efficiency, and plate-voltage swing must be achieved for best linearity of the amplifier.

The S-Curve   The perfect linear amplifier delivers a signal that is a replica of the input signal. Inspection of the plate-characteristic curve of a typical tube will disclose the tube linearity under class-AB₁ operating conditions (figure 17). The curve is usually of exponential shape, and the signal distortion is held to a small value by operating the tube well below its maximum output, and centering operation over the most linear portion of the characteristic curve.

The relationship between exciting voltage in a class-AB₁ amplifier and the r-f plate-

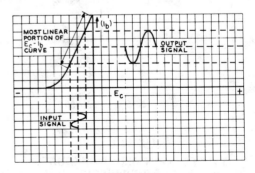

Figure 17

E_c-I_b CURVE

Amplifier operation is confined to the most linear
portion of the characteristic curve.

circuit voltage is shown in figure 18. With a small value of static plate current the lower portion of the line is curved. Maximum undistorted output is limited by the point on the line (A) where the instantaneous plate voltage drops down to the screen voltage. This "hook" in the line is caused by current diverted from the plate to the grid and screen elements of the tube. The characteristic plot of the usual linear amplifier takes the shape of an S-curve. The lower portion of the curve is straightened out by using the proper value of static plate current, and the upper portion of the curve is avoided by limiting minimum plate voltage swing to a point substantially above the value of the screen voltage.

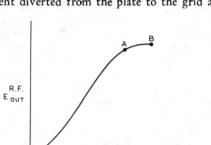

Figure 18

LINEARITY CURVE OF TYPICAL TETRODE AMPLIFIER

At point A the instantaneous plate voltage is swinging down to the value of the screen voltage. At point B it is swinging well below the screen and is approaching the point where saturation, or plate-current limiting takes place.

Operating Parameters for the Class-AB₁ Linear Amplifier The approximate operating parameters may be obtained from the constant-current curves (E_c-E_b) or the E_c-I_b curves of the tube in question (figure 19). The following example will make use of the latter information, although equivalent results may be obtained from constant current curves. An operating load line is first approximated. One end of the load line is determined by the dc operating voltage of the tube, and the required static plate current. As a starting point, let the product of the plate voltage and current approximate the plate dissipation of the tube. Assuming a 4-400A tetrode is used, this end of the load line will

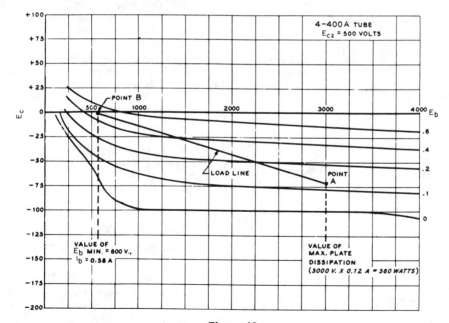

Figure 19

OPERATING PARAMETERS FOR TETRODE LINEAR AMPLIFIER ARE OBTAINED FROM CONSTANT-CURRENT CURVES.

fall on point A (figure 19). Plate power dissipation is 360 watts (300V at 120 mA). The opposite end of the load line will fall on a point determined by the minimum instantaneous plate voltage, and by the maximum instantaneous plate current. The minimum plate voltage, for best linearity should be considerably higher than the screen voltage. In this case, the screen voltage is 500, so the minimum plate voltage excursion should be limited to 600 volts. Class-AB₁ operation implies no grid current, therefore the load line cannot cross the $E_c = 0$ line. At the point $e_{b\ min} = 600$, $E_c = 0$, the maximum instantaneous plate current is 580 mA (Point B).

Each point at which the load line crosses a grid-voltage axis may be taken as a point for construction of the E_c-I_b curve, just as was done in figure 21, chapter 6. A constructed curve shows that the approximate static bias voltage is −74 volts, which checks closely with point A of figure 19. In actual practice, the bias voltage is set to hold the actual dissipation slightly below the maximum limit of the tube.

The single tone PEP power output is:

$$P_o = \frac{(E_b - e_{b\ min}) \times i_{b\ max}}{4} =$$

$$\frac{(3000 - 600) \times 0.58}{4} = 348 \text{ watts}$$

The plate current conduction angle efficiency factor for this class of operation is 0.73, and the actual plate circuit efficiency is:

$$N_p = \frac{E_b - e_{b\ min}}{E_b} \times 0.73 = 58.4\%$$

The peak power input to the stage is therefore:

$$\frac{P_o}{N_p} \times 100 = \frac{348}{58.4} = 595 \text{ watts PEP}$$

The peak plate dissipation is:

$$595 - 348 = 247 \text{ watts}$$

(Note: A 4-250A may thus be used in lieu of the 4-400A as peak plate dissipation is less than 250 watts, provided resting plate current is lowered to 70 mA.)

It can be seen that the limiting factor for either the 4-250A or 4-400A is the static plate dissipation, which is quite a bit higher than the operating dissipation level. It is possible, at the expense of a higher level of distortion, to drop the static plate dissipation and to increase the screen voltage to obtain greater power output. If the screen voltage is set at 800, and the bias increased sufficiently to drop the static plate current to 70 mA, the single-toned dc plate current may rise to 300 mA, for a power input of 900 watts. The plate circuit efficiency is 55.6 percent, and the power output is 500 watts. Static plate dissipation is 210 watts, within the rating of either tube.

At a screen potential of 500 volts, the maximum screen current is less than 1 mA, and under certain loading conditions may be negative. When the screen potential is raised to 800 volts maximum screen current is 18 mA. The performance of the tube depends on the voltage fields set up in the tube by the cathode, control grid, screen grid, and plate. The quantity of current, flowing in the screen circuit is only incidental to the fact that the screen is maintained at a positive potential with respect to the electron stream surrounding it.

The tube will perform as expected as long as the screen current, in either direction, does not create undesirable changes in the screen voltage, or cause excessive screen dissipation. Good regulation of the screen supply is therefore required. Screen dissipation is highly responsive to plate loading conditions and the plate circuit should always be adjusted so as to keep the screen current below the maximum dissipation level as established by the applied voltage.

7-7　　Grounded-Grid Linear Amplifiers

The popularity of grounded-grid (cathode-driven) linear amplifiers for SSB service is unique in the Amateur Service. Elimination of costly and bulky bias and screen power supplies make the "g-g" amplifier an economical and relatively light-weight power unit.

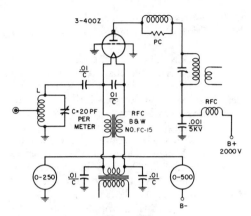

Figure 20

SIMPLE GROUNDED-GRID
LINEAR AMPLIFIER

Tuned cathode (L-C) is required to prevent
distortion of driving-signal waveform.

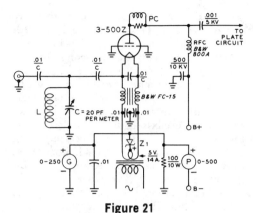

Figure 21

ZENER-DIODE BIAS FOR
GROUNDED-GRID STAGE

The resting plate current of a grounded stage
may be reduced by inclusion of a Zener diode
in the filament return circuit. At a plate potential of 3250 volts, for example, a Zener
bias of 4.7 volts reduces the resting plate
current of the 3-500Z from 160 to approximately 90 milliamperes. A 1N4551 Zener may
be used, bolted to the chassis for a heat sink.

A typical grounded-grid amplifier is shown
in figure 20. The driving signal is applied
between the grid and the cathode, with the
grid held at r-f ground potential. The control grid serves as a shield between the
cathode and the plate, thus making neutralization unnecessary at medium and high
frequencies. High-μ triodes and triode-
connected tetrodes may be used in this configuration. Care must be taken to monitor
the #1-grid current of the tetrode tubes as
it may run abnormally high in some types
(4X150A family) and damage to the tube

R-F LINEAR AMPLIFIER SERVICE FOR SSB AND CW
CATHODE-DRIVEN (GROUNDED-GRID)
CLASS-B MODE

TUBE	PLATE VOLTAGE E_b	FIL. V/A	APPROX. ZERO SIG. PLATE I_{bo} CURRENT	MAX.SIG. PLATE I_b CURRENT	MAX.SIG. GRID I_{c1} CURRENT	DRIVING IMPEDAN. R_k	PLATE LOAD R_L IMPEDAN.	MAX.SIG. DRIVING POWER W.	PEP PLATE INPUT POWER W.	USEFUL OUTPUT POWER W.	AVERAGE PLATE DISSIPAT. P_d	APPROX. 3d ORDER IMD-Db
811A	1250 1700	6.3/4	18 30	175 160	28 28	320	3600 5200	12 15	220 270	135 175	70 85	−33 −28
572 B T-160L	2400	6.3/4	20	250	45	215	4500	30	600	350	160	−28
813	2000 2500	10/5	20 30	200	50 50	270	5000 7000	10 11	400 500	270 350	130 150	−30 −33
3-400Z 8163	2000 2500 3000	5/14.5	62 73 100	400 400 333	148 142 120	120	2750 3450 4750	— — 32	800 1000 1000	445 600 655	355 400 345	−40 −35 −32
3-500Z	2000 2500 3000	5/14.5	95 130 160	400 400 370	130 120 115	115	2750 3450 5000	— — 30	800 1000 1100	500 600 750	300 400 350	−38 −33 −30
3-1000Z 8164	2500 3000 3500	7.5/21	162 175 200	800 670 750	270 220 245	65 65 65	1800 2400 2600	95 65 85	2000 2000 2600	1250 1250 1770	750 750 830	−38 −35 −30
3CX1000A7 8283	2500 3000	5/30	200 310	875 800	590 320	41 42	1100 1670	78 67	2200 2400	1000 1200	1000 1000	−32 −32
4-125A	2000 2500 3000	5/6.5	10 15 20	105 110 115	55 55 55	340 340 340	10500 13500 15700	16 16 16	210 275 345	145 190 240	65 85 100	— — —
4-400A	2000 2500 3000	5/14.5	55 60 70	265 270 330	100 100 106	160 150 140	3950 4500 5600	38 39 40	530 675 990	325 435 572	200 225 390	— — −30
4-1000A	3000 4000 5000	7.5/21	60 90 120	700 675 540	200 200 115	104 106 110	2450 2450 5500	130 105 70	2100 2700 2700	1475 1870 1900	750 730 700	−34 −34 —

Figure 22

may possibly result unless a protective circuit of the form shown in figure 21 is used.

"Zero-bias" triodes (811-A, 3-400Z and 3-1000Z) and certain triode-connected tetrodes (813 and 4-400A, for example) require no bias supply and good linearity may be achieved with a minimum of circuit components. An improvement of the order of 5 to 10 decibels in intermodulation distortion may be gained by operating such tubes in the grounded-grid mode in contrast to the same tubes operated in class-AB$_1$, grid-driven mode. The improvement in the distortion figure varies from tube type to tube type, but all so-called "grounded-grid" triodes and triode-connected tetrodes show some degree of improvement in distortion figure when cathode-driven as opposed to grid-driven service.

Cathode-Driven High-μ Triodes　High-μ triode tubes may be used to advantage in cathode-driven (grounded-grid) service. The inherent shielding of a high-μ tube is better than that of a low-μ tube and the former provides better gain per stage and requires less drive than the latter because of less feedthrough power. Resistive loading of the input or driving circuit is not required because of the constant feedthrough power load on the exciter as long as sufficient Q exists in the cathode tank circuit. Low-μ triodes, on the other hand, require extremely large driving signals when operated in the cathode-driven configuration, and stage gain is relatively small. In addition shielding between the input and output circuits is poor compared to that existing in high-μ triodes.

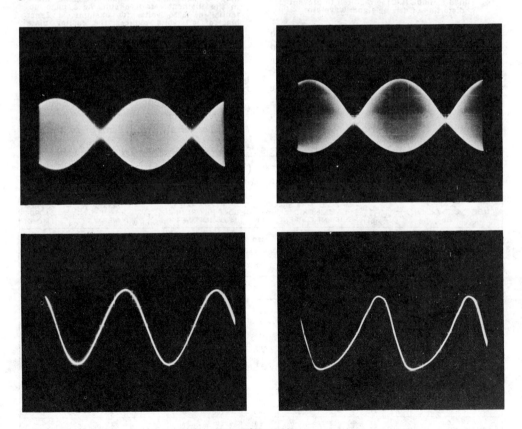

Figure 23

Waveform distortion caused by half-cycle loading at cathode of grounded-grid amplifier may be observed (right) whereas undistorted waveform is observed with tuned cathode circuit (left). Two-tone tests at 2.0 MHz proved the necessity of using a cathode tank circuit for lowest intermodulation distortion.

Bias Supplies for G-G Amplifiers Medium-μ triode tubes that require grid bias may be used in cathode-driven service if the grid is suitably bypassed to ground and placed at the proper negative dc potential. Bias supplies for such circuits, however, must be capable of good voltage regulation under conditions of grid current so that the dc bias value does not vary with the amplitude of the grid current of the stage. Suitable bias supplies for this mode of operation are shown in the *Power Supply* chapter of this Handbook. *Zener bias* (figure 21) may be used for low values of bias voltage. Approximate values of bias voltage for linear amplifier service data may be obtained from the audio data found in most tube manuals, usually stated for push-pull class-AB_1 or AB_2 operation. As the tube "doesn't know" whether it is being driven by an audio signal or an r-f signal, the audio parameters may be used for linear service, but the stated dc currents should be divided by two for a single tube, since the audio data is usually given for two tubes. Grounded-grid operating data for popular triode and tetrode tubes is given in figure 22.

The Tuned Cathode Circuit Input waveform distortion may be observed at the cathode of a grounded-grid linear amplifier as the result of grid- and plate-current loading of the input circuit on alternate half-cycles by the single-ended stage (figure 23). The driving source thus "sees" a very low value of load impedance over a portion of the r-f cycle and an extremely high impedance over the remaining portion of the cycle. Unless the output voltage regulation of the r-f source is very good, the portion of the wave on the loaded part of the cycle will be degraded. This waveform distortion contributes to intermodulation distortion and also may cause TVI difficulties as a result of the harmonic content of the wave. Use of a tuned cathode circuit in the grounded-grid stage will preserve the waveform as shown in the photographs. The tuned-cathode circuit need have only a Q of 2 or more to do the job, and should be resonated to the operating frequency of the amplifier. Various versions of cathode tank circuits are shown in figure 24.

In addition to reduction of waveform distortion, the tuned-cathode circuit provides a short r-f return path for plate current pulses from plate to cathode (figure 25). When the tuned circuit is not used, the r-f return path is via the outer shield of the coaxial line, through the output capacitor of the exciter plate-tank circuit and back to the cathode of the linear amplifier tube via the center conductor of the coaxial line. This random, uncontrolled path varies with the length of interconnecting coaxial line, and permits the outer shield of the line to be "hot" compared to r-f ground.

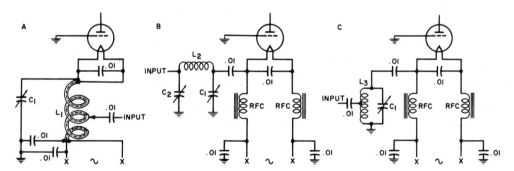

Figure 24

Tuned cathode network for cathode-driven circuit may take form of bifilar coil (A), pi-network (B), or shunt LC circuit (C). Circuit Q of at least 2 is recommended. Capacitor C_1 may be a 3-gang broadcast-type unit. Coils L_1, L_2, or L_3 are adjusted to resonate to the operating frequency with C_1 set to approximately 13 pF-per meter wavelength. Capacitor C_2 is approximately 1.5 times the value of C_1. The input taps on coils L_1 and L_3, or the capacitance of C_2 are adjusted for minimum SWR on coaxial line to the exciter.

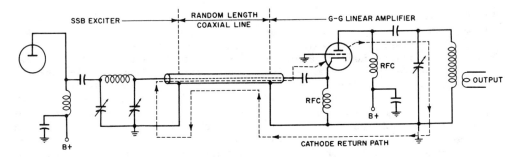

Figure 25

Untuned cathode circuit of grounded-grid amplifier offers high-impedance path to the r-f current flowing between plate and cathode of the amplifier tube. The alternative path is via the interconnecting coaxial line and tank circuit of the exciter. Waveform distortion of the driving signal and high intermodulation distortion may result from use of alternative input circuit.

7-8   Intermodulation Distortion

If the output signal of a linear amplifier is an exact replica of the exciting signal there will be no distortion of the original signal and no distortion products will be generated in the amplifier. Amplitude distortion of the signal exists when the output signal is not strictly proportional to the driving signal and such a change in magnitude may result in *intermodulation distortion* (IMD). IMD occurs in any nonlinear device driven by a complex signal having more than one frequency. A voice signal (made-up of a multiplicity of tones) will become blurred or distorted by IMD when amplified by a nonlinear device. As practical linear amplifiers have some degree of IMD (depending on design and operating parameters) this disagreeable form of distortion exists to a greater or lesser extent on most SSB signals.

A standard test to determine the degree of IMD is the *two-tone test,* wherein two radio-frequency signals of equal amplitude

are applied to the linear equipment, and the resulting output signal is examined for spurious signals, or unwanted products. These unwanted signals fall in the fundamental-signal region and in the various harmonic regions of the amplifier. Signals falling outside the fundamental-frequency region are termed *even-order products,* and may be attenuated by high-Q tuned circuits in the amplifier. The spurious products falling close to the fundamental-frequency region are termed *odd-order products.* These unwanted products cannot be removed from the wanted signal by tuned circuits and show up on the signal as "splatter," which can cause severe interference to communication in an adjacent channel. Nonlinear operation of a so-called "linear" amplifier will generate these unwanted products. Amateur practice calls for suppression of these spurious products to better than 30 decibels below peak power level of one tone of a two-tone test signal. Commercial practice demands suppression to be better than 40 decibels below this peak level.

Additional data on IMD and two-tone test techniques is given in chapter 9.

CHAPTER EIGHT

Specialized Circuitry for Semiconductors and Vacuum Tubes

Semiconductor and vacuum tube usage is not limited to the field of radio or wire communication, nor merely the generation and reception of electromagnetic signals. This chapter covers some of the more common semiconductor and vacuum tube circuits encountered in computer technology and advanced information transmission and storage applications.

8-1 Limiting Circuits

A *limiter* is a device in which some characteristic of the output signal is automatically prevented from exceeding a predetermined level (figure 1). Limiters are useful in waveshaping circuits where it is desirable to square off the peaks of the applied signal. For example, a sine wave may be applied to a limiter to produce a rectangular wave, or a peaked wave may be processed to eliminate either the positive or negative excursions from the output. Limiters are used in f-m receivers to limit signal amplitude at the detector stage and are used to reduce impulse noise on received signals. They may also be

used to limit input signals to special devices or to maintain a high average level of modulation in a transmitter.

The Diode Limiter The characteristics of a diode are such that the device conducts only when the anode is positive with respect to the cathode. A positive potential may be placed on the cathode, but the diode will not conduct until the voltage on the anode rises above an equally positive value. When the anode becomes positive with respect to the cathode, the diode conducts and passes that portion of the signal which is more positive than the cathode. Diodes may be used as either series or parallel limiters, as shown in figure 2, and the diode may be biased so that only a portion of the positive or negative signal is removed.

Peak Limiting A peak limiter, or clipper, consisting of two diode limiters may be used to limit the amplitude of an ac signal to a predetermined value to provide a high average signal level. Limiters of this general type are useful in transmitters to provide a high level of modulation without danger of overmodulation. An effective limiter for this service is the twin diode clipper shown in figure 3. This circuit is a variation of the circuits of figure 2A and 2B, the clipping level being set by the clipping level control, R_4.

The square-topped audio waves generated by a clipper are high in harmonic content, but these high-order harmonics can be

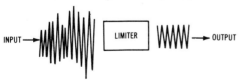

Figure 1
AMPLITUDE LIMITER ACTION

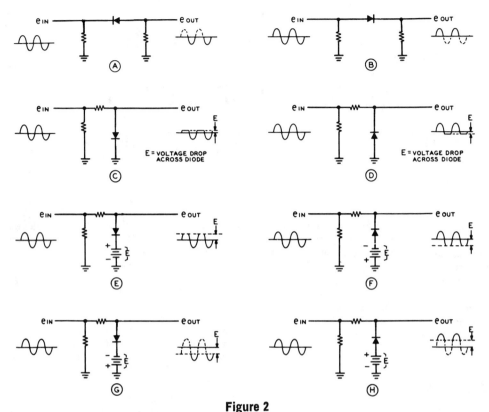

Figure 2

VARIOUS DIODE LIMITING CIRCUITS

Series diodes limiting positive and negative peaks are shown in A and B. Parallel diodes limiting positive and negative peaks are shown in C and D. Parallel diodes limiting above and below ground are shown in E and F. Parallel-diode limiters which pass negative and positive peaks are shown in G and H.

greatly reduced by a low-pass speech filter following the clipper.

8-2   Clamping Circuits

A circuit which holds either amplitude extreme of a waveform to a given reference level of potential is called a *clamping circuit* or a *dc restorer*. Clamping circuits are used after RC coupling circuits where the waveform swing is required to be either above or below the reference voltage, instead of alternating on both sides of it (figure 4). Clamping circuits are usually encountered in oscilloscope sweep circuits. If the sweep voltage

does not always start from the same reference point, the trace on the screen does not begin at the same point on the screen each time the sweep is repeated and therefore is "jittery." If a clamping circuit is placed between the sweep amplifier and the deflec-

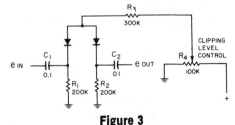

Figure 3

THE SERIES-DIODE GATE CLIPPER FOR PEAK LIMITING

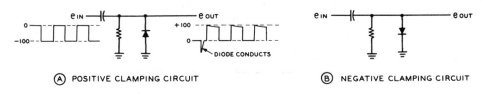

(A) POSITIVE CLAMPING CIRCUIT   **(B) NEGATIVE CLAMPING CIRCUIT**

Figure 4

SIMPLE POSITIVE AND NEGATIVE CLAMPING CIRCUITS

tion element, the start of the sweep can be regulated by adjusting the dc voltage applied to the clamping diode (figure 5).

8-3   Positive Feedback Amplifiers

Positive feedback may be employed in an amplifier to execute transitions between two distinct stable states. Amplifiers employing positive feedback are called *multivibrators*, Schmitt-triggers, "one-shots," *relaxation oscillators*, or *flip-flops*. They are used for timing, frequency division, and switching. The switching time is commonly called *regeneration time* and the time required to reach a final steady state condition is called *resolution time*.

Monostable, Bistable, and Astable Operation — Amplifiers having heavy positive feedback such as shown in this section may operate in more than one mode. A *monostable* state implies that the amplifier has one stable state. An external trigger is required to shift the circuit to a quasi-stable state for one cycle of a predetermined interval. A *bistable* state implies an amplifier having two stable operating states. The amplifier changes state for each external trigger signal. This circuit is also called a *flip-flop*. An *astable* state implies that the amplifier continuously alternates between two unstable states at a frequency determined by the circuit constants.

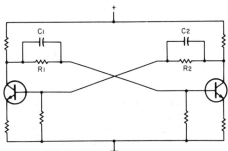

Figure 6

BASIC MULTIVIBRATOR CIRCUIT

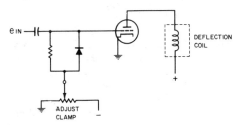

Figure 5

NEGATIVE CLAMPING CIRCUIT EMPLOYED IN ELECTROMAGNETIC SWEEP SYSTEM

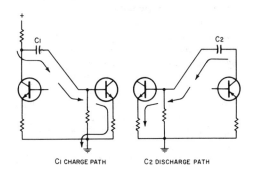

C1 CHARGE PATH   C2 DISCHARGE PATH

Figure 7

THE CHARGE AND DISCHARGE PATHS IN THE FREE-RUNNING MULTIVIBRATOR OF FIGURE 6

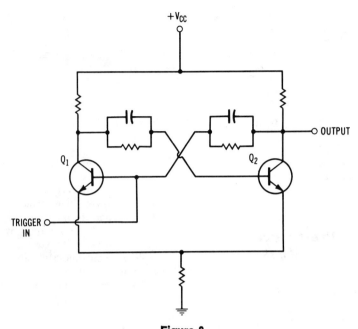

Figure 8

BISTABLE MULTIVIBRATOR

A *multivibrator* is a circuit characterized by a large degree of positive feedback which causes the circuit to operate in abrupt transitions between two blocked end-states. The transition rate may be stabilized by the introduction of a synchronizing voltage of harmonic or subharmonic frequency. A representative multivibrator is shown in figure 6. Basically, it is a two-stage resistance-capacitance coupled amplifier with the output coupled back to the input. In general, the duration of a quasi-stable state is determined by the exponential decay of charge stored in a coupling circuit time constant. Action is started by thermal agitation or miscellaneous noise and is maintained by the charge and discharge of energy in the coupling capacitors. The paths for these actions are shown in figure 7.

The output of a free-running multivibrator may be used as a source of square waves, as an electronic switch, or as a means of obtaining frequency division. Submultiple frequencies as low as one-tenth of the injected synchronizing frequency may easily be obtained.

A bistable multivibrator, or flip-flop circuit, is one that is not free running, but that has two conditions of stable equilibrium (figure 8). One condition is when Q_1 is conducting and Q_2 is cut off; the other is when Q_2 is conducting and Q_1 is cut off. The circuit remains in one or the other of these stable conditions until a trigger signal causes the nonconducting device to conduct. The two devices then reverse their functions and remain in the new condition until the next trigger signal is applied.

A monostable multivibrator is shown in figure 9. This circuit accomplishes a complete cycle when triggered by a positive pulse. Such a device is called a *one-shot* multivibrator. For initial action, Q_1 is cut off and Q_2 is conducting. A large positive pulse is applied to the base of Q_1 causing this transistor to conduct, and the voltage at its collector decreases by virtue of the voltage drop through resistor R_3. Capacitor C_1 is charged rapidly by this abrupt change in Q_1 collector voltage and Q_2 is cut off while Q_1 conducts. This condition exists until C_1 discharges, allowing Q_2 to conduct,

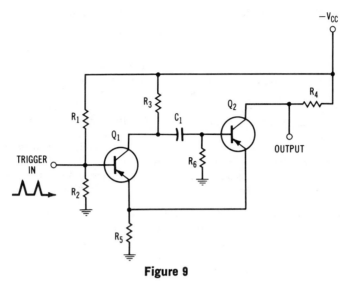

Figure 9

MONOSTABLE (ONE-SHOT) MULTIVIBRATOR

raising the emitter bias of Q_1 until it is once again cut off.

8-4 The Blocking Oscillator

The *blocking oscillator* shown in figure 10 is a common-base, unity gain current amplifier coupled to a current transformer. When the circuit is triggered, Q_1 and the transformer inject a current into the emitter circuit that is larger than the collector current. The collector-to-ground voltage drops abruptly and the collector junction appears as a short circuit. The collector current rises in an interval (t), equals the emitter current, and the transistor is cut off. The collector current then drops to zero and the energy accumulated in the inductance of the transformer windings causes a rapid increase in the collector voltage, as shown in the diagram. The "ringing" caused by the combination of capacitor C and the magnetization inductance of the transformer can be suppressed by an external damping circuit.

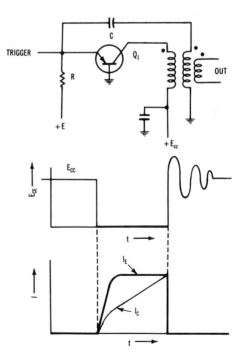

Figure 10

THE BLOCKING OSCILLATOR

8-5   The Schmitt Trigger Circuit

The Schmitt trigger circuit is similar to the monostable multivibrator but it is bi-stable in operation (figure 11). It is used as a switching device to change sinusoidal and other waveforms into square waveforms for digital computer circuitry.

In a quiescent condition, transistor Q_2 is conducting and Q_1 is cut off because there is no effective base bias path for Q_1. Transistor Q_1 remains off because of the bias voltage developed across resistor R_6.

When a sinusoidal voltage greater in magnitude that the bias voltage is applied to Q_1, the transistor begins to conduct. At the same time Q_2 begins to turn off very rapidly, primarily because the conduction current through Q_1 causes the bias voltage across R_6 to rise, thus making the emitter of Q_2 more positive. As Q_2 is cut off, Q_1 continues to turn on very rapidly because the bias voltage across R_6 is decreasing, producing a corresponding increase in Q_1 base-emitter forward bias.

When the input voltage decreases, the circuit returns to the original state with Q_2 conducting and Q_1 cut off. The level of the output signal and pulse width are controlled by varying the values of R_2 and R_3.

The multivibrator, flip-flop, one-shot, and Schmitt-trigger circuits shown in figures 6, 8, 9, and 11 are depicted as circuits made up of discrete components. Such circuits are rarely built up in this manner commercially, since it is much cheaper to use ICs that are designed for these functions. Astable multivibrators may be simply implemented with the *Signetics* NE555 or one of the similarly numbered devices made by other manufacturers. Representative circuits are shown in figure 12.

The NE555 may also be used as a one-shot, with a slightly different connection, but more often a member of one of the IC logic families is used. In the TTL family, the 74122 provides a single one-shot in an IC package (figure 13). If two one-shots per package are required a 74123 is used. In the CMOS family, a *Motorola* MC 14538 provides two one-shots in one IC package (figure 14). The 74122 and 74123 (TTL) will operate only over voltages close to $+5$ volts. The NE555 operates on any dc voltage from $+5$ to $+30$, and the CMOS version (MC 14538) operates on any voltage from $+5$ to $+15$.

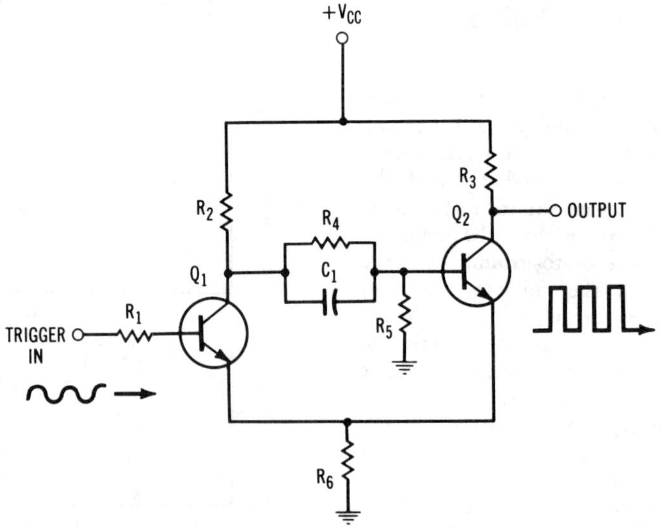

Figure 11

SCHMITT TRIGGER CIRCUIT

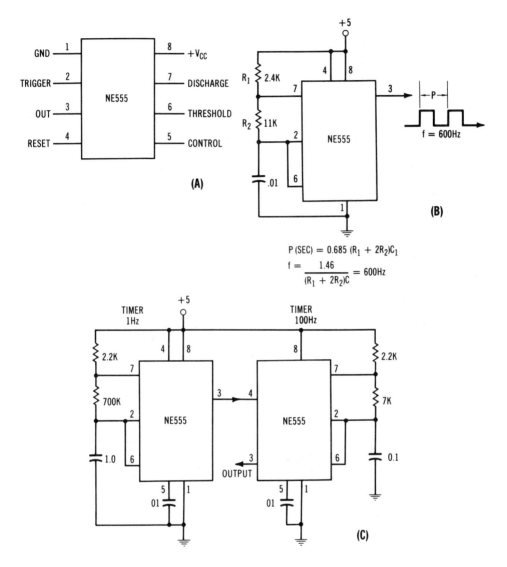

$$P\ (SEC) = 0.685\ (R_1 + 2R_2)C_1$$

$$f = \frac{1.46}{(R_1 + 2R_2)C} = 600Hz$$

Figure 12

ASTABLE MULTIVIBRATOR USING NE555 IC

The astable multivibrator using the NE555 IC produces positive and negative pulses timed by C_1, R_1 and R_2. A—Pin configuration of NE555. B—Multivibrator providing a square pulse at a 600 Hz rate. C—First pulser (1 Hz) controls second (100 Hz) by connecting output of first to input (reset) of second. Only when the first timer has high output will the second function. NE556 has two NE555 devices in one case.

In the case of the flip-flop (bistable multivibrator), suitable units are available in all the common IC logic families. In the TTL family the 7476 is a good example, offering great flexibility with both a *preset* and a *clear* input for each of its two flip-flops (figure 15). In the CMOS family, the 4027 (dual) is representative (figure 16).

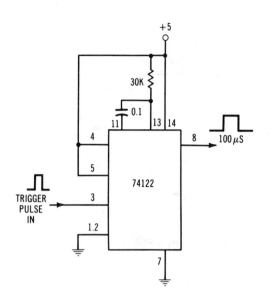

Figure 13

TTL ONE-SHOT MULTIVIBRATOR

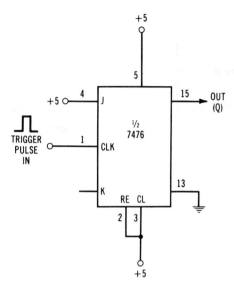

Figure 15

**TTL FLIP-FLOP BISTABLE
MULTIVIBRATOR**

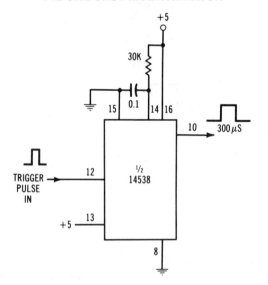

Figure 14

CMOS ONE-SHOT MULTIVIBRATOR

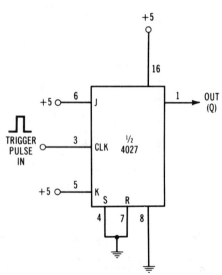

Figure 16

**CMOS FLIP-FLOP BISTABLE
MULTIVIBRATOR**

For Schmitt-trigger designs it is possible to ·use some ICs in the TTL logic family such as the 7413, and some ICs in the CMOS family such as the MC 14584 which are essentially logic gates having hysteresis (positive feedback) built into them. A flexible Schmitt-trigger is, however, best built using one of several types of IC comparators figure 17). A very versatile comparator is the LM 311 which will operate on voltages ranging from $+5$ to $+30$. Hysteresis is controlled by the value of R_1 and the trip level by the setting of potentiometer R_2.

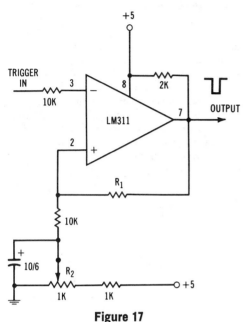

Figure 17

SCHMITT TRIGGER USING LM311 COMPARATOR

8-6   The Relaxation Oscillator

The neon lamp oscillator, although impractical for many purposes other than experimentation, illustrates the basic operation of a *relaxation oscillator* and its application for time control. In this circuit, the dc supply voltage is applied to a series RC circuit and the neon lamp serves as a load connected in parallel with the capacitor (figure 18).

The capacitor is charged from the supply through the series resistor. When the voltage across the capacitor becomes equal to the firing voltage of the lamp (approximately 60 volts), the capacitor discharges through the lamp, causing it to flash. With the capacitor discharged, the charge-discharge cycle is repeated. Time constant of the RC circuit may be determined by the information given in Chapter 2, section 2-5.

The response of the RC circuit during charge time results in a gradual increase in neon lamp voltage. However, capacitor voltage decreases abruptly through the lamp during the discharge portion of the cycle producing an output voltage that has a sawtooth waveform.

The equivalent of the neon lamp oscillator can be achieved by using solid-state active devices as shown in figure 19. Illustration A shows a 4-layer diode taking the place of the lamp. Illustration B shows a unijunction transistor in an equivalent circuit. This configuration has the added advantage of putting out pulses as well as sawtooth waveforms.

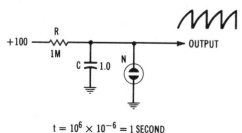

$$t = 10^6 \times 10^{-6} = 1 \text{ SECOND}$$

Figure 18

RELAXATION OSCILLATOR

Time constant is product of R (ohms) and C (farads).

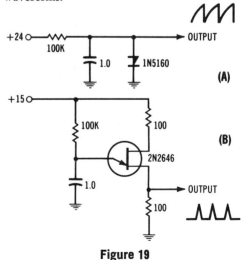

Figure 19

SOLID-STATE RELAXATION OSCILLATOR

A—Using 4-layer diode
B—Using unijunction transistor

Either of these two circuits can be modified to produce a linear sawtooth waveform by replacing the series resistor with a constant current diode (a JFET with source and gate connected together, for example). A circuit of this type is shown in figure 20.

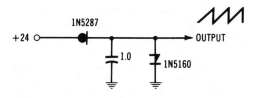

Figure 20

RELAXATION OSCILLATOR

Constant-current diode (1N5287) replaces series resistor to provide linear sawtooth waveform.

8-7   The Resistance-Capacitance Oscillator

In an *RC oscillator*, the frequency is determined by a resistance capacitance network that provides regenerative coupling between the output and input of a feedback amplifier. No use is made of a tank circuit consisting of inductance and capacitance to control the frequency of oscillation.

The *Wien-Bridge* oscillator employs a *Wien network* in the RC feedback circuit and is shown in figure 21. Since the feedback at pin 6 of the μA 741C is in phase with the input signal from the bridge at pin 3 at all frequencies, oscillation is maintained by voltages of any frequency that exist in the circuit. The bridge circuit is used, then, to eliminate feedback voltages of all frequencies except the single frequency desired at the output of the oscillator. The bridge allows a voltage of only one frequency to be effective in the circuit because of the degeneration and phase shift provided by this circuit. The frequency at which oscillation occurs is:

$$f = \frac{1}{2\,\pi\,R_1\,C_1}$$

when,

$R_1 \times C_1$ equals $R_2 \times C_2$

A lamp (R_3) is used as a thermal stabilizer of the oscillator amplitude. The variation of the resistance of the lamp bulb holds the oscillator output voltage at a nearly constant amplitude.

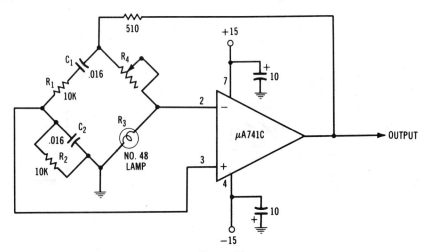

Figure 21

WIEN-BRIDGE OSCILLATOR

IC op-amp in TO-5 case is used. Lamp acts as thermal stabilizer.

The *phase-shift oscillator* shown in figure 22 is a single-transistor oscillator using a four mesh phase-shift network. Each section of the network produces a phase shift in proportion to the frequency of the signal that passes through it. For oscillations to be produced, the signal through the network must be shifted 180°. Four successive phase shifts of 45° accomplish this, and the frequency of oscillation is determined by this phase shift.

In order to increase the frequency of oscillation, either the resistance or the capacitance must be decreased by an appropriate amount.

A *bridge-type Twin-T oscillator* is shown in figure 23. The bridge is so proportioned that only at one frequency is the phase shift through the bridge equal to 180°. Voltages of other frequencies are fed back to the amplifying device out of phase with the existing input signal, and are cancelled by being amplified out of phase.

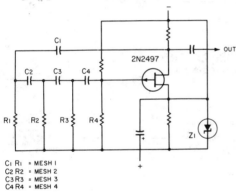

C₁ R₁ = MESH 1
C₂ R₂ = MESH 2
C₃ R₃ = MESH 3
C₄ R₄ = MESH 4

Figure 22
THE PHASE-SHIFT OSCILLATOR

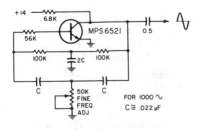

Figure 23
THE TWIN-TEE OSCILLATOR

The *Bridge-T oscillator* developed by the National Bureau of Standards consists of a two-stage amplifier having two feedback loops, as shown in figure 24. Loop 1 consists of a regenerative loop, consisting of R_1 and R_2. The bulb regulates the positive feedback, and tends to stabilize the output of the oscillator, much as in the manner of the Wien circuit. Loop 2 consists of a degenerative circuit, containing the Bridge-T.

Oscillation will occur at the null frequency of the bridge, at which frequency the bridge allows minimum degeneration in loop 2 (figure 25).

8-8 Closed-Loop Feedback

Feedback amplifiers have been discussed in Chapter 6, of this Handbook. A more general use of feedback is in automatic control and regulating systems. Mechanical feedback has been used for many years in such forms as engine-speed governors and servo steering engines on ships.

A simple feedback system for temperature control is shown in figure 26. This is a *cause-and-effect system*. The furnace (F) raises the room temperature (T) to a predetermined value at which point the sensing thermostat (TH) reduces the fuel flow to the furnace. When the room temperature drops below the predetermined value the fuel flow is increased by the thermostat control. An interdependent control system is created by this arrangement: the room temperature depends on the thermostat action, and the thermostat action depends on the room temperature. This sequence of events may be termed a *closed-loop feedback system*.

Error Cancellation   A feedback control system is dependent on a degree of error in the output signal, since this error component is used to bring about the correction. This component is called the *error signal*. The error, or deviation from the desired signal, is passed through the feedback loop to cause an adjustment to reduce the value of the error signal. Care must be taken in the design of the feedback loop to reduce

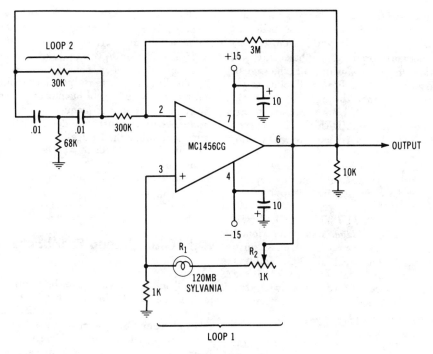

Figure 24

**NBS BRIDGE-T OSCILLATOR
WITH MC 1456 CG op-amp**

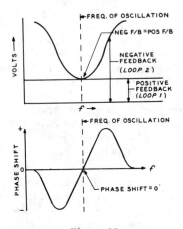

Figure 25

**BRIDGE-T FEEDBACK
LOOP CIRCUITS**

Oscillation will occur at the null frequency of
the bridge, at which frequency the bridge allows
minimum degeneration in loop 2.

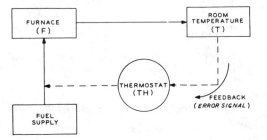

Figure 26

**SIMPLE CLOSED-LOOP
FEEDBACK SYSTEM**

Room temperature (T) controls fuel supply to
furnace (F) by feedback loop through thermostat (TH) control.

over-control tendencies wherein the correction signal would carry the system past the
point of correct operation. Under certain
circumstances the new error signal would

cause the feedback control to overcorrect in the opposite direction, resulting in *hunting* or oscillation of the closed-loop system about the correct operating point.

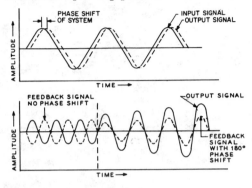

Figure 27

PHASE SHIFT OF ERROR SIGNAL MAY CAUSE OSCILLATION IN CLOSED LOOP SYSTEM

To prevent oscillation, the gain of the feedback loop must be less than unity when the phase shift of the system reaches 180 degrees.

Negative-feedback control would tend to damp out spurious system oscillation if it were not for the time lag or phase shift in the system. If the overall phase shift is equal to one-half cycle of the operating frequency of the system, the feedback will maintain a steady state of oscillation when the circuit gain is sufficiently high (figure 27). In order to prevent oscillation, the gain figure of the feedback loop must be less than unity when the phase shift of the system reaches 180 degrees. In an ideal control system the gain of the loop would be constant throughout the operating range of the device, and would drop rapidly outside the range to reduce the bandwidth of the control system to a minimum.

The time lag in a closed-loop system may be reduced by using electronic circuits in place of mechanical devices, or by the use of special circuit elements having a *phase-lead* characteristic. Such devices make use of the properties of a capacitor, wherein the current leads the voltage applied to it.

CHAPTER NINE

Single-Sideband
Transmission and Reception

Single-sideband (SSB) communication is a unique, sophisticated information transmission system well suited for wire and radio services. Although known in theory for several decades, "sideband" was sparingly used in commercial service for a number of years, and only in the last decade has it achieved popularity and general acceptance in the Amateur Service. Economical in cost, sparing of valuable spectrum space, and usable under the most trying propagation conditions, SSB is the stepping stone to a future era of better and more reliable rapid hf communication.

9-1   The SSB System

Single sideband is a recent attempt to translate human intelligence into electrical impulses capable of being economically transmitted over great distances. The general flow of information in a communication system includes a *source*, followed by a *translator* which propagates the intelligence through a conducting *medium*. A second translator is used to extract the intelligence conveyed by the medium and to make it available in a usable form. The vocal chords, vibrations in the atmosphere, and the ear drum accomplish this sequence of events for sound; the light source, the "ether," and the human eye provide the same sequence for sight.

Experiments before the turn of the century proved the existence of electromagnetic waves which could be propagated and put to use for transmission of information. When voice transmission via radio waves was successfully accomplished *circa* 1907, the concept of carrier waves and sidebands was unknown, although it was understood that "a channel separation high compared with the pitch of the sound waves transmitted" was required. An implication that a *transmission band* of frequencies was involved was apparently not grasped at the time, and the idea that intelligence could be transmitted by a single carrier wave of constant frequency and varying amplitude persisted until about 1921 at which time the sideband concept had been established by a series of discoveries, experiments, and inventions.

Early SSB experiments with single-sideband transmission were conducted by the telephone industry which was interested in transmitting electrical impulses corresponding to the human voice over long-distance telephone circuits. Since the transmission properties of wire and cable deteriorate rapidly with cable length and increasing frequency, a means of frequency conservation was desired which would permit the "stacking" of different voices in an electromagnetic package so that many voices could be sent over a single circuit. The voice impulses were mainly concentrated in the band 300—3,000 Hz and the problem at hand was to translate this voice band to a higher

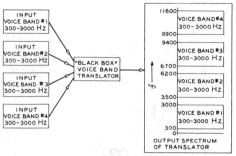

Figure 1

THE "BLACK BOX" VOICE BAND TRANSLATOR

A simple device for "stacking" voice bands in an electromagnetic "package" for transmitting many voices over a single circuit cannot be built as it is impossible to translate a band of frequencies directly to another band. Translation must be accomplished by an indirect method, making use of an auxiliary carrier wave and a mixing process termed "modulation."

band of frequencies (15,300—18,000 Hz, for example) for transmission on the telephone circuit, then to reverse the translation process at the receiving terminal to recover the original band of frequencies. Experiments proved, however, that a simple and economical apparatus for translation of the voice frequencies from one band to another was not forthcoming. No device could be built that would do the job that looked so simple when sketched on paper (figure 1). It proved possible, however, to generate a continuous electrical signal at some high frequency (15,000 Hz, for example) and to impress the voice impulses on this signal. For convenience, the continuous signal was termed the *carrier wave*, as it was assumed to "carry" the intelligence in some way or other. A suitable device at the receiving terminal detected the intelligence on the carrier, recovering the original speech frequencies impressed on the carrier at the transmitter. Mathematical analysis of this process (called *modulation*) showed that the carrier remained unchanged and additional frequencies were created lying on either side of the carrier, spaced from it by a frequency proportional to the modulation frequency (figure 2). These additional frequencies were termed *sidebands* and conclusive evidence of separate sidebands was achieved in 1915 by the use of electric filters that sep-

arated sidebands and carriers, proving their individuality.

The sideband theory was of little more than passing interest to radio engineers, but it was a matter of considerable importance to the telephone industry. The carrier wave was useless except as an operator necessary to generate and then upon which to "hang" the two sidebands, both of which carried the same information (figure 3). For economic reasons and spectrum conservation it was desirable to remove one sideband and the carrier from the translator, passing only one sideband through the conducting medium. At the receiver, a locally generated carrier wave of the correct frequency and amplitude was combined with the incoming single-sideband signal. The resulting output was a reproduction of the signal impressed on the translator. Commercial wire telephone systems using this technique were placed in operation in 1918 and the first h-f SSB telephone link was activated in 1927.

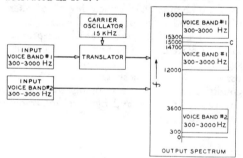

Figure 2

THE TRANSLATOR MIXER

Voice band #1 is impressed on a carrier signal in a translator (mixer) stage. Voice band #2 is unchanged. The output spectrum of the device shows that two voice bands are available, one "stacked" above the other in frequency. Addition of other translators will permit additional voice bands to be "stacked" in the frequency region between 3600 Hz and 12,000 Hz. The voice packages thus created could be sent over a single circuit. Note that the translation process creates two symmetrical voice bands from the original #1 signal, spaced each side of a carrier frequency between the bands. Elimination of carrier signal and one voice band would permit addition of another signal in this portion of the spectrum.

Practical Application of SSB The spectrum waste arising from a frequency translation process utilizing simple amplitude modulation could be eliminated by suppression of one sideband

and the carrier, and the transmission of only the remaining sideband. To date, no method exists to directly generate an SSB signal. All translation techniques involve the use of a carrier wave, and the resulting signal includes the original carrier and two auxiliary sidebands.

The post-World War II acceptance of SSB transmission for military and commercial circuits has stimulated research and development in this field and has contributed to a heightened interest in the technique by the radio amateur. Mass production of sharp-cutoff filters and stable translation oscillators, plus the use of advanced and simplified circuitry has brought SSB to the point of obsoleting simple amplitude-modulation transmission on the high-frequency amateur bands. Undoubtedly, in the years ahead, further design refinements and technical advances will make the use of SSB even more advantageous to all concerned with transmission of intelligence by electrical means.

The popularity of SSB for general amateur use has been brought about as this technique has consistently proved to allow more reliable communication over a greater range than has amplitude modulation. It has greater ability to pierce interference, static, and man-made noise than has amplitude

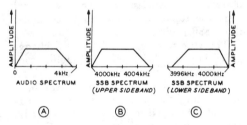

Figure 3

RELATIONSHIP OF AUDIO AND SSB SPECTRUMS

The single-sideband components are the same as the original audio components except that the frequency of each is raised by the frequency of the carrier. The relative amplitude of the various components remains the same.

modulation and is inherently resistant to propagation abnormalities that render a-m completely useless. In addition, the annoying interference caused by heterodynes between a-m carriers is completely missing in SSB service.

Basic SSB A single-sideband signal can be best be described as an audio signal *raised* (or translated) to the desired radio frequency. The translation process may not result in the inversion of the audio-frequency components in the signal, depending on the sideband selected (figure 4). For example, a single audio tone of 2000 Hz is to be translated into an SSB signal in the 455-kHz region. The tone is amplified and applied to one input of a translator stage (usually termed a *balanced modulator*). A radio-frequency *carrier* is applied to the other input terminal of the modulator. For this example, the frequency of the

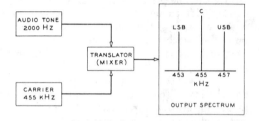

Figure 4

THE TRANSLATOR SPECTRUM

The SSB signal is an audio signal raised (mixed, or translated) to the desired radio frequency. A 455-kHz carrier signal upon which is impressed a 2-kHz audio tone in a translator stage will possess two side-bands, separated from the carrier frequency by the frequency of the tone. The carrier has been generated by the separate oscillator and the two adjacent signals (sidebands) are a product of the mixing process taking place between the audio signal and the carrier. The output spectrum pictured is of a double sideband, with carrier To produce an SSB signal, it is necessary to eliminate the carrier and one sideband.

carrier is 455 kHz. The translation process takes place in the balanced modulator; creating two *sidebands* positioned each side of the carrier, and separated from it by the modulation frequency. Thus, at least four signals are flowing within the modulator: the 2000-Hz (2-kHz) *audio signal*, the *lower sideband* (455 − 2 = 453 kHz), the *carrier* (455 kHz), and the *upper sideband* (455 + 2 = 457 kHz). The carrier, of course, has been generated by the separate local oscillator, and the two sidebands are a product of the mixing process taking place between the audio signal and the carrier.

The balanced modulator is usually designed to balance (or cancel) the carrier signal to a large degree, leaving only the two sidebands and the audio signal to appear in the output circuit. Some modulators also balance out the audio signal. Part of the job of creating an SSB signal has now been accomplished. The high-frequency components of the output signal of the balanced modulator comprise a *double-sideband, suppressed-carrier signal*. The remaining step to create an SSB signal is to eliminate one of the sidebands and to reduce to minor proportions any vestige of carrier permitted to pass through the balanced-modulator stage. A *sideband filter* accomplishes this last step. At the output of the filter is the desired SSB signal. The passband of the filter should be just wide enough to pass the intelligence without passing the carrier wave or the unwanted sideband. For voice communication, such filters usually pass a band of radio frequencies about 2 or 3 kHz wide.

The unwanted carrier and sideband that are eliminated by the filter and balanced modulator are actually absorbed by the filter and modulator and converted to heat. In order to hold the cost and size of the filter to a reasonable figure, it is necessary that the above process take place at a relatively low signal level, of the order of a volt or two, so that power dissipation is low.

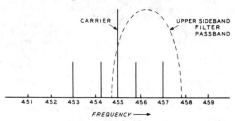

Figure 5

THE SSB SIGNAL

The SSB signal may be generated by passing a double-sideband-with-carrier signal through a filter which removes one sideband and partially suppresses the carrier. In this example, a two-tone audio signal (700 and 2000 Hz) is mixed with a 455-kHz carrier signal. The output signal from the mixer, or modulator, contains four sideband frequencies: 453 kHz, 454.3 kHz, 455.7 kHz, and 457 kHz, in addition to the carrier at 455 kHz. Additional carrier suppression may be obtained by the use of a balanced modulator.

The SSB Spectrum A single audio tone in a perfect SSB system remains a simple sine wave at all points in the system and cannot be distinguished from a

c-w signal generated by more conventional means. A voice signal, on the other hand, is a complex band of audio components having many frequencies of varying amplitudes. A simple and useful compromise signal for testing SSB equipment is the *two-tone* signal, composed of two equal and separate sine waves separated a very small percentage in frequency. If two audio tones are applied to the input circuit of the SSB exciter previously discussed, the output of the 455-kHz balanced modulator will contain *four* sideband frequencies (figure 5). Assume the audio tones are 700 and 2000 Hz. The output frequencies of the balanced modulator will be: 453 kHz, 454.3 kHz, 455 kHz (the partially suppressed carrier), 455.7 kHz and 457 kHz. The two lower frequencies represent the lower sideband, and the two higher frequencies represent the upper sideband. With a properly designed filter following the balanced modulator, both the frequencies in one sideband and the remainder of the carrier will be almost completely eliminated. If the filter completely eliminates the lower sideband and the carrier, the output of the exciter will be two radio frequencies at 455.7 kHz and 457 kHz. An observer examining these r-f signals could not tell if the signals were generated by two oscillators operating at the observed frequencies, or if the two signals were the result of two audio tones applied to an SSB exciter.

The waveform of the SSB signal changes drastically as the number of audio tones is increased, as shown in figure 6. A single-tone waveform is shown in illustration A and is simply a single, steady sine-wave r-f output. A signal composed of two audio tones is shown in illustration B. The two radio-frequency signals are separated by the difference in frequency between the audio tones and beat together to give the SSB envelope shown. The figure has the shape of half-sine waves, and from one null to the next represents one full cycle of the difference frequency. If one tone has twice the amplitude of the other, the envelope shape is as shown in illustration C. The SSB envelope of three equal tones of equal frequency spacings and at one particular phase relationship is shown in illustration D. Illustration E shows the SSB envelope of four equal tones having equal frequency spacings

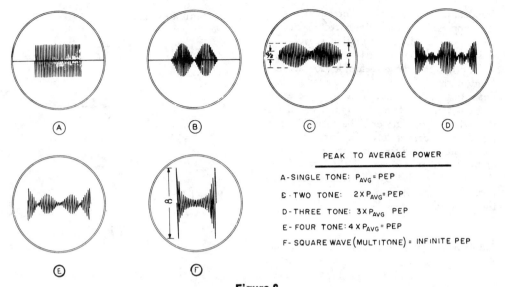

PEAK TO AVERAGE POWER
A – SINGLE TONE: P_{AVG} = PEP
B – TWO TONE: $2 \times P_{AVG}$ = PEP
D – THREE TONE: $3 \times P_{AVG}$ PEP
E – FOUR TONE: $4 \times P_{AVG}$ = PEP
F – SQUARE WAVE (MULTITONE) = INFINITE PEP

Figure 6

SSB WAVEFORMS

The waveform of the SSB signal changes with the nature of the modulating signal, and the envelope shape of the SSB wave may not be the same as the original audio waveshape. The peak power in the SSB wave is a direct function of the r-f waveform, as shown here. Peak and average power in the SSB wave will be discussed later in this chapter.

and at one particular phase relationship. Finally, illustration F shows the SSB envelope of a square wave having an infinite number of odd harmonics. A pure square wave requires infinite bandwidth, so in theory the SSB envelope requires infinite amplitude. This emphasizes the point that the SSB envelope shape may not be the same as the original audio waveshape, and usually bears no similarity to it. This is because the percentage difference between the radio frequencies is small, even though one audio tone may be several times the other in terms of frequency. Because of nonlinearity and phase shift in the practical SSB transmitter, the peak amplitude of a transmitted square wave is not so great as predicted by theory through the addition of the harmonic coefficients, making it impossible to faithfully reproduce a square wave. Speech processing in the form of heavy audio *clipping* therefore is of limited value in SSB because the SSB r-f envelopes are so different from the audio envelopes. A heavily clipped wave approaches a square wave which will have the tendency to exhibit the high amplitude peaks shown in illustration 6F, a waveform

the SSB transmitter is theoretically unable to transmit.

The Received SSB Signal In summary, if an *audio spectrum* containing many different tones (the human voice, for example) is applied to the SSB exciter, an *r-f spectrum* is generated that corresponds to the audio tones. If the audio spectrum encompasses the range of 300—3000 Hz, the output of the 455-kHz balanced modulator will be 452 to 454.7 kHz (the lower sideband), 455 kHz (the partially suppressed carrier), and 455.3 to 458 kHz (the upper sideband). An "upper-sideband" type filter having a passband of 455.3 to 458 kHz will substantially eliminate the residual carrier and lower sideband.

Listening to the output of the SSB exciter on a typical a-m receiver will divulge a series of unintelligible sounds having no apparent relation to the original speech impressed on the SSB exciter. (A low-pitched voice can be read with difficulty as the syllabic content is preserved and is apparent). Injection in the receiver of a local carrier frequency of 455 kHz (corresponding to the suppressed carrier eliminated in the exciter) will produce intelligible speech that is a replica of the original voice frequencies.

In order to transmit simple double sideband with carrier (amplitude modulation) with this SSB exciter, it is only necessary to bypass the sideband filter and unbalance the balanced modulator. The resulting a-m signal with carrier may be intelligible on the ordinary receiver without the necessity of local-oscillator injection, the latter function being fulfilled by the transmitted carrier, if it has sufficient strength relative to the sidebands.

SSB Power Rating The SSB transmitter is usually rated at *peak envelope input or output power*. Peak envelope power (PEP) is the root-mean-square (rms) power generated at the peak of the modulation envelope. With either a two-equal-tone test signal or a single-tone test signal, the following equations approximate the relationships between single-tone and two-tone meter readings, peak envelope power, and average power for class-B or class-AB linear amplifier operation:

Single tone:

DC Plate Current (Meter Reading):

$$I_b = \frac{i_{pm}}{\pi}$$

Plate Input (Watts):

$$P_{in} = \frac{i_{pm} \times E_b}{\pi}$$

Average Output Watts and PEP:

$$P_o = \frac{i_{pm} \times e_p}{4}$$

Plate Efficiency:

$$N_p = \frac{\pi \times e_p}{4 \times E_b}$$

Two equal tones:

DC Plate Current (Meter Reading):

$$I_b = \frac{2 \times i_{pm}}{\pi^2}$$

Plate Input (Watts):

$$P_{in} = \frac{2 \times i_{pm} \times E_b}{\pi^2}$$

Average Output Watts:

$$P_o = \frac{i_{pm} \times e_p}{8}$$

PEP Output Watts:

$$P_o = \frac{i_{pm} \times e_p}{4}$$

Plate Efficiency:

$$N_p = \left(\frac{\pi}{4}\right)^2 \times \frac{e_p}{E_b}$$

where,

i_{pm} equals peak of the plate-current pulse,
e_p equals peak value of plate-voltage swing,
π equals 3.14,
E_b equals dc plate voltage,
N_p equals efficiency in percent.

"Average" Speech Section 97.67 of the Amateur Radio Service Rules of the FCC indicates that the average power input of an SSB transmitter in the amateur service shall not exceed one kilowatt on modulation peaks, as indicated by a plate-current meter having a time constant of not more than 0.25 second. It is common practice among amateurs to define this as equivalent to a *peak envelope power* input of two kilowatts. This is convenient, since a two-tone test signal having a peak-to-average power ratio of two to one can thereby be employed for tuneup and adjustment purposes with the reasonable assumption that the SSB equipment will be properly adjusted for one kilowatt average power voice operation.

It is difficult to determine the ratio of peak to average power in the human voice, as the range of intensity of speech sounds may vary as much as 40 decibels. "Average" speech seems to have an intensity range of about 20 decibels and a ratio of instantaneous peak-to-average power of about 14 decibels for 99 percent of the time of speech.

Speech processing (clipping or compression) may alter this figure, bringing the peak to average power ratio closer to unity. In any event, adjustment of the amateur SSB transmitter to achieve a peak power input of twice the average power input level has proven by experience to allow sufficient peak-power capability to cover the majority of cases. In those situations where the peak capability of the equipment is exceeded at an average-power input level of one kilowatt, the average-power level must be reduced to conform with the maximum capability of the transmitter. In any case, the use of an oscilloscope is mandatory to determine the peak-power capability of an SSB transmitter.

Power Advantage of SSB over AM Single sideband is a very efficient form of voice communication by radio. The amount of radio-frequency spectrum occupied can be no greater than the frequency range of the audio or speech signal transmitted, whereas other forms of radio transmission require from two to several times as much spectrum space. The r-f power in the transmitted SSB signal is directly proportional to the power in the original audio signal and no strong carrier is transmitted. Except for a weak pilot carrier present in some commercial usage, there is no r-f output when there is no audio input.

The power output rating of an SSB transmitter is given in terms of *peak envelope power* (PEP). This may be defined as the rms power at the crest of the modulation envelope. The peak envelope power of a conventional amplitude-modulated signal at 100% modulation is four times the carrier power. The average power input to an SSB transmitter is therefore a very small fraction of the power input to a conventional amplitude-modulated transmitter of the same power rating.

Single sideband is well suited for long-range communications because of its spectrum and power economy and because it is less susceptible to the effects of selective fading and interference than amplitude modulation. The principal advantages of SSB arise from the elimination of the high-energy carrier and from further reduction in sideband power permitted by the improved performance of SSB under unfavorable propagation conditions.

In the presence of narrow-band manmade interference, the narrower bandwidth of SSB reduces the probability of destructive interference. A statistical study of the distribution of signals on the air versus the signal strength shows that the probability of successful communication will be the same if the SSB power is equal to one-half the power of one of the two a-m sidebands. Thus SSB can give from 0 to 9 dB improvement under various conditions when the *total* sideband power is equal in SSB and regular amplitude modulation. In general, it may be assumed that 3 dB of the possible 9 dB advantage will be realized on the average contact. In this case, the SSB power required for equivalent performance is equal to the power in one of the a-m sidebands. For example, this would rate a 100-watt SSB and a 400-watt (carrier) a-m transmitter as having equal performance. It should be noted that in this comparison it is assumed that the receiver bandwidth is just sufficient to accept the transmitted intelligence in each case.

To help evaluate other methods of comparison the following points should be considered. In conventional amplitude modulation two sidebands are transmitted, each having a peak envelope power equal to $1/4$ carrier power. For example, a 100-watt a-m signal will have 25-watt peak envelope power in each sideband, or a total of 50 watts. When the receiver detects this signal, the voltages of the two sidebands are added in the detector. Thus the detector output voltage is equivalent to that of a 100-watt SSB signal. This method of comparison says that a 100-watt SSB transmitter is just equivalent to a 100-watt a-m transmitter. This assumption is valid only when the receiver bandwidth used for SSB is the same as that required for amplitude modulation (e.g., 6 kHz), when there is no noise or interference other than broadband noise, and if the a-m signal is not degraded by propagation. By using half the bandwidth for SSB reception (e.g., 3 kHz) the noise is reduced 3 dB so the 100-watt SSB signal becomes equivalent to a 200-watt carrier a-m signal. It is also possible for the a-m signal to be degraded another 3 dB on the average due to narrow-band interference and poor propagation conditions, giving a possible 4 to 1 power advantage to the SSB signal.

It should be noted that 3 dB signal-to-noise ratio is lost when receiving only one sideband of an a-m signal. The narrower receiving bandwidth reduces the noise by 3 dB but the 6 dB advantage of coherent detection is lost, leaving a net loss of 3 dB. Poor propagation will degrade this "one-sideband" reception of an a-m signal less than double-sideband reception, however. Also under severe narrow-band interference conditions (e.g., an adjacent strong signal) the ability to reject all interference on one side of the carrier is a great advantage.

Advantage of SSB with Selective Fading   On long-distance communication circuits us-ing amplitude modulation, selective fading often causes severe distortion and at times makes the signal unintelligible. When one sideband is weaker than the other, distortion results; but when the carrier becomes weak and the sidebands are strong, the distortion is extremely severe and the signal may sound like "monkey chatter." This is because a carrier of at least twice the amplitude of either sideband is necessary to demodulate the signal properly. This can be overcome by using *exalted-carrier reception*

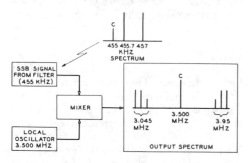

Figure 7

SSB FREQUENCY TRANSLATION

The SSB signal may be translated higher in frequency in the same manner the voice signals are translated to a radio-frequency spectrum. In this example, the 455-kHz two-tone, suppressed-carrier signal is translated (mixed) with a 3.5-MHz oscillator to provide two new sidebands, one at 3.045 MHz and the other at 3.95 MHz. If the 3.95-MHz signal is desired, filter circuits may be used to eliminate the unwanted 3.045-MHz sideband and the 3.5-MHz carrier signal from the local oscillator. The 3.95-MHz signal may now be shifted in frequency by changing the frequency of the local oscillator.

in which the carrier is amplified separately and then reinserted before the signal is demodulated or detected. This is a great help, but the reinserted carrier must be very close to the same phase as the original carrier. For example, if the reinserted carrier were 90 degrees from the original source, the a-m signal would be converted to phase modulation and the usual a-m detector would deliver no output.

The phase of the reinserted carrier is of no importance in SSB reception and by using a strong reinserted carrier, exalted-carrier reception is in effect realized. Selective fading with one sideband simply changes the amplitude and the frequency response of the system and very seldom causes the signal to become unintelligible. Thus the receiving techniques used with SSB are those which inherently greatly minimize distortion due to selective fading.

SSB Amplification and Frequency Changing   The single-sideband signal appearing at the output of the filter must be amplified to a sufficiently strong level for practical use. The amplifying stage must have low distortion and the output signal must be a faithful replica of the input signal. An amplifier meeting these requirements is called a *linear amplifier*. Any deviation from amplitude linearity produces signal distortion and spurious products which rapidly degrade the SSB signal. It is therefore impossible to pass the SSB signal through frequency doublers or class-C amplifiers without creating severe distortion, because these are inherently non-linear devices. Linear amplifier stages must be used, and if a change of frequency of the SSB signal is desired, it must be heterodyned to the new frequency by means of a mixer stage and another local oscillator (figure 7). The resulting signal may be vfo controlled by varying the frequency of the local oscillator, but the frequency at which the SSB signal is generated is held constant. Thus by means of linear amplifiers and mixer stages, a low frequency SSB signal may be amplified and converted to any other frequency desirable for communication purposes.

9-2 A Basic Single-Sideband Transmitter

The general outline of a practical SSB filter-type transmitter suitable for high-frequency operation can be assembled from the preceding information. A block diagram of such a unit is shown in figure 8. The transmitter consists of a speech amplifier, a carrier oscillator, a balanced modulator, a sideband filter, a high-frequency mixer stage and conversion oscillator, and a linear amplifier having a high-Q tuned output circuit. Incidental equipment such as power supplies and metering circuits are also necessary. Many variations of this basic block diagram are possible.

The Speech Amplifier—A typical speech amplifier consists of a microphone which converts the voice into electrical signals in the audio band, followed by one or more stages of voltage amplification. No appreciable audio power output is required making the audio system of the SSB transmitter quite different from that of the usual a-m transmitter, which requires an audio power level equal to one-half the class C amplifier power input. Included in the speech system is a *speech level* (audio volume) *control* and additional stages to allow *automatic voice operation* (VOX) of the equipment.

The Carrier Oscillator—A highly stable r-f oscillator (often crystal-controlled) is used to generate the carrier signal required in the mixing process. The choice of carrier frequency is determined by the design of

the sideband filter, and frequencies in the range of 250 kHz to 20 MHz are common. Power output is low and frequency stability is a prime necessity in this circuit.

The Balanced Modulator—The balanced modulator translates the audio frequencies supplied by the speech amplifier into r-f sidebands adjacent to the carrier generated by the carrier oscillator. In addition, the balanced modulator partially rejects the carrier which has no further use after the mixing process is completed. A *carrier-balance (null) control* is an integral part of this circuit and is adjusted for optimum carrier suppression.

The Sideband Filter—Selection of one of the two sidebands at the output of the balanced modulator is the function of the filter. A practical filter may consist of small tuned LC circuits, or it may consist of mechanical resonators made of quartz or steel. A representative passband for a sideband filter is shown in figure 9. The filter must provide a sharp cutoff between the wanted sideband and the carrier, as well as rejection of the unwanted sideband.

The Converter (Mixer) Stage and Conversion Oscillator—It is usually necessary to obtain an SSB signal at a frequency other than that of the sideband filter passband. Frequency conversion is accomplished in the same manner the voice frequencies were translated to the filter frequency region; that is, by the use of a converter stage and conversion oscillator. The process carried out in this step may be referred to as *translation, mixing, heterodyning,* or *converting.* For this example, it is desired to

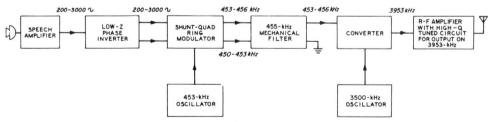

Figure 8

BLOCK DIAGRAM OF FILTER-TYPE SSB TRANSMITTER

Voice frequencies in the range of 200 to 3000 Hz are amplified and fed to a balanced modulator. Depending on the choice of frequency of the local oscillator, either the upper or lower sideband may be passed through to the mechanical filter. The carrier has, to some extent, been reduced by the balanced modulator. Additional carrier rejection is afforded by the filter. The SSB signal at the output of the filter is translated directly to a higher operating frequency. Suitable tuned circuits follow the converter stage to eliminate the conversion oscillator signal and the image signal.

convert a 455-kHz SSB signal to 3.95 MHz. The operation takes place in a second balanced-modulator circuit. One input is the 455-kHz SSB signal, and the other input signal is from an oscillator operating on 3.500 MHz. The output of the second mixer is a partially suppressed carrier (3.500 MHz), the lower sideband in the 3.045-MHz range (3.500 − 0.455 = 3.045 MHz), and the upper sideband in the 3.95-MHz range (3.500 + 0.455 = 3.95 MHz). The upper sideband is the desired one, so a simple auxiliary *image filter* is used to separate it from the unwanted sideband and the partially suppressed carrier. In most cases, this filter consists of the two or three parallel-tuned circuits normally associated with the following amplifier stages tuned to 3.95 MHz.

The Linear Amplifier—The output of the last mixer stage is usually of the order of a few milliwatts and must be amplified to a usable level in one or more *linear amplifier* stages. For lowest distortion, the output of the linear amplifier should be a nearly exact reproduction of its input signal. Any amplitude nonlinearity in the amplifier not only will produce undesirable distortion within the SSB signal, but will also produce annoying spurious products in adjacent channels. Distortion may be held to a low

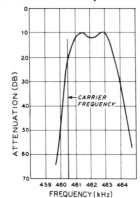

Figure 9

PASSBAND OF CRYSTAL LATTICE FILTER

A 460.5-kHz crystal-lattice filter composed of eight crystals has an excellent passband for voice waveforms. Carrier rejection is about −20 decibels, and unwanted sideband rejection is better than −35 decibels. Passband is essentially flat up to 463 kHz, providing an audio passband of about 300 to 2500 Hz.

value by the proper choice of tubes, their operating voltages and driving-circuit considerations, and by the use of external negative feedback, as discussed in Chapter Twelve.

9-3 The Balanced Modulator

The *balanced modulator* is used to mix the audio signal with that of the local carrier to produce sideband components which may be selected for further amplification. Any *nonlinear* element will serve in a modulator, producing sum and difference signals as well as the original frequencies. This phenomenon is objectionable in amplifiers and desirable in mixers or modulators. The simplest modulator is a rapid-action switch, commonly simulated by diode rectifiers for r-f service. Either semiconductors or vacuum-tube rectifiers may be employed and some of the more commonly used circuits are shown in figure 10. The simplest modulator is that of figure 10A, the two-diode series-balanced modulator. The input transformer introduces the audio signal to the balanced diode switches, which are turned off and on by the carrier voltage introduced in an in-phase relationship. If the carrier amplitude is large with respect to the audio signal, the only current flowing in the output transformer is due to the action of the audio voltage added to the carrier voltage. A properly designed DSB output transformer will filter out the switching transients, the audio component, and the carrier signal, leaving only the desired double-sideband output. A shunt version of this circuit is shown in illustration B wherein the diodes form a short-circuit path across the input transformer on alternate half-cycles of carrier switching voltage.

Four-diode balanced modulators are shown in illustrations C through E. Circuits C and E are similar to the two-diode circuits except that untapped transformers may be used to save cost. The double-balanced ring circuit of illustration D is popular as both carrier and audio signal are balanced with respect to the output, which is advantageous when the output frequency is not sufficiently different from the inputs to allow ready separation by inexpensive filters. The

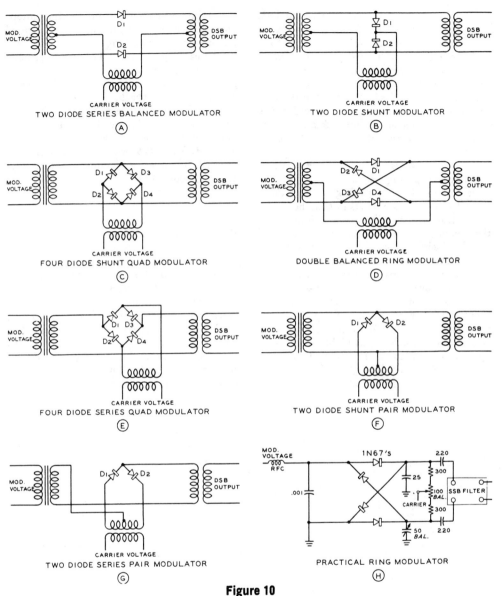

TWO DIODE SERIES BALANCED MODULATOR
(A)

TWO DIODE SHUNT MODULATOR
(B)

FOUR DIODE SHUNT QUAD MODULATOR
(C)

DOUBLE BALANCED RING MODULATOR
(D)

FOUR DIODE SERIES QUAD MODULATOR
(E)

TWO DIODE SHUNT PAIR MODULATOR
(F)

TWO DIODE SERIES PAIR MODULATOR
(G)

PRACTICAL RING MODULATOR
(H)

Figure 10

BALANCED-MODULATOR CIRCUITS

The balanced modulator is used to mix the audio signal with that of the carrier to produce sideband components. It may also be used as a converter or mixer stage to convert an SSB signal to
a higher frequency. The diodes act as an r-f driven switch and may be arranged in series or shunt
mode as shown in the illustrations. A practical diode modulator incorporating balancing circuits
is shown in illustration H.

series and shunt-quad configuration may be
adapted to two diodes as shown in illustrations F and G, substituting a balanced carrier transformer for one side of the bridge.
In applying any of these circuits, r-f chokes

and capacitors must be employed to control
the path of audio and carrier currents and
balancing capacitors are usually added to
null the carrier as shown in the circuit of
illustration H.

The double-diode circuits are useful but, in general, it is more difficult to balance a transformer at the carrier frequency, than it is to use an additional pair of diodes. Untapped transformers are desirable, thus eliminating this critical component from the circuit. Paired diodes combined with balancing potentiometers and capacitors usually provide a good compromise, permitting a high degree of carrier balance at minimum cost.

In recent years, however, the double-balanced diode ring modulator has become widely available as a package. These units are a form of figure 10D, but without the lower transformer. The carrier input and the modulation inputs are exchanged and double sideband comes out of the right-most output. The transformer balancing and diode matching is done at the component factory.

Even inexpensive models have 0.5 to 500 MHz capability at the transformer-coupled ports, and dc to 500 MHz at the remaining port. Special models are available which will operate to as high as 18 GHz. Figure 11 shows a typical double-balanced modulator made by *Watkins Johnson*.

An integrated circuit differential amplifier serves as a high-quality balanced modulator under varying voltages and temperatures (figure 12). The *bias* terminal of the IC provides a port for control voltage for cw operation, allowing the carrier to pass through the modulator stage.

Several ICs have been created especially for use as double-balanced modulators. The one most commonly used is perhaps the *Motorola MC-1496*, which is widely second-sourced by other semiconductor manufacturers (figure 13).

9-4   The Sideband Filter

The heart of a filter-type SSB exciter is the sideband filter. Conventional coils and capacitors may be used to construct a filter based on standard wave-filter techniques. Such filters are restricted to relatively low frequencies because of the rapid cutoff required between the filter passband and adjacent stopbands. The Q of the filter inductors must be relatively high when com-

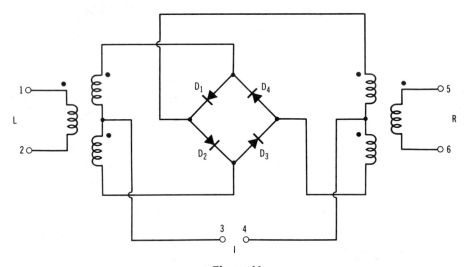

Figure 11

**WJ-M6A DOUBLE-BALANCED
MODULATOR**

This device operates up to 2000 MHz at a maximum power level of 50 mW. Noise figure is 6 dB and isolation is 45 dB.

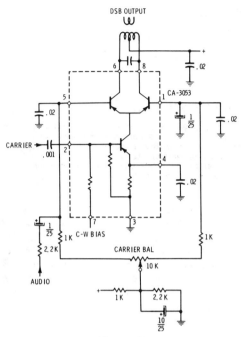

Figure 12

**DIFFERENTIAL AMPLIFIER USED
AS BALANCED MODULATOR**

pared with the reciprocal of the fractional bandwidth. If a bandwidth of 3 kHz is needed at a carrier frequency of 50 kHz, for example, the bandwidth expressed in terms of the carrier frequency is 3/50, or 6 percent. This is expressed in terms of fractional bandwidth as 1/16. For satisfactory operation, the Q of the filter inductances should be ten times the reciprocal of this, or 160.

For voice communication purposes, the lower frequency response of the sideband filter is usually limited to about 300 Hz. Frequencies above 2500 Hz or so contribute little to speech intelligence, moreover, and their elimination permits closer grouping for SSB signals. Practical filters for speech transmission, therefore, have a passband from about 300 to 2500 Hz or so, rejecting signals in the unwanted passband and those above 3000 Hz by over 40 decibels. A ten-pole LC SSB filter and the characteristic response is shown in figure 14.

Crystal Filters   Practical and inexpensive filters are designed around quartz crystal resonators well into the hf

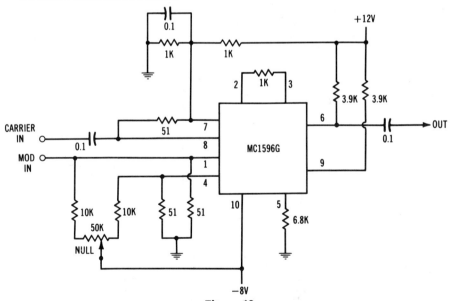

Figure 13

MC-1596G BALANCED MODULATOR

Output signal contains sum and difference frequency components and amplitude information of modulating signal.

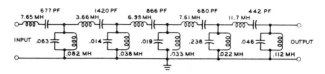

Figure 14

TEN-POLE BUTTERWORTH-TYPE SSB FILTER

The carrier frequency is 70 kHz and filter impedance is 600 ohms. Each series-resonant and parallel-resonant circuit is tuned to the carrier frequency. Using high-Q inductors, the filter passband is about 4 kHz wide at a response of −40 decibels. Nose of filter is about 2500 Hz wide. Low-frequency SSB filters of this type require two or more conversion stages to provide h-f SSB signal without troublesome images. High-frequency quartz-crystal filters, on the other hand, make possible SSB exciters capable of single conversion operation up to 50 MHz or so.

range. A representative lattice filter is shown in figure 15. Two such filters may be cascaded for additional selectivity. The frequency spread between the two filter paths determines the filter passband. A spread of 500 Hz may be used for a cw filter and a spread as great as 2 to 3 kHz for an SSB filter. In general, the greater the spread, the greater the passband ripple.

Mechanical Filters Filters using mechanical resonators have been studied by a number of companies and are offered commercially by *Rockwell-Collins*. They are available in a variety of bandwidths

at center frequencies of 250 and 455 kHz. The 250-kHz series is specifically intended for sideband selection. The selectivity attained by these filters is intermediate between good LC filters at low center frequencies and engineered quartz-crystal filters. A passband of two 250-kHz filters is shown in figure 16.

9-5   The Phasing Type SSB Exciter

An SSB signal may be generated by the phasing of two a-m signals in such a way

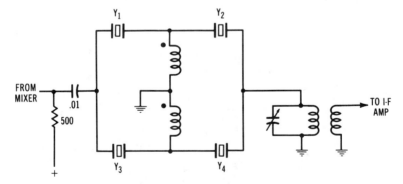

Figure 15

CRYSTAL-LATTICE FILTER

This filter provides good skirt selectivity for SSB service. Filter bandwidth depends on frequency separation between pairs Y_1-Y_2 and Y_3-Y_4. Filter must be terminated properly to achieve lowest passband ripple. Output of filter terminates in a transformer to match it to the following stage. Input is terminated by collector load resistor of mixer.

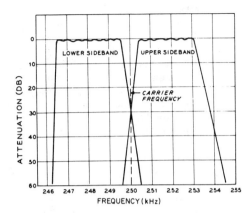

Figure 16

**PASSBAND OF LOWER- AND UPPER-
SIDEBAND MECHANICAL FILTER**

that one sideband is enhanced, and the other sideband and carrier are cancelled or balanced out. This technique is known as the *phasing system* and exchanges the problems of filter design for those of accurately controlled phase shifts. In general, the phasing transmitter is more economical in cost than is the filter-type transmitter and may be less complex. It requires adjustment of various audio and r-f balancing controls for maximum suppression of the unwanted sideband and carrier that is otherwise accomplished by bandpass-filter action in the filter-type equipment. The phasing system has

the advantage that all electrical circuits which give rise to the SSB signal can operate in a practical transmitter at the nominal output frequency of the transmitter. Thus, if an SSB signal is desired at 50.1 MHz, it is not necessary to go through several frequency conversions in order to obtain an SSB signal at the desired output frequency. The balanced modulator in the phasing transmitter is merely fed with a 50.1 MHz carrier and with the audio signal from a balanced phase splitter. Practical considerations, however, make the construction of a 6-meter SSB phasing-type exciter a challenge to the home constructor because of the closely controlled r-f phase shifts that must be achieved at that frequency.

**A Practical
Phasing Exciter** A simplified block diagram illustrating the phasing method of SSB generation is shown in figure 17. An audio signal is amplified, restricted in bandwidth by a speech filter and then split into two branches (ϕ_1 and ϕ_2) by the *audio phase network*. The resulting signals are applied independently to two balanced modulators. The audio networks have the property of holding a 90° *phase difference* between their respective output signals within the restricted range of audio frequencies passed by the speech filter and applied to their input terminals. In addition, the amplitude response of the networks remains essentially constant over this frequency range.

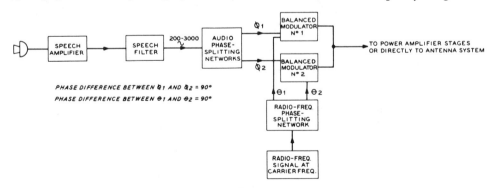

Figure 17

BLOCK DIAGRAM OF A PHASING TYPE EXCITER

The phasing method of obtaining a single-sideband signal is simpler than the filter system in regard to the number of tubes and circuits required. The system is also less expensive in regard to adjustments for the transmission of a pure single-sideband signal.

Each balanced modulator is driven by a fixed-frequency carrier oscillator whose output is also split into two branches (θ_1 and θ_2) by a 90° r-f phase shift network operating at the carrier frequency. The algebraic sum of the output signals of the two balanced modulators appears at the output of a combining circuit and is the desired single-sideband, suppressed-carrier signal. The degree of sideband suppression is dependent on the control of audio phase shift and amplitude balance through the system; a phase error of two degrees, for example, will degrade the sideband attenuation by over 10 decibels.

By way of illustration, assume that the carrier oscillator frequency is 3.8 MHz and that a single modulating tone of 2000 Hz

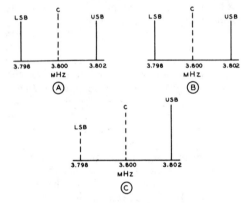

Figure 18

THE PHASING-TYPE SSB SIGNAL

Two signals having identical spectrum plots may be combined to produce an SSB signal. The signals of illustrations A and B, however, have simultaneous 90-degree phase shifts applied to the audio and carrier signals and when properly combined produce an SSB signal whose lower sidebands are out of phase and whose upper sidebands are in phase. By use of twin balanced modulators, the carrier may be suppressed and proper sideband addition and subtraction achieved (illustration C).

is used. The output from balanced modulator #1 is represented by the spectrum plot of figure 18A, in which the carrier frequency is represented by the vertical dashed line at 3.8 MHz with the symmetrical sidebands at 3.798 MHz and 3.802 MHz. The carrier frequency is balanced within the modulator and so does not appear in its out-

put. Similarly, the output of balanced modulator #2 produces a signal which has an identical spectrum plot, as shown in figure 18B. While the spectrum plots appear identical, they do not show everything about the output signals of the two modulators as addition of two identical quantities yields a result which is simply twice as great as either quantity. However, the result of the two simultaneous 90° phase shifts applied to the audio and carrier signals impressed on the modulators produces sideband signals in their respective outputs that are *in phase* for the identical upper-sideband frequency of 3.802 MHz but 180° *out of phase* for the lower-sideband frequency of 3.798 MHz as shown in figure 18C. Addition of the output signals of the two balanced modulators thus doubles the strength of the upper-sideband component while balancing out the lower-sideband component. Conversely, subtraction of the output signals of one balanced modulator from those of the other will double the strength of the lower-sideband component while cancelling the upper-sideband component. In either case, an SSB signal is created. A double-pole, double-throw reversing switch in two of the four audio leads to the balanced modulators is all that is required to switch from one sideband to the other.

The phase-shift method works not so much because the system passes a certain band of frequencies but because it is able to cancel a closely adjacent band of frequencies. The result, however, is equivalent to that obtained by the use of bandpass filters.

Filter versus Phasing? The phasing system of SSB generation does not necessarily produce a better or worse signal than does the filter-type of SSB generator. Suppression of the unwanted sideband in the phasing generator depends on the characteristic of the audio phase-shift networks and on matching the differential phase shift these networks provide to the r-f phase shift at carrier frequency. These adjustments must be accomplished by the equipment operator. On the other hand, in the filter-type SSB generator, unwanted sideband suppression depends on the built-in characteristics of the sideband filter and on the placement of the carrier relative to the filter passband. How well the job is done in each case is primarily a matter of design and cost—not one of basic superiority of one method over the other. Reduced cost of high-frequency crystal filters has dropped the price of the filter equipment to that of the previously less-expensive phasing system and most of today's commercial and amateur SSB gear makes use of the filter technique of sideband generation. Even so, for equivalent quality of components and design, it would be hard for an observer to tell whether a given SSB signal was generated by the phasing method or by the filter method.

Radio Frequency Phasing   A single sideband generator of the phasing type requires that the two balanced modulators be fed with r-f signals having a 90-degree phase difference. Figure 19 shows two MC-1596G integrated circuits used as balanced modulators. The outputs of the two balanced modulators are summed in the r-f transformer. A broadband, ferrite core transformer is used to eliminate circuit adjustment. The carrier is suppressed by about 60 dB and the second harmonic of the carrier about 30 dB. A low-pass filter should follow the succeeding amplifier stages to provide additional second harmonic attenuation.

A diagram of a representative r-f phase-shift circuit is shown in figure 20. A portion of MC-1035 serves as a flip-flop, or digital r-f phase shifter.

Audio Frequency Phasing   Audio frequency phasing provides a 90-degree phase shift over the voice range of about 150 to 3000 Hz. A representative circuit is shown in figure 21. The values of

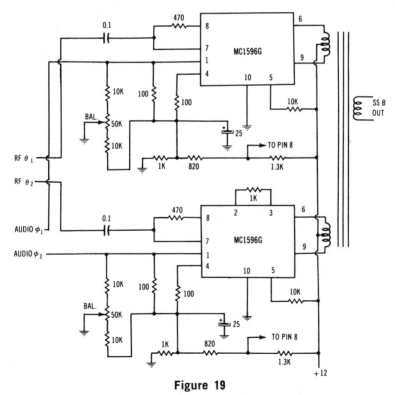

Figure 19

SSB GENERATION BY PHASING METHOD

Two Motorola MC1496G integrated circuits serve as balanced modulators. Outputs are summed in broadband ferrite-core transformer T₁. Carrier is suppressed by about 60 dB.

resistance and capacitance must be carefully held to ensure minimum deviation from a 90-degree shift over the voice spectrum.

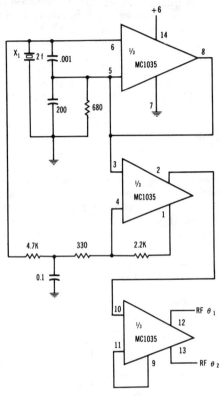

Figure 20

R-F OSCILLATOR AND PHASE SHIFTER

Motorola MC 1035 serves as digital r-f phase shifter. The outputs of the second MC 1035 section are two square waves, 180 degrees out of phase. These square waves are divided by two in the third section, a ring counter which also provides synchronization to the phase quadrature. The crystal frequency is thus always twice the desired suppressed-carrier frequency.

A passive audio phase-shift network employing no tubes is shown in figure 22. This network has the same type of operating restrictions as those described above. A solid state phase-shift network is shown in figure 23. The network is preceded by a low-pass audio filter. In addition selective bypassing of the audio amplifier helps to roll-off the low audio frequencies. Sideband

switching is accomplished by reversing the phase of one audio channel.

9-6　Single-Sideband Frequency Conversion

The output signal from the low-level SSB generator is usually at a fixed frequency and must be converted, or translated, to the desired operating frequency. This conversion is accomplished by a heterodyne process involving converter or mixer stages and suitable oscillators. Frequency multipliers cannot be used with the SSB signal since this process would alter the frequency relationships present in the original audio signal.

The heterodyne process mixes two signals in a manner to produce new signal components equal in frequency to the sum and difference of the original frequencies. One of the two products is useful and is passed by the tuned circuits of the equipment which reject the undesired products as well as the original signals. Mixing imposes many problems in keeping the output signal free from spurious products created in the mixer. Selection of mixing frequencies and signal levels is required to aid in holding the level of unwanted products within reasonable limits. A discussion of frequency-conversion problems will follow later in this chapter.

Mixer Stages　A mixer stage is commonly used to convert the SSB signal from the generated frequency to the operating frequency. A simple mixer stage is shown in figure 24. A circuit using a MOSFET which provides somewhat better isolation between the signal frequencies is shown in figure 25. A balanced mixer is shown in figure 26 which provides better than 20 dB of carrier attenuation.

The modulator stages shown earlier in this chapter may also be used as mixers for frequency-conversion techniques.

9-7　Selective Tuned Circuits

The selectivity requirements of the tuned circuits following a mixer stage often

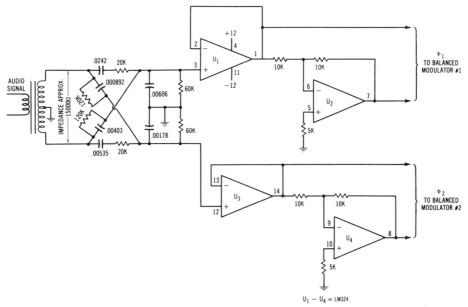

Figure 21

DOME AUDIO-PHASE-SHIFT NETWORK

This circuit arrangement is convenient for obtaining the audio phase shift when it is desired to use a minimum of circuit components.

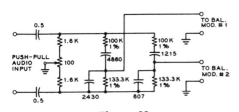

Figure 22

PASSIVE AUDIO-PHASE-SHIFT NETWORK, USEFUL OVER RANGE OF 300 TO 3000 Hz.

become quite severe. For example, using an input signal at 250 kHz and a conversion injection frequency of 400 kHz the desired output may be 4250 kHz. Passing the 4250-kHz signal and the associated sidebands without attenuation and realizing 100 dB of atenuation at 4000 kHz (which is only 250 kHz away) is a practical example. Adding the requirement that this selective circuit must tune from 2250 to 4250 kHz further complicates the basic requirement. The best solution is to cascade a number of tuned

circuits. Since a large number of such circuits may be required, the most practical solution is to use permeability tuning, with the circuits tracked together.

If an amplifier tube is placed between each tuned circuit, the overall response will be the sum of one stage multiplied by the number of stages (assuming identical tuned circuits). Figure 27 is a chart which may be used to determine the number of tuned circuits required for a certain degree of attenuation at some nearby frequency. The Q of the circuits is assumed to be 50, which is normally realized in small permeability-tuned coils. The number of tuned circuits with a Q of 50 required for providing 100 dB of attenuation at 4000 kHz while passing 4250 kHz may be found as follows:

$$\Delta f \text{ is } 4250 - 4000 = 250 \text{ kHz}$$

where,

f_r is the resonant frequency (4250 kHz), and,

$$\frac{\Delta f}{f_r} = \frac{250}{4250} = 0.059$$

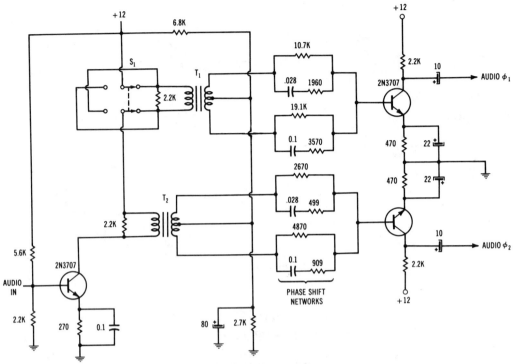

Figure 23

AUDIO PHASE-SHIFT NETWORK

Sideband is selected by switch S_1.

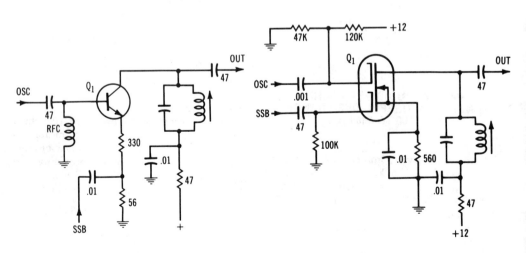

Figure 24

**MIXER STAGE FOR
SSB FREQUENCY CONVERSION**

Figure 25

**MIXER STAGE FOR
SSB FREQUENCY CONVERSION
USING A MOSFET**

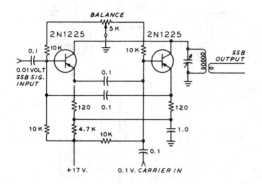

Figure 26

BALANCED MODULATOR CIRCUIT FOR SSB FREQUENCY CONVERSION

The point on the chart where .059 intersects 100 dB is between the curves for 6 and 7 tuned circuits, so 7 tuned circuits are required.

Another point which must be considered in practice is the tuning and tracking error of the circuits. For example, if the circuits were actually tuned to 4220 kHz instead of

$$4250 \text{ kHz, the } \frac{\Delta f}{f_r} \text{ would be } \frac{220}{4220} \text{ or}$$

0.0522. Checking the curves shows that 7 circuits would just barely provide 100 dB of attenuation. This illustrates the need for very accurate tuning and tracking in circuits having high attenuation properties.

Coupled Tuned Circuits When as many as 7 tuned circuits are required for proper attenuation, it is not necessary to have the gain that 6 isolating amplifier tubes would provide. Several vacuum tubes can be eliminated by using two or three coupled circuits between the amplifiers. With a coefficient of coupling between circuits 0.5 of critical coupling, the overall response is very nearly the same as isolated circuits. The gain through a pair of circuits having 0.5 coupling is only eight-tenths that of two critically coupled circuits, however. If critical coupling is used between two tuned circuits, the nose of the response curve is broadened and about 6 dB is lost on the skirts of each pair of critically coupled circuits. In some cases it may be necessary to broaden the nose of the response curve to avoid adversely affecting the frequency response of the desired passband. Another tuned circuit may be required to make up for the loss of attenuation on the skirts of critically coupled circuits.

Frequency-Conversion Problems The example in the previous section shows the difficult selectivity problem encountered when strong undesired signals appear near the desired frequency. A high-frequency SSB transmitter may be required to operate at any carrier frequency in the range of 1.7 to 30 MHz. The problem is to find a practical and economical means of heterodyning the generated SSB frequency to any carrier frequency in this range. There are many modulation products in the output of the mixer and a frequency scheme must be found that will not have undesired output of appreciable amplitude at or near the desired signal. When tuning across a frequency range some products may "cross over" the desired frequency. These undesired crossover frequencies should be at least 60 dB below the desired signal to meet modern standards. The amplitude of the undesired products depends on the particular characteristics of the mixer and the particular order of the product. In general, most products of the 7th order and higher will be at least 60 dB down. Thus any crossover frequency lower than the 7th order must be avoided since there is no way of attenuating them if they appear within the desired passband. The book *Single Sideband Principles and Circuits* by Pappenfus, McGraw Hill Book Co., Inc., N. Y., covers the subject of spurious products and incorporates a "mix selector" chart that is useful in determining spurious products for various different mixing schemes.

In general, for most applications when the intelligence-bearing frequency is lower than the conversion frequency, it is desirable that the ratio of the two frequencies be between 5 to 1 and 10 to 1. This is a compromise between avoiding low-order harmonics of this signal input appearing in the output, and minimizing the selectivity requirements of the circuits following the mixer stage.

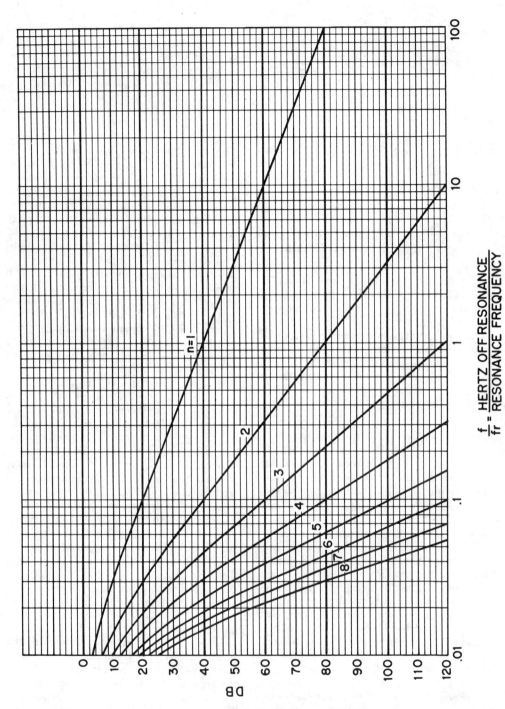

Figure 27

RESPONSE OF "N" NUMBER OF TUNED CIRCUITS,
ASSUMING EACH CIRCUIT Q IS 50

9-8 Distortion Products Due to Nonlinearity of R-F Amplifiers

When the SSB envelope of a *voice or multitone* signal is distorted, a great many new frequencies are generated. These represent all of the possible combinations of the sum and difference frequencies of all harmonics of the original frequencies. For purposes of test and analysis, a *two-tone* test signal (two equal-amplitude tones) is used as the SSB source. Since the SSB radio-frequency amplifiers use tank circuits, all distortion products are filtered out except those which lie close to the desired frequencies. These are all odd-order products; third order, fifth order, etc. The third-order products are $2p-q$ and $2q-p$ where p and q represent the two SSB r-f tone frequencies. The fifth order products are $3p-2q$ and $3q-2p$. These and some higher order products are shown in figure 28 A, B, and C. It should be noted that the frequency spacings are always equal to the difference frequency of the two original tones. Thus when an SSB amplifier is badly overloaded, these spurious frequencies can extend far outside the original channel width and cause an unintelligible "splatter" type of interference in adjacent channels. This is usually of far more importance than the distortion of the original tones with regard to intelligibility or fidelity. To avoid interference in another channel, these distortion products should be down at least 30 dB below the adjacent channel signal. Using a two-tone test, the distortion is given as the ratio of the amplitude of one test tone to the amplitude of a third-order product. This is called the *signal-to-distortion ratio* (S/D) and is usually given in decibels. The use of feedback r-f amplifiers makes S/D ratios of greater than 40 dB possible and practical.

Vacuum-Tube Nonlinearity Distortion products caused by amplifier departure from a linear condition are termed *intermodulation products* and the distortion is termed *intermodulation distortion*. This distortion can be caused by nonlinearity of amplifier gain or phase shift with respect to input level, and only appears when a multi-

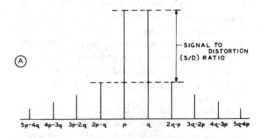

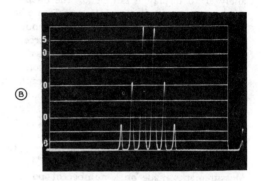

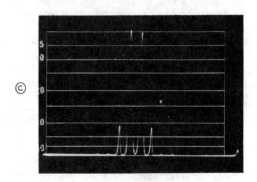

Figure 28

A shows SSB distortion products pictured up to ninth order. B shows SSB distortion products as seen on a panoramic analyzer. Third-order products are 19 decibels below two-tone test signal and fifth-order products are 32 decibels below the test signal. C illustrates that third-order products are better than 40 decibels down from test signal.

tone signal is used to drive the linear amplifier. This is the case for a voice signal which is composed of many tones, and intermodulation distortion will show up as a "gravelly" tone on the voice and will create interference to signals on adjacent channels. The main source of intermodulation distortion in a linear amplifier is the vacuum tube or transistor as these components have inherently nonlinear characteristics. Maximum linearity may be achieved by proper choice of tube or transistor and their operating conditions.

A practical test of linearity is to employ a two-tone, low-distortion signal to drive the tube or transistor and to use a spectrum analyzer to display a sample of the output spectrum on an oscilloscope (figure 28). The test signal, along with spurious intermodulation products may be seen on the screen, separated on the horizontal axis by the difference in frequency between the two tones. A reading is made by comparing the amplitude of a specific intermodulation product with the amplitude of the test signal. For convenience, the ratio between one of the test signals and one of the intermodulation products is read as a power ratio expressed in decibels below the test signal level. Measurements made on a number of power tubes have shown typical intermodulation distortion levels in the range of -20 to -40 decibels below one tone of a two-tone test signal.

The present state of the art in commercial and military SSB equipment calls for third-order intermodulation products better than -40 to -60 decibels below one tone of a two-tone test signal. Amateur requirements are less strict, running as low as -20 decibels, and may be justified on an economic basis since signal distortion, at least to the listener, is a highly subjective thing. To date, the use of inexpensive TV-type sweep tubes as linear amplifiers in amateur SSB gear has been acceptable, regardless of the rather high level of distortion inherent in these tube types.

Interpreting SSB Meter Readings The operator of an SSB transmitter should closely monitor the amplifier plate or collector current as it provides him with a quick check of equipment operation. Representative meter readings are shown in figure 29. These illustrations represent meter readings for a linear amplifier whose maximum plate current is 500 mA. In drawing A, the amplifier is loaded to the maximum value of

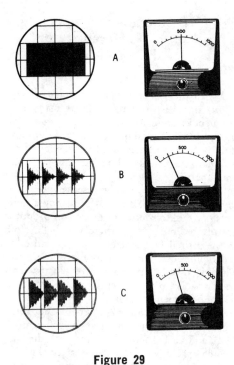

Figure 29

SSB AMPLIFIER METER READINGS

A—Carrier insertion at maximum input level.
B—Voice modulation with peaks reaching maximum level.
C—Voice modulation with audio clipping or compression. Peaks reaching maximum levels.

plate current by carrier insertion in the exciter or other means. An oscilloscope rendition of the output wave is shown to the left of the meter. Tuning and loading operations are conducted to provide maximum output from the amplifier at this value of plate current. The inserted carrier is now removed and the plate current drops to the resting (quiescent) value. Under the voice modulation, the peaks of the waveform just reach the same level of amplitude on the 'scope as was previously exhibited by the carrier (B). Because of the high peak-to-average power ratio of the human voice and the inertia of the meter, peak voice meter readings run about one half, or less, of the fully loaded condition. In this example, peak meter reading on voice runs about 200 mA.

If speech processing, or other form of compression is used, the signal peaks will remain the same on the oscilloscope (C) but the peak voice meter reading will increase. In this example, peak current is about 325 mA.

Thus, the steady state condition (A) sets the parameters for peak voice operation, as shown in B and C. Under no circumstances should the peak voice meter reading reach the steady state value shown in A or severe distortion and signal spatter will occur.

9-9   SSB Reception

Single-sideband reception may be considered the reverse of the process used in SSB transmission. The received SSB signal is amplified, translated downward in frequency, further amplified and converted into a replica of the original audio frequencies. The SSB receiver is invariably a superheterodyne in order to achieve high sensitivity and selectivity.

To recover the intelligence from the SSB signal, it is necessary to restore the carrier in such a way as to have the same relationship with the sideband components as the original carrier generated in the SSB exciter. To achieve this, it is important that the

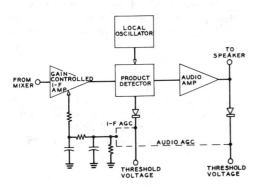

Figure 30

BLOCK DIAGRAM OF AUTOMATIC GAIN CONTROL SYSTEM

Audio or i-f derived control signal is applied to low-level gain-controlled i-f amplifier in typical SSB receiver. Agc system reduces the gain of controlled stage(s) on signal peaks to prevent receiver overload. Control voltage must be derived from the modulation envelope. Since carrier is not transmitted with voice SSB signal.

receiver oscillators have good frequency accuracy and stability.

To take advantage of the narrow bandwidth occupied by the SSB signal, selectivity characteristics of the receiver must be held to narrow limits. Excessive receiver bandwidth degrades the signal by passing unnecessary interference and noise.

SSB Receivers   In a conventional a-m receiver, the audio intelligence is recovered from the radio signal by an envelope amplitude detector, such as a diode rectifier. This technique may be used to recover the audio signal from an SSB transmission provided the amplitude of the local carrier generated by the beat oscillator is sufficiently high to hold audio distortion at a reasonable low level. Better performance with respect to distortion may be achieved if a *product detector* is used to recover the audio signal.

The characteristics of the *automatic volume control* (or *automatic gain control*) system of an SSB receiver differ from those of a conventional a-m receiver. In the latter, the agc voltage is derived by rectifying the received carrier, as the carrier is relatively constant and does not vary rapidly in amplitude. The agc system can therefore have a rather long time constant so that an S-meter may be used to indicate relative carrier amplitude.

In an SSB receiver, however, the signal level varies over a large range at a syllabic rate and a fast time-constant agc system is required to prevent receiver overload on initial bursts of a received signal. To prevent background noise from receiving full amplification when the SSB signal is weak or absent, a relatively slow agc release time is required (figure 30).

The agc system, moreover, must be isolated from the local-oscillator voltage to prevent rectification of the oscillator voltage from placing an undesired no-signal static bias voltage on the agc line of the receiver.

Thus, the SSB receiver differs from the a-m receiver in that it requires a higher order of oscillator stability and i-f bandwidth, a more sophisticated agc system, and the capability of receiving signals over a very wide range of strength without overload or cross modulation. In addition, the tuning rate of the SSB receiver should be substantially less than that of an a-m receiver; generally speaking, tuning rates of 25 to 100 kHz per dial revolution are common in modern SSB receivers.

Because of variations in the propagation path, transmitter power, and distance between stations, the input signal to an SSB receiver can vary over a range of 120 decibels or so. The receiver requires, therefore, a large dynamic range of signal-handling capability and an enhanced degree of gain-adjusting capability.

SSB Receiver Circuitry For minimum spurious response it is desired to have good selectivity ahead of the amplifier stages in the SSB receiver. This is possible to a degree, provided circuit simplicity and receiver sensitivity are not sacrificed. For the case when sensitivity is not important, an attenuator may be placed in the receiver input circuit to reduce the amplitude of strong, nearby signals (figure 31). To further reduce the generation of cross-modulation interference, it is necessary to carefully select the tube or device used in the r-f amplifier stage to determine if it will retain its linearity with the application of agc-bias control voltage. Suitable r-f stage circuits are shown in the *Radio Receiver Fundamentals* chapter of this Handbook.

Avoidance of images and spurious responses is a main problem in the design of SSB receiver mixers. Due to the presence of harmonics in the mixer/oscillator signal and

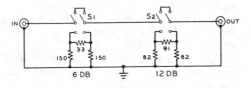

Figure 31

R-F ATTENUATOR FOR SSB RECEIVER

The dynamic signal range of an SSB receiver may be increased, and troubles resulting from overload may be decreased with the use of a simple r-f attenuator placed in the coaxial line from the antenna to the receiver. This attenuator is designed for use with either 50- or 70-ohm transmission lines and may be built in a small aluminum Minibox.

nonlinearity in the mixer, higher-order products are generated in addition to the desired mixing product. These undesired products vary in frequency as the oscillator is tuned and may fall within the received passband, creating *crossovers*, or *birdies* (spurious beatnotes which tune faster than the normal tuning rate).

The twin problems of images and crossovers can be resolved through the use of double conversion. The first (high) conversion provides adequate image rejection and the second (low) conversion may be adjusted so as to reduce crossover points to a minimum. In addition, double conversion allows the use of a crystal-controlled oscillator for the first converter stage, which can provide a higher order of stability than a tunable oscillator. The oscillator for the lower mixer stage may be made tunable, covering only a single frequency range, eliminating some of the mechanical and electrical factors contributing to receiver instability.

Choice of an intermediate frequency low with respect to signal frequency minimizes the probability of strong birdie signals within the receiver passband. The low intermediate frequency, however, may lead to image problems at the higher received frequencies.

The bandwidth of the low-frequency i-f system determines the overall selectivity of the SSB receiver. For SSB voice reception, the optimum bandwidth at the 6-dB point is about 2 kHz to 3 kHz. It is good practice to place the selective filter in the circuit ahead of the i-f amplifier stages so that strong adjacent-channel signals are attenuated before they drive the amplifier tubes into the overload region. In addition to the sideband filter, additional tuned circuits are usually provided to improve overall receiver selectivity, especially at frequencies which are down the skirt of the selectivity curve. Some types of SSB filters have spurious responses outside the passband which can be suppressed in this manner.

Desensitization, Intermodulation, and Crossmodulation   When a receiver is tuned to a weak signal with a strong signal close to the received frequency, an apparent decrease in receiver gain may be noted. This loss of gain is called *desensitization* or *blocking*. It commonly occurs when the unwanted signal voltage is sufficient to overcome the operating bias of an amplifier or mixer stage, driving the stage into a nonlinear condition. Rectified signal current may be coupled back into the gain-control system, reducing overall gain and increasing signal distortion.

Amplifier and mixer stages using transistors and vacuum tubes may generate in-band spurious products resulting from beats between the components of the desired signal in the receiver, or between two received signals. This class of distortion is termed *intermodulation distortion* and is evident in a nonlinear device driven by a complex signal having more than one frequency, such as the human voice.

Intermodulation occurs at any signal level and spurious products are developed by this action. For example, assume a signal is on 900 kHz and a second signal is on 1.5 MHz. The receiver is tuned to the 80-meter band. Intermodulation distortion within the receiver can result in a spurious signal appearing at 3.9 MHz as a result of mixing in a nonlinear stage. The product mix is: $(2 \times 1.5) + 0.9 = 3.9$ MHz.

This particular spurious signal (often termed a *spur*) is a result of a harmonic of the 1.5-MHz signal being produced in the receiver and beating against the incoming 0.9-MHz signal. Other spurious signals, composed of the sums and differences and harmonics of the fundamental signals exist in addition to the one at 3.9 MHz. Some of these products fall at: 0.3, 1.8, 2.1, 2.7, 3.0, 3.3, and 4.5 MHz. Other spurs may be generated by higher order linearities. Thus, two signals passed through a nonlinear device can create a whole range of unwanted signals. Since the radio spectrum is crowded with numerous strong signals, all of which can create spurious intermodulation products simultaneously in varying degrees of severity, it is important that high-Q circuits or a number of tuned circuits be used in the front-end of a receiver to prevent out-of-band signals from entering the receiver. In addition, the optimum choice of transistor or tube must be made for each receiver stage, and its correct operating point established.

Crossmodulation is the transfer of intelligence from an unwanted strong signal to a wanted weak one. Thus, if a receiver is tuned to a wanted signal at 3.9 MHz and a strong unwanted signal is at 3.8 MHz, the modulation on the second signal may be imposed on the wanted signal, even though the second signal is well outside the i-f passband of the receiver. Multiple signals, moreover, can produce multiple crossmodulation effects. Crossmodulation can be minimized by optimum selection of amplifying and mixing devices and by careful selection of signal levels and operating voltages in the various receiver stages.

Intermodulation, crossmodulation, and desensitization can all occur simultaneously in a receiver and the overall effect is a loss in intelligibility and signal-to-noise ratio of the desired signal. These receiver faults may be ascertained by injecting test signals of various frequencies and amplitudes into the receiver, a stage at a time.

Generally speaking, field-effect transistors and remote-cutoff vacuum tubes exhibit a significant improvement in linearity and provide enhanced rejection to these unwanted effects as opposed to bipolar transistors, which have a lower linearity figure than the other devices.

Passband Tuning An unwanted signal can be rejected in the i-f system of a receiver by means of *passband tuning* (figure 32). A notch filter is used to pro-

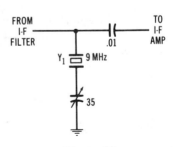

Figure 32

I-F NOTCH FILTER

This simple notch filter is used in hf amplifier strips. Adjustment of the capacitor can attenuate a narrow frequency band in the i-f passband.

vide a rejection slot in the passband at any point in the passband. The variable capacitor shifts the resonant point of the crystal across the passband. A more complicated technique making use of a *varifilter* circuit in which the i-f signal is mixed to a lower frequency and passed through two selective filters whose center frequencies are slightly different. Tuning the mixing oscillator moves the new i-f channel across the filters thus effectively varying the passband of the i-f system.

Automatic Gain Control and Signal Demodulation   The function of an *automatic gain control* system is to reduce the gain of the controlled stages on signal peaks to prevent receiver overload and hold constant audio output. Since the carrier is not transmitted in SSB, the receiver agc system must obtain its signal voltage from the modulation envelope. The agc voltage may be derived either from the i-f signal or the audio signal (figure 33). Audio-derived agc has the advantage of easier iso-

lation from the local carrier voltage, but the i-f system will function on both SSB and a-m signals in a satisfactory manner.

Product detectors are preferred for SSB reception because they minimize intermodulation distortion products in the audio signal and, in addition, do not require a large local-oscillator voltage. The product detector also affords a high degree of isolation between the carrier oscillator and the agc circuit. The undesired mixing products present in the output circuit of the detector may be suppressed by a low-pass filter placed in the audio line.

A Representative SSB Receiver   A typical SSB receiver is made up of circuits resembling those discussed in the previous section. To achieve both high stability and good image rejection, many amateur SSB receivers are double-conversion types, such as outlined in figure 34. An accurate, stable low-frequency tunable oscillator is employed, together with a standard 455-kHz i-f channel and a crystal or mechanical SSB filter. The frequency coverage of the vfo may be as high as 500 kHz

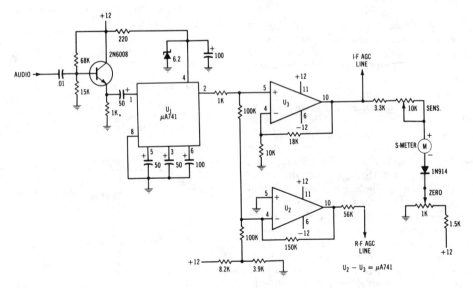

Figure 33

AUDIO DERIVED AGC CIRCUIT

Agc voltage is derived from the audio signal. Device U₁ provides "hang" period to maintain gain during speech pauses. Period is one second which is determined by capacitance value at pin 6 of U₁.

to cover all of the low-frequency amateur bands, or it may be restricted to only 100 kHz or so, necessitating the use of a multiplicity of crystals in the first conversion oscillator to achieve complete band coverage. A tunable first i-f stage covering the required passband may be ganged with the variable-frequency oscillator and with the r-f amplifier tuning circuits. The high-frequency tuning range is chosen by the appropriate high-frequency crystal.

To permit sideband selection, the bfo may be tuned to either side of the i-f passband. Proper tuning is accomplished by ear, the setting of the bfo on the filter passband slope may be quickly accomplished by experience and by recognition of the proper voice tones.

In addition to the special circuitry covered in this chapter, SSB receivers make full use of the general receiver design information given in this Handbook.

9-10 The SSB Transceiver

The SSB *transceiver* is a unit in which the functions of transmission and reception are combined, allowing single-channel semi-duplex operation at a substantial reduction in cost and complexity along with greatly increased ease of operation. The transceiver is especially popular for mobile operation

where a savings in size, weight, and power consumption are important. Dual usage of components and stages in the SSB transceiver permits a large reduction in the number of circuit elements and facilitates tuning to the common frequency desired for two-way communication.

Figure 35 shows a basic filter-type transceiver circuit. Common mixer frequencies are used in each mode and the high frequency vfo is used to tune both transmit and receive channels to the same operating frequency. In addition, a common i-f system and sideband filter are used.

The transceiver is commonly switched from receive to transmit by a multiple-contact relay which transfers the antenna and removes blocking bias from the activated stages. Transceivers are ideal for net operation since the correct frequency may be ascertained by tuning the received signal to make the voice intelligible and pleasing. With practice, the SSB transceiver may be adjusted to a predetermined frequency with an error of 100 Hz or less by this simple procedure.

Single-Band Transceivers An important development is the single-band transceiver, a simplified circuit designed for operation over one narrow frequency band. Various designs have been made available for the 50-MHz band as well as the popular h-f amateur bands. Commercial transceiver designs are usually operated on crystal-

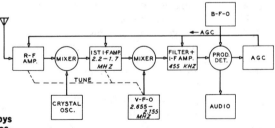

Figure 34

DOUBLE CONVERSION SSB RECEIVER

Typical double-conversion SSB receiver employs tunable first i-f and crystal-controlled local oscillator, with tunable oscillator and fixed-frequency i-f amplifier and sideband filter. This receiver tunes selected 500-kHz segments of the hf spectrum. Additional conversion crystals are required for complete coverage of the 10-meter band.

TUNING RANGE (MHZ)	CRYSTAL (MHZ)
3.5 – 4.0	5.7
7.0 – 7.5	9.0
14.0 – 14.5	16.2
21.0 – 21.5	23.2
28.5 – 29.0	30.7

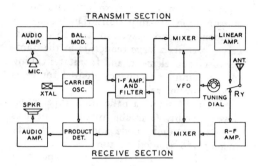

TRANSMIT SECTION

RECEIVE SECTION

Figure 35

THE SSB TRANSCEIVER

Common carrier oscillator, i-f amplifier/filter, and vfo are used in transceiver, designed to communicate on a single frequency selected by proper vfo setting. Transfer from receive to transmit is carried out by relays and by application of blocking voltage to unused stages.

controlled channels in the h-f and vhf spectrum using a crystal synthesizer for channel control. Elaborate synthesizers permit selection of discrete operating frequencies as closely separated as 100 Hz. Some units include a *clarifier* control which permits a slight frequency adjustment to place the unit exactly on the chosen operating channel (figure 36).

9-11 Spurious Frequencies

Spurious frequencies (spurs) are generated during every frequency conversion in a receiver or transmitter. These unwanted frequencies mix with the harmonics generated by the mixing oscillators to produce undesired signals that either interfere with reception of the wanted signal or can be radiated along with the desired signal from the transmitter. If the spurs are known, this information can help to determine the required r-f and i-f selectivity characteristics, the number of conversions, the allowable harmonic content of the oscillators, and the optimum intermediate frequencies.

The severity of interference from a given spur depends upon its proximity to the desired signal frequency, rather than the absolute frequency difference. For example, a simple tuned circuit has sufficient selectivity to reject a spur 4 MHz away from a 1-MHz frequency, while much more complicated means are needed to reject a spur that is 4 MHz away from a frequency of 100 MHz. Spur interference is dependent on the ratio of the spur frequency to the tuned frequency, and the lower the ratio, the more serious the problem.

Another indication of the importance of a particular spur is contained in the order of response. This order may be defined as the sum of the signal and oscillator harmonics that produce the spur. For example, a spur produced by the second harmonic of the signal and the third harmonic of the oscillator is known as a *fifth-order* spur. Lower-order spurs are more serious because higher harmonics of both input signals are easier to reject by circuit design techniques.

A Spur Chart Graphical relationships between the frequencies of the various spurious signals and the desired signal are presented by the spur chart of figure 37. A given ratio of spur to desired frequency is represented by a constant horizontal distance on the chart.

The local-oscillator frequency is represented by F_R and the relative signal frequency by F_O. The curves cover all spurious products up to the sixth order for spur-signal frequencies that fall within an octave of the signal frequency. Each line on the chart represents a normalized frequency difference of 1 for $mF_O + nF_R$ where m and n may be positive or negative integers. The heavy, central lines labeled $F_R - F_O$ and $F_O - F_R$ are plots of the desired frequency conversion when the oscillator frequency is either higher or lower than the signal frequency. Whichever line represents the desired signal, the other line represents the image spur.

To determine the spurious environment for a given conversion, first normalize the desired signal and oscillator frequencies by dividing both frequencies by the mixing output frequency. Then locate the desired point on one of the heavy lines representing either $F_O - F_R$ or $F_R - F_O$. Since the oscillator frequency does not change for spurs, simply

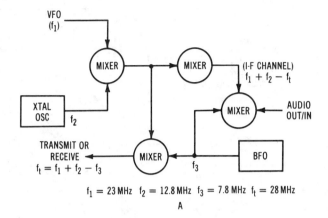

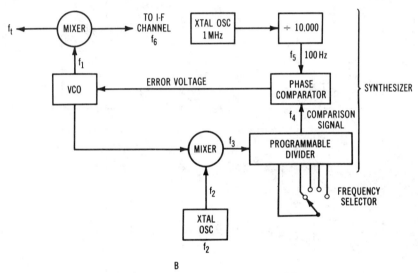

Figure 36

SYNTHESIZED SSB TRANSCEIVERS

A—Crystal synthesizer for 10-meter transceiver uses 7.8-MHz i-f channel. Vfo at 23 MHz controls operating frequency in 500-kHz ranges. Four crystals at f_2 cover the 10- meter band.

B—Simplified diagram of synthesized transceiver tunable in 100-Hz steps. Transceiver is tunable in 100-Hz increments. The output of the phase comparator is an error voltage that varies the frequency of the vco (voltage- controlled oscillator) until error voltage is zero. Good shielding and filtering is required in transceivers of this class to keep the various mixing frequencies where they belong.

trace horizontally in either direction to determine the relative frequency of the spurs.

Example: Desired signal frequency is 10 MHz.

Mixing output frequency is 2 MHz.

Oscillator frequency is 12 MHz.

Then, relative signal frequency F_O is 10 MHz/2 MHz = 5.

And, relative oscillator frequency F_R is 12 MHz/2 MHz = 6.

Since oscillator frequency is higher, we use the $F_R - F_O$ curve.

Locate the $F_O = 5$, $F_R = 6$, point on the curve. Tracing horizontally to the left, the spur lines intercepted on the F_O scale are: $3F_O - 2F_R$ at F_O of 4.35, or signal frequency

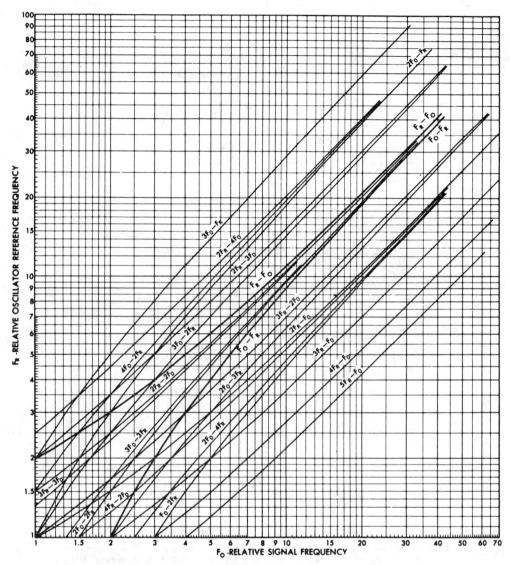

Figure 37

SPUR CHART

Curves cover all spurious mixer products that fall within an octave of the signal frequency.

that causes the spur is 8.70 MHz (2 MHz × 4.35).

$2F_R - 3F_O$ at F_O of 3.70, equivalent to a signal frequency of 7.40 MHz.

$2F_O - 2F_R$ at F_O of 3.50, equivalent to a signal frequency of 7.00 MHz.

Tracing right, nearest spur lines are:

$2F_R - 2F_O$ at F_O of 5.50, equivalent to a signal frequency of 11.0 MHz.

$3F_R - 3F_O$ at F_O of 5.70, equivalent to a signal frequency of 11.4 MHz.

$3F_O - 3F_R$ at F_O of 6.36, equivalent to a signal frequency of 12.7 MHz.

And the image frequency, $F_O - F_R$, occurs at 7.00 or 14.0 MHz.

CHAPTER TEN

Communication Receiver Fundamentals

Part I—The HF Receiver

Communication receivers vary widely in their cost, complexity and design, depending on the intended application and various economic factors. A receiver designed for amateur radio use must provide maximum intelligibility from signals varying widely in received strength, and which often have interfering signals in adjacent channels, or directly on the received channel. The practical receiver should permit reception of continuous wave (c-w), amplitude-modulated (a-m) and single-sideband (SSB) signals. Specialized receivers (or receiver adapters) are often used for reception of narrowband f-m (NBFM), radio teletype (RTTY), slow scan television (SSTV) and facsimile (FAX) signals.

The desired signal may vary in strength from a fraction of a microvolt to several volts at the input terminals of the receiver. Many extraneous strong signals must be rejected by the receiver in order to receive a signal often having a widely different level than the rejected signals.

The modern receiver, in addition, must have a high order of electrical and mechanical stability, and its tuning rate should be slow enough to facilitate the exact tuning of c-w and SSB signals. Finally, the receiver should be rugged and reliable as well as easy to service, maintain, and repair. All of these widely differing requirements demand a measure of compromise in receiver design in order to achieve a reasonable degree of flexibility.

Modern solid-state receivers can readily meet most of these requirements. In many instances the receivers are incorporated in a transceiver package but the fundamentals discussed in this chapter apply equally well to either configuration. Frequency-modulation (f-m) reception is discussed at length in Chapter 13 of this handbook.

10-1 Types of Receivers

All receivers are *detectors* or *demodulators* which are devices for removing the modulation (intelligence) carried by the incoming signal. Figure 1 illustrates an elementary receiver wherein the induced voltage from the signal is diode rectified into a varying direct current. The capacitor C_2 is charged to the average value of the rectified wave-

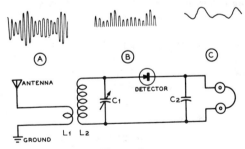

Figure 1

ELEMENTARY FORM OF RECEIVER

This is the basis of the "crystal set" type of receiver. The tank circuit (L_2-C_1) is tuned to the frequency it is desired to receive. The bypass capacitor across the phones should have a low reactance to the carrier frequency being received, but a high reactance to the modulation on the received radio signal.

form. The resulting current is passed through earphones which reproduce the modulation placed on the radio wave.

The Autodyne Detector   Since a c-w signal consists of an unmodulated carrier interrupted by dots and dashes, it is apparent that such a signal would not be made audible by detection alone. Some means must be provided whereby an audible tone is heard when the carrier is received, the tone stopping when the carrier is interrupted. Audible detection may be accomplished by generating a local carrier of a slightly different frequency and mixing it with the incoming signal in the detector stage to form a beat note. The difference frequency, or *heterodyne*, exists only when both the incoming signal and the locally generated signal are present in the mixer. The mixer (or

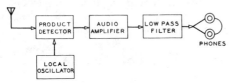

Figure 2

DIRECT DETECTOR CIRCUIT

Direct conversion receiver uses separate heterodyne oscillator to produce audio beat note signal. Passband is restricted by use of audio filter.

detector) may be made to supply the beating signal, as in the *autodyne detector*. A variation of the autodyne detector makes use of a separate oscillator and is termed a *direct conversion* receiver. A product detector may be used and signal selectivity is obtained at audio frequencies through the use of a low-pass audio filter (figure 2).

10-2 Receiver Performance Requirements

Receiver performance may be defined in terms of *sensitivity, selectivity, tuning rate, stability, spurious response, dynamic signal range*, and *gain compression*. Other factors enter into receiver specifications, but these properties are of the greatest interest to the user. A well-designed communication receiv-

er must be able to receive all modes of emission used in the band of reception while meeting minimum levels of performance in these important areas.

Sensitivity   The *sensitivity* of a receiver may be defined as the input signal required to give a signal-plus-noise output of some ratio (usually 10 dB) above the noise output of the receiver. A perfect "noiseless" receiver would generate no internal noise and the sensitivity would be limited only by the thermal noise (or "r-f smog") about the receiving location. Below 30 MHz or so, external noise, rather than internal receiver noise, is the limiting factor in weak signal reception.

Random electron motion, or *thermal agitation noise*, is proportional to the absolute temperature and is independent of frequency when the absolute bandwidth and input impedance of the receiver are constant. The noise is expressed as *equivalent noise resistance*, or that value of resistance which, if placed at the input circuit of a stage, will produce output circuit noise equivalent to the noise of the amplifying device in the stage.

The degree to which a "perfect" receiver is approached by a practical receiver having the same bandwidth is called the *noise figure* of the receiver. This is defined as the ratio of the signal-to-noise power ratio of the "perfect" receiver to that ratio of the receiver under test. The noise figure is expressed in decibels or as a power ratio, and the larger the noise figure, the noisier is the receiver.

The noise figure is defined as:

$$F = \frac{S/kTB}{S_0/N_0}$$

where,

F equals noise figure of receiver,
S equals available signal power from source,
S_0 equals available signal power from receiver,
N_0 equals noise power from receiver,
k equals Boltzmann's constant (1.38 $\times$ 10^{-23} joules per °k),

T equals absolute temperature of signal source,

B equals bandwidth of receiver.

The quantity kTB represents the available noise power from a resistor of arbitrary value at temperature T. Thus, a signal, no matter how it is generated, has associated with it a minimum amount of noise (kTB). If the signal is passed through a receiver that amplifies it without adding noise, the ratio of signal power to noise power at the output of the receiver will be the same as at the input and the noise figure (F) will be unity. If the receiver adds additional noise, F will be greater than unity. The noise figure of a good hf communications receiver runs between 5 to 15 dB below 30 MHz; a noise figure better than this is of little use considering the high atmospheric noise level. In the vhf spectrum, very low noise figures are extremely useful as the external noise level is quite low.

Noise Figure Measurement Expressed in decibels, the *noise figure* of a receiver is:

$$F = 10 \log_{10}\frac{N_2}{N_1}$$

where,

N_1 and N_2 are the noise power figures in watts and represent the output from an actual receiver, (N_2) at 290° K (63°F), divided by the noise power output from an ideal receiver (N_1) at the same temperature.

The noise figure of a receiver may be ascertained by direct measurement with a *noise generator*. The receiver input is terminated with a resistor and wideband random noise, generated by thermal agitation in a suitable generator, is injected into the input circuit of the receiver. The power output of the receiver is measured with no noise input and the generator output is then increased until the receiver noise output is doubled. The noise figure of the receiver is a function of these two levels, and may be computed from these measurements.

It is common practice to match the impedance of the antenna transmission line to the input impedance of the amplifying device of the first r-f amplifier stage in a receiver. However, when vhf tubes and transistors are used at frequencies somewhat less than their maximum capabilities, a significant improvement in noise figure can be attained by *increasing* the coupling between the antenna and first tuned circuit to a value greater than that which gives greatest signal amplitude out of the receiver. In other words, in the vhf bands, it is possible to attain somewhat improved noise figure by increasing antenna coupling to the point where the gain of the receiver is slightly reduced.

It is always possible, in addition, to obtain improved noise figure in a vhf receiver through the use of devices which have improved input-impedance characteristics at the frequency in question over conventional types.

The relationship between the sensitivity in microvolts, noise figure and audio bandwidth is shown in figure 3 which assumes an antenna input impedance of 50 ohms and room temperature of 80.5°F.

Selectivity The *selectivity* of a receiver is the ability to distinguish between the desired signal and signals on closely adjacent frequencies. The bandwidth, or passband, of the receiver must be sufficiently wide to pass the signal and its sidebands if faithful reproduction of the signal is desired. For reception of a double-sideband amplitude-modulated broadcast signal, a passband of about 10 kHz is required. SSB passband response may be as small as 2 kHz for voice reception. For cw reception, a passband less than 100 Hz is often employed. As the circuit passband is reduced, transmitter and receiver frequency stability requirements become more strict and practical bandwidth in receivers may often have to be greater than the theoretical minimum value to compensate for frequency drift of the equipment.

Receiver bandwidth is defined in terms of *skirt selectivity*, or the degree of attenuation shown to a signal received at some frequency removed from the center frequency of reception, as shown in figure 4. The bandwidth is the width of the resonance curve of the receiver and is specified at the −6dB and −60 dB points. For the example shown, the −6 dB bandwidth is about 2.8 kHz for curve A and the −60 dB bandwidth is 12 kHz.

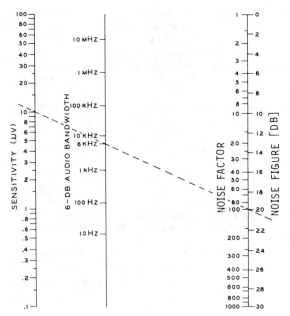

Figure 3

NOISE-FIGURE NOMOGRAPH

To find the noise figure of a receiver, a line extended between sensitivity and audio bandwidth points will intersect noise-figure line at right. Dashed line shows bandwidth of 6 kHz and sensitivity of 10 microvolts gives a noise factor of 100, or a noise figure of 20 dB.

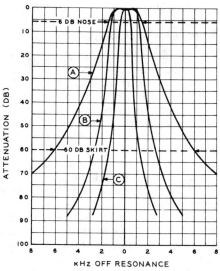

Figure 4

SKIRT SELECTIVITY

Receiver bandwidth is determined by selectivity of i-f system. Curve A shows typical response for reception of double-sideband, amplitude-modulated signal. SSB reception on a good communication receiver is shown by curve B. C-w selectivity is shown by curve C. Strong-signal selectivity is determined by bandwidth at 60-dB skirt points.

Stability The ability of a receiver to remain tuned to a chosen frequency is a measure of the *stability* of the receiver. Environmental changes such as variations in temperature, supply voltage, humidity and mechanical shock or vibration tend to alter the receiver characteristics over a period of time. Most receivers, to a greater or lesser degree, have a steady frequency variation known as *warm-up drift* which occurs during the first minutes of operation. Once the receiver components have reached operating temperature, the drift settles down, or subsides. *Long-term drift* may be apparent over a period of days, weeks or even years as components age or gradually shift in characteristics due to heat cycling or usage. Many receivers include a high-stability calibration oscillator to provide *marker signals* at known frequencies to allow rapid frequency calibration of the receiver dial. Typical short-term receiver drift is shown in figure 5.

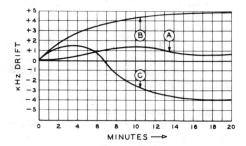

Figure 5

RECEIVER FREQUENCY STABILITY

Frequency drift of receiver depends on electrical and mechanical stability of tuned circuits. Temperature compensation (A) reduces warmup drift to a minimum. No compensation may result in long term, continual drift (B) and overcompensation can show as reversal of drift (C). Frequency compensation may be achieved by use of special capacitors having controlled temperature characteristics in critical circuits and by temperature stabilization of oscillator circuitry.

Tuning Rate A good communication receiver should have a slow *tuning rate*. That is, each revolution of the tuning control should represent only a moderate frequency change when compared to the bandwidth of reception. SSB receivers often have

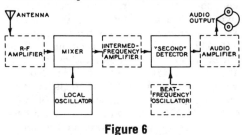

Figure 6

ESSENTIAL UNITS OF A SUPERHETERODYNE RECEIVER

The basic portions of the receiver are shown in solid blocks. Practicable receivers employ the dotted blocks and also usually include such additional circuits as a noise limiter, an agc circuit, and a bandpass filter in the i-f amplifier.

a tuning rate of 10 to 50 kHz per tuning dial revolution. Receivers intended for c-w reception may have a tuning rate as low as 5 kHz per dial revolution. The tuning rate may be determined mechanically by means of a step-down gear train or rim-drive mechanism placed between the tuning dial and the tuning control of the receiver. In some instances, electrical *bandspread* (see Section 10-4) may be employed. Regardless of the technique used, the tuning mechanism should have a smooth action and be free of mechanical or electrical backlash.

The Superheterodyne Receiver By changing the frequency of a received signal to a lower, fixed, *intermediate frequency* before ultimate detection, high gain and selectivity may be obtained with a good order of stability. A receiver that performs this frequency changing (heterodyning) process is termed a *superheterodyne* or *superhet* receiver. A block diagram of a typical superhet receiver is shown in figure 6.

The incoming signal is applied to a *mixer* consisting of a nonlinear impedance such as a vacuum tube, transistor, or diode. The signal is mixed with a locally generated variable-frequency signal, with the result that a third signal bearing all the modulation applied to the original signal but of a frequency equal to the difference between the local oscillator and the incoming signal frequency appears in the mixer output circuit. The output from the mixer is fed into a fixed-tuned intermediate-frequency amplifier, wherein it is amplified, detected, and passed on to an audio amplifier.

Although the mixing process is inherently noisy, this disadvantage can be overcome by including a radio-frequency amplifier stage ahead of the mixer, if necessary.

Advantages of the Superheterodyne The advantages of superheterodyne reception are directly attributable to the use of the fixed-tuned *intermediate-frequency* (i-f) *amplifier*. Since all signals are converted to the intermediate frequency, this section of the receiver may be designed for optimum selectivity and high amplification. High amplification is easily obtained in the intermediate-frequency amplifier, since it operates at a relatively low frequency, where conventional pentode-type tubes and transistors give adequate voltage gain.

Spurious Responses The mark of a good receiver is its ability to reject spurious signals outside of the passband of the receiver and to generate no spurious signals within the passband. While the superheterodyne receiver is universally accepted as the best combination of circuit principles for optimum reception, the device has practical disadvantages that should be recognized. The greatest handicap of this type of receiver is its susceptibility to various forms of spurious response and the complexity of design and adjustment required to reduce this response. Most of the responses, but not all, are a result of frequency conversion.

Image Interference The choice of a frequency for the i-f amplifier involves several considerations. One of these considerations concerns selectivity — the lower the intermediate frequency the greater the obtainable selectivity. On the other hand, a rather high intermediate frequency is desirable from the standpoint of *image* elimination, and also for the reception of signals from television and f-m transmitters both of which occupy a rather wide band of frequencies, making a broad selectivity characteristic desirable. Images are a peculiarity common to all superheterodyne receivers and for this reason they are given a detailed discussion later in this chapter.

While intermediate frequencies as low as 50 kHz are used where extreme selectivity is a requirement, and frequencies of 60 MHz and above are used in some specialized forms of receivers, many communication receivers use intermediate frequencies near 455 or 1600 kHz. Some receivers make use of high-frequency crystal-lattice filters in the i-f amplifier and use an intermediate frequency as high as 5 MHz or 9 MHz to gain image rejection. Entertainment receivers normally use an intermediate frequency centered about 455 kHz, while many automobile receivers use a frequency of 262 kHz. The standard frequency for the i-f channel of f-m receivers is 10.7 MHz, whereas the majority of television receivers use an i-f which covers the band between 41 and 46 MHz.

Arithmetical Selectivity Aside from allowing the use of fixed-tuned bandpass amplifier stages, the superheterodyne has an overwhelming advantage over the autodyne type of receiver (figure 2B) because of what is commonly known as *arithmetical selectivity*.

This can best be illustrated by considering two receivers, one of the autodyne type and one of the superheterodyne type, both attempting to receive a desired signal at 10,000 kHz and eliminate a strong interfering signal at 10,010 kHz. In the autodyne receiver, separating these two signals in the tuning circuits is practically impossible, since they differ in frequency by only 0.1 percent. However, in a superheterodyne with an intermediate frequency of, for example, 1000 kHz, the desired signal will be converted to a frequency of 1000 kHz and the interfering signal will be converted to a frequency of 1010 kHz, both signals appearing at the input of the i-f amplifier. In this case, the two signals may be separated much more readily, since they differ by 1 percent, or 10 times as much as in the first case.

An audio filter will provide a degree of selectivity that makes the autodyne receiver practical for use in the hf spectrum.

Images There always are *two* signal frequencies which will combine with a given frequency to produce the same difference frequency. For example: assume a superheterodyne with its oscillator operating on a higher frequency than the signal (which is common practice in many superheterodynes) tuned to receive a signal at 14,100 kHz. Assuming an i-f amplifier frequency of 450 kHz, the mixer input circuit will be tuned to 14,100 kHz, and the oscillator to 14,100 plus 450, or 14,550 kHz. Now, a *strong* signal at the oscillator frequency plus the intermediate frequency (14,550 plus 450, or 15,000 kHz) will also give a difference frequency of 450 kHz in the mixer output and will be heard also. Note that the image is always *twice* the intermediate frequency away from the desired signal. Images cause *repeat points* on the tuning dial.

The only way that the image could be eliminated in this particular case would be to make the selectivity of the mixer input circuit, and any circuits preceding it, great enough so that the 15,000-kHz signal never reaches the mixer input circuit in sufficient amplitude to produce interference.

For any particular intermediate frequency, image interference troubles become increasingly greater as the frequency (to which the signal-frequency portion of the receiver is tuned) is increased. This is due to the fact that the percentage difference between the desired frequency and the image frequency decreases as the receiver is tuned to a higher frequency. The ratio of strength between a signal at the image frequency and a signal at the frequency to which the receiver is tuned producing equal output is known as the *image ratio*. The higher this ratio is, the better the receiver will be in regard to image interference troubles .

With but a single tuned circuit between the mixer grid and the antenna, and with 400- to 500-kHz i-f amplifiers, image ratios of 40 dB and over are easily obtainable up to frequencies around 2000 kHz. Above this frequency, greater selectivity in the mixer grid circuit through the use of additional tuned circuits between the mixer and the antenna is necessary if a good image ratio is to be maintained.

Image signal reception can be confusing, especially in SSB reception, when an image

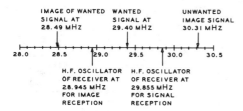

Figure 7

IMAGE SIGNAL

Relation between image signal and wanted signal when receiver local oscillator operates on high-frequency side of wanted signal. Image of 29.40 MHz signal appears at 28.49 MHz when 455 kHz i-f system is used. Unwanted signal at 30.31 MHz appears as image signal when receiver is tuned to desired signal at 29.40 MHz. Conditions are reversed for operation of oscillator on low-frequency side of signal.

signal may appear on the opposite sideband and tune "in the wrong direction" as compared to normal signals. Figure 7 illustrates the relationship between image signals when the receiver local oscillator operates on the high-frequency side of the received signal. The conditions are reversed for oscillator operation on the low-frequency side of the received signal. For reasons of economy and maximum oscillator stability, many receivers employ "low-side" oscillator operation on all but the highest frequency bands, where "high-side" operation is often used.

Phantom Signals   Any combination of desired and undesired signals, combined with the mixing oscillator and its harmonics, can produce signals at the intermediate frequency of the receiver. These, plus the spurious responses of the mixer stage are termed *birdies.*

Many spurious responses can be reduced by inclusion of adequate selectivity ahead of the mixer stage and by the use of shielding and filtering to prevent unwanted signal leak-through or pickup by later stages of the receiver.

Intermodulation distortion products are spurious signals generated in a nonlinear device. These signals are difficult to eliminate unless the frequencies for the mixing signal and the intermediate frequency are carefully chosen. Many undesired mixer products fall within the receiver passband and follow a

predictable frequency relationship as the receiver mixing oscillator is tuned. Multiples of the signal and oscillator frequency are present in a mixer stage which corresponds to the second, third, fourth, fifth, and sixth harmonics of the mixing signals. Higher order products are also present but are usually attenuated sufficiently so as not to cause any birdie problem. In the case of the lower order products, typical crossover combinations are:

Odd-order products

$$2f_o \pm f_s \qquad\qquad 2f_s \pm 2f_o$$
$$f_o \pm 2f_s \qquad\qquad 2f_o \pm 2f_s$$
$$3f_o \pm f_s \qquad\qquad 3f_o \pm 2f_s$$
$$f_o \pm 3f_s \qquad\qquad 2f_o \pm 3f_s$$
$$4f_o \pm f_s \qquad\qquad \text{. . . and so on.}$$
$$f_o \pm 4f_s$$

where,

f_o equals frequency of local oscillator,
f_s equals signal frequency.

Dynamic   The dynamic range of a re-
Signal Range   ceiver is that range over which the signal output of the receiver is a replica of the input signal. At the low-sensitivity end of the range, the limit is set by the noise and hum "floor" of the receiver. At the high end of the range, the limiting factors are *intermodulation distortion, gain compression,* and *crossmodulation.*

Dynamic measurements on a receiver are made in terms of power, specified in decibels with respect to one milliwatt—abbreviated *dBm.* Specifically, 0 dBm is one milliwatt. A typical communications receiver will have a noise floor of −140 dBm and at a signal level of −40 dBm the receiver may show indications of blocking or crossmodulation. The dynamic range of the receiver, then, is the difference between the two levels, or 100 dB. Modern high-frequency communications receivers have a dynamic range from 70 dB to better than 120 dB, as measured above the noise floor.

The dynamic range of a receiver can be specified by measuring the third order products and receiver gain for various levels of input signal. This is done with the test arrangement shown in figure 9.

Using a single generator, input power versus output power is plotted as shown in

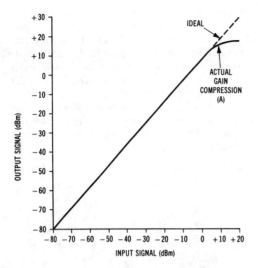

Figure 8

**Comparison of input power versus output power
for a receiver showing gain compression.**

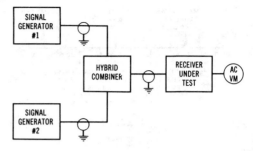

Figure 9

RECEIVER EVALUATION

**Dynamic range of a receiver can be specified
by measuring third-order mixing products from
two signal sources.**

figure 8. The input-output power relationship is linear until a certain signal level (A) is reached at which point the gain is compressed and the relationship is no longer linear. When 1 dB of gain compression occurs, the receiver has reached the level of desensitization.

A two-tone intermodulation test is now run (figure 9) to determine intermodulation distortion at the receiver. Two equal-level signals at slightly different frequencies (f_1 and f_2) are injected into the receiver. The third-order intermodulation products ($2f_1 - f_2$ and $2f_2 - f_1$) will appear in the mixer of the receiver and can be measured. For example, assume the two input frequencies are 7.0 MHz and 7.02 MHz. The third order intermodulation products, then, will be 6.98 MHz and 7.04 MHz. The intermodulation products can be detected by a spectrum analyzer or by a selective receiver coupled to the r-f circuitry of the receiver under test. As the amplitude of the two equal tones is increased, the frequencies at which the third-order products appear is monitored and the spurious signal levels plotted against the input signals.

If these two plots are combined as one graph, the intercept point of the two curves is called the *third-order intermodulation*

intercept and it defines the dynamic performance of the receiver for maximum signal levels as shown in figure 10.

The curves in this illustration show the input-output power relationship for the received signal and the third-order responses in a nonlinear system. The solid line represents the received signal and is linear with a slope of unity indicating the output power of the system changes on a dB-for-dB basis with the input power up to the overload point. The dashed line represents the output power of the intermodulation products as a function of the signal power. The curve is linear for small signals and has a slope of about three. This slope indicates that intermodulation power increases 3 dB for each dB increase in the input signal.

The intercept point of a system cannot be measured directly but can be computed from:

$$\text{Intercept point} = \frac{R_s}{2} + P_i$$

where,

R_s equals suppression in dB of third-order products,

P_i equals signal power level in dBm at which relative suppression is measured.

In figure 10 this is equal to the vertical difference between the two curves.

An important specification of an amplifier system is the output power at which the amplifier gain decreases by one dB from its small signal value. This is termed the *1-dB compression point* and occurs in the example shown in figure 10 at about the point the curves meet. A high 1-dB compression point permits amplification of high level signals with low intermodulation. The amplifier efficiency can be maximized if circuitry is chosen such that the amplifying device operates at the highest possible 1-dB compression point.

Crossmodulation is the superposition of modulation from an unwanted signal. It can be measured with the test arrangement of figure 9. A specific signal level is chosen for one signal and the second signal is injected into the receiver. The second signal has a tone modulation of 30 percent and its power level is raised until a three percent modulation level is impressed on the first signal. The power level of the second signal necessary to achieve this indicates the ability of the receiver to handle adjacent strong signals.

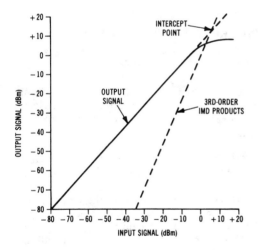

Figure 10

DYNAMIC RANGE OF A RECEIVER

The output power level at which the two curves cross each other is called the third order, intermodulation intercept point and defines the intermodulation performance of the receiver.

Additional selectivity ahead of the mixer stages will help to protect against crossmodulation. Pre-mixer crystal filters are often used in specialized receivers.

10-3 The Superheterodyne Receiver

A block diagram of a single conversion superheterodyne was shown in figure 6. This is the basic "building block" for communications receivers of all types. Frequency conversion permits r-f amplification at a fixed low frequency (the i-f stages) which provides high gain, good selectivity and a high order of stability. Improved image rejection and superior high-frequency performance can be achieved with the use of a double-conversion receiver.

Frequency Conversion   As previously mentioned, the use of a higher intermediate frequency will also improve the image ratio, at the expense of i-f selectivity, by placing the desired signal and the image farther apart. To give both good image ratio at the higher frequencies and good selectivity in the i-f amplifier, a system known as *double conversion* is sometimes employed. In this system, the incoming signal is first converted to a rather high intermediate frequency, and then amplified and again converted, this time to a much lower frequency. The first intermediate frequency supplies the necessary wide separation between the image and the desired signal, while the second one supplies the bulk of the i-f selectivity (figure 11).

The recent development of high quality, low cost crystal filters has made feasible the use of a high intermediate frequency in a single-conversion receiver to provide good selectivity and a high order of image rejection. Filters in the 5 MHz to 10 MHz range provide satisfactory selectivity for SSB reception at less than the cost of a lower-frequency filter and the attendant conversion and mixer stages.

Some specialized high-frequency receivers make use of an intermediate frequency *above* the tuning range of the receiver (40 MHz, for example). The extremely high i-f permits a high order of attenuation of image responses and allows continuous tuning of all frequencies up to 30 MHz or so (figure 11B). In other designs, dual conversion is accomplished with the use of but a single local oscillator, with the injection frequency

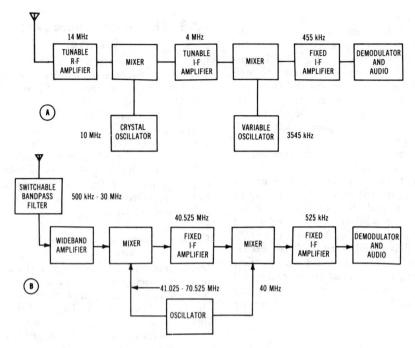

Figure 11

TYPICAL DOUBLE-CONVERSION SUPERHETERODYNE RECEIVERS

Illustrated at A is the basic circuit of a dual-conversion superheterodyne receiver. Diagram B shows use of intermediate frequency higher than the signal frequency.

chosen so that oscillator drift is automatically eliminated.

In all double-conversion receivers, the problem of spurious responses is aggravated because of the multiple-frequency signals existing within the receiver circuitry. Careful shielding and filtering of power leads must be incorporated in a receiver of this type if birdies and spurious signals are to be avoided.

The Demodulator   The *demodulator* (detector, second mixer or second detector as it is variously named) retrieves the intelligence from the incoming signal. A simple diode detector is suitable for a-m reception, and a *beat-frequency oscillator* (bfo) can provide a heterodyne note, suitable for c-w reception. For SSB reception, the demodulator must have an extremely wide dynamic range of operation, plus a bfo that provides a strong mixing

signal for low distortion reception of strong signals. A control voltage for automatic gain control may also be obtained from the demodulator stage.

Automatic Gain Control   *Automatic gain control* (agc) provides for gain regulation of the receiver in inverse proportion to the strength of the received signal. The circuitry holds receiver output relatively constant despite large changes in the level of the incoming signal. In addition to control of gain, the agc circuit can also provide signal strength indication by means of an *S-meter*, whose reading is proportional to the agc control voltage.

Audio Circuitry   The communication receiver has no need to reproduce audio frequencies outside of the required communication passband. The high-frequency response of such a receiver is usually limited by the selective i-f passband. For voice reception, the lower audio frequencies are also attenuated in order to make speech crisp and clear. An audio passband of about 200 to 2000 Hz is all that is normally required for good SSB reception of speech. For c-w reception, the audio passband can be narrowed further by peaking the response to a frequency span ranging from 100 to 1000 Hz. High-Q audio filters may be used in the communication receiver to shape the audio response to the desired characteristic. In addition, audio or i-f filters may be added to either provide a special, narrow response characteristic, or a sharp rejection notch to eliminate heterodynes or objectionable interference.

Control   Under normal circumstances, the
Circuitry communication receiver is disabled during periods of transmission. A standby control may take the form of a switch or circuit that removes high voltage from certain tubes or transistors in the receiver. Alternatively, the bias level applied to the r-f and i-f stages may be substantially increased during standby periods to greatly reduce receiver gain. This will permit use of the receiver as a monitoring device during periods of transmission. In all cases, the input circuitry of the receiver must be protected from the relatively strong r-f field generated by the transmitter. Receiver control circuitry may be actuated by the transmitter control devices through the use of suitable interconnecting relay circuits (VOX), as discussed in Chapter 18 of this Handbook.

Receiver   Communications receivers are gen-
Power   erally designed to operate from a
Supplies   120- or 240-volt, 50- to 60-Hz power source, with the possible addition of auxiliary circuitry to permit operation from a 12-volt automotive electrical system. In some instances, voltage regulation circuits or devices are added to the supply to stabilize the voltages applied to critical oscillator circuits. In all instances, the primary circuit of a well designed communications receiver is fused to protect the equipment from overload and the complete receiver is designed and built to protect the operator from accidental shock.

10-4 The R-F Amplifier Stage

Since the necessary tuned circuits between the mixer stage and the antenna can be combined with solid-state devices or tubes to form r-f amplifier stages, the reduction of the effects of mixer noise and enhancement of the image ratio can be accomplished in the input section of the receiver. The tuned input stages, moreover, provide protection against unwanted signal response but, unfortunately, may increase the susceptibility of the receiver to cross-modulation, blocking, and desensitization because of the enhanced gain level of the received signals. In all cases, receiver gain (and particularly front-end gain) should be limited to that amount necessary to only override mixer noise. Excess receiver gain usually creates more problems than it solves.

If the r-f amplifier stage has its own tuning control, it is often known as a *preselector*. Some preselectors employ regeneration to boost signal gain and selectivity at the expense of the signal-to-noise ratio, which usually is degraded in such a circuit.

Generally speaking, atmospheric and man-made noises below about 30 MHz are so high that receiver sensitivity and signal-to-noise ratio is not a serious problem. Above 30 MHz or so, noise generated within the receiver is usually greater than the noise received on the antenna. Vhf and uhf r-f amplifiers will be discussed in Section II of this Chapter.

Experience has shown that about an 8-dB noise figure is adequate for weak-signal reception under most circumstances below 30

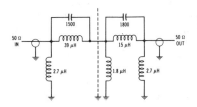

Figure 12

HIGH-PASS INPUT FILTER

This filter provides a rejection of greater than 60 dB below 1 MHz. It has an insertion loss of about 0.5 dB at 1.9 MHz (design by W6URH).

MHz. Interference immunity is very important below 30 MHz because of the widespread use of high-power transmitters and high-gain antennas and large-signal handling ability is usually more important to the hf communicator than is extreme weak-signal reception.

To minimize receiver overload from strong local signals, a variable attenuator such as the type shown in Chapter 9 may be placed in the receiver input circuit. The attenuation can be varied in 10-decibel steps and the unit is useful in dropping the signal level of strong, local transmitters.

A high-pass filter is shown in figure 12 which eliminates crossmodulation and intermodulation from local broadcast stations. Both of these devices provide good front-end protection from unwanted signals.

Image Rejection Image rejection ability of a receiver is a function of the selectivity response of the tuned circuits ahead of the mixer stage. The image rejection figure (or image response) is expressed in decibels by which the image signal is reduced below the fundamental response:

$$\text{Image rejection} = \frac{1}{\sqrt{1 + (Q_y)^2}}$$

where,

$$y = \frac{f_i}{f_o} - \frac{f_o}{f_i}$$

f_i is the image frequency,
f_o is the resonant frequency.

For example, if an a-m broadcast receiver has an i-f channel of 455 kHz and is tuned to 550 kHz by an input circuit having a Q of 60, the image rejection is:

$$\text{Image rejection} = \frac{1}{\sqrt{1 + \left(\frac{1460}{550} - \frac{550}{1460}\right)^2}} = \frac{1}{136}$$

or -42.5 dB

If an r-f stage having the same Q as the detector tuned circuit was added, the selectivity would be doubled and the image rejection would then be -85 dB.

These calculations are simplified by the use of the Universal Selectivity Curve shown in figure 13.

Greatly improved image rejection can be achieved by using an intermediate frequency which lies above the maximum tuning range of the receiver. The image frequency thus falls even higher and may easily be removed by a lowpass filter having a cutoff frequency below the frequency of the i-f amplifier. Selectivity is achieved by the use of crystal and ceramic filters in the i-f chain.

Small Signal R-F Amplifiers Typical common solid-state r-f amplifiers are shown in figure 14. A *common-base amplifier* is shown in illustration A. To overcome the possibility of oscillation at the higher frequencies, an external neutralizing circuit may be added, which consists of a neutralizing capacitor placed between the collector and the lower end of the input circuit, which is lifted above ground. If the external feedback circuit cancels both resistive and reactive changes in the input circuit due to voltage feedback, the amplifier is considered to be *unilateralized*. If only the reactive changes in the input circuit are cancelled, the amplifier is considered to be neutralized. Neutralization, then, is a special case of unilateralization. Modern silicon npn epitaxial planar type transistors are designed for vhf use up to 470 MHz and many have sufficiently low feedback capacitance so that neutralization is unnecessary.

The *common-emitter amplifier* (figure 14B) corresponds to the grounded-cathode vacuum-tube circuit and provides the highest power gain of common transistor circuitry. As the phase of the output signal is opposite to that of the driving signal, the feedback from output to input circuit is essentially negative.

Field-effect transistors may be used in *common-source, common-gate,* or *common-drain* configurations. The common-source arrangement (figure 14C) is most frequently used as it provides high input

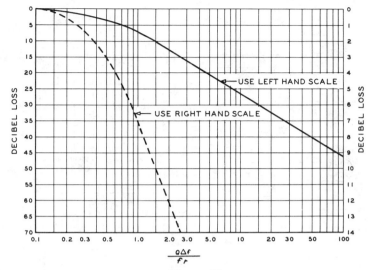

USE LEFT HAND SCALE

USE RIGHT HAND SCALE

$$\frac{Q \Delta f}{f_r}$$

Figure 13

UNIVERSAL SELECTIVITY CURVE

Image rejection capability may be determined with aid of universal curves. Selectivity required to adequately suppress the various spurious signals is provided by tuned input circuits. The number of circuits required depends upon Q, frequency, and attenuation desired. These curves are for a single tuned circuit.

impedance and medium-to-high output impedance. The first neutralized transistor drives the second connected in common-gate configuration which is used to transform from a low or medium input impedance to a high output impedance. The relatively low voltage gain of the second stage makes dual neutralization unnecessary in most cases. The two FET transistors are arranged in a cascode amplifier circuit, with the first stage inductively neutralized by coil L_N. FET amplifiers of this type have been used to provide low-noise reception at frequencies in excess of 500 MHz. A single gate MOSFET amplifier is shown in figure 14D.

A dual-gate diode-protected MOSFET r-f amplifier is shown in figure 15A. The signal input is coupled to gate 1 and the output signal is taken from the drain. Gain control is applied to gate 2 and a dc sensing current may be taken from the source to be applied to the S-meter circuit, if desired. With proper intrastage shielding, no neutralization of this circuit is required in the hf region.

An integrated circuit may be used as an r-f amplifier (figure 15B). It is connected as a differential amplifier and provides high gain, good stability and improved agc characteristic as compared to a bipolar device.

A dual-gate MOSFET device is shown in figure 15C and will be more fully discussed in the vhf section of this chapter.

The Cascode Amplifier The cascode amplifier consists of a grounded-emitter stage directly coupled to a grounded-base stage (figure 16). The bias level is set separately for each stage. When properly designed, no neutralization is required and stage gain is equivalent to that of a single grounded-emitter stage.

Vacuum-Tube R-F Amplifiers A typical hf vacuum-tube amplifier circuit is shown in figure 17. A high-gain pentode such as a 6BA6 or 6BZ6 may be used with the input circuit connected between grid and cathode. The output signal is taken from the plate circuit. Modern pentode tubes provide very high gain, combined with low grid-to-plate capacitance, and usually do not require neutralization. Remote-cut-

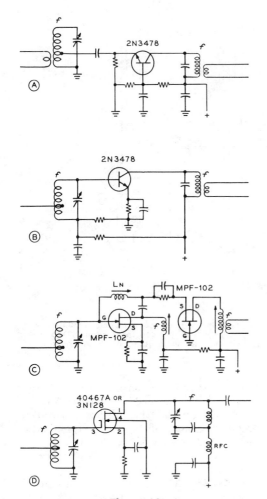

Figure 14

HIGH-FREQUENCY TRANSISTOR R-F STAGES

A—Common-base amplifier.
B—Common-emitter amplifier.
C—Cascode amplifier using FET transistors in cascode circuit.
D—Single-gate MOSFET amplifier.

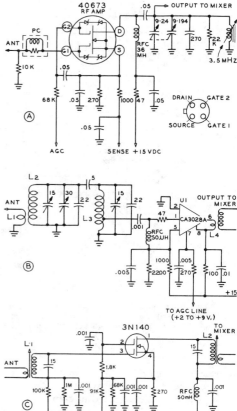

Figure 15

SOLID-STATE R-F AMPLIFIER STAGES

A—Dual-gate, diode-protected MOSFET amplifier. B—Integrated circuit differential amplifier with double-tuned input circuit. C—Dual-gate MOSFET amplifier.

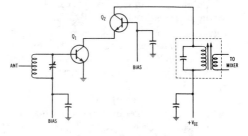

Figure 16

TYPICAL VHF CASCODE AMPLIFIER

off tubes are most often used in r-f amplifier stages because of their superior large-signal handling capability and their good agc characteristics.

Signal-Frequency Circuits   The signal-frequency tuned circuits in high-frequency superheterodyne receivers consist of coils of either the solenoid or uni-

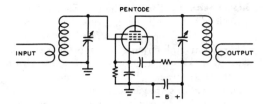

Figure 17

**TYPICAL PENTODE
R-F AMPLIFIER STAGE**

versal-wound (air or powdered-iron core) type shunted by variable capacitors. It is in these tuned circuits that the causes of success or failure of a receiver often lie. The universal-wound type coils usually are used at frequencies below 2000 kHz; above this frequency the single-layer solenoid type of coil is more satisfactory.

Impedance and Q The two factors of greatest significance in determining the gain-per-stage and selectivity, respectively, of a tuned amplifier are tuned-circuit impedance and tuned-circuit Q. Since the resistance of modern capacitors is low at ordinary frequencies, the resistance usually can be considered to be concentrated in the coil. The resistance to be considered in making Q determinations is the r-f resistance, not the dc resistance of the wire in the coil.

The latter ordinarily is low enough that it may be neglected. The increase in r-f resistance over dc resistance primarily is due to skin effect and is influenced by such factors as wire size and type, and the proximity of metallic objects or poor insulators, such as coil forms with high losses. Higher values of Q lead to better selectivity and increased r-f voltage across the tuned circuit. The increase in voltage is due to an increase in the circuit impedance with the higher values of Q.

Frequently it is possible to secure an increase in impedance in a resonant circuit (and consequently an increase in gain from an amplifier stage) by increasing the reactance through the use of larger coils and smaller tuning capacitors (higher LC ratio).

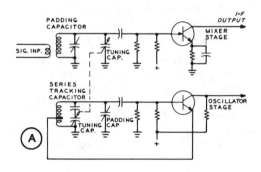

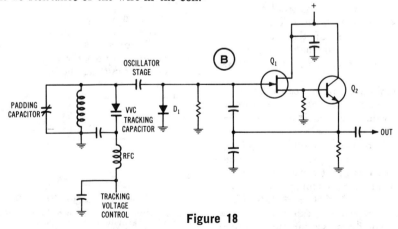

Figure 18

TRACKING SYSTEMS FOR SUPERHETERODYNE RECEIVER

A—Series tracking capacitor.
B—Varactor-tuned oscillator. Tracking is controlled electronically.

Tuning Range For a general coverage receiver the choice of oscillator frequency is a function of the range of the tuning capacitor:

$$\text{Tuning range} = \frac{C_{\max}}{C_{\min}} = \left(\frac{f_{\max}}{f_{\min}}\right)^2$$

The oscillator tuning range may either be less or more than the signal-frequency tuning range. When it is less, the relationship between the two ranges is:

$$\text{i-f} = f_s - f_o$$

When it is greater,

$$\text{i-f} = f_o - f_s$$

where,
i-f is the intermediate frequency,
f_o is the oscillator frequency,
f_s is the signal frequency.

In the case of an a-m broadcast receiver the signal-frequency range is 540 to 1600 kHz giving a frequency ratio of 1600/540 = 2.96. The tuning range, then, is $(2.96)^2 = 8.76$.

If the oscillator tuning range is higher than the signal-frequency range, then:

$$f_o = \text{i-f} + 455 \text{ to } f_o = \text{i-f} + 1600$$

For an i-f of 455 kHz, the oscillator tuning range is thus 995 to 2055 kHz.

The frequency ratio of the oscillator is 2055/995 = 2.07 and the tuning range for the oscillator tuning capacitor is $(2.07)^2 = 4.28$.

If the minimum circuit capacitance in the oscillator stage is 55 pF, the maximum capacitance value required is 55 × 4.28 = 235 pF.

The signal-frequency circuit frequency ratio is 2.96 and the tuning capacitor range is 8.76. If the minimum capacitance in the circuit is 40 pF, the maximum capacitance value required is 40 × 8.76 = 350 pF.

A dual 350-pF variable capacitor may be used with the oscillator section tuning range reduced by the addition of a series capacitance or a special two-gang capacitor having sections of 235 and 350 pF may be used.

Superheterodyne Tracking Because the tunable local oscillator in a superheterodyne operates "offset" from the other front-end circuits, it is often necessary to make special provisions to allow the oscillator to track when similar tuning capacitor sections are ganged. The usual method of obtaining good tracking is to operate the oscillator on the high-frequency side of the mixer and use a *series tracking capacitor* to retard the tuning rate of the oscillator. The oscillator tuning rate must be slower because it covers a smaller range than does the mixer when both are expressed as a percentage of frequency. At frequencies above 7000 kHz and with ordinary intermediate frequencies, the difference in percentage between the two tuning ranges is so small that it may be disregarded in receivers designed to cover only a small range, such as an amateur band.

A mixer- and oscillator-tuning arrangement in which a series tracking capacitor is provided is shown in figure 18A. The value of the tracking capacitor varies considerably with different intermediate frequencies and tuning ranges, capacitances as low as 100 pF being used at the lower tuning-range frequencies, and values up to .01 μF being used at the higher frequencies.

An electronic tracking system using a varactor (voltage-variable capacitor) is shown in figure 18B. Tracking voltage is controlled by a potentiometer ganged to the receiver tuning dial.

Superheterodyne receivers designed to cover only a single frequency range, such as the standard broadcast band, sometimes obtain tracking between the oscillator and the r-f circuits by cutting the variable plates of the oscillator tuning section to a different

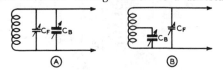

Figure 19

BANDSPREAD CIRCUITS

Parallel bandspread is shown in drawing (A). The tuning capacitor can be tapped on the coil to achieve bandspreading, as shown in (B). Padding capacitor is connected across the whole coil. Parallel bandspread is employed in inexpensive receivers to reduce complexity of bandswitch.

shape than those used to tune the r-f stage. In receivers using large tuning capacitors to cover the shortwave spectrum with a minimum of coils, tuning is likely to be quite difficult, owing to the large frequency range covered by a small rotation of the variable capacitors. To alleviate this condition, some method of slowing down the tuning rate, or *bandspreading*, must be used as shown in figure 19.

Stray Circuit Capacitance The stray circuit capacitance is that value of capacitance remaining in a circuit when all the tuning and padding capacitors are set at their minimum value.

Circuit stray capacitance can be attributed to the input capacitance of the device in the circuit and also to the capacitance to ground of the components and wiring of the stage. In well-designed receivers, every effort is made to reduce stray capacitance to a minimum since a large stray value reduces the tuning range available with a given coil and reduces the LC ratio of the circuit.

10-5 The Mixing Process

The *mixer*, or *frequency-converter*, stage of a superhet receiver translates the received signal to the intermediate frequency by means of a modulation process similar to that employed in transmitters (figure 20). The signal and local-oscillator voltages appearing in the output circuit of the mixer are rejected by selective circuits and only the

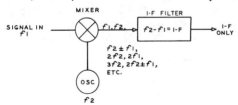

Figure 20

RECEIVER MIXER STAGE

Received signal is translated to intermediate frequency by the mixer stage. Signal and local-oscillator voltages and various mixer products are rejected by selective circuits in i-f amplifier and only the mixer product at the intermediate frequency is accepted.

mixer product at the intermediate frequency is accepted.

Any nonlinear circuit element will act as a mixer, with the injection frequencies and sum and difference frequencies appearing in the output circuit. Thus any diode, vacuum-tube or solid-state device may be used as a mixer.

The Diode Mixer Representative diode mixers are shown in figure 21. In illustration A, a single diode is used, the input signal being attenuated below the local oscillator signal by resistor R to provide low-distortion mixing action. A twin diode mixer is shown at B, the mixing signal being applied in parallel to the diode cathode while the input signal is applied in series with the diodes. Mixing produces a product of the signals, instead of sums and differences, and this circuit is termed a *product* mixer.

A double diode mixer is shown in figure 22. This circuit offers good isolation between the mixing signal and the output signal, attributable to the inherent circuit balance between the two input ports. No isolation exists between the signal port and the output (i-f) port and the circuit is not practical for vhf work as wiring capacitance, transformer winding capacitance and location of components can upset the circuit balance.

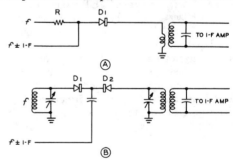

Figure 21

DIODE MIXER STAGES

A—Single diode mixer, B—Double diode mixer with input signal applied in series and mixing signal applied in parallel to diodes.

The Double-Balanced Mixer The double-balanced mixer has become a standard component in vhf and uhf communication systems and is now being used in hf single-sideband equipment because of its high degree of freedom from intermodulation, distortion and good isolation (figure 22). Balanced transformers and matched diodes are used for best results.

Mixer balance is achieved as shown in figure 23. If D_1 and D_2 and the local-oscillator transformer (T_1) are symmetrical, then voltage at point A is the same as at the transformer centertap (ground). In like manner, if D_3 and D_4 are symmetrical, voltage at point B is zero. Therefore there is no voltage across A or B and no voltage across the input or output ports. This illustrates how isolation is obtained between the three ports.

Looking at the circuit from the signal input terminal, if D_4 is equal to D_1 and D_2 equal to D_3, the voltage at point C will be equal to that at point D. Thus there is no voltage difference between C-D and no input signal will appear at the local-oscillator port. From symmetry it can be seen that the voltage at the output port is the same as the voltage at C, D, or zero; thus there is no input signal at the output port.

As with single- or double-diode mixers, unbalance and a subsequent drop in isolation will result from diode junction capacitance differences and transformer winding variations.

FET and MOSFET Mixers

Typical FET mixer circuits are shown in figure 24. These circuits are preferred over bipolar mixer circuits because the dynamic characteristics of bipolar transistors prevent them from handling high signal levels without severe intermodulation distortion. Illustration A shows a junction FET with signal and oscillator frequencies applied to the gate. Source injection is shown at B.

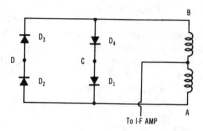

Figure 23

I-f leakage to signal and input ports is minimized in a double-balanced mixer by use of balanced transformers and matched diodes.

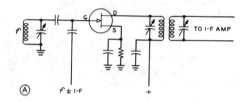

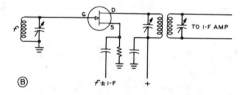

Figure 24

TYPICAL FET MIXER STAGES

A—Junction FET mixer with gate injection.
B—JFET mixer with source injection.

Both circuits can handle high input signal levels without overloading.

A dual gate MOSFET is shown in a typical mixer circuit in figure 25A. The unit shown has no internal chip protection and

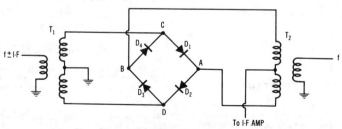

Figure 22

THE DOUBLE BALANCED MIXER

great care must be taken during installation to prevent the thin dielectric material of the gate from being punctured by static electricity. All leads should be shorted together until after the device is connected in the circuit. The MOSFET should be handled by its case and it should never be inserted or removed from a circuit when operating voltages are applied.

The dual gate MOSFET shown in illustration B has internal protection diodes that allow it to be handled with ordinary care.

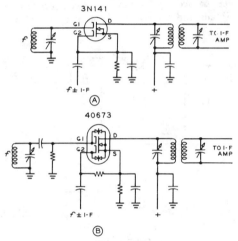

Figure 25

TYPICAL MOSFET MIXER STAGES

A—Dual-gate MOSFET mixer. B—Dual-gate MOS-FET with diode protection. Both circuits offer high conversion gain and relative immunity from cross modulation.

Both circuits offer high conversion gain, relative immunity from cross modulation, and do not load the local oscillator heavily.

A balanced mixer using JFETs is shown in figure 26. This mixer provides excellent immunity to intermodulation and cross-modulation effects while exhibiting a noise figure of about 8 dB at 150 MHz. A common-gate configuration is used, with the mixing oscillator coupled to the input circuit. Wideband ferrite-core transformers are used for good performance over the 50-MHz to 150-MHz range. A trifilar i-f output transformer is used to match the input impedance of the following stage.

Vacuum-Tube Mixers   Vacuum tubes have been used for decades as mixers and figure 27 illustrates one of the more common circuits. The *pentagrid converter* is shown. Tubes of this type are good conversion devices at medium frequencies, although their performance tends to drop off above 20 MHz or so.

Triodes and pentodes may also be used as mixer tubes, the mixing voltage being injected on the control grid, screen grid, or cathode.

10-6 The Mixing Oscillator

The exact frequency of reception of a superheterodyne receiver is controlled by the

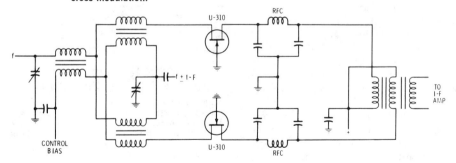

Figure 26

BALANCED MIXER USING FETs

JFET balanced mixer uses two devices operating in depletion mode. Mixer provides excellent immunity to cross modulation and overload.

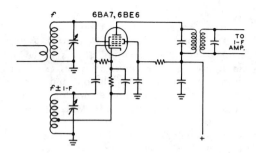

Figure 27

PENTAGRID MIXER STAGE

frequency of the mixing oscillator or oscillators. The overall stability of the receiver, moreover, is determined by the frequency stability of the oscillator. The frequency accuracy for SSB reception is rather precise when compared with most other communication systems. A frequency error of, say, 50 Hz in carrier reinsertion results in noticeable voice distortion, and intelligibility is impaired when the frequency error is 150 Hz or greater.

Oscillator stability should be relatively immune to mechanical shock and temperature rise of the receiver. A tunable oscillator should have good resetability and tuning should be smooth and accurate. Construction should be sturdy, with short, heavy interconnecting leads between components, that resist vibration. Variable capacitors should be mounted so that no strain exists on the bearings and the capacitors should be selected to have good, low-inductance wiping contacts that will resist aging.

The oscillator coil should be preferably wound on a ceramic form and the winding should be locked in position for maximum stability. Variable inductors with movable cores should be avoided if possible, because of possible movement of the core under vibration.

In case of double conversion receivers, one of the mixing oscillators is usually crystal-controlled. Information on crystal oscillators is given in chapter 11 of this handbook.

Solid-State   Transistor local-oscillator cir-
Oscillators   cuitry is employed in more modern SSB receivers. A bipolar circuit is shown in figure 28A. The base

element is near r-f ground potential and feedback is between the collector and the emitter. A JFET oscillator circuit (B) and a MOSFET circuit (C) are shown for comparison. The diode placed between gate and ground limits the level of gate bias to improve oscillator stability.

Because of the nonlinear change in the collector-base capacitance during oscillator operation, most transistor oscillators exhibit a high level of harmonic energy. A low-pass filter may be required after the oscillator to minimize spurious response in the receiver caused by mixing between unwanted signals and oscillator harmonics. In addit-

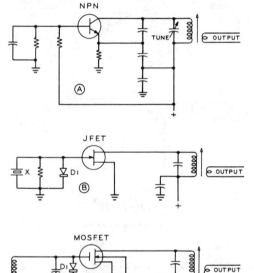

Figure 28

TYPICAL SOLID-STATE OSCILLATOR CIRCUITS

A—Bipolar transistor with emitter feedback from collector. B—JFET crystal oscillator. C—MOSFET oscillator. Diode D₁ between gate and ground limits level of gate bias to improve oscillator stability.

ion, one or more buffer stages may be required between oscillator and mixer to prevent the mixer from "pulling" the oscillator frequency when the strength of the incoming signal varies up and down.

Local-Oscillator Noise All oscillators produce phase noise but until recently this source of dynamic range reduction was overshadowed by the effects of intermodulation and overload. In modern receivers, however, the noise sidebands of the local oscillator must be reduced to avoid degrading strong signal performance of the receiver. Noise sidebands of a typical oscillator may be 80 dB below the intercept point but will nevertheless mix with strong signals out of the receiver passband to create enough noise power within the passband to mask weak signals. A very-low-noise oscillator circuit is shown in figure 29.

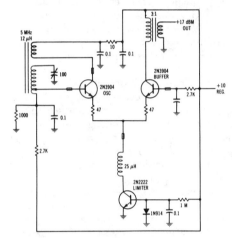

Figure 29

LOW-NOISE MIXING OSCILLATOR

(Design by DJ2LR)

The Frequency Synthesizer A higher order of accuracy of frequency control for both receiver and transmitter may be achieved by crystal control of the various conversion oscillators. Multiple-frequency operation, however, calls for an uneconomical and bulky number of crystals. These problems are solved by the use of a *frequency synthesizer* (figure 30). This is a device in which the harmonics and subharmonics of one or more oscillators are mixed to provide a multiplicity of output frequencies, all of which are harmonically related to a subharmonic of the master oscillator. A discussion of the frequency synthesizer is included in Chapter 12, "Frequency Synthesis."

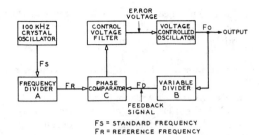

FS = STANDARD FREQUENCY
FR = REFERENCE FREQUENCY

Figure 30

FREQUENCY SYNTHESIZER

Subharmonics (F_R) of crystal oscillator are compared with divided signal (F_D) of voltage-controlled variable oscillator. Error signal corrects frequency of voltage-controlled oscillator.

10-7   The I-F Amplifier

The main voltage gain of a superhet receiver is achieved in the i-f amplifier stages. Intermediate-frequency amplifiers commonly employ bandpass circuits which can be arranged for any degree of selectivity, depending on the ultimate application of the amplifier. I-f amplifier circuitry is very similar to those circuits discussed for r-f amplifiers earlier in this chapter and the stage gain of the i-f chain may be controlled by an automatic gain control circuit actuated by the received signal.

Choice of Intermediate Frequency The intermediate frequency used is a compromise between high gain, good selectivity, and image rejection. The lower the frequency, the higher will be the gain and selectivity, and the lower the image rejection of the particular receiver. Conversely, the higher the i-f, the lower the gain and selectivity will be and the higher the image rejection. By traditional usage and international agreement, the most commonly used intermediate frequencies are 262 kHz, 455 kHz, and 1600 kHz for communication and entertainment receivers. Many sideband equipments make use of crystal-filter i-f systems in the 5-MHz to 9-MHz range and vhf equipment may have intermediate frequencies as high as 150 MHz. When a high value of i-f is employed, it is common technique to convert the signal a second time to a lower intermediate frequency in order to pick up gain and selectivity that cannot be economically achieved in the higher i-f.

I-F Transformers   Intermediate - frequency transformers commonly consist of two or more resonant circuits coupled together. The circuits are usually mounted in a metal shield. Either air-, or powdered-iron core windings may be used, the latter providing a higher Q and greater selectivity.

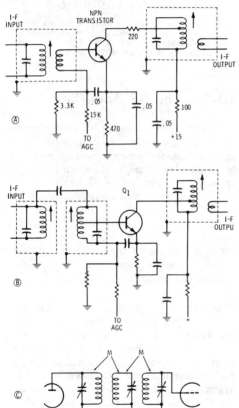

Figure 31

I-F COUPLED CIRCUITS

A—Bipolar transistor with low-impedance base drive.
B—Two tuned circuits capacitively coupled provide enhanced selectivity and better shape factor in transistor i-f stage.
C—Triple-tuned circuit provides high degree of selectivity. Center circuit acts as a sharply tuned coupler between input and output circuits.

The transformers are tuned by means of small parallel-connected capacitors, the capacitor being variable in some cases and in others the capacitors are fixed and the winding is tuned by varying the position of the slug core. Some representative examples are shown in figure 31. The circuit shown at A is the conventional i-f transformer, with inductive coupling provided between the windings. As the coupling is increased, the selectivity curve becomes broader and overcoupling the windings provides a flat-top response.

The windings of this type of i-f transformer, as well as most others used for low-frequency work, consist of small, flat, universal-wound pies mounted on either an insulated core, or on a powdered-iron core. The iron-core transformers generally have somewhat more gain and better selectivity than equivalent air-core units.

The circuit of illustration B utilizes capacitive coupling between the windings of separate transformers to improve selectivity. In some cases, three resonant circuits are used, as shown in illustration C. The energy is transferred from the input to the output winding by virtue of the mutual coupling to the center winding.

The selectivity of the i-f amplifier depends on the number of transformers used and the Q of the transformer windings. A single i-f stage operating at 455 kHz, for example, utilizing two transformers having two windings each could exhibit a response having a bandwidth of 3.5 kHz at the −6 dB points, and 16 kHz at the −50 dB points. Additional tuned circuits, of course, will sharpen the skirt selectivity of the amplifier, as discussed in the following section.

Shape Factor   It is obvious that to accept an SSB signal the i-f amplifier must pass not a single frequency but a band of frequencies. The width of this passband, usually 2 kHz to 3 kHz in a good communication receiver, is known as the *passband*, and is arbitrarily taken as the width between the two frequencies at which the response is attenuated 6 dB, or is "6 dB down." However, it is apparent that to discriminate against an interfering signal which is stronger than the desired signal, much more than 6 dB attenuation is required. The attenuation commonly chosen to indicate adequate discrimination against an interfering signal is 60 dB.

It is apparent that it is desirable to have the bandwidth at 60 dB down as narrow as possible, but it must be done without making the passband (6-dB points) too narrow for satisfactory reception of the desired signal. The figure of merit used to show the ratio of bandwidth at 6 dB down to that at 60 dB down is designated as *shape factor*. The ideal i-f curve (a rectangle), would have a shape factor of 1.0. The i-f shape factor in typical communications receivers runs from 2.0 to 5.5.

The most economical method of obtaining a low shape factor for a given number of tuned circuits is to employ them in pairs, adjusted to *critical coupling* (the value at which two resonance points just begin to become apparent). If this gives too sharp a *nose* or passband, then coils of lower Q should be employed, with the coupling maintained at the critical value. As the Q is lowered, closer coupling will be required for critical coupling.

Conversely if the passband is too broad, coils of higher Q should be employed, the coupling being maintained at critical. If the passband is made more narrow by using looser coupling instead of raising the Q and maintaining critical coupling, the shape factor will not be as good.

The *passband* will not be much narrower for several pairs of identical, critically coupled tuned circuits than for a single pair. However, the *shape factor* will be greatly improved as each additional pair is added, up to about 5 pairs, beyond which the improvement for each additional pair is not significant. The passband of a typical communication receiver is shown in figure 4.

The Crystal Filter A quartz crystal can be used as a selective filter in an i-f amplifier, providing a sharply peaked response that is suitable for c-w reception. Rejection of the audio image signal as high as 50 dB can be obtained with the proper circuitry. The skirt selectivity of a single crystal filter, however, is poor and the shape factor of the response is poor for SSB reception.

Bandpass Crystal Filters The sharply peaked response of the single crystal filter can be modified by the use of several crystals in a bandpass configuration. Typically, a good bandpass filter for SSB reception might have a passband of 2100 Hz at −6dB and 4000 Hz at −60 dB. Representative filters are shown in figure 32. A simple filter utilizing two crystals is shown in illustration A. The series resonance of the crystals differs by an amount equal to the desired bandwidth. To improve the shape factor of the passband, additional crystals may be added to the filter, as shown in B. Provided there is no leakage of signal around the filter, an extremely good shape factor can be achieved with this filter design at a center frequency in excess of 100 MHz. Vhf crystal filters, moreover, have been used in commercial communication systems.

A representative cascade crystal filter is shown in figure 33. Filter bandwidth is determined by the series-resonance frequencies of the crystals. The values of the input and output terminating resistances are determined by the make of the filter and the passband required. For SSB reception, the frequency spacing between the X_1-X_3 and X_2-X_4 pairs is about 2 kHz.

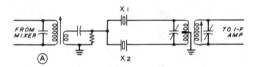

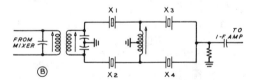

Figure 32

BANDPASS CRYSTAL FILTERS

A—Dual crystal filter. B—Multiple crystal filter improves passband response.

The Mechanical Filter The *mechanical filter* is an electromechanical bandpass device about a quarter the size of a cigarette package. As shown in figure 34, it consists of an input transducer, a resonant mechanical section comprised of

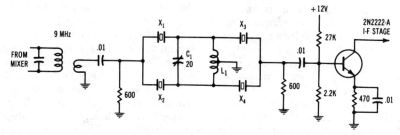

Figure 33

CASCADE CRYSTAL FILTER

Inductor L_1 is tuned to center frequency of filter by C_1.

a number of metal discs, and an output transducer.

The input and output transducers serve only as electrical-to-mechanical coupling devices and do not affect the selectivity characteristics which are determined by the metal discs. An electrical signal applied to the input terminals is converted into a mechanical vibration at the input transducer by means of *magnetostriction*. This mechanical vibration travels through the resonant mechanical section to the output transducer, where it is converted by magnetostriction to an electrical signal which appears at the output terminals.

In order to provide the most efficient electromechanical coupling, a small magnet in the mounting above each transducer applies a magnetic bias to the nickel transducer core. The electrical impulses then add to or subtract from this magnetic bias, causing vibration of the filter elements which corresponds to the exciting signal. There is no mechanical motion except for the imperceptible vibration of the metal discs.

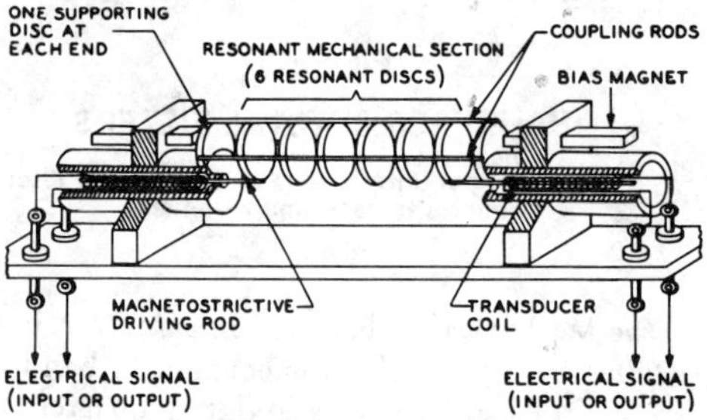

The frequency characteristics of the mechanical filter are permanent, and no adjustment is required or is possible. The filter is enclosed in a hermetically sealed case.

In order to realize full benefit from the mechanical filter's selectivity characteristics, it is necessary to provide shielding between the external input and output circuits, capable of reducing transfer of energy external to the filter by a minimum value of 100 dB. If the input circuit is allowed to couple energy into the output circuit external to the filter, the excellent skirt selectivity will deteriorate and the passband characteristics will be distorted (figure 35).

Diode Filter Switching   Two filters of different bandwidths are commonly used for SSB and c-w reception. Mechanical switching of such filters may lead to unwanted coupling between input and output, thus seriously degrading the shape factor of the filter. By using diode-controlled switching (figure 36), the switching components may be placed close to the filter terminals, thus offering a minimum of deterioration in isolation between ports. The diodes are triggered by a panel switch, and the appropriate diode pair places the desired filter in the i-f signal path. Operation of switch S_1 forward-biases a pair of diodes at a time and reverse-biases the other pair, allowing one filter to function at a time.

The Transfilter   A small mechanical resonator *(transfilter)* may be used in place of an i-f transformer in transistor i-f

Figure 34

MECHANICAL FILTER FUNCTIONAL DIAGRAM

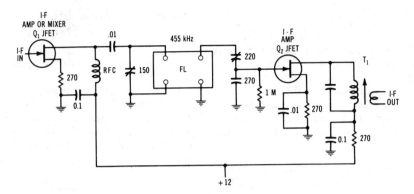

Figure 35

THE MECHANICAL FILTER IN AN I-F STRIP

circuits (figure 37A). A second transfilter resonator may be substituted for the conventional emitter bypass capacitor to enhance i-f selectivity. Transfilters may also be employed in the high-Q oscillator tuned circuits. The passband of a single transfilter i-f stage with emitter resonator is shown in figure 37B.

Bilateral Amplifier A *bilateral amplifier* is one that amplifies in two signal directions (figure 38). Such a stage is useful in SSB transceivers wherein r-f and i-f stages

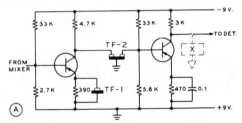

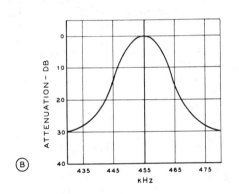

Figure 36

DIODE FILTER SWITCHING

Diode-controlled switching reduces unwanted coupling between input and output circuits of filters, thus preserving shape factor of the filter. Appropriate diode pairs are triggered by panel switch (S₁). One diode pair is forward-biased at a time, allowing proper filter to function.

Figure 37

MECHANICAL RESONATOR USED AS I-F FILTER

A—Transistorized i-f amplifier using Transfilters (TF-1, TF-2). Addition of second Transfilter (X) will sharpen selectivity. B—Passband of single Transfilter i-f stage with emitter resonator.

function in both receive and transmit modes. During the receive function, the bilateral amplifier passes the signal from the mixer to the balanced modulator and during transmit it passes the signal in the opposite direction —from the balanced modulator to the mixer. The same tuned circuits are used for both transmitting and receiving. The various injection oscillators operate continuously, supplying the local mixing signals to the proper mixer stages.

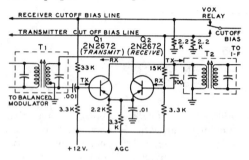

Figure 38

BILATERAL I-F AMPLIFIER FOR TRANSCEIVER

Bilateral i-f amplifier stage functions in both receive and transmit modes in SSB transceiver. Cutoff-bias lines transfer operation from transistor Q_1 to transistor Q_2 as VOX relay is actuated. Common-emitter stages are used with base-bias control.

In the circuit shown, the amplifier operates in the common-emitter configuration. In the receive mode, the 33K base-bias resistor is returned to the receiver cutoff-bias control line, disabling transistor Q_1. The 15K base-bias resistor of transistor Q_2 is returned to the transmitter bias-control circuit, which is at ground potential when the VOX relay is actuated. Thus, in the receive mode, a signal appearing at the receiver i-f transformer (T_2) will be amplified by transistor Q_2 and delivered to the i-f transformer (T_1). When the VOX circuit is activated to the transmit mode, the two bias-control lines are inverted in polarity so that transistor Q_2 is cut off and Q_1 is able to conduct. Therefore, a signal appearing at transformer T_1 is amplified by Q_1 and impressed on transformer T_2. Unilateral stages that are not required on either transmit or receive may be turned off by returning their base-bias resistors to an appropriate cutoff bias control line.

10-8   Solid-State I-F Strips

A very compact i-f/a-f strip can be built around modern ICs. The model shown is designed for SSB reception and utilizes a 9-MHz crystal lattice filter for selectivity, a product detector, and a local oscillator. Various high gain, linear ICs have been developed for i-f amplifier service, and a typical unit is the CA 3028A, shown in

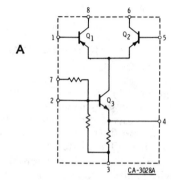

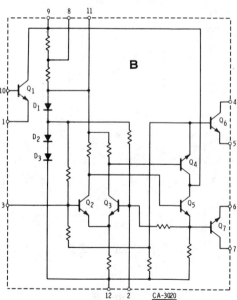

Figure 39

DIFFERENTIAL AND OPERATIONAL AMPLIFIERS

A—CA-3028A differential amplifier
B—CA-3020 operational amplifier

figure 39A. This device consists of a differential amplifier (Q_1, Q_2) with the common-emitter elements connected to the collector of Q_3. Because of the high impedance of Q_3, the sum of the emitter currents of Q_1 and Q_2 are practically independent of the operating points of Q_1 and Q_2. Transistor Q_3 is termed a *current sink*. The output of the IC device is a function of the difference between the input signals and, as such, functions as an amplifier. The 3028A can also serve as a limiter, product detector, frequency multiplier, and mixer.

The more complex 3020 integrated circuit (figure 39B) is used as an audio amplifier and agc control device. This is a high-gain, direct-coupled amplifier with cascaded stages, incorporating a separate output stage (Q_6, Q_7).

The circuit of the 9-MHz i-f amplifier chain is shown in figure 40. A 9-MHz crystal lattice filter (FL_1) is placed at the input of the amplifier to determine the overall selectivity. The input impedance of IC_1 together with the parallel-connected RC circuit form the load impedance for this particular filter. The signal is impressed on the base of transistor Q_3 in the CA 3028A device which, together with Q_2, forms a low-noise cascode amplifier. Transistor Q_1

is unused and connections 1 and 8 of IC_1 are unconnected.

Gain control voltage is fed to pin 7 of IC_1 and IC_2 to vary the base bias of transistor element Q_3. Maximum gain is achieved at maximum voltage ($+7$) and minimum gain at about $+1.7$ volts. This voltage range varies the gain of the two stages over a 45-dB range. Pin 4 of IC_1 is grounded through an RC network which permits a varying degree of negative feedback voltage to be applied to the emitter of Q_3 (figure 39). Potentiometer R_1 thus serves as a manual gain control, permitting adjustment of gain to achieve best overload characteristics.

The output signal of IC_1 is taken from the collector of Q_2 which is tapped on the interstate circuit at the proper impedance level to achieve good interstage selectivity. The second amplifier stage (IC_2) is essentially connected in the same manner as the first stage. The output circuit is an untuned r-f choke.

IC_3 forms the product detector and local oscillator. The signal is fed to the base of device Q_1 through a series of isolation circuit which prevents oscillator voltage from reaching IC_2. Device Q_3 serves as a Colpitts oscillator with crystal X_1 for emitter injection into the differential amplifier Q_1, Q_2.

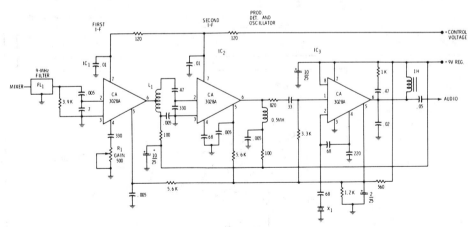

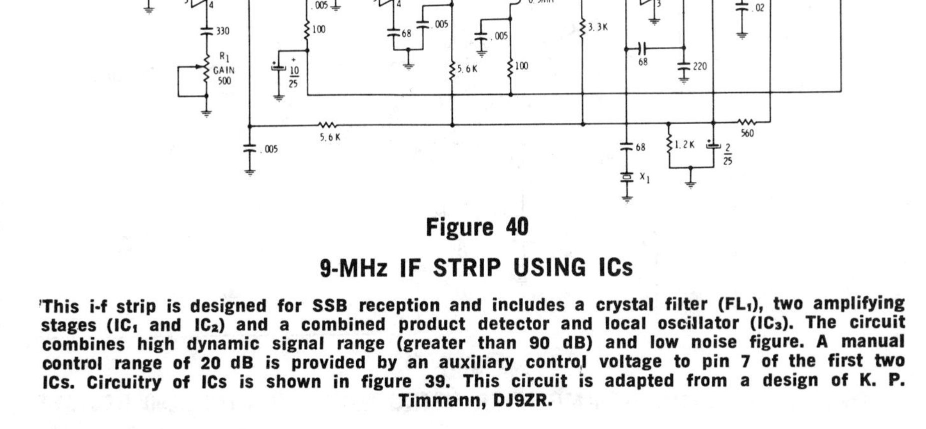

Figure 40

9-MHz IF STRIP USING ICs

This i-f strip is designed for SSB reception and includes a crystal filter (FL_1), two amplifying stages (IC_1 and IC_2) and a combined product detector and local oscillator (IC_3). The circuit combines high dynamic signal range (greater than 90 dB) and low noise figure. A manual control range of 20 dB is provided by an auxiliary control voltage to pin 7 of the first two ICs. Circuitry of ICs is shown in figure 39. This circuit is adapted from a design of K. P. Timmann, DJ9ZR.

The detected audio signal is taken from pin 6 of IC_3, with the higher frequency components filtered out by a series RC circuit. The collector to Q_2 is fed via a small audio choke from the +9-volt power line.

A 60-MHz I-F Strip Shown in figure 41 is a two stage i-f amplifier for 60 MHz. It provides a nominal gain of 80 dB and does not require neutralization. Only three tuning adjustments are required. Bandwidth is 1.5 MHz and sensitivity is 6 μV for a 10 dB signal-to-noise ratio.

The values of the resistors in series with the agc line (pin 2) were chosen so that gain reduction of the first stage is larger than that of the second to prevent overloading of the strip.

10-9 The Beat-Frequency Oscillator

The *beat-frequency oscillator (bfo)* or *carrier-injection oscillator* is a necessary ad-

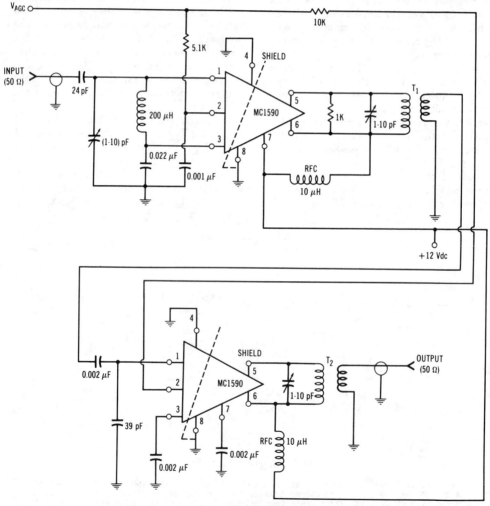

Figure 41

TWO-STAGE 60-MHz I-F AMPLIFIER USING MOTOROLA MC 1590 DEVICES

junct to the communication receiver for the reception of c-w or SSB signals.

The oscillator is coupled into or just ahead of the second detector circuit and supplies a signal of nearly the same frequency as that of the desired signal from the i-f amplifier. If the i-f amplifier is tuned to 455 kHz, for example, the bfo is tuned to approximately 454 or 456 kHz to produce an audible (1000-Hz) c-w beat note in the output of the second detector of the receiver. The carrier signal itself is, of course, inaudible. The bfo is not used for a-m reception, except as an aid in searching for weak stations.

Care must be taken with the bfo to prevent harmonics of the oscillator from being picked up at multiples of the bfo frequency. The complete bfo together with the coupling circuits to the second detector, should be thoroughly shielded to prevent pickup of the bfo harmonics by the input circuitry of the receiver. The local hf oscillator circuits shown in Section 10-6 may be used for beat-frequency oscillators, as can the various

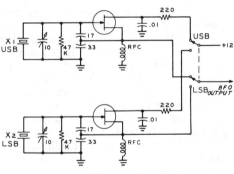

Figure 42

DUAL BFOs FOR USB AND LSB RECEPTION

oscillator circuits shown in the chapter "Generation of R-F Energy."

Many modern SSB receivers employ separate crystal-controlled beat-frequency oscillators to provide upper- and lower-sideband reception (figure 42). Dc switching is used in this particular circuit which is preferable to crystal switching. A buffer stage isolates the oscillators from the load, while increasing the bfo voltage to the proper level for the detector stage. The crystals are placed

on the correct frequencies by means of the trimming capacitors.

10-10 The Detector or Demodulator

Detection, or *demodulation,* is the process of recovering intelligence from a signal. The simplest and oldest detector is the diode, which makes use of the unilateral characteristics of a semiconductor or vacuum-tube device to produce an output voltage proportional to the modulation level (figure 43). The diode detector may be used directly on an a-m signal and can be used for SSB or c-w

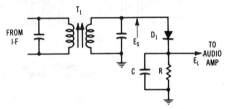

Figure 43

THE DIODE DETECTOR

RC is the load impedance across which the audio voltage is developed by the detector.

reception with a carrier oscillator. The diode loads the driving circuit and thus reduces the selectivity of the i-f system to a degree unless the transformer is designed for the low-impedance load. To minimize audio distortion on a-m signals having a high percentage of modulation, the capacitance across the diode load resistor should be as low as possible. The diode may be used as a single-ended or push-pull detector, the latter circuit having somewhat lower audio distortion than the former.

Action of a diode detector is illustrated in figure 44. When a signal is applied to the detector circuit, current flows only during that part of the cycle over which the diode conducts. The capacitor C is charged up to a potential that is almost equal to the peak of the signal voltage. Between signal peaks, a portion of the charge on capacitor C is discharged through resistor R, to be replenished by a new charge at the peak of the next signal cycle. The result is that the voltage de-

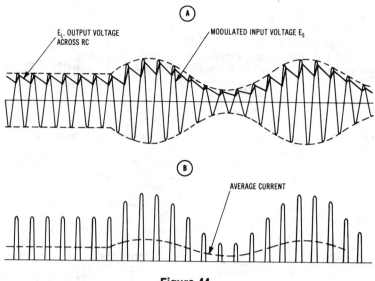

Figure 44

A—Load voltage follows modulating envelope.
B—Current flow through diode and load resistance.

veloped across the load resistor varies with respect to the modulation envelope, the value being smoothed out by capacitor C which acts as a filter for the r-f component of the signal output voltage.

SSB Demodulators   The *product detector* is a *linear demodulator* in which two signals are multiplied together to produce a resultant output audio signal. Product detectors are preferred over other detectors for SSB reception because they minimize intermodulation distortion products in the audio output signal and do not require excessively large local carrier voltage. A simple double-diode product detector is shown in figure 45A. This circuit has good large-signal handling capability and may be used with an inexpensive high impedance i-f transformer.

A diode ring demodulator is shown in figure 45B. This demodulator provides better low-signal response than the double-diode demodulator and provides a substantial degree of carrier cancellation. The i-f signal is applied to the ring demodulator in push-pull and the local carrier is applied in a par-

allel mode, where it is rejected by the push-pull output configuration.

A simple transistor sideband demodulator is shown in figure 45C. The transistor is heavily reverse-biased to a class-C condition and the two input signals are mixed in the base circuit. The audio product of mixing is taken from the collector circuit.

A source-follower product detector employing two JFETs is shown in figure 45D. Its vacuum-tube counterpart will be recognized in figure 46. The two gates provide high-impedance input for both the i-f signal and the carrier oscillator, while providing good isolation between the two signals. Both intermodulation distortion and conversion gain are low in this circuit.

A dual-gate MOSFET is used as a product detector in figure 45E. Various MOSFETs, designed for mixer applications, provide a wide dynamic operating range which permits them to handle large signal levels.

Good isolation between i-f signal and carrier signal may be obtained with simple vacuum-tube product detector circuits. A single triode product detector is shown in figure 45F. The tube is cathode-biased into the nonlinear operating region and the demodulated signal is taken from the plate

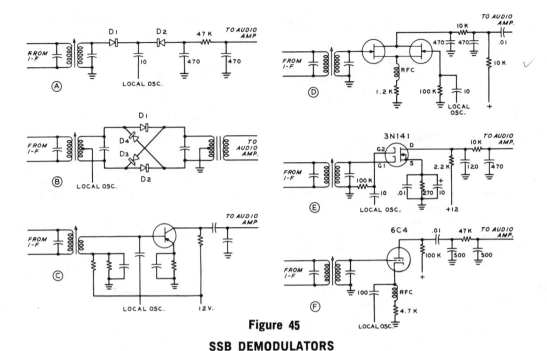

Figure 45

SSB DEMODULATORS

A—Double diode product detector. Simple RC filter is used in audio circuit to remove r-f products from output. B—Diode ring demodulator. C—Bipolar transistor demodulator. Input and local oscillator are mixed in base circuit. D—Source follower demodulator using two JFETs. E—Dual gate MOSFET product detector. F—Cathode-biased triode product detector.

circuit through a simple r-f network that filters out the unwanted r-f mixing products.

A dual-triode demodulator circuit (similar to the JFET circuit shown in figure 45D) provides excellent isolation and low intermodulation distortion (figure 46). The SSB signal from the i-f amplifier is applied to a cathode-follower stage that effectively isolates the signal source from the mixing circuit. The carrier signal is fed to the mixing tube and is amplified. The signals mix

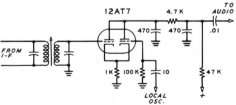

Figure 46

TYPICAL VACUUM-TUBE DEMODULATOR

Dual triode product detector provides low intermodulation distortion at high signal level.

within the tube and the product output is taken from the plate circuit of the mixer.

Sideband Detectors in General Any sideband modulator can be altered to become a demodulator by feeding in carrier and a sideband signal instead of a carrier and audio signal and changing appropriate r-f transformers to audio transformers. Generally speaking, the magnitude of the carrier signal should be from 10 to 20 times as strong as the sideband signal for lowest intermodulation distortion and highest signal overload capability. All signal components other than the desired audio signal must be filtered from the output section of the demodulator if good performance is to be achieved. Carrier injection level should be adjusted for minimum intermodulation distortion on large signals, however, care must be taken to prevent the carrier signal from reaching the i-f stages of the receiver by radiation and conduction along circuit wiring. Excessive carrier signal may also cause overloading or desensitization of the audio section of the receiver and also cripple the agc action. Stray coupling from the carrier oscillator to other portions of the receiver circuitry, then, must be carefully controlled.

10-11 Automatic Gain Control

Modern communication receivers include a control loop to automatically adjust the r-f and i-f gain level. The loop holds the receiver output substantially constant despite changes in input signal level. This system is termed *automatic gain control (agc)*. Conventional a-m automatic volume control systems are generally not usable for SSB since they operate on the level of the carrier, which is suppressed in SSB. A system must be used which obtains its information directly from the modulation envelope of the incoming signal. The control voltage derived from the agc detector is applied to a variable gain element in the receiver, usually in the r-f and i-f chain.

For optimum SSB reception, the control voltage must be applied rapidly to the variable element to avoid transient overload at the beginning portion of each word, otherwise an annoying *agc thump* will be appar-

ent at the start of the first syllable. As the syllabic envelope of the SSB signal is a replica of the original audio signal, the agc voltage must rise rapidly with the start of the syllable and then hold at a value corresponding to the average of the syllable undulations of the signal over an extended period of seconds. Too-rapid variations of the agc voltage with respect to syllabic peaks may bring up background noise in an objectionable manner termed *agc pumping*. The ideal agc action, then, exhibits a fast-attack, slow-decay time constant. Circuits having a charge time of 50 to 200 milliseconds and a discharge time of 0.5 to 3 seconds have proven successful.

The i-f signal may be used to control the agc system in a solid-state receiver, as shown in figure 47. An IC is used as an amplifier to provide gain and isolation. The resulting signal is rectified and further amplified by cascaded dc amplifiers Q_1 and Q_2. Transistor Q_1 is forward-biased by the agc voltage to provide a voltage drop across the collector load resistor. This voltage biases Q_2 more heavily in the forward direction when a large signal arrives and increases the voltage drop across the emitter resistor. This voltage varies in accord with the strength of the incoming signal and changes the bias voltage on various signal stages. The agc characteristic is determined by the agc time constant, R_1, R_2, C_1.

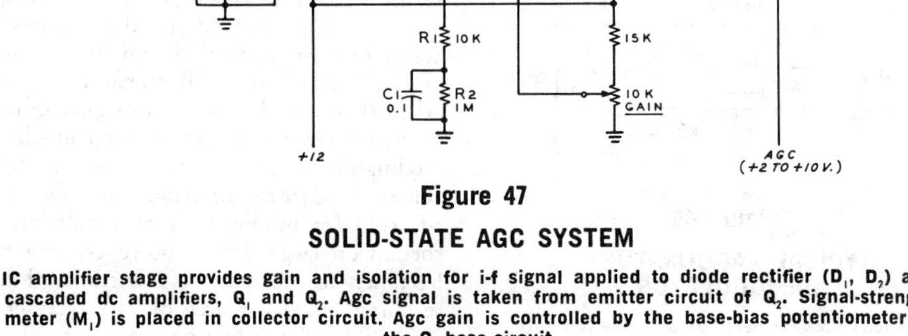

Figure 47

SOLID-STATE AGC SYSTEM

IC amplifier stage provides gain and isolation for i-f signal applied to diode rectifier (D_1, D_2) and cascaded dc amplifiers, Q_1 and Q_2. Agc signal is taken from emitter circuit of Q_2. Signal-strength meter (M_1) is placed in collector circuit. Agc gain is controlled by the base-bias potentiometer in the Q_1 base circuit.

An advanced agc system capable of handling a large dynamic signal range is shown in figure 48. This is a simplified schematic of the agc circuit of the *Heath SB-104* transceiver. An emitter-follower in the i-f circuit provides the driving power for the agc system. The i-f signal is sensed by a differential amplifier (Q_1 and Q_2). When the output level exceeds the threshold level, Q_1 conducts and pulls the base of Q_3 down on each signal peak and places positive pulses on the base of Q_4. This transistor is an integrator which converts the pulses to a dc voltage. It has separate time constants which set the agc attack and delay time constants for the agc system. Resistor R_1 and capacitor C_1 determine the attack time constant. Capacitor C_1 discharges either through resistor R_2 or resistors R_2 and R_3 in parallel depending on whether *fast* or *slow* agc is selected by switch S_1 to set the delay time constant. This voltage, whose level is a function of the i-f output level, is fed through the Darlington emitter follower (Q_5 and Q_6) where it is then applied to the integrated circuit i-f amplifier through diode D_3. The gain of the IC is thus controlled so that the output remains relatively constant for varying input levels.

Audio Derived AGC Since agc voltage follows the average SSB syllabic undulation of speech, it is possible to derive the agc voltage from the audio system of the receiver as shown in figure 49A. A portion of the audio signal is rectified and returned to the controlled stages after passing through a combination filter and delay network. Transistor Q_1 is operated without base bias so that no output is obtained until the input signal exceeds a critical peak level (0.6 volt), enough to turn on the transistor. Once this level is reached, very little additional voltage is needed to achieve full output from the agc rectifier. This results in a very flat agc characteristic.

A different audio-derived agc circuit is shown in figure 49B. A JFET serves as a source follower from the audio line, driving

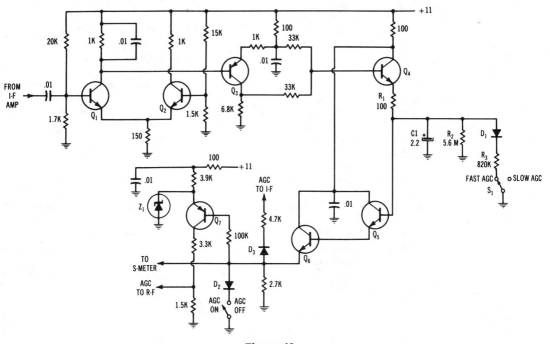

Figure 48

AN ADVANCED AGC SYSTEM WITH LARGE DYNAMIC SIGNAL RANGE CAPABILITY

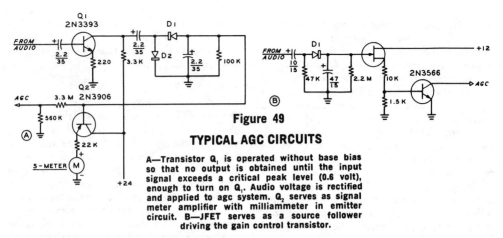

Figure 49

TYPICAL AGC CIRCUITS

A—Transistor Q₁ is operated without base bias so that no output is obtained until the input signal exceeds a critical peak level (0.6 volt), enough to turn on Q₁. Audio voltage is rectified and applied to agc system. Q₂ serves as signal meter amplifier with milliammeter in emitter circuit. B—JFET serves as a source follower driving the gain control transistor.

the gain control transistor (Q₁). The no-signal voltage at the base of Q₁ is about 0.4 volt, rising to about 0.55 volt before gain reduction starts.

A compact audio, agc and S-meter can be built using two CA 3020 integrated circuits and two transistors (figure 50). The audio signal from the product detector is fed to the base of Q₁ of IC₁ (see figure 40B). A peak limiter consisting of reverse-connected diodes D₆-D₇ is used as a peak suppressor, clipping all pulse-type interference peaks that are greater than the envelope of the audio signal. The emitter of device Q₁ (pin 1) is grounded to r-f by a parallel RC circuit while the audio signal is passed through a volume control and back into Q₂ of IC₁. The common emitter pair (Q₂, Q₃) deliver a push-pull, balanced signal to Q₄ and Q₅ which, in turn, drive the output devices, Q₆ and Q₇. IC₁ provides about ½-watt output into a 130 ohm load if a heat sink is used.

The agc control voltage is derived from the base of Q₁ of IC₁ (pin 1). Integrated circuit IC₂ provides an amplified voltage at the collector of device Q₆. The voltage is coupled to the control rectifier (D₄, D₅) which provides a positive voltage, and the rectifier network (D₂, D₃) which provides time-constant (negative) voltage. Transistor Q₁ is used as a time constant switch providing small time constants for the agc loop at low signal levels and large time constants at high signal levels. This compensates for the characteristics of IC₁ and IC₂ which have an effect on the control time constant.

The control voltage must be reversed in polarity for control of the i-f amplifier stages

and this is done by transistor Q₂. A voltage variation of 1.7 to 7.3 volts is available for control purposes. The same circuit provides control voltage for an S-meter.

10-12 The Signal Strength Indicator

Visual means of determining the relative strength of the received signal may be provided by a signal strength indicator, or S-meter. For a-m service, the S meter may consist of a high impedance dc voltmeter that registers the average agc control voltage (figure 51A). The collector current of the transistor rises as the negative agc voltage increases and this causes a corresponding rise in the meter reading.

For SSB service, a better approach to signal strength indication is by voltage amplification of the detected audio, as shown in figure 51B. The FET stage provides a high impedance to the detector and minimizes loading and distortion. The second stage (Q2) is a common-emitter voltage amplifier with a positive pulse rectifier for the meter. Full-scale meter reading is obtained with an audio signal of 60 mV (peak to peak). Meter damping is controlled by increasing or decreasing the shunt capacitor.

Calibration of the S-meter varies with the receiver gain and the actual reading, therefore, is only a relative indication of signal strength.

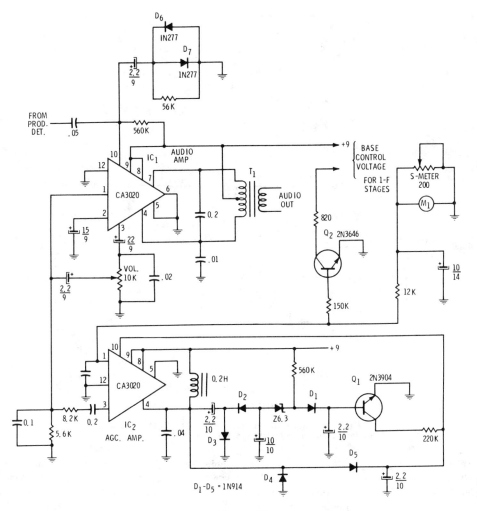

Figure 50

AUDIO AND AGC CIRCUITRY USING ICs

This compact audio and agc strip uses two ICs and two transistors. IC_1 serves as an audio amplifier, driving an external speaker via T_1. IC_2 serves as the agc amplifier and control stage. The dc control voltage is obtained from rectifiers (D_1-D_5). A variable time constant in the control voltage is achieved, whereby small time constants are obtained at low signal levels and a large time constant is achieved at a high signal level. Q_2 inverts the control voltage levels for correct polarity when applied to the i-f chain of figure 41. Refer to figure 40B for internal circuitry of IC_1, IC_2.

10-13 Impulse Noise Limiters

High-frequency reception is susceptible to interference from *impulse-type noise* generated by certain types of electronic equipment, ignition systems, switches, or like circuitry. Impulse noise, because of the short pulse duration, has a low value of energy per pulse and to cause appreciable interference, must have a peak amplitude appreciably greater than the received signal. Noise may be reduced or eliminated by reducing the receiver gain during the period of the noise pulse or by clipping the pulse to the amplitude of the received signal.

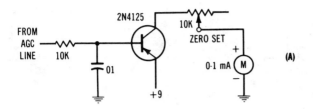

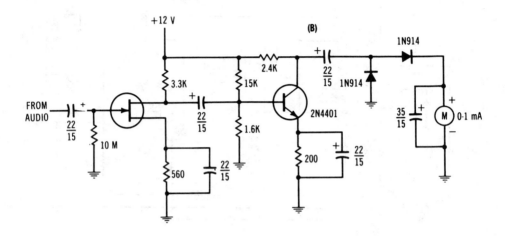

Figure 51

SIGNAL STRENGTH INDICATORS

A—High-impedance voltmeter measures average agc voltage.
B—Meter amplifier registers average value of detected audio.

If the receiver gain is reduced during the short duration of the pulse, a "hole" will be left in the signal. In some instances, the presence of the "hole" will degrade the intelligibility of the signal nearly as much as the original noise pulse. Practical *noise-blanker* circuits are able to silence the receiver without appreciably degrading signal intelligibility.

Noise reduction may be accomplished by amplitude limiting, wherein the r-f or a-f signal is clipped, or limited, at a level which substantially eliminates the noise pulse. Both blanking and limiting are most effective on short-duration noise pulses and, when the noise passes through the receiver tuned circuits, the pulse duration is increased because of the selectivity of the tuned circuits. Thus, the closer the noise reduction system is to the input of the receiver, the more effective the suppression will be.

Audio Noise Limiters Some of the simplest and most practical peak limiters for voice reception employ one or two diodes either as shunt or series limiters in the audio system of the receiver (figure 52). When a noise pulse exceeds a certain predetermined threshold value, the limiter diode acts either as a short or open circuit, depending on whether it is used in a shunt or series circuit. The threshold is made to occur at a level high enough that it will not clip modulation peaks enough to impair voice intelligibility, but low enough to limit the noise peaks effectively.

Because the action of the peak limiter is needed most on very weak signals, and these usually are not strong enough to produce proper avc action, a threshold setting that is correct for a strong voice signal is not correct for optimum limiting on very weak signals. For this reason the threshold control

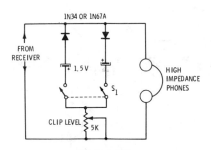

Figure 52

AUDIO NOISE LIMITER

When noise peaks exceed a predetermined voltage determined by the diode bias, the diodes conduct and shunt the noise peaks to ground. Clipping level may be increased by means of potentiometer.

is often tied in with the agc system so as to make the optimum threshold adjustment automatic instead of manual.

Suppression of impulse noise by means of an audio peak limiter is best accomplished at the very front end of the audio system (figure 53).

The amount of limiting that can be obtained is a function of the audio distortion that can be tolerated. Because excessive distortion will reduce the intelligibility as much as will background noise, the degree of limiting for which the circuit is designed has to be a compromise.

Peak noise limiters working at the second detector are much more effective when the

i-f bandwidth is broad, because a sharp i-f amplifier will lengthen the pulses by the time they reach the second detector, making the limiter less effective.

The I-F Noise Limiter   I-f noise limiting is more effective than audio limiting. A representative i-f noise limiter is shown in figure 54. This circuit clips the positive and negative noise peaks above a certain signal level, the capacitors in series with the diodes holding the diode bias constant for the duration of the heavy noise pulse. The time constant is determined by the value of the capacitors and the shunt resistance. The limiter is disabled by the 2N2222 control transistor.

10-14 The Noise Blanker

The noise blanker employs a blanking gate in the i-f system which silences the receiver in the presence of certain types of noise. Noise blanking is most effective on short duration, high amplitude, low repetition rate noise such as automobile ignition noise and make or break switching. It is less effective with long duration, high repetition noise such as lightning crashes, some power line noise and brush arcing on motors.

The reason the noise blanker is ineffective on low-amplitude, long-duration noise

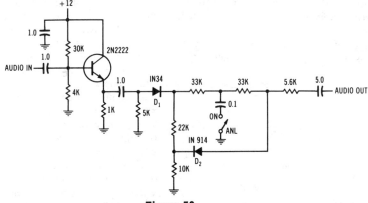

Figure 53

AUDIO NOISE LIMITER

Negative pulses disable diode D_1 and positive pulses cause diode D_2 to limit the positive pulses.

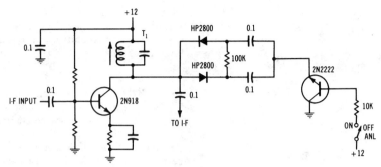

Figure 54

SELF-ADJUSTING I-F NOISE LIMITER

pulses is because the blanker operates by first sensing the pulse and then by silencing the receiver for the duration of the pulse (figure 55). Sensing requires discrimination between signal and noise and the greater the difference between the two, the easier discrimination becomes.

A simple noise blanker is shown in figure 56 which employs the signal to directly drive blanking diodes which "punch a hole" in the signal. Device Q_1 is a noise amplifier and diodes D_1 and D_2 serve as a noise pulse rectifier. The negative-going rectified noise voltage is amplified by Q_2 and applied to a switching transistor, Q_3. The switching level is set by the operator.

A more effective noise blanker is shown in figure 57. The incoming signal is split into two channels. The main signal channel passes through the noise gate and on to the rest of the receiver. In the noise channel, the noise and signal are amplified, and the noise impulses detected, with the detector output used to trigger a pulse generator.

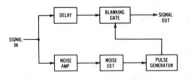

Figure 55

**BLOCK DIAGRAM OF
NOISE BLANKER**

Incoming signal is split into two paths. Delay is introduced into signal path to ensure that the two signals arrive at the blanking gate simultaneously.

The pulse generator forms a signal of proper amplitude and polarity to cut off the gate for the duration of the impulse.

The blanker is placed in the receiver at a point of low signal level and prior to the narrow bandwidth selectivity circuits. High-Q circuits will stretch the pulse unreasonably and degrade the capability of the blanker. In many receivers, the blanking is done immediately after the first mixer stage.

In this circuit Q_1 and its double-tuned drain circuit comprise a low gain bandpass amplifier that removes the remaining oscillator signal while setting the bandwidth at about 50 kHz. In the main channel, Q_2 drives the 50-ohm lowpass delay network. The network is a seven-pole Butterworth lowpass filter with a 700-kHz cutoff frequency. The phase shift is about 200 degrees at 500 kHz, therefore the delay is about 1.1 μS.

In the noise channel U_2 operates as a video amplifier driving Q_3, the pulse detector and CR_1, the agc detector. The agc time constant is set by C_1 and R_1 and is long enough to be unaffected by short noise pulses but will follow the average signal level. Operational amplifier U_1 amplifies the agc and controls the gain of U_2. This sets the point at which diode CR_1 conducts.

Detection takes place in Q_3 and the resulting positive pulses are applied to buffer U_3A which triggers the one-shot multivibrator and gates U_3B and U_3C. The remaining gates of U_3 develop the proper phase and current amplitude to operate the blanking gate.

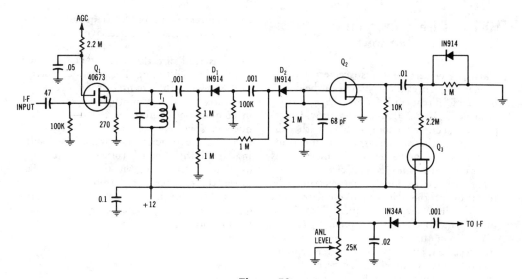

Figure 56

NOISE BLANKER FOR I-F SYSTEM

Noise blanker employs i-f signal to drive blanking diodes (D₁, D₂) which "punch hole" in signal.

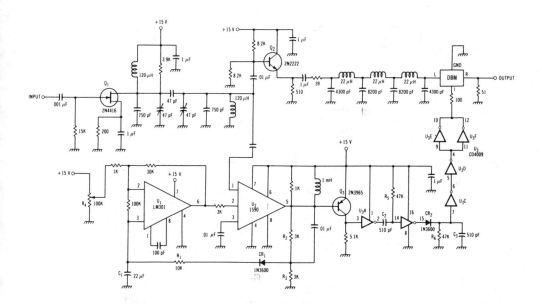

Figure 57

EFFECTIVE I-F NOISE BLANKER

CMOS IC is used as a retriggerable, one-shot multivibrator and a double-balanced modulator as a noise gate. Detection takes place in transistor Q₃ and resulting positive pulses are applied to buffer U₃A. Delay network is placed between Q₂ and double-balanced modulator. Blanker is inserted in i-f strip at point where bandwidth is about 50 kHz and signal level is a few millivolts. Design by K7CVT.

10-15 Direct Frequency Readout

Many receivers and transceivers have a frequency counter incorporated in the design to provide direct readout of the operating frequency. *Digital readout* can provide frequency accuracy comparable with the accuracy of the measuring clock, and readout to 100 Hz, or better, in the hf region is achievable with inexpensive circuitry.

The simplest readout device measures the frequency of the conversion oscillator and adds the intermediate frequency to it to obtain the operating frequency (figure 58A). A counter of this type is suitable for a-m reception, but for SSB or c-w reception, a more complex interface between the receiver or transceiver and the counter is required.

A representative counter for c-w and SSB reception is shown in figure 58B. For SSB, the counter monitors the frequency of the suppressed carrier of the received signal and for c-w, the frequency of the incoming signal is read directly without zero-beating or other special tuning. For a double-conversion receiver, the counter is connected to the three oscillators in the receiver which, in combination, determine the received frequency. The counter mixes the two hf oscillator frequencies, then mixes the resulting signal with the i-f (or beat-frequency) oscillator. Depending on the coupling between the counter and the bfo, the counter can either measure the actual tuning frequency, or the suppressed carrier frequency of an SSB signal.

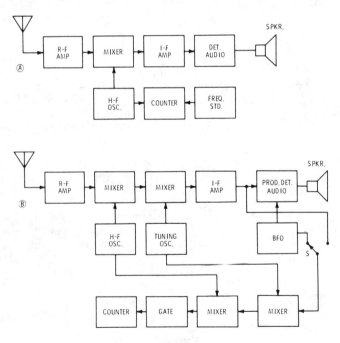

Figure 58

DIRECT FREQUENCY READOUT

A frequency counter can provide digital readout for a receiver, transmitter, or transceiver. A— Simple counter measures frequency of conversion oscillator and adds the intermediate frequency to show operating frequency. B—C-w and SSB counter adds frequency of all receiver oscillators. Counter can measure either zero-beat frequency, or carrier frequency.

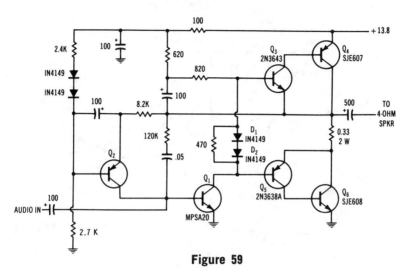

Figure 59

TWO-WATT AUDIO STAGE FOR RECEIVER

10-16 The Audio System

The audio system of many high-frequency receivers is of inferior quality. This is surprising in view of the inexpensive, high quality stereo equipment on the market. It would seem that some of this circuitry would be incorporated in the more sophisticated communication receivers. Such is not the case. Most receivers have a high level of audio distortion, marginal audio power capability and a midget, low quality speaker. The older tube-style receivers, while not having a "high fidelity" audio system in most instances still surpass most new receivers having an IC audio system.

It is possible to have good quality audio when using solid-state circuitry. Figure 59 is a representative circuit that will provide good quality audio at a power level of about 2 watts. The audio signal from the detector is fed to the base of transistor Q_1. This device drives a complementary output stage consisting of $Q_3 - Q_6$. The signal for the speaker is taken from the collector of Q_4 through a large coupling capacitor which removes the dc from the speaker. The frequency response of the amplifier is estab-

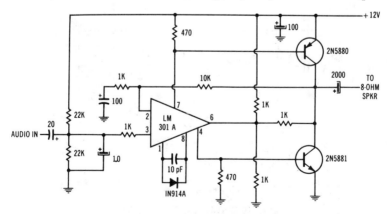

Figure 60

HIGH-QUALITY AUDIO AMPLIFIER FOR RECEIVER

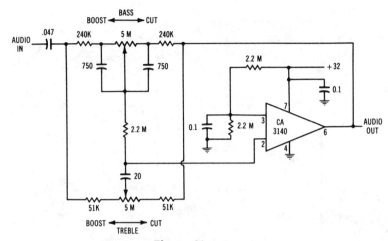

Figure 61

TONE CONTROL CIRCUIT

lished by a feedback circuit consisting of Q_2 and associated components.

A 3-watt audio system capable of quite low distortion is shown in figure 60. A complementary-symmetry output circuit is used, driven by a low cost optional amplifier.

The Audio Filter An audio filter can be used in the audio system of a receiver to improve the signal-to-noise ratio, to increase the intelligibility of a weak signal, or to tailor the audio response to meet listening conditions. While LC filters have been used extensively in this type of service, the modern *active filter* employing RC components has proven to give superior results. A simple tone-control circuit which can furnish up to 15 dB bass and treble boost or cut at 100 Hz and 10 kHz, respectively, is shown in figure 61. With controls set for flat frequency response, the circuit offers unity gain.

An active c-w filter peaked at 800 Hz is shown in figure 62. The bandwidth of this type of filter can be adjusted from a flat response to a peaked response of less than 10 Hz. In addition, it can be arranged to

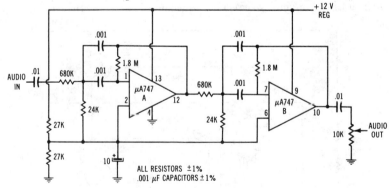

Figure 62

TWO-SECTION C-W FILTER WITH ADJUSTABLE BANDWITH

Bandwidth is set by choice of components in feedback circuit.

provide bandpass, highpass, or lowpass frequency response. Thus, this device is ideal for shaping the audio passband of a receiver for speech, RTTY or c-w reception. Filter response is determined by the feedback circuit placed around the active device. Additional discussion of feedback circuits for operational amplifiers is included in Chapter 4.

Part II—The VHF and UHF Receiver

Vhf and uhf receiver design and construction follows the same general philosophy discussed in the first part of Chapter 10 for hf receivers, but with important consequences dictated by the peculiarities of radio waves and propagation at frequencies above 30 MHz.

It should be remembered that at 50 MHz a half-wavelength is about 118 inches and at 420 MHz a half-wavelength is only 14 inches. At the latter frequency, a one-watt resistor is about .04 wavelength long and a radio tube of the audio output variety is nearly an eighth-wavelength high. Thus at vhf and uhf wavelengths, ordinary radio components approach the physical size of the radio wave and under these conditions unusual things begin to happen to radio parts that function well in the hf region. As a result, special small components designed for vhf/uhf operation are used, component layout is critical and unusual precautions must be taken to make sure that hidden circuit resonances do not alter the proper operation of the equipment.

The outstanding factor in vhf/uhf reception, as compared to reception at the lower frequencies is that the ultimate system sensitivity is primarily limited by equipment noise, rather than by noise external to the receiver. It is therefore possible to realize superior performance in terms of usable signal-to-noise ratio and sensitivity as opposed to an hf system, in which external atmospheric and manmade noise makes such receiver attributes relatively useless.

Vhf/uhf receivers are externally limited in sensitivity only by extraterrestrial (galactic) noise and some forms of man-made noise. Sophisticated receivers for this portion of the spectrum can reach the galactic noise level while rejecting manmade noise to a great degree. The state-of-the-art receiver noise figure is approximately as shown in figure 1.

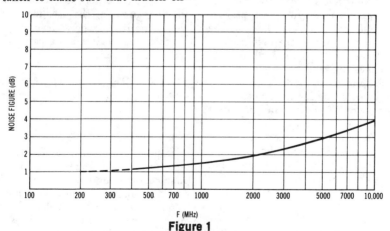

F (MHz)

Figure 1

REPRESENTATIVE RECEIVER
NOISE FIGURE

State-of-the-art receiver noise figure rises from about 1.2 dB at 450 MHz to near 4 dB at 10,000 MHz for specialized solid-state devices operating at room temperature.

10-17 VHF/UHF Noise Sources

External noise may be composed of atmospheric noise, galactic (cosmic) noise, and man-made noise as shown in figure 2. Above 30 MHz or so, external noise drops to a level that makes receiver noise of paramount importance. The development of low-noise vhf/uhf receivers is a continuing task as this portion of the spectrum becomes greater and greater importance to the modern world.

Atmospheric noise is due mainly to lightning discharges in the atmosphere which are propagated worldwide by ionospheric reflection. The noise varies inversely with frequency, being greatest at the lower frequencies and least at the higher frequencies. It also varies in intensity with time of day, weather, season of the year, and geographical location. It is particularly severe in the tropical areas of the world during the rainy seasons.

Galactic noise is caused by disturbances that originate outside the earth's atmosphere. The primary sources of such noise are the sun and a large number of "radio stars" distributed principally along the galactic plane. Galactic noise is largely blocked out by atmospheric noise at frequencies below approximately 20 MHz.

Man-made noise tends to decrease with increasing frequency, although it may peak at some discrete frequency, depending on the electrical characteristics of the noise source. It can be caused by electrical appliances of all types, television receivers, ignition systems, motors, and erratic radiation of high-frequency components from power lines. Propagation is by direct transmission over power lines and by radiation, induction, and occasionally by ionospheric reflection.

Thermal noise, or *Johnson noise*, is caused by the thermal agitation of electrons and pervades nature. It is only at absolute zero that such motion ceases. As the temperature of a conductor rises above absolute zero, the random motion of free electrons increases and this motion corresponds to a minute electric current flowing in the conductor. This "white noise" is generated over a wide band of frequencies and the portion of it falling within the passband of a receiving

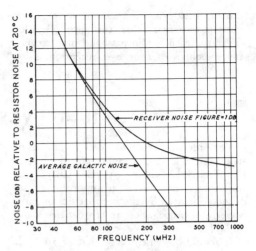

Figure 2

AVERAGE GALACTIC (COSMIC) NOISE LEVEL

Atmospheric noise predominates below 30 MHz. Galactic noise drops with increasing frequency, reaching low values at uhf. Receiver with 1-dB noise figure would have ultimate capability shown by top curve. Reduction of receiver noise figure becomes increasingly important for weak-signal reception above 100 MHz.

system will contribute to the noise output of the system. Limiting system bandwidth, therefore, will tend to limit the thermal noise. Thermal noise takes place in the receiving antenna, the feedline, and the receiver itself, the noise level of the input stage of the receiver being particularly critical as to system performance.

10-18 Receiver Noise Performance

Receiver noise figure was discussed in the first part of this chapter. The overall noise figure to a great degree is determined by the first r-f stage of the receiver.

Noise figure, or effective noise temperature, can be used to stipulate the excellence of a receiver. The noise figure is defined as the ratio of the total noise power available at the output of the receiver when the input termination is at 290°K (63°F) to that portion of the total available noise power produced by the input termination. The "perfect" receiver has, therefore, a noise figure of zero decibels and its noise contribution is zero.

The noise temperature is an apparent temperature that is representative of the internally generated noise of the receiver. It represents the number of degrees that the input noise power must be raised before the receiver noise output power reaches a new value representing the internally generated noise. The equivalent noise temperature is expressed in degrees Kelvin and is shown relative to the noise figure in figure 3.

Absolute noise figure measurement can be made using two resistive inputs for the receiver under test. One resistor is immersed in liquid nitrogen at 77.3°K and the other is in a temperature controlled oven at 373.1°K. The noise power in watts from either receiver is equal to *kBT*, where *T* is the temperature of the resistor in degrees Kelvin, *k* is Boltzmann's Constant, and B is the noise bandwidth of the receiver in hertz.

The quantities N_1 and N_2 are discussed in the first part of this chapter.

For an absolute comparison of noise figure between two receivers, both terms must be referenced to the standard temperature of 290°K. A representative test set-up is shown in figure 4.

Relative noise figures may be ascertained by direct receiver measurement with a noise generator. The receiver input is terminated with a resistor and wideband random noise is injected into the input circuit. The power output of the receiver is measured with no noise input and the generator output is then increased until the receiver noise output is doubled. The relative noise figure of the receiver is a function of these two levels, and may be computed from these measurements.

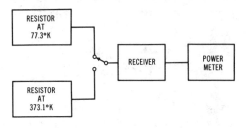

Figure 4

NOISE FIGURE MEASUREMENT

Receiver is switched between "cold" and "hot" noise sources.

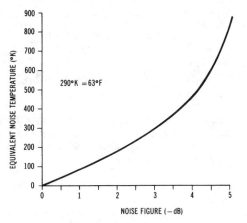

Figure 3

EQUIVALENT NOISE TEMPERATURE EXPRESSED AS NOISE FIGURE

The noise figure is:

$$F = \frac{(T_2/T_0 - 1) - Y\,(T_1/T_0 - 1)}{(Y - 1)}$$

where,

F equals noise figure
T_0 equals 290°K
T_1 equals 77.3°K
T_2 equals 373.1°K
Y equals N_2/N_1

Vacuum Tubes in VHF/UHF Receivers   The vacuum tube has been eclipsed for low-noise reception above 30 MHz by solid-state devices. Because of the hot filament within the tube, thermal agitation and noise level are excessive for weak-signal reception. Vacuum-tube noise is composed of *shot noise* (electron noise), *partition noise* (noise caused by a random division of space current between the elements of the tube), and *induced grid noise* caused by fluctuations in cathode current passing the grid element. The summation of these noises is expressed as the *equivalent noise resistance* of the vacuum tube. In addition to noise, most vacuum tubes have comparatively high input and output capacitances and a low input impedance, all of which inhibit the design of high-Q, high-impedance tuned circuits above 50 MHz or so.

Semiconductors in VHF/UHF Receivers Great advances have been made in recent years in both bipolar and field-effect devices and these improved units have pre-empted the vacuum tube in vhf/uhf operation in low-noise receiver circuitry. While the bipolar transistor exhibits circuit loading due to low input impedance and often has characteristics that vary widely with temperature, these problems are being overcome by new design and production techniques. The field-effect device, on the other hand, exhibits an input impedance equal to, or better, than vacuum tubes in the vhf/uhf region.

The better solid-state devices are superior to vacuum tubes as far as good noise factor is concerned and noise figures of 2 dB or better are possible up to 2000 MHz or so with selected transistors and field-effect devices.

10-19 VHF Receiver Circuitry

Vhf r-f receiver circuitry resembles the configurations discussed for hf receivers to a great degree. Solid-state r-f circuits specifically designed for efficient vhf operation are discussed in this section and they may be compared against the circuitry shown earlier in this chapter.

The common-base (or gate) r-f amplifier circuit (figure 5) is often used with bipolar devices in the vhf range since it is stable and requires no neutralization. Either PNP or NPN transistors may be used, with due attention paid to supply polarity. The input signal is fed to the emitter (source); the base (gate) is at r-f ground potential; and the output signal is taken from the collector (drain) circuit. Stage gain is low and two or more stages are often cascaded to provide sufficient signal level to overcome mixer noise. The input impedance of the common-base circuit is low and this configuration does not offer as much r-f selectivity as does the common-emitter (source) circuit of figure 6. This circuit often requires neutralization, accomplished by feeding energy back from the output to the input circuit in proper amplitude and phase so as to cancel the effects of spurious signal feedthrough in

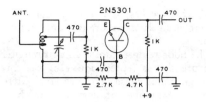

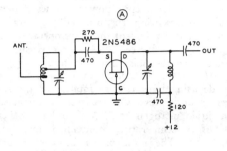

Figure 5

COMMON-BASE (GATE) R-F AMPLIFIER

Input signal is applied to emitter (A) or source (B) and output signal is taken from collector (A) or drain (B). Stage gain and input impedance are both low in this configuration.

and around the device. Tuning and neutralization are interlocking adjustments.

The cascode amplifier (figure 7) is a series-connected, ground-emitter (source), grounded-base (gate) circuit. Neutralization, while not always necessary, may be employed to achieve lowest noise figure.

A neutralized, IGFET vhf amplifier stage is shown in figure 8A. Protective diodes D_1 and D_2 (discussed in the next section) are used in the input circuit. A dual-gate, diode-protected MOSFET is employed in the amplifier circuit of figure 8B. Input and output points are tapped down the tuned circuits to reduce stage gain and to remove the necessity for neutralization, which otherwise may be necessary.

Special vacuum tubes, such as high-gain TV pentodes and low-noise triodes may be used in these typical vhf circuits and are often used in simple converters designed for 6 and 2 meters.

To optimize the noise figure of all of these circuits, the input coupling, bias level, and neutralizing adjustment (if any) are made with a weak signal source used for

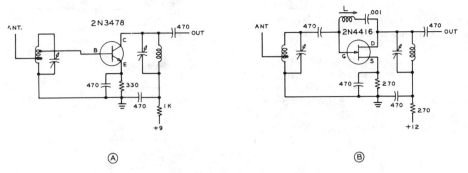

Figure 6

COMMON-EMITTER (SOURCE) R-F AMPLIFIER

Input signal is applied to base (A) or gate (B) and output signal is taken from collector (A) or drain (B). Stage gain is high and neutralization is often required to cancel signal feedthrough, as shown in (B).

alignment. Adjustment is not complicated provided proper vhf construction techniques and shielding are used in construction of the amplifier.

Amplifier Protection Vhf solid-state devices are vulnerable to burnout by accidental application of high input signal voltage to the receiver. Reverse-connected diodes (either silicon or germanium) placed across the input circuit will limit maximum signal voltage to a few tenths of a volt, providing automatic protection against damaging overload. In particular, the protection

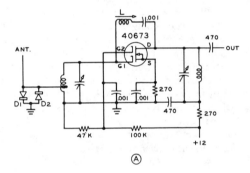

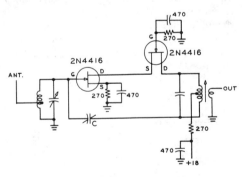

Figure 7

CASCODE R-F AMPLIFIER

Two FET devices are series-connected, the first being driven at the gate and the second at the source. Bipolar transistors or tubes are used in a similar arrangement. Neutralization is required to achieve highest overall gain and optimum noise figure.

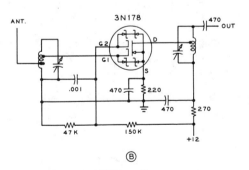

Figure 8

FETs IN VHF CIRCUITRY

A—Neutralized IGFET using 1N100 protective diodes in input (gate) circuit. B—Dual-gate, self-protected MOSFET circuit. Neutralization may be required for maximum stage gain and optimum noise figure.

diodes will absorb r-f energy that leaks around an antenna changeover relay, or that is received from a nearby transmitter.

The amplifier must also be protected from off-frequency energy, particularly that radiated by high power f-m and television transmitters. A compact cavity resonator is often incorporated in the r-f stage to reject offending signals (figure 9). Seriesed resonators can provide a front-end bandwidth of less than 200 kHz in the vhf spectrum.

VHF/UHF Mixers Diodes and bipolar transistors may be used as mixers in the vhf/uhf spectrum. Use of the transistor is limited because it lacks the large signal capability offered by other devices.

Various diodes are available for use as mixers and the *hot-carrier diode* serves as a low noise mixer for applications up to and including the uhf region (figure 10). This device (also known as a *Schottky-barrier* diode) is a planar version of a conventional point-contact microwave mixer diode. The hot-carrier diode has closely matched transfer characteristics from unit to unit and a high front-to-back ratio. In addition, it provides extremely fast switching speed combined with low internal noise figure. Input and output impedances are low, but overall conversion efficiency is high.

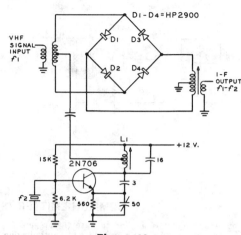

Figure 10
HOT-CARRIER DIODE MIXER

Schottky-barrier diode is a planar version of a conventional point-contact microwave mixer diode having closely matched transfer characteristics from unit to unit and high front-to-back ratio. It provides extremely fast switching time combined with low internal noise figure.

A portion of the receiver noise originates in the local oscillator and if this noise were eliminated, there would be a reduction in the overall noise figure of the receiver. The balanced mixer of figure 10 balances out local oscillator noise products, thus reducing receiver noise from this source.

The FET or MOSFET devices make good mixers in the vhf region. A popular mixer circuit using source injection to a JFET is shown in figure 11. It provides good isolation between the input circuit and the local oscillator, thus reducing unwanted oscillator radiation.

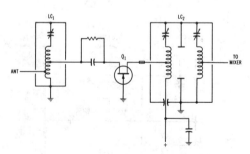

Figure 9
HELICAL RESONATORS USED IN VHF R-F STAGE

Two helical resonators are employed to provide front-end selectivity for a vhf receiver. The resonator consists of a high-Q tuned circuit mounted in a low-loss cavity. The output resonator (LC₂) is a two-pole filter. A slot or hole between the cavity sections provides coupling.

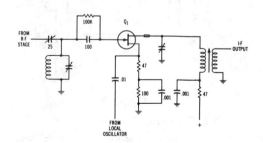

Figure 11
JFET MIXER FOR VHF SERVICE

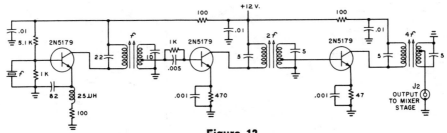

Figure 12

LOCAL OSCILLATOR "STRING" FOR VHF RECEIVER

Multiple tuned high-Q circuits between stages prevent unwanted harmonics of oscillator from reaching the mixer stage. Fundamental oscillator signal and 3rd and 5th harmonics could produce spurious responses in receiver unless suitably attenuated.

10-20   I-F Strips and Conversion Oscillators

To combine good image rejection with a high order of selectivity, double frequency conversion is normally used for vhf/uhf small-signal reception. The first intermediate frequency is usually rather high to provide adequate rejection of image signals and the second is low to provide good selectivity. Care must be used in choosing the first intermediate frequency or image problems will arise from signals in the 80- to 130-MHz range, which includes high power f-m transmitters and strong aircraft signals.

It is common practice to construct the r-f amplifier and first conversion circuits in a separate *converter* unit, the i-f output of which is fed into an hf communications receiver which serves as the low-frequency i-f strip. Choice of the first i-f channel is important, since many vhf/uhf converters provide scant selectivity at the received frequency, having bandwidths measured in hundreds of MHz. If the image ratio is unity, the image signal may be as strong as the wanted signal and the noise figure of the receiving system is degraded by 3 decibels, regardless of the noise figure of the converter. The first i-f channel, and the r-f selectivity of the converter should therefore be sufficiently high so that images are not a problem. Generally speaking a first i-f channel of 15 MHz to 30 MHz is suitable for 144-MHz and 220-MHz reception and a frequency in the region of 144 MHz is

often used as the first i-f channel for 432-MHz (and higher) reception.

In addition to attention to image problems, care must be taken to ensure that the harmonics of the local oscillator of the communications receiver used for the i-f strip do not fall within the input passband of the converter. Attention should also be given to the input circuit shielding of the communications receiver to prevent breakthrough of strong hf signals falling within the first i-f passband. Unwanted hf signals may also enter the receiver via the speaker wires or the power cord.

Spurious signals and unwanted "birdies" can be reduced to a minimum by using the highest practical injection frequency for the local oscillator. Most first conversion oscillators in vhf receiving systems are crystal controlled and high-overtone crystals are to be preferred as contrasted to lower-frequency

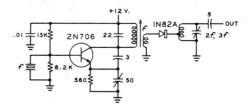

Figure 13

DIODE MULTIPLIER FOR LOCAL OSCILLATOR INJECTION AT A HIGH HARMONIC

One or more tuned circuits or traps are used after diode multiplier to attenuate unwanted harmonics of local oscillator.

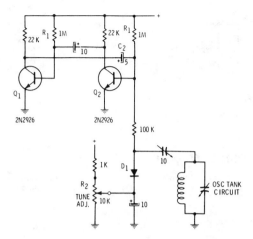

Figure 14
MULTIVIBRATOR SCANNING CIRCUIT

Sawtooth waveform from multivibrator (Q_1, Q_2) sweeps oscillator across band. The scanning rate is determined by multivibrator constants and sweep limit is set by potentiometer R_2.

crystals and a multiplier string. Unwanted harmonics generated by a multiplier string must be prevented from reaching the mixer stage by means of a high-Q trap circuit in order to avoid unwanted mixing action between received signals and the various harmonics.

When low-frequency conversion crystals are employed, the use of multiple tuned intermediate circuits in the multiplier string

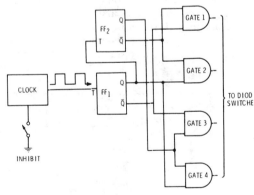

Figure 15
FOUR CHANNEL SCANNER

Scanner uses two J-K flip-flops (FF₁, FF₂) and four two-input gates which sequentially select one of four crystals. Sequence of operation is shown in figure 16.

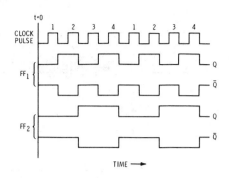

Figure 16
FLIP-FLOP WAVEFORM TO ACTIVATE DIODE SWITCHES

NAND gates produce grounded output (logic zero) when both inputs are high (logic 1). During first clock pulse, Q of FF₁ and FF₂ are high and drive gate 1 of figure 17. When clock pulse 2 arrives, Q of FF₁ and FF₂ are high and drive gate 2. This sequence continues through all four pulses of the clock, then repeats.

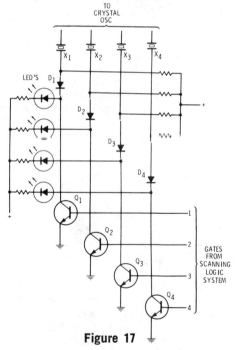

Figure 17
DIODE SWITCH SELECTS CONVERSION CRYSTALS

When the scanning logic selects a channel, a bias voltage saturates one transistor (Q_1-Q_4), resulting in a collector-emitter resistance of a few ohms. One of the diodes (D_1-D_4) is forward biased, grounding the crystal and illuminating LED indicator.

is suggested, as shown in figure 12. A simple diode multiplier may also be used in place of a tube or transistor, as shown in figure 13.

10-21   Band Scanning Receivers

Monitor (*scanning*) receivers are capable of searching many vhf channels for activity. The receiver sequentially looks at preset channels and a signal on one channel will increase the agc voltage of the receiver, causing the scanner to stop seeking and lock onto the signal.

The simplest form of scanning receiver continually sweeps a band of frequencies and the receiver is manually locked on a received signal by the operator (figure 14). A multivibrator circuit sweeps the oscillator by means of a varactor diode (D_1). Transistors Q_1 and Q_2 form the multivibrator, providing a sawtooth waveform in the base circuit. This voltage is applied to the varactor diode which sweeps the frequency of the variable oscillator. The scanning rate is determined by the values of the base resistors (R_1) and capacitor C_2.

For crystal-control service, wherein the channels are preselected by the choice of crystals, the scanning receiver selects the proper conversion crystal and also squelches the receiver between channels. A unijunction transistor is used as the timing clock supplying a series of sawtooth pulses to the pulse-shaping circuits and logic scanning circuits. An "inhibit" control circuit interrupts, on command, the series of pulses to the decade counter.

The binary coded decimal output from the counter is fed to a decoder which selects one of several output lines each time an input pulse is received. Shown in figure 15 is a typical four-channel scanner using two J-K flip-flops and four two-input gates which sequentially selects from among four crystals.

The sequence of operation is illustrated in figure 16. The NAND gates are connected to the flip-flops so that they produce a grounded output (logic zero) only when both inputs are high (logic 1), as shown in the waveforms from FF_1 and FF_2 underneath clock impulse 1. At this time only the Q outputs of FF_1 and FF_2 are high so they are used to drive gate 1. When clock pulse 2 arrives, the Q outputs of FF_1 and FF_2 are high while all others are low. They are used to drive gate 2. This sequence continues through all four pulses of the clock, then repeats.

More complex scanning receivers scan up to 8 or 16 channels. This is accomplished by dividing the crystals into two groups, which are scanned alternatively. An additional flip-flop sequentially selects these groups in an odd-even select system.

CHAPTER ELEVEN

Generation and Amplification of Radio-Frequency Energy

PART I
SOLID-STATE HF CIRCUITS

A radio communication or broadcast transmitter consists of a source of radio frequency power, or *carrier;* a system for *modulating* the carrier whereby voice or telegraph keying or other modulation is superimposed upon it; and an antenna system, including feedline, for *radiating* the intelligence-carrying radio-frequency power. The power supply employed to convert primary power to the various voltages required by the r-f and modulator portions of the transmitter may also be considered part of the transmitter.

Modulation usually is accomplished by varying either the amplitude or the frequency of the radio-frequency carrier in accord with the components of intelligence to be transmitted or by generation of an SSB signal (a form of amplitude modulation).

Radiotelegraph keying normally is accomplished either by interrupting, shifting the frequency of, or superimposing an audio tone on the radio-frequency carrier in accordance with the intelligence to be transmitted.

The complexity of the radio-frequency generating portion of the transmitter is dependent on the power, order of stability, and frequency desired. An oscillator feeding an antenna directly is the simplest form of radio-frequency generator. A modern high-frequency transmitter, on the other hand, is a very complex generator. Such equipment comprises a very stable crystal-controlled or synthesized oscillator to stabilize the output frequency, a series of frequency multi-pliers, or mixers, one or more amplifier stages to increase the power up to the level which is desired for feeding the antenna system, and a filter system for keeping the harmonic energy generated in the transmitter from being fed to the antenna system.

11-1   Self-Controlled Oscillators

The amplifying properties of a three- (or more) element vacuum tube, a bipolar transistor, or an FET give them the ability to generate an alternating current of a frequency determined by auxiliary components associated with them. Such circuits are termed *oscillators.* To generate ac power with an amplifier, a portion of the output

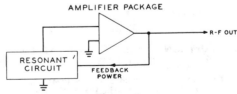

Figure 1

THREE TERMINAL OSCILLATOR

A portion of the output of a three-terminal amplifier is fed back to the input in proper phase and amplitude with the starting power which is generated initially by thermal noise. Power delivered to the load is output power less feedback power. Resonant circuit in input determines frequency of oscillation.

power must be returned or fed back to the input in phase with the starting power (figure 1). The power delivered to the load will be the output power less the feedback power.

Basic   Oscillation may be initially
Oscillators   caused in a transistor or tube circuit by external triggering, or by self-excitation. In the latter case, at the moment the dc power is applied, the energy level does not instantly reach maximum but, instead, gradually approaches it. Oscillations build up to a point limited by the normal operation of the amplifier, the

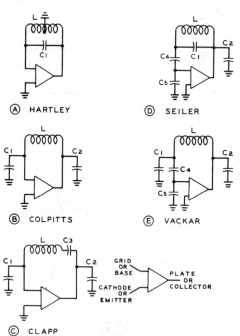

Ⓐ HARTLEY   Ⓓ SEILER

Ⓑ COLPITTS   Ⓔ VACKAR

Ⓒ CLAPP

Figure 2

COMMON TYPES OF SELF-EXCITED OSCILLATORS

The circuits are named after the inventors and are based on variations in the method of coupling and introducing feedback into oscillator tank circuit. A—Hartley circuit with inductive feedback. B—Colpitts circuit with capacitive feedback. C—Clapp circuit with capacitive feedback plus series-tuned tank. D—Seiler circuit with capacitive feedback and separate parallel-tuned tank circuit. E—Vackar circuit with capacitive feedback plus parallel-tuned tank circuit. Circuits may be used with either solid-state devices or vacuum tubes by adjustment of feedback amplitude and applied potentials.

feedback energy, and the nonlinear condition of the circuit. Practical oscillator circuits employ a variety of feedback paths, and some of the most useful ones are shown in figure 2. Either tubes, transistors, or FETs may be used in these circuits.

The oscillator is commonly described in terms of the feedback circuit. The *Hartley* oscillator (figure 2A) employs a tapped inductor in the resonant circuit to develop the proper phase relationship for the feedback voltage, while the *Colpitts* oscillator derives the exciting voltage by means of a capacitive voltage divider. The *Clapp* circuit (figure 2C) employs a series-tuned tank circuit, shunted by a large capacitive voltage divider (C_1-C_2).

The *Seiler* and *Vackar* circuits employ a voltage divider (C_4-C_5) to establish the correct feedback level for proper operation. At resonance, all circuits are versions of a pi-network in one way or another, the tuning scheme and feedback path being different for the various configurations.

The Colpitts   Of these circuits, the Colpitts
Oscillator   and Seiler configurations have proven to be the most practical. A solid-state version of the Colpitts is shown in figure 3. The frequency determining circuit is composed of inductor L plus capacitors C_2, C_3, and C_4. Feedback is determined by the ratio of C_2 to C_3 which is approximately 1:5. The calculation for determining the value of circuit capacitance is given in the drawing.

Capacitors C_2 and C_3 are large so as to shunt out any changes in the junction capacitance of the FET. Capacitor C_4 is relatively small, otherwise the LC ratio of the tuned circuit would also be small. This would result in degraded circuit stability as the inductance of leads, switches and ground returns in the circuit become a larger portion of the total inductance. The minimum value of C_4 should be used that is consistent with stable oscillation.

The Seiler   A version of the *Seiler oscillator*
Oscillator   is often chosen to overcome these circuit limitations (figure 4). The use of series tuning reduces the effect

of the shunt capacitance presented by the gate capacitors and a larger inductor may be used for a given frequency than in the circuit of figure 3.

Circulating current in the tuned circuit is quite high and the variable tuning capacitor should have a constant low-impedance path to ground through the rotor bearings.

Note that in both circuits a limiting diode (D_1) clamps the positive peaks of the gate waveform. This places a limiting value on the transconductance of the FET and also inhibits harmonic generation caused by changes in junction capacitance.

The Vackar Oscillator The *Vackar* oscillator is a variation of the basic Clapp circuit which has improved tuning range and relatively constant output combined with good stability with respect to a varying load. A practical Vackar circuit designed for 30 MHz is shown in figure 5. With the constants shown, the range is from 26.9 to 34.7 MHz, with an output amplitude change of less than −1.5 dB relative to the lower frequency. Capacitor C_1 tunes the circuit while capacitor C_2 is adjusted for optimum drive level such that the transistor is not driven to cutoff or saturation.

The output level, when properly adjusted, is about 4 volts peak-to-peak for a 9-volt supply. The emitter-bias resistor is bypassed for r-f and audio frequencies to eliminate a tendency for the circuit to oscillate at a parasitic frequency that is low in comparison to the working frequency. The value of capacitors C_3 and C_1 are approximately:

$$C \ (pF) = \frac{3000}{f \ (MHz)}$$

The frequency of oscillation is approximately:

$$f_{(osc)} = \frac{1}{2\pi \ \sqrt{L \ (C_1 + C_2)}}$$

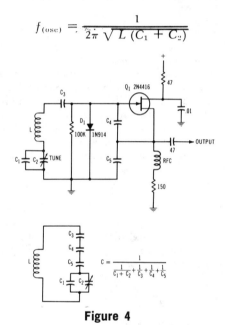

Figure 4

THE SEILER OSCILLATOR

Tank circuit has a better LC ratio than that of the Colpitts. Capacitor C_1 should be large in relation to C_2. Total circuit capacitance is given by the formula.

$$C = \frac{1}{\frac{1}{C_1} + \frac{1}{C_2} + \frac{1}{C_3} + \frac{1}{C_4} + \frac{1}{C_5}}$$

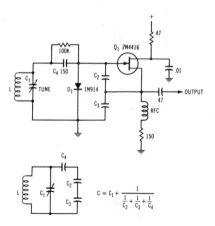

Figure 3

THE COLPITTS OSCILLATOR

Total capacitance in the tuned circuit is determined by the formula. Circuit feedback is determined by the ratio of C_2 to C_3, which is about 1:5.

$$C = C_1 + \frac{1}{\frac{1}{C_2} + \frac{1}{C_3} + \frac{1}{C_4}}$$

It is important that high-quality components be used throughout the oscillator circuit. The coil form should be ceramic with the wire turns tightly wound on the form and cemented in place with low-dielectric adhesive, such as clear acrylic liquid or spray. If a slug-tuned form must be used, choose one that has a good mechanical lock on the slug shaft so that it will not inad-

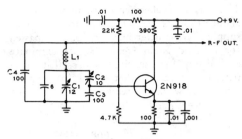

Figure 5

THE VACKAR OSCILLATOR

**Thirty-MHz oscillator for vhf frequency control.
Coil L_1 is 1.5 μH, wound on a ceramic form.
Capacitor C_2 is adjusted for optimum drive level.**

vertently move about. Adjust the coil inductance so that the slug entry into the winding is a minimum.

As mentioned before, the tuning capacitor should have dual bearings for maximum mechanical stability. The oscillator assembly should be firmly fixed to a metal plate or heavy circuit board to prevent flexing when the capacitor is tuned or when the oscillator is operated under conditions of physical vibration.

Design Summary The oscillator is fed from a voltage-regulated power supply, uses a well-designed and temperature-compensated tank circuit, is of rugged mechanical construction to avoid the effects

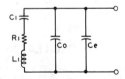

Figure 6

EQUIVALENT CIRCUIT OF A QUARTZ PLATE

The equivalent series-resonant circuit of the crystal itself is at the left, with shunt capacitance of electrodes and holder (C_o) and capacitance between electrodes with quartz as the dielectric (C_e) at right. The composite circuit may exhibit both series resonance and parallel resonance (antiresonance), the separation in frequency between the two modes being very small and determined largely by the ratio of series capacitance (C_1) to shunt capacitance.

of shock and vibration, is protected against excessive changes in ambient room temperature, and is isolated from feedback or stray coupling from other portions of the transmitter by shielding, filtering of voltage supply leads, and incorporation of one or more buffer-amplifier stages. In a high-power transmitter a small amount of stray coupling from the final amplifier to the oscillator can produce appreciable degradation of the oscillator stability if both are on the same frequency. Therefore, the oscillator usually is operated on a subharmonic or image of the transmitter output frequency, with one or more frequency multipliers or mixers between the oscillator and final amplifier.

11-2   Quartz-Crystal Oscillators

Quartz is a naturally occurring crystal having a structure such that when plates are cut in certain definite relationships to the crystallographic axes, these plates will show the *piezoelectric effect*. That is, the plates will be deformed in the influence of an electric field, and, conversely, when such a plate is deformed in any way a potential difference will appear on its opposite sides.

A quartz-crystal plate has several mechanical resonances. Some of them are at very-high frequencies because of the stiffness of the material. Having mechanical resonance, like a tuning fork, the crystal will vibrate at a frequency depending on the dimensions, the method of electrical excitation, and crystallographic orientation. Because of the piezoelectric properties, it is possible to cut a quartz plate which, when provided with suitable electrodes, will have the characteristics of a resonant circuit having a very high LC ratio. The circuit Q of a crystal is many times higher than can be obtained with conventional inductors and capacitors of any size. The Q of crystals ranges from 10,000 to several million.

The equivalent electrical circuits of a quartz-crystal plate are shown in figure 6. The shunt capacitance of the electrodes and holder is represented by C_o, and the capacitance between the electrodes with quartz as the dielectric is C_e. The series capacitance (C_1) represents the motional elasticity of the quartz, while the inductance (L_1) is a function of the mass. The series resistance (R_1) represents the sum of the crystal losses, including friction, acoustic loading, and power transmitted to the mounting structure.

Practical Quartz Crystals   While quartz, tourmaline, Rochelle salts, ADP, and EDT crystals all exhibit the piezoelectric effect, only quartz has a low temperature coefficient and exhibits chemical and mechanical stability. The greater part of the raw quartz used today for frequency control is man-made rather than natural and crystal blanks are produced in large quantities at low prices. The crystal blank is cut from a billet of quartz at a predetermined orientation with respect to the optical and electrical axes, the orientation determining the activity, temperature coefficient, thickness coefficient, and other characteristics of the crystal.

The crystal blank is rough-ground almost to frequency, the frequency increasing in inverse ratio to the oscillating dimensions (usually the thickness, but often the length). It is then finished to exact frequency by careful lapping, by etching, or

by plating. Care is taken to stabilize the crystal so frequency and activity will not change with time.

Unplated crystals are mounted in pressure holders, in which an air gap exists between the crystal and electrodes. Only the corners of the crystal are clamped. At frequencies requiring a low ratio of length to thickness (usually below 2 MHz or so) a "free" air gap is required because even the corners of the crystal move.

Control of the orientation of the blank when cut from the quartz billet determines the characteristics of the crystal. The *turning point* (point of zero temperature coefficient) may be adjusted to room temperature, usually taken as 20°C. A graph of the normal frequency ranges of popular crystal cuts is shown in figure 7. For frequencies between 550 kHz and 55 MHz, the AT-cut crystal is now widely used.

Crystal Holders   Crystals are normally purchased ready-mounted. Modern high-frequency crystals are mounted within metal holders, hermetically sealed with glass insulation and a metal-to-glass bond. Older crystal types make use of a phenolic holder sealed with a metal plate and a rubber

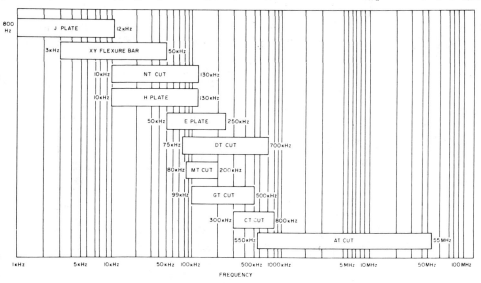

Figure 7
FREQUENCY RANGE OF CRYSTAL CUTS

gasket. A summary of crystal holders and crystal types is given in figure 8.

Precision crystals for calibrating equipment are vacuum-sealed in a glass envelope. Special vacuum-sealed crystals having a relatively constant temperature coefficient are used in high-stability frequency standards in place of the near-obsolete and expensive temperature-controlled "crystal oven."

Overtone-cut Crystals Just as a vibrating string can be made to vibrate on its overtone frequencies, a quartz crystal will exhibit mechanical resonance (and therefore electrical resonance) at overtones of its fundamental frequency. (The terms *overtone* and *harmonic* should not be used interchangeably. The overtone is a mechanical phenomenon and its frequency differs from the harmonic by virtue of the mechanical loading of the crystal. The harmonic is an electrical phenomenon and is an exact multiple of the fundamental frequency.)

By grinding the crystal especially for overtone operation, it is possible to enhance its operation as an overtone resonator. AT-cut crystals designed for optimum overtone operation on the 3rd, 5th, and even the 7th overtone are available. The 5th- and 7th-overtone types, especially the latter, require special holders and circuits for satisfactory operations, but the 3rd-overtone type needs little more consideration than a regular fundamental type. It is possible in some circuits to operate a crystal on the fundamental and 3rd overtone simultaneously and produce an audio beat between the third harmonic and the third overtone. Unless specifically desired, this operation is to be avoided in conventional circuits.

The overtone frequency for which the crystal is designed is the working frequency which is not the fundamental, since the crystal actually oscillates on this working frequency when it is functioning in the proper manner. The Q of an overtone crystal, moreover, is much higher than that of a fundamental crystal of the same frequency. As a result, overtone crystals are less prone to frequency change brought about by changes of oscillator input capacitance. Many frequency-standard crystals in the hf range, therefore, are overtone types.

Crystal Drive Level Crystal dissipation is a function of the drive level. Excessive crystal current may lead to frequency drift and eventual fracture of the blank. The crystal oscillator should be run at as low a power level as possible to reduce crystal heating. Drive levels of 5 milliwatts or less are recommended for fundamental AT blanks in HC-6/U style holders, and a level of 1 milliwatt maximum is recommended for overtone crystals or

QUARTZ CRYSTAL HOLDERS

Holder Type	Pin Spacing	Pin Diam.	Size		
			H	W	T
HC-5/U	0.812	0.156	2.20	1.82	1.60
HC-6/U	0.486	0.050	0.78	0.76	0.35
HC-10/U	(1)	0.060	1.10	—	0.56D
HC-13/U	0.486	0.050	0.78	0.76	0.35
HC-17/U	0.486	0.093	0.78	0.76	0.35
HC-18/U	(2)	—	0.53	0.40	0.15
HC-25/U	0.192	0.040	1.53	0.76	0.35
FT-243	0.500	0.093	1.10	0.90	0.40

(1)—Barrel Mount
(2)—Wire Leads 0.018 Diam.

QUARTZ CRYSTAL TYPES

Mil. Type	Holder Used	Type	Resonance
CR-15B/U	HC-5/U	Fund.	Parallel
CR-16B/U	HC-5/U	Fund.	Series
CR-17/U	HC-10/U	Overtone	Series
CR-18A/U	HC-6/U	Fund.	Parallel
CR-19A/U	HC-6/U	Fund.	Series
CR-23/U	HC-6/U	Overtone	Series
CR-24/U	HC-10/U	Overtone	Series
CR-27/U	HC-6/U	Fund.	Parallel
CR-28A/U	HC-6/U	Fund.	Series
CR-32A/U	HC-6/U	Overtone	Series
CR-52A/U	HC-6/U	Overtone	Series
CR-53A/U	HC-6/U	Overtone	Series

Figure 8
CRYSTAL HOLDERS AND TYPES

fundamental crystals above 10 MHz in HC-6/U holders. The older FT-243 style crystal is capable of somewhat greater drive levels by virtue of the larger blank size.

Series and Parallel Resonance The shunt capacitance of the electrodes and associated wiring is considerably greater than the capacitive component of an equivalent series LC circuit, and unless the shunt capacitance is balanced out, the crystal will exhibit both series- and parallel-resonance frequencies, the latter being somewhat higher than the former. The series-resonant condition is employed in filter circuits and in oscillator circuits wherein the crystal is used in such a manner that the phase shift of the feedback voltage is at the series-resonant frequency.

The only difference between crystals designed for series-resonance and those for parallel-resonance operation is the oscillator input reactance (capacitance) for which they are calibrated. A crystal calibrated for parallel resonance will operate at its calibrated frequency in a series-resonant circuit with the addition of an appropriate value of series capacitance. Thus, a crystal cannot be specified in frequency without stating the reactance with which it is to be calibrated. The older FT-243 fundamental crystals were usually calibrated with a parallel capacitance of 35 pF, while many of the new hermetic sealed crystals are calibrated with a capacitance of 32 pF.

Crystal Grinding Techniques Crystals may be raised in frequency by grinding them to smaller dimensions. Hand grinding can be used to raise the frequency of an already finished crystal and this can be accomplished without the use of special tools or instruments. In the case of the surplus FT-243 style of crystal, the blank may be raised in frequency up to several hundred kilohertz, if it is a fundamental-frequency cut.

A micrometer is required to measure the crystal thickness and grinding is done on a small sheet of optically flat glass. A piece of plate glass will suffice for the home workshop. A grinding compound composed of carborundum powder and water is required. A few ounces of #220 and #400 grits are suggested.

Before grinding is started, the crystal should be checked in an oscillator to make sure it is active. Activity of the crystal can be rechecked during the grinding process to make sure that the faces of the crystal remain parallel.

One face of the crystal is marked with a pencil as a reference face. All grinding is done on the opposite face in order to maintain a reference flat surface. A small amount of #400 grinding grit is placed on the glass disc and enough water added to make a paste. The unmarked side of the crystal is placed face down on the disc and the blank is rubbed in a figure-8 motion over the disc, using just enough pressure from the index finger to move the crystal.

After about a dozen figure-8 patterns have been traced (depending on the amount of frequency change desired), the crystal is washed with water and wiped dry. The crystal is then placed in the holder for a frequency check. The process is repeated a number of times until the crystal is gradually moved to the new frequency.

For larger movement of the crystal frequency, the #220 grit may be used. Additional grit should be added to the glass plate as the compound gradually loses its cutting power with use.

If crystal activity drops with grinding, the blank should be measured with a micrometer to determine the degree of flatness. Normally, the corners are one to three ten-thousandths of an inch thinner than the center of the blank. A thick corner will tend to reduce activity. Grinding the edge of the crystal will restore activity in some cases.

When reassembling the FT-243 holder make sure that the raised corners of the top electrode press against the blank; these are the only points of the electrode that make contact with the crystal.

11-3 Crystal-Oscillator Circuits

A crystal may replace the conventional tuned circuit in a self-excited oscillator, the crystal oscillating at its series- or parallel-resonant frequency. Basic oscillator circuits are shown in figure 9. Series mode operation of the crystal is used in these circuits.

The Colpitts Oscillator   The circuit of figure 9A places the crystal in a feedback network composed of capacitors C_2 and C_3. These are a portion of the resonant circuit formed by inductor L_1 and capacitors C_1, C_2, and C_3. The ratio of the capacitor network is about 1:4. Capacitor C_1 is adjusted for optimum oscillator stability.

The base of a bipolar transistor has a very low input impedance and makes design of a crystal oscillator difficult when parallel-resonant mode crystals are used. The series resonance mode, such as employed in this circuit, eliminates the problem.

A small inductor (L_2) is placed in parallel with the crystal to form a parallel-resonant circuit with the holder capacitance of the crystal. Balancing the holder capacitance ensures that the crystal oscillates as marked.

An FET crystal oscillator is shown in figure 9B. The ratio of capacitance between C_1 and C_2 is about 1:4. The series-connected capacitors across the crystal must be taken into account when ordering a crystal for a specific frequency. A small high-quality variable capacitor can be placed across the crystal for frequency adjustment.

The Pierce Oscillator   A representative Pierce oscillator is shown in figure 10. Few components are required and the high gate impedance of the FET results in very light crystal loading. Use of a bipolar transistor in the Pierce circuit is not recommended because of the very low base impedance of the device. If an IGFET is used in the Pierce circuit, a separate limiting diode is usually added across the gate-source circuit.

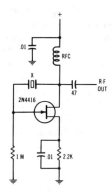

Figure 10

THE PIERCE OSCILLATOR

The high gate impedance of the FET results in light crystal loading.

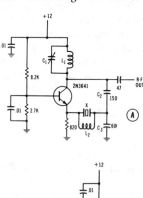

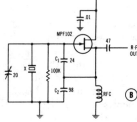

Figure 9

CRYSTAL OSCILLATOR CIRCUITS

A—Colpitts oscillator with feedback network composed of capacitors C_2 and C_3. B—FET version with network composed of capacitors C_1 and C_2.

The Overtone Oscillator   The *overtone crystal oscillator* is very useful for frequency control in the vhf region as the undesired harmonics of a low-frequency oscillator-doubler chain are eliminated

Overtone crystals make possible vhf output from crystals operating on their third, fifth, or seventh mode. Three practical overtone circuits are shown in figure 11. Circuit

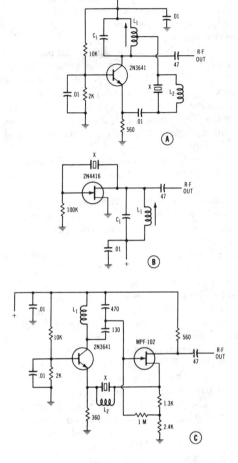

Figure 11

OVERTONE CRYSTAL OSCILLATORS

A bipolar transistor is used in circuit A. The degree of feedback is determined by position of tap on coil L_1. Coil L_2 is resonant at the overtone frequency with the capacitance of the crystal holder. An FET is used in circuit B. Crystal loading is reduced to a minimum in this configuration. A Butler circuit is shown at C. This circuit employs an npn transistor and n-channel FET and resembles an emitter-coupled multivibrator.

A is a variation of the basic Hartley configuration with a bipolar transistor. An FET version is shown in illustration B. A Butler oscillator is shown in circuit C. This employs an npn transistor and n-channel FET and resembles an emitter-coupled multivibrator.

Note that in these overtone circuits the crystal acts as a small series resistance at the overtone resonant frequency. It resembles a resonant coupling capacitor. That is, the circuit would still oscillate if the crystal were replaced by a short. To prevent the residual crystal holder capacitance from causing unwanted oscillation at a frequency other than desired, it is resonated out by a means of inductor L_2.

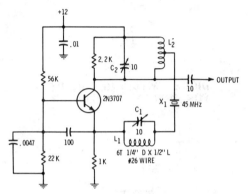

Figure 12

VARIABLE FREQUENCY CRYSTAL OSCILLATOR (VXO)

The VXO Circuit   A moderately wide frequency range of operation of a crystal oscillator may be achieved by operating the crystal below its resonant frequency and loading it with an inductance. Frequency stability is reduced by a factor of about 10, but bandwidth operation up to one or two percent of the crystal frequency may be achieved. Shown in figure 12 is a circuit for use with an overtone crystal in the 45-MHz range which provides a variation of plus or minus 20 kHz at the operating frequency. A circuit of this type is termed a *variable crystal oscillator (VXO)*.

Crystal Switching and Offsetting   Crystals may be switched to obtain more than one oscillator frequency. An electrical switching circuit is shown in figure 13. The control switch (which may be remotely located) switches diodes D_1 and D_2 to remove the unwanted crystal from the circuit. If it is desired to offset

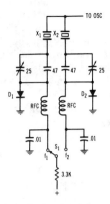

Figure 13

**DIODE SWITCHING CIRCUIT
FOR CRYSTAL OSCILLATOR**

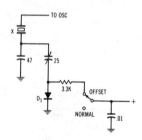

Figure 14

**DIODE OFFSET CIRCUIT
FOR CRYSTAL OSCILLATOR**

the crystal a small amount a diode is used to vary the capacitance of the shunt capacitor as shown in figure 14.

11-4 HF Power Circuits

Most high-frequency power transistors are silicon, planar, diffused npn structures having a high ratio of active to physical area. Upward of 200 watts average power at frequencies in the neighborhood of 450 MHz may be handled by modern silicon power transistors of advanced design. In the coming decade the efficiency, power gain, and temperature stability of these devices will lead to their use in many r-f amplifier applications heretofore solely reserved for electron tubes.

Circuit Considerations The power output capability of a transistor is determined by current and voltage limitations at the frequency of operation. The maximum current capacity is limited by maximum breakdown limits imposed by layer resistivity and by the penetration of the junction. The *high-frequency current gain* figure of merit (f_T) defines the frequency at which the current gain is unity, and a high value of f_T at high emitter or collector current levels characterizes a good r-f transistor.

In many cases, components and construction techniques used for vacuum tubes are not appropriate for transistor circuits. This variance in circuit considerations results mainly because of the lower circuit impedances encountered in transistor circuits. The most troublesome areas are power dissipation and unwanted oscillation. In the case of power dissipation, the levels reached under a given r-f power input are considerably higher than equivalent levels achieved under dc operating conditions, since the junction temperature is a complex function of device dissipation, which includes r-f losses introduced in the pellet mounting structure. The package, then, is an integral part of the r-f power transistor having thermal, capacitive, and inductive properties. The most critical parasitic features of the package are emitter and base lead inductances. These undesired parameters can lead to oscillations, most of which occur at frequencies *below* the frequency of operation because of the increased gain of the transistor at lower frequencies. Because transistor parameters change with power level, instabilities can be found in both common-emitter and common-base circuits. Some of the more common difficulties are listed below:

Parametric Oscillation—Parametric instability results because the transistor collector-base capacitance is nonlinear and can cause low-frequency modulation of the output frequency. This effect can be suppressed by careful selection of the bypass capacitors, and by the addition of a low-frequency bypass capacitor in addition to the high-frequency bypass capacitor (figure 15).

Low Frequency Oscillation—With transistor gain increasing at about 6 dB per

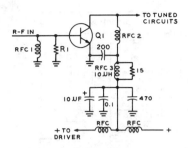

Figure 15

WIDEBAND DECOUPLING CIRCUIT FOR POWER TRANSISTOR

To suppress parametric oscillation collector by-pass circuit must be effective at very low fre-quencies. Multiple bypass capacitors and series r-f chokes provide an adequate filter when used in conjunction with regular hf and vhf filtering techniques.

octave, any parasitic low-frequency circuit can cause oscillation. Inadequate bypassing plus the use of high-Q, resonant r-f chokes can lead to this difficulty. This effect can be eliminated by placing small resistances in series with the r-f choke, or by the use of low Q chokes of the ferrite-bead variety.

Hysteresis—Hysteresis refers to discontinuous mode jumps in output power that occur when the input power or operating frequency is increased or decreased. This is caused by dynamic detuning resulting from nonlinear junction capacitance variation with change in r-f voltage. The tuned circuit, in other words, will have a different resonant frequency for a strong drive signal than for a weak one. Usually, these difficulties can be eliminated or minimized by careful choice of base bias, by proper choice of ground connections, and by the use of transistors having minimum values of parasitic capacitance and inductance. Circuit wiring should be short and direct as possible and all grounds should be concentrated in a small area to prevent chassis inductance from causing common-impedance gain degeneration in the emitter circuit. In common-emitter circuits, stage gain is dependent on series emitter impedance and small amounts of degeneration can cause reduced circuit gain at the higher frequencies and permit unwanted feedback between output and input circuits.

Thermal Considerations All semiconductor devices are temperature sensitive to a greater or lesser degree and the operating temperature and power dissipation of a given unit must be held below the maximum specified rating either by limiting the input power or by providing some external means of removing the excess heat generated during normal operation. Low power devices have sufficient mass and heat dissipation area to conduct away the heat energy formed at the junctions, but higher power devices must use a *heat sink* to drain away the excess heat.

Transistors of the 200-watt class, for example, have a chip size up to $\frac{1}{4}$ inch on a side and the excess heat must be removed from this very small area. For silicon devices, the maximum junction temperature is usually in the range of 135°C to 200°C. The heat generated in the chip is passed directly to the case through the collector-case bond.

The heat sink is a device which takes the heat from the transistor case and couples it into the surrounding air. Discrete heat sinks are available in various sizes, shapes, colors and materials. It is also common practice to use the chassis of the equipment as a heat sink. The heat dissipation capability of the heat sink is based on its *thermal resistance*, expressed in degrees per watt, where the watt is the rate of heat flow. Low power semiconductor devices commonly employ a clip-on heat sink while higher power units require a massive cast-aluminum, finned, radiator-style sink.

The interface between transistor case and sink is extremely important because of the problem of maintaining a low level of thermal resistance at the surfaces. If it is required to electrically insulate the device from the sink a mica washer may be used as an insulator and the mounting bolts are isolated with *nylon* or *teflon* washers. Some case designs may have a case mounting stud insulated from the collector so that it can be connected directly to the heat sink.

If the transistor is to be soldered into the circuit, the lead temperature during the soldering process is usually limited to about 250°C for not more than 10 seconds and the connections should not be made less than 1/32 inch away from the case.

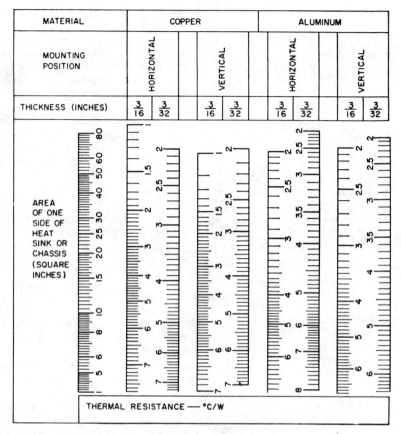

Figure 16

DIMENSIONS OF HEAT SINK AS FUNCTION OF
THERMAL RESISTANCE

The use of a thermal conductive compound such as a zinc-oxide, silicone compound (*Corning PC-4*), for example, is recommended to fill the air insulating voids between the transistor case and the sink to achieve maximum heat transfer across the interface.

Figure 16 is a nomograph for obtaining the physical dimensions of a heat sink as a function of its thermal resistance. The data pertain to a convention- and radiation-cooled sink that is unpainted.

Input Circuits Once the dynamic input impedance has been determined from published data or from measurements, the input circuit may be designed. In practice, the input circuit must provide a match between a source impedance that is high compared to the input impedance of the transistor, which may be of the order of a few tenths of an ohm. Lumped LC circuits are used in the high-frequency region and air-line or strip-line circuits are used in the vhf region, as shown in figure 17.

The reactive portion of the input circuit is a function of the transistor package inductance and the chip capacitance; at the lower frequencies the input impedance is capacitive, and at the higher frequencies it becomes inductive; at some discrete intermediate frequency, it is entirely resistive. The inductive reactance present at the higher frequencies may be tuned out by means of a line section presenting capacitive reactance to the transistor. This advantageously results in an appreciable increase in overall line length, as compared to the more common quarter-wave matching transformer (figure 17D).

At the very high frequencies, the input impedance of a power transistor is commonly inductive and the interstage network of figure 18 is often used. A representative 20-watt, 150-MHz silicon device may have a series input impedance of about $1 + j2$ ohms. Because of the low input impedance, network design and assembly is critical and care should be taken to observe the high circulating currents flowing in the final network loop, particularly through the shunt capacitance (C_3). Current values in the amperes range may flow through this capacitor at drive powers of well less than 5 watts or so. Special ceramic microwave capacitors having an extremely high value of Q and low lead inductance are available for configurations of this type. The low-loss porcelain units are expensive, but their cost is still small compared to the expensive transistors needed to produce appreciable power at the very high frequencies.

Output Circuits   In most transistor power amplifiers, the load impedance (R_L) presented to the collector is dictated by the required power output and the allowable peak dc collector voltage, and thus is not made equal to the output resistance of the transistor. The peak ac voltage is always less

than the supply voltage and the collector load resistance may be expressed as:

$$R_{1.} = \frac{(V_{CC})^2}{2 \times P_o}$$

where,
   V_{CC} equals supply voltage,
   P_o equals peak power output.

The nonlinear transfer characteristic of the transistor and the large dynamic voltage and current swings result in high-level harmonic currents being generated in the collector circuit. These currents must be suppressed by proper design of the output coupling network, which offers a relatively high impedance to the harmonic currents and a low impedance to the fundamental current (figure 19). Parallel-tuned, or pi-network circuitry may be used, with the reactive component of the output admittance tuned out by the proper design of the series choke (RFC_1). At the lower frequencies, the collector of the transistor may be tapped down the tank coil as shown in the illustration. Capacitor C_1 provides tuning, and capacitor C_2 provides load matching. If the value of the inductor is properly chosen, harmonic suppression may be adequate.

A more flexible output circuit is shown in figure 20. This is commonly used with lumped constants in the hf region and also with strip-line configuration in the vhf re-

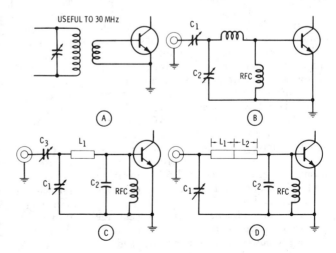

Figure 17

COMMON-EMITTER INPUT CIRCUITRY

Gain of common-emitter circuit is very dependent on emitter series impedance which should be low. Base input impedance is usually less than one ohm and a matching circuit must be provided from a source impedance that is high compared to input impedance. A low-impedance inductive circuit (A) may be used, or various tuned networks that combine impedance transformation with rejection of harmonic frequencies (B). A linear pi network is shown at C. If the input circuit is inductive, the reactance may be tuned out by means of a line section (L,) that presents a capacitive reactance to the transistor (D).

USEFUL TO 30 MHz

gion. A form of the lumped constant circuit is shown in figure 21.

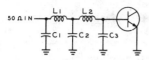

Figure 18

TRIPLE L-NETWORK INPUT CIRCUIT

Network steps down 50-ohm termination to low input impedance of base circuit. In the vhf region, the input impedance is commonly inductive, making up the missing series inductance of the third L network.

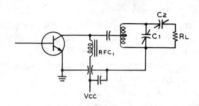

Figure 19

TRANSISTOR OUTPUT MATCHING CIRCUITRY

The reactive component of the output circuit of the transistor stage may be tuned out by proper design of the collector r-f choke (RFC₁). Tuning is accomplished by capacitor C_1 and load matching by capacitor C_2.

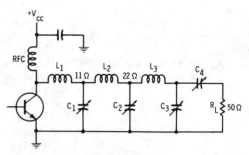

Figure 20

REPRESENTATIVE OUTPUT MATCHING NETWORK

Transistor presents series-conjugate load impedance to network. Center point design impedances are 11 and 22 ohms. Load impedance is usually given on manufacturer's data sheet in either series or parallel equivalent.

on-off (class-C) operation and the forward bias necessary to place them in a class-AB mode leaves them susceptible to *second breakdown,* a destructive phenomenon characterized by localized heating within the transistor pellet, which leads to a regenerative layer damage.

Second breakdown may be controlled by the addition of emitter resistance of low value. A compromise amount is usually chosen as excessive emitter resistance can limit power gain and output. Developmental transistors designed for linear amplifier service have emitter resistance in the chip, in amounts of a fraction of an ohm. Other transistor types may incorporate a zener diode on the chip to provide controlled, positive base voltage.

The forward bias must, in any event, be maintained over a wide temperature range to prevent an increase in idling current accompanied by a rise in chip temperature, which leads to a destructive runaway condition under maximum output conditions when transistor temperature is highest.

Mode of Operation From the stability standpoint, the common-emitter configuration provides a more stable circuit at the higher frequencies than does the common-base circuit. Collector efficiency in either case is about the same. Generally speaking, breakdown voltages under r-f conditions are considerably lower than the normal dc breakdown voltages, and the capability of the r-f power transistor to work into loads having a high value of SWR is limited. A well-designed circuit operated at low supply voltage where power gain is not excessive is found to be less prone to SWR mismatch. High values of SWR mismatch lead to excessive r-f peak voltages, poor efficiency, and instability.

Single-sideband, linear operation calls for class-AB transistor operation. Most high-frequency power transistors are designed for

Class of Operation Low-level solid state r-f amplifiers may run either in the class A, B, or C mode. The class-A mode is used when maximum linearity and high stage gain are desired. The class-B mode is often used for a linear stage when low idling current is required and linearity is not such an important factor. When linear-

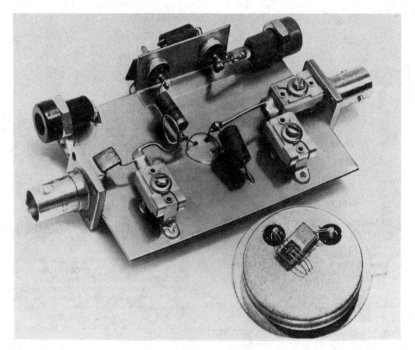

Figure 21

**HIGH-FREQUENCY TRANSISTOR AMPLIFIER SHOWING
INPUT AND OUTPUT CIRCUITS**

The input circuit is similar to that shown in figure 17B and the output circuit is a simplified
version of that shown in figure 20. A magnified view of the transistor element is shown below
the amplifier.

ity is not a consideration, class-C operation provides low idling current and high stage efficiency.

The device may be operated in either the common-emitter or common-base configuration (figure 22). Common-emitter gain increases on a 6-dB per octave slope until the frequency reaches the beta cutoff. Gain at this point may be as high as 40 dB. Common-base gain increases on a 6-dB per octave slope until the frequency reaches the alpha cutoff (see Chapter 4). Below the alpha cutoff the gain flattens out at approximately 12 to 15 dB and remains at this level.

The gains shown assume zero feedback due to any common lead inductance. If common lead inductance is added, common-emitter gain decreases as emitter inductance produces negative feedback. Common base gain increases if base inductance is added because it produces positive feedback.

Choice of Circuitry Common-base and common-emitter circuits have different stability problems. Common-emitter circuits tend to oscillate at the low frequencies where the gain is very high. Common-base circuits are more stable at lower frequencies because the gain remains at a reasonable level. The only real stability problem is regeneration due to the positive feedback. If this is minimized by holding the base inductance low, common-base configuration offers a high degree of stability.

As a linear amplifier, common-emitter circuitry is superior. It is easier to bias and negative feedback can be added to improve linearity. Because there is no phase shift from input to output negative feedback cannot be used to improve linearity of the common-base circuit.

Finally, the saturated power output of the common-emitter circuit is greater than that

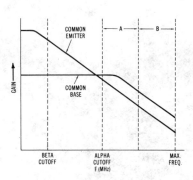

Figure 22

**COMMON-EMITTER AND
COMMON-BASE GAIN**

Common-emitter circuitry is normally used in
Region A. Common-base circuitry is normally
used in Region B.

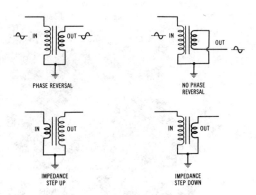

Figure 23

**CONVENTIONAL COUPLING
TRANSFORMERS**

The basic transformer can provide phase reversal or a change in impedance level. An alternative is the autotransformer which has a single
tapped winding instead of two separate windings. Ferrite core transformers of this general
design may be built having a frequency range
of 1.5 to 30 MHz.

of the common-base because the drive
power feeding through to the output increases the overall efficiency. Balanced against
this is the fact that the ability to withstand
high SWR loads is much better with the
common-base circuit.

When making the decision between common-base and common-emitter configurations, the packaging and common lead inductance of the device must be very good
before the common-base circuit becomes use-
able. Special common-base devices are available that provide very low base lead inductance and the use of these transistors is
mandatory, especially in the vhf region.

11-5　Broadband Transformers and Matching Networks

The low input and output impedances of
vhf transistors make the use of low-Q, broadband transformers attractive. The transformers may be conventional or of transmission
line configuration. The conventional transformer can accomplish a phase reversal or
not, as shown in figure 23. R-f transformers
are commonly wound on ferrite or iron-
powder core material having a permeability
between 10 and 1800.

The transmission line transformer consists
of a transmission line wrapped around a ferrite core (figure 24). The transformers may
be configured either balanced or unbalanced
with various transformation ratios, as indicated. Two units may be cascaded to provide a wider range of transformation.

A variation of the broadband transformer
is the *hybrid combiner*. This device is useful
when two or more transistors deliver power
to a common load.

Because of the unusually low port impedances of the power transistor, the design
of matching networks is important to achieve
maximum power transfer and harmonic suppression. A representative set of output networks is shown in figure 25. The transistor
manufacturer usually specifies the series load
impedance required to obtain a rated specification. For power transistors, the resistive
portion of the impedance is very low.

11-6　Power Amplifier Design

The operating parameters for linear service present severe circuit problems for the

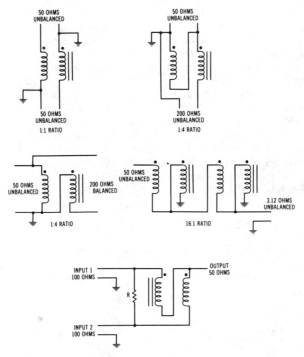

Figure 24

WIDEBAND TRANSMISSION-LINE TRANSFORMERS

Transmission-line transformers provide phase reversal and impedance transformations of 1, 4, 9, and 16 in conventional designs. The 16:1 transformer is composed of two 4:1 units in cascade. The hybrid combiner transformer converts two 100-ohm input ports to a 50-ohm output port. Any phase or amplitude imbalance is dissipated in load resistor R.

solid-state device, among which is the wide variation in the base input impedance, which may vary widely with frequency and tuning, because of the low value of impedance and the relatively large value of collector-base capacitance. A representative 50-watt transistor designed for linear service may have a series input impedance ranging from $4 - j2$ ohms at 3.5 MHz to $0.5 - j0.5$ ohms at 30 MHz.

The transistor for linear service should be chosen on the basis of good current-gain linearity at high values of collector current. A transistor having rapid h_{fe} falloff at high collector currents will generally have poor intermodulation distortion characteristics. In addition to good linearity, the device should have the ability to survive a mismatched load and maintain a low junction temperature at full power output. Transistors are available which combine these attributes, at power

levels up to 100 watts PEP output, having intermodulation distortion levels of − 30 dB for the ratio of one distortion product to one of two test tones. Power gain and linearity are shown in figure 26 for the 2N5492 *Motorola* silicon transistor, specifically designed for linear amplifier service up to 30 MHz.

Operation of a solid-state linear amplifier at reduced collector voltage drastically reduces the maximum power output for a given degree of linearity since the device must deliver correspondingly higher collector peak currents for a given power output, thus placing a greater demand upon the h_{fe} linearity at high values of collector current.

Bias Considerations   A typical class-C solid-state device is operated with both the base and emitter grounded and the transistor is cut off when no driving signal is present. The linearity of a solid-state device requires operation with forward bias, as stated previously. This implies a finite no-signal value of collector current. Optimum values of no-signal (quiescent) collector current range from 5 to 50 mA for devices in the 10- to 100-watt PEP range. Such values fall under the definition of class-B operation. Class-B operation is complicated by thermal runaway problems and large variations in the transistor base current as the r-f drive level is varied. For best linearity, the dc base bias should remain constant as the r-f drive level is varied. This is in

conflict with the conditions required to prevent thermal runaway. A representative bias circuit that meets these critical requirements is shown in figure 27. This circuit supplies an almost constant base bias by virtue of the zener diode (D_1) which is also used to temperature-compensate the transistor. The diode is thermally coupled to the transistor by mounting it on the same heat sink, thus providing temperature compensation due to its decrease in forward voltage drop with increasing temperature. Using this particular transistor, base current rises from the no-signal value of 3 mA to about 200 mA at

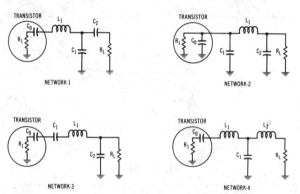

Figure 25

OUTPUT MATCHING NETWORKS

Four commonly used output networks for high-power transistors. Computer-generated tables provide component values for termination to a 50-ohm load. Complete data on these networks is found in Motorola Application Bulletin AN-267 (Motorola Semiconductor Products, Box 20912, Phoenix, AZ 85036).

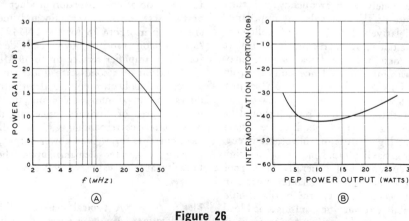

Figure 26

POWER GAIN AND LINEARITY OF 2N5492

Motorola 2N5492 power transistor is designed for linear amplifier service up to 30 MHz and has intermodulation distortion level better than −30 dB.

80 watts output with a two-tone test signal. The current through the diode at the no-signal condition is about 260 mA and when r-f drive is applied, the transistor receives its additional base current from the diode, since the voltage drop across the diode is always slightly greater than the base-emitter voltage of the transistor due to the voltage drop in choke RFC_1.

Resistor R_1 has a dual function in that it causes current to flow through RFC_1 in the no-signal condition and it also reduces the impedance from base to ground, helping to improve the stability of the amplifier.

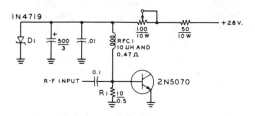

Figure 27

BASE BIAS CIRCUIT FOR 2N5070 IN LINEAR AMPLIFIER SERVICE

Zener diode D_1 is also used to temperature-compensate the transistor by mounting it on common heat sink.

Combining Power Transistors When a single transistor is not capable of providing the output power necessary, extra devices may be added to the circuit. Or it may be desirable to use multiple devices to achieve better reliability or heat distribution. Suitable combining choices for r-f work include the use of transformers, the use of hybrid coupling devices and the utilization of conventional LC networks.

Difficulties are often encountered by unequal load sharing and matching extremely low load impedance levels when power devices are connected directly in parallel. These problems are minimized through the use of signal splitting techniques in both the input and output networks.

Shown in figure 28 are two power transistors combined to provide twice the output power capability of a single device. Inductor L_1 in conjunction with capacitors C_1-C_2 provides an impedance match between the driver impedance and the very low input impedance of Q_1 and Q_2. This is a modified form of pi-network, inductor L_2 in the collector circuit divides the load between the transistors and permits the power output of each device to be combined at a higher impedance level at the common output terminating point.

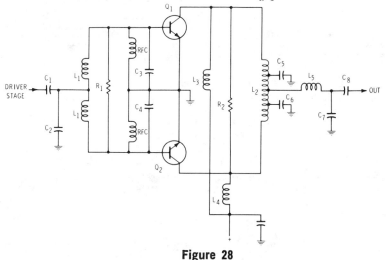

Figure 28

OUTPUT STAGE COMBINES TWO TRANSISTORS TO PROVIDE TWICE THE POWER OF A SINGLE DEVICE

Signal splitting networks and load equalizing resistors provide equal load sharing for two transistors. Conventional LC networks are used to provide circuitry. Coils L_1 and L_2 are air wound inductors. Capacitors are ceramic chip.

External capacitors have been added at or near the base of each transistor to provide an impedance match at the operating frequency and a low-impedance path to ground at the second harmonic frequency for improved efficiency. In some transistors, these capacitors are incorporated in the device.

Resistors R_1 and R_2 help compensate for differences that may occur in transistor power gains and input impedances and therefore help equalize load sharing between the two devices. This results in improved amplifier stability as collector voltage and drive levels are varied. Under symmetrical conditions, signals equal in phase and amplitude will appear on each terminal of R_1 and each terminal of R_2 and thus no current will flow through the resistors. In a practical case, a small current will flow but its effect on the matching network is minimal.

The inductors L_3 and L_4 function as r-f chokes, but also must present a low impedance at frequencies below the lowest operating frequency. This is necessary in order to assure stable operation, since the device gain is very high and the normal transmitter load is essentially removed by series coupling capacitor C_8 at these frequencies. In addition, the inductors must have a low dc resistance to permit efficient operation at the dc current levels involved.

Because of lead inductance and other parasitic effects, actual capacitance values may deviate significantly from the design values, particularly at the higher frequencies and representative capacitors should be measured at the desired operating frequency. For example, a mica capacitor having a nominal low-frequency value of 125 pF can exhibit an effective capacitance as high as 147 pF at 175 MHz.

Broadband Circuitry   The use of transmission-line type wideband transformers permits the construction of a single-ended broadband amplifier (figure 29), whose power gain versus frequency performance is shown in figure 26. The special transformers consist of a low-impedance, twisted wire transmission line wound about a ferrite toroid. These devices have a much wider frequency response than conventional core-coupled or air-coupled transformers due to the utilization of transmission-line techniques and design. A representative transformer is shown in figure 30. The characteristic impedance of the twisted line is the

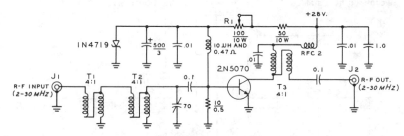

Figure 29

BROADBAND 2- TO 30-MHz LINEAR AMPLIFIER USING 2N5070

Nominal 50-ohm input is stepped down to the base impedance by series-connected 4:1 balun transformers. Single 4:1 balun transformer steps up collector impedance to 50-ohm level.

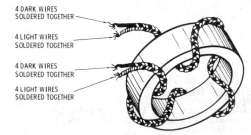

4 DARK WIRES SOLDERED TOGETHER

4 LIGHT WIRES SOLDERED TOGETHER

4 DARK WIRES SOLDERED TOGETHER

4 LIGHT WIRES SOLDERED TOGETHER

Figure 30

BROADBAND FERRITE TOROID TRANSFORMER

A short transmission line made of twisted conductors is wound on the ferrite core. In this example, each conductor consists of four wires in parallel.

geometric mean between the two impedances to be matched and the optimum length of the line is somewhat shorter than an eighth-wavelength at the highest frequency of operation. The impedance of the line is affected by the wire size, tightness of the twist (designated in *crests per inch*) and the number of wires in the line. In general, the impedance may be decreased by using larger wires, a tighter twist, or increasing the number of wires. In the transformer shown in the illustration, four small wires connected in parallel are used for each line, colored insulation being used for ready identification of wires.

The ferrite core selected for the hf transformer is material usually used at frequencies below 10 MHz. Optimum performance over the hf range is achieved with a low-frequency core, since these transformers are not core-coupled and the primary function of the core is to increase winding inductances

to improve performance at the lower end of the operating frequency range.

Transformation ratios of 4:1 or 9:1 may be achieved with the proper winding connections. Two series-connected transformers can be used to achieve greater ratios, if required. Additional information on transformer design may be obtained in Motorola Application Note AN-546, available from *Motorola Semiconductor Products, Inc.,* Box 20912, Phoenix, AZ 85036. A representative amplifier schematic utilizing these wideband transformers is shown in figure 29.

Broadband Push-Pull Circuitry Broadband push-pull transformers made up of a ferrite core stack provide hf coverage from 3 to 30 MHz (figure 31). The low-impedance primary winding consists of one turn of brass tubing soldered to printed

Figure 31

BROADBAND FERRITE-CORE TRANSFORMERS

These small transformers are used with two power transistors to provide high-frequency coverage from 3 to 30 MHz. The primary of the transformer consists of two brass tubes connected together at one end by a copper clad plate, forming a U-turn. The opposite ends of the tubes are provided with insulated terminations for direct connection to the transistors (see transformer at right). The secondary winding is made up of parallel-connected lengths of flexible hookup wire. Ferrite cores are slipped over the brass tubes to complete the assembly. The larger transformer is 1¾″ long and uses ½″ diameter ferrite cores. It is rated at 200 watts PEP input. The transformer at right is 1¼″ long and is rated at 100 watts PEP input.

Figure 32

SOLID STATE 100-WATT PEP LINEAR AMPLIFIER

The transistors are mounted to a printed-circuit board which, in turn, is fixed to the aluminum heat sink. The input and output broadband transformers are placed immediately adjacent to the transistors, with the r-f feedback circuitry grouped around the transistors. A third harmonic LC filter in the output circuitry is in the foreground.

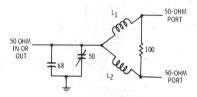

Figure 33

POWER COMBINER FOR 28-30 MHz

This lumped constant combiner serves as either an input or output combiner. Coils L_1 and L_2 are 11 turns #18 wire, 3/8″ diam. As an input combiner, R_1 is 100 ohms, 2-watt carbon. As an output combiner, R_1 is 100 ohms, 25-watt Dale type NH-25.

(Design courtesy of Communication Transistor Corp.)

circuit board end plates. Ferrite beads are slipped over the tubing. The impedance transformation ratio depends on the number of secondary turns, two turns providing a 4:1 ratio, 3 turns a 9:1 ratio, and 4 turns a 16:1 ratio. The secondary turns are made by passing insulated wire through the primary winding tubes. The simplicity and ease of construction provide a balanced transformer that is rugged, adaptable to printed-circuit board construction and relatively inexpensive. A representative amplifier utilizing these transformers is shown in figure 32.

The Power   Solid-state amplifiers may be
Combiner   connected together with a *power combiner* to provide twice the power output of one amplifier (figure 33). In this example, two 30-MHz amplifiers are

combined at a 50-ohm impedance. The network converts a nominal 50-ohm source and load impedance into two 50-ohm ports which are in phase. Any amplitude or phase imbalance causes power to be dissipated in the load resistor, thus assuring equal load sharing between the two amplifiers. A number of amplifier units may be combined in this fashion to provide very high power, solid-state amplifiers for hf or vhf operation.

Part II   Vacuum-Tube HF Power Amplifiers

Vacuum tubes are used as r-f power amplifiers as discussed in Chapters 6 and 7 of this Handbook. By and large, they have been supplanted by solid-state devices for low-power service at frequencies below 450 MHz. At high power levels and at the upper reaches of the vhf spectrum, however, the power tube is in universal use.

11-7   Power Amplifier Operation

The three classes of vacuum-tube r-f amplifiers that find widest application in modern communication equipment are the class-AB_1, class-B, and class-C types.

The angle of plate-current conduction determines the class of operation. Class B is a 180-degree conduction angle and class C is less than 180 degrees. Class AB is the region between 180 degrees and 360 degrees of conduction. The subscript "1" indicates that no grid current flows, and the subscript "2" means that grid current is present. The class of operation has nothing to do with whether the amplifier is grid driven or cathode driven (grounded grid). A cathode-driven amplifier, for example, can be operated in any desired class, within limitations imposed by the tube.

Operating Characteristics The class-AB amplifier can be operated with very low intermodulation distortion in linear amplifier service. Typical plate efficiency is about 60 percent, and stage gain is about 20 to 25 decibels. The class-B amplifier will generate more intermodulation distortion than the class-AB circuit but the distortion level is acceptable in many applications. Typical plate efficiency is about 66 percent and power gains of 15 to 20 decibels are readily achieved. The class-C amplifier is used where large amounts of r-f power are to be amplified with high efficiency. Class-C amplifiers operate with considerably more than cutoff bias, much like a limiter; therefore, this configuration cannot amplify a modulated signal without serious distortion. Class-C amplifiers are used for high-level amplitude modulation wherein the plate voltage (or plate and screen voltages for tetrodes) is modulated at an audio rate. The output power of a class-C amplifier, adjusted for plate modulation, varies with the square of the plate voltage. That is the same condition that would take place if a resistor equal to the voltage on the amplifier, divided by the plate current, were substituted for the amplifier. Therefore, the stage presents a resistive load to the plate modulator. Typical plate efficiency is 70 percent and stage gain is 8 to 10 decibels.

Grid Excitation Adequate grid excitation must be available for class-B or class-C service. The excitation for a plate-modulated class-C stage must be sufficient to produce a normal value of dc grid current with rated bias voltage. The bias voltage preferably should be obtained from a combination of grid-resistor and fixed grid-bias supply.

Cutoff bias can be calculated by dividing the amplification factor of the tube into the dc plate voltage. This is the value normally used for class-B amplifiers (fixed bias, no grid resistor). Class-C amplifiers use from 1.5 to 5 times this value, depending on the available grid drive, or excitation, and the desired plate efficiency. Less grid excitation is needed for c-w operation, and the values of fixed bias (if greater than cutoff) may be reduced, or the value of the grid-bias resistor can be lowered until normal rated dc grid current flows.

The values of grid excitation listed for each type of tube may be reduced by as much as 50 percent if only moderate power output and plate efficiency are desired. When consulting the tube tables, it is well to remember that the power lost in the tuned circuits must be taken into consideration when calculating the available grid drive. At very-high frequencies, the r-f circuit losses may even exceed the power required for actual grid excitation.

Excessive grid current damages tubes by overheating the grid structure; beyond a certain point of grid drive, no increase in power output can be obtained for a given plate voltage.

11-8   Vacuum Tube Neutralization

Because the input-to-output isolation of a vacuum tube is not perfect it is often necessary to neutralize the internal feedback of the tube, especially at frequencies above about 500 kHz. Various tetrode and pentode tubes in a grid-driven configuration may operate without neutralization in the hf region, provided the stage gain is less than the overall feedback from output to input circuit.

Neutralizing Circuits   The object of *neutralization* is to cancel the capacitive feedback path from input to output that appears to a greater or lesser degree in any vacuum tube. The method commonly used to compensate for the feedback is through the use of a capacitance bridge which balances out the effect of grid-plate capacitance (figure 1). When the bridge is balanced, a high degree of isolation is achieved between the input and output circuits of the tube. The bridge is made up of the grid-plate capacitance, the plate-filament capacitance and an external neutralizing capacitor. The capacitor is adjusted to achieve balance. Note that points A and B

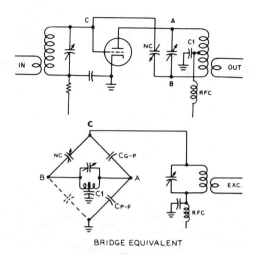

BRIDGE EQUIVALENT

Figure 1

A NEUTRALIZED HF AMPLIFIER

The equivalent neutralizing circuit is shown. This amplifier exhibits positive feedback at the operating frequency even when perfectly neutralized. If a split-stator capacitor (with rotor grounded) is used in the plate circuit and by-pass capacitor C_1 removed, the amplifier exhibits negative feedback at the operating frequency. In either case, feedback amounts to about 3 dB. The circuit may be reversed, with the split coil placed in the input circuit if a single-ended output circuit is required.

in the bridge are 180 degrees out of phase with each other by virtue of the split output coil. The centertap of the coil is at r-f ground potential by virtue of capacitor C_1. To obtain a closer balance in the vhf region a small capacitor equal to the plate-to-filament capacitance of the tube is often added to the circuit from point B to ground.

The neutralizing circuit may be reversed, placing the split coil in the grid circuit with the neutralizing capacitor returned to the plate of the tube. A balanced, or push-pull amplifier is cross-neutralized as shown in figure 2.

Energy feedback from plate to grid of a tube may also be neutralized by placing an inductor between plate and grid. If the reactance is of equal value and opposite sign to the reactance of the grid-plate capacitance, the neutralizing circuit is resonant and a very high impedance will exist from grid to plate. This technique is often used in the vhf region because lead length in the neutralizing circuit is negligible. The disadvantages of the circuit is that neutralization is frequency sensitive and a blocking capacitor must be placed in series with the circuit to isolate the plate voltage from the grid circuit.

Neutralization of Cathode-Driven Amplifiers Stable operation of the cathode-driven (grounded-grid) amplifier often requires neutralization, particularly above 25 MHz or so. Complete circuit stability requires neutralization of two feedback paths, as shown in figures 3A and B.

The first path involves the cathode-to-plate capacitance and proper neutralization may be accomplished by a shunt inductance or by a balanced-bridge technique. The bridge technique is less critical of adjustment than the shunt-inductance circuit, and a reasonable bridge balance over a wide frequency range may be achieved with a single setting of the neutralizing capacitance.

The second feedback path includes the grid-to-plate capacitance, the cathode-to-

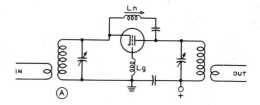

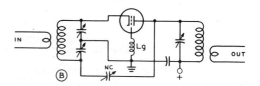

Figure 3

NEUTRALIZATION OF CATHODE DRIVEN AMPLIFIER

A—Cathode-to-plate feedback path may be neutralized by making it part of a parallel-tuned circuit by addition of neutralizing coil L_n. Series capacitor removes plate voltage from neutralizing coil. Adjustments tend to be frequency sensitive.
B—Cathode-to-plate feedback path is neutralized by introducing out-of-phase voltage from drive circuit into plate circuit by means of capacitor NC. Inductor L_g represents grid-lead inductance of vacuum tube, whose effects are not cancelled by either neutralizing circuit.

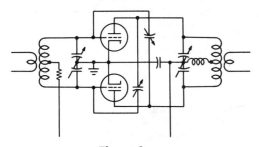

Figure 2

STANDARD CROSS-NEUTRALIZED PUSH-PULL TRIODE AMPLIFIER

grid capacitance and the series inductance of the grid-to-ground path (figure 3B). If this path is not neutralized, a voltage appears on the grid of the tube which either increases or decreases the driving voltage, depending on the values of grid inductance and internal capacitances of the tube. A certain frequency exists at which these two feedback paths nullify each other and this self-neutralizing frequency may be moved about by adding either positive or negative reactance in the grid circuit, as shown in the illustration. If the operating frequency is

above the self-neutralizing frequency a series capacitance is used to reduce the grid inductance. If the operating frequency is below the self-neutralizing frequency, the series grid inductance should be increased. For most tubes of the amateur power class, the self-neutralizing frequency lies between 50 and 150 MHz.

Neutralizing the Tetrode Amplifier A single-ended tetrode amplifier may be neutralized in the same manner as a triode amplifier, or it may be neutralized by the technique shown in figure 4. In illustration 4A the plate-grid capacitance is balanced out when capacitors C and NC bear the following ratio to the grid-plate and grid-ground capacitances of the tube:

$$\frac{NC}{C} = \frac{C_{g-p}}{C_{g-gnd}}$$

It is important to note that the grid-ground capacitance includes the socket capacitance and other circuit strays.

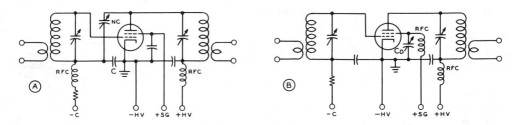

Figure 4

TETRODE NEUTRALIZATION

A—Bridge circuit for neutralization below the self-neutralizing frequency. B—Screen neutralization for use above the self-neutralizing frequency.

Illustration 4B shows a useful neutralization circuit for a vhf amplifier. A variable
screen bypass capacitor is used to series-
resonate the screen lead inductance to
ground. The circuit is frequency sensitive
and requires adjustment unless the equipment is only operated over a narrow frequency band.

Cancellation of   For each tetrode or pentode
Screen Lead   a frequency exists at which
Inductance   the screen capacitance and
                       screen lead inductance are
in a condition of series resonance. This places
the screen at a zero r-f potential with respect to ground. The frequency is called the
self-neutralizing frequency and usually falls
between 20 MHz and 200 MHz, depending
on the physical size and lead arrangement
of the tube. Above or below this frequency,
the screen will allow an amount of coupling
between input and output circuits.

Common neutralizing circuits will eliminate the effects of coupling in the hf region.
Above the self-neutralizing frequency, the
circuit shown in figure 4B is effective.

Alternatively, screen lead inductance can
be neutralized above the self-neutralizing
frequency by feeding back a small amount
of energy from plate to grid by means of
a small capacitance connected between these
tube elements. Note this capacitor is connected in such a manner as to increase the
grid-plate capacitance of the tube. This
technique, like the other neutralization configurations effective above the self-neutralizing frequency, operates over a relatively
narrow band of frequencies with one adjustment and is considered to be frequency
sensitive.

Neutralizing   Voltage feedback from output
Procedure   to input through the distrib-
                    uted constants of the vacuum
tube has a deleterious effect on amplifier
performance. The magnitude, phase and rate
of change with respect to frequency of this
feedback voltage determine the stability of
the amplifier. Control of feedback is termed
neutralization. The purpose of neutralization
of an amplifier is to make the input and
output circuits independent of each other
with respect to voltage feedback. *Proper
neutralization may be defined as the state in
which, when output and input tank circuits
are resonant, maximum drive voltage, minimum plate current, and maximum power
output occur simultaneously.*

The state of correct neutralization, therefore, may be judged by observing these
operating parameters or by observing the
degree of feedback present in the amplifier.

Passive   An amplifier may be neutral-
Neutralization   ized in the passive state with
                     the aid of a signal generator,
an r-f voltmeter, and a grid-dip oscillator.
The input and output circuits of the amplifier are resonated to the operating frequency
and a small signal from the generator is
applied to the input circuit of the amplifier.
An r-f voltmeter (or well-shielded receiver)
is connected to the output circuit of the
amplifier. Neutralizing adjustments are now
made to reduce to a minimum the feedthrough voltage reaching the receiver from
the signal generator. Adjustments may be
made with no filament or plate voltage
applied to the amplifier. Once a null adjustment has been achieved, the amplifier may

be activated and the neutralization adjustment touched up at the full power level.

Passive neutralization is a highly recommended technique since no voltages are applied to the equipment, and adjustments and circuit modifications may be made without danger to the operator of accidental shock.

11-9   Frequency Multiplication

The vacuum tube may be used as a *frequency multiplier* to provide gain and power at two, three, and four times the input frequency. Circuitry is conventional, with the plate circuit tuned to the desired harmonic. Operating efficiency is about 50 percent at the second harmonic, 35 percent at the third harmonic, and 20 percent at the fourth harmonic.

The angle of plate current flow (Chapter 7) must be quite small for a frequency multiplier and high grid bias is required so that the peak excitation voltage will exceed the cutoff voltage for only a short portion of the excitation cycle.

Figure 5 illustrates doubler action showing how the cutoff bias is overcome by the peaks of the exciting signal. The missing pulses in the plate circuit are filled in by the *flywheel effect* of the tuned circuit.

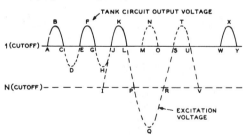

Figure 5

ACTION OF A FREQUENCY MULTIPLIER

Pulses ABC, EFG, and JKL illustrate 180-degree excitation pulses under class-B operation, the upper X-axis being cutoff bias. If the bias is increased N times (the lower dashed X-axis) and excitation increased, the drive signal is represented by the positive, alternate wave peaks. Pulses EFG and MNO illustrate the missing pulses which are filled in by the flywheel effect in the plate tank circuit.

Two tubes can be connected in push-pull to provide tripling action, or they may be connected with grids in push-pull and plates in parallel for doubler service. Such configurations are often used in high-frequency, high power multiplier circuits.

11-10   Plate Tank Circuit Design

A class-B or -C amplifier draws plate current in the form of pulses of short duration. The r-f plate current is therefore full of high frequency harmonics which can be removed from the output circuit by means of a tuned inductance-capacitance *tank circuit* which tends to smooth out the pulses by its storage, or tank, action into a sine wave of r-f output.

Tank Circuit Q   The plate tank circuit must be able to store enough r-f energy so that it can deliver current in essentially sine wave form to the load. The ability of a circuit to store energy is designated as the *effective* Q of the circuit. The Q is defined as the ratio of the energy stored to 2π times the energy lost per cycle. The energy lost is the sum of the circuit losses and the energy delivered to the load.

Tank circuit Q at resonance is equal to the parallel resonant impedance (which is resistive at resonance) divided by the inductive or capacitive reactance (both equal at resonance). Thus,

$$Q = \frac{R_L}{X_C} = \frac{R_L}{X_L}$$

where,

R_L is the resonant impedance of the tank,
X_C is the reactance of the tank capacitor,
X_L is the reactance of the tank coil.

This value of resonant impedance (R_L) is the r-f load which is presented to the power amplifier tube in a single-ended circuit such as shown in figure 6.

The value of r-f load impedance (R_L) which the class-B/C amplifier tube sees may

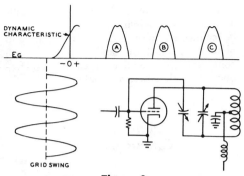

Figure 6

CLASS-C AMPLIFIER OPERATION

Plate current pulses are shown at A, B, and C. The dip in the top of the plate current waveform will occur when the excitation voltage is such that the minimum plate voltage dips below the maximum grid voltage. A detailed discussion of the operation of class-C amplifiers is given in Chapter 7.

be obtained, looking in the other direction from the tank coil, from a knowledge of the operating conditions on the class-B/C tube. This load impedance may be obtained from the following expression, which is true in the general case of any class-B/C amplifier:

$$R_L = \frac{(e_{p\,max})^2}{1.8 \times N_p \times I_b \times E_b}$$

where the values in the equation have the characteristics listed in the beginning of Chapter 7.

The expression is academic, since the peak value of the fundamental component of plate voltage swing ($e_{p\,max}$) is not ordinarily known unless a high-voltage peak ac voltmeter is available for checking. Also, the decimal value of plate circuit efficiency is not ordinarily known with any degree of accuracy. However, in a *normally operated* class-B/C amplifier the plate voltage swing will be approximately equal to 0.8 to 0.9 times the dc plate voltage on the stage, and the plate-circuit efficiency will be from 60 to 80 percent (N_p of 0.6 to 0.8), the higher values of efficiency normally being associated with the higher values of plate voltage swing. With these two assumptions as to the normal class-B/C amplifier, the expression for the plate r-f load impedance can be greatly sim-

plified to the following approximate expression, which also applies to class-AB₁ stages:

$$R_L = \frac{R_{dc}}{1.8}$$

which means simply that the resistance presented by the tank circuit to the class-B/C tube is *approximately equal to one-half the dc load resistance* which the class-C stage presents to the power supply (and also to the modulator in case of high-level modulation of the stage is to be used).

Combining the above simplified expression for the r-f impedance presented by the tank to the tube, with the expression for tank Q given in a previous paragraph we have the following expression which relates the reactance of the tank capacitor or coil to the dc input to the class-B/C stage:

$$X_C = X_L = \frac{R_{dc}}{Q}$$

The foregoing expression is the basis of the usual charts giving tank capacitance for the various bands in terms of the dc plate voltage and current to the class-B/C stage, including the chart of figure 7.

Harmonic Radia- The problem of harmonic
tion versus Q radiation from transmitters
has long been present, but it has become critical during the past decades along with the extensive occupation of the vhf range. Television signals are particularly susceptible to interference from other signals falling within the passband of the receiver, so that the TVI problem has received the major emphasis of all the services in the vhf range which are susceptible to interference from harmonics of signals in the hf or lower-vhf range.

Inspection of figure 8 will show quickly that the tank circuit of an r-f amplifier should have an operating Q of 10 or greater to afford satisfactory rejection of second-harmonic energy. The curve begins to straighten out above a Q of about 15, so that a considerable increase in Q must be made before an appreciable reduction in second-harmonic energy is obtained. Above

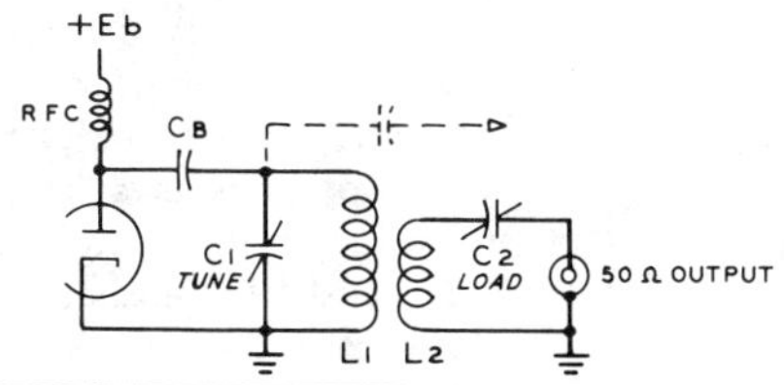

PARALLEL RESONANT
TANK CIRCUIT Q = 10

| COMPONENT | | R-F PLATE LOAD RESISTANCE (OHMS) = $\dfrac{Eb}{2\,Ib}$ |||||||| |
| | | Q = 10   Ib = AMPERES ||||||||
	F (MHz)	1000	2000	3000	4000	5000	6000	7000	8000
C1 (pF)	1.8	900	450	300	225	180	150	130	112
	3.5	450	225	150	112	90	75	65	56
	4.0	390	195	130	100	80	65	57	49
	7.0	225	112	75	66	45	38	33	28
	14.0	112	56	38	33	23	19	16	14
	21.0	75	37	25	22	15	12	11	9
	28.0	56	23	19	17	12	9	8	7
	50.0	31	13	11	9	7	5	4	3
L1 (μH)	1.8	9.0	18.0	27.0	36.0	45.0	54.0	63.0	72.0
	3.5	4.5	9.0	13.5	18.0	22.5	27.0	31.5	36.0
	4.0	4.0	8.0	12.0	16.0	20.0	24.0	28.0	32.0
	7.0	2.2	4.5	6.6	9.0	11.0	12.0	15.4	18.0
	14.0	1.1	2.2	3.3	4.5	5.5	6.0	7.7	9.0
	21.0	0.7	1.5	2.1	3.0	3.5	4.0	4.9	6.0
	28.0	0.5	1.1	1.5	2.2	2.7	3.0	3.5	4.5
	50.0	0.3	0.6	0.9	1.2	1.5	1.8	2.1	2.4

C2 (pF)	F (MHz)	C2 values	L2 (μH)
	1.8	900	9.0
	3.5	450	4.5
	4.0	390	4.0
	7.0	225	2.2
	14.0	112	1.1
	21.0	75	0.7
	28.0	56	0.5
	50.0	31	0.3

NOTE:

WHEN SPLIT–STATOR CAPACITOR IS USED IN BALANCED TANK CIRCUIT, CAPACITANCE OF EACH SECTION IS DIVIDED BY TWO AND INDUCTANCE (L1) SHOULD BE MULTIPLIED BY FOUR.

Figure 7

PARALLEL-TUNED CIRCUIT CHART

Component values listed are for a Q of 10. For other values of Q, use $Q_A/Q_B = C_A/C_B$ and $Q_A/Q_B = L_B/L_A$. Capacitance values shown are divided by four for balanced tank circuit (figure 9C) and inductance is multiplied by four. See figures B and D for split-stator circuitry.

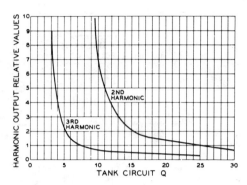

Figure 8

RELATIVE HARMONIC OUTPUT PLOTTED AGAINST CIRCUIT Q FOR PARALLEL-TUNED TANK

a circuit Q of about 10 any increase will not afford appreciable reduction in the third-harmonic energy, so that additional harmonic filtering circuits external to the amplifier proper must be used if increased attenuation of higher-order harmonics is desired. The curves also show that push-pull amplifiers may be operated at Q values of 6 or so, since the second harmonic is cancelled to a large extent if there is no unbalanced coupling between the output tank circuit and the antenna system.

Plate Tank Circuit Design Chart The chart of figure 7 shows circuit capacitance (C) required for a circuit Q of 10, generally considered to be a good compromise value for class AB, B, and C amplifier stages. The capacitance value includes the output capacitance of the tube and stray circuit capacitances. Total stray capacitance may run from perhaps 5 pF for a low-power vhf stage to as high as 50 pF for a high-power hf stage. Also included in the chart are appropriate values for the tank inductance (L_1).

While tank circuit constants are determined by the r-f load resistance, as discussed earlier, this chart has been modified to read in terms of the dc load resistance, as determined by the ratio of dc plate voltage to twice the value of the maximum (peak) dc plate current in amperes. For linear amplifier service, the maximum plate current may be taken as that noted for proper loading at resonance with full carrier injection.

If a different value of circuit Q is desired, a new Q value may be established by a simple ratio. For example, with a given value of plate voltage to plate current ratio, revised values of constants for a Q of 12 may be found by multiplying the capacitance by 12/10 and the inductance by 10/12. When a split tank circuit is used (figure 9B, D), the capacitance value may be reduced as shown and the inductance raised, while still maintaining a constant value of circuit Q.

At the higher frequencies, stray circuit capacitance may be larger than the value determined for a Q of 10. In this case, the Q must be raised to a higher value. Circuit

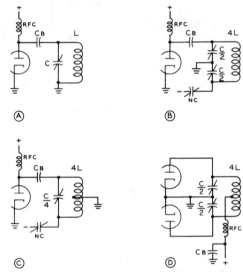

Figure 9

PARALLEL-TUNED TANK CIRCUITS

A—Single ended, use chart of figure 7 for values of L and C. B—Single-ended, split tank. Multiply values of L by four. Each section of split-stator capacitor is ½ value listed in figure 7. C—Split tank with single-section capacitor. Capacitor value is ¼ value listed in figure 7. D—Push-pull circuit with split-stator capacitor. Each section of capacitor is ½ value indicated in figure 7.

Q values of 15 to 50 are often unavoidable and commonly used in the vhf range because of high stray circuit capacitance.

At the lower frequencies, on the other hand, circuit Q may be decreased to as low as 3 to reduce the cost of the tank tuning capacitor and to reduce circuit selectivity to

USUAL BREAKDOWN RATINGS OF COMMON PLATE SPACINGS	
Air-gap in inches	Peak voltage breakdown
.030	1000
.050	2000
.070	3000
.100	4000
.125	4500
.150	5200
.170	6000
.200	7500
.250	9000
.350	11,000
.500	15,000
.700	20,000

Recommended air-gap for use when no dc voltage appears across plate tank capacitor (when plate circuit is shunt fed, or when the plate tank capacitor is insulated from ground).

D C plate voltage	CW/SSB	Plate mod.
400	.030	.050
600	.050	.070
750	.050	.084
1000	.070	.100
1250	.070	.144
1500	.078	.200
2000	.100	.250
2500	.175	.375
3000	.200	.500
3500	.250	.600

Figure 10

Spacings should be multiplied by 1.5 for same safety factor when dc voltage appears across plate tank capacitor.

eliminate sideband clipping. The increased harmonic content of the output waveform, in this instance, is reduced by placing a suitable harmonic filter in the transmission line from amplifier to antenna.

The tank circuit operates in the same manner whether the tube driving it is a pentode, triode, or tetrode; whether the circuit is single-ended or push-pull; or whether it is shunt-fed or series-fed. The prime factor in establishing the operating Q of the tank circuit is the ratio of the loaded resonant impedance across its terminals to the reactance of the coil and capacitor which make up the circuit.

Effect of Loading on Q   The Q of a circuit depends on the resistance in series with the capacitance and inductance. This series resistance is very low for a low-loss coil not loaded by an antenna circuit. The value of Q may be from 100 to 400 under these conditions. Coupling an antenna circuit has the effect of increasing the series resistance, though in this case the power is consumed as useful radiation by the

antenna. Mathematically, the antenna increases the value of R in the expression $Q = \omega L/R$ where L is the coil inductance in microhenrys and ω is the term $2\pi f$ (f being in MHz).

The coupling from the final tank circuit to the antenna or antenna transmission line can be varied to obtain values of Q from perhaps 3 at maximum coupling to a value of Q equal to the unloaded Q of the circuit at zero antenna coupling. This value of unloaded Q can be as high as 400, as mentioned in the preceding paragraph. However, the value of $Q = 10$ will not be obtained at values of normal dc plate current in the class-C amplifier stage unless the C-to-L ratio in the tank circuit is correct for that frequency of operation.

Tuning Capacitor Air Gap   To determine the required tuning capacitor air gap for a particular amplifier circuit it is first necessary to estimate the peak r-f voltage which will appear between the plates of the tuning capacitor. Then, using figure 10, it is possible to estimate the plate spacing which will be required.

The instantaneous r-f voltage in the plate circuit of a class-C amplifier tube varies from nearly zero to nearly twice the dc plate voltage. If the dc voltage is being 100 percent modulated by an audio voltage, the r-f peaks will reach nearly four times the dc voltage.

These rules apply to a loaded amplifier or buffer stage. If either is operated without an r-f load, the peak voltages will be greater and can *exceed* the dc plate supply voltage. For this reason no amplifier should be operated without load when anywhere near normal dc plate voltage is applied.

If a plate blocking capacitor is used, it must be rated to withstand the dc plate voltage plus any audio voltage. This capacitor should be rated at a dc working voltage of at least *twice the dc plate supply in a plate-modulated amplifier,* and at least *equal to the dc supply* in any other type of r-f amplifier.

Inductive Coupling to a Coaxial Line   The chart of figure 7 provides data for coupling the resonant tank circuit to a low-impedance coaxial transmission line. To achieve proper coupling the coupling coil should be series-resonated to the tank frequency. The inductance of the link coil is such that its reactance at the operating frequency is equal to the characteristic

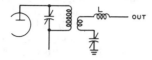

Figure 11

AUXILIARY LOADING COIL (L) USED IN SERIES-TUNED ANTENNA CIRCUIT TO ACHIEVE MAXIMUM COUPLING

impedance of the transmission line. The circuit Q of the link-capacitor combination may be as low as 2. In such case, the value of series capacitance is quite large and the value may be reduced to a more practical amount by placing an auxiliary inductance (L) in series with the link coil as shown in figure 11.

11-11 L, Pi, and Pi-L Matching Networks

Various types of networks are used to transform one impedance to another and network types known as *L*, *pi*, and *pi-L* are commonly used in transmitter circuitry for this purpose. The reason these networks are able to complete a transformation is that,

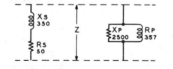

$$① \quad Zs = \sqrt{Rs^2 + Xs^2} \qquad Zp = \frac{Rp \, Xp}{\sqrt{Rp^2 + Xp^2}} \quad ③$$

$$② \quad Q = \frac{Xs}{Rs} \qquad\qquad Q = \frac{Rp}{Xp} \quad ④$$

$$\text{AND} \quad ⑤$$

$$\frac{Rp}{Rs} = Q^2 + 1$$

Figure 12

SERIES TO PARALLEL IMPEDANCE CONVERSION

for any series circuit consisting of a series reactance and resistance, there can be found an equivalent parallel network which possesses the same impedance characteristics (figure 12). Such networks are used to accomplish a match between the tube or device of an amplifier and a transmission line.

The L-Network The *L-network* is the simplest of the matching networks and may take either of the two forms of figure 13. The two configurations are equivalent, and the choice is usually made on the basis of other components and circuit considerations apart from the impedance matching characteristics. The circuit shown in illustration (B) is generally preferred because the shunt capacitor (C) provides a low impedance path to ground for the higher harmonic frequencies.

The L-network is of limited utility in impedance matching since its ratio of impedance transformation is fixed at a value equal to $(Q^2 + 1)$. The operating Q may be

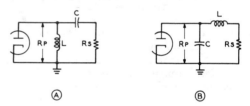

Figure 13

TWO EQUIVALENT L-NETWORKS

A—Inductance in parallel leg, capacitance in series leg. B—Capacitance in parallel leg, inductance in series leg. Impedance values for both circuits are given in figure 12.

relatively low (perhaps 3 to 6) in a matching network between the plate *tank circuit* of an amplifier and a transmission line; hence impedance transformation ratios of 10 to 1 and even lower may be attained. But when the network also acts as the plate tank circuit of the amplifier stage, as in figure 14, the operating Q should be at least 10 and preferably 15. An operating Q of 15 represents an impedance transformation of 225; this value normally will be too high even for transforming from the 2000- to 10,000-ohm plate impedance of a class B/C amplifier stage down to a 50-ohm transmission line.

However, the L-network is interesting since it forms the basis of design for the pi-network. Inspection of figure 14 will show that the L-network in reality must be considered as a parallel-resonant tank circuit in which R_A represents the coupled-in load resistance; only in this case the load resistance is directly coupled into the tank circuit rather than being inductively coupled as in the conventional arrangement where the load circuit is coupled to the tank circuit by means of a link. When R_A is shorted, L and C comprise a conventional parallel-resonant tank circuit, since for proper operation L and C must be resonant in order for the network to present a resistive load to the class-C amplifier.

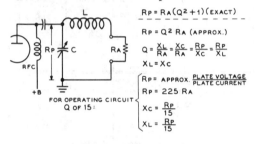

Figure 14

THE L-NETWORK IMPEDANCE TRANSFORMER

The L-network is useful with a moderate operating Q for high values of impedance transformation, and it may be used for applications other than in the plate circuit of a tube with relatively low values of operating Q for moderate impedance transformations. Exact and approximate design equations are given.

The Pi-Network   The *pi-network* can be considered as two back-to-back L-networks as shown in figure 15. This network is much more general in its application than the L network since it offers greater harmonic attenuation and since it can be used to match a relatively wide range of impedances, while still maintaining any desired operating Q. The values of C_1 and L_1 in the pi-network of figure 15 can be thought of as having the same values of the L network in figure 14 for the same operating Q, but, what is more important from the comparison standpoint these values will

be about the *same as in a conventional tank circuit.*

The value of the capacitance may be determined by calculation with the operating Q and the load impedance which should be reflected to the plate of the class-C amplifier as the two known quantities—or the actual values of the capacitance may be obtained for an operating Q of 10 by reference to the chart of figure 16.

The inductive arm in the pi-network can be thought of as consisting of two inductances in series, as illustrated in figure 15.

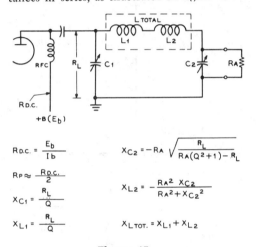

Figure 15

THE PI-NETWORK

The pi-network is valuable for use as an impedance transformer over a wide ratio of transformation values. The operating Q should be at least 10 when the circuit is to be used in the plate circuit of a class-C amplifier. Design equations are given above. Inductor L_{Tot} represents a single inductance, usually variable, with a value equal to the sum of L_1 and L_2.

The first portion of this inductance (L_1) is that value of inductance which would resonate with C_1 at the operating frequency —the same as in a conventional tank circuit. However, the actual value of inductance in this arm of the pi-network, L_{Tot} will be greater than L_1 for normal values of impedance transformation. For high transformation ratios L_{Tot} will be only slightly greater than L_1, for a transformation ratio of 1.0, L_{Tot} will be twice as great as L_1. The amount of inductance which must be added to L_1 to restore resonance and maintain circuit Q

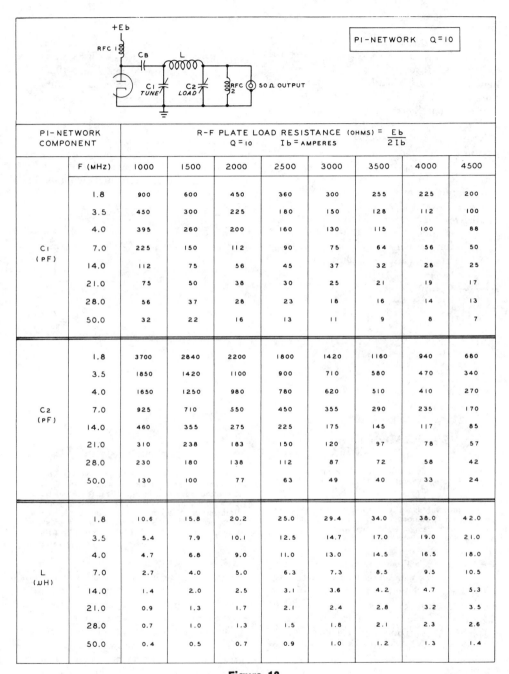

PI-NETWORK COMPONENT	R-F PLATE LOAD RESISTANCE (OHMS) $= \dfrac{Eb}{2Ib}$ $Q=10$ $Ib=$ AMPERES							
F (MHz)	1000	1500	2000	2500	3000	3500	4000	4500
	1.8	900	600	450	360	300	255	225	200
	3.5	450	300	225	180	150	128	112	100
	4.0	395	260	200	160	130	115	100	88
C1 (PF)	7.0	225	150	112	90	75	64	56	50
	14.0	112	75	56	45	37	32	28	25
	21.0	75	50	38	30	25	21	19	17
	28.0	56	37	28	23	18	16	14	13
	50.0	32	22	16	13	11	9	8	7
	1.8	3700	2840	2200	1800	1420	1160	940	680
	3.5	1850	1420	1100	900	710	580	470	340
	4.0	1650	1250	980	780	620	510	410	270
C2 (PF)	7.0	925	710	550	450	355	290	235	170
	14.0	460	355	275	225	175	145	117	85
	21.0	310	238	183	150	120	97	78	57
	28.0	230	180	138	112	87	72	58	42
	50.0	130	100	77	63	49	40	33	24
	1.8	10.6	15.8	20.2	25.0	29.4	34.0	38.0	42.0
	3.5	5.4	7.9	10.1	12.5	14.7	17.0	19.0	21.0
	4.0	4.7	6.8	9.0	11.0	13.0	14.5	16.5	18.0
L (µH)	7.0	2.7	4.0	5.0	6.3	7.3	8.5	9.5	10.5
	14.0	1.4	2.0	2.5	3.1	3.6	4.2	4.7	5.3
	21.0	0.9	1.3	1.7	2.1	2.4	2.8	3.2	3.5
	28.0	0.7	1.0	1.3	1.5	1.8	2.1	2.3	2.6
	50.0	0.4	0.5	0.7	0.9	1.0	1.2	1.3	1.4

Figure 16

Pi-NETWORK CHART

Component values listed are for class-AB/B service for a Q of 10. For other values of Q, use $Q_A/Q_B = C_A/C_B$ and $Q_A/Q_B = L_A/L_B$. When plate load resistance is higher than 3000 ohms, or for class-C service, it is recommended that components be selected for a circuit Q between 12 and 15. For 70-ohm termination, multiply values of capacitor C_2 by 0.72.

is obtained through use of the expression for X_{L1} and X_{L2} in figure 15.

The peak voltage rating of the main tuning capacitor (C_1) should be the normal value for a class-C amplifier operating at the plate voltage to be employed. The inductor (L_{Tot}) may be a plug-in coil which is changed for each band of operation, or some sort of variable inductor may be used. A continuously variable slider-type variable inductor may be used to good advantage if available, or a tapped inductor may be employed. However, to maintain good circuit Q on the higher frequencies when a variable or tapped coil is used on the lower frequencies, the tapped or variable coil should be removed from the circuit and replaced by a smaller coil which has been especially designed for the higher frequency ranges.

The peak voltage rating of the output or loading capacitor (C_2) is determined by the power level and the impedance to be fed. If a 50-ohm coaxial line is to be fed from the pi-network, receiving-type capacitors will be satisfactory even up to the power level of a plate-modulated kilowatt amplifier. In any event, the peak voltage which will be impressed across the output capacitor is expressed by:

$$e_p = \sqrt{2 \times R_a \times P_o}$$

where,

e_p is the peak voltage across the capacitor,
R_a is the value of resistive load which the network is feeding,
P_o is the maximum value of the average power output of the stage.

The harmonic attenuation of the pi-network is greater than that of the simple L-network but is not considered great enough to meet the FCC transmitter requirements for harmonic attenuation. The attenuation to second harmonic energy is approximately -35 dB for the pi-network for a transformation ratio of 40, and increases to -40 dB when the operating Q is raised from 10 to 15.

The Pi-L The *pi-L network* is made up of
Network three L-networks and provides a greater transformation ratio and higher harmonic suppression than do either of the simpler networks (figure 17). Because the loading capacitor is placed at the *image impedance* level (R_1), which is usually of the order of 300 to 700 ohms, the peak voltage across the capacitor ($C_{2A} + C_{2B}$) will be higher than that across the output capacitor of an equivalent pi-network, and the value of the pi-L capacitor will be appreciably less than that of the equivalent pi-network loading capacitor. A formal calculation of the pi-L circuit parameters is given in the article "*The Pi-L Plate Circuit in Kilowatt Amplifiers,*" QST, July, 1962. A free reprint of this article may be obtained by writing to: Amateur Service Department, EIMAC division of Varian, San Carlos, CA 94070. Typical components for pi-L network design for the various hf amateur bands is given in the chart of figure 18.

For a transformation ratio of 40 the attenuation to second harmonic energy is about -52 dB for a pi-L network having a Q of 10 and an image impedance of 300 ohms, rising to -55 dB for a Q of 15 (figure 19).

Network Data In this decade new amateur
for the New bands will become author-
Amateur Bands ized near 10.1, 18.1, and 24.9 MHz. To assist in designing equipment for these frequencies, the pi and pi-L network values are summarized in figures 20 and 21. An image resistance of 300 ohms is chosen for the pi-L configuration.

11-12   The "Grounded-Grid" Amplifier

The cathode-driven or *grounded-grid* amplifier is well suited for linear service in the hf and vhf region. "Grounded grid" implies cathode drive, but in such a circuit the grid may not be at dc ground potential (figure 22), especially with respect to screen voltage. The design of the plate circuit, however, applies equally well to this class of amplifier and the data given in the previous section is correct for the design of a grounded-grid stage.

The grounded-grid amplifier requires considerably more excitation than if the same tube were employed in a grid-driven circuit and the waveform of the drive signal must

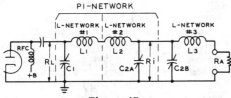

Figure 17

Pi-L NETWORK IS MADE UP OF THREE L-NETWORKS IN SERIES

Pi-L network provides greater transformation ratio and higher harmonic suppression than do either the L- or the pi-networks. Loading capacitor (C_2) is common to networks 2 and 3 and is placed at image impedance level (R_1) which is usually of the order of 300 to 700 ohms.

be maintained. Since the amplifier normally operates in the class-AB or class-B mode, heavy grid current is drawn on the peak of the positive drive cycle. Unless sufficient circuit Q is maintained, the waveform will be distorted and intermodulation products in the amplifier will rise. In addition, many modern solid-state exciters require a load having reasonable SWR and circuit Q. Unless an input matching circuit is used between the grounded-grid amplifier and the exciter, these requirements will not be met, and exciter loading problems may arise.

A pi-network in the cathode circuit of the amplifier will provide an impedance match between the 50-ohm output circuit of the driver-exciter and the input impedance of the cathode-driven amplifier (figure 23). The input impedance of a cathode-driven tube is related to the ratio of the peak cathode signal voltage to the peak cathode current (sum of grid and plate currents), and is commonly given in the tube data sheet. For the 3-500Z high-mu triode operating at 2500 volts, for example, it is about 110 ohms. And for two tubes in parallel it is about 55 ohms only over the operating cycle.

It is tempting to jump to the conclusion that if the amplifier input impedance is about 55 ohms and the output circuit impedance of the exciter is 50 ohms, that no cathode impedance-matching circuit is required. The omission is poor engineering practice as circuit Q at this point *is* required. A Q value of 2 is sufficient, and a simple rule-of-thumb is that the network circuit capacitance at resonance should be about 20

pF per meter of wavelength for a one-to-one impedance transformation.

A computer-derived chart is given in figure 24 covering cathode input impedances ranging from 20 to 250 ohms for the amateur bands. Circuit values for other frequencies may be found by interpolation from this data.

11-13 Bias and Screen Voltage

Radio frequency amplifiers often require an external grid bias supply for proper operation. The bias places the grid at a negative potential with respect to the cathode. Special, "zero-bias" tubes usually require no external bias voltage to establish the correct operating point.

The amount of grid bias depends on the individual tube characteristics and the mode of operation. Amplitude-modulated class-C amplifiers are operated with a bias slightly more than twice that value which will cut off the plate current under the operating plate voltage. Cw, RTTY, and f-m transmitters can be operated with a bias value as low as cutoff for class-C service, or less than cutoff value for class-B operation.

Linear amplifier operation requires that the bias supply have good regulation. Some so-called "zero-bias" tubes actually require a small amount of bias to reduce the resting plate current to a reasonable value. Since resting plate current of a linear amplifier depends on the plate voltage it is possible for the tube to exceed the maximum plate dissipation rating in standby condition at very high values of plate voltage. Addition of a small amount of external bias will solve this problem. Grid bias may be obtained in a number of ways. The most practical techniques follow.

Self-Bias   A resistor can be connected in the grid circuit of a class-C amplifier to provide *self-bias*. This resistor (R_1 in figure 25) is part of the dc path in the grid circuit.

The r-f excitation applied to the grid circuit of the tube causes a pulsating direct current to flow through the bias supply lead,

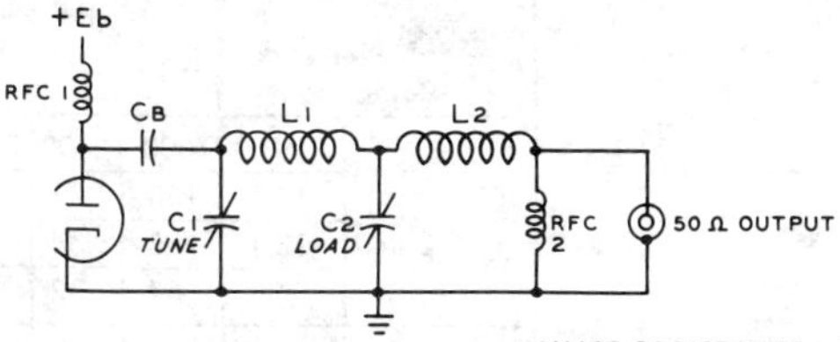

PI-L NETWORK Q = 10

IMAGE RESISTANCE = 350 Ω

PI-L NETWORK COMPONENT	R-F PLATE LOAD RESISTANCE (OHMS) = $\frac{Eb}{2\,Ib}$ Q = 10 Ib = AMPERES							
F (MHz)	1000	1500	2000	2500	3000	3500	4000	4500
C1 (PF) 1.8	900	600	450	360	300	256	224	200
3.5	450	300	225	180	150	128	112	100
4.0	395	260	200	160	130	115	100	88
7.0	225	150	112	90	75	64	56	50
14.0	112	75	56	45	37	32	28	25
21.0	76	50	38	30	25	21	19	17
28.0	56	37	28	23	18	16	14	13
50.0	32	22	16	13	11	9	8	7
C2 (PF) 1.8	2160	1880	1690	1600	1500	1440	1380	1320
3.5	1080	940	845	800	750	720	690	660
4.0	940	820	740	690	650	620	600	570
7.0	540	470	422	400	375	360	345	330
14.0	270	235	211	200	187	180	175	165
21.0	180	155	140	130	125	120	117	110
28.0	135	117	105	100	93	90	87	82
50.0	75	66	59	56	53	50	48	46
L1 (µH) 1.8	14.2	20.0	25.2	31.0	36.0	41.0	46.0	51.0
3.5	7.1	10.0	12.6	15.5	18.0	20.5	23.3	25.5
4.0	6.3	8.8	11.0	13.5	15.5	18.0	20.0	22.5
7.0	3.6	5.0	6.3	7.6	9.0	10.3	11.5	12.7
14.0	1.8	2.5	3.2	3.8	4.5	5.2	5.7	6.3
21.0	1.2	1.6	2.1	2.5	3.0	3.4	3.8	4.2
28.0	0.9	1.3	1.6	1.8	2.2	2.6	2.8	3.1
50.0	0.5	0.7	0.9	1.0	1.2	1.4	1.6	1.8

F (MHz)	1.8	3.5	4.0	7.0	14.0	21.0	28.0	50.0
L2 (µH)	11.2	5.6	4.8	2.7	1.4	0.9	0.7	0.4

Figure 18
Pi-L NETWORK CHART

Component values are listed for class AB/B/C service for a Q of 10. For other values of Q, use conversion transformations listed in figures 15 and 17. Image impedance of 300 ohms is used for calculations.

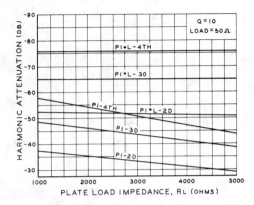

Figure 19

HARMONIC ATTENUATION OF Pi-AND Pi-L NETWORKS

Second, third and fourth harmonic levels are shown relative to fundamental signal. Pi-L configuration provides improved attenuation to all higher harmonics as compared to pi-network.

Pi-L Network Component	R-F Plate Load Resistance (Q = 10)					
	f(MHz)	1000	2000	3000	4000	5000
C_1 (pF)	10.1	158	79	53	39	32
	18.1	88	44	29	22	18
	24.9	64	32	21	16	13
C_2 (pF)	10.1	402	315	276	252	236
	18.1	224	176	154	141	131
	24.9	163	128	112	102	96
L_1 (μH)	10.1	2.40	4.29	6.09	7.84	9.56
	18.1	1.34	2.40	3.40	4.38	5.33
	24.9	0.98	1.74	2.47	3.18	3.88
f(MHz)	10.1	18.1	24.9	Image Resistance = 300 ohms		
L_2 (μH)	1.76	0.98	0.71			

Figure 21

Pi-L NETWORK COMPONENTS FOR 10.1, 18.1, AND 24.9 MHz

Pi-Network Component	R-F Plate Load Resistance (Q = 10)			
	f(MHz)	1000	2000	3000	4000
C_1 (pF)	10.1	158	79	53	39
	18.1	88	44	29	22
	24.9	64	32	21	16
C_2 (pF)	10.1	634	389	261	161
	18.1	354	217	145	90
	24.9	257	158	106	65
L_1 (μH)	10.1	1.87	3.51	5.07	6.56
	18.1	1.05	1.96	2.83	3.66
	24.9	0.76	1.42	2.06	2.66

Figure 20

Pi-NETWORK COMPONENTS FOR 10.1, 18.1, AND 24.9 MHz

due to the rectifying action of the grid, and any current flowing through R_1 produces a voltage drop across that resistor. The grid of the tube is positive for a short duration of each r-f cycle, and draws electrons from the filament or cathode of the tube during that time. These electrons complete the circuit through the dc *grid return*. The voltage drop across the resistance in the grid return provides a *negative bias* for the grid.

Self-bias automatically adjusts itself over fairly wide variations of r-f excitation. The

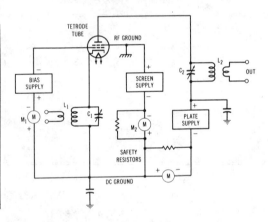

Figure 22

THE GROUNDED-GRID AMPLIFIER

Diagram of the so-called "grounded-grid" amplifier. The grid and screen elements are bypassed to ground as far as rf is concerned, but each element has normal operating voltages applied and is "above ground" as far as dc is concerned. Metering is inserted in the supply return leads to dc level. This eliminates the screen bypass capacitor, a tricky component that often causes circuit instability at the higher frequencies.

value of grid resistance should be such that normal values of grid current will flow at the maximum available amount of r-f excitation. Self bias cannot be used for grid-modulated or linear amplifiers in which the

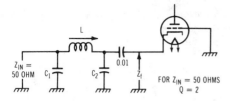

Figure 23

PI-NETWORK CIRCUIT FOR GROUNDED GRID AMPLIFIER

Zt is the input impedance of the amplifier.

Safety Bias Self bias alone provides no protection against excessive plate current in case of failure of the source of r-f grid excitation. A well-regulated low-voltage bias supply can be connected in series with the grid resistor. This fixed protective bias will protect the tube in the event of failure of grid excitation. "Zero-bias" tubes do not require this bias source, since their plate current will drop to a safe value when the excitation is removed.

Cathode Bias A resistor can be connected in series with the cathode or center-tapped filament lead of an amplifier to se-

average dc current is constantly varying with modulation.

Cathode Z_t (Ω)	Band	C1(pF)	C2(pF)	L(μH)	Cathode Z_t (Ω)	Band	C1(pF)	C2(pF)	L(μH)
20	160	3300	4100	2.50	75	160	3300	2870	3.81
	80	1700	2120	1.34		80	1700	1540	2.05
	40	900	1120	0.68		40	900	770	1.03
	20	440	560	0.33		20	440	380	0.51
	15	300	370	0.22		15	300	250	0.34
	10	220	275	0.16		10	220	180	0.25
30	160	3300	3900	2.84	100	160	3300	2520	4.20
	80	1700	2100	1.52		80	1700	1350	2.26
	40	900	1050	0.77		40	900	680	1.14
	20	440	520	0.38		20	440	330	0.56
	15	300	350	0.25		15	300	220	0.38
	10	220	258	0.19		10	220	160	0.28
40	160	3300	3360	3.01	150	160	3300	2100	4.81
	80	1700	1800	1.62		80	1700	1130	2.59
	40	900	910	0.82		40	900	570	1.30
	20	440	450	0.40		20	440	280	0.66
	15	300	300	0.27		15	300	180	0.43
	10	220	220	0.20		10	220	138	0.32
50	160	3300	3300	3.33	200	160	3300	1800	5.32
	80	1700	1700	1.79		80	1700	980	2.86
	40	900	900	0.90		40	900	490	1.44
	20	440	440	0.45		20	440	245	0.71
	15	300	300	0.30		15	300	164	0.48
	10	220	220	0.22		10	220	120	0.35
60	160	3300	3100	3.53	250	160	3300	1640	5.78
	80	1700	1670	1.90		80	1700	880	3.11
	40	900	840	0.96		40	900	440	1.57
	20	440	417	0.47		20	440	220	0.78
	15	300	275	0.32		15	300	140	0.52
	10	220	205	0.23		10	220	100	0.38

Figure 24

CATHODE CIRCUIT VALUES FOR GROUNDED-GRID AMPLIFIER

This chart provides approximate values for the components of the cathode circuit. Capacitors should be 1-kV silver mica or equivalent. Inductors can be wound on a slug-tuned form. Value of C_2 should take into account the cathode-grid (ground) capacitance of the tube which appears in parallel with C_2.

cure *automatic bias*. The plate current flows through this resistor, then back to the cathode or filament, and the voltage drop across the resistor can be applied to the grid

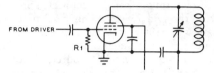

Figure 25
SELF-BIAS

The grid resistor on an amplifier or multiplier stage may also be used as the shunt feed impedance to the grid of the tube when a high value of resistor (greater than perhaps 20,000 ohms) is used. When a lower value of grid resistor is to be employed, an r-f choke should be used between the grid of the tube and the grid resistor to reduce r-f losses in the grid resistance.

circuit by connecting the grid bias lead to the grounded or power-supply end of resistor R, as shown in figure 26.

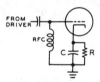

Figure 26
R-F STAGE WITH CATHODE BIAS

Cathode bias sometimes is advantageous for use in an r-f stage that operates with a relatively small amount of r-f excitation.

The grounded (B-minus) end of the cathode resistor is negative relative to the cathode by an amount equal to the voltage drop across the resistor. The value of resistance must be so chosen that the sum of the desired grid and plate current flowing through the resistor will bias the tube for proper operation.

Separate Bias Supply An external supply often is used for grid bias, as shown in figure 27. The bleeder resistance across the output of the filter can be made sufficiently low in value that the grid current of the amplifier will not appreciably change the amount of negative grid-

bias voltage. Alternately, a voltage-regulated grid-bias supply can be used. This type of bias supply is used in class-B audio and

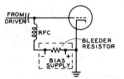

Figure 27
SEPARATE BIAS SUPPLY

A separate bias supply may be used for triodes or tetrodes. Bias is applied across a low-resistance bleeder. Grid current (if any) flowing through bleeder will boost bias voltage over nominal value of supply. Bias supply for AB, linear amplifier, even though no grid current is encountered, must still have low-resistance bleeder to help overcome rise in bias due to collection of primary electrons on grid of tube.

class-B r-f linear amplifier service where the voltage regulation in the bias supply is important.

Zener Bias A few volts of bias may be needed to reduce the zero-signal plate current of a "zero-bias" triode. A low-impedance bias source is required and the simplest way of obtaining well-regulated bias voltage is to place a zener diode in the fila-

Figure 28
ZENER CATHODE BIAS

Zener diode may be used to obtain a few volts of well-regulated cathode bias. This circuit may be used to reduce zero-signal plate current of high-μ triodes in cathode-driven (grounded-grid) service.

ment or cathode return circuit (figure 28). The 1N4551, for example, has a nominal voltage drop of 4.7 volts and an impedance of 0.1 ohm, making it ideal for this service. At this value of bias, the zero-signal plate current of a 3-500Z at a plate potential of 3250 volts is reduced from 160 to approximately 90 mA.

The 1N4551 diode may be bolted directly to the chassis which will act as a heat sink.

Screen Voltage   In addition to plate and bias voltages, tetrode tubes require screen voltage. Screen voltage is critical in that a change in plate potential or grid bias will usually require a change in screen voltage. Screen current, moreover, is a function of grid excitation and plate loading and screen dissipation can quickly be exceeded if either of these two parameters vary outside the design limits. In particular, if plate voltage is removed or reduced under excitation, the resulting voltage and current surges in the screen circuit are apt to permanently damage the tube.

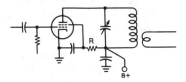

Figure 29

DROPPING-RESISTOR SCREEN SUPPLY

The Series Screen Supply   A simple method of obtaining screen voltage is by means of a dropping resistor from the high-voltage plate supply; as shown in figure 29. This circuit is recommended for use with low power tetrodes (6146, 5763, etc.) in class-C service. Because of poor regulation with varying screen current it should not be used in a linear amplifier stage. Since the current drawn by the screen is a function of the exciting voltage applied to the tetrode, the screen voltage will rise to equal the plate voltage under conditions of no exciting voltage. If the control grid is overdriven, on the other

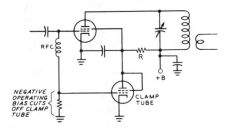

Figure 30

CLAMP-TUBE SCREEN SUPPLY

hand, the screen current may become excessive. In either case, damage to the screen and its associated components may result. In addition, fluctuations in the plate loading of the tetrode stage will cause changes in the screen current of the tube. This will result in screen voltage fluctuations due to the inherently poor voltage regulation of the screen series dropping resistor. These effects become dangerous to tube life if the plate voltage is greater than the screen voltage by a factor of 2 or so.

The Clamp Tube   A clamp tube may be added to the series screen supply, as shown in figure 30. The clamp tube is normally cut off by virtue of the dc grid bias drop developed across the grid resistor of the tetrode tube. When excitation is removed from the tetrode, no bias appears across the grid resistor, and the clamp tube conducts heavily, dropping the screen voltage to a safe value. When excitation is applied to the tetrode the clamp tube is inoperative, and fluctuations of the plate loading of the tetrode tube could allow the screen voltage to rise to a damaging value. Because of this factor, the clamp tube does not offer complete protection to the tetrode.

The Separate Screen Supply   A low-voltage screen supply may be used instead of the series screen-dropping resistor. This will protect the screen circuit from excessive voltages when the other tetrode operating parameters shift. However, the screen can be easily damaged if plate or bias voltage is removed from the tetrode, as the screen current will reach high values and the screen dissipation will be exceeded.

In linear amplifier service, tuneup is commonly accomplished with carrier injection and the drive level can be advanced slowly from zero as the stage is loaded. This will protect the screen from the unfortunate combination of heavy drive and light plate loading. Some commercial transmitters have a current-limiting circuit in the screen supply lead, or other means of indicating excessive screen current.

Screen Protection In designing equipment using high-power tetrodes, consideration must be given to control of secondary emission from the screen element of the tube. The screen is normally operated at a relatively low potential to accelerate the electrons emitted from the cathode. Not all of the electrons pass through the screen grid on the way to the plate, some of them being intercepted by the grid. In the process of striking the screen grid, other electrons are emitted, some of which may be attracted by the higher potential of the plate. The result is a flow of electrons from the screen to the plate. It is possible that more electrons will leave the screen than will arrive and a screen meter will indicate a reverse electron flow, or negative screen current, under this condition. A low-impedance path to ground must be provided for this flow, otherwise the screen voltage will attempt to rise to the value of the plate voltage, by virtue of the voltage drop created by the negative screen current flowing across the high-impedance screen circuit. As the screen voltage rises, the plate current of the tetrode increases and the tube is in a runaway condition. The addition of a resistor from screen to ground will compensate for the effect of negative screen current. The value of this resistor will be such that the bleeder current will run from 20 mA to as high as

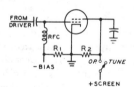

Figure 31

SCREEN CONTROL CIRCUIT

The dc return path to ground for screen of a tetrode should not be broken. Resistor R₂ completes the circuit and screen high-voltage lead may be open to reduce stage gain for tuneup purposes.

70 mA, depending on the tube type. Tube data sheets normally state the amount of bleeder current required to counteract the emission current.

A correct circuit for the screen supply of a linear amplifier, including a "tune-operate" switch is shown in figure 31. In the "tune"

position, screen voltage is removed, permitting adjustments to be made to the circuit at a very low power level for tuneup purposes.

Grid Protection The impedance of the grid circuit must be considered, particularly in class AB₁ amplifiers wherein a regulated bias source is required. Primary grid emission can cause trouble if the impedance of the grid circuit is too high. The dc resistance to ground of the bias supply should be sufficiently low (below 1000 ohms or so) to prevent appreciable reverse bias from being developed by the flow of emission current through the internal resistance of the bias supply. The reverse bias produced by this effect tends to subtract from the grid bias, causing a runaway condition if not controlled.

Arc Protection Modern transmitting tubes have very close internal spacing between elements to achieve high power gain and good performance at very high operating frequencies. Components, too, tend toward more compact sizes to allow high-density construction in modern equipment. Under these conditions, flashovers or arcing between high- and low-potential points in the circuit or tube may possibly occur. The impedance of an arc is very low, of the order of an ohm or so, and extremely high values of *fault current* flow during the flashover. Fault current flowing through a small resistance or impedance creates a high voltage drop in unexpected places and may result in damaged equipment. A flashover in a dc plate circuit, for example, can discharge the power-supply filter capacitor in a fraction of a second and allow thousands of amperes of current to pass through the arc and any components in series with the discharge path.

A sparking gap (G₁) may be placed at a critical point, as shown in figure 32, to protect tube and components against transient arc voltages and a high-voltage, quick-action fuse can be placed in series with high capacity filter circuits to prevent damaging fault currents from flowing through delicate metering circuits or zener diodes. Meters may

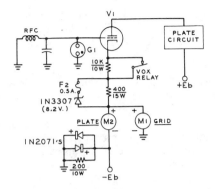

Figure 32

FLASHOVER PROTECTION

Equipment can be protected from flashover and high flashover currents by placing spark gap (G₁) from grid to ground, zener diode fuse in cathode lead and reverse-connected diodes placed across metering circuit. Spark gap arcs over at a predetermined voltage to provide low-impedance path from grid to ground, thus protecting r-f choke and grid bypass capacitor. Cathode fuse opens under heavy arc current, protecting zener diode, while shunt resistor provides path to ground for fault current. Reverse-connected diodes across plate and grid meters provide low-impedance shunt when voltage across meters reaches level of forward voltage drop across the diodes (about 0.4 to 0.8 volt, depending on diode temperature). Filter capacitors in the power supply may also be series-connected with a high-voltage quick-action fuse to prevent discharge through fault circuit in the equipment.

be protected from overload by placing reverse-connected silicon diodes, across them to carry the fault current, as shown in the illustration.

The fault current may be limited to provide tube protection by the inclusion of a 25-ohm resistor in series with the B-plus lead to the amplifier. If a 2-watt composition resistor is used it will usually disintegrate under high fault current, thus opening the circuit and protecting the tube. If a high-wattage wirewound resistor is used it will reduce the fault current to a lesser value and provide time for an overload relay to open, removing plate voltage from the tube. In either case, the tube is protected from an excessively high level of fault current.

Filament Inrush Current   The cold resistance of a thoriated tungsten filament is about one-tenth the hot resistance. As a result, a high level of fila-

ment inrush current exists when the tube is first turned on. The abnormal current can damage the vacuum seal of the filament leads of the tube, or the magnetic field accompanying the current can actually distort the filament structure leading to eventual filament-to-grid shorts within the tube.

A *step-start* circuit is suggested that will limit filament inrush current by applying a fraction of the filament voltage for a period of a few seconds. This allows the filament temperature to rise and for filament resistance to approach near-normal value. In general, holding filament voltage to sixty percent of normal for a period of two seconds will provide inrush protection for most small- and medium-size power tubes. Filament voltage may be brought up by means of a variable transformer in the primary side of the filament circuit or may be boosted to normal after an interval by a time delay relay which shorts out a series resistor in the filament primary circuit.

11-14   R-F Feedback

Comparatively high gain is required in single-sideband equipment because the signal is usually generated at levels of one watt or less. To get from this level to a kilowatt requires about 30 dB of gain. High gain tetrodes may be used to obtain this increase with a minimum number of stages and circuits. Each stage contributes some distortion; therefore, it is good practice to keep the number of stages to a minimum. It is generally considered good practice to operate the low-level amplifiers below their maximum power capability in order to confine most of the distortion to the last two amplifier stages. *R-f feedback* can then be utilized to reduce the distortion in the last two stages. This type of feedback is no different from the common audio feedback used in high-fidelity sound systems. A sample of the output waveform is applied to the amplifier input to correct the distortion developed in the amplifier. The same advantages can be obtained at radio frequencies that are obtained at audio frequencies when feedback is used.

R-F Feedback Circuits R-f feedback circuits have been developed by the Collins Radio Group of *Rockwell-International* for use with linear amplifiers. Tests with large receiving and small transmitting tubes showed that amplifiers using these tubes without feedback developed signal-to-distortion ratios no better than 30 dB or so. Tests were run employing cathode-follower circuits, such as shown in figure 33A. Lower distortion was achieved, but at the cost of low gain per stage. Since the voltage gain through the tube is less than unity, all gain has to be achieved by voltage step-up in the tank circuits. This gain is limited by the dissipation of the tank coils, since the circuit capacitance across the coils in a typical transmitter is quite high. In addition, the tuning of such a stage is sharp because of the high-Q circuits.

The cathode-follower performance of the tube can be retained by moving the r-f ground point of the circuit from the plate to the cathode as shown in figure 33 B. Both ends of the input circuit are at high r-f potential so inductive coupling to this type of amplifier is necessary.

Inspection of figure 33 B shows that by moving the top end of the input tank down on a voltage-divider tap across the plate tank circuit, the feedback can be reduced from 100%, as in the case of the cathode-follower circuit, down to any desired value. A typical feedback circuit is illustrated in figure 34. This circuit is more practical than those of figure 33, since the losses in the input tank are greatly reduced. A feedback level of 12 dB may be achieved as a good compromise between distortion and stage gain. The volt-

age developed across C_2 will be three times the grid-cathode voltage. Inductive coupling is required for this circuit, as shown in the illustration.

The circuit of figure 35 eliminates the need for inductive coupling by moving the r-f ground to the point common to both tank circuits. The advantages of direct coupling between stages far outweigh the disadvantages of having the r-f feedback voltage appear on the cathode of the amplifier tube.

In order to match the amplifier to a load, the circuit of figure 36 may be used. The ratio of X_{L1} to X_{C1} determines the degree of feedback, so it is necessary to tune them in unison when the frequency of operation is changed. Tuning and loading functions are accomplished by varying C_2 and C_3. L_2 may also be varied to adjust the loading.

Feedback Around a Two-Stage Amplifier The maximum phase shift obtainable over two simple tuned circuits does not exceed 180 degrees, and feedback around a two-stage amplifier is possible. The basic circuit of a two-stage feedback amplifier is shown in figure 37. This circuit is a conventional two-stage tetrode amplifier except that r-f is fed back from the plate circuit of the PA tube to the cathode of the driver tube. This will reduce the distortion of both tubes as effectively as using individual feedback loops around each stage, yet will allow a higher level of overall gain. With only two tuned circuits in the feedback loop, it is possible to use 12 to 15 dB of feedback and still leave a wide margin for stability. It is possible to reduce the distortion by nearly as many dB as are used in feedback. This

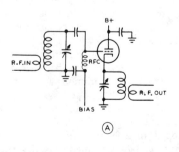

(A)

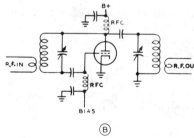

(B)

Figure 33

SIMILAR CATHODE FOLLOWER CIRCUITS HAVING DIFFERENT R-F GROUND POINTS

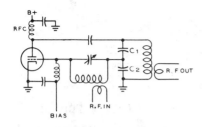

Figure 34

SINGLE STAGE AMPLIFIER WITH R-F FEEDBACK CIRCUIT

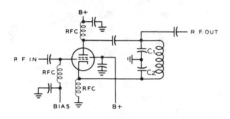

Figure 35

SINGLE STAGE FEEDBACK AMPLIFIER WITH GROUND RETURN POINT MODIFIED FOR UNBALANCED INPUT AND OUTPUT CONNECTIONS

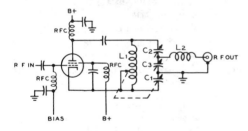

Figure 36

R-F AMPLIFIER WITH FEEDBACK AND IMPEDANCE MATCHING OUTPUT NETWORK

Tuning and loading are accomplished by C_2 and C_3. C_1 and L_1 are tuned in unison to establish the correct degree of feedback.

circuit has two advantages that are lacking in the single-stage feedback amplifier. First, the filament of the output stage can now be operated at r-f ground potential. Second, any conventional pi output network may be used.

R-f feedback will correct several types of distortion. It will help correct distortion

caused by poor supply regulation, too low grid bias, and limiting on peaks when the plate voltage swing becomes too high.

Neutralization and R-F Feedback   The purpose of neutralization of an r-f amplifier stage is to balance out effects of the grid-plate-capacitance coupling in the amplifier. In a conventional amplifier using a tetrode tube, the effective input capacity is given by:

Input capacitance $= C_{in} + C_{gp}(1 + A \cos \theta)$

where,

C_{in} equals tube input capacitance,
C_{gp} equals grid-plate capacitance,
A equals grid-to-plate voltage amplification,
θ equals angle of load.

In a typical unneutralized tetrode amplifier having a stage gain of 33, the input capacitance of the tube with the plate circuit in resonance is increased by 8 pF due to the unneutralized grid-plate capacitance. This is unimportant in amplifiers where the gain (A) remains constant but if the tube gain varies, serious detuning and r-f phase shift may result. A grid or screen-modulated r-f amplifier is an example of the case where the stage gain varies from a maximum down to zero. The gain of a tetrode r-f amplifier operating below plate-current saturation varies with loading so that if it drives a following stage into grid current the loading increases and the gain falls off.

The input of the grid circuit is also affected by the grid-plate capacitance, as shown in this equation:

$$\text{Input resistance} = \frac{1}{2\pi f \times C_{gp} (A \sin \theta)}$$

This resistance is in shunt with the grid current loading, grid tank circuit losses, and driving source impedance. When the plate circuit is inductive there is energy transferred from the plate to the grid circuit (positive feedback) which will introduce negative resistance in the grid circuit. When this shunt negative resistance across the grid circuit is lower than the equivalent positive resistance of the grid loading, circuit losses, and driving source impedance, the amplifier will oscillate.

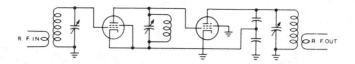

Figure 37

BASIC CIRCUIT OF TWO-STAGE AMPLIFIER WITH R-F FEEDBACK

Feedback voltage is obtained from a voltage divider across the output circuit and applied directly to the cathode of the first tube. The input tank circuit is thus outside the feedback loop.

When the plate circuit is in resonance (phase angle equal to zero) the input resistance due to the grid-plate capacitance becomes infinite. As the plate circuit is tuned to the capacitive side of resonance, the input resistance becomes positive and power is actually transferred from the grid to the plate circuit. This is the reason that the grid cur-

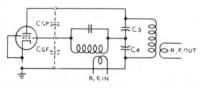

Figure 38

SINGLE STAGE R-F AMPLIFIER WITH FEEDBACK RATIO OF C_3/C_4 to C_{gp}/C_{gf} DETERMINES STAGE NEUTRALIZATION

rent in an unneutralized tetrode r-f amplifier varies from a low value with the plate circuit tuned on the low-frequency side of resonance to a high value on the high-frequency side of resonance. The grid current is proportional to the r-f voltage on the grid which is varying under these conditions. In a tetrode class-AB_1 amplifier, the effect of grid-plate feedback can be observed by placing an r-f voltmeter across the grid circuit and observing the voltage change as the plate circuit is tuned through resonance.

If the amplifier is over-neutralized, the effects reverse so that with the plate circuit tuned to the low frequency side of resonance, the grid voltage is high, and on the high frequency side of resonance, it is low.

Amplifier Neutralization Check A useful "rule of thumb" method of checking neutralization of an amplifier stage (assuming that it is nearly correct to start with) is to tune both grid

and plate circuits to resonance. Then, observing the r-f grid current, tune the plate circuit to the high-frequency side of resonance. If the grid current rises, more neutralization capacitance is required. Conversely, if the grid current decreases, less capacitance is needed. This indication is very sensitive in a neutralized triode amplifier, and correct neutralization exists when the grid current peaks at the point of plate current dip. In tetrode power amplifiers this indication is less pronounced. Sometimes in a supposedly neutralized tetrode amplifier, there is practically no change in grid voltage as the plate circuit is tuned through resonance, and in some amplifiers it is unchanged on one side of resonance and drops slightly on the other side. Another observation sometimes made is a small dip in the center of a broad peak of grid current. These various effects are probably caused by coupling from the plate to the grid circuit through other paths which are not balanced out by the particular neutralizing circuit used.

Feedback and Neutralization of a One-Stage R-F Amplifier Figure 38 shows an r-f amplifier with negative feedback. The voltage developed across C_4 due to the divider action of C_3 and C_4 is introduced in series with the voltage developed

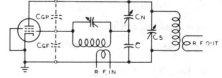

Figure 39

NEUTRALIZED AMPLIFIER AND INHERENT FEEDBACK CIRCUIT

Neutralization is achieved by varying the capacity of C_N.

across the grid tank circuit and is in phase-opposition to it. The feedback can be made any value from zero to 100% by properly choosing the values of C_3 and C_4.

For reasons stated previously, it is necessary to neutralize this amplifier, and the relationship for neutralization is:

$$\frac{C_3}{C_4} = \frac{C_{gp}}{C_{gf}}$$

It is often necessary to add capacitance from plate to grid to satisfy this relationship.

Figure 39 is identical to figure 38 except that it is redrawn to show the feedback in-

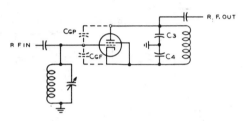

Figure 40

UNBALANCED INPUT AND OUTPUT CIRCUITS FOR SINGLE-STAGE R-F AMPLIFIER WITH FEEDBACK

herent in this neutralization circuit more clearly. C_N and C replace C_3 and C_4, and the main plate tank tuning capacitance is C_5. The circuit of figure 39 presents a problem in coupling to the grid circuit. Inductive coupling is ideal, but the extra tank circuits complicate the tuning of a transmitter which uses several cascaded amplifiers with feedback around each one. The grid could be coupled to a high source impedance such as a tetrode plate, but the driver then cannot use feedback because this would cause the source impedance to be low. A possible solution is to move the circuit ground point from the cathode to the bottom end of the grid tank circuit. The feedback voltage then appears between the cathode and ground (figure 40). The input can be capacitively coupled, and the plate of the amplifier can be capacitively coupled to the next stage. Also, cathode type transmitting tubes are available that allow the heater to remain at ground potential when r-f is impressed on the cathode. The output voltage available with capacity coupling, of course, is less than the plate-cathode r-f voltage developed by the amount of feedback voltage across C_4.

Part III Vhf Power Amplifiers

The representative circuits shown in Parts I and II of this chapter apply equally well to the lower part of the vhf portion of the spectrum as they do to the lower frequencies. Above approximately 100 MHz, however, the clear distinction between external lumped circuit parameters and the amplifying device becomes indistinct and different design techniques are required to achieve proper circuit efficiency.

11-14 Vacuum-Tube Limitations

The vacuum tube becomes progressively less efficient as the frequency of operation is raised, requiring more drive power for a given power output level. At the same time, the input impedance of the tube drops as does the maximum impedance realizable in the plate circuit. *Lead inductance* of tube and socket creates undesirable r-f voltage drops so that the available driving voltage does not appear across the tube elements (figure 1A). In addition, the interelectrode capacitance of the tube approaches a large fraction of the capacitance required to establish circuit resonance with the result that the tank circuit may "disappear" within the tube (figure 1B). The combination of lead inductance and interelectrode capacitance of the tube will cause an internal resonance in the upper vhf region, possibly leading to parasitic oscillation and instability.

Cathode Lead Inductance   Tube gain is adversely affected by *cathode lead inductance* which, in conjunction with grid-cathode capacitance, causes a resistive load to appear across the input of the tube. This load results from a voltage drop across the cathode lead inductance which drives the cathode as in a grounded grid amplifier stage. A portion of the drive signal thus appears in the output circuit (termed *feedthrough* power) which must be supplied by the driving stage. As the frequency of operation is raised, input loading due to cathode lead inductance rises, roughly as the square of the increase in frequency. Thus, input loading is nine times as great at 432 MHz as it is at 144 MHz for a given tube.

The cathode lead inductance may be neutralized by choosing a value of cathode bypass capacitance such that the total lead inductance (tube, socket, and stray circuit inductance) is approximately series-resonant at the operating frequency, as shown in figure 2.

Cathode lead inductance may also be neutralized by placing an inductance (L_s)

in series with the screen-to-ground circuit as shown in figure 3 or by utilizing the grid structure of the tube as a screen and placing the exciting signal on the cathode (figure 4). The cathode lead inductance is now a part of the input tuned circuit and the *grid lead inductance* (while having a voltage drop across it) usually is of much smaller magnitude than cathode lead inductance in a well designed vhf tube.

The grid lead inductance can either cause instability and a loss of drive voltage or it may provide a method of neutralizing the amplifier, as discussed in the previous part of this chapter.

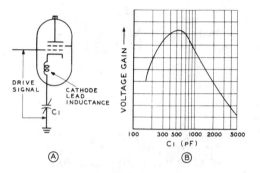

Figure 2

CATHODE LEAD INDUCTANCE

A—Cathode lead inductance is neutralized by series-resonant cathode circuit. B—Voltage gain of the tube may be peaked by adjustment of cathode bypass capacitor.

Screen Lead Inductance   *Screen lead inductance* may help or hinder the operation of the tube. Below the self-neutralizing frequency of the tube (see Part II, Section 11-8) screen lead inductance is detrimental to amplifier stability as r-f current flowing through the inductance will cause an unwanted r-f voltage to be developed on the screen element. At operating frequencies above the self-neutralizing frequency, a variable screen-bypass capacitor is sometimes added to allow the self-neutralizing frequency to be moved up to the operating frequency.

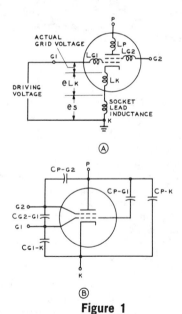

Figure 1

LEAD INDUCTANCE AND INTERNAL CAPACITANCE

A—Interelectrode capacitances of the tube may approach a large fraction of the capacitance required to establish circuit resonance. B—Lead inductance of the tube and socket creates voltage drops so that only a portion of the drive voltage appears between grid and cathode.

Input Capacitance   The *input capacitance* of a grid-driven tetrode is the sum of the grid-cathode and grid-screen capacitances. The larger the input capacitance the lower the reactance and the greater the exciting current needed to charge the capacitance. The driving stage must supply the current to charge this capacitance. Stray input capacitance external to the tube must be held to the minimum value, and peak driving voltage should be limited by operating with low bias to reduce the effects of charging current and accompanying waste of drive power. The charging current can cause heating of the tube seals and expansion and detuning of the resonant circuits.

The cathode-driven amplifier has a lower input capacitance for a given tube than the grid-driven equivalent since the input capacitance consists only of the cathode-grid capacitance, and its use is widespread in vhf equipment.

Feedback Capacitance   The *feedback capacitance* in a grid-driven amplifier is the grid-plate capacitance of the tube, which becomes a larger factor in circuit design as the frequency of operation is raised. The cathode-driven amplifier minimizes feedback capacitance since the cathode-plate capacitance is usually quite small in most vhf tetrode tubes, with the grid (or grids) shielding the output from the input circuit.

Regardless of circuitry, the higher the operating frequency is, the greater are the chances for amplifier instability due to r-f feedback from the output through the feedback capacitance of the tube to the input circuit.

Circuit and Tube Losses   The power losses associated with tube and circuit all tend to increase with frequency. In the vhf region all r-f current flows in the surface layers of a conductor because of *skin effect*. Resistance and r-f losses in a conductor increase with the square root of the frequency, since the layer in which the current flows decreases in thickness as the frequency of operation increases. Additional circuit losses will accrue due to *radiation of energy* from wires and components carrying r-f current. The power radiated from a short length of conductor increases as the square of the frequency.

Dielectric loss within insulating supports in the tube and in external circuitry increases directly with frequency and is due to the molecular movements produced within the dielectric by the electric field. Both dielectric and radiation loss contribute to a general reduction in tube and circuit efficiency as the frequency of operation is raised.

Transit-Time Effect   *Transit time* is the finite time an electron takes in passing from the cathode to the grid of a tube and is a function of the grid-to-cathode spacing and grid-to-cathode voltage, increasing as the frequency of operation is increased. If transit time is an appreciable fraction of one operating cycle, electrons in transit will be "out of step" with instantaneous grid potential, and the resulting plate current pulses are not as sharp and defined as the current pulses liberated

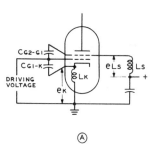

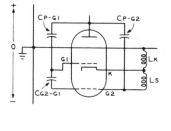

(A)   (B)

Figure 3
VHF SCREEN NEUTRALIZATION

A—Cathode lead inductance may be neutralized by placing inductance in series with screen-to-ground circuit. B—Cathode and screen lead inductances form bridge with grid-to-screen and grid-to-plate capacitances. Bridge balance places grid and cathode at same voltage level as far as internal feedback is concerned. Bridge is balanced by adjustment of screen inductor L_s.

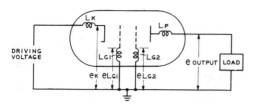

Figure 4

CATHODE-DRIVEN VHF AMPLIFIER

Cathode lead inductance is part of the input circuit and a degenerative signal now appears across grid-to-ground inductance. Grid inductance (L_G) may be used for neutralization of the stage when proper phase shift is present.

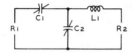

$$R_1 > R_2$$

$$X_{L_1} = Q_L R_2$$

$$X_{C_1} = R_1 \sqrt{\frac{R_2(Q_L^2+1)}{R_1} - 1}$$

$$X_{C_2} = \frac{R_2(Q_L^2+1)}{Q_L} \cdot \frac{1}{1-\left(\frac{X_{C_1}}{Q_L R_1}\right)}$$

$$Q_L = \text{LOADED Q OF NETWORK}$$

Figure 5

T-NETWORK FOR
CATHODE-DRIVEN AMPLIFIER

Simple T-network can be used for step-down or step-up transformation between cathode impedance and nominal 50-ohm termination. In this circuit, R_1 is greater than R_2. Network Q of 2 to 5 is commonly used.

from the cathode. This increases the conduction angle of operation and reduces the plate efficiency of the tube.

11-15 Input and Output Circuitry

Single-ended vhf amplifiers make use of linear versions of parallel-tuned or network circuits in the input and output configurations. A practical and simple input circuit for a cathode-driven amplifier is the version of the T-network shown in figure 5. For the

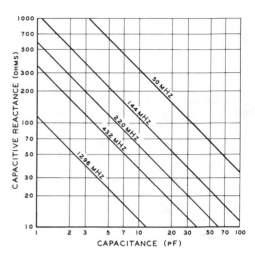

Figure 6

REACTANCE CHART FOR VHF BANDS

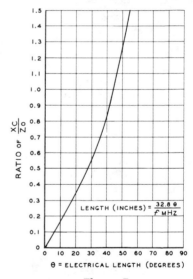

Figure 7

ELECTRICAL LENGTH OF LINE AS
FUNCTION OF X_c/Z_o

lower portion of the vhf region the network can be made up of lumped constants.

The output circuitry, in addition to matching the tube to the transmission line may also be called upon to dissipate the anode heat of the tube. In order to do this, and to prevent rapid detuning of the circuit with rising temperature, the circuit Q

Figure 8

HALF-WAVELENGTH STRIPLINE PLATE CIRCUIT

Tuning capacitor is placed at the high-impedance end of the line away from the tube. Inductive
output coupling loop is placed at a low-impedance on the line, near the center.

should be as low as practicable. The strip-line technique (see section 11-18) is often used since it provides a large thermal capacity and requires a minimum of machine work, as compared to a coaxial cavity.

The stripline (or cavity) can operate in the 1/4-, 1/2-, or 3/4-wave mode, with increasing loaded Q, increasing impedance, and decreasing bandwidth as the electrical length is increased. The impedance of the output circuit is limited by tube and stray output capacitance:

$$X_c = Z_0 \times \tan l$$

where,

X_c = tube and stray output capacitance,

Z_0 = characteristic impedance of line or cavity,

l = length of line or cavity in electrical degrees.

For minimum loaded Q and greatest bandwidth, the ratio X_c/Z_0 should approximate 0.5 for a quarter-wave circuit and 0.83 for a half-wave or three-quarter-wave circuit.

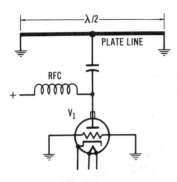

Figure 9

**HALF-WAVELENGTH PLATE CIRCUIT
WITH TUBE AT CENTER**

Stripline or coaxial circuit design may be aided by the charts of figures 6 and 7. For example, a 3CX1000A7 high-mu triode in grounded-grid configuration has an average output capacitance (plate-to-grid) of 15 pF. Circuit stray and tuning capacitance are estimated to total 15 pF. At 144 MHz, X_c is about 35 ohms for the total value of 30 pF. For an X_c/Z_o ratio of 0.5 and given the X_c value of 35 ohms the line impedance should be about 70 ohms. From figure 7, the point $X_c/Z_o = 35$ is found and the line length noted to be 27 electrical degrees, or about 6⅛ inches. This is the total physical length of the stripline and includes the path through the tube anode cooler and tuning capacitor. If this short a line poses coupling problems, the experimenter may go to a longer half wavelength line, with the attendant problems of increased circuit Q for the longer length.

The line, in any event, resonates with a fixed value of capacitance and decreasing line impedance increases the electrical length, whereas increasing line impedance decreases the electrical length.

| **The Half-Wavelength Line** | The half-wavelength line or cavity is useful when the capacitance of the tube is appreciable and the use of a |

quarter-wavelength line places the low impedance end of the line close to the tube socket terminals. A single ended, half-wave stripline circuit is shown in figure 8 with the tuning adjustment placed at the high-

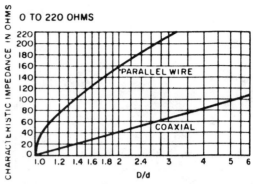

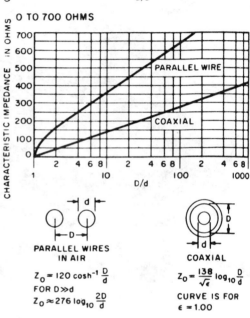

Figure 10

**CHARACTERISTIC IMPEDANCE OF
PARALLEL AND COAXIAL LINES
HAVING AIR DIELECTRIC**

impedance end of the line at the point of low impedance and minimum r-f voltage. The whole circuit, including the output capacitance of the tube, becomes an electrical half wavelength, capacitively loaded at one end by the tube, and at the other by the tuning capacitor.

Alternatively, the tube may be placed at the center of a half-wavelength line, as shown in figure 9. Both ends of the line are at r-f ground potential. The line is adjusted to resonance by means of a small

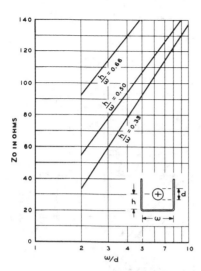

Figure 11

CHARACTERISTIC IMPEDANCE OF OPEN TROUGH LINE FOR VARIOUS HEIGHT-TO-WIDTH RATIOS

variable capacitor placed near the plate of the tube.

This arrangement distributes the heavy r-f current flowing in the plate circuit more evenly around the anode of the tube than does the single-ended arrangement shown in figure 8. R-f output may be taken by means of a variable "flipper" capacitor placed at the high-potential of the line.

Tank Circuit Impedance The characteristic impedance of the transmission line making up the resonant tank circuit must be known in order to determine the physical attributes of the configuration. The characteristic impedance of parallel and coaxial lines having an air dielectric are given in figure 10. The impedance of an open *trough line* having height to width ratios of 0.33, 0.50 and 0.66 may be determined from the graph of figure 11. The characteristic impedance of a strip line having various height to width ratios can be calculated with the aid of the nomograph of figure 12.

11-16   Operating Parameters

When operating a power tube in the vhf region it is recommended that minimum

Table 1
Comparison of HF and VHF Operating Parameters for Two Power Tubes

TUBE: 4CX250B TYPICAL CLASS-C OPERATION	150 MHz	500 MHz
Plate Voltage	2000	2000
Screen Voltage	250	300
Grid Voltage	−90	−90
Plate Current (mA)	250	250
Screen Current (mA)	20	10
Grid Current (mA)	26	10
Plate Input Power (W)	500	500
Plate Output Power (W)	390	300
Heater Voltage	6.0	5.5
Efficiency (%)	78	60
TUBE: 6146 TYPICAL CLASS-C OPERATION	**50 MHz**	**175 MHz**
Plate Voltage	750	400
Screen Voltage	160	190
Grid Voltage	−62	−54
Plate Current (mA)	120	150
Screen Current (mA)	11	10
Grid Current (mA)	3	2
Plate Input Power (W)	90	60
Plate Output Power (W)	70	35
Efficiency	78	58

drive power and maximum plate loading be achieved. This will provide good efficiency and hold circulating r-f currents to a minimum.

A minimum amount of control grid bias should be employed and, in the case of a tetrode, a high value of screen voltage should be used. Screen current should be monitored so that maximum screen dissipation is not exceeded.

A minimum of r-f excitation should be used that will allow the required level of plate efficiency, even though the dc grid current is considerably lower than the value expected at lower frequencies.

Generally speaking, for a given power input, it is advisable to run the lowest plate voltage possible and make up the power level by an increased value of plate current. Apparently, the use of lower r-f voltages in the plate circuit is desirable. Fortunately, this condition reduces driving power and

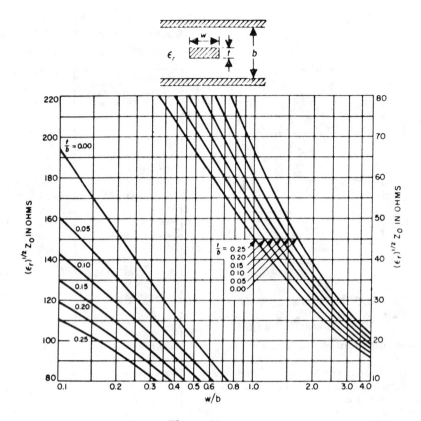

Figure 12

PLOT OF STRIP TRANSMISSION LINE Z₀ VERSUS w/b FOR VARIOUS VALUES OF t/b.

For lower left family of curves, refer to left-hand ordinate values; for upper right curves, use right-hand scale.

screen current (in the tetrode) and improves life expectancy of the tube (Table 1).

Tube Cooling   A tube operated in the vhf region is subject to greater heating action than one operated at a lower frequency. This results from the flow of larger r-f charging currents in the tube capacitances, by dielectric losses and through the tendency of electrons to bombard parts of the tube structure other than the grid and plate. Greater cooling is therefore required at the higher frequencies. Even if no cooling is specified for a particular tube type, ample free space for air circulation is required or else air must be forced past the tube.

Filament Voltage Derating   At high frequencies the filament voltage of a power tube should be maintained at the operating voltage plus or minus five percent. At frequencies above 250 MHz, transit-time effects begin to influence the cathode temperature. The amount of drive power diverted to cathode heating will depend on frequency, plate current, and drive level. When the tube is driven to maximum input as a class-C amplifier, cathode back-heating is a maximum and the filament voltage should be reduced. For example, the 4CX250B filament voltage is reduced to 5.75 volts (from a normal value of 6.0 volts) at frequencies between 300 and 400 MHz and to 5.5 volts at frequencies between 400 and 500 MHz. Further reduction in filament voltage may be needed in pulse service above 500 MHz. Filament derating information on other tube types can be obtained from the manufacturer.

11-17 Solid-State VHF Circuitry

Power transistors are available that provide up to 150 watts power output to over 200 MHz and up to 100 watts power output to 500 MHz for class-C service. Experimental transistors can provide upward of 50 watts in class-C operation at frequencies in excess of 1000 MHz. These devices make practical, low cost solid-state power amplifiers for amateur f-m service up through 432 MHz.

Vhf power transistors are tailored for operation over certain popular frequency ranges (25-80 MHz, 100-200 MHz, or 200-600 MHz, for example) and the power capability and reliability require that the user operate the device within the intended range, since the ruggedness of the vhf power transistor is a function of *both* voltage and frequency. A transistor rated for operation near 175 MHz will be less rugged at 100 MHz and may be too delicate for use at 30 MHz. In addition, the device must be operated well within the manufacturer's rating and due attention paid to the standing-wave ratio appearing on the transistor output load network.

For f-m service, the vhf transistor is operated in the zero bias, class-C mode and stripline circuitry is commonly employed.

Transistor Service Classes   Solid state devices are classified as to the stage mode of operation and efficiency much in the same manner as vacuum tubes. The classes are:

Class A—Bias and drive signal are adjusted to allow continuous output current at all times. Current passing through the load resistor generates the output voltage. Efficiency is low, in practice running about 25 percent. This class of service is generally restricted to audio and linear r-f applications.

Class B—Bias and drive signal effect a 50-percent output current duty cycle for

each element in the circuit, which are generally arranged in push pull. Improved efficiency is achieved with good linearity. When collector voltage is at its maximum value, current is zero and when collector voltage is minimum, current is maximum. Input power is proportional to average load current and efficiency in practice runs about 55 percent. Amplitude of the output is independent of the supply voltage up to amplifier saturation.

Class C—Bias is adjusted so that drive signal voltage produces output current for less than half the operating cycle. The device normally operates in a saturated condition and is relatively insensitive to drive variations. Modulation can be achieved by variations in the collector supply voltage. Linearity is very poor and efficiency in practice runs about 75 percent. Use is generally restricted to r-f applications.

Class D—Configuration is push-pull operating in a switching mode. A square wave is delivered to a tuned circuit that passes only the switching (fundamental) frequency. High linearity and very high efficiency (approaching 100 percent) are possible. Use is generally restricted to audio and low frequency r-f service.

Class E—A modified switching-mode design using only one device to combine switching action with the transient response of the tuned output circuit to achieve high efficiency. Useful in the medium-frequency range.

Class F—Similar to other switching modes except that the tuned output circuit introduces a third-harmonic component properly phased to improve output power capability. Efficiency in practice runs as high as 90 percent.

Class G—Two class-B amplifiers having different voltages are combined. Small-amplitude signals are boosted by the device operating at the lower voltage, resulting in high efficiency for audio signals.

Class H—A class-B amplifier wherein the collector voltage is varied by an external circuit so that it remains just above the minimum value required to prevent saturation. This provides high efficiency for audio signals.

Class S—A pulse-width modulation technique wherein the devices are switched by

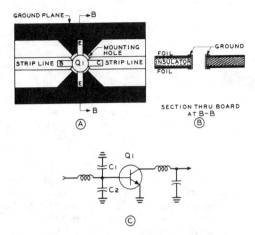

Figure 13

VHF TRANSISTOR MOUNTED IN STRIPLINE CONFIGURATION

(A) Two emitter leads of transistor are connected to ground plane. Base and collector leads are soldered to resonant striplines. Dual-surface board is used with top and bottom ground planes connected together with straps under each emitter lead (B). Small ceramic chip capacitors are often placed in parallel at base terminal to form portion of input matching network (C). Extremely low impedance to ground is required at this point because current flowing in capacitors is heavy.

a control frequency several times higher than the signal frequency being amplified. Efficiency in practice runs about 90 percent.

Circuit Techniques   Transistor input and output impedances are extremely low and stray circuit inductance and ground current return paths play a large role in circuit design. Impedance levels of one ohm, or less, are common and lead lengths in r-f circuitry of 0.1 inch or so become quite critical. Special vhf ceramic capacitors having ribbon leads may be used in impedance matching circuits and uncased mica/porcelain chip capacitors used for high r-f current paths. The technique of

grounding the r-f components becomes a very critical aspect of the circuit design as a result of the very low impedance characteristics of the transistor.

The common-base or common-emitter lead should be grounded *at the body* of the transistor for proper performance. With the stripline package, the device may be mounted to a ground plane (such as a printed-circuit board) as shown in figure 13. Dual-surface board is used, with the top and bottom ground planes connected together using straps under each emitter lead. Capacitors in the input matching net-

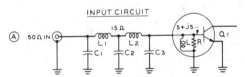

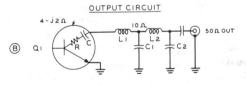

Figure 15

INPUT AND OUTPUT MATCHING NETWORKS

(A) Input impedance of vhf transistor, typically, is inductive. Two-section network with center impedance of 15 ohms matches 50-ohm input to the base circuit of the transistor. (B) Output impedance presents a low value of series reactance. Two-section network with center impedance of 10 ohms provides proper match to 50-ohm termination. Circuit Q of networks is held to 2 or 3 for optimum bandwidth.

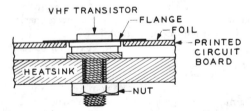

Figure 14

STUD-MOUNTED TRANSISTOR IS BOLTED TO HEAT SINK

Flange connections of transistor should not be twisted or bent. Printed-circuit board is elevated above the heat sink so that flange leads are not stressed and provide shortest possible connection to the stripline. Silicone grease is used on the stud to lower thermal resistance between transistor and heat sink.

work require a good ground and extremely low inductive impedance. Two small chip capacitors are often used in parallel at this point, as shown in the illustration.

The stud-mounted transistor should be mounted on a flat surface (figure 14) for proper heat transfer. The flange connections should not be twisted or bent, and should not be stressed when the transistor is torqued to the heat sink. Silicone grease should always be used on the stud to lower the thermal resistance between transistor and sink.

The transistor user should remember that the vhf power transistor will not tolerate overload as the thermal time constant of the small chip is very fast, thus, the allowable dissipation rating of the transistor must be capable of handling momentary overloads. Generally speaking, for class-C operation, the r-f output level of the vhf power transistor should be held to about 50 percent of the power dissipation rating.

VHF Circuit Design   Vhf transistor circuitry involves impedance matching networks and dc feed systems. It is common practice to make networks up of simple, cascaded L-sections which provide low-pass filter characteristics and ample impedance transfer. If the Q of each

step of the network is held to a low figure (2 or 3), the bandwidth of the amplifier will be wide enough to cover any of the vhf amateur bands. Representative two-section networks for input and output terminations are shown in figure 15.

The transistor input impedance in the vhf range is usually inductive and a shunt capacitor (circuit A, capacitor C_3) is used to cancel the reactive portion of the impedance. Two series-connected L-sections are used, the first matching the 50-ohm input impedance down to 15 ohms and the second matching down from 15 ohms to the 5-ohm impedance level of the transistor. The intermediate impedance point is often chosen as the mean value between the output and input impedance levels. If a stripline configuration is used, line impedance may be taken as the mean value to simplify calculations.

The vhf transistor generally has a capacitive reactance and the proper load impedance is usually given by the manufacturer. A series inductance (circuit B, inductor L_1) equalizes the series capacitance of the device and two series-connected L-sections step the transistor impedance level up to 50 ohms.

A representative three-stage, vhf amplifier using conventional tuned circuits is shown in figure 16.

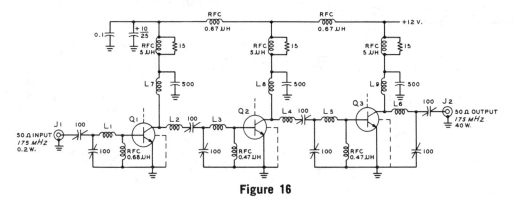

Figure 16

40-WATT, 175-MHZ THREE STAGE AMPLIFIER

L_1—#16 wire, ½-inch long
L_2, L_4—#16 wire about ¾-inch long formed into "U"
L_3, L_5—¼" × ⅛" strap, .005" thick about ½" long
L_6—½" × ¼" strap, .005" thick about ½" long
L_7—8 turns #16 e., ¼" diam.

L_8—6 turns, as L_7
L_9—5 turns, as L_7
Q_1—CTC type B3-12
Q_2—CTC type B12-12
Q_3—CTC type B40-12
Note: 100-pF capacitors are mica compression type. (All transistors by Communications Transistor Corp.)

DC Feed Systems Design The dc feed network permits the operating voltages to be applied to the transistor without interfering with the r-f circuitry. Voltages may be fed to the transistor via r-f chokes, which must be carefully designed in order to prevent low-frequency parasitic oscillations. Transistor gain increases rapidly with decreasing signal frequency and a figure of 40 dB is not uncommon for low-frequency gain. The dc feed network therefore must present a load impedance which will not sustain low-frequency oscillation. This may be done by using as small r-f chokes as possible consistent with the operating frequency and

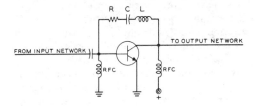

Figure 17

NEGATIVE COLLECTOR FEEDBACK DECREASES LOW-FREQUENCY STAGE GAIN

impedance level and large bypass capacitors. In addition negative collector feedback can be used to decrease the stage gain below the design frequency (figure 17).

Frequency Multipliers Although single-transistor frequency multipliers are most common, it is possible to use the push-pull multiplier for high order odd multiples and the push-push multiplier

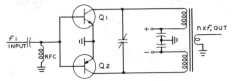

Figure 18

COMPLEMENTARY BASE-DRIVEN MULTIPLIER

Circuit may be considered to be either push pull or push push depending on phasing of the collector windings. Only one winding need be reversed to change mode of operation.

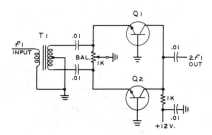

Figure 19

BROADBAND PUSH-PUSH DOUBLER

Balancing potentiometer permits attenuation of fundamental and third harmonic levels when circuit is used as a frequency doubler.

for high order even multiples of the fundamental frequency.

It is possible to build multipliers using bipolar transistors that are impossible to realize with tubes, because both npn and pnp types of active devices are available.

Figure 18 shows a complementary base-driven frequency multiplier. It may be considered to be either a push-pull or a push-push configuration depending upon the phasing of the collector windings. Only one winding need be reversed to change from one design to the other since it is the balance of the circuit, in addition to the selectivity of the output tank, that attenuates adjacent harmonics in the output. A broadband hf push-push doubler is shown in figure 19. In this configuration, the amplitudes of the fundamental and third-harmonic signals are respectively 28 dB and 32 dB below the level of the second harmonic output signal.

A second mechanism that may be used for frequency multiplication makes use of the base-collector depletion capacitance and is called *parametric multiplication* (figure 20). A number of idler circuits are used to reflect undesired harmonics back to the collector-base capacitance.

11-18 Stripline Circuitry

Stripline, or *microstripline*, circuitry is universally used for solid-state vhf and uhf amplifiers. The microstrip line most com-

Figure 20

PARAMETRIC FREQUENCY MULTIPLIER

Bipolar transistor makes use of base-collector depletion capacitance to work as frequency multiplier. Idler circuits are used to reflect undesired harmonics back to collector-base capacitance.

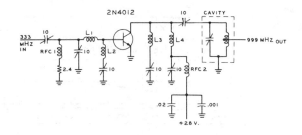

monly employed is a single conductor supported by a low loss, high dielectric material (figure 21). The dielectric material is affixed to a ground plane and has the ability to reduce the physical size of the line for

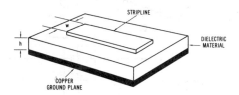

Figure 21

MICROSTRIP LINE

The microstrip line consists of a conductor above a ground plane. The strip is bonded to a dielectric sheet, the other side of which is bonded to the metallic ground plane. The physical length of the stripline is a function both of the electrical length, the characteristic impedance of the line, and the dielectric.

any operating frequency. Propagation along the line is the same as for a coaxial line; both propagate in the TEM mode. The combination of the dielectric and the ground plane is called a *substrate*.

The characteristic impedance of a stripline is determined by its width (W), the thickness of the dielectric material (h), and the *dielectric constant* of the material (ε).

The power handling capability of the stripline circuit is limited by dielectric heating and breakdown. Heating is due to resistive loss in the strip conductor and r-f power loss in the dielectric. The cross section of the conductor and the ambient temperature of the dielectric also influence the power handling capability.

For general use up to 150 MHz, at power levels up to 100 watts or so, G-10 epoxy-

glass board ($\varepsilon = 4.8$) is commonly used. The losses in this material increase rapidly with frequency, and for work above 150 MHz teflon fiberglass board ($\varepsilon = 2.55$) is preferred.

Because of the stripline's asymmetrical construction, the *relative dielectric constant* (ε_R) differs from that of the dielectric substrate and also changes over the impedance range and with strip width. Generally speaking, the wider the strip, the closer the relative dielectric constant is to the constant of the material. The relative dielectric constants of several materials are given in figure 22 and the characteristic impedance of various microstrip lines is given in figure 23.

Once the stripline design is formulated a choice of board must be made. When the line impedance is known, the width/to height ratio can be determined from figure 23 for a particular board material. Then the relative dielectric constant is found from figure 22.

For example, a 50-ohm line is required on a 1/16" (.0625") thick G-10 epoxy board. Using figure 23, the 50-ohm line is followed across until it intersects with the ε curve for epoxy-glass material. At this point, drop down to the W/H scale and read the width/height ratio as 2.3.

The height (thickness) of the board is .0625", therefore the required width of the strip is .0625 × 2.3 = .1438" to a first approximation.

Referring to figure 22, it will be noted that for a W/H ratio of 2.3 and an ε of 4.8, the *relative dielectric constant* is about 3.5. Referring back to figure 23, it is noted that the W/H ratio must be modified slightly to accommodate the 50-ohm impedance. Since a curve for the new ε_R is not shown, an interpolation must be made between the closest curves (2.55 and 4.8). The revised width/height ratio is now about 2.2 and

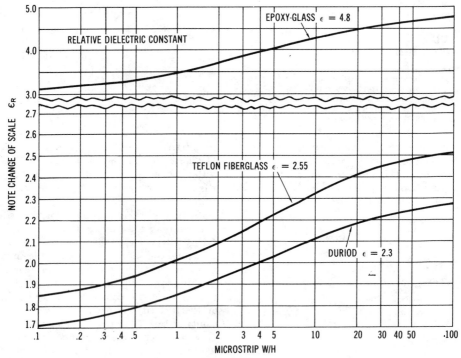

Figure 22

RELATIVE DIELECTRIC CONSTANT OF SUBSTRATES

The wider the stripline, the closer the relative dielectric constant is to the constant of the material.

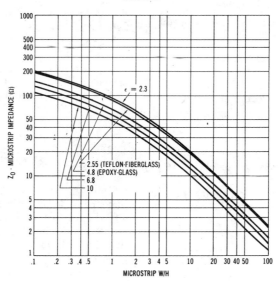

Figure 23

IMPEDANCE VS. WIDTH/HEIGHT RATIO FOR STRIPLINE

the revised strip width is $.0625 \times 2.2 = .1375''$.

Stripline Components Circuit elements may be modeled in stripline configuration. The normal lumped elements may be approximated by means of distributed lines (figure 24). Note that in all cases the length of the distributed element is less than a quarter-wavelength. The effective dieletric constant used to determine line lengths is:

$$\lambda = \frac{\lambda_o}{\sqrt{\varepsilon_R}}$$

where,

λ equals line length in centimeters,
λ_o equals wavelength in free space,
ε_R equals relative dielectric constant.

For wide lines with large width/height ratios the effective dielectric constant is nearly equal to the actual dielectric constant of the material.

Circuit Design Transistor power amplifier circuit design primarily involves impedance matching networks and dc feed networks. The impedance matching networks are usually constructed of L-net-

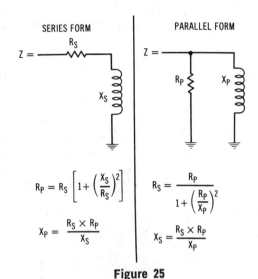

Figure 25

SERIES-PARALLEL CONVERSION EQUATIONS

SERIES FORM

$$Z = \quad R_S \quad X_S$$

$$R_P = R_S \left[1 + \left(\frac{X_S}{R_S} \right)^2 \right]$$

$$X_P = \frac{R_S \times R_P}{X_S}$$

PARALLEL FORM

$$Z = \quad R_P \quad X_P$$

$$R_S = \frac{R_P}{1 + \left(\frac{R_P}{X_P} \right)^2}$$

$$X_S = \frac{R_S \times R_P}{X_P}$$

works. In addition to matching, the L-sections provide a low-pass filter for input and output to reduce troublesome harmonics. If the Q of each matching step is held low (Q of 2 to 3), the bandwidth of the resulting amplifier will be greatest. The values for the L, C, and Q required are usually determined using a Smith Chart.

Figure 24

STRIPLINE CIRCUIT ELEMENTS

A—Width of stripline is reduced to simulate inductance. B—High impedance line shunted to ground simulates a shunt inductance. C—Break in stripline simulates series capacitance. For large values of series capacitance a chip capacitor is inserted in the line. D—Shunt capacitor is simulated by widening line to form a low-impedance section which increases capacitance to ground. Complete information on this circuitry is given in "Stripline Circuit Design," by H. Howe, Artech House, Dedham, MA (1964).

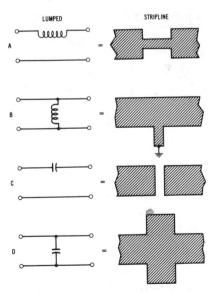

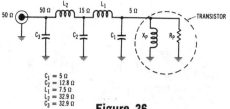

Figure 26

INPUT MATCHING NETWORK FOR VHF POWER TRANSISTOR

Network is made up of a series of L sections.
The transistor input impedance is usually inductive due to lead inductance inside the package.

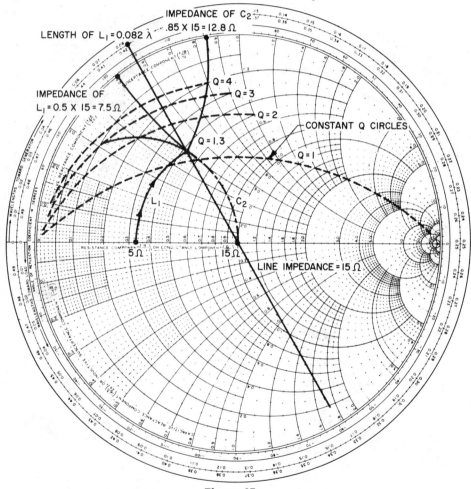

Figure 27

REACTANCE PORTION OF IMMITANCE CHART

Constant-Q circles are added to the chart. For this example the chart is normalized at 15 ohms.
Complete chart with admittance circles is shown in figure 30.

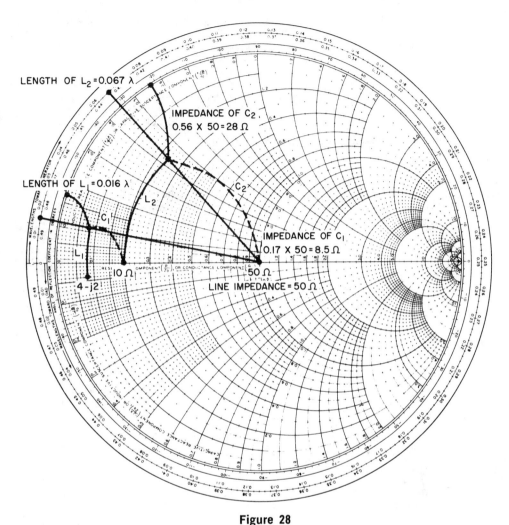

Figure 28

OUTPUT MATCHING NETWORK

Reactance portion of Immittance Chart used in determining components of output matching network. Complete chart is shown in figure 30.

Input and load impedances for transistors are usually given on the data sheet in either series or parallel equivalents which are mathematically related and may be converted back and forth (figure 25).

In designing a matching network, it is common to work *from the transistor to the 50-ohm termination*. If the first matching component is to be a shunt element, the parallel-equivalent impedance is used. The series equivalent impedance is used when the first matching element is a series element.

The Input Matching Network   A representative input matching network is shown in figure 26, with typical impedance levels marked. The transistor input impedance is usually inductive due to lead inductance inside the package. In order to maintain the lowest loaded Q for the first matching step, the first component used should be a shunt capacitor equal in impedance to the input impedance (X_p). This makes the impedance at the input of the transistor purely resistive and equal to R_p. If the input capacitance reactance is less than about 8 ohms, it is best to use two capacitors in parallel back to each emitter lead to minimize inductance and equalize ground currents.

If the parallel input impedance is higher than 15 ohms, then the matching network may require only one section. However if it is quite low (as low as 2 ohms, in some cases), then two L sections probably will be required. If two sections are required, the intermediate impedance point should be the geometric mean of the input and output impedances of the network.

Values for the stripline components are determined with the use of a special Smith Chart that has *admittance circles* overprinted on it. A suitable chart, *Normalized Impedance and Admittance Coordinates,* Smith Chart form ZY-01-N may be obtained from Analog Instruments Co., New Providence, NJ 07974. The impedance portion of the chart is printed in red and the admittance part in green (figure 30).

Somewhat easier to work with is an oversize *Immittance Chart* (form 2308) available from Cincinnati Electronics Co., 2630 Glendale-Milford Rd., Cincinnati, OH 45241.

Using the appropriate chart, an intermediate impedance point is chosen for the network (figure 27). The best choice of stripline impedance for easiest calculation is equal to this value—in this case 15 ohms. The Smith Chart is normalized to 15 ohms to make the calculations.

Start at the transistor impedance (5 ohms) on the chart and progress clockwise on a circular path, with the chart center as origin, until an admittance circle is reached which also passes through the desired output impedance for this portion of the network (15 ohms). Note that the values of L_1 and C_1 are read directly on the outer wavelength perimeter of the chart. The circuit Q is interpreted between the dashed constant-Q arcs which can be added to the chart. The additional L section is calculated in the same manner.

(Note: The chart shown in figure 27 does not indicate the green admittance circles which do not reproduce in reprinting.)

The Output Matching Network   The transistor manufacturer often specifies the series load impedance required to obtain rated specifications. Assume it is $4 + j2$ ohms. The network is started from the conjugate of this value on the impedance-admittance Smith Chart (figure 28). In this case, the complex conjugate figure is $4 - j2$. The same technique as used on the input matching network is followed and the final value obtained is that value seen "looking into" the network. Since this is a single section design, the impedance chosen for the microstrip is 50 ohms and the chart is normalized at this value.

Note that in this case, as well as the input matching network, the length of the stripline inductances in terms of wavelengths may be read around the perimeter of the chart.

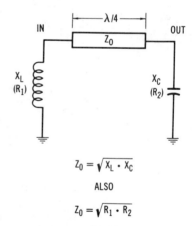

$$Z_0 = \sqrt{X_L \cdot X_C}$$

ALSO

$$Z_0 = \sqrt{R_1 \cdot R_2}$$

Figure 29

QUARTER-WAVE STRIPLINE TRANSFORMER

Matching Transformers   Impedance matching with striplines may take the form of quarter- and eighth-wave matching transformers or stubs. A quarter-wave transmission line transformer can invert and transform a reactance (figure 29). Inductive reactance at the input port will

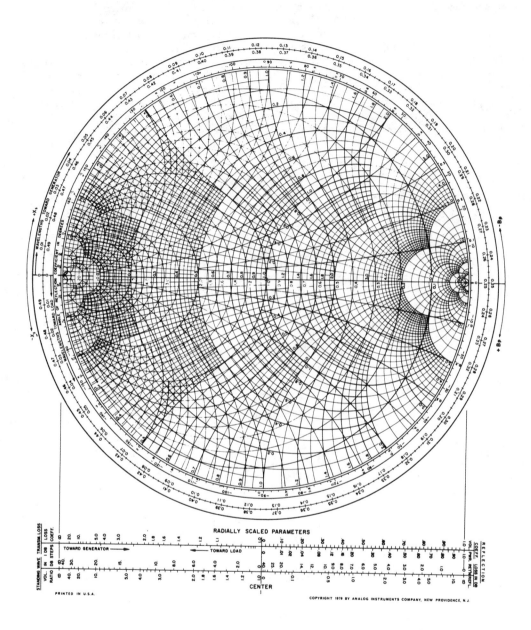

Figure 30

IMMITANCE CHART

The immitance version of the Smith Chart combines normalized impedance configuration with overlay of admittance circles.

appear as a capacitive reactance at the output port, and vice-versa. At the same time, the magnitude of the transformation can be modified by adjusting the characteristic impedance of the quarter-wave transformer section.

An eighth-wave matching section has useful properties when used as a shunt matching element. If the line is terminated in an open circuit, the reactance is capacitive and when the line is terminated in a short, the reactance is inductive. In either case, the reactance value is equal to the characteristic impedance of the line. One use of an eighth-wave line is to replace the shunt capacitor on the input of a transistor, as shown. A second useful property of an open eighth-wave line is that it appears as a short at the second harmonic frequency thus simplifying output filtering circuits.

CHAPTER TWELVE

Frequency Synthesis

Frequency synthesis is a technique that has gained widespread use in recent years, especially in radio communication services which are channelized. In Class D Citizens Radio Service and others, channelization is dictated by law; in vhf amateur repeaters the channelization is by mutual agreement among the users. But regardless of how the channel assignment occurs, this arrangement to operate at discrete (equally spaced) frequencies makes the use of frequency synthesis attractive.

The task of frequency synthesis may be accomplished in several ways, but the fundamental concept is to use one or more reference oscillators and combine their outputs to produce a multiplicity of output frequencies that are different from the reference frequencies.

referred to as the *direct method* of frequency synthesis and most of the early frequency synthesizers used this technique. The direct synthesis method has the advantage of being extremely fast-switching compared to other techniques.

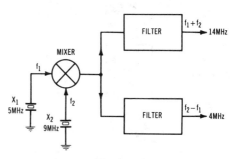

Figure 1

SIMPLE MIXER SYNTHESIS

Two frequencies are mixed to produce sum and difference frequencies.

12-1 Synthesis Techniques

The first and most straightforward method of synthesis is by mixing two reference frequencies together and filtering the output to exclude one of the two resultant mixer outputs (figure 1). As an example, 5 MHz and 9 MHz reference signals may be applied to a mixer stage and the output filtered to pass the *difference frequency* of 4 MHz; or the output may be filtered to pass the *sum frequency* of 14 MHz. This technique is often

Figure 2 illustrates a simple frequency synthesizer utilizing eleven crystals to produce 23 channels in the 27-MHz region. The drawing shows how channel #1 is synthesized and the frequencies involved in generating the remaining 22 channels. This approach reduces the number of crystals for a 23-channel transmitter by more than one half.

The technique can be extended to the point where one reference crystal frequency can be used to generate frequencies every 1.0 Hz (or finer) over the entire band of 1.0 Hz to 10 MHz.

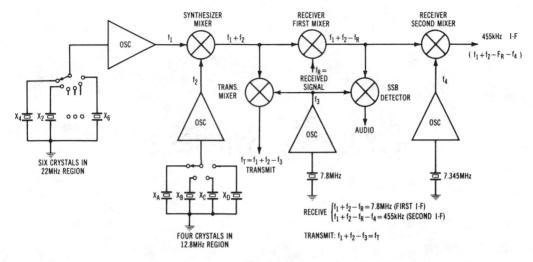

Figure 2

CRYSTAL SYNTHESIS FOR SSB TRANSCEIVER

Twelve crystals are used to provide 23 channels in the 27-MHz region.

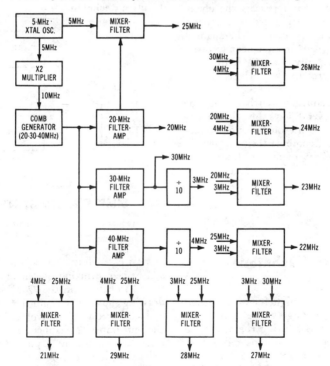

Figure 3A

"BUILDING BLOCK" SYNTHESIZER

Synthesizer creates signals at 21, 22, 23, 24, 26, 27, 28, and 29 MHz from 5-MHz reference oscillator. These frequencies are mixed as shown in figure 3B.

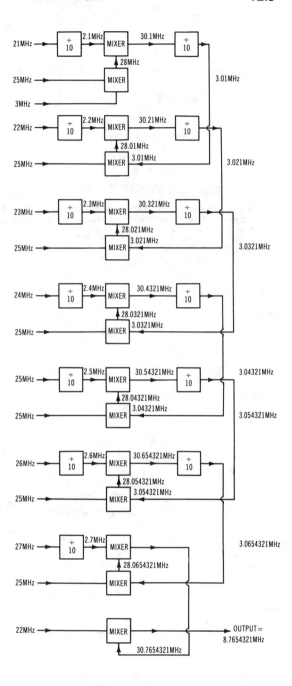

Figure 3B

FIFTEEN MIXERS PROVIDE DESIRED FREQUENCY

Frequencies derived from the "building block" generator of figure 3A are again mixed to provide specimen frequency of 8.7654321 MHz. Note that readout becomes more fine in each succeeding mixing process. (Read top to bottom of diagram.)

Single Crystal Synthesizer A single crystal synthesizer is shown in figure 3. A stabilized 5-MHz reference crystal oscillator creates a group of "building block" frequencies from the reference frequency by frequency multiplication and division (illustration A) and then mixes them together in eight modules that each contain one or two mixers and one or two divide-by-ten counters (B) (see Chapter 4). This technique al-

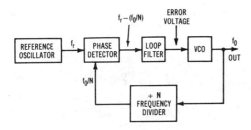

Figure 4

BASIC PLL SYNTHESIZER

If the frequency of the VCO is divided by ten and compared to a reference frequency which is one-tenth the VCO frequency, a phase comparison can be made. The output of the comparator (phase detector) is filtered and used to control the VCO frequency.

lows the eight mixer-divider modules to be nearly identical. The diagram may seem complex but such designs are being used and work reliably using solid state components.

The advantage of this direct method of frequency synthesis is that frequency changes can be made in milliseconds or less which allows rapid frequency hopping, essential in special modes of transmission. These special modes are generally under computer control.

The PLL Synthesizer　　The *phase-locked loop* (PLL) system of frequency synthesis is the most used method in amateur equipment. A basic circuit is shown in figure 4. The use of a *voltage-controlled oscillator* (VCO) at the output allows direct output at the desired frequency. The VCO has relatively low inherent frequency stability; in fact the VCO is voltage-sensitive and varying the dc voltage input to its control port varies the output frequency in a predictable way.

If the frequency of the VCO (f_o) is divided by 10, for example, and compared to a *reference frequency* (f_r) which is approximately one-tenth of f_o, a phase comparison between f_r and $f_o/10$ can be made in a phase comparator. If f_o is fairly close to ten times f_r, the output of the phase comparator will be a low frequency that is the difference between f_r and $f_o/10$. This phase comparator output frequency is filtered by a simple RC low-pass filter (the *loop filter*)

and is used to control the VCO at its control port. If the loop is properly designed, the VCO will lock-in in such a way as to put f_r and $f_o/10$ in a phase-quadrature relationship. That is, f_r and $f_o/10$ will be 90° out of phase with each other, but on the same frequency. This signifies that f_o will be forced to be exactly ten times f_r and the output frequency then acquires the same long-term stability as the reference frequency. By making f_r a high stability "standard frequency" and making N_1 a programmable counter (N_1 can be any integer) many output frequencies are available, each selectable by means of N_1 and each having the stability of f_r. The frequencies will be $N_1 \times f_r$ over the range that is within the voltage variable range of the VCO. The channel frequency spacing will be f_r.

If channel spacing smaller than 100 kHz (which is about the lowest frequency of "standard frequency" crystals) is desired, then the standard frequency is divided down by a fixed N_2 frequency divider, as shown in figure 5. The best frequency standards operate in the 1 to 5 MHz range and thus most phase-locked loop synthesizers use this technique, otherwise channel spacing would be too coarse. The fixed N_2 division of the standard can be carried to extremes; for instance suppose a 1 Hz channel spacing is desired with a 1 MHz standard frequency. The fixed N_2 would be 10^6 and the inputs to the phase

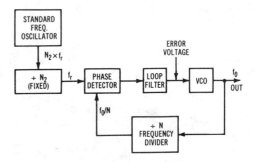

Figure 5

SINGLE LOOP PLL WITH CHANNEL SPACING SMALLER THAN STANDARD FREQUENCY

Best standard frequency oscillators work in the 1- to 50-MHz range and frequency must be divided by a fixed N_2 divider; otherwise channel spacing is too coarse.

comparator would be at 1 Hz. Since the phase comparator is essentially a mixer, it would show both the sum and difference of the input signals in its output port. For this example, the loop filter would have to have an RC time constant of at least tens of seconds to discriminate against the sum output, thus the loop would be very slow to acquire and lock.

For this reason, phase-comparison is generally not done at frequencies below about 1 kHz, but multiple phase-locked loops are used instead. An example of a multiple loop, phased-locked synthesizer is shown in figure 6. It is considerably more difficult to design than a single PLL device. Merely deciding which values of N to use in the various dividers can be a tedious task and computer solutions are generally used for this sort of design.

12-2 Synthesizer Building Blocks

A *voltage controlled oscillator* (VCO) is shown in figure 7. The VCO is a stable r-f oscillator circuit, such as the Hartley or Col-

pitts (or some of their variations such as the Clapp, Vackar, etc.), with one of the tuning capacitors replaced by a varicap. The varicap (also called a varactor) is a back-biased diode which has a capacitance which varies with the amount of back-bias voltage. Varicaps are usually specified as to capacity with −4 volts applied. (The large MV-series of *Motorola* and the VC-series of *TRW* are representative of varicaps used in VCOs as well as for other purposes.)

The VCO shown in figure 7 uses a pair of varicaps in a 400-MHz circuit designed around a junction FET. The use of back-to-back varicaps is a common practice in VCO design as it allows larger r-f voltage to be used in the oscillator without the danger of forward-biasing the varicaps, with resulting nonlinearity.

There are numerous other ways of making an oscillator voltage-variable: saturable reactors in the inductance of the resonant circuit, dependence upon the voltage variable capacitance of the base-collector junction of a bipolar transistor, the FET equivalence of a reactance tube, and the use of a voltage-sensitive astable RC flip-flop circuit. An astable flip-flop VCO is shown in figure 8.

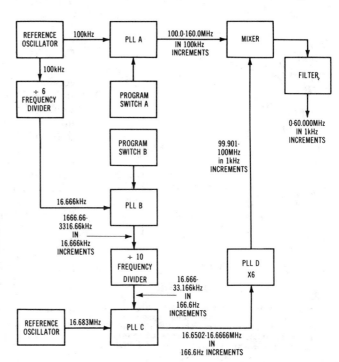

Figure 6

FREQUENCY SYNTHESIZER FOR 0-60 MHz

Four PLLs can be programmed to provide 1-kHz resolution between 0 and 60 MHz. Switching response time is about 2 ms.

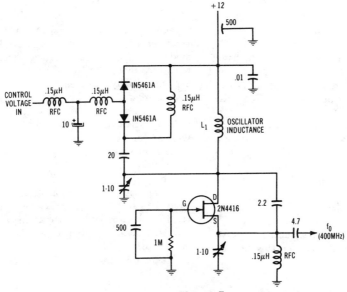

Figure 7

REPRESENTATIVE VCO FOR 400 MHz

Back-to-back varicaps are used for best linearity with respect to control voltage.

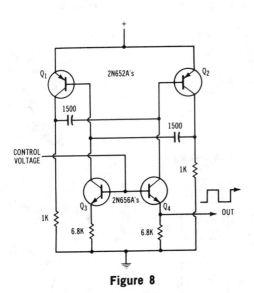

Figure 8

F-M MODULATOR

In this astable multivibrator emitter followers Q₃ and Q₄ replace conventional base resistors of Q₁ and Q₂. Input control voltage is applied to base circuit. Linear collector current-input voltage relationship provides a linear frequency variation.

It is intended for rather low-frequency operation.

The IC VCO In recent years, several manufacturers have introduced IC versions of the VCO. The *Signetics* NE566 and the *Motorola* MC4024 are good examples. The NE566 is a square wave/triangular wave VCO operating on +10 to +24 volts and the MC4024 is a dual VCO operating on +5 volts and producing rectangular wave output at TTL level. Both of these VCOs are RC types and are rather limited in upper frequency capability. The NE566 will operate up to about 100 kHz and the MC4024 up to about 25 MHz.

Motorola also makes a VCO in their ECL family of logic, the MC1648. This device will operate up to 200 MHz and uses an external coil and varicap to determine frequency and provide frequency/voltage dependance. Because the MC1648 uses a coil, it has higher equivalent Q in the oscillator than most other IC VCOs. This means that the output frequency has less near-carrier noise.

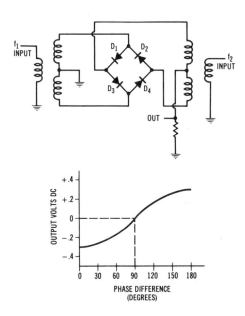

Figure 9

TYPICAL PHASE DETECTOR

Output voltage waveform as a function of phase difference is sinusoidal. With two square-wave inputs the output voltage is linear. With identical frequencies injected into the ports, a dc output related to the phase difference between the signals will appear at the output port.

The Phase The *phase detector* is another cru-
Detector cial block in a PLL frequency
synthesizer. While phase detectors are not common in normal communication electronics, they are used in disguised forms. Most mixers and product detectors are phase detectors and have the same function, that is, to multiply two signals together and produce the difference frequency. The phase detector always has a dc coupled output, however, so that an average dc level can be delivered to the VCO control port. The common double-balanced diode mixer has a good phase detector characteristic, as shown in figure 9.

It should be noted that a phase detector is quite different from a' frequency detector, such as a discriminator, or ratio detector. The frequency detector has a tuned circuit built into it and provides a dc level that is a function of the frequency difference between the built-in frequency "standard" and *one* external input frequency. *Frequency-locking* of a VCO can be reversed in sense, that is, the loop can be miswired so that locking is discriminated against. *Phase-locking* does not function that way, and it is impossible to reverse the feedback in a PLL provided that only phase control is in operation. Other forms of loop instability can prevent a loop from locking, however.

Phase detectors are available in IC form, just as VCO ICs are. The *Motorola* MC4044 is a digital phase detector which is TTL compatible and specifically made to operate with the MC4024 VCO. The two devices can operate together in a TTL-PLL circuit as shown in figure 10. Note that the MC-4044 is both a frequency and a phase detector, and so its sense can be reversed, making locking impossible. The *Motorola* MC-12040 is also a phase and frequency detector, but for use at frequencies up to 80 MHz it is part of the ECL family and is generally used with the MC1648 VCO.

Frequency The dividers used in a PLL fre-
Division quency synthesizer are usually
IC devices since building digital frequency dividers any other way is extremely expensive. Programmable dividers (usually decade types) are the rule. The *Motorola* MC4016 and the *Texas Instrument* SN-74190N are representative types. These ICs can divide up to one decade per package at speeds of 10 MHz and 20 MHz respectively. More recently, even larger-scale programmable dividers have become available in MOS ICs; the *Mostek* MK50398N is representative of a six decade divider that can be *loaded* (programmed) and which will count up or down. The *Mostek* device is only good to 1 MHz, but similar ICs with higher frequency capability are on the horizon.

The programming of a digital counter is done by loading a number into the counter and then allowing the device to count down to zero or up to its maximum design count number. This loading must be done each time the counter counts up or down. This means, for a decade-down counter, that each time there is a 0000 to 1001 transition, the counter must be reloaded. For a decade-up

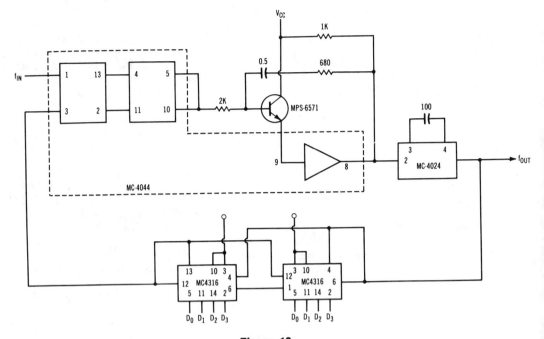

Figure 10

DIGITAL PHASE DETECTOR

The MC-4044 is a digital phase detector and the MC-4024 is a VCO.

counter, reloading must occur each time a 1001 to 0000 transition occurs.

The loop filter, as mentioned previously, is a simple RC low-pass circuit which is very important to the operation of the PLL. The basic form of the loop filter is shown in figure 11A, but it is usually implemented in the form shown in 11B so that the output has a lower impedance in order to drive the VCO. In general, R_2 in the loop filter is considerably smaller than R_1. Note that R_1 and C establish the cutoff frequency (f_c) of the filter as shown in figure 12. The presence of R_2 causes the filter rolloff to stop its 6-dB/octave rate and flatten out at some frequency f_t. This action occurs because at the higher frequencies, C_1 is a short circuit and the filter becomes a voltage divider consisting of R_1 and R_2. The design of a PLL is fairly complex and the stability of it depends in a large part on the loop filter once the VCO sensitivity (in MHz/volt) and the phase detector sensitivity (in volts/radian) are chosen.

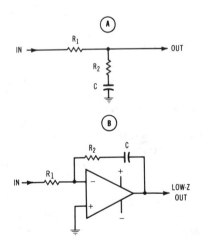

Figure 11

PLL LOOP FILTER

A—Basic RC filter.
B—Operational filter has low output impedance to drive a VCO.

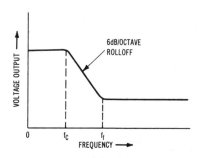

Figure 12

RESPONSE OF LOOP FILTER

In practice, an f_c compatible with the lock-up time of the loop is picked (being careful to make the time constant long enough so that the loop filter will ade-

quately attenuate the two input frequencies and the sum frequency from the phase detector). Next, the value of R_2 is raised enough to cause the loop to be stable. This value of R_2 is rarely more than one-tenth of R_1. Any attempt to make the filter roll off at greater than 6-dB/octave by the use of a sharper cutoff design will probably cause the loop to become unstable.

12-3 A VHF Synthesizer

The synthesizer building blocks can be put together to form a 114-178 MHz synthesizer suitable for the amateur and mobile service bands. This unit provides outputs every 30 kHz (figure 13). The system

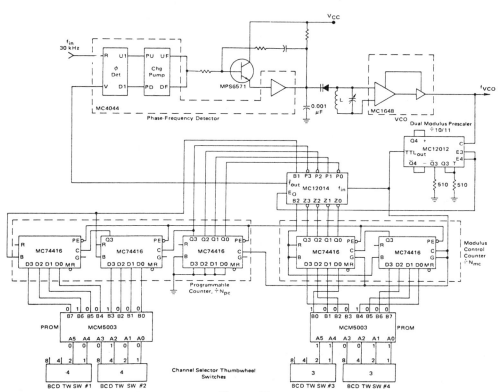

Figure 13

144- TO 178-MHz FREQUENCY SYNTHESIZER

This device provides 30-kHz channel spacing across a range of 34 MHz in the vhf region. The dual modulus prescaler (MC 12012) divides by either 10 or 11 in a pulse swallowing technique discussed later in this chapter. Channels are selected by four thumbwheel switches shown at bottom of diagram.

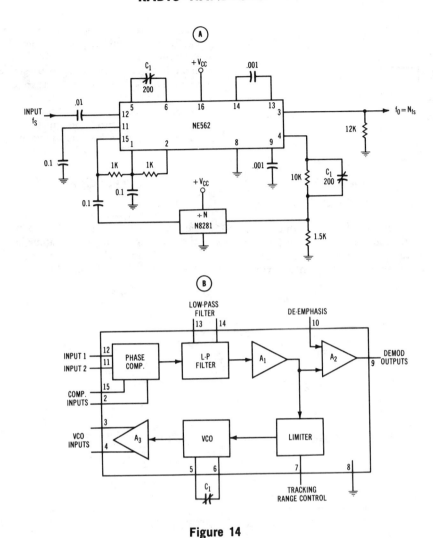

Figure 14

A—Frequency Synthesizer and programmable divider.
B—Synthesizer on a chip.

uses both ECL and TTL logic ICs in a mixed package. Details on this system are given in the *Motorola* brochure "MC4016/MC74416 Data Sheet." Motorola Semiconductor Co., Box 20912, Phoenix, AZ 85036. *Motorola* application brochures AN-564 and AN-594 are also helpful in understanding the details of this circuit.

In addition to PLLs made up of a number of IC blocks, as shown in figure 13, there are some ICs available that are complete phase-locked loops or complete synthesizers. Neith-

er of these categories of ICs are really complete, but they do offer an increased level of on-chip capability for synthesis. The NE560 series of *Signetics* devices are an example of PLL and related ICs. Figure 14 shows a simple synthesizer using a NE562 and N8281 (TTL programmable divider). The only additional item required is the reference frequency input. The NE565 is another PLL chip with a lower maximum operating frequency (500 kHz) but which is available from other manufacturers.

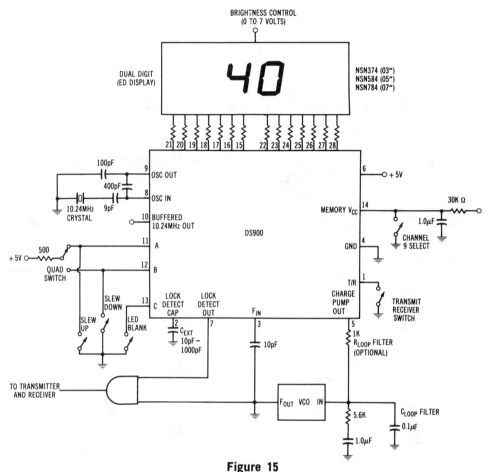

Figure 15

40-CHANNEL SYNTHESIZER ON A CHIP

Great capability is achieved at the expense of great specialization.

Complex PLLs Since the first IC PLLs were introduced about 1970, there has been considerable development toward more and more on-chip complexity. The programmable counters were included on the chip, the reference crystal oscillator put on the chip, and numerous other functions added. The *National* DS8900 (figure 15) demonstrates how a chip can have so many features added to it that it is only applicable to a single use. In this case, 40 channel CB transceivers. Note that great capability is achieved at the expense of great specialization, which is a help to the equipment manufacturer, but of little use to the experimenter. *National* also makes a series of chips that are not so complex (the 55100 series) which offer various features which should be checked for a particular usage. Other manufacturers also make similar synthesizer ICs, including *Micro-Power Systems, Nitron*, and several Japanese firms.

Synthesizer Considerations There are considerations and techniques that are important in certain synthesizer designs. It has been stated that the long term stability of the VCO in a PLL is the same as that of the reference frequency. This does not mean that the short-term stability is as good as the reference. "Short-term" in this instance means frequencies plus or minus

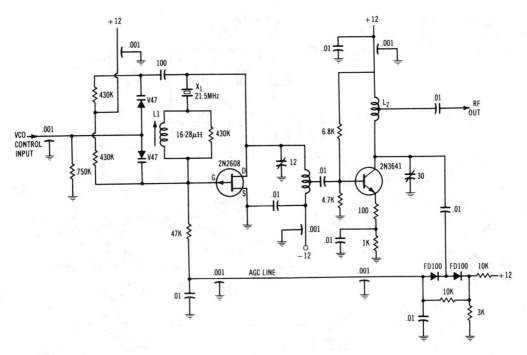

Figure 16

VOLTAGE-CONTROLLED CRYSTAL OSCILLATOR

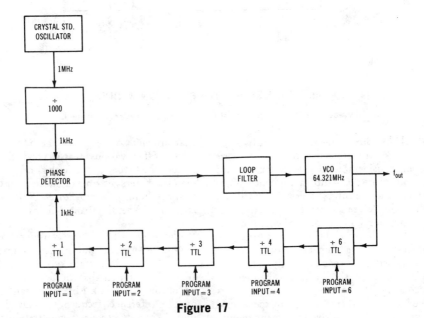

Figure 17

VHF PLL SYNTHESIZER

Channel Spacing = 1 kHz.

the VCO output that are high enough so that the loop filter does not pass them. Thus, a VCO with rather poor inherent stability will have only its average frequency stabilized by the feedback of the loop. Short term instability (showing up as phase noise) around the VCO output frequency will be present.

Two things can be done to improve short-term stability. First, the VCO must be made more stable by improving the Q of the VCO and mechanical stability problems must be solved. Second, a shorter time constant must be used in the loop filter. Increasing the Q of the VCO ultimately leads to the use of a crystal as the resonant element. Such a circuit is termed a *Voltage-Controlled Crystal Oscillator* (VCXO). While improving stability, the VCXO severely limits the range of output frequencies. A representative VCXO is shown in figure 16.

Shortening of the time constant of the loop filter can only be done if the frequency of the phase comparison is high enough so that f_r/N_2 is not passed by this filter. In short, attempts to clean up the phase noise or otherwise improve the short term stability of the synthesizer introduce additional conditions to be compromised in the total PLL design.

Prescaling A synthesizer for vhf or uhf operation is complicated by the fact that programmable dividers are most available for frequencies of 100 MHz, or lower. Thus, a different method of dividing down the VCO output frequency must be used when the VCO operates between 100 and 1000 MHz. Since ECL dividers are available that can operate up to about 1000 MHz, the straightforward approach is to divide first by ten in ECL, then make an ECL/TTL conversion and use programmable TTL dividers for lower frequencies. This is comparable to using a fixed decade prescaler on a frequency counter and has the same effect, that is, the least significant digit of frequency resolution is lost. Figure 17 shows an example of synthesizing 64.321 MHz. A PLL using TTL programmable di-

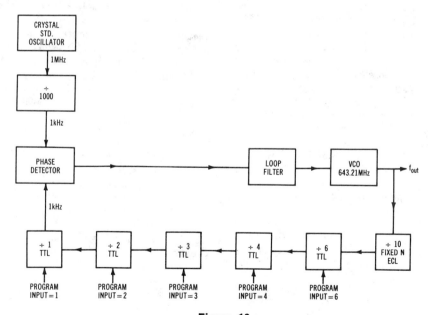

Figure 18

UHF PLL SYNTHESIZER

Channel Spacing = 10 kHz.

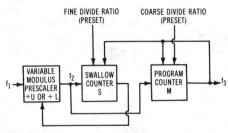

FINE DIVIDE RATIO (PRESET)   COARSE DIVIDE RATIO (PRESET)

Figure 19

PULSE-SWALLOWING COUNTER

The prescaler usually has a divide ratio of 10/11 or 5/6. U and L differ only by one and changing S by a certain amount changes N by the same amount. The prescaler and swallow counter thus act as a single, high speed, programmable divider.

viders is used. Also a method of synthesizing 643.21 MHz is shown in figure 18 using an ECL fixed decade followed by TTL dividers. Since it is decided that phase comparison will not be done below 1 kHz for economic reasons, the uhf synthesizer must be limited to having 10 kHz channel spacing instead of 1 kHz.

There is a solution to the fixed prescaler problem, at least for frequencies up to 650 MHz. By using one of the divide by 10/11 units such as the *Fairchild* 11C90 (or the 95C90 for frequencies up to 250 MHz) the technique of *pulse swallowing* can be used. As the 11C90 device can be controlled by either 10 or 11, it can count by 10 for a number of counts and then count by 11 for a number of counts. This is illustrated in figure 19. It can be shown that the total count (or "divide by") number N is given by:

$$N = (U\text{-}L)S + LM$$

where,

U is the upper count (11 in this case),
L is the lower count (10 in this case),
S is the divide ratio of the swallow counter,
M is the divide ratio of the program counter.

Thus, $N = S + 10M$ and the design has achieved the equivalent of a high-frequency programmable divider by using the pulse-swallowing technique.

12-4 A HF SSB Synthesizer Transceiver

Many modern hf SSB transceivers are fully synthesized and can provide discrete frequencies in the range of 1.8 to 30 MHz in 100-Hz steps. Upper and lower sideband, plus c-w and FSK modes, are available in units providing up to 100 watts PEP output. A representative transceiver is discussed in this section (figure 20). The unit is a *Sunair* GSB-900DX designed for fixed or mobile operation from plug-in, modularized power supplies.

General Operation   Figure 21 is a block diagram of the transceiver. The synthesizer consists of six function blocks which are built up on separate printed-circuit boards: the spectrum generator, the low-digit generator, the translator, the vhf divider, the VCO, and the synthesizer master board.

The *synthesizer* generates the three local-oscillator injection frequencies needed to determine the operating frequency of the transceiver. The synthesizer input is the 5-MHz precision reference signal from the frequency standard. The three frequencies are obtained by a combination of direct synthesis and digital phase-lock techniques. The frequency accuracy of the transceiver is thus solely determined by the accuracy of the frequency standard. Frequency stability is $\pm 1 \times 10^{-6}$ over a temperature range of $-30°C$ to $+65°C$ and under 100% humidity at 50°C.

The *third local oscillator* (10.5 MHz) is derived by direct synthesis and this signal is used for product detector injection on receiver and as a carrier generator on transmit. The *second local oscillator* consists of a crystal oscillator (80.750 MHz) and this signal is used in the vhf mixer assembly to convert the first i-f of 91.250 MHz to the second i-f of 10.5 MHz. Because of the mixing technique, any frequency error in this oscillator appears on the first local-oscillator frequency and is therefore cancelled at the output of the vhf mixer.

The *first local oscillator* is a voltage-controlled oscillator (VCO), phase-locked to cover the range of 91.25 to 121.2499 MHz in 100-Hz steps. The exact frequency range of the oscillator is:

$$F_1 = 91.250 + F_0 + e$$

where,

F_1 equals first local-oscillator frequency (MHz),

F_0 equals the dialed frequency (MHz),

e equals second local-oscillator error frequency (MHz).

On receive, the first local oscillator converts the incoming signal up to the first i-f channel (91.25 MHz). On transmit, it is used to convert the transmitted signal at the first i-f channel down to the final operating frequency.

The *spectrum generator* block diagram is shown in figure 22. It generates the fixed reference frequencies needed in the synthesizer. The input is the 5-MHz reference frequency which is amplified by U_1 and formed into a short pulse by pulse generator U_2. The fourth harmonic (20 MHz) is filtered by double-tuned bandpass circuit and amplified by U_3. The output signal is applied to the 17-MHz mixer and also to the buffer amplifier (U_8) at a low impedance level.

The Reference Generators The 5-MHz pulse from U_2A is fed to U_4, a divide-by-five counter. The resultant 1-MHz signal is fed to three stages of divide-by-ten counters (U_5, U_6 and U_7) to produce 100-kHz and 1-kHz output signals.

A 1-MHz pulse from U_4 is passed through a circuit tuned to 3 MHz to derive the third harmonic. This signal is amplified by Q_5, filtered, and applied as a mixing signal to mixer Q_6. The resultant 17-MHz signal is filtered by a double-tuned circuit and matched to a 50-ohm output by a complementary emitter follower (Q_7, Q_8).

To derive the 21-MHz mixing signal, the 1-MHz pulse from U4 is passed through a tuned circuit and through an emitter follower (Q_1, Q_2) to match the low-impedance input of the balanced mixer, CR_4-CR_7. The 20-MHz reference signal from U_3 is amplified by U_8 and applied to the mixer. The resultant signal at 21 MHz is passed through a filter and amplified and transformed to a low impedance value by U_9.

The 21-MHz signal from U_9 is also fed to flip-flop U_{10} which generates a 10.5-

Figure 20

100-WATT OUTPUT SYNTHESIZED HF SSB TRANSCEIVER

The GSB-900DX transceiver covers the range of 1.600 to 29.999 MHz in 100-Hz frequency steps. Two sets of frequency select dials are provided to set in two channels. Frequency of operation is shown in a six-digit LED display. An antenna coupler and SWR meter are incorporated as an option. Continuous tuning between the 100-Hz increments is also provided. The transceiver is fully modularized and can operate from both ac and dc power sources. (Sunair Electronics, Inc., Ft. Lauderdale, FL).

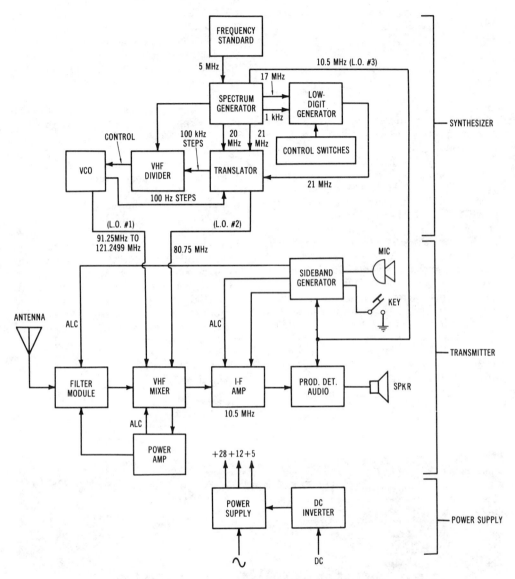

Figure 21

BLOCK DIAGRAM OF GSB-900DX TRANSCEIVER

This circuitry is representative of some of the synthesized hf SSB transceivers in use. The upper blocks are the synthesizer elements and the lower blocks are the rf generation and detection elements. The synthesizer provides two local oscillator signals for the double conversion transceiver, plus bfo injection for SSB or cw service. Frequency control is accomplished by a 5 MHz frequency standard. First i-f channel is 91.25 MHz which provides high spurious signal rejection. Broadband power amplifier provides 100 watts cw or PEP with automatic level control (ALC).

MHz squarewave which is filtered to a sine wave and then matched to 50 ohms by emitter follower Q₄.

The flip-flop is disabled in the a-m receive mode to prevent a beat note from appearing in the audio signal. A blanking pulse

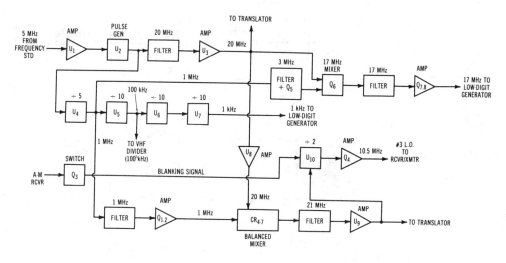

Figure 22

THE SPECTRUM GENERATOR

The spectrum generator provides 284,000 synthesized hf channels having 100-Hz spacing. Three injection frequencies are obtained from a 5-MHz reference by a combination of direct synthesis and digital phase-lock techniques. Frequency accuracy of the transceiver is solely determined by the accuracy of the frequency standard. Unwanted signals from the mixing process are removed by use of high attenuation filters.

from the local-oscillator blanker circuit also disables the flip-flop for about 100 ms when the setting of the 1-MHz frequency switch on the front panel is changed.

The Low Digit Generator The *low digit generator* (figure 23) generates the 100-Hz, 1-kHz, and 10-kHz synthesized frequency steps. The inputs are the 17-MHz reference and the 1-kHz reference from the spectrum generator, BCD frequency control lines from the frequency control switches and the coarse steering voltage from the 10-kHz frequency control switch on the front panel. Output of the low-digit generator is 1.500 to 1.5999 MHz in 100-Hz steps and is fed to the translator as a mixing reference.

The *voltage-controlled oscillator* (VCO) (Q_1) covers the range of 15.0 to 15.999 MHz. Coarse frequency tuning is provided by action of the coarse steering voltage on two varactor diodes. *Fine frequency control* is provided by the phase detector (U_2) acting through the loop filter and a 1-kHz notch filter on a second set of varactor

diodes. The oscillator output is coupled to isolation amplifier U_1.

The mixer (Q_3) transforms the VCO frequency to 2.00 MHz to 1.001 MHz to place the signal in the range of the preset counters. The inputs to the mixer consist of the VCO signal and the 17-MHz reference. The output of the mixer is filtered by a 2.5-MHz low-pass filter and amplified by U_6. Quad NAND gate U_7 acts as a monostable multivibrator, forming the signal into a short pulse to drive the preset counter.

The preset counter (U_8-U_{11}) has a division ratio controlled by the frequency-control switches on the panel. During the normal counting interval, the counter functions as a divide-by-2000 counter. During the preset interval, the clock is disabled and the counter preset to a count determined by the frequency-control switches. The frequency-control information is entered in binary-coded decimal (BCD) format and the division ratio (D) is determined by:

$$D = 2000 - (100 N_{10} + 10 N_1 + N_{100})$$

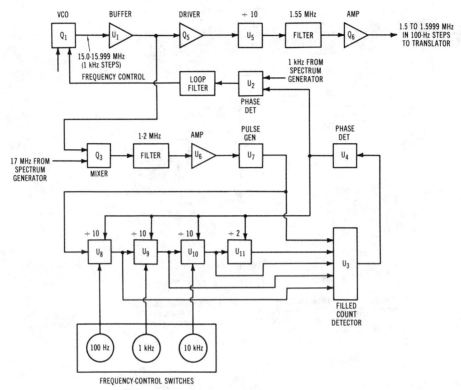

Figure 23

FREQUENCY READOUT AND LOW-DIGIT GENERATOR

This circuitry generates the 100-Hz, 1-kHz, and 10-kHz synthesized frequency steps. The inputs are the 17-MHz reference and the 1-kHz reference from the spectrum generator and the BCD frequency control lines from the panel frequency-control switches. An additional input is the "coarse" steering voltage from the 10-kHz control switch on the panel.

where,

N_{10} equals the setting of the 10-kHz frequency dial,

N_1 equals the setting of the 1-kHz frequency dial,

N_{100} equals the setting of the 100-Hz frequency dial.

Representative dial settings are shown in figure 24.

The preset generator applies a short pulse to the data strobe inputs of the preset counter when a full count is detected. A "look ahead" technique is employed to eliminate miscounting due to the propagation delays in the counter. When the counter has reached a count of 1999, the inputs to NAND gate will be in a "one" state. As

soon as the clock input to the gate returns to a "one" state, the input of U_3 will go to a "zero" state, thereby triggering monostable multivibrator U_4. Then U_4 presets the counters by applying a "zero" to their data strobe inputs for approximately 100 ns. The output of U_4 will return to a "one" state before the beginning of the next clock pulse.

The Phase Detector   The *phase detector* (U_2) compares the frequency of the output of the preset counter (U_4) with that of the 1-kHz reference signal from the spectrum generator. If the VCO frequency is high, for example, the output frequency of mixer Q_3 will be low. The output frequency of the preset counter,

DIAL SETTINGS			PRESET	COUNT (D)
10 kHz	1 kHz	100 kHz		
0	0	0	000	2000
0	0	1	001	1999
0	0	2	002	1998
0	1	1	011	1989
1	9	9	199	1801
9	9	9	999	1001

Figure 24

DIAL SETTINGS FOR PRESET COUNTER

The frequency control information is entered in binary-coded-decimal (BCD) format. During the normal counting interval the counter functions as a divide-by-2000 counter. During the preset interval, the clock is disabled and the counter is loaded (or preset) to a count determined by the settings of the frequency-control switches.

therefore, will also be low. The phase detector output voltage will decrease until the frequency error is corrected. If there is no frequency error, the output voltage of the phase detector will remain constant.

The loop filter removes any 1-kHz components in the phase detector output and also determines the transient response of the loop. The 1-kHz frequency components are further attenuated by a twin-T notch filter following the loop filter.

The action of the phase-lock loop is to make the VCO frequency follow the relationship:

$$F_{VCO} = 17.000 - D \text{ (in MHz)}$$

where D is the count ratio. The VCO will therefore vary from 15.000 to 15.999 MHz in-1 kHz steps.

The output from buffer U_1 is further amplified by Q_5 and fed to divide-by-ten counter U_5. The output of U_5 is filtered to a sine wave and fed to emitter follower Q_6 which matches the output to 50 ohms. The output from the low-digit generator is 1.5000 to 1.5999 MHz in 100-Hz steps and follows the relationship:

$$F_{OUT} = 1.5000 \text{ MHz} + N \text{ (kHz)}$$

where,

N equals the knob setting of the 10-kHz, 1-kHz, and 0.1-kHz dials.

The Translator A block diagram of the *translator* package is shown in figure 25. This unit combines the signals from the *low-digit generator* and VCO and generates a signal which, after subsequent frequency division in the vhf divider (figure 26), is used to phase-lock the VCO to the proper frequency. The second local oscillator (Q_7) and vfo signals are also generated in this assembly. The inputs to this assembly are: 20- and 21-MHz references from the spectrum generator; first local oscillator from the VCO; 1.5000 to 1.5999 MHz from the low-digit generator; and the vfo control and vfo on/off signals from the front panel. The output is the 10.0- to 39.9-MHz reference signal which is fed to the vhf divider. In the vfo mode the internally generated 21-MHz vfo is substituted for the 21-MHz reference from the spectrum generator.

Since the second local oscillator is a free-running crystal oscillator and is not referenced to the frequency standard, a small frequency error can exist. However, because of the mixing technique employed, both the first and second local oscillators will have the same frequency error which can be cancelled in the vhf mixer assembly.

The vfo (Q_1) is a crystal oscillator covering the range of 20.995 to 21.005 MHz, thereby providing approximately ± 5-kHz tuning adjustment around the dialed frequency. The vfo control voltage is applied to varactor diodes in series with the crystal. The oscillator output is amplified by U_2 when in the vfo mode. The 21-MHz reference from the spectrum generator is amplified by U_1 when the vfo mode is not selected. The second local oscillator (Q_7) is crystal controlled at 80.75 MHz. A portion of this signal is amplified by Q_9 and converted to a 50-ohm level for injection in the receiver portion of the device.

The 100.75-MHz mixer (Q_8) heterodynes the second local-oscillator signal from Q_7 and the 20-MHz reference signal from the spectrum generator. A triple-tuned bandpass filter at 100.75 MHz selects the desired sum frequency and rejects the 80.75 MHz and 60.75 MHz components.

Depending on the mode selected, the balanced mixer (CR_1-CR_4) heterodynes the 1.500- to 1.5999-MHz output of the

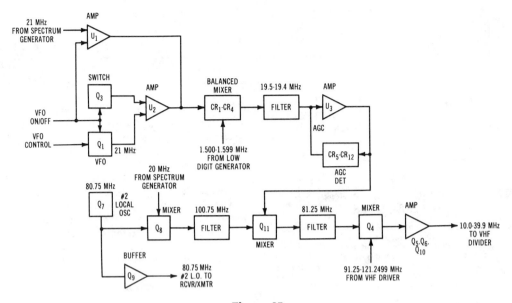

Figure 25

THE TRANSLATOR

The translator combines the signals from the low-digit generator and VCO and generates a signal which, after subsequent frequency division in the vhf divider, is used to phase-lock the VCO to the proper frequency. The second local-oscillator and vfo signals are also generated in this assembly.

low-digit generator and either the 21-MHz reference or vfo signal. The 19.45-MHz bandpass filter selects the desired difference frequency. The output is fed to U_3 for further amplification and filtering. Automatic gain control (agc) is provided at this point to ensure a proper signal level to the 81.25-MHz mixer (Q_{11}). This mixer combines the 19.5000- to 19.4001-MHz signal from Q_3 and the 100.750-MHz signal from mixer Q_8 to produce the difference frequency of 81.2500 to 81.3499 MHz. A bandpass filter selects the desired difference frequency.

The output mixer (Q_4) heterodynes the 81.25-MHz mixer output and the VCO signal. The output signal is filtered by a 10- to 50-MHz bandpass filter and then transformed to a low impedance by emitter follower Q_{10}. Negative feedback around the amplifier provides flat gain well beyond 50 MHz as well as a constant input impedance and low output impedance.

The VHF Divider   The *vhf divider* is shown in figure 26. This unit contains a divide-by-400 high-speed preset counter which forms the 10-MHz, 1-MHz, and 100-kHz frequency steps. A phase detector compares the frequency and phase of the output of this counter with that of the 100-kHz reference from the spectrum generator and develops a "fine" steering correction voltage which is fed back to control the frequency of the VCO. This phase-lock loop, by controlling the frequency of the VCO, forces the input to the vhf divider to follow the relationship:

$$F_{IN} = 10.0 + 10N_{10} + N_1 + 0.1N_{100}$$
$$(\text{in MHz})$$

where,

N_{10} equals the 10-MHz digit,
N_1 equals the 1-MHz digit,
N_{100} equals the 100-kHz digit.

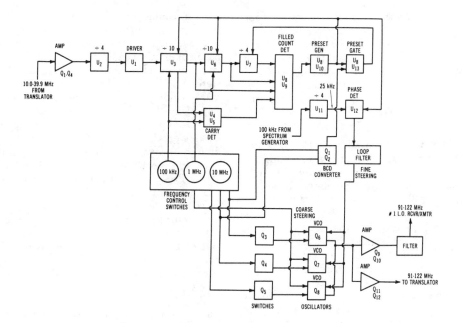

Figure 26

THE VHF DIVIDER

This unit contains a divide-by-400 high speed preset counter which forms the 10-MHz, 1-MHz, and 100-kHz frequency steps. A phase detector develops a "fine" steering correction voltage which is fed back to control the frequency of the VCO.

The input frequency therefore varies from 10.0 to 39.9 MHz in 100-kHz steps. The 10-MHz input corresponds to dial settings of "000" whereas the 39.9-MHz input corresponds to dial settings of "299" on the 10-MHz, 1-MHz, and 100-kHz dials respectively.

The inputs to the vhf divider are: the 100-kHz reference from the spectrum generator, the output signal from the translator, the frequency control lines from the 1-MHz and 100-kHz switches on the front panel, and the 10-MHz preset lines from the VCO. The output is the "fine" steering voltage which is fed back to the VCO.

The broadband input amplifier consists of a two-stage feedback amplifier (Q_1, Q_2) followed by a complementary-emitter follower (Q_3, Q_4). A negative feedback network is placed around the first two stages. The output of the emitter follower provides a low impedance driving source for the subsequent high speed prescaler and also establishes the proper logical "zero"

and "one" levels for the following TTL logic ICs.

The prescaler (U_2) is a high speed dual flip-flop connected in a divide-by-four configuration. The output is buffered by NAND gate a1c so as not to place excessive loading on U_2.

The preset counters (U_3, U_6, and U_7) consist of two stages of preset decade counters (U_3, U_6) followed by a preset divide-by-four dual flip-flop (U_7). Device U_7 is preset by the two-input NAND gate (U_{13}) and gate U_8. During the normal counting mode the data strobe lines on U_3 and U_6 are held in a "one" state by preset flip-flop U_{10}. This permits these counters to function in their normal divide-by-ten mode. Similarly, the 10-MHz preset bus is held in a "zero" state by U_{10}. This forces the output of gates U_8A and U_8B and the preset inputs to dual flip-flop U_7 to be in a "one" state. In addition, the outputs of U_{13}C and U_{13}D and the "clear" inputs to U_7 are forced to a "one" state. U_7, therefore, counts in its normal divide-by-four mode.

During the preset interval, the data strobe lines to U_3 are held in a "zero" state by U_{10} and the inputs to the preset 10-MHz gate are held in a "one" state by U_{10}. The clock pulse to the counters is inhibited and the preset information from the ten frequency control lines is entered into the counters (U_3, U_6, and U_7).

The "Carry" Generator   When all four 100-kHz preset lines are programmed to a "zero" state by the front panel switches, corresponding to a dial setting of "0" on the 100-kHz frequency control, a special "carry" signal must be generated to program the counters to the correct division ratio.

Mathematically, this is necessary because a dial setting of zero requires the input counter (U_3) to divide-by-zero, an impossible operation. The count is corrected by programming U_3 to divide-by-ten in this state and then subtracting one count from the next decade counter.

Quad two input NAND gate U_4 is connected as a quad inverter with a common output. One of the four 100-kHz input lines is connected to each section of the gate. This special gate enables all four outputs to be connected together. The output of U_4 is inverted by U_5A. If all four inputs to U_4 are "zero" (dial set to "0" on the 100-kHz switch), the output of U_4 will be in a "one" state and the U_5A output will be a "zero." If any of the 100-kHz inputs are in a "one" state, the U_5A output will also be a "one."

The Preset Generator   During the normal counting interval, the Q output of flip-flop U_{10} is in a "one" state, the preset bus is in a "one" state and the 10-MHz preset bus is in a "zero" state. In order to count properly, the presetting must occur between input clock pulses. A "look-ahead" technique is therefore employed to eliminate the propagation delays through the various counters.

First assume that the 100-kHz dial is not in the "0" position (the output of U_5A is in a "one" state). When the preset counter has reached a count of 399 (that is, one count from being filled); counter U_3 will

have a count of "9" (binary 1001), U_6 will have a count of "9" and U_7 will have a count of "3" (binary 11). The output of eight-input NAND gate U_9 will sense this unique state and will go to a "zero" state. U_8C inverts this output to a "one" state, making the input to master/slave flip-flop U_{10} a "one." On the next transition of the U_3 input clock to a "zero" state, the output of U_{10} will toggle to a "zero" state and the preset bus will also be in a "zero" state and the 10-MHz preset bus in a "one" state. Presetting will therefore occur. On the next transition of the U_3 input clock back to a "one" state, the output of U_8D will transition from a "one" to a "zero" state, applying a "zero" to the present input of U_{10}, thus forcing the output of U_{10} back to a "one" state. This terminates the preset cycle, and normal counting sequence is restored.

Finally, if the 100-kHz dial is set in the "0" position, the U_5A output will be in a "zero" state. The output of carry gate U_5B will therefore always be in a "one" state and will not follow the output of U_6. Flip-flop U_{10} will now be armed at the 389th counter state instead of at the 399th state. The desired carry of ten counts will therefore occur.

The Phase Detector   The 100-kHz reference signal from the spectrum generator is divided in frequency by four to 25 kHz by dual flip-flop U_{11}. In the U_{12} *phase detector* the frequency and phase of the output of the preset counter is compared with that of the 25-kHz reference and a "fine" steering voltage correction is fed back to control the frequency of the VCO. This voltage changes in the correct direction to bring the VCO into phase lock. The phase detector operates in the following manner: If the frequency of the preset counter output is greater than that of the 25-kHz reference, the phase-detector output will decrease in voltage. If the frequency of the preset counter output is less than that of the reference, the output will increase in voltage. If the two frequencies are exactly the same, the phase-detector output will remain constant.

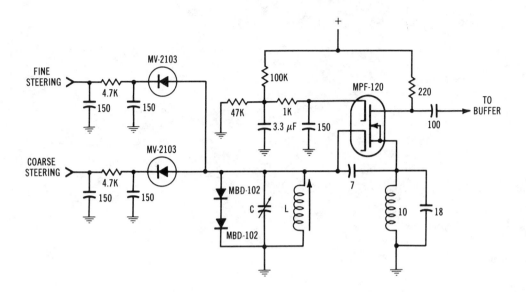

Figure 27

THE VOLTAGE CONTROLLED OSCILLATOR

Three separate oscillators are used, each covering a 10-MHz frequency range in the 91.25- to 121.499-MHz region. A typical oscillator is shown in this drawing. "Coarse" and "fine" steering voltages are applied by means of varactor diodes.

The Voltage-Controlled Oscillator The VCO generates the variable frequency first local oscillator signal that controls the operating frequency of the transceiver (figure 27). Three separate oscillators are used, each covering a 10-MHz frequency range and are selected by the 10-MHz switch on the panel of the radio. The three ranges are 91.25 to 101.2499 MHz, 101.250 to 111.2499 MHz, and 111.250 to 121.499 MHz. The exact frequency of each oscillator is controlled by two dc voltages, designated "coarse" and "fine" steering. Each steering voltage is applied to a varactor diode across the tank circuit. The "coarse" voltage is derived from a precision voltage divider located on the 1-MHz frequency control switch on the panel. This voltage sets the oscillator frequency within the acquiring range of the phase-lock loop. The "fine" steering voltage is derived from the phase detector on the vhf divider. This voltage is the dc feedback within the loop which forces the oscillator to the correct frequency.

Oscillator tracking is provided by adjustment of the inductor at the low-frequency end of the band and by the padding capacitor at the high-frequency end. Logic switching applies +12 volts to the appropriate oscillator circuit when the appropriate band control line is grounded by the 10-MHz switch on the front panel. The LED frequency display is driven by digital control signals from the frequency dials.

CHAPTER THIRTEEN

Frequency Modulation and Repeaters

Exciter systems for f-m and single-sideband transmission are basically similar in that modulation of the signal in accordance with the intelligence to be transmitted is normally accomplished at a relatively low level. Then the intelligence-bearing signal is amplified to the desired power level for ultimate transmission. True, amplifiers for the two types of signals are basically different; linear amplifiers of the class-A or class-B type being used for SSB signals, while class-C or nonlinear class-B amplifiers may be used for f-m amplification. But the principle of low-level modulation and subsequent amplification is standard for both types of transmission.

13-1 Frequency Modulation

Early frequency-modulation experiments were conducted by Major Edwin H. Armstrong of Columbia University based on the belief that noise and static were amplitude variations that had no orderly variations in frequency. In 1934 Armstrong conducted his classic f-m transmissions in the old 2½ meter amateur band in conjunction with W2AG in Yonkers, N.Y. Subsequent amateur experiments in 1936 showed that f-m promised excellent prospects for static-free, reliable, mobile communication in the vhf bands.

Postwar vhf development centered around amplitude modulation in the amateur bands for over two decades, aided by the flood of surplus military vhf gear, and it was not until the "mid-sixties" that amateur interest in f-m was stimulated by a quantity of obsolete commercial mobile f-m gear available on the surplus market at modest prices.

Vhf commercial two-way mobile radio is now standardized on channelized frequency-modulation techniques which provide superior rejection to random noise, interference, and fading as compared to conventional a-m systems. When the amplitude of the r-f signal is held constant (limited) and the intelligence transmitted by varying the frequency or phase of the signal, some of the disruptive effects of noise can be eliminated. In addition, audio squelch circuits silence noise peaks and background effects in the receiver until an intelligible signal appears on the frequency. The combination of noise rejection and squelch control provides superior range for a given primary power, as compared to an equivalent a-m power allocation.

Amateur vhf f-m techniques are based on the channel concept. Transmitters and receivers are mainly crystal controlled on a given frequency and random tuning techniques common to the lower frequency amateur bands are absent. F-m channels on the 10-meter band are standardized by common agreement at 40 kHz separation, starting at 29.55 MHz. A national calling channel is reserved at 29.60 MHz. On the 6-meter band the f-m channels start at 52.50 MHz, with 52.525 MHz reserved as a national calling frequency. Channel spacing is 40 kHz beginning at 52.60 MHz. F-m channels are spaced 30 kHz apart on the 2-meter band, beginning with 146.01 MHz, the repeater output channels being 600 kHz higher than the input channels up to 146.97 MHz. Above this frequency, the repeater channels are inverted, with the input channels starting at 147.99 MHz and running down to 147.60 MHz. The output channels run from 147.39 MHz to 147.00 MHz. Simplex channels fall in the regions of 146.40 MHz to 146.58 MHz and 147.42 MHz to 147.57 MHz.

On the 220 MHz band, the f-m channels start at 222.30 MHz, with 40 kHz separation. The repeater input channels begin at 222.30 MHz, with the outputs 1.6 MHz higher in frequency. Simplex channels begin at 223.42 MHz and the national calling frequency is 223.50 MHz.

On the 420 MHz band, channel spacing is 50 kHz, with the f-m channels beginning at 438.05 MHz. Repeater inputs or outputs begin at 442.00 MHz, with the input or the output channel 5.0 MHz higher (or lower) in frequency. Simplex channels begin at 445.00 MHz, with a national calling frequency on 446.00 MHz.

In this chapter various points of difference between frequency-modulation and amplitude-modulation transmission and reception will be discussed and the advantages of frequency modulation for certain types of communication pointed out. Since the distinguishing features of the two types of transmission lie entirely in the modulating circuits at the transmitter and in the detector and limiter circuits in the receiver, these parts of the communication system will receive the major portion of attention.

Modulation   *Modulation* is the process of altering a radio wave in accord with the intelligence to be transmitted. The nature of the intelligence is of little importance as far as the process of modulation is concerned; it is the *method,* by which this intelligence is made to give a distinguishing characteristic to the radio wave which will enable the receiver to convert it back into intelligence, that determines the type of modulation being used.

Figure 1 is a drawing of an r-f carrier amplitude-modulated by a sine-wave audio voltage. After modulation the resultant modulated r-f wave is seen still to vary about the zero axis at a constant rate, but the strength of the individual r-f waves is proportional to the amplitude of the modulation voltage.

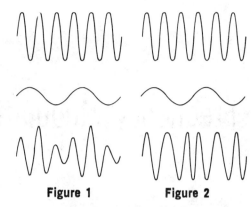

Figure 1   **Figure 2**

A-M AND F-M WAVES

Figure 1 shows a sketch of the scope pattern of an amplitude-modulated wave at the bottom. The center sketch shows the modulating wave and the upper sketch shows the carrier wave.

Figure 2 shows at the bottom a sketch of a frequency-modulated wave. In this case the center sketch also shows the modulating wave and the upper sketch shows the carrier wave. Note that the carrier wave and the modulating wave are the same in either case, but that the waveform of the modulated wave is quite different in the two cases.

In figure 2, the carrier of figure 1 is shown frequency-modulated by the same modulating voltage. Here it may be seen that modulation voltage of one polarity causes the carrier frequency to decrease, as shown by the fact that the individual r-f waves of the carrier are spaced farther apart. A modulating voltage of the opposite polarity causes the frequency to increase, and this is shown by the r-f waves being compressed together to allow more of them to be completed in a given time interval.

Figures 1 and 2 reveal two very important characteristics about amplitude- and frequency-modulated waves. First, it is seen that while the amplitude (power) of the signal is varied in a-m transmission, no such variation takes place in frequency modulation. In many cases this advantage of frequency modulation is probably of equal or greater importance than the widely publicized noise-reduction capabilities of the system. When 100 percent amplitude modulation is obtained, the average power output of the transmitter must be increased by 50 percent. This additional output must be supplied either by the modulator itself, in the high-level system, or by operating one or more of the transmitter stages at such a low output level that they are capable of producing the additional output without distortion in the low-level system as is commonly done in SSB—a form of amplitude modulation. On the other hand, a frequency-modulated transmitter requires an insignificant amount of power from the modulator and needs no provision for increased power output on modulation peaks. All of the stages between the oscillator and the antenna may be operated as high-efficiency class-B or class-C amplifiers or frequency multipliers.

Carrier-Wave   The second characteristic of
Distortion   f-m and a-m waves revealed
                   by figures 1 and 2 is that both
types of modulation result in distortion of the r-f carrier. That is, after modulation, the r-f waves are no longer sine waves, as they would be if no frequencies other than the fundamental carrier frequency were present. It may be shown in the amplitude-modulation case illustrated, that there are only two additional frequencies present, and these are

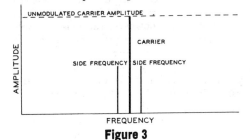

Figure 3

A-M SIDE FREQUENCIES

For each a-m modulating frequency, a pair of side frequencies is produced. The side frequencies are spaced away from the carrier by an amount equal to the modulation frequency, and their amplitude is directly proportional to the amplitude of the modulation. The amplitude of the carrier does not change under modulation.

the familiar *side frequencies,* one located on each side of the carrier, and each spaced from the carrier by a frequency interval equal to the modulation frequency. In regard to frequency and amplitude, the situation is as shown in figure 3. The strength of the carrier itself does not vary during modulation, but the strength of the side frequencies depends on the percentage of modulation. At 100 percent modulation the power in the

side frequencies is equal to one-half that of the carrier.

Under frequency modulation, the carrier wave again becomes distorted, as shown in figure 2. But, in this case, many more than two additional frequencies are formed. The first two of these frequencies are spaced from the carrier by the modulation frequency, and the additional side frequencies are located out on each side of the carrier and are also spaced from each other by an amount equal to the modulation frequency. Theoretically, there are an infinite number of side frequencies formed, but, fortunately, the strength of those beyond the frequency *swing* of the transmitter under modulation is relatively low.

One set of side frequencies that might be formed by frequency modulation is shown in figure 4. Unlike amplitude modulation, the strength of the component at the carrier frequency varies widely in frequency modulation and it may even disappear entirely under certain conditions. The variation of strength of the carrier component is useful in measuring the amount of frequency modulation, and will be discussed in detail later in this chapter.

One of the great advantages of frequency modulation over amplitude modulation is the

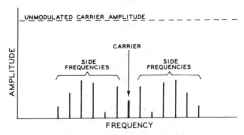

Figure 4

F-M SIDE FREQUENCIES

With frequency modulation, each modulation frequency component causes a large number of side frequencies to be produced. The side frequencies are separated from each other and the carrier by an amount equal to the modulation frequency, but their amplitude varies greatly as the amount of modulation is changed. The carrier strength also varies greatly with frequency modulation. The side frequencies shown represent a case where the deviation each side of the "carrier" frequency is equal to five times the modulating frequency. Other amounts of deviation with the same modulation frequency would cause the relative strengths of the various sidebands to change widely.

reduction in noise at the receiver which the system allows. If the receiver is made responsive only to changes in frequency, a considerable increase in signal-to-noise ratio is made possible through the use of frequency modulation, when the signal is of greater strength than the noise. The noise-reducing capabilities of frequency modulation arise from the inability of noise to cause appreciable frequency modulation of the noise-plus-signal voltage which is applied to the detector in the receiver.

F-M Terms   Unlike amplitude modulation, the term *percentage modulation* means little in f-m practice, unless the receiver characteristics are specified. There are, however, three terms, *deviation, modulation index,* and *deviation ratio,* which convey considerable information concerning the character of the f-m wave.

Deviation is the amount of frequency shift each side of the unmodulated carrier frequency which occurs when the transmitter is modulated. Deviation is ordinarily measured in kilohertz, and in a properly operating f-m transmitter it will be directly proportional to the amplitude of the modulating signal. When a symmetrical modulating signal is applied to the transmitter, equal deviation each side of the resting frequency is obtained during each cycle of the modulating signal, and the total frequency range covered by the f-m transmitter is sometimes known as the *swing.* If, for instance, a transmitter operating on 1000 kHz has its frequency shifted from 1000 kHz to 1010 kHz, back to 1000 kHz, then to 990 kHz, and again back to 1000 kHz during one cycle of the modulating wave, the *deviation* would be 10 kHz and the *swing* 20 kHz.

The *modulation index* of an f-m signal is the ratio of the deviation to the audio modulating frequency, when both are expressed in the same units. Thus, in the example above if the signal is varied from 1000 kHz to 1010 kHz to 990 kHz, and back to 1000 kHz at a rate (frequency) of 2000 times a second, the modulation index would be 5, since the deviation (10 kHz) is 5 times the modulating frequency (2 kHz).

The *deviation ratio* is similar to the modulation index in that it involves the ratio between a modulating frequency and deviation. In this case, however, the deviation in question is the peak frequency shift obtained under full modulation, and the audio frequency to be considered is the maximum audio frequency to be transmitted. When the maximum audio frequency to be transmitted is 5000 Hz, for example, a deviation ratio of 3 would call for a peak deviation of 3 × 5000, or 15 kHz at full modulation. The noise-suppression capabilities of frequency modulation are directly related to the deviation ratio. As the deviation ratio is increased, the noise suppression becomes better *if* the signal is somewhat stronger than the noise. Where the noise approaches the signal in strength, however, low deviation ratios allow communication to be maintained in many cases where high-deviation-ratio frequency modulation and conventional amplitude modulation are incapable of giving service. This assumes that a narrow-band f-m receiver is in use. For each value of r-f signal-to-noise ratio at the receiver, there is a maximum deviation ratio which may be used, beyond which the output audio signal-to-noise ratio decreases. Up to this critical deviation ratio, however, the noise suppression becomes progressively better as the deviation ratio is increased.

For high-fidelity f-m broadcasting purposes, a deviation ratio of 5 is ordinarily used, the maximum audio frequency being 15,000 Hz, and the peak deviation at full modulation being 75 kHz. Since a swing of 150 kHz is covered by the transmitter, it is obvious that wide-band f-m transmission must necessarily be confined to the vhf range or higher, where room for the signals is available.

In the case of television sound, the deviation ratio is 1.67; the maximum modulation frequency is 15,000 Hz, and the transmitter deviation for full modulation is 25 kHz. The sound carrier frequency in a standard TV signal is located exactly 4.5 MHz higher than the picture carrier frequency. In the *intercarrier* TV sound system, which is widely used, this constant difference between the picture carrier and the sound carrier is employed within the receiver to obtain an f-m subcarrier at 4.5 MHz. This 4.5 MHz subcarrier then is demodulated by the f-m detector to obtain the sound signal which accompanies the picture.

Narrowband
F-M Transmission
Narrowband f-m transmission has become standardized for use by the mobile services such as police, fire, and taxicab communications, and is also authorized for amateur work in portions of each of the amateur radiotelephone bands. A maximum deviation of 15 kHz has been standardized for the mobile and commercial communication services, while a maximum deviation of 3 kHz is authorized for amateur nbfm hf communication. For a maximum audio frequency of 3000 Hz, the maximum deviation ratio is 1.0. For vhf f-m, the deviation ranges from 3 kHz to 15 kHz for a deviation ratio of up to 5.0.

The new channelized f-m concept for amateur communication has standardized on 5 kHz deviation on 10 meters and 6 meters, 5 to 15 kHz deviation on 2 meters, and 40 to 50 kHz deviation on the higher vhf bands. F.C.C. amateur regulations limit the bandwidth of f-m to that of an a-m transmission having the same audio characteristics below 29.0 MHz and in the 50.1 to 52.5 MHz frequency segment. Greater bandwidths are allowed above 29 MHz and above 52.5 MHz.

F-M Sidebands
Sidebands are set up when a radio-frequency carrier is frequency modulated. These sidebands differ from those resulting from a-m in that they occur at integral multiples of the modulating frequency; in a-m a single set of sidebands is generated for each modulating frequency. A simple method of determining the amplitude of the various f-m sidebands is the family of *Bessel curves* shown in figure 5. There is one curve for the carrier and one for each pair of sideband frequencies up to the fourth.

The Bessel curves show how the carrier and sideband frequency pairs rise and fall with increasing modulation index, and illustrate the particular values at which they disappear as they pass through zero. If the curves were extended for greater values of modulation index, it would be seen that the carrier amplitude goes through zero at modulation indices of 5.52, 8.65, 11.79, 14.93, etc. The modulation index, therefore, can

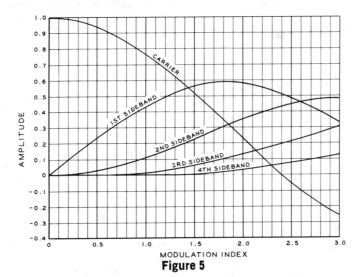

Figure 5

BESSEL CURVES SHOW VARIATION IN CARRIER AND SIDEBAND AMPLITUDE AS MODULATION INDEX IS INCREASED

The carrier and sideband frequency pairs rise and fall with increasing modulation index and pass through zero at certain values. Carrier drops to zero at modulation index of 2.40. The negative amplitude of the carrier above the 2.40 index indicates that the phase is reversed as compared to the phase without modulation.

be measured at each of these points by noting the disappearance of the carrier.

The relative amplitudes of carrier and sideband frequencies for any modulation index can be determined by finding the y-axis amplitude intercept for the particular function. Representative spectrum plots for three different values of modulation index are shown in figure 6. The negative amplitude in the Bessell curves indicate that the phase of the particular function is reversed as compared to the phase without modulation. In f-m, the energy that goes into the sideband frequencies is taken from the carrier; the total power in the overall composite signal remains the same regardless of the modulation index.

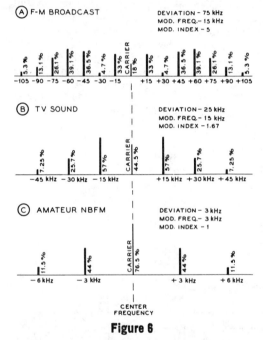

Figure 6

EFFECT OF F-M MODULATION INDEX

Showing the side-frequency amplitude and distribution for the three most common modulation indices used in f-m work. The maximum modulating frequency and maximum deviation are shown in each case.

It might be thought that the large number of side frequencies thus formed might make the frequency spectrum produced by an f-m transmitter prohibitively wide. However, the additional side frequencies are of small amplitude, and, instead of increasing the

bandwidth, modulation by a complex wave actually reduces the effective bandwidth of the f-m spectrum. This is especially true when speech modulation is being used, since most of the power of voice sounds is concentrated in the lower audio frequencies.

13-2 Direct F-M Circuits

Frequency modulation may be obtained either by the direct method, in which the frequency of an oscillator is changed directly by the modulating signal, or by the indirect method which makes use of phase modulation. Phase-modulation circuits will be discussed in the following section.

A successful frequency-modulated transmitter must meet two requirements: (1) The frequency deviation must be symmetrical about a fixed frequency, for symmetrical modulation voltage. (2) The deviation must be directly proportional to the amplitude of the modulation, and independent of the modulation frequency. There are several methods of direct frequency modulation which will fullfill these requirements. Some of these methods will be described in the following paragraphs.

Reactance Modulators   One of the most practical ways of obtaining direct frequency modulation is through the use of a *reactance modulator*. In this arrangement the modulator output circuit is connected across the oscillator tank circuit, and made to appear as either a capacitive or inductive reactance by exciting the modulator with a voltage which either leads or lags the oscillator tank voltage by 90 degrees. The leading or lagging input voltage causes a corresponding leading or lagging output current, and the output circuit appears as capacitive or inductive reactance across the oscillator tank circuit. When the transconductance of the modulator is varied by varying one of the element voltages, the magnitude of the reactance across the oscillator tank is varied. By applying audio modulating voltage to one of the elements, the transconductance (and hence the frequency) may be varied at an audio rate. When properly designed and operated, the reactance modulator provides linear frequency modulation, and is capable of producing large amounts of deviation.

There are numerous possible configurations of the reactance modulator circuit. The difference in the various arrangements lies principally in the type of phase-shifting circuit used to provide an input voltage which is in phase quadrature with the r-f voltage at the output of the modulator. A representative tube circuit showing four phase-shift arrangements is shown in figure 7.

A simple reactance modulator is shown in figure 8. An FET is coupled through a capacitor to the "hot" side of the oscillator tank circuit. The phase-shift network consists of the blocking capacitor (C_1), resistor R_1, and the input conductance of the FET (C_2). The value of resistor R_1 is made large in comparison with the reactance of capacitor C_2 at the oscillator frequency, and the current through the series circuit will be nearly in phase with the voltage across the tank circuit. Thus, the voltage across capacitor C_2 will lag the oscillator tank voltage by almost 90 degrees. The result of the lagging voltage is as though an inductance were connected across the oscillator tank circuit, thus raising the oscillator frequency. The increase in frequency is proportional to the amplitude of the lagging current in the reactance modulator stage.

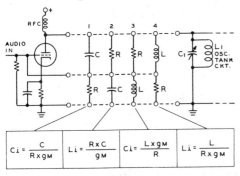

$$C_i = \frac{C}{R \times g_M} \qquad L_i = \frac{R \times C}{g_M} \qquad C_i = \frac{L \times g_M}{R} \qquad L_i = \frac{L}{R \times g_M}$$

Figure 7

FOUR POSSIBLE LOAD ARRANGEMENTS FOR REACTANCE MODULATOR

Stabilization Due to the presence of the reactance-tube frequency modulator, the stabilization of an f-m oscillator in regard to voltage changes is considerably more involved than in the case of a simple self-controlled oscillator for transmitter frequency control. If desired, the oscillator itself may be made perfectly stable under voltage changes, but the presence of the frequency modulator destroys the beneficial effect of any such stabilization. It thus becomes desirable to apply the stabilizing arrangement to the modulator as well as the oscillator.

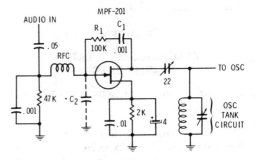

Figure 8

REACTANCE MODULATOR FOR DIRECT F-M

Phase-shift network consists of blocking capacitor C_1 plus R_1 and C_2 (the input conductance of the FET).

Linearity Test It is almost a necessity to run a static test on the reactance modulator to determine its linearity and effectiveness, since small changes in the values of components, and in stray capacitances will almost certainly alter the modulator characteristics. A frequency-versus-control voltage curve should be plotted to ascertain that equal increments in control voltage, both in a positive and a negative direction, cause equal changes in frequency. If the curve shows that the modulator has an appreciable amount of nonlinearity, changes in bias, electrode voltages, r-f excitation, and resistance values may be made to obtain a straight-line characteristic.

Figure 9 shows a method of connecting two batteries and a potentiometer to plot the characteristic of the modulator. It will be necessary to use a zero-center voltmeter to measure the voltage, or else reverse the voltmeter leads when changing from positive to negative grid voltage. When a straight-line characteristic for the modulator is obtained by the static test method, the capacitances of the various bypass capacitors in the circuit must be kept small to retain this characteristic when an audio voltage is used to vary the frequency in place of the dc volt-

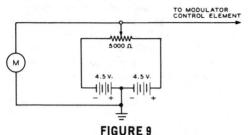

FIGURE 9

REACTANCE-MODULATOR LINEARITY CHECKER

age with which the characteristic was plotted.

The Diode Modulator   When a resistor and a capacitor are placed in series across an oscillator tank circuit, the current flowing in the series circuit is out of phase with the voltage. If the resistance or capacitance is made variable, the phase difference may be varied. If the variation is controlled at an audio rate, the resultant current can be used to frequency-modulate an oscillator (figure 10). The *diode modulator* may be a vacuum tube acting as a variable resistance or a solid-state voltage-variable capacitor whose capacitance varies inversely as the magnitude of the reverse bias. The variable element is placed in series with a small capacitance across the tank circuit of an oscillator to produce a frequency-modulated signal. The bias voltage applied to the diode should be regulated for best results.

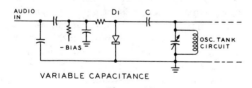

Figure 10

THE DIODE MODULATOR

13-3   Phase Modulation

By means of *phase modulation* (pm) it is possible to dispense with self-controlled os-

cillators and to obtain directly crystal-controlled frequency modulation. In the final analysis, phase modulation is simply frequency modulation in which the deviation is directly proportional to the modulation frequency. If an audio signal of 1000 Hz causes a deviation of 0.5 kHz, for example, a 2000-Hz modulating signal of the same amplitude will give a deviation of 1 kHz, and so on. To produce an f-m signal, it is necessary to make the deviation independent of the modulation frequency, and proportional only to the modulating signal (figure 11). With phase modulation this is done by including a frequency-correcting network in the transmitter. The audio-correction network must have an attenuation that varies directly with frequency, and this requirement is easily met by a very simple resistance capacitance network.

The only disadvantage of phase modulation, as compared to direct frequency modulation such as is obtained through the use of a reactance modulator, is the fact that very little frequency deviation is produced directly by the phase modulator. The deviation produced by a phase modulator is independent of the actual carrier frequency on which the modulator operates, but is dependent only on the phase deviation which is being produced and on the modulation frequency. Expressed as an equation:

$$F_d = M_p \times \text{modulating frequency}$$

where,

F_d is the frequency deviation one way from the mean value of the carrier,

M_p is the phase deviation accompanying modulation expressed in radians (a radian is approximately $57.3°$).

Thus, to take an example, if the phase deviation is $\frac{1}{2}$ radian and the modulating frequency is 1000 Hz, the frequency deviation applied to the carrier being passed through the phase modulator will be 500 Hz.

It is easy to see that an enormous amount of multiplication of the carrier frequency is required in order to obtain from a phase modulator the frequency deviation of 75 kHz required for commercial f-m broadcasting. However, for amateur and com-

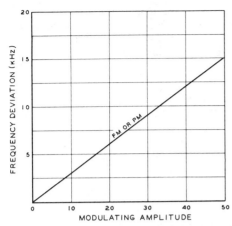 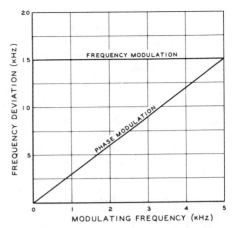

Figure 11

RELATIONSHIP BETWEEN FREQUENCY AND PHASE MODULATION

Frequency deviation is a function of amplitude and frequency of modulating signal for phase modulation (left) and a function of the amplitude only of modulating signal for frequency modulation (right). Most modern f-m transmitters use phase modulation as it may be easily applied to a crystal-controlled circuit.

mercial f-m work only a quite reasonable number of multiplier stages are required to obtain a deviation ratio of approximately one.

Many vhf f-m transmitters employ crystal control with the crystal frequency one twenty-fourth or one thirty-second of the carrier frequency. A deviation of 15 kHz at 144 MHz, for example, is equivalent to a deviation of 0.625 kHz at a crystal frequency of 6 MHz, which is well within the linear capability of a phase modulator. Some high-frequency f-m gear for the 30-MHz region employs crystals in the 200- to 500-kHz region to achieve sufficient frequency multiplication for satisfactory phase modulation at the crystal frequency.

Odd-harmonic distortion is produced when frequency-modulation is obtained by the phase-modulation method, and the amount of this distortion that can be tolerated is the limiting factor in determining the amount of phase modulation that can be used. Since the aforementioned frequency-correcting network causes the lowest modulating frequency to have the greatest amplitude, maximum phase modulation takes place at the lowest modulating frequency, and the amount of distortion that can be tolerated at this frequency determines the maximum deviation that can be obtained by the p-m method.

For high-fidelity broadcasting, the deviation produced by phase modulation is limited to an amount equal to about one-third of the lowest modulating frequency. But for nbfm work the deviation may be as high as 0.6 of the modulating frequency before distortion becomes objectionable on voice modulation. In other terms this means that phase deviations as high as 0.6 radian may be used for amateur and commercial nbfm transmission.

The Phase Modulator A change in the phase of a signal can be produced by passing the signal through a network containing a resistance and a reactance. If the series combination is considered to be the input, and the output voltage is taken from across the resistor, a definite amount of phase shift is introduced, the amount depending on the frequency of the signal and the ratio of the reactance to the resistance. When the resistance is varied with an applied audio signal, the phase angle of the output changes in direct proportion to the audio signal amplitude and produces a phase-modulated signal.

A representative phase modulator is shown in figure 12. The basic RC phase-shift network is composed of the resistance represented by the FET and the capacitor placed between input and output terminals of the modulator. The modulator is placed after the crystal oscillator and before the frequency multiplier stages. Phase modulation occurs as the modulator, in effect, detunes the amplifier tank circuit and thus varies the phase of the tank current to achieve phase modulation. The degree of phase shift that occurs during the detuning process depends upon the Q of the circuit, the higher the Q the smaller amount of detuning required to secure a given number of degrees of phase shift. With a Q of 10, for example, the relation between phase shift and the degree of detuning in kHz either side of the resonant frequency is substantially linear over a phase-shift range of nearly 25 degrees.

Since frequency deviation increases with the modulating frequency in phase modulation, as contrasted to frequency modulation, it is necessary to attenuate the higher frequencies to reduce the unnecessary sidebands that could be generated at frequencies far-removed from the carrier.

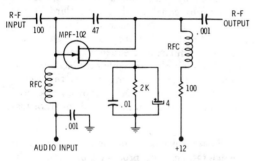

Figure 12

SOLID-STATE PHASE MODULATOR

Modulator stage is placed between crystal oscillator and the following amplifier or multiplier stages.

Shown in figure 13 is a simple phase modulator which employs a varactor diode to vary the phase of a tuned circuit. The modulator is installed between the oscillator and the subsequent frequency multiplier stage.

A phase modulator capable of a greater degree of modulation is shown in figure 14. This configuration is often used in vhf crystal-controlled f-m transmitters. In general a FET is used as a crystal oscillator, followed by a second FET as a phase modu-

lator, with the modulating network in the gate circuit. Two inexpensive silicon diodes used as varactors across a phasing coil are driven by the modulating voltage. The r-f output of the 2N5459 is about 30 milliwatts.

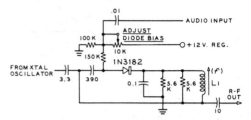

Figure 13

PHASE MODULATOR EMPLOYING VARACTOR DIODE

Audio voltage applied to varactor diode varies the phase of the tuned circuit. Diode bias is adjusted for largest phase shift consistent with linearity.

The F-M Transmitter   The various direct and indirect methods of producing f-m involve changing either the frequency or the phase of an r-f carrier in accordance with the modulating signal. The f-m signal is then raised to the operating frequency by passing it through a series of frequency multipliers. When the frequency is multiplied, the frequency deviation is multiplied by a like amount.

Inexpensive and highly stable crystals are available in the 3- to 10-MHz range and many popular f-m transmitters in the vhf region use such crystals, multiplying the crystal frequency by a factor of 12, 18 or 24. Because the amplitude of an f-m signal is constant, the signal may be amplified by nonlinear stages such as doublers and class-C amplifiers without introducing signal distortion. Actually, it is an advantage to pass an f-m signal through nonlinear stages, since any vestige of amplitude modulation generated in the phase modulator may be smoothed out by the inherent limiting action of a class-C amplifier.

Measurement   When a single-frequency mod-
of Deviation   ulating voltage is used with an f-m transmitter the relative amplitudes of the various sidebands and the carrier vary widely as the deviation is varied by increasing or decreasing the amount

of modulation. Since the relationship between the amplitudes of the various sidebands and carrier to the audio modulating frequency and the deviation is known, a simple method of measuring the deviation of a frequency-modulated transmitter is possible. In making the measurement, the result is given in the form of the modulation index for a certain amount of audio input.

The measurement is made by applying a sine-wave audio voltage of known frequency to the transmitter, and increasing the modulation until the amplitude of the carrier component of the frequency-modulated wave reaches zero. The modulation index for zero carrier may then be determined from the table below. As may be seen from the table, the first point of zero carrier is obtained when the modulation index has a value of 2.405—in other words, when the deviation is 2.405 times the modulation frequency. For example, if a modulation frequency of 1000 Hz is used, and the modulation is increased until the first carrier null is obtained, the deviation will then be 2.405 times the modulation frequency, or 2.405 kHz. If the modulating frequency happened to be 2000 Hz, the deviation at the first null would be 4.810 kHz. Other carrier nulls will be obtained when the index is 5.52, 8.654, and at increasing values separated approximately by π. The following is a listing of the modulation index at successive carrier nulls up to the tenth:

Zero carrier point no.	Modulation index
1	2.405
2	5.520
3	8.654
4	11.792
5	14.931
6	18.071
7	21.212
8	24.353
9	27.494
10	30.635

The only equipment required for making the measurements is a calibrated audio oscillator of good wave form, and a communication receiver equipped with a narrow passband i-f filter, to exclude sidebands spaced from the carrier by the modulation frequency. The unmodulated carrier is accurately tuned on the receiver. Then modulation from the audio oscillator is applied to the transmitter, and the modulation is increased until the first carrier null is obtained. This carrier null will correspond to a modulation index of 2.405, as previously mentioned. Successive null points will correspond to the indices listed in the table.

A heterodyne deviation meter is shown in figure 15. This device provides a quick and easy means of "netting" an f-m transmitter. A diode mixer is used in conjunction with a local oscillator to provide an audio signal which is amplified and clipped in an operational amplifier, IC_1. The resulting signal is a square wave which is applied to a rectifier

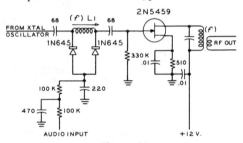

Figure 14

FET PHASE-MODULATED IN GATE CIRCUIT

Two silicon diodes are used as varactors across a phasing coil (L_1). R-f output of 2N5459 is about 30 milliwatts. Circuit permits a small degree of amplitude modulation which is limited out by succeeding stages of f-m exciter.

and indicating meter. The squarewave signal is passed through an adjustable coupling capacitor which allows calibration for the meter ranges of 1, 10, and 20 kHz. The meter reads average rectified current which is proportional to frequency.

The deviation meter is calibrated by applying a low level audio signal to pin 2 of U_1. The frequency of the applied signal is set at the indicated frequencies and the appropriate trimmer capacitor adjusted for full-scale deflection. As the audio frequency is varied, the meter reading should correspond with the frequency over the greater portion of the range.

The crystal is chosen so as to produce a harmonic signal at the carrier frequency of the f-m channel in use. Sine-wave modula-

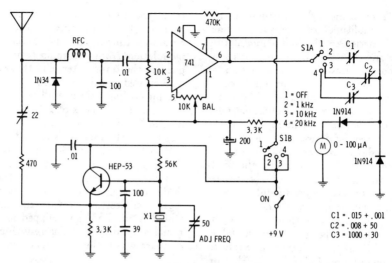

Figure 15

F-M DEVIATION METER

Simple direct-conversion receiver is coupled to a meter whose reading is proportional to the frequency of the applied audio signal. In this case, the audio signal is produced by the beat between the crystal harmonic frequency and the observed frequency.

tion is applied to the transmitter under test and the deviation level adjusted for the amount desired, as indicated on the meter of the instrument.

Modulation Limiting Deviation in an f-m transmitter can be controlled by a circuit that holds the audio level within prescribed limits. Simple audio clipping circuits may be used, as well as more complex deviation control circuits. Diode limiting circuits, such as discussed in Chapter 9 are commonly used, followed by a simple audio filter which removes the harmonics of the clipped audio signal. A representative clipping and filtering circuit is shown in figure 16.

13-4　Reception of F-M Signals

A conventional communications receiver may be used to receive narrow-band f-m transmission, although performance will be much poorer than can be obtained with an nbfm receiver or adapter. However, a receiver specifically designed for f-m reception must be used when it is desired to receive high deviation f-m such as used by

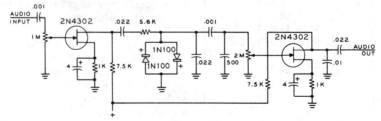

Figure 16

MODULATION LIMITING

Deviation in an f-m transmitter can be controlled by a clipping circuit which holds peak audio level within prescribed limits. Simple audio filter removes higher harmonics of clipped signal.

f-m broadcast stations, TV sound, and mobile communications.

The f-m receiver must have, first of all, a bandwidth sufficient to pass the range of frequencies generated by the f-m transmitter. And since the receiver must be superheterodyne if it is to have good sensitivity at the frequencies to which frequency modulation is restricted, i-f bandwidth is an important factor in its design.

The second requirement of the f-m receiver is that it incorporate some sort of device for converting frequency changes into amplitude changes, in other words, a detector operating on frequency variations rather than amplitude variations. Most f-m equipment operates in the vhf region, and at these frequencies it is not always possible to obtain optimum performance at reasonable cost with a single-conversion superheterodyne receiver. When good adjacent-channel selectivity is necessary, a low i-f channel is desirable; this, however lowers the image rejection ability of the receiver. Similarly, if good image rejection is desired, a high i-f channel should be used, but this is not compatible with good adjacent-channel rejection unless an expensive i-f filter is employed.

These difficulties are compromised by the use of a double-conversion receiver, such as the one shown in the block diagram of figure 17. In many receiver designs, the high i-f channel is chosen so that a harmonic of the mixing oscillator used for the second mixer may be used with the first mixer to reduce the number of crystals in the receiver. In other cases, a frequency synthesizer is used to generate the proper mixing frequencies.

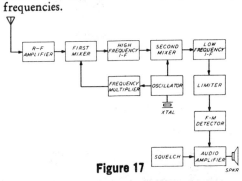

Figure 17

**DOUBLE-CONVERSION RECEIVER
FOR VHF F-M RECEPTION**

The third requirement, and one which is necessary if the full noise-reducing capabilities of the f-m system of transmission are desired, is a limiting device to eliminate amplitude variations before they reach the detector.

The Frequency Detector The simplest device for converting frequency variations to amplitude variations is an "off-tune" resonant circuit, as illustrated in figure 18. With the carrier tuned in at point A, a certain amount of r-f voltage will be developed across the tuned circuit, and, as the frequency is varied either side of this frequency by the modulation, the r-f voltage will increase and decrease to point C and B in accordance with the modulation. If the voltage across the tuned circuit is applied to an ordinary detector, the detector output will vary in accordance with the modulation, the amplitude of the variation being proportional to the deviation of the signal, and the rate being equal to the modulation frequency. It is obvious from figure 18 that only a small portion of the resonance curve is usable for linear conversion of frequency

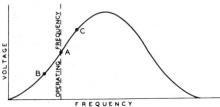

Figure 18

SLOPE DETECTION OF F-M SIGNAL

variations into amplitude variations, since the linear portion of the curve is rather short. Any frequency variation which exceeds the linear portion will cause distortion of the recovered audio. It is also obvious by inspection of figure 18 that an a-m receiver used in this manner is vulnerable to signals on the peak of the resonance curve and also to signals on the other side of the resonance curve. Further, no noise-limiting action is afforded by this type of reception.

Double-Tuned Discriminator A better frequency detector or *discriminator*, is shown in figure 19A. In this arrangement two tuned circuits are used, one tuned on each side of the i-f amplifier frequency, and with their resonant frequencies spaced slightly more than the expected transmitter swing. Their outputs are combined in a differential rectifier so that the voltage across series load resistors R_1 and R_2 is equal to the algebraic sum of the individual output voltages of each rectifier. When a signal

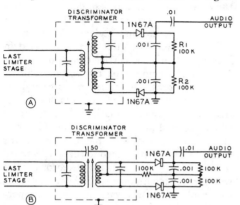

Figure 19

THE F-M DETECTOR

A—The double-tuned discriminator uses two secondary windings on the detector transformer, one tuned on each side of the i-f amplifier center frequency. On either side of center frequency a voltage of polarity and magnitude proportional to direction and magnitude of frequency shift is developed. B—Foster-Seeley discriminator employs a single, tapped secondary winding. Vector diagram of summed output voltages is shown in figure 20B-C.

at the i-f midfrequency is received, the voltages across the load resistors are equal and opposite, and the sum voltage is zero. As the r-f signal varies from the midfrequency, however, these individual voltages become unequal, and a voltage having the polarity of the larger voltage and equal to the difference between the two voltages appears across the series resistors, and is applied to the audio amplifier. The relationship between frequency and discriminator output voltage is shown in figure 20A. The separation of the discriminator peaks and the linearity of the output voltage-versus-frequency curve depend on the discriminator frequency, the Q of the tuned circuits, and the value of the diode load resistors.

Foster-Seeley Discriminator   A popular form of discriminator is that shown in figure 19B. This type of discriminator yields an output voltage-versus-frequency characteristic similar to that shown in figure 20B. Here, again, the output voltage is equal to the algebraic sum of the voltages developed across the load resistors of the two diodes, the resistors being connected in series to ground. However, this *Foster-Seeley* discriminator requires only two tuned circuits instead of the three used in the previous discriminator. The operation of the circuit results from the phase relationships existing in a transformer having a tuned

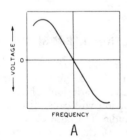
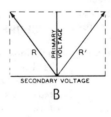
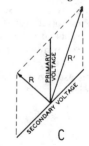

Figure 20

DISCRIMINATOR CHARACTERISTICS

A—Discriminator of figure 19A produces zero voltage at the center frequency. On either side of this frequency it gives a voltage of a polarity and magnitude which depend on the direction and amount of frequency shift.

B—Vector diagram of discriminator of figure 19B. Signal at the resonant frequency will cause secondary voltage to be 90 degrees out of phase with the primary voltage and the resultant voltages (R and R') are equal.

C—If the signal frequency changes, the phase relationship changes and the resultant voltages are no longer equal. A differential detector is used to provide an output voltage proportional to the difference between R and R'.

secondary. In effect, as a close examination of the circuit will reveal, the primary circuit is in series for r-f, with each half of the secondary to ground. When the received signal is at the resonant frequency of the secondary, the r-f voltage across the secondary is 90 degrees out of phase with that across the primary. Since each diode is connected across one half of the secondary winding and the primary winding in series, the resultant r-f voltages applied to each are equal, and the voltages developed across each diode load resistor are equal and of opposite polarity. Hence, the net voltage between the top of the load resistors and ground is zero. This is shown vectorially in figure 20B where the resultant voltages R and R' which are applied to the two diodes are shown to be equal when the phase angle between primary and secondary voltages is 90 degrees. If, however, the signal varies from the resonant frequency, the 90-degree phase relationship no longer exists between primary and secondary.

The result of this effect is shown in figure 20C where the secondary r-f voltage is no longer 90 degrees out of phase with respect to the primary voltage. The resultant voltages applied to the two diodes are now no longer equal, and a dc voltage proportional to the difference between the r-f

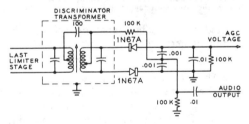

Figure 21

THE RATIO DETECTOR

This detector is inherently insensitive to amplitude modulation and does not require the use of a limiter ahead of it. Automatic volume control voltage is provided for controlling gain of r-f and i-f stages ahead of the detector.

voltages applied to the two diodes will exist across the series load resistors. As the signal frequency varies back and forth across the resonant frequency of the discriminator, an ac voltage of the same frequency as the original modulation, and proportional to the

deviation, is developed and passed on to the audio amplifier.

Ratio Detector A third form of f-m detector circuit, called the *ratio detector* is diagrammed in figure 21. The input transformer can be designed so that the parallel input voltage to the diodes can be taken from a tap on the primary of the transformer.

The circuit of the ratio detector appears very similar to that of the more conventional discriminator arrangement. However, it will be noted that the two diodes in the ratio detector are polarized so that their dc ouput voltages add, as contrasted to the Foster-Seeley circuit wherein the diodes are polarized so that the dc output voltages buck each other. At the center frequency to which the discriminator transformer is tuned, the voltage appearing at the top of the 100K resistor will be one-half the dc voltage appearing at the agc *output* terminal, since the contribution of each diode will be the same. However, as the input frequency varies to one side or the other of the tuned value (while remaining within the passband of the i-f amplifier feeding the detector) the relative contributions of the two diodes will be different. The voltage appearing at the top of the 100K resistor will increase for frequency deviations in one direction and will decrease for frequency deviations in the other direction from the mean or tuned value of the transformer. The audio output voltage is equal to the ratio of the relative contributions of the two diodes, hence the name ratio detector.

The ratio detector offers several advantages over the simple discriminator circuit. The circuit does not require the use of a limiter preceding the detector since the circuit is inherently insensitive to amplitude modulation on an incoming signal. This factor alone means that the r-f and i-f gain ahead of the detector can be much less than the conventional discriminator for the same overall sensitivity, further, the circuit provides agc voltage for controlling the gain of the preceding r-f and i-f stages. The ratio detector is, however, susceptible to variations in the amplitude of the incoming signal as in any other detector circuit except the discriminator *with* a limiter preceding it,

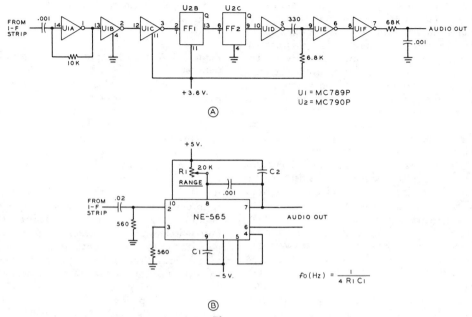

Figure 22

UNUSUAL F-M DETECTORS MAKE USE OF INTEGRATED CIRCUITS

A—Pulse counting detector uses two small ICs and provides quieting and linear detection over wide frequency ranges. First three stages provide limiting and produce a pulse train which is fed to a "divide-by-four" pair of flip-flops. Low-frequency pulses trigger a multivibrator (U_{1D}) whose repetition rate varies in direct proportion to frequency variation of i-f signal. Pulses are converted to audio signal by RC de-emphasis network at output of detector. B—Single IC performs as phase-locked loop detector for f-m. Error voltage proportional to frequency deviation is applied to voltage-controlled oscillator, locking it to incoming signal. Error voltage is replica of frequency shift on incoming signal.

so that agc should be used on the stage preceding the detector.

The Pulse-Counting Detector Shown in figure 22A is a compact detector that provides inherent quieting and linear detection over wide frequency ranges. Two ICs (RTL logic) provide the functions of a limiter and discriminator. The first inverter serves as a signal amplifier and the following two stages provide limiting to produce a pulse train at the intermediate frequency. This train is fed to a "divide-by-four" circuit composed of flip-flops FF_1 and FF_2. The low-frequency signal triggers a monostable multivibrator (U_{1D}), whose period is about 0.5 that of the i-f signal. The output pulses of the multivibrator have a repetition rate which varies in direct proportion to the frequency vaiation of the i-f signal. The pulses are amplified by two inverter stages and con-

verted to an audio signal by the RC de-emphasis network at the output of U_{1F}.

The Phase-Locked Loop Detector The phase-locked loop, discussed in Chapter 11 is now available in a single IC package or in separate building block ICs. The PLL consists of a phase detector, a filter, a dc amplifier, and a voltage-controlled oscillator which runs at a frequency close to that of an incoming signal. The phase detector produces an error voltage proportional to the difference in frequency between the oscillator and the incoming signal, the error voltage being applied to the voltage-controlled oscillator. Any change in frequency of the incoming signal is sensed, and the resulting error voltage readjusts the oscillator frequency so that it remains locked to the incoming signal. As a result, the error voltage is a replica of the audio variations originally used to shift the frequency of the f-m signal, and the PLL functions directly as an f-m detector. The functional bandwidth of the system is determined by a filter placed on the error voltage line. The *Signetics* NE565 is especially designed for this service (figure 22B).

The Quadrature Detector   The *quadrature detector* (figure 23) demodulates an f-m signal by combining two versions of the i-f signal which are in quadrature (a phase difference of 90 degrees).

The input stages in the representative IC f-m quadrature detector are wideband limiting amplifiers which remove the a-m component of the wave and pass on a clipped, squarewave series to a signal splitter which feeds a portion of the signal to an external, 90-degree phase-shift network (illustration B). The shifted signal is fed to one input port of the synchronous detector. The gated detector integrates the pulsed signals to extract the audio signal.

Alignment of the quadrature detector requires that the phase-shift coil be adjusted

for maximum audio level, or the coil may be adjusted to null the noise level on an unmodulated signal.

Limiters   The limiter of an f-m receiver using a conventional discriminator serves to remove amplitude modulation and pass on to the discriminator a frequency-modulated signal of constant amplitude; a typical circuit is shown in figure 24.

Up to a certain point the output of the limiter will increase with an increase in signal. Above this point, however, the limiter becomes overloaded, and further large increases in signal will not give any increase in output. To operate successfully, the limiter must be supplied with a large amount of signal, so that the amplitude of its output will not change for rather wide variations in amplitude of the signal. Noise, which causes little frequency modulation but much amplitude modulation of the received signal, is virtually wiped out in the limiter.

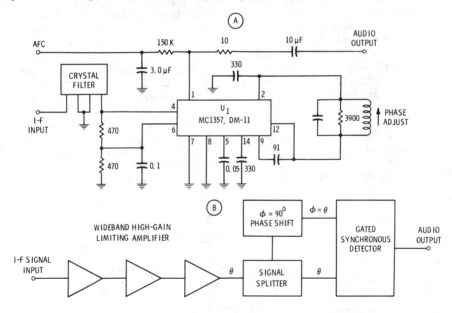

Figure 23
THE QUADRATURE DETECTOR

A—F-m quadrature detector using MC 1357P integrated circuit
B—Block diagram of the MC 1357P quadrature detector

Proper limiting action calls for a signal of considerable strength to ensure full clipping, typically several volts for tubes and about one volt for transistors. Limiting action should start with an r-f input of 0.2 μV, or less, at the receiver antenna terminals, consequently a large amount of signal gain is required between antenna and the limiter stages. Typically 100 dB to 140 dB gain is used in modern f-m receivers, most of this gain being achieved in the i-f amplifier chain. The high gain level amplifies internal and external noise and an annoying blast of noise emits from the speaker of the f-m receiver unless some form of *audio squelch* is provided, as discussed later in this chapter.

Receiver Bandwidth One of the most important factors in the design of an f-m receiver is the frequency swing which it is intended to handle. It will be apparent from figure 20 that if the straight portion of the discriminator circuit covers a wider range of frequencies than those generated by the transmitter, the audio output will be reduced from the maximum value of which the receiver is capable.

In this respect, the term *modulation percentage* is more applicable to the f-m receiver than it is to the transmitter, since the modulation capability of the communication system is limited by the receiver bandwidth and the discriminator characteristic; full utilization of the linear portion of the characteristic amounts, in effect, to 100 percent modulation. This means that some sort of standard must be agreed on, for any particular type of communication, to make it unneccessary to vary the transmitter swing to accommodate different receivers.

Two considerations influence the receiver bandwidth necessary for any particular type of communication. These are the maximum audio frequency which the system will handle, and the deviation ratio which will be employed. For voice communication, the maximum audio frequency is more or less fixed at 3000 to 4000 Hz. In the matter of deviation ratio, however, the amount of noise suppression which the f-m system will provide is influenced by the ratio chosen, since the improvement in signal-to-noise ratio which the f-m system shows over amplitude modulation is equivalent to a constant *multiplied by the deviation ratio*. This assumes that the signal is somewhat stronger than the noise at the receiver, however, as the advantages of wideband frequency modulation in regard to noise suppression disappear when the signal-to-noise ratio approaches unity.

As mentioned previously, broadcast f-m practice is to use a deviation ratio of 5. When this ratio is applied to a voice-communication system, the total swing becomes 30 to 40 kHz. With lower deviation ratios, such as are most frequently used for voice work, the swing becomes proportionally less, until at a deviation ratio of 1 the swing is equal to twice the highest audio frequency. Actually, however, the receiver bandwidth must be greater than the expected transmitter swing, since for distortionless reception the receiver must pass the complete band of energy generated by the transmitter, and this band will always cover a range somewhat wider than the transmitter swing.

Figure 24

TWO-STAGE F-M LIMITER

F-m limiter circuit serves to re-move amplitude variations of incoming f-m signal. Limiter satu-rates with small signal and further increases in strength of incoming signal will not give any increase in output level. Noise, which causes little f-m but much a-m, is virtually eliminated in effective limiter stages.

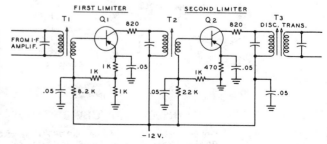

On the other hand, a low deviation ratio is more satisfactory for strictly communication work, where readability at low signal-to-noise ratios is more important than additional noise suppression when the signal is already appreciably stronger than the noise.

Deviations of 15, 5, and 2.5 kHz are common on the amateur vhf bands and are termed wideband, narrowband, and sliver band, respectively. Bandwidth required in an f-m receiver is about 2.4 times the deviation: 36 kHz for wideband reception and 13 kHz for narrowband reception.

The proper degree of i-f selectivity may be achieved by using a number of over-coupled transformers or by the use of a ceramic or crystal filter. Shown in figure 25 is a transistorized i-f strip using a packaged filter for adjacent channel selectivity and four stages of resistance-coupled amplification to provide adequate gain. The stages are paired in regard to the supply voltage, with the paired transistors placed in series so that each has half the supply voltage. I-f filters for vhf f-m service generally have a center frequency of 455 kHz, 9.0, 10.7, or 21.5 MHz with bandwidths ranging from 12 kHz to 36 kHz.

Pre-Emphasis and De-Emphasis Standards in f-m broadcast and TV sound work call for the pre-emphasis of all audio modulating frequencies above about 2000 Hz, with a rising slope such as would be produced by a 75-microsecond RL network. Thus the f-m receiver should include a compensating de-emphasis RC network with a time constant of 75 microseconds so that the overall frequency response from microphone to speaker will approach linearity. The use of pre-emphasis and de-emphasis in this manner results in a considerable improvement in the overall signal-to-noise ratio of an f-m system. Appropriate values for the de-emphasis network, for different values of circuit impedance are given in figure 26.

Squelch Circuits Squelch circuits are used to mute the audio of an f-m receiver when no signal is present. In a high-gain receiver, speaker noise can be very annoying to the operator who must monitor a channel for a long period. When the receiver is squelched, no background noise is heard; when an r-f signal comes on, squelch is turned off and the audio system becomes operative. Squelch circuits may be carrier operated or noise operated.

A solid-state squelch circuit is shown in figure 27. Audio voltage is amplified and rectified and applied to the gate of a JFET which acts as a series audio gate. Squelch level is controlled by varying the signal gate voltage of the MPF-103 device. The output impedance of the MPS-A10 amplifier is quite low and suitable for running into an audio line, if required.

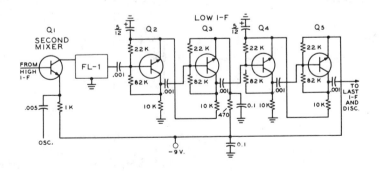

Figure 25

TRANSISTOR I-F STRIP USES CASCODE CIRCUIT

Transistors in pairs (Q₁-Q₃ and Q₄-Q₅) are placed in series in regard to the supply voltage in the manner of a cascode amplifier so that each transistor of a pair has half the dc voltage across it. A crystal or mechanical filter provides good adjacent-channel selectivity.

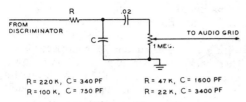

Figure 26

75-MICROSECOND DE-EMPHASIS CIRCUIT

The audio signal transmitted by f-m and TV stations has received high-frequency pre-emphasis, so that a de-emphasis circuit should be included between the output of the f-m detector and the input of the audio system.

A single IC carrier operated squelch circuit is shown in figure 28. A squelch voltage greater than $+4$ volts turns the audio stage on. The squelch sensing voltage is taken from the rectified carrier.

A Simple F-m Adapter Many transceivers (and CB equipment converted to the 10-meter amateur band) can be adapted for f-m reception using the circuit shown in figure 29. This adapter works with any receiver having a 455-kHz i-f system and requires a single IC and a few parts. The device is designed for tv sound service and functions as an i-f amplifier, limiter, f-m detector, and audio driver.

The driving signal is acquired from the last 455-kHz amplifier of the receiver and the audio output signal is coupled back into the existing audio stage of the receiver.

The adapter can be built up on a small piece of vector board and mounted within the receiver. The input is connected to the base of the last i-f amplifier transistor and the audio output is connected to the top of the audio volume control in the receiver. The existing lead to this point must be removed or switched off so as not to receive f-m and the existing receiver mode simultaneously.

The slug of the transformer should be adjusted for maximum audio response. A small capacitor (C_1) may be required to achieve resonance. If the audio signal is too large, a 100K resistor may be inserted between pin 8 of the IC and the top of the volume control.

FM Stereo Two program channels are transmitted. Additional signal processing in the receiver is required to recover the separate programs. The detector output with a stereo signal is a composite signal that contains the basic 50 to 15,000 Hz audio band (which is in mono form and consists of the sum of the two channels, or $L + R$), a 19-kHz pilot carrier, and a double-sideband signal about a 38-kHz suppressed subcarrier. This signal contains the difference of the stereo channels ($L - R$).

After processing in the multiplex demodulator, the $L + R$ and $L - R$ signals are recovered and can be combined in a resistive matrix. The addition and subtraction of these signals results in separation of the left

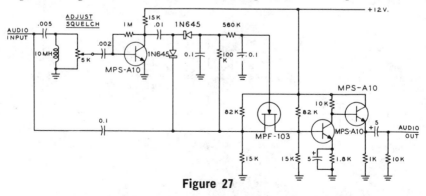

Figure 27

AUDIO OPERATED SOLID-STATE SQUELCH CIRCUIT

Audio voltage is amplified, rectified, and applied to the gate element of a JFET which acts as a series audio gate. Squelch level is controlled by varying the signal gate voltage of MPF-103 squelch amplifier stage.

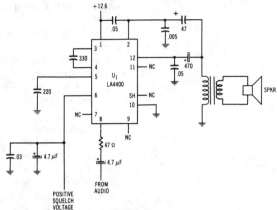

Figure 28

SINGLE IC CARRIER-OPERATED SQUELCH CIRCUIT

IC chip is LA 4400. Audio stage is cut off when pin 6 drops to low dc level. Squelch voltage greater than 4 volts turns on the audio stage. Squelch sensing voltage is taken from rectified carrier signal.

and right program channel signals. Each signal is then de-emphasized by a simple RC network that has a 75 μs time constant that rolls off the response at a 6 db/octave rate above 2100 Hz (complementing a similar boost at the transmitter) to yield a flat overall frequency response.

13-5   THE F-M Repeater

Since radio transmission in the vhf region is essentially short range, a form of radio relay station termed a *repeater* may be employed to expand the communication range of base or mobile stations over an extended distance. Various types of relays are in use in the United States, their operation depending on the requirements of the commmunications circuit.

The *relay unit* is a fixed repeating station whose specific purpose is to extend station-to-station communication capability. The user's transmitter is on the input frequency while his receiver is on the output frequency of the relay (figure 30). When desired, direct communication between stations may take place by using a closely spaced frequency domain and a two-frequency transmitter.

The *remote base* is a form of relay unit whose location has a height or tactical

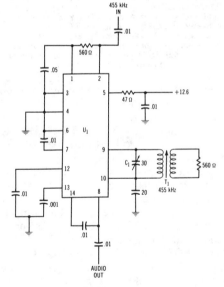

FIGURE 29

SIMPLE F-M ADAPTER FOR A COMMUNICATIONS RECEIVER

IC chip is MC1358, CA3065, or ECG712. Capacitor C_1 is a 30-pF mica compression unit.

advantage. Means must be provided to control such an installation which in amateur service most often is working in conjunction with a pair of frequencies—input and

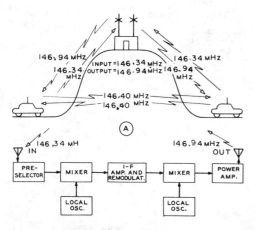

Figure 30

**F-M REMOTE REPEATER
FOR MOBILE SERVICE**

Radio relay station serves as a repeater to
extend the range of base or mobile f-m stations.
Communication between units may be achieved
either directly, or through repeater. The repeater consists of a back-to-back receiver and
transmitter having a common i-f and remodu-
lator system. Most repeaters are limited to a
single channel, but multiplex operation permits
simultaneous transmission of different information forms on the channel.

output. In so doing, remote bases serve on
common frequencies by which individual
groups operating their own installation can
cross-communicate. Frequencies above 220
MHz or direct-wire lines must be used for
remote control.

Simplex communication, on the other
hand, refers to communication between individual units operating on a common transmit and receive frequency. Thus simplex
operation can be interfaced with relay operation, using either a local or remote base.
Remote base operation must take place
under FCC license to a responsible controlling authority and each application for such
service is judged individually on the merits
of the case.

Repeater   There are two basic categories of
Types   repeaters: *open* and *closed*. The
                  open repeater is one which has
been installed for the benefit of all who
wish to use it for communications; the
closed repeater is one which is designed to

selectively benefit a specific group of users.
Both types are in widespread use throughout the United States and many foreign
countries. Early repeaters were a-m open
types, which later gave way to the f-m
open and closed repeaters. The open repeater
is virtually always carrier operated, switching to the transmitting mode only with an
incoming signal.

The closed repeater, as the name implies,
gives the benefits of repeater coverage to a
select group of subscribers or users. Special
selective circuits are used on the repeater to
reject all signals other than those for which
the system was designed. This function is
almost universally achieved with a system of
access tones, whereby a specific tone on the
incoming signal is a prerequisite to being
automatically relayed to the repeater output. One technique calls for a continuous
low-frequency tone (below 120 Hz) to be
transmitted. A decoding device is employed
at the repeater that responds only to signals
bearing this tone. This is termed a *continuous tone squelched private line* (PL) system.
A second technique requires that the incoming signal be accompanied by a short high-
frequency *tone burst* of a few milliseconds.
The decoder at the repeater allows the transmitter to be energized only when the signal
bears the proper tone. This access approach
is called the *single-tone,* or "whistle-on"
system, since it may be activated by an
operator with a good ear for tone and a
talent for whistling!

Many repeaters make use of a *transmission
limiter*, which consists of a timer which disables the repeater when input time exceeds
3 minutes or so. The repeater is reactivated
when the input signal is removed. More
complicated control techniques exist, too,
which make use of channelized tones
between 1500 and 1650 Hz.

Control   The basic control element of
Techniques   most amateur repeaters is the
                   carrier-operated relay (COR),
a squelch-responding circuit that provides a
relay closure (K_1) with each signal that
occupies the channel (figure 32). When the
repeater is at a remote location, functional
control may be exerted over a wire (telephone line) or by a uhf radio link. The

Figure 31

TYPICAL REMOTE REPEATER INSTALLATION

A vhf amateur remote repeater installation at a commercial facility atop 8500-foot Blueridge
Summit in California.

control scheme is based upon the transmission of specific and precise audio frequency signals which activate turn-on and shut-down systems, frequency selections, and automatic time-out devices. The audio frequencies are generated by a tone generator termed an *encoder* and the responding device is called a *decoder*. Multiple functions

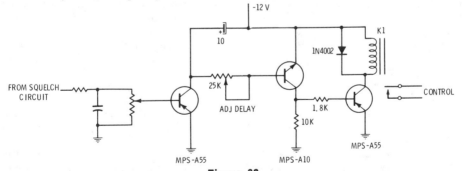

Figure 32

CARRIER-OPERATED RELAY

Adjustable delay circuit permits repeater to remain on the air for a few seconds after being
keyed off.

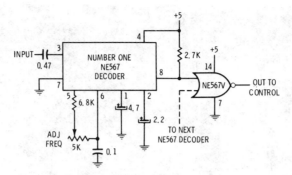

Figure 33

PLL DECODER BUILDING BLOCK

Phase-lock loop (PLL) basic building block. Seven NE 567 ICs are used in the decoder section, each centered on a particular tone frequency. Four TTL quad gates are used in the logic section. When a tone is received, the output goes from a logic high to a logic low. The output of the gate can be used to drive a relay or a function decoder. Proper interconnections to the logic sections can provide for encoded pair of tones, two low inputs switching the gate to a high level, for example.

may be achieved through the use of a single decoder by the use of tone filters and phase-locked loops (figure 33).

One of the most promising tone-control techniques makes use of the multitone (*Touchtone*) technique. *Touchtone* command signals are generated with a conventional *Touchtone* telephone dial which has an integral multitone encoder. The system makes use of eight discrete tone frequencies arranged in two groups of four tones each (a high group and a low group). Sixteen digits can then be represented by the combination of one tone from the high group with one tone from the low. The individual frequencies and various combinations are shown in figure 33, which is a schematic of the standard 25A3 10-button *Touchtone telephone pad*. The supply voltage is fed to the pad over the same path as the output of the tones.

The *Touchtone* encoder pad can be connected directly into the microphone amplifier of an f-m transmitter for transmission the tones over the air to the decoder unit at the repeater site.

The *Touchtone* signal can be decoded by separating the two-tone combination via bandpass and band-elimination filters into groups so that each tone can be regulated, limited, and applied to the desired control circuit.

Other tone systems exist, including the dual-tone (*Secode*) system and the single-tone approach. The latter may be used with a telephone dial pulsing system, as shown in figure 35. Control pulses are sent serially, at a rate of about 10 pulses per second to initiate a command function at the repeater.

The Repeater   The repeater is a receiver-transmitter combination capable of duplex operation. That is, the receiver must be capable of functioning regardless of whether the transmitter is activated or not. Since the repeater equipment must run continuously (probably in a remote spot without air conditioning) it must be well ventilated. Most repeaters have air continuously circulated about within the cabinet or enclosure by means of exhaust and intake fans as shown in figure 36.

Transmitter Noise—Broadband noise may be radiated by any r-f generating equipment as the result of random noise components generated and amplified in the driver stages, which are amplified and passed on to the antenna through the relatively broad selectivity of the amplifier output circuitry. Enough noise may be radiated to degrade the performance of a nearby receiver operating several MHz away (figure 37A). Transmitter noise is bothersome as "off-channel"

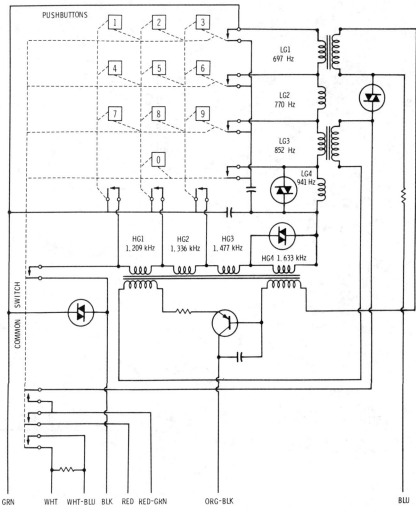

Figure 34

TOUCHTONE PAD

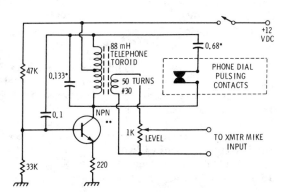

Figure 35

SECODE-TYPE 600/1500-Hz OSCILLATOR

* TO ADJUST TONE FREQUENCY, FIRST OPEN THE TELEPHONE DIAL PULSING CONTACTS. SELECT THE 0.133-μF CAPACITOR FOR 1500-Hz OUTPUT. CLOSE THE DIAL PULSING CONTACTS, AND SELECT THE 0.68 μF CAPACITOR FOR 600-Hz OUTPUT.

** ANY NPN TRANSISTOR WITH h_{FE} OF 50 to 100.

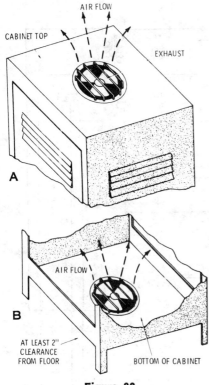

Figure 36

**METHODS FOR MOUNTING
VENTILATING FANS**

A—Top-mounted exhaust fan. B—Bottom-
mounted forced-air type.

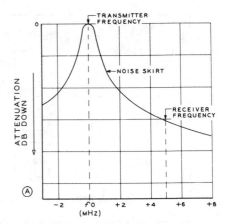

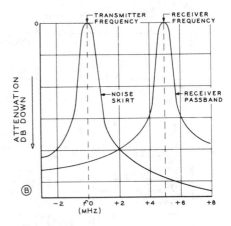

Figure 37

A—Broadband noise is radiated by an f-m transmitter as result of random noise components
amplified and passed to antenna through relatively broad selectivity of output circuitry.
Enough noise may be radiated to degrade performance of nearby receiver operating several
MHz away from transmitting frequency. B—
Bandpass cavity on output of transmitter and
input of receiver provides sufficient attenuation and rejection of off-channel noise to protect
receiver from desensitization.

noise which cannot be filtered out at the receiver, competing with the desired signal and reducing effective receiver sensitivity.

Receiver Desensitization—This form of interference is the result of a strong off-channel frequency signal entering the front-end of the receiver, upsetting critical voltage and current levels, and reducing receiver gain.

Intermodulation—Intermodulation is the generation of spurious frequencies in a non-linear circuit element. The undesired frequencies correspond to the sum and differences of the fundamental and harmonics of two or more frequencies passing through the element, as discussed in Chapter 16.

Intermodulation interference may occur from signals outside the normal operating range of the equipment to produce a product which can interfere with a desired signal.

Receiver   Sufficient electrical isolation
Protection   between receiver and transmitter
                 at a repeater site will protect the
receiver from desensitization, intermodula-

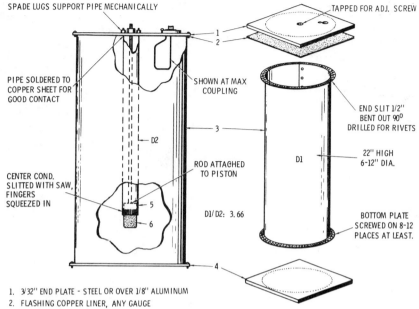

1. 3/32" END PLATE - STEEL OR OVER 1/8" ALUMINUM
2. FLASHING COPPER LINER, ANY GAUGE
3. ALUMINUM CYLINDER (0.032" OR THICKER)
4. 3/32" STEEL OR ALUMINUM END PLATE
5. COPPER PIPE - DIA: 1/3.66 x OUTSIDE DIA. OF CAVITY (NOT CRITICAL)
6. TUNING PISTON - ANY MATERIAL WITH FLASHING COPPER WRAPPED ON OUTSIDE.
   LENGTH TO ALLOW TRAVEL MAKING TOTAL CENTER CONDUCTOR VARIABLE FROM 17" TO 21".

NOTE: FOR PISTON ROD SCREW, USE 5/16-18 THREADED ROD. SECURE AT TOP WITH LOCKNUT.

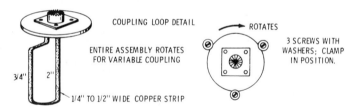

Figure 38

DESIGN DETAILS OF THE 144- TO 148-MHz CAVITY

tion, and spurious transmitter noise. Receiver protection may be brought about by physically separating the receiver and transmitter antennas in space and by the use of a high-Q bandpass cavity at the input of the receiver to reject frequencies outside of the cavity passband (figure 38). The cavity resonator is placed in the antenna circuit in such a way as to pass the received frequency and reject the transmitted frequency. A second cavity on the output of the transmitter will reduce off-frequency transmitter noise passing to the antenna, as shown in figure 37B.

CHAPTER FOURTEEN

Specialized Amateur Communications Systems and Techniques

Electromagnetic communication includes various modulation techniques and propagation modes that lie afield from the more common voice and code modulation systems and ionospheric reflection propagation used by the majority of radio amateurs. Great strides have been made in recent years by small, dedicated groups of radio amateurs operating in the forefront of technology, exploring new methods and techniques of intercommunication.

Chief among these interesting, new modes and techniques are *satellite communication, earth-moon-earth communication, radio teletype, slow-scan television, broadband television, facsimile, and radio control of models.* Of these new modes and techniques, satellite communication and earth-moon-earth (moonbounce) have excited the greatest interest, both in the United States and abroad as they have pointed the way to a more extensive utilization of the vhf bands for long distance communication.

The very nature of amateur radio is such that from its beginning more than 70 years ago, it has not only kept pace with the development of other radio services, but it has often been well in the vanguard. It is not surprising, therefore, that the radio amateur should be among the first to utilize new, specialized techniques and modes of communication. This chapter will cover some of the more interesting developments.

14-1 Amateur Space Communication

Radio amateurs have been interested in space communication ever since the first *Sputnik* was placed in orbit in the fall of 1957. Thousands of amateurs monitored the 20-MHz signal and shortly thereafter some of them began to discuss the exciting prospect of constructing a satellite of their own.

The first space experiments consisted of monitoring telemetry signals from satellites launched in other services. In 1959, however, a group of radio amateurs in California formed the *Project Oscar Association,* Oscar being an acronym for Orbiting Satellite Carrying Amateur Radio. The objective was to design, build, and launch an amateur radio space satellite. The satellite would operate in a band allocated to the amateur service and would permit radio amateurs everywhere to make useful contributions to the new field of space communications. The task was enormous, but the Project Oscar group completed their first satellite in about a year of spare time work. The satellite contained a simple 100-milliwatt radio beacon transmitting on 144.98 MHz.

Amateur radio entered the space age on December 12, 1961 when OSCAR-1 was successfully launched as ballast aboard a scheduled research vehicle of the U.S. Air Force (figure 1). Before the historic flight

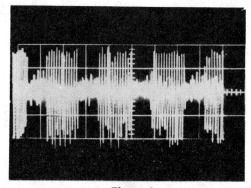

Figure 2

OSCAR TELEMETRY SENDS "HI" SIGNAL TO AMATEURS WORLDWIDE

Early OSCAR satellites sent the Morse letters "HI" in the form of telemetry on the 144-MHz band. This photograph of the OSCAR-3 signal was recorded by F3NB near Paris, France in 1965.

over 700 different amateur stations throughout the world.

By 1962, then, the first two satellites were successful in introducing radio amateurs to space communications. The telemetry beacons provided useful propagation data as well as continuous observations of the satellites' behavior, thus paving the way for OSCAR-3, amateur radio's first active communications satellite.

Figure 1

OSCAR-1, AMATEUR RADIO'S FIRST SPACE SATELLITE

Fifty years after Marconi sent the letter "S" across the North Atlantic, amateur radio operators entered the space age with the launch of OSCAR-1. This tiny space satellite was launched from California and transmitted a telemetry signal in the amateur 144-MHz band. Radio amateurs in all continents and 28 countries filed more than 5000 telemetry reports with Project OSCAR headquarters. OSCAR-1 operated for about 3 weeks before batteries expired.

ended three weeks later, the beacon signal had been tracked and logged by amateurs in all continents and 28 countries, and more than 5000 telemetry reports were received by the Project from interested amateurs (figure 2).

Amateur radio's second satellite, OSCAR-2, was launched in June, 1962. It consisted of a 144-MHz telemetry beacon and gave amateurs further training in this new and exciting aspect of amateur radio. More than 6000 reception reports were received from

Satellite History Made in 1965   OSCAR-3 made telecommunications history. By being launched a month before *Early Bird* (the first International Telecommunications Satellite Consortium INTELSAT) it holds the distinction of being the world's first free-access communications satellite (figure 3). In many instances amateur communication through OSCAR-3 marked the first time that a space communication project had been conducted in overseas countries. Over 400 amateurs in 16 countries communicated through the satellite repeater during the two week life of the device. The Atlantic Ocean was bridged twice with contacts logged between the United States and Germany and Spain, and California amateurs heard Hawaiian signals through OSCAR-3. The first Asia-Europe contact was logged between Israel

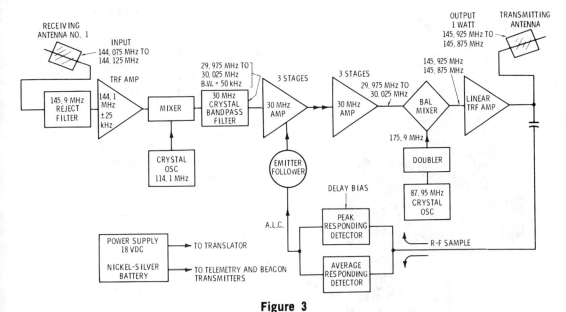

Figure 3

BLOCK DIAGRAM OF OSCAR-3 SATELLITE

OSCAR-3 was a frequency-translating satellite that received a 50-kHz segment of the two-meter band, amplified it, and translated it to another portion of the band for retransmission. Maximum transmitter power was 1 watt, PEP. There was no detection and remodulation, and within the bandwidth limitations of the system, any mode of communication was possible. This was the first multiple-access device ever launched. Input signals were received on a separate antenna, filtered, and passed to conventional amplification and mixer stages. The intermediate frequency was 30 MHz. A second mixer converted the i-f passband back to 144 MHz for further amplification. A second antenna was used to radiate the received signals. A special ALC circuit acted to reduce overload from strong signals.

and Bulgaria and Alaskan amateurs heard signals from the United States via the satellite.

Continuing the program, OSCAR-4 was launched in December, 1965 (figure 4). This communication satellite featured an uplink in the 144-MHz band and a downlink in the 432-MHz band. The goal was to place the 3-watt repeater in a semisynchronous orbit, about 18,000 miles above the earth. At this altitude, the satellite would move with the speed of the earth's rotation, and thus hang steady over the northern tip of Brazil, providing vhf communication over the American hemisphere for radio amateurs.

While the satellite equipment functioned, the desired orbit was not achieved, the satellite being placed in a highly elliptical orbit, tumbling rapidly as it revolved about the earth. Nevertheless, a number of successful contacts were made through the repeater, in-cluding the first two-way satellite contact between the U.S.A. and the U.S.S.R.

Australis Demonstrating the worldwide na-
OSCAR-5 ture of Project Oscar, the fifth amateur satellite was designed and constructed by students at Melbourne University in Australia, under the auspices of the Wireless Institute of Australia. Working with the *Radio Amateur Satellite Corporation* (AMSAT), a Washington, D.C. based international organization of radio amateurs, the satellite was prepared and qualified for launch by NASA in early 1970. It was carried as a secondary payload on the *Itos-1* weather satellite mission. OSCAR-5 included a two-band beacon on 144 MHz and 29.45 MHz, the latter incorporating a command control permitting it to be turned on and off from the ground tracking stations. This was an important demonstration that the emissions from the amateur satellite could

Figure 4

**OSCAR-4 SATELLITE WITH
432- TO 144-MHz REPEATER**

OSCAR-4 was a translator device having a 144-MHz up-link and a 432-MHz down-link. Solar cells covered the tetrahedron-shaped vehicle. Designed as a semisynchronous satellite for an 18,000- mile orbit, OSCAR-4 was placed in a highly elliptical orbit when one of the launch stages failed to ignite. First satellite contact between the United States and the Soviet Union was made through OSCAR-4.

be controlled in the event interference developed, thus greatly enhancing the practicality of operating amateur satellites in those amateur bands shared with other services.

OSCAR-5 was the first amateur satellite to transmit in the hf as well as the vhf spectrum, permitting propagation studies to be made at two distinctly different frequency ranges. A significant number of propagation anomolies were reported, such as over-the-horizon and antipodal reception of the 10-meter beacon.

OSCAR-6   OSCAR-6, launched in late 1972, was a far more elaborate satellite than the previous models. This AMSAT device included two beacon trans-

mitters and a 144 MHz to 28 MHz repeater. A block diagram of the satellite is shown in figure 5. A command receiver is incorporated in the package which accepts pulsed commands from the ground control station and converts them to level commands which turn on and off the 435.1-MHz beacon transmitter. In a similar manner, the control logic converts ground commands to change the modulation modes of the beacon transmitters. Either Morse Code telemetry or the Codestore system can be commanded to key the beacons. Additional commands control the 24-channel telemetry system incorporated in the satellite. A block diagram of the 144 MHz to 28 MHz repeater is shown in figure 6.

The Linear   The *linear repeater*, or *frequency*
Repeater   *translator*, is the heart of the repeater satellite. This device receives a segment of one band and retransmits the segment on another frequency. The transmitted band may or may not be in the same band as the input spectrum. Many separate signals can be accommodated within the spectrum and all signals received by the satellite in the spectrum are translated and rebroadcast simultaneously.

As more signals appear in the passband, the output power of the translator is divided between the signals, so that an ultimate limit is reached when the translator is saturated with signals. In a similar manner, a strong signal can overload the translator circuitry and cause weaker signals to be suppressed in signal strength. Improved circuitry is constantly being developed to overcome the limitations of translator devices, especially those designed to accept random signals.

In the case of OSCAR-6 the translator is designed to receive amateur signals in the frequency range of 145.9 to 146.0 MHz, relaying them in the down-link frequency range of 29.45 to 29.55 MHz. The repeater makes use of input and output filters in order to reduce spurious responses and to prevent the repeater from listening to the "white noise" signal of the transmitter. All stages, except the mixers, operate in the linear mode and the output of the repeater is an exact replica of the input. This device differs from the more commonly known f-m

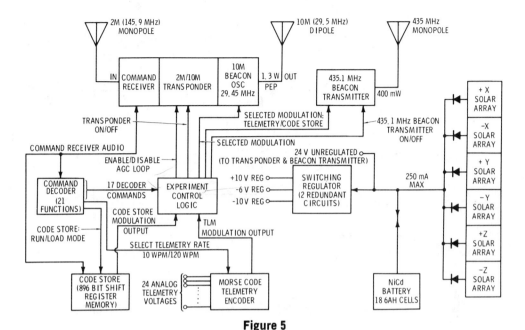

Figure 5

THE OSCAR-6 REPEATER SATELLITE

Block diagram of the repeater showing command and control circuitry. OSCAR-6 incorporated a digital decoder activated by an up-link command signal. Twenty one command functions are available. Codestore system uses a reprogrammable shift-register memory to transmit binary messages loaded on ground command. Repeater block diagram is shown in figure 6.

repeater in that the satellite repeater reproduces a frequency spectrum which may contain a multitude of separate signals. In some instances, spectrum "translation" is inverted, as shown in figure 7.

Using the Repeater The spacecraft repeater of OSCAR-6 is typical and its use will be described briefly. While the repeater will handle most forms of narrow-band modulation, SSB and c-w are recommended as they make the most efficient use of the repeater because a number of users can operate simultaneously, each taking different proportions of the repeater's power capability at a given moment.

Most amateurs engaging in repeater contacts monitor their own down-link signals which enables them to hear their signal as others hear it. This requires that a separate receiver and antenna be available for down-link reception while up-link transmission is being accomplished. This type of operation

makes break-in, or duplex, contacts possible and the power level and frequency of each station can be adjusted for best performance. If a transmitter is vfo controlled, its frequency can be continually adjusted to keep the apparent down-link frequency constant in the presence of Doppler shift, which can be as much as ± 4.5 kHz during an overhead pass.

Experience with satellite-repeated signals leads to operating expertise and various techniques have been developed to assist the operator in making the best use of a particular satellite. Additional information on the subject may be obtained from *AMSAT*, Box 27, Washington DC 20044.

OSCAR-7 The OSCAR-7 spacecraft was launched in late 1974, while OSCAR-6 was still functioning. For the first time, amateur radio operators had two operable communication satellites in orbit at the same time. OSCAR-7 was many mag-

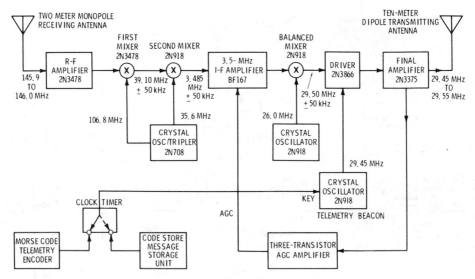

Figure 6

BLOCK DIAGRAM OF OSCAR-6 144-MHz TO 29-MHz REPEATER

The linear repeater "listens" over the range of 145.9 MHz to 146.0 MHz, converting received signals to the first i-f of 39.1 MHz and the second i-f of 3.5 MHz. After 35 dB of amplification, the passband is up-converted to 29.45 to 29.55 MHz. Maximum power output is 1.3 watts, PEP. The power source is a 24-volt Nicad battery charged by solar cells. The repeater also contains a beacon oscillator on 29.45 MHz. Input and output filters are used to reduce spurious responses and to eliminate television-band signal interference with the repeater.

nitudes more complex than the previous amateur satellites and was designed for long life. It was built in an octahedral configuration to allow sufficient surface area for a powerful solar cell power supply (figure 8). The satellite contained two repeaters and two tracking beacons and both Morse Code and teletype telemetry encoders. Down-link

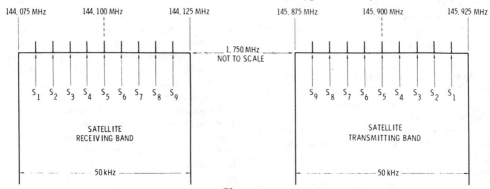

Figure 7

OSCAR-3 FREQUENCY TRANSLATION

The frequency translation satellite receives a 50-kHz segment falling between 144.075 MHz and 144.125 MHz. All energy in this spectrum will be processed by the translator. This device inverts the spectrum, that is, signals at the low-frequency end of the input band (S$_1$, S$_2$) appear at the high-frequency end of the output band. A station transmitting upper-sideband SSB will be retransmitted as a lower-sideband SSB signal. Other satellites may not necessarily invert the spectrum, but the principle of translation still applies.

Figure 8

AMSAT OSCAR-7

This is an applications communications space-craft designed for noncommercial public service and educational use by the amateur radio community. OSCAR stands for Orbiting Satellite Carrying Amateur Radio. Developed by the Radio Amateur Satellite Corporation (AMSAT), a nonprofit scientific corporation headquartered in Washington, D.C., the purpose of the spacecraft is to have students around the world work with their teachers to get a direct understanding of space science by actually participating in demonstrations through local radio amateur operators anywhere in the world. OSCAR-7 was launched by the National Aeronautics and Space Administration—the third OSCAR to be launched by NASA as a piggy-back spacecraft aboard a Delta rocket from Vandenburg Air Force Base, California. (Photograph courtesy of NASA)

telemetry and stored message data could be routed to either of the beacons. It was thus possible, for example, to receive Morse Code on the 10-meter beacon and *Codestore* information on the 435-MHz beacon at the same time, using two ground receivers.

The satellite normally alternated between a 144 MHz to 10 meters repeater and a 432

MHz to 146 MHz repeater, switching every 24 hours. The timer could be ground controlled so that the mode change could be conducted at approximately the same time each day.

OSCAR-7 contained automatic power-supply monitoring circuitry so that if the battery voltage dropped below a predetermined level, the spacecraft would switch to a low-power condition for recharge from the solar cells.

While many amateurs communicated via OSCAR-7, some of the most meaningful contacts took place via satellite repeater, using both OSCAR-6 and -7 satellites working together. Thus, a 432-MHz ground station could be repeated by OSCAR-7 on the 144-MHz band to OSCAR-6 which would re-repeat the signal on the 10-meter band.

OSCAR 8   The OSCAR-8 satellite was launched in early March, 1978. It was built by radio amateurs in the United States, Canada, West Germany and Japan. This satellite carries transponders for two modes of operation. There is a conventional 145.9- to 29.4-MHz transponder and a new 145.9- to 435.1-MHz transponder, a similar frequency combination that was pioneered by the OSCAR-4 spacecraft in 1966. Six channels of telemetry are provided to monitor the onboard status of the spacecraft.

The principal objective of OSCAR-8 is the educational use of a low orbiting satellite. It is to provide a means for the use of such a satellite as an educational tool in schools and other institutions. Other objectives include the continuation of communications demonstrations by amateurs, and of the feasibility of using satellites with small amateur terminals for "bush" communication, emergency communication and satellite-to-home broadcasting to amateur receivers.

Satellite Operation   The OSCAR-8 satellite contains two communications transponders and command and telemetry systems. The spacecraft is solar powered, weights 60 pounds, and is a 15 inch rectangular solid 13 inches high (figure 9). Its anticipated useful life is three years.

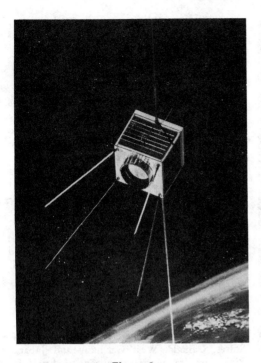

Figure 9

AMSAT OSCAR 8 SATELLITE

This is the eighth of a series of satellites built by radio amateurs that began with the launch of OSCAR 1 in 1961. The spacecraft was built by AMSAT (Radio Amateur Satellite Corporation) a Washington, D.C.-based international organization of amateurs and was provided to NASA at no cost for launch on a noninterference basis. (Photo by NASA).

The *Mode A* transponder is a two-to-ten device similar to the one on OSCAR-7 and with the same frequency passband (input passband of 145.85 to 145.95 MHz, and output frequency passband between 29.40 and 29.50 MHz). A 250 mW telemetry beacon provides telemetry data in Morse code at a frequency of 29.402 MHz. Approximately -95 dBm is required at the transponder input terminals for an output of one watt. This corresponds to an *effective radiated power* (ERP) from the ground of 80 watts for a distance to the satellite of 1200 miles and a polarization mismatch of 3 dB. The transponder translation frequency (input frequency minus output frequency) is 116.458 MHz. Thus, the relationship between the uplink (f_u) and downlink (f_d) is:

$$f_d = f_u - 116.458 \pm \text{Doppler shift}$$

where both f_d and f_u are in MHz.

As in the recent satellites, the passband is not inverted and upper sideband uplink signals become upper sideband downlink signals. Output power is one to two watts.

The *Mode J* transponder uses the same receiving antenna as the Mode A device, a canted turnstile fed by a hybrid and matching network so as to develop circular polarization. Left-hand circular polarization can be used by amateurs in the Northern Hemisphere and right-hand circular polarization by users in the Southern Hemisphere. The Mode A ten meter dipole is oriented perpendicular to the stabilization magnet in the spacecraft. The Mode J 435 MHz downlink antenna is a monopole, linearly polarized, on top of the spacecraft.

The spacecraft contains solar panels on its four sides and on the top. The solar cells, combined with a 12-cell, six ampere-hour battery should be adequate to power the spacecraft for several years. A battery charge regulator is also contained which converts from the 28-30 volt solar array voltage to the 12-16 volts required by the battery.

OSCAR-8 was launched from the NASA Western Test Range as a secondary payload with the LANDSAT-C earth resources technology satellite on a two-stage Thor-Delta 2910 launch vehicle. The orbital parameters were programmed to be:

Apogee: 577 statute miles
Perigee: 549 statute miles
Period: 103 minutes
Inclination: 99.0 degrees
Time of descending mode: 9:30 AM

The orbit is sun-synchronous, with passes repeating at the same time each day on a one-day cycle. Because the satellite operates in a 560 statute mile orbit, at just about half the altitude of the 910 statute mile orbit of OSCAR-7, the communication range will be less. The usable time on an overhead pass will be about 18 minutes and the horizon range will be about 2000 miles. The schedule of orbits for OSCAR-8 is a feature in *QST* magazine.

In this rapid and exciting fashion the science of amateur space communication has advanced over the few short years between

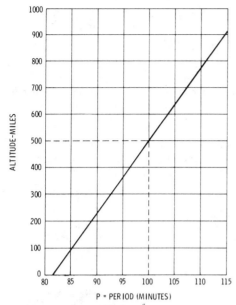

Figure 10

PERIOD OF SATELLITE AND ALTITUDE RELATIONSHIP

The Orbital period of the satellite is related to the altitude as shown in this graph for a circular orbit. For example, if the period is 100 minutes, the altitude is 500 miles. If the period is below approximately 85 minutes, the satellite quickly falls back to earth after a few orbits.

OSCAR-1 and the modern, sophisticated space satellite of today.

OSCAR Satellite Tracking   To communicate through an OSCAR repeater satellite, it is necessary to know the location, orbit, and orbital time of the spacecraft in addition to the parameters of the onboard repeater.

In general, communication satellites are launched in a circular orbit about the earth. *Orbital height* and *period of orbit* are related to each other, as shown in figure 10. If the period drops below 85 minutes, the satellite will not remain in orbit, but will plunge back to earth. Once the satellite's height has been determined from the orbital period, the maximum *ground range* (range to the horizon from a point on earth beneath the satellite) may be determined, as shown in figure 11.

Satellite Range   Unlike earth-moon-earth (moonbounce) communication, space satellites in orbit relatively close to the earth's surface appear to move rapidly across the sky from horizon to horizon. It is confusing to picture yourself on a stationary earth with the satellite whirling overhead on various erratic passes, sometimes going north to south and other times going south to north. A much clearer picture may be gained by visualizing the satellite as rotating about the earth in a fixed plane, with the earth revolving inside the satellite orbit (figure 12). Thus, when a satellite passes over a ground station on one orbit, the rotation of the earth will cause the satellite to pass over a different spot, lying to the west of the ground station, on the next orbit. This is termed *progression*.

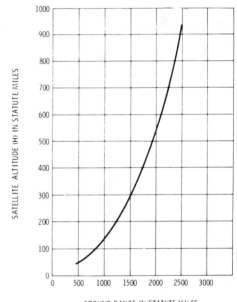

Figure 11

GROUND RANGE AS A FUNCTION OF SATELLITE ALTITUDE

The ground range is the distance measured along the surface of the earth from the ground station to a point on the earth directly below the satellite. At an altitude of 500 miles, for example, the ground range is nearly 2000 miles and two stations broadside to the satellite path and 4000 miles apart could theoretically communicate with each other through a repeater satellite orbiting at that height.

Each successive orbit will progressively cross the earth's equator farther west from the original point of observation and, to the observer at the ground station, each successive orbit has moved further west from his point of observation. In reality, the observer has moved east with the earth's rotation, and the orbit of the satellite has remained fixed in the sky. When the ground station's position has rotated 180 degrees (12 hours), the observer is looking at the reverse side of the orbit and if he was watching north-to-south passes, he is now watching south-to-north passes (figure 13).

How long the satellite will remain within range of a ground station is dependent on two factors: the distance it will be at the *point of closest approach* (PCA) to the station and the altitude of the satellite. The longest duration at any altitude will occur on orbits that pass directly over the station location, and the duration of the pass will decrease for orbits that pass further away

from the station (figure 14). For example, a satellite in a 1000 mile high orbit would be within line of sight range of a ground station for about 25 minutes on an overhead pass, about 20 minutes when it comes within 1000 miles of the ground station and only about 10 minutes with a 2000-mile distance of closest approach. When using a satellite for two-way communication, it is therefore necessary to take into consideration the length of time the satellite will be within the simultaneous range of all ground stations involved. Communication will be possible with any other ground station having the satellite within its range at the same time, but the length of time of contact will vary with the position of the ground station relative to the satellite, as shown in figure 14.

The higher a satellite is, the greater the effective range of a ground station using it will have. Since higher satellites are further away from the ground station, signal strengths will be less due to path losses unless either more powerful transmitters or higher gain receiving systems are used. The

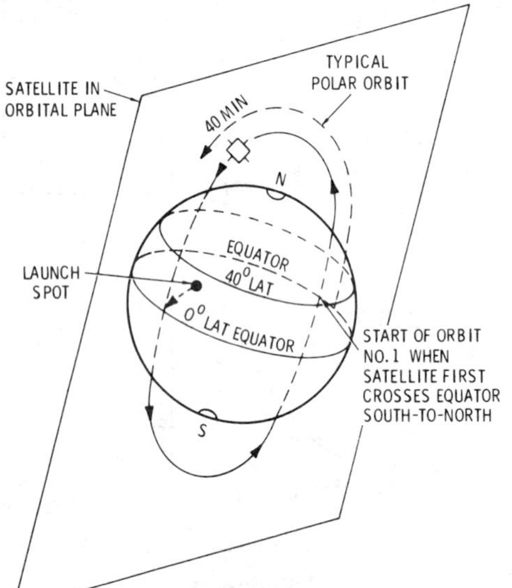

SATELLITE IN ORBITAL PLANE
TYPICAL POLAR ORBIT
40 MIN
N
EQUATOR 40° LAT
LAUNCH SPOT
0° LAT EQUATOR
START OF ORBIT NO. 1 WHEN SATELLITE FIRST CROSSES EQUATOR SOUTH-TO-NORTH
S

Figure 12

EARTH ROTATES WITHIN SATELLITE ORBITAL PLANE

As the earth rotates within the orbit of the satellite, all areas on the earth's surface will pass beneath the satellite if it is in a polar orbit. If the orbital plane is tipped, areas of high longitude will lie outside the orbit of the satellite. Orbit number 1 starts when the satellite first crosses the equator in a south-to-north direction.

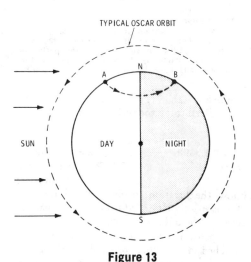

Figure 13

EARTH REVOLVES WITHIN SATELLITE ORBIT

Satellite orbit remains fixed in space while the earth revolves inside it. For example, if a space satellite is launched from California, in a southward direction, all future daytime passes will be in a south-north direction. For daytime passes the observer is at point A and for nighttime passes at point B.

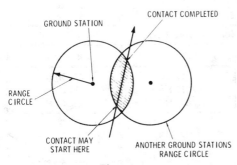

Figure 14

GROUND RANGE AND POINT OF CLOSEST APPROACH

Ground range of two stations overlap when satellite passes between them. The longest duration of pass occurs on an orbit that passes directly over a ground station, but maximum ground range occurs on a pass to the side of the ground station. In this example two ground stations are within range of the satellite and the time of communication is shown by the shaded area.

greater the line-of-sight path distance between user and satellite, the more circuit gain will be required to maintain adequate signal levels. Therefore, although high altitude satellites will allow contacts with more distant stations, a more elaborate ground station will be required, or else a satellite with larger transmitters and antennas.

Doppler Shift The movement of a satellite relative to the ground station results in a change of frequency of signals received in either direction. This change, known as *Doppler Shift*, can be determined from the following formula:

$$f_{(ds)} = 5.4\,(f_u - f_d)\,V$$

where,

f_{ds} = shift on either side of the center frequency (MHz),

f_u = frequency of ground station (up-link) in MHz,

f_d = frequency of the satellite-repeated signal (down-link) in MHz,

V = speed of the satellite in miles/second (a function of the altitude).

Maximum Doppler Shift will occur on overhead passes. It can be seen in the formula that Doppler Shift is a function of frequency as well as speed and is greater at the higher frequencies. Table 1 indicates the total shift that may be expected at various altitudes and frequencies.

This shift in frequency of course must be taken into account when tuning receivers and transmitters for satellite communication. The frequency of a satellite transmitter moving toward a ground station will appear *higher* than the actual satellite transmission frequency and will drop as the satellite approaches until at the exact point of closest approach, when it will be on the true frequency. Past this point, the received signal will continue to drop *lower* in frequency as the satellite moves away from the ground station.

Problems of tuning transmitted and received frequencies are reduced when the satellite receiver and transmitter frequencies are sufficiently far apart to permit the ground station to monitor its down-link while it is transmitting, as for example, when the up-link is 2 meters and the down-link is 10 meters. This allows maximum efficiency of spectrum use since mutual interference between stations on the same frequency can be immediately detected.

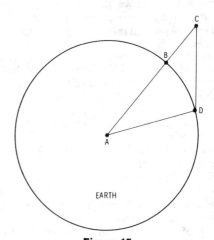

Figure 15

POSITION OF THE SATELLITE

Line ABC is the distance from the earth's center to a satellite in orbit (C). The spot where this line intersects the earth (B) is the sub-satellite point. Point D is the location of a ground station and distance CD is the slant range to the satellite. Arc BD is the distance along the earth's surface between the station and the subsatellite point. An observer at B would see the satellite directly overhead.

TABLE 1.

Total Doppler Shift for Overhead Pass			
Altitude	29 MHz	145 MHz	436 MHz
100 st. mi.	1510 Hz	7550 Hz	22,720 Hz
500 st. mi.	1440 Hz	7210 Hz	21,680 Hz
1000 st. mi.	1360 Hz	6834 Hz	20,550 Hz

Satellite Position In determining the position of a satellite and predicting its future location some knowledge of terms and orbital relationships is useful. Satellite distances and speeds may be expressed in different ways. Nautical and statute miles, as well as meters and kilometers are commonly used. A summary of these terms is shown in Table 2.

Figure 15 summarizes satellite position. Line ABC is the distance from the center of the earth to a satellite in orbit (point C). The spot where this line intersects the earth's surface is called the *sub-satellite point* (B). Point D is the location on the earth of an amateur station. Line CD is called the *slant range* to the satellite. Arc BD is the distance along the earth's surface between the station and the *sub-satellite point*. This distance can be plotted on a map of the earth to locate the satellite to see if it is within range of a particular ground station. The distance may be expressed as a distance of angular degrees (one degree on the earth's surface being equal to 69.09 statute miles, 59.97 nautical miles, or 111.14 km). All points on the diagram except points A and D are continually changing as the earth rotates and the satellite moves. A station at point B would observe the satellite directly overhead.

All factors in a satellite's orbit are interrelated and much can be determined from a few known facts. For instance, the velocity of a satellite through space is a function of altitude and the period and time of one revolution are a function of altitude and velocity. Rough orbital predictions can be made if three pieces of information are available: the altitude or period of the satellite, the time and longitude of any equator-

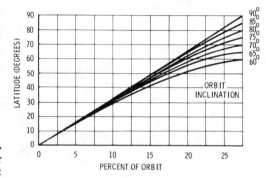

Figure 16

ORBITAL TRAVEL TIME FROM EQUATOR

The travel time of a satellite to reach a given latitude, as expressed as a percent of total orbital time. For example, if the observer is located at altitude 40 degrees, and the satellite has an orbital inclination of 80 degrees, it takes about 12 percent of total period time to travel from the equator to that latitude.

ial crossing and the angle at which the satellite crossed the equator.

Informational broadcasts are commonly given during an OSCAR flight by the ARRL Headquarters station, W1AW, and selected OSCAR stations. These broadcasts include tracking data for the satellite and provide the predicted times of south-to-north equatorial crossings in GMT, the points of crossing in degrees of west longitude and the time of pass over major cities on the earth. Once you have heard a satellite in your vicinity on one orbit, all that is really needed to predict when it will again be within your range is the orbital period, the progression per orbit and the time it takes to reach your location from the equatorial crossing point (figure 16).

TABLE 2. Conversion Table

1 st. mile = 0.868 naut. mile = 1609.344m = 1.609344 km
1 kilometer = 0.6214 st. mile = 0.5396 naut. miles = 1000 meters.

Ground Station Antennas The most recent OSCAR satellites have operated in the 432-MHz, 144-MHz, and 10-meter amateur bands. For 10-meter reception of satellite signals a 10-meter rotary beam is satisfactory. When mounted well in the clear it provides a low angle of radiation which is desirable for maximum communication range when the satellite is just over the horizon. For passes close to the ground station. a dipole antenna mounted at a lower height (having a high angle of radiation) is useful.

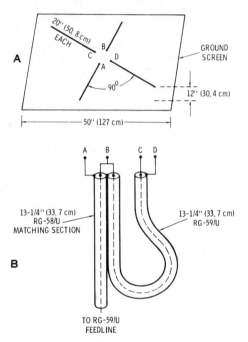

TURNSTILE ANTENNA WITH SCREEN
REFLECTOR FOR SATELLITE
COMMUNICATION AT 144 MHz

This simple antenna provides an omnidirectional, high-angle pattern suitable for satellite reception above the horizon. Full coverage is provided at angles above 20 degrees. Crossed dipoles are mounted above a ground screen measuring 50 inches on a side. The screen reflector may be made of galvanized wire having openings less than one-half inch square. Window screen can serve as a substitute. The turnstile antenna is placed 12 inches above the screen. A phasing harness (B) provides the correct 90-degree phase difference between the dipoles. Antenna is designed to be fed with a 70-ohm coaxial line.

For transmitting to the satellite on the vhf bands, many amateurs use a simple high gain, Yagi antenna. Experience with the OSCAR satellites, however, has shown that rapid fading of the signals repeated by the satellite is partially due to the radiation pattern nulls of the vhf beam antenna. In addition, because of the random positioning of the satellite in the orbital path, cross polarization of the transmitting and receiving antennas can contribute to observed fading.

Cross-polarization fading can be reduced by using circular polarization at the ground station and radiation pattern nulls can be compensated for by using either a null-free antenna or a continuously tracking antenna that holds the satellite at the center of the radiation pattern.

A simple turnstile antenna mounted about a quarter-wavelength above a reflector screen will provide a circularly polarized pattern. The maximum lobe of the radiation pattern is vertical, providing a broad lobe that is effective at all elevation angles above approximately 40 degrees. A practical turnstile antenna array for 144 MHz is shown in figure 17.

Crossed-Yagi antennas can be used to provide circular polarization and details on the construction of such an antenna is contained in the *VHF Handbook For Radio Amateurs*, available from Radio Publications, Inc., Box 149, Wilton, Conn., 06897. Additional information on satellite techniques may be found in *Specialized Communications Techniques*, published by the American Radio Relay League, Newington, Conn., 06111. A quarterly newsletter covering amateur satellite activity is published by AMSAT, Box 27, Washington, DC 20044.

14-2 EME (Moonbounce) Communication

The moon presents a good radio target when it rides high in the sky and by the end of World War II circuits and techniques were available to use it as a passive reflector for radio signals. The first instance of amateur moon-reflected signals was the reception of W4AO's 144-MHz signals by W3GKP in mid-1950. In 1960, the first two-way moonbounce contact took place on the 1296-MHz band between W6HB and

W1BU using dish antennas and experimental, 1-kilowatt vhf klystron tubes in the transmitters. From these early tests, moonbounce communication has grown rapidly, as interested vhf operators turned to this new and exciting mode of communication. Today, moonbounce activity is taking place on the various vhf bands, with the major interest concentrating on the 144- and 432-MHz bands (figure 18).

Figure 18

**THE 144-MHz MOONBOUNCE
ANTENNA OF W6PO**

The array consists of 160 elements arranged in 32 Yagi beams formed into eight 20-element collinear assemblies, stacked four wide. Overall antenna size is 33 feet wide, 24 feet high, and 8 feet deep. Gain is estimated to be approximately 23 decibels. Similar arrays are also in use on the 432-MHz band by active "moonbouncers".

The EME The moon is about 2160 miles in
Circuit diameter and orbits the earth at a
distance that varies from 221,463 miles to 252,710 miles. The orbital period is 28 days and because the orbit is somewhat eccentric, the moon travels along a somewhat different path each night of the lunar month.

As a target for radio reflection, the moon subtends an arc of about one-half degree when seen from the earth. The reflection coefficient of the moon's surface is about 7 percent so the remaining 93 percent of the signal striking the moon is absorbed. The portion of the signal that is reflected is dif-

fused all over space and only a minor portion of it is returned to earth. A smaller fraction of the returned signal is captured by the receiving antenna, which is small compared to the earth surface area facing the moon; about 98,470,000 square miles. Thus, the EME path loss is quite high and moonbounce communication at the maximum amateur power level is a challenge to the best talents of many of the world's most skillful radio amateurs.

Radio signals travelling through space are attenuated as the square ratio of the frequency. Consequently the EME path loss is about 8.3 times (9 dB) greater on 144 MHz than on 50 MHz, and a similar increase in path attenuation takes place between the 144-MHz and 432-MHz bands. In addition, transmitter efficiency tends to decrease and receiver noise figure and transmission line loss increase with increasing frequency.

On the other hand, the power gain of a directive antenna of a given size increases by the same ratio that the path loss increases and, because the antenna gain is real-

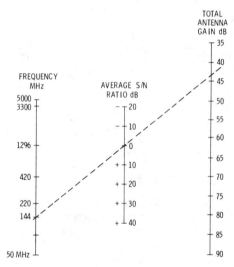

Figure 19

**ANTENNA GAIN REQUIREMENT
FOR EME CIRCUIT**

This graph is based on 600 watts transmitter power output, a zero-dB receiver noise figure, and 100-Hz receiver bandwidth. At 144 MHz, for example, for an average signal-to-noise ratio of zero decibels, a total antenna gain is about 42 dB. Two 21-dB antennas should be satisfactory.

ized during both transmission and reception at each end of the circuit, there is a net signal gain with increase in frequency, notwithstanding the increased circuit losses.

The free space loss for the EME circuit varies about 2 dB depending on whether the moon is at perigee or apogee. Typically, the circuit loss at perigee (the point of closest approach of the moon) is 216 dB for 50 MHz, 225 dB for 144 MHz, 235 dB for 432 MHz, and 244 dB for 1296 MHz. The nomograph of figure 19 illustrates antenna gain and average signal-to-noise requirements as a function of frequency for an average path loss. This graph and the circuit losses are based on a transmitter power output of 600 watts to the antenna, a zero-decibel noise figure and a receiver bandwidth of 100 Hz. As an example, at 144 MHz, for an average signal-to-noise ratio of zero decibels, the total antenna gain should be about 42 decibels. Thus, two 21-dB antennas are required, one at each end of the path. If the gain of one antenna is higher than this, the gain of the other may be correspondingly lower to achieve the same signal-to-noise ratio.

Under the best of conditions, then, using the maximum legal power, the most sensitive receiver and the largest possible antenna array, two-way amateur communication via the moon is a marginal means of communication. Even so, the number of successful c-w and SSB contacts via moon reflection speak well for the experimenters doing this fascinating, space-age means of world-wide vhf communication.

Faraday Rotation During the passage of a radio signal to and from the moon, it may rotate in polarization several times. This effect is called *Faraday Rotation* and is thought to be produced by the the effect of the earth's magnetic field on the signal. Faraday Rotation produces a cyclic fading in the signal received, as the path length between the earth and the moon is constantly changing. The fade is quite rapid at the lower frequencies and the period increases with frequency until it ceases to be significant above 1000 MHz. Special antennas can be used to combat Faraday Rotation, at a loss in signal gain, but most experimenters accept the slow fade and work

around it, especially on 144 MHz, where the fade period is rather long, typically 20 minutes between signal peaks.

A second fading phenomena known as *libration fading* of moon reflected signals is caused by a rocking motion in the movement of the moon in orbit. The fading is characterized by a rapid flutter in the received signal.

The EME Reporting System Because of the weakness and unpredictability of moon-reflected signals, special reporting systems have been devised by experimenters to provide quick and reliable confirmation of a valid contact. Each of the vhf bands has its own unique system, the majority of which convey information with a series of dashes, since dots have a low energy content and tend to disappear in the noise.

On 144 MHz, for example, the *TMO Report System* is used. The letter *T* is sent repeatedly when the signal can be heard but no intelligence can be detected. The letter *M* is sent when portions of call letters can be copied, and the letter *O* is sent when a complete call set is copied. Once contact is established, and the signals are loud enough, normal amateur procedure is commonly used. At 144 MHz, where the Faraday Rotation is long, the usual moonbounce calling sequence is 2 minutes, whereas at 50 MHz, where the Faraday Rotation is rapid, the calling sequence is 30 seconds. In all cases, the sequence is agreed to beforehand and synchronized with time signals from WWV.

For more information about moonbounce experiments and activity, write to Amateur Service Department, EIMAC division of Varian, 301 Industrial Way, San Carlos, CA 94070 and ask for their free bulletin series AS-49 (*Almost Everything You Want to Know About Moonbounce*).

14-3 Radioteletype Systems

Teleprinting is a form of communication based on a simple binary (on-off) code designed for electromechanical transmission. The code consists of dc pulses generated by a special electric typewriter, which can be reproduced at a distance by a separate machine. The pulses may be transmitted from one machine to another by wire or by a radio circuit. When radio transmission is used, the system is termed *radioteletype* (RTTY). The name *teletype* is a registered trademark of *Teletype Corporation* and the term *teleprinter* is used in preference to the registered term.

Although the first, teleprinter machine was put in service in the United States before World War I, radio amateur RTTY experiments did not start until about 1946 using make-and-break (c-w) keying on the 80-meter band and audio keying on the 144-MHz band. *Frequency-shift keying* (FSK) was permitted on the hf bands in 1953 using a shift of 850 Hz. Since that date, interest in RTTY has grown rapidly among radio amateurs, particularly with the advent of solid-state devices, microprocessor systems, etc. in some cases eliminating the need for "teleprinters" entirely unless paper copy is required.

Radioteleprinter Systems   The dc pulses that comprise the teleprinter signal may be converted into three basic forms of emission suitable for radio transmission.

These are: (1) *frequency-shift keying* (FSK), designated as F1 emission; (2) *make-break keying* (MBK), designated as A1 emission; and (3) *audio frequency-shift keying* (AFSK), designated as F2 emission.

Frequency-shift keying is achieved by varying the transmitted frequency of the radio signal a fixed amount (usually 170 Hz using tones of 2125 and 2295 Hz) during the keying process. The shift is accomplished in discrete intervals designated *mark*

and *space*. Both types of intervals convey information to the teleprinter. *Make-break keying* is analogous to simple c-w transmission in that the radio carrier conveys information by changing from an *on* to an *off* condition. Early RTTY circuits employed MBK equipment, which is now considered obsolete since it is less reliable than the frequency-shift technique. *Audio frequency-shift keying* is the most popular method, particularly when an SSB transmitter is used and tones are introduced into the microphone jack. The transmitter is modulated by an audio tone which is shifted in frequency according to the RTTY pulses. Other forms of information transmission may be employed by a RTTY system which also encompass translation of binary pulses into r-f signals.

The Teleprinter Code   The teleprinter code (*Baudot*) consists of 26 letters of the alphabet and additional characters that accomplish machine nonprinting functions, such as line feed, carriage return, bell, and upper- and lower-case shift. These special characters are required for the complete automatic process of teleprinter operation in printing received copy. Numerals, punctuation, and symbols may be taken care of in the case shift, since all transmitted letters are capitals (figure 20).

The teleprinter code is made up of spaces and pulses, for transmission at 60, 67, 75, or 100 words per minute. Each character (at 60 w.p.m.) is made up of five elements, plus a 22 millisecond *start space* and a 31 millisecond *stop pulse*. All characters are equal

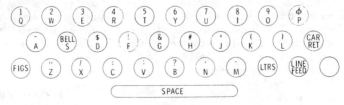

Figure 20
TYPICAL TELEPRINTER KEYBOARD

Shown here is the Western Union keyboard. The lower case is identical to that of an American-style typewriter, with the exception of the auxiliary keys, which control line feed, carriage return, and figures. Various types of upper-case keyboards exist, including the Bell System (TWX), The weather system, the American Communications Keyboard, and the CCITT (European) styles. Only three rows of keys are used instead of the four as on a normal typewriter. All printed letters are capitalized.

in total transmission time to 163 milliseconds duration to achieve machine synchronization at both ends of the RTTY circuit. Timing is usually accomplished by the use of synchronous motors in the equipment, locked to the ac line frequency. The sequence of mark and space pulses for the letter *R* is shown in figure 21. The start space provides time for synchronization of the receiving machine with the sending machine. The stop pulse provides time for the sending mechanism as well as the receiving mechanism to properly position themselves for transmission of the following character.

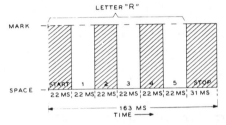

Figure 21

THE TELEPRINTER CODE

Teleprinting is based on a simple binary code made up of spaces and pulses, each of 22 milliseconds duration. Normal transmission is at the rate of 60 w.p.m. The sequence of mark and space pulses for the letter R are shown here. Start space provides time for machine synchronization and stop pulse provides time for sending and receiving mechanisms to position themselves for transmission of the following character.

The keying system normally employs the higher radio frequency as the mark and the lower frequency as the space. This relationship holds true in the AFSK system also. The lower audio frequency is 2125 Hz and the higher audio frequency 2295 Hz, giving a frequency difference or shift of 170 Hz.

The Teleprinter and Keyboard The older style *teletypewriter* (keyboard) is an electro-mechanical device that resembles a typewriter in appearance having a keyboard, a type basket, a carriage, and other familiar appurtenances. The keybord, however, is not mechanically linked to the type basket or printer. When a key is pressed on the keyboard of the sending apparatus a whole code sequence for that character is generated in the form

of pulses and spaces. When this code sequence is received on a remote machine, a type bar is selected and made to print the

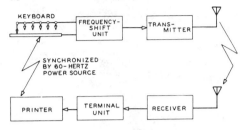

Figure 22

BLOCK DIAGRAM OF ONE-WAY-RTTY CIRCUIT

The teleprinter generates code sequence in the form of on-off pulses for the alphabet and additional special characters. Teleprinter code is transmitted at rate of 60 w.p.m. by means of frequency-shift technique. The receiving apparatus drives a mechanical printer that is usually synchronized with the keyboard by the common 60-Hz power source.

letter corresponding to the key pressed. Synchronization of machines is accomplished by means of start and stop pulses transmitted with each character. An electromechanical device driven by the motor of the teleprinter is released when a key is pressed and transmission of the complete character is automatic.

The receiving apparatus operates in reverse sequence, being set in operation by the transmitter mechanism. While each character is sent at the speed of 60 w.p.m., actual transmission of a sequence of characters may be much slower, depending on the typing speed of the operator. A simplified diagram of a one-way RTTY circuit is shown in figure 22. Many amateurs have obtained these machines from surplus channels and, although obsolete, they are still in use on the amateur bands.

The modern keyboard, however, is completely solid-state in construction and uses no mechanical linkages between the keyboard and the pulse generator, type basket, or printer. A typical device is shown in figure 23. This dual-mode keyboard permits the operator to send either the teleprinter pulse code or Morse code. In the RTTY mode, transmission at standard data rates of 60, 66, 75, or 100 w.p.m. is available. In

Figure 23

DUAL MODE, SOLID-STATE KEYBOARD

The HAL DKB-20-10 electronic keyboard transmits both RTTY and Morse codes. It includes a station identifier which automatically transmits the station call sign at the touch of a key and a buffer memory that stores characters typed for transmission at a constant rate. Extra keys are used for double characters commonly used, such as SK, AS, AR, KN, and BT. A tune key overrides the keyboard and keys the transmitter on for adjustment. The device uses 57 ICs, 12 transistors, and 125 diodes. Keying speeds are controlled by three precision crystals. (Photo courtesy HAL Communications Corp.)

the c-w mode, transmission at speeds between 8 and 60 w.p.m. is possible. In either mode, a built-in sidetone oscillator allows the operator to monitor the transmission. A block diagram of the dual-mode keyboard is shown in figure 24.

The ASCII Code The *ASCII* (American Standard Code for Information Interchange) Code was adopted in 1968. It uses seven binary data bits which have 128 possibilities. An eighth bit is used for error checking. Thirty-two of these bits are reserved for control of the printing mechanism or other aspect of message handling. Of the remaining 96 bits, all but one represent a character. The last bit is a space. ASCII code is authorized for transmission by radio amateurs on frequencies presently authorized for RTTY use between 3.5 and 21.45 at a maximum rate of 300 baud. On RTTY frequencies between 28 and 225 MHz, F_1, F_2 and A_2 emission are permitted, with a baud rate up to 1200. Above 420

MHz, F_1, F_2 and A_2 are permitted, with a rate up to 19.6 kilobaud.

The *baud* is a unit of signalling speed equal to the number of discrete conditions or signal events per second. A baud equates to the reciprocal of the data pulse width.

The ASCII code is summarized in Table 3. The names of the control characters are tabulated. Note that each lower case letter is only one bit different than the upper case letter or character. Finally, if only the least significant four bits are retained, the numerals are already coded into BCD language.

Figure 25 shows the scanning principle used by many electronic keyboards. At each intersection of the 8 × 8 matrix, a key switch is placed. A free running six-bit counter has its least significant three bits driving a decoder which grounds each horizontal wire, one at a time. The most significant three bits drive a data selector which examines each vertical wire, one for each full scan of the horizontal wires. If a key is pressed, the data selector notes the ground and makes an output. This stops the counter and generates a pulse.

The keyboard output may be transmitted over seven wires, or the bits may be sent over one circuit one at a time (serial asynchronous communication). Figure 26 shows two ways to connect a keyboard and a receiving device.

By convention, the resting state of the current loop is "one", or mark. When the current is interrupted, the state is "zero", or space.

Character Transmission To transmit a character, the resting state is reversed for one bit period (figure 27). Seven data bits follow, the least significant first. This sequence is followed by a parity bit for error checking. The parity bit is generated so that there is always an even number of "ones" in the eight bits (even parity) or always an odd number (odd parity). The receiving device can then total the number of "ones" and tell if one of the bits was distorted during transmission. On short wire circuits, the parity bit is not used in this fashion and is always left in a mark or space condition. At the end of the char-

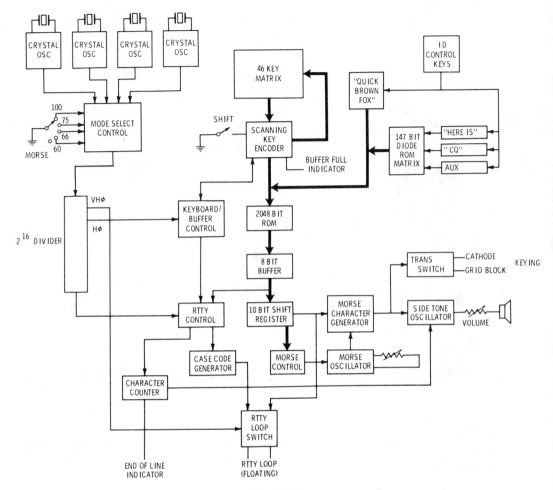

Figure 24

BLOCK DIAGRAM OF DUAL-MODE KEYBOARD

Keying speeds are controlled by four crystal oscillators. The 46-key matrix (driven by the keyboard) is connected to a large-scale IC key encoder whose output is a seven-bit digital code which is the ASCII equivalent of the character to be transmitted. The digital code is applied to the eight address inputs of the ROM (read-only memory) and addresses a particular eight-bit address in the memory. In that location is stored the correct code for the character, which appears at the ROM output. The ASCII code is thus converted to the Morse or RTTY bit pattern. The ROM output is fed to an eight-bit buffer memory which stores the code until the shift register is cleared and ready to accept new data. The ten-bit shift register now transforms the information from a parallel to a series mode. Clock pulses are applied to the register, causing the bits to appear in sequence at the register output. Depending on the setting of the mode switch, the register output activates either the Morse character generator or the RTTY loop-switching circuit. For Morse transmission, the bits are converted to pulses of unequal length, forming dots and dashes. The generator output activates the Morse keying transistor and the sidetone oscillator.

acter, the circuit returns to its resting state for one- or two-bit periods.

Speed of transmission is limited by system bandwidth or the response of the printing mechanism, or both. The signalling rate is the baud. A system using ten bits per character (one start, seven data, one parity, and one stop) running at 1200 baud has a data

BIT POSITION									
7 6 5	7 6 5	7 6 5	6 7 5	7 6 5	7 6 5	7 6 5	7 6 5		
0 0 0	0 0 1	0 1 0	0 1 1	1 0 0	1 0 1	1 1 0	1 1 1	4 3 2 1	
NUL	DLE	SP	0	@	P	`	p	0 0 0 0	
SOH	DC1	!	1	A	Q	a	q	0 0 0 1	
STX	DC2	"	2	B	R	b	r	0 0 1 0	
ETX	DC3	#	3	C	S	c	s	0 0 1 1	
EOT	DC4	$	4	D	T	d	t	0 1 0 0	
ENQ	NAK	%	5	E	U	e	u	0 1 0 1	
ACK	SYN	&	6	F	V	f	v	0 1 1 0	
BEL	ETB	'	7	G	W	g	w	0 1 1 1	
BS	CAN	(	8	H	X	h	x	1 0 0 0	
HT	EM	)	9	I	Y	i	y	1 0 0 1	
LF	SUB	*	:	J	Z	j	z	1 0 1 0	
VT	ESC	+	;	K	[	k	{	1 0 1 1	
FF	FS	,	<	L	\	l	;	1 1 0 0	
CR	GS	−	=	M	]	m	}	1 1 0 1	
SO	RS	.	>	N	∧	n	~	1 1 1 0	
SI	US	/	?	O	—	o	DEL	1 1 1 1	

NUL	Null, or all zeros
SOH	Start of heading
STX	Start of text
ETX	End of text
EOT	End of transmission
ENQ	Enquiry
ACK	Acknowledge
BEL	Bell, or alarm
BS	Backspace
HT	Horizontal tabulation
LF	Line feed
VF	Vertical tabulation
FF	Form feed
CR	Carriage return
SO	Shift out
SI	Shift in
DLE	Data link escape
DC1	Device control 1
DC2	Device control 2
DC3	Device control 3
DC4	Device control 4
NAK	Negative acknowledge
SYN	Synchronous idle
ETB	End of transmission block
CAN	Cancel
EM	End of medium
SUB	Substitute
ESC	Escape
FS	File separator
GS	Group separator
RS	Record separator
US	Unit separator
SP	Space
DEL	Delete

TABLE 3

ASCII Code and names of control characters. The ASCII code contains many more symbols than does the older Baudot code. The keyboard is arranged in the same manner as a typewriter with the extra symbol keys arranged to the right, or around the main bank of characters. The typewriter keyboard is an array of single-pole switches arranged in the standard pattern. A binary number is assigned to each key on the board.

rate of 0.7 × 1200 = 840 data bits per second. All-electronic systems may use one of the following rates: 300, 600, 1200, 2400, 4800, or 9600 baud. Radio amateur hf transmissions are commonly at a 45-baud data rate.

When a circuit has separate paths for simultaneous send and receive it is called *full duplex*. If a single two-way path is used, only one device can talk at a time and it is termed *half duplex*. A one-way path is called *simplex*.

The parallel output of a keyboard is converted to serial by loading a 10-bit shift register and taking the information out in a series mode. This may be accomplished in a single IC *UART* (Universal Asynchronous Receiver-Transmitter) having eight parallel inputs and a series output, plus other inputs and outputs to register change of state. The UART is commonly clocked at 16 times the baud rate. The output may go to a video display monitor that draws lines and writes in the fashion of a television screen. Raster scanning is used, with the screen driven by a character display circuit (see Section 14-5.)

14-4 RTTY Transmission

The pulsed dc voltage generated by the teleprinter is used to operate a keyer circuit

in the radio transmitter to shift the carrier frequency back and forth in accord with the mark and space signals of the RTTY code. *Audio frequency-shift keying* (AFSK) is generally used on the amateur bands. For many years the frequency shift was 850 Hz (equal to an audio shift of 2125 Hz to 2975 Hz). The newer systems employ a closer shift, 170 Hz being commonly used, with tones of 2125 Hz and 2295 Hz comprising the audio shift. In AFSK, the nominal transmitter frequency is chosen as the mark and the shift condition is chosen as the space signal.

Frequency shift keying (FSK) may be accomplished by varying the frequency of the transmitter oscillator in a stable manner between the mark and the space frequencies. The amount of shift must be held within close tolerances as the shift must match the frequency difference between the selective circuits in the receiving unit. The degree of frequency shift of the transmitting oscillator is, of course, multiplied by any factor of multiplication realized in succeeding multiplier stages of the transmitter. In most simple heterodyning systems, there is no frequency multiplication so the oscillator shift is equal to the desired mark/space relationship. However, depending on which side of the carrier the mixing process occurs,

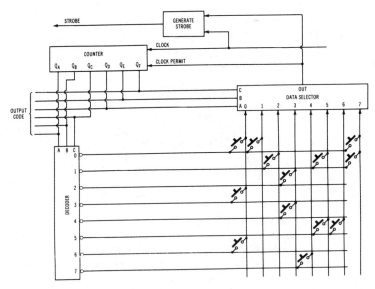

Figure 25

BASIC KEYBOARD

The keyboard has 64 keys in an 8 × 8 matrix. The counter drives a decoder, which grounds each horizontal wire, one at a time. The data selector interrogates each vertical wire for a closed key. When one is found, the data selector is grounded and generates a pulse.

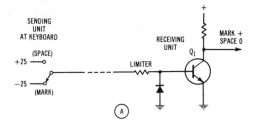

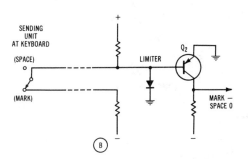

Figure 26

SERIAL ASYNCHRONOUS TRANSMISSION

A—Serial voltage interface.
B—Serial current loop interface.

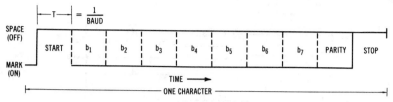

Figure 27

TIME SEQUENCE OF ASYNCHRONOUS SERIAL TRANSMISSION

the shift may be inverted on one or more bands.

Frequency Shift Circuitry A widely used FSK device is the diode switch (figure 29). Upon receiving a pulse from the teleprinter, the diode conducts and places the open terminal of the shift capacitor at ground, thus lowering the frequency of the oscillator. The series-connected choke and associated bypass capacitors remove the r-f from the keying leads. C-w identification is provided by an auxiliary key, the series potentiometer permitting the operator to adjust the amount of frequency shift used for identification.

To invert the keying, the diode is biased to conduct with a small auxiliary supply, the teleprinter pulses removing the bias during the keying cycle (figure 30).

Audio shift keying (AFSK) is primarily used by radio amateurs for compatibility with SSB equipment. An audio oscillator is employed to generate the mark (2125 Hz) and space (2295 Hz) tones when driven by the teleprinter or by a tape unit. The audio signal is then applied to the microphone jack of the transmitter and the resulting frequency-shifted signal is detected and put to use by an audio converter. A simple AFSK keying oscillator is shown in figure 31.

RTTY Duty Cycle The *duty cycle* during an RTTY transmission is unity; that is, the average-to-peak power ratio is one. Most amateur equipments, particularly SSB equipments, are designed with a speech duty cycle in mind and must be derated for RTTY service. Generally speaking, the duty cycle for RTTY equals 2 × the plate dissipation rating of the tube or tubes (or the collector dissipation of the transistor or transistors) in the amplifier stage. Thus, if the amplifier has a pair of, say, 6LQ6 tubes having a combined plate

Figure 28

HOMEBUILT KEYBOARD

Keyboard under construction by K4TWJ. All circuitry except power supply and keyboard interface is contained on the single printed circuit-board designed by W8OZA.

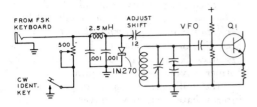

Figure 29

DIODE KEYER FOR FREQUENCY-SHIFT KEYING OF VFO

A simple diode switch may be used to vary the frequency of the transmitter in a stable manner between two chosen frequencies. The amount of shift must match the frequency difference between the selective filters in the receiving demodulators unit.

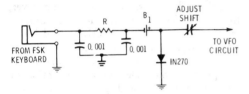

Figure 30

DIODE KEYER FOR INVERTED KEYING

The diode is biased to conduct, the teleprinter pulses removing the bias during the keying cycle.

dissipation of 60 watts for continuous service, the maximum input to the amplifier for RTTY service is limited to about 120 watts.

Auxiliary RTTY Equipment   RTTY transmission by prepunched tape is made possible by means of a *transmitter-distributor* (T-D) unit. This is an electromechanical device which senses perforations in a teleprinter tape and translates this information into electrical impulses of the five-unit teleprinter code at a constant speed. The information derived from the punched tape by contact fingers is transmitted in the proper time sequence by a commutator-distributor driven at a constant speed by a synchronous motor (figure 32). Used in conjunction with the T-D is a *tape perforator* which punches the teleprinter code in a paper tape. The perforator operates mechanically from a teleprinter keyboard for originating messages. A *reperforator* may be connected to receiving equipment to "tape" an incoming message for storage or retransmission.

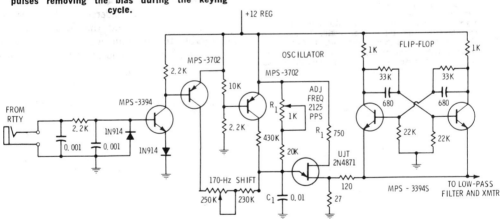

Figure 31

SIMPLIFIED SCHEMATIC OF AFSK KEYING OSCILLATOR

A portion of the Mainline AK-1 AFSK unit. Keyboard providing a positive voltage for the space character is required. The frequency of the unijunction pulse generator is set by R_1-C_1. A mylar capacitor is used for maximum frequency stability. The shift is set by a selector switch. UJT generator runs at 4250 pulses per second and flip-flop divides by two to provide 2125 pulses per second. Flip-flop also squares pulses. Audio pulses are then passed through a low-pass filter to remove all harmonics above 3000 Hz, changing the square wave into a sine wave. Since the UJT generator does not have an LC circuit to determine the frequency, keying transients are minimized when shifting from mark to space.

14-5   RTTY Reception

The RTTY receiving system must respond to a sequence of pulses and spaces transmitted by wire or radio. Frequency-shift keying may be demodulated by a beat-frequency technique, by means of a discriminator as employed in f-m service or by a pulse counting technique. The received signal is converted into dc pulses which are used to operate the printing mechanism in

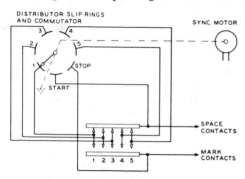

Figure 32

TRANSMITTER DISTRIBUTOR (T-D) UNIT

T-D unit is an electromechanical device which senses perforations in a teleprinter tape and translates this information into the electrical impulses of the teleprinter code. Information derived from the tape by contact fingers is transmitted in proper time sequence by a commutator-distributor driven by a constant-speed motor.

the teleprinter. Conversion of RTTY signals into proper pulses is accomplished by a *receiving converter* (*terminal unit*, abbreviated TU, or *demodulator*). RTTY converters may be either i-f or audio units, the former having been used quite extensively by the military. A block diagram of an intermediate-frequency converter is shown in figure 33A. The RTTY signal in the i-f system of the receiver is considered to be a carrier frequency-modulated by a 22.8-Hz square wave having a deviation of plus and minus 85 Hz (for 170-Hz shift). Amplitude variations in the signal are removed by the limiter stage and the discriminator stage converts the frequency shift into a 22.8-Hz waveform, applied to the teleprinter by means of an electronic keyer. In its simplest form, the i-f demodulator requires that adequate selectivity and interference rejection be achieved by the i-f system of the receiver. I-f demodulators do not provide good selectivity or rejection of interfering signals and they are not well suited for operation in the crowded amateur bands.

The Audio RTTY Demodulator   The audio converter, or demodulator, is generally considered to be superior to the i-f device, and the former unit is preferable for amateur work. A block diagram of a simple audio-frequency demodu-

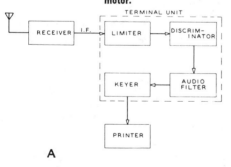

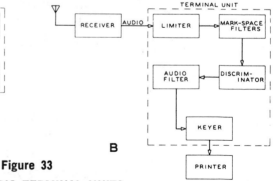

Figure 33

I-F AND AUDIO TERMINAL UNITS

A shows a block diagram of an i-f terminal unit employing f-m discriminator technique. I-f converter requires that selectivity and interference rejection be achieved by means of selective tuned circuits of the receiver. B shows a block diagram of audio-frequency terminal unit. Mark and space filters are used ahead of audio discriminator, followed by a low-pass audio filter. Beat oscillator of the receiver is used to provide audio beat tones of 2125 and 2295 Hz required for nominal 170-Hz shift system.

lator is shown in figure 33B. An audio limiter is followed by mark-frequency and space-frequency filters placed ahead of the discriminator stage. A low-pass filter and electronic keyer provide the proper signal required by the teleprinter. The beat oscillator of the receiver may be used to provide the beat tones of 2125 and 2295 Hz required in the 170-Hz shift systems. Either frequency may be used for either mark or space, and the signal can be inverted by tuning the beat oscillator to the opposite side of the i-f passband of the receiver.

The demodulator may ignore one tone and concentrate on the other tone, the space tone generally being used to actuate the printer, which is biased to rest on the mark tone. It is more reliable, however, to take advantage of both tones, providing negative keying voltage for one tone and positive voltage for the other, as is done in the more sophisticated converters.

High-frequency RTTY signals often exhibit severe fading, with the mark and space frequencies fading independently as skywave reflection varies. Selective fading can often obliterate one frequency and then the other in a random sequence and even the demodulation of both tones will often not permit proper copy during a prolonged fade period, but with properly designed circuitry, normal operation of the demodulator and

teleprinter will continue even during periods of severe fading.

A representative audio frequency RTTY demodulator is shown in figure 34. This simple unit works with 2125-2295 mark and space tones required by SSB receivers. Two small op-amps and a 300-volt rated transistor are used, along with nine diodes.

The first op-amp is a high gain limiter. Reverse-connected zener diodes in the input circuit protect the amplifier against an excessive signal level. The 25K *balance* potentiometer compensates for a small degree of offset input voltage.

The output of the op-amp is fed to the discriminator filters which use surplus 88-mH toroidal inductors (T_1, T_2). Full-wave rectification and a simple RC low-pass filter remove the audio component of the signal as the shifting audio tones are converted into dc pulses in a *slicer* stage. This op-amp takes the small voltages from the tuned filters and changes them to $+10$ volts for *mark* and -10 volts for *space*. Overall gain is sufficient so that the unit will operate with shifts as low as a few cycles.

The keyer transistor (Q_1) has a 300-volt collector-emitter rating and will pass the 60 mA loop current required for teleprinters. A simple RC network in the collector-emitter circuit protects the transistor from the back-emf developed by the inductance

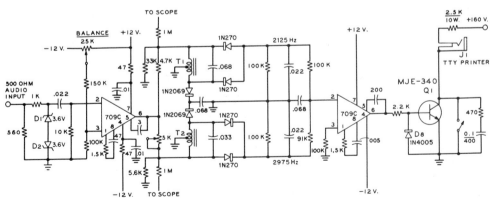

Figure 34

REPRESENTATIVE RTTY DEMODULATOR (CONVERTER)

This solid-state audio RTTY demodulator is based on a design by W6FFC (the Mainline ST-5). It uses two 709C operational amplifiers, one as an audio limiter, and the other as a trigger stage to drive the keyer transistor, which has a 300-volt collector-emitter rating. Reverse-connected zener diodes limit the drive signal to the demodulator unit and the mark and space tones are separated by tuned filters, which are built around surplus 88-mH toroid inductors (T_1, T_2).

of the selector magnets in the printer. The teleprinter keyboard may be connected in series with the printer magnets, both seriesed through jack J_1, if desired.

An Advanced RTTY Demodulator The *Mainline ST-6* demodulator, designed by W6FFC, is a popular unit and provides many advantages over the more simple circuits. The *ST-6* accepts frequency-shifted audio tones from the station receiver and converts them into dc pulses to operate a teleprinter or a video display (figure 35).

The ST-6 is designed to accept various shifts, the most widely used of which are 850 Hz and 170 Hz. Bandpass filters at the input of the device provide a high order of selectivity to eliminate interfering signals. The filters are followed by a limiter having a dynamic range of about 90 dB to correct for signal fading. The output signal from the limiter is fed to a discriminator and detector stage which provides the low-frequency switching pulses. A lowpass active filter after the discriminator/detector provides over 50 dB attenuation to transients normally encountered above the keying speed in service.

The filter is followed by a threshold corrector which provides symmetry to the pulses and corrects the effects of the lowpass filter, which tends to change the desired square wave into a sine wave. The processed signal then passes to a slicer which is a low-frequency amplifier compensated for proper response to the control signals. The output of the slicer drives the keyer stage which provides a mark-hold signal to the teleprinter when there is no input from the slicer. Auxiliary equipment includes a loop supply for the teleprinter, automatic start control, and an antispace circuit that locks the printer to mark-hold when a non-RTTY signal in the space channel tends to activate the printer. A tuning meter is provided to allow the operator to correctly tune the receiver to "straddle" the RTTY signal.

Additional features of the ST-6 are a normal-reverse switch for copying stations having inverted mark/space characteristic and an optional limiterless operation wherein copy may be made from mark-only or space-only signals.

RTTY Video Display A recent development in RTTY apparatus is the *video display generator* which converts the output of a demodulator unit into RTTY readout which may be fed to a TV monitor or to a standard TV receiver (fig-

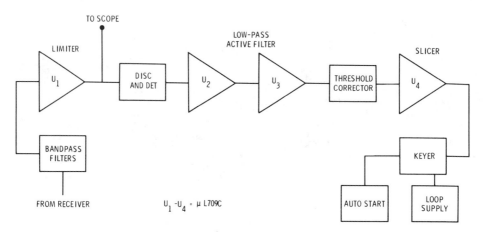

Figure 35

BLOCK DIAGRAM OF ST-6 RTTY DEMODULATOR

The *Mainline ST-6* demodulator accepts frequency-shifted tones from the station receiver and converts them into dc pulses to operate a teleprinter or video display. A commercial version of the ST-6 is produced by HAL Communications Corp.

ure 36A). The display generator block diagram is shown in figure 36B. The RTTY characters are shown as white letters on a black background and are made up as a 5 × 7 dot matrix. There are 40 characters per line and 25 lines per page, displaying 1000 characters per screen. Characters are continually on the screen and new information is written letter by letter as it is received on the bottom line of the display, much in the manner of a typewritten page. When the screen is filled, the top line is pushed off the screen by the next bottom line of display.

Video signal bandwidth is about 4 MHz, the line rate is 15,750 kHz and the field rate is 60 Hz. Frame rate is 30 Hz, with 262.5 lines per field and two fields per frame, with interlaced lines. This provides a compatible signal with U.S. television standards.

Diversity Reception At best, RTTY communication over an hf path is subject to fading. Various methods of diversity reception and signal processing have been utilized to combat this problem. The methods include space diversity with the attendant requirement of multiple antennas and frequency diversity with the required multiple transmitters, antennas, and receivers.

Another less common approach is the *in-band diversity* technique whereby use of the redundant information inherent to the mark/space FSK signal provides an improvement over a nondiversity system. A

A

Figure 36

RTTY VIDEO DISPLAY GENERATOR

The *HAL RVD* 1005 display generator converts the output of RTTY demodulator into readout which is fed to a standard TV receiver or monitor. RTTY characters are shown as white letters on a black background. The generator takes the output of an RTTY demodulator and converts the pulsed signals into impulses compatible with any television receiver. The unit works with speeds of 60, 66, 75, and 100 words per minute, at 40 characters per line and 25 lines per page. (Photo courtesy *HAL Communications Corp.*).

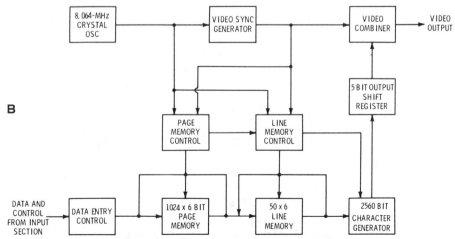

B

Figure 37

DIVERSITY TERMINAL UNIT

The Dovetron MPC-1000C multipath diversity terminal unit provides special circuits for correcting signal distorted by multipath reflections. Block diagram of unit is shown in figure 38.

recent implementation of the in-band diversity method is the *multipath diversity terminal unit* (figure 37). This design also provides special circuits for correcting signals distorted by multipath reflections. The block diagram of the system is shown in figure 38. A detailed functional block diagram of the basic terminal unit is shown in figure 39.

14-6   Slow-Scan Television

Slow-scan television (SSTV) is a narrow-band system for transmitting video images approved by the FCC for use in various amateur bands. Signal bandwidth of an SSTV image is limited to 3 kHz. This transformation is commonly accomplished by converting the video information to a vary-

ing tone which is fed into the audio system of an amateur transmitter. Either a-m, SSB, or f-m transmission may be used. SSB is used for SSTV on the hf bands and f-m on the vhf bands. Because of the restricted bandwidth, the video signal may be received on a communication receiver and may be preserved on an audio tape recorder running at 3¾ inches per second, or more.

The first experiments with SSTV were conducted by WØORX in the early 50's on the then-available 11-meter band, the only portion of the hf spectrum where emissions of this type were permitted. As a result of these early experiments, the FCC granted permission for SSTV transmissions on an experimental basis on the 14- and 28-MHz bands. Since 1958 SSTV has been permitted in the Advanced and Extra Class portions of all hf bands, and in the General Class portion (phone) of the 10- and 6- meter bands, as well as in the vhf bands. Inde-

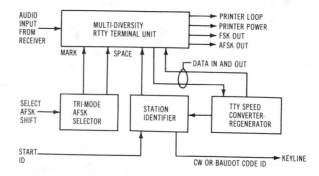

Figure 38

BLOCK DIAGRAM
REGENERATIVE RTTY TERMINAL SYSTEM

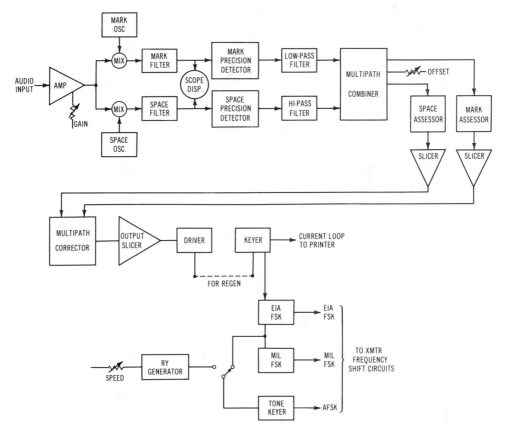

Figure 39

BLOCK DIAGRAM OF BASIC TERMINAL UNIT

pendent sideband transmission is permitted, with picture information in one sideband and voice in the other sideband.

SSTV Transmission A representative SSTV signal consists of a 1500-Hz tone which is shifted down to 1200-Hz for sync information and modulated upward to 2300-Hz for video (picture) information. The 1500-Hz frequency represents the *black level* and the 2300-Hz frequency is the *white level*, with tones in between giving shades of gray. The sync pulse durations are 5 milliseconds for the horizontal and 30 milliseconds for the vertical. The scanning sequence is left to right and top to bottom. Normally, 120 lines are scanned per frame, with an aspect ratio of 1:1. For 60-Hz areas, the horizontal sweep rate is 15 Hz and the vertical sweep rate is 6 to 8 seconds. Since picture transmission time is only a few seconds, it permits rapid alternation of the voice and picture transmission over the same circuit. See Table 4 for a summary of SSTV standards.

A representative SSTV installation is shown in figure 40. The input of the SSTV monitor connects in parallel with the receiver speaker and the audio input circuit of the transmitter is switched to select either video or audio inputs. A good quality cassette or reel-to-reel tape recorder may also be used to record either incoming or outgoing SSTV material.

The P7 Monitor—The simplest and least expensive method of getting started with SSTV involves the use of a P-7 cathode-ray tube type monitor. Such units may be purchased commercially or homebuilt from readily available components. A working

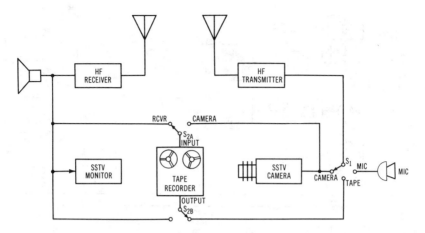

Figure 40

REPRESENTATIVE SSTV STATION

Tape recorder should not be the type that records and plays back simultaneously.

knowledge of audio and dc circuitry is the only prerequisite for building this type monitor.

Stripped of extras, the operational concept of all P-7 monitors is basically the same. The block diagram of a hypothetical unit is shown in figure 41. Incoming SSTV signals are applied to an audio amplifier and limiter stage, producing a constant amplitude fm signal which can be applied to video and sync discriminators. The frequency-sensitive video discriminator produces minimum output for black (1500 Hz) and maximum output for white (2300 Hz). These signals are then detected, amplified and applied to the picture tube as instantaneous cutoff bias (video). During this same time, 1200 Hz sync pulses are removed from the signal by the sync discriminator. These pulses are then detected, amplified, and used to trigger ramp generating circuits in the horizontal and vertical sweep stages. Since the vertical frequency is much lower than the horizontal frequency, a timing circuit is used to eliminate false triggering caused by interference or noise on the signal. If infrablack sync pulses (1200 Hz) get through the video discriminator they cannot be seen as they do not develop any picture tube voltage. A simple high-Q tuned circuit can be used for the sync discriminator while the video discriminator can be a lower-Q tuned circuit resonant at 2300 Hz.

The video amplifier is usually a single transistor stage capable of linearly driving the picture tube from cutoff (black) to saturation (white). Likewise, the sync amplifier amplifies sync pulses to the proper level for accurately triggering the horizontal and vertical sawtooth generating circuits. The output of these sweep circuits can be amplified with complementary-symmetry solid-state circuits, if necessary.

An SSTV Camera The cameras used for SSTV operation fall into two general categories: *plumbicon* units and conventional *vidicon* units. The plumbicon tube employs a form of cesium oxide target which has sufficient time lag to permit its direct use at slow scan rates. Conventional SSTV circuitry is employed in such homebuilt units. Plumbicon tubes are expensive and often TV station "pull-outs" are logical sources for these devices. Vidicon units, on the other hand, have a short time lag and thus must be operated at fast-scan rates and their signal output sampled at the appropriate times. Each of the video samples is then used to modulate a voltage-controlled SSTV oscillator. A simplified block diagram of a sampling camera is shown in figure 42.

Two sets of sync signals are employed in this unit: one set for fast scan and one for slow scan. The fast scan circuitry drives the

TABLE 4. SSTV Standards

Item	60 Hz line frequency	50 Hz line frequency
Horizontal sweep rate	15 Hz	16⅔ Hz
Vertical sweep rate	8 sec.	7.2 sec.
Scanning lines	120	120
Picture aspect ratio	1:1	1:1
Direction of horizontal scan	L-R	L-R
Direction of vertical scan	Top to bottom	Top to bottom
Horizontal sync pulse	5 ms	5 ms
Vertical sync pulse	30 ms	30 ms
Subcarrier frequencies:		
Sync	1200 Hz	
Black	1500 Hz	
White	2300 Hz	

vidicon and triggers the SSTV sync generator. The vidicon output signal is passed to the video amplifier during the time each raster line is scanned. Then, during predetermined times of each line, the comparator senses proper levels of fast and slow scan sweep and sends a "load pulse" to the sample and hold circuit. This permits certain video picture elements (*pixels*) to move into storage and their consequent levels modulate the SSTV voltage-controlled oscillator. The SSTV sync generator is periodically shifted to a preset level for driving the SST VCO to sync frequency.

Digital Scan Conversion One of the most recent innovations to SSTV has been the application of digital scan conversion techniques. These concepts permit unmodified fast scan equipment to be used directly in a slow scan mode. Two modes of digital scan conversion are possible: fast to slow and slow to fast. Essentially, a scan converter may be considered as a time buffer which stores incoming information, accelerating or decelerating it approximately 1000 times, providing output at the required data rate. Approximately 65,000 bits of memory are required for

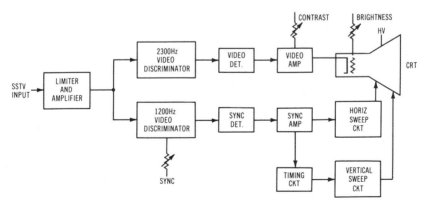

Figure 41

BLOCK DIAGRAM OF TYPICAL SSTV MONITOR

Figure 42

SAMPLING CAMERA FOR SSTV

Vidicon camera tube is scanned with a 4000-Hz vertical rate and a 15-Hz horizontal rate. A timing and sampling circuitry picks individual picture elements out of the fast-scan picture. Elements are then stretched in time and form to form a baseband slow-scan signal. Diagram is of the *Robot 70A Monitor.* (Photo courtesy *Robot Research, Inc.*).

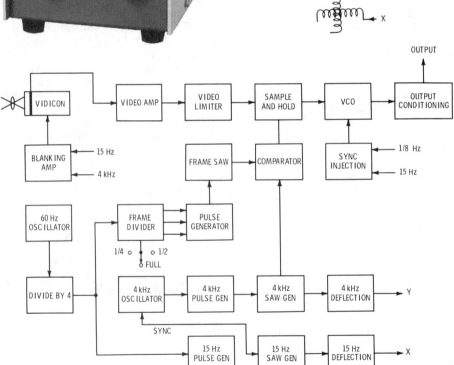

these operations. Due to the relatively large memory requirements and the time consuming task of constructing such scan converters commercial units are commonly used, the *Robot* Model 400 being a representative device.

In addition to converting scan rates both ways (fast to slow and slow to fast) the unit also functions as a complete SSTV station control, gray-scale test generator and auxiliary storage device for any single fast or slow scan picture. The concepts as-

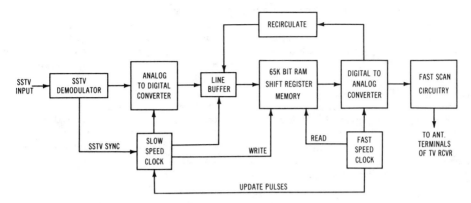

Figure 43

BLOCK DIAGRAM OF DIGITAL SCAN CONVERTER

Slow to fast mode. Approximately 65,000 bits of memory are required to resolve sixteen shades of gray in reproduced pictures.

sociated with digital scan conversion techniques are basically identical, thus a simplified description of slow to fast conversion will suffice (figure 43).

An incoming slow scan picture is initially received and impressed on an SSTV demodulator which functions in a manner similar to the input circuitry of many P7 monitors. The resulting video voltages from the demodulator are then converted to digitalized equivalents by an analog to digital converter and stored in the single-line buffer. An example of the operation performed by the A to D converter is shown in figure 44.

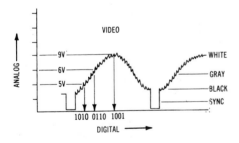

Figure 44

A-TO-D CONVERSION

Analog to digital conversion is accomplished by taking approximately 256 samples of the voltage in each line of the picture, then producing a binary coded equivalent signal. Usually four bits are utilized to describe sixteen shades of gray.

After each line of SSTV video has been received, an SSTV sync pulse is extracted and used to trigger the slow speed clock. This clock then directs data from the line buffer to the main memory during the next available recirculate/update pulse. This operation permits one full line of SSTV to enter the high speed 65,000-bit main memory. A complete SSTV picture is loaded in the memory after approximately 128 of these load functions have been implemented (about 8 seconds). As the fast-speed clock is continuously directing data in the main memory to recirculate a fast scan rate while also updating with new data from the line buffer, the necessary scan rate conversion is performed.

The next step is to release this data. The digital to analog converter performs this function by monitoring data from the main memory and delivering voltage equivalents to the fast-scan circuitry. The fast-speed clock sends preset (sync) levels to the D to A converter at the end of each fast scan line. Finally, the video and sync voltages applied to the fast-scan circuitry are used to modulate a vhf oscillator which is attached to the antenna terminals of a conventional television receiver. Meanwhile, additional incoming SSTV signals continue to refresh the main memory and are consequently displayed on the TV screen. The resulting pictures (which bear a striking resemblance to ordinary black and white TV pictures) can be viewed in a brightly illuminated room.

Living Color SSTV   The first techniques of real-time color SSTV employed a modified concept used in commercial fast scan television. A 500-Hz subcarrier was modulated in quadrature with blue and red video information while green video, which contains 59 percent of the composite video information, modulated the regular SSTV carrier. Frequency-interleaving concepts were also investigated during this early period.

Three digital memories were required for this system: one for red, one for blue, and

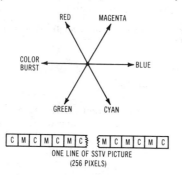

ONE LINE OF SSTV PICTURE
(256 PIXELS)

Figure 45

VECTOR ANALYSIS OF COLOR BURST

Picture element multiplexing is shown below.

one for green. Because of the complexity and cost, a more flexible technique was developed. The resulting system, which is presently gaining popularity was primarily developed by Dr. Don Miller, W9NTP. There are only two colors used in this system (figure 45). The colors are *cyan* and *magenta*. The logic of using these colors may be visualized by reference to the color burst shown in the illustration. Cyan is the complement (inverse) of red and magenta is the complement of green. Additionally, blue may be synthesized by combining noninverted cyan and magenta. Conventional black-and-white SSTV installations receiving these cyan and magenta color multiplexed pictures display them as conventional slow-scan television. Meanwhile, a color SSTV installation demodulates the

signals using the condensed technique illustrated in figure 46.

Incoming color-multiplexed SSTV information is demodulated and A to D converted, then demultiplexed and alternately loaded into the proper 65K bit memory. Then, at the proper fast-scan times, the one line keyer alternately extracts information from the memories as required to reconstruct each line of the color picture. Next, the D to A converter and fast-scan circuitry encode this information into conventional Y, I, and Q signals which can be applied to a regular fast-scan color television receiver.

Y, I, and Q are descriptive terms for the constant luminance principle used in conventional (fast-scan) color TV systems. This principle is necessary for full black-and-white and color compatibility. Essentially, this means that all brightness variations (gray scale) are transmitted in the regular luminance, or Y, channel while color-difference signals containing the specific shades of colors (hue) and their amounts (saturation) are conveyed by the color subcarrier. The luminance, or Y, channel is amplitude modulated with black-and-white video, and the subcarrier is modulated in phase and amplitude with color information. The subcarrier's color modulating signals are either in Phase (I) or 90 degrees out of phase (Q) with the luminance (Y) signals. Black and white TV's receive and detect only "Y" information. Color TV's receive and detect I, Q, and Y information. Since only "color information" is contained in the I and Q signals, the color TV combines this information with "Y" to determine all variations of color brightness.

This interesting two-color SSTV system provides full black-and-white compatibility and picture element multiplexing also allows the maximum amount of full color information to be displayed in fine detail. The system simplicity is a virtue and many digital scan converters commonly in use can be modified for live color slow scan by the addition of a second 65K bit memory board.

During black-and-white television operation, the additional memory board can also perform these interesting operations: *Interference-Processing, 3-D* and *limited motion SSTV.* If each of the 65K memories is loaded with partial or distorted black-and-

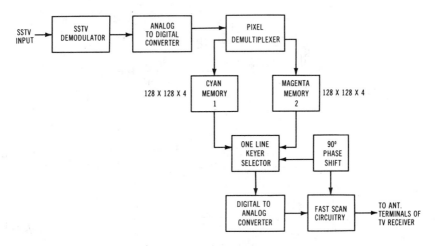

Figure 46

TWO COLOR SSTV SYSTEM
USING PIXEL MULTIPLYING TECHNIQUE

This system will reproduce full color, real-time television pictures from SSTV.

white SSTV information during periods of interference, and pixel averaging techniques are used, a relatively high quality picture can be produced. Likewise, 3-D (three-dimension) SSTV can be produced by loading each memory with color-keyed eye viewing information, and viewing the resultant pictures through color-polarized glasses. These color-keyed 3-D pictures may initially be produced by loading two 65K memories from one camera, which is moved approximately four inches between frames and its lens covered with the appropriate color-key filter.

Limited motion SSTV may be accomplished by initially loading the first of a "motion series" of Slow Scan pictures into one memory, then allowing the next frame of "difference information" (the motion) to load the second memory. Consequent SSTV frames are then used to update each memory in a leap-frog manner. Special sync and memory-select signals are required for this system.

Video Sampling Techniques   One of the most outstanding means of securing the highest possible video resolution from a transmitted SSTV signal involves video sampling. This technique, which may be applied to any SSTV moni-

tor, is similar to that used in modern digital voltmeters. Sampling operates on the principle of time detection rather than frequency detection, that is, a sine (or square) wave for 1500 Hz will have a longer time rate than a wave for 2300 Hz. Thus each incoming alteration of video is compared with a 150 microsecond ramp (frequency = 1/time) and a corresponding voltage for each alternation of video is delivered to the video amplifier. The basic concept of video sampling is shown in figure 47.

Incoming slow-scan signals are initially amplified and limited in amplitude, producing square waves at point A. Positive video alternations of these waves then "fire" the top Schmidt trigger (point B) while negative transitions "fire" the bottom Schmidt trigger (point C). The top single impulse produces an output pulse for each input pulse, and the bottom single impulse does likewise. The time between these pulses (point D and E) is determined by the time, or frequency, of each incoming video alternation. The gated ramp generator produces a sawtooth of voltage each time an *initiate* pulse hits the input (point D). The ramp will increase in amplitude for 150 microseconds, unless it is stopped by an incoming pulse from the bottom single impulse (point E). If the incoming frequency is 1500 Hz,

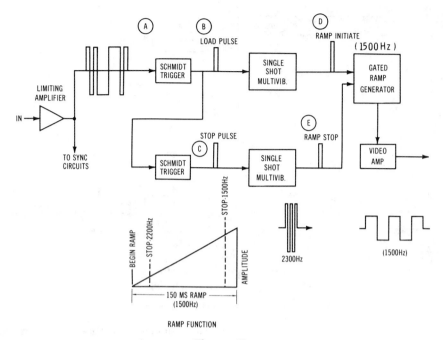

Figure 47

BASIC VIDEO SAMPLING TECHNIQUE

the ramp builds in amplitude for 150 microseconds. Voltage developed by the ramp generator is then amplified and applied to the cathode-ray tube a bias —high bias for black picture elements and low bias for white picture elements. Incoming video can also be converted to digitalized equivalents using this technique. For example, the "initiate" pulse at point *D* can start a counter and the *"stop"* pulse at point E can stop the counter. These digital counts then represent precise shades of gray for each picture element. The sampling process restarts with each new alteration of incoming video. The use of a coincidence detector circuit before the ramp generator also permits this sampling concept to be used for amplitude-modulated (fast-scan) video systems.

OSCAR SSTV   Operating SSTV via an OS-CAR satellite can be a challenging and rewarding experience. The array of experiments via this medium is only limited by the imagination. Initially, the operator should operate cw and SSB via the satellite to gain knowledge in working with

variables such as satellite tracking, Doppler shift, roll rate, etc. Then, after successfully communicating via the satellite, slow-scan TV may be included in future experiments (figure 48).

A typical OSCAR SSTV experiment is shown in figure 49. Separate tape recorders are used for transmitting and receiving SSTV, permitting maximum operational flexibility and reducing the many variables which could inhibit operation. Transmitting a prerecorded program during a satellite pass is very desirable, as it allows the operator to concentrate on such aspects as signal levels, antenna positions and the possibility of monitoring the transmitted signal returned via the satellite. Once SSTV operations have begun during a pass, the receiver's tape recorder should run continuously to enable the operator to review the exercise once the satellite has passed from radio range.

One of the most important considerations in satellite SSTV work is the use of low transmitter power. The 100 percent duty cycle of SSTV can heavily tax the sat-

Figure 48

SSTV VIA OSCAR SATELLITE

Slow-Scan Television picture transmitted by WA9UHV on 145.98 MHz and received by W9NTP on 29.53 MHz via OSCAR 4. An amateur radio "first" (October, 1972). Picture interference is caused by other transmissions through the satellite.

ellite transponder. Returned signals should never be allowed to be as strong as the satellite's beacon.

Operating SSTV Transmitting and receiving SSTV pictures via the hf and vhf amateur bands can be an enjoyable experience provided the operator takes time to plan and organize an efficient video setup (figures 50, 51). Several high quality audio cassettes, prerecorded with station identification, operator views, etc., will ease the operating workload during periods of high activity. A cart rack or cassette "lazy susan" are ideal for holding these cartridges. In addition, operating controls and lights should be placed so that the operator does not have to reach over the equipment to make adjustments during a transmission. One widely used setup technique involves placing all equipment in a desired location, then simulating operation before permanently connecting all gear for operation. An optimum location for the SSTV camera would allow it to view a wall-mounted pegboard covered with an appropriate pattern for quick focus, then by turning the tripod handle, shift the view to the station and operator. An incandescent light dimmer for controlling camera floodlights is

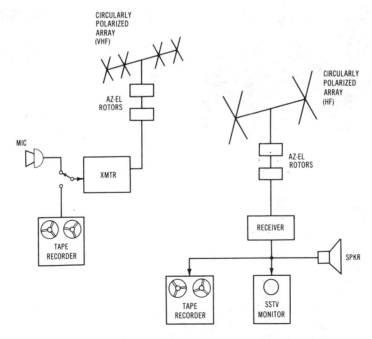

Figure 49

REPRESENTATIVE SSTV STATIONS FOR OSCAR OPERATIONS

Figure 50

SSTV INSTALLATION AT K6SVP

Camera and tape deck are to the left of the transceiver. Fast scan monitor is at right of desk atop digital scan converter.

Figure 51

SSTV INSTALLATION AT N6WQ

SBE Scanvision P7 monitor features built-in cassette tape deck. A second recorder above monitor permits additional station flexibility. SSTV keyboard is home-built.

handy and can be placed at the operating desk to eliminate frequent *f-stop* adjustments to the camera.

The transmitter output power should be reduced to one-half the SSB rating when transmitting SSTV. This can be accomplished by decreasing microphone gain. Never detune the exciter or amplifier, as this places undue strain on the output tubes or transistors.

As an example, assume the station wattmeter registers 400 watts SSB output on voice peaks. The transmitter gain control should be reduced until the wattmeter reads not over 200 watts on SSTV transmission. An extra fan directed toward the final amplifier tubes of the exciter and amplifier should be used during the time of transmission. Half-power signal reduction at the transmitter will result in approximately equal video and audio levels at the receiving station.

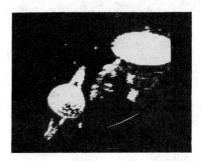

Figure 52

PLANET SATURN VIA SSTV

SSTV picture of Voyager 1 approaching Saturn. Picture was transmitted to K4TWJ by K6SVP.

The majority of SSTV activity on the amateur bands at the time of writing will be found approximately ± 10 kHz of the following frequencies: 3845 kHz, 7171 kHz, 14,230 kHz, 21,340 kHz and 28,680 kHz. Amateurs desiring additional information or assistance with special problems should look for the International Slow Scan Net which meets each Saturday at 1800Z on 14,230 kHz.

14-7 Amateur Facsimile

Facsimile (FAX) is the process whereby graphic or photographic information is either transmitted or recorded by electronic means. Commercial use of FAX includes transmission of weather maps, drawings, and photographs.

FAX transmission is permitted in the United States above 50.1 MHz on the 6-meter band, above 144.1 MHz on the 2-

meter band and on all amateur frequencies above 220 MHz. F-m facsimile is permitted above 220 MHz.

FAX Transmission In general, a facsimile image is created by photoelectric scanning of a printed image (figure 53). The most common technique is to wrap the material to be transmitted around

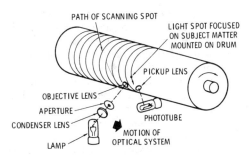

Figure 54

FACSIMILE SCANNING SYSTEM

Material to be transmitted is placed on a revolving drum. A scanning spot of light explores the area of the subject material. The light is focused on the drum and the reflection is picked up by a photocell. The optical system moves along the axis of the revolving drum to provide coverage of the subject by the scanner.

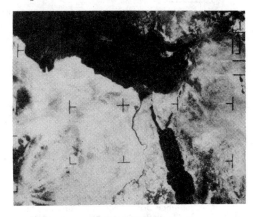

Figure 53

FACSIMILE PHOTOGRAPH VIA SATELLITE

This facsimile photograph was received from the ITOS-1 satellite showing the Middle East area of the Red Sea, Nile and Delta, Dead Sea, Cyprus, etc. The light sandy terrain of North Africa shows up as near-white. (Photo courtesy of *Science Department, Ambassador College.*)

The voltage output of the photoelectric device is called the *baseband* which consists of varying dc levels representing the range of contrast from white to black. Maximum output may be taken to be either white or black. The baseband signal is then used to control the frequency of a voltage-controlled oscillator to generate a subcarrier in which the shades of black and white are represented by a band of frequencies.

The FAX transmission is synchronized with reception by the use of synchronous motors locked to the 60-Hz line frequency. In addition, a series of phasing pulses sent by the FAX transmitter control the start of each line scan so that the receiving unit starts each line of reproduction at the same point on the page.

In general, drum writing speed is 120 lines per minute, with a scan density of 96 lines per inch. Drum speed, and other specifications, vary greatly between equipments of different manufacture and no universally accepted standards are in effect, at least as far as amateur facsimile is concerned.

a cylinder which is rotated about its axis while a light spot is projected on the image. The light reflected from the image is focused on a photomultiplier tube whose output is a function of the varying light intensity reflected from the image on the drum. As the drum is turned, the photoelectric tube is moved slowly by a lead screw causing a slight separation of the scanning lines, much in the manner of operation of a stereo pickup head on a record (figure 54).

A second technique is to use a "flying-spot" scanner, similar to that process discussed in the previous section. Scanning, in either case, is the same as the normal reading process: from left to right and top to bottom.

FAX Reception FAX may be received on a communications receiver, the signal being detected and demodulated. The resulting signal has a varying dc component which corresponds to the light shades in the transmitted subject material. The transmission process, in effect, is reversed.

Sensitized paper is placed on a revolving drum in contact with a stylus which advances along the paper in unison with the movement of the photoelectric device transmitting the picture. A current is passed through the stylus onto the paper on the drum, which is treated with a special electrolyte. The variations in stylus current cause a variation in the darkness of the paper. In some machines, a lamp replaces the stylus and photosensitive paper is used. After exposure, the paper is developed, in the manner of a photographic negative.

14-8 Amateur Television

Amateur television (ATV) transmissions first took place in the prewar 160-meter band using primitive scanning-disc techniques. Electronic television transmissions were experimentally run in the prewar 112-MHz amateur band, but it was not until after 1950 that amateurs used the present 432-MHz band for wideband picture transmission. ATV transmission is growing in popularity, with video transmission in the 432-MHz band and audio transmission in the 144-MHz band.

ATV Transmission   The amateur television transmitter employs the same standards as commercial television. In the United States, this consists of 525 lines per picture at 30 frames per second. The video channel is 4.25 MHz wide and negative modulation is used. The line frequency is 15.75 kHz (525 lines per frame × 30 frames per second). Other standards are in use in other countries.

The *video modulator* of a television transmitter must pass up to 3.5 MHz for black and white service. While the r-f portion of a television transmitter is conventional, the video modulator is unique, and a representative grid-modulation system is shown in figure 55. High-frequency response is enhanced by reducing shunt capacitances and by using series or shunt peaking circuits.

The video signal to be transmitted consists of: (1) impulses corresponding to the brightness of the scanned picture elements conveyed by the camera signal; (2) the

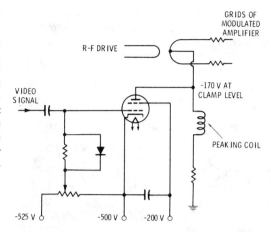

Figure 55

VIDEO MODULATOR FOR ATV TRANSMISSION

The video modulator can transmit a dc component. Clamping diode provides dc restoration for maximum brightness at the peak of the sync signal. Video modulator plate potential is −170 volts with respect to ground, with screen at −200, cathode at −500 and control grid biased to −525 volts. Actual plate and screen voltages are 330 and 300 volts.

blanking of the scanning signal at the receiver during the retrace motions, by the blanking level, or *pedestal* of the signal; and (3) the synchronization of the scanning signal by the vertical and horizontal synchronization signals. When the video signal is imposed on the carrier wave, the envelope of the modulated carrier constitutes the video signal waveform.

The portion of the carrier envelope below the black level is called the *camera signal* and polarity of transmission is negative, that is, increased light on the camera results in a decrease in carrier amplitude. The maximum white level is 15 percent or less of maximum carrier amplitude.

The synchronizing pulses are above the black level (in the *infrablack* region) and do not produce light in the received image. The synchronizing signals contain horizontal impulses for initiating the motion of the scanning spot along each horizontal line and vertical impulses for initiating motion of the scanning spot vertically at the beginning of each field.

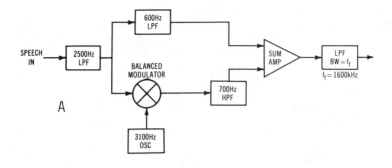

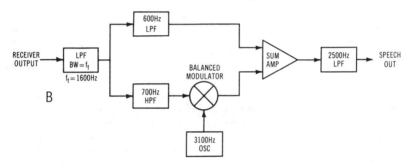

Figure 56

BLOCK DIAGRAM OF FREQUENCY COMPANDOR SYSTEM

A—The frequency compressor is placed between the microphone and the transmitter. The speech is divided into two voice bands and the higher frequency voice band is inverted by the balanced modulator and "folded" down into the lower speech band.

B—The reverse process takes place at the receiver with the frequency expandor placed between the receiver and the speaker. The "folded" speech is "unfolded" in a continuous manner and the natural balance of the voice is restored. (This diagram is based upon the compandor design of VBC, Inc. of San Mateo, Calif.).

ATV Reception Since ATV standards are the same as commercial TV, the least expensive reception technique is to make use of a conventional black-and-white TV receiver, in conjunction with a 432-MHz converter. Tunable converters are in general use, as opposed to a crystal-controlled converter, as it is desirable to be able to tune off to one side of the ATV carrier to obtain the clearest picture consistent with local interference and the shape of the receiver passband. Since amateur TV transmits both sidebands, instead of one as is done in commercial practice, it is convenient to be able to tune to either sideband for best reception.

14-9 Narrowband Voice Modulation (NBVM)

The increasing demand for hf and vhf spectrum space has brought about various methods for reducing channel separation in commercial and military service and for allowing more amateurs to operate comfortably in the narrow frequency bands assigned to them.

Narrowband voice modulation has been under investigation for many years and various solutions to this problem have been advanced. The demand for spectrum space will continue to increase as new uses for radio develop. This is especially true in the vhf region because this band has particularly good characteristics for mobile use.

Two modern technologies have been recently introduced that permit significant improvement in signal-to-noise ratio and required band width for a communication circuit. These are the *amplitude compandor* which compresses signal amplitude on transmission and expands it on reception and the *frequency compandor* which compresses bandwidth on transmission and expands it on reception.

The amplitude compandor is well known in amateur terms as a "speech compressor," or "speech processor" which compresses the amplitude on transmission. The companion device which expands the signal back to its original proportions on reception is not as well known. Frequency compression and expansion, on the other hand, are relatively new concepts. Both systems modify the basic voice signals, one in the amplitude domain and the other in the frequency domain, to make more efficient use of a frequency channel.

The Compandor   The *amplitude compandor* works by reducing the amplitude of loud syllables and increasing the amplitude of weak ones to achieve a transmitted signal more even (compressed) in power level. After transmission and reception, the signal is restored (expanded) to its original form (compressor plus expandor equals "compandor"). The result is that noise occurring during quiet passages is greatly reduced; noise during loud passages is increased, but it is masked by the loudness of the passage itself. For the average listener, FCC tests have shown the amplitude compandor results in an apparent 15 dB improvement in signal-to-noise ratio. That is the compandored signal sounds as good as a normal signal that has 15 dB less noise power. The 15 dB improvement is obtained for amplitude compression giving 1 dB output variation for every 2 dB of input variation.

In the second technique, *frequency companding*, the voice frequencies are compressed prior to transmission and expanded upon reception. One system takes advantage of the tendency in human speech for either

the lower frequencies to be voiced (on vowel sounds) or the higher frequencies to be voiced (on consonant sounds), but not together. This system "folds" the higher frequencies down over the lower frequencies and transmits both together. Expansion at the receiver gives a high-quality signal that, in this particular system, has been transmitted in about 0.60 of the bandwidth normally required for intelligible voice.

These two techniques are applied to the basic voice signal prior to modulation of the transmitter and at the receiver before the speaker, or reproducer. Alteration of the basic transmitter and receiver is not required.

Shown in figure 56 is the block diagram of a practical frequency compandor. Illustration A outlines the frequency compressor system used at the transmitter. Distortion that may reduce speech intelligibility is reduced by the use of a 2500-Hz audio filter before the compandor. By the use of additional filters the speech is separated into two bands, one from dc to 600 Hz and the second from f_1 to 2500 Hz. The frequency f_1 is chosen based upon the characteristics of the second lowpass filter cutoff frequency. A practical frequency for f_1 is 1600 Hz.

The frequency range of dc to 600 Hz is passed essentially straight through the system. The characteristics of the SSB exciter generally limits the audio frequencies below 250 Hz. The other range, however, is inverted by the balanced modulator and local oscillator and down-converted. The resulting speech range is then added to the first one and passed through a 1600-Hz lowpass filter. Although a gap does exist in the final output spectrum, the speech is of high intelligibility and has high recognizability. The process is reversed for reception (illustration B).

Acoustically, voice consonants are emphasized and folded, electrically, into blank spaces not occupied by vowels as speech occurs. This is possible since vowels and consonants do not interfere in time domain because a vowel and a consonant cannot occur at the same time. On receiving the "folded" speech, the vowels and consonants are "unfolded" in a continuous manner and the natural balance between vowels and consonants restored.

CHAPTER FIFTEEN

Amplitude Modulation and Audio Processing

A listener to the amateur bands would conclude that *amplitude modulation* is extinct and that all communication is carried on by single sideband, RTTY or c-w on the high-frequency bands and by frequency modulation and SSB on the very-high frequency bands.

While it is true that the Amateur Radio Service has "outmoded" amplitude modulation, the greater bulk of everyday radio communication in the United States (and throughout the world) is still conducted by amplitude modulation (a-m).

Over 4500 a-m broadcast stations exist in the United States, together with over 240,-000 a-m stations in the Aeronautical Radio Service. And of the 8,000,000 CB transmitters in existence, an estimated 80-percent of these are amplitude modulated.

As far as SSB goes, it too is basically an amplitude-modulated signal whose carrier and one sideband have been removed.

Basically, then, amplitude modulation is the heart of the modern communication system and the details of this fundamental means of superimposing intelligence on a radio-frequency carrier are discussed in this chapter.

15-1   Sidebands

Modulation is essentially a form of mixing, or combining, already covered in a previous chapter. To transmit intelligence at radio frequencies by means of a-m, the intelligence is converted to radio-frequency *sidebands*. The sidebands appear symmetrically above and below the frequency of the unmodulated carrier signal, as shown in figure 1.

Even though the amplitude of radio-frequency voltage representing the composite signal (resultant of the carrier and sidebands, called the *envelope*) will vary from zero to twice the unmodulated signal value during full modulation, the amplitude of the *carrier* component does not vary. Also, as long as the amplitude of the modulating voltage does not vary, the amplitude of the sidebands will remain constant. For this to be apparent, however, it is necessary to measure the amplitude of each component with a highly selective filter. Otherwise, the measured

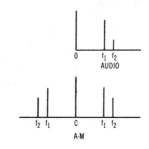

Figure 1

FREQUENCY SPECTRUM

Comparison of two-tone signal in the audio spectrum and resulting amplitude-modulated waveform. Unmodulated carrier is at C.

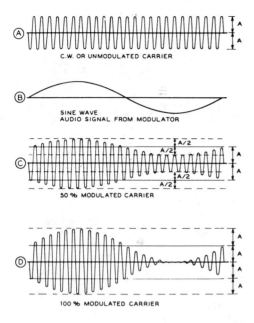

C.W. OR UNMODULATED CARRIER

SINE WAVE
AUDIO SIGNAL FROM MODULATOR

50 % MODULATED CARRIER

100 % MODULATED CARRIER

Figure 2

AMPLITUDE-MODULATED WAVE

Top drawing A represents an unmodulated carrier wave; B shows the audio output of the modulator. Drawing C shows the audio signal impressed on the carrier wave to the extent of 50 percent modulation; D shows the carrier with 100 percent amplitude modulation.

power or voltage will be a *resultant* of two or more of the components, and the amplitude of the resultant will vary at the modulation rate.

If a carrier frequency of 5000 kHz is modulated by a pure tone of 1000 Hz, or 1 kHz, two sidebands are formed: one at 5001 kHz (the sum frequency) and one at 4999 kHz (the difference frequency). The frequency of each sideband is independent of the amplitude of the modulating tone, or *modulation percentage*; the frequency of each sideband is determined only by the frequency of the modulating tone. This assumes, of course, that the transmitter is not modulated in excess of its linear capability.

When the modulating signal consists of multiple frequencies, as is the case with voice or music modulation, two sidebands will be formed by each modulating frequency (one on each side of the carrier), and the radiated signal will consist of a *band* of frequencies. The *bandwidth*, or *channel*, taken

up in the frequency spectrum by a conventional double-sideband amplitude-modulated signal, is equal to twice the highest modulating frequency. For example, if the highest modulating frequency is 5000 Hz, then the signal (assuming modulation of complex and varying waveform) will occupy a band extending from 5000 Hz below the carrier to 5000 Hz above the carrier.

Frequencies up to at least 2000 Hz, and preferably 2500 Hz, are necessary for good speech intelligibility. If a filter is incorporated in the audio system to cut out all frequencies above approximately 2500 Hz, the bandwidth of an a-m signal can be limited to 5 kHz without a significant loss in intelligibility. However, if harmonic distortion is introduced subsequent to the filter, as would happen in the case of an overloaded modulator or overmodulation of the carrier, new frequencies will be generated and the signal will occupy a band wider than 5 kHz.

15-2   Mechanics of Modulation

A c-w or unmodulated r-f carrier wave is represented in figure 2A. An audio-frequency sine wave is represented by the curve of figure 2B. When the two are combined or "mixed," the carrier is said to be amplitude modulated, and a resultant similar to 2C or 2D is obtained. It should be noted that under modulation, each half cycle of r-f voltage differs slightly from the preceding one and the following one; therefore at no time during modulation is the r-f waveform a pure sine wave. This is simply another way of saying that during modulation, the transmitted r-f energy no longer is confined to a single radio frequency.

It will be noted that the *average* amplitude of the peak r-f voltage, or modulation envelope, is the same with or without modulation. This simply means that the modulation is symmetrical (assuming a symmetrical modulating wave) and that for distortionless modulation the upward modulation is limited to a value of twice the unmodulated carrier wave amplitude because the amplitude cannot go below zero on downward portions of the modulation cycle. Figure 2D illustrates the maximum obtainable distortionless modulation with a sine modulating wave, the r-f voltage at the peak of the r-f cycle varying from zero to twice the unmodulated value, and the r-f power varying from zero to four times the unmodulated value (the power varies as the square of the voltage).

While the average r-f *voltage* of the modulated wave over a modulation cycle is the same as for the unmodulated carrier, the average *power* increases with modulation. If the radio-frequency power is integrated over the audio cycle, it will be found with 100 percent sine-wave modulation the average r-f power has increased 50 percent. This additional power is represented by the sidebands, because, as previously mentioned, the carrier power does not vary under modulation. Thus, when a 100-watt carrier is modulated 100 percent by a sine wave, the total r-f power is 150 watts—100 watts in the carrier and 25 watts in each of the two sidebands.

Modulation Percentage So long as the *relative proportion* of the various sidebands making up voice modulation is maintained, the signal may be received and detected without distortion. However, the higher the average amplitude of the sidebands, the greater the audio signal produced at the receiver. For this reason it is desirable to increase the *modulation percentage,* or degree of modulation, to the point where maximum "negative" peaks just hit 100 percent. If the modulation percentage is increased so that the peaks exceed this value, distortion is introduced, and if carried very far, bad interference to signals on nearby channels will result.

Modulation Measurement The amount by which a carrier is being modulated may be expressed either as a *modulation factor,* varying from zero to 1.0 at maximum modulation, or as a percentage. The percentage of modulation is equal to 100 times the modulation factor. Figure 3A shows a carrier wave modulated by a sine-wave audio tone. A picture such as this might be seen on the screen of a cathode-ray oscilloscope with sawtooth sweep on the horizontal plates and the modulated carrier impressed on the vertical plates. The same carrier without modulation would appear on the oscilloscope screen as figure 3B.

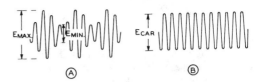

Figure 3

GRAPHICAL DETERMINATION OF MODULATION PERCENTAGE

The procedure for determining modulation percentage from the peak voltage points indicated is discussed in the text.

The percentage of modulation of the positive peaks and the percentage of modulation of the negative peaks can be determined separately from two oscilloscope pictures such as shown.

The modulation factor of the positive peaks may be determined by the formula:

$$M = \frac{E_{max} - E_{car}}{E_{car}}$$

The factor for negative peaks may be determined from the formula:

$$M = \frac{E_{car} - E_{min}}{E_{car}}$$

In the above two formulas E_{max} is the maximum carrier amplitude with modulation and E_{min} is the minimum amplitude; E_{car} is the steady-state amplitude of the carrier without modulation.

If the modulating voltage is symmetrical, such as a sine wave, and modulation is accomplished without the introduction of distortion, then the percentage modulation will be the same for both negative and positive peaks. However, the distribution and phase relationships of harmonics in voice and music waveforms are such that the percentage modulation of the negative modulation peaks may exceed the percentage modulation of the positive peaks, or vice versa. The percentage modulation when referred to without regard to polarity is an indication of the average of the negative and positive peaks.

Modulation Capability The *modulation capability* of a transmitter is the maximum percentage to which that transmitter may be modulated before spurious sidebands are generated in the output or before the distortion of the modulating waveform becomes objectionable. The highest modulation capability which *any* transmitter may have on the *negative* peaks is 100 percent. Overmodulation on negative peaks causes clipping of the wave at the zero axis and changes the envelope wave shape to one that includes higher-order harmonics which appear as additional side frequencies, showing up in a receiver as sideband "splatter" and distortion of the imposed signal intelligence.

Overmodulation on upward modulating peaks does not cause distortion, within the linearity limit of the transmitter. In the United States, an increase in positive peak modulation to 125 percent is permitted in the a-m broadcast service.

Speech Waveform Dissymmetry   The manner in which the human voice is produced by the vocal cords gives rise to a certain dissymmetry in the waveform of voice sounds when they are picked up by a good quality microphone. This is especially pronounced in the male voice, and more so on certain voice sounds than on others. The result of this dissymmetry in the waveform is that the voltage peaks on one side of the average value of the wave will be considerably greater, often two or three times as great, as the voltage excursions on the other side of the zero axis. The *average* value of voltage on both sides of the wave is, of course, the same.

As a result of this dissymmetry in the male voice waveform, there is an optimum polarity of the modulating voltage that must be observed if maximum sideband energy is to be obtained without negative peak clipping and generation of *splatter* on adjacent channels.

The use of the proper polarity of the incoming speech wave in modulating a transmitter can allow a useful increase in the average level of intelligence that may be placed on the signal. If the modulating amplitude is adjusted so that the peak downward (negative) modulation is held to 100 percent, or less, the peak upward (positive) modulation may reach a greater value. If the modulation envelope reproduces the waveform of the modulating signal, there is no distortion.

Overmodulation   If the peak negative modulation level is too great, a period of time will exist during which the instantaneous voltage applied to the modulated stage is zero, or negative, and the stage is cut off. The shape of the modulation envelope is then no longer accurately reproduced and the modulation is distorted. This condition is called *overmodulation* and results in the creation of new, additional side frequencies generated on both sides of the carrier. These spurious frequencies widen the sidebands of the signal and can cause severe adjacent channel interference termed *splatter*.

The splatter is a direct consequence of clipping the r-f waveform at the zero axis during peaks of negative modulation. A neg-

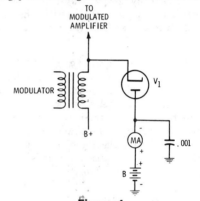

Figure 4

NEGATIVE PEAK OVERMODULATION INDICATOR

The milliammeter will show a reading on modulation peaks that carry the instantaneous voltage on the plate-modulated amplifier below zero. Bias voltage (B) may be adjusted to provide indication of negative modulation peaks of any value below 100 percent.

ative peak modulation indicator (figure 4) can be used to monitor this form of clipping. The effect of modulation beyond 100 percent of both positive and negative peaks is illustrated in figure 5.

15-3   Audio Processing

Speech waveforms are characterized by frequently recurring high-intensity peaks of very short duration. These peaks will cause overmodulation if the average level of mod-

Figure 5

SPEECH-WAVEFORM AMPLITUDE MODULATION

Showing the effect of using the proper polarity of a speech wave for modulating an a-m transmitter. A shows the effect of proper speech polarity on a transmitter having an upward modulation capability of greater than 100 percent. B shows the effect of using proper speech polarity on a transmitter having an upward modulation capability of only 100 percent. Both these conditions will give a clean signal without objectionable splatter. C shows the effect of the use of improper speech polarity. This condition will cause serious splatter due to negative-peak clipping in the modulated-amplifier stage.

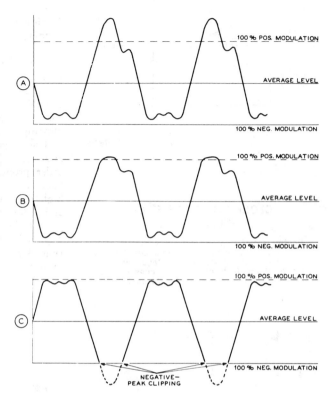

ulation on loud syllables exceeds approximately 30 percent. Careful checking into the nature of speech sounds has revealed that these high-intensity peaks are due primarily to the vowel sounds. Further research has revealed that the vowel sounds add little to intelligibility, the major contribution to intelligibility coming from the consonant sounds such as *v*, *b*, *k*, *s*, *t*, and *l*. Measurements have shown that the power contained in these consonant sounds may be down 30 dB or more from the energy in the vowel sounds in the same speech passage. Obviously, then, if we can increase the relative energy content of the consonant sounds with respect to the vowel sounds it will be possible to understand a signal modulated with such a waveform in the presence of a much higher level of background noise and interference. Experience has shown that it is possible to accomplish this desirable result by *audio processing* which builds up the effective level of the weaker sounds without overmodulation of the carrier. Various systems exist that accomplish this goal without

appreciable audio distortion. Among these systems are *dynamic compression* and *amplitude limiting*.

Dynamic Compression   Dynamic compression of the audio signal may be used to maintain a high level of modulation over a large range of audio input to the modulating system. This is accomplished by taking a control voltage from the output voltage of the system and using it to control system gain so that the output voltage is virtually constant.

A practical dynamic compressor rectifies and filters the audio signal as it passes through the amplifier and applies the dc component to a gain control element in the amplifier (figure 6). Simple compressors exhibit an attack time of 300 milliseconds or longer. A compression range of the order of 20 to 35 dB is realizable, corresponding to the dynamic range of the human voice. Reverberation and background noise usually limit the practical compression range to about 15 dB. A representative solid-state

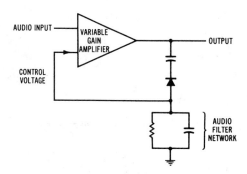

Figure 6

BLOCK DIAGRAM OF AUDIO COMPRESSOR

Control signal is taken from the output of the compressor, rectified and filtered and fed back to a low-level gain-controlled stage. Time constants of the filter network are chosen to prevent oscillation and distortion.

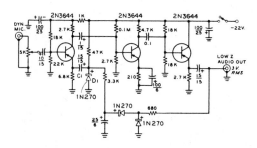

Figure 7

SOLID-STATE COMPRESSOR AMPLIFIER FOR DYNAMIC MICROPHONE

Compression is brought about by variation of emitter bypass capacitor C_1 in the first-stage transistor. Variable-resistance network is driven by two 1N270 diodes as a voltage doubler of output signal taken from emitter of the third stage emitter follower.

compressor for voice waveform is shown in figure 7.

The main disadvantage of a simple audio compressor is that in the intervals between words the compressor gain rises and background noise appears to rise also. If the time constant of the audio filter is fast enough to follow fast speech sounds then the possibility exists of undesired clipping on initial sounds with consequent distortion. A slow time constant, on the other hand, means that initial sounds can overmodulate before the system can compensate for them.

Amplitude Limiting   Limiting may take place in either the audio or r-f systems of a transmitter. An audio limiter can take the form of a *peak clipper* that passes signals up to a certain amplitude but limits all signals greater than this level (figure 8). The net effect of this is to "flat-top" the wave envelope, which at an extreme clipping level, can approach a square wave.

The high order products produced by audio clipping can cause splatter and the low order products fall within the audio passband and cause distortion of the signal. A high pass audio filter following the clipper and reduction of low frequency audio components before the filter can allow a higher clipping level for a given degree of distortion. A representative audio clipper is shown in figure 9.

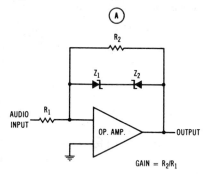

$$GAIN = R_2/R_1$$

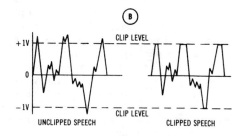

Figure 8

AUDIO CLIPPER

A—Block diagram of audio clipper.
B—Unclipped and clipped speech.

R-F Clipping   Once the audio signal is transposed into an rf SSB signal, clipping and filtering may be employed.

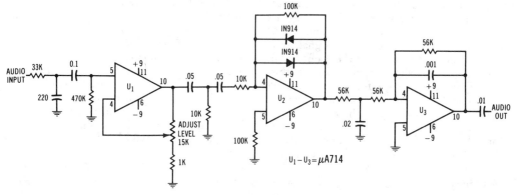

Figure 9

AN AUDIO CLIPPER

A preamplifier (U₁) incorporates r-f filtering and high-frequency audio cutoff in the output circuit. A high impedance microphone should be used. U_1 is an adjustable gain amplifier which sets the input level to limiting amplifier U_2. U_2 limits because of the nonlinear resistance characteristics of the back-to-back diodes which supply increasingly heavier negative feedback as the output amplitude of U_2 increases. A lowpass filter (U_3) follows the compressor. Frequencies above 2.8 kHz that are generated in U_2 are removed in this stage.

This has the advantage that fewer in-band distortion products are created for a given degree of clipping than in an equivalent audio clipper. This results in a higher quality signal, provided an r-f filter maintains the original circuit bandwidth (figure 10). With 15 dB of clipping, an increase in speech intelligibility of nearly 8 dB may be achieved. Generally speaking, the distortion produced by r-f envelope clipping is less objectionable than that caused by an equivalent amount of audio clipping.

R-F Compression R-f compression (often termed *automatic load control*, or *ALC*) may take the form shown in figure 11. Operation is very similar to the i-f stage of a receiver having automatic gain control. Control voltage is obtained from the amplifier output circuit and a large delay (threshold) bias is used so that

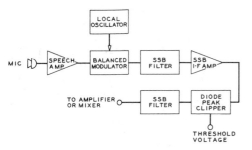

Figure 10

BLOCK DIAGRAM OF R-F ENVELOPE CLIPPER

An r-f clipper may be placed in the i-f portion of the SSB transmitter to limit amplitude of SSB signal. The clipper is followed by an r-f filter to remove harmonics and out-of-band products caused by clipping action. Clipping level is controlled by threshold voltage.

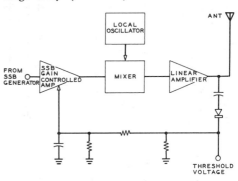

Figure 11

BLOCK DIAGRAM OF R-F COMPRESSOR

R-f compression (automatic load control) is similar to automatic gain control circuit of a receiver. Control voltage is obtained from rectified output signal of final linear amplifier stage and is applied to low level gain-controlled stage. Threshold bias is set so that no gain reduction takes place until output signal is nearly up to the maximum linear signal capability of the amplifier.

no gain reduction takes place until the output signal is nearly up to the maximum linear signal capability of the amplifier. At this level, the rectified output signal overcomes the delay bias and the gain of the preamplifier is reduced rapidly with increasing signal level. Peak r-f compression levels of up to 15 decibels are commonly used in SSB service, providing an increase in average-to-peak power of up to 5 decibels. Speech *intelligibility* may be improved only by about one decibel by such a technique.

A Comparison of Processing Techniques Outboard speech-processing adapters incorporated into existing equipment are becoming quite popular, but should be viewed with caution, since the equipment in question may have inherent limitations that preclude the use of a driving signal having a high average-to-peak ratio. Excessive dissipation levels may be reached in amplifier tubes, or low-level stages may be overloaded by the intemperate use of speech processing equipment. In any case, the output spectrum of the transmitter should be carefully examined for out-of-passband emissions.

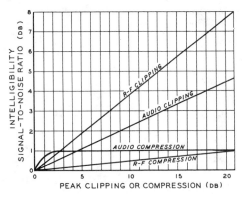

Figure 12

COMPARISON OF SPEECH-PROCESSING TECHNIQUES

In terms of overall speech intelligibility, r-f clipping has an advantage of several decibels over other systems. R-f clipping up to 10 decibels or so may be used with many SSB transmitters without objectionable distortion. Use of add-on speech processing of any type should be done with caution since the user has no knowledge of limitations of the transmitter, which may preclude drastic changes in peak-to-average ratio of driving signal.

Figure 12 shows a comparison of the four different methods of speech processing used in SSB work. R-f envelope clipping has an advantage of several decibels over the other systems. All techniques increase transmitted average-to-peak power to a degree, thereby improving the overall speech intelligibility. Use of two speech-processing systems, however, is not directly additive, and only the larger improvement factor should be considered.

A Practical R-F Envelope Clipper Shown in figure 13 is a block diagram of a practical r-f envelope clipper which employs the phase-shift method of generating and processing a low-frequency SSB signal. The circuit provides up to 20 dB clipping with low distortion and results in an improvement in intelligibility signal-to-noise ratio of better than 6 dB.

The schematic of the clipper is shown in figure 14. Inexpensive ICs made by *RCA, National Semiconductor* and others are used. Three LM-741 operational amplifiers are used in the audio phase-shift networks driving two CD-4013/4016s as the balanced modulator. A CD-4007 dual complementary pair plus inverter serves as an r-f amplifier and limiter stage. The local oscillator (80 kHz) is a CD-4007 which also serves as the product detector. A two-stage amplifier provides a low-impedance audio output and drives level meter M_1.

The clipper is easily adjusted using the ALC meter of the transmitter. The audio level of the transmitter is set so that a continuous tone provides a small reading on the ALC meter. The clipper is now added to the circuit and the input gain adjustment of the clipper increased until the same audio tone drives the clipper level meter to 0 dB with the same ALC reading. On a monitor scope, this adjustment produces the same *peak* signal level in either case but increases the average output level several times.

Power-Supply Requirements The power load of an SSB transmitter can fluctuate between the zero-signal value and that required for maximum signal power output. For a class-B stage, this may represent a current ratio of 10 to 1, or more. The rate and amount of current fluctuation are

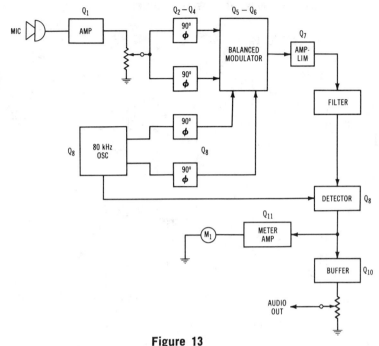

Figure 13

BLOCK DIAGRAM OF PHASING-TYPE R-F SPEECH PROCESSOR

related to the envelope of the SSB signal and the frequency components in the supply current variation may be much lower and higher than the frequency components of the driving signal. For voice modulation, supply current fluctuations corresponding to syllabic variations may be as low as 20 Hz and high-order distortion products of non-linear stages may produce fluctuations higher than 3000 Hz. The power supply for an SSB transmitter, therefore, must have good *dynamic regulation,* or the ability to absorb a sudden change in the load without an abrupt voltage change. The most effective means of achieving good dynamic regulation in the supply is to have sufficient filter capacity in the supply to overcome sudden current peaks caused by abrupt changes of signal level. At the same time, *static regulation* of the supply may be enhanced by reducing voltage drops in the power transformer, rectifier, and filter choke, and by controlling transformer leakage reactance.

15-4   Systems of Amplitude Modulation

The simplest form of modulation uses a single diode driven by two signals (figure 15). The carrier to be modulated is normally at a high frequency compared with the modulating frequency. The current at point A consists of positive pulses passed by the diode and at point B, because of the tuned circuit, a double sideband, a-m signal is produced. To hold distortion to a low value, modulation of the carrier is limited to about 10 percent in this circuit.

Modulator Classifications   The following discussion concerns modulation systems employing vacuum tubes. A later section will cover solid-state modulators. There are many different systems and methods for amplitude-modulating a carrier, but most may be grouped under three general classifications: (1) *variable efficiency sys-*

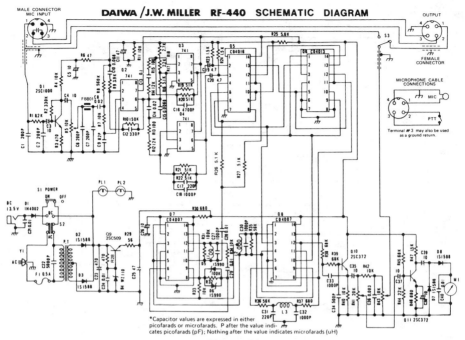

Figure 14

R-F SPEECH PROCESSOR

The Daiwa/Miller processor employs the phase-shift method of generating and processing a low frequency SSB signal. Inexpensive CMOS IC devices are used. The processor is connected between the microphone and the transmitter.

tems in which the average input to the stage remains constant with and without modulation and the variations in the efficiency of the stage in accordance with the modulating signal accomplish the modulation; (2) *constant-efficiency* systems in which the input to the stage is varied by an external source of modulating energy to accomplish the modulation; and (3) so-called *high efficiency* systems in which cir-

cuit complexity is increased to obtain high plate-circuit efficiency in the modulated stage without the requirement of an external high-level modulator. The various systems under each classification have individual characteristics which make certain ones best suited to particular applications.

Variable-Efficiency Modulation   Since the *average* input remains constant in a stage employing variable-efficiency modulation, and since the average power output of the stage increases with modulation, the additional average power output from the stage *with* modulation must come from the plate dissipation of the tubes in the stage. Thus, for the best relation between tube cost and power output, the tubes employed should have as high a plate dissipation rating per unit cost as possible.

The plate efficiency in such an amplifier is doubled when going from the unmodulated condition to the peak of the modula-

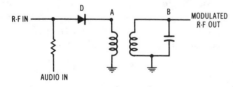

Figure 15

DIODE MODULATOR

The current at A consists of positive pulses passed by diode D, and at point B, because of the tuned circuit, a double sideband a-m signal is produced.

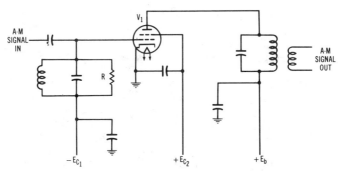

Figure 16

CLASS-B GRID-DRIVEN LINEAR AMPLIFIER

Swamping resistor R is included in the grid circuit to reduce effects of grid impedance
variation on the driving stage when grid current is drawn.

tion cycle. As a result, the unmodulated efficiency of such an amplifier *must always be less than 40 percent,* since the maximum peak efficiency obtainable in a conventional amplifier is in the vicinity of 80 percent. Since the peak efficiency in certain types of amplifiers will be as low as 60 percent, the unmodulated efficiency in such amplifiers will be in the vicinity of 30 percent.

There are many systems of efficiency modulation, but they *all* have the general limitation discussed in the previous paragraph—so long as the carrier amplitude is to remain constant with and without modulation, the efficiency at carrier level must be not greater than one-half the peak modulation efficiency, if the stage is to be capable of 100-percent modulation.

The Class-B Grid Driven Linear Amplifier This is the simplest practicable type amplifier for an amplitude-modulated wave or a single-sideband signal. The system requires that excitation, grid bias, and loading must be carefully controlled to preserve the linearity of the stage. Also, the grid circuit of the tube, in the usual application where grid current is drawn on peaks, presents a widely varying value of load impedance to the source of excitation. It is thus necessary to include some sort of *swamping resistor* to reduce the effect of grid-impedance variations with modulation (figure 16). If such a swamping resistance across the grid tank is not included, or is too high in value, the positive modulation peaks of the incoming

modulated signal will tend to be flattened with resultant distortion of the wave being amplified.

Since a class-B a-m linear amplifier is biased to *extended* cutoff with no excitation (the grid bias at extended cutoff will be approximately equal to the plate voltage divided by the amplification factor for a triode, and will be approximately equal to the screen voltage divided by the grid-screen μ factor for a tetrode or pentode) the plate current will essentially flow in 180-degree pulses. Due to the relatively large operating angle of plate current flow the theoretical peak plate efficiency is limited to 78.5 percent, with 65 to 70 percent representing a range or efficiency normally attainable.

The carrier power output from a class-B linear amplifier of a normal 100-percent modulated a-m signal will be about one-half the rated plate dissipation of the stage, with optimum operating conditions. The peak output from a class-B linear, which represents the maximum-signal output as a single-sideband amplifier, or peak output with a 100 percent a-m signal, will be about twice the plate dissipation of the tubes in the stage. Thus the carrier-level input power to a class-B linear should be about 1.5 times the rated plate dissipation of the stage.

Screen-Grid Modulation Amplitude modulation may be accomplished by varying the screen-grid voltage in a class-C amplifier which employs a pentode, beam

15.12   **RADIO HANDBOOK**

tetrode, or other type of screen-grid tube. The modulation obtained in this way is not especially linear as the impedance of the screen grid with respect to the modulating signal is nonlinear. However, *screen-grid modulation* does offer other advantages and the linearity is quite adequate for communications work.

A screen-grid modulated r-f amplifier operates as an efficiency-modulated amplifier the same as does a class-B linear amplifier and a grid-modulated stage. The *plate circuit* loading is relatively critical as in any efficiency-modulated stage, and must be adjusted to the correct value if normal power output with full modulation capability is to be obtained. As in the case of any efficiency-modulated stage, the operating efficiency at the peak of the modulation cycle will be between 70 and 80 percent, with efficiency at the carrier level (if the stage is operating in the normal manner with full carrier) about half of the peak-modulation value.

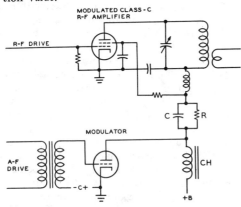

Figure 17

HEISING PLATE MODULATION

This type of modulation was the first form of plate modulation. It is sometimes known as "constant-current" modulation. Because of the effective 1:1 ratio of the coupling choke, it is impossible to obtain 100 percent modulation unless the plate voltage to the modulated stage is dropped slightly by resistor R. The capacitor (C) merely bypasses the audio around R, so that the full a-f output voltage of the modulator is impressed on the class-C stage.

Suppressor-Grid Modulation   Still another form of efficiency modulation may be obtained by applying the audio modulating signal to the suppressor

grid of a pentode class-C r-f amplifier. Basically, *suppressor-grid modulation* operates in the same general manner as other forms of efficiency modulation; carrier plate-circuit efficiency is about 35 percent, and antenna coupling must be rather heavy.

The suppressor grid is biased negatively to a value which reduces the plate-circuit efficiency to about one-half the maximum obtainable from the particular amplifier, with antenna coupling adjusted until the plate input is about 1.5 times the rated plate dissipation of the stage.

Audio signal is applied to the suppressor grid. In the normal application the audio voltage swing on the suppressor will be somewhat greater than the negative bias on the element. Suppressor-grid current will flow on modulation peaks, so that the source of audio signal voltage must have good regulation.

15-5   Input Modulation Systems

Constant-efficiency variable-input modulation systems operate by virtue of the addition of external power to the modulated stage to effect the modulation. There are two general classifications that come under this heading; those systems in which the additional power is supplied as audio-frequency energy from a modulator (usually called *plate-modulation systems*) and those systems in which the additional power to effect modulation is supplied as direct current from the plate supply.

Modulation systems coming under the second classification have been widely applied to broadcast work. There are quite a few systems in this class. Two of the more widely used are the *Doherty* linear amplifier, and the *Terman-Woolyard* high-efficiency grid-modulated amplifier. Both systems operate by virtue of a carrier amplifier and a peak amplifier connected together by electrical quarter-wave lines. They will be described later in this section.

Plate Modulation   Plate modulation is the application of the audio power to the *plate circuit* of an r-f amplifier. The r-f amplifier must be operated class C for this type of modulation in order to obtain a radio-frequency output which changes in exact accord with the variation in plate voltage. *The r-f amplifier is 100 percent modulated when the peak ac voltage from the modulator is equal to the dc voltage applied to the r-f tube.* The positive peaks of audio voltage increase the instantaneous plate voltage on the r-f tube to *twice* the dc value, and the negative peaks reduce the voltage to zero.

The instantaneous plate *current* to the r-f stage also varies in accord with the modulating voltage. The peak alternating current in the output of a modulator must be equal to the dc plate current of the class-C r-f stage at the point of 100 percent modulation. This combination of change in audio voltage and current can be most easily referred to in terms of *audio power in watts.*

By properly matching the plate impedance of the r-f tube to the output of the modulator, the ratio of voltage and current swing to dc voltage and current is automatically obtained. The modulator should have a peak voltage output equal to the average dc plate voltage on the modulated stage. The modulator should also have a *peak* power output equal to the dc plate input power to the modulated stage.

Heising modulation is the oldest system of plate modulation, and usually consists of

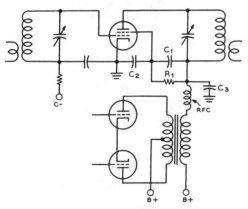

Figure 18

CLASS-B PLATE MODULATION

This type of modulation is the most flexible in that the loading adjustment can be made in a short period of time and without elaborate test equipment after a change in operating frequency of the class-C amplifier has been made.

a class-A audio amplifier coupled to the r-f amplifier by means of a modulation choke, as shown in figure 17.

Class-B Plate Modulation   High-level class-B plate modulation is the least expensive method of plate modulation. Figure 18 shows a conventional class-B plate-modulated class-C amplifier.

The statement that the modulator output power must be one-half the class-C input for 100 percent modulation is correct only if the waveform of the modulating power is a *sine wave.* Where the modulator waveform is unclipped speech waveforms, the average modulator power for 100 percent modulation is considerably less than one-half the class-C input.

Power Relations in Speech Waveforms   It has been determined experimentally that the ratio of peak-to-average power in a speech waveform is approximately 4 to 1 as contrasted to a ratio of 2 to 1 in a sine wave. This is due to the high harmonic content of such waveform, and to the fact that this high harmonic content manifests itself by making the wave unsymmetrical and causing sharp peaks of high energy content to appear. Thus for unclipped speech, the *average* modulator plate current, plate dissipation, and power output are approximately one-half the sine wave values for a given *peak* output power.

For 100 percent modulation, the *peak* (instantaneous) audio power must equal the class-C input, although the average power for this value of peak varies widely depending on the modulation waveform, being greater than 50 percent for speech that has been clipped and filtered, 50 percent for a sine wave, and about 25 percent for typical unclipped speech tones.

Plate-and-Screen Modulation   When *only* the plate of a tetrode tube is modulated, it is difficult to obtain high-percentage linear modulation under ordinary conditions. The plate current of such a stage is not linear with plate voltage. However, if the screen is modulated simultaneously with the plate, the instantaneous screen voltage drops in proportion to the drop in the plate voltage, and linear modulation can then be obtained (figure 18).

The screen r-f bypass capacitor (C_2) should not have a greater value than 0.005 μF, preferably not larger than 0.001 μF. It should be large enough to bypass effectively all r-f voltage without short-circuiting high-frequency audio voltages. The plate bypass capacitor can be of any value from 0.002 μF to 0.005 μF. The screen-dropping resistor (R_1) should reduce the applied high voltage to the value specified for operating the particular tube in the circuit.

15-6 The Doherty and the Terman-Woodyard Modulated Amplifiers

These two amplifiers will be described together since they operate on very similar principles. Figure 19 shows a greatly simplified schematic diagram of the operation of both types. Both systems operate by virtue of a *carrier tube* (V_1 in both figures 19 and 20), which supplies the unmodulated carrier, and whose output is reduced to supply negative peaks, and *a peak tube* (V_2), whose function is to supply approximately half the positive peak of the modulation cycle and whose additional function is to lower the load impedance on the carrier tube so that it will be able to supply the other half of the positive peak of the modulation cycle.

The peak tube is able to increase the output of the carrier tube by virtue of an impedance-inverting line between the plate circuits of the two tubes. This line is designed to have a characteristic impedance of one-half the value of load into which the carrier tube operates under the carrier conditions. Then a load of one-half the characteristic impedance of the quarter-wave line

is coupled into the output. It is known that a quarter-wave line will vary the impedance at one end of the line in such a manner that the geometric mean between the two terminal impedances will be equal to the characteristic impedance of the line. Thus, if a value of load of *one-half* the characteristic impedance of the line is placed at one end, the other end of the line will present a value of *twice* the characteristic impedance of the lines to carrier tube V_1.

This is the situation that exists under the carrier conditions when the peak tube merely floats across the load end of the line and contributes no power. Then as a positive peak of modulation comes along, the peak tube starts to contribute power to the load until at the peak of the modulation cycle it is contributing enough power so that the impedance at the load end of the line is equal to R, instead of the $R/2$ that is presented under the carrier conditions. This is true because at a positive modulation peak (since it is delivering full power) the peak tube subtracts a negative resistance of $R/2$ from the load end of the line.

Now, since under the peak condition of modulation the load end of the line is terminated in R ohms instead of $R/2$, the impedance at the *carrier-tube* will be *reduced* from $2R$ ohms to R ohms. This again is due to the impedance-inverting action of the line. Since the load resistance on the carrier tube has been reduced to half the carrier value, its output at the peak of the modulation cycle will be doubled. Thus the necessary condition for a 100 percent modulation peak exists and the amplifier will deliver four times as much power as it does under the carrier conditions.

On negative modulation peaks the peak tube does not contribute; the output of the carrier tube is reduced until, on a 100 percent negative peak, its output is zero.

The Electrical Quarter-Wave Line   While an electrical quarter-wave line (consisting of a pi network with the inductance and capacitance units having a reactance equal to the characteristic impedance of the line) does have the desired impedance-inverting effect, it also has the undesirable effect of introducing a 90° phase shift across such a line. If the shunt elements

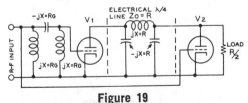

Figure 19

DIAGRAMMATIC REPRESENTATION OF THE DOHERTY LINEAR

are capacitances, the phase shift across the line lags by 90°; if they are inductances, the phase shift leads by 90°. Since there is an undesirable phase shift of 90° between the plate circuits of the carrier and peak tubes, an equal and opposite phase shift must be introduced in the exciting voltage of the grid circuits of the two tubes so that the resultant output in the plate circuit will be in phase. This additional phase shift has been indicated in figure 19 and a method of obtaining it has been shown in figure 20.

Comparison Between The difference between
Doherty and Terman- the Doherty linear am-
Woodyard Amplifiers plifier and the Terman-
Woodyard grid - modu-
lated amplifier is the same as the difference between any linear and grid-modulated stages. Modulated r-f is applied to the grid circuit of the Doherty linear amplifier with the carrier tube biased to cutoff and the peak tube biased to the point where it draws substantially zero plate current at the carrier condition.

In the Terman-Woodyard grid-modulated amplifier the carrier tube runs class-C with comparatively high bias and high plate effi-

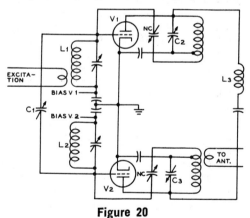

Figure 20

SIMPLIFIED SCHEMATIC OF A "HIGH-EFFICIENCY" AMPLIFIER

The basic system, comprising a "carrier" tube (V₁) and a "peak" tube (V₂) interconnected by lumped-constant quarter-wave lines, is the same for either grid-bias modulation or for use as a linear amplifier of a modulated wave.

ciency, while the peak tube again is biased so that it draws almost no plate current.

Unmodulated r-f is applied to the grid circuits of the two tubes and the modulating voltage is inserted in series with the fixed bias voltages. From one-half to two-thirds as much *audio* voltage is required at the grid of the peak tube as is required at the grid of the carrier tube.

Operating The resting carrier efficiency
Efficiencies of the grid-modulated amplifier
may run as high as is obtainable in any class-C stage—80 percent or better. The resting carrier efficiency of the linear will be about as good as is obtainable in any class-B amplifier—60 to 70 percent. The overall efficiency of the bias-modulated amplifier at 100 percent modulation will run about 75 percent; of the linear—about 60 percent.

In figure 20 the plate tank circuits are detuned enough to give an effect equivalent to the shunt elements of the quarter-wave "line" of figure 19. At resonance, coils L_1 and L_2 in the grid circuits of the two tubes have each an inductive reactance equal to the capacitive reactance of capacitor C_1. Thus we have the effect of a pi network consisting of shunt inductances and series capacitance. In the plate circuit we want a phase shift of the same magnitude but in the opposite direction; so our series element is inductance L_3 whose reactance is equal to the characteristic impedance desired of the network. Then the plate tank capacitors of the two tubes (C_2 and C_3) are increased an amount past resonance, so that they have a capacitive reactance equal to the inductive reactance of the coil L_3. It is quite important that there be no coupling between the inductors.

Other High-Efficiency Many other high-effi-
Modulation Systems ciency modulation
systems have been described since about 1936. The majority of these, however, have received little application either by commercial interests or by amateurs. Nearly all of these circuits have been published in the *Proceedings of the IRE* (now IEEE) and the interested reader can refer to them in back copies of that journal.

Pulse-Duration Modulation   A recent innovation in high-level plate modulation is the *pulse-duration modulation* technique wherein the modulator tube is operated in a saturated switching mode and is placed in series with the r-f power tube.

The plate modulator in a conventional a-m transmitter operates in a linear mode that may be compared to an analog system. In the pulse-duration modulator, the modulator operates in a switching mode that may be compared to a digital computer, having two conditions: *off* and *on*. Audio information is contained in the duration of the *on* pulse.

Audio amplitude is determined by the duty cycle of the modulator tube. A square-wave signal of about 70 kHz is pulse-width modulated by the audio signal, whose amplitude causes the symmetry of the square

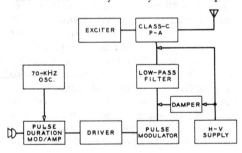

Figure 21

GATES PULSE-DURATION MODULATION SYSTEM

The audio signal is combined with a 70-kHz square-wave signal and processed to produce a modulated pulse-width modulated train which is amplified and applied to the cathode of the class-C r-f amplifier through a low-pass filter that removes the 70-kHz signal and its sidebands, thereby recovering the original audio. The modulator tube acts like a variable resistance whose value varies with the amplitude and frequency of the applied audio signal. The driver stages for the modulator are simple "on"-"off" switches. A damper diode is connected between the output of the modulator and the high-voltage supply to conduct when the modulator does not.

wave to vary. The audio signal is imposed on the 70-kHz square wave train at a low level and the resulting signal is amplified to the modulating level. The square-wave component is then filtered out to leave the amplified audio voltage, plus a dc component that is the modulated plate voltage

for the class-C amplifier. This technique eliminates the need of a modulation transformer and modulation choke.

A block diagram of the *Gates* VP-100 pulse-duration modulated a-m transmitter is shown in figure 21.

15-7   Spread-Spectrum Modulation

In conventional communications a bandwidth is generally used that is just wide enough to transmit the information involved. The *spread-spectrum* technique, on the other hand, uses a much larger transmission bandwidth than the information bandwidth being communicated. The spread-spectrum system thus makes use of some function other than information bandwidth to establish the transmitted signal bandwidth. Current spread-spectrum systems use a transmitted bandwidth up to a million times the information bandwidth.

One of the immediate advantages of spread-spectrum distribution of a signal over a great bandwidth is that power density (watts per Hz) is lowered by the same amount that the spectrum is widened. This interchange of power density for spectrum space can reach a point where signals can be transmitted and received while hidden many decibels below the background noise. Obviously, such low-density signals can reduce the problem of message interception, while at the same time preventing interference to other circuits. For civilian as well as military networks, spread-spectrum systems allow many users to share a single channel.

Information to be transmitted by spread-spectrum techniques is first converted into digital data to provide a primary modulation of the carrier. A secondary modulation of much wider bandwidth is then applied to the carrier to spread the spectrum of the primary modulation (figure 22). A pseudo-random noise generator (PRN) is one method of establishing the spectrum spread. Frequency-hopping may also be used.

The total energy expended is the same both in the spread and the conventional unspread signals. An important difference is that the power density in the former system

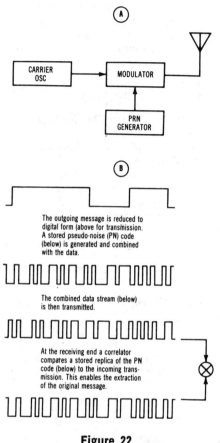

Figure 22

SPREAD-SPECTRUM MODULATION

A—Block diagram of spread-spectrum
transmitter.
B—PRN modulation and demodulation.

is distributed over a wider area of the frequency spectrum. Various space satellites rely on spreading transmissions over wide bandwidths to provide high resistance to jamming, security, and multiple access.

15-8   A-M Stereo Transmission

Many a-m broadcast stations have seen a steady erosion of their audience as the interest in f-m stereo has grown. The added dimension of stereo might recover some of the lost audience.

A-m stereo was first demonstrated in 1925 but before it became practical, interest shifted to f-m and later to stereo f-m. Only recently has interest in stereo a-m been revived.

A variety of techniques exist to generate stereo a-m. One of the simplest systems is shown in figure 23. In this composite modulation system, the constant frequency a-m signal carries the $L+R$ channel combination and the variable frequency (f-m) signal within the a-m envelope carries the $L-R$ signal. Channel bandwidth is about 12 KHz to accommodate the significant sidebands arising from the composite modulation stereo system.

Reception is accomplished with a special receiver having both an a-m and an f-m detector to derive the left and right channel information.

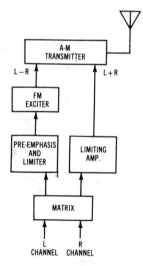

Figure 23

BLOCK DIAGRAM OF A COMPATIBLE A-M STEREO TRANSMITTER

15-9   Practical High Level Modulation

A High-Power Modulator with Beam Tetrodes Listed in Table 1 are representative operating conditions for various tetrode tubes providing power levels up to 1500 watts of audio. Complete operating data on these tube types may be obtained from the manufacturer.

Because of the power level involved and the design of the external-anode tube, the 4X150A/4CX250B tubes must be forced-air cooled in this application. It is recommended that the 4-250A, and 4-400A tubes be convection cooled with a small fan.

Triode Class-B Modulators High level class-B modulators can make use of triodes such as the 811A or 810 tubes with operating plate voltages between 750 and 2500 (Table 2). Because of the grid driving power required, a low-impedance driving source is necessary for this class of service. Push-pull, low-μ triodes, such as the 2A3 type are commonly used as the

driver stage. A well-regulated bias supply for the class-B stage is also required, as the triodes draw heavy grid current when the grids are driven into the positive region.

15-10 Solid-State Modulation Circuitry

Transistors and other solid-state devices are useful in amplitude-modulated service and the great majority of CB and Aeronautical Radio Service equipment is a-m, solid state.

The modulation requirements of transistor r-f power amplifiers are similar to those for tube amplifiers. For good modulation

Table 1. Typical Operating Conditions for Class AB, Tetrode Modulator

Tubes	Plate Volts	Screen Volts	Grid Bias	P-P Load (Ohms)	Plate Current (mA)	Sine Wave Power Output (Watts)
4-125A	2500	600	—96	20,300	235	330
4-250A	3000	600	—116	15,000	420	750
4-400A	4000	750	—150	14,000	580	1540
4CX250B	2000	350	—55	9,500	500	600

Table 2. Typical Operating Conditions For General-Purpose Modulator

Class-B Tubes	Plate Voltage	Grid Bias (Volts)	Plate Current (mA)	Plate-To-Plate Load (Ohms)	Sine Wave Power Output (Watts)
811-A	750	0	30-350	5,100	175
811-A	1000	0	45-350	7,400	245
811-A	1250	0	50-350	9,200	310
811-A	1500	—4.5	32-315	12,400	340
810	2000	—50	60-420	12,000	450
810	2500	—75	50-420	17,500	500

linearity, class-C operation of the device is commonly used. In this class of service, the collector conduction angle is less than 180 degrees. This corresponds to the same mode of operation in a vacuum tube class-C amplifier.

For class-C operation, the base-emitter junction of the transistor must be reverse-biased so that the collector quiescent current is near-zero during zero signal conditions. This may be accomplished by the application of a reverse dc bias to the base, or the reverse bias may be obtained by the flow of dc base current through a resistance (figure 24A) or by the use of self-bias developed across an emitter resistor (figure 24B). Because no external base resistance is added in the latter circuit, the collector-emitter

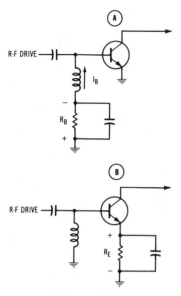

Figure 24

**CLASS-C MODULATION OF
SOLID-STATE AMPLIFIER**

A—Base-emitter junction bias is obtained by IR
drop through base resistance.
B—Self-bias is developed across emitter resistor.

breakdown voltage is not affected. The base
bias resistor must be carefully bypassed to
provide a low impedance r-f path to ground
to prevent degeneration of stage gain at the
signal frequency.

Simple amplitude modulation of the collector supply of a transistor output stage
does not result in full modulation as a portion of the r-f drive feeds through the transistor and is unmodulated. Better modulation
characteristics can be obtained by modulation of the collector supply of the last two
stages in the transmitter chain (figure 25).

On the downward portion of the modulation cycle, drive from the preceding modulated stage is reduced, and less feedthrough
power appears in the output. Flattening of
the r-f output on the upward portion of the
modulation cycle is reduced because of the
increased drive from the modulated lower-
level stage.

The modulated stages must be operated at
half their normal voltage levels to avoid
high collector voltage swings that may exceed transistor collector-emitter breakdown
ratings.

Amplitude modulation of a solid-state
transmitter may also be obtained by modulation of a lower-level stage and operation
of the higher-level stages in a linear mode.

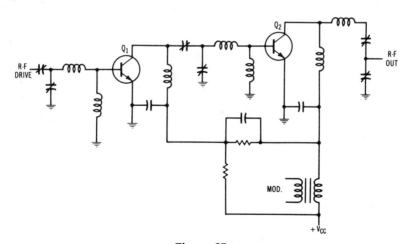

Figure 25

**FULL MODULATION IS ACCOMPLISHED BY MODULATION OF THE
COLLECTOR SUPPLY OF TWO STAGES IN THE TRANSMITTER CHAIN.**

CHAPTER SIXTEEN

Radio Interference (RFI)

The radio amateur may be the cause, or the victim, of *radio frequency interference* (RFI). Equally toublesome is the fact that he may be accused of creating RFI for which he is not responsible.

In 1980 the Federal Communications Commission received over 60,000 complaints of RFI. The greater percentage of these involved home-entertainment equipment of which a large portion had no provision for protection from nearby r-f energy. Basic design deficiencies in most equipment of this type, therefore, are a cause of a great deal of the RFI that is reported.

Even while only a small proportion of the population lives in the vicinity of a radio amateur, the tremendous growth in radio communications over the past decade has resulted in a high density of radio transmitters in urban and suburban areas. In addition to radio amateurs, there are over one million transmitters operating in the Citizens Radio Service, in addition to hundreds of thousands of transmitters in the Land Mobile Service and the television and broadcast service. In addition there are thousands of transmitters in the military, microwave-repeater, and maritime services, all of which could be potential sources of radio frequency interference to a poorly designed electronic device.

A second type of prevalent RFI is *radio noise*. Impulse noise generated by a spark discharge or by solid-state switching devices creates an annoying type of interference that can be transmitted for many miles by conduction and radiation. A serious form of impulse noise is power line interference, with appliance interference as an additional source of widespread radio noise.

Many of the problems associated with RFI could be alleviated if there was some control over spurious r-f emissions and if technical standards were set for the protection of electronic equipment against unwanted radiation. Unfortunately, this is not being done at the present time. The burden of RFI, then, falls mainly upon the radio amateur, as he is a visible source of FRI to his neighbors and—at the same time—uniquely qualified to assist his neighbors in understanding and correcting RFI problems.

16-1   Television Interference

Television interference (TVI) is an annoyance to many viewers. More likely than not, TVI is often blamed on the amateur, regardless of the cause. Over the years, amateur transmitting equipment has been designed with the idea in mind of reducing TVI-causing harmonics and spurious emissions and, as a result, moden SSB equipment is relatively TVI-free. The FCC reports that over 90 percent of all TVI complaints can be cured only at the TV receiver. If your own TV set is free of interference from your station, it is likely that interference to a more distant TV receiver at your neighbors' home is not the fault of your equipment. All amateur equipment, however, is not TVI-free and certain precautions must be taken to make sure that your station does not cause interference to nearby television receivers.

Types of TV Interference   There are two main causes of TVI which may occur singly or in combination as caused by emissions from an amateur transmitter. These causes are:

1. Overload of the television receiver by the fundamental signal of the transmitter.

2. Impairment of the TV picture by either spurious emissions or harmonic radiation from the transmitter.

In the first instance, the television receiver can be protected by the addition of a high-pass filter in the antenna feed line, directly at the receiver. In the second instance, filtering of transmitter circuits and/or circuit modifications to the transmitting equipment are called for.

TV Receiver Overload   Even if the amateur transmitter were perfect and had no harmonic or spurious emission whatsoever, it could still cause overloading to a TV receiver whose antenna is within a few hundred feet of the transmitting antenna. The overload is caused by the fact that the field intensity in the immediate vicinity of the transmitting antenna is sufficiently high so that the selective circuits of the TV receiver cannot reject the signal which is greater than the dynamic range the receiver can accept. Spurious responses are then generated within the television receiver that cause severe interference. A characteristic of this type of interference is that it will always be eliminated when the transmitter in question is operated into a dummy antenna. Another characteristic of this type of overloading is that its effects are substantially continuous over the entire frequency range of the TV receiver, all channels being affected to approximately the same degree.

The problem, then, is to keep the fundamental signal of the amateur transmitter out of the affected receiver. (Other types of interference may or may not show up when the fundamental signal is eliminated, but at least the fundamental signal must be eliminated first).

Elimination of the fundamental signal from the television receiver is normally the only operation performed on or in the vicinity of the receiver. After this has been accomplished, work may then begin on the transmitter toward eliminating this as the cause of the other type of interference.

Removing the Fundamental Signal   A strong signal, out of the passband of the television receiver, can cause objectionable interference to either the picture or the audio signal, or both. The interference may be caused by crossmodula-

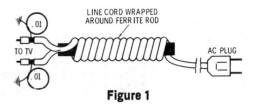

Figure 1

LINE FILTER FOR TV RECEIVER

The line cord of the TV receiver is bypassed at the chassis with two .01-μF, 1.6-kV ceramic capacitors and a portion of the line is wound around a ½-inch diameter ferrite rod to form a simple r-f choke. Cord may be held in position about the rod with vinyl electrical tape.

tion within the receiver, interference with the audio or i-f circuitry, or mixing of the local signal with other strong nearby signals. The interfering signal, or signals, can enter the TV receiver via the antenna circuit or via the power line. It is possible to install suitable filters in these leads to reduce, or eliminate, the interfering signal.

The Power Line Filter—The power line can act as an antenna, picking up a nearby signal and radiating it within the sensitive circuits of the TV receiver. If the interference continues when the antenna is removed from the television receiver, it is probable that the signal is entering the set via the ac power line. A filter of the type used to suppress electric shavers, vacuum cleaners, etc., placed in the power line *at the receiver* may remove this interference. Alternatively, the power line should be bypassed to the chassis of the receiver as shown in figure 1 and the line cord formed into an r-f choke by winding the cord around a ferrite antenna rod. Make sure the capacitors are rated for continuous operation under ac conditions.

The Antenna Filter—Fundamental overloading can be prevented by reducing the nearby signal to such a level that the selective circuits of the television receiver can reject it. A high-pass filter in the antenna lead of the TV set can accomplish this task, in most cases. The filter, having a cutoff frequency between 30 MHz and 54 MHz is installed *at the tuner input terminals* of the receiver. Design data for suitable filters are given in figure 2. The filters should preferably be built in a small shielded box for highest rejection, although "open-air" filters work quite well if maximum rejection is not required. The series-connected capacitors are

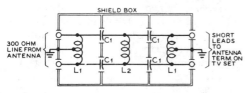

(A) FOR 300-OHM LINE, SHIELDED OR UNSHIELDED

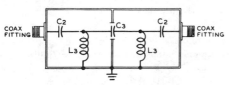

(B) FOR 50-75 OHM COAXIAL LINE

Figure 2

HIGH-PASS TRANSMISSION LINE FILTERS

The arrangement of A will stop the passing of all signals below about 45 MHz from the antenna transmission line into the TV set. Coils L_1 are each 1.2 microhenrys (17 turns No. 24 enam. closewound on ¼-inch dia. polystyrene rod) with the center tap grounded. It will be found best to scrape, twist, and solder the center tap before winding the coil. The number of turns each side of the tap may then be varied until the tap is in the exact center of the winding. Coil L_2 is 0.6 microhenry (12 turns No. 24 enam. closewound on ¼-inch dia. polystyrene rod). The capacitors should be about 16.5 pF, but either 15- or 20-pF ceramic capacitors will give satisfactory results. A similar filter for coaxial antenna transmission line is shown at B. Both coils should be 0.12 microhenry (7 turns No. 18 enam. spaced to ½ inch on ¼-inch dia. polystyrene rod). Capacitors C_2 should be 75-pF midget ceramics, while C_1 should be a 40-pF ceramic.

mounted in holes cut in the interior shields of the box, if such an assembly is used. Various commercial filters are also available. Input and output terminals of the filter may be standard TV connectors, or the inexpensive terminal strips usually employed on "ribbon" lines.

Operation on the 50-MHz amateur band in an area where TV channel 2 is in use imposes a special problem in the matter of receiver blocking. High-pass filters of the normal type simply are not capable of giving sufficient protection to channel 2 from a strong 50-MHz signal whose frequency is so close to the necessary passband of the filter. In this case, a resonant circuit element, such as shown in figure 3 is recommended to trap out the transmitter signal at the input of the television set. The stub

is selective and therefore protects the television receiver only over a small range of frequencies in the 50-MHz band. The stub is trimmed for minimum TVI while the transmitter is tuned to the most-used operating frequency.

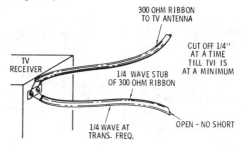

Figure 3

RESONANT STUB FOR 50-MHz PROTECTION

A ¼-wave open stub will provide protection against a local 50 MHz transmitter. The stub is placed in parallel with the 300-ohm ribbon line at the antenna terminals of the TV set. Using line with a velocity of propagation of 0.84, the line should be about 50 inches long. It is trimmed a quarter-inch at a time for minimum TVI. If it is too short, it will affect reception of TV channel 2.

Transmission Line Pickup—In most cases, the "ribbon line" connecting the antenna to the television receiver is longer in terms of wavelengths than the TV antenna is, especially at the high frequencies represented by the amateur bands up through 6 meters. Thus, the transmission line will actually pick up more energy from a nearby amateur transmitter than will the TV antenna.

The induced currents flowing in the TV line flow in parallel and in phase, the two-wire line acting as a single wire antenna. Most TV input circuits respond strongly to such parallel currents and the nearby signal at the input circuit of the tuner is much stronger than if the interference were only picked up by the relatively small TV antenna.

A solution to this form of overload is to use a shielded transmission line from the antenna to the television receiver. Balanced, *twinax* 300-ohm line is readily available, or coaxial line may be used for an unbalanced feed system. In either case, the outer shield of the line should be grounded to the TV receiver chassis.

16-2 Harmonic Radiation

After any condition of blocking at the TV receiver has been eliminated, and when the transmitter is completely free of transients and parasitic oscillations, it is probable that TVI will be eliminated in many cases. Certainly general interference should be eliminated, particularly if the transmitter is a well-designed affair operated on one of the lower frequency bands, and the station is in a high-signal TV area. But when the transmitter is to be operated on one of the higher frequency bands, and particularly in a marginal TV area, the job of TVI-proofing will just have begun. The elimination of harmonic radiation from the transmitter is a job which must be done in an orderly manner if completely satisfactory results are to be obtained.

First it is well to become familiar with the TV channels presently assigned, with the TV intermediate frequencies commonly used, and with the channels which will receive interference from harmonics of the various amateur bands. Figures 4 and 5 give this information.

Even a short inspection of figures 4 and 5 will make obvious the seriousness of the interference which can be caused by harmonics of amateur signals in the higher frequency bands. With any sort of reasonable precautions in the design and shielding of the transmitter it is not likely that harmonics higher than the 6th will be encountered. For this reason, the most frequently found offenders in the way of harmonic interference will almost invariably be those bands above 14 MHz.

Nature of Harmonic Interference Investigations into the nature of the interference caused by amateur signals on the TV screen, assuming that blocking has been eliminated as described earlier in this chapter, have revealed the following facts:

1. An unmodulated carrier, such as a c-w signal with the key down or an a-m signal without modulation, will give a crosshatch or herringbone pattern on the TV screen. This same general type of picture also will occur in the case of a narrow-band f-m signal either with or without modulation.

2. A relatively strong a-m or SSB signal will give in addition to the herringbone

TRANSMITTER FUNDAMENTAL	2ND	3RD	4TH	5TH	6TH	7TH	8TH	9TH	10TH
7.0–7.3		21–21.9 TV I.F.			42–44 TV I.F.		56–58.4 CHANNEL ②	63–65.7 CHANNEL ③	70–73 CHANNEL ④
14.0–14.35		42–43 TV I.F.	56–57.6 CHANNEL ②	70–72 CHANNEL ④	84–86.4 CHANNEL ⑥	98–100.8 F-M BROADCAST			
21.0–21.45		63–64.35 CHANNEL ③	84–85.8 CHANNEL ⑥	105–107.25 F-M BROADCAST				189–193 CHANNELS ⑨ ⑩	210–214.5 CHANNEL ⑬
28.0–29.7	56–59.4 CHANNEL ②	84–89.1 CHANNEL ⑥			168–178.2 CHANNEL ⑦	196–207.9 CHANNELS ⑩ ⑪ ⑫			
50.0–54.0	100–108 F-M BROADCAST		200–216 CHANNELS ⑪ ⑫ ⑬				450–486　500–540 POSSIBLE INTERFERENCE TO UHF CHANNELS		

Figure 4

HARMONICS OF THE AMATEUR BANDS

Shown are the harmonic frequency ranges of the amateur bands between 7 and 54 MHz, with the TV channels (and TV i-f systems) which are most likely to receive interference from these harmonics. Under certain conditions amateur signals in the 1.8- and 3.5-MHz bands can cause interference as a result of direct pickup in the video systems of TV receivers which are not adequately shielded.

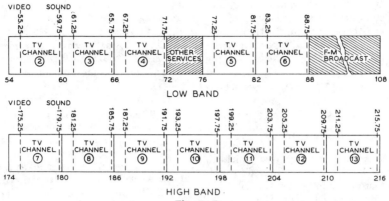

Figure 5

FREQUENCIES OF THE VHF TV CHANNELS

Showing the frequency ranges of TV channels 2 through 13, with the picture carrier and sound carrier frequencies also shown.

a very serious succession of light and dark bands across the TV picture.

3. A moderate strength c-w signal without transients, in the absence of overloading of the TV set, will result merely in the turning on and off of the herringbone on the picture.

To discuss condition 1 above, the herringbone is a result of the beat note between the TV video carrier and the amateur harmonic. Hence the higher the beat note the less obvious will be the resulting crosshatch. Further, it has been shown that a much stronger signal is required to produce a discernible herringbone when the interfering harmonic is as far away as possible from the video carrier, without running into the sound carrier. Thus, as a last resort, or to eliminate the last vestige of interference after all corrective measures have been taken, operate the transmitter on a frequency such that the interfering harmonic will fall as far as possible from the picture carrier. The worst possible interference to the picture from a continuous carrier will be obtained when the interfering signal is very close in frequency to the video carrier.

Isolating the Source of the Interference Throughout the testing procedure it will be necessary to have some sort of indicating device as a means of determining harmonic field intensities. The best indicator, of course, is a nearby television receiver. The home receiver may be borrowed for these tests. A portable "rabbit ears" antenna is useful since it may be moved about the transmitter site to examine the intensity of the interfering harmonics.

The first step is to turn on the transmitter and check all TV channels to determine the extent of the interference and the number of channels affected. Then disconnect the transmitting antenna and substitute a shielded dummy load, noting the change in interference level, if any. Now, remove excitation from the final stage of the transmitter, and determine the extent of interference caused by the exciter stages.

In most cases, it will be found that the interference drops materially when the transmitting antenna is removed and a dummy load substituted. It may also be found that the interference level is relatively constant, regardless of the operation of the output stage of the transmitter. In rare cases, it may be found that a particular stage in the transmitter is causing the interference and corrective measures may be applied to this stage. The common case, however, is general TVI radiating from antenna, cabinet, and power leads of the transmitter.

The first corrective measure is to properly bypass the transmitter power leads before they leave the cabinet. Each lead should be bypassed to chassis ground with a .01-μF, 1.6-kV ceramic capacitor, or run through a 0.1-μF, 600-volt feedthrough (*Hypass*) capacitor. If possible, the transmitter chassis should be connected to an external ground.

The next step is to check transmitter shielding. Paint should be removed from mating surfaces wherever possible and the cabinet should be made as "r-f tight" as possible in the manner discussed in Chapter 33.

16-3 Low Pass Filters

After the transmitter has been shielded, and all power leads have been filtered in such a manner that the transmitter shielding has not been rendered ineffective, the only remaining available exit for harmonic energy lies in the antenna transmission line. Thus, the main burden of harmonic attenuation will fall on the low-pass filter installed between the output of the transmitter and the antenna system.

Experience has shown that the low-pass filter can best be installed externally to the main transmitter enclosure, and that the transmission line from the transmitter to the low-pass filter should be of the coaxial type.

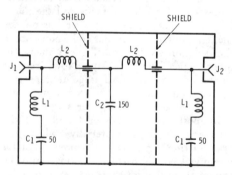

Figure 6

SIMPLE LOW-PASS FILTER FOR 1.8- TO 30-MHz TRANSMITTER

This filter is suitable for high frequency transmitters of up to 2 kW PEP power operating up to 30 MHz. Capacitors designated C₁ are 50-pF, 5-kV ceramic units (Centralab type 850S-50Z) Capacitor C₂ is composed of two 75-pF, 5-kV units connected in parallel (Centralab type 850S-75N). Coils designated L₁ are 4 turns of #12 enamel wire, ½-inch inside diameter, ½-inch long. Coils designated L₂ are 7 turns wound as same as L₁ and about 1 inch long. Coils L₁ and L₂ are mounted at right angles to each other. The filter is designed for use in a 50-ohm coaxial line. Receptacles J₁ and J₂ are matching units, such as SO-239 for type PL-259 plugs.

As a result, the majority of low-pass filters are designed for a characteristic impedance of 50 ohms, so that RG-8/U cable (or RG-58/U for a small transmitter) may be used between the output of the transmitter and the antenna transmission line or the antenna tuner.

Transmitting-type low-pass filters for amateur use usually are designed in such a manner as to pass frequencies up to about 30 MHz without attenuation. The nominal cutoff frequency of the filters is usually between 38 and 45 MHz, and *m*-derived sections with maximum attenuation in channel 2 usually are included. Well-designed filters capable of carrying any power level up to one kilowatt are available commercially from several manufacturers. Alternatively, filters in kit form are available from several manufacturers at a somewhat lower price. Effective filters may be home constructed, if the test equipment is available and if sufficient care is taken in the construction of the assembly.

Construction of Low-Pass Filters Shown in figure 6 is a simple low-pass filter suitable for home construction. The filter provides at least 30 dB attenuation to all frequencies above 54 MHz when properly built and adjusted. The filter is built in a small aluminum utility box measuring 2¼" × 2¼" × 5". Two aluminum partitions are installed in the box to make three compartments. Small holes are drilled in the partition to pass the connecting leads.

The coils are self-supporting and wound of #14 enamel or *formvar* covered copper wire. The ceramic capacitors are bolted to the side of the box. Since appreciable r-f current flows through the capacitors, heavy-duty ceramic units of the type specified must be used. In the case of the center capacitor, two units connected in parallel by a ⅜-inch wide copper strap are used. The capacitors are placed side by side so that minimum strap length is achieved. The coils are connected between capacitor terminals and the coaxial fittings mounted on the end walls of the box.

Once the filter is complete, it is adjusted before the lid of the box is bolted in place. To check the end sections, the coaxial connectors are shorted out on the inside of the case with short leads and the resonant frequency of the end sections is checked with the aid of a grid dip meter. The coils L_1 should be squeezed or spread until resonance occurs between 56 and 57 MHz. The shorts are now removed and the cover placed on the box.

A High Performance Low-Pass Filter Figure 7 shows the construction and assembly of a high performance low-pass filter designed for a 50-ohm transmission line. The filter is built in a slip-cover aluminum box measuring 17″ × 3″ × 2⅝″. Five aluminum baffle plates have been bolted in the box to make six shielded sections.

The filter is designed for a nominal cutoff frequency of 45 MHz, with a frequency of maximum rejection of 57 MHz.

Either high power or low power components may be used in the filters. Using small zero-coefficient ceramic capacitors, power levels up to 100 watts ouput may be used provided the filter is terminated in a load having a low value of SWR. For higher power levels, *Centralab type 850S and 854S* capacitors, or equivalents, have proven suitable for power levels up to 2 kW, PEP at standing wave ratios less than 3 to 1.

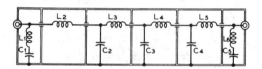

Figure 7

LOW-PASS FILTER

The filter uses m-derived terminating half sections at each end, with three constant-k mid-sections.
C_1, C_5—41.5 pf (40 pF will be found suitable.)
C_2, C_3, C_4—136 pf (130 to 140 pF may be used.)
L_1, L_6—0.2 µH; 3½ t. No. 14
L_2, L_5—0.3 µH; 5 t. No. 12
L_3, L_4—0.37 µH; 6½ t. No. 12

Capacitors C_1, C_2, C_4, and C_5 can be standard manufactured units with normal 5 percent tolerance. The coils for the end sections can be wound to the dimensions given (L_I and L_6). Then the resonant frequency of the series-resonant end sections should be checked with a grid-dip meter, after the adjacent input or output terminal has been

shorted with a very short lead. The coils should be squeezed or spread until resonance occurs at 57 MHz.

The coils in the intermediate sections of the filter (L_2, L_3, L_4, and L_5) may be checked most conveniently outside the filter unit with the aid of a small ceramic capacitor of known value and a grid-dip meter.

Using Low-Pass Filters The low-pass filter connected in the output transmission line of the transmitter is capable of affording an enormous degree of harmonic attenuation. However, the filter must be operated in the correct manner or the results obtained will not be up to expectations.

In the first place, all direct radiation from the transmitter and its control and power leads must be suppressed. This subject has been discussed in the previous section. Secondly, the filter must be operated into a load impedance approximately equal to its design characteristic impedance. The filter itself will have very low losses (usually less than 0.5 dB) when operated into its nominal value of resistive load. But if the filter is not terminated correctly, its losses will become excessive, and it will not present the correct value of load impedance to the transmitter.

If a filter being fed from a high-power transmitter is operated into an incorrect termination it may be damaged; the coils may be overheated and the capacitors destroyed as a result of excessive r-f currents. Thus, it is wise when first installing a low-pass filter, to check the standing-wave ratio of the load being presented to the output of the filter with a standing-wave bridge.

The Half-Wave Filter A *half-wave filter* is an effective device for TVI suppression and is easily built. It offers the advantage of presenting the same value of impedance at the input terminal as appears as a load across the output terminal. The filter is a single-band unit, offering high attenuation to the second- and higher-order harmonics. Design data for high-frequency half-wave filters is given in figure 8.

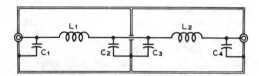

Figure 8

SCHEMATIC OF THE TWO-SECTION HALF-WAVE FILTER

The constants given below are for a characteristic impedance of 50 ohms, for use with RG-8/U and RG-58/U cable. Coil L_1 should be checked for resonance at the operating frequency with C_1, and the same with L_2 and C_4. This check can be made by soldering a low-inductance grounding strap to the lead between L_1 and L_2 where it passes through the shield. When the coils have been trimmed to resonance with a grid-dip meter, the grounding strap should of course be removed. This filter type will give an attenuation of about 30 dB to the second harmonic, about 48 dB to the third, about 60 dB to the fourth, 67 to the fifth, etc., increasing at a rate of about 30 dB per octave.

C_1, C_2, C_3, C_4—Silver mica or small ceramic for low power, transmitting type ceramic for high power. Capacitance for different bands is given below.

160 meters—1700 pF
 80 meters— 850 pF
 40 meters— 440 pF
 30 meters— 330 pF
 20 meters— 220 pF
 10 meters— 110 pF
  6 meters— 60 pF

Miniductor for power levels below 250 watts, or of No. 12 enam. for power up to one kilowatt. Approximate dimensions for the coils are given below, but the coils should be trimmed to resonate at the proper frequency with a grid-dip meter as discussed above. All coils except the ones for 160 meters are wound 8 turns per inch.

160 meters—4.2 μH; 22 turns No. 16 enam. 1" dia. 2" long
80 Meters—2.1 μH; 13 t. 1" dia. (No. 3014 Miniductor or No. 12 at 8 t.p.i.)
40 meters—1.1 μH; 8 t. 1" dia. (No. 3014 or No. 12 at 8 t.p.i.)
30 meters—0.8 μH; 8 t. ¾" dia. (No. 3010 or No. 12 at 8 t.p.i.)
20 meters—0.55 μH; 7 t. ¾" dia. (No. 3010 or No. 12 at 8 t.p.i.)
10 meters—0.3 μH; 6 t. ½" dia. (No. 3002 or No. 12 at 8 t.p.i.)
6 meters—0.17 μH; 4 t. ½" dia. (No. 3002 or No. 12 at t.p.i.)

A High-Power Filter for Six Meters The second and higher harmonics of a six-meter transmitter fall directly into the f-m and uhf and vhf television bands. An effective low-pass filter is required to adequately suppress unwanted transmitter emissions falling in these bands. Described in this section is a six-meter TVI filter rated at the two-kilowatt level which

provides better than 75 decibels suppression of the second harmonic and better than 60 decibels suppression of higher harmonics of a six-meter transmitter (figure 9). The unit is composed of a half-wave filter with added end sections which are tuned to 100 MHz and 200 MHz. An auxiliary filter ele-

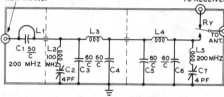

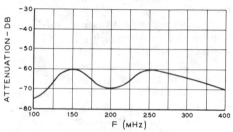

Figure 9

SIX METER TVI FILTER

C_1—50-pF Centralab 8505-50Z. Resonates with L_1 to 200 MHz.
C_2, C_7—4-pF piston capacitor. JFD type VC-4G.
C_3, C_4, C_5, C_6—60 pF. Three 20-pF capacitors in parallel. Centralab 853A-20Z.
L_1—Copper strap, ½" wide, 2¼" long, 1⅞" between mounting holes, approximately 0.01" thick. Strap is bent in U-shape around capacitor and bolted to capacitor terminals.
L_2—11 turns #18 enam. wire, ¼" diameter, ¾" long, airwound. Resonates to 100 MHz with capacitor C_2.
L_3, L_4—3 turns 3/16" tubing, 1¼" i.d., spaced to occupy about 2½". Turns are adjusted to resonate each section at 50 MHz.
L_5—6 turns #18 enam. wire, ¼" diameter, ⅝" long, airwound. Resonates to 200 MHz with capacitor C_7.

ment in series with the input is tuned to 200 MHz to provide additional protection to television channels 11, 12, and 13.

The filter (figure 10) is built in an aluminum box measuring 4" $\times$ 4" $\times$ 10" and uses *type-N* coaxial fittings. The half-wave filter coils are wound of 3/16-inch diameter copper tubing and have large copper lugs soldered to the ends. The 60-pF capacitors are made up of three 20-pF, 5kV ceramic units in parallel. A small sheet of copper is cut in triangular shape and joins the capacitor terminals and a coil lug is attached to the center of the triangle with heavy brass bolts.

The parallel-tuned 200-MHz series filter element at the input terminal is made of a length of copper strap shunted across a 50-pF, 5kV ceramic capacitor. In this particular filter, the parallel circuit was affixed to the output capacitor of the pi-network tank circuit of the transmitter and does not show in the photograph.

The filter is adjusted by removing the connections from the ends of the half-wave sections and adjusting each section to 50 MHz by spreading the turns of the coil with a screwdriver while monitoring the resonant frequency with a grid-dip oscillator. The next step is to ground the top end of each series-tuned section (C_2, L_2 and C_7, L_5) with a heavy strap. The input section is tuned to 100 MHz and the output section to 200 MHz. When tuning adjustments are completed, the straps are removed and the top of the filter box is held in place with sheet-metal screws.

A Two-Meter Most filter construction tech-
Lowpass Filter niques that are usable in the high-frequency spectrum have serious shortcomings when used above 100 MHz. Normal capacitor lead lengths reduce the effectiveness of the capacitor and at medium power levels the loss in the capacitor can be excessive.

The requirement that the harmonic output of a vhf amplifier be held to 60 dB below the signal level is a difficult one to meet with conventional construction techniques. The filter described in this section was developed to combine simple construction with sufficient attenuation to comply with the FCC standard.

The common three-pole filter can provide about 18 dB rejection to the second harmonic using standard components (figure 11). This is insufficient to meet modern requirements. Substitution of quarter wavelength open coaxial sections for the capacitors, however, provides superior rejection of the second harmonic, as shown in figure 12. For power levels up to several hundred watts, RG-188/U line may be used for the coaxial sections. When higher power is used, the small coaxial line should be replaced with a copper line having air dielectric, where the capacitance per inch is selected to achieve one-quarter wavelength at the second harmonic. The resultant air line has an impedance of 57.3 ohms with conductor diameter ratios of 2.6:1. As an example, the outer conductor would have an inner diameter of 7/8 inch (2.22 cm) and the inner conductor would have a diameter of 5/16-inch (0.793 cm).

The construction of a simple low-pass filter for 2 meters is shown in figure 13. All that it requires is a printed-circuit board, two lengths of coaxial line, and a small coil. The board provides termination for the line and coil. The far ends of the line are soldered to the board. The open end shields are soldered to small "islands" cut in the board which act as support points. The length of

Figure 10

INTERIOR VIEW OF SIX-METER FILTER

The input compartment of the filter is at the left. The series coil is wound of copper tubing, and the connection to the output section (right) is made by a length of tubing which passes through a hole in the center shield. Series elements carry less current and employ wirewound coils. At right is antenna relay, with power leads bypassed as they leave filter compartment. Filter is set to correct frequency by adjusting the inductance of the tubing coils.

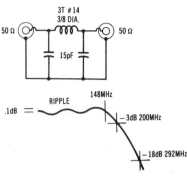

Figure 11

ATTENUATION OF 3-POLE VHF FILTER

18 dB protection is provided for 146-MHz signal for filter with cutoff frequency of 148 MHz.

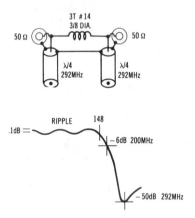

Figure 12

VHF FILTER WITH COAXIAL CAPACITORS

50 dB protection is provided for 146-MHz signal when coaxial capacitors are used.

the lines may be trimmed slightly to place maximum attenuation at a specific point in the spectrum.

16-4   Stereo-FM Interference

With the growing popularity of imported, solid-state stereo f-m equipment the problem of interference to these devices has become severe in the past few years. Most of this

home-entertainment equipment has little or no effective filtering to prevent RFI and is "wide open" to nearby, strong signals. Unfortunately, the prospective purchaser of such a device has little or no knowledge of the subsceptibility to RFI of the various imports and the burden falls on any nearby amateur to convince the neighbor that the set, and not the amateur, is at fault when RFI shows up.

RFI rejection in stereo f-m equipment is especially poor when the device is solid state and uses printed-circuit boards wherein a good, r-f ground is almost impossible to maintain. This description covers the majority of home entertainment devices sold today.

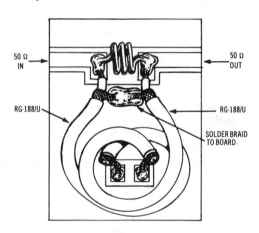

Figure 13

LAYOUT OF COAXIAL VHF FILTER

Filter is built on printed-circuit board. Two lengths of RG-188/U (teflon dielectric) cable, each 6⅜" (16.9 cm) long are used. The braid at free end of cable is soldered to "island" on circuit board. Coil is 3 turns #14 enameled wire, ⅜-inch (0.95 cm) in diameter. Filter designed by K6KBE.

Reduce External Pickup   Most stereo f-m units have long leads running between the speakers and the set, with additional leads running to the changer and/or auxiliary equipment. These leads make excellent antennas and are the major path for unwantd r-f energy to enter the equipment. The first step, then, in trying to eliminate the RFI path is to remove the input leads to the equipment, one at a time, and note which one reduces or eliminates the interference. The speaker leads can be disconnected and a pair of low impedance earphones with short leads substituted for the interference tests.

If interference is still present with the leads disconnected, the interference may be entering the equipment via the power line, or else is picked up by the internal wiring of the equipment. A power-line filter, such as described for a television receiver in an earlier Section of this chapter is recommended in the first case. Power line-type interference can be checked by pulling the plug out of the wall receptacle while the interference is manifesting itself. If the RFI is entering the equipment via the ac line it will disappear the instant the plug is pulled; if it is being picked up by the internal wiring of the stereo equipment it will slowly fade away as the power supply filter capacitors discharge.

If the interference seems to be arriving via the speaker leads, the leads should be made as short as possible and each lead bypassed to the chassis (ground) of the equipment by a .01-μF disc ceramic capacitor. If interference is still present to a degree, the speaker leads may be wrapped around a ferrite rod, or core, at the equipment. About 20 turns around the core will suffice. Leads to the pickup may be treated in the same manner using a small ferrite core. An extra ground lead between the changer pickup and the stereo chassis may also be of assistance in reducing r-f pickup.

Equipment Problems   R-f interference to solid-state amplifiers is caused primarily by the rectifying action of the transistor junction which demodulates a strong, nearby signal. A small ceramic capacitor should be connected between the emitter-base junction (figure 14). A ferrite bead in series with the base lead may also be of benefit. An additional ferrite bead on the feedback line is recommended.

In spite of shielded patch cords being used in modern stereo gear, the cords are poor shields as far as r-f energy goes. In many cases, the "shield" consists of a spirally wrapped wire partially covering the main

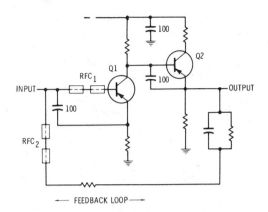

Figure 14

RFI SUPPRESSION IN STEREO EQUIPMENT

A small ceramic bypass capacitor is placed between the base-emitter junction in the first stages of the amplifier. Ferrite beads can also be used in the input and feedback circuits to further suppress RFI. The collector supply is also bypassed with a ceramic disc capacitor.

lead. Substituting coaxial cable (RG-59/U, for example) for the original leads will also help in stubborn cases of RFI.

If it is apparent that the interference is entering the equipment via the f-m antenna, installation of a TV-type high-pass filter will attenuate the interfering signal. Only as a last resort should shielding of the stereo equipment itself be attempted as many units have floating ground circuits. It is possible, however, to make small shields out of aluminum foil that may be clipped or fastened in place around critical circuits.

Each piece of stereo equipment must be handled as a special case, but if these broad guidelines are followed, it should be possible to suppress the majority of RFI cases. The techniques outlined in this section also apply to electronic organs or other home entertainment devices.

In many cases the equipment manufacturer has special service guides to aid in the suppression of RFI. This information should be obtained by writing directly to the *manufacturer* of the equipment.

16-5   Broadcast Interference

Interference to broadcast signals in the 540- to 1600-kHz band is a serious matter to those amateurs living in a densely populated area. Although broadcast interference (BCI) has been overshadowed by TVI and stereo problems, BCI still exists, especially for amateurs working the lower frequency bands.

Blanketing This is not a tunable effect, but a total blocking of the receiver. A more or less complete "washout" covers the entire receiver range when the carrier is switched on. This produces either a complete blotting out of all broadcast stations, or else knocks down their volume several decibels—depending on the severity of the interference. Voice modulation causing the blanketing will be highly distorted or even unintelligible. Keying of the carrier which produces the blanketing will cause an annoying fluctuation in the volume of the broadcast signals.

Blanketing generally occurs in the immediate neighborhood (inductive field) of a powerful transmitter, the affected area being directly proportional to the power of the transmitter. Also, it is more prevalent with transmitters which operate in the 160-meter and 80-meter bands, as compared to those operating on the higher frequencies.

The great majority of "modern" broadcast receivers employ a loopstick antenna concealed within the receiver cabinet. Loopstick pickup at the higher frequencies is quite restricted and the receiver may be physically oriented for minimum pickup of the interfering signal. In addition, bypassing each side of the receiver power line to the chassis or negative return bus with a pair of .01-μF, 1.6-kV ceramic disc capacitors is recommended. The remedies applicable to the stereo receiver circuits, previously discussed, also apply to a broadcast receiver.

Phantoms With two strong local signals applied to a nonlinear impedance, the beat note resulting from cross modulation between them may fall on some frequency within the broadcast band and will be audible at that point. If such a "phantom" signal falls on a local broadcast frequency, there will be heterodyne interference as well. This is a common occurrence with broadcast receivers in the neighborhood

of two amateur stations, or an amateur and a broadcast station. It also sometimes occurs when only one of the stations is located in the immediate vicinity.

As an example: an amateur signal on 3514 kHz might beat with a local 2414 kHz carrier to produce a 1100-kHz phantom. If the two carriers are strong enough in the vicinity of a circuit which can cause rectification, the 1100-kHz phantom will be heard in the broadcast band. A poor contact between two oxidized wires can produce rectification.

Two stations must be transmitting simultaneously to produce a phantom signal; when either station goes off the air the phantom disappears. Hence, this type of interference is apt to be reported as highly intermittent and might be difficult to duplicate unless a test oscillator is used "on location" to simulate the missing station. Such interference cannot be remedied at the transmitter, and often the rectification takes place some distance from the receivers. In such occurrences it is most difficult to locate the source of the trouble.

It will also be apparent that a phantom might fall on the intermediate frequency of a simple superhet receiver and cause interference of the untunable variety if the manufacturer has not provided an i-f wavetrap in the antenna circuit.

This particular type of phantom may, in addition to causing i-f interference, generate harmonics which may be tuned in and out with heterodyne whistles from one end of the receiver dial to the other. It is in this manner that *birdies* often result from the operation of nearby amateur stations.

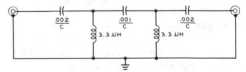

Figure 15

HIGH-PASS FILTER FOR AMATEUR RECEIVER

This simple filter attenuates signals below 1600 kHz to reduce overload caused by strong nearby broadcast stations. Filter is designed to be placed in series with coaxial line to receiver. Filter should be built in small shield box with appropriate coaxial fittings. J. W. Miller ferrite choke 74F336AP may be used for 3.3 μH inductor.

When one component of a phantom is a steady unmodulated carrier, only the intelligence presence on the other carrier is conveyed to the broadcast receiver.

Phantom signals almost always may be identified by the suddenness with which they are interrupted, signaling withdrawal of one party of the union. This is especially baffling to the inexperienced interference locater, who observes that the interference suddenly dissapears, even though his own transmitter remains in operation.

If the mixing or rectification is taking place in the receiver itself, a phantom signal may be eliminated by removing either one of the contributing signals from the receiver input circuit.

In the case of phantom crosstalk in an amateur-band receiver, a simple high-pass filter designed to attenuate signals below 1600 kHz may be placed in the coaxial antenna lead to the receiver (figure 15). This will greatly reduce the strength of local broadcast signals, which in a metropolitan area may amount to fractions of a volt on the receiver input circuit.

Ac/dc Receivers Inexpensive tube-type ac/dc receivers are particularly susceptible to interference from amateur transmissions. In most cases the receivers are at fault; but this does not absolve the amateur of his responsibility in attempting to eliminate the interference.

In cases of interference to inexpensive receivers, particularly those of the ac/dc type it will be found that stray receiver rectification is causing the trouble. The offending stage usually will be found to be a high-μ triode as the first audio stage following the second detector. Tubes of this type are quite nonlinear in their grid characteristic, and hence will readily rectify any r-f signal appearing between grid and cathode. The r-f signal may get to the tube as a result of direct signal pickup due to the lack of shielding, but more commonly will be fed to the tube from the power line as a result of the series heater string.

The remedy for this condition is simply to ensure that the cathode and grid of the high-μ audio tube (usually a 12AV6 or equivalent) are at the same r-f potential.

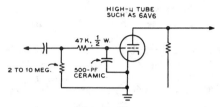

Figure 16

CIRCUIT FOR ELIMINATING AUDIO-STAGE RECTIFICATION

This is accomplished by placing an r-f bypass capacitor with the shortest possible leads directly from grid to cathode, and then adding an impedance in the lead from the volume control to the grid of the audio tube (figure 16).

In many ac/dc receivers there is no r-f bypass included across the plate-supply rectifier for the set. If there is an appreciable level of r-f signal on the power line feeding the receiver, r-f rectification in the power rectifier of the receiver can cause a particularly bad type of interference which may be received on other broadcast receivers in the vicinity in addition to the one causing the rectification. The soldering of a 0.01-μF, 1.6-kV disc ceramic capacitor directly from anode to cathode of the power rectifier (whether it is of the vacuum-tube or silicon-rectifier type) usually will bypass the r-f signal across the rectifier and thus eliminate the difficulty.

Image Interference In addition to those types of interference already discussed, there are two more which are common to superhet receivers. The prevalence of these types is of great concern to the amateur, although the responsibility for their existence more properly rests with the broadcast receiver.

The mechanism whereby image production takes place may be explained in the following manner: when the first detector is set to the frequency of an incoming signal, the high-frequency oscillator is operating on another frequency which differs from the signal by the number of kHz of the intermediate frequency. Now, with the setting of these two stages undisturbed, there is another signal which will beat with the high-frequency oscillator to produce an i-f signal. This other signal is the so-called *image*, which is separated from the desired signal by twice the intermediate frequency.

Thus, in a receiver with a 175-kHz intermediate frequency tuned to 1000 kHz; the hf oscillator is operating on 1175 kHz, and a signal on 1350 kHz (1000 kHz plus 2 × 175 kHz) will beat with this 1175-kHz oscillator freqency to produce the 175-kHz i-f signal. Similarly, when the same receiver is tuned to 1450 kHz, an amateur signal on 1800 kHz can come through.

The second variety of superhet interference is the result of harmonics of the receiver high-frequency oscillator beating with amateur carriers to produce the intermediate frequency of the receiver. The amateur transmitter will always be found to be on a frequency equal to some harmonic of the receiver hf oscillator, *plus or minus the intermediate frequency.*

As an example: when a broadcast superhet with 465-kHz intermediate frequency is tuned to 1000 kHz, its high-frequency oscillator operates on 1465 kHz. The third harmonic of this oscillator frequency is 4395 kHz, which will beat with an amateur signal on 3930 kHz to send a signal through the i-f amplifier. The 3930 kHz signal would be tuned in at the 1000-kHz point on the dial.

Insofar as remedies for image and harmonic superhet interference are concerned, it is well to remember that *if* the amateur signal did not in the first place reach the input stage of the receiver, the annoyance would not have been created. It is therefore good policy to try and reduce or eliminate it by the means discussed in this chapter. However, in some solid-state equipments, it is almost impossible to make the necessary circuit changes, or the situation does not allow the amateur to work on the equipment. In either case, if this form of interference exists, the only alternative is to try and select an operating frequency such than neither image nor harmonic interference will be set up on favorite stations in the susceptible receiver.

16-6   Other Forms of Interference

Telephone   The carbon microphone of the
Interference telephone, as well as varistors
                in the compensation networks
incorporated therein may serve as efficient

rectifiers of nearby r-f energy, injecting the modulation of the signal on the telephone circuit. The first step to take when this form of interference develops is to contact the repair department of your local telephone company, giving them the details. Depending upon the series nomenclature of the phone in use, the company is able to supply various types of filters to suppress or reduce the interference. The widely-used *series 500* phones require the replacment of the existing compensation network with a type 425J network (supplied by the phone company). This device has the varistors replaced with resistors in the network. In addition, a .01-μF ceramic capacitor should be placed across the carbon microphone and also across the receiver terminals. The older *series 300* phones require only a .01-μF ceramic capacitor placed across the microphone.

The newer ("touchtone") phones, which include *series 1500, 1600* and *1700, require* the same modification as the series 500 units, except that the replacement network is a type 4010E.

In addition to the modification devices for the telephone instrument the phone company can also supply a type 40BA line filter capacitor which acts to bypass the drop wire coming into the telephone and also a type 1542A r-f inductor which is placed at the connector block. All of these items are available, upon request, from your local telephone company, in most cases.

Power-Line   Power-line interference may
Interference reach a radio receiver by trans-
                mission along the line or by
direct radiation. Typical sources of power-line interference are spark and electrostatic discharge. Spark discharge from brush-type motors, heaters for fish aquariums, thermostats on sleeping blankets, and heating pads are prolific sources of such interference. If the interfering unit can be located, bypass capacitors on the power line directly at the unit will usually suppress the noise. The noise may often be located by using a portable radio as a direction finder, homing in on the noise source. Direct power-line noise, caused by leaky insulators or defective hardware on high-voltage transmission lines is harder to pinpoint, as the noise may be carried for a considerable distance along the line.

Standing waves of noise are also apparent on power lines, leading to false noise peaks that confuse the source. Many power companies have a program of locating interference and it is recommended that the amateur contact the local company office and register a complaint of power-line interference rather than to try and find it himself, since the cure for such troubles must be applied by the company, rather than the amateur.

Electrostatic discharge may be caused by intermittent contact between metallic objects in a strong electric field. Guy wires or hardware on power poles are a source of this form of interference. In addition, loose hardware on a nearby TV antenna, or the tower of the amateur antenna may cause this type of interference in the presence of a nearby power line. This type of interference is hard to pinpoint, but may often be found with the aid of a portable radio. In any event, suspected power-line interference originating on the power-line system should be left to the power-company interference investigator.

Interference from TV Receivers   The sweep oscillator of a modern TV receiver is a prolific generator of harmonics of the 15.75-kHz sweep signals. Harmonics of high amplitude are observed as high as 50 MHz from inadequately shielded receivers. Sweep oscillator radiation may take place via the power line of the TV set, from the antenna or directly from the picture tube and associated sweep circuit wiring. Most cases of nearby interference use a combination of all three paths.

Oscillator radiation along the power line can be reduced by the use of a power-line filter or by wrapping the line around a ferrite rod. Radiation from the TV antenna can be substantially reduced by the use of a high-pass filter installed at the receiver and/or the use of a shielded lead-in.

Radiation from the sweep-circuit wiring itself is difficult to suppress and modifications to the television receiver are not recommended. However, it should be pointed out that radiation of this type, if of sufficient intensity to cause serious interference to another radio service, falls under Part 15 of the FCC Rules and Regulations. When such interference is caused and is reported, the user of the receiver is obligated to take steps to eliminate it. The owner of the receiver is well advised to contact the manufacturer of the receiver for information concerning the alleviation of the radiation.

Light Dimmers   Inexpensive wall-receptacle light dimmers are a prolific source of r-f interference which resembles a high buzz which increases in strength at the lower frequencies. These devices make use of an expensive silicon controlled rectifier (SCR) which is a high speed unidirectional switch. When the SCR conducts, it creates a very steep wavefront, which is rich in harmonic energy. More expensive dimmer controls are available having r-f harmonic suppression built in the case, and the easiest way to get rid of this annoying source of RFI is to replace the offending unit with a model incorporating the suppression circuit.

16-7 Help in Solving TVI

Some TV set manufacturers will supply high-pass TV filters at cost for their receivers or provide information on TVI reduction upon request. When writing to the manufacturer about TVI problems, supply complete details, including model and serial number of the TV set involved; the name and address of the TV set owner; the name, address, and call letters of the amateur involved; and particulars of the interference problem (channels affected, frequency of amateur transmitter, sound or picture affected, etc.). The following manufacturers can supply information and assistance:

Manufacturer Service Representatives
(The information contained in this listing has been supplied by the American Radio Relay League (ARRL), Newington, Connecticut.)

Admiral
No longer in business. For parts, tel. 800-447-8361.

Akai America

Akai products include audio tape recorders, video tape recorders, a-m/f-m receivers, speaker systems and related accessory products. Inquiries related to RFI should be addressed to the Customer Service Department, 800 W. Artesia Blvd., Compton, CA 90220, or to P. O. Box 6010, Compton, CA 90224, tel. 213-537-3880. "Upon receipt of these inquiries, we will investigate the situation and, to our utmost, try to resolve the customer's problems."

Allen Organ Company

When a complaint is received via the dealer, Allen Organ Co. sends the dealer an informational service bulletin on RFI and sufficient components to cover all amplifiers in the affected instrument. This service is offered at no extra cost to the customer. Refer RFI problems to the local Allen dealer. Inquiries may be made to Mr. David L. George, National Service Manager, Macungie, PA 18062, tel. 215-966-2200.

Altec Lansing International

Customer RFI problems are referred to the authorized Altec warranty stations located nationwide and denoted by an information card furnished with each piece of equipment. Unusual situations are, at the option of the warranty station, referred to Altec Customer Service, 1515 W. Katella Ave., Anaheim, CA 92803, tel. 714-774-2900, or to the Engineering Department, 1515 S. Manchester Ave., Anaheim, CA 92803, Attention: Chief Engineer, Electronics.

Apple Computer, Inc.

"Our products include business, professional, educational, scientific, industrial and home computers, peripheral devices, and software. These products are designed to be compliant with the FCC guidelines covering Class A and Class B computer devices. Inquiries related to RFI should be addressed to any of our more than 800 dealer-operated Level One service centers. If the service technicians there are unable to solve the situation, they will contact our Corporate Engineering Services Group."

Arvin Industries, Inc., Consumer Electronics Division

Customer problems involving RFI should be referred to Mr. John Currey, Manager, Engineering Support Group, E. 15th St., Columbus, IN 47201, tel. 812-372-7271.

Audio Research Corporation

In the event of an RFI problem, the customer may write to Mr. Richard Larson, Chief Engineer, 6801 Shingle Creek Pkwy., Minneapolis, MN 55430, tel. 612-566-7570.

Baldwin Piano and Organ Company

"RFI complaints are usually handled by the local Baldwin service technician. Factory personnel are available to assist a technician when needed. Baldwin maintains its own staff of technical representatives who travel in the field and may be called upon to assist a dealer technician with difficult problems, including RFI. Several Baldwin Technical Manual Supplements are available with specific instructions for RFI suppression on specific models. This information is readily available upon request. Inquiries may be directed to Mr. Gilbert C. Carney, Manager, Organ Technical Service, Baldwin Piano and Organ Co., 1801 Gilbert Ave., Cincinnati, OH 45202, tel. 513-852-7838."

Bogen Division of Lear Siegler, Inc.

"Bogen Division manufactures professional, commercial and industrial sound equipment. In the event of an RFI problem with any Bogen unit, write for the division's free Field Service Bulletin No. 59 about RFI signal interference, or contact Allen Guthman, Service Manager, Bogen Division/LSI, Box 500, Paramus, NJ 07652, tel. 201-343-5700."

Carver Corporation, Inc.

Carver Corporation manufactures high-fidelity components. "Problems pertaining to RFI should be directed to our service manager, Mr. Philip Fenner, P. O. Box 664, 14304 N.E. 193rd Pl., Woodinville, WA 98072, tel. 206-487-3483."

Conn Keyboards, Inc.

RFI complaints should be referred to the local Conn dealer, whether instrument is in or out of warranty. Factory assistance is available to the dealers who are unable to correct the RFI. RFI problems encountered within the term of instrument warranty are usually corrected by the selling dealer without cost to the organ owner. Contact Mr. Thomas A. Umbaugh, National Service Manager, 350 Randy Rd., Carol Stream, IL 60187, tel. 312-653-4330.

Crown International

"Crown International is the manufacturer of high-end audio products. RFI suppression is incorporated in the design of the product. If a customer should encounter an RFI problem, he may contact the Technical Services Department of Crown International, 1718 W. Mishawaka Rd., Elkhart, IN 46517."

Curtis Mathes

Curtis Mathes products include color TVs and stereos (100% solid state) in portable, console and combination configurations. Customer complaints involving RFI should first be resolved at the retail-dealer level. If not satisfied, then the complaint should be made in writing to the Consumer Relations Department giving all details of the problem, along with the model information, serial number, date of sale, dealer and service history. Each complaint will be handled individually. Write to Curtis Mathes Manufacturing Co., Curtis Mathes Pkwy., Athens, TX 75751, tel. 800-527-7646, Texas only tel. 800-492-9543.

Delco Electronics, Division of GM Corporation (see GM Corp.)

Dumont (see Emerson Quiet/Kool Corp.)

Electra Company, Division of Masco Corporation of Indiana

Electra Co. asks that RFI problems with "Bearcat," its automatic scanning radio, be referred to its service department at 300 E. County Line Rd., Cumberland, IN 46229, tel. 317-844-1440.

Emerson Quiet/Kool Company

Mr. Jerome Roth reports that his company has not made TVs or audio devices since 1972. As a continuing gesture of goodwill, however, Mr. Roth suggests that customers may refer RFI problems with equipment previously marketed by Emerson Quiet/Kool Co. to him for recommendations, at the mailing address below. *Do not confuse* this company with Emerson Radio Corp., which is an entirely different publicly owned corporation. Contact Emerson Quiet/

Kool Co., P. O. Box 300, Woodbridge, NJ 07095, tel. 201-381-7000.

Emerson Radio Corporation

Customers may refer RFI inquiries related to Emerson Radio Corp. TV and radio problems to Mr. Dave Buda. Emerson Radio does not supply filters. The new address is: Emerson Radio Corp., One Emerson Way, Secaucus, NJ 07094, tel. 201-865-4343.

Epicure Products, formerly Elpa Marketing Industries, Inc.

"Complaints are handled with respect to parts and labor on an individual basis. Necessary modifications for RFI are made on a no-charge basis for parts and labor during the term of instrument warranty. Beyond warranty, modification parts are available free of charge. The customer then pays for labor involved in the installation of the parts. Refer RFI problems to Mr. John F. King, National Service Manager, 25 Hale St., Newburyport, MA 01950, tel. 800-225-7932."

Fisher Corporation

Fisher Corporation asks that RFI problems involving a Fisher product be handled as follows: request assistance from the local selling dealer or request assistance from the local Fisher authorized service station (a list is packed with every Fisher unit). Contact with local Fisher agencies is the preferred method of handling. Fisher's service coordination group maintains close communications with Fisher authorized service stations and Fisher's Engineering Department, and works under the supervision of the office of the National Service Manager. If the problem cannot be solved at the first two service levels, contact Service Coordination, 21314 Lassen St., Chatsworth, CA 91311, tel. 213-998-7322.

Garrard/Plessey Consumer Products

Garrard advises the consumer on methods that may eliminate RFI. In unusual cases where the suggestions are ineffectual, customers should refer the RFI problem to Mr. Al Pranckevicus, National Service Manager, 85 Sherwood Ave., Farmingdale, NY 11735, tel. 516-293-2400.

General Electric Company

RFI problems involving G.E. television receivers should be referred to the nearest

General Electric Customer Care Service Operation. If G. E. Customer Care Service is unable to correct the RFI, the customer should refer the problem to General Electric Co., Mr. J. F. Hopwood, Manager of Consumer Affairs, Appliance Park, Louisville, KY 40225, tel. 502-452-3754. All RFI problems involving G. E. radios, record players and other audio products should be referred to Manager of Consumer Counseling, Mrs. Patricia C. Cleary, Electronics Park, Bldg. 5, Syracuse, NY 13221, tel. 315-456-3388.

General Motors Corporation

"From time to time you may have questions concerning the electromagnetic compatibility of mobile transmitters when installed on General Motors vehicles. To help avoid such questions from arising, it is urged that care be taken to follow any applicable GM service procedures. The local GM Service Manager for the Car or Truck Division whose vehicle is involved should be contacted for information about such service procedures. If you are unable to obtain such assistance locally or if questions nevertheless arise, we have established a central contact point for all such inquiries. Accordingly, you should direct your inquiries to: Mr. Henry J. Lambertz, GM Service Research (GMSR), Service Development Center, 30501 Van Dyke, Warren, MI 48090, tel. 313-492-8448. He will direct your inquiries to the appropriate divisions or staff within GM and follow up to see that appropriate action is taken."

Gulbransen, Division of CBS Musical Instruments, Inc.

Gulbransen cooperates with dealers and customers in offering suggested solutions to RFI. Gulbransen does not reimburse the consumer for servicing. When extreme cases are encountered because of the proximity of the transmitter and relative power, however, the dealer may sometimes absorb the cost of servicing RFI problems. Customers should refer RFI problems to the local dealer. Inquiries may be directed to Mr. J. A. Iacono, Consumer Service Supervisor, 100 Wilmot Rd., Deerfield, IL 60015, tel. 800-323-1814.

Hammond Organ Company

"RFI difficulties are usually handled by the local Hammond dealer service technician.

Hammond maintains a staff of technical service representatives who travel in the field and may be called upon to assist local dealer technicians with difficult or unusual service problems, including RFI." Hammond states that the services of the Engineering and Technical Field Service Departments under its control are provided to consumer and dealer without charge. RFI problems should be referred to the local Hammond dealer. Inquiries may be directed to the Hammond Technical Service Department, 4200 W. Diversey Ave., Chicago, IL 60639, Attention: Jerry J. Welch.

Harman/Kardon, Inc.

RFI problems should be directed to Harman/Kardon at 240 Crossways Park West, Woodbury, NY 11797, tel. 516-496-3406, Attention: Customer Relations Dept.

Heath Company

Heath Co. suggests that, for fastest service on matters related to RFI regardless of the product line involved, customers may now reach the Technical Consultation Department by either writing directly to that department at Heath Co., Benton Harbor, MI 49022, or by using a new direct-line telephone system to the department by calling 616-982-3302. Do not write to an individual.

Hitachi Sales Corporation of America

"Our primary products are TVs, radios, tape recorders, hi-fi components and video tape recorders. Hitachi Sales Corp. of America attempts to cure each RFI problem on an individual basis. Customers should provide model number and information concerning the nature of the problem. RFI problems should be referred to the nearest Hitachi Regional Office." *Eastern Regional Office*, 1200 Wall St. West, Lyndhurst, NJ 07011, tel. 021-935-8980, Attention: Service Dept. *Mid-Western Regional Office*, 1400 Morse Ave., Elk Grove Village, IL 60007, tel. 312-593-1550, Attention: Service Dept. *Western Regional Office*, 612 Walnut, Compton, CA 90220, tel. 213-537-8383, Attention: Service Dept. *Southern Regional Office*, 510 Plaza Dr., College Park, GA 30349, tel. 404-763-0360, Attention: Service Dept.

J. C. Penney Company, Inc.

J. C. Penney Company asks that customers with RFI problems contact their nearest J. C. Penney store for personal assistance. J. C. Penney Company, Inc., 1301 Avenue of the Americas, New York, NY 10019.

Kenwood Electronics, Inc.

Kenwood asks that customers with RFI problems take the affected unit to an authorized service center where an adjustment will be made at no cost to the customer if the product is properly registered with Kenwood and is within warranty. It is suggested that prior authorization for the return be obtained from Mr. Toshi Furutsuki, 1315 E. Watsoncenter Rd., Carson, CA 90745, tel. 213-518-1700.

Lafayette Radio Electronics Corporation

"Customers should refer RFI problems involving Lafayette products to the local dealer. If the dealer cannot alleviate the problem, the customer may contact Mr. Charles Tanner, Vice President Administration, 111 Jericho Tpk., Syosset, NY 11791, tel. 516-921-7700."

Lowrey Division of Norlin Music, Inc.

"Lowrey customers should refer RFI problems to the local Lowrey dealer or certified Lowrey technician. Lowrey provides all technicians with technical literature regarding RFI and will provide assistance to local service organizations through its staff of field technical representatives when needed. Inquiries may be directed to Mr. Larry R. Thomas, Director of Product Service, 707 Lake Cook Rd., Deerfield, IL 60015."

Magnavox Consumer Electronics Company

"RFI problems are usually handled by the local Magnavox Authorized Service Center. Technical assistance in resolving such problems is provided by the Magnavox Field Service Staff through four Area Service Offices. Technicians or customers may refer unusual RFI problems involving Magnavox products to their nearest Area Service Center." In the *New York area* contact Magnavox Consumer Electronics Co., 161 E. Union Ave., East Rutherford, NJ 07073. In the *Chicago area* contact Magnavox Consumer Electronics Co., 7510 Frontage Rd., Skokie, IL 60077. In the *Atlanta area* contact Magnavox Consumer Electronics Co., 1898 Leland Dr., Marietta, GA 30067. In the *Los Angeles area* contact Magnavox Consumer Electronics Co., 2645 Maricopa St., Torrance, CA 90503.

Marantz (see Superscope)

McIntosh Laboratory, Inc.

"McIntosh has a number of authorized service agencies located throughout the country. Customers will be assisted to receive prompt help. RFI and other service-related problems can be directed to Mr. John Behory, Customer Service Manager, 2 Chambers St., Binghamton, NY 13903, tel. 607-723-3512."

MGA Mitsubishi Electric Sales America, Inc.

MGA is the new sales and service representative for the Mitsubishi Electric Corp. RFI reports from the field, beyond the dealer's capability to resolve and in which MGA becomes involved, are handled on an individual basis, as in the past. "All attempts will be made to give customer satisfaction." MGA suggests that requests for assistance be addressed to 3030 E. Victoria St., Compton, CA 90221, or the Service Department may be contacted by telephone, toll free, at 800-421-1132. Mr. Ken Kratka is the new National Service Manager.

Midland International Corporation

Midland policy remains the same. If any RFI problems are encountered with Midland portable black-and-white and color TVs or audio and radio products, individuals should contact Mr. Dennis Oyer, Vice President Customer Service, P. O. Box 1903, Kansas City, MO 64141, or at 1690 N. Topping, Kansas City, MO 64120, tel. 816-241-8500.

Montgomery Ward

Service for RFI should be obtained from the nearest Montgomery Ward location. If service is not obtainable locally, the customer may write to: Customer Service Product Manager, Corporate Offices, Montgomery Ward Plaza 4-N, Chicago, IL 60671. The Montgomery Ward field service organization can call upon factory and corporate engineering talent for assistance in handling difficult RFI problems.

Morse Electro Products Corporation

"RFI complaints related to Morse entertainment products may be referred to Mr. Phillip Ferrara, Service and Parts Dept., 3444 Morse Dr., Dallas, TX 75221, tel. 214-337-4711 or 800-527-6422."

Nikko Audio

"Nikko's line of products includes stereo receivers, tuners, amplifiers, combination preamp and main-amp pairs, tape decks and signal processors. For information and assistance with any Nikko products, inquiries should be made to Mr. Robert Fontana, National Service Manager, Service Dept., 320 Oser Ave., Hauppauge, NY 11787, tel. 516-231-8181."

North American Phillips Corporation

This corporation no longer manufactures its own RFI-prone products. (See Sylvania.)

Nutone Division

"Refer RFI problems to Mr. Norman W. Aims, Field Service, Scovil Housing Products Group, Madison and Red Bank Rds., Cincinnati, OH 45227, tel. 513-527-5415."

Panasonic Company

When instances of RFI occur, the customer should contact Panasonic at the following address: Panasonic Co., Division of Matsushita Electric Corp. of America, One Panasonic Way, Secaucus, NJ 07094, Attention: Supervisor of Quality Assurance Group, tel. 201-348-7000. The customer should provide model number, serial number and information concerning the problem. Upon review of the problem, the customer will be contacted and advised where to return the unit for corrective repair. "Panasonic will absorb both parts and labor costs in these instances."

Phase Linear Corporation

"RFI problems should be directed to Phase Linear Service Dept., Rick Bernard, Service Manager, 20121 48th Ave. West, Lynnwood, WA 98036, tel. 206-774-8848. In-house articles regarding RFI cures are available upon request at no charge."

Quasar Company (Matsushita Corporation of America)

For a high-pass filter, the consumer should contact Quasar Co., Consumer Relations Manager, Mr. George Datillo, 9401 W. Grand Ave., Franklin Park, IL 60131, tel. 312-451-1200. Model and serial number of the receiver and frequency of the interfering signal, if known, should be included with the written request, as well as whether sound or picture or both are affected. The Quasar distributor serving the local area should be contacted relative to any other interference problem that is unique to Quasar products.

Radio Shack

"Customers who encounter unique interference problems involving Radio Shack audio products may write to Mr. Dave Garner or Mr. Al Zuckerman, Product Development Engineers, National Headquarters, 1100 One Tandy Center, Fort Worth, TX 76102, tel. 817-390-3205."

RCA Consumer Electronics

"RFI problems involving both TV and audio products may be referred to Mr. J. J. Sanchez, 600 N. Sherman Dr., Indianapolis, IN 46201, tel. 317-267-6448. Requests for filters should include model number and serial number of the RCA television receiver. Filter installation charges will be the customer's responsibility."

Rodgers Organ Company, Division of CBS Musical Instruments, Inc.

RFI problems involving the Rodgers Organ may be referred to Custom Organ Test Department, 1300 N. East 25th Ave., Hillsboro, OR 97223, tel. 503-648-4181.

Rotel of America, Inc.

Stereo receivers, amplifiers, tuners and tape decks are made by Rotel. RFI problems should be referred to Michael Gregory, National Service Manager, 13528 S. Normandie Ave., Gardenia, CA 90249. "RFI problems will be handled according to the terms of our limited warranty."

Sansui Electronics Corporation

"RFI problems should now be directed to Mr. Frank Barth, Vice President Frank Barth, Inc., 500 5th Ave., New York, NY

10110, tel. 212-398-0820. Frank Barth, Inc. is the new advertising and public relations agency representing Sansui. Mr. Barth will direct the customer to an appropriate Sansui Service Center." A Sansui representative has previously stated that all Sansui products are carefully checked prior to final engineering commitments for susceptibility to RFI. "Units are often taken to high rf-level areas such as New York City to determine any design flaws."

Sanyo Electric, Inc.

"In the event an RFI problem should occur, the customer is requested to take the set to the nearest Sanyo authorized repair station. Transportation to and from the shop is the responsibility of the customer. Should the shop not alleviate the problem, either the customer or the shop should contact Mr. Brad Coulter, Consumer Relations Manager, Sanyo Electric, Inc., Electronics Division, 1200 W. Artesia Blvd., Compton, CA 90220, tel. 213-537-5830."

Scientific Audio Electronics, Inc.

"Refer RFI inquiries to Mr. Michael L. Joseph, National Marketing Manager, or contact Mr. Robert Hunt, National Service Manager, 701 E. Macy St., Los Angeles, CA 90012, tel. 213-489-7600."

H. H. Scott, Inc.

This manufacturer offers a simple instruction sheet to aid customers in resolving problems involving rf pickup. The information includes suggestions about suitable equipment grounding, power-line bypassing and hints and suggestions on how to determine where rf is entering the equipment. "Customers should refer any RFI problems to Mr. D. F. Merryman, Engineering Dept., 20 Commerce Way, Woburn, MA 01801, tel. 617-933-8800."

Sears, Roebuck and Company

Sears asks that customers with an RFI problem involving a Sears product contact the nearest Sears service department for assistance. Inquiries may be directed to Mr. R. C. Good, Manager Marketing Communications, Home Appliances, Dept. 703, Sears Tower, Chicago, IL 60684, tel. 312-875-8366.

Sharp Electronics Corporation

"Sharp Electronics will, with proof of purchase, supply customers with a Drake TV-300 high-pass filter at no cost. Audio rectification problems are handled on an individual basis by the Service Department. Refer RFI problems to Service Manager, 2 Keystone Pl., Paramus, NJ 07652, tel. 201-262-9000."

Sherwood, Division of Inkel Corporation

Customers with interference problems should contact Mr. David Daniels, Vice President Marketing, 17107 Kingsview Ave., Carson, CA 90746, tel. 213-515-6866.

Shure Brothers, Inc.

The manufacturer recommends the use of balanced-line, low-impedance microphones and cables. If an RFI problem persists after the above measures have been taken, the customer should contact Shure Brothers, Inc. with specifics so that they may be able to help solve the problem. Refer RFI problems to Customer Services Dept., 222 Hartrey Ave., Evanston, IL 60204, tel. 312-866-2553.

Sony Corporation of America

"Our primary products are color television, black-and-white television, video tape recorders, stereo equipment, audio components and word-processing equipment. RFI assistance is provided through regional service managers of Sony Factory Service Centers through the Customer Care Dept. An RFI booklet is available from the company on request. Sony Corp., 47-47 Van Dam St., Long Island City, NY 11101, tel. 212-361-8600."

Sound Concepts

"We handle all RFI complaints at our main laboratories at 27 Newell Rd., Brookline, MA 02146, tel. 617-566-0110. We request that the offending unit be accompanied by a description of the nature of the RFI; there is no charge for this service."

Soundesign Corporation

"Soundesign Corp./Acoustic Dynamics requests that all service problems relating to nonstereo merchandise be referred to Mr. Thomas R. Greene, Administrative Vice President, 34 Exchange Pl., Jersey City, NJ

07302, tel. 201-434-1050. All service problems on stereo merchandise are to be referred to our authorized service centers. The nearest one can be found by calling toll free in the continental U.S., 800-631-3092."

Superscope/Marantz Corporation

Superscope/Marantz manufacturers a-m/f-m receivers, tuners, amplifiers, tape recorders, record players and audio systems. In the event of special RFI cases resulting from extremely high fields, contact the Technical Services Dept. at Superscope corporate offices. "Modifications necessary to resolve such RFI problems are provided to customers on an individual basis." Superscope/Marantz Corp., 20525 Nordhoff St., Chatsworth, CA 91311, tel. 213-998-9333. For Service Dept., call toll free, 800-423-5224, Attention: Mr. Albert Almeida, Technical Service Manager.

Sylvania/Philco, Division of North American Phillips Corporation

Sylvania policy remains as follows: "Factory field service and field engineering personnel work together to solve many of the TVI and audio rectification problems. If the consumer has an interference condition, he should contact his local dealer. He is in touch with the manufacturer's services that will help resolve it." Consumers should contact the dealer and work through his services first. RFI problems are handled on an individual basis. Sylvania has available for their technicians an excellent pictorial TVI training manual titled, *Diagnosis, Identification and Elimination of TVI.* Sylvania/Philco, Mr. Jack Berquest, Manager Service Training, Consumer Electronics Division, 700 Ellicott St., Batavia, NY 14020, tel. 716-344-5000.

Tandberg of America, Inc.

When RFI occurs in Tandberg products, the manufacturer suggests that the unit be returned to them. "We will do any modification possible to eliminate the RFI." Authorization should be obtained from Mr. Tor Sivertsen prior to return of the unit. Mr. Tor Sivertsen, Technical Vice President, Labriola Ct., Armonk, NY 10504, tel. 914-273-9150.

Thomas International Electronic Organs, Division of Whirlpool Corporation

"RFI is usually resolved at the dealer level. If the manufacturer's field service is made aware of a consumer complaint regarding RFI, they contact the seller and advise him on how to eliminate the problem." Thomas has six field service engineers. In the event of a call for assistance, an engineer personally contacts the consumer by telephone and makes an appointment to visit the home of the consumer to correct the RFI condition, with or without the dealer's technician. "We do not charge the consumer for this service. Refer RFI complaints to the dealer. Inquiries may be directed to Mr. Daniel E. Hofer, Manager Field Service, 7300 Lehigh Ave., Chicago, IL 60648, tel. 312-647-8700 or 800-323-4301.

Toshiba America, Inc.

Customers should contact the nearest regional office, an updated listing of which appears below, for obtaining assistance in solving RFI problems involving Toshiba televisions, radios, tape products, amplifiers, tuners and receivers. Mr. Stanley Friedman, National Service Manager, 82 Totowa Rd., Wayne, NJ 07470, tel. 201-628-8000. Mr. Sy Rosenthal, *Eastern Regional* Service Manager, 82 Totowa Rd., Wayne, NJ 07470, tel. 201-628-8000. Mr. Ray Holich, *Mid-West Regional* Service Manager, 2900 MacArthur Blvd., Northbrook, IL 60062, tel. 312-564-5110. Mr. C. B. Monroe, *Southwest Regional* Service Manager, 3300 Royalty Row, Irving, TX 75062, tel. 214-438-5814. Mr. S. Ito, *Western Regional* Service Manager, 19515 S. Vermont Ave., Torrance, CA 90502, tel. 213-538-9960.

U.S. JVC Corporation

"Inquiries related to RFI involving JVC products may be referred to Mr. T. Sadato, Chief Engineer, 41 Slater Dr., Elmwood, NJ 07407, tel. 800-526-5308."

U.S. Pioneer Electronics Corporation

"Contact: Mr. Andrew Adler, *Eastern Region,* 75 Oxford Dr., Moonachie, NJ 07074; Mr. John Noa, *Southern Region,* 1875 Walnut Hill Ln., Irving, TX 75062; Mr. Clarence Skroch, *Western Region,* 4880 W. Rosecrans Ave., Hawthorne, CA 90250;

Mr. Daniel Brostoff, *Mid-West Region*, 737 Fargo Ave., Elk Grove Village, IL 60007."

Wells-Gardner Electronics Corporation

"Wells-Gardener is a private-label manufacturer of consumer products. Inquiries related to RFI should be referred to our private-label customers whose address appears on the model-number label attached to the product. Special problems which may be encountered by private-label customers are usually referred to Wells-Gardner, Mr. Harry McComb, Service Manager, 2701 N. Kildare Ave., Chicago, IL 60639, tel. 312-252-8220."

Wurlitzer Company

"The Wurlitzer Company makes available a toll-free telephone line, 800-435-2930, to assist any technician or customer in any and all needs pertaining to the Wurlitzer product. The Wurlitzer company maintains a staff of field service managers who can assist should an RFI problem arise." Wurlitzer Co., 403 E. Gurler Rd., DeKalb, IL 60015.

Yamaha International Corporation

The Yamaha organization attempts to cure each RFI problem on an individual basis. Yamaha supplies all necessary technical information at no charge. If interference is caused by design error, Yamaha takes steps at its own expense to remedy the problem. Refer RFI problems to the local dealer. The dealers are kept well informed and current on RFI countermeasures. Inquiries may be directed to Mr. William Perkins, Electronic Service Manager, Electronic Service Dept., P.O. Box 6600, Buena Park, CA 90622, tel. 714-522-9351.

Zenith Radio Corporation

"Zenith gives consideration to handling and providing relief for RFI problems on a case-by-case basis. RFI problems should be referred to Service Division, 11000 W. Seymour Ave., Franklin Park, IL 60131, tel. 312-671-7550. RFI referrals should include model and serial numbers of the affected unit. Customers with a unique, difficult problem may direct a letter to Mr. Richard Wilson, National Service Manager, at the same address."

Other Manufacturers

Ms. Sally Browne, Director of Consumer Affairs, Consumer Electronics Group, Electronic Industries Association, 2001 Eye St., N.W., Washington, DC 20006, tel. 202-457-4900, may be contacted for assistance or recommendations in the handling of RFI problems involving manufacturers not listed here, or for assistance when the product is no longer manufactured.

Note: This list has been compiled by Harold W. Richman, W4CIZ, a former FCC Engineer in Charge and a member of the ARRL RFI Task Group. Additional RFI information appears from time to time in QST magazine, the monthly publication of the ARRL.

CHAPTER SEVENTEEN

Equipment Design

The performance of communication equipment is a function of the design, and is dependent on the execution of the design and the proper choice of components. This chapter deals with the study of equipment circuitry and the basic components that go to make up this circuitry. Modern components are far from faultless. Resistors have inductance and reactance, and inductors have resistance and distributed capacitance. None of these residual attributes show up on circuit diagrams, yet they are as much responsible for the success or failure of the equipment as are the necessary and vital bits of resistance, capacitance, and inductance. Because of these unwanted attributes, the job of translating a circuit on paper into a working piece of equipment often becomes an impossible task to those individuals who disregard such important trivia. Rarely do circuit diagrams show such pitfalls as ground loops and residual inductive coupling between stages.

Parasitic resonant circuits are seldom visible from a study of the schematic. Too many times radio equipment is rushed into service before it has been entirely checked. The immediate and only too apparent results of this enthusiasm are receiver instability, transmitter instability, difficulty of neutralization, r-f wandering all over the equipment, and a general "touchiness" of adjustment.

Hand in glove with these problems go the more serious ones of receiver overload, TVI, keyclicks, and parasitics. By paying attention to detail, with a good working knowledge of the limitations of the components, and with a basic concept of the actions of ground currents, the average amateur will be able to build equipment that will work "just like the book says."

The twin problems of TVI and parasitics are an outgrowth of the major problem of overall circuit design. If close attention is paid, to the cardinal points of circuitry design, the secondary problems of TVI and parasitics will in themselves be solved.

17-1   The Resistor

A *resistor* is a device which impedes the flow of current and dissipates electrical energy as heat. The range of available resistors is great, ranging from less than one ohm to many million ohms.

Two fundamental types of resistors exist: fixed and variable. Fixed resistors are commonly either carbon composition, wirewound, or film. Film types may be either carbon, metal, or nonmetal film.

The *carbon composition resistor* is composed of carbon held in a suitable binder and fired within a ceramic jacket. Resistance range is from 10 ohms to 22 megohms, with power ratings of $\frac{1}{4}$, $\frac{1}{2}$, 1, and 2 watt being most in demand. Resistance tolerances are typically $\pm 20\%$, with $\pm 10\%$ and $\pm 5\%$ units available. Most units have tinplated axial leads.

The *wirewound resistor* consists of resistance wire wound around an insulating form and fired with a ceramic jacket (figure 1). These units are used where temperature stability is a prerequisite. Units are available with resistance ratings of less than a fraction of an ohm to several hundred thousand ohms. General tolerance is ± 2% and the temperature coefficient of a typical resistor is about ± 100 ppm/°C. Power ratings of wirewound resistors run from 2 watts to as much as 250 watts, or higher.

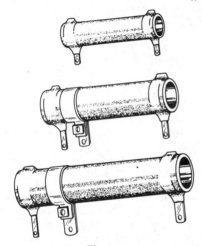

Figure 1

WIREWOUND RESISTORS

Resistors are wound with nichrome wire on a ceramic form. Inductive reactance becomes a problem when these resistors are used in high-frequency applications. Special spirally wound, noninductive resistors are used to cancel out the inductive effects at the higher frequencies.

The basic construction of a wirewound resistor involves a winding of nichrome wire and is by nature an inductance. Inductive reactance becomes a problem when these resistors are used in high-frequency applications. Special spirally wound (noninductive) resistors are often used to cancel the inductive effects at the higher frequencies.

Wirewound resistors are available with either radial or axial leads and often have an uninsulated area so that contact may be made to the body of the resistor at a random point.

The *film-type* resistor is made of a thin conductive film deposited and fired on an aluminum oxide, or glass, mandrel. The film may be nickel chromium, tin oxide, or a powdered precious metal mix (*cermet*).

Resistance value of the metal film resistor is set after the film has been fired on the mandrel. A spiral groove is ground or cut around the mandrel to set the desired value. The metal film resistor is finished by fitting end caps with leads over the ends. The unit is protected with a molten plastic dip.

Metal film resistors commonly available are in the ⅛- and ¼-watt power capacity with tolerances of ± 1%. Resistance values up to 200 megohms are available with a typical temperature coefficient of 100 ppm/°C.

The *variable resistor* (often called a *rheostat*, or *potentiometer*) is a unit whose resistance value may be changed by the user. The rheostat is primarily considered to be a power handling device, with ratings often in excess of 1000 watts. Rheostats are used for control of generator fields, motor speed, lamp dimming, and like services. The rheostat is commonly disc shaped and controlled by a rotating shaft. The resistance element is wound on an open ceramic ring and is welded at each end to a terminal band having connection points. The wound core is covered (except for an exposed track) with a fired enamel coating. The control arm is insulated from the moving contact assembly.

The contact brush, carried by the movable arm, is generally a powdered-metal compound (copper-graphite) which is connected by a flexible stranded shunt to a slip ring which rubs against a center lead supported by the rheostat framework.

Wattage rating of a common rheostat is based on a maximum attained temperature of 340°C measured at the hottest point on the enamel coating. The maximum hot-spot temperature varies with the percentage of the rheostat winding in use.

The general purpose *wirewound potentiometer* is available in resistance ranges from 0.5 ohm to about 150,000 ohms. The most common ratings are 1.5, 2, 4, 5, and 10 watts with a resistance tolerance of ± 10%. The great majority of potentiometers have a linear resistance winding, but special units are available wherein the resistance change is not constant throughout the shaft rotation.

An important property of the wirewound potentiometer is *resolution*. With such a device, the resistance change, as the slider moves from one extreme of rotation to the

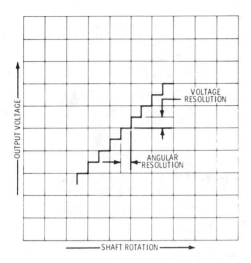

Figure 2

RESOLUTION OF WIREWOUND POTENTIOMETER

The resistance of a wirewound potentiometer varies in a step-like progression as the slider moves from one turn of wire to the next. Resolution is expressed either as angular or voltage resolution.

other, does not occur as a straight line but rather as a step-like progression, as the slider moves from one turn of wire to the next. Resolution is expressed as either angular or voltage resolution (figure 2). Precision potiometers having high tolerance and good resolution provide a resistance value that is proportional to shaft rotation to better than $\pm 1\%$. The precision devices may be either single turn, rotary; multiturn, rotary; or linear motion designs.

The *composition potentiometer* is widely used in all types of electronic equipment. Power ratings range from 1/10 watt to 4 watts, while resistances from 20 ohms through 10 megohms are commonly available. Various taper characteristics are shown in figure 3. The most common taper is the audio taper which provides 10% resistance at 50% rotation.

The resistance element may be carbon film, carbon-ceramic or molded carbon. More expensive potentiometers make use of *cermet* material. The composition potentiometer is available in a number of tolerances ranging from $\pm 40\%$ for commercial carbon-film devices to $\pm 5\%$ for high quality

cermet units. Ambient temperature rating for commercial units is 55°C.

For high resistance values, the maximum voltage rating across the end terminals of the potentiometer is an important factor. At a value of resistance defined as the critical value, the potentiometer is operating at

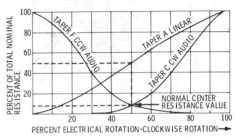

Figure 3

TAPER CHARACTERISTICS OF COMPOSITION POTENTIOMETERS

The linear (A) taper provides 50 percent of the resistance value at 50 percent of the clockwise rotation. The tapers C and F provide 10 percent of the resistance value at 50 percent of the rotation. Taper F is counterclockwise and taper C is clockwise.

maximum voltage and power at the same time. Above this value, the wattage of the unit must be derated. Most potentiometers have a maximum terminal potential of 500 volts.

The *trimming potentiometer* is a "set and forget" device that is not intended for dynamic control. These units are quite small in size and often have a very limited rotational life of less than 1000 cycles. Once set, they are not normally readjusted except as part of a regular maintenance or calibration program.

Common trimmers are packaged as either rectangular, multiturn units or single-turn, round units (figure 4). Resistance values of standard products range from 10 ohms to 50,000 ohms, with a usual tolerance of $\pm 10\%$. Power rating of the common units is $\frac{1}{4}$ to $\frac{3}{4}$ watt at a maximum temperature rating of 70°C.

Inductance of Resistors Every resistor because of its physical size has in addition to its desired resistance, less desirable amounts of inductance and distributed capacitance. These quantities are illus-

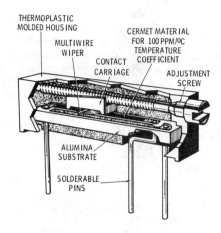

THERMOPLASTIC
MOLDED HOUSING
MULTIWIRE
WIPER
CERMET MATERIAL
FOR 100 PPM/ºC
TEMPERATURE
COEFFICIENT
CONTACT
CARRIAGE
ADJUSTMENT
SCREW
ALUMINA
SUBSTRATE
SOLDERABLE
PINS

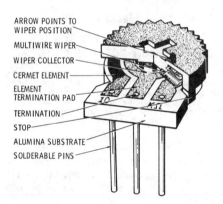

ARROW POINTS TO
WIPER POSITION
MULTIWIRE WIPER
WIPER COLLECTOR
CERMET ELEMENT
ELEMENT
TERMINATION PAD
TERMINATION
STOP
ALUMINA SUBSTRATE
SOLDERABLE PINS

Figure 4

THE TRIMMING POTENTIOMETER

The trimming potentiometer is a "set and forget" device that is not intended for dynamic
control. The top unit is a multiturn unit that
offers infinite resolution. The lower unit is a
single-turn design having a universal adjustment slot that accepts either a blade or Phil-
lips-type screwdriver. Both units have pin terminals for circuit board mounting.

trated in figure 5A, the general equivalent
circuit of a resistor. This circuit represents
the actual impedance network of a resistor
at any frequency. At a certain specified frequency the impedance of the resistor may be
thought of as a series reactance (X_s) as
shown in figure 5B. This reactance may be
either inductive or capacitive depending on
whether the residual inductance or the distributed capacitance of the resistor is the
dominating factor. As a rule, skin effect
tends to increase the reactance with frequency, while the capacitance between turns

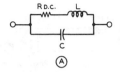

EQUIVALENT CIRCUIT OF A RESISTOR

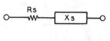

EQUIVALENT CIRCUIT OF A RESISTOR
AT A PARTICULAR FREQUENCY

Figure 5

of a wirewound resistor, or capacitance between the granules of a composition resistor
tends to cause the reactance and resistance to
drop with frequency. The behavior of various types of composition resistors over a

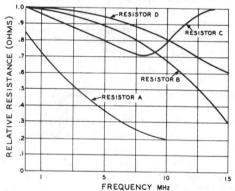

Figure 6

FREQUENCY EFFECTS ON SAMPLE
COMPOSITION RESISTORS

large frequency range is shown in figure 6.
By proper component design, noninductive
resistors having a minimum of residual reactance characteristics may be constructed.
Even these have reactive effects that cannot
be ignored at high frequencies.

Wirewound resistors act as low-Q inductors at radio frequencies. Figure 7 shows
typical curves of the high-frequency characteristics of cylindrical wirewound resistors.
In addition to resistance variations wirewound resistors exhibit both capacitive and
inductive reactance, depending on the type

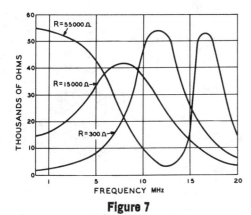

Figure 7

CURVES OF THE IMPEDANCE OF WIRE WOUND RESISTORS AT RADIO FREQUENCIES

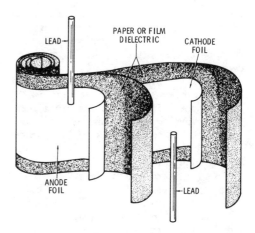

Figure 8

THE CAPACITOR

The capacitor is a device which stores and releases electrical energy. In its simplest form it consists of a layer of insulation (dielectric) sandwiched between two metallic plates, or foils. Leads are attached to the foils for external connections. The inexpensive bypass capacitors use either paper or film for the dielectric.

of resistor and the operating frequency. In fact, such resistors perform in a fashion as low-Q r-f chokes below their parallel self-resonant frequency.

17-2   The Capacitor

A *capacitor* is a device which stores and releases electrical energy. In its simplest form it consists of a layer of insulation or dielectric sandwiched between two metallic plates, or foils. The plates are oppositely charged and the electrical energy is stored in the polarized dielectric (figure 8).

The property of capacitance depends directly on the area of the plates, or foils, a product of dielectric constant and area, and is inversely proportional to the separation of the plate surfaces. Capacitance changes with temperature, frequency and dielectric age.

The two basic capacitor designs are fixed and variable units. Fixed capacitors are classified according to their dielectric material. *Mica* is a natural dielectric and has a dielectric constant averaging about 6.85. High quality mica fixed capacitors have very high dielectric strength and a sheet having a thickness of .001 inch has a breakdown potential of about 2000 volts. Mica capacitors are commonly used in high power r-f applications. Most fixed mica capacitors are planar devices with the mica

sandwiched between foil; in others the mica is metallized (silver mica).

Glass is an important dielectric and is superior to mica in many ways. The quality can be controlled more closely and there are no irregularities in a good glass dielectric. Layers of aluminum foil and glass can be interleaved and fused to form a monolithic capacitor having excellent resistance to moisture. *Vitreous enamel* is sometimes employed as a substitute dielectric for glass.

Inexpensive bypass capacitors use *paper* as a dielectric. The paper is often impregnated with mineral oil to improve the insulation and breakdown characteristics.

Organic film capacitors provide better and more reliable operation than do the older paper capacitors and these units are replacing the paper units in most applications. The film capacitors provide better insulation and can operate at higher temperatures than the paper counterparts. *Polyester film* (Mylar) is a standard dielectric which can handle peak voltages up to 1000 volts. *Polycarbonate film* is used in precision capacitors which require very high insulation resistance and a low temperature coefficient. *Polystyrene, polypropylene, poly-*

sulfone and *teflon* are also used as thermoplastic dielectrics in special capacitors.

Mylar is the least expensive and most commonly used film. It has a dielectric constant between 2.8 and 3.5, but this parameter varies widely with temperature. In addition, mylar working voltage must be derated above 85°C. Polystyrene has a linear negative temperature coefficient of about 120 ppm/°C and is often used in temperature compensating capacitors. Maximum operating temperature is 85°C. Polysulfone has high temperature capability but is expensive and unproven in regard to reliability. Teflon works well up to 250°C and has a linear temperature characteristic but suffers from a low dielectric constant.

Ceramic dielectric capacitors are widely used in audio and rf circuitry. The inexpensive disc ceramic capacitor is made of *barium titanate* with a silver paste screened on the ceramic wafer to form the electrodes. Firing fixes the electrode to the ceramic and after leads are attached the unit is encapsulated. The general purpose ceramic capacitors have a temperature-capacitance curve that is generally positive below 25°C and negative above that point. *Temperature compensated* ceramic capacitors are available with a wide range of temperature coefficients. *P-types* have a positive temperature change, while *N-types* exhibit negative change. The *NPO* type exhibits virtually no capacitance change over the temperature range of −25°C to +85°C. *Temperature stable* ceramic capacitors are refinements of the NPO type, extended out to wider temperature limits.

Layer-built, ceramic *monolithic capacitors* are composed of alternate layers of thin ceramic dielectric and noble metal thick films (figure 9). The structure is fired into a homogeneous block. After firing, the block is cut up to form capacitors. Some are less than a tenth of an inch on a side. These small units are called *chip capacitors* and common varieties are available in capacitances as high as 0.1 μF at 100 volts. The chips are leadless and unencapsulated and are designed to be attached to circuit substrates by solder reflow technique or thermal compression bonding.

The *electrolytic capacitor* is a polarized device consisting of two metallic electrodes separated by an electrolyte. A thin film of

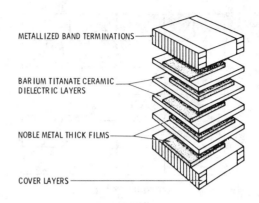

METALLIZED BAND TERMINATIONS

BARIUM TITANATE CERAMIC DIELECTRIC LAYERS

NOBLE METAL THICK FILMS

COVER LAYERS

Figure 9

THE MONOLITHIC LAYER-BUILT CAPACITOR

This capacitor is composed of alternate layers of barium titanate ceramic dielectric and noble metal thick films. The structure is fired into a homogeneous block which is cut up to form individual capacitors. The outer layers are metallized to allow solder connections to the unit.

oxide on the electrodes is produced by chemical (electrolytic) action to form the dielectric (figure 10).

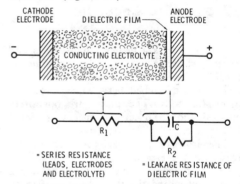

CATHODE ELECTRODE   DIELECTRIC FILM   ANODE ELECTRODE

CONDUCTING ELECTROLYTE

R_1

C

R_2

= SERIES RESISTANCE (LEADS, ELECTRODES AND ELECTROLYTE)

= LEAKAGE RESISTANCE OF DIELECTRIC FILM

Figure 10

BASIC CELL AND SIMPLIFIED EQUIVALENT CIRCUIT FOR A POLAR ELECTROLYTIC CAPACITOR

Among capacitors, the electrolytic device has the highest possible capacitance per unit volume. Common types are the *aluminum foil* and the *dry tantalum slug* versions, but there are also wet tantalum foil and slug types available. Foil units, regardless of the base metal, contain a liquid or gel electrolyte between the foil anode and the case that is in continuous contact with the oxide layer and participates in its formation. The solid, or slug-type capacitor employs a solid semi-conducting electrolyte in place of the liquid or gel, and the anode is a sponge-like porous metal slug. For dry tantalum capacitors, manganese dioxide is used as the electrolyte.

Electrolytic capacitors are classified as either *aluminum oxide* or *tantalum oxide* capacitors. While aluminum foil capacitors are widely used in power supply, high energy storage and smoothing applications, tantalum slug units are used in miniaturized circuits where space is a premium.

The electrolytic capacitor element consists of two foils separated by a dielectric and wound convolutely and sealed in an aluminum can. In order to reduce the series resistance of the capacitor, multiple, parallel connected leads are attached to each foil, reducing the ohmic path to the terminals. Computer grade (*energy storage*) capacitors employ low inductance leads for minimum series resistance and charge/discharge capability.

Dc leakage is a significant factor in the life of an electrolytic capacitor. As the capacitor ages, and leakage increases, internal gasses form which are vented off through a special seal. Reverse voltage also causes excessive gassing. In either case, gassing drives the electrolyte out of the winding, causing a loss of capacitance and an increase in the internal resistance of the capacitor. The useful life of the electrolytic capacitor can be extended by operating the voltage below the maximum rated level, operation at a low temperature and positioning of the unit to permit adequate heat dissipation.

The miniature, epoxy-dipped solid electrolyte *tantalum capacitor* provides long operating life and is hermetically sealed against moisture (figure 11). Outgassing does not occur with this type of device. These compact capacitors are available in ratings up to 680 μF in a voltage range of 3 to 50 volts. The capacitance tolerance is $\pm$ 20%.

Capacitor Inductance The inherent *residual* characteristics of capacitors include series resistance, series inductance and shunt resistance, as shown in figure 12. The series resistance and inductance depend to a large extent on the physi-

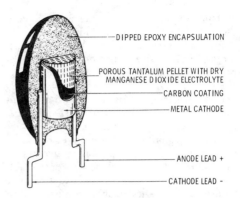

DIPPED EPOXY ENCAPSULATION

POROUS TANTALUM PELLET WITH DRY MANGANESE DIOXIDE ELECTROLYTE

CARBON COATING

METAL CATHODE

ANODE LEAD +

CATHODE LEAD −

Figure 11

MINIATURE EPOXY-DIPPED TANTALUM CAPACITOR

This dry electrolytic is hermetically sealed and is designed for insertion in printed-circuit boards.

cal configuration of the capacitor and on the material from which it is composed. Of great interest to the amateur constructor is the series inductance of the capacitor. At a certain frequency the series inductive reactance of the capacitor and the capacitive reactance are equal and opposite, and the capacitor is in itself series resonant at this frequency. As the operating frequency of the circuit in which the capacitor is used

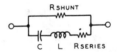

R SHUNT

C   L   R SERIES

Figure 12

EQUIVALENT CIRCUIT OF A CAPACITOR

is increased above the series-resonant frequency, the effectiveness of the capacitor as a bypassing element deteriorates until the unit is useless.

When considering the design of transmitting equipment, it must be remembered that while the transmitter is operating at some relatively low frequency (for example, 7 MHz), there will be harmonic currents flowing through the various bypass capacitors of the order of 10 to 20 times the operating frequency. A capacitor that behaves properly at 7 MHz however, may offer considerable impedance to the flow of these harmonic currents. For minimum harmonic generation and radiation, it is obviously of greatest importance to employ bypass capacitors having the lowest possible internal inductance.

Mica-dielectric capacitors have much less internal inductance than do most paper capacitors. Figure 13 lists self-resonant frequencies of various mica capacitors having various lead lengths. It can be seen from inspection of this table that most mica capacitors become self-resonant in the 12- to 50-MHz region. The inductive reactance they would offer to harmonic currents of 100 MHz, or so, would be of considerable magnitude. In certain instances it is possible to deliberately series-resonate a mica capacitor to a certain frequency somewhat below

CAPACITOR		LEAD LENGTHS	RESONANT FREQ.
.02	μF MICA	NONE	44.5 MHz
.002	μF MICA	NONE	23.5 MHz
.01	μF MICA	¼"	10 MHz
.0009	μF MICA	¼"	55 MHz
.002	μF CERAMIC	⅝"	24 MHz
.001	μF CERAMIC	¼"	55 MHz
500	pF BUTTON	NONE	220 MHz
.0005	μF CERAMIC	¼"	90 MHz
.01	μF CERAMIC	½"	14.5 MHz

Figure 13

SELF-RESONANT FREQUENCIES OF VARIOUS CAPACITORS WITH RANDOM LEAD LENGTH

its normal self-resonant frequency by trimming the leads to a critical length. This is sometimes done for maximum bypassing effect in the region of 40 to 60 MHz.

The *button-mica* capacitors shown in figure 14 are especially designed to have extremely low internal inductance. Certain types of button-mica capacitors of small physical size have a self-resonant frequency in the region of 600 MHz.

Ceramic-dielectric capacitors in general have the lowest amount of series inductance per unit of capacitance of these three universally used types of bypass capacitors. Typical resonant frequencies of various ceramic units are listed in figure 13. Ceramic capacitors are available in various voltage and capacitance ratings and different physical configurations. Standoff types such as shown in figure 14 are useful for bypassing socket and transformer terminals. Two of these capacitors may be mounted in close proximity on a chassis and connected together by an r-f choke to form a highly effective r-f

filter. The inexpensive *disc* type of ceramic capacitor is recommended for general bypassing in r-f circuitry, as it is effective as a bypass unit to well over 100 MHz.

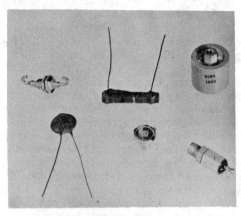

Figure 14

TYPES OF CERAMIC AND MICA CAPACITORS SUITABLE FOR HIGH-FREQUENCY BYPASSING

The Centralab 858 S (1000 pF) is recommended for screen and plate circuits of tetrode tubes.

The large TV *doorknob* capacitors are useful as by-pass units for high voltage lines. These capacitors have a value of 500 pF, and are available in voltage ratings up to 40,000 volts. The dielectric of these capacitors is usually titanium dioxide. This material exhibits piezoelectric effects, and capacitors employing it for a dielectric will tend to "talk-back" when a-c voltages are applied across them.

An important member of the varied line of capacitors is the *coaxial*, or *Hypass*, type of capacitor. These capacitors exhibit superior bypassing qualities at frequencies up to 200 MHz and the bulkhead type is especially effective when usd to filter leads passing through partition walls between two stages.

Variable Air Capacitors Even though air is the perfect dielectric, air capacitors exhibit losses because of the inherent resistance of the metallic parts that make up the capacitor. In addition, the leakage loss across the insulating supports may become of some consequence at high frequencies. Of greater concern is the inductance of the capacitor at high frequencies. Since the capacitor must be of finite size, it will have tie rods, metallic braces, and end plates; all of which contribute to the inductance of the unit. The actual amount of the inductance will depend on the physical size of the capacitor and the method used to make contact to the stator and rotor plates. This inductance may be cut to a minimum value by using as small a capacitor as is practical, by using insulated tie rods to prevent the formation of closed inductive loops in the frame of the unit, and by making connections to the centers of the plate assemblies rather than to the ends as is commonly done. A large transmitting capacitor may have an inherent inductance as large as 0.1 microhenry, making the capacitor susceptible to parasitic resonances in the 50- to 150-MHz range of frequencies.

The question of optimum C/L ratio and capacitor plate spacing is covered in Chapter Eleven. For all-band operation of a high-power stage, it is recommended that a capacitor just large enough for 40-meter operation be chosen. (This will have sufficient capacitance for operation on all higher-frequency bands.) Then use fixed padding capacitors for operation on 80 meters. Such padding capacitors are available in air, ceramic, and vacuum types.

Specially designed variable capacitors are recommended for uhf work; ordinary capacitors often have "loops" in the metal frame which may resonate near the operating frequency.

17-3 Wire and Inductors

Wire Leads Any length of wire, no matter how short, has a certain value of inductance. This property is of great help in making coils and inductors, but may be of great hindrance when it is not taken into account in circuit design and construction. Connecting circuit elements (themselves having residual inductance) together with a conductor possessing additional inductance can often lead to puzzling difficulties. A piece of No. 10 copper wire ten inches long (a not uncommon length for a plate lead in an amplifier) can have a self-inductance of 0.15 microhenry. This inductance and that of the plate tuning capacitor together

with the plate-to-ground capacity of the vacuum tube can form a resonant circuit which may lead to parasitic oscillations in the vhf regions. To keep the self-inductance at a minimum, all r-f carrying leads should be as short as possible and should be made out of as heavy material as possible.

At the higher frequencies, solid enameled copper wire is most efficient for r-f leads. Tinned or stranded wire will show greater losses at these frequencies. Tank-coil and tank-capacitor leads should be of heavier wire than other r-f leads.

The best type of flexible lead from the envelope of a tube to a terminal is thin copper strip, cut from thin sheet copper. Heavy, rigid leads to these terminals may crack the envelope glass when a tube heats or cools.

Wires carrying only audio frequencies or direct current should be chosen with the voltage and current in mind. Some of the low-filament-voltage transmitting tubes draw heavy current, and heavy wire must be used to avoid voltage drop. The voltage is low, and hence not much insulation is required. Filament and heater leads are usually twisted together. An initial check should be made on the filament voltage of all tubes of 25 watts or more plate dissipation rating. This voltage should be measured right at the tube sockets. If it is low, the filament-transformer voltage should be raised. If this is impossible, heavier or parallel wires should be used for filament leads, cutting down their length if possible.

Coaxial cable may be used for high-voltage leads when it is desirable to shield them from r-f fields. RG-8/U cable may be used at dc potentials up to 8000 volts, and the lighter RG-58/U may be used to potentials of 3000 volts. Spark plug-type high-tension wire may be used for unshielded leads, and will withstand 10,000 volts.

If this cable is used, the high-voltage leads may be cabled with filament and other low-voltage leads. For high-voltage leads in low-power exciters, where the plate voltage is not over 450 volts, ordinary radio hookup wire of good quality will serve the purpose.

No r-f leads should be cabled; in fact it is better to use enameled or bare copper wire for r-f leads and rely on spacing for insulation. All r-f joints should be soldered, and the joint should be a good mechanical junction before solder is applied.

The Inductor The *inductor* is an electric coil that stores and releases magnetic energy in the field about the coil. When the flow of current through the coil is varied, the resulting change in the magnetic field about the coil induces a voltage in the coil which opposes the supply voltage. This results in the coil having self inductance. The amount of inductance of a coil depends upon the number, size and arrangement of the turns forming the coil and the presence or absence of magnetic substances in the core of the coil.

Coils are classified according to the coil material and the type of winding. The *solenoid*, or *single-layer* winding is the simplest device, whereas a *multilayer wound* coil provides more inductance per unit of volume as compared to the solenoid. The *pi*, or *universal winding* provides a larger value of inductance per unit of volume. The coil material, in any case, may be either magnetic or nonmagnetic. *Adjustable inductors* are made by the addition of a moveable core which can be inserted or withdrawn from the inductor body. When inductance range is important, *ferrite* or other high permeability powdered core material is used, when stability is more important, a lower permeability core material is used. *Ceramic* core material is often used to approximate an *air-core inductor*, providing an electrically stable winding platform.

Air-core inductors are used for r-f chokes and tuned circuits in modern communication equipment. Coil specification is difficult because of the fact that inductors, unlike resistors, capacitors and transistors, cannot be labeled as producing a particular electrical characteristic when placed in a circuit as the frequency at which a coil is tested affects its inductance as well as its Q, or figure of merit. Also, the inductor has a great many independently variable characteristics, such as *distributed capacitance,* resistance, impedance, etc. In the main, the inductor is evaluated for Q at the chosen frequency of operation and when placed in its operating position.

Physically small inductors can be coated with a waxlike substance to protect the

TABLE 1 AIRWOUND INDUCTORS

AIRWOUND INDUCTORS									
COIL DIA. INCHES	TURNS PER INCH	B & W	AIR DUX	INDUCTANCE μH	COIL DIA. INCHES	TURNS PER INCH	B & W	AIR DUX	INDUCTANCE μH
1/2	4	3001	404T	0.18	1 1/4	4	—	1004	2.75
	6	—	406T	0.40		6	—	1006	6.30
	8	3002	408T	0.72		8	—	1008	11.2
	10	—	410T	1.12		10	—	1010	17.5
	16	3003	416T	2.90		16	—	1016	42.5
	32	3004	432T	12.0	1 1/2	4	—	1204	3.9
5/8	4	3005	504T	0.28		6	—	1206	8.8
	6	—	506T	0.62		8	—	1208	15.6
	8	3006	508T	1.1		10	—	1210	24.5
	10	—	510T	1.7		16	—	1216	63.0
	16	3007	516T	4.4	1 3/4	4	—	1404	5.2
	32	3008	532T	18.0		6	—	1406	11.8
3/4	4	3009	604T	0.39		8	—	1408	21.0
	6	—	606T	0.87		10	—	1410	33.0
	8	3010	608T	1.57		16	—	1416	85.0
	10	—	610T	2.45	2	4	—	1604	6.6
	16	3011	616T	6.40		6	—	1606	15.0
	32	3012	632T	26.0		8	3900	1608	26.5
1	4	3013	804T	1.0		10	3907-1	1610	42.0
	6	—	806T	2.3		16	—	1616	108.0
	8	3014	808T	4.2	2 1/2	4	—	2004	10.1
	10	—	810T	6.6		6	3905-1	2006	23.0
	16	3015	816T	16.8		8	3906-1	2008	41.0
	32	3016	832T	68.0		10	—	2010	108.0
NOTE: COIL INDUCTANCE APPROXIMATELY PROPORTIONAL TO LENGTH. I.E., FOR 1/2 INDUCTANCE VALUE, TRIM COIL TO 1/2 LENGTH.					3	4	—	2404	14.0
						6	—	2406	31.5
						8	—	2408	56.0
						10	—	2410	89.0

winding from damage and encapsulation of the inductor in a plastic case resembling a composition resistor is common. The less expensive small inductors are machine wound on a plastic form, with an exposed winding. Open windings have the least environmental protection and more expensive units are either encapsulated or hermetically sealed and metal-encased. Temperature coefficients for air inductors generally vary from 150 to 300 ppm/°C. Inductance tolerances are commonly ± 20% for values up to 1 μH and ± 10% above this value. The more expensive moulded inductors have a tolerance as close as 1%.

Application By Frequency Low frequency (below 100 kHz) inductors are commonly wound with solid wire, often on a laminated iron core. However above 10 kHz fine, stranded (Litz) wire is often used to improve coil Q by reducing the series dc resistance. In the medium frequency region (100 kHz to 3 MHz) solid wire is used for the majority of small inductors and ferrite cores are employed to achieve high Q in a small volume. Above 3 MHz, inductors are generally space wound with solid wire to achieve a high order of Q. Ferrite core material is often used, as discussed later. Above 30 MHz, it is common practice to use nonferrous core material, such as brass or copper, with a silver plating to reduce r-f losses. This type of core permits adjustment of the inductance but introduces losses similar to those caused by a shorted turn.

Radio-Frequency Chokes R-f chokes may be considered to be special inductances designed to have a high value of impedance over a large range of frequencies. A practical r-f choke has inductance, distributed capacitance, and resistance. At low frequencies, the distributed capacitance has little effect and the electrical equivalent circuit of the r-f choke is as shown in figure 15A. As the operating frequency of the choke is raised the effect of the distributed capacitance becomes more evident until at some particular frequency the distributed capacitance resonates with the inductance of the choke and a parallel-resonant circuit is formed. This point is shown in figure 15B. As the frequency of operating is further increased the overall reactance of the choke becomes capacitive, and finally a point of series resonance is reached (figure 15C). This cycle repeats itself as the operating frequency is raised above the series-resonant point, the impedance of the choke rapidly becoming lower on each successive cycle. A chart of this action is shown in figure 16. It can be seen that as the r-f choke approaches and leaves a condition of series resonance, the performance of the choke is seriously impaired. The condition of series resonance may easily be found by shorting the terminals of the r-f choke in question with a piece of wire and exploring the windings of the choke with a grid-dip oscillator. Most commercial transmitting-type chokes have series resonances in the vicinity of 11 or 24 MHz.

High Power R-F Chokes By observing the series-resonant frequency of the choke, a homemade, high power r-f choke may be made very inexpensively. Representative designs are listed in Table 2. The first choke covers the 7.0- to 30-MHz frequency region with the first series resonance at 43 MHz. The choke is rated for an operating potential of 5 kV and a maximum dc current of 2 amperes. The second choke covers the 3.5- to 30-MHz region, with the

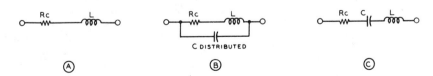

Figure 15

ELECTRICAL EQUIVALENT OF R-F CHOKE AT VARIOUS FREQUENCIES

exception of the series-resonance frequency near 25 MHz. The choke is rated for 3 kV at 1 ampere. The third choke is designed for the 21- to 54-MHz region with a series resonance near 130 MHz. It has the same voltage and current ratings as the second choke.

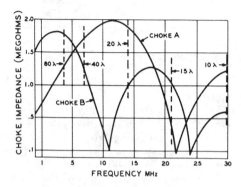

Figure 16

FREQUENCY-IMPEDANCE CHARACTERISTICS FOR TYPICAL PIE-WOUND R-F CHOKES

Ferrite "Beads"   Small, hollow sections of ferrite material can serve as an effective r-f choke when slipped over a conductor (figure 17). Unwanted, harmonic currents create a magnetic field about the conductor and, as the field cuts the ferrite material, the local impedance rises rapidly, creating the effect of an r-f choke in that immediate area (figure 18). At the lower frequencies, where the per-

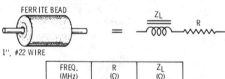

FREQ. (MHz)	R (Ω)	Z_L (Ω)
50	53	+ j45
100	95	+ j50
200	230	+ j80
250	350	+ j120

Figure 17

THE FERRITE BEAD INDUCTOR

The ferrite bead, slipped over a wire, acts as an r-f choke to harmonic currents flowing on the wire. The equivalent series impedance of a ferrite bead placed on a #22 wire is shown above. Bead is Ferroxcube K5-001-00/3B.

meability of the bead is low, there is almost no impedance to the flow of current. By stringing one or more ferrite beads on a conductor, good high frequency isolation between stages is easily achieved.

Electrically equivalent to an r-f choke, these tiny devices offer a convenient, simple and inexpensive method of obtaining effective r-f decoupling at the higher frequencies.

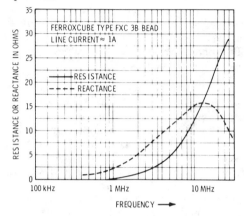

Figure 18

IMPEDANCE OF FERRITE BEAD AS A FUNCTION OF FREQUENCY

Ferrite bead functions as effective r-f choke in low-impedance circuits in the hf and vhf regions. One or more beads can be strung on a conductor to achieve isolation of harmonic currents.

17-4   Insulators

An insulator is a substance which has high resistivity to electric current flow. The characteristics of an insulator include dielectric strength, dielectric constant, dissipation factor, resistivity, and arc resistance.

The *dielectric strength* is the voltage gradient across the material at which electric failure results. It is expressed in volts per unit of thickness and failure results in an excess flow of current and possible destruction of the material.

The *dielectric constant* is the ratio of the capacitance formed by the plates of a capacitor with some insulating material between them and the capacitance when the

Table 2. HF Radio-Frequency Chokes for Power Amplifiers

4000-Watt Peak Rating	
7-30 MHz:	90 turns #18 Formex, close-wound, about 4⅛" long on ¾" diam. ✕ 6½" long teflon form. Series resonant at 43 MHz (32μH).
14-54 MHz:	43 turns #16 Formex space-wound wire diameter, about 4⅛" long on ¾" diam. ✕ 6½" long Teflon form. Series resonant at 96 MHz (15μH) It is suggested that the form be grooved on a lathe for ease in winding.
2000-Watt PEP Rating	
3.5-30 MHz:	110 turns #26e., space-wound wire diameter, about 4" long on 1" diam. ✕ 6" long ceramic form. Series resonant at 25 MHz. (78μH).
21-54 MHz:	48 turns #26e., space-wound wire diameter, about 1½" long on ½" ✕ 3" long ceramic form. Or Air-Dux 432-T (B & W 3004) on wood form. Series resonant near 130 MHz. (75μH.)

plates are in a vacuum. The *dissipation factor* is the ratio between the parallel resistance and the parallel reactance, (see figure 15, Chapter 3) and is directly related to the energy dissipation in the capacitor.

The *resistivity* is the ratio between the voltage applied to an insulator and the current flowing in it. Resistivity is sensitive to temperature, humidity, surface cleanliness, and surface contour. *Arc resistance* is a measure of insulator surface breakdown caused by an arc that tends to form a conducting path.

Thermal properties of insulators are important in electronic equipment that often operates at an elevated temperature. Generally speaking, resistivity decreases as temperature increases. Dissipation factor and dielectric constant increase under this condition. In the case of some plastics, physical stability decreases with increasing temperature. Physical and electrical stability are often indicated by a continuous-use temperature figure given for the insulator.

17-5 Relays

A relay is an electrically operated switch which permits current flow as a result of contact closure, or prohibits the flow during the open-contact state. Relays are also used as protection devices and for time-delay or multiple circuit operation.

A basic relay is shown in figure 19. A pivoted armature is held in position by a

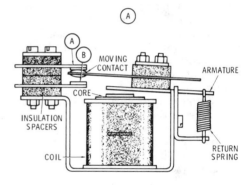

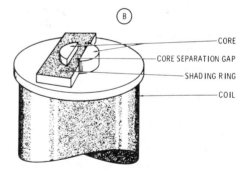

Figure 19

THE BASIC RELAY

A—The relay consists of a pivoted armature held in position by a spring. When the magnetic coil is energized, the armature moves toward the magnet, transferring the electrical circuit from the upper contact (A) to the lower contact (B). B—Relay designed for alternating current is equipped with a copper shading ring mounted above the coil to eliminate hum and chatter caused by current variations through the coil.

spring, holding the armature contact in the normally open position. When the magnetic coil is energized, the moving contact is pressed against the lower contact and the circuit is closed. Standard relays range from single pole, single throw to eight pole, double throw. The usual relay breaks the upper contact before it makes the lower contact, however, certain designs provide make-before-break sequence.

The size and material of the relay contacts are determined by the circuit requirements, usually the amount of current that will pass through the contacts. The possibility of contacts sticking is greater when making than when breaking. An inductive load persents problems, as there may be high values of current flowing during the make and break relay sequence. When the inductive load is switched off, for example, the sudden collapse of the magnetic field around the inductive load produces a very high transient voltage which can cause excessive sparking at the relay contact. A capacitive load can produce high current surges which may cause pitting and burning of the relay contacts. Tungsten carbide, mercury-wetted, or silver contacts are often used to allow good contact life.

The relay coil uses a relatively small current and coils are generally designed for 6, 12, 24, 48, or 115 volts dc and ac. Ac relay coils come equipped with copper shading ring to eliminate hum, a problem encountered because of current variations through the coil (figure 19B).

The miniature *reed relay* has recently been introduced into amateur equipment, particularly in keying circuits. The basic reed relay is a normally open switch consisting of two ferromagnetic reeds, each of which is sealed in the end of a glass tube. The reeds are positioned with their ends overlapping about $1/16$ in and are separated by a gap of about .01-inch. When a magnetic field is introduced to the switch, the reeds become flux carriers, the overlapping ends assume opposite magnetic polarities, and attract each other, making electrical contact.

The amount of power required to actuate a reed relay is typically 125 milliwatts. The more power that is applied the faster the reeds will close, until the saturation point of the reeds is reached. Maximum operating time is about 0.8 ms. Contact bounce is increased when the reed relay is driven hard, so speed is dependent on permissible bounce. Standard contact material is gold, with the more expensive relays having mercury-wetted contacts. Relay operation over a temperature range of $-65\,°C$ to $+85\,°C$ is common.

The *static relay* has no moving parts to perform the switching function. This device utilizes solid-state components to provide isolation between the signal and load circuits and provides a high ratio of off to on impedance in the controlled circuit. The static relay eliminates the mechanical problems of the electromagnetic relay but does not provide the degree of isolation between input and output circuits inherent in the older device. In addition, static relays often produce electromagnetic interference and can be temperature sensitive.

17-6   Grounds

At frequencies of 30 MHz and below, a chassis may be considered as a fixed ground reference, since its dimensions are only a fraction of a wavelength. As the frequency

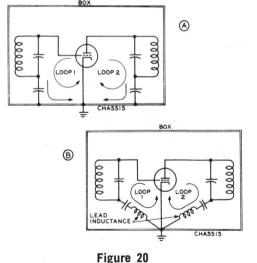

Figure 20

GROUND LOOPS IN AMPLIFIER STAGES

A. Using chassis return
B. Common ground point

is increased above 30 MHz, the chassis must be considered as a conducting sheet on which there are points of maximum current and potential. However, for the lower amateur frequencies, an object may be assumed to be at ground potential when it is affixed to the chassis.

In transmitter stages, two important current loops exist. One loop consists of the input circuit and chassis return, and the other loop consists of the output circuit and chassis return. These two loops are shown in figure 20. It can be seen that the chassis forms a return for both the input and output circuits, and that *ground currents* flow in the chassis toward the cathode circuit of the stage. For some years the theory has been to separate these ground currents from the chassis by returning all ground leads to one

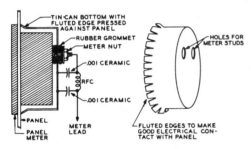

Figure 21

SIMPLE METER SHIELD

point, usually the cathode of the tube for the stage in question. This is well and good if the ground leads are of minute length and do not introduce cross couplings between the leads. Such a technique is illustrated in figure 20B, wherein all stage components are grounded to the cathode pin of the stage socket. However, in transmitter construction the physical size of the components prevent such close grouping. It is necessary to spread the components of such a stage over a fairly large area. In this case it is best to ground items directly to the chassis at the nearest possible point, with short, direct grounding leads. The ground currents will flow from these points through the low inductance chassis to the cathode return of the stage. Components grounded on the top of the chassis have their ground currents flow through holes to the cathode circuit which

is usually located on the bottom of the chassis, since such currents travel on the surface of the chassis. The usual "top to bottom" ground path is through the hole cut in the chassis for the tube socket. When the gain per stage is relatively low, or there are only a small number of stages on a chassis this universal grounding system is ideal. It is only in high gain stages (i-f strips) where the "gain per inch" is very high that circulating ground currents will cause operational instability.

Intercoupling of Ground Currents It is important to prevent intercoupling of various different ground currents when the chassis is used as a common ground return. To keep this intercoupling at a minimum, the stage should be completely shielded. This will prevent external fields from generating spurious ground currents, and prevent the ground currents of the stage from upsetting the action of nearby stages. Since the ground currents travel on the surface of the metal, the stage should be enclosed in an electrically tight box. When this is done, all ground currents generated inside the box will remain in the box. The only possible means of escape for fundamental and harmonic currents are imperfections in this electrically tight box. Whenever we bring a wire lead into the box, make a ventilation hole, or bring a control shaft through the box we create an imperfection. It is important that the effect of these imperfections be reduced to a minimum.

17-7   Holes, Leads, and Shafts

Large size holes for ventilation may be put in an electrically tight box provided they are properly screened. Perforated metal stock having many small, closely spaced holes is the best screening material. Copper wire screen may be used provided the screen wires are bonded together every few inches. As the wire corrodes, an insulating film prevents contact between the individual wires, and the attenuation of the screening suffers. The screening material should be carefully soldered to the box, or bolted with a spacing

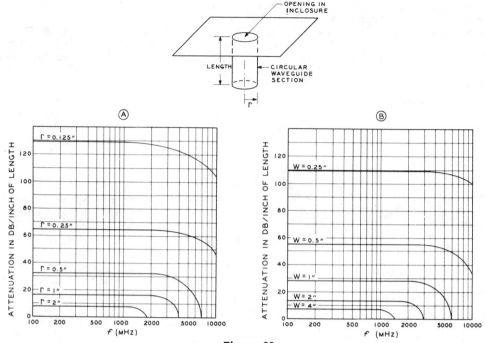

Figure 22

WAVEGUIDE-BEYOND-CUTOFF ENCLOSURE OPENINGS

Waveguide section at enclosure opening can provide improved shielding efficiency. Air passes through the waveguide but r-f attenuated to a greater degree than a simple opening can provide. Chart (A) provides attenuation in decibels/inch for circular waveguide. Chart (B) provides attenuation for rectangular waveguide for TE_{10} mode. All curves continue horizontally down to 10 MHZ.

of not less than two inches between bolts. Mating surfaces of the box and the screening should be clean.

A screened ventilation opening should be roughly three times the size of an equivalent unscreened opening, since the screening represents about a 70 percent coverage of the area. Careful attention must be paid to equipment heating when an electrically tight box is used.

Commercially available panels having half-inch ventilating holes may be used as part of the box. These holes have much less attenuation than does screening, but will perform in a satisfactory manner in all but the areas of weakest TV reception. If it is desired to reduce leakage from these panels to a minimum, the back of the grill must be covered with screening tightly bonded to the panel.

Doors may be placed in electrically tight boxes provided there is no r-f leakage around the seams of the door. Electronic weatherstripping or metal "finger stock" may be used to seal these doors. A long, narrow slot in a closed box has the tendency to act as a slot antenna and harmonic energy may pass more readily through such an opening than it would through a much larger circular hole.

Variable-capacitor or switch shafts may act as antennas, picking up currents inside the box and re-radiating them outside of the box. It is necessary either to ground the shaft securely as it leaves the box, or else to make the shaft of some insulating material.

A two- or three-inch panel meter causes a large leakage hole if it is mounted in the wall of an electrically tight box. To minimize leakage, the meter leads should be bypassed and shielded. The meter should be en-

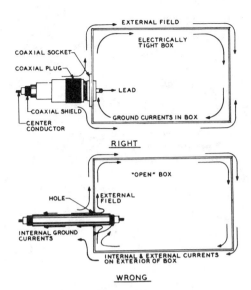

Figure 23

Use of coaxial connectors on electrically tight box prevents escape of ground currents from interior of box. At the same time external fields are not conducted into the interior of the box.

cased in a metal shield that makes contact to the box entirely around the meter. The connecting studs of the meter may project through the back of the metal shield. Such a shield may be made out of the end of a tin or aluminum can of correct diameter, cut to fit the depth of the meter. This complete shield assembly is shown in figure 21.

Openings for shafts, meters, and ventilation are sources of r-f leakage, and this spurious radiation may be reduced by designing the aperture through which leakage occurs as a waveguide-type attenuator.

A cutoff frequency for any waveguide is the lowest frequency at which propagation occurs without attenuation. Below cutoff, attenuation is a function of guide length and frequency. When an aperture is designed as a *waveguide below cutoff*, shielding efficiencies of a high order are achieved.

Figure 22A shows a set of design curves for circular waveguides ranging from 0.125″ to 2″ in radius and figure 22B shows curves for rectangular guides up to 4″ in width. When the diameter or width of the opening is known, select the maximum frequency at which r-f suppression is desired. Select the appropriate curve from either chart and read

attenuation in decibels per inch of length. Making the length of the waveguide three times the diameter for 100 dB of attenuation, and 80 dB with rectangular guides is a useful design shortcut.

As an example a 1″ diameter hole is required in an enclosure and 100-dB transmission attenuation through the hole is desired at 100 MHz. From figure 23A attenuation is 32 dB per inch at 100 MHz for radius = ½″. The required length is 100/32 equals 3.13 inches.

Pass-Through Leads   Careful attention should be paid to leads entering and leaving the electrically tight box. Harmonic currents generated inside the box can easily flow out of the box on power or control leads, or even on the other shields of coaxially shielded wires. Figure 23 illustrates the correct method of bringing shielded cables into a box where it is desired to preserve the continuity of the shielding.

Unshielded leads entering the box must be carefully filtered to prevent fundamental and harmonic energy from escaping down the lead. Combinations of r-f chokes and low-inductance bypass capacitors should be used in power leads. If the current in the lead is high, the chokes must be wound of

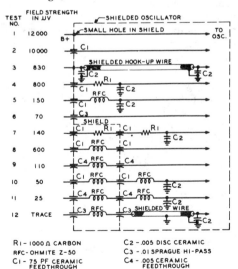

Figure 24

LEAD LEAKAGE WITH VARIOUS LEAD FILTERING SYSTEMS

large-gauge wire. Composition resistors may be substituted for the r-f chokes in high-impedance circuits. Bulkhead or feedthrough-type capacitors are preferable when passing a lead through a shield partition. A summary of lead leakage with various filter arrangements is shown in figure 24.

Internal Leads Leads that connect two points within an electrically tight box may pick up fundamental and harmonic currents if they are located in a strong field of flux. Any lead forming a closed loop with itself will pick up such currents, as shown in figure 25. This effect is enhanced if the lead happens to be self-resonant at the frequency of the existing energy. The solution for all of this is to bypass all internal power leads and control leads at each end, and to shield these leads their entire length. All filament, bias, and meter leads should be so treated. This will make the job of filtering the leads as they leave the box much easier, since normally "cool" leads within the box will not have picked up spurious currents from nearby "hot" leads.

Chassis Material From a point of view of electrical properties, aluminum is a poor chassis material. It is difficult to make a soldered joint to it, and all grounds must rely on a pressure joint. These pressure joints are prone to give trouble at a later date because of high resistivity caused by the formation of oxides from electrolytic action in the joint. However, the ease of working and forming the aluminum material far outweighs the electrical shortcomings, and aluminum chassis and shielding may be used with good results provided care is taken in making all grounding connections. Cadmium and zinc plated chassis are preferable from a corrosion standpoint, but are much more difficult to handle in the home workshop.

17-8 Parasitic Resonances

Filament leads within vacuum tubes may resonate with the filament bypass capacitors at some particular frequency and cause instability in an amplifier stage. Large tubes

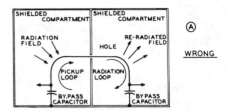

Illustration of how a supposedly grounded power lead can couple energy from one compartment to another

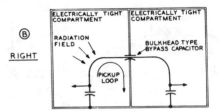

Ilustration of lead isolation by proper use of bulkhead bypass capacitor

Figure 25

of the 4-1000A and 3-1000Z type are prone to this spurious effect. In particular, an amplifier using .001-μF filament bypass capacitors had a filament resonant loop that fell in the 7-MHz amateur band. When the amplifier was operated near this frequency, marked instability was noted, and the filaments of the tubes increased in brilliance when plate voltage was applied to the amplifier, indicating the presence of r-f in the filament circuit. Changing the filament bypass capacitors to .01 μF lowered the filament resonance frequency to 2.2 MHz and cured this effect. A 1-kV mica capacitor of .01 μF used as a filament bypass capacitor on each filament leg seems to be satisfactory from both a resonant and a TVI point of view. Filament bypass capacitors smaller in value than .01 μF should be used with caution.

17-9 Parasitic Oscillation in Vacuum-Tube Amplifiers

Parasitics (as distinguished from self-oscillation on the operating frequency) are undesirable oscillations either of very-high or very-low frequency which may occur in r-f amplifiers, regardless of operating frequency or power level.

A parasite may cause spurious signals, splatter outside the normal passband of the equipment, TVI, key clicks, voltage break-down or flashover, circuit instability, and shortened life or failure of the tubes. The oscillation may be continuous or be triggered by keying or modulation and is usually a result of unwanted resonant circuits external to the tube coupled in such a way as to allow spurious oscillation or feedback. Normal neutralization circuits usually do not suppress parasitic oscillations.

The cure for parasites is twofold: The oscillatory circuit is damped (loaded) until sustained oscillation is impossible, or it is detuned until oscillation ceases.

Low-Frequency Parasites One type of parasite is created when r-f chokes in input and output circuits of an amplifier are resonant and coupled through the interelectrode capacitance of the tube (figure 26). This type of parasite is generally at a much lower frequency than the operating frequency and causes spurious signals to appear, spaced ten to several hundred kHz on either side of the main signal. The parasitic circuit is composed of the r-f chokes and bypass capacitors. The neutralizing circuit no longer provides out-of-phase feedback at the parasite frequency but actually enhances the low-frequency oscillation. A neon bulb held near the oscillatory circuit will glow *yellow* and the parasite may be heard in a nearby receiver tuned to the frequency of oscillation.

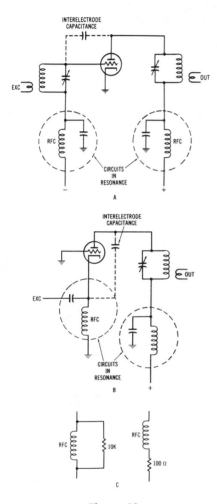

Figure 26

LOW-FREQUENCY PARASITIC CIRCUITS IN VACUUM-TUBE AMPLIFIERS

A—Low frequency parasitic circuit formed by grid and plate r-f chokes and bypass capacitors in grid-driven amplifier. B—Parasitic circuit in cathode-driven amplifier. C—Parasitic circuits are de-Qed by addition of either series or parallel resistance until circuit will not sustain parasitic oscillation.

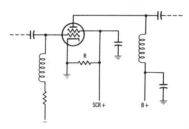

Figure 27

SCREEN-RETURN RESISTOR FOR TETRODE TUBE

The tetrode tube can draw negative screen current under certain operating conditions, leading to an R-C oscillation in the screen power supply. A resistor (R) placed between screen and ground that draws about twice the maximum positive value of screen current will suppress the oscillation.

The parasitic circuit may be broken up by changing the inductance of one of the chokes, or removing one choke and replacing it with a wirewound resistor of sufficient wattage to carry the current in the circuit. If the choke is to remain in the circuit, it may be shunted with a high value of resistor to de-Q it, or a series resistor may be added before the bypass capacitor. In some cases the value of the capacitor must be changed.

In the case of a high-gain tetrode, low-frequency oscillation can take place because of the negative resistance that exists under certain portions of the operating cycle (figure 27). A bleeder resistor from screen to ground will suppress this type of oscillation.

17-10 Elimination of VHF Parasitic Oscillations

Vhf parasitic oscillations are often difficult to locate and difficult to eliminate since their frequency often is only moderately above the desired frequency of operation. But it may be said that vhf parasitics always may be eliminated if the operating frequency is appreciably below the upper frequency limit for the tubes used in the stage. However, the elimination of a persistent parasitic oscillation on a frequency only moderately higher than the desired operating frequency will involve a sacrifice in either the power output or the power sensitivity of the stage, or in both.

Beam-tetrode stages, particularly those using 6146 or TV-style sweep tubes, will almost invariably have one or more vhf parasitic oscillations unless adequate precautions have been taken in advance. Many of the units described in the constructional section of this edition had parasitic oscillations when first constructed. But these oscillations were eliminated in each case; hence, the expedients used in these equipments should be studied. Vhf parasitics may be readily identified, as they cause a neon lamp to have a purple glow close to the electrodes when it is excited by the parasitic energy.

Parasitic Oscillations with Triodes   In the case of triodes, vhf parasitic oscillations often come about as a result of inductance in the neutralizing leads. This is particularly true in the case of push-pull amplifiers. The cure for this effect will usually be found in reducing the length of the neutralizing leads and increasing their diameter. Both the reduction in length and increase in diameter will reduce the inductance of the leads and tend to raise the parasitic oscillation frequency until it is out of the range at which the tubes will oscillate. The use of straightforward circuit design with short leads will assist in forestalling this trouble at the outset.

Vhf parasitic oscillations may take place as a result of inadequate bypassing or long bypass leads in the filament, grid-return, and plate-return circuits. Such oscillations also can take place when long leads exist between

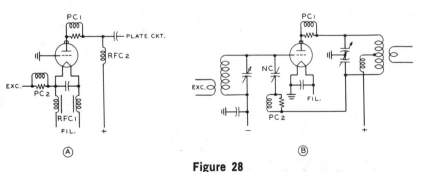

Figure 28

PARASITIC SUPPRESSION CIRCUITS

A—Plate parasitic suppressor is used in grounded-grid circuit. Filament suppressor may be added if secondary parasitic is present. B—Plate parasitic suppressor is used for grid-driven circuit, with second suppressor added in neutralizing circuit, if necessary.

the grid and the grid tuning capacitor or between the plate and the plate tuning capacitor. Sometimes parasitic oscillations can be eliminated by using iron or nichrome wire for the neutralizing lead. But in any event it will always be found best to make the neutralizing lead as short and of as heavy conductor as is practicable.

To increase losses at the parasitic frequency, the parasitic coil may be wound on 100-ohm 2-watt resistors. The "lossy" suppressor should be placed in the plate or grid lead of the tube close to the anode or grid connection, as shown in figure 28.

Parasitics with Where beam-tetrode tubes are
Beam Tetrodes used in the stage which has
been found to be generating the parasitic oscillation, all the foregoing suggestions apply in general. However, there are certain additional considerations involved in elimination of parasitics from beam-tetrode amplifier stages. These considerations involve the facts that a beam-tetrode amplifier stage has greater power sensitivity than an equivalent triode amplifier, such a stage has a certain amount of screen-lead inductance which may give rise to trouble, and such stages have a small amount of feedback capacitance.

Beam-tetrode stages often will require the inclusion of a neutralizing circuit to eliminate oscillation on the operating frequency. However, oscillation on the operating frequency is not normally called a parasitic oscillation, and different measures are required to eliminate the condition.

When a parasitic oscillation is found on a very high frequency, the interconnecting leads of the tube, the tuning capacitors and the bypass capacitors are involved. This type of oscillation generally does not occur when the amplifier is designed for vhf operation where the r-f circuits external to the tube have small tuning capacitors and inductors. Without tuning capacitors, the highest frequency of oscillation is then the fundamental frequency and no higher frequencies of resonance exist for the parasitic oscillation.

The vhf oscillation commonly occurs in hf amplifiers, using the capacitors and associated grid and plate leads for the inductances of the tuned circuit. The frequency

of unwanted oscillation is generally well above the self-neutralizing frequency of the tube. If the frequency of the parasitic can be lowered to or below the self-neutralizing frequency, complete suppression of the parasitic will result. It is also possible to suppress the oscillation by loading the circuit so that the circuit is "lossy" at the parasitic frequency. This may be done by the use of a parasitic choke in the plate and/or grid lead of the stage in question. A parallel coil and resistor combination operates on the principle that the resistor loads the vhf circuit but is shunted by the coil for the lower fundamental frequency. The parasitic choke (figure 29) is usually made up of a noninductive resistor of about 25 to 100 ohms, shunted by three or four turns of wire, approximately one-half inch in diameter and frequently wound over the body of the resistor.

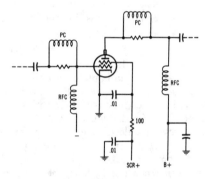

Figure 29

PARASITIC SUPPRESSOR

RL-type parasitic suppressor is placed in plate or grid lead of tetrode and pentode tubes. Screen circuit is isolated from r-f entering power lead by two bypass capacitors and series 100-ohm resistor.

In the process of adjusting the resistor-coil combination, it may be found that the resistor runs too hot. The heat is usually caused by the dissipation of fundamental power in the resistor, which is an indication of too many turns in the suppressor coil. Just enough turns should be used to suppress the parasitic oscillation, and no more. Once the circuit is properly loaded and the parasitic suppressed, no parasitic power will be present and no power other than primary power will be lost in the resistor of the suppressor.

For medium power levels, a plate suppressor may be made of a 22-ohm, 2-watt Ohmite or Allen-Bradley composition resistor wound with 4 turns of No. 18 enameled wire. For kilowatt stages operating up to 30 MHz, a satisfactory plate suppressor may be made of three 220-ohm, 2-watt composition resistors in parallel, shunted by 3 or 4 turns of No. 14 enameled wire, ½ inch diameter and ½ inch long.

The parasitic suppressor for the plate circuit of a small tube such as the 6146, 6LQ6, or similar type normally may consist of a 47-ohm composition resistor of 2-watt size with 4 turns of No. 18 enameled wire wound around the resistor. However, for operation above 30 MHz, special tailoring of the value of the resistor and the size of the coil wound around it will be required in order to attain satisfactory parasitic suppression without excessive power loss in the parasitic suppressor.

Tetrode Screening   Isolation between the grid and plate circuits of a tetrode tube is not perfect. For maximum stability, it is recommended that the tetrode stage be neutralized. Neutralization is *absolutely necessary* unless the grid and plate circuits of the tetrode stage are each completely isolated from each other in electrically tight boxes. Even when this is done, the stage will show signs of regeneration when the plate and grid tank circuits are tuned to the same frequency. Neutralization will eliminate this regeneration. Any of the neutralization circuits described in the chapter *Generation of R-F Energy* may be used.

17-11   Checking for Parasitic Oscillations

It is an unusual transmitter which harbors no parasitic oscillations when first constructed and tested. Therefore it is always wise to follow a definite procedure in checking a new transmitter for parasitic oscillations.

Parasitic oscillations of all types are most easily found when the stage in question is running by itself, with full plate (and screen) voltage, sufficient protective bias to limit the plate current to a safe value, and no excitation. One stage should be tested at a time, and the complete transmitter should never be put on the air until all stages have been thoroughly checked for parasitics.

To protect tetrode tubes during tests for parasitics, the screen voltage should be applied through a series resistor which will limit the screen current to a safe value in case the plate voltage of the tetrode is suddenly removed when the screen supply is on. The correct procedure for parasitic testing is as follows (figure 30):

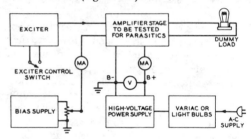

Figure 30

SUGGESTED TEST SETUP FOR PARASITIC TEST

1. The stage should be coupled to a dummy load, and tuned up in correct operating shape. Sufficient protective bias should be applied to the tube at all times. For protection of the stage under test, a lamp bulb should be added in series with one leg of the primary circuit of the high-voltage power supply. As the plate-supply load increases during a period of parasitic oscillation, the voltage drop across the lamp increases, and the effective plate voltage drops. Bulbs of various sizes may be tried to adjust the voltage under testing conditions to the correct amount. If a *Variac* or *Powerstat* is at hand, it may be used in place of the bulbs for smoother voltage control. Don't test for parasitics unless some type of voltage control is used on the high-voltage supply! When a stage breaks into parasitic oscillations, the plate current increases violently and some protection to the tube under test *must* be used.

2. The r-f excitation to the tube should now be removed. When this is done, the grid, screen, and plate currents of the tube should drop to zero. Grid and plate tuning capacitors should be tuned to minimum capacity. No change in resting grid, screen, or plate current should be observed. If a parasitic is present, grid current will flow, and there will be an abrupt increase in plate current. The size of the lamp bulb in series with the high-voltage supply may be varied until the stage can oscillate continuously, without exceeding the rated plate or screen dissipation of the tube.

3. The frequency of the parasitic may now be determined by means of an absorption wavemeter, or a neon bulb. Low-frequency oscillations will cause a neon bulb to glow yellow. High-frequency oscillations will cause the bulb to have a soft, violet glow.

4. When the stage can pass the above test with no signs of parasitics, the bias supply of the tube in question should be decreased until the tube is dissipating its full plate rating when full plate voltage is applied, with no r-f excitation. Excitation may now be applied and the stage loaded to full input into a dummy load. The signal should now be monitored in a nearby receiver which has the antenna terminals grounded or otherwise shorted out. A series of rapid dots should be sent, and the frequency spectrum for several MHz each side of the carrier frequency carefully searched. If any vestige of parasitic is left, it will show up as an occasional "pop" on a keyed dot. This "pop" may be enhanced by a slight detuning of the input or output circuit.

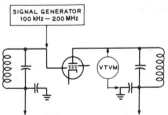

Figure 31

PARASITIC GAIN MEASUREMENT

Grid-dip oscillator and vacuum tube voltmeter may be used to measure parasitic stage gain over 100 kHz-200 MHz region.

5. If such a parasitic shows up, it means that the stage is still not stable, and further measures must be applied to the circuit. Parasitic suppressors may be needed in both screen and grid leads of a tetrode, or perhaps in both grid and neutralizing leads of a triode stage. As a last resort, a 10,000-ohm 25-watt wirewound resistor may be shunted across the input circuit of a high powered stage. This strategy removed a keying "pop" that showed up in a commercial transmitter, operating at a plate voltage of 5000.

Test for Parasitic Tendency in Tetrode Amplifiers In most high-frequency transmitters there are a great many resonances in the tank circuit at frequencies other than the desired operating frequency. Most of these parasitic resonant circuits are not coupled to the tube and have no significant tendency to oscillate. A few, however, are coupled to the tube in some form of oscillatory circuit. If the regeneration is great enough, oscillation at the parasitic frequency results. Those spurious circuits existing just below oscillation must be found and suppressed to a safe level.

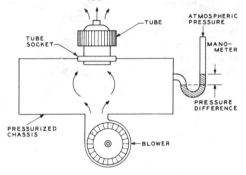

Figure 32

FORCED-AIR COOLING SYSTEM

Centrifugal blower pressurizes plenum chamber (air-tight chassis) and air is exhausted through the tube socket and anode cooler of vacuum tube. Pressure difference between plenum chamber and atmosphere is measured with manometer tube.

One test method is to feed a signal from a grid-dip oscillator into the grid of a stage and measure the resulting signal level in the plate circuit of the stage, as shown in figure 31. The test is made with all operating voltages applied to the tubes. Class-C stages should have bias reduced so a reasonable amount of static plate current flows. The grid-dip oscillator is tuned over the range of 100 kHz to 200 MHz, the relative level of the r-f voltmeter is watched, and the frequencies at which voltage peaks occur are noted. Each significant peak in voltage gain in the stage must be investigated. Circuit changes or suppression must then be added to reduce all peaks by 10 dB or more in amplitude.

17-12 Forced Air Cooling

A large percentage of the primary power drain of a transmitter is converted to heat emitted by tubes and components. The resulting temperature rise must be held within reasonable limits to ensure satisfactory life for the equipment.

Forced-air-cooled systems may be used to remove excess heat. A typical system consists of an *air blower,* a *conduit* to guide the air to the tube or component, a *heat radiator* on the component, and an air *exhaust exit.* The resistance to the air passage through such a system is termed system *back pressure, pressure drop,* or *static pressure.* Air requirements are normally expressed as a pressure drop defined in *inches of water* (as measured by a manometer) with a corresponding volumetric air flow defined in

cubic feet per minute (c.f.m.). A typical air-cooling system is shown in figure 32. Cooling requirements for most transmitting tubes are provided on the data sheet and air requirements and blower data for some popular tubes are given in figure 33.

Adequate cooling of the tube envelope and seals is one of the factors leading to long tube life. Deteriorating effects increase directly with the temperature of the tube envelope and seals. Even if no cooling air is specified by the technical data sheet for a particular tube, ample free space for circulation of air about the tube is required, or else air must be forced past the tube.

As the frequency of operation of the tube is extended into the vhf region, additional cooling is usually required because of the larger r-f losses inherent in the tube structure.

Temperature-sensitive paint or crayons may be used to monitor the temperature of a tube under operating conditions. If the paint is applied to the tube envelope in a *very* thin coat, it will melt and virtually disappear at its critical temperature. After subsequent cooling, it will have a crystalline appearance indicating that the surface with which it is in contact has exceeded the critical temperature. Temperature-sensitive tapes and decals are also available to measure envelope temperature of transmitting tubes.

TUBE TYPE	AIR CFM	BACK PRESSURE	BLOWER SIZE	RPM	SOCKET CHIMNEY
3-400Z 3-500Z	13	0.20	3	1600	SK410 SK416 SK406
3-1000Z	25	0.64	3 3/4 2 1/2	3000 6000	SK510 SK516
4-1000A	25	0.64	3 3/4 2 1/2	3000 6000	SK510 SK506
4CX250B	6.4	1.12	2 1/2	6000	SK600 SK606
4CX1000A 4CX1500B	22	0.3	3	3100	SK800 SK806
5CX1500A	47	1.12	3	6000	SK840 SK806

Figure 33

COOLING REQUIREMENTS FOR TRANSMITTING TUBES

Air-system sockets and chimneys are required for high-power transmitting tubes. Complete air-cooling data for these types may be obtained from Application Engineering Department, Eimac Division of Varian, San Carlos, Calif. 94070.

17-13 Conduction Cooling

The anode power dissipation density in a modern transmitting tube is extremely high and *conduction cooling* is often used to remove the heat from the tube structure.

A conduction cooling system comprises the heat source (the power tube), a *thermal link* to transfer the heat, and a *heat sink,* where the heat is removed from the system. The thermal link has the dual properties of a thermal conductor and an electrical insulator. *Beryllium oxide* (BeO) combines these properties and is generally used for the thermal link. The BeO link may be brazed to the tube or be a detachable accessory (figure 34).

Most conduction-cooled tubes have an output capacitance which is higher than con-

Figure 34

CONDUCTION-COOLED TUBE WITH INTEGRAL THERMAL LINK

Experimental-type Y-406 tetrode makes use of beryllium oxide thermal link to transfer anode heat to an external heat sink. Link is pressed against the sink, with mating surfaces coated with silicone grease to improve interface thermal resistance. The heat sink transfers excess system heat to the surrounding atmosphere.

ventional air-cooled tubes due to the added capacitance between the tube anode and the heat sink, typically 6 to 10 pF. The capacitance is caused by the BeO dielectric. Below about 150 MHz, this added capacitance causes little difficulty since it can be included in the matching network design. Above 150 MHz, care in network design still permits successful operation up to the frequency limit of the tube, but attention must be given to bandwidth and efficiency requirements and the physical length and configuration of the required resonating inductance as the added capacitance of the thermal link will limit the value of reasonating inductance.

Normal use of electron tubes having beryllium oxide is safe. However, BeO dust or fumes are highly toxic and breathing them can be injurious to health. Never per-

form work on any ceramic part of a power tube utilizing this material which could possibly generate dust or fumes. At the end of the useful life of the tube or heat sink, the BeO material should be returned prepaid to the manufacturer with written authorization for its disposal.

17-14 Transient Protection

Circuits that are switched on and off can produce transients because of inductance in the circuits. The inductance may be a desired portion of the circuit or it may be the residual inductance of the circuit wiring and components. Transients can range as high as five to ten times the nominal circuit voltage and may damage equipment and components associated with the circuit.

Transient protection will reduce the damaging effects of over-voltage and various forms of *transient suppressors* are available to do this work. The simplest form of transient protection takes the form of a voltage sensitive gap which trips, or fires, at a given value of peak voltage. Devices are made that trip from 550 volts to 20 kilovolts. Vacuum encapsulated gaps fire in a matter of microseconds when the trip voltage is exceeded and are capable of passing peak current pulses of thousands of amperes while preserving infinite resistance up to the trip voltage. Solid state suppressors are less costly but may exhibit a trip voltage that is dependent upon the rate at which the voltage is applied.

The life of a voltage suppressor can be approximated in terms of the cumulative charge, in coulombs, that can be passed through the device without changing its trip voltage by more than 10 percent. As an example, a vacuum-type protector may have a life of 3000 discharges under a given set of circumstances.

When the protection device fires, the near-infinite resistance drops to a near short. Thus, the follow-on current is limited only by circuit impedance. This current must be limited and finally interrupted to allow arrester recovery .

CHAPTER EIGHTEEN

Transmitter Keying and Control

Information is imparted to a radio wave by the process of modulation, which implies that the radio signal is changed in amplitude, frequency or phase. *On-off* (c-w) keying is a simple type of amplitude modulation and is a basic form of communication among radio amateurs.

Keying is usually accomplished in a low power stage of a transmitter so that the controlled power is small. The amplifiers following the keyed stage must be designed so that their power consumption remains within a safe limit when the drive signal is cut off during keying.

In certain styles of operation, it is convenient for the operator to listen through his transmission so that the station at the other end of the circuit can *break-in* while the first operator is transmitting. This requires that the sending station avoid generating an interfering signal, or *back wave*, in the local receiver when the transmitter is keyed off.

In simple on-off keying, the carrier is broken into dots and dashes of the Morse Code for transmission. The carrier signal is of constant amplitude when the key is closed, and is entirely removed when the key is open. If the change from key-up to key-down condition occurs too rapidly, the rectangular pulse which forms the keying character contains high-frequency components which take up a wide frequency band as sidebands and are heard as *key clicks* on the signal.

To be capable of transmitting code characters and at the same time not be causing unnecessary adjacent channel interference, the c-w transmitter must meet two important specifications:

1. The transmitter must have no parasitic oscillations either in the stage being keyed, or in any preceding or following stage.
2. The transmitter must have filters in the keying circuit capable of shaping the leading and trailing edge of the waveform.

Both of these specifications must be satisfied before the transmitter is capable of meeting the FCC regulations concerning spurious emissions. Merely turning a transmitter carrier on and off by the haphazard insertion of a telegraph key, or keyer, in some power lead is an invitation to trouble.

Shown in this chapter are keying circuits and keyers capable of keying a transmitter to provide clean, clickless keying at high speed and which keep the keying circuit at ground potential so that no danger of shock exists to the operator.

18-1 Keying Requirements

The transmitter keying circuit must provide fast, clickless keying with no frequency variation or *chirp* in the keyed wave. Key click elimination is accomplished by preventing a too rapid make and break of power in the keyed circuit, thus rounding off the keying characters so as to limit the sidebands to a value which does not cause interference in adjacent channels. The optimum keying characteristic is a highly subjective thing and "on the air" checks are questionable, since many amateurs hesitate to be truly critical of another amateur's signal unless it is causing objectionable interference.

Various keying characteristics are shown in figure 1. Illustration A shows a keyed wave with the envelope rising from zero to full value in 10 microseconds (μs). The leading edge of the signal has the same shape as one modulation cycle of an r-f signal modulated with a frequency of 100 kHz. Sidebands 100 kHz on each side of the carrier are therefore generated by this waveform. Up to a keying speed of 100 words per minute, a rise time as slow as 5 milliseconds can be used (illustration B), reducing the sidebands to 200 Hz. Suitable filter circuits in the keying system reduce the rise and decay times of the keyed characteristic to conservative values, thus decreasing the keyed bandwidth of the signal.

Poor power supply regulation can alter an otherwise perfect keyed waveform (illustration C). Insufficient filter capacitance permits the power in the keyed wave to sag during the long dash, adding an unusual-sounding characteristic to the signal.

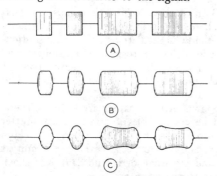

Figure 1
C-W KEYING CHARACTERISTICS
A—Abrupt rise and decay time of dot character leads to severe key clicks on make and break. B—Simple keying filter rounds dot character reducing transition time between key-open and key-closed condition. C—Poor power-supply regulation can distort keying waveform and add "yoop" to the signal.

With high power equipment, transmitter keying can affect power line regulation and possibly make the lights blink in the vicinity of the transmitter. The variation in line voltage may affect the regulation of certain power supplies in the equipment, or make a slow variation in filament voltage, that will change the keying characteristic of the transmitted signal.

Location of Keyed Stage   If the transmitter is keyed in a stage close to an oscillator, the change in r-f load to the oscillator may cause the oscillator to shift frequency with keying. This will cause the oscillator to have a distinct chirp. If the oscillator is followed by frequency multiplication, the chirp will be multiplied as many times as the frequency is multiplied. The oscillator itself may be keyed but it is common to employ differential keying, as described later, to eliminate the frequency instability caused by turning the oscillator off and on.

In a heterodyne system, a mixer stage is often keyed with the frequency-generating stage protected by one or two class-A intermediate buffer-amplifiers. When done at a low level, the keyed stage is commonly followed by a class-A amplifier stage as the keyed waveform is retained in a linear amplifier. Linear stages are employed up through the output stage in an SSB transmitter, which are capable of preserving the keyed waveform.

Class-C (nonlinear) stages that follow the keyed stage sharpen the keyed characters, introducing sharp leading and trailing edges to the character and thus spoil an otherwise well-keyed signal. The class-C stage acts as a peak clipper, tending to square up the rounded keying impulse, and the cumulative effect of several such stages is sufficient to alter the keyed waveform to the point where bad clicks are reimposed on a clean signal.

Differential Keying   Oscillator keying is tempting since it permits break-in operation, permitting the operator to listen to the other station between keyed characteristics. The use of *differential keying* permits break-in, as the oscillator is turned on quickly by the keying sequence, a moment before the rest of the transmitter stages are energized, and remains on a moment longer than the other stages (figure 2). The chirp, or frequency shift, associated with abrupt switching of the oscillator is thus removed from the emitted signal. In addition, the differential keyer can apply waveshaping to the amplifier section of the transmitter, eliminating the click caused by rapid keying of the later stages.

The ideal differential keying sequence is shown in the illustration. When the key is closed, the oscillator reaches maximum output almost instantaneously. The following stages reach maximum output in a fashion determined by the waveshaping circuits of the keyer. When the key is opened, the output of the amplifier stages starts to decay in a predetermined manner, followed shortly

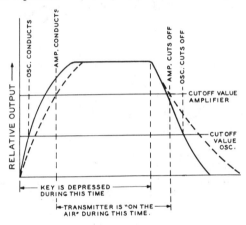

Figure 2

DIFFERENTIAL KEYER TIME SEQUENCE

When differential keying is used, the oscillator is turned on quickly by the keying sequence, a moment before the rest of the transmitter is energized (at left of illustration). The oscillator remains on a moment longer than the rest of the transmitter (at right of illustration). Any chirp, or frequency shift associated with abrupt oscillator switching is thus removed from the emitted signal.

by cessation of the oscillator. The end result of this sequence is to provide relatively soft make and break to the keyed signal, meanwhile preventing oscillator frequency shift during the active keying sequence.

The rate of charge and decay in a representative RC keying circuit may be varied independently by the blocking diode system shown in figure 3. Each diode permits the charging current of the timing capacitor to flow through only one of the two adjustable potentiometers, thus permitting independent

adjustment of the make and break characteristics of the keying system.

18-2   Transmitter Keying

The problems of keying a vacuum tube are somewhat different from keying a solid-state circuit. The vacuum tube may be keyed in the grid, cathode or screen circuit and the tube element may be either blocked with a negative voltage or opened with respect to ground or the positive potential of the supply.

The transistor may be keyed in a similar fashion by varying the base or emitter voltage as described in Chapter 4. The transistor can readily be adjusted to either cutoff or saturation by controlling the base-to-emitter potential. The potential of the keyed waveform will depend on the polarity of the keyed voltage and the choice of either an NPN or PNP keying transistor.

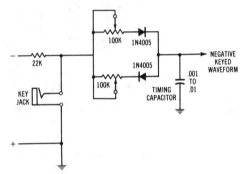

Figure 3

WAVEFORM SHAPING CIRCUIT

Reverse-connected diodes vary time constant of "make" and "break" characteristics of keyed stage.

Keying the Vacuum Tube The vacuum tube may be keyed either in the cathode or grid circuit (figure 4). Cathode keying opens both the plate and grid dc return circuits, thus blocking the grid at the same time the plate return circuit is opened (A). Voltage exists across the keying contacts and an electronic switch or relay should be used for shock protection.

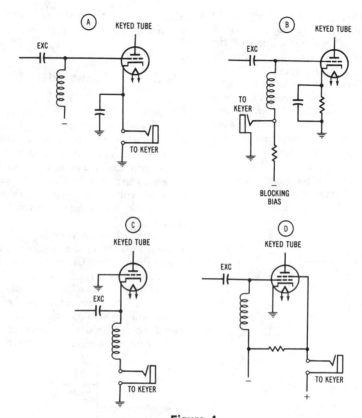

Figure 4

VACUUM TUBE KEYING

A—cathode keying. B—Blocked-grid keying. C—Blocked-grid keying with zero-bias triode.

D—Screen keying with tetrode tube.

Blocked-grid keying, wherein the exciting voltage is overridden by a negative blocking voltage, can be used with tetrode and high-mu triode tubes (B). The keyed waveform may be determined by an RC constant in the grid circuit. Grid current flows through the keying circuit and voltage exists across the keying contacts.

Self-blocking keying in the grid circuit may be achieved with high-mu triodes such as the 811A and 3-500Z which automatically cut themselves off when the grid return circuit is opened (C).

In the case of the tetrode, the screen circuit may be keyed, with a blocking voltage applied to the screen in the key-up condition to reduce the backwave caused by r-f leakage through the grid-plate capacitance of

the tube (D). Voltage exists across the keying contacts.

18-3   Break-in Keying

Break-in c-w operation permits information to be transmitted back and forth between two stations at will. For true break-in, each station must be able to listen to the other during the key-up period, while the receiver remains mute (or operates at reduced gain) during the key-down period. Thus, one operator can "break" the other at any time between the dots and dashes of a single letter.

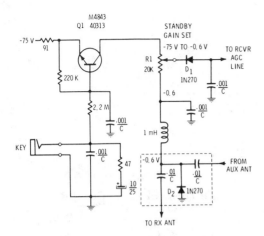

Figure 5

BREAK-IN CIRCUIT FOR TRANSMITTER AND RECEIVER CONTROL

In key-up position, or receive, the auxiliary receiving antenna is connected to the receiver through a simple T-R switch and the receiver agc system functions normally. In the key-down position, or transmit, the receiving antenna is shorted to ground and a negative voltage is applied to the receiver agc line. The keying constants are adjusted by an RC network placed across the key.

In order to achieve break-in capability, the receiver must be protected against overload from the nearby transmitter during the key-down period and must be able to recover full sensitivity in the key-up periods.

Simple break-in technique calls for the use of a separate receiving antenna, as the ordinary antenna relay cannot respond fast enough to follow high speed keying. The separate antenna, in most instances, may be a random length of wire run at right-angles to the main station antenna to reduce transmitter pickup. A more complex technique makes use of an electronic transmit-receive switch (*T-R switch*) which offers automatic protection to the receiver from the transmitter power.

Shown in figure 5 is a representative break-in circuit that provides gain reduction and receiver input circuit protection during the key-down period of the transmitter. In the key-up, or receive, position, the auxiliary receiving antenna is connected to the receiver through a simple T-R switch and the receiver agc circuit functions normally. In

the key-down position, transistor Q_1 conducts and the collector assumes a negative potential. A negative voltage is thus applied to diode D_2 which conducts, effectively shorting the receiver antenna circuit to ground. An adjustable negative voltage is taken from potentiometer R_1 and applied to the receiver agc line, silencing the receiver. Diode D_1 prevents shorting the agc line to ground during key-up condition. The keying characteristic may be achieved by a simple R-C network placed across the key terminals.

A more complex break-in circuit is shown in figure 6. Transistors Q_1-Q_4 form a complementary switch that controls transmitter bias. The three control circuits are near zero

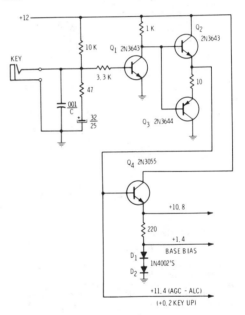

Figure 6

TRANSMITTER BREAK-IN CIRCUIT

Transistors Q_1-Q_4 control transmitter bias on two stages of solid-state circuitry. When the key is closed, the circuits go to the positive voltages indicated. The 1.4-volt line is used to bias-on the base circuits of the r-f driver stages of the transmitter. Agc/alc control voltage is derived from transistors Q_2-Q_3.

potential during key-up periods. When the key is closed, the circuits go to the positive voltages indicated in the diagram. The potential of the +1.4-volt line is determined by the two diodes D_1 and D_2. This voltage is used to bias-on the base circuits of the r-f section of the transmitter. The agc/alc control voltage is derived from transistors Q_2-Q_3.

A break-in circuit utilizing relay control for antenna switchover is shown in figure 7. Transistors Q_1 and Q_2 conduct when the key is closed. Transistor Q_1 provides collector voltage to the low-level stages of the exciter and Q_2 provides driving voltage for switch transistor Q_3 through an RC waveshaping circuit. The transistor (Q_3) is normally cut off by the two diodes in the emitter circuit when the key is up.

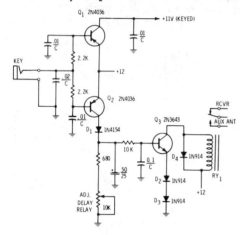

Figure 7

BREAK-IN CIRCUIT FOR RELAY CONTROL

Transistors Q_1-Q_2 conduct when the key is closed. Q_1 provides collector voltage to a low level r-f stage of the exciter and Q_2 provides driving voltage for switch transistor Q_3 through a waveshaping circuit. The relay is controlled by Q_3.

A companion *sidetone* generator is shown in figure 8 which provides an audio tone for the receiver when the transmitter is keyed.

A break-in circuit designed around the NE-555 timing IC is shown in figure 9. This device contains a set-reset flip-flop as well as an output buffer and has a wide range of operation control. The 555 timer may be used in electronic keyers or in break-in circuitry as well as automatic T-R (transmit-receive) switching. In the circuit illustrated, device U_1 is the main tim-

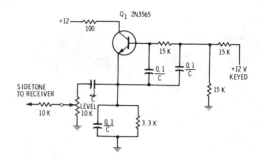

Figure 8

SIDETONE GENERATOR FOR BREAK-IN KEYING

ing element. When the key is closed U_1 switches on, activating the antenna relay through Q_1 and muting the receiver by means of Q_2. After a short delay which allows the antenna relay to operate, U_2 turns on a low-level mixer stage. At the same time these operations take place, the keyed stages of the transmitter are activated by Q_4, an emitter follower. The combination R_1-C_1 provides proper keying characteristics.

In the case of a vacuum-tube amplifier driven by a solid-state device, the circuit of figure 10 may be used. Driver Q_2 is keyed through transistor Q_1 and the amplifier tube V_1 is blockgrid keyed directly through a waveform shaping circuit (R_1-C_1).

The PIN Diode Keyer The PIN diode discussed in Chapter 4 is a current-controlled resistor at r-f and microwave frequencies and can switch a large amount of r-f power with a low value of dc control voltage. When the PIN diode is at zero or reverse bias the diode appears as a capacitor shunted by a parallel resistance. The resistance value is proportional to voltage and inversely proportional to frequency. In most r-f applications, the resistance is much higher than the resistance of the capacitance (figure 11A).

When the PIN diode is forward biased the diode appears as a low value of resistance in series with a small value of inductance. Various diode switching circuits are shown in figure 11B.

When used as a transmit-receive switch the PIN diode connects the antenna to the

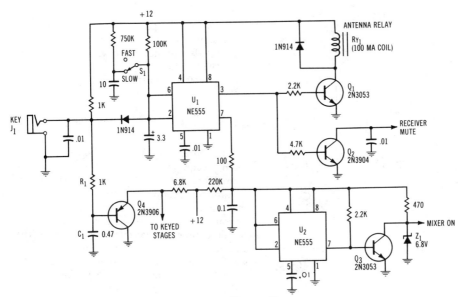

Figure 9

TRANSMITTER CONTROL CIRCUIT

The timer V₁ controls antenna relay and receiver-mute timing. Timer V₂ turns on mixer stage and transistor Q₄ controls keyed stages. Time delay is determined by switch S₁.

transmitter in the transmit mode and to the receiver during the receiving mode. The basic circuit for such a T-R switch is shown in figure 12 and consists of a PIN diode connected in series with the transmitter and a shunt diode connected a quarter-wavelength away from the antenna in the direction of the receiver. Lumped elements or a coaxial line can be used to simulate the quarter-wave section.

In the transmit mode each diode is forward biased. The series diode appears as a low impedance to the transmitter circuit and the shunt diode effectively shorts the receiver antenna circuit to prevent overload. With proper diodes, insertion loss during transmit is less than 0.2 dB and greater than 30 dB isolation to the receiver during the transmit mode can be expected.

In the receive mode, the diodes are at zero or reverse bias and present essentially a low capacitance which creates a direct low-insertion-loss path between the antenna and the receiver. The transmitter is isolated by the high impedance series diode.

The amount of power this r-f switch can handle depends on the power rating of the

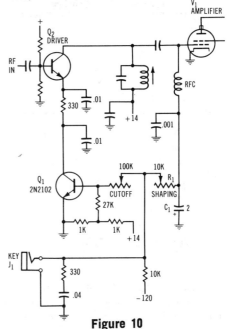

Figure 10

TWO STAGE KEYER CIRCUIT

Blocking bias to amplifier V₁ is keyed via a shaping circuit R₁-C₁. The driver is keyed by transistor Q₁.

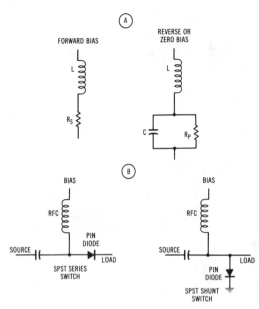

Figure 11

PIN DIODE R-F SWITCHES

A—Diode electrical model under forward- and
reverse-bias conditions.
B—R-f switches, series and shunt connected.

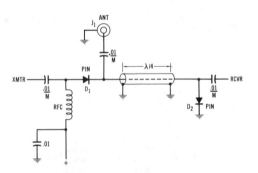

Figure 12

**QUARTER-WAVE ANTENNA SWITCH
USING PIN DIODES**

PIN diode and the diode resistance. In a 50-ohm antenna system where the condition of a mismatched antenna must be considered, the power capability of the PIN switch is,

$$P_{(watts)} = \frac{P_D \cdot Z_o}{R_S} \left(\frac{a+1}{2a} \right)^2$$

where,
 P_D equals the power rating of the diode,
 R_S equals the diode resistance,
 a equals the maximum value of SWR on the system,
 Z_o equals 50 ohms.

18-4 The Electronic Key

The *International Morse Code* used in radio telegraphy is made up of three elements: the *dot*, the *dash*, and the *space* (see Chapter 1, Section 4). Intelligence can be transmitted at high rates of speed by using various combinations of these elements. A standard time relationship exists between the elements and between the space between words. The dot is a unit pulse and one pulse per second is termed one *baud*. The dot has a duty cycle of fifty percent, thus making the space equivalent in length to a unit pulse. The dash has a duty cycle of seventy-five percent, or three unit pulses in length. The space between words is seven unit pulses in length.

These fixed relationships between the code elements make it possible to use digital techniques to generate the timing characteristics used in an automatic electronic keying device, or *keyer*.

The representative keyer is actuated by the operator who keys at approximately correct times, the keyer functioning at precisely correct times determined by the *clock circuit* of the device.

In most keyers either an astable multivibrator or a pulse generator is used as a clock to create precise dots and dashes. The latter are made by filling in the space between two dots. Latching (memory) circuits are used so that an element, or code character, will be completed once it is initiated by the keyer paddle, or lever.

Since the transmitter following the keyer has wave-shaping circuits and possibly relay closure delay, a *weight control* may be incorporated in the keyer to vary the dot-to-space ratio.

Modern electronic keyers make use of solid-state circuitry which is admirably suited to on-off position. A basic electronic key uses a single or dual key lever, movable in a horizontal plane and having two side contacts, much in the style of the mechanical key, or *bug*. Moving the keying paddle to the right produces a uniform string of dots and moving the paddle to the left produces a uniform string of dashes. A more sophisticated keyer makes use of a dual *squeeze* paddle having double paddles, levers, and contacts, one set for dots and one for dashes. In one version of this *squeeze keyer* (the *iambic* keyer), closing both paddles at once produces a string of sequential dots and dashes. This simplifies the sending of the letters having this sequence, such as C, Q, A, L, X, R, and K. Other versions of the squeeze keyer produce a string of dots or dashes when both paddles are closed. The keyer may be modified to send dots over dashes or dashes over dots when one paddle is closed after another. This action is termed *override*. Automatic dot completion is achieved by incorporating a *memory* circuit in the keyer.

A Basic Keyer   The logic functions of a typical keyer are performed by silicon integrated circuits (figure 13). The pulse (dot) generator, or *clock*, is a free running multivibrator made up of two inverters (IC_{1A}, IC_{1B}) with the pulse speed controlled by potentiometer R_1. The free running, astable multivibrator allows precise spacing between the code elements as the space will always be one dot long, regardless of the sending speed. A dual flip-flop (IC_{2A}, IC_{2B}) is used as a character generator. Grounding the dot contact of the two-contact key triggers the *set* (S) input of the dot flip-flop (IC_{2A}) which then sends precise square-wave dots as long as the dot contact is closed. If the dot contact is opened before the completion of a dot, the element will be completed (dot memory).

Grounding the dash contact of the key triggers the *set* input of the dash flip-flop (IC_{2B}) and also grounds the *set* input of the dot flip-flop through diode D_1. The dot flip-flop starts a dot, the dash flip-flop is triggered, and a second dot is initiated completing the dash element at the end of the second dot. The outputs of the flip-flops are added in a summing gate (IC_3). Once a character has started, it is impossible to alter it with the paddle and characters are self-completing.

The transmitter is actuated by a keying transistor (Q_1) employing a fast-operating relay in the collector circuit. In many instances, a *reed relay* is used. This type of relay has operate and release times of less than one millisecond and can allow good keying up to 100 words per minute. Some keyers eliminate the relay in favor of a keying transistor having a high collector-to-emitter voltage rating and a large collector

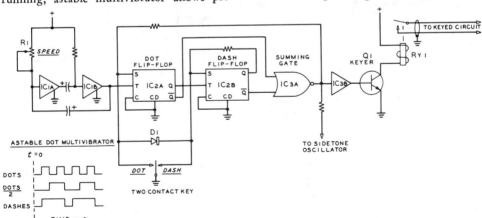

Figure 13

LOGIC FUNCTIONS OF ELECTRONIC KEYER

Astable multivibrator (IC_1) generates string of pulses (dots) with speed controlled by potentiometer R_1. Dot flip-flop sends precise square-wave dots when key contact is closed. Dash flip-flop adds long pulse to dot, forming 3-baud dash at output of summing gate. Amplifier and keying transistor drive a reed relay which controls the transmitter circuit. Dot memory, sidetone monitor, and iambic characteristic may be added to the basic keyer, if desired.

current rating, thus permitting the transistor to be used to directly key cathode or grid circuits carrying up to several hundred milliamperes with an open-key voltage up to 300.

A *sidetone oscillator* or keying monitor can be driven by the keyer to provide the operator with an audible indication of the keying process.

Variation in the control logic and the use of a double paddle key permits conversion of the basic keyer to iambic keying whereby grounding either the dot or the dash contact and then immediately grounding the other produces alternating dots and dashes. Another version will produce a dot or dash override sequence whereby closing both contacts simultaneously, only dots (or dashes) are generated.

A representative keyer is shown in figure 14. This unit employs a dual flip-flop for dot and dash generation at a three-to-one ratio. The IC-3 (NE-555) serves as a clock generator whose speed is set by control potentiometer R_1. A second timer serves as a sidetone generator. The transmitter is keyed by means of transistor Q_1 and a reed relay which isolates the keyer circuit from the transmitter voltages.

18-5   The COSMOS Keyer—Mark II

This compact and reliable keyer is an up-to-date version of the popular W9TO keyer that has appeared in various versions, revisions, and modifications over the past decade. The latest design uses the newest and the best adapted IC logic form: CMOS (figure 16). A recently developed IC does it all, the *Curtis 8044*. Rather than building up a keyer using several small-scale-integration type IC's, the builder can use only one *8044* and have dot and dash memory, variable weight, and even iambic (squeeze keying) mode.

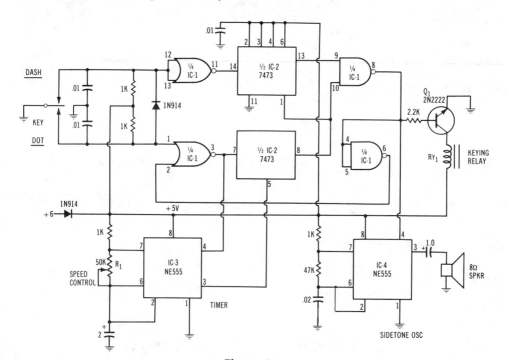

Figure 14

ELECTRONIC KEYER

Device IC-1 is a 7400 Quad, two input NAND gate. RY₁ is a reed relay.

Figure 15

THE COSMOS
INTEGRATED-CIRCUIT KEYER—
MARK II

The COSMOS keyer uses CMOS logic with a single IC, the Curtis 8044. This device provides dot and dash memory, variable weight, and iambic (squeeze keying) mode. The device works either from an internal 3-volt battery or from 120 volts 60 Hz. The unit is built in a Moduline cabinet measuring 5" wide, 3½" high and 5" deep, exclusive of controls. The small speaker is mounted in the removable lid of the box. Pitch and weight controls are in line across the top, with volume and speed controls across the bottom. Rubber feet are placed at the bottom of the box to prevent scratching the operating table. Box color is gray, with an off-white panel.

Because CMOS is inherently capable of operating from a wide range of supply voltages, the 8044 can operate on $+5$ to $+12$ volts dc. Since 9-volt transistor radio batteries are cheap and common, that voltage was chosen for this keyer. Either battery or ac operation is selectable by the front panel power switch. Since the keyer consumes only about 50 μA "key-up" and 50 mA "key-down," leaving the power switch on in the battery mode causes little drain.

Keyer Circuitry   The circuit of the COSMOS keyer is shown in figure 18, as arranged for cathode keying of a tube-type transmitter. Note that transistor Q_3 is a type capable of withstanding $+300$ volts in the "key-up" condition and 200 mA in the "key-down" position.

The transmitter may be turned on for tuning, by closing the *tune* switch on the keyer. Also, the keyer may be used for code practice, without keying the transmitter by closing the *self-test* switch.

The ac power supply uses a four terminal regulator, the *Fairchild* 78 MG. This regulator is very much like the 3-terminal types having fixed voltages, but has a fourth terminal by which the output voltage may be adjusted. The power supply circuit is shown

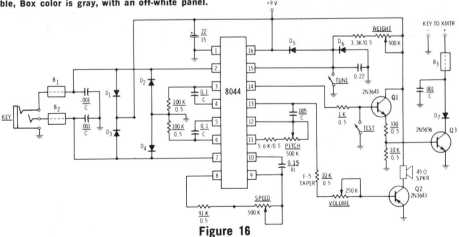

Figure 16

SCHEMATIC OF THE COSMOS KEYER—MARK II

Diodes D_1-D_4 are germanium (1N270). Diode D_1 is a 1N4006. All capacitors may be ceramic except the 0.15-μF unit between pins 9 and 10 of the IC, which is a mylar unit. Ceramic beads (B_1-B_2) are placed on key lines to reduce r-f feedback into the keyer. A set of the major components and a glass-epoxy circuit board for the keyer may be ordered from: Curtis Electro Devices, Box 4090, Mountain View, CA 94040.

in figure 17. Since the regulator is fully adjustable, it can be set to $+9$ volts, the same as the nominal battery voltage.

Note that all leads passing in and out of the keyer cabinet (*Moduline P-355*) are r-f decoupled. The two keyer paddle leads and the transmitter keying line are choked, using a ferrite bead and a 1000-pF feedthrough capacitor in each lead, forming a simple L-network. These RFI precautions may not always be required because the *Curtis 8044* IC is relatively insensitive to r-f, being CMOS. It is safest to put it in as the keyer is built, rather than having to add it on later if trouble does develop. The keyer is assembled on a peg board, as shown in the interior view of figure 18.

Using the Keyer   The only adjustment to be made once the keyer is wired and checked is to set the regulator (figure 17) for $+9$ volts. The keyer should be tested on the external supply and then on the internal battery. A Ni-cad battery is recommended for longest life. If desired a small LED may be placed across the supply to indicate when the keyer is turned on.

The *Curtis 8044* comes with an IC socket and instruction manual. The manual shows how the IC may be used to key a transmitter having a negative "key-up" potential. This method uses a high-voltage PNP transistor as a saturated switch, in much the same manner as the NPN (Q_3) device was used to key a positive voltage ("key-up") transmitter. The circuit shown in the *Curtis* man-

ual, however, places the keyer paddle common at -9 volts relative to the transmitter chassis. If complete isolation is desired, an inexpensive reed relay, offering millisecond response and minimum bounce may be used in the circuit of figure 19.

The keyer provides self-completing dots, dashes, and spaces. Once a character (or space) is commenced, there is no way to prevent it from being completed. The self-completing function of an electronic keyer can cause dots to get lost because the operator tends to "lead" the keyer. Since dashes are held longer, they seldom get lost. To prevent lost dots, the 8044 CMOS device employs a memory to remember when a dot is called for and to insert it at the proper time. The dot memory also aids in "squeeze" keying where a tap on the dot paddle will insert a dot into a series of dashes. When the dot paddle is pressed, a continuous string of dots is produced. When the dash paddle is pressed, continuous dashes are produced. When both paddles are closed, an alternating series of dots and dashes (iambic) is produced. The series can be started with either a dot or a dash depending upon which paddle is closed first. Iambic operation allows "squeeze" keying if desired by using a twin-lever paddle. A single-lever paddle allows the "non-squeeze" mode.

The keyer provides a speed range of 8 to 50 w.p.m. Resistor R_{12} sets the upper end of the speed range and may be decreased in value for higher speed keying.

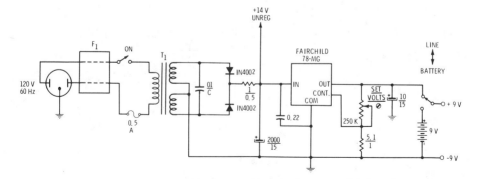

Figure 17

SCHEMATIC OF KEYER POWER SUPPLY

F_1—Line receptacle and filter. CORCOM 6EF-1. T_1—10-0-10 volts at 60 mA. Signal PC-20-60.

Figure 18

INTERIOR VIEW OF KEYER

The keyer is built upon a prepunched terminal board, P pattern with 0.042″ diameter holes and type T45-4 terminals (Vector). The 9-volt transistor battery is mounted to the side of the cabinet with a clip. At the rear of the board is the IC voltage regulator, with the power transformer and filter capacitor to the side. The Curtis IC is near the center of the board, mounted in a 16-pin socket. Connections between the terminals is made on the under side of the terminal board. The test and tune switches, along with the RFI-proof power receptacle and terminal board for keying connections are mounted on the rear wall of the cabinet.

Figure 19

REED RELAY KEYING CIRCUIT FOR COSMOS KEYER

B₁, B₂—Ferrite bead. RY₁—Reed relay, 12-volt coil. Potter & Brumfield JR-M-1009.

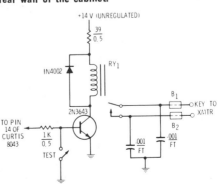

18-6   The Keyboard Keyer

Use of a *keyboard*-style keyer is growing, especially among radio amateurs interested in very high speed c-w, from 50 to 90 w.p.m., at which speed accurate manual transmission is very difficult. Keyboards are also used by lower speed operators interested in accurate c-w independent of physical dexterity.

A keyboard keyer consists of a keyboard, usually arranged similar to a typewriter, an *encoding system* for the keys, a *converter* for obtaining the Morse code characters with proper element spacing, a *sidetone monitor* and an *output section* for keying the transmitter. The *Curtis KB-4200 Morse Keyboard* is shown in figure 20, and a block diagram of the device is given in figure 21.

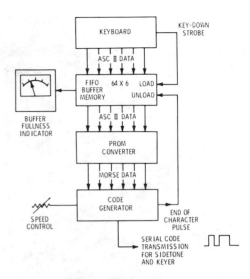

Figure 21

THE CURTIS KB-4200 KEYBOARD KEYER

Keyboard provides standard ASCII code of six parallel lines to a 64-character FIFO (first-in, first-out) buffer memory. A fullness meter indicates the amount of storage in use. As each character is withdrawn from the memory, the stack falls by one character. The ASCII data is routed to a PROM code converter where Morse equivalents are generated. A sidetone monitor and keyer are run by the code generator.

Figure 20

THE CURTIS KEYBOARD

Standard typewriter format is used in the keyboard. At upper left is the buffer status meter, with the speed control, calibrated in words per minute at the center. Volume, pitch, and weight controls are at upper right.

Keyboard Operation   Although there are several ways of implementing a keyboard keyer, the machines fall into two general classes; those with a *buffer memory* and those without. This difference has a large effect upon the sequence of operation of the device. On keyboards without buffers, character and word spacing is provided by the operator and is variable as a result. On units having buffers, the operator types a few code characters ahead of the actual transmission. The buffering circuitry supplies character spacing, and by depressing a *space bar* on the keyboard, the operator inserts standard word spaces into the message.

Buffer memory sizes range from one character to as many as 128. A buffer storage of 64 characters is more than adequate for normal operation. Buffered keyboards are normally designed to produce only one character per key depression similar to a typewriter, whereas certain of the unbuffered designs send a continuous stream of characters on key depression. While helpful in sending words with rapid, repetitive letters, such as "keep," the key must be released very quickly to avoid sending unwanted duplicates of short letters at high keying speeds. Also, on an unbuffered device, the rhythm of key depression is tied to the rhythm of the Morse transmission, that is, some letters are short and some very long. On buffered units, the operator is free to type independent of transmission speed once he has a few characters stored in the buffer.

A Buffered Keyboard The diagram of figure 21 illustrates a buffered keyboard. This device makes use of a standard computer terminal keyset and associated electronics to prevent key de-bouncing and two-key roll often caused by overlapping key depressions by the operator. The output of this section is the standard *ASCII* code (American Standard Code for Information Interchange) for alphanumeric characters consisting of six parallel lines. A strobe output indicates when a key has been depressed and the key data is valid.

The ASCII information is routed to a 64 character *FIFO* (first-in, first-out) buffer memory (using two FSC 3341 ICs), where it is stacked up, ready for transmission. As each character is withdrawn for transmission the stack falls one character. Operation of the FIFO memory is similar to an old-time trolley car conductor's coin changer where the rate of coins deposited and extracted is completely independent. In the KB-4200

keyboard, buffer fullness is indicated by a panel meter, calibrated in characters, as an operating convenience.

Data for each character exciting the FIFO buffer memory is routed to a code converter (using two Signetics 8223 PROMS, 32 × 8, each) where the ASCII representation is changed to a Morse code representation. A convenient and compact Morse representation consists of eight parallel bits. The first five describe the character elements (*dit* or *dah*) and the last three contain a binary count of the number of elements in the character. Calling a dit "1" and a dash "0," some examples are:

Letter	Element	Count
E	10000	001
A	10000	010
V	11100	100
6	01111	101

Six-element characters can be accommodated with this system using advanced electronic circuitry.

The parallel Morse representation is routed to a code generator which makes up the actual Morse character in a serial form ready for transmission.

18-7 VOX Circuitry

A form of VOX (*voice-operated transmission*) is often employed in SSB operation. A representative VOX system is shown in figure 22. The VOX signal voltage is taken

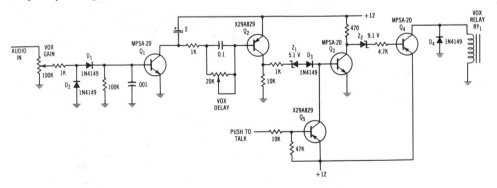

Figure 22
TRANSISTOR VOX CIRCUIT

from the speech amplifier stages of the transmitter and adjusted to the proper signal level by means of the VOX gain potentiometer. The audio signal is rectified by diodes D_1 and D_2 and amplified by transistor Q_1. The resulting voltage pulsations are amplified by Q_2 and Q_3 after the time constant of the dc waveform is determined by the VOX delay potentiometer in the base circuit of Q_2. The VOX voltage or push-to-talk voltage is applied to Q_3 after the levels have been set and long time constant pulses are passed to Q_4, which serves as a switch to drive the transmit-receive relay.

VOX Bias Control   It is desirable to completely disable a high-power linear amplifier during reception for two reasons: first, the amplifier consumes standby power unless it is biased to cutoff and, second, many amplifiers will generate "white noise" when in a normal standby condition. The white noise, or diode noise, may show up in the receiver as a loud hiss interfering with all but the loudest signals.

The circuit of figure 23 provides an automatic cutoff-bias system for a VOX-controlled amplifier stage. The resting plate current of the amplifier is passed through a 50K resistor in the filament return circuit, creating a voltage drop that is applied as cutoff bias to the tube(s). The filament circuit is raised to a positive voltage with respect to the grid, thus leaving the grid in a negative, cutoff condition. On activation of the VOX relay, a separate set of con-

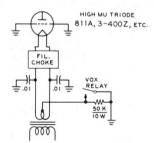

Figure 23

VOX BIAS CONTROL

Cutoff bias for grounded-grid triode may be obtained from cathode bias resistor. Action of VOX relay shorts out resistor, restoring amplifier to normal operating conditions.

tacts short out the bias resistor, restoring the amplifier stage to normal operating condition.

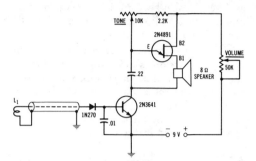

Figure 24

SCHEMATIC OF R-F ACTUATED KEYING MONITOR

18-8   An R-F Operated Keying Monitor

For proper sending and clean code transmission it is mandatory for the operator to monitor his signal. This may be done by copying the output of an audio oscillator that is simultaneously keyed with the transmitter. The oscillator shown in figure 24 is triggered by r-f picked up from the transmitter and thus provides an accurate replica of the keyed signal.

A unijunction transistor (2N4891) serves as a simple relaxation oscillator whose tone and volume are controlled by two potentiometers. The oscillator runs a small speaker and is enabled by grounding the junction of the 0.22-μF capacitor and the speaker. This is accomplished by a keying transistor (2N3641) which is forward-biased by a small r-f voltage developed by pickup coil L_1 and rectified by a diode.

The keying monitor may be built on a perforated circuit board and placed within an aluminum utility box. It is powered by a 9-volt transistor radio battery. The r-f pickup coil is introduced into the transmitter, in the vicinity of the tank coil of the final amplifier stage, and the trigger voltage level adjusted by moving the coil away from, or closer to, the tank inductor.

18-9 The Phone Patch

The *phone-patch* is an electrical interconnection between the amateur station and the telephone line. Effective in 1959, the Bell System responded to an FCC order covering interconnection of the System with privately owned facilities, which legalized phone-patching. Accordingly, most telephone companies will provide a unit called a *voice coupler* which is a connecting device to be attached to a telephone set, along with a switch to connect and disconnect the coupler. The coupler isolates the station equipment from the telephone line and provides an impedance match and level control between the line and the station equipment. The coupler is connected in parallel with the telephone set when a phone patch is in progress.

To effect a phone patch, the average voice level to be applied to the phone line is restricted by the telephone company and the audio power in various a-f bands is specified, in particular, the band from 2450 Hz to 2750 Hz, which if present, must not exceed a prescribed level. This band is used for signaling.

Modern SSB equipment uses VOX and antivox circuitry, and provisions for voice control are helpful for full phone-patch service. In order for this to be accomplished correctly, a *hybrid* circuit is included in the phone patch. This is a network which resembles a bridge and prevents the receiver audio signal from reaching the audio circuitry of the transmitter (figure 25). The signal-level loss of this circuit is approximately 10 dB. In some patches, a 2600-Hz filter is added in the line from the receiver to prevent unwanted disconnections resulting from heterodynes or interference on the received signal falling in that audio-frequency range. Such a filter is helpful on long distance phone calls but is usually not required for local calls.

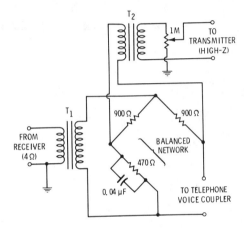

Figure 25

REPRESENTATIVE PHONE PATCH

T₁—1000 ohms to 4 ohms (reversed)
T₂—1000 ohms to 5000 ohms
Adjust balancing network for minimum signal feedthrough between receiver and transmitter.

CHAPTER NINETEEN

Mobile and Portable Equipment

Mobile operation is permitted on all amateur bands. Tremendous impetus to this phase of the hobby was given by the suitable design of compact mobile equipment. Complete mobile installations may be purchased as packaged units, or the whole mobile station may be home built, according to the whim of the operator.

The problems involved in achieving a satisfactory two-way installation vary somewhat with the band, but many of the problems are common to all bands. For instance, ignition noise is more troublesome on 10 meters than on 80 meters, but on the other hand an efficient antenna system is much more easily accomplished on 10 meters than on 80 meters.

Compact mobile equipment is available for f-m operation on the vhf bands and this popular mode has flourished, at the expense of mobile operation on the hf bands. The use of fixed f-m repeaters placed on elevated locations has done much to enhance vhf mobile communication.

The majority of high-frequency mobile operation takes place on single sideband. The low duty-cycle of SSB equipment, as contrasted to the heavy power drain of conventional a-m gear, has encouraged the use of relatively high-power sideband equipment in many mobile installations.

Portable operation is extremely popular on all hf and vhf bands and specialized equipment for this mode of operation is available, using battery power as a primary source. To conserve battery drain, solid-state devices are commonly used and power input is limited for the same reason. Some amateurs employ gasoline driven power generators for portable and emergency service. ventional a-m gear, has encouraged the use In all cases, however, the power source is critical since even mobile power sources are limited in their ultimate capacity.

19-1 Mobile and Portable Power Sources

A small transistor converter for casual listening may be run from a 9-volt battery, but larger mobile receivers, transmitters, and transceivers require power from the electrical system of the automobile. SSB equipment, with its relatively light duty cycle, is ideally suited for mobile use and demands the least primary power drain for a given radiated signal of all the common types of amateur transmission.

F-m, on the other hand, is universally used for vhf mobile service. In any case, a total equipment power drain of about 250 watts for SSB or f-m is about the maximum power that may be taken from the electrical system of an automobile without serious regard to discharging the battery when the car is stopped for *short* periods of mobile operation.

With many SSB mobile-radio installations now requiring 100 to 250 watts peak power from the automotive electrical system, it is usually necessary to run the car engine when the equipment is operated for more than a few minutes at a time to avoid discharging the battery. Fortunately, a majority of automobiles have a 12-volt *alternator system* as standard equipment and as a result, most SSB transceivers may be run directly from the automotive electrical system without undue strain on the battery during the course of normal driving.

The Alternator A typical alternator circuit is shown in figure 1. The alternator differs from the classic generator in that it uses a rotating field to which dc is supplied through the slip rings and carbon brushes. Field current is quite low, of the order of 3 amperes or so, for many alternators. The rotating field usually has six

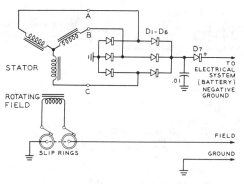

Figure 1

THREE-PHASE AUTOMOBILE ALTERNATOR

Three-phase output voltage is converted to dc by full wave rectifier D_1, - D_6. Rectifier D_7 protects rectifier assembly from transients and voltage surges in electrical system of auto.

pairs of poles, and the output of one stator winding represents six electrical cycles for each revolution of the field. The output frequency in cycles per second is one-tenth the shaft speed expressed in revolutions per second.

The high output current of the alternator is supplied directly from the fixed stator windings in the form of three-phase current. The stator is usually connected in a *wye* (Y) configuration to an internal rectifier assembly made up of six silicon diodes which provide full-wave rectification. The ripple frequency is six times the frequency developed in one winding. Thus, at a shaft speed of 4000 rpm, the nominal voltage is 14, output frequency will be 400 Hz, and the ripple frequency is 2400 Hz.

The diode assembly (D_1, D_6) may be mounted on or behind the rear end-bell of the alternator, in conjunction with an isolation diode (D_7) which protects the rectifier assembly from voltage surges and helps to suppress radio noise.

The output voltage of the alternator system is a function of the shaft speed to about 5000 rpm or so. Above this speed, output voltage tends to stabilize because of hysteresis losses. In any case, the alternator output is regulated through adjusting

the current in the field by a mechanical voltage regulator or by a solid-state regulator. Because the reverse current through the rectifier diodes is small, the alternator is usually connected directly to the battery without the use of a cutout relay.

Batteries   The voltage available at the terminals of a battery is determined by the chemical composition of the cell. Many types and sizes of batteries are available for portable radio and communication equipment. The inexpensive *carbon-zinc* cell provides a nominal 1.5 volts and, unused, will hold a charge for about a year. The current capacity of the cell depends on the physical size of the electrodes and the composition of the electrolyte. A battery may be made up of a number of cells connected in series, providing good life under intermittent service.

Next to the carbon-zinc cell, the most commonly used unit is the *alkaline cell* (1.2 volts) which has about twice the total energy capacity per unit size as compared to the carbon-zinc cell. This cell is capable of a high discharge rate over an extended period of time and provides longer life in continuous service than does the carbon-zinc cell.

The *mercury cell* (1.34 volts) is more expensive than the previously mentioned cells, but it has an extremely long working life. In addition, the mercury cell maintains full rated voltage until just before expiration, then the voltage drops sharply. Shelf life of the mercury cell is excellent and it may be stored for long periods of time.

These three types of batteries may be recharged to some extent by reversing the chemical action by application of a reverse current to the cell. For best results, the current should be low and should have a small ac component to provide a more even redeposit of material on the negative electrode. Recharged cells have an uncertain operating life, and the recharging cycle may vary from cell to cell.

The *nickel-cadmium (Nicad)* cell (1.25 volts) is the most expensive cell in terms of initial costs, but it may be recharged at a slow rate a number of times in reliable cycles of operation (figure 2).

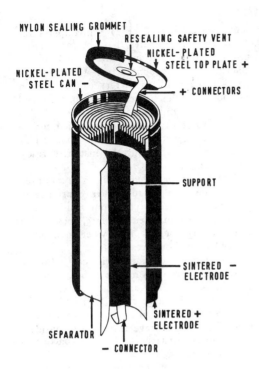

NYLON SEALING GROMMET
RESEALING SAFETY VENT
NICKEL-PLATED
STEEL TOP PLATE +
NICKEL-PLATED
STEEL CAN −
+ CONNECTORS
SUPPORT
SINTERED −
ELECTRODE
SINTERED +
ELECTRODE
SEPARATOR
− CONNECTOR

Figure 2

CONSTRUCTION OF TYPICAL NICKEL-CADMIUM CELL

The Nicad battery is a sealed device. Oxygen produced during operation is recycled so there is no loss of electrolyte. Most cells have a safety vent that enables cell to release gas under heavy load and then to reseal automatically. Chemical action of cell causes a temperature rise and it is necessary to limit charging current to prevent overheating and overcharge.

The wet cell (*lead-acid*) storage battery is in near-universal use in automotive equipment. The cell delivers about 2.1 volts and is rechargeable. The lead-acid cell is made of coated lead plates immersed in a solution of sulphuric acid and water. The acid content of the dielectric varies with the state of charge, which may be determined by measuring the specific gravity of the electrolyte. Generally speaking, a hydrometer reading of 1.27 indicates a fully charged cell, whereas a reading of 1.15 or below indicates the cell is in need of charging. The wet cell may be fast-charged as high as 40 amperes for a 12-volt battery, provided that care is taken to let escaping gases free themselves and

provided that electrolyte temperature is held below 125° Fahrenheit.

The Nickel-Cadmium Cell   The nickel-cadmium (Nicad) cell is a high-efficiency cell capable of being recharged hundreds or thousands of times in the proper circumstances. The cell has a positive nickel electrode and a negative cadmium electrode immersed in a solution of potassium hydroxide at a specific gravity of 1.300 at 72° F. The common and popular lead-acid battery does not equal the recharge ability of the nickel-cadmium battery and use of the latter is common in mobile and portable equipment and other devices where small cell size and high recharge capability are an asset.

There are two common types of nickel-cadmium batteries classified as *vented* and *nonvented*. The nonvented cell is a hermetically sealed unit which resembles a conventional dry cell in appearance. The vented cell resembles a lead-acid cell and often has a removable plug which covers a port for gas venting during the charging process.

The terminal voltage of a nickel-cadmium cell varies with the state of charge and normally runs between 1.25 and 1.30 volts on open circuit. Exact terminal voltage depends on the state of charge, the charging current, and the time of charge. The specific gravity of the electrolyte, moreover, does not change appreciably between charge and discharge, as is commonly done with lead-acid cells. At end of charge, nickel-cadmium cell voltage may drop as low as a fraction of a volt and it is possible under heavy discharge for a cell to show a negative or reversed voltage, indicating a state of extreme discharge. A terminal voltage of 1.1 volts is usually considered to be a state of complete discharge, for all practical purposes and should not be exceeded.

For standby service the nickel-cadmium cell can be maintained on a trickle charge, with the charger adjusted to maintain a terminal potential of 1.36 to 1.38 volts per cell. Following a substantial discharge, a regular charge should be given, after which the cell is placed back on trickle charge. While the overcharge tolerance is good and the cell may be left on charge for long periods of time, severe overcharge must be avoided because the cell may be destroyed by accumulation of gases within the container.

The nickel-cadmium cell may be charged by a *constant-potential process* whereby charger current is continually adjusted to maintain a constant potential of 1.55 volts across the cell. This requires a charger designed for such service, as very high current occurs at the start of charge, tapering rapidly as the charge progresses. A fully discharged cell can be completely recharged by this method in an hour or so.

The nickel-cadmium cell may also be charged by the *constant-current* process. This technique requires a charging source having an ammeter and control rheostat in the charging circuit. The cell is charged at a constant current rate. To maintain constant current, the rheostat requires adjustment during the charge period as the counter-emf of the cell rises.

The practical value of charging current varies from cell to cell and is usually specified by the manufacturer. If the extent of discharge is not known, the cell may be charged at a constant current rate until the cell voltage ceases to rise. Reasonable overcharge is not harmful as long as the electrolyte level is above the plate tops and the electrolyte temperature does not exceed 125°F.

When charging at a high rate, the nickel-cadmium cell will gas rather vigorously when approaching full charge. This gassing will cause the electrolyte level to rise above the limit line. This apparent excess electrolyte should not be removed as the level will drop back after the cell stands on open circuit following the charge. Charging disassociates water from the electrolyte which forms this gas.

The energy capability of a nickel-cadmium cell is usually rated in milliampere-hours. for small cells and ampere-hours for large ones. The rating is based on cell capability to a specific end point (usually 1.1 volts per cell) over a 10-hour period. This figure is used as the capacity of the cell and depends upon the rate of discharge. Generally speaking, the charging current is held to 10 percent of the milliampere-hour rating of a small cell and the time of charge is set at 150 percent of the time required to re-establish the maximum milliampere-hour rating of the cell. Thus a 250 milliampere-

hour cell is charged at 25 milliamperes for 15 hours. This ensures that the lost energy is restored and various other losses and inefficiencies are accounted for. With a simple charger the standard battery can be left on extended trickle charge (at less than 10 percent of the milliampere-hour rating) for years. This constant current extended charge feature has value in standby applications where the battery must be instantly ready to operate.

Nonvented, or sealed, cells can be mounted in any position because their construction prevents the electrolyte from spilling out. Since they are maintenance free, these sealed cells are frequently totally encased in a molded plastic or metal housing.

The nickel-cadmium cell can also be stored for years with no significant degradation in performance and then, after just a few charge-discharge cycles, can be brought back to the point where it will be good as new. This long storage feature has considerable value in situations where the battery is only used occasionally.

The following precautions are recommened to users of Nicad cells or batteries:

1. Do not dispose of batteries in a fire.
2. Do not attempt to solder directly to a sealed cell because the seal can be damaged by too much heat.
3. Do not place a charged cell in your pocket. If you have keys, coins, or other metal objects in your pocket, the cell may be shorted and produce extreme heat.

19-2 Transistor Supplies

The vibrator-type of mobile supply achieves an overall efficiency in the neighborhood of 70%. The vibrator may be thought of as a mechanical switch reversing the polarity of the primary source at a repetition rate of 120 transfers per second. The switch is actuated by a magnetic coil and breaker circuit requiring appreciable power which must be supplied by the primay source.

One of the principal applications of the transistor is in switching circuits. The transistor may be switched from an "off" condition to an "on" condition with but the

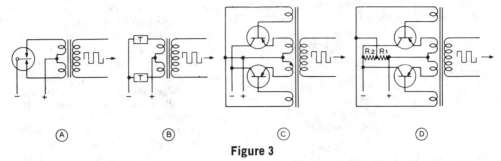

Figure 3

TRANSISTORS CAN REPLACE VIBRATOR IN MOBILE POWER SUPPLY SYSTEM

A—Typical vibrator circuit.
B—Vibrator can be represented by two single-pole single-throw switches, or transistors.
C—Push-pull square-wave "oscillator" is driven by special feedback windings on power transformer.
D—Addition of bias in base-emitter circuit results in oscillator capable of starting under full load.

application of a minute exciting signal. When the transistor is nonconductive it may be considered to be an open circuit. When it is in a conductive state, the internal resistance is very low. Two transistors properly connected, therefore, can replace the single-pole, double-throw mechanical switch representing the vibrator. The transistor switching action is many times faster than that of the mechanical vibrator and the transistor can switch an appreciable amount of power. Efficiencies in the neighborhood of 95 percent can be obtained with 28-volt primary-type transistor power supplies, permitting great savings in primary power over conventional vibrators and dynamotors.

Transistor Operation The transistor operation resembles a magnetically coupled multivibrator, or an audio-frequency push-pull square-wave oscillator (figure 3C). A special feedback winding on the power transformer provides 180-degree phase-shift voltage necessary to maintain oscillation. In this application the transsistors are operated as on-off switches; i.e., they are either completing the circuit or opening it. The oscillator output voltage is a square wave having a frequency that is dependent on the driving voltage, the primary inductance of the power transformer, and the peak collector current drawn by the conducting transistor. Changes in transformer turns, core area, core material, and feedback turns ratio have an effect on the frequency of oscillation. Frequencies in common use are in the range of 120 Hz to 3500 Hz.

The power consumed by the transistors is relatively independent of load. Loading the oscillator causes an increase in input current that is sufficient to supply the required power to the load and the additional losses in the transformer windings. Thus, the overall efficiency actually increases with load and is greatest at the heaviest load the oscillator will supply. A result of this is that an increase in load produces very little extra heating of the transistors. This feature means that it is impossible to burn out the transistors in the event of a shorted load since the switching action merely stops.

Transistor Power Rating The power capability of the transistor is limited by the amount of heat created by the current flow through the internal resistance of the transistor. When the transistor is conducting, the internal resistance is extremely low and little heat is generated by current flow. Conversely, when the transistor is in a cut-off condition the internal resistance is very high and the current flow is extremely small. Thus, in both the "on" and "off" conditions the transistor dissipates a minimum of power. The important portion of the operating cycle is that portion when the actual switching from one transistor to the other occurs, as this is the time during which the transistor may be passing through the region of high dissipation. The greater the rate of switching, in general, the faster will be the rise time of the square wave (figure 4) and the lower will be the internal losses of the transistor. The average transistor can switch about eight times the power rating of class-A operation of the unit. Two switching transistors having 5-watt class-A power output rating can therefore switch 80 watts of power when working at optimum switching frequency.

Self-Starting Oscillators The transistor supply shown in figure 3C is impractical because oscillations will not start under load. Base bias of the proper polarity has to be momentarily introduced into the base-emitter circuit before oscillation will start and sustain itself. The addition of a bias resistor (figure 3D) to the circuit results in an oscillator that is capable of starting under full load. R_1 is usually of the order of 10 to 50 ohms while R_2 is adjusted so that approximately 100 milliamperes flow through the circuit.

The current drawn from the battery by this network flows through R_2 and then divides between R_1 and the input resistances of the two transistors. The current flowing in the emitter-base circuit depends on the value of input resistance. The induced voltage across the feedback winding of the trans-former is a square wave of such polarity that it forward-biases the emitter-base diode of the transistor that is starting to conduct collector current, and reverse-biases the other transistor. The forward-biased transistor will have a very low input impedance, while the input impedance of the reverse-biased transistor will be quite high. Thus, most of the starting current drained from the primary power source will flow in R_1 and the base-emitter circuit of the forward-biased transistor and very little in the other transistor. It can be seen that R_1 must not be too low in comparison to the input resistance of the conducting transistor, or it will shunt too much current from the transistor. When switching takes place, the transformer polarities reverse and the additional current now flows in the base-emitter circuit of the other transistor.

The Power Transformer The power transformer in a transistor-type supply is designed to reach a state of maximum flux density (saturation) at the point of maximum transistor conductance. When this state is reached the flux density drops to zero and reduces the feedback voltage developed in the base winding to zero. The flux then reverses because there is no conducting transistor to sustain the magnetizing current. This change of flux induces a voltage of the opposite polarity in the transformer. This voltage turns the first transistor off and holds the second transistor on. The transistor instantly reaches a state of maximum conduction, producing a state of saturation in the transformer. This action repeats itself at a very fast rate. Switching time is of the order of 5 to 10 microseconds, and saturation time is perhaps 200 to 2000 microseconds. The collector waveform of a typical transistor supply is shown in figure 4. The rise time of the wave is about 5 microseconds, and the saturation time is 500 microseconds. The small "spike" at the leading edge of the pulse has an amplitude of about 2.5 volts and is a product of switching transients caused by the primary leakage reactance of the transformer. Proper transformer design can reduce this "spike" to a minimum value. An excessively large "spike" can puncture the transistor junction and ruin the unit.

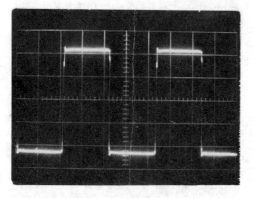

Figure 4

EMITTER-COLLECTOR WAVEFORM OF SWITCHING CIRCUIT

Square waveshape produces almost ideal switching action. Small 2-volt, "spike" on leading edge of pulses may be reduced by proper transformer design. Pulse length is about 1000 microseconds and rise time is 10 microseconds.

A 35-Watt Supply   The 35-watt power unit uses two inexpensive 2N2870 power transistors for the switching elements and four silicon diodes for the high-voltage rectifiers. The complete schematic is shown in figure 5. Because of the relatively high switching frequency only a single 20-μF filter capacitor is required to provide pure direct current.

Regulation of the supply is remarkably good. No-load voltage is 310 volts, dropping to 275 volts at maximum current drain of 125 milliamperes.

The complete power package is built on an aluminum chassis-box measuring 5¼″ × 3″ × 2″. Paint is removed from the center portion of the box to form a simple heat sink for the transistors. The box therefore conducts heat away from the collector elements of the transistors. The collector of the transistor is the metal case terminal and in this circuit is returned to the negative terminal of the primary supply. If the negative of the automobile battery is grounded to the frame of the car the case of the transistor may be directly grounded to the unpainted area of the chassis. If the positive terminal of the car battery is grounded it is necessary to electrically insulate the transistor from the aluminum chassis, yet at the same time permit a low *thermal* barrier to exist between the transistor case and the power–supply chassis. A simple method of accomplishing this is to insert a thin mica sheet between the transistor and the chassis. *Two-mil* (0.002″) mica washers for transistors are available at many large radio supply houses. The mica is placed between the transistor and the chassis deck, and fiber washers are placed under the retaining nuts holding the transistors in place. When the transistors are mounted in place, measure the collector-to-ground resistance with an ohmmeter. It should be 100 megohms or higher in dry air. After the mounting is completed, spray the transistor and the bare chassis section with plastic *Krylon* to retard oxidation. Several manufacturers produce anodized aluminum washers that serve as mounting insulators. These may be used in place of the mica washers, if desired.

An 85-Watt Supply   Figure 6 shows the schematic of a dual-voltage transistor mobile power supply. A bridge rectifier permits the choice of either 250 volts or 500 volts, or a combination of both at a total current drain that limits the secondary power to 85 watts. Thus, 500 volts at 170 milliamperes may be drawn, with correspondingly less current as additional power is drawn from the 250-volt tap.

The supply is built on an aluminum box chassis measuring 7″ × 5″ × 3″, the layout closely following that of the 35-watt supply. HEP-231 or SK3012 transistors are used as the switching elements and eight silicon diodes form the high-voltage bridge rectifier.

The transistors are affixed to the chassis in conjunction with a homemade aluminum heat sink formed from two pieces of aluminum sheet bent into channels, as shown in figure 7. Silicone grease is spread

Figure 5

SCHEMATIC, TRANSISTOR POWER SUPPLY FOR 12-VOLT AUTOMOTIVE SYSTEM

T₁—Transistor power transformer. 12-volt primary, to provide 275 volts at 125 mA. Stancor DCT-1
D₁-D₄—1N4005 with .01 μF and 100K across each diode
Use 6 amp. fuse in +12-volt lead.

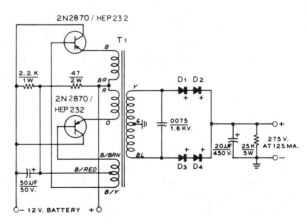

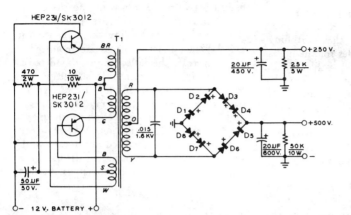

Figure 6

SCHEMATIC, 85-WATT TRANSISTOR POWER SUPPLY FOR 12-VOLT AUTOMOTIVE SYSTEM

T_1—Transistor power transformer. 12-volt primary to provide 275 volts at 125 mA. Stancor DCT-2.
D_1-D_4—1N4005 with .01 μF and 100K across each diode.

thinly between the transistors, heat sinks, and the chassis to permit better heat transfer between the various components of the assembly.

A 270-Watt Transceiver Supply   SSB transceivers suitable for mobile service are capable of PEP power inputs up to 250 watts or more. Shown in figure 8 is a compact triple-voltage supply capable of running many transceivers from

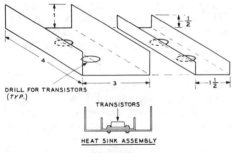

DRILL FOR TRANSISTORS (*TYP.*)

TRANSISTORS

HEAT SINK ASSEMBLY

Figure 7

HOMEMADE HEAT SINK FOR POWER TRANSISTOR

a 12 volt dc supply. The unit provides 900 volts at 300 milliamperes, 275 volts at 180 milliamperes, and an adjustable bias voltage of -15 to -150. Additionally, -150 volts at 40 milliamperes is available for VOX standby circuitry in auxiliary equipment.

Two heavy-duty switching transistors are used, driven by base feedback from a winding of oscillator transformer T_1. The transistors are forward-biased by a voltage divider circuit and are protected from voltage spikes by the two 1N4719 diodes. Two zener diodes (1N4746) provide transient suppression in the primary circuit of transformer T_1. A power transformer (T_2) is driven by the squarewave pulses provided by the switching circuit based on transformer T_1.

The supply is built on an aluminum chassis measuring $12'' \times 6'' \times 3''$. The main components are mounted atop the chassis with the heat sinks mounted on one side, with the fins in a vertical position. To improve thermal conductivity, the heat sinks are bolted to a $\frac{1}{8}$-inch thick copper plate (measuring $12'' \times 6''$) affixed to the side of the chassis. The transistors are insulated from the chassis by thin insulators coated with silicone grease.

All primary leads to the power transistors, transformer T_1, and the input terminals are wired with #6 conductors, with the negative primary circuit grounded at one point in the supply. Heavy $\frac{1}{4}$-inch battery leads run from the supply to the automobile battery. The supply should be mounted close to the battery to reduce primary voltage drop to a minimum.

A DC to AC Inverter For the Car or Boat   Radio and electrical equipment of all kinds up to about 200 watts intermittent power consumption may be run from this compact dc to ac power inverter. Designed for use with 12-volt automotive systems, the inverter provides a nominal 115-volt, 60-Hz square-wave

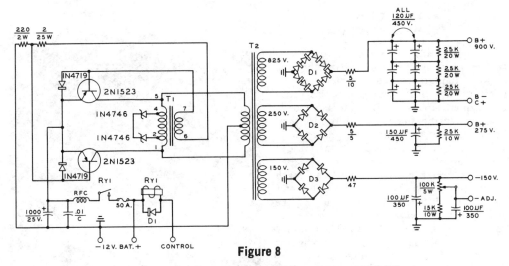

Figure 8

270-WATT MOBILE TRANSCEIVER POWER SUPPLY

D_1-D_3—Use 1N4005 diodes. Two diodes in series are used in each leg of D_1. Place 470K 1-watt resistor and 0.1, 1.6-kV disc across each diode

RY_1—SPST contactor, 60 ampere, with 12-volt coil. Potter-Brumfield MB-3D

RFC—10 turns #10 enamel wire on 1" form

T_1—Oscillator transformer (1000 Hz). Osborne 6784 (Osborne Transformer Co. 3834 Mitchell Ave., Detroit, Michigan)

T_2—Power transformer, Osborne 21555

Heat sink—One for each 2N1523. Thermalloy 6421B, or Delco 7281366

Use Delco insulator kit 7274633 for transistors

output, suitable for transformer-powered equipment, lights, or motors.

The inverter construction is straightforward, and assembly is on an aluminum chassis measuring 8″ × 6″ × 2″. A standard heat sink for the transistors is specified, however, the sink shown in figure 7 may be used. A grounded-collector circuit is used (negative ground) so the transistors need not be insulated from the heat sink or chassis. Silicon grease should be placed between the transistor, sink sections, and chassis to ensure good thermal conductivity between the units. The low-voltage primary circuit should be wired with heavy-duty flexible line cord, or stranded #12 hookup wire.

This supply is designed to start under full load, and should be turned on loaded, since unloaded operation (especially starting and stopping) may give rise to transients which may endanger the transistors.

The supply is capable of 100 watts continuous power and about twice this amount in intermittent service. Because of the square-wave output, additional line filtering

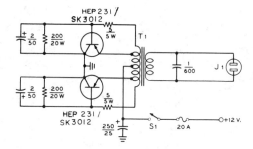

Figure 9

DC TO AC INVERTER FOR THE CAR

T_1—Inverter transformer. 12-volt dc, tapped primary, 115-volt ac, tapped secondary (Triad TY-75A)

Line Filter—J. W. Miller 5521 choke, 4 μH at 20 amperes, bypassed with 0.1-μF capacitors on each side (12-volt circuit). J. W. Miller 7818 (115-volt circuit)

Heat Sink—Wakefield NC 623A for each transistor

may be necessary in the power line to the equipment, and a suitable line filter is tabulated in the parts list of figure 9.

19-3   Antennas for Mobile Operation

The mobile antenna is the key to success-ful operation on any amateur band. Because of space limitations on the vehicle and the sweep of the vehicle body panels, the verti-cal whip antenna is the most popular mobile antenna, regardless of the band of operation. For hf service, the whip takes the form of a flexible, tapered steel rod with a threaded base fitting.

Unless the whip is a resonant length (common only on the vhf, 6- and 10-meter bands) it is brought into resonance by the addition of a *loading coil* which makes up for the missing antenna length. The coil may be placed either at the base of the whip, or near the center. Overall antenna efficiency is generally a function of the Q or circuit efficiency of the loading coil, and every effort should be made to design and use a high-Q coil, well removed from the body of the vehicle.

Antenna Mounts   High-frequency whip an-tennas, because of their height, are usually mounted low on the vehicle, often on the rear bumper or fender as shown in figure 10. Chain or strap-type mounts are available; they clamp directly over the edges of the bumper without the need of drilling mounting holes in the ve-hicle. The antenna is held in position by an insulated adapter bolted to the top bracket of the mount. Sometimes a heavy spring is included in the mount to absorb the road shock.

The whip antenna must remain free and clear of the body of the vehicle. Use of a bumper mount on station wagons, trucks and vans is not recommended because the whip passes too close to the upper metal body panels of the vehicle and severe de-tuning of the antenna may result. In this situation, a shorter antenna mounted higher on the body or roof is recommended.

A ball mount and spring (figure 11) can be used to mount the whip antenna at an angle on the vehicle so that the antenna it-self is in a vertical plane, regardless of the plane of the mount. Usual placement in-cludes the rear deck, the side or top of the fender or (for short antennas) the top, flat

Figure 10

MULTIBAND MOBILE WHIP USING HIGH-Q AIRWOUND COIL

Heavy base section provides support for adjust-able loading coil. Antenna may be used over a range of about 15 kHz on 80 meters without re-tuning and correspondingly larger ranges on the higher frequency bands. Coil is mounted well clear of automobile body. Outer braid of coax line is grounded to bumper and to auto frame at base of antenna.

portion of the roof. In the latter case, care must be taken to make sure the antenna does not strike overhead electrical wires and tree limbs.

The ball mount requires that a mounting hole be drilled in the skin of the vehicle on a relatively flat surface. Once the mount is in place, the whip is inserted in the socket and the rotary ball joint adjusted to align the whip in a vertical position.

Many amateurs hesitate to drill holes in their vehicle and are interested in an an-tenna mount that will not scar the body of the automobile. The trunk lip mount is a device that meets this need. The adjustable antenna mount is slipped beneath the edge of the trunk lid and bolted firmly to the groove of the car body. Enough clearance

Figure 11

ADJUSTABLE BASE MOUNT
FOR MOBILE WHIP

Mount may be placed on automobile panel and then adjusted so that whip is vertical regardless of position of panel. Jumper wire inside spring ensures that inductance of spring does not become part of the antenna.

exists around the edge of most trunk lids to permit the user to bring a small coaxial cable (RG-58/U) through the gap and up to the antenna mount as shown in the illustration. Some trunk mounts fasten to the trunk lid as shown in figure 12.

A vhf whip may be clamped to the rain gutter of the vehicle by means of a gutter clamp. The mount is affixed to the outer rim of the gutter, taking care to be sure that the clamp breaks through the enamel coating of the gutter to make a good electrical contact to the body of the vehicle. Scraping off the paint at this point is a good idea. The mount is adjustable to permit placing the antenna in a vertical position.

Vhf Antennas   In areas where vertical polarization is predominant, the vertical whip antenna is used for mobile operation. The most logical place for a vhf whip is at the center of the vehicle roof since this provides a relatively large ground-plane area and nearly omnidirectional coverage. The next best location is at or near the

center of the trunk lid at the rear of the vehicle. Field-strength tests have shown that trunk-lid mounting of a 144-MHz whip antenna provides an omnidirectional pattern that is only 1 decibel less in signal strength than the same antenna in a roof-mount position.

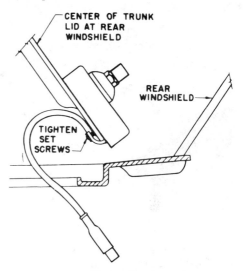

Figure 12

TRUNK-LID ANTENNA MOUNT

Antenna mount is bolted to underside of trunk lid so that auto body is not damaged by mounting holes.

The Vhf Whip Antenna   By far the most popular and inexpensive antenna for vhf mobile service is the quarter-wave whip, which uses the automobile body as a ground plane. Nominal whip length is 55″ (140 cm) for the 50-MHz band, 19″ (48.5 cm) for the 144-MHz band, 12½″ (32 cm) for the 220-MHz band and 6½″ (16.5 cm) for the 430-MHz band. The radiation resistance of the whip is about 30 ohms when mounted on the car body and overall length of the whip may be adjusted for lowest value of SWR on the coaxial feed system.

A popular antenna for 50-MHz and 144-MHz operation is a 55″ (140 cm) whip which operates as a ¼-wavelength radiator on the lower band and as a ¾-wavelength radiator on the higher band. A collapsible whip can be adjusted for minimum SWR on either band since the resonant points for each band are only a few inches apart.

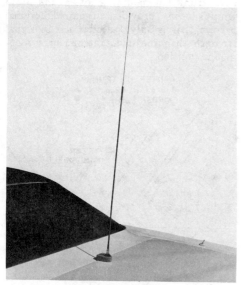

Figure 13

**VHF EXTENDED WHIP EQUALS
ROOF-MOUNTED GROUND PLANE**

Five-eighths wave antenna mounted on rear trunk area of vehicle provides equivalent performance to quarter-wavelength ground plane mounted at center of vehicle roof. Base coil is 6 turns #18 wire, ½″ diam., 1″ long.

A typical ⅝-wavelength whip for the 2-meter band is shown in figure 13. The whip is reduced in length to 47″ (119.3 cm) and is base-loaded with a small coil which is mounted in the base assembly mount. Whip length is adjusted a quarter-inch at a time for lowest SWR on the transmission line to the antenna.

Hf Whip At frequencies lower than 28 **Antennas** MHz, the common mobile whip antenna is appreciably shorter than a quarter-wavelength. As the length of the whip decreases with respect to the wavelength of operation, the radiation resistance of the whip drops sharply. The antenna thus requires some kind of matching system to match the 50-ohm nominal output impedance of most transmitting equipment. If the matching device were 100 percent efficient, the whip antenna performance would compare favorably with a full size antenna.

However, the short whip, combined with the imperfect ground system in a mobile installation is a very lossy device, whose efficiency drops as the operating frequency is lowered. Depending on the length of the antenna and other factors, the radiation resistance of a whip antenna may be as low as one ohm at 80 meters, with a capacitive reactance component as high as 3500 ohms.

In addition to the radiation resistance, the loss resistance of the matching network must be recognized as well as the ground loss resistance, the sum of which comprise the total resistive component of the impedance appearing at the base of the antenna. The loss resistance, taken in total, is usually much greater than the radiation resistance, especially at the lower operating frequencies (figure 14). In this example of an 80-meter whip, the radiation resistance is 1 ohm, the loading coil resistance is 10 ohms and the ground loss is 9 ohms. The overall radiating efficiency is 5 percent, representing a transmitter power loss of about 12 dB. In spite of such inefficiency, mobile whip antennas are used to good advantage on the 80- and 160-meter bands for short range, ground-wave communication.

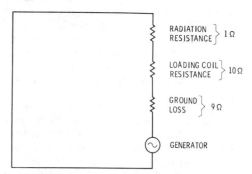

RADIATION RESISTANCE } 1 Ω

LOADING COIL RESISTANCE } 10 Ω

GROUND LOSS } 9 Ω

GENERATOR

Figure 14

**80-METER MOBILE WHIP HAS
LOW EFFICIENCY**

A representative 80-meter mobile whip center loaded, has an overall radiation loss of 19 ohms compared to a radiation resistance of about 1 ohm. Efficiency is about 5 percent, representing a transmitter power loss of 12 decibels.

10-Meter Mobile The most popular mobile **Antennas** antenna for 10-meter operation is a rear-mounted whip approximately 8 feet long, fed with coaxial line. This is a highly satisfactory antenna, but a few remarks are in order on the subject of feed and coupling systems.

The feed-point resistance of a resonant quarter-wave rear-mounted whip is approx-

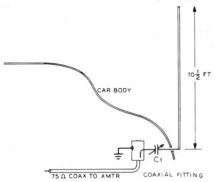

Figure 15

5/16-WAVE WHIP RADIATOR FOR 10 METERS

If a whip antenna is made slightly longer than one-quarter wave it acts as a slightly better radiator than the usual quarter-wave whip, and it can provide a better match to the antenna transmission line if the reactance is tuned out by a series capacitor close to the base of the antenna. Capacitor C_1 may be a 100-pF midget variable.

imately 20 to 25 ohms. While the standing-wave ratio when using 50-ohm coaxial line will not be much greater than 2 to 1, it is nevertheless desirable to make the line to the transmitter exactly odd multiples of one-quarter wavelength long electrically at the center of the band. This procedure will minimize variations in loading over the band.

A more effective radiator and a better line match may be obtained by making the whip approximately $10\frac{1}{2}$ feet long and feeding it with 75-ohm coax (such as RG-11/U) via a series capacitor, as shown in figure 15. The relay and series capacitor are mounted inside the trunk, as close to the antenna feedthrough or base-mount insulator as possible. The $10\frac{1}{2}$-foot length applies to the overall length from the tip of the whip to the point where the lead-in passes through the car body. The leads inside the car (connecting the coaxial cable, relay, series capacitor and antenna lead) should be as short as possible. The outer conductor of both coaxial cables should be grounded to

the car body at the relay end with short, heavy conductors.

A 100-pF midget variable capacitor is suitable for C_1. The optimum setting for lowest SWR at the transmitter should be determined experimentally at the center of

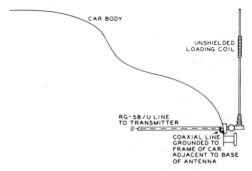

Figure 16

THE CENTER-LOADED WHIP ANTENNA

The center-loaded whip antenna when provided with a tapped loading coil or a series of coils, may be used over a wide frequency range. The loading coil may be shorted for use of the antenna on the 10-meter band.

the band. This setting then will be satisfactory over the whole band.

If an all-band center-loaded mobile antenna is used, the loading coil at the center of the antenna may be shorted out for operation of the antenna on the 10-meter band. The usual type of center-loaded mobile antenna will be between 9 and 11 feet long, including the center-loading inductance which is shorted out. Thus such an antenna may be shortened to an electrical quarter wave for the 10-meter band by using a series capacitor as just discussed. If a pi network is used in the plate circuit of the output stage of the mobile transmitter, any reactance presented at the antenna terminals of the transmitter by the antenna may be tuned out with the pi network.

The All-Band Center-Loaded Mobile Antenna The great majority of mobile operation on the 14-MHz band and below is with center-loaded whip antennas. These antennas use an insulated bumper or body mount, with provision for coaxial feed from the base of the antenna to the transmitter, as shown in figure 16.

The center-loaded whip antenna must be tuned to obtain optimum operation on the desired frequency of operation. These antennas will operate at maximum efficiency over a range of perhaps 20 kHz on the 75-meter band, covering a somewhat wider range on the 40-meter band, and covering the whole 20-meter phone band. The procedure for tuning the antenna is as follows:

The antenna is installed, fully assembled, with a coaxial lead of RG-58/U from the base of the antenna to the place where the transmitter is installed. The rear deck of the car should be closed, and the car should be parked in a location as clear as possible of trees, buildings, and overhead power lines. Objects within 15 or 20 feet of the antenna can exert a considerable detuning effect on the antenna system due to its relatively high operating Q. The end of the coaxial cable which will plug into the transmitter is terminated in a link of 3 or 4 turns of wire. This link is then coupled to a grid-dip meter and the resonant frequency of the antenna determined by noting the frequency at which the grid current fluctuates. The coils furnished with the antennas normally are too large for the usual operating frequency, since it is much easier to remove turns than to add them. Turns then are removed, *one at a time*, until the antenna resonates at the desired frequency. If too many turns have been removed, a length of wire may be spliced on and soldered. Then, with a length of insulating tubing slipped over the soldered joint, turns may be added to lower the resonant frequency. Or, if the tapped type of coil is used, taps are changed until the proper number of turns for the desired operating frequency is found. This procedure is repeated for the different bands of operation.

Ground loss resistance in the automobile and capacitance of the car body to ground have been measured to be about 20 ohms at 3.9 MHz. These radiation and loss resistances, plus the loss resistance of a typical loading coil may bring the input impedance of a typical 80-meter center-loaded whip to about 25 to 30 ohms at the resonant frequency. Overall radiation efficiency is about two to five percent and operational bandwidth (for a 3/1 SWR on the transmission line) is about 25 kHz when the antenna is properly matched.

The relatively low efficiency of the loaded whip antenna at the lower frequencies indicates that attention must be paid to all details of the antenna installation. The loading coil must be of the highest possible Q and all joints in the antenna system must be low resistance. To properly match the 25-ohm antenna load to a typical 50-ohm transmission line, the matching system of figure 18 may be used. The loaded whip antenna forms a portion of a network whose input impedance over a small frequency range is close to 50 ohms. The antenna is made a part of an equivalent parallel-reso-

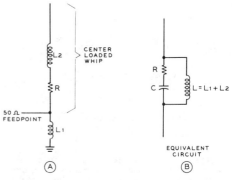

Figure 17

CENTER-LOADED WHIP ANTENNA

A—Center-loaded whip represents large loss resistance (R) which is inverse function of coil Q. High-Q coil (300 or better) provides minimum losses consistent with practical coil design. B—Equivalent circuit provides impedance match between whip antenna and 50-ohm feedpoint.

nant circuit in which the radiation resistance appears in series with the reactive branch of the circuit. The input impedance of such a circuit varies nearly inversely with respect to the radiation resistance of the antenna, thus the very low radiation resistance of the whip antenna may be transformed to a larger value which will match the impedance of the transmission line.

The radiation resistance of the whip antenna can be made to appear as a capacitive reactance at the feedpoint by shortening the antenna. In this case, this is done by slightly reducing the inductance of the center-loading coil. The inductive portion of the tuned network (L_1) consists of a small coil placed across the terminals of the antenna as shown in figure 17A. The LC ratio of antenna and matching coil determine the transformation ratio of the network when the LC product is parallel resonant at the operating frequency of the antenna.

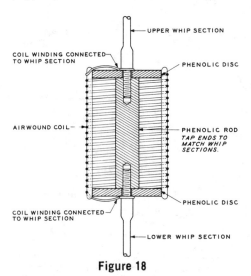

Figure 18

HIGH-Q MOBILE LOADING COIL

Efficient loading coil is assembled from section of air-wound coil stock (i-Core or B-W). 2½″ diam coil is recommended. Approximate inductance for various bands, when used in center of 8-foot whip: is 160 meters, 700 μH; 80 meters, 150 μH; 40 meters, 40 μH; 20 meters, 9 μH; 15 meters, 2.5 μH. Complete antenna is grid-dipped to operating frequency and number of turns in coil adjusted for proper resonance.

Typically, coil L_1 at the base of the center-loaded whip may be about 5μH for operation on the 80-meter band. The turns are shorted out for operation on the higher-frequency bands. A coil consisting of 13 turns of #12 wire, 2½″ diameter and 4″ long will be satisfactory.

The antenna system is grid-dipped to the operating frequency and the coaxial line is then tapped on the base coil at a point which provides a satisfactory impedance match, which may be determined with the aid of a SWR meter in the line to the transmitter.

Construction of a high-Q center loading coil from available coil stock is shown in figure 18.

Top Loading   A *capacity hat* may be added to a loaded whip antenna figure 19) to improve the efficiency at the expense of the wind resistance. The capacitance added above the loading coil requires a reduc-

tion in the number of turns in the coil to reestablish resonance. Since the loss resistance of the coil is proportional to the inductance,

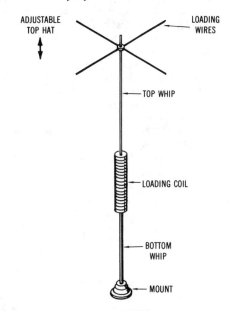

Figure 19

CAPACITY HAT LOADING FOR MOBILE WHIP

A "top hat" is made of stiff wires attached to a ferrule which may be slid up and down the top whip section. It is held in place with a setscrew. Whip tuning may be achieved over a small frequency range by adjusting the position of the top hat for lowest SWR on the transmission line.

any reduction in the number of turns for a given antenna is beneficial.

The hat may be made out of lengths of hard copper wire and hat diameters of several feet have been used with success for 80- and 160-meter operation. The larger the hat, in terms of surface area, the greater the capacitance and the fewer the turns needed in the loading coil.

An SWR Meter   This simple reflectometer is
for Mobile Use   designed to be used with mobile equipment over the 3- to 30-MHz range at power levels up to 500 watts. It may be placed in the 50-ohm coaxial transmission line to the antenna and mounted under the dash of the automobile to provide a constant check of transmitter

power output and antenna operation. It is
also useful for tuneup purposes, since the
transmitter stages may be adjusted for maximum forward-power reading of the instrument. The circuit is bidirectional; that is,
either terminal may be used for either input
or output connection.

The SWR meter is constructed in an
aluminum utility box measuring 4" × 4"
× 2" and the circuit is shown in figure 21.
The heart of the device is a 4¾" long pickup line made of the inner conductor of a
length of RG-58A/U coaxial line and a
piece of ¼-inch copper tubing, which
makes a close slip fit over the polyethylene
inner insulation of the line.

To assemble the pickup line, the outer
jacket and braid are removed from a length
of coaxial line. Before the line is passed within the tubing, the insulation is cut and removed at the center point, which is tinned.
A small hole is drilled at the center of the
copper-tubing section so that a connection
may be made to the inner line. The line is
passed through the tubing, and one lead
of a 51-ohm, ½-watt composition resistor

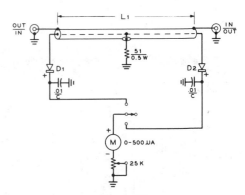

Figure 21

SCHEMATIC, MINI-SWR METER

D₁, D₂—1N34A
L₁—See text
M—0-500 μA, dc, Simpson 1212

the ends of the tubing are affixed to the
coaxial connectors, as shown in figure 22.

Sensitivity of the SWR meter is controlled
by the variable resistance in series with the
meter. To check the instrument, power is
fed through it to a matching dummy load
and the meter switch set to read forward
power. On reversal of the switch, the meter
will read reflected power. In the case of
a good load match, the reflected reading will
be near zero, increasing in value with the
degree of mismatch of the load.

19-4   Construction of
Mobile Equipment

The following measures are recommended
for the construction of mobile equipment,
either transmitting or receiving, to ensure
trouble-free operation over long periods:

Use only a stiff, heavy chassis unless the
chassis is quite small.

Use lock washers or lock nuts when
mounting components by means of screws.

Use stranded hookup wire except where
r-f considerations make it inadvisable (such
as for instance the plate tank circuit leads
in a vhf amplifier). Lace and tie leads wherever necessary to keep them from vibrating
or flopping around.

To facilitate servicing of mobile equipment, all interconnecting cables between

Figure 20

**MINI-SWR METER FOR
MOBILE EQUIPMENT**

Inexpensive reflectometer is built in 4" × 4" ×
2" aluminum utility box and many be used over
3- to 30-MHz range at power levels up to 500
watts or so.

is soldered to the line at this point. The
pickup line is then bent into a semicircle and

Figure 22

INTERIOR, MINI-SWR METER

Pickup line is bent in semicircle and tubing is soldered to loops of wire which connect to center pin of SO-239 coaxial receptacles. Center conductor of line is attached to diodes D_1, D_2.

units should be provided with separable connectors on at least one end.

Finally, it should be remembered that the interior of the vehicle can get very hot when it is left in the sun for a period of time. Excessive heat may possibly damage solid-state devices and some crystal microphones. Try and place the mobile equipment where it will not be exposed to such heat. Excessive cold, on the other hand, may render solid-state equipment inoperative as the transistorized power supply may refuse to start.

Control Circuits The send-receive control circuits of a mobile installation are dictated by the design of the equipment, and therefore will be left to the ingenuity of the reader. However, a few generalizations and suggestions are in order.

Do not attempt to control too many relays, particularly heavy-duty relays with large coils, by means of an ordinary push-to-talk switch on a microphone. These contacts are not designed for heavy work, and the inductive "kick" will cause more arcing than the contacts on the microphone switch are designed to handle. It is better to

actuate a single relay with the push-to-talk switch and then control all other relays, including the heavy-duty contactor for the transistor power pack with this relay.

A recommended general control circuit, where one side of the main control relay is connected to the hot 12-volt circuit, but all other relays have one side connected to the ground, is illustrated in figure 23.

Microphones and Circuits The standardized connections for a majority of hand-held microphones provided with push-to-talk switch are shown in figure 24.

The high-impedance *dynamic* microphone is probably the most popular with the *ceramic*-crystal type next in popularity. The conventional crystal type is not suitable for mobile use since the crystal unit will be destroyed by the high temperatures which

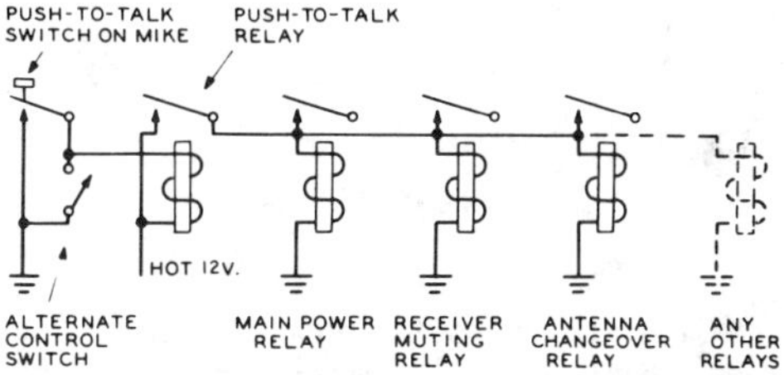

Figure 23

RELAY CONTROL CIRCUIT

Simplified schematic of the recommended relay control circuit for mobile transmitters. The relatively small push-to-talk relay is controlled by the button on the microphone or the communications switch. Then one of the contacts on this relay controls the other relays of the transmitter; one side of the coils of all the additional relays controlled should be grounded.

can be reached in a closed car parked in the sun in the summer time.

The use of low-level microphones in mobile service requires careful attention to the elimination of common-ground circuits in the microphone lead. The ground connection for the shielded cable which runs from the transmitter to the microphone should be made at only one point, preferably directly adjacent to the input of the first tube or transistor in the speech amplifier.

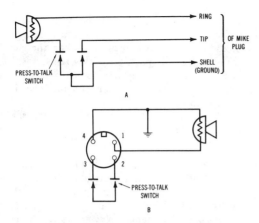

Figure 24

CONNECTION OF PUSH-TO-TALK SWITCH ON HAND-HELD MICROPHONES

(A)—Two-circuit microphone plug.
(B)—Standard microphone plug.

19-5   Vehicular Noise Suppression

Satisfactory communication on hf and vhf channels usually requires noise suppression measures. The measures vary with the mode of communication and the frequency range involved. Vhf f-m reception, on the one hand, usually requires little noise suppression whereas hf SSB reception requires substantial noise suppression in most vehicles equipped with an internal combustion engine having an ignition system.

In addition to the ignition noise generated by the gasoline engine, most vehicles also contribute noise generated by their electrical circuits and additional radio noise may be created by the movement of the vehicle through the atmosphere.

Most of the various types of common noise present in a vehicle may be broken down into the following categories:

1. Ignition noise.
2. Generator or alternator noise.
3. Voltage regulator noise.
4. Instrument noise.
5. Wheel or tire static.
6. Intermittent ground contacts.

Identifying the Noise Each type of noise you hear on a mobile receiver provides a clue as to its identity by its characteristic sound. Ignition noise is a steady popping sound that increases in tempo with higher engine speed. It stops instantly when the ignition key is turned off at fast idle. Generator or alternator noise is a high pitched whine that increases in frequency with higher engine speeds. Voltage regulator noise is a ragged, rasping sound that occurs at an irregular rate. Instrument noise is a hissing, crackling sound that occurs irregularly as the gauges operate. Wheel and tire static is an irregular popping or rushing sound that occurs in dry weather at high speeds.

Ignition Noise The ignition system furnishes a high-voltage spark to ignite the gas-air mixture in the cylinders of the engine. The distributor breaker points select the voltage for the proper plug and an interrupted dc voltage is provided to the ignition coil by a separate set of breaker points driven by the engine.

To reduce the radio noise, it is necessary to make sure the metal ignition coil case is grounded to the vehicle. Scrape the paint around the bolt holes and use lock washers under the nuts to make a firm ground connection. Next, install a .005 μF, 1.6-kV ceramic disc capacitor at the coil *distributor* terminal to ground. Finally, install a 0.1 μF coaxial capacitor at the *battery* terminal of the coil. This is connected in line with the ignition switch. Do not use a conventional capacitor at this point as it is ineffective in the hf-vhf region.

The next step is to install a sparkplug suppressor on each plug or else substitute resistor plugs. Wires to the plugs can be removed and resistance ignition cable substituted which contains a resistive conductor instead of wire. In severe cases of radiation both resistor plugs and suppressor cable must be used. The plugs and cables are often combined in a shielded ignition kit. The kit must be purchased for a specific engine.

Generator or Alternator Noise To reduce generator or alternator noise, the leads to the unit must be filtered. In the case of the alternator, install a 0.5

μF coaxial capacitor at the output terminal. Ground the capacitor to the alternator frame. Two capacitors are required for the dual terminals of a heavy duty alternator.

In the case of the generator, the factory installed capacitor is removed and replaced with a 0.5 μF coaxial capacitor at terminal A (armature). *Do not* connect a capacitor to the field terminal (F). Finally, make sure the body of the device is securely grounded to the frame of the vehicle.

Voltage Regulator Noise Little or no regulator noise is caused by the regulator on newer vehicles equipped with a solid-state ignition system. The older mechanical regulator, however, can produce a crackling noise during operation. To reduce it, a 0.1 μF coaxial capacitor is placed in the battery (B) lead at the regulator and a second capacitor is placed in the armature (A) lead. The field lead (F) is *not* bypassed as this may cause damage to the regulator.

Instrument or Accessory Noise The various instruments or lights sometimes require noise suppression. The rasping noise heard from the gas gauge, for example, can be suppressed by installing a 0.1 μF coaxial capacitor at the gauge. In some cases a small *hash choke* must be placed in series with the line. A suitable choke can be made of 15 turns of #18 enamel wire on a ¼-inch diameter form. A similar capacitor or choke may be required on the windshield wiper motor.

Wheel Static Wheel static is either static electricity generated by rotation of the tires and brake drums, or is noise generated by poor contact between the front wheels and the axles (due to the grease in the bearings). The latter type of noise seldom is caused by the rear wheels, but tire static may of course be generated by all four tires.

Wheel static can be eliminated by insertion of grounding springs under the front hub caps, and by inserting "tire powder" in all inner tubes. Both items are available at radio parts stores and from most auto radio dealers.

Body Static Loose linkages in body or frame joints anywhere in the car are potential static producers when the car is in motion, particularly over a rough road. Locating the source of such noise is difficult, and the simplest procedure is to give the car a thorough tightening up in the hope that the offending poor contacts will be caught up by the procedure. The use of braided bonding straps between the various sections of the body of the car also may prove helpful.

Miscellaneous There are several other potential noise sources on a passenger vehicle, but they do not necessarily give trouble and therefore require attention only in some cases.

At high car speeds under certain atmospheric conditions, corona static may be encountered unless means are taken to prevent it. The receiving-type auto whips which employ a plastic ball tip are so provided in order to minimize this type of noise, which is simply a discharge of the frictional static built up on the car. A whip which ends in a relatively sharp metal point makes an ideal discharge point for the static charge, and will cause corona trouble at a much lower voltage than if the tip were hooded with insulation. A piece of *Vinylite* sleeving slipped over the top portion of the whip and wrapped tightly with heavy thread will prevent this type of static discharge under practically all conditions. An alternative arrangement is to wrap the top portion of the whip with *Scotch* brand electrical tape.

In many cases, the control rods, speedometer cable, etc., will pick up high-tension noise under the hood and conduct it up under the dash where it causes trouble. If so, all control rods and cables should be bonded to the fire wall (bulkhead) where they pass through using a short piece of heavy flexible braid of the type used for shielding.

In some cases it may be necessary to bond the engine to the frame at each rubber engine mount in a similar manner. If a rear mounted whip is employed, the exhaust tail pipe also should be bonded to the frame if it is supported by rubber mounts.

19-6   A Portable 40-Meter Receiver

This simple and compact receiver covers the 40 meter band and may be used on other bands by adding a converter. It is ideal for portable or Field Day operation. The receiver is completely solid state and operates from a nominal 12-volt power supply. A single 5.5 MHz i-f filter provides good selectivity for either c-w or SSB reception. The receiver was designed and built by W6XM.

The Receiver Circuit   A block diagram of the receiver is shown in figure 25. An MPF-102 (Q_1) is used as a tuned r-f stage to provide selectivity and image rejection. A 40673 dual-gate MOSFET is used for the mixer with the signal applied to gate 1 and the local oscillator to gate 2. The mixing oscillator is an MPF-102 (Q_3) which covers the range of 12.5 to 12.8 MHz. Bandspread can be set by choosing the value of capacitor C_1. Two 2N2222 transistors are used as a buffer stage between the oscillator and the mixer to assure good stability.

Receiver selectivity is determined by the i-f filter. Either a c-w or SSB filter may be used at this point. A single 40673 is used as an i-f amplifier (Q_6) and a second 40673 serves as a product detector to achieve good

signal gain and to provide good overload capability. A 2N2222 audio driver (Q_8) powers a LM380 N IC which provides up to 2 watts audio power into a small speaker. The bfo is an MPF-102 (Q_9) followed by a 2N4123 buffer stage.

The R-F Section—The front-end schematic of the receiver is shown in figure 26. Receiver impedance is a nominal 50 ohms. The r-f coil is wound on a small ferrite core. Circuit resonance is checked using a dip-oscillator with a loop around its coil coupled to a loop around the toroid. The r-f stage and mixer circuits are separately peaked by small capacitors mounted side by side on the front panel of the receiver.

Oscillator construction is as described elsewhere in this Handbook. All components are firmly mounted to reduce vibration and movement. The slug-tuned oscillator coil is adjusted for proper frequency coverage and the slug then fixed in place with a drop of epoxy cement. Drive level to the mixer stage is adjusted to 1.5 volts rms at the emitter of the last buffer stage by varying the value of the emitter resistor.

The I-F, Product Detector, and Audio Section (figure 27)—Two 46073 devices are used in the i-f and product detector stages. The bfo is crystal-controlled by a crystal that matches the passband of the i-f filter. A single 2N4123 serves as a buffer and couples the bfo signal to gate 2 of the product detector. An audio gain control is

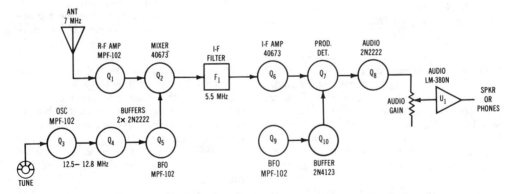

Figure 25

BLOCK DIAGRAM OF PORTABLE 40-METER RECEIVER

This compact, portable receiver tunes the 40-meter band for SSB or c-w reception. It operates from a nominal 12-volt power supply. I-f selectivity is provided by a crystal filter. The complete schematic of the receiver is given in figures 26 and 27.

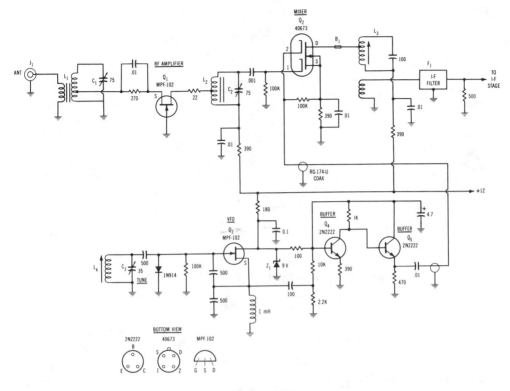

Figure 26

FRONT END OF PORTABLE 40-METER RECEIVER

L_1—40 turns #28 enamel wire on T-37-6 core. Primary 4 turns #28 e. Note: T-37-6 power iron core (SF material) has 3/8-inch outside diameter, rated at 30 μH/100 turns (A_1=30). Yellow dot, mu=8.

L_2—Same as L_1. Tap 4 turns from ground end.

L_3—25 turns #30 e. on 3/8-inch diameter ceramic slug-tuned form. Secondary 13 turns #30 e. at bottom.

L_4—7 turns #26 e. on 3/8-inch diameter slug-tuned form.

F_1—5.5 MHz filter, SSB or c-w passband. Swan Electronics Filtronix model, 437-006 or equivalent.

B—Ferrite bead. Amidon FB-75B-101 or equivalent.

C_3—35 pF air variable capacitor.

Note: All resistors 1/2 watt.

provided in the input circuit of the LM380N audio amplifier.

Receiver Construction   The receiver may be built on a perforated board or a printed-circuit board. If the latter assembly is chosen, a simple technique is to use black PC drafting tape to lay out the board, which is then etched in ferric chloride. The receiver can be made quite compact and it is suggested that the vfo assembly be made up and tested separately.

A PC box arrangement will provide good frequency stability. The receiver is built starting from the audio end and tested a stage at a time. An r-f voltmeter is useful in tuning the oscillators. Final adjustment can be easily made by listening to on-the-air signals.

19-7 A Portable Amateur Band Receiver

The availability of low priced solid-state devices and integrated circuits makes fea-

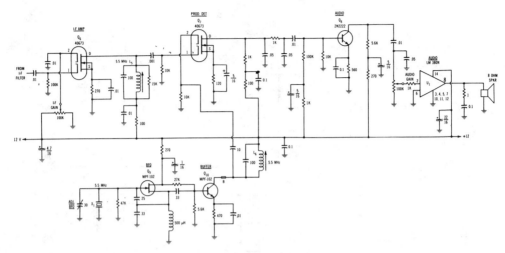

Figure 27

I-F AND AUDIO PORTION OF PORTABLE RECEIVER

L_5, L_6—25 turns #30 enamel wire on 3/8-inch diameter slug-tuned coil form.
X_1—Crystal to match passband of mechanical filter.
Note: All resistors 1/2 watt.

sible the design of a compact, completely solid-state amateur band receiver for c-w and SSB reception that performs as well as or better than an equivalent receiver using conventional vacuum tubes. The advanced receiver described in this section (figure 28) is completely solid state, making use of improved MOSFET and IC devices, and covers the amateur bands between 80 and 10 meters in 500-kHz segments. The design goal was to produce a compact receiver of top-notch performance, but one not so small as to be difficult to assemble and wire, or to operate. The receiver may be run from a battery power supply or from an ac supply so it is well suited for either portable or fixed service. This receiver was designed and built by VE3GFN.

The Receiver Circuit A block diagram of the complete solid-state receiver is shown in figure 29. The circuit is basically a four-band crystal-controlled front-end converter, followed by a tunable i-f receiver which covers the fifth band (80 meters). The bandswitching front-end, or converter, is shown in detail in figure 30. This separate assembly covers the amateur bands between 7 MHz and 29 MHz, with allowance in design for out-of-band

coverage, as well as coverage as high as 30 MHz, or more. Using a *Motorola* 2N5459 high-frequency MOSFET device in the tunable r-f amplifier stages results in high gain and good circuit stability. The r-f amplifier circuitry does not require neutralization, while permitting agc (automatic-gain-control), voltage to be applied to the front end, a feature very necessary in solid-state receivers. The dual-gate feature of the MFE-3006 allows a separation of these functions, the incoming signal being applied to gate 1 of the MOSFET and the agc control voltage to gate 2 of the device.

The R-F Section—The tuned circuits in the high-frequency portion of the receiver are basically 20-meter circuits, which are made resonant in the other high-frequency bands by means of appropriate shunt impedances brought into the circuit by the bandswitch. For 40-meter operation, the basic tuned circuit is padded to a lower resonant frequency by means of capacitor C_1 (figure 31). For 15- and 10-meter operation, the inductance of the tuned circuit is shunted by parallel inductors (L_2 and L_3) thus effectively raising the resonant frequency of the new circuit formed by the auxiliary inductors. These tuned circuits are designed to have an essentially flat response

Figure 28

A SOLID-STATE AMATEUR BAND RECEIVER

This advanced communication receiver covers all amateur bands between 80 and 10 meters. It uses 3 MOSFETS, 5 FETs, 5 transistors, 2 ICs, and 3 hot carrier diodes. Measuring only 10″ × 4″ (panel size) and 7″ deep, the solid-state receiver provides excellent reception of SSB and c-w signals, combined with exceptional strong signal overload capability.
Panel controls (l. to r.) are: Sideband selector switch (S_1); bandswitch; peak preselector (C_1) power switch (S_3); agc switch (S_2); phone jack (J_2) insulated from the panel; r-f gain potentiometer (R_2); audio gain control (R_1); and signal-strength meter (M).
The main tuning dial is calibrated every 100 kHz, with 5-kHz markers and is made of a panel mask (figure 37). The pointer window is cut from a piece of ¼-inch aluminum stock and has a plastic window insert epoxied to the underside of the frame. The cursor line is scratched on the rear of the window.

over 500 kHz of the band in use, making a peaking control unnecessary. The 10-meter tuned circuits can be adjusted to pass any 500-kHz segment of the 10-meter band, allowing the receiver to cover the complete band, by the proper choice of local-oscillator conversion crystal and auxiliary inductor tuning.

Maximum gain is obtained from the MOSFETs in the r-f amplifier stages when gate 2 has +12 volts applied to it; however, this amount of gain has a tendency to overload the i-f system on any strong signal. Hence, provision has been made in the design of the agc system to limit the positive swing of the front-end agc input, eliminating this problem.

The Mixer-Oscillator—A 2N5459 FET is used as a common-source mixer with local oscillator and received signals applied to the gate element. The crystal controlled local oscillator is capacitively coupled to the gate and the incoming signal is inductively cou-

pled through transformer L_1. The converter oscillator employs a 2N4124 bipolar transistor and uses an r-f choke as a broadband collector load on the lower frequencies (RFC_3). Series-connected parallel-tuned circuits provide properly selective collector loads on the two higher-frequency bands.

The schematic of the tunable 80-meter stages and low-frequency i-f section is shown in figure 32. The front end of this section of the receiver has two stages of r-f amplification using MFE-3006 MOSFETs to provide needed sensitivity and image rejection. The tuned circuits for these stages are adjustable from the panel of the receiver and provide a preselector function (*PEAK*). Good electrical isolation between the stages is necessary as the gain of this cascade circuitry is considerable. To avoid crossmodulation and overload, these stages are followed by a 2N5459 FET mixer (Q_3), using a common-gate circuit proven to be tolerant of high input levels.

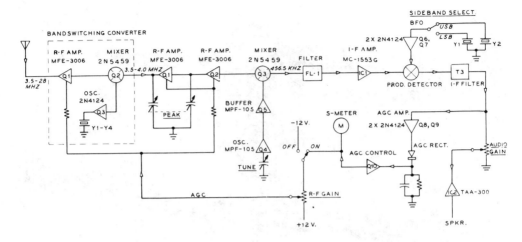

Figure 29

BLOCK DIAGRAM OF THE SOLID-STATE COMMUNICATION RECEIVER

The main portion of the receiver covers the 80-meter band (3.5 - 4.0 MHz) and serves as an i-f section for a bandswitching converter covering the 40-, 20-, 15-, and 10-meter bands in 500-kHz segments. The high-frequency converter unit is crystal controlled and the low-frequency variable oscillator in the 80-meter section is not switched permitting a high degree of electrical and mechanical stability to be achieved. I-f gain is provided by an integrated circuit module (MC-1553G) and suitable SSB selectivity is achieved by a mechanical filter. Audio agc is provided for the various r-f stages and front-end gain may be separately controlled, if desired. The complete schematic of the receiver is given in figures 30 and 32.

The intermediate frequency of the receiver is 455 kHz and the frequency response of the i-f system is largely established by a mechanical filter having a passband (2.1 kHz) suitable for SSB reception. Intermediate-frequency gain is provided by a *Motorola* integrated circuit element (MC-1553G), matched to the mechanical filter by a simple transformer and resistance network.

The Product Detector—A product detector is used to provide good linearity, low insertion loss, and a minimum of beat-oscillator leakthrough into the audio system. One-half of a diode quad is used for the detector, employing 1N2970 hot-carrier diodes, resulting in excellent circuit balance. Closely matched 1K load resistors ensure minimum leakthrough while a simple low-pass audio filter (T_3) placed after the product detector attenuates all residual high-frequency products. The filter is a parallel-tuned circuit at 455 kHz offering high impedance to the intermediate frequency, and a low impedance to audio frequencies.

The local oscillator (bfo) consists of separate crystal-controlled oscillators with the

outputs selected by switch S_2, feeding the input of the product detector through transformer T_2. A switch on the panel of the receiver (*SIDEBAND SELECT*) turns on one oscillator or the other for upper- or lower-sideband reception. The specified oscillator crystals should be as close to the target frequency as possible, since reduced detector output will result if one or the other of the crystals is misplaced on the slope of the filter passband.

The audio system is a second integrated-circuit package (TAA-300) delivering almost a watt of audio power with a 10-milli-volt driving signal. Speakers of 3 to 30 ohms impedance may be used, and the receiver will drive an efficient 10-inch diameter speaker with impressive results. A jack is provided on the panel for use with low-impedance earphones.

The AGC System—The agc network is novel in that the agc lines swing from positive to negative potential with increasing input signal level (figure 24). The three control lines are terminated at the arm of the R-F GAIN control potentiometer (R_2). One

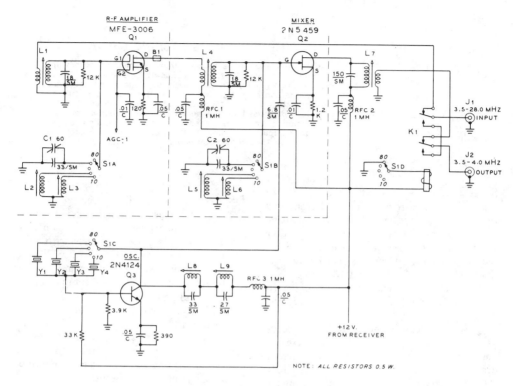

Figure 30

CONVERTER PORTION OF COMMUNICATION RECEIVER

B₁—Ceramic bead (Ferroxcube K5-001-03B or Stackpole 7D)
C₁, C₂—10 to 60-pF piston capacitors (Voltronics TM-60C, or equiv.)
J₁, J₂—Type BNC receptacles, UG-657/U
K₁—Dpdt relay, crystal-can style, 12-volt coil (Potter-Brumfield SC-11DB or equiv.)
L₁, L₄—24 turns #32 enameled wire, closewound on ¼" diameter form. Approx. 4 µH (Q = 50).
   Use J. W. Miller 4500-2 (red) form, powdered iron core. Link winding is 5 turns #42 e. around
   "cold" end of coil
L₂, L₅—(15 meters). 20 turns #32 e., closewound on ¼" diam. form. Approx. 3.4 µH. J. W. Miller
   4500-3 (green) form, powdered-iron core
L₁, L₆—(10 meters). 11 turns #32 e., as L₂. Approx. 1.4 µH
L₇—40 turns #32 e., closewound on ¼" diameter form. J. W. Miller 4500-3 (green) form, pow-
   dered iron core. Tunes to 3.9 MHz. Link winding is 10 turns #32 e. around "cold" end of coil
L₈—10 turns #32 e., closewound on ¼" diameter form. J. W. Miller 4500-2 (red) form, powdered
   iron core. Resonates to 24.5 MHz.
L₉—15 turns #32 e., as L₈. Resonates to 17.5 MHz
RFC₁, ₃—1 millihenry. J. W. Miller 9350-44 or equiv.
S₁A, D—4 pole 6 position ceramic switch. Centralab 2021 or equiv.
Y₁—3.500 MHz crystal, HC-6/U holder
Y₂—10.500 MHz, as Y₁
Y₁—17.500 MHz, as Y₁
Y₄—24.500 MHz, as Y₁

end of the potentiometer (max) is connected to the +12-volt supply line, and the other end (min) to about −3 volts when the agc switch (S₂) is off. When agc is on, the control is switched to the drain circuit of an agc control FET (Q₁₀). With no input signal, the gate of the control FET is near zero potential and the FET conducts, placing the negative end of the r-f gain control potentiometer close to ground potential. The agc lines, therefore, are at some positive potential between ground and +12 volts, depending on the setting of the potentiometer, allowing maximum receiver gain to

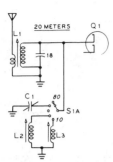

Figure 31

SIMPLIFIED R-F SWITCHING CIRCUIT

The external antenna is coupled to a resonant LC circuit for
20-meter reception. When the bandswitch is changed to 40
meters, the 20-meter circuit is padded to the lower frequency
by the addition of piston capacitor C₁, placed in the circuit by
switch section S₁ₐ. On 15 meters, the inductance of 20-meter
coil L₁ is decreased by the added shunting action of coil Lₜ.
On 10 meters, coil L₃ is switched in the circuit. Alignment of
the tuned circuit must first be done on 20 meters before the
15- and 20-meter bands are adjusted.

be established, if desired. When a higher
input signal level requires reduced front-end
receiver gain, rectified audio of a positive
polarity from the agc amplifiers (Q_8, Q_9)
is applied to the gate of the control FET,
reducing its conduction. Accordingly, the
drain element of the FET drops toward
-12 volts, taking the agc lines along with
it, thus reducing front-end gain of the
receiver.

A signal-strength meter is incorporated
as part of the agc system. The meter is
connected so as to measure the current
drawn by the control FET. The *METER-
ADJUST* control (R_3) is set so the meter
indicates full-scale current when the antenna
input terminals are grounded. In operation,
the *RF-GAIN* control (R_2) is set so that
a small deflection of the meter (toward zero
current) takes place with antenna connected
but without signal input. At this point, the
agc system will control receiver front-end
gain in the proper manner, between near
cutoff and maximum usable gain.

Power and Switching Circuits—The receiver is operated from a $+12$-volt 200 mA
supply. In addition, -12 volts is required
for agc action. The drain of the -12 volt
section is only 20 milliamperes and series
connected "penlite" cells may be incorporated in the receiver, if desired, for this
function.

The converter portion of the receiver is
switched in and out by means of a small
crystal-can relay (K_1, figure 30) operated
by the bandswitch. The relay is normally
unenergized in all band positions except 80
meters. On this band, the relay removes the
converter from the circuit and bypasses
the antenna connections around the converter portion of the receiver.

Receiver Construction A multiband receiver such as
this is a complex device and its
construction should only be
undertaken by a person familiar with solid-
state devices in general and MOSFETs in
particular, and who has built and aligned
equipment approaching this complexity.

The solid-state receiver is built on a
chassis within a wrap-around metal cabinet
measuring $10'' \times 7'' \times 4''$. The cabinet
assembly specified comes complete with
panel, chassis, and rubber mounting feet.
Other cabinets of the same general configuration, of course, may be used.

General receiver assembly may be seen in
the photographs and drawings. The high-
frequency converter covering 40 through
10 meters is the most complex assembly and
the most compact (figure 34). This unit is
built in an aluminum box measuring $4'' \times
2'' \times 2\frac{3}{4}''$ and is mounted to the left rear
of the main chassis. The converter band-
switch (S_1) is panel driven by means of an
extension shaft as seen in the top-view
photograph. Power and control leads are
brought out through miniature feedthrough
insulators mounted on the side of the box.

The variable-frequency oscillator is a
second subassembly built within an alumi-
num box measuring $3\frac{1}{4}'' \times 2\frac{1}{8}'' \times 1\frac{5}{8}''$.
The tuning capacitor used (C_3) is a high-
quality unit having full ball-race bearings
front and back and a controlled torque. This
unit provides minimum drag on the geared
dial.

The first step in construction of the solid-
state receiver is to lay out the chassis, panel,
tuning dial, and other major components in
a "mockup" assembly to ensure that the
receiver will go together without a physical

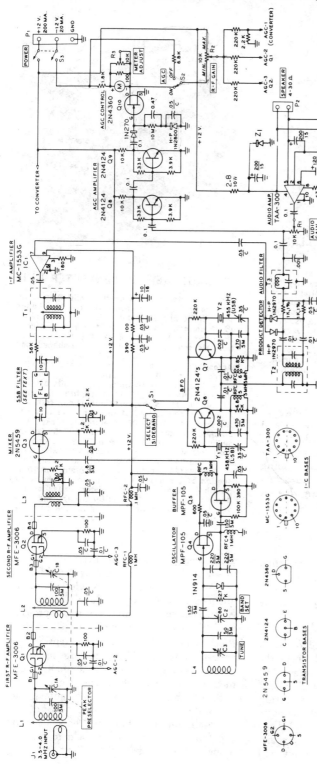

Figure 32

SCHEMATIC OF SOLID-STATE RECEIVER

B_1-B_4—Ceramic bead. Ferroxcube K5-001-003B or Stackpole 7D.

C_{1A}, $_B$—7-110 pF per section. Miniature two section broadcast-type mica compression capacitor. Mitsumi PVC-2Z or equiv.

C_2—10- to 60-pF piston capacitor. Voltronics TM-60C or equiv.

C_3—8- to 45-pF air capacitor. Jackson Bros. 804-50 or equiv. (Obtainable from: M. Swedgal, 258 Broadway, New York, N. Y. 10007).

FL_1—Mechanical Filter 455-kHz center frequency, 2.1-kHz bandpass. Toyo 455-2.4C. Collins Radio Co. amateur-type filter may be used by substituting 150 pF variable mica compression capacitors for 10 pF resonating capacitors.

T_1-T_2—Double-tuned miniature i-f transformer, 455 kHz. Armaco TR-104, or J. W. Miller 8807 or equiv.

Y_1, Y_2—Sideband-selection crystals to match filter characteristics. Type HC-6/U

Z_1—10 volt, 7-watt zener diode on 1½" square heat sink

1N2800, 1N2970—Hewlett Packard hot-carrier diodes

Teflon terminals—Sealectro FT-SM1 or equiv.

Cabinet—Hammond 1426-G (10" X 7" X 4") or equiv.

Meter—0-200 dc microammeter

Note: All resistors ¼ watt unless otherwise specified.

Also, Kokusai MF-455-15 mechanical filter may be used. See filter data for full application information.

IC_1—Integrated-circuit module. Motorola MC-1552G or equiv.

IC_2—Integrated-circuit module. Phillips/Amperex TAA-300 or equiv.

L_1-L_3—40 turns #32 e. wire on ¼" diameter form. Approx. 11 µH. J. W. Miller 4500-2 (red) form, powdered iron core. Link is 10 turns #32 e. on "cold" end

L_4—40 turns #32 e., as L_1. Approx. 15 µH (see text)

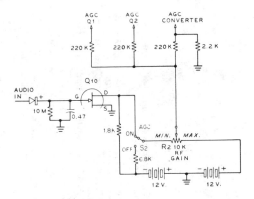

Figure 33

SIMPLIFIED AUDIO-CONTROLLED AGC SYSTEM

The three agc lines (Q_1, Q_2 and converter) are terminated at the arm of r-f gain control R_2. When agc switch S_2 is off, control voltage may be varied between +12 and −3 volts. When the agc system is on, control is switched to the drain circuit of FET Q_{10}. Agc voltage is now proportional to the audio input signal, varying between zero and +12 volts under normal conditions. A strong signal will drive the agc towards −12 volts, sharply reducing receiver gain. Maximum gain is controlled by the potentiometer.

Figure 34

OBLIQUE VIEW OF CONVERTER UNIT

The converter section of the solid-state communications receiver covers the amateur bands between 80 and 10 meters and has an i-f output of 80 meters. The unit is built in a small aluminum box (4″ × 2″ × 2¾″) with the major components mounted on the inner, U-shaped box section.

Across the rear of the assembly are the slug tuned r-f coils (l. to r.): 20-, 15- and 10-meter coils. The 15- and 10-meter mixer coils are immediately to the right. In the righthand corner of the box is the mixer output coil (L_1).

Along the center line of the converter unit are (l. to r.): The MFE-3006 r-f amplifier socket, the 20-meter mixer coil, and the 2N5459 mixer socket. At the front of the unit are the conversion crystals (l. to r.): 3.5 MHz, 10.5 MHz, 17.5 MHz, and 24.5 MHz. To the right of the crystals is the 2N4124 oscillator socket. Along the front section of the assembly are (l. to r.): the relay feedthrough terminal and piston capacitor C_1, bandswitch S_1, piston capacitor C_2, agc and voltage feedthrough terminals, and (at the extreme right) oscillator collector coils L_8 and L_9.

conflict between the components. Figure 35 shows placement of the converter and oscillator assemblies and the i-f filter. The exact location of the vfo box behind the panel and the height of the main tuning capacitor on the side of the box are determined by the position of the tuning dial on the main panel.

Receiver Wiring The receiver should be wired in an orderly manner, a stage at a time. To reduce r-f ground currents, all grounds for a single stage should be returned to that stage, preferably to a common ground point at or near the transistor socket.

It is suggested that the r-f stages of the main receiver section be wired first, followed by the oscillator assembly, and then the product detector and the audio stage. The agc system, S-meter, and power wiring may be done last. A very small pencil soldering iron, miniature solder, and small diameter (No. 22) hookup wire are recommended for ease in assembly. The various tuned circuits are wired and grid-dipped to frequency and

the interstage shields are made up and cut to fit (a "nibbling" tool is handy here) as the work progresses. A closeup of the under-chassis r-f stages is shown in figure 36. A two-section variable mica compression-tuning capacitor is used for C_1 (*PEAK PRESELECTOR*) and has an extension shaft press-fit onto the short tuning stub. The capacitor is supported from a small bracket mounted directly behind the panel.

Small shields are mounted across each MOSFET socket. The shields are cut of scrap aluminum or brass and have a mounting foot on them which is held in place by a nearby 4-40 bolt. The first r-f stage

Figure 35

TOP VIEW OF RECEIVER ASSEMBLY

Placement of the major receiver components may be observed in this view. The h-f crystal-controlled converter assembly is at the left with the bandswitch extension shaft running to the front panel. At the center of the main chassis are the mechanical filter and the variable oscillator for the 80-meter portion of the receiver. Directly behind the oscillator are the i-f amplifier and the bfo stage with the associated sideband-selection crystals. At the right is the audio IC stage (with heat sink) and the "meter-adjust" potentiometer. The agc stages are in the right front corner of the receiver, with the 80-meter r-f section located at the front left corner of the chassis.

MOSFET socket (Q_1) is at the left of the photograph with the small coaxial line from the converter unit visible at the lower edge of the assembly. To the right is the second r-f stage MOSFET socket (Q_2), with the FET mixer socket above and to the right. The injection line from the vfo passes through a *Teflon* feedthrough insulator mounted in the chassis immediately behind the tuning dial and runs to the gate terminal of the FET socket.

The main tuning dial is made up of a reduction drive, a home-made pointer, and a calibrated scale etched on a piece of copper-plated circuit board of the glass-epoxy variety. The mask for the negative of the board is reproduced in figure 37. It may be photocopied from the page and used to make a negative for direct reproduction.

The Converter Assembly—The general layout of the converter assembly is shown in figures 38 through 40. The MOSFETs and conversion crystals are mounted in sockets placed atop the converter box, with the various slug-tuned coils mounted at the rear of the assembly. Figure 38 shows the rear of the box with the cover removed. Note that several *Teflon* feedthrough insulators are mounted in the L-shaped shield partition to pass power leads between the stages within the box. An oblique view of the r-f compartment is shown in figure 39.

An end view of the converter assembly is shown in figure 40. The relay is held in

Figure 36

UNDER-CHASSIS VIEW OF SOLID-STATE COMMUNICATIONS RECEIVER

The 80-meter r-f amplifier and mixer stages are seen in the upper left corner of the chassis. The two-section variable mica compression capacitor ($C_{1A, B}$) is mounted to the chassis by means of a small aluminum bracket affixed behind the main panel. The capacitor is driven by a short extension shaft. An intrastage shield is placed across the first r-f amplifier MOSFET socket (Q_1) and a second similar shield is placed across the second r-f amplifier socket. The shields may be made of copper-plated circuit board, aluminum, or thin copper shim stock. The audio circuit and agc components are placed along the right-hand edge of the chassis, with the bfo, detector and i-f components strung along the rear of the chassis area (bottom of the photograph). The two 35-pF capacitors used to adjust the frequency of the bfo crystals are supported below the chassis by their leads.
Note: The cutout at the front of the chassis is to provide room for the gear-reduction drive mounted to the panel.

position with a small aluminum U-clamp over the body, and the opposite side of the L-shaped intrastage shield is visible.

The Variable Oscillator—The vfo is the only other separate subassembly. Layout of parts (aside from placement of the main tuning capacitor, mentioned earlier) is not critical. The components are self-supported around the capacitor using short, direct leads to prevent vibration. It is possible to build the unit in a much smaller box, but the good drift characteristic (100-Hz total warmup drift) makes the larger box worthwhile. Both FET sockets are mounted on the vertical front surface of the box, with the oscillator coil (L_4) mounted to one end; and the bandset capacitor (C_2) mounted to the other end of the box.

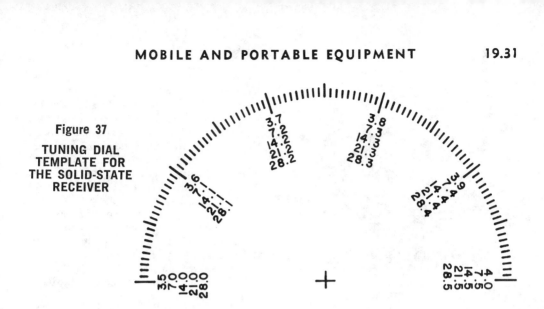

Figure 37

TUNING DIAL TEMPLATE FOR THE SOLID-STATE RECEIVER

Figure 38

REAR VIEW OF CONVERTER ASSEMBLY

The r-f amplifier and bandswitch are seen at the right of the internal shield partition. R-f coils are (l. to r.): 20, 15 and 10 meters. Note Teflon feedthrough terminals mounted in the intrastage partition. The mixer stage and crystal can relay (K_1) are at the left of the partition. Mixer coils are (l. to r.): 15 and 10 meters. I-f output coil L_7 is at the extreme left.

Figure 39

THE R-F AMPLIFIER

The r-f amplifier coils are in the foreground with the bandswitch and piston capacitor (C_1) at the right. The coaxial leads run to the crystal-can relay. The outer shields of the various coaxial lines are grounded to a common point near the relay and also at the free ends in the receiver assembly. Note that coils and bandswitch have been arranged for shortest possible lead lengths.

Receiver Alignment   Alignment of the receiver is not difficult if done in a systematic manner and may be done by ear alone. A quicker and better job may be achieved, however, with the use of proper instruments. The main receiver chassis is aligned first, so that a proper output indicator will be available for subsequent alignment of the converter. All alignment is done with the agc switched off. Before beginning the alignment and before power is applied to the receiver, the tuning meter should be disconnected to prevent its possible damage due to accidental overcurrent. The builder should also note the information in the transistor chapter of this Handbook regarding the handling procedures to be used with the MOSFET transistors, which are inserted toward the end of the alignment operation.

Figure 40

SIDE VIEW OF THE CONVERTER UNIT

The crystal-can relay is in the lower foreground with the 10-meter oscillator coil at the top left and the 15-meter oscillator coil at the bottom left. The internal shield (also seen in figure 38) is L-shaped and isolates the oscillator coils from the mixer coils located at the rear of the chassis deck.

The audio portion of the receiver is tested first. A heat sink is placed over the audio IC (TAA-300) before tests are begun. A 1000-Hz, 10-millivolt sine-wave audio signal is applied at the arm of the *AUDIO-GAIN* potentiometer (R_1) and should result in a signal in the speaker when primary power is applied to the receiver, indicating the audio stage is working. Check the voltage at the drain of the 2N4360 agc control transistor (Q_{10}). It should be close to -12 volts. Removing the audio signal should cause it to drop to almost zero volts. This indicates that the complete agc system is working.

Next, set the *METER-ADJUST* potentiometer (R_3) for zero resistance (short circuit) and reconnect the tuning meter. With the audio signal applied again as before, adjust the meter current for minimum deflection (minimum reading). Removing the audio signal should cause the meter current to increase to a full-scale value. Although the agc is off, the system still controls the meter and it can now be used as an indicator of input signal level to the receiver. Advance the *R-F GAIN* control (R_2) fully clockwise to *Max* position. Apply a 456.5-kHz modulated signal of 1-millivolt

level to the input (pin 1) of the IC i-f amplifier (MC-1553G). If the amplifier, the bfo, and the product-detector stages are working, an audio signal should be heard in the speaker. Adjust the detector filter circuit (T_3) for minimum hiss in the speaker when the audio modulation is turned *off*. Now, adjust the *AUDIO-GAIN* control (R_1) back and forth to make sure it functions properly. Apply the same r-f signal to the input of the mechanical filter and adjust i-f transformer T_1 for maximum signal in the speaker. Varying the input signal frequency above and below 456.5 kHz will provide an indication of the intermediate-frequency passband response of the receiver. Switch the bfo *SELECT-SIDEBAND* switch (S_1) to both positions to ensure that both oscillator circuits are working. Crystal alignment on the filter passband is accomplished by adjustment of the series capacitors.

The next step is to test the variable tuning oscillator. The transistors are inserted in their sockets and the oscillator tuned circuit should be adjusted to tune over the range of 3043.5 kHz to 3543.5 kHz between the extreme positions of the dial. The bandset capacitor (C_2) may be used for this adjustment, along with the slug adjustment of coil L_1. After the slug position has been determined, it should be fastened in place with a drop of cement to prevent vibration.

The tuned circuits in the r-f stages and the mixer should be adjusted to track across the 80-meter band when the *PEAK-PRESELECTOR* control is adjusted. Preliminary alignment should be done with a grid-dip oscillator with transistors Q_1, Q_2, and Q_3 *removed from their sockets*. When MOSFETs Q_1 and Q_2 are inserted in their respective sockets, a ferrite bead is slipped over the gate and drain leads of each device to suppress any tendency toward vhf parasitic oscillations. Place the peaking control (C_1) at half capacitance and apply a 10 microvolt, 3750-kHz signal at the input terminal (J_1) of the main receiver. Tune the receiver to the signal and adjust the three tuning slugs in coils L_1, L_2, and L_3 for maximum signal output. The receiver may now be used for 80-meter reception.

Converter Alignment—The high-frequency converter should now be attached to the main chassis and the various leads connected. Before the MOSFETs are placed in the sockets, the converter tuned circuits should have been grid-dipped to the approximate working frequencies. Now, the converter bandswitch is set to the 20-meter position and the main tuning dial of the receiver set to 14.250 MHz. A 10-microvolt signal at this frequency is applied to the converter input circuit, making sure that the relay K_1 is properly activated. Adjust the slug of the mixer coil (L_4) for maximum output signal, followed by adjustment of r-f coil L_1. These adjustments will not be critical due to the large bandwidth of these circuits. The converter must be first aligned on 20 meters since the tuned circuits are basically tuned to that band. Once they are aligned, do not touch them further.

The bandswitch is now placed in the 40-meter position and a 7.2-MHz signal applied to the receiver. Capacitors C_1 and C_2 are adjusted for maximum signal level. In the same fashion, a midband signal is applied to the converter for the 15- and 10-meter bands, aligning them by the slugs in the shunt coils, as before, mixer circuit first. Finally, adjust the 10-meter oscillator circuit (L_8) for best received signal on that band, then adjust the 15-meter oscillator circuit (L_9) for *minimum* received signal when a 20-meter signal is injected into the receiver. This completes alignment of the receiver.

19-8   A QRP 40-Meter Transceiver

This low power transceiver is designed for 40-meter c-w operation and is suitable for portable, mobile, or fixed operation from a nominal 12.6 volt dc power supply (figure 41). The transceiver is vfo controlled and provides better than 1-watt power output. The receiver is a direct-conversion design having a wide dynamic range and excellent c-w selectivity (figure 42). The transceiver was designed and built by WB5DJE.

Figure 41

QRP 40-METER C-W TRANSCEIVER

This compact transceiver provides 1-watt power output in the 40 meter band. It features wide dynamic range in a direct conversion receiver and an 800 Hz filters having a bandwidth of 200 Hz for optimum c-w reception. The detector rejects a-m signals for improved reception. In this front view, the mode switch (S_2) and phone receptacle are at left of the main dial and the filter bandwidth switch (S_1) and audio gain controls are at the right. The main dial has a reduction gear and is directly calibrated in kilohertz. Panel and dust cover are homemade.

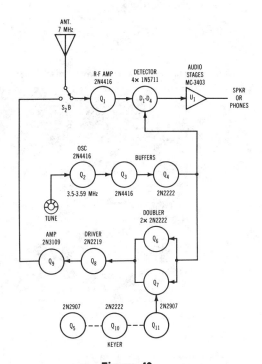

Figure 42

BLOCK DIAGRAM OF 40-METER TRANSCEIVER

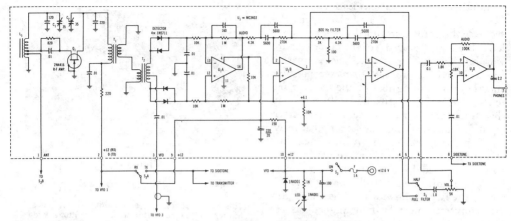

Figure 43

FRONT END OF 40-METER RECEIVER SECTION

L_1—3.5 µH. 33 turns #28 enamel wire on T-37-6 core. Tap 3 turns from ground.
T_1—2.2 µH primary. 25 turns #28 e. on T-37-6 core. Secondary is 6 turns #28 e.
T_2—1.3 µH primary. 5 trifilar-wound turns #26 e. on FT-37-1 core.
Note: T-37-6 powder-iron core (SF material) has ⅜-inch outside diameter, rated at 30 µH per 100 turns (A_1=30). Yellow dot, mu=8. FT-37-1 ferrite core (Q_1 material) has ½-inch outside diameter, rated at 510 µH/100 turns (A_1=510).
C_1, C_2—35 pF ceramic trimmer. Erie 538-002D-9-35 or equivalent.
Note: All resistors ¼ watt.

The Receiver Circuit   The receiver front end (figure 43) has a grounded-gate FET r-f amplifier (Q_1) which has good sensitivity and a low noise level.

The response is about 1.5 microvolt input level for a 10 dB signal-plus-noise to signal ratio. The r-f stage is inductively coupled to the detector and isolated from the vfo to

Figure 44

RECEIVER CIRCUIT BOARD

The board is viewed from the top of the receiver. The front end circuits and the four-diode detector are at the rear of the board, along with the input circuit tuning capacitors. The integrated circuit that serves as filter and audio amplifier is at the center of the board. Switch S_1 is mounted to the front panel above the board and the variable tuning capacitor for the oscillator is seen at the bottom of the photograph. Boards are mounted above an aluminum chassis frame by 4-40 hardware.

prevent interaction between the two stages.

The detector is composed of four diodes which act as r-f switches. It is driven by the vfo at *one-half* the received frequency. Each set of diodes operates differentially and amplitude-modulated signals having double sidebands cancel out in each diode pair. This provides good rejection against the a-m broadcast stations that infest the 40-meter band (figure 44).

Detector output is d-c coupled to a differential audio amplifier (U_1) which provides about 46 dB signal gain. The second sections of U_1 are 800-Hz filters with a bandwidth of 200 Hz and a gain of 30 dB. The Q of the filters is selected to prevent ringing. The final section of U_1 provides good audio gain, picks up the transmitting sidetone, and drives the headphones or a small speaker.

The VFO The schematic of the vfo is shown in figure 45. It provides output between 3.5 MHz and 3.59 MHz to the transmitter and receiver. This provides a tuning range of 7.0 to 7.18 MHz. On re-

ceive the vfo frequency is used directly but on transmit it is doubled. Also, in the receive mode the frequency is offset and a station worked will be shifted approximately 800 Hz, so that the beat note falls in the audio passband.

The vfo is a Seiler type using a 2N4416 FET followed by a FET buffer and output amplifier. Device U_2 in the vfo is a 5-volt, three-terminal regulator biased to provide +7 volts which is set by the grounding resistor value. Voltage regulation of this device is far superior to that of a zener diode and the frequency of the oscillator holds within 10 Hz for input supply voltage variations between 9 and 15 vdc (figure 46).

The oscillator is very stable although not temperature compensated. The coil is a powdered-iron toroid and polypropylene capacitors are used since they have better drift characteristics than do silver-mica capacitors. The inductor is mounted to the board with a coating of polystyrene cement to secure the turns and to stabilize the inductance.

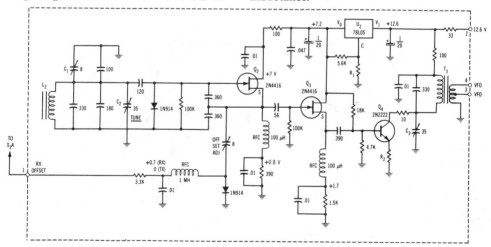

Figure 45

TRANSCEIVER VFO CIRCUIT

L_2—3.0 μH. 30 turns #28 e. on T-37-6 core. Adjust number of turns and spacing to set center frequency and range. Coat with coil dope when adjustments completed.

T_3—6.4 μH primary. 48 turns #28 e. on T-37-6 core. Secondary is 6 turns #26 e. tapped at 3 turns.

C_1—8 pF ceramic trimmer. Erie 538-002A-2-8 or equivalent.

C_2—35 pF air variable capacitor.

Select R_2 for 150 mV rms at terminal 4 of T_3; range value is 33 to 68 ohms.

Select R_1 for 7.2-volt output of U_2; range value is 330 to 1600 ohms.

Note: Fixed capacitors in the oscillator tuned circuit are polypropylene. All resistors are ¼ watt. See figure 43 for core data.

Figure 46

VFO CIRCUIT BOARD

The vfo board is mounted directly behind the main tuning capacitor. The oscillator coil is seen at the right of the board with the connection to the tuning capacitor made through a short length of coaxial cable. Output transformer T_3 is at the left of the photograph.

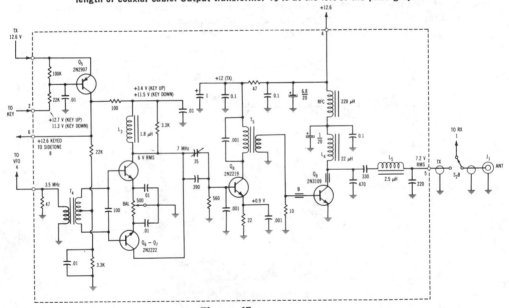

Figure 47

TRANSMITTER PORTION OF TRANSCEIVER

L_3—1.8 μH. 23 turns #26 e. on T-37-6 core.
L_4—22 H. 20 turns #24 e. on FT-37-1 core.
L_5—2.5 μH. 27 turns #26 e. on T-37-6 core.
T_4—1.3 μH. 5 trifilar-wound turns #26 e. on FT-37-1 core.
T_5—0.53 μH. 13 turns #26 e. on T-37-6 core. Secondary is 4 turns #24 e.
B—Ferrite bead. Amidon FB-75B-101 or equivalent.
Note: All resistors ¼ watt. See figure 43 for core data.

Figure 48

TRANSMITTER CIRCUIT BOARD

The transmitter board is mounted to the left of the main tuning capacitor. The output jack of the transceiver is to the rear of the board and the mode switch is mounted to the front panel above the board. Note that output transistor at rear of board has a heat sink snapped over it.

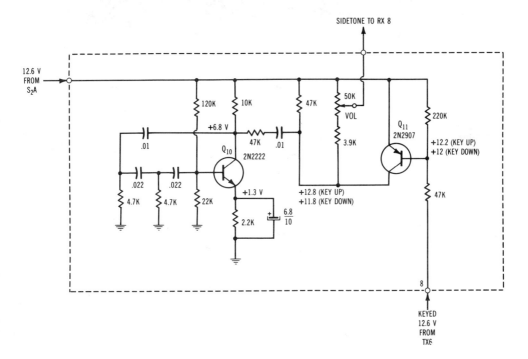

Figure 49

TRANSCEIVER KEYER CIRCUIT

This is a part of the receiver board. All resistors are ¼ watt.

Figure 50

TOP VIEW OF TRANSCEIVER

Transmitter board is at left of assembly, vfo board at center and receiver board at right. Main tuning capacitor is supported from the front panel by means of a subpanel spaced out on studs. Output jack is mounted to the rear panel of the enclosure. The panel-mounted terminal strip supports the dial LED.

The vfo output stage is operated class A and provides energy to receiver and transmitter continuously. The output circuit (T_3) is peaked in the center of the tuning range and output level is about 150 mV at receiver output terminal 3. The transmitter doubler is designed to operate at a level of 320 mV, provided by the second output (4).

The Transmitter The transmitter portion is shown in figures 47 and 48. Transistor keying is employed and an amplifier (Q_3, Q_4) follows the vfo. A doubler, driver and power amplifier proved a 1-watt output signal. Keying is accomplished by Q_{11} which is turned on when the key is closed. This device keys the voltage to the frequency-doubler stage as the driver and final amplifier do not require keying.

Transistors Q_6 and Q_7 are configured as a push-push doubler with push-pull base feed from the trifilar-wound transformer (T_4). The balance potentiometer in the emitter circuits allows the fundamental signal to be balanced out so that the output waveform contains very little 80-meter energy.

The doubler output circuit is peaked in the center of the band. The capacitor and balance control are adjusted for a stable 7-MHz output that contains a minimum of fundamental component.

Driver Q_8 is operated in the class-C mode with some self-bias which provides a clean output signal. The power amplifier is a 2N3109 (or equivalent 5-watt, 1-ampere device). The transistor should have an f_t of about 100 MHz. A higher value of f_t may provide higher harmonic value with corresponding TVI problems. With the values shown, the second harmonic of the transmitter is 40 dB down from the fundamental signal and other harmonics down 50 dB, or better.

Dial Calibration Before the coil is coated, the vfo tuning range is set to calibrate the dial with capacitor C_1 in midposition. The number of turns and spacing on the coil are adjusted and final calibration is done with C_1. The tuning range of 7000 to 7180 kHz was chosen and the dial provides about 10-kHz per 10 degrees of rotation. The dial can be laid out with a protractor and inked when calibration is established.

Transceiver Construction The transceiver is laid out on printed-circuit boards as shown in figures 44, 46, and 48. Coil data is given in the captions. Fifty-ohm

miniature coaxial cable is used to interconnect the r-f circuits between the boards. As built, the vfo requires no shielding. The boards are placed in a homebuilt enclosure as shown in figure 50. The vfo dial is made from a 2-inch diameter piece of *plexiglas*. A thin sheet of *mylar* is cut to fit and dry transfers are used for the lettering. An adhesive is sprayed on the front of the *mylar* and it is attached to the back of the *plexiglas* dial. The pointer is cut from an aluminum plate. A slit is cut in the plate with a saw blade. The plate is mounted behind the dial and illuminated with a green LED which shines through the slot. The dial is mounted to a 6:1 reduction drive unit.

Transceiver Operation The vfo board should be tested first and aligned with the help of an auxiliary receiver or frequency counter. The receiver board is tested next and the r-f amplifier can be aligned on an incoming signal. The transmitter adjustments are done last, with the transmitter section coupled into a 47-ohm 2-watt resistor used as a dummy load. Foil and lay-

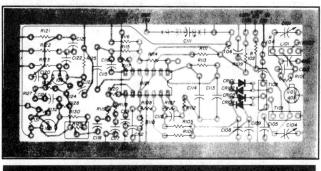

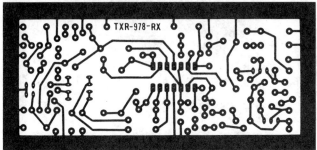

Figure 51

RECEIVER BOARD FULL-SIZE PATTERN

At top: parts layout. At bottom: foil side of the board.

out sides of the three printed-circuit boards are shown in figures 51, 52, and 53.

19-9 A 160-Watt 144-MHz Amplifier

This high-power, solid-state amplifier is intended for f-m service in the 2-meter band. With the addition of proper biasing circuitry the amplifier may be used for SSB service as well. Drive power is 5 to 10 watts as supplied by a typical transceiver and power output is 160 watts, or better, when a 13.8-volt power supply is used. Once adjusted, the amplifier will exhibit broadband performance over the entire band (144–148 MHz) (figure 54). The amplifier is free of spurious outputs under all operating conditions with the exception of harmonic energy which is greatly suppressed by the amplifier design. The amplifier is reproducible and employs a minimum number of parts commensurate with performance, size, and cost.

Amplifier Circuitry The amplifier is a two stage configuration, with the schematic shown in figure 55. Input power is applied to the base of the CD 4024 transistor (Q_1) through a broadband-tuned microstrip matching network composed of C_1, L_1, and C_2. A low value of loaded Q is maintained to optimize circuit efficiency and bandwidth, as well as a low input SWR across the band. The output of the driver is coupled through a matching network (C_4, C_5, L_3) to a common feedpoint where the power is divided and delivered to the base circuits of two parallel-connected BM-80 transistors (Q_2, Q_3) by individual microstrip circuits. Independent dc return paths are provided for each base to enhance the isolation between the transistors already established by the microstrip sections. To reduce any circuit tendency toward push-pull oscillation, a suppressor resistor has been added across the base and collector circuits (R_2 and R_3). These resistors introduce significant loss only during conditions of

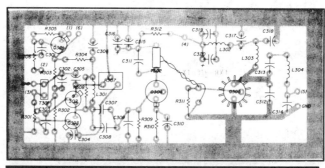

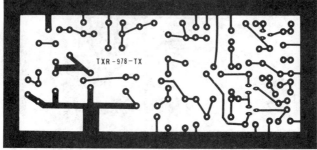

Figure 52

FULL-SIZE RENDITION OF TRANSMITTER BOARD LAYOUT

At top: parts layout. At bottom: foil side of the board.

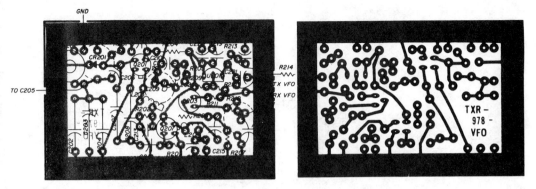

Figure 53

FULL-SIZE VFO BOARD LAYOUT

At left: parts layout. At right: foil side of the board.

phase unbalance within the circuit, providing a balancing effect, thus suppressing out-of-phase modes within the circuit. Each collector is matched with an individual microstrip structure to a common point. Isolation efforts similar to those of the base circuits are realized as separate dc power-supply lines are provided for each collector.

Each supply has separate high- and low-frequency decoupling to further isolate each side of the parallel-connected amplifier. The common point in the output circuit is then

Figure 54

BROADBAND 160-WATT AMPLIFIER FOR 144 MHz

This two stage, solid-state amplifier delivers 160 watts power output over the 2-meter band when driven by a 10-watt exciter. It is designed for mobile service and can accept wide supply voltage variations. The input stage is at the left with the parallel-connected amplifier stage at the right. Printed-circuit techniques are used and feedback is employed around the amplifier stage for greatest stability. Individual microstrip networks are used for the two output devices. The double-sided circuit board is bolted to a matching heat sink for proper heat dissipation.

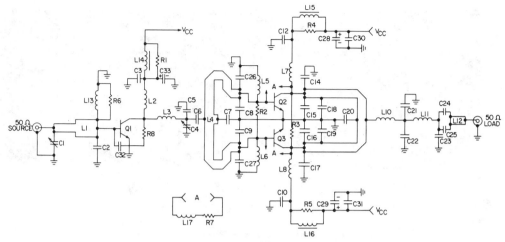

Figure 55

SCHEMATIC OF TWO-METER, TWO STAGE AMPLIFIER

C₁, C₄—5-45 pF mica compression capacitor. ARCO C-4103/OX or equivalent.
The following are metal-case mica capacitors. See supplier list at end of the chapter.
C₂, C₂₀—200 pF
C₃, C₁₀, C₁₂—1000 pF
C₅—68 pF
C₆, C₂₄, C₂₅—750 pF
C₇—100 pF
C₈, C₉, C₁₄, C₁₅, C₁₆, C₁₇, C₂₆, C₂₇—300 pF
C₁₈, C₁₉—250 pF
C₂₁, C₂₂—47 pF
C₂₃—33 pF
The following are tantalum electrolytic capacitors.
C₂₈, C₂₉—50 μF, 50 volts
C₃₃—1 μF, 50 volts
The following are disc ceramic capacitors.

C₃₀, C₃₁—0.1 μF
C₃₂, C₃₄—.01 μF
The following are carbon resistors.
R₁, R₆—15 ohm, ½ watt
R₂, R₃—15 ohm, 2 watt
R₄, R₅—15 ohm, 1 watt
R₇—82 ohm, 2 watt
R₈—47 ohm, ½ watt
See figure 58 for coil data.
Transistors.
Q₁—CD 4024 (CTC)
Q₂, Q₃—BM-80 (CTC)

Figure 56

TOP VIEW OF AMPLIFIER

The stripline circuitry and placement of capacitors can be seen in this view. Input amplifier is at the left and parallel-connected output stage at the right. Voltage terminals are to either side of the output connector on the bracket at right. Compare schematic of figure 55 with this view.

matched to the 50-ohm output port with a low loss multisection filter structure (L_{10}, L_{11}, C_{23}–C_{25}, and L_{12}). This circuit provides the necessary impedance transformation and has a low-pass characteristic which reduces the harmonic level in the output circuit significantly.

Amplifier Construction The amplifier is constructed on a piece of G10 glass-epoxy board having 2-ounce copper on both sides (figure 56). The board is .062-inch thick. The board is affixed to an aluminum heat sink measuring about 9.4" × 4.25". End plates are added to hold the r-f and dc connectors.

The top side of the circuit board is etched while the bottom is not etched and serves as a ground plane. The top ground-plane areas are tied to the lower plane at the transistor cutout areas by means of thin copper shim strips, as shown in figure 57.

The board is mounted directly to the heat sink with screws. Only drilling and tapping are required on the sink for assembly. After drilling the board for the transistor mounting holes (4-40 size) the board is fastened down. Mount all components, giving special attention to the capacitors placed at the emitter leads on the BM 80 devices. Be sure to place these capacitors as shown in the illustrations and solder them in place, carefully avoiding any shorts. Proper placement is critical to good amplifier performance. All but a few of the components may be mounted prior to mounting the transistors. Prior to mounting, two precautions should be observed: (1) A flush, smooth interface between the heat sink and the transistor flange is mandatory for proper heat transfer. Inspect the surface of the heat sink for any blemishes that would disturb this interface. (2) Prior to transistor mounting, apply an ample amount of heat transfer compound. Use only a high quality, oxide-loaded compound such as those supplied by *Dow-Corning* and *General Electric*. Finally mount the transistors and remaining components as shown in the illustrations and figure 58. Again, be sure that the shunt capacitors in the networks are positioned as shown.

Amplifier Adjustment Upon completion, the amplifier is connected to an exciter, a dummy load and a metered 12.6-volt source capable of supplying 300 watts dc power. Apply about 5 to 8 watts of drive power at 144 MHz. Adjust input capacitor C_1 for minimum reflected power to the driver. Change frequency to 146 MHZ and adjust capacitor C_4 for maximum output power. If all components were mounted correctly, no additional adjustments need be made to achieve the typical performance shown in figure 59.

Variations in component tolerance and printed-circuit board characteristics sometimes occur and slight adjustment in the collector matching circuit may be necessary. The components having the greatest impact on overall power gain are capacitors C_{18} and C_{19}. Make slight adjustments to the position of these capacitors along the line, making adjustments to both simultaneously. The value of the capacitors need not be changed.

The recommended output for this amplifier is 160 watts at a primary potential of 13.8 volts. The amplifier is capable of output power in excess of 200 watts, but this level should be avoided to prevent overload and damage to the BM 80 devices. Forced-air cooling of the heatsink is necessary and a very low SWR load is mandatory for excess power levels.

Linear Operation To reduce distortion when amplifying an SBB signal, linear amplification is necessary and may be accomplished by adding forward bias to each stage of the amplifier. This bias establishes an idling current in each stage under no signal conditions. The bias supply must be capable of supplying necessary base current while maintaining a constant output voltage. This requires a low impedance bias source. The quality of regulation will directly influence the degree of linearity of the amplifier. A separate bias supply must be used for each stage and no interaction between bias supplies should be allowed. Finally, note that applying bias to an optimized class-C design usually requires slight retuning, resulting in higher overall power gain with slightly reduced efficiency. Make sure that the r-f input level does not drive

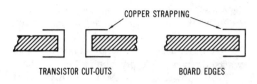

Figure 57

COPPER-STRAPPING DETAIL

Top side ground plane areas are tied to the lower plane at transistor cutout areas.

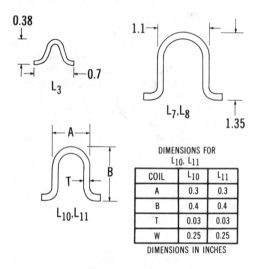

DIMENSIONS FOR
L_{10}, L_{11}

COIL	L_{10}	L_{11}
A	0.3	0.3
B	0.4	0.4
T	0.03	0.03
W	0.25	0.25

DIMENSIONS IN INCHES

Figure 58

COIL DATA FOR TWO-METER AMPLIFIER

the amplifier into saturation. Proper linearity is usually maintained if the amplifier is operated at a peak power output level slightly below typical class-C c-w power output.

A half-size circuit board layout is shown in figure 60.

19-10   A Solid-State 10-Watt Linear Amplifier for 420 MHz

This inexpensive 10-watt *linear* stripline amplifier is designed and built by WB6QXF for mobile use, or fixed station service using either SSB or f-m modes (figure 61). With a nominal 12.6-volt supply, the amplifier provides 10 watts PEP output with a 10 dB, or better, power gain. With a simple modification, the amplifier is converted to class-C mode for f-m service, providing the same power output.

Many amateurs find solid-state stripline amplifiers difficult and expensive to build. The special teflon-glass board is hard to find and costly, and the printed stripline circuits become quite critical to make, especially in the 450-MHz region. This amplifier overcomes these problems. It is designed around low cost *G-10 glass filled epoxy* board and employs stripline circuitry made of short lengths of flashing copper. No intricate circuit board work is required.

Amplifier Circuitry   The amplifier schematic is shown in figure 62. A base-driven circuit is used, with a simple L-network in the base circuit. A pi-L network is used in the collector output circuit to provide a good match to a nominal 50-ohm load impedance. A *CTC CM10-12* power transistor is used. This device was developed for land mobile service and is inexpensive and rugged.

For linear service, the power transistor is forward biased by the use of *byistor* (Q_2). This device consists of a diode and a silicon resistor in one package. It is physically coupled to the heat sink of the amplifier and tracks the power amplifier thermally, assuring that thermal runaway problems are minimized by automatically adjusting the forward bias of the transistor to compensate for changing heat sink temperature.

Special low impedance, high current, vhf-type *Underwood* capacitors are placed directly at the base and collector terminals of the transistor to achieve a proper impedance match to the input and output networks. Low frequency oscillations are suppressed in the power circuits by means of r-f chokes and bypass capacitors.

Amplifier Construction   The amplifier is built on a piece of epoxy circuit board, copper plated on both sides. To make a good ground plane, the entire outside edge of the board is lined with thin copper foil making an electrical connection from the top to the bottom of the board

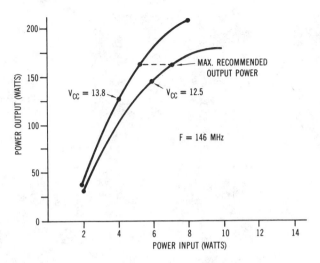

Figure 59

POWER OUTPUT VERSUS POWER INPUT

**Maximum output at 13.8 volts drops to 120 watts
at 140 MHz and 160 watts at 150 MHz.**

(figure 63). Similarly, short, narrow pieces of foil are cut and soldered at the four edges of the transistor mounting hole, as shown in the illustration.

Once the board has been prepared around the edges, it is placed on the heatsink and secured in position with four 4-40 bolts whose matching holes are then drilled and tapped in the aluminum sink. The byistor mounting hole is also drilled at this time.

The circuit board is removed from the sink and a large diameter drill is used to cut a space through the fins of the heat sink so that the bolt may be placed on the byistor stud, which projects through the heat sink. Be careful not to drill the clearance hole through the base of the heat sink.

The next step is to solder the four special *Underwood* vhf capacitors to the board as close as possible to the transistor mounting hole. The cases are soldered to the board in such a way as to allow the leads to overlap each other, as shown in figure 64. The overlap provides base and collector connections to the transistor. The leads of the transistor are now trimmed to size and silicone grease (*GE Insulgrease* or equivalent) is placed on the mounting flanges and bottom of the transistor. The transistor must be bolted in

position, to the heat sink before the leads are soldered in place to prevent the transistor case from being strained.

After the transistor is mounted, the islands are cut in the copper foil of the circuit board for the ends of the striplines. An *Exacto* knife or razor blade is used for this operation. The input island is one inch (2.54 cm) away from the overlapped leads of the input capacitors, as measured from the edge of the cutout. The island area is $5/16''$ (0.8 cm) square in the center of a cutout $1/2$ inch (1.3 cm) square. The collector island is the same size and is located $1\frac{1}{16}''$ (1.74 cm) away from the overlapped leads of the capacitors. Smaller islands are cut for the byistor supplier and injector leads.

The remaining components are placed as shown in the photograph. Placement of parts is not critical, except for placement of the base circuit bypass capacitors at the termination of RFC_1. Since the choke is only a short length of wire, the capacitors are positioned at the terminal point of the wire.

Testing the Amplifier—Temporarily disconnect the collector dc choke (RFC_2) from the stripline. Insert a 0-500 dc milli-

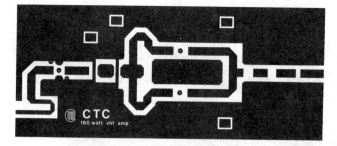

Figure 60

AMPLIFIER BOARD

Half-size rendition of layout.
Board is 9½ inches long.

Figure 61

10-WATT LINEAR AMPLIFIER FOR 420-450 MHz

This broadband, solid-state linear amplifier is designed for either SSB or f-m service. Power gain is better than 10 dB. The amplifier is built on inexpensive, 2-side epoxy circuit board. The edges of the board are trimmed with copper shim that provides a conductive path between the top and bottom surfaces. The CTC power transistor and associated "UNELCO" uhf capacitors are located near the center of the board. The inexpensive mica compression tuning capacitors are in the input and output circuits. A stripline configuration is used, with the lines cut from thin copper stock. The lines are mounted about ³⁄₃₂" above the board and held in position by soldering the free ends to isolated islands cut in the copper board material. Input and output connectors are mounted to aluminum plates bolted to the ends of the heat sink.
Choke RFC_1 is visible as a single length of wire running between the center point of the input line (at left) and the associated bypass capacitors, hidden beneath the series-connected resistors that make up R_1. The output circuit and RFC_2 are to the right of the power transistor.

ammeter in series with the byistor at point X in the schematic. Gradually apply low voltage to the byistor circuit, gradually increasing it toward 12.6 volts while observing the injector current. At 12.6 volts, the current should be in the range of 300 mA to 350 mA. If not, the value of series resistor R_1 should be adjusted until this level is achieved. The resistor may be composed of two units in series to aid in adjustment.

Once the injector current of the byistor is set at the proper level, the collector choke RFC_2 is reconnected to the stripline. Short base resistor R_2 with a 1-ohm resistor and apply 12.6 volts to the amplifier, with the

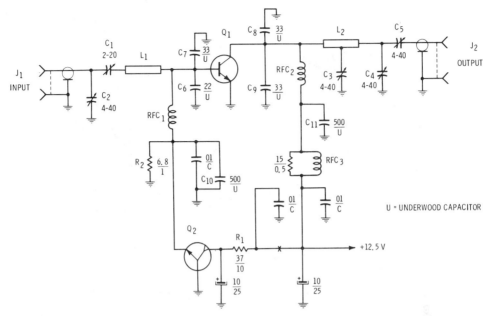

Figure 62

SCHEMATIC OF 450-MHz AMPLIFIER

C_1—Mica trimmer 2-20 pF (ARCO T51113-1 or equivalent)

C_2,—C_5—Mica trimmer, 4-40 pF (ARCO T51213-1 or equivalent)

C_6—22 pF Underwood (UNELCO) uncased mica capacitor. Do not substitute.

C_7,—C_9—33 pF Underwood (UNELCO) uncased mica capacitor. Do not substitute.

C_{10},—C_{11}—500 pF Underwood (UNELCO) uncased mica capacitor. Do not subst tute.

L_1—Copper shim strap, 0.175" (0.45 cm) × 1.1" (2.8 cm), including bend at input end of line.

L_2—Copper shim strap, 0.175" (0.45 cm) × 1.2" (3.0 cm) including bend at end of line. Tap at midpoint for capacitor C_3

RFC_1—1.5" (3.8 cm) straight piece of #20 wire

RFC_2—Two turns #20, 0.3" (0.8 cm) diameter (wound around pencil eraser)

RFC_3—5 turns #20 around 0.5" (1.3 cm) diam. toroid (#6 material) in parallel with 15-ohm, 0.5-watt resistor

Q_1—CTC type CM10-12

Q_2—CTC type BYI-1

Heatsink—6" × 4" (15.3 × 10.0 cm.) Wakefield or equivalent.

milliammeter connected to read total current drain. The meter now indicates byistor current plus the quiescent current of the power amplifier. Increase the value of R_2 in one-half ohm steps until an idling current of 50 mA to 60 mA is achieved. It may be necessary to parallel two resistors to reach the correct value of idling current. The meter will now read approximately 350 mA, of which about 300 mA is byistor current and 50 mA is amplifier current. The amplifier is now ready for tuneup.

Adjusting the Amplifier Preliminary tuneup can most easily be accomplished with an SWR bridge, power meter, and dummy load (figure 65). An accurate power meter is necessary for proper adjustment.

Voltage is applied to the amplifier and the power meter checked for zero power. Any indication of power at this point indicates amplifier oscillation. (No such oscillations were detected in four amplifiers built to these specifications. For additional information on amplifier instability, refer to Chapter 11). Next, apply about 200 mW of drive power and tune output capacitors C_3, C_4, and C_5 for maximum output. Tune capacitors C_1 and C_2 for minimum input SWR. Increase the drive level to about 600 mW and repeat the adjustments. The power meter should now show a reading of about 10 watts. If a two-tone drive signal

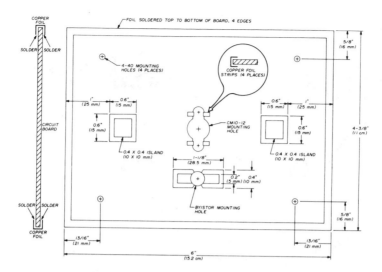

Figure 63

LAYOUT AND CONSTRUCTION OF CIRCUIT BOARD

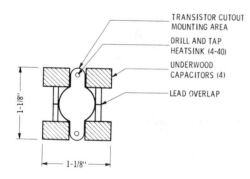

Figure 64

The four Underwood capacitors are soldered to the circuit board with their leads overlapping, as shown. The transistor is placed in the mounting hole and attached to the heat sink with 4-40 bolts run into holes tapped in the heat sink. The four transistor leads are soldered to the cases of the capacitors. The striplines are soldered to the overlapping inner terminals. See photograph for details. Whole assembly is 1⅛" (2.86 cm) square.

is used, and the power meter is an average-reading device (i.e., *Bird #43*), the reading will be approximately one-half of the actual PEP output.

Amplifier Performance—A well regulated 12.6-volt power supply is necessary for lin-ear operation. Since current drain is less than 2 amperes, a simple series regulator will be satisfactory. An oscilloscope should be used to check for flattopping of the waveform under voice modulation. The collector current should rise to about 500 mA to 600 mA under proper drive with a voice signal.

Care must be taken to operate the amplifier into a load having a low value of SWR. Although the transistor will survive an infinite SWR, sustained operation into a load having a high value of SWR is not recommended. Performance data for the amplifier is listed in Table 1.

For f-m service, the byistor circuit can be removed and the bottom end of base choke RFC_1 returned to ground. This removes the forward bias from the transistor and allows class-C operation. Tuneup adjustments for this class of service are as described previously.

19-11   Two Solid-State Linear Amplifiers for Mobile SSB

Described in this section are two solid-state, broadband class-B linear amplifiers

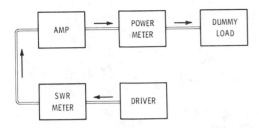

Figure 65

TEST SETUP FOR 450-MHz AMPLIFIER

A Bird #43 power meter may be used for the SWR meter as well as the output measuring device. A 20-watt dummy load having a low value of SWR at 150 MHz should be employed. When a two-tone test signal is used, the wattmeter will read approximately one-half the actual PEP output.

TABLE 1

Performance Data for
450-MHz Amplifier

SINGLE TONE	TWO TONE
Supply Volts = +12.6	Supply Volts = +12.6
Input Power = 15 W	Input
Output Power = 10 W	Power = 600 MW PEP
I_c = 1.2 A	Output
	Power = 10 W PEP
	I_c = 750 mA
	IMD = −29 dB, 3rd
	Order Products
	At 10W PEP

Note: Meter will read 300 mA High Because of Byistor Current.

covering the 1.5-MHz to 30-MHz spectrum. They are suitable for mobile operation with a nominal 12.6 volt dc power supply. The amplifiers are untuned and provide outputs of 25 watts PEP and 100 watts PEP, respectively. They exhibit intermodulation distortion product levels of better than −30 dB below one tone of a two-tone test signal at full output level.

The 25-Watt Amplifier   The 25-watt PEP output amplifier is shown in figures 66 through 70. It requires only 0.4 watt PEP drive at 30 MHz for full output, having a power gain of about 18 decibels. Amplifier efficiency is about 55 percent under c-w (carrier) conditions. Even-order harmonics are better than −35 decibels below peak power output. The level of the

odd-order harmonics is such that a harmonic filter should be incorporated after the amplifier to suppress the 3rd, 5th, and 7th order harmonics which are attenuated less than −30 decibels below peak power output in the amplifier.

Amplifier Circuitry—The schematic of the 25-watt amplifier is shown in figure 67. Two TRW type PT 5740 epitaxial silicon NPN power transistors, specially designed for hf SSB service are used (Q_1, Q_2). The transistors incorporate temperature-compensating emitter resistors on the chip and are designed to work into an infinite SWR load without damage at a maximum collector potential of 16 volts.

The PT 5740 devices are connected in a push-pull configuration with broadband, ferrite-loaded transformers used in the input and output circuits to match unbalanced terminations. A simple RLC compensation network is placed across the input winding of transformer T_1 to equalize amplifier gain across the operating range.

The input impedance of a PT 5740 power transistor is below 5.5 ohms and is capacitive over the operating range of the amplifier. The output impedance is of the order of 4 ohms. As a result, special r-f transformers must be built to match these very low impedance levels to 50 ohms.

The push-pull collectors of the transistors are connected to a balanced feed transformer (T_2) and to a matching output transformer (T_3) to provide single-ended output at a nominal impedance value of 50 ohms. The push-pull configuration is used since the amplifier covers five octaves of bandwidth, and suppression of even harmonics is of major importance since the harmonics are a function of the ratio of the cutoff frequency to the operating frequency and the selectivity of the output matching network.

Bias Stability—One of the most demanding aspects of solid-state linear amplifier design is the bias network and the associated temperature stability of the transistors. Factors influencing the bias value and network include: (1) Large signal r-f amplifiers generally rectify a portion of the input signal and if the base-emitter resistance is high the amplifier will be biased class AB for small signals, but will self-bias to class-

Figure 66

25-WATT PEP OUTPUT SOLID-STATE HF LINEAR AMPLIFIER

Two TRW type PT 5740 transistors in a broadband circuit provide high performance over the 1.5- to 30-MHz range. The amplifier is built on an etched-circuit board with ferrite-loaded input and output transformers. The input transformer is at the left with the two NPN power transistors at center. The multiple output transformer and feed transformer are at the right of the assembly. Transistors are heat-sinked to an aluminum radiator beneath the circuit board. Ground points atop the board are jumpered to the copper foil on the underside of the printed-circuit board.

C operation under large signal conditions. This shift in operating point seriously increases intermodulation distortion. The bias source resistance, therefore, must be held to a low value, typically 0.5 to 1 ohm. (2) Intermodulation distortion is usually minimum over a relatively narrow range of resting collector current. The devices used in this amplifier have a large safe operating current range and the resting collector current may be set high enough to achieve the lowest value of intermodulation. (3) Under small-signal conditions transistor dissipation is low and junction temperature is low. However, under conditions of peak power dissipation the junction temperature rises. Using a constant-voltage bias source with a

device having a negative temperature coefficient for emitter-base voltage change can lead to thermal destruction of the chip unless thermal equilibrium is established by proper transistor design and use of the proper heat sink.

In both of these amplifiers, the design of the PT 5740 power transistor and the accompanying circuitry solves the important bias, temperature and collector current problems.

Impedance Matching—Broadband, ferrite-loaded input and output transformers are used in this amplifier to achieve the required frequency response. The ferrite material used has an initial permeability of 800 which remains above 200 at 30 MHz. Losses in the

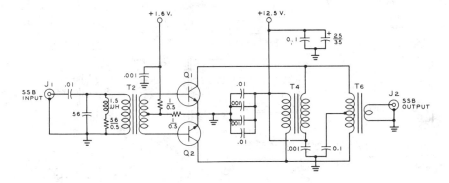

Figure 67

SCHEMATIC, 25-WATT AMPLIFIER

T₂, T₄, T₆—See text and figures 68-70
Q₁, Q₂—TRW type PT 5740 r-f power transistors
Circuit-board material—Glass-filled epoxy board, G-10, 0.060" thick
Heat sink—Wakefield 620 or equivalent
NPO chip capacitors—UNELCO (Underwood Electronics)

Figure 68

**FERRITE-LOADED INPUT
TRANSFORMERS T₂ and T₃**

Each transformer consists of six ferrite cores in two stacks of three each, epoxied between end plates made of p.c. board material. Each transformer consists of a single-turn winding of two pieces of 0.190" diameter brass rod, each 0.80" long. The pieces are connected together at one end by the foil of one p.c. end board thus forming a one turn loop. For the 25-watt amplifier, the primary consists of 4 turns #20 e. wire. For the 100-watt amplifier, the primary consists of 5 turns #18 e. wire.

ferrite material are quite low and ferrite temperature rise is less than 20°C in any transformer at full power output at any frequency in the operating range.

Input transformer T₁ is shown in figures 68 and 69. The unit consists of a very low impedance, split secondary made of two short brass tubes mounted between end plates made of printed-circuit board (foil on one side). One end board serves as the terminal end connections for the tubes and the other acts as a connecting strap and center-tap point between the tubes. Two stacks of three ferrite cores are slid over the tubes which are then soldered in position between the boards and the assembly is epoxied for rigidity. The high impedance wire winding (50 ohms) is threaded in continuous fashion through the tubes.

The dc feed transformer (T₂) and the output matching transformer (T₃) are mounted side by side between two printed-circuit board end plates, in the manner described for the input transformer. Each transformer consists of two stacks of six ferrite cores. The assembly is shown in figure 69. The end plates are soldered to the two brass tubes that make up the low impedance winding of transformer T₃ and the whole assembly of ferrites and end plates is epoxied for rigidity. The secondary winding of transformer T₃ and the twisted-pair dual winding of T₂ are then wound back

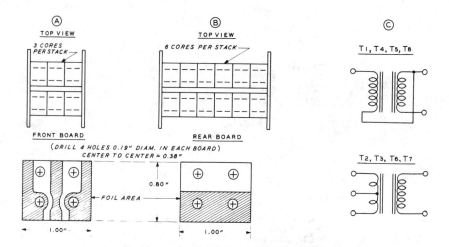

Figure 69

DETAILS OF FERRITE-LOADED TRANSFORMERS

(A)—Top view of input transformer stack of 3 ferrite cores showing assembly and view of front p.c. board. Foil areas provide terminations for brass tubes and connections to main circuit board.

(B)—Top view of transformer assembly of output and feed transformers. Each transformer is made up of two stacks of six cores each. Brass tubes are connected to foil on p.c. board at front and rear.

(C)—Schematic of ferrite transformers. Transformers T_1 and T_8 are identical to T_4 and T_5 but are not mounted on p.c. board frame.

and forth through the ferrite stacks as shown in the photograph. When completed, the transformers are soldered to the copper foil of the circuit board. The low impedance (brass tube) winding ends are soldered directly to the foil of the end boards and the foil to that of the master board.

Amplifier Assembly and Testing—The amplifier is assembled on an etched-circuit board measuring $4\frac{1}{2}" \times 2"$ and mounted to an aluminum heat sink. The sink ends are trimmed to fit the board.

Upon completion, the amplifier is connected to an exciter, a dummy load, and a metered 12.5-volt source capable of supplying 5 amperes. Base bias is supplied from a well-regulated source and is adjusted for a resting collector current of 150 mA. With full carrier insertion, the collector current will rise to nearly 5 amperes, and will approximate 2.5 ampere peaks under voice modulation. The third harmonic is -13 dB below the fundamental signal level and a suitable harmonic filter should be used before the antenna to reduce this emission.

(Note: the unfiltered waveform is essentially a square wave. Output power measurement should be made with a calorimetric power meter or other thermal sensing instrument. Power meters using a diode detector will read low by a factor of 0.785).

The 100-Watt Amplifier   The 100-watt PEP output amplifier is shown in figures 71 and 72. The unit requires 3 watts PEP drive power at 30 MHz for full output, having a power gain of about 15 decibels. It may be easily driven by the amplifier described in the previous section.

Amplifier Circuitry—The schematic of the amplifier is shown in figure 72. Two pairs of TRW-type PT 5741 transistors are operated push-pull and then combined with zero-degree hybrid transformers (T_1 and T_8) which convert the nominal 50-ohm source and load impedances to two 100-ohm ports which are in phase. Any amplitude or phase unbalance causes power to be dissipated in resistors R_1 and R_2. As in the smaller amplifier previously described, an

Figure 70

OUTPUT AND FEED TRANSFORMER ASSEMBLY T₄, T₆ and T₅, T₇

The feed transformer is shown atop the output transformer in a four-stack assembly. Twenty-four ferrite cores are used, stacked between two p.c. end plates. The feed transformer has two, one-turn windings made of #18 enamel wire, twisted at 5 twists per inch. Hybrid transformers T_1 and T_8 are identical but are not mounted on p.c. end boards (see figure 71). Transformers T_6 and T_7 are composed of copper-tube windings, similar to T_2 and T_3 except that six stack cores are used and the tubes are 1.375″ long. The output winding consists of 4 turns #18 enamel copper wire.

Figure 71

100-WATT PEP OUTPUT SOLID-STATE HF LINEAR AMPLIFIER

Four TRW type PT 5741 transistors are used in a combined, push-pull configuration to cover the 1.5- to 30-MHz range. The four transistors are in line across the middle of the printed-circuit board. At the right are the two input transformers T_2 and T_3 with the hybrid transformer T_1 between them. At the left are the two output-feed transformer assemblies with the hybrid transformer T_8 between them. Ground points atop the board are jumpered to the copper-foil ground plane on the underside of the p.c. board.

RLC compensation network is placed across the input winding of transformer T_1 to equalize amplifier gain across the operating range.

The collector-feed transformers (T_4, T_5) combine with the output matching transformers (T_6, T_7) to form a modified 180° hybrid combiner. Differences in phase or amplitude that would otherwise exist at the collectors are minimized by allowing the difference current to be bypassed to ground. The resulting output currents in the two transformers are highly balanced and provide good second harmonic rejection. Any minor amplitude or phase unbalance is dissipated in resistor R_2. The port impedance is transformed to an unbalanced value of about 50 ohms by transformer T_8.

Amplifier Assembly and Testing—Data for the various ferrite transformers is given in figures 68, 69, and 70 and the amplifier layout is shown in figure 71. The unit is assembled on an etched-circuit board measuring 4½″ × 4″ in size. Placement of the four output transistors is critical in that the connection between the collectors and the brass-tubing winding of the output transformers should be extremely short, being composed of the copper foil on the mating circuit boards. Multiple bypass capacitors at the "cold" end of the windings contribute to the low impedance collector path to ground.

Using a 12.5-volt source capable of supplying 16 amperes, the amplifier is adjusted to draw a resting collector current of 0.5 ampere by varying the base bias potential. With full carrier insertion, the collector current will rise to nearly 16 amperes, and will approximate 7-ampere peaks under voice modulation. As in the case of the smaller amplifier, a suitable harmonic filter should be used between the amplifier and the antenna to suppress odd-order harmonics.

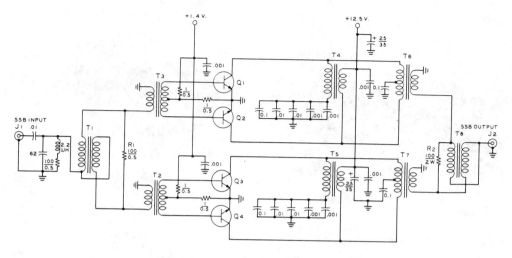

Figure 72

SCHEMATIC, 100-WATT AMPLIFIER

T_1-T_8—See text and figures **68-70**
Q_1-Q_4—TRW type **PT 5741** power transistors
Circuit-board material—Glass-filled epoxy board, type G-10, 0.060" thick
Heat sink—Wakefield, or equivalent
NPO chip capacitors—UNELCO (Underwood Electronics)

(Note additional information on the amplifiers, a circuit-board template, and data on the TRW transistors may be obtained by requesting Application Notes CT-122-71 and CT-113-71 from the *Semiconductor Division, TRW Inc.,* 14520 Aviation Blvd., Lawndale, CA, 90260.)

NOTE: The metal-cased capacitors specified in various units in this chapter may be obtained from:

Underwood Electric Co.
148 South 8th Ave.
Maywood, IL 60153

VHF Engineering Co.
320 Water St.
Binghamton, NY 13902

SEMCO
South Walnut St.
Wauregan CT 06387

The CTC transistors may be obtained from:

R. F. Gain, Inc.
100 Merrick Rd.
Rockville, NY 11570

National Electro Sales
12063 W. Jefferson Blvd.
Culver City, CA 90230

Richardson Electronics
3030 No. River Rd.
Franklin Park
Franklin Park, IL 60131

CHAPTER TWENTY

Receivers and Exciters

Equipment construction has just about become a lost art. Aside from the many excellent kits, the average amateur has a difficult time building his own gear. Reliable communication equipment is available, ready to go, at moderate prices and the home-built equipment is further handicapped because it has no resale value. Finally, the job of finding the desired components is a difficult one, and many frustrating hours can be spent searching for one or two inexpensive components that have held up a home building project.

On the other hand, the purchaser of ready-made equipment pays a penalty for convenience. The c-w operator must often pay for the SSB operator's wide passband and S-meter that he never uses, and the SSB operator must pay for the c-w operator's narrow filter. For one amateur, the receiver or transceiver has too little bandspread or power, for the next, too much. The design of the equipment is often compromised for economy's sake and for ease of alignment, low-Q circuits are often found where high-Q circuits are called for, making the receiver a victim of overloading from nearby signals. Inexpensive transistors are used in the interest of economy, leaving the receiver wide open to crossmodulation and desensitization. Rarely does the purchaser of commercial equipment realize the manufacturing trade-offs encountered considering the results he might achieve if he built his own equipment to his desired specifications.

The ardent experimenter, however, needs no such arguments. He builds his equipment for the enjoyment of construction and creation, and the thrill of using a product of his own manufacture.

It is hoped that the equipment described in this, and the succeeding chapters, will awaken the experimenter's instinct in the reader, even in those fortunate individuals owning expensive commercial equipment. These lucky amateurs have the advantage of comparing their home-built product against the best the commercial market has to offer. Sometimes such a comparison is surprising.

Check Your Equipment When the builder has finished the wiring of his equipment it is suggested that he check his wiring and connections carefully for possible errors before any voltages are applied to the circuits. If possible, the wiring should be checked by a second party as a safety measure. Some transistors can be permanently damaged by having the wrong voltages applied to their electrodes. Electrolytic capacitors can be ruined by hooking them up with the wrong voltage polarity across the capacitor terminals. Transformer, choke, and coil windings may be damaged by incorrect wiring of the high-voltage leads.

The problem of meeting and overcoming such obstacles is just part of the game. A true radio amateur should have adequate knowledge of the art of communication. He should know quite a bit about his equipment (even if purchased) and, if circumstances permit, he should build a portion of his own equipment. Those amateurs who do such construction work are convinced that half of the enjoyment of the hobby may be obtained from the satisfaction of building and operating their own receiving and transmitting equipment.

Figure 1

COMPONENT NOMENCLATURE

CAPACITORS:

1— VALUES BELOW 999 PF ARE INDICATED IN UNITS.
   EXAMPLE: 150 PF DESIGNATED AS 150.

2— VALUES ABOVE 999 PF ARE INDICATED IN DECIMALS.
   EXAMPLE: .005 µFD DESIGNATED AS .005.

3— OTHER CAPACITOR VALUES ARE AS STATED.
   EXAMPLE: 10 µFD, 0.5PF, *ETC.*

4— TYPE OF CAPACITOR IS INDICATED BENEATH THE VALUE
   DESIGNATION.
   SM = SILVER MICA
   C = CERAMIC
   M = MICA
   P = PAPER

   EXAMPLE: $\frac{250}{C}$, $\frac{.01}{P}$, $\frac{.001}{M}$

5— VOLTAGE RATING OF ELECTROLYTIC OR "FILTER"
   CAPACITOR IS INDICATED BELOW CAPACITY DESIGNATION.

   EXAMPLE: $\frac{10}{450}$, $\frac{20}{600}$, $\frac{25}{10}$

6— THE CURVED LINE IN CAPACITOR SYMBOL REPRESENTS
   THE OUTSIDE FOIL "GROUND" OF PAPER CAPACITORS,
   THE NEGATIVE ELECTRODE OF ELECTROLYTIC CAPACITORS,
   OR THE ROTOR OF VARIABLE CAPACITORS.

RESISTORS:

1— RESISTANCE VALUES ARE STATED IN OHMS, THOUSANDS
   OF OHMS (K), AND MEGOHMS (M).
   EXAMPLE: 270 OHMS = 270
              4700 OHMS = 4.7 K
            33,000 OHMS = 33 K
           100,000 OHMS = 100 K OR 0.1 M
        33,000,000 OHMS = 33 M

2— ALL RESISTORS ARE 1-WATT COMPOSITION TYPE UNLESS
   OTHERWISE NOTED. WATTAGE NOTATION IS THEN INDICATED
   BELOW RESISTANCE VALUE.

   EXAMPLE: $\frac{47 K}{0.5}$

INDUCTORS:

MICROHENRIES = µH
MILLIHENRIES = MH
HENRIES = H

SCHEMATIC SYMBOLS:

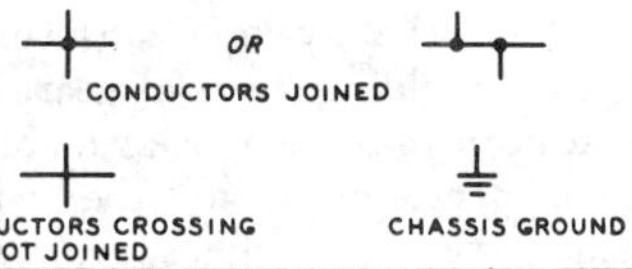

OR

CONDUCTORS JOINED

CONDUCTORS CROSSING   CHASSIS GROUND
BUT NOT JOINED

Circuitry and Components It is the practice of the editors of this Handbook to place as much usable information in the schematic illustration as possible. In order to simplify the drawing, the component nomenclature of figure 1 is used in all the following construction chapters.

The electrical value of many small circuit components such as resistors and capacitors is often indicated by a series of colored bands or spots placed on the body of the component. Several color codes have been used in the past, and are being used in modified form at present to indicate component values. The most important of these color codes for resistors, capacitors, power transformers, chokes, i-f transformers, etc. can be found in Chapter 35 of this Handbook.

20-1 A Deluxe, Solid-State Amateur Band Receiver— Mark II

This is the latest up-to-date version of a popular receiver previously described in this handbook. The receiver, including the present modified design, is a creation of VE3GFN. It provides good sensitivity ($1\mu V$ for 20 dB signal-plus-noise to noise ratio, or better) with good overload capability (figure 2). Features such as IC power-supply regulation, varactor diode tuning of front-end circuits and variable oscillators, integrated double-balanced modulator mixers, diode switching of filters and tuned circuits, and a solid-state digital readout counter, are included in the design.

Modular construction is used as much as possible. Most of the circuits are built as separate, shielded modules, and are tested and aligned as such, completely independent of the receiver system (figure 3). This technique makes system modification easy, simplifies testing and alignment, and contributes greatly to freedom from spurious mixing products and circuit radiation. Input/output specifications are provided for each module, allowing the receiver to be duplicated module by module; by meeting the various module requirements, the builder is assured that his system will function properly when assembled.

The detailed description of this receiver is along modular lines as well, with the description of each module including necessary circuit theory, construction details, and electrical specifications.

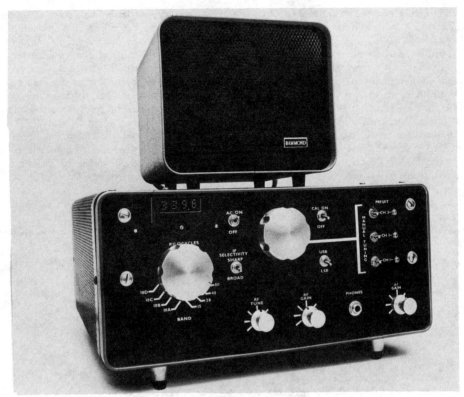

Figure 2

SOLID-STATE DELUXE AMATEUR BAND RECEIVER

This advanced receiver covers the amateur bands between 80 and 10 meters in 500-kHz segments. Featuring direct readout, varactor diode tuning, integrated circuit double-balanced modulators, and diode switching, the modularized receiver is an ideal construction project for the advanced amateur. The direct read-out escutcheon is at the upper left of the panel, with the KILOCYCLES-BAND switch directly below it. Readout is to 100 Hz. The large knob to the right is the tuning control, with the three pre-set channel switches at the right of the panel, the R-F TUNE, R-F GAIN, and A-F GAIN controls and earphone jack are along the lower edge of the panel. To the left of the main tuning control are the AC ON switch and the I-F SELECTIVITY switch. Two crystal filters provide optimum selectivity for SSB and c-w modes. A separate speaker sits atop the receiver. Construction is simplified by building the receiver in modules, each of which may be tested independently before the receiver is assembled.

The Receiver Circuit The receiver is single conversion on all amateur bands, 80 through 10 meters; coverage of the entire 10-meter band is included (figure 4). For good stability and to avoid tracking problems, the local-oscillator injection voltage is derived from the mixing product of a 5.0- to 5.5-MHz *variable-oscillator module* (A) with a *crystal-oscillator module* (B), the frequency of which is changed for each band. On 20 meters only, the variable oscillator is not mixed with a crystal oscillator, but drives the signal-path mixer directly. The frequency of the variable oscillator is counted by the *digital counter and display module* (D), to 100Hz resolution, and displayed as "kHz above the band edge." While a display of exact frequency may be more convenient from the operator's standpoint, the system used is simpler, and enables the digital counter to be built and tested as separate a module as any other, completely independent of the bandswitch.

Band changing is accomplished by a rugged rotary switch built into the *front-end mixer module* (C). Extra wafers on this switch control the switching of the heterodyne mixer output circuits, and the variable oscillator output, through or around the heterodyne mixer system.

The antenna signal first passes through a PIN diode, agc controlled, r-f attenuator which provides 30 dB of agc range. Signal input from the antenna is amplified by a dual-gate MOSFET r-f amplifier (Q_1) which is tuned from the front panel (*RF Tune*) by controlling the bias of varactor diodes D_1 and D_2 in the input and output tuned circuits (figure 5). The amplified antenna signal then passes through the signal-path

mixer (U_1), a double-balanced IC modulator. Local-oscillator injection for this mixer comes from the *heterodyne mixer module* (E), and is the sum of the variable-oscillator frequency, and the frequency of one of the heterodyne crystal oscillators. The heterodyne mixer (U_4) has diode-gated tuned circuits in the output to control the mixing frequency. The variable oscillator (Q_5) is a JFET circuit, varactor-tuned, of high stability. Injection to the signal-path mixer from the heterodyne mixer or the variable oscillator is controlled by diode gates through the bandswitch.

The output of the signal-path mixer (U_1) is the 9.0-MHz i-f signal, which passes through two crystal filters; a 2.4-kHz filter

Figure 3

TOP VIEW OF RECEIVER CHASSIS

Placement of the major modules may be seen in this view. The VFO module (A) is at the right, rear of the chassis with the heterodyne mixer module (E) in the left rear corner. The front-end module (C) is hidden under the digital counter board (D). At the right, next to the vfo module is the bfo module (F). The varactor diode tuning assembly is centered behind the National PW-0 dial mechanism which is set back from the panel to allow room for the various control switches. The heterodyne oscillator module (B) is at the rear, center of the chassis. Switches and "trimpots" for the preset, manual tuning channels are on the right, front corner of the panel. All modules are pretested before mounting on the receiver chassis

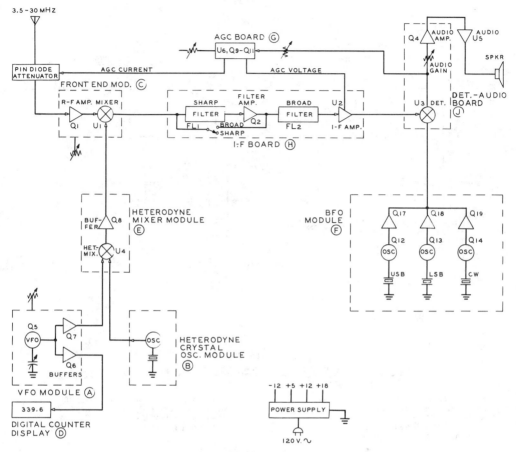

Figure 4

BLOCK DIAGRAM OF DELUXE AMATEUR BAND RECEIVER

The receiver is built and described in modules. The mixing signal is derived from a heterodyne mixer module (E). The mixing frequency is changed for each band. On 20 meters only, the variable oscillator (A) is not mixed with the heterodyne crystal oscillator (B), but drives the mixer (U_1) directly. The frequency of the variable oscillator is counted by the digital counter (D) to 100-Hz resolution and displayed as "kHz above the band edge." For 20 meters, the frequency shown is 14,339.6 kHz. Bandchanging is accomplished by a rotary switch in the mixer module. Separate i-f filters provide c-w and SSB selectivity and switchable bfo crystals provide upper and lower sideband.

for SSB, (FL_2), and a 500-Hz filter for c-w (FL_1). The choice of filter is made by the i-f *Selectivity Broad/Sharp* panel switch, which controls diode gates that direct the i-f signal to the filters.

The sharp filter has a dual-gate MOSFET amplifier (Q_2) after its output, which is adjusted for equal system gain when using either filter; the sharp filter has more insertion attenuation, making this necessary. The sharp filter and the compensation amplifier are shorted out by a diode gate for ssb reception.

The i-f signal is amplified by an IC amplifier (U_2) common to both filters, providing up to 20 dB i-f gain.

The second detector is a double-balanced modulator (U_3), used in a product-detector configuration, obtaining beat-frequency injection from the *beat-frequency oscillator* module (F), employing crystal oscillators whose frequency is selected by the *USB/LSB* panel switch, or the i-f *Selectivity* switch, depending on the choice of c-w or SSB reception. The detector output drives a high gain bipolar audio amplifier (Q_4).

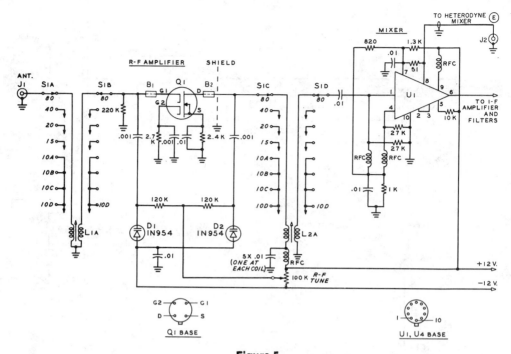

Figure 5

SCHEMATIC, FRONT-END BANDSWITCHING MODULE

B₁, B₂—Ferrite beads
J₁, J₂—BNC connector, UG-185/U
L₁, L₂—See Table 1
Q₁—RCA 40673 or Fairchild FT 0601
Note: All resistors ¼ watt

RFC—1 millihenry, 35-mA, J.W. Miller 10F-103A1
S₁—3 deck heavy duty rotary switch, 6 pole. 8
position. Centralab JV-9037, or equivalent
U₁—Motorola MC 1596G or Fairchild μA 796

The audio output stage (U_5) is a six-watt IC amplifier, with its own power supply regulator.

Frequency readout is obtained from LED devices (light emitting diodes) in the *digital counter display module* (D) which are driven by a highly stable time base and decade counters. A 100-kHz crystal is used as a standard for the count. The frequency of the variable oscillator is read to 100 Hz. A separate 100-kHz oscillator is provided for receiver calibration checks (figure 7B).

General Construction Technique Most of the circuits in this receiver are hand wired on G-10 glass epoxy printed-circuit board subchassis, using teflon press-fit terminals at the interconnection points. Ground connections are soldered di-

rectly to the copper foil. Solid-state devices are soldered directly into the circuits, with no device sockets used except in the digital counter, the only module where printed-circuit technique was necessary to simplify duplication. In a few cases (the i-f strip, for example) the main aluminum chassis was used as the construction base.

This assembly technique is ideal for high-frequency circuit work; it is quickly and easily modified, and short-lead construction is easy. Employing the copper-board sub-chassis utilizes most of the advantages of printed-circuit construction, but eliminates the extra time needed for artwork and board fabrication. When the final design has been completed, it can be easily adapted to the usual printed circuits if desired.

PIN Diode Attenuator—The PIN diode attenuator is designed to be inserted between

Figure 6

**FRONT VIEW OF INSTALLED
FRONT-END MODULE**

Front-end module (C) is installed in cutout in
main receiver chassis. The module is held in
position by angle brackets on the sides. Module
assembly is made of two aluminum chassis
mounted back-to-back. Connecting terminals
are on sides of the module. Digital counter
board (D) mounts on top plate of module. R-f
coils (L_1 series) are mounted to front of lower
module chassis (left to right): 10-, 15-, 20-, 40-,
and 80-meter coils.

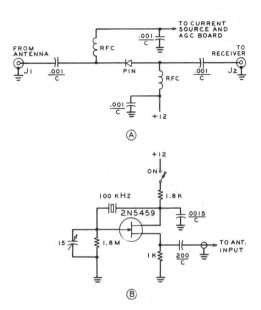

Figure 7

**PIN ATTENUATOR AND
CALIBRATOR**

A—The PIN attenuator PIN diode functions as
variable attenuator driven from agc circuit and
provides up to 40 dB of control.
PIN diode: Hewlett Packard 5082-3379
RFC = 1 mH, 100 mA.
B—100-kHz calibrator.

the antenna and the antenna input port of
the receiver. The PIN diode has a very low
impedance when conducting a relatively
high bias current and a very high impedance when the bias current is small (figure
7A). While most PIN diodes are designed
to be used above 100 MHz, certain *Hewlett-Packard* diodes are useful down to 1
MHz. The attenuator itself is built in a
separate shielded enclosure with coaxial receptacles provided on each end. Depending
upon construction, this attenuator can provide up to 40 dB attenuation (and therefore an equal agc range) between 3 and 30
MHz when terminated in a 50-ohm impedance.

The circuit layout of the diode attenuator is critical. Attenuation can be compromised by stray capacitance so all leads should
be kept as short as possible. Only disc
ceramic capacitors should be used. The attenuator uses a piece of printed-circuit
board as a chassis with ground connections
soldered directly to the board. The enclosure is also made of circuit-board stock, with
the entire assembly soldered together after
the final tests are completed.

*Construction of Front-End/Bandswitch-
ing Module* (C)—The front-end module
contains the r-f amplifier (Q_1), the signal-
path mixer (U_1), associated tuned circuits,
and the receiver bandswitch. The schematic
is shown in figure 5.

The r-f amplifier is a dual-gate MOSFET,
providing up to 20 dB of r-f gain.

The r-f input and output circuits are
tuned by means of varactor diodes D_1 and
D_2, the bias (capacitance) of which is controlled by the *R-F Tune* panel control. The
amplifier is stable on all bands without neutralization. Ferrite beads on the MOSFET
input and output leads contribute to inherent stability.

The signal-path mixer (U_1) uses an IC as a double-balanced modulator, which provides great attenuation to undesired mixing products. This IC device is used throughout the receiver for all signal-translation applications.

The local-oscillator injection for the signal-path mixer is obtained from the heterodyne mixer module (E) on all bands except 20 meters, where the variable-frequency oscillator module drives it directly. Input to U_1 at J_2 from the mixer module should be 50-300 mV p-p of sinusoidal waveform.

Additional wafers of bandswitch S_1 control the heterodyne crystal oscillators, the gating of the output tuned circuits in the heterodyne mixer module, and the diode gating of the variable-frequency oscillator module output.

The output of the front-end module (with the first i-f transformer connected)

should be a sinusoidal waveform at 9.0 MHz, of a level about 40 dB greater than the antenna signal level, with the r-f amplifier adjusted for maximum gain. This gain figure is only a nominal one. Views of the front-end module are shown in figures 6, 8, and 9.

The bandswitching module contains the bandswitch, the r-f amplifier (Q_1) and the signal-path mixer U_1. It is the most complex and compact of the receiver modules, and its assembly will be simplified if

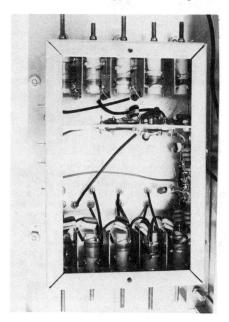

Figure 9

**UNDER-CHASSIS VIEW OF
FRONT-END MODULE**

The horizontal shield across the enclosure contains the r-f amplifier FET and associated components. At the bottom of the compartment are the r-f input coils, with the detector coils at the top (rear) of the compartment. The signal-path mixer (U.) is mounted on a small circuit board on the side of the enclosure. The shield sections between the coils are soldered to p.c. boards mounted on the front and back sides of the module chassis. Mounting holes for the coils are drilled through board and chassis. Note the feedthrough terminals from the bandswitch enclosure protruding through the clearance holes in the deck of the chassis.

Figure 8

**INTERIOR OF FRONT-END
MODULE**

Main bandswitch is centered in compartment with press-fit feedthrough insulators grouped near switch terminals. Wires running between switch points and underside of module (for control of crystal oscillators and heterodyne mixer gating) are kept as short as possible and laid out so as not to interfere with switch action.

the following step-by-step procedure (used in the construction of the prototype) is followed. See Table 1 for coil data.

The bandswitching mixer module (C) is built in two aluminum chassis, each 6" × 4" × 2" mounted back-to-back, as seen in figure 6. The bandswitch is installed in the top chassis, and the solid-state circuits and coils in the bottom chassis. The contacts of the switch are wired to press-fit feed-through terminals mounted in the bottom of the switch chassis; these terminals protrude through clearance holes drilled in the circuit chassis, and the coils and proper circuits are wired to them. Thus the switch is shielded from the r-f circuitry, yet leads

Table 1. R-F Amplifier and Mixer Coils (L₁, L₂)

Band (Meters)	L₁ and L₂
80	50 turns #29 e. on Cambion 1534/2/1 form, closewound. Inductance = 17 µH, Q = 55. Link = 10 turns #27 close-wound on "cold" end. 100 pf connected across primary, 80 meters only.
40	35 turns # 29 e., as above. Inductance = 10 µH, Q = 85. Link = 7 turns #27, as above.
20	15 turns #29 e., as above. Inductance = 4.4 µH, Q = 70. Link = 4 turns, as above.
15	12 turns #27 e. on Cambion 1534/3/1 form, closewound. Inductance = 1.8 µH, Q = 115. Link = 3 turns as above.
10	8 turns #27 e., closewound, as above. Inductance = 0.8 µH, Q = 140. Link = 3 turns, as above.

are kept short. The bandswitching module is constructed as a separate assembly and mounted in a slot in the main chassis. Assembly of the Module is as follows:

Step 1. Cut ⅜" clearance hole in switch chassis front, center, to mount bandswitch. Do not mount the switch.

Step 2. Drill 9 64" mounting holes in all four corners of the switch chassis, allowing room for a 6-32 nut to cover the hole and clear the chassis corner. Place the two chassis back-to-back, and mark the centers for the mounting holes in the circuits chassis, using the drilled switch chassis as a template.

Step 3. Mount and secure the band-switch in its chassis. Refer to the bot-tom-view photograph of the switch chassis (figure 8) for the feedthrough terminal layout. Note where the common switch arm of each wafer is on each deck, as its location requires more than a casual glance. Mark centers for the feedthrough terminals close to each wafer of the switch, being very careful of clearances when marking the terminals for the inner wafers. Now, remove the switch from its chassis, center punch the marked hole centers, bolt the two chassis together, and drill a centering hole through both chassis. Separate the chassis. The feedthrough terminals require a 9/64" hole, and the clearance holes should be enlarged to ¼".

Step 4. Install the switch in the switch chassis. Now examine the location of the feedthrough terminals, and the switch contacts to which they must be wired. The terminals for the inner wafers are almost covered by the switch, and are virtually inaccessible. These feedthrough terminals should be pre-wired (before the switch is installed) using 4" lengths of bare wire. As the switch is installed, these wires can be drawn up to the proper contacts, and wired to them, after sliding a length of insulating tubing over each lead. The switch contacts that are accessible (front and rear) should be pre-wired in a similar manner, and these wires run to the proper feedthrough teminals after the switch is installed. This completes the wiring of the switch chassis.

Step 5. Make up two coil-shield partition assemblies as shown in the under-chassis photograph (figure 9), using the following procedure: Mount a pre-cut printed-circuit board on the front and back ends of the circuit chassis. Mark the centering holes for the five coil forms (L₁ series and L₂ series) on the outside of the chassis end pieces. Center punch and drill through both chassis and p.c. boards. Then enlarge the holes to the required size. Remove the boards and temporarily mount the coil forms, then mark the locations of the brass shield partitions. Remove the coil forms and solder the partitions into place. This method ensures proper clearance for the coil forms after the shield partitions are installed.

Step 6. Bolt the two chassis together, install all coils, and wire them to the terminals of the bandswitch.

Step 7. The r-f amplifier stage is built on the aluminum shield section which separates the r-f amplifier coils (L_1 series) from the mixer section. Install the wired r-f amplifier shield section, then install the mixer subchassis. Wire these to the proper switch terminals, and to the terminals on the side of the chassis for input, output, gain control, tune voltage, and supply lines. This completes the assembly of the Bandswitching Mixer Module.

Construction of Heterodyne Crystal Oscillator Module (B) — The seven crystal oscillators for the heterodyne mixer are built as separate units to avoid the bandswitching complexities and design compromises necessary in one oscillator covering 7.5 to 33.5 MHz (figures 10 and 11).

Figure 10

TYPICAL HETERODYNE CRYSTAL OSCILLATOR ASSEMBLY

The heterodyne crystal oscillator board (B) shown contains one of the seven crystal oscillator stages. The other three boards each have two oscillator stages on them and are visible in the main under-chassis view of the receiver. The oscillator slug-tuned inductor is adjustable from beneath the receiver.

The output of each oscillator should be a reasonably undistorted sinusoid, of 200-500 millivolts p-p amplitude, measured at the input (50 ohms impedance) of the heterodyne mixer (U_1). The series output at-

tenuator circuit (10 pF, 1K) prevents oscillator loading and eliminates any problems due to the oscillator signal being routed through the bandswitch and around the chassis.

Table 2. Heterodyne Oscillator Module—Circuit Details

Band (Meters)	C_1 (pF)	X_1 (MHz)	L_1	Output mV (P-P)
80	100	7.500	35 turns #29 on Cambion 1536/2/1 $L = 7.5\ \mu H$	600 into 50 Ω
40	32	11.000	Same as above	550 into 50 Ω
15	45	25.000	12 turns #29 on Cambion 1536/3/1 $L = 1.5\ \mu H$	450 into 50 Ω
10A	—	32.000	7 turns #29 on Cambion 1536/3/1 $L = 0.55\ \mu H$	450 into 50 Ω
10B	—	32.500	Same as above	same
10C	—	33.000	Same as above	same
10D	—	33.500	Same as above	same

The output of each oscillator should be measured after the series attenuator network, using an oscilloscope of at least 150-MHz bandwidth capability as an instrument of lesser bandwidth will not reveal harmonic distortion in the output. The frequency of each oscillator should be accurately checked as overtone crystals often have a penchant for operating on their second harmonic.

If a crystal does not oscillate, or if there is distortion in the output, the output tuned circuit probably requires adjustment. See Table 2 for coil and capacitor data.

The crystal oscillator module construction is straightforward using an enclosure $3'' \times 2\frac{3}{8}'' \times 6''$ in size.

Construction of Heterodyne Mixer Module (E) — The heterodyne mixer module (figures 13, 14, and 15) consists of an IC double-balanced modulator (U_4) and a JFET buffer (Q_8) to

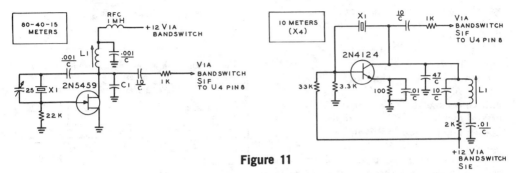

Figure 11

HETERODYNE CRYSTAL
OSCILLATORS

See Table 2 for component details

enable the mixer to drive a 50-ohm load, plus an output filter made up of seven tuned circuits and seven diode switches. The module is completely self-contained, with inputs being supplied through 50-ohm coaxial cables, and bandchanging accomplished

Figure 12

UNDERCHASSIS VIEW OF RECEIVER

At the right is the front-end module (C), recessed in a hole cut in the chassis. To the left of it, at center, is the i-f strip (H) and behind it is the small detector-audio board (J) with the large heat sink on the audio device. Behind the front-end module are the r-f attenuator and the vfo diode gating board. At the rear of the chassis is the crystal calibrator board and along the side of the chassis is the detector-audio board (J). Note that decoupling capacitors are installed at the power connector at the rear of the chassis.

by using one deck of the bandswitch (S_1E) to supply twelve volts to the appropriate diode switch.

The mixer/buffer output signal is 200-300 millivolts peak-to-peak of sinusoidal waveform into a 50-ohm load. The vfo injection signal (pin 1, U_4) must be 200 millivolts peak-to-peak (or less) and the crystal-oscillator injection signal (pin 8, U_4) must be 300 millivolts peak-to-peak (or less) of as sinusoidal a waveform as possible. Distortion in the input or output sine waveform increases the possibility of spurious frequencies occurring in the receiver system. The heterodyne mixer/buffer is not used on 20 meters.

The mixer module output filter consists of seven tuned circuits (L_4A to L_4G), each of which is resonant at midband of one of the necessary injection spectrum frequencies required by the signal-path mixer as listed in Table 3.

Only one coil must be switched in the mixer output circuit at any time, and this coil completes the mixer dc output circuit. The gating diode in series with the coil in use is forward-biased, and completes the circuit to pin 6 of the mixer (U_4),

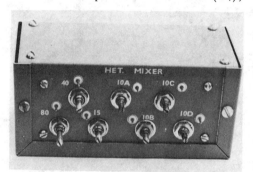

Figure 13

THE HETERODYNE MIXER MODULE

The heterodyne mixer module (E) is built in an aluminum minibox. The seven coils (L_4 series) and gating-control press-fit feedthrough terminals are mounted on the side of the box. The four corner screws secure the component chassis board inside the module.

which is at an 11.3-volt level. The cathodes of the remaining switching diodes are at this same positive level, while the anodes are close to ground potential; and so are reverse-biased. The *Amperex (Philips)* BA-

Table 3. Heterodyne Mixer Module (E) Frequencies

Band (Meters)	Heterodyne Mixer Output Range (MHz)	Heterodyne Oscillator (MHz)
80	12.5 - 13.0 MHz	7.5 MHz
40	16.0 - 16.5 MHz	11.0 MHz
20	—	—
15	30.0 - 30.5 MHz	25.0 MHz
10A	37.0 - 37.5 MHz	32.0 MHz
10B	37.5 - 38.0 MHz	32.5 MHz
10C	38.0 - 38.5 MHz	33.0 MHz
10D	38.5 - 39.0 MHz	33.5 MHz

COIL DATA

80	L_4A—22 turns #29 on Cambion 1534/2/1 form. $L = 5\ \mu H$. $Q = 75$. Note: 18-pf capacitor connected across coil.
40	L_4B—25 turns as above. $L = 6\mu H$. $Q = 75$. Resonates with circuit capacitance.
15	L_4C—10 turns #27. $L = 1.5\ \mu H$. $Q = 75$. Cambion 1534/3/1.
10A—D	L_4D–L_4G—5 turns #27. $L = 1\ \mu H$. $Q = 100$. Cambion 1534/3/1. Note: 18-pf trimmer capacitor connected across each of the 10-meter coils.

182 diodes (D_1—D-) were selected for their low forward impedance and their small reverse capacitance.

A silicon diode (D_s) in series with the dc current path to the rest of the mixer corrects the output dc level imbalance caused by the diode junction voltage drop of the switching diodes. The series circuit of an r-f choke and 1K resistor across the mixer output (pin 6) drains additional current though the switching diodes.

The heterodyne mixer module is built into the "U" shaped portion of a minibox $2\frac{1}{4}" \times 2\frac{1}{4}" \times 5"$. Because of the large number of components that must be installed in this module, it is assembled using "layered" construction.

The heterodyne mixer consists of three separate sections; a chassis board, containing the components of the mixer itself (figure 15), the U-shaped section of the minibox, the sides of which have been drilled to ac-

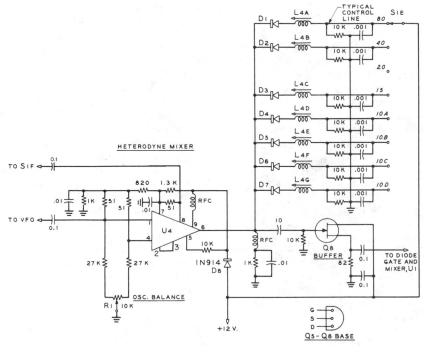

Figure 14

SCHEMATIC, HETERODYNE MIXER MODULE

Q₈—2N5459
U₄—Motorola MC 1496G or Fairchild μA 796
D₁-D₇—Amperex Phillips BA 182

L₄ₐ-L₄g—See Table 3
RFC—1 millihenry, 35-mA, J. W. Miller 10F-103A1
Note: All resistors ¼ watt.

cept the seven coils, and another chassis board, drilled to match the side of the minibox, on which components connected directly to the coils are mounted. Figure 13 showing the completed mixer module illustrates how the side of the minibox must be drilled for the coils; board containing the coil components is simply cut to match the dimensions and holes of the minibox side.

The mixer component board is wired, keeping all leads short, and using sufficient heat on the ground connections. When it is completed, mount it on the bottom of the U-section of the minibox. The board containing the coil components is then cut to size and installed on the side of the mini-box. The centers of the holes for the coils and the feedthrough terminals are then

Figure 15
HETERODYNE MIXER COMPONENT BOARD

Integrated circuit U₄ is at center with the oscillator null potentiometer (R₁) at the extreme left of the board. The assembly is built as compact and flat as possible to allow clearance for the board below the coils, once the module is assembled. The buffer FET (Q₈) is at the right of the board.

marked, and the holes drilled through both the minibox side and the coil component board.

The component board is now removed from the side of the minibox, and the coil bypass capacitors and the gating-diode biasing resistors are installed on the board. The board is then reinstalled on the side of the minibox and the coils are now installed in the module and wired in to both of the

Figure 16

VFO MODULE

Counter output from buffer stage Q_6 is at left (BNC connector) and oscillator inductor L_1 is atop the chassis, along with calibrating capacitor, C_1.

component boards. If the coils mounted nearest the bottom of the minibox are installed first, wiring will be easy. It is important to keep the diode leads as short as possible. For this reason, the 10-meter coils are mounted nearest the output pin of the mixer component board.

To align the heterodyne mixer, inject a 5.0- to 5.5-MHz signal at the correct amplitude into one port (pin 1), and the speci-

fied frequency and amplitude to simulate the crystal oscillators into the other port (pin 8). Terminate the module output in 50 ohms, and check with a high-frequency oscilloscope and digital counter for the correct mixer-output frequency, and a clean sinusoidal waveform, as each tuned circuit is gated into the output by applying +12 volts to each of the gating terminals. Make sure, as each band is checked, that the correct crystal-oscillator frequency is being injected. Check each band for uniform output over the 5.0-5.5 MHz vfo range. On the 10-meter band, the output tuned circuits may interact to a certain extent, and the tuning process may have to be repeated several times. For each band, the idea is to obtain maximum output, and uniform output amplitude, at the correct frequency.

When testing the mixer, use the mixer module (pin 6) to drive the oscilloscope; do not try to probe the output circuits of the mixer itself with the oscilloscope, as even the small input capacitance of a high-frequency oscilloscope will load the tuned circuits.

Construction of Variable-Frequency Oscillator Module (A)—The vfo module consists of a voltage-controlled oscillator (Q_5), two buffer stages (Q_6 and Q_7) for the two necessary oscillator outputs, and a regulated supply for the frequency-control potentiometer that is derived from the positive and negative 12-volt rails (figures 16, 17, and 18).

The vfo itself covers 5.0 to 5.5 MHz, and is a Colpitts circuit using a JFET as the oscillating device, and varactor diodes D_1, D_2) as the bandspread tuning capacitor. A piston trimmer (C_1) across the varactor diode circuit enables the limits of the tuning range to be accurately calibrated, once

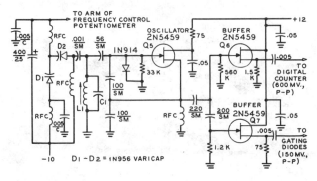

Figure 17

SCHEMATIC, VFO-BUFFER MODULE

L_1—7μH. Cambion 2419-2
C_1—Piston capacitor, 60 pF. MC-606Y
RFC—1 mH, 35 mA. J. W. Miller 10F-103A1

the inductance has been established. Elimination of the large plate-type variable capacitor often used in such circuits allows the vfo to be built in a much smaller enclosure than usual, and the JFET oscillator is inherently quite thermally stable.

The vfo covers a wide frequency range with the circuit constants provided, so the voltage range of the varactor diode is limited by the regulator circuits of the frequency-control potentiometer as shown in figure 19. The regulators also serve the purpose of stabilizing the tuning voltage, which directly affects the stability of the oscillator.

The tuned circuits for the vfo are wired with solid wire, with leads as short as possible; the entire vfo should be built mechanically stable and vibration proof. After construction, the output ports of the buffer

Figure 18

INTERIOR VIEW OF VFO MODULE

Components of vfo Module (A) are securely mounted to p.c. board bolted to one half of minibox. Tuned circuit (L_1-C_1) is at right, with buffer FETs just behind the BNC coaxial connector. Oscillator leads are short and heavy.

stages should be checked for a sinusoidal output, of amplitudes approximating those indicated in figure 17.

The frequency range of the oscillator can be adjusted with the aid of an accurate digital frequency counter on the high-level output; the inductance adjustment will affect the width of the frequency variation, and the piston trimmer is used to set the lower-frequency limit of the tuning range. Adjusting each of these in turn, and checking the frequency range with each adjust-

ment, should result in proper calibration being attained in short order.

The oscillator circuitry is built into a minibox measuring $2\frac{1}{4}'' \times 2\frac{1}{4}'' \times 4''$. The Oscillator Module is thermally coupled to the main chassis by cleaning the paint from the bottom of the module and covering it with a layer of silicone grease before mounting it. By heat-sinking the module to the main chassis, excellent thermal stability is attained, even though the oscillator enclosure is quite small. No electrical temperature compensation is required.

The potentiometer used to control the frequency of the receiver is a matter of choice for the builder. Since a *National PW-O* gear reduction drive is used to drive the control potentiometer, the tuning of the receiver is very smooth, and reasonably slow even with a single-turn carbon unit of linear taper (figure 19). Using a single-turn potentiometer, a tuning rate of 35 to 60 kHz per tuning knob revolution is obtained, depending on the total coverage adjustment of the oscillator. This tuning rate enables the operator to cover the band at a fairly rapid rate, and is ideal for SSB operation. It is a bit fast, however, for tuning the band with the very selective 500-Hz c-w filter. Using a 10-turn helipot, (the *Amphenol* 2151B-104 is recommended for installation in the same space) a tuning rate of 3 to 5 kHz per knob revolution is easily attained. While it is tedious to tune the entire band at this rate, even with a "spinner" knob, it is perfect for use with the sharp filter.

By varying the total coverage adjustment of the oscillator, and by varying the voltage limits between which the tuning potentiometer wiper moves (a maximum of + and −12 volts), a wide variation in tuning rates can be attained.

Variable-Frequency Oscillator Output Gating Detail— The output of the vfo drives the heterodyne mixer module on all bands except 20 meters. The heterodyne mixer is driven by one of the seven crystal oscillators as well, and it drives the local-oscillator injection port of the signal-path mixer (U_1, pin 8) on all bands except 20 meters.

On 20 meters alone, the various crystal oscillators are disabled, and the vfo drives

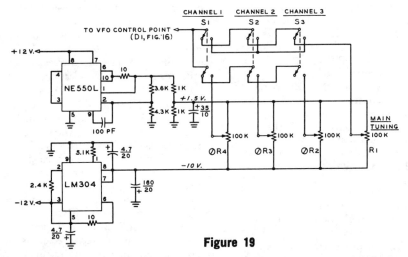

Figure 19

FREQUENCY-CONTROL SYSTEM AND REGULATED SUPPLY

R_1—100K Amphenol 215B-104
R_2-R_4—100K "Trimpot" Amphenol 2750SL or Bourns 3052-S
Note: All switches shown in "Main tuning" position. If more than one switch at a time is set to "Preset Channel" position, only one preset control is active. All wiring done with shielded wire to prevent noise pickup on varactor control leads.

the signal-path mixer directly, while the heterodyne mixer input and output circuits are completely isolated from the system.

The circuit isolation and signal gating are accomplished by means of diode gates, which **only change conduction states when the bandswitch is in the 20-meter position** (figure 20). In this position, none of the

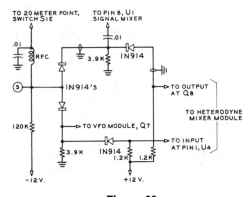

Figure 20

VFO GATING CIRCUIT

RFC—1 millihenry, 35 mA. J. W. Miller 10F-103A1
Note: All resistors ¼ watt

crystal oscillators are supplied with dc power; the oscillator outputs are not connected to the heterodyne mixer through the bandswitch, and the +12 volt line which normally powers the crystal oscillators now changes the status of the vfo output gates and the heterodyne mixer gates.

Diode gate status change is accomplished by changing the level at point S from −12 volts through a high impedance to +12 volts through a low impedance (the dc resistance of the isolating choke) by means of the bandswitch. This forces the diode gates between the vfo and signal-path mixer to conduct, while at the same time reverse-biasing the diode gates at the heterodyne mixer input and output circuits.

VFO Module Frequency-Control System —The use of a varactor-diode tuned circuit in the variable oscillator allows several modes of frequency control not seen in the usual amateur receiver.

The usual mode, employing a front-panel knob ganged to a tuning element in the oscillator, is here approximated by the ten-turn potentiometer which controls the varactor bias and which is driven through a gear-reduction drive from the front panel *Main Tuning Control.*

Because the frequency of the oscillator is controlled by a varying dc voltage, this receiver is also equipped for *channel* operation; by switching one of three switches provided, one of three preset frequencies may be selected (figure 19). These preset channels are adjustable from the front panel, by means of three panel-mounting *trimpots*. In case more than one channel switch is activated at the same time, the circuit is arranged so only one channel is actually enabled, even though all three switches may be in the channel position. Only when all three switches are in the *Main Tuning* position is frequency control by means of the panel tuning control possible. The *trimpot* channel adjusts may be easily set by a small screwdriver, while watching the digital counter; they will adjust the vfo frequency independently of band changing, of course. Once set, it is impossible to knock them out of adjustment accidentally. This feature will be found useful for net operation, for DX-ing, and for cross-frequency operation.

Remote Operation—This receiver is easily adapted for remote operation: For mobile work, the receiver can be located in the rear of the vehicle. A small control box near the driver would require only a speaker and a multiturn potentiometer to replace the main tuning potentiometer (it should be wired into the vfo module in exactly the same way). In addition, a second control to adjust the *R-F Tune* voltage is necessary, as the tuning of these circuits is quite sharp.

It is also possible to adapt this receiver to automatic tuning; a dc sawtooth voltage of approximately one cycle/30 seconds, and ranging from -10 to $+1.5$ volts applied to the varactor control point in place of voltage from a tuning potentiometer will cause the receiver to scan its complete range, quickly return to the starting point, and scan again. Because of the very sharp tuning of the r-f circuits scanning over a small range is probably most practical.

I-F Strip—The i-f strip uses an RCA CA3002 chip (U_2) which provides 30 dB of gain at 9 MHz and is specified as having

Figure 21

THE I-F BOARD

The two filters are mounted on the reverse side of the board. Small metal shields are placed across the filter to isolate input and output terminals. These shields are seen on-edge and are difficult to locate. One is at top, left of the board just above Q_2 and the second is to the left of the device U_2. The gain compensation helipot is at the bottom of the board.

80 dB of agc control range (figure 21). Used in conjunction with the PIN attenuator, agc range in excess of 70 dB was measurable and the range is thought to be better than 120 dB, far beyond most instrumentation to confirm.

The agc control port of U_2 is driven through Q_3 from the PIN attenuator agc system (figure 22). Circuit values are used that provide a nominal value of attenuation by U_2, until the attenuation limits of the PIN attenuator are approached; after this level attenuation in the i-f system increases very rapidly.

In front of the i-f IC (U_2), two KVG 9-MHz crystal filters are installed such that in normal operation the two filters are cascaded. The sharp filter (500 Hz) drives a MOSFET amplifier (Q_2), which has its own gain control. This amplifier is adjusted so as to have a gain which exactly compensates for the insertion loss of the sharp filter. This ensures that the i-f strip has the same gain whether the sharp filter is in the circuit, or not. When a wider i-f bandwidth is desired, the sharp filter and its compensation amplifier are shorted out by activating a diode gate. The switch that controls

the gate also switches in the proper bfo crystal when the sharp filter is used. The input circuit of the i-f board is designed to supply the dc operating voltage to the mixer which drives it.

An optional muting circuit can be provided by adding a front panel switch to connect *pin 1* of the i-f amplifier device to —6 volts. Receiver muting is complete and recovery is immediate. A jack on the rear apron allows an external transmitter to perform the mute-switching function.

The Detector and Audio Stages—The detector is a double-balanced modulator IC (U_3), used in a product-detector configuration (figure 23A). It requires 50 to 300 millivolts p-p injection from the beat-frequency oscillator, applied to pin 8.

The detector drives a high-gain, bipolar transistor a-f preamplifier stage (Q_4), the output of which drives the output a-f amplifier.

To test this system, connect an oscilloscope to the a-f preamplifier transistor collector. Inject 8999.0 kHz at 300 millivolts p-p from an oscillator into pin 8 of the detector. Inject 9.0 MHz at 100 millivolts p-p from a second oscillator into the input

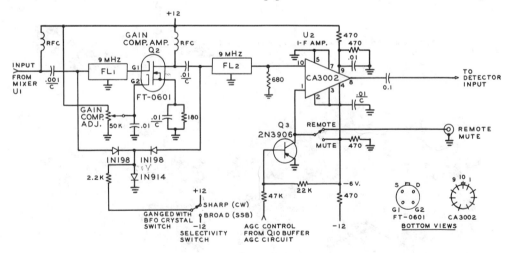

Figure 22

SCHEMATIC, I-F BOARD

Cascaded crystal filters are used to obtain optimum bandwidth for SSB and C-W operation.

FL₁—Crystal filter, KVG type XL-10M (500 Hz bandwidth)
FL₂—Crystal filter, KVG type XL-9B (2.1 to 3 kHz bandwidth)
RFC—1 mH, 3 mA. J. W. Miller 1741
Q₂—Fairchild FT-0601
U₁—RCA CA3002

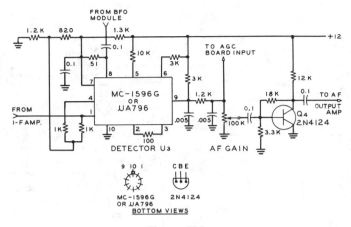

Figure 23A

SCHEMATIC, DETECTOR-AUDIO BOARD

U₃—Motorola MC-1596G or Fairchild μA-796

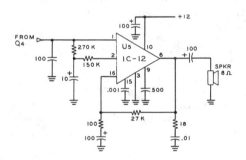

Figure 23B

SCHEMATIC, AUDIO AMPLIFIER

U₅—Sinclair IC-12 or equivalent

of the i-f strip. Adjust the gain-compensation amplifier gain control for equal audio output when either one or both of the i-f filters are used.

The audio output stage is a *Sinclair* IC12 (figure 23B) amplifier (U_5), which delivers up to six watts into an 8-ohm speaker. The output connectors may be arranged so that a pair of low-impedance (stereo) headphones can be plugged into a front panel jack, disabling the speaker.

Construction of Beat Frequency Oscillator Module (F)—The bfo module contains three beat-frequency oscillators and three source followers which transform both the oscillator output amplitude and impedance

to the level required by the detector (figure 24).

The bfo module frequencies are crystal controlled, with three-frequency capability necessary for operating c-w and SSB (both upper and lower sideband). One of the three oscillator-follower combinations is selected by applying a positive 12-volt level to the appropriate power supply line by means of front panel switches *I-F Selectivity* and *USB/LSB* (S_2B). When the *I-F Selectivity* switch is set to the *Sharp* position, the bfo frequency is automatically set to 8999.0 kHz for use with the sharp filter. When the switch is in the *Broad* position, bfo frequency is selected by the *USB/LSB* switch.

The frequency of each oscillator can be adjusted over quite a range by means of the trimmer capacitor across each crystal. No interaction of adjustments will be noticed. The module is designed so that the trimmers may be adjusted while the receiver is operating, without it being necessary to disassemble the module to gain access to the trimmers.

The bfo output should be a reasonably sinusoidal waveform of about 300 millivolts p-p amplitude when driving a 50-ohm load. Once the module is built, the frequencies may be adjusted by connecting the module output to a digital counter.

Construction of AGC and PIN Board—The PIN current source, buffer and agc

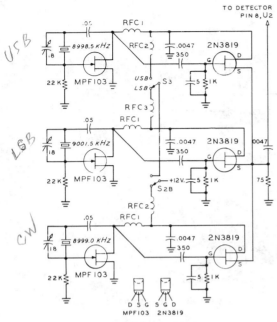

Figure 24

SCHEMATIC, BFO MODULE

RFC₁, RFC₂—1 millihenry, 35 mA. J. W. Miller 10F-103A1 or equivalent

Crystals—KVG. 8999.0 kHz, type XF-903. 8998.5 kHz, type XF-901. 9001.5 kHz, type XF-902 (Spectrum International, Box 1084, Concord, MA 01742), or equivalent

circuit are built on a separate board (figure 25). An NPN transistor is used as a current source, providing in excess of 100 mA to the PIN diode. The current-source transistor is driven from the agc circuit through a JFET buffer (Q_{10}), which prevents the low impedance of the current source Q_{11} from loading the agc line and affecting the time constant. The time constant is determined by R_1 C_1, the values shown being a compromise between slow (SSB) and fast (c-w) agc.

The agc voltage is audio derived. The signal is taken from the a-f gain control and is amplified and rectified; 200 mV rms at the input of U_6 is sufficient to cause maximum attenuation. The centertap of transformer T_1 can be grounded, but if it is tied to the arm of a potentiometer connected between ground and the −12 volt bus, manual control of attenuation level provides an r-f gain control function while the automatic control feature is still maintained.

Construction of Digital Counter Module (D)—The digital counter is the single module built on a printed-circuit board; the only practical way of doing it. By following a board layout and figures 26, 27, and 28 (use the schematic as a reference, while following the board layout) it should be possible to duplicate the counter easily. Since the board

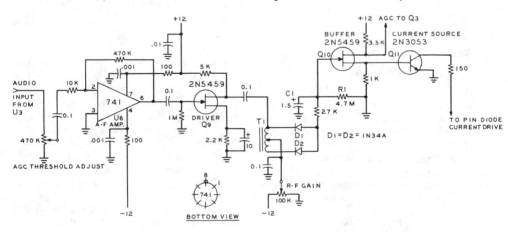

Figure 25

AGC AND PIN CURRENT SOURCE BOARD

T₁—Small audio transformer with center-tap winding (200 mW)

Figure 26

DIGITAL COUNTER AND DISPLAY BOARD (D)

Frequency count-out of receiver is displayed on panel-viewing LEDs (light-emitting diodes) seen at the front (right) of this assembly. Time base is established by 1000.00-kHz crystal at opposite end of board. ICs and other components are mounted to printed-circuit board. A full-size template of the p.c. board may be obtained by writing the publishers of this Handbook, enclosing 25 cents to cover cost of mailing.

layout drawings and photo are made directly from the board itself, and the schematic does not show all the many decoupling capacitors in detail, the layout and photo should guide the builder through any points of confusion. It is important that small components be used where indicated. A print of the board layout and a component layout drawing may be obtained by sending one dollar to cover the cost of mailing, to the Editor of this Handbook.

The frequency counter may be divided into two parts, a *time-base* and a *divider-display*.

The Time Base—The time-base consists of a 1-MHz clock oscillator, the output of which is shaped and divided down to give the needed timing pulses. The oscillator is a FET tuned-drain oscillator, crystal-controlled. The output is buffered by a FET source follower, and shaped by two NAND gates into a squarewave suitable for the inputs of the chain of decade dividers. The

time-base output is a series of 5-Hz pulses, which drive the *enable/disable count* line, resulting in 5 frequency sample/updates per second of the display. The display does not blink during the count, nor can it be seen to "run up."

During the *disable* cycle, *read-in* and *clear* signals are generated using additional NAND gates, to control the operation of the *memory* circuits of the display, located inside the *readout* chips.

The Divider/Display—If this section seems to be a bit sparse, it is because the *LED Readout* integrated-circuits contain their own decoder drivers and memory circuits, making .the divide/display circuitry relatively simple to build.

The input signal to the counter is buffered by an emitter follower, squared by a *Schmitt Trigger,* and then counted down by decade counters in a manner similar to the reference clock. The decade counters count the input signal for a time determined by the time-base control, then the counter outputs are read into the memories (in binary-coded-decimal) of the readout chips. The BCD is decoded into decimal inside the readouts, and the appropriate digit is indicated on the display.

The counter module has three connection points at the rear of the board:

(1) 5V—to power supply.
(2) GD—chassis ground.
(3) SIGNAL IN—input signal from the Vfo buffer, Q6.

Construction of the   The power supply is built
Power Supply   on a chassis separate from the main receiver chassis, and is housed in the speaker cabinet. An *on/off* switch is provided on the supply chassis for testing purposes; this is connected in parallel (through the power cable) with a power control switch on the receiver panel.

The supply provides three dc rails; plus and minus twelve volts, and plus five volts

(all with respect to chassis ground) and all rails are extremely well regulated, using IC regulators (figure 29).

The negative supply is regulated by means of a positive regulator, by isolating the regulator from ground; for this reason, the negative supply regulator is not mounted/heat-sinked to the chassis; however, little current is drawn from this supply, making this procedure quite safe.

The unregulated input to the five-volt supply would be as high as eighteen volts, were it not for the series five-ohm resistors, which gradually decrease the input to the regulator with increasing load, thus minimizing the dissipation of the regulator. The load on the five-volt supply is close to one ampere and will not vary; the input voltage to the regulator under these conditions is about eight volts.

The supply is built on a 6″ × 4″ × 1″ chassis and is housed in a 8″ × 8″ × 6″

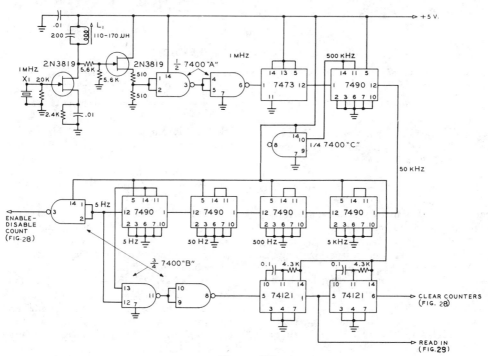

Figure 27

SCHEMATIC, COUNTER TIME-BASE MODULE

L₁—Cambion 3338-21 or equivalent
Integrated Circuits—National Semiconductor DM 7400, Quad, two input NAND. National Semiconductor DM 7490 Decade counter; National Semiconductor DM 7473 Dual J-K flip-flop; National Semiconductor DM 74121 Monostable multivibrator

speaker cabinet. Separating the power supply from the main chassis eliminates possible hum problems, prevents component crowding, and facilitates portable operation of the receiver. If battery operation is contemplated and the digital readout is to be powered as well, it will be necessary to provide a means of shutting off the readout, except during actual frequency measurement; the high current drain of the readout will otherwise run down the supply battery quite rapidly.

It is all-important that the dc supply lines be as free as possible from ac ripple. Not only will audio hum rise with increasing ripple, but residual f-m hum in the output of the vfo will create intolerable audio distortion on received signals. Keep in mind that if a 10-volt change in potential across the vfo varactor tuning diode causes the frequency to change by 500 kHz, one millivolt of ripple on the varactor control line will cause 50 Hz of frequency shift, which can be easily heard in the audio, particularly when listening to a c-w signal!

Careful attention must be paid to power-supply lead lengths, lead size, and grounding when wiring the power-supply rectifier circuits; particularly the positive rectifier, which has the highest load current. The filter capacitors are mounted directly above the rectifiers, with the capacitor terminals close to the chassis, and short, heavy leads are run from the terminals directly to the bridge rectifiers. The positive rectifier is grounded directly to a solder lug, and the ground line in the receiver power cable is grounded to the same lug, to prevent ground loops. Additional decoupling capacitors are installed across the dc input lines, directly at the receiver power connector.

It may be necessary to add a separate heavy ground strap between the power supply and the receiver chassis to eliminate unwanted ground loops. With all these efforts, additional supply regulations inside the vfo module itself will be necessary, to reduce residual f-m on the vfo output to an acceptable level (details are included in the section on the vfo module).

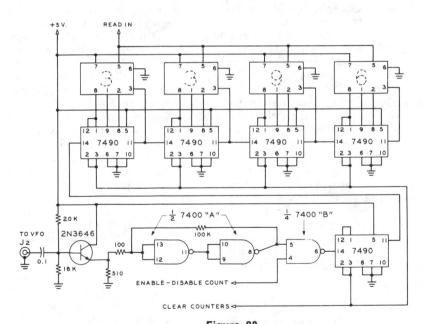

Figure 28

SCHEMATIC, DISPLAY MODULE

Integrated Circuits—National Semiconductor DM 7400, Quad, two input NAND. National Semiconductor DM 7490 Decade Counter; Hewlett-Packard 5082-7300 LED Digital Readout.

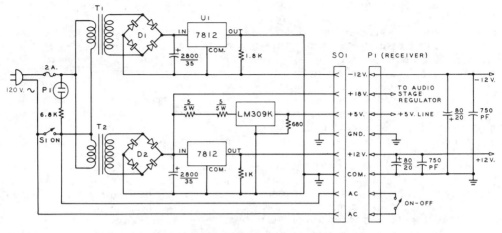

Figure 29

SCHEMATIC, RECEIVER POWER SUPPLY

T₁—12.6 volts, 0.3 ampere
T₂—12.6 volts, 2 amperes
D₁—Silicon bridge rectifier, 200 volts, p.i.v., 0.5 ampere
D₂—Silicon bridge rectifier, 200 volts p.i.v., 2 amperes
P₁—Neon pilot lamp
Integrated Circuits—Fairchild 7812; National LM 309K

System Alignment and Test Once the individual modules and circuit boards have been built, tested, and aligned, and the complete receiver system has been wired, the following procedure may be used to test the receiver as a system.

(1) Before connecting the receiver to the power supply, turn on the supply and check the + and −12, +5, and +18 volt lines for the proper voltage. Using a low-frequency oscilloscope, verify that ripple and noise on all lines (except the unregulated 18-volt line) is 2 millivolts, p-p or less. Now, connect the supply to the receiver, and repeat the supply tests. Note that ripple on the lines will likely increase somewhat with the increase in load current.

(2) With a 150-MHz oscilloscope, confirm that there is output from each crystal oscillator, measured at the common terminal of the oscillator output switch deck, as the bandswitch is rotated through its range.

(3) With a digital counter and oscilloscope, check for output from the vfo module of the proper waveform, amplitude, and frequency at the input to the digital counter in the receiver. With the bandswitch in the 80-, 40-, 15-, and 10-meter positions, confirm that there is vfo injection to the heterodyne mixer module. With the bandswitch in the 20-meter position, check for vfo injection directly at the signal-path mixer input, and no vfo injection to the heterodyne mixer module.

(4) With the bandswitch in the 80-, 40-, 15-, and 10-meter positions, check for oscillator injection to the signal-path mixer from the heterodyne mixer module. Check the waveform with a high-frequency oscilloscope and digital counter, and tune the vfo through its range while making this check. Output from the heterodyne mixer module should be sinusoidal, and of reasonably constant amplitude across the range on each band.

(5) With oscilloscope and counter, check for proper injection to the detector from the bfo module, while switching from one i-f filter to the other, and from USB to LSB.

(6) With a one-microvolt unmodulated signal at the antenna input, check for a comfortable audio level in headphones or speaker (*A-F Gain* at maximum) while tuning across the signal. Confirm that the digital counter in the receiver indicates the proper frequency; and keep in mind that any errors in calculation are the result of inaccuracies in the frequency being generated by the heterodyne crystal oscillator in use. During prototype testing, the receiver was able to copy 0.5-μV signals on all bands, using the speaker.

If instability is noted on any band, it may be necessary to alter wiring layout, lead dress, or add ferrite beads and/or "losser" resistors to the inputs and/or outputs of high-frequency amplifiers in the receiver. With the receiver properly shielded and grounded, no instability should be observed.

(7) On the prototype it was found necessary to do extensive decoupling of the +12 and —12 volt rails using 0.1 μF and 1 μF capacitors where the feedpoints entered the various modules. The rails should be checked at every module, using a high frequency oscilloscope. No signals should appear on the rails.

20-2   An Advanced, Solid-State HF Communications Receiver

The high-frequency communications receiver described in this section was built by W8ZR. It is designed to meet the problems of both weak- and strong-signal reception with overload or crossmodulation by making use of modern circuit techniques and high performance solid-state devices (figure 30).

The receiver covers the amateur bands, 80 through 10 meters, plus WWV, in eight 500-kHz segments. The dial calibration is linear and, because of the conversion technique used, none of the bands tune backwards. Provisions are made for three crystal filters for c-w, SSB and a-m. Incremental tuning may be used in the event the receiver vfo is used for transceiver operation. A summary of receiver performance is given in Table 4.

Receiver performance is quite in line with the best modern practices. The signal/noise ratio with a 0.17-μV input signal is better than 20 dB up to 14.4 MHz, dropping to 15

Table 4.

Receiver Performance Data

1. **I-f Rejection** (9 MHz signal into antenna terminals. Level adjusted until 1 μV i-f breakthrough signal detected):

Band	9 MHz level
80	2 volts
40	100 mV
20	600 mV
15	2 volts
10	2 volts

2. **Dynamic Range:** Agc Threshold: 2 μV. No overload detected with 3-volt signal.

   Agc off: I-f overload begins at 20 μV with r-f gain open.

3. **Image Rejection:** Signal injected at image frequency.

Band	Image Signal	Detected Signal
80	3 Volts	None
40	3 Volts	0.5 μV
20	3 Volts	none
15	3 Volts	none
10	3 Volts	none

4. **Frequency Stability:** (14 MHz)

Time	Frequency Shift (Hz)
1 min.	+100
30 min.	+250
1 hr.	+305
2 hrs.	+320
3 hrs.	+310

5. **Sensitivity:** (0.17 μV signal)

Band	S/N
80	> 20 dB
40	> 20 dB
20	> 20 dB
15	> 15 dB
10	> 10 dB

6. **Spurious Responses:** 7,196 kHz (1 μV)
   21,099 kHz (0.3 μV)

   Also responses less than 0.1 μV noted at 14,032; 14,082; 14,316 and 21,189 kHz

dB at 21 MHz and 10 dB at 29 MHz. Frequency stability is better than 250 Hz after 30 minutes operation and spurious responses are under 0.1 μV except for one spur at 7196 kHz which is approximately 1μV. The dynamic range is such that no overload is detected at agc threshold on a 2-μV signal when a nearby signal has an amplitude of 3 volts. With the agc off, the i-f system begins to overload at 20 μV with the r-f gain control wide open. These figures equal, or exceed, the specifications of the best communications receivers on the amateur market.

The Receiver Design   In order to minimize spurious responses (birdies) and to hold image rejection high, only one frequency conversion is used in the main signal path of the receiver. A 9-MHz intermediate frequency is used, while the vfo tunes 6.0 to 6.5 MHz and is premixed to provide an injection signal which is 9 MHz higher than the signal frequency (figure 31). Although the cost of two crystals could have been saved by operating the vfo over the 5.0- to 5.5-MHz range and by injecting the vfo signal directly into the mixer for 80- and 20-meter coverage, one of the bands would then have tuned backward on the dial, and image rejection would have suffered on both bands.

To minimize crossmodulation effects in the mixer, the r-f amplifier stage operates at a low gain level and is triple-tuned to provide a good measure of adjacent channel preselectivity (Table 4). Toroid inductors are used in the front end tuned circuits to reduce interstage coupling and to maintain high circuit Q. A sophisticated audio-derived hang agc circuit is used to control the gain of both the r-f and i-f stages; the agc loop is very "tight," resulting in a dynamic range well in excess of 120 dB and making the receiver virtually immune to overload prob-

Figure 30

THE W8ZR HIGH PERFORMANCE COMMUNICATIONS RECEIVER

This solid state hf communications receiver combines both weak- and strong-signal reception with protection from overload and cross modulation. The coverage includes all amateur bands 80 through 10 meters, plus WWV, in 500-kHz band segments. The dial is directly calibrated and readable to 1 kHz. Three crystal filters are employed for SSB, c-w and a-m reception. The OFFSET (Receiver Incremental Tuning) control is at the upper left of the panel, with the PRESELECT alignment capacitor beneath it. Directly below is the R-F GAIN control potentiometer.

To the right of the gain control is the AGC ON-OFF switch and the bandswitch. At the upper right of the panel is the S-METER, calibrated in microvolts and S-units and below it is the OFF-STANDBY-ON-CALIBRATE switch. Directly below this switch is the A-F GAIN control potentiometer.

The receiver is built in a homemade wraparound cabinet and is completely self-contained.

lems. A low distortion 5-watt audio ouput stage is used which results in very crisp sounding c-w and smooth sideband copy, particularly when using high quality stereo earphones.

The Receiver Circuit   A block diagram of the complete receiver is shown in figure 31. The circuit of the r-f amplifier is shown in figure 32. The incoming signal is inductively coupled by two tuned circuits to the r-f amplifier MOSFET (Q_1). This device has reverse agc voltage applied to its second gate, and on strong signals provides more than 30 dB of signal attenuation. After passing through a third tuned circuit in the drain connection, the signal is applied to source-follower Q_2 (figure 33), which provides an impedance match to the doubly-balanced diode ring mixer, consisting of transformers T_1, T_2, and hot-carrier diodes D_1-D_1. This mixer has a conversion loss of about 8 dB, but is superior to a dual-gate MOSFET with respect to intermodulation and overloading. The primary of T_2 is tuned

to the 9-MHz intermediate frequency. The i-f output from T_2 is coupled through one of the crystal filters (FL$_1$-FL$_3$) to the i-f amplifier consisting of devices U_1 and U_2 (figure 34). To assure stability, the tuned circuits in the i-f strip (L_5-C_6) and (L_7-C_7) are intentionally "de-Q'ed" by resistors R_1 and R_2 which also serve to limit the overall i-f strip gain to about 60 dB. Forward agc voltage is applied to both U_1 and U_2 resulting in a control range for the i-f amplifier of about 100 dB.

The output from the i-f amplifier is coupled via a low impedance line to the product detector (U_3, figure 36A-B). The bfo is crystal controlled and consists of an oscillator (Q_3) and a buffer (Q_1) (figure 37). A conventional diode detector is used for a-m reception.

The potentiometer at the output of U_3 (placed at the arm of switch S_2G) controls the audio level to the audio board. This is not used as an audio gain control, but rather to set the agc threshold value, as it adjusts the input level to transistor Q_7 in the agc

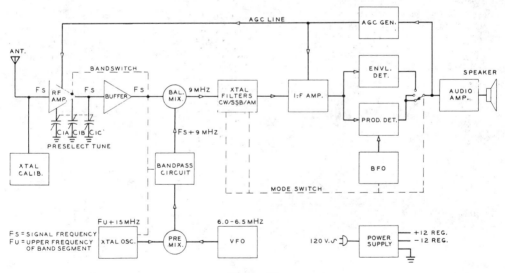

Figure 31

BLOCK DIAGRAM OF SOLID-STATE HF COMMUNICATIONS RECEIVER

This eight-band solid-state receiver covers the vhf amateur bands plus WWV (15 MHz). The receiver is built in modules on small circuit boards. The r-f amplifier uses a JFET for low noise figure and good dynamic signal range. Three tuned circuits precede the mixer to reduce image response and unwanted signal pickup. The mixing signal is derived from a vfo and crystal oscillator combination which is followed by a premixer and a simple bandpass circuit to attenuate unwanted "birdies". Three degrees of selectivity are provided by crystal i-f filters and a product detector and amplified agc provide good overload characteristics. The dial is direct reading to 1 kHz and each amateur band is covered in a 500-kHz segment.

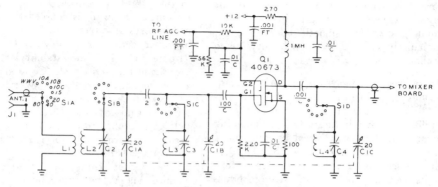

Figure 32

R-F AMPLIFIER SECTION OF RECEIVER

C_1A-B-C—Three section capacitor, 20 pF per section. Miller 1460, or equivalent
C_2, C_3, C_4—Arco compression trimmers. See Table 5
S_1—Bandswitch assembly consisting of eight, 2-12 pole ceramic switch sections (Centralab PA-1) mounted on index assembly (Centralab PA-302). Eight switch positions are used
Notes: All resistors, unless otherwise specified, are metal film, ¼-watt, 2% tolerance. Corning C-4, or equivalent. Feedthrough capacitors are Centralab FT-1000. R-f chokes are Miller 70F103A1. Tuning dial for C_1 is Bourns H-510-2 turns-counting dial. Main tuning dial is Eddystone 898 assembly

system. The control is placed beneath the chassis and can be seen in the upper corner of figure 36B. Audio gain is controlled by the *a-f gain* potentiometer placed at the input of device U_3, the audio preamplifier stage.

The audio system is shown in figure 38A-B. It consists of preamplifiers U_1 and

U_5 and a 5-watt output stage (Q_5-Q_6). It is desirable, though not essential, that Q_5 and Q_6 have matched *beta* characteristics. The agc voltage is derived from the audio signal by U_8, a sophisticated device which provides a "hang" period to maintain the receiver gain during speech or c-w pauses, but which will smoothly follow a fading signal and at

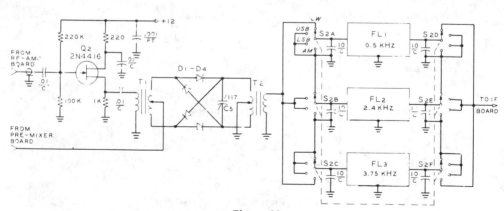

Figure 33

SCHEMATIC, MIXER AND I-F FILTER SECTION

C_5—87 pF to 117 pF. 82-pF silver mica parallel with 35-pF compression trimmer
D_1-D_4—Matched Quad. Hewlett-Packard HPA 5082-2830 or equiv.
FL_1, FL_2, FL_3—KVG XF9-M (0.5 kHz), KVG XF9-B (2.4 kHz), KVG XF9-C (3.75 kHz). Spectrum International, Box 1084, Concord, MA 01742

T_1—Primary: 10 turns #26 e. wound on T-50-2 core (Amidon). Secondary is 20 turns #26. Coils are trifilar wound.
T_2—Primary: 20 turns #20 e., wound on T-50-2 core (Amidon). Secondary is 24 turns #26. Coils are trifilar wound.

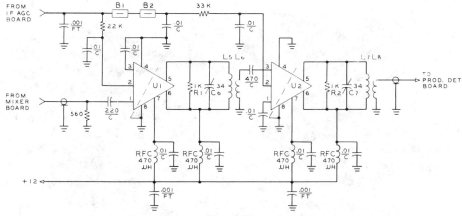

Figure 34

SCHEMATIC, RECEIVER I-F AMPLIFIER

B₁, B₂—Ferrite beads. Amidon
C₆, C₇—34 pF. Johnson 189-506-5, or equivalent in parallel with 18-pF silver mica capacitor
L₅, L₇—42 turns #28 e. on Amidon T-50-2 core
L₆, L₈—11 turns #28 teflon insulated wire wound over "cold" ends of L₅, L₇

U₁, U₂—MC-1590G integrated circuit. Motorola
Notes: R-f chokes are Miller 70F474A1. All resistors, unless otherwise specified, are metal film, ¼-watt, 2% tolerance. Corning C-4, or equivalent

the same time suppress impulse noise. "Hang" periods of 1 second or 0.25 second selected by switch S_3, are used for SSB and c-w reception, respectively. Devices U_7 and U_8 are used to convert the output of U_6 to the voltage levels required by the r-f and i-f amplifiers.

The vfo uses a variation of the Vackar circuit (figure 39) and is extremely stable. Stability is important in the local oscillator which the narrow bandwidth (500-Hz) filter is used for c-w reception. Polystyrene padding capacitors are used for the fixed elements in the oscillator, as opposed to silver-mica capacitors, because the former are sealed against moisture and have a superior temperature coefficient. The silver-mica units, particularly those of the larger values tend to have erratic drift characteristics, especially with regard to temperature and this effect shows up markedly when the narrowband filter is in use.

The oscillator is followed by a buffer-amplifier stage (Q_9) which boosts the output of the oscillator and at the same time isolates it from the mixer stage. The mixing signal is pure and oscillator noise is very low, compared to the peak oscillator voltage.

Electrical *dial correction* is provided by varicap diode VC_1 which is panel-controlled

by a small knob placed next to the main tuning dial. This adjustment brings the dial into calibration on each band, and provides precise calibration at any point on the dial. *Incremental tuning* (for use with an external transceiver or exciter) is provided by varicap VC_2. This control is a 10-turn potentiometer which is cut in or out of the circuit either by switch S_5 (*RIT on/off*) or from the external exciter by means of relay RY_2, which removes the RIT bias voltage when it is energized.

This control is helpful in chasing DX as the receiver may be set to the transmitting frequency with the RIT control and, when this is defeated by RY_2, the receiver is instantly returned to the frequency of the DX station, which is set on the main tuning dial.

The vfo output is mixed by U_9 with crystal oscillator Q_{10}-Q_{11} (figure 40 A-B) and injected via a tuned circuit and low-impedance link (L_{11}, L_{12}, C_{16}) into the main receiver mixer. Although a bandpass coupler might ordinarily be used at this point instead of a simple tuned circuit, it was not required because of the inherent suppression of fundamental and even-order mixer products provided by the double balanced design of mixer U_9.

Figure 35

TOP VIEW OF RECEIVER CHASSIS

Placement of the major parts atop the chassis may be seen in this view. The vfo module is at center, driven by the main panel dial. At the right is the r-f preselector alignment capacitor, with the 10-turn potentiometer for incremental tuning directly above it. To the left of the vfo is the audio-agc board with the S-meter controls on a small bracket above it. The controls are: Zero set and sensitivity.

In front of the vfo compartment are the KVG i-f filters, with the power supply components at the corner of the chassis. The individual antenna coaxial receptacles for each band are along the rear apron of the chassis with the slugs of the hf oscillator coils projecting through the chassis near the center of the assembly. The line cord receptacle, primary fuse, speaker connections, and muting terminals are at the left corner of the chassis apron.

Chassis and panel are joined with two angle support strips on the sides and an angle plate is fastened above the dial assembly to provide a connection to the slip-on cabinet.

Receiver muting is accomplished by solid-state switches Q_{13}-Q_{15} (figure 41), and operates smoothly without pops or thumps. A reed relay (RY_1) shorts the antenna terminals of the receiver to ground during standby periods but can be omitted if this task is already accomplished elsewhere, for example in a transceiver or linear amplifier.

The power supply (figure 42A-B) provides regulated +12 and —12 volts at 1 ampere. Although a less complicated design could be used without impairing the performance of the receiver, this particular circuit has exceptionally good regulation as well as adjustable current limiting. This last feature is particularly valuable if the opera-

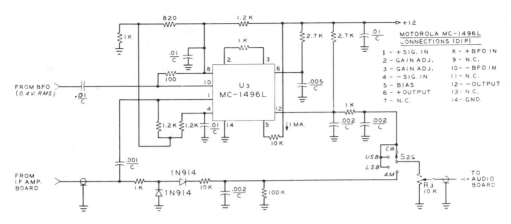

Figure 36A

SCHEMATIC OF PRODUCT AND A-M DETECTORS

U₃—MC-1496L integrated circuit. Motorola
Notes: All resistors, unless otherwise specified, are metal film, 2% tolerance. Corning C-4, or equivalent, Potentiometer is Trimpot.

Figure 36B

ARRANGEMENT OF OSCILLATOR AND PRODUCT DETECTOR BOARD

This board is located in the upper right corner of the underchassis, as observed in figure 43. At the left are the two sideband crystals and padding capacitors plus an extra crystal for a-m reception when the receiver is used as an i-f strip for a vhf converter. Selector switch S_1 is adjacent to the crystals, with devices Q_3 and Q_4 immediately to the right. Near the right-hand edge of the board is U_1, the product detector. This board comprises the circuitry shown in figures 36 and 37. The 10K sensitivity control is mounted to the aluminum flange that supports switch S_2.

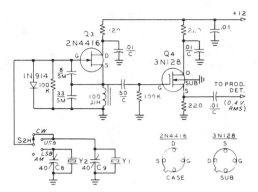

Figure 37

SCHEMATIC, BEAT-FREQUENCY OSCILLATOR

C_8, C_9—40 pF. Arco 403 compression trimmer, or equivalent

Y_1—8998.5 kHz. KVG XF-901

Y_2—9001.5 kHz. KVG XF-902. Crystals by Spectrum International, Box 1084, Concord, MA 01742

Note: All resistors, unless otherwise specified, are metal film, ¼-watt, 2% tolerance. Corning C-4, or equivalent

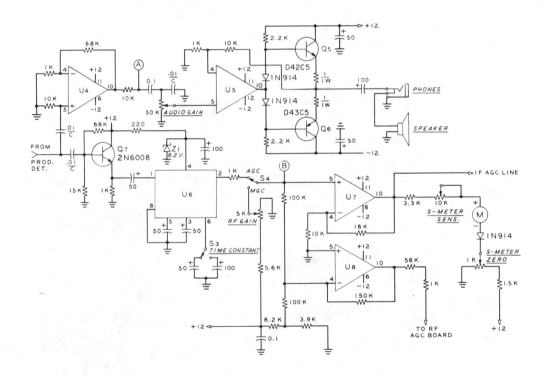

Figure 38A

SCHEMATIC, AUDIO AND AGC CIRCUITRY

Q_5, Q_6—NPN and PNP 5-watt power transistors. General Electric

U_6—Plessey SL-621 AGC generator

U_4, U_5, U_7, U_8—μA741 operational amplifiers. Fairchild U6A774193 pin connections for DIP configuration

M_1—0-1 dc mA. Simpson "Century", Model 8122 panel meter

Z_1—Zener diode. 1N753A

Notes: All resistors, unless otherwise specified, are metal film, ¼-watt, 2% tolerance. Corning C-4, or equivalent. Circled letters refer to connections to the muting circuit. Speaker is 8 ohms.

Figure 38B

AUDIO AND AGC AMPLIFIER ASSEMBLY

This board includes the audio stages and agc system. At lower right is the small bracket holding the S-meter controls. The four ICs are mounted in sockets affixed to the board. The board is held in place above the chassis by means of ¼" spacers and 4-40 hardware.

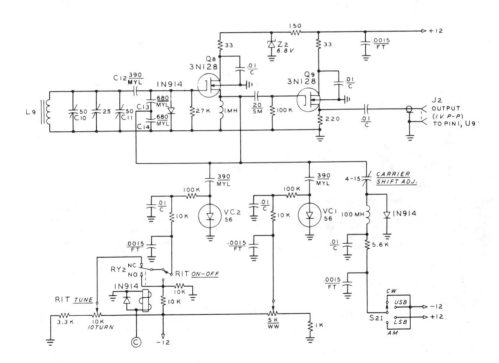

Figure 39

SCHEMATIC, VARIABLE FREQUENCY OSCILLATOR

C_{10}—50 pF. Millen 23050MK, or equivalent
C_{11}—50 pF, 750 NPO
C_{12}, C_{13}, C_{14}—Mylar capacitors
L_9—Wound on Millen 69046 coil form, ½" diameter, orange core. Wind with 17 turns #20 enamel or formvar wire
RY_2—Spdt reed relay. Magnecraft W103-MX2, or equivalent
VC_1, VC_2—HEP R2504 variable capacitor diodes. Motorola
Notes: RIT tune control is 10 turn Helipot (surplus). All resistors, unless otherwise specified, are metal film, ¼-watt, 2% tolerance. Corning C-4, or equivalent. RF chokes are Miller 70F104A1. Feedthrough capacitors are Centralab FT-1500

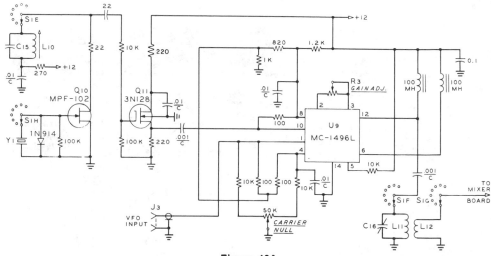

Figure 40A

SCHEMATIC, HF CRYSTAL OSCILLATOR AND PREMIXER

C_{15}, C_{16}—See Table 6
L_{10}, L_{11}, L_{12}—See Table 6. Toroid cores by Amidon, or equivalent
U_9—MC-1496L integrated circuit. Motorola, or equivalent
Y_1-Y_8—Sentry SGP-6 crystals. See Table 6
Trimmer capacitors—Johnson 189-506-5
Note: All resistors, unless otherwise specified, are metal film, ¼-watt, 2% tolerance. Corning C-4, or equivalent

Figure 40B

HF OSCILLATOR AND PREMIXER ASSEMBLY

This view covers the assembly shown in figure 40. At the front are the eight crystals which cover the 80 through 15 meter bands, plus three segments on 10 meters and WWV. The circuit board behind the crystals contains the components for crystal oscillator Q_{10} and buffer stage Q_{11}. At the left of the board is the premixer (U_9). Bandswitch segment S_1A-E is in the foreground. The slug-tuned coils (series L_{10}) are immediately behind the circuit board. Coil assemblies L_{11}-L_{12} are wound on small ferrite cores which are mounted to the rear of the circuit board immediately behind the small Johnson air-variable capacitors at the rear of the photograph. The coaxial receptacle for vfo input is mounted to the right of the compartment.

tor ever short-circuits the power supply accidentally during troubleshooting or alignment periods.

General Construction Technique The circuitry of the receiver is constructed on thirteen double-sided, glass epoxy etched circuit boards. The vfo, audio circuitry, filters, power transformer, pre-

selector tuning capacitor, and S-meter circuitry are mounted atop the chassis, as shown in figure 35. The r-f and i-f stages, mixers, crystal oscillators, and product detector are mounted below the chassis, as seen in figure 43.

The most vexing mechanical consideration in a multiband receiver is the problem of locating all of the tuned circuits close to the

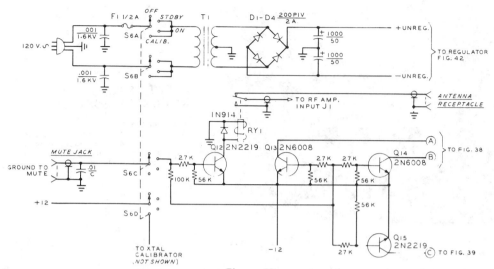

Figure 41

SCHEMATIC, RECEIVER POWER SUPPLY

D₁-D₄—200-volt piv, 2-ampere bridge rectifier. Sarkes-Tarzian S-6211
RY₁—Spdt reed relay. Magnecraft W103MX-2
S₆—Two sections, 4 poles, 2-6 position ceramic rotary switch. Centralab 2011

T₁—25.2 volts, 2 amperes. Stancor P-8357
Note: All resistors, unless otherwise specified are metal film, ¼-watt, 2% tolerance. Corning C-4, or equivalent

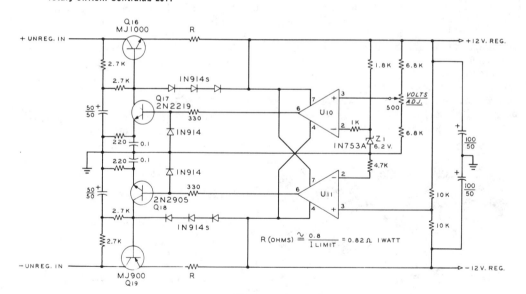

$$R \ (OHMS) \cong \frac{0.8}{I \ LIMIT} = 0.82 \ \Omega \quad 1 \ WATT$$

Figure 42A

SCHEMATIC, REGULATOR CIRCUIT

Q₁₆, Q₁₉—Motorola power Darlington transistors.
U₁₀, U₁₁— μA 741 operational amplifiers. Fairchild Semiconductor
Z₁—Zener diode, 6.2 volt
Notes: All resistors, unless otherwise specified, are metal film, ¼-watt, 2% tolerance. Corning C-4, or equivalent. Resistors marked R are proportional to the limiting value of current, as indicated.

Figure 42B

POWER SUPPLY AND REGULATOR ASSEMBLY

This view covers the assembly shown in figure 42. At the left are the filter capacitors for the power supply, with the dual regulator for the +12 volt and −12 volt power supply on the circuit board at the center.

bandswitch while at the same time avoiding unwanted interstage coupling. In this design all of the bandswitched circuits were constructed on $2\frac{1}{4}'' \times 5\frac{3}{4}''$ circuit boards. A representative board is shown in figure 44. The components are mounted directly to the glass-epoxy boards, including the switch section and the boards are mounted on edge underneath the chassis. The bandswitch shaft is then inserted from the front panel through all of the boards, and finally the detent assembly attached to the shaft with a shaft coupling. Aluminum shield plates, the same size as the circuit boards, are used to isolate the various stages of the bandswitch assembly from each other.

The inclosure is homemade, with the side pieces milled out of thick aluminum stock to provide guides for the various boards. It would be difficult to duplicate this with-

out machine shop facilities, but there are a variety of commercial circuit board guides available, such as the *Vector SR-1* or *SR-2* *"Frame-Loc Rail"* series.

Referring to figure 43 the boards (from the front to the back of the receiver) are: *Board 1:* holds reed relay RY_1 and bandswitch segment S_1A. *Board 2:* Input circuit L_1, L_2 and bandswitch segment S_1B. *Board 3:* Input circuit L_3, C_3 and switch segment S_1C. A shield plate separates board 3 from the next board. *Board 4:* R-f amplifiers Q_1 and Q_2. *Board 5:* Output circuits L_4, C_4 and switch segment S_1D. *Board 6:* Doubly-balanced mixer and transformers T_1 and T_2. *Board 7:* Coupler L_{11}-L_{12} and switch segments S_1F and S_1G.

Immediately adjacent to board 7 is the area holding the crystal oscillator and premixer assembly and switch segments S_1E

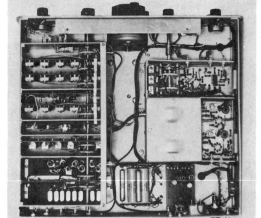

Figure 43

UNDERCHASSIS VIEW OF RECEIVER

General placement of circuit boards may be seen in this view. At the right are the r-f boards, mounted vertically about the bandchange switch. A common ground strap is soldered to the top corner of the r-f boards, running from the front to the back of the assembly. The conversion crystals are at the upper right corner of the chassis.

On the opposite front portion of the chassis is located the sideband oscillator components and the sideband selector switch. At the center is the shielded box containing the i-f filter switch sections. The i-f amplifier board is to the left of this inclosure. Behind the inclosure are the power supply and voltage regulator boards. A large flange is placed across the inside front panel of the receiver to allow it to be firmly fastened to the cabinet.

and S_1H. Finally, at the rear of the inclosure are the high-frequency crystals. Extra shield partitions are placed between boards 5 and 6, boards 6 and 7, and boards 7 and 8. The whole assembly measures 11" deep by 6" wide by 2¼" high. It is assembled and tested, a board at a time, before inclusion within the receiver chassis.

The receiver chassis measures approximately 13" deep by 15" wide by 2½" high. It is made up of a top plate (which includes the back lip), two side plates that act as panel support brackets (see figure 35) and a panel. After all main holes are drilled, the chassis is sandblasted to give it a matte finish; etching the chassis in a caustic lye solution would have a similar effect. A double panel measuring 16" × 7½" is used to avoid unsightly screw holes. The inner panel is countersunk to accept mounting screws and is then concealed by an outer panel covered with black vinyl (obtained from an upholstery shop).

Epoxy cement is used to attach the vinyl to the aluminum; other cements should not be used as they will not make a good bond. Press-on labels are used to label the panel which is finally sprayed with a thin coat of matte-finish plastic spray. The same letters are used to label the dial and the panel meter. The cabinet is also homemade; the curved sections made by bending the sheet aluminum around a piece of pipe. Important dimensions for the chassis assembly are given in figure 45.

The vfo assembly is shown in figure 46. To ensure good mechanical stability, the vfo is constructed on ⅛-inch thick glass-epoxy circuit board and is mounted inside of a homemade aluminum housing whose side panels are ¼ inch thick. Recessed edges are milled in the panels to accept ⅛-inch thick mating panels resulting in an r-f tight inclosure with battleship rigidity. A die-cast aluminum box *(Bud CU-347)* would make an acceptable substitute, although an unreinforced "minibox"-type inclosure would not have the required rigidity. It was found necessary to add a temperature compensating capacitor (C_{11}) to the vfo to reduce a slow, gradual drift resulting from the heat generated by the power transformer. Until this was done, the vfo drifted about 2 kHz before stabilizing; with the temperature compensation the warmup drift is less than 300 Hz in the first hour and 10 to 20 Hz per hour afterwards.

The crystal i-f filters are mounted atop the receiver, with their terminals projecting into an aluminum box visible from the underside of the receiver. The interior of the box, showing the switching mechanism is shown in figure 47. A grounded shield plate isolates the input and output sections of the filters. The *mode* switch shaft runs into this box and has extra sections on each side of the

Figure 44

REPRESENTATIVE CIRCUIT BOARD OF RF ASSEMBLY

This board contains the input tuned circuits shown in figure 32. Padding capacitors C_7 for each band are across the top of the board, with the ferrite core inductors (series L_1, L_2) directly below them. At the right is bandswitch segment S_1A-B. The board for the second coupled circuit (L_3-C_3) is similar to this one. The 2-pF coupling capacitor is connected between the boards, which are mounted vertically in the upper left-hand corner of the bottom view (figure 43).

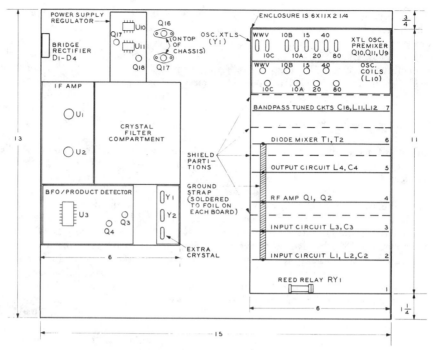

Figure 45

UNDERCHASSIS LAYOUT OF MAJOR COMPONENTS

Refer to photograph of figure 43 for comparison

Figure 46

INTERIOR VIEW OF VFO

The vfo schematic is given in figure 39. The box is constructed of heavy aluminum with the tuning capacitor (C_{10}) bolted to the front wall of the box. Once inductor L_9 is properly adjusted, the slug is held in position with a drop of epoxy cement to prevent it from moving or vibrating within the coil form. The adjustable padding capacitors are next to the tuned circuit. The coaxial connector for vfo output is at the right side of the box.

shield plate for shorting the terminals of the unused filters to ground. If this precaution is not taken, the ultimate stopband attenuation of the receiver is likely to be limited by leakage around the filters rather than by the rejection characteristics of the filters themselves.

Receiver Alignment and Test As with any receiver, construction and alignment should begin with the simpler stages (audio, product detector, and power supply) and proceed backward towards the r-f stages. After the audio sections are working, the bfo injection into the product detector should be checked to make sure that it is about 0.4 volt rms. The bfo crystals can then be trimmed with their padders to the correct frequency, placing them about 20 decibels down the skirt of the SSB crystal filter. Next, the i-f amplifier tuned circuits are brought to resonance; this adjustment is not critical since the Q of these circuits is relatively low.

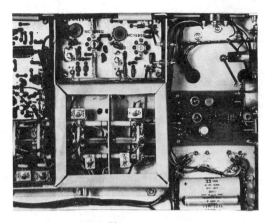

Figure 47

CLOSEUP VIEW OF CRYSTAL FILTER SWITCH AND I-F AMPLIFIER

The filter switch S_2 runs through the filter box at left. Filters FL_1 and FL_2 are installed atop the chassis with the alignment capacitors mounted on the switch deck terminals. At right is the i-f amplifier with the two Motorola ICs adjacent to toroid inductors L_6 and L_7. A shield plate is mounted across U_1 with the air trimmer capacitors immediately adjacent to the toroid inductors. These compartments correspond with figures 33 and 34.

Once the audio and detector stages are operating correctly, the vfo should be calibrated so that it covers the range of 6.0 to 6.5 MHz with an output of about 0.3 volt rms. Do not try to get the calibration exact at this time since the job must be repeated when the vfo is linearized. Next, adjust the tuned circuits in the crystal oscillator stage until all the crystals start reliably; the proper

Table 5. R-F Tuned Circuits

Band	L_2, L_3, L_4	L_1	Amidon Core	C_2, C_3, C_4 (pF)
80	73t. #28e.	4t. #26	T-68-2	4-60
40	37t. #24e.	3t. #24	T-68-2	4-60
20/WWV	23t. #22e.	3t. #22	T-50-6	4-40
15	17t. #20e.	3t. #20	T-50-6	4-40
10 A,B,C	13t. #22e.	2t. #22	T-37-6	1.5-20

Notes: L wound over "cold" end of L_2
capacitors are compression-type

point is just after the oscillator output begins to drop off on the high side of resonance.

Now comes an important part of the alignment procedure; setting the gain of the premixer (U_9). The gain of this stage is adjusted with potentiometer R_8 and should be set to as low a value as possible consistent with adequate drive into the diode ring mixer (about 0.2 to 0.3 volt rms). If the drive level is too low, mixer performance and noise figure will be degraded, while if it is too high U_9 will not operate in a linear mode, and spurious signals (birdies) will be generated. The mixer will also be prone to overload and desensitization on strong signals. The tuned circuits in the r-f amplifier stage are now brought into resonance and the dc voltage on gate 2 of Q_1 checked to be about 0.5 volt with the antenna terminals shorted to ground.

At this point, the overall gain distribution of the receiver should be checked. With a 50-ohm composition resistor connected across the antenna terminals, a definite increase in speaker hiss should be noted when the preselector control is swept through resonance; if this is not heard, it probably means that the i-f gain is too high. Agc voltage should begin to be developed with an input signal of about 2 μV, and should keep all the amplifier stages operating in their linear range up to an input signal of several volts.

The final part of the alignment procedure is the linearization of the vfo. The uncorrected vfo is not more than about 15 kHz away from linearity at any point in its range. However, it is not a difficult procedure to reduce this error, if a frequency counter is at hand, and it does permit the vernier window on the dial to be used for direct 1-kHz readout.

Begin by adjusting vfo coil L_9 and trimmer capacitor C_{10} so that the vfo tunes from 6.5 MHz to approximately 6.0 MHz. Don't worry about setting the low frequency limit precisely at this time. Next, make a graph of the exact deviation from linearity of the oscillator, keeping in mind that the vfo should be at 6.5 MHz when the dial reads zero. The horizontal scale of the graph is labelled in dial divisions (0 to 500) and the vertical scale in kHz (0 to —40). Now, if at a dial reading of 100, the vfo frequency is actually 6390 kHz, a point is placed at

—10 on the graph, indicating that the vfo is actually 10 kHz too low at the correct dial setting.

After the initial calibration curve is obtained, return the vfo to its high frequency limit and grind a piece 1/64-inch deep by 1/2-inch long off the outside edge of each of the rotor plates, toward the end of the assembly which will first enter the stator. Use a small hand-held grinding tool (such as the *Dremel Roto-tool*) with a fine grain grinding wheel, and exercise care to avoid breaking the solder bond which holds the plates to the rotor shaft. Then, run a new calibration curve and repeat the grinding process until the calibration curve is essentially linear.

Because a variable capacitor is inherently nonlinear at the ends of its range, it is difficult to obtain a linear calibration over about a 10-kHz range at one end of the dial. After the calibration is completed to your satisfaction, remove the variable capacitor from the vfo and wash off all brass filings which may have collected between the plates. Finally, adjust the carrier-shift capacitor so that the frequency of the receiver does not change when shifting between upper- and lower-sideband.

20-3   An Advanced Six-Band Solid-State SSB Exciter

The SSB exciter described in this section was designed and built by K5JA. It is a state-of-the-art device capable of exceptionally good efficiency and low intermodulation distortion (IMD) over the range of 3.5 MHz to 54 MHz (figure 48). Power output is in excess of 5 watts PEP on all bands except the 50-MHz band where the output is 1 watt PEP. The IMD is better than —33 decibels below one tone of a two-tone test signal on the lower bands and —45 dB on the 50-MHz band. Operating convenience has not been overlooked as provision is made for VOX operation and/or push-to-talk. In addition, a frequency-spotting switch for split operation and a carrier-insertion circuit for linear amplifier tuneup have been incorporated. No tuning of the exciter is required when changing frequency or bands as the output circuits are broadbanded over the full 3.5- to 54-MHz frequency range.

Also incorporated in this exciter is front-panel control of both audio and r-f clip-

Table 6. HF Oscillator Tuned Circuit Details

Band	C_{15} (pF)	L_{10}	Y_1 (MHz)	C_{16} (pF)	L_{11}	L_{12}
80	22	17½t. #26	19.000	4-25	27t. #24	7t. #24 T-50-6 core (Amidon)
40	22	14t. #26	22.500	4-25	21t. #20	6t. #24 T-50-6 core (Amidon)
20	22	10t. #24	29.500	4-25	16t. #20	3t. #20 T-50-10 core (Amidon)
15	18	9t. #24	36.500	4-25	11t. #20	2½t. #20 T-50-10 core (Amidon)
10	12 each	8½t. #24 each	A=43.500 B=44.000 C=44.500	4-25 each	11t. #20 each	2t. #20 T-37-10 core each (Amidon)
WWV (15 MHz)	22pF	9½t. #24	30.000	20-meter circuits used		

Notes: 1. L_{10} uses J. W. Miller 4500-2 form (0.26 × 0.86) 1.0-20 MHz, red core
    2. L_{12} wound over "cold" end of L
    3. C_{15} are silver-mica capacitors
    4. C_{16} is Johnson 189-509-5 or equivalent

ping (variable from zero to 20 dB of clipping). This allows the operator to tailor his signal to meet the existing conditions; clipping may be reduced for local ragchews or turned up for more audio punch in DX pileups. An audio speech compressor adjustable from the panel is also incorporated in the design. All of these features add up to provide a very potent SSB exciter for the advanced amateur who has had experience with the sophisticated components and circuitry used in this unit.

Circuit Description The exciter and power supply are completely solid state and wideband circuitry is employed to simplify tuning and adjustment. Special, switchable filters are used in the low-level stages to eliminate unwanted mixing signals, and dual crystal filters are used in the r-f processing circuitry. A phase-locked loop synthesizer is used to generate the conversion signal. This results in an exceptionally clean signal, free of the spurious problems often associated with a premixer and also provides the same tuning rate and degree of frequency stability on all bands. The master reference oscillator tunes over the range of 3.21 to 3.71 MHz, providing excellent stability on all bands. Provision is made for coverage of four 500-kHz bands in the 10-meter range and four 500-kHz bands in the six-meter range, although this combination may be changed, if desired. Operation on nonamateur frequencies is also possible (with some exceptions) by the proper choice of crystal and tuned-circuit components. The 3.21- to 3.71-MHz oscillator tuning range was chosen by careful consideration of all mixing products up to the tenth order with the aid of a digital computer. During several months of on-the-air operation no spurious problems have been observed.

For best spurious rejection, the mixing frequency is 9 MHz *above* the desired operating band, which places the mixing frequency quite high for 6-meter operation.

Figure 48
SOLID-STATE SIX-BAND SSB EXCITER

This compact, solid-state SSB exciter delivers over 5 watts PEP output over the range of 3.5 to 29.7 MHz and provides over 1-watt PEP output on the six-meter band. Audio and r-f clipping circuits provide good audio "punch." A phase-locked-loop synthesizer is used for the conversion oscillator and r-f circuits are broadbanded over the full operating range. The main tuning dial is at the right, with the phase-lock light above it. Across the bottom of the panel are (left to right): Audio level, Audio compression (gain and recovery time), carrier insertion, VOX (gain, delay, and antivox gain), and the VOX override switch. The general purpose multimeter and switch are at the upper left of the panel and to the right are the bandswitch and drive-level controls, with the sideband selector switch centered between them. The multimeter has two ranges: 0 to 30 volts and 0 to 900 milliamperes. The +28, +12, and −12 volt supplies are monitored, as well as amplifier current.

However, the use of the frequency synthesizer provides stable frequency control from a low-frequency oscillator of good stability.

If the exciter is used to drive the antenna directly, a half-wave low-pass filter such as described in Chapter 16, Section 3 should be used between the exciter and the antenna to attenuate the harmonics of the fundamental signal. If a linear amplifier with high-Q tuned circuits is used after the exciter, however, the low-pass filter may not be required since the tuned circuits of the linear amplifier will attenuate the harmonics. If desired, an extra switch section could be added on the exciter bandswitch to remotely select the appropriate low-pass filter automatically.

Exciter Circuitry — *Audio and VOX Circuits—* Shown in figure 49 is a block diagram of the audio and VOX circuits. The schematic for these circuits is shown in figures 50 and 51. An FET device (Q_1) provides a high input impedance for the microphone and drives the first IC audio amplifier (U_1, figure 50) and the VOX amplifier (U_1, figure 51). The *AUDIO GAIN* control in the source circuit of Q_1 allows the drive level of U_1 and U_2 to be set for optimum operation of the compressor

circuit consisting of U_1, U_2, Q_2, and Q_3. The *COMPRESSOR GAIN ADJUST* control varies the amount of compression and the *RECOVERY TIME CONSTANT* adjustment varies the time required for the circuit to return to maximum gain after a large signal is removed from the input. The *COMPRESSION LEVEL ADJUST* is an internal control which is set to give 3 volts rms at pin 6 of device U_2 with a large input signal. The *AUDIO CLIPPING ADJUST* control then allows the clipping to be varied from zero to 20 decibels. Transistors Q_4 and Q_5 are used as reverse-connected diodes to provide clipping and they function much better than ordinary diodes in this circuit. Integrated circuit U_3 and associated components form an active three-pole low-pass filter with a 3-kHz cutoff frequency which removes the higher-frequency components generated by the clipping circuit.

Referring to figure 51, integrated circuit U_1 amplifies the signal from the microphone by 40 dB and the *VOX GAIN* control varies the signal level applied to U_2. The output signal of U2 is rectified and the positive voltage coupled to the base of transistor Q_3, turning it on when an audio signal is generated by the microphone. This causes Q_1 and Q_2 to turn on and the VOX

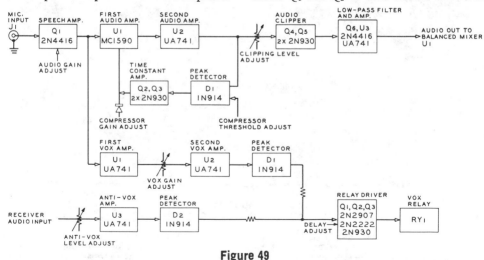

Figure 49

BLOCK DIAGRAM OF AUDIO AND VOX CIRCUITS

Audio clipping and compression are included in the speech amplifier of this versatile exciter. Compression gain and recovery time are adjustable. An audio filter follows the clipper to remove higher order harmonics. Vox gain and delay are adjustable permitting the operating time and hold-in time to be varied at the operator's preference.

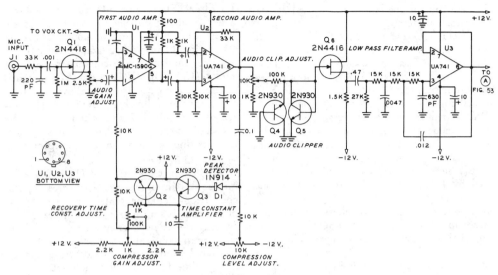

Figure 50

SCHEMATIC, AUDIO CIRCUITRY OF SSB EXCITER

U₁—Motorola MC 1590G
U₂, U₃—Fairchild μA 741
Note: All resistors ¼ watt. All potentiometers audio taper

relay (RY₁) to close. Integrated circuit U₃ amplifies the signal from the receiver output circuit, which is rectified, and the resulting negative voltage also is applied to the base of Q₃. Adjusting the *ANTIVOX* control prevents the speaker output picked up by the microphone from closing the VOX relay. The *DELAY* adjustment allows the

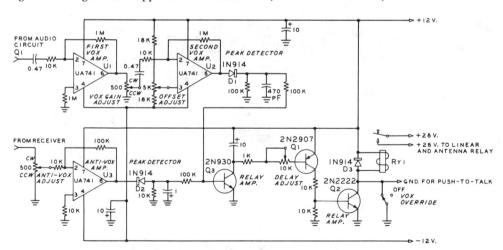

Figure 51

SCHEMATIC, VOX CIRCUITRY

U₁-U₃—Fairchild μA 741
RY₁—Crystal can relay or reed relay. Potter Brumfield JMF 1080-61
Note: All resistors ¼ watt. All potentiometers audio taper

hold-in time of the relay to be varied at the operator's preference.

The RF Circuitry—The block diagram of the rest of the exciter is shown in figure 52, including the phase-locked synthesizer. Schematic diagrams of the r-f circuits are shown in figures 53, 54, and 55. The balanced mixer (U_1 in figure 53) generates a DSB signal from the processed audio (Q_3, Q_4). Diode switches are used to remotely select crystals for upper- or lower-sideband operation. The first 9-MHz crystal filter (FL_1) selects one sideband which is amplified by FET device Q_1 and applied to the r-f clipping circuit. Diodes D_1 and D_2 are inexpensive ultrafast switching diodes and perform almost perfect clipping of the signal. The amount of clipping is adjustable by varying the gain of Q_1 over the range of zero to 20 dB of r-f clipping. The clipped signal is then passed through a second crys-

tal filter (FL_2, figure 54) to remove high-order products outside the passband of the filter. The clipped signal, now restored to its original bandwidth, is amplified by driver Q_3 and applied to the conversion balanced mixer (U_2). The *DRIVE ADJUST* potentiometer in the #2 gate of Q_3 allows the drive level to the following circuits to be adjusted as required. Drive is not adjusted by the audio circuits as is done in conventional exciters due to the various clipping circuits in this design.

The Conversion Mixer—The conversion mixer (U_2 in figure 54) has three inputs: conversion-oscillator injection from the phase-locked synthesizer; a 9-MHz signal from Q_3; or the carrier-insertion signal from the circuit consisting of diodes D_3, D_4, and associated components. The diodes are long-storage-time PIN devices which act as variable resistors (instead of diodes) at this fre-

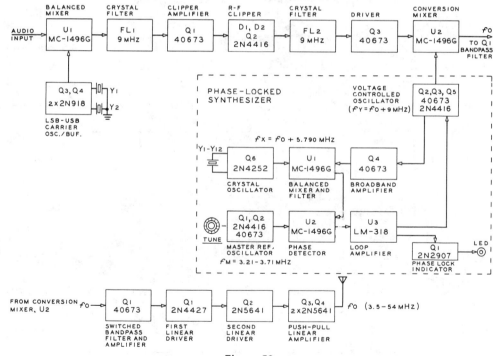

Figure 52

BLOCK DIAGRAM, R-F CIRCUITRY OF SSB EXCITER

The conversion frequency (F_x) is 9 MHz above the signal frequency. The master reference oscillator tunes the range of 3.21 MHz to 3.71 MHz. The SSB signal is passed through a switched bandpass filter (lower left) before being amplified by the three-stage linear amplifier. Operation of phase-locked loop is indicated by light-emitting diode (LED).

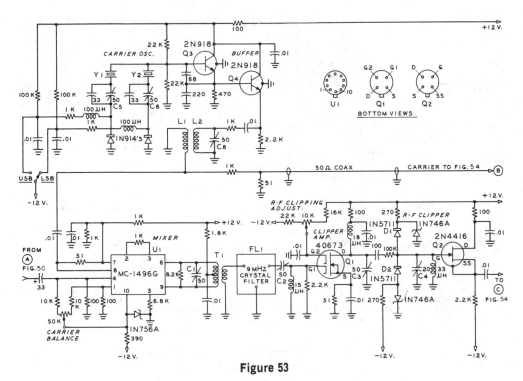

Figure 53

SCHEMATIC, SSB GENERATOR AND R-F CLIPPER

D_1, D_2—Hewlett-Packard HPA 5082-2800 (1N5711)
C_1-C_3—5-50 pF. Johanson 9305
C_4—20 pF, Johanson 9-02
C_5, C_8—5-50 pF, Johanson 9305
L_1, L_2—19 turns #28 on CTC 1536-6-2 form. Link is 5 turns #28 closewound on "cold" end of L_2

FL$_1$—9-MHz filter with 2.4-kHz bandwidth. KVG XF-9A (Spectrum International, Box 1084, Concord, Mass. 01742)
Y_1—8998.5 kHz. KVG XF-901 (see above)
Y_2—9001.5 kHz, KVG XF 902 (see above).
T_1—Primary: 19 turns #30 bifilar wound, secondary: 8 turns #24 e. Wound on .437 × .250 × .187 Carbonyl SF toroid

Note: All Resistors ¼ watt

quency. This allows a variable amount of carrier signal to be inserted by a front-panel control when the *PUSH TO SPOT* switch is depressed. In normal operation of the exciter these diodes are biased open to prevent the carrier from appearing at the output of U_2. Depressing the switch allows the bias to be adjusted by the *CARRIER INSERTION* potentiometer, causing the diodes to act as a variable attenuator, controlling carrier level as desired.

Transformer T_2 at the output of mixer U_2 is a broadband device (balun) which matches the mixer output impedance to the low-impedance coaxial cable interconnection to Q_1 in figure 55. The output of this device contains a double-tuned filter circuit which passes only the desired mixer product

to the high-gain, three-stage linear amplifier. Switch S_1 selects the proper filter for the band in use and may be eliminated if a single-band exciter is desired. In that case, the proper filter is wired directly into the circuit.

The Linear Amplifier—The linear amplifier (figures 56 and 57) consists of two class-A driver stages (Q_1, Q_2) followed by a push-pull class-AB power output stage (Q_3, Q_4). All stages are broadbanded across the 3.5- to 55-MHz range and the power gain is essentially flat to 30 MHz, decreasing to about 6 dB at 50 MHz. The resistive attenuator at the input to Q_1 is necessary to ensure stable operation on all bands. Devices Q_2, Q_3, and Q_4 are vhf power transistors with balanced emitter construction; the

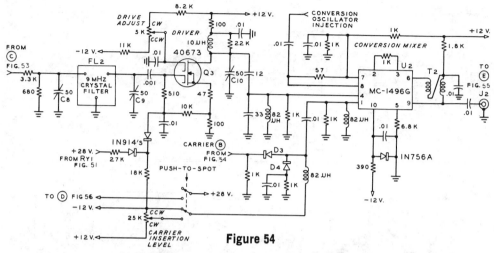

Figure 54

SCHEMATIC, FILTER AND CONVERSION MIXER

D$_3$, D$_4$—Hewlett-Packard HPA 5082-3081
C$_8$-C$_{10}$—5-50 pF. Johanson 9305
T$_2$—9 turns #28 insulated bifilar wound on CF-102 core (Indiana General)
Note: All resistors ¼ watt. See figure 53 for filter data

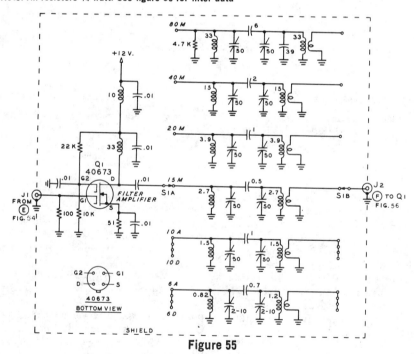

Figure 55

SCHEMATIC, SWITCHED FILTER

J$_1$, J$_2$—Subminiature coaxial receptacle
S$_1$—2-pole, 12-position ceramic switch, 2 decks
Note: All inductance values in microhenries. All inductors are J. W. Miller 9200 series or equivalent
with 2 turn link of #28 insulated wire wound on ground end. All variable capacitos are 5-50 pF
(Johanson 9305) except for 6 meters. All resistors ¼ watt

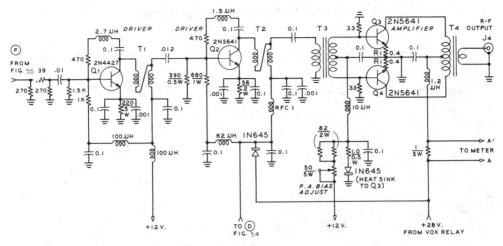

Figure 56

SCHEMATIC, LINEAR AMPLIFIER STAGES

Q₁-Q₄—Motorola transistors (see text)
T₁, T₂—Four-to-one wideband transformer. 8 turns of #28 twisted pair, 8 twists per inch wound on
  CF-102 core (Indiana General)
T₃, T₄—See figure 57 and text
R₁—Three 1.2-ohm, ½-watt carbon resistors in parallel
RFC₁—Ferroxcube VK200-10/3B
Note: All resistors ¼ watt unless otherwise specified. Dual emitter leads of Q₂ bypassed with .001
µF and .1 µF on each lead. All inductance values in microhenrys

Motorola type 2N5641 was found to combine excellent linearity and ruggedness. Other manufacturer's 2N5641s were not tested and unless linearity testing equipment is available, the *Motorola* devices should be used. A 1N645 diode provides temperature compensation for the bias of Q_3 and Q_4 and should be thermally connected to one of these transistors with heatsink thermal compound (*Dow Corning* 340 or equivalent). As mentioned previously, a low-pass filter for the band of operation should follow the linear amplifier to suppress the r-f harmonics of the signal if the amplifier is connected directly to an antenna.

The Power Supply—The circuit of the power supply is shown in figure 58. It utilizes IC power regulators to provide plus and minus regulated 12 volts. Both positive and negative full-wave rectifier circuits are connected to the secondary of transformer T_1. The +28 volts is used to drive the linear power-amplifier circuits directly and is also connected to regulator U_1, which delivers +12 volts at a maximum current of 500 mA. Regulator U_2 is connected to the negative supply and delivers −12 volts at up to 500 mA. The metering circuit allows the power-supply voltages to be measured as well as allowing the operator to monitor the power amplifier supply current.

The Master Reference Oscillator—Shown in figure 59 is the circuit of the master reference oscillator. The circuit is of the *Seiler* type and gives excellent frequency stability. A box made of ¼″ thick aluminum plate is used for the assembly and is mounted on the rear of a *National NPW-0* dial mechanism to achieve the required mechanical rigidity. Drift of the unit shown is less than 50 Hz during the first five minutes of operation at 20°C and less than 10 Hz per hour thereafter at a given temperature. With the temperature compensation shown, frequency change is less than 200 Hz over the range of 0° to 50°C.

A buffer amplifier (Q_2) provides isolation between the oscillator and the load and is partially responsible for the excellent performance of the circuit. Variable capacitor C_3 allows the output level of the unit to

Table 7. Oscillator Components

	VCO VALUES ($f_Y = f_0 + 9$ MHz)					XTAL OSC. VALUES					
BAND (MHz)	f_0	R_4	L_1	C_5 (PF)	C_7 (PF)	C_9 (PF)	C_{10} (PF)	C_{11} (PF)	L_2 (µH)	Y_1 (MHz)
3.5 – 4.0	80 M	4.3 K	19 T. N°28	—	120	10	27	270	6.8	9.290
7.0 – 7.5	40	4.3 K	15 T. N°28	1 – 15	82	10	33	250	2.7	12.790
14.0 – 14.5	20	4.3 K	11 T. N°28	1 – 15	47	10	200	18	1.5	19.790
21.0 – 21.5	15	4.3 K	8 T. N°24	1 – 15	33	22	62	—	1.2	26.790
28.0 – 28.5	10 A	6.2 K	6 T. N°22	—	30	18	62	—	0.82	33.790
28.5 – 29.0	10 B	—	6 T. N°22	—	30	18	62	—	0.82	34.290
29.0 – 29.5	10 C	—	6 T. N°22	—	30	18	62	—	0.82	34.790
29.5 – 30.0	10 D	—	6 T. N°22	—	30	18	62	—	0.82	35.290
50.0 – 50.5	6 A	—	2.5 T. N°18	—	17	12	62	—	0.33	55.790
50.5 – 51.0	6 B	—	2.5 T. N°18	—	17	12	62	—	0.33	56.290
51.0 – 51.5	6 C	—	2.5 T. N°18	—	17	12	62	—	0.33	56.790
51.5 – 52.0	6 D	—	2.5 T. N°18	—	17	12	62	—	0.33	57.290

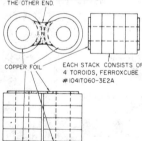

FOIL EXTENDS FROM END TO END INSIDE THE STACK

COPPER FOIL

EACH STACK CONSISTS OF 2 TOROIDS, FERROXCUBE #266T125-3E2A

OUTPUT TRANSFORMER T4

SOLDER TOGETHER ON ONE END, INSULATE FROM EACH OTHER ON THE OTHER END.

COPPER FOIL

EACH STACK CONSISTS OF 4 TOROIDS, FERROXCUBE #104IT060-3E2A

INPUT TRANSFORMER T3

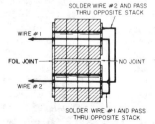

SOLDER WIRE #2 AND PASS THRU OPPOSITE STACK

WIRE #1

FOIL JOINT — NO JOINT

WIRE #2

SOLDER WIRE #1 AND PASS THRU OPPOSITE STACK

TYPICAL TRANSFORMER WINDING (CROSS SECTION THRU TRANSFORMER)

Figure 57

CORE STACK FOR WIDEBAND R-F TRANSFORMERS

Transformers T_3 and T_4 in the linear amplifier are wideband devices made up of stacks of ferrite cores. The stacks are held together by a cylinder of copper foil with adhesive on one side (available Newark Electronics part 38F1301 or 38F1222). Roll the foil around a drill shank of proper size, adhesive side out, to form a cylinder. Slide the toroids on the cylinder. Remove the drill and cut the foil so it is 1/8″ longer than the stack of toroids on each end. Make 4 to 6 slits in the extended foil and flare out flat against the toroids. Fill in the gaps with small pieces of foil tape and carefully solder in place. Trim even with the edge of the core. Place two stacks side by side and tape together with paper tape. Solder the foil on the end of one stack to the foil on the end of the other stack. This junction forms the center tap of one winding. Solder a short piece of #24 insulated wire to the foil on the other end of each stack and pass the two wires through the adjacent toroid stack. This completes one turn on either side of the center tap. Wind on the remaining turns of the center-tap winding. Finally, wind on the second winding so that the ends of the winding extend from the opposite end of the assembly from the center-tap connection. (Ferroxcube cores available from: Ferroxcube Corporation, 5635 Yale Blvd., Dallas, Texas). See Lowe, QST, December, 1971 for additional transformer data.

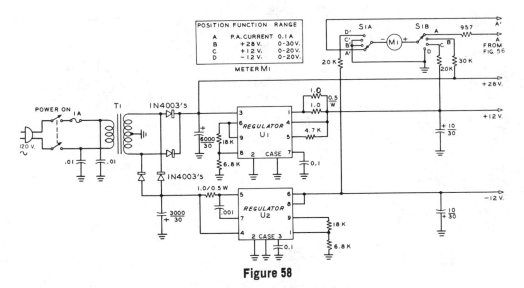

Figure 58

SCHEMATIC, POWER SUPPLY

T_1—Stancor TP-4. Use green, yellow, and red secondary leads
U_1—MC 1461 R
U_2—MC 1463 R
M_1—0-1 mA dc (Simpson or Weston)

be adjusted to 100 millivolts, rms, to drive the following circuit.

The Phase Lock Synthesizer—Shown in figures 60 and 61 are the schematic diagrams of the phase-lock synthesizer. Component values for the oscillators are given in Table 7. Referring to figure 60, integrated circuit U_1 is a balanced mixer with two inputs: one from the crystal oscillator (figure 62) and the other from the voltage-controlled oscillator (VCO) and buffer amplifier shown in figure 61. The crystal frequency is chosen to be below the voltage-controlled oscillator (VCO) frequency so that the difference frequency between the two oscillators falls in the range of 3.21 to 3.71 MHz, determined by the exact frequency the VCO is tuned to. The output of mixer U_1 is filtered by a three-pole bandpass filter (figure 60) which removes unwanted mixer products before they reach the input of the phase detector U_2. The phase detector compares the phase of the signal (and consequently the frequency) to the phase of the master reference oscillator, shown in figure 59, and generates an output signal proportional to the phase difference between the two input signals. This reference signal is dc coupled to the input of the loop amplifier (U_3, figure 60) after passing through the loop filter (R_1, C_1). This filter shapes the gain-frequency response of the loop and

Figure 59

SCHEMATIC, MASTER REFERENCE OSCILLATOR AND BUFFER

C_1—82-pF silver mica with 54-pF, N220 capacitor in parallel

C_2—6 to 78 pF Polar C341- 20/016 (Jackson Bros.)

C_3—5 to 50 pF, Johanson 9305

L_1—51 turns #28 e. on CTC 3354-6 coil form

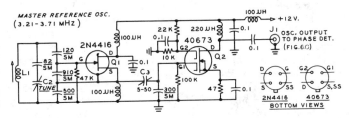

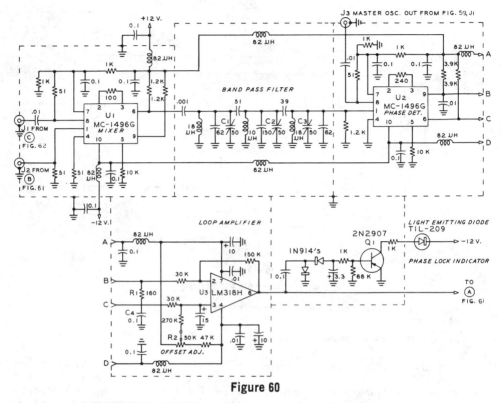

Figure 60

**SCHEMATIC, MIXER, BANDPASS FILTER, PHASE DETECTOR,
LOOP FILTER, AND LOCK INDICATOR**

C_1-C_3—5 to 50 pF. Johanson 9035
Note: All inductance values in microhenries. All inductors J. W. Miller 9200 series. All resistors 1/4 watt

is very important for proper operation of the synthesizer. The values are chosen so that the loop has a 100-kHz pull-in range; that is, if the frequency difference between the master oscillator and the output of U_1 is less than 100 kHz, the loop will lock-up and remain locked. Thus, the VCO will have the same stability as the master reference oscillator.

The output of U_3 is connected to the varicap diode (D_2) in the VCO circuit (figure 61) and also to the lock indicator circuit (Q_1, figure 60). When the loop is locked, only a dc voltage is present at the output of U_3 and Q_1 is turned off, preventing current from flowing through the light-emitting diode (LED) placed on the exciter panel above the main tuning dial. Should the loop become unlocked, however, a large ac

voltage is developed at the output of U_3, which is rectified by the diodes, thus turning on Q_1. This causes the LED to light, signaling the loop is unlocked. On-the-air operation of the exciter should never be attempted if the loop is unlocked because in this condition the VCO output consists of many frequencies instead of one.

The Voltage-Controlled Oscillator (figure 61)—Another *Seiler* circuit similar to the one used for the master reference oscillator is used as a voltage-controlled oscillator. Two varicap diodes are used to tune the frequency; the first (D_1) is driven from a potentiometer (*coarse tune*) which is mechanically coupled to the dial shaft of the master reference oscillator. This coupling causes the frequency of the VCO to be approximately tuned to the desired frequency

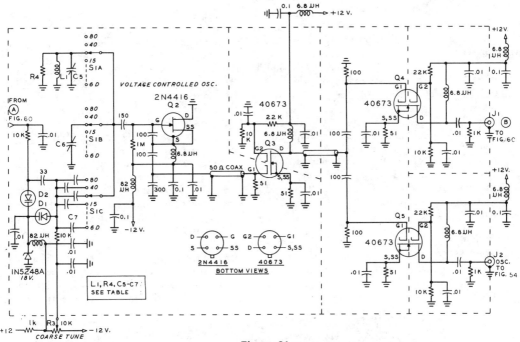

Figure 61

SCHEMATIC, VOLTAGE-CONTROLLED OSCILLATOR AND BUFFERS

D_1, D_2—1N5148A
S,—6-pole, 12-position switch
Note: For coil and capacitor data, See Table 7. All resistors ¼ watt. All inductance values in microhenrys

selected by the reference oscillator. The second diode (D_2) driven by the loop amplifier, readjusts the frequency slightly so that the loop will lock-up.

Component values for the frequency determining circuit of the VCO (Table 7)

are selected to allow the circuit to tune the proper frequency range for the bands shown. Other bands may be covered after considering the mixer products. Devices Q_3, Q_4, and Q_5 (figure 61) are broadband amplifiers which isolate the VCO from the

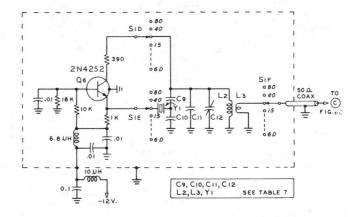

Figure 62

SCHEMATIC CRYSTAL OSCILLATOR

Note: For coil and crystal data, see Table 7. All resistors ¼ watt loads and the loads from each other. The output of Q_4 is used to drive the phase-locked loop and the output of Q_5 drives the conversion mixer (U_2, in figure 54).

The Crystal Oscillator (figure 62)—The crystal oscillator consists of a grounded-base Colpitts circuit with the crystal in the feedback path. These crystals have a series-resonant frequency as listed in Table 7. Coil L_2 is a subminiature choke about the size of a ¼-watt resistor. Link L_3 consists of 1½ turns of #28 insulated wire wound on the ground end of L_2. The output of this oscillator connects to U_1 in figure 60 to complete the phase-lock synthesizer circuit.

Exciter Construction The exciter is built in several modules which are mounted on an aluminum chassis measuring 9″ × 14″ × 2″. The unit is housed within a *Bud Shadow Cabinet* (SB-2142), as shown

in figure 48. Placement of the modules is shown in the rear-view photograph (figure 63). The chassis is mounted to the panel with two end brackets. A small gap is left at the rear of the chassis to aid cooling and the rear panel of the cabinet is replaced with a sheet of perforated aluminum. The linear-amplifier module is built on a finned heat sink with all but the end fins removed on one side. It is mounted in a vertical position at the rear of the chassis with the remaining fins projecting beyond the chassis into the gap between chassis and cabinet.

Type BNC connectors and miniature co-axial cable are used to interconnect the various modules. Immediately in front of the linear-amplifier module are the switched-filter module (figure 55) and the synthesizer module (figures 59 through 62). Adjacent to the linear amplifier is the driver module (part of figure 54) which includes the *P-A BIAS ADJUST* potentiometer.

Figure 63

TOP VIEW OF SOLID-STATE SSB EXCITER

Exciter is built in modules that may be tested and aligned one at a time. In the upper left corner is the reduction dial drive and the rugged aluminum box for the master reference oscillator. At center of the chassis are the power transformer and filter capacitors. Directly behind the oscillator is the synthesizer assembly containing the circuits of figures 59, 60, and 61. The bandswitch passes out the end of the module and is ganged with the bandswitch of the switched filter (figure 45). Across the rear of the chassis (foreground of photo) are the driver stages and bias potentiometer (left) and the push-pull linear amplifier (right). Power plug, fuse, and antenna reseptacle are on bracket at right. Microphone and VOX inputs are at lower left.

These modules are tested individually then bolted together and mounted as one unit to the rear of the main chassis.

The aluminum box containing the master reference oscillator is behind the *National* dial drive assembly, with the power supply centered on the chassis.

The enclosures for the switched filter and the synthesizer are built from rectangles cut from double sided 0.06″ *fiberglas* p.c. board material and are soldered together. The synthesizer box measures 7″ × 4″ ×

Figure 65

INTERIOR OF SWITCHED FILTER

Filter schematic is shown in figure 55. Filter components are mounted to printed-circuit board placed between the switch decks. Input and output coaxial receptacles are on the ends of the box.

Figure 64

OBLIQUE VIEW OF SYNTHESIZER MODULE

Module is built of double-sided fiberglas printed-circuit board cut into rectangles and soldered together. The bandswitch passes through the two compartments of the assembly. At the left is the crystal oscillator and crystals with the bandpass filter (figure 59) at the upper right. Across the top of the assembly are compartments containing (left to right): mixer, bandpass filter, phase detector, loop filter, and buffer stages (Q_4, Q_5, figure 60). Ten, 15, and 20 meter oscillator coils are at the right.

2″ and the filter box measures 3″ square. A view of the interior of the synthesizer is shown in figure 64. The enclosures are assembled in a similar manner. Threaded brass spacers ¼″ long are soldered in the corners to add strength to the box and to secure the covers. The pieces of circuit board used for the center dividers in the synthesizer box should have the copper soldered together along the exposed edge to provide the best grounds. This was done by wrapping a narrow strip of .001″ copper shim stock over the edge of the dividers and soldering on both sides. A good fit between the box panels is obtained by sawing the parts to a slightly large size and then filing the pieces to exact size. After the boxes are soldered together the exposed edges where the covers fit are

ground flat with a piece of fine emery cloth placed on a flat surface. This results in a neat enclosure which is strong and compact.

The switches for both units are assembled from *Centralab* PA-1 ceramic decks with a PA-302 index assembly used in the synthesizer and a PA-301 assembly used in the switched filter. An interior view of the switched filter is shown in figure 65.

The remainder of the exciter circuitry is mounted on two pieces of circuit board, each measuring 4″ × 10″, mounted below the chassis on ¼″ spacers (figure 66). The board nearest the front panel contains the audio processing circuits, VOX, and anti-VOX circuitry. The rear board contains the r-f circuitry including the two crystal filters, r-f clipper, and conversion mixer. Input and output terminations are made with BNC coaxial fittings, and each board is tested and aligned before it is placed in the chassis. Small standoff terminals are soldered directly to the copper foil to provide tie points (since no holes are drilled). This is a very fast and convenient method to build the circuits and provides a good ground plane since all grounds may be soldered directly to the copper. Circuit changes or modifications can be done easily and quickly, should the need arise.

Miniature components are used throughout the exciter. The resistors are ¼-watt carbon units, the inductors are approximately the same size (*J W Miller* 9200 series or equivalent), and the bypass capacitors are miniature ceramic units. The small capacitors are *El Menco* DM-5 type mica units. The power-supply components except for the IC regulators are mounted on a vertical p.c. board between the power transformer and the master reference oscillator. A right-angle drive is used to drive the bandswitch from the front panel. When wiring the switches remember that one switch rotates in a direction opposite that of the other when viewed from the front of the switch. The IC regulators are mounted on either end of the chassis to distribute the power dissipation.

Exciter Adjustment   Exciter tuneup is not complicated if all modules have been pretested before installation on the chassis. An electronic voltmeter with an r-f probe is required. as well as an audio generator and an oscilloscope. A frequency counter is desirable, but not mandatory.

After checking the units and the power-supply voltage, connect a 5-watt, 50-ohm load to the output terminal. Connect the audio generator through a variable attenuator to the microphone input receptacle and adjust the *COMPRESSION LEVEL ADJUST* control to provide 3 volts rms at pin 6 of U_2 (figure 50) when the *AUDIO GAIN* and *COMPRESSOR GAIN ADJUST* controls are at mid-setting. The output at pin 6 should remain constant over a signal input range of 40 decibels from the threshold point to the point where waveform distortion becomes visible on the oscilloscope. The *AUDIO GAIN* control can be used to adjust the input level to the compressor to compensate for different microphones and the *COMPRESSOR GAIN* adjustment used to determine how much, if any, compression is used.

To adjust the VOX controls, set the *ANTIVOX* and *VOX GAIN* controls at minimum, turn down the receiver output and turn up the *VOX GAIN* until the VOX relay closes when speaking into the microphone in a normal manner. Now, turn up the speaker to normal output and adjust the

Figure 66

UNDER-CHASSIS VIEW OF SSB EXCITER

The two circuit boards are mounted beneath the aluminum chassis. The board at the rear of the chassis contains the r-f circuitry and the two i-f crystal filters. The board adjacent to the front panel contains the audio circuitry. At the left is the microphone-input coaxial receptacle and at the right is the audio-output coaxal fitting. The r-f output to the switched filter atop the chassis is at the left of the rear circuit board.

Figure 67

SSB EXCITER CHASSIS AND INTERIOR VIEW OF VFO

The synthesizer module has been removed for this photograph to show the linear amplifiers mounted across the rear of the chassis. The VOX reed relay is directly behind the power-supply filter capacitors. The lid of the vfo module has been removed to show the tuning capacitor and the 1:1 gear drive to the "coarse-tune" potentiometer.

ANTIVOX control until the relay does not close on loud signals. The *VOX DELAY* control can now be adjusted for proper hold-in time, as desired.

As a final check, connect the oscilloscope to the source of Q_6 (figure 50). No clipping of the waveform should be observed when the *AUDIO CLIP* control is at minimum and clean clipping of the waveform should be visible at maximum clipping setting.

RF Alignment—To align the r-f circuits first adjust capacitors C_5 and C_6 (figure 53) to midrange and peak capacitor C_8 in the emitter circuit of the buffer stage (Q_1) for a maximum reading on the electronic voltmeter with the r-f probe connected to the top of coil L_1. Indicated voltage should be about 100 millivolts rms and may be adjusted, if necessary, by changing the value of the 1K resistor connected to L_2 and C_8.

Now, apply 300 millivolts rms of 1-kHz audio signal to pin 4 of U_1 (figure 53) and peak the r-f output at the source of Q_2, the

buffer FET, by tuning capacitors C_1, C_2, C_3, and C_4 in the first crystal filter stage, the clipper amplifier, and the buffer stage. Set the audio frequency to 2.7 kHz and adjust capacitors C_5 (or C_6, depending upon the sideband selected) in the oscillator stage for maximum response. Continue tuning the capacitor until the output decreases 3 to 6 decibels. Repeat this procedure with the other capacitor for the opposite sideband. Next, vary the frequency of the audio generator from 300 to 3000 Hz and note the ripple in the filter passband and the upper frequency at which the output has fallen off by 6 decibels. The ripple should be less than plus or minus one decibel across the band. If it is greater than this, adjust capacitors C_1 and C_2 slightly. In an extreme case, it may be necessary to alter the number of secondary turns of transformer T_1.

Next, disconnect the +28-volt line from the VOX relay to the linear amplifier and turn on the VOX OVERRIDE switch to remove the cutoff bias applied to Q_3 in the standby mode. Connect the r-f probe to pin 4 of U_2 (figure 54) and adjust capacitors C_8, C_9, and C_{10} (filter FL_2 and the driver transistor) for maximum response with the *DRIVE ADJUST* potentiometer at mid-setting. Again, check the passband ripple and realign capacitors C_8 or C_9, if necessary.

Synthesizer Alignment—Apply +12 volts to the master reference oscillator (figure 67) and adjust coil L_1 and capacitor C_1 so the oscillator tunes the range of 3.185 to 3.735 MHz. Adjust the potentiometer coupled to the shaft of capacitor C_2 so that it is at the clockwise end of its rotation when the oscillator is tuned to 3.735 MHz. Rotate the shaft of the potentiometer back about 10 degrees before locking in place to eliminate the nonlinear portion of rotation next to the stop. Adjust capacitor C_3 to provide 100 millivolts rms output when the oscillator is connected to a 50-ohm load through the subminiature 50-ohm coaxial line and connectors. Apply —12 volts to the crystal oscillator (figure 62) and tune capacitor C_{12} for maximum output on each band. Adjust the coupling between coil L_2 and link L_3 by sliding L_3 up or down the form until the oscillator output is about 100 millivolts on each band, using a 50-ohm load.

Disconnect pin 6 of U_3 (figure 60) from the 10K resistor and varicap diodes (figure 61) and ground the open end of the resistor. (Do *not* ground pin 6 of U_3). Place all padding capacitors in the VCO (figure 61) to midrange and tune coil L_1 until the output frequency (as measured at the drain terminal of FET Q_1) is nearly correct for each band, starting with 80 meters and working up in frequency. (Remember the frequency you are measuring is 9 MHz higher than the desired band).

With the electronic voltmeter connected to pin 6 of U_3 and power disconnected from the crystal oscillator, adjust the *OFFSET ADJUST* potentiometer (figure 60, loop amplifier U_3) for a reading of zero volts, dc. Turn off the power and reconnect the 10K resistor to pin 6 of U_3.

Next, tune the master reference oscillator to 3.185 MHz and set the bandswitch to *80 meters*. Connect the electronic voltmeter and oscilloscope to pin 6 of U_3, being careful not to short this point to ground. Turn on power and adjust coil L_1 (80 meters) for zero volts dc at pin 6. The phase-lock indicator should be off and the oscilloscope should indicate no ac voltage present. Repeat the tuning of L_1 for each band in sequence, leaving the oscillator at 3.185 MHz. Now set the master reference oscillator to 3.735 MHz and adjust capacitor C_5 or C_6 for the appropriate band for a zero volt dc reading. Again, start with the 80-meter band and work up in frequency. It probably will be necessary to repeat the procedure twice to get all bands properly tuned. As a final check, tune completely across each band to make sure the loop does not become unlocked at any frequency. If the loop unlocks, the voltage at pin 6 will rise, possibly as high as 10 volts. If this happens, readjust the oscillator capacitor and inductor slightly for a different L/C ratio. For conditions of lock, the voltage at any point in the band should remain between zero and 6 volts.

If a frequency counter is available, the above procedure can be speeded up. Break the line that connects one end of the *coarse tune* potentiometer to the 1N5248A diode and insert a switch in the line. Tune the master reference oscillator to 3.185 MHz with the switch closed and adjust coil L_1. Open the switch and the VCO will be tuned

Figure 68

LINEAR AMPLIFIER STAGES

The push-pull amplifier is on the left chassis, with the driver stages on the right chassis. The chassis have been removed from the main chassis and tilted forward for this picture. The "bias adjust" potentiometer is mounted to the plate connecting the two assemblies together.

to the high end of the band, even though the master oscillator is still tuned to the low-frequency end. Now capacitors C_5 and C_6 may be aligned as indicated by the counter which is connected to pin 1 of the mixer (U_1, figure 60). Remember that the counter reads a frequency that is 9 MHz above the desired band.

Switched Filter Alignment—A 50-ohm load is connected to the linear amplifier and power is applied to all stages. The *DRIVE ADJUST* potentiometer is set for minimum drive and the *VOX OVERRIDE* switch is turned on. The idling current to the power output stage (as read on the panel meter) is set to 20 mA by adjusting the *PA BIAS ADJUST* control. Connect the audio generator and inject a 1-kHz tone into the exciter, advancing the *DRIVE ADJUST* control (figure 54) to mid-setting. Set the master reference oscillator to a midband frequency and tune the capacitors in each filter section (starting with 80 meters) for a peak current reading on the meter, adjusting the drive control as necessary so as not to exceed 400 mA. The higher bands will require more drive than the lower bands, and output on the 50-MHz band is drive-limited. With 400 mA indicated current power output will be in excess of 5 watts. If a two-tone source is available, maximum current drain should be limited to 300 mA for 5 watts PEP output.

20-4 A Very Low Noise Preamplifier for 144 MHz

The preamplifier shown in this section was designed by W2AZL and built by W6PO for 144-MHz moonbounce communication. The unit is easy to build and get working and provides a noise figure of 1.5 dB with an unselected JFET device. It is designed for placement at the antenna, or for use directly at the station receiver. The preamplifier is self-contained except for an external 12 Vdc power source (figure 69).

The Preamplifier Circuit The circuit of the device is shown in figure 70. A *Siliconix* U-310 or 2N5397 is used in a common-gate circuit. The preamplifier provides about 10 dB gain and does not require neutralization. Gate bias is provided by a resistor in the source cir-

Figure 69

LOW NOISE PREAMPLIFIER FOR 2 METERS

On the left is the low noise preamplifier built in a cast aluminum box. Input and output connectors are located on the printed-circuit chassis board. The hole in the side of the box is to allow adjustment of capacitor C_1. A similar hole on the opposite side of the box is for adjustment of capacitor C_4. On the right is the power distribution unit used when the preamplifier is located at the antenna site.

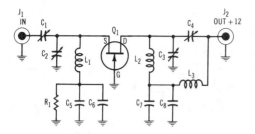

Figure 70

SCHEMATIC OF 2-METER LOW-NOISE PREAMPLIFIER

C_1-C_4—0.5 to 10 pF. (Johanson)
C_5, C_7—500 pF disc capacitor
C_6, C_8—.01 μF disc capacitor
Box—Pomona 2417
L_1, L_2—5 turns #18 e., 1/4" inside diam.
  Space diameter of wire
L_3—1 μH r-f choke
Q_1—Siliconix U-310 or 2N5397
R_1—Chosen for drain current of 10 mA. About 180
  ohms
J_1, J_2—BNC receptacles

cuit to ground and both source and drain return circuits are bypassed both for vhf as well as high frequency paths. Input and output impedances are 50 ohms.

If it is desired to operate the preamplifier remotely at the antenna site, the circuit of figure 83 is used in conjunction with the preamplifier to allow the +12 volts to be fed to the unit using the center conductor of the coaxial cable connecting the preamplifier to the converter.

Preamplifier   The preamplifier is designed
Construction   to fit within a cast aluminum
                     box (figure 69). A double
sided printed-circuit board is used for the chassis, as shown in figure 72. Components are mounted directly to the board and a small shield (figure 73) is placed over the JFET socket to provide isolation between input and output circuits.

The JFET is mounted in a very small transistor socket (figure 74) and all leads made as short as possible between the various components. If desired, the output adjustment capacitor C_4 may be made a fixed 2 pF silver mica unit, thus reducing the cost of the preamplifier without detracting from the gain or noise figure. Wiring should be done with a small soldering iron with a pointed tip.

Layout for the power distribution box chassis is shown in figures 74B and 75. Again, very short interconnecting leads should be used in this unit.

Preamplifier   Once the preamplifier is com-
Alignment   pleted and checked it should
                     be connected to a suitable
receiver and a noise generator. The out-

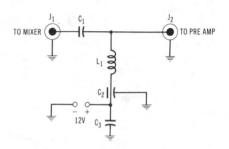

Figure 71

POWER DISTRIBUTION CIRCUIT

C_1—500 pF disc capacitor
C_2—500 pF feedthrough capacitor
C_3—.01 disc capacitor
J_1, J_2—BNC receptacles
L_1—1 μH rf choke
Box—Pomona 2417

put circuit is adjusted for maximum gain and the input circuit adjusted for best noise figure. These adjustments should be undertaken only after the preamplifier is placed within the aluminum box.

Certain precautions should be taken with this preamplifier to protect the JFET from the transmitted signal. A typical JFET will experience degraded noise figure or destruction when subjected to an r-f signal level exceeding 100 mW (+20 dBm). A high-power transmitter may have a power output of 800 watts (+69 dBm). Thus at least 49 dB of isolation is required between the receiving antenna and the preamplifier when in the transmitting mode.

One solution to this problem is to use an additional antenna relay on the preamplifier which will switch its input circuit to a 50-

Figure 72

INTERIOR VIEW OF PREAMPLIFIER

On the left is the preamplifier showing placement of the tuned circuits and the interstage shield mounted across the JFET socket. On the right is the power distribution unit.

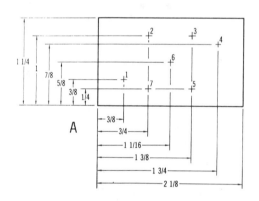

A

#1 & 4-3/8" DRILL-BNC CONNECTORS
#2 & 5-#1 DRILL-JOHANSON CAPACITORS
#3 & 7-#27 DRILL-TEFLON STANDOFF
#6-HOLE FOR SOCKET

MATERIAL-1/16 DOUBLE SIDED P.C. BOARD

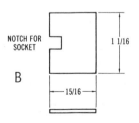

NOTCH FOR SOCKET

B

MATERIAL-1/16 DOUBLE SIDED P.C. BOARD

Figure 73

DRILLING DATA FOR PREAMPLIFIER CHASSIS AND SHIELD

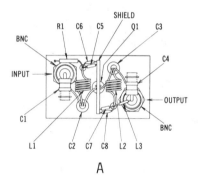

A

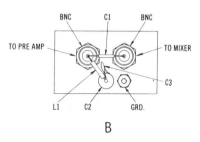

B

Figure 74

COMPONENT LAYOUT

A—Preamplifier parts placement
B—Power distribution deck parts placement

ohm load resistor during transmit. The preamplifier input should not be short- or open-circuited while transmitting, since this may encourage oscillation in the preamplifier.

Since no mechanical relay switches instantaneously, there can exist an instant

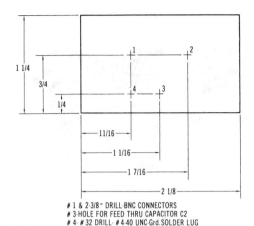

1 & 2-3/8" DRILL-BNC CONNECTORS
3-HOLE FOR FEED THRU CAPACITOR C2
4- # 32 DRILL- # 4-40 UNC-Grd.SOLDER LUG

MATERIAL-1/16 DOUBLE SIDED P.C'BOARD

Figure 75

**DRILLING DATA FOR POWER
DISTRIBUTION UNIT**

when transmitter power may be present on the line before the preamplifier is adequately isolated from the antenna. An effective way to handle this problem is to use a delay relay in the final power amplifier, preferably in the high-voltage line. This will allow antenna and preamplifier relays to switch before transmitter power is applied to the antenna. In addition to protecting the preamplifier, this will lessen the chances of burning the contacts of the antenna relay.

20-5 GaAsFET Low-Noise Preamplifiers for 144, 220 and 432 MHz

The family of *gallium arsenide field-effect transistors* (GaAsFET) provides a noise figure range and performance that is

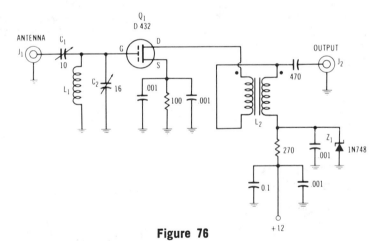

Figure 76

GaAsFET PREAMPLIFIER FOR 144-220-432 MHz

C$_1$—10 pF miniature variable air capacitor. Johanson 5751 or equivalent.
C$_2$—16 pF as above, Johanson 5451 or equivalent.
L$_1$—(144 MHz): 6 turns #14 wire, 1/4" (0.64 cm) inside diam., 1/2" (1.27 cm) long.
    (220 MHz): 4 turns #14, as above.
    (432 MHz): Copper strap 2 1/4" (5.72 cm) long by 0.6" (1.52 cm) wide spaced 0.171" (0.43 cm) above ground plane.
L$_2$—(144 MHz): 12 turns of twisted pair of #24 enameled wire on 0.375" diam. iron-powder toroid, permeability=1. Micrometals T37-0 or equivalent. Connected as a 4-to-1 transformer.
    (220 MHz): 14 turns, as above on 0.307" diam. iron-powder toroid, permeability=1. Micrometals T30-0 or equivalent.
    (432 MHz): 5 turns of twisted pair of #30 enamel wire on 0.250" diam. iron-powder toroid, permeability=1. Micrometals T25-0 or equivalent.
Q$_1$—D-432 GaAsFET. Order from Device Group, Dexcel, Inc., 2285 C Martin Ave., Santa Clara, CA 95050.
Feedthrough capacitors—1000 pF "solder-in." Stettner-Thrush BDZK-5-48 or equivalent.
J$_1$, J$_2$—(144 and 220 MHz): BNC connectors. (432 MHz): SMA connectors. All resistors 1/4 watt.

Figure 77

LAYOUT OF THE 144 and 220 MHz PREAMPLIFIER

The copper plated circuit board is bolted to the lid of the box by a single bolt, which also serves as a ground connection on the exterior of the assembly. Coaxial receptacles also hold the board to the lid. Capacitor C_2 is adjustable from the outside of the box and C_1 may be adjusted through a hole drilled in the end of the box. Note that ferrite transformer L_2 and input coil L_1 are mounted at right angles to each other.

better than other available devices in the vhf range. Shown in this section are three GaAsFET preamplifiers for 144, 220, and 432 MHz. Circuits of the amplifiers are identical and the units are built up in die-cast aluminum enclosures. The noise figure is 0.5 dB, or lower, for all three amplifiers. The amplifiers were designed and built by W6PO.

The Preamplifier Circuit The general circuit for all three amplifiers is shown in figure 76. A simple network matches the nominal 50-ohm input

Figure 78

LAYOUT OF THE 432 MHz PREAMPLIFIER

Stripline circuit L_1 occupies most of the circuit board area. It is soldered to the circuit board at one end and is supported to capacitor C_2 at the opposite end. Capacitor C_1 is adjusted through a hole drilled in the side of the box. All components are grouped closely around the transistor. Output transformer L_2 is self-supported by the leads.

impedance to the gate of the transistor. The unloaded Q of this circuit must be very high to ensure a low noise figure. As an example, at 432 MHz using a coil and capacitor circuit in the gate, the noise figure would not go below 0.7 dB. With the high-Q stripline, the noise figure dropped to 0.47 dB.

The GaAsFET drain impedance is in the range of 100 to 200 ohms and a 4-to-1 ferrite transformer matches this value to a nominal 50-ohm output termination.

Gain of the preamplifier is 18 to 20 dB on 432 MHz and 20 to 24 dB on 144 and 220 MHz. All amplifiers are stable in or out of the shielded enclosure. Stability was checked with a spectrum analyzer, noise-figure meter, and on-the-air checks.

Preamplifier Construction   Each amplifier is built in a die-cast enclosure measuring 3.63″ × 1.5″ × 1.06″ (*BUD CU-123*, or equivalent). Layout of the 144 and 220 MHz amplifiers is shown in figure

77. A double-sided printed-circuit board is used as a ground plane and is held in place by the BNC receptacles and the 4-40 grounding screw which projects through the lid of the box. No input-to-output shield is required in this design.

All components including the feed-through insulators for the power leads are soldered to one side of the board. The sink leads of the transistor are attached to two feedthrough capacitors soldered to the board. The leads of the capacitors are trimmed short on the opposite side of the board with the tips projecting through matching holes drilled in the lid of the box. The 100-ohm sink resistor is soldered between one feedthrough and the chassis ground plane.

The ferrite output transformer is self-supported between the 1000 pF bypass capacitor and the output receptacle. One lead of the transformer is soldered to the transistor. The gate lead of the device is attached to capacitor C_2, at which point the end of coil L_1 terminates. Capacitor C_1 is connected between the self-supported end of L_1 and the input receptacle. Very short interconnecting leads should be used in this unit.

Layout of the 432 MHz amplifier is shown in figure 78. The copper strap used for inductor L_1 occupies most of the ground plane area. It is soldered to the plane at the far end with the input receptacle and associated capacitors grouped closely around the gate terminal of the transistor. The miniature output circuit is to one side of the assembly.

Preamplifier Alignment Good alignment and a near-perfect noise figure can be accomplished by ear. The input circuit of the receiver following the pre-amplifier is peaked for maximum noise from the preamplifier and the input capacitors of the preamplifier are peaked for best sounding signal-to-noise ratio on a very weak signal. This simple procedure has been compared directly to alignment with a noise generator and the resulting noise figure alignment in each case was identical. The amplifier is completely stable with no signs of unwanted instability or oscillation.

20-6 A Tuneable 2-Meter Receiver With Digital Readout

A modern tuneable two meter f-m receiver has to have excellent dial readout,

Figure 79

SOLID-STATE 2-METER RECEIVER WITH DIGITAL READOUT

This sensitive 2-meter receiver is designed for f-m service in today's crowded band. It covers the range of 146.0 MHz to 148.0 MHz in four bands. Dial readout is to one kilohertz and electronic tuning system provides 50 kHz per turn. Receiver measures 7¼" long by 6" deep by 2¾" high. At left is the band switch, with the primary power switch directly beneath it. Volume and squelch controls are at right. In the center is the tuning control directly beneath the frequency readout window. Press-on decals are used for panel labeling.

crystal filter selectivity, and a high order of freedom fom overload and intermodulation distortion. Only a few years ago these characteristics were difficult to obtain in a high-frequency receiver, much less one designed for 144-MHz service.

The receiver shown in this section was designed and built by K2BLA for use in today's crowded 2-meter band. It is compact, easy to build and get working and is not complex (figure 79). The receiver works from a 12-volt supply and is intended for either fixed or mobile operation.

The Receiver Circuit   A popular receiver design for the vhf enthusiast is the combination of a crystal-controlled converter feeding a tuneable i-f system (usually the station hf receiver). This concept is satisfactory for SSB or c-w but leaves much to be desired for f-m service. However, this concept can be adapted to f-m operation as shown in this com-

pact receiver. The design features a 144-MHz converter working into a packaged f-m i-f system. A dial accuracy of one kHz is achieved, along with 0.3 μV sensitivity, good f-m limiting and excellent audio response for voice communication. Dual i-f filters are used and the design is specially tailored for amateur construction. It is possible to build and align this receiver with equipment no more sophisticated than a signal generator and a high-impedance voltmeter with an r-f probe.

A block diagram of the receiver is shown in figure 80. A crystal-controlled converter is built up of modules A and C. The r-f amplifier is a common-gate connected JFET (Q_1) having about 12 dB of gain. The transistor used is a J-308. As a substitute, the 2N4416 may be substituted without circuit change but with a slight decrease in performance. The first mixer (Q_2) is a dual-gate MOSFET, type 40673, which provides about 12 dB of conversion gain.

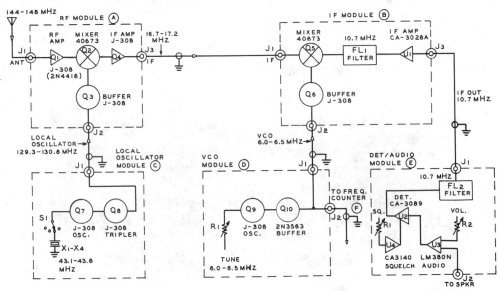

Figure 80

BLOCK DIAGRAM OF 2-METER RECEIVER

The receiver is built and described in modules. The r-f module (A) and local oscillator module (C) form a converter with an i-f output frequency of about 17 MHz. Four crystals are used to cover the f-m range of the two meter band. The i-f module (B) contains a second mixer with variable frequency injection from the VCO module (D). The second intermediate frequency is 10.7 MHz. The VCO is tuned by a varactor diode and the frequency is read in a simple counter (figure 84). The detector/audio module contains a second i-f filter and a crystal discriminator. An adjustable squelch control quiets the receiver between signals.

The first local oscillator (Q_7) uses third-overtone crystals in the 43-MHz range and is followed by a tripler (Q_8) to the 129-MHz region. Multiple tuned circuits in the oscillator chain ensure a clean local-oscillator signal since unwanted harmonic energy in the mixer would result in spurious responses. These circuits are necessary as the r-f amplifier module has only two tuned circuits for the sake of simplicity. A more exotic front-end would provide greater rejection to off-frequency signals but would require a sweep generator for proper alignment. In keeping with the design philosophy of simplicity, a less-complex front-end was adopted at the expense of image rejection. However, the first intermediate frequency was carefully chosen so that the image frequencies fall in the aeronautical navigation band. Transmitters in this band are intended to be received by airborne receivers and when heard on the ground, the signals are usually very weak. If this receiver is used in close proximity to a navigation aid station that causes interference, a simple parallel-tuned trap inserted in the antenna lead and tuned to the frequency of the interference will solve the problem.

The first i-f signal band is at 17 MHz and a coaxial cable connects the r-f module (A) to the i-f module (B). The second mixer is a dual-gate MOSFET (Q_5) which provides about 15 dB conversion gain. As in the first oscillator chain, the second oscillator chain is filtered to eliminate unwanted harmonics and only a single tuned circuit is necessary to remove unwanted responses. The second mixer feeds an eight-pole monolithic crystal filter (FL_1). A second eight-pole filter (FL_2, module E) is used before the detector to provide an excellent overall bandpass response. The second filter is necessary in any case for the proper operation of the CA-3089 i-f subsystem. A crystal-filtered discriminator is used in conjunction with the CA-3089 i-f subsystem (module E) and a simple audio amplifier (U_3) provides up to two watts of power which is sufficient for mobile or fixed operation. Rapid and accurate squelch control is provided by device U_4, a CA-3140 integrated circuit.

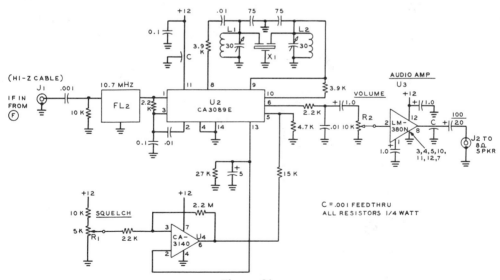

Figure 81

SCHEMATIC, I-F DETECTOR/AUDIO MODULE-E

FL_2—8-pole, 10.7 MHz filter. Piezo-Technology, Inc. Module 1475
X_1—Crystal discriminator. Piezo Technology 2283
L_1, L_2—2.2 μH molded choke. J. W. Miller 9360-02 or equivalent

The frequency of the second local oscillator (module D, Q_9) is adjusted by the main tuning control. The oscillator tunes fom 6.0 MHz to 6.5 MHz, allowing the two-meter band to be covered in 500 kHz segments. The local oscillator is electronically tuned with a varactor diode and a 10-turn potentiometer. This system provides a very smooth, backlash-free tuning system allowing 50 kHz per turn of resolution.

Figure 82

DETECTOR-AUDIO MODULE

This module is built on a copper-clad board. Volume and squelch controls are at left. Crystal discriminator and adjustable alignment capacitors are at the center of the board. Filter FL₂ is at right end of board, away from the panel controls. Integrated circuit U₂ is next to the filter. The module box is bolted in a vertical position to the side of the counter module.

Since the actual frequency is measured by an electronic frequency counter (figure 84) dial accuracy and tracking is no problem, and the received frequency may be set within one kHz with ease.

The frequency counter (module F) displays the last three kilohertz digits of the second local oscillator. For example, if the oscillator is at 6.050 MHz, the counter will read 050. Likewise, if the oscillator is at 6.485 MHz, the counter will read 485. The actual received frequency is determined by the first local-oscillator crystal frequency plus the second local-oscillator frequency as displayed on the counter. For example, if the first local oscillator is set to the 146.5-MHz position and the counter indicates 485, the received frequency is 146.5 + 0.485 = 146.985 MHz.

The receiver operates from a power source within the range of 12 to 14 volts with a 0.3 to 0.5A drain. An internal regulator provides +5 volts for the counter circuits.

Receiver Construction and Wiring The receiver is built in separate modules for easy construction and testing, as well as to achieve good shielding. Excellent shielding and good power lead bypassing techniques are absolute necessities

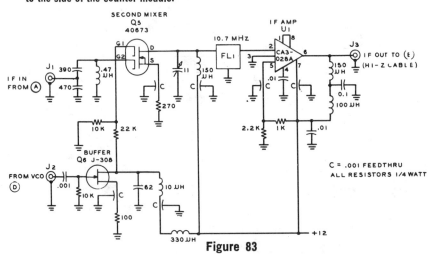

Figure 83

SCHEMATIC, I-F SECOND MIXER MODULE-B

FL₁—8-pole, 10.7 MHz Filter. Piezo-Technology, Inc. Model 1475
Inductors—J. W. Miller 9340-9350 series molded chokes
Small capacitors are silver mica or NPO ceramic

if internally generated spurious signals are to be kept to a minimum. If the receiver is constructed and tested in the suggested order, the entire unit may be aligned with simple test equipment. It is necessary to use some of the completed modules to test and align other portions of the receiver.

The I-F Subsystem and Detector/Audio Module—This is the first module to build as it will be used to test other modules. The assembly is built up on perforated board. The assembly is designated *module E* and the schematic is shown in figure 81. A photograph of this module is shown in figure 82. The unit is built within an aluminum minibox measuring $2\frac{1}{8}'' \times 1\frac{1}{8}'' \times 3\frac{1}{4}''$ (*Bud CU-2117A*).

Make sure to keep lead lengths short, particularly aound U_2, the CA-3089. The *Squelch* and *Volume* controls are located on one end of the box and the power leads on the other. After the unit has been wired and checked, a speaker and a 12-volt supply

are connected to it. A signal generator is connected to input jack J_1 though a 2.2K resistor. The signal generator is set to 10.7 MHz and tuned through this range while listening to the background noise. The noise should be the lowest when the signal generator is tuned to 10.7 MHz. Alignment of these circuits is not critical and if the signal generator can be heard, they may be left alone until the receiver is finished. Then they are peaked for best audio quality while listening to a local f-m signal. Squelch action can also be tested on a local signal or with the signal generator.

A 0-500 μA in series with an adjustable 5K potentiometer can be placed between pins 7 and 10 of device U_2. This will serve as a deviation meter and is quite useful during tune-up as well as for everyday operation.

The Second Mixer I-F Module—The second mixer module schematic is shown in figure 83. The module is assembled on a

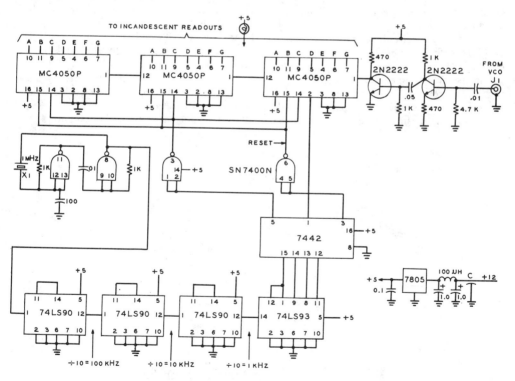

Figure 84

FREQUENCY COUNTER SCHEMATIC MODULE-F

double-sided copper clad board. Feedthrough capacitors (C) are used liberally throughout this module (and others) not only for the purpose of bypassing various circuits, but also to mechanically support the circuitry. When laying out the components, remember that these capacitors are used for support as well as for electrical connection and place the capacitors as close to the components as possible. Larger components, such as the filters, are soldered directly to the copper clad by their case. This type of construction can achieve higher density than most printed-circuit boards and, if interconnecting leads are short, makes a very strong and rigid assembly. The minibox used for this module is the same size as for module E.

After this module is built and the wiring checked, connect the i-f output (J_3) to the i-f input of module E. Apply power to both modules and connect the signal generator directly to the mixer input (J_1). The background noise from the speaker should be considerably louder than before. Tune the generator through 10.7 MHz. Even though the mixer input circuit is tuned to about 17 MHz, sufficient signal will be passed into the 10.7 MHz i-f amplifier (U_1) to make the necessary alignment adjustments. Tune the input circuit to the filter for maximum reduction of background noise with a weak input signal. A final adjustment may be

made to this circuit with a weak off-the-air signal after the receiver has been completed.

The Frequency Counter and VCO—A schematic of the frequency counter is shown in figure 84. The unit is built within a minibox measuring $\frac{1}{8}'' \times 3'' \times 5\frac{1}{4}''$ (*Bud CU-2106A*). It is assembled on a perforated board. The VCO voltage regulator and main tuning potentiometer are also mounted in this box for convenience. The counter may be tested by applying a 0.5 volt rms signal to receptacle J_1 from the signal generator at about 6.0 MHz and observing the display.

The counter will be required to adjust the second local oscillator which is the next item to be constructed. If the oscillator and buffers are functioning properly, there should be some reading on the frequency counter as the main tuning dial is rotated. Turn the main tuning control from full clockwise to full counterclockwise and the frequency span shown should be about 5.98 MHz to 6.52 MHz. The oscillator must tune from 6.0 to 6.5 MHz with some overtravel on each end. If the span is incorrect, the slug in the oscillator coil (figure 85) may be adjusted and R_A and R_B changed in value. Any overtravel between 20 and 50 kHz is acceptable.

Receiver Assembly and Alignment—The remaining sections of the receiver are the first local oscillator (module C, figure 86) and the r-f circuitry (module A, figure 87).

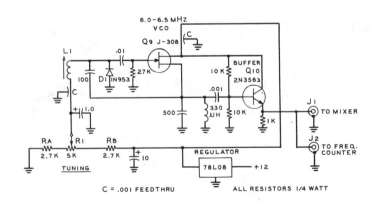

Figure 85

SCHEMATIC, SECOND OSCILLATOR MODULE-D

D_1—Varactor, 1N953 (100 pF)
L_1—14 turns, $\frac{3}{8}''$ diam. #20 e. Slug-tuned form. J. W. Miller 42A000CBI, or equivalent
R_1—5000 ohm, 10 turn potentiometer. Bourns or Beckman

A side view of the assembly showing modules C and D in position (figure 88) is helpful in the final receiver assembly. The view of the r-f module A is shown in figure 89. These units are built in accord with good construction practices.

The crystal oscillator (module C) is built in a separate minibox measuring 2¼″ × 1⅜″ × 1⅜″ (LMB-CR211) because of space considerations. The crystal oscillator and tripler do not have to be physically separate from the front-end circuitry, but since the oscillator contains a panel mounted switch, the circuitry was separated from the rest of module A to conserve front panel space.

Once modules A and C are completed, the first i-f amplifier should be aligned. If a sweep generator is available, the most accurate alignment can be accomplished by coupling the generator through a 0.01 μF capacitor to gate #2 of the first mixer (Q_2). If a sweep generator is not available, connect the signal generator in the same manner and connect an r-f probe to the i-f output connector (J_3).

Using the sweep generator, tune L_3, L_4, and L_5 for a flat response over the i-f spectrum of 16.7 MHz to 17.2 MHz. If a sweep generator is not used, a weak signal is injected into the module and inductor L_3 is adjusted for maximum response at 16.7 MHz. The generator is then tuned to 17.2 MHz and inductor L_5 is adjusted in similar fashion. Inductor L_4 is adjusted to smooth out peaks in the i-f passband.

Once the crystal oscillator (module C) is finished, the bandswitch should be set to a midfrequency crystal. The r-f probe is touched to the gate of oscillator Q_7 and the slug of inductor L_1 adjusted for maximum reading. The probe is next moved to the tripler coil (L_2) and the circuit capacitor adjusted for maximum output. The buffer stage in module A is next resonated for maximum signal. The r-f voltage at gate #2 of mixer Q_2 should be between 0.5 and 1.5 volts, rms.

The last adjustment is to peak the r-f amplifier circuits at about 146 MHz. This may be done with the signal generator. The circuits should be tuned for minimum receiver background noise while reducing generator output so that the generator barely quiets the receiver. Sensitivity, typically, is less than 0.5 microvolt. Final adjustment should be done on a weak signal as most generators leak enough signal so as to make this adjustment difficult.

When alignment is complete, the various modules can be placed in a small cabinet. This unit shown is in a *LMB-OH743* cabinet measuring 7¼″ × 6″ × 2¾″. Placement of the modules can be seen in figure 90.

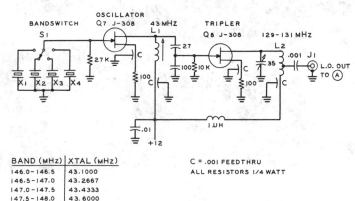

BAND (MHz)	XTAL (MHz)
146.0 – 146.5	43.1000
146.5 – 147.0	43.2667
147.0 – 147.5	43.4333
147.5 – 148.0	43.6000

C = .001 FEEDTHRU
ALL RESISTORS 1/4 WATT

Figure 86

SCHEMATIC, FIRST OSCILLATOR MODULE-C

L_1—9 turns #20 e. on ¼″ diam. slug tuned form. J. W. Miller 4500

L_2—5 turns #20 e., ¼″ diam. airwound, tap at 1½ turns from bottom

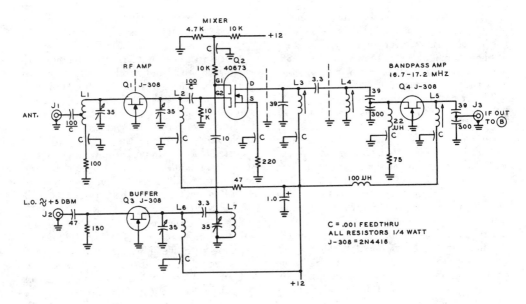

Figure 87

SCHEMATIC, R-F MODULE - A

L_1—5 turns #20 e., ¼″ diam. spacewound. Tap 3 turns from ground
L_2—Same as L_1, no tap
L_3, L_4-L_5—24 turns #22 e. on ¼″ diam. slug-tuned form. J. W. Miller 4500
L_6, L_7—6 turns #22 e. spacewound

Figure 88

SIDE VIEW OF RECEIVER

Modules C (local oscillator) and D (VCO) are exposed in this view. Slug of the VCO coil is adjustable from top of module box. The local oscillator module is panel mounted with the crystal switch projecting through the panel. At the rear of the assembly is the r-f module, A. Power switch is at lower corner of panel.

Figure 89

R-F MODULE OF 144 MHZ RECEIVER

Circuit is built up on copper plated board. The various stages have shields placed between them. The shields are made of board stock, cut to size and soldered to the copper overlay. End-view of slug-tuned coils is shown with all components supported by their leads. Module is bolted to rear of receiver cabinet.

If more of the two-meter band is desired, crystals may be added. The crystal frequency is determined by the following formula:

$$X_1 = \frac{F_1 - 16.7}{3}$$

where,

X_1 is the crystal frequency in MHz,
F_1 is the lowest frequency of the desired band segment.

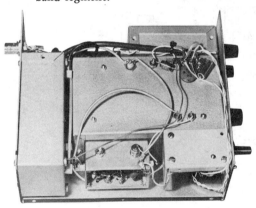

Figure 90

TOP VIEW OF RECEIVER ASSEMBLY

Receiver modules are mounted in the receiver cabinet. R-f module is at the rear, right, with other modules bolted to center counter module. Oscillator modules are in foreground.

The receiver offers interesting possibilities for other vhf bands. By simply changing the front end module (A), the receiver can be modified for use on 50-, 220-, or 450-MHz. Any converter having an output spectrum between 16.7 and 17.2 MHz may be used. Those commercial converters having output in the 20-meter band can be easily modified for use in this receiver design.

Another possibility, since tuning is accomplished by varying a dc voltage, is to use a ramp generator to automatically scan a band segment, the scan stopping when the squelch opens. Preset channels could be set by means of individual potentiometers and a selector switch. A tuning meter and signal-strength meter may also be added.

20-7 A Variable Active Audio Filter

Audio filters have proven their worth in c-w and weak signal reception of all types. Most popular filters have been fixed-frequency devices having a narrow bandwidth determined by high-Q LC circuits. These filters have the disadvantage of ringing; that is, the tuned filter circuit can oscillate when excited by a signal, resulting in a ringing noise which sounds very much like the original signal, making it difficult to copy a very weak signal. In addition, being fixed-tuned, the LC filter cannot be easily adjusted to optimize either the passband or the center frequency.

The disadvantages of the LC filter have been overcome by the *active filter* which can provide variable bandwidth and variable center frequency. In addition, the active filter uses no LC elements, eliminating the annoying ringing effect of the high-Q filter circuit.

The Active Filter An active audio filter can be built using three operational amplifiers in a stable, negative-feedback circuit (figure 91). The bandwidth of this circuit can be adjusted for flat, bandpass, highpass, or lowpass response. Such a device is ideal for such uses as speech filters, notch filters, tone decoders, RTTY filters, and c-w filters.

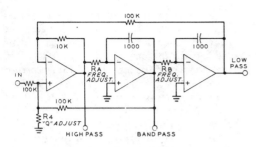

Figure 91

ACTIVE AUDIO FILTER

An active audio filter can be built using three ICs in a stable, negative-feedback circuit which may be adjusted for flat, bandpass, highpass, or lowpass response. The three ICs are packaged as one unit in the KTI FX-60 device.

The filter described in this section utilizes the newly developed *Universal Active Filter* produced by the *Kinetic Technology* division of *Baldwin Electronics, Inc.* The filters are packaged in a 14-pin IC configuration and utilize three optional amplifiers in a variation of the basic circuit. The filter tunes from 300 Hz to 1800 Hz, with an adjustable bandwidth of 50 Hz to 1200 Hz. In the *lowpass* mode, the filter is useful for SSB reception, removing the annoying high-frequency sounds from the received signal. In the *bandpass* mode, the filter is useful for c-w reception, and various filters patterned after the one shown herewith are used for moonbounce communication on the 144-MHz and 432-MHz amateur bands.

The schematic of the active filter is shown in figure 92. The active filter is a *KTI* model FX-60 device and is followed by a *National* LM-380N audio amplifier, which provides up to 2 watts of audio power. The filter is designed to be plugged into the low-impedance headphone jack of a receiver and the filter output impedance matches either low-impedance earphones or a speaker.

Filter Construction   The active filter shown here is constructed by K6HCP on a printed-circuit board and is mounted in a small aluminum cabinet (figure 93). The input signal level is adjusted by resistor R_1, if required. Device Q_f is the active filter element. The bandwidth of the filter is adjusted by potentiometer R_4 and dual potentiometer R_3A-B adjusts the center frequency of the filter. Switch S_1 selects the available outputs.

A bias network sets the voltage at pin 4 of Q_1 and the resistor between pin 6 and pin 7 allows the three outputs to hold the same level. The 5.6K resistor in series with pin 6

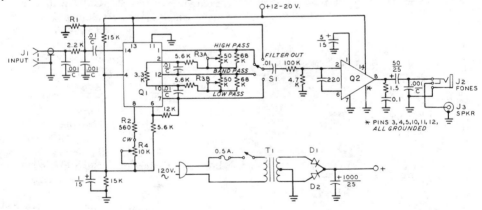

Figure 92

SCHEMATIC OF ACTIVE AUDIO FILTER

D_1, D_2—1N4001
Q_1—FX-60 universal active filter. Kinetic Technology.
Q_2—LM-380N. National Semiconductor.

R_3—50K-50K dual potentiometer. Counter clockwise log taper. Allen-Bradley 70CIN048-503B
R_4—10K potentiometer. Linear taper
T_1—24 volt, 180 mA. Signal PC24-180

Figure 93

THE ACTIVE AUDIO FILTER

This adjustable filter is built in a small aluminum channel box and has its own power supply.
The bandwidth and center frequency controls
are at center of the panel, with the function
switch immediately below.

Figure 94

LAYOUT AND CONSTRUCTION
OF FILTER

The filter is assembled on a printed-circuit
board which is held to the bottom section of
the cabinet by 4-40 hardware. The output receptacle is at lower left. At the rear of the p-c
board is the built-in power supply.

sets the widest bandwidth limit and resistor R_2 sets the narrowest limit.

The capacitors connected between pins 2 and 12, and 7 and 10 on device Q_1 set the frequency range of the filter. Resistor R_2 is selected to set the narrowest bandwidth. To adjust this, potentiometer R_4 is set clockwise for zero resistance and R_2 is selected until the circuit just goes into oscillation. The correct value of R_2 is one that prevents cir-cuit oscillation.

The audio level to the LM-380 is set by the two resistors at pin 2. The 220-pF ca-pacitor provides a high-frequency rolloff at 4 kHz.

Using the Filter The filter is usually set to the lowpass mode for SSB reception and to the bandpass mode for c-w reception. It is peaked on the c-w signal, but the center frequency setting and bandwidth will vary from person to person. The ear-brain combination is capable of acting like a variable-bandwidth, variable-frequency filter itself and the operator, with experience, can copy signals buried in noise or interference. Signals that are as low as 10 dB below the noise level can be copied, as shown by lis-tening tests conducted under controlled conditions.

A relatively low beat frequency is sug-gested for weak signal c-w reception because the signal is easier to copy in the presence of interference. If, for example, the signal has a pitch of 1000 Hz, and there is an inter-fering signal 100 Hz away, the difference between the desired and undesired signals is 10 percent. If the pitch of the signal is changed to 500 Hz, and the interfering sig-nal is 100 Hz away, the difference is now 20 percent.

CHAPTER TWENTY-ONE

HF and VHF Power Amplifier Design

A *power amplifier* is a converter that changes dc into r-f output. Chapter Seven of this Handbook discussed the various classes of r-f power amplifiers and Chapters Eleven and Seventeen covered the calculation of input and output circuit parameters. This chapter covers power-amplifier design and adjustment.

Modern hf amateur transmitters are capable of operating on c-w, SSB and often RTTY and SSTV on one or more amateur bands between 3.5 MHz and 29.7 MHz. Very few pieces of commercially built amateur equipment have amplitude-modulation capability, other than some gear designed for 6- and 2-meter operation, since the changeover from a-m to SSB is now complete. On the other hand, expansion of 160-meter privileges in the past years has brought about the inclusion of that band in most amateur equipment.

The most popular and flexible amateur hf transmitting arrangement usually includes a compact bandswitching exciter or transceiver having 100 to 250 watts PEP input on the most commonly used hf bands, followed by a single linear power-amplifier stage having 1 kW to 2 kW PEP input capacity. In many instances, the exciter is an SSB transceiver unit capable of mobile operation, while the amplifier may be a compact table-top assembly. The amplifier is usually coupled to the exciter by a coaxial cable and changeover relay combination, permitting the exciter to run independently of the amplifier, if desired, or in combination with the amplifier for maximum power output. For c-w or RTTY, the amplifier is usually operated in the linear mode, since conversion to class-C operation is not required.

These practical designs are a natural outgrowth of the importance of vfo operation and the use of SSB and c-w modes in amateur practice. It is not practical to make a rapid frequency change when a whole succession of stages must be retuned to resonance, or when bandswitching is not employed.

Power-Amplifier Design   Power amplifiers are classified according to operating *mode* and *circuitry*. Thus, a particular amplifier mode may be class AB_1, class B, or class C; the circuitry can be either single-ended or push-pull; and the unit may be grid- or cathode-driven. Mode of operation and circuit configuration should not be confused, since they may be mixed in various combinations, according to the desire of the user and the characteristics of the amplifier tube.

High-frequency silicon power transistors are used in amateur and commercial equipment designs up to the 300-watt PEP power level or so. Undoubtedly solid-state devices will become of increasing importance in hf power amplifiers in the coming decade.

Either triode or tetrode tubes may be used in the proper circuitry in hf and vhf power amplifiers. The choice of tube type is often dependent on the amount of drive power available and, in the case of home-made gear, the tube at hand. If an exciter of 100 to 200 watts PEP output capacity is to be used, it is prudent to employ an amplifier whose drive requirement falls in the same power range as the exciter output. Triode or tetrode tubes may be used in cathode-driven (grounded-grid) circuitry which will pass along an excess of exciter power in the form of feedthrough power to the antenna circuit. The tubes may also be grid-driven in combination with a power absorption network that will dissipate excess exciter power not required by the amplifier.

On the other hand, if the power output of the exciter is only a few watts PEP, either low-drive, high-gain tetrodes must be used in grid-driven configuration, or an intermediate amplifier must be used to boost the drive to that level required by triode tubes. Thus, the interface between the exciter and the amplifier in terms of PEP level must be reconciled in the design of the station transmitting equipment.

21-1 Triode Amplifier Design

Triode tubes may be operated in either grid- or cathode-driven configuration, and may be run in class-AB_1, class-AB_2, class-B or class-C mode. Plate dissipation and amplification factor (μ) are two triode characteristics which provide the information necessary to establish proper mode and circuitry and to evaluate the tube for linear-amplifier or class-C service.

Plate dissipation is important in that it determines the ultimate average and peak power capabilities of the tube. Linear amplifiers commonly run between 55- and 65-percent plate efficiency, with the majority of the remainder of the power being lost as plate dissipation. Class-C service often runs at about 70- to 75-percent plate efficiency. Knowing the plate dissipation rating of the tube, the approximate maximum power input and output levels for various modes of service may be determined by the methods outlined in Chapter 7.

Amplification Factor (μ) of a triode expresses the ratio of change of plate voltage for a given change in grid voltage at some fixed value of plate current. Values of μ between 10 and 300 are common for triode transmitting tubes. High-μ tubes (μ greater than about 30) are most suitable in cathode-driven (grounded-grid) circuitry as the cathode-plate shielding of a high-μ tube is superior to that of a comparable low-μ tube, and because a high-μ tube provides more gain and requires less driving power than a low-μ tube in this class of service. Low-μ triodes, on the other hand, are well suited for grid-driven class-AB_1 operation since it is possible to reach a high value of plate current with this type of tube, as opposed to the high-μ equivalent, without driving the grid into the power-consuming, positive region. Even though a large value of driving voltage is required for the low-μ tube, little drive power is required for class-AB_1 service, since the grid always remains negative and never draws current.

As a rule-of-thumb, then, a triode tube to be used for linear r-f service in a power amplifier should have a large plate-dissipation capability, and the output power to be expected from a single tube will run about twice the plate-dissipation rating. High-μ triodes, generally speaking, perform better in cathode-driven, class-B circuitry; whereas medium- and low-μ triodes are to be preferred in grid-driven, class AB_1 circuitry. Circuit neutralization may often be disposed with in the first case (at least in the hf region), and is usually necessary in the second case, otherwise the circuits bear a striking similarity.

Grid-Driven Circuitry   Representative grid- and cathode-driven triode circuits are shown in figure 1. The classic grid-driven, grid-neutralized circuit is shown in illustration A. The drive signal is applied to a balanced grid tank circuit (L_1, C_1) with an out-of-phase portion of the exciting voltage fed through capacitor NC to the plate circuit in a bridge neutralization scheme. A pi network is employed for the plate output coupling circuit. The plate inductor (L_2) may be tapped or otherwise variable and is normally adjustable from the amplifier panel, eliminating the necessity of plug-in coils and access openings into the shielded amplifier inclosure. The grid circuit may also be switched or varied in a similar manner.

A high ratio of capacitance to inductance (high-C) is required in the tuned grid circuit in order to preserve the phase relationship in the neutralizing circuit under

TRIODE CIRCUITRY

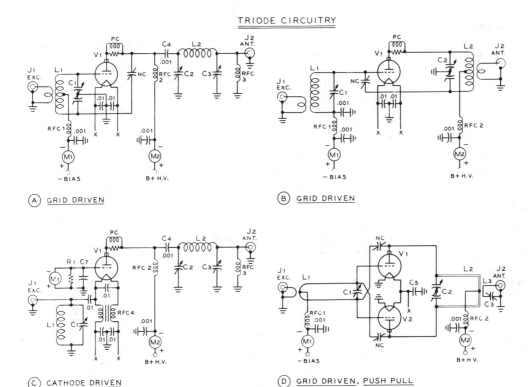

(A) GRID DRIVEN

(B) GRID DRIVEN

(C) CATHODE DRIVEN

(D) GRID DRIVEN, <u>PUSH PULL</u>

Figure 1

REPRESENTATIVE TRIODE AMPLIFIER CIRCUITS

Circuits A, B and C are for the 3–54 MHz region. Circuit D is intended for the 50–500 MHz region. Note that one filament leg is grounded in circuit D to reduce inductance of filament return circuit.

C₁—Input tuning capacitor. Typically, 3 pF per meter of wavelength. Spacing 0.03″ for power level up to 2 kW, PEP

C₂—Output tuning capacitor. Refer to plate-circuit design data in Chapter 11.

C₃—Loading capacitor. Typically, 20 pF per meter of wavelength. Refer to Chapter 11.

C₄—Plate-blocking capacitor. Typically, 500 pF to 1000 pF, 5 kV

C₅, C₇—Low-inductance mica or ceramic capacitor, series resonant near operating frequency. See Chapter 17

M₁—Grid-current meter

M₂—Plate-current meter

RFC₁—Grid choke, receiving type rated to carry grid current. Typically, 1 to 2.5 mH for 3- to 30-MHz range

RFC₂—Plate choke, transmitting type, solenoid. Rated to carry plate current. Typically, 800 μH. See Chapter 17

RFC₃—Receiving-type choke. 2½ mH for 3- to 30-MHz range.

RFC₄—Bifilar windings, 15 turns each #12 wire on ½-inch diameter ferrite core, 3″ long for 3- to 54-MHz range

PC—Plate parasitic suppressor. Typically, 3 turns #18 enamel, ½-inch diameter, ½-inch long, in parallel with 50-ohm 2-watt composition resistor. See Chapter 17

operating conditions. If a low-C grid circuit is used, grid loading will unbalance the neutralizing network, the r-f voltage at the grid dropping and the voltage at the neutralizing end of the grid circuit rising. A high-C circuit tends to alleviate this problem.

Plate circuit neutralization (figure 1B) does not exhibit such a degree of unbalance under load and is to be preferred, especially for operation at the higher frequencies. A split plate-tank circuit is required in place of the split grid circuit, making the use of a single ended pi-network output circuit impractical. Theory and adjustment of grid and plate neutralizing circuits are covered in Chapter 11. In either configuration, care must be taken in construction to make sure that a minimum of stray coupling exists between grid and plate tank circuits. Whenever possible, the grid and plate coils should be mounted at right angles to each other, and should be separated sufficiently to reduce coupling between them to a minimum. Unwanted coupling will tend to make neutralization frequency-sensitive, requiring that the circuit be reneutralized when a major frequency change is made.

Cathode-Driven Circuitry   A representative cathode-driven (grounded-grid) triode circuit is shown in figure 1C. A pi-network plate circuit is used, and excitation is applied to the filament (cathode) circuit, the grid being at r-f ground potential. If the amplification factor of the triode is sufficiently high so as to limit the static plate current to a reasonable value, no auxiliary grid bias is required. A parallel-tuned cathode input circuit is shown, although pi-network circuitry may be used in this position. Filament voltage may either be fed via a shunt r-f choke as shown, or applied through a bifilar series-fed cathode tank coil.

While nominally at r-f ground, the grid of the triode may be lifted above ground a sufficient amount so as to insert a monitoring circuit to measure dc grid current. The grid to ground r-f impedance should remain very low, and proper attention must be paid to the r-f circuit. A considerable amount of r-f current flows through the grid bypass capacitor (C_7) and this component should be rated for r-f service. It should be shunted with a low value of resistance (of the order of 10 ohms or less) and the dc voltage drop across this resistor is monitored by the grid voltmeter, which is calibrated in terms of grid current. Both resistor and capacitor aid in establishing a low-impedance path from grid to ground and should be mounted directly at the socket of the tube. If multiple grid pins are available, each pin should be individually bypassed to ground. Control of the grid-to-ground impedance in the cathode-driven circuit establishes the degree of intrastage feedback, and an increase in grid impedance may alter stage gain, leading to possible uncontrolled oscillation or perhaps making the stage difficult to drive. At the higher frequencies, stage gain may be controlled by

the proper choice of the grid-to-ground impedance.

From a practical standpoint, it is suggested that the cathode tank circuit be made fixed-tuned and peaked at the middle of the amateur band in use. This form of construction is suggested because if the cathode circuit is inadvertently tuned too far off-frequency, it will turn the cathode-driven amplifier into a robust oscillator! The user might suspect instability, or a possible parasitic oscillation, which is not the case. It is merely that the circuit constants are such that a phase shift may be unintentionally created between cathode and plate which will sustain oscillation. The use of a fixed-tuned, or slug-tuned, cathode circuit will prevent this, as it cannot be adjusted sufficiently far off frequency to sustain oscillation.

Push-Pull Circuitry   A push-pull triode amplifier configuration is shown in figure 1D. This circuit design is now rarely used in the hf region because of the mechanical difficulties that ensue when a large frequency change is desired. In the vhf region, on the other hand, where operation of an amplifier is generally restricted to one band of frequencies, linear push-pull tank circuits are often employed. Lumped-inductance tank coils are usually avoided in the vhf region since various forms of parallel-line or strip-line circuitry provide better efficiency, higher Q, and better thermal stability than the coil-and-capacitor combination tank assemblies used at the high frequencies. Push-pull operation is of benefit in the vhf region as unavoidable tube capacitances are halved, and circuit impedances are generally higher than in the case of single-ended circuitry. At the higher vhf regions, parallel- and strip-line circuitry give way to coaxial tank circuits in which the tube structure becomes a part of the resonant circuit.

The output coupling circuit may be designed for either balanced or unbalanced connection to coaxial or twin-conductor transmission line. In many cases, a series capacitor (C_3) is placed in one leg of the line at the feed point to compensate for the inductance of the coupling coil.

Common hf construction technique employs plug-in plate and grid coils which necessitate an opening in the amplifier inclosure for coil-changing purposes. Care must be taken in the construction of the door of the opening to reduce harmonic leakage to a minimum. While variations in layout, construction, and voltage application are found, the following general remarks apply to hf amplifiers of all classes and types.

Circuit Layout The most important consideration in constructing a push-pull amplifier is to maintain electrical symmetry on both sides of the balanced circuit. Of utmost importance in maintaining electrical balance is the control of stray capacitance between each side of the circuit and ground.

Large masses of metal placed near one side of the grid or plate circuits can cause serious unbalance, especially at the higher frequencies, where the tank capacitance between one side of the tuned circuit and ground is often quite small in itself. Capacitive unbalance most often occurs when a plate or grid coil is located with one of its ends close to a metal panel. The solution to this difficulty is to mount the coil parallel to the panel to make the capacitance to ground equal from each end of the coil, or to place a grounded piece of metal opposite the "free" end of the coil to accomplish a capacity balance.

All r-f leads should be made as short and direct as possible. The leads from the tube grids or plates should be connected directly to their respective tank capacitors, and the leads between the tank capacitors and coils should be as heavy as the wire that is used in the coils themselves. Plate and grid leads to the tubes may be made of flexible tinned braid or flat copper strip. Neutralizing leads should run directly to the tube grids and plates and should be separate from the grid and plate leads to the tank circuits. Having a portion of the plate or grid connections to their tank circuits serve as part of a neutralizing lead can often result in amplifier instability at certain operating frequencies.

Filament Supply The amplifier filament transformer should be placed right on the amplifier chassis in close proximity to the tubes. Short filament leads are

necessary to prevent excessive voltage drop in the connecting leads, and also to prevent r-f pickup in the filament circuit. Long filament leads can often induce instability in an otherwise stable amplifier circuit, especially if the leads are exposed to the radiated field of the plate circuit of the amplifier stage. The filament voltage should be the correct value specified by the tube manufacturer when measured *at the tube sockets*. A filament transformer having a tapped primary often will be found useful in adjusting the filament voltage. When there is a choice of having the filament voltage slightly higher or slightly lower than normal, the lower voltage is preferable.

Filament bypass capacitors should be low internal inductance units of approximately .01 μF. A separate capacitor should be used for each socket terminal. Lower values of capacitance should be avoided to prevent spurious resonances in the internal filament structure of the tube. Use heavy, shielded filament leads for low voltage drop and maximum circuit isolation.

Plate Feed The series plate-voltage feed shown in figure 1D is the most satisfactory method for push-pull stages. This method of feed puts high voltage on the plate tank inductor, but since the r-f voltage on the inductor is in itself sufficient reason for protecting the inductor from accidental bodily contact, no additional protective arrangements are made necessary by the use of series feed.

The insulation in the plate supply circuit should be adequate for the voltages encountered. In general, the insulation should be rated to withstand at least four times the maximum dc plate voltage. For safety, the plate meter should be placed in the cathode return lead, since there is danger of voltage breakdown between a metal panel and the meter movement at plate voltages much higher than one thousand.

Parallel plate feed, such as shown in figures 1A and 1B, is commonly used for single-ended pi-network amplifier configurations. The plate r-f choke is a critical component in this circuit, and a discussion of choke design is covered in Chapter 17. The plate-blocking capacitor (C_1) should be rated to withstand the peak r-f plate current (usually about three to four times the dc plate current) and the peak r-f voltage (up to twice the dc plate voltage.)

In the case of the push-pull stage, the amplifier grid and plate circuits should be symmetrically balanced to ground. In some instances, a small differential capacitor is placed in the grid circuit to effect balance, and the grid current of each tube is monitored individually to ascertain correct balance. The rotor of the split-stator plate-tuning capacitor is usually ungrounded, permitting the plate tank circuit to establish its own r-f balance.

The various filament, grid, and plate by-pass capacitors are often vhf coaxial types which have inherently low inductance well into the vhf region. These capacitors should be checked to make sure that their internal self-resonant frequency is well above the operating frequency of the amplifier.

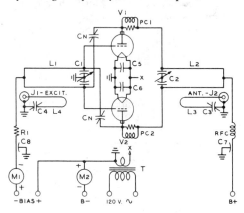

Figure 2

TYPICAL PUSH-PULL VHF TRIODE AMPLIFIER CIRCUIT

C₁, C₂—Low-capacitance, balanced split-stator capacitor. Typically, 10 pF per section for 144 MHz.

C₃, C₄—Loading capacitor. Capacitance chosen to series resonate at operating frequency with coupling loop

C₅, C₆—Low inductance mica or ceramic capacitor, series resonant near operating frequency. See Chapter 17

C₇, C₈—Low-inductance feedthrough capacitor. See Chapter 17

C_N—Neutralizing capacitor. Approximately equal to grid-plate capacitance of triode tube.

M₁—Grid-current meter

M₂—Plate-current meter

R₁—Wire resistor (100-500 ohms) to act as low-Q r-f choke

RFC—Vhf choke rated to carry plate current. See text

In most cases, the push-pull amplifier may be cross-neutralized in the normal manner. At the higher frequencies (above 150 MHz or so) it is common practice to operate the triode tubes in cathode-driven configuration which usually eliminates the need for neutralization if proper shielding is used.

Plate parasitic suppressors may or may not be necessary depending on the operating frequency of the amplifier and the natural parasitic frequency of the input and output circuits. Both grid- and plate-tuning capacitors should be located close to the tube elements and not tapped down the tuned lines, otherwise unwanted parasitic circuits may be created. If oscillations are encountered, they may possibly be suppressed by placing noninductive carbon resistors across a portion of the plate (and grid) lines as shown in figure 2.

The plate choke (RFC) should be mounted at right angles to the plate line and care should be taken that it is not coupled to the line. In particular, the choke should not be mounted within the line, but rather outside the end of the line, as shown. A resistor (R₁) is used to take the place of a grid choke, thus eliminating any possibility of resonance between the two chokes, with resulting circuit instability.

In order to prevent radiation loss from the grid and plate lines, it is common practice to completely inclose the input and output circuits in "r-f tight" inclosures, suitably ventilated to allow proper cooling of the tubes.

The plate parasitic suppressor (PC₁) is a critical component. The suppressor is designed to present a load to the amplifier tube at the parasitic frequency only, leaving the fundamental frequency component undisturbed. In theory, the inductor short-circuits the loading resistor at the fundamental frequency and acts as a high impedance at the parasitic frequency which, in most cases, is higher than the fundamental frequency. In the vhf region, the shunt inductor of the suppressor must have a very low value of inductance to prevent too much fundamental power from being dissipated in the parallel-connected resistor. In the 2-meter band, it is common practice to connect the parasitic resistor across a section of the plate lead which usually is a copper strap. The amount of lead shunted by the resistor constitutes the inductor and determines the degree of coupling between the fundamental signal and the parasitic suppressor.

When large tubes are used in the vhf region, the parasitic frequency of the circuit may fall near, or at, the fundamental operating frequency. If this is so, parasitic suppression is unnecessary as the conventional cross-neutralization circuit will also inhibit parasitic oscillation.

21-2 Tetrode Amplifier Design

As in the case of triode tubes, tetrodes may be operated in either grid- or cathode-driven configuration and may (within certain limits) be run in class-AB_1, -AB_2, -B, or class-C mode. Much of the information on circuit layout and operation previously discussed for triode tubes applies in equal context to tetrodes. Other differences and additional operational data will be discussed in this section.

Tetrode tubes are widely used in hf and vhf amplifiers because of their high power gain and wide range of simple neutralization. Tetrode circuitry resembles triode circuitry in that comparable modes and circuit configurations may be used. Various popular and proven tetrode circuits are shown in figure 3. Illustration A shows a typical single-ended neutralized tetrode circuit employing a pi-network output circuit and a bridge neutralization scheme. Tetrode neutralization techniques are discussed in detail in Chapter 11.

Tetrode plate current is a direct function of screen voltage and means must be employed to control screen voltage under all conditions of operation of the tetrode. In particular, if the *dc screen-to-ground* path is broken, the screen voltage may rise to equal the plate potential, thus damaging the tube and rupturing the screen bypass capacitor. It is dangerous, therefore, to reduce screen voltage for tuneup purposes by simply breaking the screen power lead unless a protective screen bleeder resistor (R_2) is placed directly at the tube socket, as shown in the illustrations of figure 3. If this resistor is used, the screen supply may be safely

broken at point X for tuneup purposes, or for reduced-power operation. The value of screen bleeder resistance will vary depending on tube characteristics, and a typical value is generally specified in the tube data sheet. For tubes of the 4CX250B family, the value of resistance is chosen to draw about 15 to 20 mA from the screen power supply. The 4CX1000A, on the other hand, requires a screen bleeder current of about 70 mA.

In any case, regardless of whether the screen circuit is broken or not, the use of a screen bleeder resistor in the circuit at all times is mandatory for those tetrodes which produce reverse screen current under certain operating conditions. This is a normal characteristic of most modern, high-gain tetrodes and the screen power supply should be designed with this characteristic in mind so that correct operating voltages will be maintained on the screen at all times.

With the use of a screen bleeder resistor, full protection for the screen may be provided by an overcurrent relay and by interlocking the screen supply so that the plate voltage must be applied before screen voltage can be applied.

Power output from a tetrode is very sensitive to screen voltage, and for linear service a well-regulated screen power supply is required. Voltage-regulator tubes or a series-regulated power supply are often used in high-power tetrode linear-amplifier stages.

A tetrode neutralizing circuit suitable for the lower portion of the vhf region is shown in figure 3B. When the operating frequency of the tetrode is higher than the self-neutralizing frequency, the r-f voltage developed in the screen circuit is too great to provide proper voltage division between the internal capacitances of the tube (see Chapter 11). One method of reducing the voltage across the screen lead inductance and thus achieving neutralization is to adjust the inductive reactance of the screen-to-ground path so as to lower the total reactance. This reactance adjustment may take the form of a variable series capacitor as shown in illustration B. This circuit is frequency sensitive and must be readjusted for major changes in the frequency of operation of the amplifier.

Balanced input and output tuned circuits are used in the configuration of figure 3B.

TETRODE CIRCUITRY

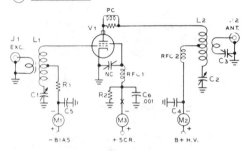

(A) GRID DRIVEN

(B) GRID DRIVEN

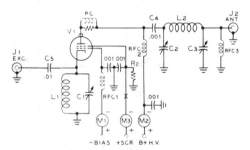

(C) CATHODE DRIVEN

Figure 3

REPRESENTATIVE TETRODE AMPLIFIER CIRCUITS

Circuit B is intended for operation above the self-neutralizing frequency of the tetrode. Above 30 MHz or so, the screen bypass capacitor of circuits A and C is often chosen so as to be self-resonant at the operating frequency of the amplifier.

C_1, L_1—Input tuned circuit. Typically, 3 pF per meter of wavelength for circuits A and B. 20 pF per meter of wavelength for circuit C

C_2, C_3, L_2—Pi-network plate circuit. Refer to plate-circuit design data in Chapter 11

C_4—Plate-blocking capacitor. Typically, 500 pF to 1000 pF at 5 kV

R_1—Wirewound resistor (100-500 ohms) to act as low-Q r-f choke

R_2—Screen resistor to carry negative screen current and complete screen-to-ground circuit. See tube data sheet for details

PC—Plate parasitic suppressor. See Chapter 17 and figure 1 of this chapter. For vhf operation, suppressor may consist of composition resistor shunted across a short portion of the plate lead

RFC_1—Grid choke, receiving type. Typically, 2.5 mH for 3- to 30-MHz range. Vhf-rated choke for 50 MHz and 144 MHz

RFC_2—Plate choke, transmitting type, solenoid. Rated to carry plate current. Typically, 800 μH for 3- to 30-MHz range. Vhf-rated choke for 50 MHz and 144 MHz

RFC_3—Receiving-type choke. 2.5 mH for 3- to 30-MHz range

M_1—Grid-current meter

M_2—Plate-current meter

M_3—Screen-current meter

In the grid circuit, the split capacitance is composed of variable capacitor C_1 and the grid-cathode input capacitance of the tube. The coil (L_1) is chosen so that C_1 approximates the input capacitance. The same technique is employed in the plate circuit, where a split tank is achieved by virtue of capacitance C_2 and the output capacitance of the tetrode tube.

A cathode-driven tetrode amplifier is shown in illustration C. Many tetrodes do not perform well when connected in class-B grounded-grid configuration (screen and grid both at ground potential). These tubes are characterized by high perveance, together with extremely small spacing between the grid bars, and between the grid structure and the cathode. Tubes of the 4-65A, 4X150A/4CX250B, and 4CX1000A family are in this class. For proper operation of these high-gain tubes, the screen requires much larger voltage than the control grid. When the electrodes of these tubes are tied together, the control grid tends to draw heavy current and there is risk of damaging the tube. Lower-gain tetrodes, such as the 813, 4-400A, and 4-1000A have a more balanced ratio of grid to screen current and may be operated in zero-bias, grounded-grid mode. The best way to employ the higher-gain tetrode tubes in cathode-driven service is to ground the grid and screen through bypass capacitors and to operate the elements at their rated class AB_1 dc voltages. In all cases, grid and screen current should be monitored, so as to keep maximum currents within ratings.

Tetrode Amplifier Circuitry The most widely used tetrode circuitry for hf use is the single-ended pi-network configuration, variations of which are shown in figure 4.

A common form of pi-network amplifier is shown in figure 4A. The *pi* circuit forms the matching system between the plate of the amplifier tube and the low-impedance, unbalanced, antenna circuit. The coil and input capacitor of the *pi* may be varied to tune the circuit over a 10 to 1 frequency range (usually 3,0 to 30 MHz). Operation over the 20- to 30-MHz range takes place when the variable slider on coil L_2 is adjusted to short this coil out of the circuit. Coil L_1 therefore comprises the tank inductance for the highest portion of the operating range. This coil has no taps or sliders and is constructed for the highest possible Q at the high-frequency end of the range. The adjustable coil (because of the variable tap and physical construction) usually has a lower Q than that of the fixed coil.

The degree of loading is controlled by capacitors C_1 and C_5. The amount of circuit capacitance required at this point is inversely proportional to the operating frequency and to the impedance of the antenna circuit. A loading capacitor range of 100 to 2500 pF is normally ample to cover the 3.5- to 30-MHz range.

The *pi* circuit is usually shunt-fed to remove the dc plate voltage from the coils and capacitors. The components are held at ground potential by completing the circuit to ground through the choke (RFC_1). Great stress is placed on the plate-circuit choke (RFC_2). This component must be specially designed for this mode of operation, having low interturn capacitance and no spurious internal resonances throughout the operating range of the amplifier.

Parasitic suppression is accomplished by means of chokes PC_1 and PC_2 in the screen, grid, or plate leads of the tetrode. Suitable values for these chokes are given in the parts list of figure 4. Effective parasitic suppression is dependent to a large degree on the choice of screen bypass capacitor C_1. This component must have extremely low inductance throughout the operating range of the amplifier and well up into the vhf parasitic

range. The capacitor must have a voltage rating equal to at least twice the screen potential (four times the screen potential for plate modulation). There are practically no capacitors available that will perform this difficult task. One satisfactory solution is to allow the amplifier chassis to form one plate of the screen capacitor. A "sandwich" is built on the chassis with a sheet of insulating material of high dielectric constant and a matching metal sheet which forms the screen side of the capacitance. A capacitor of this type has very low internal inductance but is very bulky and takes up valuable space beneath the chassis. One suitable capacitor for this position is the *Centralab* type *858S-1000*, rated at 1000 pF at 5000 volts. This compact ceramic capacitor has relatively low internal inductance and may be mounted to the chassis by a 6-32 bolt. Further screen isolation may be provided by a shielded power lead, isolated from the screen by a .001-μF ceramic capacitor and a 100-ohm carbon resistor.

Various forms of the basic pi-network amplifier are shown in figure 4. The A circuit uses coil switching in the grid circuit, bridge neutralization, and a tapped pi-network coil with a vacuum tuning capacitor. Figure 4B shows an interesting circuit that is becoming more popular for class-AB_1 linear operation. A tetrode tube operating under class-AB_1 conditions draws no grid current and requires no grid-driving power. Only r-f voltage is required for proper operation. It is possible therefore to dispense with the usual tuned grid circuit and neutralizing capacitor and in their place employ a noninductive load resistor in the grid circuit across which the required excitation voltage may be developed. This resistor can be of the order of 50 to 300 ohms, depending on circuit requirements. Considerable power must be dissipated in the resistor to develop sufficient grid swing, but driving power is often cheaper to obtain than the cost of the usual grid-circuit components. In addition, the low-impedance grid return removes the tendency toward instability that is often common to the circuit of figure 4A. Neutralization is not required of the circuit of figure 4B, and in many cases parasitic suppression may be omitted. The price that must be paid is the additional excitation

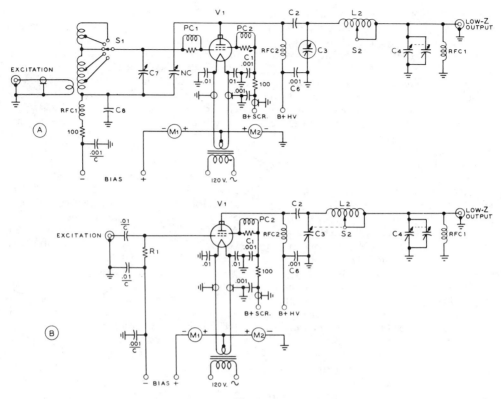

Figure 4

TYPICAL PI-NETWORK CONFIGURATIONS

A—Tapped grid and plate inductors are used with "bridge-type" neutralizing circuit for tetrode amplifier stage. Vacuum tuning capacitor is used in input section of pi-network
B—Untuned input circuit (resistance loaded) and plate inductor ganged with tuning capacitor comprise simple amplifier configuration. R_1 is usually 50-ohm, 100-watt carbon resistor.
PC_1, PC_2—50-ohm, 2-watt composition resistor, wound with 3 turns #12 enam. wire
Note: Alternately, PC_2 may be placed in the lead.

that is required to develop operating voltage across grid resistor R_1.

The pi-network circuit of figure 4B is interesting in that the rotary coil (L_2) and the plate tuning capacitor (C_3) are ganged together by a gear train, enabling the circuit to be tuned to resonance with one panel control instead of the two required by the circuit of figure 4A. Careful design of the rotary inductor will permit the elimination of the auxiliary high-frequency coil (L_1), thus reducing the cost and complexity of the circuit.

The Grounded-Screen Configuration   For maximum shielding, it is necessary to operate the tetrode tube with the screen at r-f ground potential. As the screen has a dc potential applied to it (in grid-driven circuits), it must be bypassed to ground to provide the necessary r-f return. The bypass capacitor employed must perform efficiently over a vast frequency spectrum that includes the operating range plus the region of possible vhf parasitic oscillations. This is a large order, and the usual bypass capacitors possess sufficient inductance to introduce regeneration into the screen circuit, degrading the grid-plate shielding to a marked degree. Nonlinearity and self-oscillation can be the result of this loss of circuit isolation. A solution to this problem is to eliminate the screen bypass capacitor, by grounding the screen terminals of the tube by means of a low-inductance strap. Screen voltage is then applied to the tube by grounding the positive terminal of the screen supply, and "floating" the negative of the screen and bias supplies below ground potential as shown in figure 5. Meters are placed in the separate-circuit cathode return leads, and each meter reads only the current flowing in that particular circuit. Operation of this grounded-screen circuit is normal in all respects, and it may be applied to any form of grid-driven tetrode amplifier with good results.

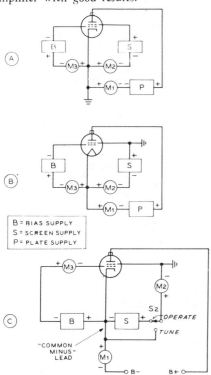

Figure 5

GROUNDED-SCREEN-GRID CONFIGURATION PROVIDES HIGH ORDER OF ISOLATION IN TETRODE AMPLIFIER STAGE

A—Typical amplifier circuit has cathode return at ground potential. All circuits return to cathode.
B—All circuits return to cathode, but ground point has been shifted to screen terminal of tube. Operation of the circuit remains the same, as potential differences between elements of the tube are the same as in circuit A.
C—Practical grounded-screen circuit. "Common minus" lead returns to negative of plate supply, which cannot be grounded. Switch S_2 removes screen voltage for tune-up purposes.

The Inductively Tuned Tank Circuit The output capacitance of large transmitting tubes and the residual circuit capacitance are often sufficiently great to prevent the plate tank circuit from having the desired value of Q, especially in the upper reaches of the hf range (28- to 54-MHz). Where tank capacitance values are small, it is possible for the output capacitance of the tube to be greater than the maximum desired value of tank capacitance. In some cases, it is possible to permit the circuit to operate with higher-than-normal Q, however this expedient is unsatisfactory when circulating tank current is high, as it usually is in high-frequency amplifiers.

A practical alternative is to employ *inductive tuning* and to dispense entirely with the input tuning capacitor which usually has a high minimum value of capacitance (figure 6). The input capacitance of the circuit is thus reduced to that of the output capacitance of the tube which may be more nearly the desired value. Circuit resonance is established by varying the inductance of the tank coil with a movable, shorted turn, or loop, which may be made of a short length of copper water pipe of the proper diameter. The shorted turn is inserted within the tank coil by a lead-screw mechanism, or it may be mounted at an angle within the coil and rotated so that its plane travels from a parallel to an oblique position with respect to the coil. The shorted turn should be silver plated and have no joints to hold r-f losses to a minimum. Due attention should be given to the driving mechanism so that unwanted, parasitic shorted turns do not exist in this device.

Push-Pull Tetrode Circuitry Tetrode tubes may be employed in push-pull amplifiers, although the modern trend is to parallel operation of these tubes. A typical circuit for push-pull operation is shown in figure 7. The remarks concerning the filament supply, plate feed, and grid bias in Section 21-1 apply equally to tetrode stages. Because of the high circuit gain of the tetrode amplifier, extreme care must be taken to limit intrastage feedback to an absolute minimum. It must be remembered with high-gain tubes of this type that almost full output can be obtained with practically zero grid excitation. Any minute amount of energy fed back from the plate circuit to the grid circuit can cause instability or oscillation. *Unless suitable precautions are incorporated in the electrical and mechanical design of the amplifier, this energy feedback will inevitably occur.*

Fortunately these precautions are simple. The grid and filament circuits must be isolated from the plate circuit. This is done by placing these circuits in an "electrically tight" box. All leads departing from this box are bypassed and filtered so that no r-f energy can pass along the leads into the box. This restricts the energy leakage path between the plate and grid circuits to the residual plate-to-grid capacity of the tetrode tubes. This capacity is of the order of 0.25 pF per tube, and under normal conditions is sufficient to produce a highly regenerative condition in the amplifier. Whether or not the amplifier will actually break into oscil-

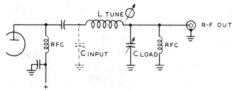

Figure 6

INDUCTIVE TUNING ELIMINATES INPUT TUNING CAPACITOR

lation is dependent upon circuit loading and residual lead inductance of the stage. Suffice to say that unless the tubes are actually neutralized a condition exists that will lead to circuit instability and oscillation under certain operating conditions.

Parasitic suppression is required with most modern high-gain tetrodes and may take place in either the plate or screen circuit. In some instances, suppressors are required in the grid circuit as well. Design of the suppressor is a cut-and-try process: if the inductor of the suppressor has too few turns, the parasitic oscillation will not be ade-

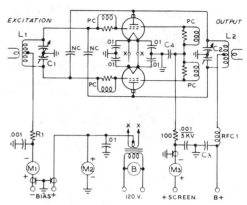

Figure 7

REPRESENTATIVE PUSH-PULL TETRODE AMPLIFIER CIRCUIT

The push-pull tetrode amplifier uses many of the same components required by the triode amplifier of figure 2. Parasitic suppressors may be placed in grid, screen, or plate leads. A low-inductance screen capacitor is required for proper amplifier operation. Capacitor C_4 may be .001 μF, 5 kV. Centralab type 858S-1000. Strap multiple screen terminals together at socket with $\frac{3}{8}$-inch copper strap for operation below 30 MHz and attach PC to center of strap. Blower required for many medium- and high-power tetrode tubes to cool filament and plate seals.

quately suppressed. Too many turns on the suppressor will allow too great an amount of fundamental frequency power to be absorbed by the suppressor and it will overheat and be destroyed. From 3 to 5 turns of #12 wire in parallel with a 50-ohm, 2-watt composition resistor will usually suffice for operation in the hf region. At 50 MHz, the suppressor inductor may take the form of a length of copper strap (often a section of the plate lead) shunted by the suppressor resistor.

VHF Push-Pull Tetrode Amplifiers   The circuit considerations for the vhf triode amplifier configuration apply equally well to the push-pull tetrode circuit shown in figure 8. The neutralization techniques applied to the tetrode tube however, may vary as the frequency of operation of the amplifier varies about the *self-neutralizing* frequency of the tetrode tube. At or near the upper frequency limit of operation, the inductance of the screen-grid lead of the tetrode cannot be ignored as it becomes of importance. Passage of r-f current through the screen lead produces a potential drop in the lead which may or may not be in phase with the grid voltage impressed on the tube. At the self-neutralizing frequency of the tube, the tube is inherently neutralized due to the voltage and current divisions within the tube which place the grid at the filament potential as far as plate-circuit action is concerned (see **Chapter 11**, Section 6). When the tetrode tube is operated below this frequency, normal neutralizing circuits apply; operation at the self-neutralizing frequency normally does not require neutralization, provided the input and output circuits are well shielded. Operation above the self-neutralizing frequency (in the range of 25 MHz to 100 MHz for large glass tubes, and in the range of 120 MHz to 600 MHz for ceramic, vhf tubes) requires neutralization, which may take the form of a series screen-tuning capacitor, such as shown in the illustration.

Neutralization is frequency sensitive and the amplifier should be neutralized at the operating frequency. Adjustment is conducted so as to reduce the power fed from the grid to the plate circuit. The amplifier may be driven with a test signal (filament and dc voltages removed) and the signal in the plate tank circuit measured with an r-f voltmeter. The neutralizing capacitors are adjusted in unison until a minimum of fed-through voltage is measured. A good null will be obtained provided that intrastage feedback is reduced to a minimum by proper shielding and lead-bypassing techniques.

Sweep Tubes in Linear Service   Listed in figure 9 are intermittent voice operation ratings for various TV sweep tubes when used for linear operation in the amateur service. While the plate dissipation of these tubes is of the order of 30 to 35 watts, the intermittent nature of amateur transmission and the high ratio of peak to average power in the human voice allow a good balance between peak power input, tube life, and tube cost to be achieved. For lower levels of intermodulation distortion,

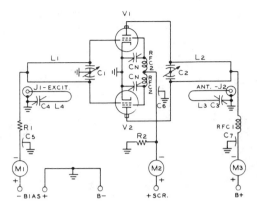

Figure 8

REPRESENTATIVE VHF PUSH-PULL TETRODE AMPLIFIER CIRCUIT

Tuned lines are used in grid and plate tank circuits in place of lumped inductances. Each screen circuit is series resonated to ground by neutralizing capacitor C_N. Wirewound resistor (R_1) is used in the grid-return circuit and frequency-rated r-f chokes in the plate and screen power leads. Screen resistor is included to complete screen-to-ground circuit, as discussed in text. Vhf type feedthrough capacitors are used for maximum suppression of r-f currents in power leads.

the user must shift to transmitting-type tubes rated for linear service, and which are designed to have low intermodulation distortion characteristics.

The owner of sweep-tube equipped SSB gear is cautioned that when the tubes are replaced, they should be of the same brand name as the original set, and the new tubes should be matched for equal values of resting plate current. Different manufacturers often have slightly different assembly techniques in matters such as lead length within the tube envelope. These minor construction differences do not affect operation in sweep circuits but may vastly alter the neutralization technique when the tube is used in r-f service. Certain brands of sweep tubes, moreover, have the internal connection between cathode and base pin taken from the top of the tube structure. This results in an extremely long cathode lead whose inductance is so high that it is impossible to secure sufficient grid drive at 28 MHz for efficient operation in linear amplifier service.

Because of electrical variations from tube-to-tube, it is suggested that sweep tubes be matched for identical values of resting plate current when they are used in parallel connection. One tube may be tested at a time in the amplifier and two tubes chosen for use whose resting plate currents are approximately equal at the same bias level. When unbalanced tubes are used, one will tend to draw more plate current than the other, thus leading to shorter tube life and increased intermodulation distortion on the signal.

21-3 Cathode-Driven Amplifier Design

The *cathode-driven,* or *grounded-grid* amplifier has achieved astounding popularity in recent years as a high-power linear stage for sideband application. Various versions of this circuit are illustrated in figure 10. In the basic circuit the control grid of the tube is at r-f ground potential and the exciting signal is applied to the cathode by means of a tuned circuit. Since the grid of the tube is grounded, it serves as a shield between the input and output circuits, making neutralization unnecessary in many instances. The very small plate-to-cathode capacitance of most tubes permits a minimum of intrastage coupling below 30 MHz. In addition, when zero-bias triodes or tetrodes are used, screen or bias supplies are not usually required.

Feedthrough Power A portion of the exciting power appears in the plate circuit of the grounded-grid (cathode-driven) amplifier and is termed *feedthrough* power. In any amplifier of this type, whether it be triode or tetrode, it is desirable to have a large ratio of feedthrough power to peak grid-driving power. The feedthrough power acts as a swamping resistor across the driving circuit to stabilize the effects of grid loading. The ratio of feedthrough power to driving power should be about 10 to 1 for best stage linearity. The feedthrough power provides the user with added output power he would not obtain from a more conventional circuit. The driver stage for the grounded-

			R-F LINEAR AMPLIFIER SERVICE FOR SSB AND CW										
			GRID DRIVEN, CLASS AB₁ MODE										
TUBE	FIL $\frac{V}{A}$	BASE	PLATE VOLTAGE Eb	SCREEN VOLTAGE Ec2	GRID VOLTAGE Ec1	ZERO SIG. PLATE CUR. Ibo	MAX.SIG. PLATE CUR. Ib	MAX.SIG. SCREEN CUR. Ic2	PL.LOAD IMPEDAN. Rp-Ω	PLATE INPUT PWR. W.	USEFUL POWER OUT. Po	AVERAGE PLATE DISSIP. Pd	3d ORDER IMD Db
6146	$\frac{6.3}{1.2}$	7CK	600	200	-46	25	103	9	3570	61	41	16	-25
			750	200	-51	25	118	7	2825	88	55	28	-22
6146B			800	290	-69	30	125	10	3620	100	59	35	-24
			800	290	-77	25	180	13	2300	145	91	45	-19
807	$\frac{6.3}{0.9}$	5AW	600	300	-34	18	70	8	4300	42	28	12	-23
			750	300	-35	15	70	8	5200	53	36	14	-23
6550	$\frac{6.3}{1.6}$	7S	680	340	-39	48	140	20	3010	95	67	26	-32
			800	290	-33	45	127	15	3920	102	70	29	-30
6DQ5	$\frac{6.3}{2.5}$	8JC	500	150	-46	48	170	17	1800	85	54	27	-28
			500	150	-46	48	182	13	1625	91	56	29	-26
			700	150	-49	35	182	11	2210	127	78	41	-23
			800	180	-67	30	250	13	1710	200	121	70	-19
6GB5	$\frac{6.3}{1.38}$	9NH	600	200	-41	23	192	14	1900	115	80	30	-18
6GE5	$\frac{6.3}{1.2}$	12BJ	600	200	-45	30	132	15	2500	79	51	23	-22
			800	250	-61	25	172	18	2750	138	90	39	-19
6HF5	$\frac{6.3}{2.25}$	12FB	500	140	-46	40	133	5	1900	67	35	29	-27
			800	125	-45	30	197	7	2170	158	100	48	-21
6JE6A	$\frac{6.3}{2.5}$	9QL	500	125	-44	40	110	4	2300	55	30	24	-26
			750	175	-63	27	218	15	1850	163	102	51	-20
6LQ6	$\frac{6.3}{2.5}$	9QL	750	175	-60	25	215	9	1850	161	102	49	-18
			800	200	-69	25	242	13	1850	197	124	60	-18

Figure 9

SWEEP TUBE DATA FOR CLASS AB₁ LINEAR AMPLIFIER SERVICE

Data for the 6LQ6 also applies to the 6MJ6

grid amplifier must, of course, supply the normal excitation power plus the feed-through power. Many commercial sideband exciters have power output capabilities of the order of 70 to 100 watts and are thus well suited to drive high-power grounded-grid linear amplifier stages whose total excitation requirements fall within this range.

Distortion Products Laboratory measurements made on various tubes in the circuit of figure 10A show that a distortion reduction of the order of 5 to 10 decibels in odd-order products can be obtained by operating the tube in cathode-driven service as opposed to grid-driven service. The improvement in distortion varies from tube type to tube type, but some order of improvement is noted for all tube types tested. Most amateur-type transmitting tubes provide signal-to-distortion ratios of −20 to −30 decibels at full output in class-AB₁ grid-driven operation. The ratio increases to approximately −25 to −40

decibels for class-B grounded-grid operation. Distortion improvement is substantial, but not as great as might otherwise be assumed from the large amount of feedback inherent in the grounded-grid circuit.

A simplified version of the grounded-grid amplifier is shown in figure 10B. This configuration utilizes an untuned input circuit.

It has inherent limitations, however, that should be recognized. In general, slightly less power output and efficiency is observed with the untuned-cathode circuit, odd-order distortion products run 4 to 6 decibels higher, and the circuit is harder to drive and match to the exciter than is the tuned-cathode circuit of figure 10A. Best results are obtained when the coaxial line of the driver stage is very short —a few feet or so. Optimum linearity requires cathode-circuit Q that can only be supplied by a high-C tank circuit.

Since the single-ended class-B grounded-grid linear amplifier draws grid current on only one-half (or less) of the operating

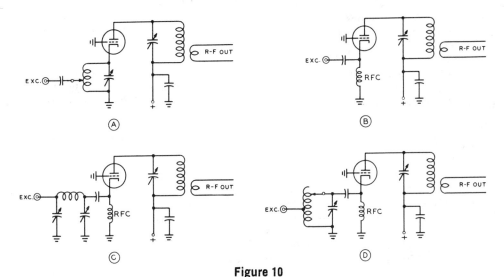

Figure 10

THE CATHODE-DRIVEN AMPLIFIER

Widely used as a linear amplifier for sideband service, the cathode-driven (g-g) circuit provides economy and simplicity, in addition to a worthwhile reduction in intermodulation distortion. A—The basic g-g amplifier employs tuned input circuit. B—A simplified circuit employs untuned r-f choke in cathode in place of the tuned circuit. Linearity and power output are inferior compared to circuit of figure A. C—Simple high-C pi-network may be used to match output impedance of sideband exciter to input impedance of grounded-grid stage. D—Parallel-tuned, high-C circuit may be employed for bandswitching amplifier. Excitation tap is adjusted to provide low value of SWR on exciter coaxial line.

cycle, the sideband exciter "sees" a low-impedance load during this time, and a very high-impedance load over the balance of the cycle. Linearity of the exciter is thereby affected and the distortion products of the exciter are enhanced. Thus, the *driving signal* is degraded in the cathode circuit of the grounded-grid stage unless the unbalanced input impedance can be modified in some fashion. A high-C tuned circuit, stores enough energy over the operating r-f cycle so that the exciter "sees" a relatively constant load at all times. In addition, the tuned circuit may be tapped or otherwise adjusted so that the SWR on the coaxial line coupling the exciter to the amplifier is relatively low. This is a great advantage, particularly in the case of those exciters having fixed-ratio pi-network output circuits designed expressly for a 50-ohm termination.

Finally, it must be noted that removal of the tuned cathode circuit breaks the amplifier plate-circuit return to the cathode, and r-f plate-current pulses must return to the cathode via the outer shield of the driver coaxial line and back via the center conductor! Extreme fluctuations in exciter loading, intermodulation distortion, and TVI can be noticed by changing the length of the cable between the exciter and the grounded-grid amplifier when an untuned-cathode input circuit and a long interconnecting coaxial line are used.

Cathode-Driven Amplifier Construction Design features of the single-ended and push-pull amplifiers discussed previously apply equally well to the grounded-grid stage. The g-g linear amplifier may have either configuration, although the majority of the g-g stages are single ended, as push-pull offers no distinct advantages and adds greatly to circuit complexity.

The *cathode circuit* of the amplifier is resonated to the operating frequency by means of a high-C tank (figure 10A). Resonance is indicated by maximum grid current of the stage. A low value of SWR on the driver coaxial line may be achieved by adjusting the tap on the tuned circuit, or by varying the capacitors of the pi-network

(figure 10C). Correct adjustments will produce minimum SWR and maximum amplifier grid current at the same settings. The cathode tank should have a Q of 2 or more.

The cathode circuit should be completely shielded from the plate circuit. It is common practice to mount the cathode components in an "r-f tight" box below the chassis of the amplifier, and to place the plate circuit components in a screened box above the chassis.

The *grid (or screen) circuit* of the tube is operated at r-f ground potential, or may have dc voltage applied to it to determine the operating parameters of the stage (figure 11A). In either case, the r-f path to ground must be short, and have extremely low inductance, otherwise the screening action of the element will be impaired. The grid (and screen) therefore, must be bypassed to ground over a frequency range that includes the operating spectrum as well as the region of possible vhf parasitic oscillations. This is quite a large order. The inherent inductance of the usual bypass capacitor plus the length of element lead within the tube is often sufficient to introduce enough regeneration into the circuit to degrade the linearity of the amplifier at high signal levels even though the instability is not great enough to cause parasitic oscillation. In addition, it is often desired to "unground" the grounded screen or grid sufficiently to permit a metering circuit to be inserted.

One practical solution to these problems is to shunt the tube element to ground by means of a 1-ohm composition resistor, bypassed with a .01-μF ceramic disc capacitor. The voltage drop caused by the flow of grid (or screen) current through the resistor can easily be measured by a millivoltmeter whose scale is calibrated in terms of element current (figure 11B).

The *plate circuit* of the grounded-grid amplifier is conventional, and either pi-network or inductive coupling to the load may be used.

Tuning the Grounded-Grid Amplifier Since the input and output circuits of the grounded-grid amplifier are in series, a certain proportion of driving power appears in the output circuit. If

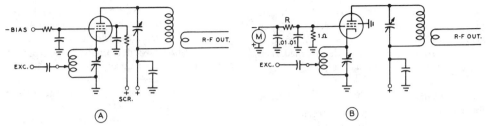

Figure 11

TETRODE TUBES MAY BE USED IN CATHODE-DRIVEN AMPLIFIERS

A—Tetrode tube may be used in cathode-driven configuration, with bias and screen voltages applied to elements which are at r-f ground potential. B—Grid current of grounded-grid tube is easily monitored by RC network which lifts grid above ground sufficiently to permit a milli-voltmeter to indicate voltage drop across 1-ohm resistor. Meter is a 0-1 dc milliammeter in series with appropriate multiplier resistor.

full excitation is applied to the stage and the output circuit is opened, or the plate voltage removed from the tube, practically all of the driving power will be dissipated by the grid of the tube. Overheating of this element will quickly occur under these circumstances, followed by damage to the tube. Full excitation should therefore never be applied to a grounded-grid stage unless plate voltage is applied beforehand, and the stage is loaded to the antenna.

Tuneup for sideband operation consists of applying full plate voltage and sufficient excitation (carrier injection) so that a small rise in resting plate (cathode) current is noted. The plate loading capacitor is set near full capacitance and the plate tank capacitor is adjusted for resonance (minimum plate current). Drive is advanced until grid current is noted and the plate circuit is loaded by decreasing the capacitance of the plate loading capacitor. The drive is increased until about one-half normal grid current flows, and loading is continued (re-resonating the plate tank capacitor as required) until loading is near normal. Finally, grid drive and loading are adjusted until PEP-condition plate and grid currents are normal. The values of plate and grid current should be logged for future reference. At this point, the amplifier is loaded to the maximum PEP input condition. In most cases, the amplifier and power supply are capable of operation at this power level for only a short period of time, and it is not

recommended that this condition be permitted for more than a minute or two.

The exciter is now switched to the SSB mode and, with speech excitation, the grid and plate currents of the cathode-driven stage should rise to approximately 40 to 50 percent of the previously logged PEP readings. The exact amount of meter movement with speech is variable and depends on meter damping and the peak to average ratio of the particular voice. Under no circumstances, however, should the voice meter readings exceed 50 percent of the PEP adjustment readings unless some form of speech compression is in use.

To properly load a linear amplifier for the so-called "two-kilowatt PEP" condition, *it is necessary for the amplifier to be tuned and loaded at the two-kilowatt level*, albeit briefly. It is necessary to use a dummy load to comply with the FCC regulations, or else a two-tone test signal should be used, as discussed in Chapter 9.

For best linearity, the output circuit of the grounded-grid stage should be over-coupled so that power output drops about 2-percent from maximum value. A simple output r-f voltmeter is indispensable for proper circuit adjustment. Excessive grid current is a sign of antenna undercoupling, and overcoupling is indicated by a rapid drop in output power. Proper grounded-grid stage operation can be determined by finding the optimum ratio between grid and plate current and by adjusting the drive level and loading to maintain this ratio. Many manufacturers now provide grounded-grid operation data for their tubes, and the ratio of grid to plate current can be determined from the data for each particular tube.

Choice of Tubes for G-G Service Not all tubes are suitable for grounded-grid service. In addition, the signal-to-distortion ratio of the suitable tubes varies over a wide range. Some of the best g-g performers are the 811A, 813, 4-400A, and 4-1000A. In addition, the 3-400Z, 3-500Z, 8873, 8877 and 3-1000Z triodes are specifically designed for low distortion, grounded-grid amplifier service.

Certain types ot tetrodes, exemplified by the 4-65A, 4X150A, 4CX300A, and 4CX-1000A should not be used as grounded-grid amplifiers unless grid bias and screen voltage are applied to the elements of the tube (figure 11A). The internal structure of these tubes permits unusually high values of grid current to flow when true grounded-grid circuitry is used, and the tube may be easily damaged by this mode of operation.

The efficiency of a typical cathode-driven amplifier runs between 55- and 65-percent, indicating that the tube employed should have plenty of plate dissipation. In general, the PEP input in watts to a tube operating in grounded-grid configuration can safely be about 2.5 to 3 times the rated plate dissipation. Because of the relatively low average-to-peak power of the human voice it is tempting to push this ratio to a higher figure in order to obtain more output from a given tube. This action is unwise in that the odd-order distortion products rise rapidly when the tube is overloaded, and because no safety margin is left (particularly in terms of grid dissipation) for tuning errors or circuit adjustment.

21-4 Neutralization of the Cathode-Driven Stage

A basic cathode-driven amplifier is shown in figure 12. The grid of the tube is at r-f ground potential and excitation is applied to the cathode, or filament. Instantaneous plate voltage is developed in series and in phase with the exciting voltage and the driver and amplifier may be thought of as operating in series to deliver power to the load. A tuned circuit is used in the input of the cathode-driven amplifier to enhance the regulation of the driver stage and to provide a proper termination for the driver over the operating cycle of the amplifier.

As the driver and amplifier are in series, the output current of the amplifier passes through the load resistance of the driver, causing a voltage drop across that resistance which opposes the original driving voltage. This indicates that inverse feedback is inherent in the cathode-driven amplifier to some degree if the driver has appreciable load resistance.

Most high-frequency, cathode-driven amplifiers are not neutralized, that is, no external neutralizing circuit is built in the amplifier. As the frequency of operation is raised, however, it will be found that intrastage feedback exists and the amplifier may exhibit signs of instability. The instability is due to voltage feedback within the amplifier tube (figure 13).

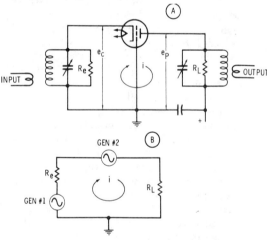

Figure 12

THE CATHODE-DRIVEN AMPLIFIER

A—Driving voltage e_c is applied to the cathodegrid circuit of the amplifier. Output voltage e_p appears across the plate load impedance.
B—The driver (generator 1) and the cathodedriven amplifier (generator 2) are in series with respect to the amplifier voltages. Cathode current of the amplifier (i) flows through the load resistance of the driver (R_e), contributing a degree of feedback to the system.

When a cathode-driven amplifier is operated at the higher frequencies, the internal capacitances and the inductance of the grid structure of the tube (or tubes) contribute to the degree of feedback. To achieve stability, the various feedback paths through the distributed constants in the tube strucure must be balanced out, or nulled, by neutralizing techniques. Proper neutralization is defined as the state in which: with plate and cathode tank circuits resonant and with maximum cathode voltage, minimum plate current and maximum power output occur simultaneously. This implies that input and output circuits are independent of each other with respect to common reactive currents, and that the tuning of the circuits reveals no interaction.

This definition provides the user of a cathode-driven ("grounded-grid") amplifier a quick and easy means of checking amplifier stability. When the amplifier is properly loaded and tuned with carrier insertion,

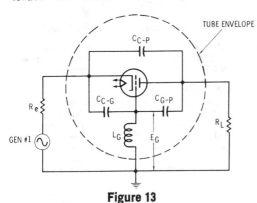

Figure 13

FEEDBACK PATHS WITHIN
CATHODE-DRIVEN AMPLIFIER

Cathode-plate, cathode-grid, and grid-plate capacitances, together with grid lead inductance (L_g) make up feedback paths that must be neutralized for proper stability of the amplifier, particularly in the vhf region. Two feedback paths enter the picture: the direct path from plate to cathode via capacitance C_{c-p}, and a more indirect path via the series capacitors (C_{c-g} and C_{g-p}) and grid inductance L_g.

maximum grid current and minimum plate current should appear at the same setting of the plate circuit tuning capacitor. If this does not happen, the amplifier is not neutralized in the strict sense of the word.

Neutralizing Circuits Stable operation, particularly at the higher frequencies, often calls for the cathode-driven amplifier to be neutralized. Complete circuit stability requires neutralization of two feedback paths, for which separate techniques are required. The first feedback path involves the cathode-plate capacitance (C_{c-p}). Although the capacitance involved is small, the path is critical and may require neutralization. This is accomplished either by a shunt inductance or by a balanced capacitive bridge. The first technique consists of connecting an inductance from plate to cathode of such magnitude as to pass back to the cathode a current equal in value but opposite in phase to the current passing through the cathode-plate capacitance (figure 14). This is a version of the well-known inductive neutralization circuit used in conventional grid-driven amplifiers to balance out the effects of grid-plate capacitance. The inductive neutralizing circuit is frequency sensitive as the inductor and cathode-plate capacitance of the tube form a frequency-sensitive resonant circuit at the operating frequency. Consequently, as the operating frequency is moved, the neutralizing circuit must be readjusted to resonance.

Bridge Neutralization The second neutralizing technique is a variation of the bridge neutralizing circuit used in grid-driven circuits (figure 15).

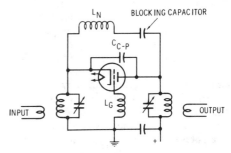

Figure 14

INDUCTIVE NEUTRALIZATION

Cathode-plate feedthrough capacitance is neutralized by making the capacitor part of a parallel-resonant circuit tuned to the operating frequency by the addition of inductor L_N. Blocking capacitor is added to remove dc plate voltage from the circuit.

The balanced input circuit provides equal out-of-phase voltage to which the cathode of the tube and the neutralizing capacitor are attached. The voltages are balanced in the output circuit when neutralization is achieved. Both capacitances are quite small, and the series lead inductance is relatively unimportant, consequently the bridge remains in balance over a wide frequency range.

Either neutralizing circuit can be properly adjusted even though the grid of the tube may not be at actual r-f ground potential because of the internal grid lead inductance L_g. Intrastage feedback resulting from this inductance requires a separate solution, apart from the neutralizing techniques just discussed.

Grid-Inductance Neutralization   A second feedback path exists in the cathode-driven amplifier which includes the grid-plate and grid-cathode capacitances and the series grid lead inductance (figure 16). These paths result in an apparent r-f leakage through the tube that may be many times greater than predicted. If the path is not neutralized, a voltage (e_L) appears on the grid which either increases or decreases the driving voltage, depending upon the values of internal tube capacitances and the value of the grid inductance. Oscillation may occur, even though the cathode-plate feedback path discussed earlier is completely neutralized.

The voltage (e_L) on the grid of the cathode-driven stage is determined by reaction between the total cathode-plate capacitance and a separate low-Q circuit composed of a capacitive voltage divider (C_{c-g} and C_{g-p} in series) together with grid inductance L_g. A certain frequency at which these two feedback paths nullify each other is the *self-neutralizing frequency* (f_1) of the tube. This frequency usually falls in the lower portion of the vhf spectrum for small transmitting tubes. All the elements composing the neutralizing circuit are within the tube, but connecting the tube into a circuit by wiring or socketing will alter the frequency.

The self-neutralizing phenomenon comes about because of a frequency-sensitive voltage balance that takes place within this net-

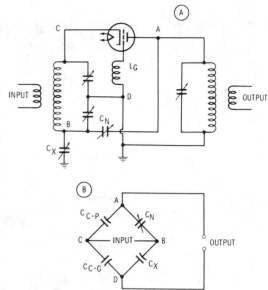

Figure 15

EQUIVALENT BRIDGE CIRCUIT

A—Cathode-plate bridge neutralizing circuit for cathode-driven amplifier. Balanced input tank provides equal, out-of-phase voltages at B and C.

B—Equivalent bridge circuit. Bridge is balanced except for C_x, which represents residual capacitance from point B to ground. If the balanced input circuit is high-C in comparison to the electrode capacitances, C_{c-g} and C_x are swamped out and bridge may be considered to be balanced. A capacitor from point B to ground provides exact balance.

work (figure 16A) and which may be explained by a vector diagram (figure 16B). The r-f plate voltage (e_p) causes a current (i) to flow through C_{g-p} and L_g. If the reactance of L_g is small in comparison with the reactance of C_{g-p} (as would be the case below the self-neutralizing frequency), the current (i) will lead the plate voltage (e_p) by 90 degrees. In flowing through L_g this current develops a grid voltage (e_L) which is 180 degrees out of phase with e_p and also the voltage (e_{fb}) fed back to the cathode via C_{c-p} and series-connected C_{c-g} and C_{g-p}.

At some frequency the voltage (e_L) developed across L_g will just equal the voltage fed back through the interelectrode capacitances (e_{fb}). The frequency at which (e_L is

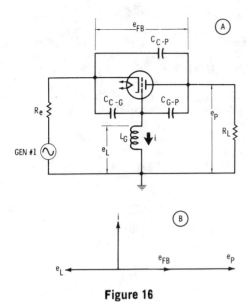

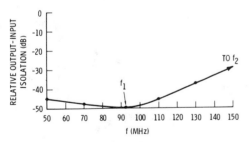

Figure 16

GRID-INDUCTANCE NEUTRALIZATION

A—Three terminal representation of cathode-driven amplifier showing internal capacitances of tube and grid-lead inductance.
B—Vector equivalent of feedback voltages in above circuit.

equal to e_{fb} is the self-neutralizing frequency. A second, somewhat higher, frequency at which the complex grid configuration is in a series-resonant state with respect to intrastage isolation is called the *grid series resonant frequency* (f_2) of the tube.

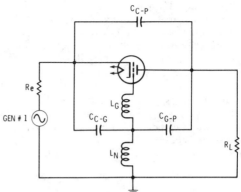

Figure 18

LOW-FREQUENCY NEUTRALIZING CIRCUIT

Below the self-neutralizing frequency of the tube, the point of self-neutralization may be adjusted by the addition of an inductance (L_N) in series with the grid-to-ground return of the tube.

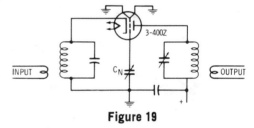

Figure 19

VHF NEUTRALIZING CIRCUIT

Cathode-driven amplifier is neutralized above the self-neutralizing frequency by placing a series capacitance in one grid lead. Neutralization adjustment is frequency sensitive and must be peaked for maximum intrastage isolation at the operating frequency.

The self-neutralizing characteristic of a 3-400Z type tube is shown in figure 17. A signal is applied to the grid of the tube and the transmission voltage through the tube measured at the plate. The test is conducted with the filament cold and no voltage applied to the tube. Above the self-neutralizing frequency, the intrastage isolation deteriorates as the test frequency approaches the series-resonant frequency f_2. Near the latter, tube operation is impractical, being further complicated by transit-time effects.

Figure 17

INPUT-OUTPUT ISOLATION OF 3-400Z IN CATHODE-DRIVEN CIRCUIT

Self-neutralizing frequency of 3-400Z is about 92 MHz. Tube is mounted in a special, shielded socket and measured in "cold" condition with filament unlit. Relative isolation is given since impedance of input and output circuits is not established.

Below the self-neutralizing frequency, the tube can be neutralized by the addition of a small inductor in the grid-to-ground path (figure 18). Above this frequency, the tube can be neutralized by the addition of a series capacitance in one of the grid leads (figure 19). The original self-neutralizing frequency (f_1) was little changed by the addition of the auxiliary circuit.

In the lower portion of the vhf spectrum only one neutralizing technique may be needed for a cathode-driven amplifier, at least as far as amplifier stability goes. As the frequency of operation is raised, however, both feedback circuits require attention to allow the amplifier to be properly neutralized. In the hf region, the cathode-driven amplifier, particularly when using well shielded, low capacitance tubes, probably will not require neutralization if the construction of the amplifier is such so that feedback between input and output circuits does not take place due to lack of shielding or feedback through the various power leads.

CHAPTER TWENTY-TWO

HF and VHF Power Amplifier Construction

Part I   HF Amplifiers

Construction of amateur SSB and vhf equipment is difficult at best because of the problems involved in obtaining many of the components. In addition, costly and complex test equipment is often required, making the task of checking and testing the equipment a formidable one for the amateur working on a slim budget.

On the other hand, dispensing with the streamlined cabinet in place of a homemade enclosure and making the equipment a single-band device, instead of a multiband one, can save money for the home builder who can spare the time to construct his equipment. This is especially true with power amplifiers for hf and vhf service which can be built at moderate cost and with a minimum of test equipment. Best of all, many of the components for these units still seem to be available at electronic surplus outlets and some of the major distributors of electronic equipment.

Shown in this chapter are amplifiers of varying complexity that are representative of current amateur construction practice, and that are relatively foolproof in construction and operation. While complete layout plans are not given, the experienced amateur should have no difficulty in building the equipment, providing the layout follows accepted engineering practices as outlined in this Handbook.

The first part of this chapter covers schematic diagrams of popular hf amplifiers that have been requested by readers of this Handbook. In order to conserve space and yet permit the maximum amount of information to be given, only a short description of each unit is provided. The more complex vhf units are described in detail in the second part of this chapter.

22-1   Amplifier Safety Summary

The amateur builder must remember that the equipment described in this chapter operates with extremely high voltages present and that consequently he should take precautions to protect himself from shock. The equipment should never be worked on when primary power is applied. This warning is doubly imperative to the solid-state experimenter, who often plunges his hand into equipment operating at a source supply of 12 volts, or less. *Voltages encountered in high power transmitting equipment are deadly and the equipment should never be turned on unless the operator is well clear of the circuitry involved.*

It is urged that a *shorting stick* be used to short out the high-voltage circuitry in equipment such as described in this chapter before work is done on it. The shorting stick is a dry, wood dowel rod having a metal point on the end. The point is connected to ground by means of a flexible, insulated wire jumper. Before work is started, the jumper is grounded to the negative of the power supply and the high-voltage terminal of the equipment shorted to ground by means of the stick. The wire-side of the shorting stick may be permanently hooked to the negative side of the power supply and mounted at the side of the workbench or operating table for quick use.

Before the equipment is placed on the air, it should be thoroughly bench-checked for low- and high-frequency parasitic oscillations as discussed in Chapter 11. It is then run at full input into a dummy load, of the type described in Chapter 31. In short, it is the responsibility of the builder and user of the equipment to make sure that it is working properly before it is put on the air in order to make certain that interference is not caused to other amateurs or other communication services.

Warning—Radiation Hazard

Avoid exposure to strong r-f fields, even at relatively low frequency. Absorption of r-f energy by human tissue is dependent on frequency. Under 30 MHz, most of the energy will pass completely through the human body with little attenuation or heating effect. Public health agencies are concerned with this hazard, however, even at these frequencies.

Many power tubes are specifically designed to generate or amplify radio-frequency power. There may be a relatively strong r-f field near the power tube and its associated circuitry—the more power involved and the larger the tube, the stronger the r-f field. Proper enclosure design and efficient coupling of r-f energy to the load will minimize the r-f field in the vicinity of the r-f power amplifier unit itself.

The dangers of r-f radiation are most severe at uhf and microwave frequencies. At these frequencies, extreme caution must be taken to avoid even brief exposures to strong r-f energy levels. In this range, the absorption of energy by human tissue is progressively greater and produces a heating effect which can be very rapid and destructive, particularly to sensitive tissue such as the eyes. Exposure of the human body to microwave radiation in excess of 10 milliwatts per square centimeter may be unsafe and may cause serious personal injury. Human eyes are particularly vulnerable to low-energy microwave radiation and blindness can result from overexposure. Exposure to high-energy microwaves can be fatal. Uhf and microwave energy must be contained properly by shielding and transmission lines. Arrangements should be made to prevent exposure of personnel to r-f fields in the vicinity of uhf and microwave tubes and in front of antenna systems.

Hazardous Operation of Power Tubes The operation of power tubes involves one or more of the following hazards, any one of which, in the absence of safe operating practices and precautions, could result in serious harm to personnel:

1. *High voltage.* Normal operating voltages can be deadly.
2. *R-f radiation.* Exposure to uhf or microwave radiation may cause serious bodily injury, possibly resulting in blindness or death. Cardiac pacemakers may be affected.
3. *Glass explosion.* Many electron tubes have glass envelopes. Breaking the glass can cause an implosion, which will result in an explosive scattering of glass particles. Handle glass tubes carefully.
4. *Hot Surfaces.* Surfaces of air-cooled radiators and other parts of tubes can reach temperatures of several hundred degrees and cause serious burns if touched.

22-2 Amplifier Schematics

Shown in this section are schematics of popular amplifier designs of interest to experimenters. Construction follows conventional techniques and where plate circuit components are not specified, the reader is referred to Chapter 11, which provides tables for pi- and pi-L networks based on the r-f plate impedance of the amplifier tube, or tubes. It is suggested that the equipment be built in shielded enclosed cabinets, or ventilated metal boxes to reduce the problem of TVI, yet at the same time provide operator safety and adequate ventilation so that tubes and components run at reasonable temperatures.

High frequency amplifiers employing a pi-network output circuit shown in this chapter are designed to have a second harmonic suppression of over 40 dB as referred to the mean power output. The third harmonic suppression should be better than 50 dB. Minor changes in lead length, component placement, and assembly may alter these figures. If more harmonic attenuation is desired, the reader is referred to Chapter 11 for the discussion and attenuation data for pi-L output networks. Substitution of a pi-L network for a pi-network will increase harmonic attenuation substantially.

A 700-Watt PEP Amplifier Using Sweep Tubes Shown in figure 1 is the schematic for an inexpensive, single-band linear amplifier. Four heavy duty 6MJ6 TV-type sweep tubes are used in cathode-driven service, operating at a plate potential of 800 volts. Plate current requirements is about 800 mA peak, or 400 mA average.

The four sweep tubes are parallel-connected, with the #1 grid tied to the cathode. This reduces the resting plate current and prevents excessive grid dissipation, which

occurs at peak power level when all grids are strapped together. The #2 and #3 grids are at r-f ground potential but have a small amount of negative bias applied by virtue of rectified filament voltage supplied by diode D_1.

The r-f plate impedance of this amplifier is of the order of 525 ohms and the plate tank circuit constants for that value are given in Table 1. The amplifier incorporates a *tune-up* switch (S_1) which permits adjustment with extra cathode bias in the circuit to reduce amplifier input during tuning.

Amplifier tuneup is straightforward. An SWR or power output meter in the antenna circuit is recommended for observation of amplifier operation. A low drive level is

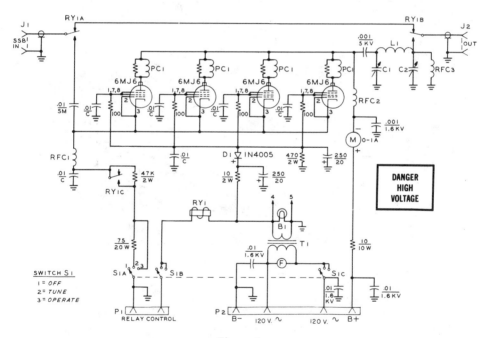

Figure 1

SCHEMATIC OF MULTIPLE SWEEP-TUBE AMPLIFIER

B_1—Red pilot lamp, 6.3-volt bulb

C_1—Tuning capacitor, 2 kV working voltage. See Table 1

C_2—Loading capacitor, 500 V working voltage. See Table 1

F—Fan. Ripley SK-4125 cooling fan, or equivalent

J_1, J—Coaxial receptacle, SO-239 or equivalent

L—Plate inductor. See Table 1

M—0-1 dc ammeter

PC_1—50-ohm, 2-watt composition resistor wound with 5 turns #14 wire spaced to length of resistor

RFC_1—7μH, 1 ampere. 60 turns #20 enamel close-wound on 1/2" diam form

RFC_2—200 μH, 800 mA. Miller RFC-3.5 or Miller 4534

RFC_3—2.5 mH, 50 mA

RY_1—3-pole, double-throw relay with 6.3-Vac coil

S_1A,B,C—3-pole, 3-position rotary switch

T_1—6.3 volts at 10 amperes

TABLE 1 PI-NETWORK VALUES FOR 50-OHM LOAD			
BAND (meters)	C_1(pF)	C_2(pF)	L_1(μH)
160	1800	7800	21.0
80	900	3400	10.6
40	440	1850	4.7
20	220	925	2.7
15	150	700	1.8
10	80	470	1.4

160- and 80-Meter coils wound with #12 wire. Other coils wound with $\frac{3}{16}$" copper tubing.

applied with switch S_1 in the *tune* position and the tuning and loading controls adjusted

for maximum output. Plate current should be held to 250 mA, or less. The switch is now turned to *operate* and the drive level increased for a plate current reading of about 500 mA. The amplifier controls are again adjusted for maximum output. Drive level is increased until maximum output is obtained at a resonant plate current of approximately 800 mA. This will occur with about 100 to 125 watts of drive power. Do not allow maximum continuous plate current to flow for more than 30 seconds, or the amplifier tubes may be damaged. If longer tuning time is required, switch to *tune* for a minute and reduce drive power, and then switch back to *operate* for another 30 sec-

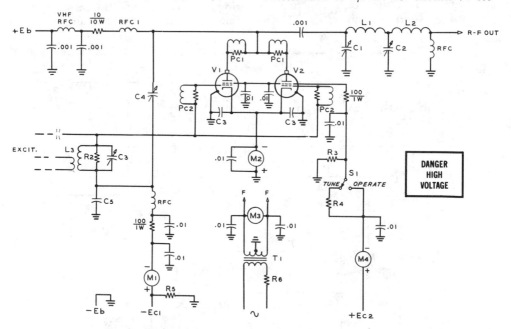

Figure 2

REPRESENTATIVE SCHEMATIC FOR GRID-DRIVEN TETRODE AMPLIFIER

C_1—Tuning capacitor. Working voltage equal to 1.5 times dc plate voltage

C_2—Loading capacitor. Working voltage equal to 0.3 times dc plate voltage

C_3—Grid tuning capacitor, receiving type. Capacitance about 2 pF/meter

C_4—Neutralizing capacitor. Working voltage equal to 2 times dc plate voltage

L_1, L_2—See chapter 11, Section 12 for data on pi-L network components

L_3—Resonates to operating frequency with given value of C_3. Excitation may be coupled by link or through capacitor, as indicated

M_1-M_4—See tube data sheet for representative meter ranges

PC_1—Four turns #16 spaced around 47-ohm, 2-watt composition resistor

PC_2—Four turns #20 spaced around 47-ohm, 1-watt composition resistor

RFC_1—Solenoid-type r-f choke. Approximately 200 μH

T_1—Filament transformer. Use unit having minimum wattage capacity to limit filament inrush current

R_6—Inrush current limiting resistor. Short out with VOX relay (see text)

Note: Check tube data sheet to determine if forced air cooling is required. See text for additional circuit details and component ratings.

onds. With experience in tuning, it will be found that 30 seconds is more than enough time.

Under voice conditions, with no speech clipping or compression, plate current will peak between 350 mA and 400 mA for full output. It is easy to produce higher meter readings but flattopping and distortion will result. For operation under c-w conditions, the function switch may be placed in the *tune* position, or amplifier loading and drive reduced for a current reading of approximately 500 mA. The amplifier is actuated for VOX operation by shorting the relay control terminals. (Note: While somewhat less rugged, type 6JE6 or 6LQ6 may be directly substituted for the type 6MJ6).

A Universal Grid-driven Tetrode Amplifier Shown in figure 2 is the generalized schematic for a parallel-connected, grid driven tetrode amplifier. The design is suitable for power tubes ranging from the 6146B to the 4CX1000A. Electrode voltages and currents should be derived from the manufacturer's data sheet for the specific tube type used.

The circuit is straightforward. A Pi-L plate circuit is chosen to provide maximum harmonic rejection. Using the data given in Chapter 11, the network can be designed to match the plate load resistance presented by the parallel-connected tubes. For two tubes, the data given in the charts for a particular value of load impedance is correct provided the total plate current figure is used to derive the load resistance.

Lead filtering and bypassing is accomplished as discussed elsewhere in this handbook. If the tube(s) have multiple cathode terminals, each one must be bypassed to ground separately (C_3). All components are chosen with regard to the operating potentials and currents.

A 10-ohm, 10-watt resistor is placed in series with the plate r-f choke to lower the Q of the choke and to serve as a surge suppressor in case of an inadvertent arc in the plate circuit. Individual parasitic suppressors are used in the plate and grid leads and the stage is neutralized by the bridge technique discussed in chapter 11. The value of the neutralizing capacitor (C_4) and the grid bypass capacitor (C_5) are chosen so as to permit the neutralization bridge to be bal-

anced with as large a value of C_5 as possible. Resistor R_2 is chosen to swamp out excessive drive power and should be a composition type device.

Resistors R_3 and R_5 are included to ensure that the screen and grid power supplies present a low dc impedance path to the cathode. This is especially critical in the case of high-gain tetrodes, such as the 4CX1000A and 4CX1500B. The grid resistor may be placed in the power supply and should hold the grid-to-ground resistance to less than 10,000 ohms. The series-connected screen resistor (R_4), permits the amplifier to be tuned up at reduced screen voltage and reduced input. Screen voltage is dropped to about half-value, or less, by means of the resistor which is brought into the circuit in the *tune* position of switch S_1. Resistor R_3 is chosen to hold the dc impedance-to-cathode path of the screen supply to between 10,000 to 50,000 ohms, depending on tube type regardless of the setting of switch S_1.

If the screen circuit of many high-gain power tetrodes is broken without a low-impedance path to ground present, the screen element of the tube will instantaneously assume the plate potential due to the electron flow within the tube. This can damage the screen and the screen bypass capacitor before the plate voltage can be removed. A permanent dc return path, provided by resistor R_3, prevents this from happening.

Plate-current metering is accomplished in the cathode circuit for safety reasons. The meter reads combined plate and screen current. The latter must be subtracted from the reading in order to ascertain the true plate current.

Some large tetrodes exhibit negative screen current over a portion of the operating cycle. The screen meter (M_4) should therefore incorporate an elevated zero point so that negative current can be read. Alternatively, the meter can be placed before the screen stabilizing resistor (R_3) so that it reads screen current (sometimes negative) and bleeder current (always positive).

In a Class-AB$_1$ amplifier, the grid is never driven positive and grid current never flows. Any appreciable grid current noted on modulation peaks indicates the amplifier is being driven into the distortion region.

For maximum stability and freedom from phase modulation of the SSB signal the amplifier should be neutralized even though it seems stable without the neutralizing circuit. Neutralization can be checked by driving the amplifier with a signal generator and measuring the r-f voltage in the plate tank circuit. No voltages are applied to the amplifier for this test.

Filament transformer T_1 should have no greater power capacity than that required for the job. This will restrict filament in-rush current, which can be as high as ten times normal current if an overly large filament transformer is used. Over a period of time, high in-rush current can crack the

tube seals or otherwise render the tube inoperative.

The easiest way to limit filament in-rush current is to use the smallest filament transformer practicable. In addition, a series resistor (R_6) placed in the primary circuit will tend to reduce the in-rush current to a great extent. The resistance value is chosen so as to reduce the filament voltage fifty percent when twice normal filament current is drawn from the secondary winding. The resistor is shorted out of the circuit by a time-delay relay after the filaments have

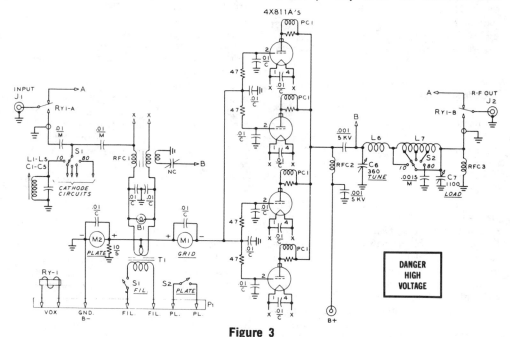

Figure 3

SCHEMATIC, 811A LINEAR AMPLIFIER

C_1—200-pF mica capacitor, 500-volt
C_2, C_3—470-pF mica capacitor, 500-volt
C_4, C_5—1000-pF mica capacitor, 500-volt
C_6—360-pF, 2.5 kV Johnson 154-2
C_7—1100-pF three-section receiving capacitor (broadcast type)
C_8—.0015-μF, 1250-volt mica capacitor. Sangamo type H
NC—Neutralizing capacitor. Approx. 15 pF, 1.25 kV. Use 100-pF midget capacitor with alternate plates removed to leave six stator plates and seven rotor plates.
L_1, L_2—(0.15 μH) 4 turns #16 on ½-inch diameter form.
L_3, L_4—(0.3 μH) 6 turns #16 on ½-inch diameter form. Coil L_3 is airwound, coil L_4 wound on form with powdered-iron slug. Both coils closewound.
L_5—(1.3 μH) 13 turns #16 on ½-inch diameter form, closewound. Dip all tuned circuits to center of band.
L_6—(10 meters): 4½ turns #10 e., 1½" diam., 1½" long

L_7—(80-15 meters): 13 turns #12 e., 2½" diam., 2" long. Resonates as follows: 80 meters: C_6—300 pF, C_7—1600 pF; 40 meters: C_6—150 pF, C_7—800 pF; 20 meters: C_6—75 pF, C_7—400 pF; 15 meters: C_6—50 pF, C_7—350 pF; 10 meters: C_6—35 pF; C_7, 200 pF. Above capacitances include output capacitance of tubes. Adjust coil taps accordingly.
RFC$_1$—16-ampere choke. 20 turns #12 enamel wire, bifilar wound (2 windings, 20 turns each) on ferrite core (½-inch diameter, Indiana General type CF-501, Q-1 material. Cut and break rod to length). Neutralizing coil: Seven turns #18 closewound around tube end of choke in same direction as bifilar winding.
RFC$_2$—200 μH, 1 ampere. B & W 800
RFC$_2$—2 mH, 100 mA. National R-100 or equiv.
PC$_1$—4 turns #18 wound over 47-ohm 2-watt composition resistor. Turns spaced the length of the resistor
T_1—6.3 volts at 16 amperes

reached operating temperature (about 0.5 second).

Plate and screen voltage should never be applied to oxide-coated, cathode type power tubes (4CX250B, 4CX1000A, etc.) until the filament has reached operating temperature, otherwise damage to the tube cathode may result. Delay time is commonly stipulated on the tube's data sheet. Thoriated tungsten type tubes (4-400A, 4-1000A, etc.) have a quick-heating filament and plate voltage may be applied to the tube at the same time the filament voltage is switched on.

The power gain of grid-driven tetrode tubes can approach 25 dB under some circumstances. This means that very little driving power is required: only a sufficient amount to overcome grid circuit losses. Swamping the grid circuit will increase the driving power and lower the stage gain. If maximum stage gain is desired, care must be taken to make sure that output power from the stage cannot return to the input circuit via the power leads of the amplifier and the power leads *of the driver*. In many instances, power is inadvertently introduced into these stages via unshielded wires. Care must be taken in amplifier construction to make sure that output power does not feed back into the driver circuitry.

External feedback can often be neutralized out by adjustment of neutralizing capacitor C_4. However, such an adjustment is frequency-sensitive and the stage must be re-neutralized if the frequency of operation is shifted. It is better to isolate the power leads and properly shield the amplifier to achieved neutralization over the frequency range of the amplifier.

It is prudent to monitor the filament voltage of the amplifier tubes. Filament voltage should be adjustable to within the limits established by the manufacturer. It is better to err on the under-voltage side, rather than the over-voltage side. Filament voltage should be monitored with a meter having a scale accuracy of one percent. The common rectifier type of meter in wide usage should not be relied upon for ac filament voltage measurement. Only an rms-responding type of meter of known accuracy should be used.

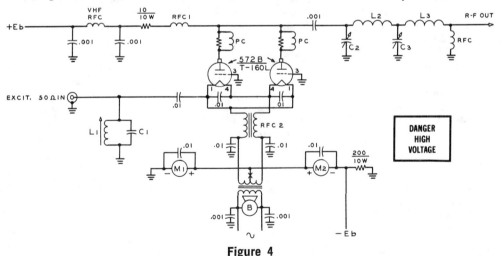

Figure 4

REPRESENTATIVE SCHEMATIC FOR GROUNDED-GRID TRIODE AMPLIFIER

C_1—Cathode tuning capacitor See table 2 for values
C_2—Tuning capacitor. Working voltage equal to 1.5 times dc plate voltage
C_3—Loading capacitor. Working voltage equal to 0.3 times dc plate voltage
L_1—Cathode tuning coil. See table 2
L_2, L_3—See chapter 11, Section 12 for data on pi-L network components
M_1—0-100 mA
M_2—0-500 mA

PC—four turns #16 spaced around 47-ohm, 2-watt composition resistor
RFC_1—Solenoid-type r-f choke. Approximate 200 μH
RFC_2—Bifilar winding. Each coil is 14 turns #12 e., on ferrite core, 5" long, ½" diameter. (Indiana-General CF-503)
T—Filament transformer. 6.3 volts at 8 amperes
Blower—Rotary fan, 4½" diameter impeller

An Inexpensive 811A Linear Amplifier This simple and inexpensive linear amplifier is designed for service on the 3.5- to 29.7-MHz hf amateur bands. It is capable of running 1-kW PEP input in SSB service when used with a plate supply providing 1500 volts at a peak plate current of 650 milliamperes. Plate load impedance is 1250 ohms. The schematic of the amplifier is shown in figure 3.

The four 811As are cathode-driven with provisions for neutralization. Drive requirement is 80 watts PEP for full input. Each grid of the 811A combination is at r-f ground and dc grid return is completed through a simple circuit that permits grid current measurement. Plate current is metered in the B-minus lead, with the negative lead of the power supply returned to the chassis ground of the amplifier. A built-in VOX relay provides antenna changeover for transceiver operation.

The amplifier is built on a chassis measuring 10" × 17" × 3" and fits within an aluminum inclosure made of perforated material bolted to the back of a standard relay rack-size panel. A bottom plate is made of the perforated material. Layout of parts is not critical provided reasonable care is taken to provide short, direct leads. The tubes are grouped at the corners of a square at one end of the chassis and quarter-inch holes are drilled around each socket to allow convection air currents to flow from beneath the chassis to help cool the tubes. The input circuits and filament transformer are mounted below deck. The tube sockets are recessed about an inch below the deck to conserve vertical space.

With the tubes in their sockets, the input circuits are dipped to the middle of each amateur band. The plate circuit inductor is dipped to frequency by setting the network capacitors to the given values and trimming the coil to establish resonance.

Injecting carrier, the amplifier is loaded to a peak plate current of 650 mA with a maximum grid current of 100 to 120 mA. The amplifier is neutralized on the 10-meter band. Starting with capacitor NC open (and with the top cover and bottom plate installed) the amplifier tuning capacitor should be swung out of resonance while grid and plate current readings are observed. The neu-

tralizing capacitance is gradually increased, using an insulated tool, until maximum grid current and minimum plate current are noted at the same setting of the tuning capacitor. Adjustment is not critical.

To limit plate dissipation, tuneup should be limited to 15 second periods of time every 30 seconds. Under SSB modulation, grid and plate currents for full input should kick up to about one half the carrier value.

A 572B/T-160L Multiband Amplifier A pair of 572B/T-160L high-mu triode tubes are used in this cathode-driven grounded-grid amplifier intended for multiband operation (figure 4). This amplifier is capable of 1-kW PEP input in SSB service when operated with a plate potential of 2500 volts at a peak plate current of 400mA. Plate load impedance is 3200 ohms and the component values for the plate tank circuit can be found in chapter 11 of this Handbook. A pi-L circuit is chosen to provide maximum harmonic attenuation but substitution of the simpler pi-network is discretionary.

The cathode circuit is fixed-tuned to the center of the amateur band in use (Table 2). Both cathode and plate circuits can be switched to provide rapid bandchange, if desired. If the coaxial lead from the exciter to the amplifier is short, the cathode tuned circuit may be omitted, provided the exciter has a variable load control adjustment (most solid-state exciters do not).

Table 2. Input Network Details

Band	L₁	C₁
	Circuit Component Values (Q ≅ 1)	
80	13 turns #18 e. on ½" diam. form. (1.3 µH) closewound	1000 pF, 1kV Mica
40	6 turns #14 e. on ½" diam. form, spaced.	1000 pF, 1kV Mica
20	6 turns #14 e. on ½" diam. form, spaced	470 pF, 1 kV Mica
15	4 turns as above adjust to resonance	200 pF, 1 kV Mica
10	4 turns as above adjust to resonance	200 pF, 1 kV Mica

The two tubes draw about 50 mA resting plate current. This may be reduced to near zero by placement of a zener diode at point X in the filament return circuit. A 3.9-volt 10-watt device (HEP-Z3500 or equivalent) should be used with the positive terminal connected to the centertap of the filament transformer.

Metering is accomplished in the negative power leads and the negative circuit of the high voltage supply is raised above ground by a 200-ohm 10-watt resistor placed across the meter circuitry. The potential difference between B-minus and ground is less than a volt but the negative circuit of the power supply must not be grounded, otherwise the meter circuitry will be shorted out.

Once the amplifier is wired and checked the tuned cathode circuit (L_1-C_1) should be adjusted to midband frequency with the aid of a dip meter. Plate circuit resonance can be roughly established in this fashion

Figure 5

TOP VIEW OF 4CX250B LINEAR AMPLIFIER

Center compartment contains the main r-f components. At the rear are the two 4CX300A tubes mounted on a small chassis adjacent to the blower. To the right of the tubes is the small drawn aluminum case containing the output reflectometer. Plate loading and tuning capacitors are mounted at the right of the compartment on the front subpanel. Central area contains the three plate-circuit inductors and the bandswitch. Low- and medium-frequency inductors are mounted to the sides of the compartment with small ceramic standoff insulators, and the high-frequency coil is supported by bandswitch and tuning capacitor. The plate r-f choke is placed vertically at the rear of the compartment with the plate-blocking capacitor atop it. The blower, filament transformer, and auxiliary components are mounted to the left of the r-f compartment. The circuit breaker overload potentiometer (R_3) is mounted to the outer wall of the inclosure. Electron tuning tube is mounted to the front panel by a bracket which encircles the tube.

when loading capacitor C_3 remains set at maximum capacitance.

Initial tuneup should be done at reduced plate voltage, say, 1500 volts. Drive is applied and tuning capacitor C_2 is adjusted for maximum power output as indicated by an external wattmeter or SWR meter. Capacitor C_3 (load) is then adjusted for maximum output.

Plate voltage is now raised to the operating value and drive power advanced until plate current is near 400 mA. The tuning and loading capacitors are adjusted for maximum power output and minimum plate current, which should be about 400 mA. Tuning adjustments should be limited to periods of less than 30 seconds in order to allow the tubes to cool. During normal operation, the anodes of the tubes will approach a dull red color on voice peaks.

It is important that proper ventilation be maintained about the tubes. The small axial fan is positioned to blow across the glass envelopes and the warm air is exhausted out the top of the amplifier cabinet.

Because these tubes are running at near maximum allowable input, the use of speech compression or clipping is not recommended.

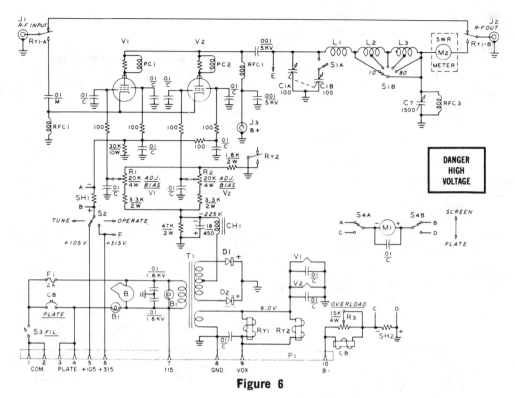

Figure 6

SCHEMATIC, 4CX250B AMPLIFIER

C_{1A}, $_B$—Split stator, 100-pF per section, .07" spacing.

C_2—1500 pF, .03" spacing. Jackson 4595/3/380 (LE-3).

L_1—(10-meter coil) 4 turns, $\frac{3}{16}$" copper tubing, 2¼" inside diam., 3" long

L_2—(15-20 meter coil) 6½ turns, ⅛" copper tubing, 2" inside diam., 3" long. Tap at center

L_3—(40-80 meter coil) 20 turns #14 (6 t.p.i.), 2¼" diam., 3¼" long. Tap at center

RFC_1, RFC_2—(84 μH) rated at 600 mA. J.W. Miller RFC-14

RFC_3—2.5 mH. J. W. Miller 4537

RY_1—Dpdt, ceramic insulation, 6.3-volt coil

RY_2—Spst, 6.3-volt coil

PC_1, PC_2—3 turns #16 spaced around 47-ohm, 2-watt composition resistor

CH_1—12 H, 30 mA. Stancor C-2318

T_1—460-volt, center tapped at 50 mA, 6.3 volts at 5 amp, Stancor P-8155. Remove turns from filament winding to provide 6.0 volts under load

B—Ripley 81 (left hand) with 2¾" impeller, 3100 r.p.m., 58 c.f.m.

J_1, J_2—coaxial receptacle, SO-239

S_{1A}, $_B$—2-pole, 6-position. Radio Switch Corp. (Marlboro, N. J.) Model 86 with two style A Rotors (standard)

SH_1—10 ohms, 1-watt

SH_2—500-mA shunt to match meter movement

CB—500-mA circuit breaker (Heinemann), 15-amp service.

M_1—0-50 dc milliamperes. Calibrate scale for 0-50 and 0-500 mA ranges

D_1, D_2—1N4005, 600-volt PIV, 1 amp

Sockets—Eimac SK-760 for 4CX300A. Eimac SK-640 plus SK-606 chimney for 4X150A and 4CX250B

4CX250B Linear Amplifier for 3.5-29.7 MHz Two 4CX250Bs, 4X150As, or 4CX300As are operated in the cathode-driven mode in this 1-kW PEP linear amplifier (figure 5). The amplifier is designed for continuous service and may be run at full input for RTTY or SSTV service.

The two high-gain tetrodes are run in class AB_1 mode with drive applied to the cathodes and normal dc operating potentials applied to the screen and grid elements from an external power supply (figures 6 and 7). Individual bias potentiometers (R_1, R_2) are provided to electrically balance the tubes to draw equal values of resting plate current.

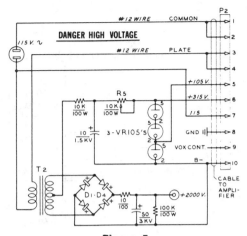

Figure 7

4CX250B PLATE AND SCREEN
POWER SUPPLY

T_2—1600-volt center tap, 500-mA secondary. Center tap insulated for 3kV. 117-volt primary
D_1-D_4—Diode bridge. Each leg requires six 1N4005 silicon diodes, 500-volt PIV at 1 ampere in series. Each diode is shunted by a .01-μF ceramic capacitor and a 470K, 1-watt resistor

A built-in bias supply provides -225 volts and the VOX relay permits plate current cutoff in the receive mode.

An electron-ray peak indicator (figure 8) is incorporated in the amplifier which samples the instantaneous r-f plate voltage, a portion of which is used for ALC voltage. The electron-ray tuning tube is used to establish proper plate loading. With no drive signal, the pattern of the tube is open, gradually closing with increased signal voltage until at the optimum plate load condition the pattern is closed, showing a solid green bar in the viewing portion of the tube.

In the standby mode, the linear amplifier is biased to cutoff by relay RY_2, permitting the use of an intermittent voice service-rated power supply (see Power Supply chapter).

The amplifier is built on a chassis measuring 14″ × 10″ × 3″ and fits within a shielded inclosure. The main bandswitch and pi-network loading capacitor are contained in cutout areas in the chassis. The tubes are mounted in a small box at the chassis rear which measures about 5¼″ × 3¼″ × 2″ high. Sockets and auxiliary components are placed in the box, one end of which has a hole cut in it to match the opening of the

blower. Cooling air is exhausted through the sockets and chimneys. The three sections of the plate tank coil are placed in the center area of the chassis behind the bandswitch. The electron-ray tube is mounted horizontally at the rear of the panel behind a thin cutout.

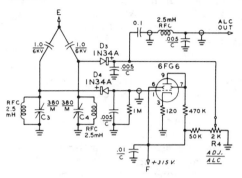

Figure 8

ELECTRON-EYE PEAK INDICATOR
AND ALC CIRCUITRY

The 6FG6/EM-84 tuning indicator is used for an r-f peak-level indicator in the linear amplifier. R-f voltage is sampled, rectified, and applied to the gate (pin 1) of the indicator. The pattern is formed between the deflection elements (pins 6 and 7) and appears as a horizontal line. Amplitude of indication is adjustable by means of mica compression capacitor C_1. ALC control voltage is taken from plate circuit and magnitude established by capacitor C_4. Control point may be set by adjusting diode bias voltage with "Adjust ALC" potentiometer R_4.

Filament voltage is checked at 6.0 volts. The amplifier inclosure is closed and high voltages are applied. One tube at a time is run with no drive signal and the bias adjusted for a resting plate current of 100 mA. Carrier is now inserted and the amplifier loaded and tuned for a maximum peak plate current of 500 mA. Screen current will be 20 to 30 mA, which includes the bleeder current flowing through the 30K screen resistor. Power output will run about 650 watts on all bands.

Once the amplifier is operating properly, the electron-ray tube is adjusted to completely close at maximum PEP power input by adjustment of capacitor C_3. Once set, voice peaks will just cause the eye to close. The magnitude of the ALC voltage is set by adjustment of capacitor C_1 and potentiometer R_1 (which controls the threshold voltage). For c-w operation, the amplifier is loaded to a current of 500 mA.

A 4-1000A Grounded- The 4-1000A tetrode
grid Amplifier makes an excellent grounded-grid triode when the grid and screen are strapped together (figure 9). At a plate potential of 3.5 kV to 4 kV the 4-1000A will operate at 2 kW PEP input with a driving power of only 110 watts, PEP.

An L-network is used to match the input impedance of the tube (about 100 ohms) to a 50-ohm drive source. Data for the network is given in Table 3. Heavy duty, transmitting-type mica capacitors are suggested for cathode capacitor C_3.

Gride and plate current metering is done in the ground return circuits. If cutoff bias is desired during standby periods, a 25 K, 50-watt resistor may be placed in the filament return circuit at X. The resistor is then shorted out by contacts on the VOX relay during transmissions.

A pi-L plate circuit is used for maximum harmonic attenuation. The plate load impedance of the 4-1000A is 3000 ohms at a plate potential of 3.5 kV and 3500 ohms for a potential of 4 kV. Data to design the plate circuit is given in chapter 11 of this Handbook. Typical operating values for the amplifier are given in Table 3.

22-3 The KW-1 Mark III Linear Amplifier Using the 8875

This compact desktop linear amplifier, is a third generation descendant of the popular 1000-watt PEP amplifier featured in various forms in the last three editions of this Handbook. This new version operates on all amateur bands between 3.5 MHz and 29.7 MHz with good efficiency. The KW-1 amplifier features a single 8875 ceramic high-μ power triode with a 300-watt anode dissipation rating operating in a class-B, cathode-driven configuration. Peak power input is 1000 watts for SSB voice operation, 800 watts for intermittent c-w operation, and 500 watts for continuous RTTY service.

The 8875 anode has a transverse cooler requiring forced-air cooling directed cross-

Table 3.

Band	Cathode Network (Dip to center of amateur band)		
	L$_3$	C$_3$	
80	(2.3 μH) 20 turns #14, ¾" diam. 2¼" long	500 pF XMTG Mica	
40	(1.2 μH) 12 turns #10, ¾" diam. 1½" long	250 pF XMTG Mica	
20	(0.6 μH) 5 turns #10, 1¼" diam. 1½" long	125 pF XMTG Mica	
15	(0.4 μH) 4 turns #10, 1½" diam. 1¼" long	90 pF XMTG Mica	
10	(0.3 μH) 4 turns #10, 1½" diam. 1½" long	60 pF XMTG Mica	
Typical Operating Characteristics			
Plate Voltage	3.0	3.5 kV	
Resting Plate Current	100	110 mA	
Single Tone Plate Current	700	650 mA	
Single Tone G$_1$ +G$_2$ Current	275	230 mA	
Single Tone Drive Power	120	110 watts	

ways. Maximum dissipation is realized with ducted air to the cooler from a small, low noise blower mounted near the tube.

The 8875 is rated for 250 mA dc continuous anode current. In intermittent voice service or keyed c-w operation where short term duty does not exceed 50%, the dc anode current may be 500 mA during the "on" time. During very short test periods, the tube may be operated at the full 500 mA value but care must be taken to keep the "on" time as short as possible, with sufficient "off" time to allow for tube cooling.

The KW-1, Mark III linear amplifier is small enough to be placed on the operating table next to an SSB transceiver or exciter (figure 10). At 2500 volts anode potential, third-order products are better than −30 decibels below one tone of a two-tone test signal.

The Amplifier The schematic of the KW-1
Circuit amplifier is shown in figure 11. The 8875 is operated in a cathode-driven mode using switchable

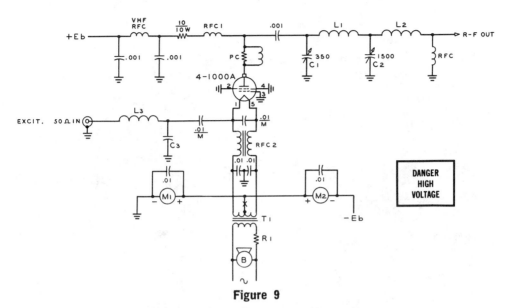

Figure 9

SCHEMATIC, 4-1000A G-G AMPLIFIER

C₁—Tuning capacitor. 350 pF at 6 kV
C₂—Loading capacitor. 1500 pF at 1 kV
C₃—Input tuning capacitor. See table 3
L₁, L₂, L₃—See table 3
M₁—Grid meter.0-500 mA
M₂—Plate meter. 0-1000 mA
PC—4 turns #16, ½″ diameter around 47-ohm 2-watt resistor
RFC₁—(Approx. 60 µH.) 90 turns #26 e., space-wound wire diameter, 3⅜″ long, ¾″ diam. on ceramic or teflon form. Series resonant at 26

MHz. (Use B&W 800 choke with 10 turns removed from top).
RFC₂—Bifilar winding. Each coil is 14 turns #10 e. on ferrite core, 5″ long, ½″ diam. (Indiana-General CF-503).
T₁—7.5 volts, 21 amperes. Chicago-Stancor R-6457
Socket for 4-1000A—SK-510
Chimney for 4-1000A—SK-506
Blower—20 cu. ft./min. Dayton 1C-180 or Ripley LR-81

cathode input transformers for each band (see Table 4) and a tapped pi-network output circuit. A small degree of r-f feedback is incorporated in the design by the choice of the 200-pF grid bypass capacitors on the tube, placing the grid above r-f ground by

Table 4.

Cathode Transformers, T_1-T_5 Wound on ⅜″ diameter forms, slug-tuned.			
T₁-(80 Meters)	24 turns #16e.	C6 (omitted).	C5= 470pF.
T₂-(40 Meters)	17 turns #16e.	C6= 510pf.	C5= 310pF.
T₃-(20 Meters)	9½ turns #18e.	C6= 360pf.	C5= 200pF.
T₄-(15 Meters)	4½ turns #18e.	C6 (omitted).	C5= 75pF.
T₅-(10 Meters)	3½ turns #18e.	C6 (omitted).	C5= 68pF.

the small voltage drop created across a divider formed by the plate-grid and grid-ground capacitances.

The power gain of the 8875 is quite high and—even with the r-f feedback—only 25 watts PEP drive power is required. A resistive T-pad is included in the input circuit which raises the drive level to about 100 watts PEP to accommodate some of the higher power SSB exciters. The pad may be omitted if a lower driving level is desired.

Because the grid of the 8875 is not at ground potential, a safety gap (surge arrestor) is placed from grid to ground (SG₁), which will ionize and "fire" when the grid potential exceeds the breakdown voltage of the gap. This protects the grid and cathode of the tube from transient voltages that may develop in the circuit.

Since the 8875 has a separate cathode, the filament may be isolated from the input circuit. It is not necessary in the hf region,

Figure 10

THE KW-1, MARK III LINEAR AMPLIFIER

This amplifier covers all hf amateur bands between 80 and 10 meters using an 8875 ceramic, high-μ power triode. A cathode-driven circuit is employed and the amplifier is capable of 1000 watts PEP input for SSB. The unit is housed in an aluminum cabinet and is self-contained except for the power supply. At the top of the panel are the multimeter and plate meter, with the plate tuning control at the center left and the loading control at the right. The plate bandswitch is at center, with the cathode bandswitch at the lower right. The amplifier cabinet is light gray with a dark gray panel. After the lettering is applied, the panel is sprayed with clear Krylon enamel to protect the lettering.

but a special trifilar filament choke is used to permit the cathode to be returned to dc ground, as shown in the schematic.

Resting plate current of the 8875 is set by the *Adjust Bias* potentiometer. A built-in bias supply also provides control voltage for the transmit relay, RY_1. A series connected diode in the control circuit serves to keep the relay transient voltage from upsetting the bias circuit. A separate filament transformer is used for the 8875 and a primary potentiometer allows the voltage to be set at 6.3 volts at the socket of the tube.

The control circuit is designed to prevent application of r-f drive without plate voltage and a 60-second time delay unit (TD) prevents plate voltage from being applied before the cathode of the tube reaches operating potential.

A single 50-μA dc meter is used to monitor grid current, high voltage, and relative power output. Grid current is read across a 5-percent resistor in the grid-bias return lead. Plate voltage is read indirectly across the last resistor in the power-supply bleeder string. The full-scale meter readings are 50 mA and 5000 volts for grid current and plate voltage respectively.

The KW-1 Mark III amplifier plate circuit is a conventional pi-network arrangement with additional plate tuning capacitance (C_2) added to the circuit on the 80-meter band by means of switch S_2. The plate coil is divided into two sections; the smaller, air-wound coil being used for 10 and 15 meters and the larger coil for 20, 40, and 80 meters. The network is designed to match a nominal 50-ohm load having an SWR of 3

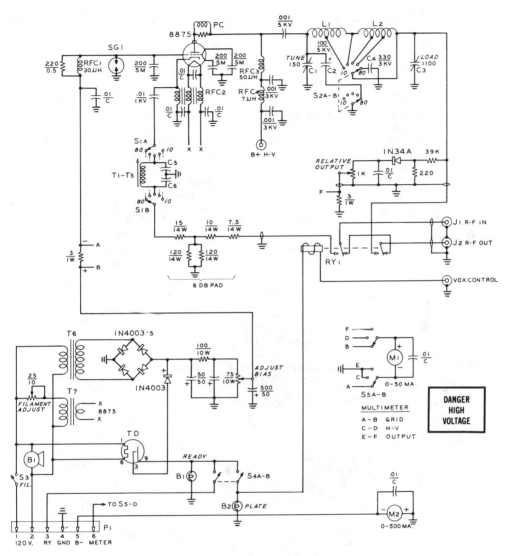

Figure 11

SCHEMATIC, KW-1 LINEAR AMPLIFIER

C_1—150 pF, 3 kV. Johnson 154-8 (26-96)

C_2—Centralab type 858S (21-109)

C_3—1100 pF, (26-97, 1700 pF may be used)

L_1—10-15 meter coil. 9 turns #10 wire, 1⅛" inside diameter, 2¼" long. 10-meter tap 4 turns from plate end (40-596)

L_2—20-40-80 meter coil. 1¹³⁄₁₆" diameter form, 4" long. Wound with #16 wire at 9 turns per inch. 20-meter section, 4 turns; 40-meter section, 7 turns; 80-meter section, 10 turns. Space between sections is ¼".

RFC_1—30 μH (45-18)

RFC_2—Trifilar choke. 20 turns #14 e. on ½" diameter ferrite core, 2¼" long (Indiana General). Interwind with third winding of #22

insulated wire. (45-60 with interwound winding of #22 insulated wire)

RFC_3—50 μH (45-61), or Ohmite Z-14

SG_1—Surge arrestor, 230-volt peak. Signalite CG-230L, Siemens B1-A230 or Reliable Electric SR-P17170

T_6—24 volts, 1 ampere

T_7—6.3 volts, 5 amperes

TD—Time delay relay, 60 seconds. Amperite 115C60T

B_1—Dayton 2C782. 3160 rpm, 2¼" wheel

PC—4½ turns #16 around 50-ohm, 1-watt composition resistor

Note: Heath part numbers given in parenthesis.

or less. An additional loading capacitance (C_4) is automatically switched into the circuit for 80-meter operation.

Amplifier Construction The amplifier is built on an aluminum chassis measuring 12″ × 8″ × 2½″. Inclosure height is 7″. Front and back panels of the box are cut from ⅛″ aluminum and the U-shaped cover is made of thin aluminum sheet. A 6″ × 3″ perforated aluminum plate is riveted in a cutout in the top of the cover to allow cooling air to escape from the inclosure. Angle stock is bolted around the top and side edges of the front and rear panels as a mounting surface for the cover.

The two meters are inclosed in a cut down minibox which serves as an r-f shield and an L-shaped bracket shields the filament transformer and antenna relay from the amplifier output circuitry.

Placement of the major components may be seen in figure 12. The 8875 is positioned carefully in front of the orifice of the blower and about one inch away. Six quarter-inch holes are drilled in the chassis around the tube socket to allow under-chassis air to be drawn up by convection to cool the base of the tube.

The cathode tuned circuits (T_1, T_5) and the time delay relay are mounted on an under-chassis shield plate, as seen in figure 13. The resistors making up the input attenuator are mounted immediately to the rear of this plate on two phenolic terminal strips.

Figure 12

INSIDE VIEW OF THE KW-1 AMPLIFIER

The 8875 tube is at the left with the blower positioned to force air across the anode cooler. Six holes are drilled in the chassis under the 8875 to allow air to escape from under the chassis by convection, thus cooling the tube base. The 80-40-20 meter plate coil is bolted vertically to the chassis at center with the high-frequency air-wound coil supported between the tuning capacitor and the bandswitch. The bias-control potentiometer is mounted on the shield plate behind the loading capacitor.

Figure 13

UNDER-CHASSIS VIEW OF AMPLIFIER

The tuned cathode circuits are in the partitioned area at the upper right with the input attenuator pad directly behind it. At center are the glass encapsulated time-delay relay and the bias power supply. The 8875 socket and filament choke are at lower left.

Many of the components used in this amplifier are replacement parts for the *Heath SB-200* linear amplifier and were ordered directly from the *Service Department, Heath Co.*, Benton Harbor, Michigan 49022 under the identification number given in the parts list. Other similar components will work as well as the particular parts used in this amplifier.

Transmitter Power Supply The schematic of the KW-1 Mark III power supply is shown in figure 15. A multi-conductor cable connects the supply to the amplifier along with the high voltage lead, which is run in RG-59/U coax. The *filament* switch on the panel of the amplifier

controls the primary power circuit and the time delay relay and *plate* switch activate the *transmit* relay control circuitry. The power supply is energized by grounding the *VOX control* terminal on the rear of the amplifier chassis. The power supply provides approximately 2500 volts under no-signal conditions and 2100 volts at a peak plate current of 450 milliamperes. The dynamic characteristics of the power supply allow the amplifier to develop about 20% greater peak SSB envelope power for a given level of c-w input. The power supply utilizes a voltage doubler circuit and incorporates high voltage metering. Supply voltage is checked with a meter of known accuracy and the *meter calibrate* potentiometer is adjusted to provide the same reading on the panel meter of the amplifier.

Figure 14.

CLOSEUP OF 8875 SOCKET WIRING

To the right of the socket is the small glass-encapsulated spark gap connected between grid terminals and the chassis. The trifilar filament choke is in the foreground.

Amplifier Tuning and Adjustment   Wiring should be completely checked before power is applied. The approximate settings of the plate tank circuit should be determined for each band with the aid of a grid-dip oscillator. The

slug cores of the cathode transformers are adjusted to midband resonance for each position of the bandchange switch.

The *adjust bias* potentiometer is set for maximum grid bias and filament voltage is applied to the 8875 and checked at the socket. *Caution:* The cabinet cover should now be bolted in place as high voltage points are exposed in the amplifier.

An exciter and dummy load are attached to the amplifier and high voltage applied. The VOX circuit should be energized by grounding the VOX terminal. The amplifier is now ready to be tuned up. After the time-delay relay has closed, the bias potentiometer is adjusted for a resting plate current of about 25 mA. A small amount of carrier is applied to the amplifier as a tuning signal until about 150 mA of plate current is indicated. The amplifier is tuned to resonance and peaked for maximum reading on the output meter. Once resonance is established, the tuning and loading controls are adjusted for maximum output as the driving signal is gradually increased. The loading capacitor should be near full capacitance for 80 and 40 meters, about 60 percent meshed for 20 meters and slightly less for 15 and 10 meters. Maximum carrier signal plate current is 450 mA and corresponding grid current is 30 mA.

The last step is to peak the input transformers for maximum grid current on each band, retarding the excitation so as not to overdrive the amplifier.

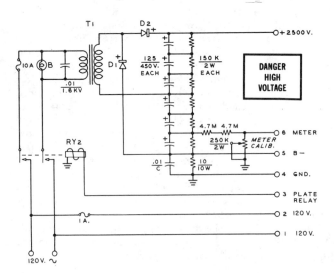

Figure 15

POWER SUPPLY, KW-1 AMPLIFIER

T₁—117-volt primary. 820-volt, 0.5-ampere secondary (54-151)
D₁, D₂—Each leg: Five 1N4005 diodes. Place .01 μF, 1.6-kV disc capacitor and 100K, 1-watt resistor across each diode
RY₂—24-volt dc coil, DPDT

Carrier is now removed and voice modulation applied. A maximum of 1000 watts PEP input is achieved with peak voice current of about 210 milliamperes. For c-w operation, carrier insertion is used and the amplifier is loaded to a plate current of 400 mA.

22-4   The 500Z 2-kW PEP Linear Amplifier for 10 thru 80 Meters

Two 3-400Z or 3-500Z high-μ triode tubes form the basis for this compact, multiband, high-power desk-top linear amplifier.

Heavy-duty design combined with rugged components permit the amplifier to be run at full legal power level for SSB or c-w service. Measuring only $16'' \times 8'' \times 13''$ deep the amplifier is small enough to be placed on the operating table adjacent to the SSB transceiver or exciter.

Auxiliary circuitry permits the exciter to bypass the amplifier, if desired, for low-power operation, and the unit incorporates automatic load control (ALC) for optimum voice efficiency in SSB operation. At maximum input level, the third-order intermodulation products are better than -33 decibels below one tone of a two-tone test signal, attesting to the high degree of linearity attained without the use of auxiliary feedback

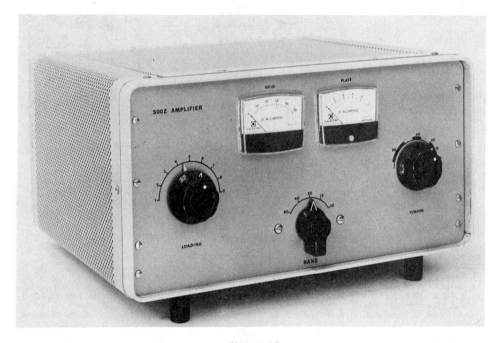

Figure 16

TWO-KILOWATT PEP INPUT IS FEATURED IN THIS COMPACT AMPLIFIER USING ZERO BIAS TRIODES

This desk-top amplifier allows maximum PEP input on all high-frequency amateur bands. Two zero-bias 3-500Z triodes are used in a cathode-driven, grounded-grid circuit. ALC is included as well as a high-efficiency, low-noise fan cooling system.
The amplifier is housed in a perforated aluminum case and is entirely self-contained, except for the power supply. At the top of the front panel are the grid and plate meters. The antenna loading control (C_7) is at the left and the plate tuning control (C_6) at the right. Both capacitors are driven through small precision planetary vernier drives. The bandswitch is centered at the lower portion of the panel.
The amplifier cabinet is gray, with light-green panel. After the lettering is applied, the panel is sprayed with clear Krylon enamel to protect the lettering. The unit is elevated above the desk top on rubber feet to permit good movement of air about the under-chassis area.

circuitry. Peak drive power is of the order of 90 watts, and the amplifier may be driven by any SSB exciter capable of this power output.

The Amplifier Circuit This 2000-watt PEP linear amplifier employs two zero-bias triode tubes connected in cathode-driven, grounded-grid configuration.

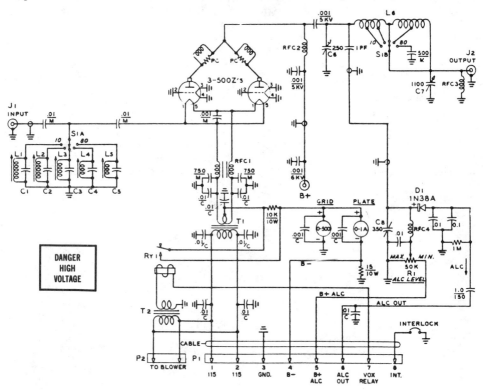

Figure 17

SCHEMATIC OF 500Z LINEAR AMPLIFIER

C_1—200 pF, 1-kV mica
C_2, C_3—470 pF, 1-kV mica
C_4, C_5—1000 pF, 1-kV mica

C_6—250 pF, 3-kV, .075″ spacing. Johnson 154-9
C_7—1100 pF, 3-section. Jackson Bros. LE3-4595-380
C_8—350-pF mica compression capacitor

L_1, L_2—(0.15 μH) 4 turns #16 on ½-inch diameter form.
L_3, L_4—(0.3 μH) 6 turns on ½-inch diameter form. Coil L_3 is airwound, coil L_4 wound on powdered-iron slug. Both coils closewound
L_5—(1.3 μH) 13 turns #16 on ½-inch diameter form, closewound. Dip all tuned circuits to center of band.
L_6—(10-15-20 meters) 10½ turns #8 wire, 2″ diam., 3½″ long. 10-meter tap is 5¼ turns from plate end; 15-meter tap, 7¼ turns. (40-80 meters) 16 turns #10 wire, 2½″ diameter, 4″ long. 40-meter tap is 8 turns from "hot" end. Coil wound on lucite plate with edges grooved for proper spacing of turns
RFC_1—Bifilar winding. Each coil is 14 turns #10 e., on ferrite core, 5″ long, ½″ diam. (Indiana General CF-503)
RFC_2—(Approx. 60 μH) 90 turns #26 e., spacewound wire diameter, 3⅜″ long, ¾″ diam. on ceramic or Teflon form. Series resonant at 26 MHz
RFC_3, RFC_4—2.5 mH
T_1—5 volts, 30 amp. Stancor P-6492
T_2—6.3 volts, 1 amp. Stancor P-8389
RY_1—Spst, 6.3-volt coil
Fan—Ripley SK-4125 or equivalent
Meters—Calectro
PC_1—Three 100-ohm, 2-watt resistors in parallel. 3½ turns #18 spacewound about one resistor
S_{1A}—Single-pole 11-position ceramic switch, 30° index. Centralab PA-6001
S_{1B}—Single-pole, 11-position ceramic switch, 30° index. Radio Switch Corp. Model 86-A
Sockets—Johson 122-275-1
Dials—General Radio with Jackson Bros. 4511-DAF Planetary Ball Drive Unit

A pi-network output circuit is used, capable of matching 50-ohm or 70-ohm coaxial antenna circuits. For improved linearity and ease of drive, a simple tuned-cathode input circuit is ganged to the pi-network amplifier bandswitch. Separate grid and plate meters are used and a variable ALC circuit is provided for connection to the exciter. The amplifier is designed for operation over a plate voltage range of 2000 to 2700 volts and a plate potential of 2500 volts is recommended.

Amplifier Circuitry—The schematic of the linear amplifier is shown in figure 17. Two 3-400Z or 3-500Z tubes are connected in parallel. Each of the three grid pins of the tube sockets is grounded, and the driving signal is applied to the filament circuit of the tubes, which is isolated from ground by a bifilar r-f choke. Neutralization is not required because of the excellent circuit isolation provided by the tubes and by the circuit layout.

The driving signal is fed in a balanced manner to the filament circuit of the two tubes. Mica capacitors suitable for r-f service are used to properly distribute the driving signal to the tuned-cathode circuit and the filaments of the tubes. Ceramic-disc capacitors are not recommended for use in this portion of the circuit because the peak r-f current under full amplifier input may be as high as 6 amperes or so. The tuned-cathode circuits (L_1-L_5) are fixed-tuned to the center of each amateur band and may be forgotten.

The Plate Circuit—Plate voltage is applied to the tubes through a heavy duty r-f choke bypassed at the B-plus end by a low-inductance, ceramic capacitor. In addition, the high voltage passes through a length of shielded cable to the high-voltage connector at the back of the chassis, and is further bypassed to ground at that point. A single .001-μF, 5-kV ceramic capacitor is used for the high-voltage plate-blocking capacitor and is mounted atop the plate r-f choke. The pi-network coil is divided into two parts for highest efficiency and ease in assembly. The first portion covers 10, 15, and 20 meters, and an additional section is added to the first to cover operation on 40 and 80 meters. Both coils are homemade and air wound at a minimum cost. The bandswitch is a *Radio Switch Corp.* high-voltage ce-

ramic-insulated unit mounted to the front panel of the amplifier

A typical circuit Q of 10 was chosen to permit a reasonable value of capacitance to be used at 80 meters and the number of turns in the plate coils was adjusted to maintain this value of Q up thorugh 15 meters. At 10 meters, the Q rises to about 18 and is largely determined by the minimum circuit capacitance achieved at this frequency. The pi-network output capacitor is a three-section, ceramic insulated 1100-pF unit. It is sufficiently large for proper operation of the amplifier on all bands through 40 meter For 80-meter operation, an additional 500-pF heavy duty mica capacitor is switched in parallel with the variable unit to provide good operation into low-impedance antenna systems commonly found on this band. The capacitor is connected to the unused 80-meter position of the bandswitch.

The instantaneous r-f plate voltage is sampled by a capacitive voltage divider and applied to a reverse-biased rectifier (D_1). Bias level is set by means of an adjustable potentiometer (*ALC Level*). When the r-f voltage exceeds the bias level, an ALC pulse is applied to the ALC control circuit of the exciter. The r-f level applied to the control circuit is set by adjustment of capacitor C_8 and the voltage is determined by the ratio of this capacitor to the 1-pF capacitor coupling the ALC circuit to the plates of the amplifier tubes. Aa a plate potential of 2500 or so, the nominal value of r-f plate voltage swing is about 1800 volts. If the ratio of the capacitive divider is 1:200, then about 90 volts of peak pulse is applied to the diode. Under normal operation, the diode is biased to about +30 volts and ALC pulses of about one-half this value are normal. Thus, the r-f voltage at the diode should be not more than 45 volts or so, calling for a capacitance ratio of about 1:300. This ratio is well within the range of the mica compression capacitor used for C_8.

The Metering Circuit—It is dangerous practice to place the plate-current meter in the B-plus lead to the amplifier unless the meter is suitably insulated from ground and isolated behind a protective panel so that the operator cannot accidentally receive a shock from the zero-adjustment fixture. If the meter is placed in the cathode return circuit, it will read the cathode

Figure 18

TOP VIEW OF LINEAR AMPLIFIER

Two 3-500Z tubes are placed at the rear of the amplifier chassis. The spacing of sockets and blower are shown in figure 21. The plate loading and tuning capacitors are mounted to each side of the pi-network coil assembly. The three stator sections of the output capacitor are connected in parallel by short lengths of copper strap. Directly below the plate coils is the aluminum box containing the cathode tuned circuits, with the adjustment slugs of the coils projecting through the top of the box. The 500-pF auxiliary 80-meter loading capacitor is placed above the bandswitch, directly in front of the 80-40 meter coil. At the left, the 1-pF coupling capacitor is attached directly to the rotor of the main tuning capacitor (see figure 20).

The filament transformer for the two 3-500Z tubes is at the rear, right corner of the chassis. The portion of the transformer facing the tubes is painted white to reflect the infra-red radiation from the tubes, which run a cherry-red color at full plate dissipation level. The cooling fan is mounted to the rear of the cabinet, and is not seen in this view.

current which is the sum of the grid and plate currents. A better idea is to place the plate meter in the B-minus lead between the cathode return circuit and the negative terminal of the power supply. The negative of the supply must thus be left "floating" above ground, or the meter will not read properly (figure 17). A protective resistor is placed across the meter circuit to ensure that the negative side of the power supply remains close to ground potential. A separate ground lead is then run between the chassis ground of the amplifier and that of the supply. Grid current is measured between grid and cathode return as shown in the simplified schematic, with the grid pins of the tubes directly connected to chassis ground.

The Cooling System—It is necessary to provide cooling air about the plate seal and filament seals of either the 3-400Z or 3-500Z tubes. Sufficient air is required to maintain the plate seal at a temperature below 225°C

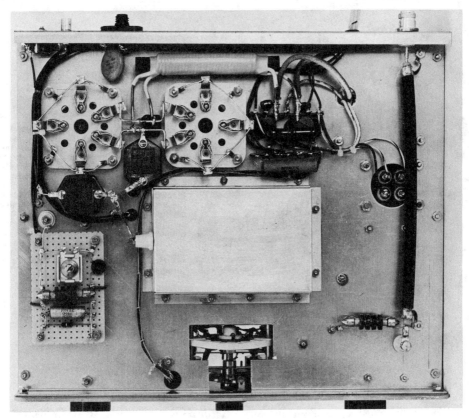

Figure 19

UNDER-CHASSIS VIEW OF AMPLIFIER

The cathode circuit box is at the center of the chassis, with the connecting load passing through a feedthrough insulator at the left. The shaft of switch S_{1A} passes through the wall of the upper section of the box, only about 1/16-inch above the level of the chassis and is joined to the plate bandswitch (S_{1B}) with a brass coupling.

The three grid pins of the tube sockets are grounded to the mounting bolts. The sockets are lowered below the chassis by means of spacers to permit cooling air to flow about the base of the tubes. The two .01-μF mica coupling capacitors are placed adjacent to the left-hand tube socket, with the ferrite-core filament choke running parallel to the rear of the chassis. Directly to the right of the sockets are placed two phenolic terminal strips which support the filament wiring, the 10K VOX resistor and the 15-ohm meter safety resistor. The bypass capacitors for the "cold" end of the filament choke are also located on one terminal strip.

At the left side of the chassis is a small phenolic board that holds mica compression capacitor C_8 and the components associated with the ALC circuit. The ALC level potentiometer is a small ¾-inch diameter control mounted on the rear lip of the chassis, adjacent to input receptacle J_1. To the right of J_1 is the high-voltage connector, with the .001-μF, 6-kV disc capacitor mounted behind it. The antenna output circuitry is at the right end of the chassis. The connection from the plate-loading capacitor passes through a ceramic feedthrough insulator near the panel, and the connection to the coaxial receptacle (J_2) at the rear of the amplifier is made via a short length of RG-8/U coaxial line. The outer braid of the line is grounded to the chassis at each end.

and the filament seals at a temperature below 200°C. Common practice calls for the use of special air-system sockets and chimneys, in conjunction with a centrifugal blower to maintain air flow requirements to meet these temperature limitations. Considerable

difficulty with conventional cooling techniques has arisen, caused by the noise created by the blower motor and the movement of air through the cooling system. Extensive tests have shown that for c-w and SSB operation at the legal power limits (1-kW c-w

Figure 20

OBLIQUE VIEW OF PLATE CIRCUIT

The Eimac HR-6 anode connectors are used on the 3-500Z tubes, with the parasitic suppressor mounted close to the connector. The plate leads are made of lengths of flexible copper braid. Both leads terminate at the plate-blocking capacitor which is mounted to a small bracket bolted to the stator terminal of the plate-tuning capacitor. At the far side of the tuning capacitor is the 1-pF ALC coupling capacitor, made of two 1-inch diameter copper discs, spaced about ¼-inch apart. The upper disc is affixed to the stator terminal of the capacitor and the lower disc is supported by the feed-through insulator mounted directly beneath it on the chassis deck.

input and 2-kW PEP voice input on SSB) either the 3-400Z or 3-500Z may be adequately cooled by a lateral air blast blown against the tube by a small rotary fan, properly spaced from the tube. A drawing of such an installation is shown in figure 21.

The Johnson 122-275-1 ceramic tube socket is used, which permits a minimum amount of lateral pressure to be exerted on the glass base of the tube. The socket is mounted below the chassis deck about 1/16″ to provide an air path around the base of the tube through which under-chassis air is drawn by convection. The rotary fan is mounted between the tubes, in line with the center of the glass envelope and blows cooling air across the envelope and plate caps. Under these conditions, maximum plate dissipation of about 350 watts per tube is achieved for the 3-400Z and 450 watts per tube for the 3-500Z. While maximum dissipation rating is not achieved with either tube, the allowable dissipation is sufficiently high so that the maximum amateur power input may be run in either case with

adequate safety factor. If it is desired to operate the amplifier under steady-state conditions (RTTY, for example), the power input will have to be reduced to about 850 watts in the case of the 3-500Z's or 750 watts for the 3-400Z's. The alternative is to install a forced-air cooling system to boost the plate dissipation capability to the maximum limit specified in the instruction sheet for the tube type in question. The air cooling system shown, however, is entirely adequate for c-w and SSB operation under normal operating conditions for extended periods of time.

The perforated metal cabinet provides maximum ventilation and, when the lid is closed, provides good r-f inclosure. In order to permit the air to be drawn into the bottom of the amplifier chassis, rubber "feet" are placed at each corner of the cabinet, raising it about one inch above the surface on which it sits. The top surface of the cabinet should be kept clear to permit the heat to freely escape from the amplifier when it is in use.

Amplifier Construction The over all dimensions of the perforated, wraparound cabinet housing the amplifier are 16" wide, 8" high, and 13" deep. The amplifier is built on a shallow chassis bent from a single sheet of aluminum and measures 15¼" wide, 12½" deep and has a 1" lip at the rear. Clearance under the chassis is 1¼" to the bottom of the cabinet. An oblique view of the chassis and cabinet, including the placement of the major components is shown in figure 20. The cooling fan is mounted to the rear of the cabinet and forces air against the two transmitting tubes through a 4"

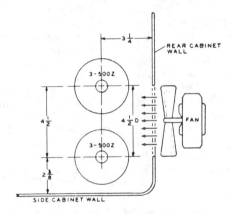

Figure 21

AIR-SYSTEM LAYOUT

The Ripley fan (Ripley Co., Inc., Middletown, Conn.) is bolted to the rear of the cabinet behind a 4½-inch diameter hole, covered with ¼-inch mesh wire screen. The air blast passes across the tube envelopes and the warm air is exhausted out the perforated top of the amplifier cabinet. The tube sockets are located with respect to the fan to permit maximum cooling air to envelop the tubes.

diameter hole cut in the rear panel of the cabinet. The hole is covered with a piece of wire mesh having ¼" squares.

Placement of the major components may be seen in the photographs. Because of the small depth of the chassis, placement of the bandswitch and tuned cathode assembly is critical. The various cathode tuned circuits and bandchange switch are mounted in an inclosed box placed near the center of the chassis, in line with the main band change switch. The cathode inclosure box is made

up of two small aluminum chassis (5" × 3½" × 1") placed back to back, one atop and the other underneath the chassis. The flanges of the chassis are cut off, and substitute flanges are attached to the outside of the chassis lips, permitting the two units to be bolted together, as shown. The various coils and bandswitch are mounted to the top chassis box, in line with the main switch and connected to it with a shaft coupler. The cathode coils and capacitors are assembled and mounted in a vertical position within the box. The cathode tank-coil assembly may be wired and the tuned circuits grid-dipped to the center of each amateur band before the chassis box is bolted to the corresponding cutout in the chassis.

The pi-network coil assembly is seen in the top view photographs. The 10-15-20 meter coil is wound of No. 8 solid copper wire. Ordinary plastic-covered house wire is used, the plastic coating stripped off before the coil is close wound on a suitable form. Once the winding is completed, the coil is spaced and the taps are soldered in place. Thin, ⅛" wide copper strap is used for the tap leads. Each lead is pretinned at the end

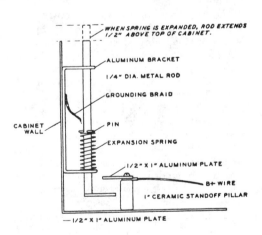

Figure 22

HOME-MADE HIGH-VOLTAGE SAFETY SHORTING SWITCH

and wrapped around the proper coil turn and soldered in place with a large iron. A good connection is important at this point as the r-f current flowing through the joint is high. Once the coil is cut to size, and the tap leads soldered in place, the coil end connections are trimmed to length and adjusted to the proper position. The coil lead to the tuning capacitor terminates in a copper soldering lug and the opposite end is flattened in a vise to make a glove fit with the proper 20-meter tap point on the bandswitch. Once all leads are properly trimmed, the coil is removed and silver plated.

The 40-80 meter coil is wound and tapped in the same fashion. Once completed, it is threaded on a strip of lucite or plastic material that has been grooved along both edges to fit the spaced winding of the coil. The grooves may be easily cut with a small triangular file. The lucite plate is supported by two plastic posts, cut to size and mounted to the chassis behind the bandswitch.

The plate parasitic suppressors for each tube are made of three composition resistors wired in parallel, with a small inductor wound around one resistor. The suppressors are placed immediately adjacent to the anode connectors of each tube, and flexible leads made of copper braid are run from the suppressors to a common terminal of the plate coupling capacitor mounted atop the plate r-f choke.

The placement of the major components beneath the chassis is shown in figure 19. A T-shaped opening is cut in the forward area of the chassis to clear the plate bandswitch, and an opening is cut in the center of the chassis for the cathode tank assembly. The tube sockets are mounted beneath the chassis by 6-32 hardware, with several washers placed on each mounting bolt beneath the chassis to lower the socket about $\frac{1}{16}$ inch, providing additional air passage around the base seal of the tube. The grid pins are grounded to the adjacent socket bolts. The large filament choke is mounted from a phenolic terminal strip to the parallel-connected filament pins of the tubes. The mica coupling capacitors are placed in close position to the filament wiring and the ceramic feedthrough insulator mounted in the sidewall of the input coil compartment.

At the side of the under-chassis area a small perforated circuit board supports the various components of the ALC circuit and the connecting lead to the 1-pF air capacitor mounted on the main tuning capacitor passes through a ceramic feedthrough insulator in the chassis deck.

The connection from the pi-network output capacitor to the coaxial receptacle mounted on the rear lip of the chassis is made via a short length of 50-ohm coaxial cable, the outer shield of the cable being grounded at both ends to nearby chassis points.

The filament transformer is mounted atop the chassis in a rear corner as seen in the photographs. The bottom area of the transformer is cleaned of paint so that the end bells make a good ground connection to the chassis to partially shield the windings from the r-f field atop the chassis. The end bell of the transformer nearest the tubes is painted white to reflect the infrared radiation emitted from the tubes, permitting the transformer to run much cooler than otherwise would be the case if the end bell was left black. The remainder of the transformer is left black so as to radiate the heat generated within the transformer.

The VOX relay and auxiliary transformer are mounted in a small shield box placed in front of the filament transformer. Sufficient room exists in this area so the box may be enlarged to also hold a rectifier and filter capacitor should it be desired to substitute a dc relay for the ac unit specified.

A shield plate measuring $6'' \times 2''$ is affixed to the rear of the meters to shield the movements from the intense r-f field surrounding the plate coils. The shield is held in position by the meter studs, each stud passing through a rubber grommet mounted in the shield plate. The plate is grounded in each corner by a short, direct lead to the meter mounting bolts.

Amplifier Adjustment   Before the tubes are inserted in the amplifier, the main bandswitch should be set to the various bands and the plate tank assembly tuned for resonance on each band when the loading capacitor is set to about $\frac{2}{3}$ maximum value. The approximate settings should be logged for future reference. The two tubes are now inserted in their sockets and filament voltage applied to the amplifier. Voltage at the tube sockets should run between 4.8 and 5.1 volts, as measured with an accurate meter. The amplifier is now placed in the cabinet and the cooling fan connected so that it runs whenever the filament circuit is energized. An interlock switch atop the cabinet should be immediately wired so that it opens the high-voltage control relay in the power supply. In addition, a high-voltage shorting switch, such as shown in the illustration (figure 22), is suggested as an integral part of the amplifier, since lethal voltages are exposed when the lid of the cabinet is raised unless precautions are taken.

Typical operating voltages and currents for the 3-500Z tube are tabulated in Table 5. An operating plate potential of 2500 is recommended with an intermittent-service power supply capability of 800 milliamperes.

Initial adjustment is greatly facilitated with the aid of an SWR meter or other

be achieved with the minimum drive level and maximum antenna load level possible.

Under voice modulation, the plate current will kick to about 440 mA and grid current will kick to about 130 mA. For c-w operation at 2500 volts, plate loading and grid drive are decreased until 400 mA plate current and 125 mA grid current are noted on the meters. As with all grounded-grid amplifiers, grid drive should never be applied before plate voltage, or damage to the tubes may result.

22-5   A Two-Stage High-Gain Amplifier Using The 3-1000Z

Table 5			
Typical Operating Data, 3-500Z			
R-F Linear Amplifier Service, Class-B			
(one tube)			
DC Plate Voltage	1500	2000	2500
Zero signal Plate Current (ma)	65	95	130
Single Tone DC Plate Current (ma)	400	400	400
Single Tone DC Grid Current (ma)	130	130	120
Two Tone DC Plate Current (ma)	260	270	280
Two Tone DC Grid Current (ma)	80	80	70
PEP Useful Output Power (watts)	330	500	600
Resonant Load Impedance (ohms)	1600	2750	3450
Intermodulation Distortion Products (db)	—46	—38	—33

output indicating device. Plate voltage is applied to the amplifier and the resting plate current is noted. A small amount of grid drive is introduced into the amplifier and resonance established in the plate circuit. Drive and loading are gradually increased, holding a ratio of about 3:1 between indicated plate and grid current. In the case of the 3-500Zs, maximum indicated grid current should be about 240 mA for a plate current of 800 mA. This ratio should

This sturdy amplifier (figure 23) is designed to operate at the 2-kW PEP input level when driven by an SSB signal of not more than 500 milliwatts PEP level. Amplifier gain is better than 33 decibels, and operation is stable under all normal conditions. The amplifier is designed for single-band operation at any frequency between 3.5 MHz and 30 MHz, and specific data is included for operation on any one of the amateur bands between 80 and 10 meters. Tank circuits are designed for a coverage of 500 kHz at the low end of the range of operation, and for 1.5-MHz coverage at the high end of the range. Used for heavy-duty service, the amplifier is capable of key-down (RTTY) service at the 2-kW power input level. Choice of rugged components and an efficient cooling system assure reliable, trouble-free, around-the-clock service.

The amplifier consists of a two-stage circuit, employing a 4CX250B ceramic tetrode operating class AB_1 to drive a 3-1000Z grounded-grid, class-B linear stage. For those amateurs having an SSB exciter capable of about 70 watts PEP output, the driver stage may be eliminated, and the 3-1000Z stage can be driven directly by the exciter. This may be accomplished by breaking the interconnecting coaxial cable between the stages at point X (figure 24). The 4CX250B stage may then be omitted, or a switch installed at this point to allow the amplifier to be used at two widely different drive levels.

Figure 23

TWO-STAGE LINEAR AMPLIFIER WITH 3-1000Z

This rugged, high-gain linear amplifier is designed for continuous-service operation at the 2-kW power level. Less than 1-watt PEP drive is required for full input. The amplifier is designed for single-band operation on any range of frequencies between 3.5 and 30 MHz.

The amplifier uses two tubes; a 4CX250B tetrode driver operating class AB_1 and a 3-1000Z high-μ triode in grounded-grid circuitry. For use with exciters having a power output of 100 watts PEP or so, the driver stage may be omitted. The amplifier is designed for mounting in a standard 19″ cabinet. The top of the shielded inclosure is removable, with top, sides and back being perforated to allow proper circulation of air within the amplifier.

Panel meters are (l. to r.): Multimeter M_1, grid-current meter M_2, and plate-current meter M_3. In a vertical position below the left-hand meter are the input and output tuning controls for the driver stage, with the ALC Adjust, Adjust output, and meter-switch knobs to the right. Primary filament and plate circuit switches and pilot lamps are at the bottom of the panel. At the right are the plate-tuning control (top) and antenna-loading control (bottom). The panel is painted a hammertone gray and lettering is placed in position, then panel is given a spray coat of clear Krylon to protect the finish.

When operated under normal conditions, the third-order intermodulation distortion figure of the two-stage amplifier is better than −33 decibels below one tone of a two-tone test signal.

Amplifier Circuitry   The circuit of the two-stage, high-gain linear amplifier is shown in figures 24 and 27. The 3-1000Z high-μ triode is operated in cathode-driven, grounded-grid service in the zero-bias mode. A pi-L network (C_5, C_6, L_4, L_5) is used in the plate circuit to achieve maximum harmonic suppression. The network is designed to match a 50-ohm load having a maximum SWR figure of 3/1, or less. To restrict overload and "flattopping," a portion of the instantaneous r-f plate volt-

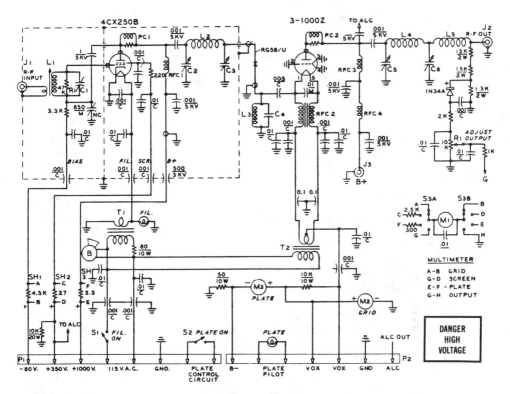

Figure 24

SCHEMATIC OF TWO-STAGE HIGH-GAIN AMPLIFIER

Note: See tables 6 and 7 for coil and capacitor data.

J₁—UG-290/U, type BNC connector
J₂—UG-58A/U, type N connector
J₃—UG-496B/U, type HN connector
PC₁—3 turns #14, spaced around a 47-ohm, 2-watt compositon resistor
PC₂—(80-40 meters): 7 turns #12, ¾" diam., around 50-ohm, 5-watt resistor. Ohmite P-300
(20-15 meters): 3 turns #12, as above. Ohmite P-300 with turns removed
(10 meters): 2 turns #12, as above. Ohmite P-300 with turns removed
RFC₁—(80 meters): 1-mH, 600-mA.
(40-10 meters): 44-μH, 600-mA. Ohmite Z-14.
RFC₂—B&W FC-30A. Home-made substitute: 14 bifilar turns #10 e. wire on ferrite core, ½" diam., 3" long. (Indiana General CF-503 core.) Notch core with file and snap to break to length.
RFC₃—(80-20 meters): 200 μH, 800 mA. B&W 800. Home-made substitute: 180 turns #20, ¾-inch diam., spaced 4½" long, 40 t.p.i. Series resonant at 24 MHz (15 meters): 130 turns, as

above, 3¼" long. Series resonant at 28 MHz. (Remove turns from B&W 800) (10 meters): 70 turns #16, ¾-inch diam., 3" long. Series resonant at 40 MHz.
RFC₄—2μH. Ohmite Z-144
T₁—6.3 volts, 3 amps. Chicago-Stancor P-6466
T₂—7.5 volts, 21 amps. Chicago-Stancor P-6457
Nc—850-pF, mica compression capacitor. ARCO 306M
Blower—20 cu. ft./min. Dayton 1C-180 or Ripley LR-81
SH₁, SH₂, SH₃—Meter shunts. Wind resistance wire around 47K, 1-watt resistor to provide proper meter ranges, as shown above
M₁—0-1 dc milliammeter, Simpson 1327
M₂—0-500 dc milliammeter, Simpson 1327
M₃—0-1000 dc milliammeter, Simpson 1327
Socket for 4CX250B—SK-600
Chimney for 4CX250B—SK-606
Socket for 3-1000Z—SK-510
Chimney for 3-1000Z—SK-516

age is sampled and rectified for use as ALC control, and applied to the exciter.

The 3-1000Z is coupled to the driver by a short length of coaxial line. The driver, a 4CX250B tetrode, is bridge neutralized for proper stability and the grid circuit is loaded by a resistor (R₁) to establish the system drive level at about 1.5 watt PEP. Without the resistor, the typical drive level is about 500 milliwatts PEP for full output of the two-stage amplifier at 3.8 MHz.

The 4CX250B has a relatively high-Q plate-tank circuit that is designed to work into a 50-ohm load. To combine high gain with maximum stability, the driver grid and plate circuits are carefully shielded from each other. In addition, the chassis is arranged to isolate the input and output of circuits within the inclosure by the use of multiple bypass capacitors and proper shielding of the power and metering leads. The majority of small components are removed

Figure 25

TOP VIEW OF HEAVY-DUTY AMPLIFIER

An interior view of the 40-meter amplifier. Inclosed 4CX250B stage is at the side of the assembly, with centrifugal blower directly behind it. Pi-L plate-circuit components are at the left, with 3-1000Z tube and chimney on center line of chassis.

On the rear apron of the inclosure are (l. to r.): antenna, high-voltage, power, and input receptacles. The last three connectors are inclosed in a small aluminum box beneath the chassis to shield the leads from the r-f circuitry. Layout of major components is identical for all amplifiers.

from the r-f inclosure and are mounted on phenolic terminal boards between the inclosure and the amplifier panel.

Amplifier operation is monitored by three panel meters. Meter M_1 is a multimeter which reads grid, screen, or plate current of the 4CX250B, in addition to monitoring relative power output of the amplifier. Meter M_2 measures grid current of the 3-1000Z, and meter M_3 measures plate current in the B-minus return lead to the power supply.

Both the 4CX250B and the 3-1000Z require forced-air cooling at 25 c.f.m. A single centrifugal blower provides this air flow, at a back pressure of about 0.4 inch of water.

Amplifier Construction   Proper interstage shielding in this amplifier contributes to the high degree of stability. The unit is built within an aluminum inclosure measuring 18″ wide, 12″ high, and 15″ deep. Sides and back of the inclosure are perforated to provide proper ventilation, as is the area of the top plate over the 3-1000Z. The inclosure is bent out of flat plate and riveted together with "pop" rivets. The centrifugal blower is mounted atop the chassis in a corner and draws air in through the rear of the inclosure and exhausts it into the under-chassis area, which serves as a plenum chamber. The under-chassis pressurized air is exhausted through the 3-1000Z air-system socket, and also passes into the driver box, providing proper cooling for the 4CX250B tube. Air chimneys are used with both the 3-1000Z and 4CX250B tubes to direct the flow of cooling air over the tube seals and anodes.

Regardless of the operating frequency, amplifier layout follows the arrangement shown in the photographs (figure 25 and 26). The 3-1000Z socket is near the center of the chassis deck, toward the front of the inclosure, with the plate-circuit components and coils to one side of the tube (figure 28). The driver inclosure is on the opposite side of the 3-1000Z. To the side of the chassis, the amplifier coils are directly behind the tank capacitors which are affixed to the front of the inclosure, and the filament transformer is mounted directly behind the 3-1000Z socket.

The exciter inclosure measures 8″ × 8″ × 3½″ in size and is bolted in position atop the chassis deck of the amplifier. Power leads pass through feedthrough capacitors mounted in the front wall of the inclosure.

Figure 26

UNDER-CHASSIS VIEW OF 10-METER AMPLIFIER

The 3-1000Z socket is near chassis center, with the filament choke directly below it, and the cathode tuned circuit at one side. To the right of the socket is the electrical conduit and shield box for the power receptacles and wiring. Air inlet from the blower is seen at lower left, with exit hole for passing cooling air to the 4CX250B buffer stage at the upper left. The air opening is covered with screening. Filament transformer for the 4CX250B is at extreme left, with primary-circuit terminal strip adajacent to it. The shaft of 4CX250B loading capacitor projects through the bottom of the chassis directly above the the transformer. Filament "Hypass" capacitors for the 3-1000Z are at the right of the filment choke, and the short coaxial lead for high voltage passes toward the back of the amplifier at the right side. Power wiring to the panel is extra length so that the panel may be removed for test purposes.

The 4CX250B socket mounts on an L-shaped bracket that incloses one-quarter of the internal area of the box. The grid-circuit components are contained in this area, as shown in the photograph of figure 29. Both sides of the box are removable for ease in wiring the stage. The portion of the box to the rear of the bracket holds the various plate-circuit components of the 4CX250B.

Cooling air is introduced into the box through a 1¼" hole in the bottom of the box which aligns with a similar hole cut in the deck of the main inclosure. The sides and top of the box are perforated to permit the air to pass out of the box after its passage through the socket and anode cooler of the 4CX250B.

The 3-1000Z cathode tuned circuit and filament choke are mounted under the amplifier deck, as is the 4CX250B filament transformer. Power and metering connectors are placed on the rear apron of the chassis and the various leads pass through the under-chassis area to the front-panel controls and components via a short length of ½-inch diameter electrical conduit pipe, grounded at both ends. A solid bottom plate completes the r-f shielding and also pressurizes the under-chassis area. The small joints, seams and holes in the chassis are filled with caulking compound to make the plenum chamber air tight.

The complete amplifier assembly is supported from the front panel by means of two U-channels made of aluminum. The intervening 2" space holds the circuit boards for various small components; and the panel meters, switches, and controls recess into this area.

The pi-network loading capacitor for the driver stage (C_3) may be set for a 50-ohm load and forgotten. Accordingly, it is not brought out to the panel, but is mounted in a vertical position, with the shaft projecting into the under-chassis area (figure 26). It may be adjusted, if desired, by placing an adjustment hole in the bottom chassis plate, and covering the hole with a

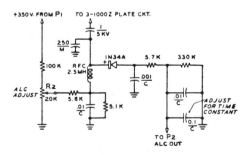

Figure 27

ALC CIRCUIT FOR AMPLIFIER

Figure 28

OBLIQUE VIEW OF 10-METER PI-L PLATE-TANK CIRCUIT

The rear frames of the pi-L capacitors are connected together with an aluminum strap and the front frames are attached to the panel with ½-inch metal spacers. The plate coil is attached to both stator terminals of each capacitor to evenly distribute the current through capacitor frame. Two .001-pF 5-kV ceramic capacitors are placed in series for the plate-blocking capacitor in this particular amplifier. The ALC capacitor (1 pF) is shown as a ceramic unit. It was replaced at a later date with a capacitor made up of two 1-inch diameter copper discs spaced about ⅜-inch apart, because several ceramic units failed in the 10-meter unit over a period of time. The 250-pF bridge capacitor (figure 27) is a mica feed-through "button" capacitor mounted to the wall of the amplifier inclosure. Note that the plate parasitic suppressor of the 10-meter amplifier has been reduced to two turns of the wire around the resistor.

Figure 29

INTERIOR VIEW OF DRIVER STAGE

The 4CX250B driver is mounted within a separate compartment atop the main amplifier chassis. The sides and top panel of the compartment are removable. The lower inclosure holds the driver grid circuit components, with the variable compression mica neutralizing capacitor in the foreground. The pi-network plate circuit of the 4CX250B is at the right.

snap bottom when it is not in use. The 4CX250B neutralizing capacitor (NC, figure 29) is adjustable through a hole drilled in the side panel of the subassembly. This adjustment hole, too, is covered when not in use to maintain the proper pressurized air system.

Amplifier Tuning and Adjustment The 4CX250B driver stage should be adjusted separately on the bench before it is placed within the amplifier. Temporary

cooling air may be applied to the unit, and it can be operated into a dummy load. The 4CX250B requires 1000 volts at 220 mA for the plate, 350 volts (regulated) at 50 mA for the screen and −80 volts (adjustable) at 10 mA for the bias supply. Bias, plate, and screen voltages to the 4CX250B are applied in that order. Bias is adjusted for a resting plate current of 100 mA.

Before voltages are applied, the 4CX250B should be neutralized according to the procedures outlined in Chapter 11. Operating voltages are then applied through the metering circuits of the amplifier, making sure the 10K screen bleeder resistor is in the circuit. When properly loaded and driven, the plate current of the 4CX250B will run about 200 to 250 mA, screen current about 15 mA or less, and grid current should be less than one division on the 0-1 mA meter range. Proper neutralization is indicated by maximum power output, maximum screen current and minimum plate current all coinciding at one setting of tuning capacitor C_2. Once the amplifier has been properly adjusted, it may be permanently placed in the amplifier compartment and wired in place. Power output should be 100 to 120 watts, single tone with less than 1.5 watts driving power.

Voltage requirements for the 3-1000Z are: 3000 volts at 670 mA (SSB) or 2000 volts at 500 mA (c.w.). For tune-up purposes, the VOX terminals on power plug P_2 should be shorted together. Plate voltage is applied, with the amplifier first being adjusted for c-w operation. Using a dummy load, excitation is slowly raised until the 3-1000Z draws 500 mA plate current at resonance at a grid current of about 200 mA. Excitation and antenna loading are interlocking, and are varied until this ratio of currents is achieved. The output circuit of the 4CX250B stage is re-resonated (and perhaps loading adjusted slightly) until the proper drive level is reached at the desired input to the 3-1000Z. Power output of the amplifier will be 650 watts, or better at 1-kW input.

In order to go from a one-kilowatt state to a 2-kW PEP state under the same load conditions, it is necessary to raise the plate voltage of the tube to 3000. At this plate potential, resting plate current should be about 220 mA when the VOX terminals are shorted. When the terminals are open, the bias provided by the 10K cathode resistor will drop the plate current to a few milliamperes. At the 3-kV potential, then, excitation is gradually raised to achieve a plate current of 665 mA at a grid current of 180 to 200 mA. Tuning adjustments need not be changed from the c-w condition.

If it is found that grid current to the 3-1000Z is too low when plate current reaches 665 mA at resonance, it is an indication that antenna loading is too heavy for the degree of grid drive. Conversely, if grid current is too high, it is an indication that excitation is too high for the amount of antenna loading. A proper balance of drive level and antenna loading will permit the proper ratio of grid current-to-plate current to be achieved. When the proper ratio is met, it will be found that when plate voltage is dropped to 2000 for c-w operation, the power input will automatically drop to 1 kW, and the only adjustment necessary to

Table 6.
4CX250B Driver-Circuit Data

Band	L_1	C_1	L_2	C_2	C_3
80	(11 μH) J. W. Miller 43A105CBI, 1/2" diam. Primary = 15t #22 e.	200 pF	(13 μH) 28 turns # 12, 1¼" diam., 2" long	200 pF 1.6 kV	1500 pF
40	(6.5 μH) J. W. Miller 42A68CBI, 1/2" diam. Primary = 10t #22 e.	100 pF	(7.5 μH) 14 turns # 12 1" diam., 2" long	100 pF 2 kV	1000 pF
20	(3.3 μH) J. W. Miller 42A336CBI, 1/2" diam. Primary = 5t #22 e.	25 pF	(3.7 μH) 8 turns # 12, 1" diam., 1½" long	75 pF 2 kV	600 pF
15	(2.2 μH) J. W. Miller 42A226CBI, 1/2" diam. Primary = 4t #22 e.	15 pF	(2.0 μH) 6 turns # 10, 2" diam., 3" long	50 pF 2 kV	300 pF
10	(1.2 μH) 8 turns #18, 5/8" diam. 1/2" long. Pri. = 4 turns #18, 5/8" diam., 1/2" long	15 pF	(1.5 μH) 6 turns # 10, 1¼" diam., 1⅛" long	35 pF 2 kV	200 pF

Table 7. 3-1000Z Circuit Data

Band	L_3	C_4	L_4	C_5	L_5	C_6
80	12 turns #10, ¾" diam., 1½" long (1.25 µH)	1600 pF XMTG. MICA	(9 µH) 14 turns #6, 3½" diam., 5" long	500 pF 3.5 kV Johnson 153-6	(6.5 µH) 20 turns #12, 1¼" diam., 2" long	1500 pF
40	4 turns #10, 1⁵⁄₁₆" diam., 1½" long (0.6 µH)	1000 pF XMTG. MICA	(4.5 µH) 8 turns #6 3½" diam., 3½" long	150 pF 4.5 kV Johnson 153-12	(3.2 µH) 11 turns #12, 1¼" diam., 1" long	1000 pF
20	4 turns #10, 1⅛" diam., 1½" long (0.3 µH)	400 pF XMTG. MICA	(2.2 µH) 10 turns ¼" tubing, 1⅞" diam., 4¼" long	100 pF 7 kV Johnson 153-14	(1.5 µH) 15 turns #12, ¾" diam., 2" long	500 pF
15	3 turns #10, 1" diam., 1" long (0.2 µH)	250 pF XMTG. MICA	(1.3 µH) 6 turns ³⁄₁₆" tubing, 2" diam., 4" long	75 pF 4.5 kV Johnson 154-13	(0.9 µH) 10 turns #10, ½" diam., 1½" long	250 pF
10	4 turns #12, ½" diam., ¾" long (0.15 µH)	200 pF XMTG. MICA	(1.2 µH) 6 turns ³⁄₁₆" tubing, 1½" diam., 3" long	35 pF 4.5 kV Johnson 154-11	(0.6 µH) 4½ turns #10, ⅝" diam., 1⅛" long	250 pF

Figure 30

3-400Z LINEAR AMPLIFIER PACKS KILOWATT PUNCH FOR THE SIX-METER OPERATOR

This compact kilowatt linear amplifier is suited for SSB, c-w or a-m service in the 50-MHz band. Utilizing the 3-400Z in a grounded-grid circuit, the amplifier requires neither bias nor screen voltage. The homemade cabinet is an "r-f tight" inclosure which helps to reduce TVI problems. Meters are shielded and are in the circuit at all times so no extra switching circuits are required. Panel size is only 8¾" × 13". Panel components are (l. to r.): Grid meter, r-f output meter, and plate meter. In the line below the meters are: r-f output calibration control, plate tuning, and antenna loading. Across the bottom of the panel are: input tuning, filament pilot light, and high-voltage pilot light.

the amplifier may be a slight "touchup" of the driving level from the auxiliary exciter.

Once proper operation at 2000 and 3000 volts has been completed with a single-tone driving signal, the amplifier may be driven with a voice signal. Because of meter inertia and the relatively low power in the human voice, peak grid and plate current readings will average about one-half of the single-tone readings. Proper peak conditions for SSB may be monitored with an oscilloscope. Operation at 1-kW dc input at 3000 volts plate potential is not recommended because efficiency is low due to the limited r-f plate-voltage swing.

22-6 A Kilowatt Linear Amplifier for Six Meters

Described in this section is a high-power amplifier expressly designed for six-meter operation. It is capable of 1-kilowatt PEP input for sideband and c-w service, and will deliver a fully modulated carrier of about 200 watts as an a-m linear amplifier. A single Eimac 3-400Z (or 3-500Z) zero-bias triode is used in this efficient, compact unit which is capable of delivering full output from an exciter providing 35 watts peak drive (or 15 watts carrier, amplitude-modulated). The cathode-driven (grounded-grid) configuration is utilized and neutralization is unnecessary.

The Amplifier The schematic of the six-
Circuit meter amplifier is shown in figure 31. A tuned-cathode circuit (L_1-C_1) is used to preserve the waveform of the driving signal and to reduce harmonic distortion that may cause TVI.

The plate circuit of the amplifier utilizes a *pi-L* network to achieve a high order of harmonic suppression and a simple diode voltmeter is used to monitor the r-f output voltage. An antenna relay (RY) is incorporated in the amplifier, and an alternative

circuit is shown for using the linear amplifier with a transceiver (figure 32).

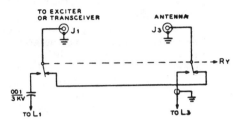

Figure 32

SUGGESTED ANTENNA-RELAY CIRCUIT FOR USING AMPLIFIER WITH TRANSCEIVER

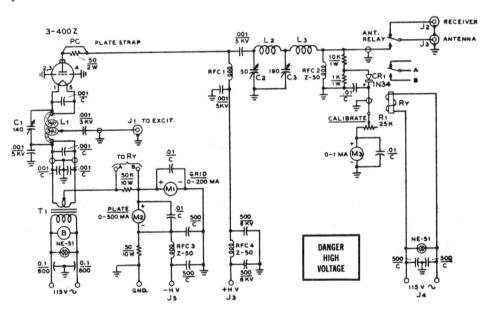

Figure 31

SCHEMATIC OF SIX-METER LINEAR AMPLIFIER

B—Blower, 13 cubic feet per minute at 0.13 inches of water. Dayton 2C-782 or equivalent
C_1—140 pF Bud 1856
C_2—50 pF, 0.07" spacing. Hammarlund MC-505X
C_3—190 pF. Bud 1858
J_4—TV-type chassis-mount cord socket
L_1—Bifilar coil. 3 turns, 1/8-inch diameter copper tubing spaced to 2", tapped 3/4 turn from grounded end. Inner conductor is No. 12 insulated or formvar wire (see text)
L_2—Pi-section coil. 5 turns, 3/16-inch copper tubing, spaced to 3". Inside diameter is 1 1/8".
L_3—L-section coil. 4 turns, 1/8-inch tubing, 3/4-inch inside diameter, spaced to 2 1/2"
RFC$_1$—3 μH choke. 48 turns No. 16 formvar wire closewound on 1/2" diameter standoff insulator.
RFC$_{2, 3, 4}$—Ohmite Z-50 choke
RY—Coaxial antenna relay.
T_1—5 volts at 15 A. Stancor P-6433
Note: 0.1 μF, 600-volt feedthrough capacitors are Sprague 80P-3. Meters are Simpson Wide-Vue.

Metering and It is necessary to measure
Suppression Circuits both grid and plate current in a cathode-driven
amplifier to establish the proper ratio of grid
to plate current. At the higher frequencies
it is desirable to directly ground the grid of
the amplifier tube and not to rely on questionable bypass capacitors to insure that the
grid remains at ground potential. Grid current, therefore, is measured in the cathode-
return circuit of the amplifier by meter M_1.

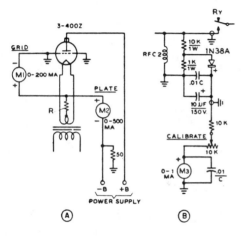

Figure 33

METERING CIRCUITS FOR KILOWATT AMPLIFIER

A—Dc meter circuit showing grid and plate
meters placed in low-potential return leads.
The B-minus of the power supply "floats"
above ground by virtue of the 50-ohm resistor, which may be placed in the power
supply, if desired.
B—Peak-responding voltmeter circuit useful for
adjusting linear for a-m service.

Plate current is measured in the B-minus lead
to the power supply by meter M_2. A simplified metering circuit is shown in figure 33.
This amplifier was checked for parasitics
and it was found that the usual plate parasitic choke was not required for stable operation. A variation in circuit layout, however, or changes in ground-return currents
may allow weak parasitic oscillation to take
place. If this condition is found, placement
of a parasitic choke in the plate lead will
suppress the unwanted parasitic. A practical
parasitic choke is shown in the schematic
and is made by merely shunting a portion of
the plate strap with a composition resistor.

Amplifier The amplifier is inclosed in an
Construction "r-f tight" cabinet measuring
$13'' \times 8\frac{3}{4}'' \times 10''$. A standard $12'' \times 10'' \times 3''$ aluminum chassis is
used, along with an $8\frac{3}{4}'' \times 13''$ panel (cut
from a standard aluminum relay rack pan-
el). The cabinet is made by bending a sheet
of light aluminum ($31'' \times 11''$) to fit
around the panel. It is riveted to $13'' \times 11''$
bottom plate. The rear of the cabinet is a
sheet of perforated aluminum fastened to
the cabinet with $\frac{1}{2}$-inch aluminum angle
stock. Additional angle stock is cut to length
and fastened to the front edge of the cabinet to secure the panel. A 4-inch hole is cut
in the cabinet directly above the 3-400Z and
is covered with a small sheet of perforated
aluminum. This shielded vent permits the
heated air from the tube to escape from the
inclosure.

A meter shield is used to protect the
panel meters from the r-f field of the plate
circuit and to suppress r-f leakage from
the cabinet via the meter face. The box-
like shield is attached to the panel by means
of aluminum angle stock which is held to
the panel by the meter mounting bolts. All
paint is removed from the rear of the panel
to provide a good ground connection to the
meter shield and to the chassis and cabinet.

The 3-400Z tube requires forced air cooling during operation and a blower (B) is
mounted on the chassis and activated with
application of filament voltage. An *Eimac*
SK-410 air-system socket and *SK-416* air
chimney are used to achieve proper air flow
around the filament and plate seals of the
tube.

Layout of the major components may be
seen in the photographs. The air-system
socket is mounted on the underside of the
chassis in a $3\frac{1}{2}$-inch diameter cutout. The
spring clips that hold the chimney in place
fasten with the same bolts used to mount
the socket, which is oriented so that filament
pins 1 and 5 are facing the front of the
chassis. The cathode tuning capacitor (C_1)
is mounted on the front apron of the chassis
with insulated washers as the rotor is above
ground by the amount of the filament voltage. The cathode coil is a dual winding, made
of copper tubing having an insulated center
conductor. A section of $\frac{1}{8}$-inch soft-drawn
copper tubing about a foot long is needed
to make the coil. Before the coil is wound,

the ends of the tubing are smoothed with a file and a length of No. 12 cotton-covered (or *formvar*-insulated) wire is passed through the tubing. The coil is then wound about a 3/4-inch diameter wood dowel rod used as a temporary form, spacing the three turns to a length of two inches. The tubing is trimmed, and the inner wire is left projecting about ten inches from each end. The coil is mounted close to the tube socket (figure 35) with one end supported by the filament pins of the tube socket. The inner conductor is trimmed to length and soldered to one filament pin, and the tubing is connected to the other filament pin by means of a short length of copper strap about 1/8-inch wide, cut from copper "flashing" material. The end of the coil is equidistant from the filament pins. The strap encircles

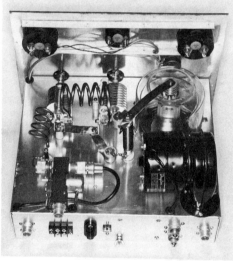

Figure 34

TOP VIEW OF 3-400Z LINEAR AMPLIFIER FOR 50 MHz

Placement of the major components above the chassis may be seen in this photograph. The meter shield has been removed for the photograph. Leads to the meter compartment are shielded, and bypass capacitors are mounted at the meter terminals.
Across the rear apron of the chassis (l. to r.) are: receiver receptacle (J₂); terminal strip (J₅); Millen high-voltage connector (J₃); Sprague feedthrough capacitors and r-f exciter receptacle (J₁). At the bottom edge of the chassis are a ground connection and the relay voltage terminal (J₄).
The copper ground strap between the plate-circuit tuning capacitors may be seen just behind the antenna relay.

one end of the tubing and is soldered in place, with the other end soldered to the pin. The filament bypass capacitor is soldered directly between the filament pins of the socket. A second short length of copper strap jumpers the first strap to the stator of the cathode tuning capacitor.

The opposite end of the cathode coil is bypassed to ground by a ceramic capacitor which also supports the coil. The inner conductor is bypassed to the outside tubing at this point, and a length of copper strap makes a connection to the rotor of the tuning capacitor. The inner conductor continues over to the filament transformer and a second length of No. 12 wire is run from the copper tubing to the second transformer terminal.

The three grid pins of the 3-400Z socket are grounded by passing a 1/4-inch wide copper strap through the slot in the socket adjacent to each grid pin and soldering the strap directly to the flat tab on the pin. The straps are then bolted to the chassis just clear of the socket.

The Plate-Circuit Assembly Layout of the components above the chassis are shown in figure 34. The plate tuning and loading capacitors (C₂ and C₃) are mounted on 1/2-inch ceramic insulators. The tuning capacitor is rotated 90 degrees on its side and held in position with small aluminum brackets. A common ground connection made of a length of 1/2-inch wide copper strap connects the rear rotor terminals of the capacitors. In addition, the capacitor rotor wipers are connected to the common ground strap.

A second strap grounds the rotors to a common ground point on the chassis under the stud of the high-voltage bypass capacitor at the lower end of the plate r-f choke. The shafts of of the variable capacitors are driven with insulated couplers to prevent ground-loop currents from flowing through the shafts into the panel.

The plate r-f choke is homemade, and is wound on a 1/2-inch diameter ceramic insulator. A commercial choke may be used, if desired. The base of the choke screws on the bolt of the high-voltage feedthrough insulator on the chassis, and is bypassed at this point with a ceramic capacitor.

The coaxial antenna relay is mounted on the top of the chassis positioned so the output lead from the L-section of the tank circuit can be connected directly to the input receptacle. The connection is made by trimming down a coaxial connector and soldering a short length of #10 wire to the center terminal to make the connection to the coil. The *antenna* receptacle of the relay extends beyond the rear apron of the chassis and through the rear of the cabinet. The *receive* receptacle is fed with a length of RG-58/U coaxial cable which terminates at the coaxial receptacle on the rear apron of the chassis. An auxiliary set of contacts on the relay are used to short out the 50K self-bias resistor in the cathode circuit of the 3-400Z when transmitting. The resistor serves to bias the tube to near cutoff during periods of reception to prevent noise being generated which may interfere with reception of weak signals and also to reduce the standby drain on the power supply. The relay is actuated by the control or VOX circuit of the exciter, and the relay coil should be chosen to match the voltage delivered from the exciter control circuit.

A diode r-f voltmeter is mounted beneath the chassis in a small aluminum box positioned over the r-f feedthrough insulator which supports the end of the L-network above the chassis. The lead from the voltmeter circuit to the *calibrate* potentiometer on the panel is run in shield braid, as are the leads from the center tap of the filament transformer. Tight rubber grommets are used in all chassis holes to restrict air leaks.

Amplifier   When the amplifier has been
Adjustment   wired and inspected, it is ready
for initial checks. Air is directed into the tube socket by means of a temporary bottom plate (cardboard) taped to the chassis. Filament voltage is applied and the blower motor should start. A strong blast of air out of the tube chimney should be noted. Tube filament voltage should be adjusted to 5.0 volts at the socket with an accurate meter. Filament voltage is now removed and the input and output coaxial receptacles are temporarily terminated in 50-ohm, 1-watt composition resistors, which may be soldered across the receptacles for

this test. A grid-dip meter is tuned to 50 MHz and brought near the cathode coil (the 3-400Z being in the socket. The meter

Figure 35

UNDER-CHASSIS VIEW OF LINEAR AMPLIFIER

The cathode circuit is mounted on the filament terminals of the air system socket (upper left), with the tuning capacitor (C_1) insulated from the panel. Filament leads run from the tuned circuit to the filament transformer mounted to the rear apron of the chassis. At the right is the small aluminum box (cover removed) holding the components for the r-f output voltmeter. The blower outlet is at the left corner of the chassis next to the feedthrough capacitors.

should show resonance with the cathode tuning capacitor about two-thirds meshed. The plate tank circuit is now tested, with the tuning capacitor about one-half meshed and the loading capacitor about two-thirds meshed. Grid-dip resonance at these settings for 50 MHZ may be achieved by slight alterations in the spacing of the pi-network coil. The L-section should also show a dip around 50 MHz.

Once resonance of the tank circuit has been verified, the 50-ohm resistors are removed and the amplifier attached to the exciter and coaxial antenna lead. A separate ground lead is run from the amplifier to the power supply. A plate potential of 2500 volts is recommended as a maximum (keydown) value, and good operation can be obtained down to 2000 volts. At the higher potential, the resting plate current will be about 80 mA. Random variations in resting plate current, or a show of grid current when the controls are tuned (with no grid drive) is an indication of parasitic oscillation and a plate parasitic choke should be installed.

After plate voltage is applied, grid drive is slowly injected until a plate current of about 150 mA is noted. The cathode circuit is resonated for maximum grid current and the plate tuning capacitor adjusted for plate-current dip. Grid drive is increased and loading adjustments made in the normal manner for pi-network operation to achieve a *single tone (carrier)* plate current of 400 mA at a grid current of about 140 mA. Proper loading is indicated by the ratio of plate current to grid current, which should be about 3:1.

For operation as a linear amplifier for SSB, carrier injection is used as described for tuning and loading. The relative-voltage output meter is very useful in the tuning process and provides a continuous check on proper operation as it increases in proportion to grid current. Maximum carrier input conditions are as stated above, and under these conditions, the anode of the 3-400Z will be a cherry red in color. With carrier removed and SSB voice modulation applied, drive is advanced until voice peaks reach about 200 mA plate current and about 70 mA grid current. For c-w operation, the full 400 mA plate current value may be run.

A-M Linear Operation The amplifier may be used for a-m linear service when properly adjusted. The amplifier efficiency at the peak of the modulation cycle is about 66 percent and efficiency under carrier conditions (no modulation) is about 33 percent. As maximum plate dissipation is 400 watts, the total a-m carrier input to the 3-400Z is limited to about 600 watts (2500 volts at 240 mA). In order to *properly* load the amplifier to this condition for a-m linear service, an oscilloscope and peak-responding voltmeter are necessary. The r-f output voltmeter in the amplifier may be converted to a peak-responding instrument as shown in figure 33B. In addition, a simple 1000-Hz audio oscillator is used for the following adjustments.

For preliminary tuneup, the a-m driver is modulated 100 percent with the 1000-Hz tone. A driver capable of about 15 watts

carrier is required. The 3-400Z amplifier is loaded and drive level adjusted to 600 watts input under this condition. Amplifier output is monitored with the peak-responding voltmeter, which is adjusted to *full-scale* reading at the 600-watt input level. Grid current will run about $\frac{1}{4}$ the plate-current value, or approximately 60 mA. Once this condition is reached, the modulation of the driver is removed, leaving only carrier excitation. If the linear amplifier is properly adjusted, the indication of the peak-responding voltmeter should drop to *one-half scale*, corresponding to an output drop to one-quarter power.

If the peak-voltage drop when modulation is removed is less than one-half, the plate circuit loading and grid-drive level of the linear amplifier must be adjusted to provide the correct ratio. This is an indication that antenna loading is too light for the given grid drive. If this process is monitored with an oscilloscope, the point of flat-topping can be noted and drive and loading adjusted to remove the distortion on the peaks of the signal. Under voice modulation, plate and grid current will flicker a small amount upward.

The combination of a peak-responding voltmeter, an oscilloscope, and an audio oscillator used with tune-up under 100 per cent single-tone modulation of the exciter affords a relatively easy and accurate method of achieving proper a-m linear amplifier service.

As with any cathode-driven amplifier, drive should never be applied to the amplifier in the absence of plate voltage, as damage to the grid of the tube may result. The proper sequence is to always apply plate voltage before drive, increasing the drive level slowly from a minimum value as tuning adjustments are made.

22-7 A Compact 2 kW PEP Linear Amplifier With the 8877

This rugged linear amplifier is designed around a single 8877 high mu, ceramic power triode and provides maximum legal

Figure 36

THE GNQ-1000 LINEAR AMPLIFIER FOR 80-10 METERS

The compact, rugged linear amplifier runs full legal input for SSB, RTTY or SSTV service on the high frequency bands. An 8877 ceramic power triode is used in a cathode driven circuit. The plate load control is at the upper left of the panel, balanced by the counter dial for the plate tuning capacitor at upper right. Centered on the panel are the grid and plate current meters. The band-switch is bottom, center with a pilot lamp above it. At the left is the grid bandswitch, with the filament switch at the lower right of the panel. The top cover plate (removed for photographs) is perforated aluminum material to allow good air circulation within the amplifer when the blower is running.

power input on all amateur bands between 3.5 MHz and 29.7 MHz. It was designed and built by Dick Drevo, W3GNQ. Amplifier power gain is high and drive requirements are less than 100 watts PEP on all bands (figure 36). The amplifier features ALC control of the exciter and a pi-network input circuit. Grid and plate currents are metered and electronic bias switching lowers the standby plate current of the amplifier during standby periods, reducing tube dissipation, heat and the power bill. The amplifier is designed to operate at a plate potential of 3 kV. Third order intermodulation

products at maximum power output are about 40 dB below one tone of a two-tone test signal.

The Amplifier Circuit   The schematic of the amplifier is shown in figure 37. The 8877 is operated in cathode-driven mode, with the grid at dc and r-f ground potential. The drive signal is coupled through a bandswitching pi-network to the cathode of the tube. The Q of the network is high enough to preserve the waveform of the driving signal yet low enough so that the networks need not be retuned when the

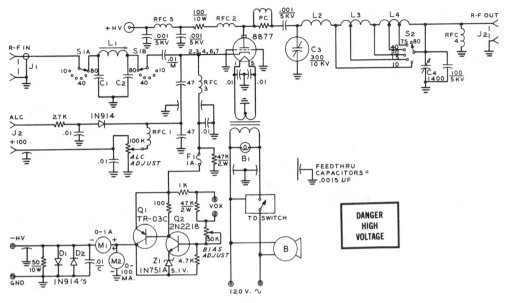

Figure 37

SCHEMATIC, 8877 AMPLIFIER

C_1, C_2—See table 8.
C_3—300 pF, 10 kV. ITT-Jennings Co.
C_4—1400 pF
L_1—See table 9.
M_1—0-1 Adc. Triplett 320G
M_2—0-100 mAdc. Triplett 320G
PC—3 turns #12 e. around 47-ohm, 2-watt resistor
Q_1—TR-03C. International Rectifier Co.
RFC_1, 4—2.5 mH, 100 mA.
RFC_2—160 turns #24 e. wound on ceramic insulator ¾" diam., 4" long.
RFC_3—2.5 mH, 300 mA

RFC_5—10 turns #14 e., ¼" diam. 1" long.
S_1A, B—2 pole, 6-position. Centralab PA-2045
S_2—1 pole, 6 position. Millen 51001 or equivalent. Radio Switch Corp.
T_1—5 volts, 15 amperes. Stancor P-6433
Blower—Dayton 1C180 or 2C781
Chimney—SK-2216
Counter dial—R. H. Bauman TC-3S
Socket—SK-2210
Time delay switch—180 seconds.
Amperite 115-N0180

frequency of operation is moved about within an amateur band. A summary of the network components is given in Table 8. Grid current metering is accomplished in the bias return circuit.

During standby, the 8877 is biased to cutoff by the electronic bias switch Q_1, Q_2. Operating bias is adjustable. A fuse in the cathode circuit protects the tube from excessive plate current and a small series resistor in the plate circuit provides protection from inadvertent flashovers in the plate circuit.

Metering Circuitry—Grid current flows from cathode to ground through meter M_2 and cathode (plate) current flows through meter M_1 which is located in the negative dc return lead to the power supply. Meter protection is provided by reverse-connected diodes D_1-D_2.

The Plate and ALC Circuitry—A pi-network output circuit is provided which is designed for a plate load impedance of about 1800 ohms with a loaded Q of 10. Data for the plate network is given in Table 9.

ALC voltage is obtained by sampling the signal in the cathode circuit through a capacitive divider. The reference level for ALC is set by means of potentiometer R_2.

Amplifier Cooling—Maximum dissipation of the 8877 in this circuit at full legal input is about 800 watts. To hold tube temperature below 250°C with 50°C ambient temperature at least 20 c.f.m. of air at a pressure drop of 0.2" is required. A *Dayton* model 1C-180 (or equivalent) blower will satisfy this requirement. The air is drawn into the underchassis area of the amplifier and exhausted through the anode of the tube and out the perforated metal top of the cabinet.

Table 8. Input Network Details

Circuit Component Values (Q = 1)

Band	C_1, C_2	L_1
80	820pF	(2.36 μH) 14 turns #24 e, 3/8" diam. closewound
75	750pF	(2.07 μH) 14 turns #24 e, 3/8" diam. closewound
40	430pF	(1.18 μH) 10 turns #24 e, 3/8" diam. closewound
20	220pF	(0.59 μH) 7 turns #16 e, 3/8" diam. closewound
15	150pF	(0.39 μH) 5 turns #16 e, 3/8" diam. closewound
10	100pF	(0.30 μH) 4 turns #16 e, 3/8" diam. spaced wire diam.

Capacitors dipped mica: DM15-J series
Coil forms: J. W. Miller 4400-0 with
red slugs (30-201-2 DP)

TVI Suppression—The amplifier is contained within an r-f tight metal cabinet for maximum containment of harmonics. All power leads are fully filtered and a screen is placed over the mouth of the blower to block this area from r-f leakage. The liberal use of feedthrough-type bypass capacitors greatly assists the reduction of radiated harmonics from the inclosure.

Table 9. Plate Circuit Details

Band	Plate Inductance
10	4½ turns ¼" thin wall copper tubing, 2" diameter, 1¾" long
15-20	5 turns ¼" thin wall copper tubing, 3" diameter, 3¾" long
40-75-80	12 turns #10, 2½" diam., 6 turns/inch

1—15-meter tap 2⅛ turns from plate end of L_3
2—40-meter tap 5 turns from plate end of L_4
3—75-meter tap 8¾ turns from plate end of L_4

The Power Supply—The high-voltage supply for this amplifier provides 3000 volts at a peak plate current of about 700 mA. Average current drain is about one-half this value. In addition, an ALC supply voltage of +100 volts at 10 mA is required. The high-voltage supply has a time delay circuit in the primary side of the transformer which delays application of plate voltage for 180 seconds after the filament is turned on.

Amplifier Construction   The amplifier is built into a metal enclosure measuring 10" × 13" × 17". This is composed of a chassis 4" high and a plate circuit box 6" high. A solid bottom plate makes the bottom chassis airtight and a perforated top plate allows cooling air to escape from the plate compartment (figure 38).

The 8877 7-pin septar socket is sub-mounted below the chassis deck with 3/8-inch metal spacers while the grid ring of the tube is electrically grounded to the chassis by means of four grounding clips on the socket assembly. The air chimney sits atop the chassis, held in position by the air cooler of the tube.

The plate circuit bandswitch is mounted to the chassis plate, with the shaft projecting into the underchassis area. The switch is panel driven through a sturdy right-angle drive unit and an extension shaft (figure 39). The cathode circuits are mounted on a small subchassis bolted to the rear wall of the enclosure. They, too, are panel-driven by means of a switch and extension shaft.

Looking at the front of the amplifier (figure 36), the panel layout is symmetrical, with the center lines of the meters 2¼" from the center line of the panel. The grid selector switch and power switch are located 4" from the center line. Plate tuning and loading controls are spaced 3⅝" in from the outer edges of the panel and on a line 3" down from the top edge of the panel.

Looking at the interior view of the amplifier (figure 38) the tube socket is centered at a point 3" behind the front of the enclosure and 7⅞" in from the left side. The center line of the main tank coil runs parallel to the chassis edge and 6" away from it. The bandswitch is centered on the chassis and 4¼" from the rear edge. Placement of the other parts in relation to these can be ascertained from the photographs.

The PNP transistor (Q_1) is mounted on the amplifier chassis for a heat sink and insulated from it by a mica washer. Transistor Q_2 is placed on a small printed-circuit board along with the associated resistors and zener diode. The ALC components are mounted on a similar board.

The vacuum variable plate-tuning capacitor is mounted on an aluminum bracket

Figure 38

TOP VIEW OF GNQ-1000 AMPLIFIER

The 8877 tube is centered on the chassis near the front panel. To the right are the plate r-f choke and coupling capacitor. The variable vacuum plate tuning capacitor is mounted on a bracket at the far right. The bandswitch is chassis mounted in line with the tube. Above it, to the right and left, are the tank circuit components. Coils L_1 and L_2 are wound of copper tubing and run from the coupling capacitor back to the bandswitch. The coils are supported on ceramic insulators. To the left of the bandswitch is coil L_3, adjacent to the loading capacitor. The chassis is supported three inches behind the panel to allow area for meters, wiring, etc.

placed behind the enclosure wall. The network output capacitor is fastened to the side wall of the enclosure while the pi-network coils are mounted to the bandswitch and ceramic insulators bolted to the chassis. The plate r-f choke is fastened to the top terminal of the plate bypass capacitor. The plate lead from the choke to the high-voltage connector on the rear of the chassis is a short length of RG-8/U coaxial cable.

For highest efficiency the plate inductor is made of three coils. Inductor L_2 is for 10 meters only, inductor L_3 covers 15 and 20 meters when added to L_2 and inductor L_4 is for 40, 75 and 80 meters when added in the circuit. Splitting the 80-meter band in two sections helps maintain a good L/C ratio, aids in loading, and reduces harmonic content in the output signal. Leads from the tap points on the coils to the bandswitch

are made with ¼-inch wide copper strap, silver soldered to the coils. The coils and leads are silver plated before final assembly.

On the rear of the amplifier chassis are the bias control, the ALC control and a large terminal strip for external wiring. The slugs of the various cathode coils also project through the rear panel of the chassis.

At the front of the chassis, meter leads and accessory wiring are brought out via a shielded cable, all leads being bypassed to ground at the panel as well as at the meter terminals.

Amplifier Alignment—Once the amplifier is wired and checked, the input network coils are adjusted to the midpoint of each band with the aid of a dip-meter. This is done with the 8877 in the socket. Final alignment takes place once the amplifier is in operation.

Figure 39

UNDERCHASSIS VIEW
OF AMPLIFIER

The bottom of the 8877 socket is centered on the chassis, with the filament transformer toward the rear corner of the chassis. The grid input circuit can be seen grouped around the grid bandswitch at the rear of the chassis, next to the compartment holding filter capacitors for the power leads. The shaft of the main bandswitch protrudes through the chassis deck and is panel driven by means of a right-angle drive unit seen behind the tube socket.

Amplifier Tuning and Adjustment The first step is to check filament voltage on the 8877 at the socket. It should be measured with a 1-percent voltmeter to determine that it is within the allowed range of 4.75 to 5.25 volts. Operation at the lower end of this range is recommended. The amplifier plate circuit should be roughly resonated with the aid of a dip meter and, after the time delay relay has activated, plate voltage may be applied to the 8877. The bias should be adjusted to provide a resting plate current of about 160 mA to 180 mA. Check to make sure the blower is operating properly and that a free flow of air exists through the anode cooler.

Drive power is applied and the amplifier is tuned and loaded for resonance. When properly loaded, grid current will run about 40 mA. The input circuits can be touched up by inserting an SWR meter in the coaxial line between the exciter and the amplifier and tuning the coil slug on each band for minimum SWR at the center frequency in each band. This adjustment should be done with the amplifier running at full power input.

Note: *The amplifier cover should not be removed during operation as high-voltage points will be exposed.*

22-8 A Modern 3-1000Z Linear Amplifier for 80-10 Meters

This compact and rugged linear amplifier is designed for continuous duty operation at the maximum legal input power on c-w, SSB or RTTY. Designed and built by Jerry Pittenger, K8RA, the unit uses a single 3-1000Z high-mu triode in a cathode driven circuit. "Grounded grid" service is especially attractive as maximum input may be run with a plate potential as low as 2500 volts, yet the power gain of the tube is high enough to allow sideband exciters of the "100-watt" class to drive it to full output. Neutralization is unnecessary up to 30 MHz as the excellent internal shielding of the 3-1000Z reduces intrastage feedback to a minimum. Distortion products of this amplifier are better than 35 dB below one tone of a two-tone test signal at maximum PEP level. A tuned cathode tank is used for greatest linearity and power output. Special attention has been given in the construction of the amplifier to protective shielding and lead filtering to reduce TVI-producing harmonics to a minimum (figure 40). A simple, solid-state power supply is also included in the equipment.

The Amplifier Circuit   The 3-1000Z amplifier covers all amateur bands between 3.5 and 29.7 MHz with generous overlaps. Bandswitching circuits are used and the amplifier is designed to operate into a coaxial antenna system of 50 to 70 ohms having an SWR of less than 3. The schematic of the amplifier is given in figure 41.

The r-f deck is shown in figure 42A-B. The driving impedance of the 3-1000Z is approximately 55 ohms, providing a close match to either a 50- or 70-ohm coaxial system. The tuned cathode circuit prevents input waveform distortion caused by the half-cycle loading of the amplifier, which operates in a near class-B mode. Filament voltage is fed to the 3-1000Z through a conventional bifilar, ferrite-core r-f choke.

Plate current metering is accomplished in the B-minus power lead to remove dangerous anode potentials from the meter circuit. The resting plate current of the tube is reduced by means of a 7.5-volt, 50-watt zener diode in the cathode circuit, and for standby operation the cathode voltage is raised by a bias resistor which is inserted in the circuit by the *in/out* relay, RY_1C. The relay shorts out the resistor to allow normal operation of the stage when actuated by the VOX circuit. The grid terminals of the 3-1000Z are

Figure 40

3-1000Z LINEAR AMPLIFIER FOR 80-10 METERS

This deluxe amplifier runs full legal input for SSB, RTTY, or SSTV service on the high-frequency bands. A single 3-1000Z high-μ triode is used in cathode-driven service. The amplifier (with cover removed) sits atop the power supply pedestal. On the panel of the amplifier are (left to right): Plate, grid, and r-f output meters; the tune and load controls; and the bandswitch. Below the bandswitch are the primary and in/out control switches.

The pedestal contains the variable voltage transformer at center, the plate voltmeter and the main pilot lamp assembly at right. Oversize casters permit the operator to move the amplifier about with ease.

directly grounded and grid current is measured in the cathode return circuit.

A pi-network plate tank circuit is used, with an additional loading capacitor switched in for 80-meter operation into low values of load impedances. In addition, a diode voltmeter is included to monitor the relative power level of the amplifier.

The amplifier plate coil is a modified commercial unit retapped to provide a loaded circuit Q of approximately 10 on all bands with a plate potential of 3200 volts and a plate current of 600 milliamperes. A suitable value of Q is maintained on c-w by lowering the plate potential to 2500 volts by means of a variable voltage transformer in the primary circuit of the high-voltage power supply.

The 3-1000Z requires forced-air cooling to maintain the base seals at a temperature below 200°C and the plate seal at a temperature below 225°C. When using an *Eimac SK-510* socket and *SK-506* air system chimney, a minimum air flow of 25 cubic feet per minute is required at a back pressure of 0.25 inches of water. Cooling air must be supplied to the tube as long as the filament is lit.

Cooling is accomplished by a 3½-inch diameter impeller running at a speed of 3200 rpm. The squirrel cage blower is mounted on the bottom plate of the chassis and extends downward into the power-supply pedestal. The chassis is pressurized, and the air is exhausted through the socket and chimney and past the anode of the tube, which is equipped with an *Eimac HR-8* heat dissipating anode connector.

Amplifier Power Supply and Controls— The control circuit of the 3-1000Z amplifier is shown in figure 43. Primary power enters the control circuitry via terminals 7 and 8. Switch S_1 is the *on/off* control switch and switch S_2 (amp. in/out) activates the 25-Vdc supply which is used to energize relay RY_1 through the auxiliary VOX contacts. The filament transformer for the 3-1000Z and the air blower come on simultaneously when S_1 is thrown. The cutoff bias resistor in series with the cathode zener diode is shorted out by the VOX circuit and relay contacts RY_1C.

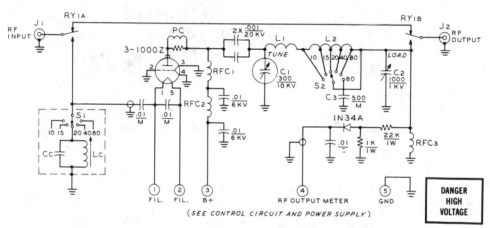

Figure 41

SCHEMATIC OF 3-1000Z LINEAR AMPLIFIER

C_1—10—300-pF, 10kV vacuum variable capacitor. Jennings UCS-300

C_2—30-1000-pF, 1-kV. (A substitute unit, 1100-pF, 3-section capacitor Jackson LE3-4595-380 may be used.)

C_3—500-pF mica, transmitting type. Sangamo type H

L_1—5 turns of ¼" copper tubing, 1¾" diam., 3" long.

L_2—Modified B-W 850A (see text). See chapter 11 for pi-network data for plate load impedance of 2600 ohms

PC —Three 120-ohm, 2-watt composition resistors in parallel. Short across ½" copper strap bent into U-shape, 2" long, 1" wide

RFC_1—(Approx. 60 μH. 90 turns #26 e., spacewound wire diameter, 3⅜" long, ¾" diam. on ceramic or teflon form. Series resonant at 26 MHz.) B-W type 800.

RFC_2—10 turns, ½" diameter, 1" long.

RFC_3—1 mH, 0.3 ampere

S_1—Single-pole, 11-position ceramic deck, 30° index. Centralab PA-6001

S_2—Single-pole, 11-position ceramic switch, 30° index. Radio Switch Corp. Model 86-A

Cathode circuit:

C_c—Silver mica. 10 meters, 200 pF; 15 meters, 470 pF; 20 meters, 470 pF; 40 meters, 1000 pF; 80 meters, 1000 pF

L_c—All coils wound on ½" diameter form, powdered-iron slug. Slug removed on 10- and 15-meter coils. 10 meters, 4 turns; 15 meters, 4 turns; 20 meters, 6 turns; 40 meters, 7 turns; 80 meters, 13 turns. Wind with #16 e. wire. Grid-dip each coil to center of appropriate band.

The power supply for the amplifier is shown in figure 44. A 240-volt primary circuit is recommended, although the amplifier could operate from a well-regulated 120-volt circuit. Relay RY₃ is a *step-start* device which allows the charging current of the capacitor bank to be reduced by virtue of the primary resistor (33 ohms, 60 watts), which is shorted out of the circuit after a few milliseconds.

A 10-ohm, 10-watt safety resistor is included in the B-plus circuit from the supply to the amplifier. In case of a flashover in the amplifier, the resistor will absorb the surge and protect the rectifier bridge and the amplifier components from the heavy short current.

Amplifier Construction   The r-f deck of the amplifier is built on an aluminum chassis measuring 12″ × 17″ × 3″ and uses a dual front panel. The main panel is 19″ wide and 14″ high and is spaced 2½″ away from the amplifier inclosure. The under-chassis area is divided into two compartments by a vertical shield. One compartment contains all wiring necessary for the 3-1000Z socket. The other compartment contains the input circuitry, power line filters and small, auxiliary components. The dual front panels allow space for the meters, power control wiring and facilitate structural support. The tube, filament transformer, antenna switching relay, and pi-network components are mounted atop the chassis. All electrical wiring from one compartment to another passes through 1000-pF feedthrough capacitors. All cables entering or leaving the r-f deck pass through pi-section r-f filters. The majority of wiring utilizes shielded cables.

Atop the chassis, the antenna switching relay (RY₁) is inclosed in a small aluminum utility box at the rear corner of the chassis. The box is insulated on the interior with ¼-inch thick cork tile, and the relay is mounted on small rubber grommets. The cork tile, plus the rubber mounting are very effective in eliminating relay noise and buzz. This relay switches the amplifier in and out of the antenna circuit and also removes the standby bias during operation.

The variable vacuum tuning capacitor and counter dial are mounted on the center-line of the assembly, directly above the loading capacitor. Placement of the other components may be seen in figure 42A-B.

The pi-network inductor incorporates its own switch and the input bandswitch is ganged to the plate bandswitch by means of a chain drive system mounted in the space between the front panels. The plate inductor is a *Barker-Williamson 850A* modified to obtain optimum efficiency. The 10-meter strap inductor is discarded and a new 10-meter coil wound using ¼-inch copper tubing. The coil has an inner diameter of 1¾″ and consists of 5 turns equally spaced out to 3 inches. The coil is silver plated.

As purchased, the *850A* unit provides too much inductance on 40 and 80 meters. Accordingly, four turns are removed from the far end of the 80-meter wire portion of the inductor and the 40-meter meter tap is moved three turns closer to the tubing portion of the inductor. Connections to the inductor are made with ½-inch wide silver plated copper strap.

Beneath the chassis, the ferrite r-f filament choke is supported at one end by the filament terminals of the air socket and at the other end by the mica bypass capacitors, which are held to the chassis deck by means of heavy angle straps (figure 45). The bottom flange of the socket is cut off to allow better air flow and the grid terminals are grounded using short lengths of copper strap passed through the socket slots near each grid pin.

Power Supply Construction   The r-f deck is mounted on a pedestal which contains the components of the high-voltage power supply. Pedestal height is 24″. Because of the weight of the components, the pedestal is constructed of ½-inch angle aluminum welded together in the form of a rectangle with a sloping top which provides a slight tilt for the r-f deck. A piece of ¾-inch thick plywood is placed at the bottom of the frame to support the power-supply components (figure 46).

The sides and front of the pedestal are covered with wrinkled aluminum sheet, available at many large hardware stores. The aluminum is held to the frame with sheet-metal screws and the front corners covered with ½-inch angle aluminum. The rear panel

The chain drive mechanism for the input circuit is seen in the foreground between the front panels. At the front of the chassis are the plate inductor, with the pi-network capacitors at the center. To the left is the filament transformer, with the 3-1000Z behind it. The antenna changeover relay is in the inclosure behind the bandswitching inductor.

Figure 42A

RIGHT-OBLIQUE VIEW OF POWER AMPLIFIER

The 3-1000Z with its glass chimney are at left. The plate blocking capacitors are supported on a small metal bracket secured to the top of the plate r-f choke.

Figure 42B

LEFT-OBLIQUE VIEW OF POWER AMPLIFIER

of the pedestal is made of $\frac{1}{16}$-inch aluminum sheet. A small jig was drilled in a piece of scrap steel and used to drill the ventilating holes in the rear panel, as shown in figure 47.

The amplifier is covered with a shield made out of perforated aluminum bent into a U-shape. The outer, decorative shield is cut from $\frac{1}{16}$-inch aluminum and a large ventilating hole is cut in it over the top of the tube. The hole is covered with perforated aluminum stock.

The front panel of the amplifier is covered with common black leather (plastic) upholstery material. A thin film of white glue holds the material to the panel. After drying thoroughly, all holes are cut in the plastic with a razor blade. Dry transfer lettering is then applied directly to the panel.

Amplifier Adjustment   After wiring is completed and checked, the cathode and plate circuits should be resonated to frequency using a dip-meter. The cathode circuits may also be resonated by temporarily wiring a 60-ohm composition resistor on the socket from a grid terminal to ground with the tube removed. With a small signal applied, the coil slugs can be adjusted for minimum SWR on the driving line on each band.

The next step is to place the tube in the socket and activate the blower and light the

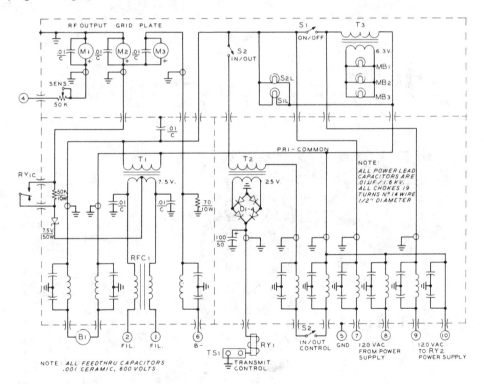

Figure 43

CONTROL CIRCUITRY FOR 3-1000Z AMPLIFIER

B_1—20 cu. ft./min. Dayton 1C-180 or Ripley LR-81 or equiv.
D_1-D_4—100-volt, 0.5 ampere diode
M_1—0-1 milliampere dc meter
M_2—0-0.3 ampere dc meter
M_3—0-1.0 ampere dc meter
MB_{1-3}—6.3-volt ac lamps in meters M_1, M_2, M_3
S_1, S_2—SPST pushbutton switches with lamps

S_1L, S_2L—120-vac lamp built into S_1, S_2
RFC_1—30-ampere bifilar filament choke. Each coil is 14 turns #10 e., on ferrite core, 5" long, $\frac{1}{2}$" diam. (Indiana General CF-503) or equivalent
T_1—7.5 volts at 22 amperes, tapped primary winding.
T_2—25 volts at 0.5 ampere

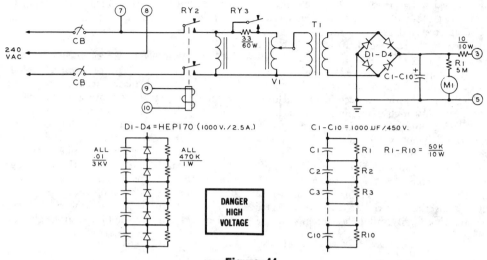

Figure 44

SCHEMATIC OF POWER SUPPLY

CB—20 ampere circuit breaker
RY₂—Double-pole, double-throw relay with 120-volt coil. 25-ampere contacts
RY₃—Single-pole, single-throw relay with 120-volt coil. 5-ampere contacts
T₁—3600 volt rms secondary winding at 1 ampere. 240 volt primary
M₁—0-1 dc milliammeter for use with 5-megohm multiplier (R₁)

Figure 45

UNDERCHASSIS VIEW OF 3-1000Z LINEAR AMPLIFIER

The tube socket and filament r-f choke are at left, with the zener diode mounted to a small heat sink immediately above the socket. At the right is the small compartment holding the tuned input circuits. The coil slugs are adjustable through holes drilled in the cover. Connections to the tuned circuit are made via lengths of coaxial line and a T-fitting. The filter networks for the primary power leads are in the foreground.

filament. The filament voltage should be checked at the tube socket. If the filament transformer primary is tapped, the tap should be set for the correct voltage, allow-

ing for the voltage drop in the filament choke.

Amplifier adjustment is initially done into a dummy load, such as the one described in

Figure 46

REAR VIEW OF PEDESTAL AND POWER SUPPLY

The power-supply components are mounted on a sheet of plywood at the bottom of the pedestal. The filter capacitors and diode rectifier assembly are at left, with the relay controls in the foreground.

Figure 47

REAR VIEW OF THE 3-1000Z AMPLIFIER AND PEDESTAL

The pedestal has a sloping top which provides a slight tilt for the r-f deck.

the Electronic Test Equipment chapter, to satisfy the requirements of the FCC. With filament and operating plate voltage applied, grid- and plate-current meters should read zero when the amplifier is switched out of the line. Shorting the *in/out* control terminals (figure 43) engages the antenna relay. In this mode, grid current with no drive should be zero. Plate current should read appoximately 120 milliamperes (the exact value varying from tube to tube). If grid current is observed, it is probably a sign of parasitic oscillation, and the circuit should be examined for parasitic resonances and the plate parasitic choke checked before additional tests are run.

Drive power can now be slowly increased to raise the plate current to about 250 mA. Tuning and loading controls are adjusted for maximum power output as indicated on the relative output power meter. Excitation and loading are now increased until approximately 600 mA plate current and 200 mA grid current are achieved, with maximum output indicated. When the above conditions have been met, the loading is increased slightly to ensure proper linearity. The amplifier is now properly tuned for 2-kW PEP operating conditions. Under speech conditions, peak plate current with no clipping will kick to about 300 mA and grid current will be approximately one-third this value.

22-9 A 4CX1500B 2-kW PEP Linear Amplifier

The linear amplifier described in this section is a deluxe 2-kW PEP, class AB₂ grid-

Figure 48

THE K5JA-2-kW PEP AMPLIFIER FOR 80-10 METER OPERATION

Designed for 2-kW PEP operation on all hf bands,   the amplifier features an Eimac 4CX1500B tetrode in a passive grid circuit. A pi-L output tank circuit provides maximum harmonic attenuation. Maximum drive signal required for full output is 5 watts, PEP. The amplifier is contained within a standard tabletop cabinet (removed for photograph). The multimeter (left) and plate current meters are at the right, top area of the panel, with the tuning and loading counter dials below the meters. The bandswitch is centered between the meters. At the left of the panel are the electron-ray tuning tube mounted in a horizontal position behind a panel cutout, the multimeter selector switch and the control switches and indicator lights. The panel is spray painted and decals are applied before a final coat of clear Krylon plastic spray is added.

driven amplifier using the low distortion 4CX1500B tube (figure 48). This is a ceramic-metal, forced-air cooled tetrode having a maximum plate dissipation of 1500 watts. It is designed for exceptionally low intermodulation in SSB service. Typically, at a plate potential of 2750 volts and a plate current of 730 mA (2-kW PEP input) the third-order intermodulation distortion products are better than 40 decibels below one tone of a two-tone test signal. Under these conditions, the useful power output is better than 1100 watts, allowing for normal tank circuit losses.

This amplifier is designed and built by John Ehler, K5JA, as a companion unit to the low power, solid-state exciter described in chapter 20 of this Handbook. The amplifier features very high power gain and very stable operation on all amateur bands between 3.5 MHz and 29.7 MHz. Well suited for use with a solid-state driver, the amplifier will deliver full output with less than 5 watts PEP drive signal.

The Amplifier Circuit   The 4CX1500B is used in a passive-grid circuit of the type shown in chapter 21. The schematic of the amplifier is shown in figure 49. In order to achieve maximum stability, the screen element of the tube is run at r-f ground potential. Screen voltage is applied to the tube by grounding the positive terminal of the screen power supply and "floating" the screen and bias supplies below ground. A special socket is used for the 4CX1500B which provides a low-inductance screen to ground path.

The Input Circuit—The grid drive requirement of the 4CX1500B is about 1.5 watts PEP for full output. The 5-watt input signal is fed to the tube through a four-to-one wideband ferrite transformer which steps up the impedance from 50 to 200 ohms. Five 1.2K, two-watt composition resistors in parallel (R_1) plus the 1.2K resistor in the bias line provide a 200-ohm terminating load. The relatively high input capacitance of the 4CX1500B is resonated on each

Table 10.
Coil and Tuning Data

Band	Grid Circuit L_1	Plate Circuit L_4 Tap	L_5 Tap
80	22 μH RFC	All	All
40	4.7 μH RFC	11½ Turns	10½ Turns
20	21 turns #30 on 1-watt resistor	7½ Turns	5½ Turns
15	14 turns #24 on 1-watt resistor	4½ Turns	4½ Turns
10	10 turns #24 on 1-watt resistor	2½ Turns	3½ Turns

L_4—7½ turns of 3/16″-diam. copper tubing spaced 0.3″ plus 13 turns #8 wire spaced 0.25″, 3″ inside diameter.

L_5—6 turns #10 spaced 0.25″ plus 10 turns #10 spaced 0.18″, 1½″ inside diameter.

Band	Plate Tuning	Plate Load
80	9.65 Turns	0.15 Turns
40	17.0 Turns	0.61 Turns
20	22.8 Turns	0.85 Turns
15	25.0 Turns	0.91 Turns
10	22.0 Turns	0.92 Turns

Note: Above turns data measured from maximum capacitance for 50-ohm load.

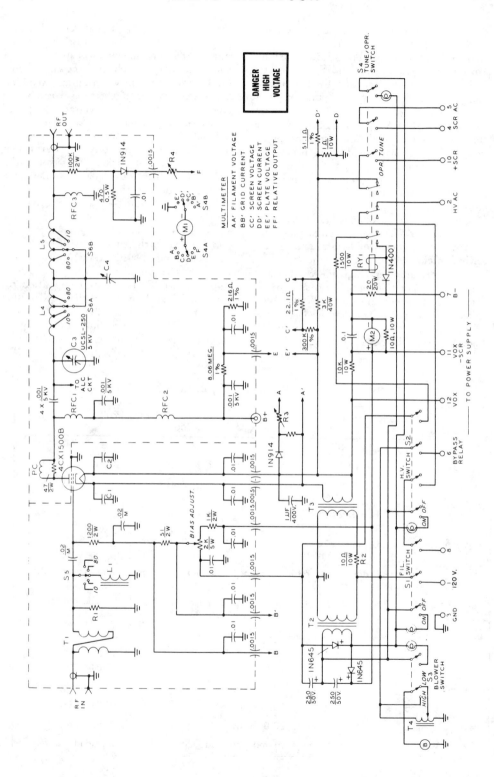

Figure 49

SCHEMATIC, R-F SECTION OF LINEAR AMPLIFIER

C_1, C_2—.001 μF and .01 μF, 500-Vdc silver mica capacitors in parallel, see text

C_3—250 pF, 5 kV. Jennings variable vacuum capacitor UCSL-250-5

C_4—1000 pF, 1 kV. Johnson 154-30 or equivalent

L_1—Grid coils. See table 10

L_4, L_5—Plate coils. See table 10

M_1—0-1 milliampere, dc. Triplett 320GLB

M_2—0-1 ampere, dc. Triplett 320GLB

PC—47-ohm, 2-watt composition resistor connected across 2" of plate line

R_1—Five 1200-ohm, 2-watt composition resistors in parallel

R_2—10-ohm, 10-watt wirewound

R_3, R_4—25K, 1-watt wirewound potentiometer

RFC_1—105 turns #22 enamel, closewound on 1" diameter teflon rod

RFC_2—22 turns #22 enamel on 100K, 2-watt composition resistor

RY_1—3-pole, double-throw relay, 24-Vdc coil. Potter-Brumfield KL-14D or equiv.

S_1, S_4—3-pole, double-throw toggle switch

S_2—4-pole, double-throw toggle switch

S_3—Double-pole, double-throw toggle switch

S_5—Single-pole, 6-position ceramic wafer deck. Centralab PA-1

S_6A,B—2-pole, 5-position ceramic switch. Radio Switch Corp. Type 86, one B-section

T_1—Four-to-one broadband transformer. 14 turns #24 bifilar wound on Indiana General CF-111-Q1 ferrite core

T_2—25 volts, 1 ampere. Stancor P-6469

T_3—6 volts, 11 amperes. Triad F-20U

T_4—Variable voltage transformer. Staco 171 or equivalent

Blower—18 c.f.m. at 0.23 inch backpressure. Dayton 4C004 or equivalent

Socket—EIMAC Y131A or Y149A

Chimney—EIMAC SK-806

band by a small inductor switched by S_5. The Q of the tuned circuit is quite low and complete coverage of each amateur band is possible without retuning.

Grid bias is applied to the 4CX1500B in shunt with the tuned circuit and provisions are incorporated for monitoring the grid current as well as for setting the zero-signal plate current of the tube (the *bias adjust* potentiometer).

With the screen element of the tube placed at dc ground, the cathode circuit and the negative side of the plate supply are connected to the negative side of the screen supply. Thus, the cathode is 225 volts negative with respect to ground and the grid, by virtue of the grid bias supply, is approximately 260 volts negative with respect to ground under normal operation. When the VOX relay contacts are open, additional negative grid bias is developed across the 10K cathode resistor, allowing the tube to draw only a few milliamperes of cathode current.

The Output Circuit—A pi-L network is used in the output circuit as it provides about 15 dB more harmonic attenuation than does the conventional pi-network. A variable vacuum capacitor is used for the plate tuning capacitor because the very small minimum capacitance (5 pF) permits the circuit Q to be held to a reasonable value on 10 meters. A design Q of 10 is used, rising to 12 at the high end of the 10-meter band.

Table 10 lists the design values and coil data for the output circuit.

Monitoring Circuit—Complete metering of the amplifier operation is provided by two meters and an electron-ray tuning tube (figure 50). The instantaneous r-f plate voltage of the amplifier is sampled by a capacitive voltage divider. A portion of the voltage is rectified and may be used for automatic load control voltage for the exciter. A second sample of voltage is used to energize the 6FG6 electron-ray tube mounted on the front panel of the amplifier. The tube is used to establish proper plate circuit loading. Under no-signal conditions, the pattern of the tube is open, gradually closing with increasing signal voltage until at maximum voltage the pattern is closed, showing a solid green bar in the viewing portion of the tube. This indication corresponds to maximum amplifier PEP input. The sensitivity of both the ALC and electron-ray tube circuits is adjustable by means of the capacitive divider.

A 0-1 dc ammeter is used to register plate current and a 0-1 dc milliameter with the movement reworked to show zero at 30 percent of full scale is used for the multimeter.

Control Circuits—Overcurrent protection is provided by a 3-pole, double-throw relay (RY_1) whose 24-volt dc coil is placed in series with the negative plate supply return. A 20-ohm resistor in parallel with the coil causes the circuit to latch-off at a plate current of 0.9 ampere. The plate supply pri-

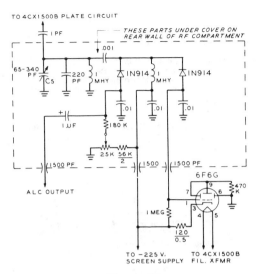

TO 4CX1500B PLATE CIRCUIT

ALC OUTPUT

TO −225 V.
SCREEN SUPPLY

TO 4CX1500B
FIL. XFMR

Figure 50

ELECTRON RAY TUBE PEAK INDICATOR AND ALC CIRCUITRY

The 6FG6/EM-84 indicator is used as an r-f peak-level indicator in the amplifier. R-f voltage is sampled, rectified, and applied to the gate (pin 1) of the indicator. The pattern is formed between the deflection elements (pins 6 and 7) and appears as a green line on the screen. Amplitude of indication is adjustable by means of mica compression capacitor C_5. ALC control voltage is taken from a separate diode and level of control is set by the "Adjust ALC" potentiometer. Electron-ray tube anode voltage is taken from screen power supply, whose positive terminal is grounded.

mary switch must be turned off and then back on to return the amplifier to normal operation.

Amplifier screen voltage is removed for tuneup by means of switch S_4 (*tune-operate*). The positive lead of the screen supply is broken, however the screen to cathode path is maintained by the bleeder resistor. *Inclusion of this resistor is imperative*, since the screen voltage must be maintained positive for any value of screen current that may be encountered. As with most high gain tetrode tubes, the 4CX1500B exhibits negative screen current under certain operating conditions (notably, when plate loading is insufficient). Dangerously high values of plate current may flow if the screen voltage rises under conditions of low or negative screen current. The bleeder resistor stabilizes the screen potential even when the screen

lead to the supply is broken and external screen voltage is zero.

The rated heater voltage for the 4CX1500B is 6.0 volts and the voltage, as measured at the tube socket, should be maintained between 5.8 and 6.0 volts by adjustment of resistor R_2 in series with the filament transformer primary. In no case should the voltage be allowed to exceed 5 percent above or below the rated value for maximum tube life. The cathode and one side of the filament are connected internally within the tube.

Power Supply Circuitry   The schematic for the power supply is shown in figure 51. Fusing of the low- and high-voltage circuits is provided and an inrush current limiting circuit (RY_3) is used in the primary circuit of the high-voltage supply. The circuit holds the charging current to a safe value when the supply is first turned on. An auxiliary plug and receptacle are provided for the connection of a variable voltage transformer to allow lower plate voltage for c-w operation. The high-voltage rectifier is a conventional full-wave, voltage doubler with series-connected electrolytic capacitors used for the filter section. The screen power supply is a full-wave configuration with choke input filter. A low resistance choke, such as the one listed, should be used for best screen voltage regulation.

The two filament windings on the screen supply power transformer (T_6) may be connected to either aid or oppose the primary winding to adjust the screen potential to 225 volts.

Amplifier Construction   The linear amplifier is built to fit within a *Bud Prestige* cabinet. The r-f section is in an inclosure having a perforated cover, and the panel is a standard $8\frac{3}{4}$" high rack panel. Views of the amplifier assembly are shown in figures 52 and 53. The r-f circuitry is assembled in a separate inclosure measuring $12" \times 11\frac{1}{2}" \times 8\frac{1}{2}"$ which is spaced $2\frac{1}{4}"$ behind the front panel to allow space for the panel meters and counter dials. The plate tuning capacitor is driven off-center from its counter dial by a miniature gear and roller chain combination providing a

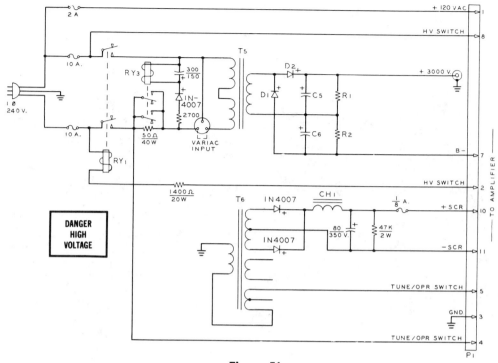

Figure 51

POWER SUPPLY FOR LINEAR AMPLIFIER

C_5, C_6—Each consists of five, series-connected 300 μF, 450-volt electrolytic capacitors, Mallory HC-45003, or equivalent
CH_1—6 henrys, 90 ohms. UTC S-29
D_1, D_2—Rectifier stack totalling 5000 volts PIV at 1 ampere with one cycle surge rating of 30 amperes.
R_1, R_2—Five 47K, 10-watt resistors series connected across filter capacitors
RY_1—Double-pole, double-throw relay, 25 amperes. Potter-Brumfield PR-11AYO
RY_3—Double-pole, double-throw relay, 12 amperes
T_5—1100-volt, 1.2-kW, ICAS rating. 120/240 volt primary. Berkshire Transformer Corp., Kent, Conn. type BTC-4905B
T_6—520 volts, center-tapped at 90 mA. Stancor PC-8404

2-to-1 reduction ratio. The loading capacitor, whose shaft is lower than that of the counter dial, is also driven by a chain and gear reduction system to provide correct alignment and to allow better resolution when presetting the loading after changing bands. The frames of the counter mechanisms are mounted to the front of the r-f inclosure on $1\frac{3}{4}''$ threaded shafts and the control shafts extend through holes in the front panel.

Within the r-f compartment, the 4CX1500B socket is mounted on a subchassis measuring $4\frac{3}{4}'' \times 8\frac{1}{4}'' \times 3\frac{1}{4}''$ placed at the rear. The plate tuning capacitor is on top of, and the loading capacitor beneath,

the extended top surface of the subchassis. The main bandswitch is mounted to the front wall of the subchassis, with its shaft extending through the wall into the subchassis to drive the smaller wafer switch for the input circuit. Flats are filed on this shaft to align the switch sections. Connection from the switch shaft to the panel knob is made by a flexible coupling.

The pi-section of the plate tank coil is supported by a strip of $\frac{3}{8}$-inch thick *teflon* sheet with notches cut in the edge to position the turns. The notches are made by drilling holes in the material and then cutting through the center line of the holes. The coil is wound from $\frac{3}{16}$-inch copper

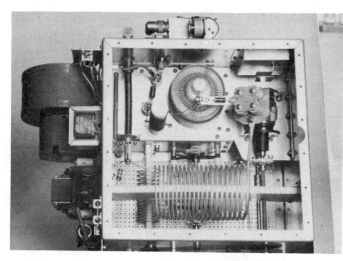

Figure 52

TOP VIEW OF AMPLIFIER

The 4CX1500B and auxiliary plate circuit components are mounted on a small chassis at the left, rear of the enclosure. The plate tuning capacitor is mounted on a bracket fastened to an extension of the top plate of the chassis. At the rear right of the amplifier is the sampling capacitor for the electron-ray tube and the shield for the circuit components for the r-f voltage divider.

The pi-L coils are in the open area at the front of the tube chassis, directly above the bandswitch. The high-voltage meter resistor is at the left of the enclosure, along with a short section of coaxial line that joins the output circuit to the coaxial connector at the rear of the amplifier. At the side of the amplifier are the squirrel-cage blower, filament transformer and control circuitry.

Top and bottom of the enclosure are made of perforated metal plate to provide ample cooling of the tube and components. The coaxial antenna relay is mounted on the rear of the box.

Figure 53

UNDERCHASSIS VIEW OF AMPLIFIER

The bandswitch assembly, coil support and pi-network capacitor are shown, with the subchassis for the 4CX1500B socket at upper left. The panel has been partially removed to show the mounting of the two counter dials.

tubing and #8 copper wire. Copper strap ⅜-inch wide is used for the tap connections on the coil and all joints are silver soldered before plating. The L-section of the plate inductor is similarly constructed and mounted at right angles underneath the main coil.

An underview of the chassis and closeup of the grid circuit compartment are shown in figures 53 and 54. All grid circuit components shown inside the dotted line at the left side of the amplifier schematic diagram are mounted inside the small subchassis. Each of the three tabs for the heater and heater/cathode socket terminals is bypassed with a .001-μF mica capacitor placed in parallel with a .01-μF mica capacitor. Additional low frequency filtering of the cathode lead is provided by the 1-μF, 400-volt *mylar* capacitor mounted outside the grid compartment.

All other grid circuit components are within the inclosure, and the various auxiliary components are mounted on the outside, left wall of the inclosure. All leads passing from the inside to the outside are bypassed by means of a 1500-pF ceramic feedthrough capacitor in parallel with a .01-μF, 1-kV disc capacitor.

The blower inlet is covered with a screen made from a small piece of ½-inch thick aluminum "honeycomb" material. This, in addition to the careful bypassing of all power leads, result in a "clean" amplifier, free of harmonics and interference problems (figure 55).

The sampling components for the ALC and electron-ray tube are located on the rear wall of the r-f inclosure with feedthrough capacitors used for all interconnections. The 1-pF sampling capacitor is made of two 1-inch square aluminum plates spaced about ¼-inch apart. A shield having a cutout for the capacitor connection to pass through covers these circuits and protects them from the strong r-f field of the plate circuit. The electron-ray tuning indicator and the rest of the components associated with it are mounted on a bracket behind the front panel.

The control toggle switches are mounted in a row across the lower left portion of the panel with a 28-volt indicator lamp above each switch. Power for these indicators is taken from the bias transformer (T_2) and an extra pole on each switch is used to turn on the indicator.

Power Supply Construction Power supply construction is straightforward. As shown in figures 56 and 57, the plate transformer, screen supply transformer, filter choke and filter capacitor are mounted atop a steel chassis measuring 11" × 17" × 3". All other components are mounted under the chassis except the high-voltage capacitors which are mounted to a 0.125" thick *fiberglas* printed circuit board having the interconnection pattern on the bottom side and the equalizing resistors on the top side. A similar piece of phenolic board is placed under the capacitor bank to insulate it from the chassis. Placement of the components beneath the chassis is not critical, provided the high-voltage circuits are sufficiently insulated from the rest of the components and wiring.

The solid-state rectifiers are mounted on a large, phenolic board near the center of the chassis. Any rectifiers, or stack of rectifiers, can be used as long as they have a one-cycle surge rating of 30 amperes, or better, and will handle 1 ampere forward current at a peak inverse voltage rating of 5000 volts. A high PIV rating is desirable

Figure 54

UNDERSIDE OF TUBE COMPARTMENT

The bandswitch segment and small grid inductors are mounted on the wall of the compartment beside the 4CX1500B tube socket. Left of the socket are the four-to-one toroid transformer and the grid load resistors, which are soldered between two thin copper plates. Note the multiple bypass capacitors on each socket terminal.

Figure 55

SIDE OF THE AMPLIFIER SHOWING BLOWER

The blower motor, impeller housing, and 25-volt transformer are to the left, with the filament transformer and associated power wiring at center. The control relay is next to the transformer.

Figure 56

TOP VIEW OF POWER SUPPLY

The filter capacitor bank and compensating resistors are at center, with the screen supply components to the side. Steel handles on the ends of the power-supply chassis assist the operator to move the heavy unit about. Front and back panels are attached to the chassis, which is covered with a U-shaped aluminum plate.

for protection against voltage surges on the power line.

Type HN coaxial connectors and RG-11A/U coaxial cable are used to interconnect the high voltage to the r-f unit and 12-pin connectors are used for the control lines. An air conditioner extension cord with the receptacle end removed provides connection to the primary power source.

Amplifier Adjustment   Before power is applied to the amplifier, zero the counter dials with the capacitors set at maximum capacity value. Then, preset the dials to the approximate values given in Table 11 for the band in use.

Remove the fuses from the plate transformer primary circuit and remove the 4CX1500B from its socket. After carefully checking all wiring and interconnections, apply filament power. Adjust the bias potentiometer for +45 volts cathode to grid, as measured at the socket. Next, check the filament open-circuit voltage at the socket; it should be about 7 volts. Replace the high voltage fuses but leave the high-voltage cable disconnected from the amplifier. Short the VOX relay terminals together. Turn on the *filament* and *high-voltage* switches and place the *tune/operate* switch in the *operate* position. Measure the screen-to-cathode voltage at the socket; it should be close to 225 volts provided the auxiliary filament windings on the screen power transformer are properly connected to either aid or oppose the primary winding. Move the switch to *tune* and check to see that the screen-to-cathode voltage drops to zero. The *blower*

Figure 57

UNDERCHASSIS VIEW OF POWER SUPPLY

Rectifier bank is mounted on a phenolic plate spaced away from the chassis. Power relay, step-start relay and various primary circuit components are mounted to the wall of the chassis. Inner conductor of RG-59/U cable is used for high-voltage wiring.

switch should be set to *high* for all tuning under full power.

Carefully replace the 4CX1500B in the socket and apply filament power. Set the filament voltage to within 5.8 to 6.0 volts as measured at the socket with an accurate meter. Set the front panel multimeter switch and adjust resistor R₃ to read a convenient reference on the meter. Reconnect the high-voltage cable and apply high voltage to the amplifier, after waiting for 3 minutes for the tube to warm up.

Adjust the *bias* potentiometer for 250 mA resting plate current with the VOX relay contacts shorted and the *tune/operate* switch in the *operate* position.

The amplifier is now ready for final tuning adjustments. Place the switch in the *tune* position and apply a single-tone signal of a few milliwatts PEP to the amplifier, adjusting the level to produce about 0.5 mA of grid current. Readjust the plate tuning for resonance, as indicated by a rise in output power and plate current; both will be small at this point. Place the switch in the *operate* position and readjust tuning and loading controls to obtain 670 mA of plate current at resonance and −15 mA screen current, holding the grid current to less than 1 mA. Power output under these conditions will be better than 1150 watts, with a plate potential of 2950 volts (see Table 11).

With carrier removed and voice modulation applied, the plate current will rise to about 350 mA and screen current will peak at about −8 mA. Grid current will be less than 0.03 mA. When the VOX relay opens,

Table 11

4CX1500B Typical Operation, Class-AB₂ R-F Linear Amplifier		
Dc plate voltage	2750	2900 volts
Dc screen voltage	225	225 volts
Dc grid voltage	−34	−34 volts
Zero-signal dc plate current	300	300 mA
Single-tone dc plate current	755	710 mA
Two-tone dc plate current	555	542 mA
Single-tone dc screen current	−14	−15 mA
Two-tone dc screen current	−11	−11 mA
Single-tone dc grid current	0.95	0.53 mA
Two-tone dc grid current	0.20	0.06 mA
Peak r-f grid voltage	45	41 volts
Driving power	1.5	1.5 watts
Resonant load impedance	1900	2200 ohms
Useful output power	1100	1100 watts

resting plate current will drop to a few milliamperes, as sufficient bias is added to produce a near-cutoff condition.

Operation of the amplifier should now be monitored with an oscilloscope to make sure than "flattopping" does not occur at maximum input level. When the maximum level has been established, adjust the capacitor on the back panel of the r-f unit until the electron-ray pattern just touches. In normal SSB voice operation, the indicator will barely reach this point at 2 kW PEP input, depending on the exact waveform of the driving signal.

22-10   A High Power Linear Amplifier With the 8877

The linear amplifier described in this section is built by Jim Garland, W8ZR. It is designed for continuous duty operation at the 2-kW PEP power level on all bands between 10 and 80 meters (figure 58). The use of a single 8877 high-mu, ceramic power power triode in a class AB_2 cathode-driven (grounded-grid) configuration provides excellent efficiency and linearity with a peak drive power requirement of about 50 watts. The amplifier and power supply are self-contained in a single console and the design features a built-in r-f wattmeter for monitoring forward and reflected power, ALC control of the exciter, sequenced relay switching and several protective features to safeguard the 8877 and power supply components against malfunction or improper use.

At a plate potential of 3 kV, the third-order intermodulation products at maximum power output are 40 decibels below one tone of a two-tone test signal.

The Amplifier Circuit   The schematic of the r-f deck of the amplifier console is shown in figure 59. The 8877 is operated with the grid at r-f ground potential with bias supplied in the cathode return circuit. The drive signal is coupled through relay RY_1A and a pi-network circuit to the cathode of the 8877. The input network (C_1, L_1 and C_2) operates at a Q of 1, which is sufficient to preserve the waveform of the input signal during the variation of the 8877 cathode impedance during each r-f cycle. An r-f choke is used to isolate the r-f drive from the grid-metering and bias circuitry.

The filament of the 8877 is internally insulated from the cathode and can be placed at, or near, ground potential, thus eliminating the expensive bifilar filament choke; only a small cathode r-f choke is required.

During standby, the 8877 is biased to near cutoff by a 10K resistor in the cathode return lead. During amplifier operation, the resistor is shorted by a set of contacts on relay RY_1, while a zener diode (Z_1) provides

Figure 58

HF LINEAR AMPLIFIER WITH 8877

This high power linear amplifier covers 80-10 meters at 2 kW PEP input. Using a single 8877 high-mu ceramic-metal triode in a grounded-grid circuit, the unit provides maximum output with a peak drive requirement of 50 watts. Amplifier and power supply are self-contained in a roll-about console. Controls and meters are (top): Plate current meter, plate tuning control, and plate loading control. Below are: Multimeter and selector switch, power and in/out switches, bandswitch, and power reset button. Console is made up of prefabricated channel stock and aluminum panels. The rack is painted black with dark gray panels.

7.5 volts of operating bias. A 1-ampere fuse in the cathode return protects the tube from excessive plate current, while a 500-ohm safety resistor prevents the voltage at the cathode from soaring if either the fuse or the zener diode fails.

Metering Circuitry—Grid current flows from the cathode of the 8877 to the grid (ground) through a 3.3-ohm resistor (R_1)

and the voltage developed across this resistor is used to obtain grid current metering and also to provide a reference voltage for the grid protection circuit. During normal amplifier operation the voltage across R_1 is insufficient to permit transistor Q_1 to conduct. If the grid current of the 8877 rises to about 180 mA, however, then Q_1 conducts and returns to ground one side of the grid overload relay (RY_3), which then latches itself

closed, illuminating the front panel *reset* button and interrupting the VOX line. Pressing the *grid reset* button unlatches the relay and permits normal operation of the amplifier to resume. A 10-μF capacitor across the coil of the relay prevents tripping of the circuit on instantaneous grid current "spikes."

The grid meter (M_1) also functions as a multimeter, monitoring the forward and

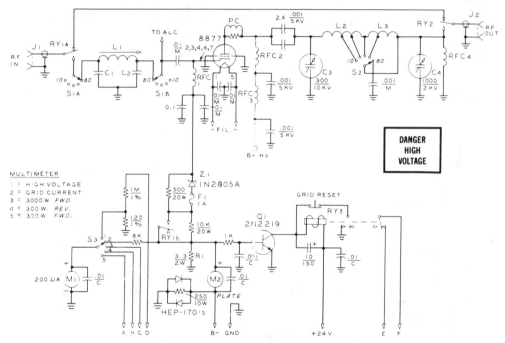

Figure 59

SCHEMATIC OF 8877 LINEAR AMPLIFIER

C_1, C_2—See Table 7

C_3—300 μF, 10-kV variable vacuum capacitor. Jennings UCS-300. An air capacitor may be substituted for this unit.

C_4—1000 μF, 2-kV variable vacuum capacitor. Jennings UCSL-1000. An air capacitor may be substituted for this unit.

L_1—See Table 12

M_1—0 to 1.5 ampere dc meter. Simpson 1327 with 1253 bezel

M_2—0 to 200 μA dc meter. Simpson 1327T with 1253 bezel

PC—Three 150-ohm, 2-watt composition resistors in parallel shunted across 1 turn of plate strap

RFC_1—110 turns #24 e., on ½" diameter teflon rod. 28 μH

RFC_2—170 turns #24 e., on 1" diameter teflon rod. 180 μH

RFC_3—12 turns #14 e., on ⅜" rod, spacewound. 1 μH

RY_1—Input relay. Potter-Brumfield KHP-17-D11, or equivalent

RY_2—Output relay. Jennings vacuum relay RF1-d, or equivalent

RY_3—Grid overload relay. Potter-Brumfield KHP-17-D11

S_1—2-pole, 6-position ceramic switch. Centralab PA series

S_2—1-pole, 6-position high voltage switch. Millen 51001

S_3—1-pole, 6-position switch. Centralab PA series

Counter dials—Bauman TC3-S

Socket—Eimac SK-2210

Chimney—Eimac SK-2216

Note: All wires passing from r-f compartment (except the high voltage lead) are bypassed with 1500-pF feedthru capacitors (not shown on drawing)

See Table 12 for plate circuit data

reverse power levels indicated by the r-f wattmeter. Plate current is monitored by a separate meter (M_2) in the B-minus return lead to the high voltage power supply; a pair of reverse-connected diodes are used to protect the meter in the event of a flashover in the plate circuit, while a 250-ohm safety resistor prevents the B-minus voltage from soaring above ground to a dangerous level if the plate meter should open up.

The Plate and ALC Circuits—The amplifier uses a conventional pi-network output circuit which is designed for a plate load impedance of 1800 ohms with a loaded Q of 12; the values of the plate circuit components for each band are given in Table 12. Circuit Q rises at 10 meters due to the output capacitance of the tube and the stray circuit capacitances (a total of about 25 pF) but circuit efficiency remains high.

ALC voltage is obtained by sampling the r-f voltage at the cathode of the 8877 with a capacitive voltage divider consisting of capacitors C_5 and C_6 (figure 60). Peak r-f drive voltage in excess of the dc reference voltage set by potentiometer R_2 (*ALC adjust*) is rectified and filtered, to appear at the *ALC output jack* for control of the exciter power level.

The broadband r-f wattmeter (figure 62) uses a conventional circuit; the amplifier output power is sampled by toroid L_4, while capacitor C_7 provides the reflected power *null adjustment*. Potentiometers R_3-R_5 allow calibration of the instrument to provide full scale meter readings of 3 kW and 300 W forward power and 300 W reverse power.

Amplifier Cooling—The 8877 requires 20.3 cfm of air at a pressure drop of 0.23″ for 1000 watts anode dissipation at sea level. A squirrel cage blower provides proper cooling. For full 1500 watts dissipation, 38.0

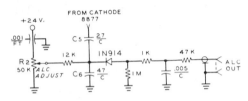

Figure 60

ALC CONTROL CIRCUIT

Figure 61

TOP VIEW OF R-F DECK

The variable vacuum capacitors are attached to the inner front panel, allowing space for the counter dial mechanisms and meters behind the main panel. The 8877 tube is at left, with the plate parasitic suppressor between it and the plate choke. The suppressor is made of three 150-ohm 2-watt composition resistors in parallel shunted across two turns of the ¼″ wide copper plate strap. Immediately behind the tube socket are the input inductors for the cathode circuit. The 8877 is mounted on a small subchassis which sits atop the bottom plate of the inclosure.

Tank coil L_2 is mounted in a vertical position behind the bandswitch, which is hidden by the inner front panel. To the right is coil L_3, immediately behind the 1000-pF loading capacitor. The plate r-f choke is mounted to the rear vertical wall of the inclosure.

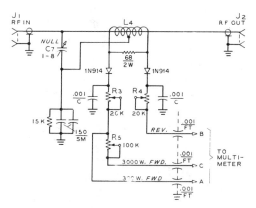

Figure 62

THE R-F WATTMETER

The pickup coil (L₄) is wound with 20 turns #22, centertap on an Amidon T-50-6 ferrite toroid. A 68-ohm 2-watt resistor is shunted across the winding at the terminals. The null capacitor is a small variable ceramic type.

cfm of air is required at a pressure drop of 0.60″. In all cases, sufficient cooling air must be supplied to hold tube temperature below 250°C with 50°C ambient temperature at sea level. A *Dayton model 1C-180* blower will satisfy the 1000-watt requirement of the 8877 under almost all operating condi-

Table 12. Input Network Details

Circuit component values (Q = 1)		
Band	C_1, C_2	$L_1(\mu H)$
80m	820 pF	2.15
40m	430 pF	1.20
20m	220 pF	0.60
15m	150 pF	0.40
10m	100 pF	0.30

Note: C_1, C_2 are made up of two paralleled silver mica capacitors.

L_1 Coil Winding Data			
Band	No. Turns	Wire Size	Inductance Range(μH)
80m	16	20	2.00-2.70
40m	10	18	0.92-1.30
20m	8	14	0.54-0.70
15m	6	14	0.35-0.48
10m	4	14	0.22-0.37

Note: Coil forms are ½″ diameter ceramic forms (Miller 69046-orange core)

tions at a low ambient noise level. For operation at 10,000 feet, or above, or for extended contest operation in a high temperature environment, a *Dayton model 2C-782* may be substituted with only moderate increase in noise level.

Power Supply and Control Circuitry— Primary power to the amplifier is applied through *control switch* S_1 to the filament transformer, blower, and time delay relay RY_4 (figure 63). After 180 seconds, the time delay relay closes and power may be applied via *plate switch* S_2 to relay RY_5, a mercury plunger solenoid relay.

The initial charging current of the filter capacitor bank in the high voltage plate supply is limited by two 15-ohm resistors in the primary circuit of plate transformer T_2. As the filter capacitors become charged, the voltage at the primary of T_3 rises because of decreased voltage drop across the resistors, eventually permitting surge-limit relay RY_6 to close. The response time of the relay is about 0.25 second and is determined by the time constant of the filter in the 24-volt dc low-voltage power supply. This supply also provides power for the VOX and antenna changeover relays and reference voltage for the ALC circuit.

A sequencing network consisting of a 150-ohm resistor and a 50-μF capacitor in series delay the closing of RY_2, the antenna changeover relay, until about 7 msec after the VOX line is actuated (figure 63). This prevents "hot-switching" the antenna relay, thus protecting the relay contacts and the plate circuit components from the high peak voltages arising from a momentarily unloaded condition. The diode across the relay coil prevents the capacitor charge from holding the relay closed after the VOX line is opened. Discharge time of the capacitor is about 100 msec through the back resistance of the parallel-connected diode.

The high voltage power supply employs a voltage doubler circuit and provides about 3000 Vdc under full load (figure 68). The diode banks are protected by RC suppressors across each diode and by *thyrector* surge suppressors (Z_1, Z_2, figure 63) across the primary winding of the plate transformer. These devices throw a low impedance short across the transformer in the presence of a high voltage transient on the primary circuit. The amplifier may be operated on either

Figure 63

POWER SUPPLY CONTROL CIRCUIT

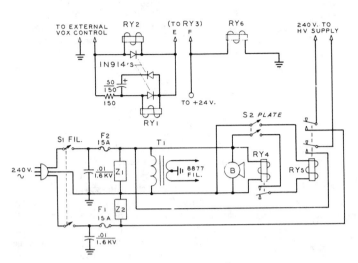

RY$_1$ — See figure 59 parts list
RY$_2$ — See figure 59 parts list
RY$_4$—180 second time-delay relay. Amperite 115-NO-180
RY$_5$—Mercury plunger relay, 2-pole, normally open. Ebert A-11
RY$_6$—Power relay, double-pole, double-throw. Potter-Brumfield PR-11-DY
S$_1$, S$_2$—Lighted Switch Assembly. Arrow-Hart actuator AH-83504 with lens AH-70 and AH-71, and contact block AH-83500
T$_1$—5 volts at 10 amperes. Stancor P-6135
Z$_1$, Z$_2$ — Thyractor. General Electric 6RS20-SP4B4
Blower—60 cfm. Dayton 1C180

240 volts or 120 volts primary power by changing taps on the plate transformer. In the interest of good regulation, operation at the higher line voltage is preferred.

Amplifier Construction   The amplifier is built into a console as shown in the various photographs. The assembly measures 15″ wide × 16″ deep × 27″ high and is mounted on heavy duty furniture casters. The aluminum framework is made of preformed material manufactured by the *Dorlec Corp.*, Box 182, Cherry Hill, NJ 08002. The framework requires only a hacksaw and a tape measure to assemble. The high voltage supply is located in the lower compartment of the console (figure 64), with the r-f deck at the top. The sides and back of the console are cut from aluminum sheet, but the sides of the amplifier section are made of perforated stock to allow adequate ventilation of the r-f components. The blower is mounted on a removable plate which fastens to the underside of the r-f chassis, while air flow to the blower intake port is through a 6-inch diameter perforated cutout in the rear of the power supply compartment.

The photographs show the location of the major power supply components. The plate transformer is bolted directly to the ⅛-inch thick aluminum baseplate while the filter capacitors are mounted on an insulating frame made of two 8-inch square plexiglass sheets (figure 65). The high voltage rectifiers are mounted on small ceramic terminals on a small plexiglass sheet placed in front of the filter capacitors.

The R-F Deck   The layout of the top portion of the r-f deck is shown in figures 61 and 66. The top portion is a cubical inclosure with the socket for the 8877 mounted on a subchassis placed at the left, rear portion of the inclosure. The chassis measures 11¾″ × 5½″ × 3½″. The 7-pin septar socket is submounted below the chassis deck with ⅜-inch metal spacers. The grid ring of the 8877 is electrically grounded to the chassis by means of four grounding clips on the socket assembly. The air chimney is held in place atop the chassis.

The filament transformer for the 8877, the antenna relay, the cathode input circuit and various auxiliary components are located inside the chassis and all electrical connections into this chassis are decoupled by 1500-pF feedthrough capacitors. The cathode input circuit switch has an extension

Figure 64

SIDE VIEW OF AMPLIFIER CONSOLE

The side panels have been removed to show placement of parts. The 8877 tube is visible in the r-f inclosure at the top, with the squirrel cage blower mounted immediately below it in the power supply compartment. The air intake vent for the blower is in the rear panel. The main plate transformer is at the rear of the lower deck, with the mercury primary relay immediately beside it. The filter capacitor bank is to the right, with the auxiliary relay controls in the foreground.

Figure 65

FILTER CAPACITOR ASSEMBLY

The computer grade filter capacitors are sandwiched between insulating plates and mounted in a horizontal position. A third plate holds the diode assembly and RC network capacitors. The inner conductor of RG-8/U cable is used for high voltage wiring. All exposed terminals are taped after assembly to prevent accidental contact.

shaft which extends out the front of the chassis through a panel bearing to the control knob. The slug-tuned coils are adjustable through the top of the chassis. The bottom plate is drilled to receive the blower.

Location of the major plate circuit components may be seen from the photographs. The two vacuum variable capacitors are mounted on a reinforced aluminum subpanel recessed 3″ behind the main panel. The plate circuit bandswitch is also mounted on the subpanel; the switch being ganged to the cathode switch by means of two brass pulleys located in the space between the panels.

Heavy gauge piano wire is used to join the pulleys.

The plate r-f choke is fastened to the rear wall of the inclosure atop the bypass capacitor and the plate end of the choke is connected to the plate r-f blocking capacitors by an angle plate made of thin copper stock. The plate inductor consists of two coils; inductor L_2 is made of $\frac{1}{4}$-inch copper tubing and is used for 10, 15, and 20 meters. It is suspended between a flange attached to the variable vacuum capacitor and a ceramic standoff insulator mounted to the bottom plate of the inclosure. Inductor L_3 is made of $\frac{1}{8}$-inch copper tubing and provides additional inductance for 40- and 80-meter operation. Leads from the tap points on the coils to the bandswitch are made with $\frac{1}{4}$-inch copper strap; coils and straps are silver plated before final assembly. Complete data for the plate circuit is given in Table 13.

The r-f wattmeter components are mounted on a small printed-circuit board placed

Figure 66

CLOSEUP OF R-F DECK

The side panel has been removed to show the 1000-pF vacuum variable loading capacitor and the plate coil inductors. The bandswitch is at left, mounted to the front panel. Connections are made to the coils with silver plated copper strap. The 8877 tube is hidden in the rear behind the loading capacitor. Visible at the right is the r-f wattmeter board, adjacent to the output receptacle.

adjacent to antenna coaxial receptacle J_2. The connection between the receptacle and the bandswitch is made with a short length of RG-8A/U cable.

Calibration The values of capacitance and **and Alignment** inductance for the plate circuit pi-network are given in Table 13. The values of the input tuning capacitance include about 15 pF of tube and stray circuit capacitance. The positions

Table 13. Plate Circuit Details

Band	C_3 Input Cap.(pF)	C_4 Output Cap.(pF)	Plate Inductance (μH)
80m	279	1460	7.33
40m	145	762	3.82
20m	74	386	1.94
15m	49	259	1.30
10m	45	252	.778

Notes: (1) Input capacitance values include 15-20 pF of tube and stray circuit capacitance.

(2) Network designed for plate impedance of 1820Ω, load impedance of 50Ω and operating Q of 12 (Q = 15 on 10 meters).

(3) Calculated reactances at resonance: (80m-15m):

$$X_C \text{ in} = 152\Omega$$
$$X_C \text{ out} = 29\Omega$$
$$X_L = 172.65\Omega$$

(10 m)
$$X_C \text{ in} = 121.6\Omega$$
$$X_C \text{ out} = 22\Omega$$
$$X_L = 139.45\Omega$$

of the taps on the plate inductors may be found by first setting capacitors C_3 and C_4 to the correct values and then adjusting the appropriate coil tap until circuit resonance is achieved, as indicated by a calibrated dip-meter. The capacitors themselves can be calibrated by the dip-meter and a known inductance; this is most easily done by constructing a graph of capacitance values for different settings of the turns-counter dial.

The input pi-network coils are aligned by inserting an SWR meter in the coaxial line between the exciter and the amplifier and tuning the coil slug on each band for minimum SWR at the center of the band. This adjustment should be done with the amplifier operating at full power input. *Since high-voltage components are in very close proximity to the slug adjustment screws, a nonmetallic screwdriver should be used for these adjustments.*

Calibration of the r-f wattmeter is done after the amplifier is finally tested. Capacitor C_7 is adjusted to null the reflected power with a 50-ohm dummy load connected to the output of the amplifier. Potentiometers R_3-R_5 are adjusted to obtain full-scale meter readings at the desired power output levels using a calibrated r-f wattmeter for comparison.

Amplifier Tuning After the amplifier wiring **and Alignment** has been completed and inspected it is ready for initial checks. The filament voltage should be measured at the socket with a 1-percent meter to determine that it is within the

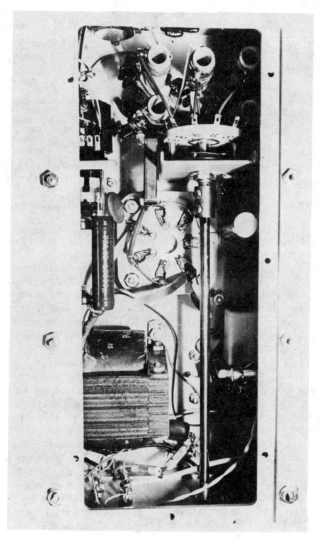

Figure 67

INTERIOR OF R-F CHASSIS

The underside of the 8877 chassis is shown in this view. The tuned cathode circuits are adjacent to the socket, with the filament transformer in a corner of the compartment. The input and overload relays are mounted to the walls of the chassis. The tube socket is recessed below the deck to permit passage of air about the tube.

allowed range of 4.75 to 5.25 volts. Operation at the lower end of this range will prolong tube life. After the 90 second time-delay relay activates, plate voltage may be applied to the 8877 and the dc resting plate current of the tube should be about 160 mA to 180 mA. At this time, check to make sure

the blower is operating properly and a free flow of air is escaping from the anode of the tube. The plate circuit controls should be set to the values determined previously.

Apply a small amount of drive power from the exciter and tune the plate circuit controls to resonance. Maximum grid cur-

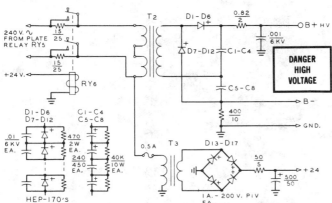

Figure 68

SCHEMATIC OF POWER SUPPLIES

D_1-D_{12}—Diode network consisting of: 2.5 amnere, 1000 piv diode (HEP-170), .01 μF, 1.6 kV disc capacitor and 470k, 1-watt resistor

C_1-C_8—240 μF, 450 WVdc capacitor. Mallory 241T450D1

RY_6—See figure 63 parts list

T_2—120/240 volt primary, 1100-volt secondary 1.2 kW rating.

T_3—26.8 volts at 1 ampere. Stancor P-8609

D_1, D_4—Diode Bridge. 2 amperes, 200 volts piv.

rent, maximum r-f output, and minimum plate current should occur at the same setting of the plate tuning capacitor. When properly loaded to 2 kW input with carrier, grid current will run about 35 mA to 40 mA, corresponding to a drive power of about 40 watts. Operation of the grid protection circuit can be checked by temporarily reducing the loading capacitor two or three turns and whistling briefly into the microphone, the grid overload relay should trip, illuminating the grid reset button and locking the amplifier into the standby mode. If the grid relay trips during subsequent operation of the amplifier, it is usually a sign of improper loading, a badly mismatched antenna, or excessive drive power. In any case, the difficulty should be remedied before resuming operation of the amplifier.

As a final check of amplifier linearity, and to establish the correct ALC threshold, the amplifier output should be monitored on an oscilloscope trapezoid display. Single-tone plate current will run 660 mA, with a grid current of about 40 mA. Voice modulation, without audio clipping or compression, will run about one-half these values on voice peaks. Power output is about 1100 watts, peak.

22-11 A 1-kW PEP Linear Amplifier for 10 Through 160 Meters

The rising popularity of 160 meters has brought a demand for a multiband amplifier

that covers the "top band," as well as the higher frequency bands. Shown in this section is a compact design using a single 3-500Z high-mu power triode that is capable of 1200 watts PEP input on the 160-, 80-, 40-, 20-, 15-, and 10-meter bands. This is a desk-top amplifier with a separate power supply that may be hidden under the operating table (figure 70). The amplifier features a tuned cathode input circuit, a pi-L plate circuit for greatest harmonic attenuation and can operate either with a transceiver or a receiver-exciter combination. In addition, the amplifier may be bypassed for low power operation.

At maximum power input the third-order intermodulation products are better than 40 dB below one tone of a two-tone test signal, without the use of auxiliary feedback. Peak drive power is of the order of 50 watts, and the amplifier may be driven by any exciter capable of providing this power level. In most cases, the measured intermodulation distortion level of the amplifier-driver combination is mainly that of the driver, as the amplifier distortion level is very low. The amplifier was designed and built by K60PZ.

The Amplifier Circuit The schematic of the linear amplifier is shown in figure 71. A single 3-500Z tube is used with excitation applied to the filament circuit via a pi-network matching configuration (C_1, C_2, L_1). There are three r-f

Figure 70

THE K6OPZ LINEAR AMPLIFIER FOR 160 THROUGH 10 METERS

This compact linear amplifier uses a single 3-500Z high-μ power triode and is capable of 1200 watts PEP input. Plate tuning and loading controls are centered vertically on the panel with the 3-500Z mounted to the left behind the grill. The meters are (left to right): grid current, plate current and plate voltage. Across the bottom of the panel are (left to right): filament power switch, input circuit bandswitch, filament pilot, plate power switch and (at the extreme right) plate circuit bandswitch. The panel is spray-painted gray with an overcoat of clear epoxy. The amplifier is raised off the desk by four rubber feet which allow cooling air to pass under the unit.

connectors on the rear apron of the amplifier, one for a receiver (if used), one for a transmitter (if used) and one for a transceiver. When used with a transceiver, the switching contacts of relay RY_1 in the input circuit are bypassed. The 3-500Z is normally biased by means of a 6.8-volt zener diode but in standby the linear is cut off by the VOX relay (RY_2) and the 20K cathode circuit bias resistor.

The Plate Circuit—The plate circuit uses a pi-L network for maximum harmonic suppression. This provides about 20 dB more harmonic suppression than the conventional pi-network. The network is switched by means of a four-section high-voltage ceramic switch (S_4). Two sections select the proper coil inductances and the other two sections add padding capacitors for 160- and 80-meter operation. Plate voltage is applied through a 90 μH r-f choke. The choke has an inductive reactance of 1200 ohms with a parallel capacitance of 83 pF at this frequency and extra capacitance in tuning capacitor C_1 is required to compensate for the

shunting action of the choke. This is accounted for in the design of the network.

The pi-network coil (L_2) is divided into sections to provide the highest efficiency and ease in assembly. A Q of 10 was chosen for 160 meters; 15 for 80, 40, and 20 meters; and 18 for 10 meters. This provides good efficiency and harmonic suppression on all bands.

The Metering Circuit—Grid and plate currents are monitored and the meters are placed at a low potential point in the circuit. The negative of the high-voltage supply "floats" a few volts above ground to permit return-lead metering. A protective resistor is placed from B-minus to ground to ensure that the negative side of the power supply remains close to ground potential. A separate ground lead is then run from the amplifier to the power supply. Grid current is measured between grid (ground) and cathode return, with the grid pins of the 3-500Z connected to chassis ground.

The Cooling System—Forced air cooling is required to maintain the seals and en-

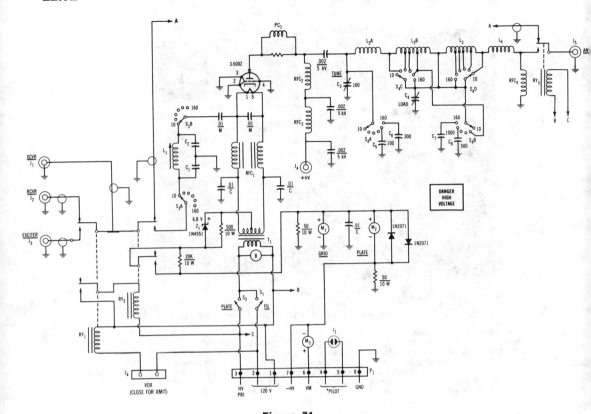

Figure 71

SCHEMATIC, 3-500Z LINEAR AMPLIFIER

C₁, C₂—See coil table.
C₃—100 pF, 0.125″ gap (Johnson 154-14 or equivalent).
C₄—480 pF, 0.045″ gap (Johnson 154-3 or equivalent).
C₅—100 pF, 5 kV mica capacitor.
C₆—300 pF, 5 kV mica capacitor.
C₇—1000 pF, 2.5 kV mica capacitor.
C₈—500 pF, 2.5 kV mica capacitor.
I₁—Pilot light, NE51-H (red).
J₁-J₃, J₅—Coaxial receptacle, SO-239.
M₁—0-200 mAdc.
M₃—0-500 mAdc.
M₃—1 mAdc. With power supply series resistor reads 5 kV full scale.
PC₁—Copper strap, ½″ wide formed into hairpin loop 1″ high × ½″ wide shunted by 47 ohm, 2
    watt composition resistor.
RFC₁—28 bifilar turns # 12e. wire on ½″ diam. ferrite rod 6½″ long. Permeability 125. (Amidon
    R61-050-750 or equiv.). Notch with file and break to length.
RFC₂—90 μH. Barker & Williamson 800 or equiv.
RFC₃—1 mH. Johnson 102-752 or equiv.
RY₁, RY₂, RY₃—dpdt power relay, 10A contacts. Parallel contacts on RY₃.
B—100 cfm blower, 120 volt. Rotron Sentinel 747 or equiv.
S₃—Two pole, six position ceramic switch. Centralab 2551 or equiv.
S₄—Four pole, six position, four deck ceramic switch. Model 86. Radio Switch Corp., Marlboro,
    NJ.
T₁—5 V, 13A. Triad F-9A or equiv.
Socket—Johnson 122-275-100 or equiv.
Plate cap—Eimac HR-6.

velope of the tube below the maximum recommended temperatures. In this design, a
100 cubic-feet-per-minute fan (*Sentinel*

Series *747* by *Rotron*) is used, with the face
of the fan mounted 2⅝″ from the vertical
axis of the tube. A ceramic socket is used,

mounted about $\frac{1}{16}$ inch below the chassis deck to provide an air path around the base of the tube through which underchassis air is drawn by convection. The air is exhausted through the perforated metal lid of the cabinet and also through a metal grill mounted on the rear wall of the box.

Maximum plate dissipation is about 450 watts using this cooling technique, sufficiently high so that the amplifier may be run at 1 kW for c-w operation, or 1200 watts PEP for SSB service. For RTTY operation, amplifier input should be reduced to about 800 watts to provide adequate protection against overheating. The top surface of the cabinet should be kept clear to permit the heat to freely escape from the amplifier when it is in use.

Amplifier Construction   The amplifier is built within an aluminum enclosure measuring 15" wide, 11" deep and 8¾" high. Front and rear panels are cut from ⅛-inch sheet aluminum and the remaining panels are cut from $\frac{1}{16}$-inch sheet. One-half-inch-wide angle stock is used at the corners for bracing. The assembly is held together with 6-32 screws tapped into the angle stock. The tube socket and small input components are mounted on an 11" × 7" × 2" aluminum subchassis bolted to the left side of the enclosure. The plate tank circuit occupies the righthand end of the enclosure as seen in the photographs.

Layout of the major components can be seen in the top view photograph (figure 72). The subchassis is at the left and contains the filament transformer, 3-500Z tube and plate r-f chokes. The solenoid-wound choke (RFC₂) is mounted in a vertical position, in the clear, with the auxiliary choke (RFC₃) mounted parallel to the chassis. The fan is mounted to the wall of the box in line with the center of the tube.

At the right are the plate circuit components. The tuning and loading capacitors

Figure 72

TOP VIEW OF THE MULTIBAND AMPLIFIER

The 3-500Z subchassis with the tube, filament transformer, and blower is at left. The plate tuning capacitor sits on an L-shaped bracket that is fastened to the side of the subchassis. At the right are the plate circuit coils. The 10-15-20 meter portion of the assembly is parallel to the front panel and the 40-80 portion is toward the rear. The meters used are shielded surplus types with the terminals bypassed with .01 μF ceramic disc capacitors. Leads to the meters are run in shielded cables. The high-voltage coaxial cable passes out the rear of the enclosure and is clamped at this point by an electrical BX box cable clamp. The cooling fan is bolted to the side of the cabinet by long bolts which pass through rubber washers on each side of the wall. This helps deaden blower noise.

are mounted to an L-shaped bracket which is bolted to the side of the subchassis. The tuning capacitor (C_3) is shown in the top view. It is panel-driven from a vernier dial, through an insulated coupling which prevents the formation of a ground loop, with consequent r-f currents flowing in the front panel.

Between the capacitor and the front panel are mounted the hf sections of the plate coil. The 10-meter portion is wound of $\frac{3}{16}$-inch copper tubing and the 15- and 20-meter section with #8 copper wire. One end of the tubing is attached to a copper strap which makes the connection to tuning capacitor (C_3). The other end of the tubing is drilled out slightly to accept the copper wire. The joint is then soldered. The 10-meter tap (a section of copper strap) is also soldered at this junction.

The 10-meter coil is supported by its terminals and the 15- and 20-meter portion is wound around a grooved piece of *Teflon* stock which is affixed to the end of the enclosure with a small angle bracket.

The 40-, 80-, and 160-meter portion of the plate inductor is mounted behind the hf coil, parallel to the end wall of the box. It is composed of two sections of commercial coil stock and is supported by its leads and a small ceramic insulator at the far end of the coil. Most of these components can be seen in figure 73.

Note that the bottom plate of the enclosure is perforated in order to achieve maximum convection cooling. The plate is removed for the underchassis view (figure 74). The area around the tube socket contains the bandswitch and associated circuits, the filament choke, two relays, and most of the small components associated with the input and control circuits. All power leads are run in shielded wire, with the shield grounded at each end of the leads.

In the plate circuit area, the loading capacitor (C_4) is bolted to the underside of the L-shaped mounting bracket and beside it is the bandswitch whose shaft projects through the front panel. Behind the switch is the L-section of the plate circuit. All

Figure 73

OBLIQUE VIEW OF AMPLIFIER

Viewed from the rear corner, the air inlet for the fan is seen on the side wall. The plate tuning capacitor is panel driven through an insulated coupling to reduce circulating ground currents. The plate circuit parasitic suppressor is connected directly to the heat-dissipating anode connector of the 3-500Z. The high-voltage cable and various connectors are mounted on the rear panel of the box. No air chimney is required for the 3-500Z with this cooling technique.

connections in the pi-section of the plate circuit are made with copper strap to reduce r-f loss.

Amplifier Adjustment and Tuning The amplifier wiring should be checked and the plate circuit adjusted for proper operation on all bands. Coil information is quite precise, but it is a good idea to check the resonant circuits on each band with the aid of a dip meter (see Tables 14 and 15). The tube should be in the socket for this test but no voltages are applied to the amplifier. The cathode circuits should be adjusted by means of the slug-tuned coils to resonate at the center of each band (this may be done before installing them in the amplifier, if desired). The plate circuit is checked by setting the loading capacitor to about three-quarters meshed and tuning capacitor (C_3) for resonance.

Filament voltage is now applied and the blower motor should start. Filament voltage should be 5.0 volts at the socket pins. The filament transformer is rated for a continuous duty current of 13 amperes but runs cool under the 14.6 ampere tube load. The transformer was selected to do the job with-

Table 14.
Input Network Details
Circuit Component Values
Design Q=4

Band	C_1 (pF)	C_2 (pF)	L_1
160	3000	2000	22t. #22e. (5.8 μH)
80	1500	1000	18t. #22e. (3 μH)
40	750	560	11t. #18e. (1.5 μH)
20	390	270	7¾t. #18e. (0.8 μH)
15	270	200	5t. #18e. (0.5 μH)
10	200	150	4t. #18e. (0.3 μH)

Capacitors: 600V dipped mica. DM15-J series
Coil forms: (160-20 meters) ½" diam.
     slug tuned (red core).
     J. W. Miller 66A022-2
     (15-10 meters) ½" diam.
     slug tuned (green core).
     J. W. Miller 66A022-3

Table 15.
Plate Circuit Details

Band	Inductor L_2A-L_2B
10	5¼ turns, ³⁄₁₆" tubing, 1¹¹⁄₁₆" i.d., 2⅝" long.
15-20	6¼ turns #8 wire, 2¹³⁄₁₆" i.d., 2" long. Tap 1¾ turns from junction with 10 meter coil.
40-80	13½ turns #12 wire, 3" i.d., 6 turns/inch. Tap 6½ turns from junction with 20 meter coil.
160	16 turns #14 wire, 3" i.d. 10 turns/inch.

L-Section Inductor L_3
23 turns #12 wire, 2" i.d. Tap from output end: 80m-15 turns, 40m-9 turns, 20m-4 turns, 15m-2¼ turns.

Inductor L_4
4 turns #12 wire, 2" i.d. ¾" long

out excess capacity, as this limits the filament inrush current when the tube is cold.

Once the control circuits have been verified, the amplifier is connected to the power supply. A separate ground lead is run between the amplifier and the supply. A plate potential between 2000 and 2500 may be used. When the VOX circuit is activated, the resting plate current of the amplifier will be between 50 and 65 mA.

The amplifier is now turned off, all shields are screwed in place, and a suitable dummy load is connected to the antenna terminals through an SWR meter. Grid drive is applied slowly until a plate current of about 150 mA is observed. The plate circuit is now adjusted for maximum power into the dummy load, consistent with minimum resonant plate current. Grid drive is increased and loading adjustments are made in a normal manner until a single-tone plate current of 400 mA is achieved. Grid current should be about 100 to 120 mA. Adjustments should be conducted to provide maximum output at 400 mA plate current within the grid current limitations.

The last step is to touch up the input circuits by inserting the SWR meter in the coaxial line between the exciter and the amplifier and tuning the coil slug on each band for minimum SWR at the center fre-

Figure 74

UNDERCHASSIS VIEW OF THE AMPLIFIER

In the plate compartment the ceramic bandswitch is directly behind the front panel with the L-portion of the plate network behind it, mounted in a vertical position. The mica plate circuit padding capacitors for 80 and 160 meters are mounted to the walls of the enclosure and connections to these units are made with copper strap. The antenna changeover relay RY₃ is mounted immediately behind the L-section coil. In the input compartment, the cathode circuit coils and capacitors are grouped around the bandswitch, which is panel mounted. The two input circuit relays and filament choke are to the rear. Long, interconnecting power leads are run in shielded wire. Bottom plate (removed for photograph) is perforated metal for maximum ventilation. The panel-driven loading capacitor is at center, mounted to the underside of the L-bracket that supports the plate tuning capacitor.

quency of each band. This adjustment should be done with the amplifier running at full power input.

The Power Supply Views of the power supply are shown in figures 75 and 76. The schematic is given in figure 77. The supply is built up in an aluminum box measuring 12″ × 10″ × 8″. Using the listed transformer, the supply provides 400 mA for intermittent voice service at over 2500 volts.

The primary control circuit of the supply is actuated by the *filament power* switch on the amplifier. When the *plate* switch is closed, relay RY_1 in the supply is activated. After the time delay relay RY_2 closes, full voltage is applied to the power transformer, which is wired for a 120-volt source. During the time the delay relay is open, reduced voltage is applied to the transformer which permits the filter capacitors to slowly charge up, reducing the diode inrush current to a minimum.

In the secondary circuit, the negative return of the supply is above ground by virtue of a 50-ohm wirewound resistor. This allows the metering circuit in the amplifier ground circuit. Protection against flashover in the amplifier is provided by a 10-ohm, 20-watt resistor in series with the B-plus lead after the filter capacitors. The resistor does not affect the operation of the supply but absorbs a tremendous amount of energy if a short occurs in the amplifier. It is inexpensive and reliable protection for the tube and meters in the amplifier.

Power supply construction can be seen in the photographs. The high-voltage diodes and RC transient suppressors are mounted to a phenolic board supported above the transformer. The filter capacitors are

Figure 75

**POWER SUPPLY FOR
LINEAR AMPLIFIER**

The power supply is self-contained in an aluminum box having perforated metal top and sides. The high-voltage receptacle is in the upper corner, behind the handle. At the lower edge of the side panel are the primary cable, the interconnecting power cable to the amplifier and the primary fuse. No high-voltage circuits are exposed in this design.

mounted between two fiberglass boards held in position with threaded rod (10-32) on each corner. Holes are drilled in the top board for the capacitor terminals. Power relays and the time-delay relay are mounted in the space adjacent to the transformer.

The high-voltage lead is a section of coaxial cable and the various interconnection leads to the amplifier are made from #18

shielded wire. The cover for the supply is made of perforated sheet metal.

An external variable voltage transformer can be used with the supply, if desired. It is placed immediately after the surge suppression circuit.

22-12 An Advanced H-F Commercial Linear Amplifier

The linear amplifier described in this section was designed and built for commercial service by Jim Garland, W8ZR. It uses a single 8877 power triode in a cathode-driven circuit and is capable of 4 kW PEP input in SSB and c-w service and 2.5 kW input in SSTV and RTTY service.

This rugged and dependable amplifier consists of a tabletop r-f deck (figures 78, 79), a remote high-voltage power supply (figure 80), and offers features which meet or exceed those available in many commercially available amplifiers. These features include:

*Tuned input and output circuits for maximum linearity and a high order of harmonic suppression.
*Industrial grade components, including vacuum variable capacitors and relays, ball-bearing blower, mercury-plunger power relay, and custom designed plate transformer.

Figure 76

**INTERIOR OF THE
POWER SUPPLY**

With perforated cover removed, the main components of the supply are visible. The plate transformer and diode rectifier board are at the left with the series-connected filter capacitors at the right. Primary power and time-delay relays are in the foreground. After initial testing, the power supply should never be run without the protective cover in place.

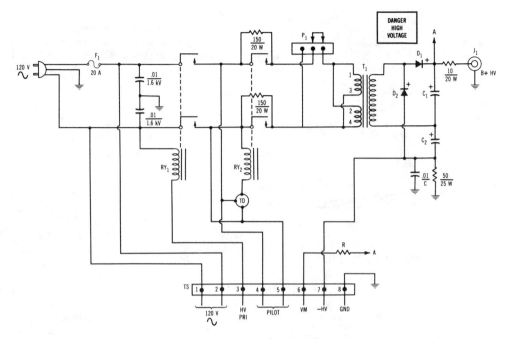

Figure 77

SCHEMATIC OF POWER SUPPLY

C_1, C_2—Each four 240 μF, 450 V electrolytic capacitors in series. Parallel each capacitor with 25K, 25 watt resistor.

D_1, D_2—Each five 2.5A, 1 kV diodes in series (HEP-170 or equivalent). Parallel each diode with 470K, ½-watt resistor and .01 μF, 1.4 kV disc ceramic capacitor.

J_1—High voltage coaxial jack UG-560/U (Amphenol 82-805) and matching plug UG-59B/U (Amphenol 82-804). Fill jack with silicone grease.

R—Five, 1-megohm, 1-watt resistors in series for meter multiplier.

RY_1, RY_2—20 A contact dpst relay, 120 Vac coil.

T_1—1100 V rms at 0.55A, 120/240 volt primary. Berkshire BTC-6181. Berkshire Transformer Co., Kent, CT.

TD—Thermal time-delay relay, spst, normally open contacts, 120-volt heater. 10 second delay.

*Advanced control and safety circuits which include automatic warmup and cool-down time delays, grid current and plate current overload protection, sequenced cold-switching of transfer relays, and surge and transient protection for the high-voltage power supply.

*Full metering of the grid and plate currents, plate voltage, forward and reflected r-f power, and temperature of the exhaust air.

*Extensive r-f screening, including double shielding of the r-f enclosure, and complete filtering of all leads entering or leaving an r-f environment.

*Ducted, forced-air cooling for minimum blower noise.

Note: Because this amplifier operates at high power and voltage levels, and as it incorporates circuitry which is more complex than is customary in homebuilt amplifiers, its construction should not be attempted by inexperienced builders.

Circuit Description   As shown in figure 81, the circuitry of the amplifier is contained in three modular interconnected units: an r-f deck which contains the main amplifier and components, control circuits, and low-voltage power supply; a detachable front panel containing the control switches, indicator lamps, and meters; and a remote high-voltage power supply. The front panel connects electrically

Figure 78

THE W8ZR HIGH FREQUENCY 4-kW PEP LINEAR AMPLIFIER

This deluxe amplifier is designed for commercial service and uses an 8877 in a cathode driven circuit. Maximum drive level is about 70 watts PEP. A number of interesting circuit features are incorporated to safeguard the tube against overload or improper operation. At the left of the main panel are the plate current meter (above) and the multimeter (below). The scales of the multimeter read: 0 to 5 kV, 0 to 100 (used for grid current), 0 to 300 (used for forward and reverse r-f power readings) and 70°C to 90°C for anode temperature measurement. To the right of the meters are the tune and load counter dials, with the bandswitch between them. Across the lower portion of the panel are the multimeter switch, the ALC control and LED, the power and operate switches and the illuminated pushbutton switches for warmup, plate reset and grid reset.

Figure 79

REAR VIEW OF RF AMPLIFIER

The various connectors on the rear panel are for r-f input, r-f output, relay control from the exciter, ALC output to the exciter, control of the remote power supply, and high voltage. A terminal for external ground connection is at the right. A BNC connector is used for r-f output and a type N connector for r-f input. The high-voltage connector is a special BNC type (see text). The air blower is at the left with the screened air intake at center and the exhaust air vent at the right.

Figure 80

THE REMOTE POWER SUPPLY

The supply provides 4 kV at 1A dc for the 8877 and is easily moved about on casters, even though it weighs 100 pounds. The power supply, which operates from 240 Vdc, is controlled entirely from the table top r-f amplifier. A small fan circulates air over the internal components. On the end of the enclosure are the control fuse, the control lamp and service/normal switch, the control plug, and the high voltage connector. Primary power cable is at the right. Note that no dangerous dc voltages are exposed.

to the r-f deck through three connectors (J_1, J_2, J_3). The high-voltage supply connects to the r-f deck through a control connector (J_4) and a high-voltage connector (J_5).

The R-F Deck—As shown in figure 82, r-f drive from the exciter is coupled through relay RY_3 to the input pi-network circuit and to the cathode of the 8877. The input network (C_1, C_2, L_1) operates with a Q of 2 and matches the 50-ohm drive impedance to the 42-ohm input impedance of the tube. The Q value is high enough to smooth out the fluctuations in cathode impedance during each r-f cycle, yet low enough to permit broadband operation. Two .01 μF capacitors provide an r-f path between cathode and filament of the 8877, thus eliminating r-f heating of the filament and reducing the possibility of r-f breakdown between cathode and filament. Two chokes (RFC_3 and bifilar-wound RFC_4) isolate the r-f voltage from the cathode and filament power circuits (figure 83).

During standby, the 8877 is biased to cut-off by the voltage developed across the 25K 10-watt resistor in the cathode lead. During operation of the amplifier, this resistor is shorted out by relay RY_3 and operating bias for the tube is supplied by Z_1, a 12-volt zener diode. Fuse F_1 in the cathode circuit is to protect the tube from excessive cath-

ode current. The fuse is a backup to the overcurrent protection circuit. An 800-ohm resistor shunted across the zener circuit keeps the cathode voltage from rising to dangerous levels should the fuse open.

Plate current is monitored by a 1.5A dc meter in the B-minus lead from the cathode to the power supply. The voltage developed across a 10-ohm resistor is used by three circuits to monitor grid current of the 8877. The first circuit consists of a metering arrangement which provides a direct indication of grid current on panel meter M_2. (Other positions of the *meter switch* (S_5) monitor the forward and reflected r-f power, the plate voltage, and the temperature of the exhaust air of the 8877 cooler.)

The other two monitoring circuits (consisting of transistors Q_3 and Q_4 and their associated components) provide grid overcurrent protection for the 8877 (figure 84). Transistor Q_3 begins to conduct when the grid current reaches about 150 mA. The conduction level is set by control R_6 (*Grid Trip Set*) and latches relay RY_8 closed. This relay illuminates a front panel indicator and switches the amplifier into the standby mode. The amplifier remains in this mode until the *Grid Reset* button is pressed by the operator. Transistor Q_4 operates in a similar manner to illuminate a red LED indicator when the peak grid current exceeds the normal operating value of about

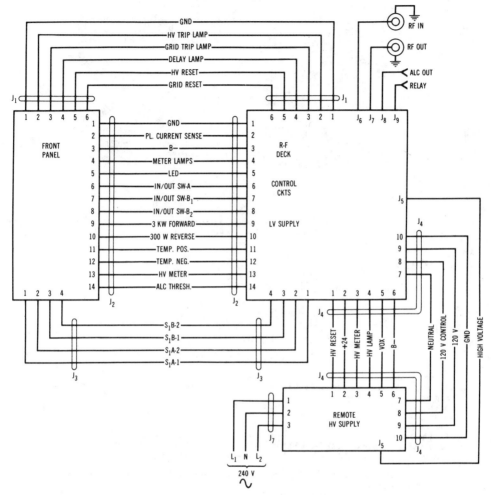

Figure 81

LINEAR AMPLIFIER INTERCONNECTIONS

J_1—Connector, 6 pin. Waldron-Molex 03-09-2061 (plug), 03-09-1061 (receptacle), or equivalent.
J_2—Connector, 15 pin. Waldron-Molex 03-09-2151 (plug), 03-09-1151 (receptacle), or equivalent.
J_3—Connector, 4 pin. Waldron-Molex 03-09-2041 (plug), 03-09-1041 (receptacle), or equivalent.
J_4—Connector, 10 pin. Amphenol MS series, Waldron-Molex 03-09-2101, or equivalent.
J_5—High voltage connector, 1 pin. Kings KV-59-03, or equivalent.
J_6—BNC-type connector.
J_7—N-type connector.
J_8—"Phono" jack.
J_9—2-pin "Jones"-type connector.

75 mA, the threshold being adjusted by potentiometer R_5 (*LED Set*).

Output Monitoring Circuits—The ALC and wattmeter circuits, shown in figure 85, are conventional designs. R-f voltage is sampled by a capacitive voltage divider (C_1, C_2) for the ALC control. Peak voltage

in excess of the dc reference voltage, set by the panel control *ALC Adjust*, is rectified, filtered and appears at the ALC output jack for control of the exciter output level.

In the wattmeter circuit, the amplifier output circuit is sampled by toroid L_1, while adjustable capacitor C_7 provides the reflect-

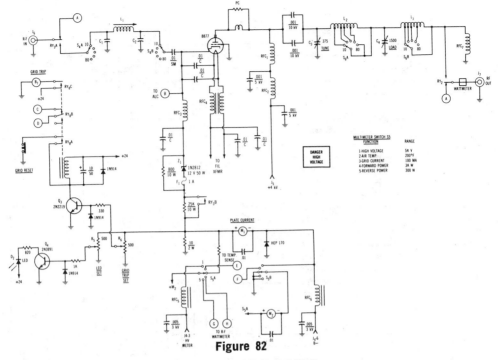

Figure 82

SCHEMATIC OF R-F DECK

C_1, C_2—See Table 16.

C_3—375 pF, 10 kV variable vacuum capacitor. Jennings UCS-375 or equivalent, with Millen 10031 counter dial.

C_4—1500 pF, 3 kV vacuum variable capacitor. Jennings UCSL-1500 or equivalent, with Millen 10031 counter dial.

L_1—See Table 16.

L_2, L_3—See Table 17.

M_1—0 to 1.5A dc. Simpson 1327-T-02650 with 01253 bezel and 01165 illuminator kit.

M_2—Multimeter. Simpson 1327-T-04401 (200 μA, taut band movement), with bezel and illuminator kit as above.

PC—Three 150 ohm, 2 watt composition resistors in parallel with 1½ turns, 1" diameter of ½-inch copper strap.

RFC_1—110 turns No. 24e. on 1" diam. ceramic rod (180 μH).

RFC_2—12 turns No. 14e. on ½" diam. form (1 μH).

RFC_3—69 turns No. 24e. on ½" diam. form (18 μH).

RFC_4—40 bifilar turns No. 12e. on ½" diam. ferrite rod (Permeability 125). Amidon R33-050-400 or equiv.

RFC_5, RFC_6—10 turns No. 14e. on ½" diam. ferrite rod.

RY_2—Vacuum antenna relay with 24 Vdc coil. Jennings RF3-A or equiv.

RY_3, RY_8—4 PDT. Potter & Brumfield KHP-17-D11 with 24 Vdc coil, and 27E006 socket.

S_4—2 pole, 6 position ceramic switch ganged with S_6. Centralab PA-2003.

S_5—Same as S_4.

S_6—2 pole, 6 position ceramic high voltage switch. Radio Switch Corp. Model 86, rotor style A, 60 degree detent.

Socket—EIMAC SK-2210. Chimney—EIMAC—SK-2216.

Note: All wires passing from r-f compartment are bypassed with r-f filters. Sprague 5JX3502 or equiv. Not shown on drawing.

ed power null adjustment. Potentiometers R_1 and R_2 are adjusted to set the forward and reverse full-scale meter readings to 3000 W and 300 W, respectively.

   The Output Circuit—The amplifier output circuit, shown in figure 86, is a pi-L configuration consisting of capacitors C_3 *(tune)* and C_4 *(load)* and inductors L_2 and L_3. High voltage is isolated from this circuit by parallel-connected .001 μF blocking capacitors. In the event one of these capacitors should short, choke RFC_7 provides a dc path to ground, instantly causing plate overload relay RY_7 in the power supply to latch the amplifier into the standby mode. Chokes RFC_1 and RFC_2 and the associated

Figure 83

THE INPUT CIRCUIT OF THE AMPLIFIER

The cathode and filament chokes are on each side of the ceramic 8877 socket, with the input bandswitch immediately in front of the socket. The H-shaped aluminum bracket holds the five slug-tuned inductors used in the cathode pi-network. To the right of this bracket is the antenna input relay and the ALC circuit board. The exhaust vent for the blower is in the floor of the chassis at the right.

Figure 84

BOTTOM VIEW OF THE AMPLIFIER WITH SHIELD PLATE IN PLACE

The cathode tuned circuits, ALC circuit and 8877 cathode and filament chokes are housed in the L-shaped area which is pressurized by the blower. Cathode circuit coil slugs can be adjusted through holes in cover plate which are normally closed by small plugs which act as an air seal. The circuit board at the left contains the grid overcurrent and LED circuitry. The three break-away plugs connecting the front panel components to the main chassis are seen at the bottom of the photograph, along with some of the components external to the main enclosure.

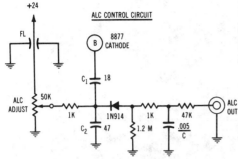

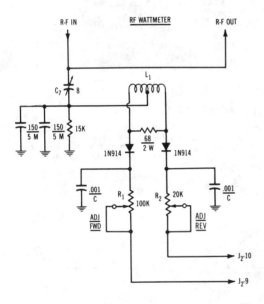

Figure 85

**ALC CONTROL CIRCUIT
AND R-F WATTMETER**

FL_1—R-f filter. Sprague 5JX3502 or equivalent.
L_1—20 turns center-tapped No. 22e. on iron-powder toroid core, 1/2″ diam. (yellow), permeability 8 (Amidon T-50-6 or equiv.).
C_7—8 pF air piston capacitor.
Note: "R-f in, r-f out" wire passes through center hole of toroid core.

bypass capacitors isolate the r-f voltages in the plate circuit from the high-voltage supply.

Control and Timing Circuits—The low-voltage dc supply, shown in figure 87, con-

sists of a 24 V, 2.8A regulated supply, fused by F_2. When the main *power switch* S_1 is closed, warmup relay RY_1 is initially open so that *warmup delay* lamp B_4 on the front panel is lighted, the *in/out* lamp (B_5) is illuminated at half-brilliance (if in/out switch S_2 is in the *in* position), and the amplifier is locked in the standby mode. During the warmup period, the 24 V supply charges capacitor C_5 through resistor R_7 and, after about 180 seconds, Darlington transistor Q_2 conducts and closes relay RY_1, allowing the amplifier to be placed in the normal operating mode.

Main power switch S_1 also activates a 120 Vac control line (J_{4-8}) to the remote high-voltage supply, closes thermal time-delay relay RY_8, energizes the filament of the 8877 through transformer T_4, and powers the blower through transformer T_3. This transformer is wired so that it reduces the voltage to 90 V across the blower to minimize noise while still maintaining adequate back pressure for the 8877. Time-delay relay RY_8 keeps the blower running for about two minutes after the amplifier has been switched off in order to exhaust warm air from the cabinet.

The *in/out* switch (S_2) enables the relay line of the amplifier to be closed by the VOX or keying circuit of the exciter. Grounding the line activates the r-f input relay (RY_3) and vacuum antenna relay RY_2 in the r-f deck. These relays are sequenced by the network C_6, R_8, D_9, and D_{10} so that relay RY_3 does not close until about 7 ms after RY_2 is closed. Cold switching RY_2 in this manner avoids contact welding which can occur if the relay is switched with power on the contacts.

The Front Panel Circuit—The front panel circuit (figure 88) consists of lamps B_4, B_6, and B_7 which indicate the *warmup plate overcurrent*, and *grid overcurrent* conditions, and their *reset* push buttons S_3 and S_4. The front panel also contains *on/off* switch S_1, *in/out* switch S_2, meters M_1 and M_2, and the *multimeter* switch (S_5).

The plate current meter is protected against circuit flashover of the high voltage by an HEP-170 power diode. Resistors R_9 and R_{10} comprise half of a Wheatstone bridge circuit which, in conjunction with a thermistor placed in the exhaust air from

Figure 86

**TOP VIEW OF THE
AMPLIFIER WITH
SHIELD PLATE
REMOVED**

The center enclosure contains the 8877 and plate circuit components. The blower at the rear of the chassis pressurizes the underchassis area around the 8877 socket. The aluminum duct pipe at upper left vents warm exhaust air from the 8877 out the rear of the amplifier. The blower, r-f wattmeter circuit board, vacuum antenna relay, and filament transformer are mounted on the L-shaped subchassis at the rear of the cabinet. The two variable vacuum capacitors and main bandswitch are mounted to the front wall of the shielded compartment. The lid of the compartment makes a good r-f seal because of the finger stock that lines the top edge of the enclosure. In the lower foreground are relays mounted on the outer wall of the enclosure.

the 8877 anode cooler, provides an indication of the heat dissipated by the tube.

High-Voltage Power Supply—The high-voltage supply (figure 89) is turned on as soon as *on/off switch* S_1 is closed (that is, when 120 Vac from control line connector J_1 is applied to the coil of mercury plunger plate relay RY_4). The *service switch* (S_3) must be set to the *normal* position for the high-voltage supply to be energized; if S_1 is set to the *service* position, lamp B_1 is lighted and the high-voltage power supply disabled. All other amplifier functions operate normally, however. This feature allows the r-f deck to be serviced under operating conditions without exposing the operator to dangerous voltage levels.

When relay RY_4 is first closed, current flow through the primary of plate transformer T_1 is limited by two 15 ohm 25 W resistors until surge-limit relay RY_5 closes, bypassing the resistors. This step-start feature (which takes about 0.25 second to complete) limits the surge current through the high-voltage diode banks (figure 90) to a safe value until the filter capacitor bank is charged. Additional protection of the diode bank is provided by transient suppressors Z_1 and Z_2 which clamp voltage spikes on the 240 Vac primary line to a safe value. The ac voltage from the secondary of the plate transformer is rectified by the voltage doubler bank and filtered by computer-grade electrolytic capacitors connected in series. The effective capacitance of C_1–C_{10} is 43 μF with a working voltage of 4500 Vdc. A voltage divider circuit is used to measure the voltage across the bottom capacitor in the string. The resistors are chosen to provide a 5kV dc full scale meter reading for multimeter M_1. Resistor R_{13} clamps the B-minus lead to within a few volts of chassis gound.

Resistor R_{14} (figure 89) monitors the current drawn from the high voltage supply. The voltage drop is applied to a resistor/diode network of the overcurrent protection circuit (figure 91). This network lights the internal LED in the opto-isolator IC_1 when plate current exceeds approximately one ampere and clamps the LED current to a safe value for short circuit currents as might momentarily occur if the capacitor bank was accidentally discharged through

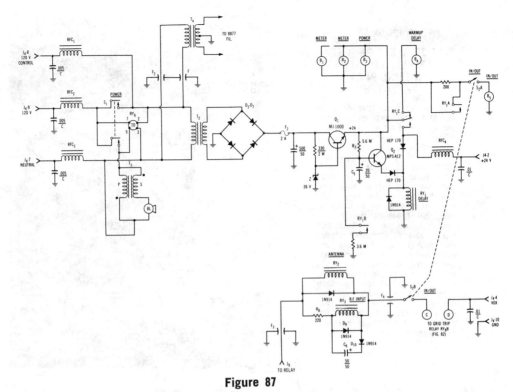

Figure 87

LINEAR AMPLIFIER CONTROL CIRCUIT

BL$_1$—Blower (see text). Rotron VS-37A2-A1.
D$_3$-D$_7$—5A, 600 V peak inverse voltage diode bridge.
F$_1$-F$_4$—R-f filter. Sprague 5JX-3502 or equivalent.
B$_1$-B$_5$—Light switch assembly, Arrow-Hart, or equiv.
   Actuator—AH-83504
   Lens—AH70 or AH84
   Contact block—AH-83500-30
   Bulb—T-1¾ (GE-334)
RFC$_1$-RFC$_4$—10 turns No. 14e. on ¼″ diam. ferrite rod (Permeability 125). Amidon R33-050-400
   or equiv.
RY$_1$, RY$_3$—4 pdt. Potter & Brumfield KHP-17-D11 with 24-volt coil and 27E006 socket.
T$_2$—25 V, 2.8A. Stancor P-8388.
T$_3$—36 V, 1A. Stancor P-8671. Wire to provide 90 Vac at blower.
T$_4$—5 V, 10A. Stancor P-6135.
Q$_1$—Motorola Power Darlington, MJ-1000.
Q$_2$—Motorola Darlington, MPS-A12.
TD—Thermal time-delay relay. Amperite 115-NO-180 or equiv. Coil to pins 2, 3. Contacts to pins
   5, 7.

a short circuit in the high-voltage line. When plate current exceeds 1.3 amperes (as determined by control R$_{16}$), transistor Q$_1$ conducts, latching control relay RY$_7$ closed and opening the high-voltage vacuum relay (RY$_6$) in the power supply. Relay RY$_7$ places the amplifier in the standby mode and illuminates *HV overcurrent* lamp B$_6$. A special wirewound resistor in the B-plus line is used as a high-voltage fuse

to back up this circuit; only the exact resistor type specified should be used for this purpose (figure 89).

Amplifier
Construction
The general internal layout of the r-f deck is shown in figures 84 and 86. The cabinet is manufactured by the *Buckeye Stamping Co.*, 555 Marion Rd., Columbus, OH 43207 (model DII-105-4-20). The cabinet con-

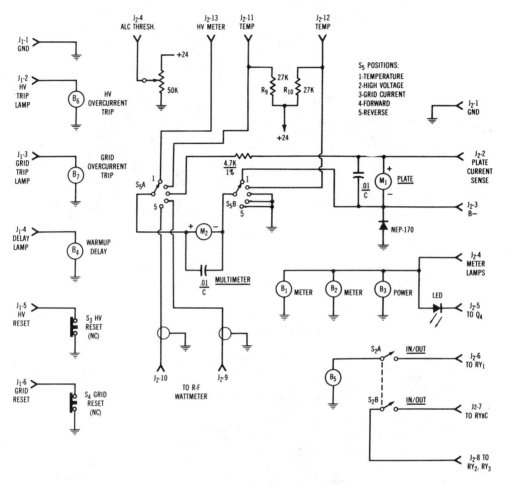

Figure 88

PANEL WIRING

Major components have been specified in other drawings or in the text. Shielded wires are run
from multimeter switch to r-f wattmeter assembly mounted on amplifier deck.

tains two subchassis. The first, which holds
the 8877 socket is 5″ wide, 3″ high and
4½″ deep and is located immediately below
an air duct made of sheet aluminum (figure 93). The second subchassis is 12″ wide,
3″ high and 4½″ deep, is located at the
rear of the cabinet and supports the filament transformer, vacuum antenna relay
and blower. All of the sheetmetal parts,
including the two chassis, were sandblasted
after fabrication to remove the glossy
sheen and scratches.

The blower draws air from a rectangular
section of perforated aluminum at the rear
of the amplifier cabinet. The air flows over
the plate circuit components before it is
sucked into the blower intake. Both of the
subchassis are pressurized by the blower
with the only air vent being through the
cooling fins and chimney of the 8877. A
cylindrical sheet of ¹⁄₁₆″ thick neoprene
rubber funnels air from the 8877 into the
air duct. The neoprene is secured to the
8877 chimney and to the flange on the

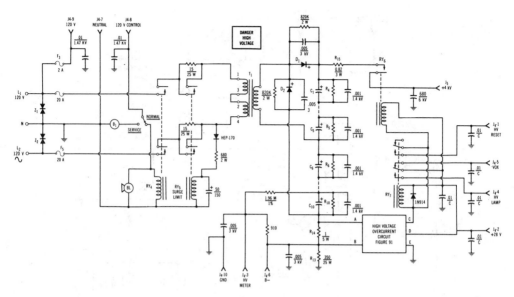

Figure 89

HIGH VOLTAGE POWER SUPPLY

BL—Rotron Muffin Fan, 7 watts.
C_1-C_{10}—425 μF, 450 V. Cornell Dubilier FAH425-450-83 or equivalent.
D_1, D_2—2.5A, 12.5 kV PIV. Eight each HEP-170 per leg in parallel with 820K, 2 W resistor and .005 μF, 3 kV disc ceramic capacitor.
R_4-R_{10}—25K, 25 watts.
R_{15}—0.82 or 1 ohm, 3 watt wirewound resistor. Ohmite 995-3A or equiv.
RY_4—dpst, 35A contacts, Mercury solenoid. Magnecraft WM35AA-120A or equiv.
RY_5—dpst, 15A contacts. Potter & Brumfield PRD-11DYO, 120 Vdc coil.
RY_6—spst vacuum relay. Jennings RB-1 or equiv. 120 Vdc coil.
RY_7—dpdt Potter & Brumfield KHP-17-D11 with socket 27E006.
T_1—Berkshire BTC-11421. 1550 Vac, 2.5 kV CCS.

Figure 90

THE HIGH VOLTAGE DIODE BANK

The high-voltage rectifier diodes are built up with a series combination of 2.5 A, 1.5 kV PIV silicon diodes mounted on a plexiglass plate. Each diode is shunted by an 820K resistor and a .005 μF capacitor to equalize the reverse voltage across the diodes and to eliminate "white noise" generated by the diodes which can appear as sideband noise on the transmitted signal.

bottom of the air duct with nylon *ty-raps*. The hot air in the duct is then exhausted from the rear of the amplifier cabinet through a length of 3″ diameter aluminum tubing. (Note that this tubing acts as a waveguide operated beyond its cutoff frequency so that r-f leakage through it is greatly attenuated.) This ducting system holds the blower and air noise to a minimum. Furthermore, warm air may be kept out of the operating area by venting the air outdoors through a length of plastic dryer tubing.

The Plate-Circuit Components—The major plate-circuit components (figure 94) are located in the center of the r-f deck in a compartment whose top is covered by a sheet of perforated aluminum. This compartment, together with the cabinet, provides double shielding of the strong r-f field in the amplifier. The front wall of the

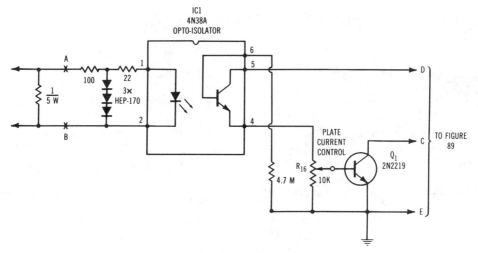

Figure 91

OPTO-ISOLATOR OVERCURRENT
PROTECTION CIRCUIT

Power-supply current is monitored across a resistor in the B-minus return. The developed
voltage is applied through a network to a
4N38A opto-isolator. Control level is set by the
potentiometer and latching control relay RY_7
(figure 89) closes, opening the high-voltage vacuum relay in the power supply.

Figure 92

POWER SUPPLY CLOSEUP

The overcurrent relay is mounted on a perf board in the center of the view, to the right of the
diode banks. The step-start relay is in the front, with the vacuum overcurrent relay on the L-
shaped bracket at the top, right. Note that each pin on the control connector plug is bypassed
with .001 μF disc capacitors.

Figure 93

CHASSIS ASSEMBLY FOR RF DECK

A view of the sheet-metal work before components are mounted. The 8877 air duct is at the left side on the small chassis and the subpanel to which the bandswitch and vacuum capacitors are mounted is at the top. Note that the top of the enclosure, as well as the rear wall, are screened.

compartment comprises a subpanel to which the bandswitch and two vacuum tuning capacitors are attached; their location and that of the other major components may be seen in the photographs. The plate-circuit bandswitch is ganged to the tuned cathode bandswitch by a chain and sprocket assembly (figure 95). The front panel of the amplifier may be easily removed by unplugging three connectors and loosening the setscrews on the shaft couplings. This allows access to the components mounted on the subpanel.

The Front Panel—The front panel contains the two meters, the turns-counter dial for plate tuning and loading, and lighted push button switches. The lettering on the panel and on the faces of the meters was applied with press-on letters obtainable from an art or engineering graphics supply store. Two coats of matte-finish plastic spray were applied to the panel after the letters were attached. A rear view of the panel showing the wiring is given in figure 96.

Plate Circuit Details—The pi-L circuit consists of two vacuum capacitors and inductors, L_2 and L_3. Inductor L_2 consists of two separate windings; section L_2A is

wound of ¼-inch diameter copper tubing and is used for the 10-, 15-, and 20-meter bands. Section L_2B is wound of ⅛-inch tubing and provides additional inductance for the 40- and 80-meter bands. The L-section inductor L_3 consists of No. 12 tinned copper wire wound on a toroidal core made of three, 2-inch diameter ferrite cores tightly wrapped with several layers of fiberglass tape. The wire is covered with teflon tubing to prevent arcing to the core. All copper tubing and interconnecting copper straps are gold plated before final assembly. (Silver plating is satisfactory and less expensive, although the silver will eventually tarnish.)

The High-Voltage Power Supply   The remote power supply measures 12" wide × 9¾" high by 21¾" deep. The enclosure is constructed of ⅛-inch thick aluminum plate bolted onto a framework of ½-inch-square aluminum stock (figure 97). Casters are mounted on the underside of the enclosure to allow it to be moved easily around the room. The power supply is cooled by a radial fan which draws air through two 3½-inch diameter screened

Figure 94

CLOSEUP VIEW OF PLATE CIRCUIT ASSEMBLY

The 80-40 meter inductor is shown beneath the right variable vacuum capacitor (load control), while the 10-15-20 meter inductor is located behind the ceramic bandswitch. The L-section inductors are located beneath the left vacuum capacitor (tune). The plate r-f choke and dc blocking capacitors are behind the tuning capacitor, with the one-turn parasitic suppressor between the anode of the tube and the plate choke. The 8877 is covered by a neoprene shroud that carries exhaust air into the aluminum air duct.

openings on the side of the enclosure. The plate transformer is bolted directly to the ⅛-inch aluminum baseplate.

The electrolytic capacitors are mounted on ½-inch-thick plexiglass sheets (figure 98) which also support the bleeder resistors, diode rectifier blocks and overcurrent protective circuit (figure 92). Layout of components in the power supply is not critical so long as adequate consideration is given to the very high voltage present.

There are two cables which connect the supply to the r-f deck. The first is an eleven conductor shielded control cable, and the second is a length of red-jacketed RG-59/U coaxial line which is used as the high-voltage cable. This cable is terminated in a special high voltage BNC connector (*Kings Electronics KV-59-03*); ordinary BNC connectors do not have sufficient insulation for this purpose. As shown in figure 92, all conductors leaving the power-supply cabinet are bypassed for r-f, as are all wires entering the r-f deck.

Figure 95

AMPLIFIER WITH FRONT PANEL REMOVED

Components attached to the front of the r-f enclosure are visible. The transformer on the right and its associated components make up the 24V dc power supply which powers the control circuits and the indicator lamps. The transformer on the left reduces the voltage supplied to the blower motor. The three relays on the upper left shelf are (l. to r.): the "after shutoff" relay, the warmup delay relay, and the grid overcurrent relay. The chain and sprocket system connects the input and output tuned circuits to the bandswitch. The disconnect plugs for the panel assembly are in the foreground.

Amplifier Calibration and Test The values of capacitance and inductance for the input and output tuned circuits are given in Tables 16 and 17. The value of the input tuning capacitance includes about 15 pF of stray circuit capacitance. The taps on the output inductors are most easily determined by making a mockup of the output circuit which includes the inductors and the bandswitch (figure 99). After the proper tap positions on the inductors are determined, the copper straps to the bandswitch may be soldered in place and dressed into position. Once the taps are in place, the entire coil and switch assembly can be removed from the mockup and placed in the amplifier.

Table 16.
Input Network Details

Band	C_1, C_2	L_1
80	820 pF	(2.2 μH) 16 turns #20e., 1/2" diam. closewound
40	430 pF	(1.2 μH) 10 turns #18e., 1/2" diam. closewound
20	220 pF	(0.6 μH) 8 turns #14e., 1/2" diam. closewound
15	150 pF	(0.4 μH) 6 turns #14e., 1/2" diam. closewound
10	100 pF	(0.3 μH) 4 turns #14e., 1/2" diam. closewound
10 MHz	330 pF	(0.9 μH) 9 turns #18e., 1/2" diam. closewound
18 MHz	180 pF	(0.5 μH) 7 turns #14e., 1/2" diam. closewound
24 MHz	120 pF	(0.35 μH) 5 turns #14e., 1/2" diam. closewound

Coil forms: Millen 69046-orange core
Capacitors: Dipped Mica DM-15J series C_1 and C_2 each made up of two parallel capacitors to distribute current.

The cathode input pi-network circuits are adjusted by inserting an SWR meter in the line between exciter and amplifier and tuning the coils for minimum reflected power at the center of each band. The amplifier

Table 17.
Plate Circuit Details

Band	C_3(Tune) pF	C_4(Load) pF	L_2	L_3
80	244	1132	11.0 μH	4.5 μH
40	104	460	6.2 μH	2.5 μH
20	53	233	3.2 μH	1.3 μH
15	35	156	2.1 μH	0.8 μH
10	26	116	1.6 μH	0.6 μH
10 MHz	69	284	5.0 μH	1.9 μH
18 MHz	38	158	2.8 μH	1.1 μH
24 MHz	29	118	2.0 μH	0.7 μH

Coil L_2 (10-15-20 meters): 9 turns, 3" inside diam. 15 meter tap 4 7/8 turns, 10 meter tap 2 7/8 turns.
L_2 (40-80 meters): 14 turns, 2 3/4" inside diam. 40 meter tap 4 3/4 turns.
Coil L_3 (10-15 meters): 6 1/4 turns, 1" inside diam. 10 meter tap 4 3/4 turns.
Coil L_3 (20-40-80 meters): Toroid, wound with #12e. wire. 80 meters 12 turns, 40 meters 6 turns, 20 meters 2 turns.
Notes: (1) Input capacitance values include 15 pF of tube and stray circuit capacitance.
(2) Network designed for plate impedance of 2250-ohm, load impedance of 50 ohms and operating Q of 12. (Q = 15 on 28 MHz).

should be operating at maximum power input while these adjustments are made. Access to the coils are through small holes in the bottom shield of the r-f compartment.

The r-f wattmeter is calibrated by operating the amplifier into a dummy load and nulling the reflected power indication with capacitor (C_7). The forward- and reverse-power potentiometers are set with the aid of a calibrated wattmeter.

Grid current threshold potentiometer (R_5) should be adjusted so that the front panel LED illuminates when the grid current exceeds 75 mA. Potentiometer R_6 is set to trip the grid overcurrent relay at a current of about 150 mA. The plate-current threshold

Figure 96

THE BACK SIDE OF THE MAIN PANEL

Front panel is removed by unplugging three connectors and loosening the setscrews on the turns counter dials and the bandswitch shaft extension. Small circuit boards are mounted on the meter terminals and hold multiplier resistors and potective diodes. The counter dials are at upper left.

potentiometer (R_6) located in the remote power supply is set to cause the protection circuit to trigger at a plate current in excess of 1.3 ampere.

Amplifier Tuneup—After the wiring has been checked the amplifier is ready for a trial run. First, set the *Service/Normal*

switch on the power supply to the *service* position and disconnect the high-voltage cable from the r-f deck. Turn on the amplifier and verify that the blower starts up and that air reaches the socket (figure 100). Check that the filament voltage at the tube socket is between 4.75 and 5.25 volts. The *warmup* lamp on the front panel should be illuminated. After about 180 seconds, the lamp should go out and a *very small* amount of r-f drive should be applied to the amplifier. Because there is no plate voltage on the 8877 the grid current will rapidly rise. Verify that the LED on the front panel lights at about 75 mA grid current and that the grid overcurrent relay closes at about 140 mA. If these checks are satisfactory, remove power from the amplifier, reconnect the high-voltage cable and set the *service/normal* switch to *normal*.

Next apply power once again to the amplifier and check that the high-voltage reading on the multimeter is about 4300 volts. Some variation is to be expected due to line voltage fluctuations. After the *warmup* lamp is extinguished, ground the relay line, and verify that the idle plate current of the 8877 is about 150 mA. If everything appears satisfactory, a small amount of r-f drive can

Figure 97

INTERIOR VIEW OF REMOTE POWER SUPPLY

The plate transformer is at center, left with the mercury plunger power relay on the side wall to the left. The capacitor bank is mounted in front of the transformer, in a horizontal position. This assembly also supports the rectifier diodes and the overcurrent protection circuit. The vacuum overcurrent relay is mounted on an L-shaped bracket on the right-hand end of the supply box, next to the cutout for the fan. The step-start components are in front of the capacitor bank.

Figure 98

THE HIGH VOLTAGE CAPACITOR BANK

The high-voltage filter capacitor consists of ten 425 μF computer grade electrolytic capacitors in series, each shunted by a 25K, 25 W wirewound resistor. The capacitors are mounted in a plexi-glass frame to make up an extremely compact unit. The capacitor bank is rated at 43 μF and 4.5 kV dc working voltage.

be applied to the amplifier and the plate-circuit controls adjusted for resonance and maximum r-f output as read on a watt-meter in the output circuit. Drive and loading can be slowly increased, with the plate *tune* control resonated for minimum plate current when proper loading is achieved. If adjustments are correct, maximum grid current, minimum plate current, and maximum power output should occur at the same setting of the plate tuning capacitors; any other condition is strong evidence of feedback or a parasitic oscillation.

At a maximum PEP input of 4 kW, grid current will run about 65 to 75 mA, corresponding to a drive power of about 70 watts. At full power, plate voltage drops to about 4 kV. Peak power output under these conditions is about 2500 watts.

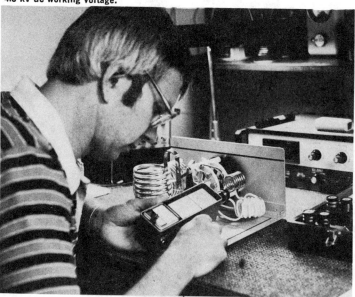

Figure 99

THE PLATE CIRCUIT MOCKUP

W8ZR determines the position of the taps on the plate-circuit coils. The assembly is built up on an aluminum mockup panel and positioned exactly as it will be in the final amplifier. A small 10 pF mica capacitor is temporarily installed to simulate the output capacitance of the 8877. Once the taps are properly positioned using a dip meter, the entire assembly is mounted in the r-f enclosure of the amplifier.

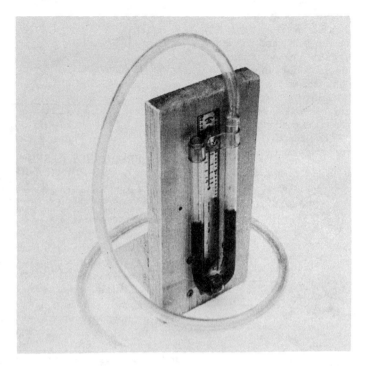

Figure 100

The HOMEMADE MANOMETER

Air pressure is checked with this homemade manometer which consists of a piece of wood, a U-shaped section of glass tubing and a six-inch scale. It is used to check the underchassis air pressure when the blower is running. Adequate cooling is provided for the 8877 when the manometer indicates at least 0.6 inch of water pressure. The glass tube contains water with a few drops of food coloring.

Part II   VHF Amplifiers

Much of the hf construction data discussed in part I of this chapter applies to vhf amplifiers. However, because the wavelength is small enough in the vhf spectrum to approach the size of the components, various design techniques (discussed in some detail in Chapter 11, part II) uncommon to hf amplifiers are commonplace in the vhf spectrum. The physical appearance of the vhf amplifier is unusual, as compared to the hf amplifier, as tank circuit components are measured in nanohenries and small values of picofarads.

Tank circuit Q usually runs quite high in vhf amplifiers as the output capacitance of the tube becomes a large portion of the tuning capacitance. As a result of the high circuit Q, circulating currents are extremely high and the tank circuit must be designed so as to safely conduct these currents (figure 1). Circuit Q values as high as 50 or 100 are not uncommon in large vhf amplifiers.

In addition to high circulating currents in the tank circuit and through the leads of the tube, tube efficiency tends to drop off in the vhf spectrum due to lead inductance, transit time, and insulator loss (figure 2). Power output, power gain, and efficiency decrease as the frequency of operation rises and grid drive and grid losses increase. Depending on the physical size of the tube, the arrangement of leads and internal geometry, most gridded tubes designed for vhf use will provide decreasing efficiency up to about 900 MHz. Above that frequency, the use of a negative grid tube is open to question.

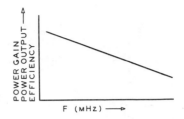

Figure 2

FREQUENCY RESPONSE OF VHF POWER TUBE

Power output, power gain, and efficiency slowly drop off as the frequency of operation is raised. Upper frequency limit of most vhf-type gridded tubes is in 900-MHz region.

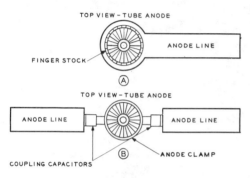

Figure 1

REPRESENTATIVE VHF TANK CIRCUITS

A—Quarter-wave stripline circuit. Anode connect-or encircles tube so that high circulating currents are distributed equally around anode.
B—Half-wave plate line with tube at high current point. Equal currents flow in each part of the plate line.

Tuned Circuit Design   In the case of a cathode-driven, grounded-grid amplifier a T-network such as the type shown in figure 3 is often used. The example shown matches a 50-ohm line to a cathode impedance of 42 ohms. Loaded Q is 2. This circuit functions well in the vhf region as the r-f cathode current flows through the cathode-to-ground capacitance of the tube and circuit. Under full plate current, the input match is adjusted to a 1-to-1 SWR because only under full-load conditions will the tube input resistance be at the design value.

Stripline or coaxial circuits are commonly used in the plate circuit of many vhf amplifiers. The tank impedance will be quite low, usually below 100 ohms. The higher the impedance of this part of the circuit,

the less will be the Q in the line, consequently bandwidth will be greater (figure 4). A wide bandwidth is not necessary to pass modulation frequencies, but it is desirable if the amplifier is to be moved about in frequency without the necessity of retrimming. In addition, a large amount of heat must be dissipated by the plate circuit and if the circuit Q is low, the tank circuit will not be detuned by expansion as rapidly as it would with a high-Q circuit.

There is a practical limit as to how high the plate line impedance can be made. Tube output capacitance, plus stray capacitance, will limit the impedance and length of the plate line:

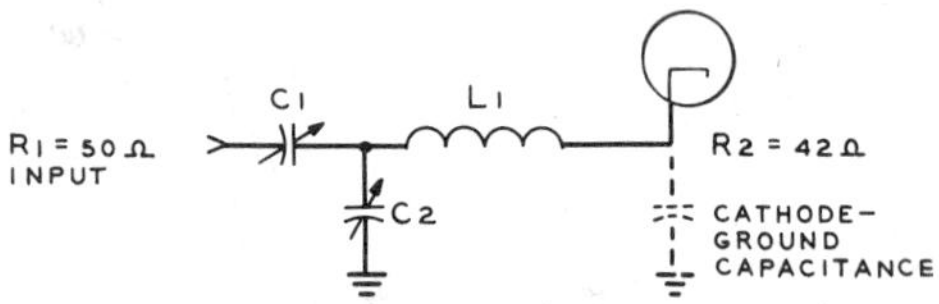

$R_1 = 50\,\Omega$ INPUT C_1 L_1 $R_2 = 42\,\Omega$
C_2 CATHODE–GROUND CAPACITANCE

$$R_1 > R_2$$
$$X_{L1} = Q_L \cdot R_2$$
$$X_{C1} = R_1 \sqrt{\frac{R_2\,(Q_L^2+1)}{R_1}-1}$$
$$X_{C2} = \frac{R_2\,(Q_L^2+1)}{Q_L} \cdot \frac{1}{1-(X_{C1}/Q_L R_1)}$$

Q_L = LOADED Q OF NETWORK

Figure 3

T-NETWORK FOR CATHODE-INPUT CIRCUIT

This circuit matches a 50-ohm source to a cathode impedance less than 50 ohms. Input match is adjusted for lowest SWR under full-load conditions of the amplifier.

$$X_C = Z_0 \tan l$$

where,

X_C is the tube output reactance,
Z_0 is the characteristic impedance of the plate line,
l is the line length in electrical degrees.

For a quarter-wave line, $Z_0/X_C = 2$ is adequate.

Using proper circuit design, most vhf tubes in the amateur service will work well in the 450-MHz band. Unfortunately, most of these tubes will not perform in the next higher band (1215-1296 MHz). Aside from the popular planar triode design (3CX100A5, etc.) which has extremely close interelectrode spacing and very low lead inductance, the majority of vhf-rated tubes have an upper frequency limit below 1000 MHz.

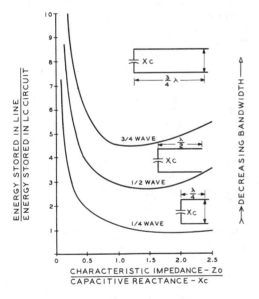

Figure 4

LONG LINEAR TANK CIRCUIT PROVIDES HIGHER Q AND LESS OPERATIONAL BANDWIDTH

Because of high output capacitance of some tubes a half- or three-quarter-wavelength plate line must be used. This reduces operational bandwidth of the amplifier. Highest possible plate line impedance provides greatest bandwidth.

Shown in this section are various amplifier designs for the 144-, 220-, and 432-MHz

bands. Aside from the use of linear tank circuits, these amplifiers follow the same general electrical design as their low-frequency counterparts. Notice that care is taken to make sure that the r-f currents in the output circuit flow from all areas of the tube anode. In the 144-MHz amplifier, for example, the plate circuit is split into two sections (L_5 and L_6) which divide the r-f plate current equally between them. The tube, in effect, is in the middle of a half-wave line with equal currents flowing in both line sections. In the 450-MHz amplifier the tube is centered in a cavity which is proportioned so that r-f currents flowing from the anode of the tube pass through all the walls of the cavity in like amounts.

Planar Tubes for VHF/UHF   Planar triodes are well suited for amateur use in the uhf/vhf spectrum. Various surplus versions, such as the 2C40, 446B, and 2C39 have been used at frequencies up to 2400 MHz. The 3CX100A5, a ceramic version of the 2C39, will provide improved service as an amplifier, doubler, or tripler in the range from 1000 to 3000 MHz.

Because of the high gain of the 3CX-100A5, grounded-grid circuitry is desirable, since intercoupling between the input and output circuits is reduced to a minimum and neutralization is not required in amplifier service.

At 1296 MHz, for example, the 3CX100-A5 is capable of about 47 watts output at an efficiency of 47 percent. Power gain is 8 dB, indicating a required drive level of about 7.5 watts as measured at the input of the cathode cavity. At 2400 MHz, power output (at 100 watts input) drops to 25 watts, and grid driving power increases to 10 watts. As a frequency doubler to 1300 MHz, the efficiency of the 3CX100A5 is about 27 percent with a power gain of 5.3 dB.

A coaxial or cavity circuit is generally employed in the 1300 MHz region. The resonant frequency of a coaxial tank, capacitively loaded at the open end by a tube is given by:

$$\frac{1}{2\pi fC} = Z_0 \tan\frac{2\pi l}{\lambda}$$

where,

f equals frequency in MHz,
C is the loading capacitance in pF,
l is the line length in centimeters,
λ is the wavelength in centimeters,
Z_0 is the characteristic impedance of the line.

The characteristic impedance is given by:

$$Z_0 = 138 \log_{10} \frac{R_1}{R_2}$$

where,

R_1 equals the inside radius of the outer conductor,

R_2 equals the outside radius of the inner conductor.

22-13 A High Performance 2-Meter Power Amplifier

This compact, high performance amplifier is rated for continuous duty at the 2-kW peak power level. It combines reliable service with good linearity and efficiency. Designed and built by W6PO, the amplifier has been used for moonbounce communications with Europe on many occasions.

The amplifier uses an 8877 high-μ ceramic power triode in a cathode-driven circuit. A half-wave plate line is employed, along with a lumped-constant T-network input circuit. The amplifier is fully shielded and built to fit on a standard 19-inch relay rack panel (figure 5A). The amplifier requires no neutralization, is completely stable and free of parasitics, and very easy to tune and operate.

The amplifier is designated for continuous duty at the 1-kW input level as well as at the 2-kW level for SSB operation. For the high power operation, plate voltage should be between 2500 and 3000 volts; under this condition the amplifier will deliver 1240 watts output. Stage gain is about 13.8 decibels and amplifier efficiency is 62 percent.

The Amplifier Circuit A schematic of the amplifier circuit is shown in figure 5B.
The 8877 is operated with the grid at dc and r-f ground potential. The grid ring at the base of the tube provides a low inductance path between the grid element and the chassis. Plate and grid currents are measured in the cathode-return lead and

a 12-volt, 50-watt zener diode is placed in series with the negative return to set the proper value of zero-signal plate current. Two diodes are reverse-connected across the instrument circuit to protect the meters.

Standby plate current of the 8877 is reduced to a very low value by the 10K cathode resistor which is shorted out when the VOX relay is activated, permitting the tube to operate in normal fashion.

A 200-ohm safety resistor ensures that the negative power lead of the amplifier does not rise above ground potential if the positive side of the high-voltage supply is accidentally grounded. A second safety resistor across the zener diode prevents the cathode potential from soaring if the zener should accidentally burn open.

The Input Circuit—The cathode input matching circuit is a T-network which matches the 50-ohm nominal input impedance of the amplifier to the input impedance of the 8877 which is about 54 ohms in parallel with 36 pF. The network consists of two series-connected inductors and a shunt capacitor. One inductor and the capacitor are variable so the network is able to cover a wide range of impedance transformation. The variable inductor (L_1) is mounted to the rear wall of the chassis and may be adjusted from the rear of the amplifier. The input tuning capacitor (C_2) is adjustable from the front panel. When the network has been properly tuned, no adjustment is then required over the 4-MHz range of the 2-meter band.

The Plate Circuit—The amplifier plate circuit is a transmission-line-type resonator. The line (L_5 plus L_6) is a half wavelength long with the tube placed at the center (figure 6). This circuit, while having less operational bandwidth than an equivalent quarter-wavelength line, is chosen because standard water pipe can be used as the center conductor of the line and the overall length of the line is long enough to be practical. In addition, the heavy r-f current that flows on the tube seals and control grid would, in the process of charging up the output capacitance to the peak plate voltage swing, tend to concentrate on one side of the tube if a single-ended, quarter-wavelength line were used. This current concentration would cause localized heating of the tube. The

Figure 5 A

THE HIGH-PERFORMANCE 2-METER POWER AMPLIFIER

This amplifier will operate at the 2-kW PEP input level for heavy-duty performance. The amplifier is built upon a 10½″ relay rack panel. The counter dial for the plate-tuning capacitor is at the center of the panel with the grid-tuning control directly beneath it. Grid and plate meters are at the left and right of the panel. The top of the r-f enclosure is covered with perforated aluminum sheet to allow the cooling air to escape from around the tube.

best tuned-circuit configuration to minimize this effect is a symmetrical, cylindrical coaxial cavity with the tube at the center. That arrangement is complex and difficult to build. A practical compromise is to use two quarter-wavelength lines connected to opposite sides of the tube. Note that each of the two quarter-wavelength lines used in this design are physically longer than if only one quarter-wavelength line were used. This is because only one-half of the tube output capacitance loads each of the two lines.

Resonance is established by a moving plate capacitor (C_5) and antenna loading is accomplished by a second capacitor (C_6)

placed at the anode of the 8877. Output power is coupled through the series capacitor into a 50-ohm output circuit. In the top-view photograph (figure 7) tuning capacitor (C_5) is at the front of the compartment; variable loading capacitor (C_6) is at the rear. The plate r-f choke is visible in the front corner.

Amplifier Construction   The 2-meter power amplifier is built in an enclosure measuring 10¼″ × 12″ × 6¼″. The 8877 socket is centered on a 6″ × 6″ subchassis plate. A squirrel-cage blower forces cooling air into the under-chassis area

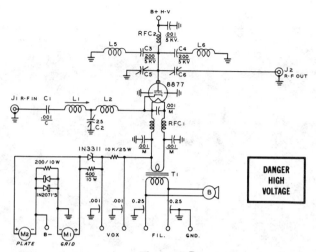

Figure 5 B

SCHEMATIC, 2-METER POWER AMPLIFIER

C₁—Centralab 858S-1000
C₂—25 pF Hammarlund HFA-25B
C₃, C₄—Each made up of two parallel connected 100 pF 5 kV ceramic capacitors. Centralab 850S-100
C₅—Plate tuning (see text)
C₇—Plate loading (see text)
L₁—5 turns #14, ¾″ long on ½″ diameter form (white slug). CTC 1538-4-3
L₂—4 turns #14, ¾″ diam., ¾″ long

L₃, L₄—(RFC₁)—Two windings; 10 turns #12 enamel each, bifilar wound, ⅝″ diameter
L₅, L₆—Plate lines (see text)
RFC₂—7 turns #14, ⅝″ diameter, 1⅜″ long
T₁—5 volts, 10 amperes. Chicago-Standard
M₁—0-100 mA dc
M₂—0-1 amp dc
Socket—Eimac SK-2210
Chimney—Eimac SK-2216

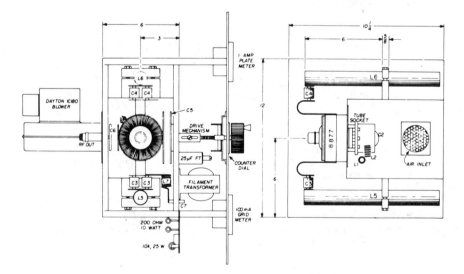

Figure 6

ASSEMBLY OF 2-METER AMPLIFIER

Structural details of the amplifier show relative size and position of the various components. Enclosure is made of aluminum panels. Bottom panel is solid and top panel is perforated to allow cooling air to escape.

Figure 7

TOP VIEW OF 2-METER AMPLIFIER

The perforated plate is removed from the plate compartment showing the 8877 tube at center. Plate-blocking capacitors and plate lines are at either side of the tube, with the plate r-f choke in the upper right corner of the enclosure. The two-plate tuning capacitor is shown just above the tube, with one of the plates attached to the anode strap of the tube. The other plate is driven in and out by means of a simple rotary mechanism driven by the counter dial. At the bottom (rear) of the amplifier the variable output coupling capacitor is seen just above the blower motor. The filament transformer and filament feedthrough capacitors are mounted to the front of the enclosure and a small plate at the right holds the various power resistors, diodes, etc.

and the air escapes through the 2⅝″ diameter socket hole.

The plate-tuning mechanism is shown in figure 8. This simple apparatus will operate with any variable plate capacitor, providing a back-and-forth movement of about one inch. It is driven by a counter dial and provides a quick, inexpensive and easy means of driving a vhf capacitor. The ground-return path for the grounded plate is through a wide, low-inductance beryllium-

copper or brass strip which provides spring tension for the drive mechanism.

The variable output coupling capacitor is located at the side of the 8877 anode. The type-N coaxial fitting is connected to the moveable plate of the coupling capacitor. The fitting is centered in a tubular assembly which allows the whole connector to slide in and out of the chassis, permitting the variable plate of the coupling capacitor to move with respect to the fixed plate mounted on

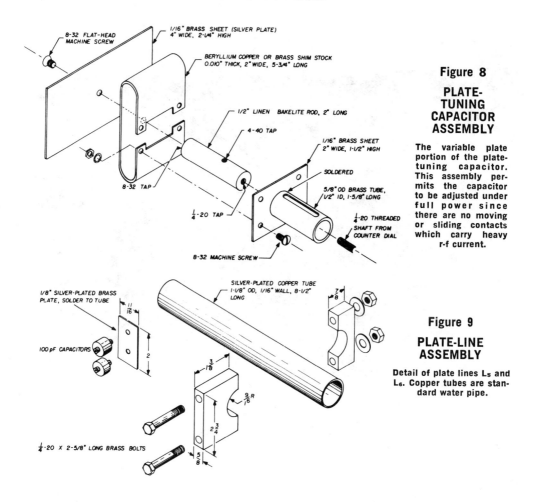

Figure 8

PLATE-TUNING CAPACITOR ASSEMBLY

The variable plate portion of the plate-tuning capacitor. This assembly permits the capacitor to be adjusted under full power since there are no moving or sliding contacts which carry heavy r-f current.

Figure 9

PLATE-LINE ASSEMBLY

Detail of plate lines L₅ and L₆. Copper tubes are standard water pipe.

the tube anode clamp (figure 6). When the final loading adjustment has been set, the sliding fitting is clamped by means of a small cable clamp passed around the tubular assembly, as shown in figure 7.

The length of the plate-line inductors (L₅, L₆) is adjusted by means of two dural blocks placed at the shorted ends of the lines (figure 9). The position of the blocks is determined by setting plate-tuning capacitor (C₅) at its lowest value and adjusting line lengths so that the plate circuit resonates at 148 MHz with the 8877 tube in the socket.

The plate r-f choke is mounted between the junction of one plate strap and a pair of the dual blocking capacitors and the high-voltage feedthrough capacitor is mounted

to the front wall of the plate circuit compartment. The r-f blocking capacitors are rated for r-f service and the substitution of TV-type capacitors at this point is not recommended.

Not observable in the photographs is a short chimney to direct cooling air from the socket through the anode of the 8877. It is made from thin, sheet *Teflon* and is clamped in place between the chassis and the anode strap.

Underchassis layout is shown in figure 10. The cathode input circuit is in the center compartment. The slug-tuned coil (L₁) is mounted on the rear wall. Air-wound filament chokes are placed in front of the socket. The cathode-heater choke coils are near the top edge of the enclosure. All of the cathode leads of the socket, plus one heater pin (pin 5) are connected in parallel and driven by the input matching network.

The ceramic socket for the 8877 is mounted one-half inch below chassis level by spacers to permit passage of cooling air to the anode. Four pieces of brass shim stock (or beryllium copper) are formed into grounding clips to make contact to the con-

8877 in the socket. An SWR meter should be placed in the input line so the input network may be adjusted for lowest SWR.

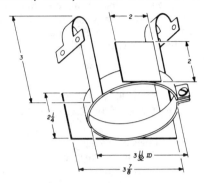

Figure 11

ANODE CLAMP ASSEMBLY

Figure 10

UNDERCHASSIS VIEW OF 2-METER AMPLIFIER

The cathode input circuit is in the center compartment, with the filament choke just above the tube socket. The socket is mounted below the chassis deck to permit cooling air to escape up around the tube anode. The dural blocks holding the ends of the plate lines are bolted to the side walls of the inner chassis. The walls are slotted to permit the blocks to be moved up and down the lines to establish resonance.

trol-grid ring. The clips are mounted between the spacers and the chassis. The aluminum clamps holding the ends of the plate lines are visible in the side compartments. The filament transformer and dial mechanism are placed in the area between the main enclosure and the panel.

Amplifier Tuning and Adjustment As with all grounded-grid amplifiers, excitation should never be applied when plate voltage is removed from the amplifier.

The first step is to grid-dip the input and output circuits to near resonance with the

Tuning and loading follow the same sequence as with any lower-frequency grounded-grid amplifier. Connect an SWR meter and dummy load to the output circuit. Plate voltage is applied, along with a very low drive level. The plate circuit is tuned for resonance and the cathode circuit is peaked for maximum grid current. Final adjustment of the cathode circuit should be done at full power input because the input impedance of a cathode-driven amplifier is a function of the plate current of the tube.

R-f drive is increased in small increments along with output coupling until the desired power level is reached. By adjusting

TABLE 1. Operating Data for 8877 for 2-kW PEP and 1-kW Conditions

Plate Voltage	3000	2500	2500V
Plate Current (peak) (single tone)	667	800	400 mA
Plate Current (no-signal)	54	44	44 mA
Grid Voltage	−12	−12	−12 V
Grid Current (single tone)	46	50	28 mA
Power Input	2000	2000	1000 W
Power Output	1240	1230	680 W
Drive Power	47	67	19 W

drive and loading together it is possible to attain the operating conditions given in Table 1. Always tune for maximum plate efficiency; that is, maximum output power for minimum input power. Do not overload and underdrive as plate efficiency will drop drastically under these conditions.

22-14 A 1-kW Power Amplifier for 220 MHz

The amplifier described in this section is well suited for the serious vhf operator interested in tropo-scatter or moonbounce work. It is intended for 50-percent duty operation at the 1000-watt dc input level and can develop 1000 watts PEP input for SSB operation. Power output in either case is about 580 watts and drive power is 30 watts.

An 8874 high-mu ceramic triode is used in a cathode-driven circuit. The tube has a plate dissipation rating of 400 watts and is well suited to linear amplifier operation as the intermodulation distortion is very low. The unit (figure 12) was designed and built by W6PO.

The Amplifier Circuit   A grounded-grid, cathode-driven circuit is used (figure 13). Plate and grid currents are measured in the cathode return lead and an 18-volt zener diode sets the bias for the proper value of zero-signal plate current. The diode is fused for protection and a resistor shunt holds the cathode near ground

potential if the fuse blows. Protective diodes are placed across the meters to protect them in case of an inadvertent short circuit or heavy current overload.

Standby plate current of the 8874 is reduced to a low value by the 10K cathode resistor which is shorted out when the VOX relay is activated.

The Input Circuit—The cathode input matching circuit is a T-network which matches a 50-ohm termination to the input impedance of the tube (about 94 ohms in parallel with 27 pF). The network consists of one series capacitor (C_1) and two series inductors (L_1 and L_2) with one shunt capacitance (C_2). Capacitor C_1 in series with L_1 allows the input inductive reactance to be varied. Inductor L_1 is larger than necessary and by placing a variable capacitor in series with it, the correct value of inductive reactance can be obtained.

The cathode of the 8874 is electrically insulated from the filament; however, filament chokes (L_3, L_4) are required because the filament/cathode structure is an appreciable fraction of a wavelength long at the operating frequency and an r-f potential appears in the filament circuit to a degree.

The Plate Circuit—The plate circuit of the amplifier is a half-wave open tranmission line (L_5). The line is made wide at the open end (figure 14) to allow sufficient area so that enough capacitance can be obtained in the tuning and loading capacitors without reducing the plate-to-plate spacing which would degrade the voltage hold-off capability. A low impedance half-wave line is too long to fit into the 17-inch enclosure;

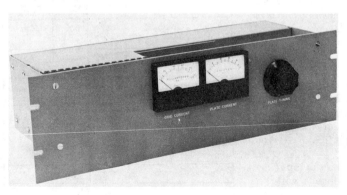

Figure 12

HIGH POWER LINEAR AMPLIFIER FOR 220 MHz

This high performance kilowatt amplifier uses an 8874 high-mu, ceramic power triode in a grounded grid circuit. The amplifier is built upon a 5¼" relay rack panel. Grid and plate meters are centered on the panel and the plate tuning control is at the right. The amplifier enclosure is spaced behind the panel to provide space for metering circuitry and auxiliary components.

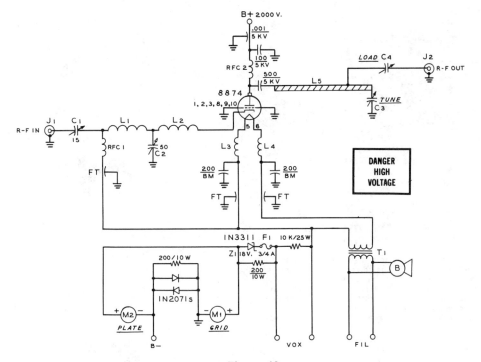

Figure 13

SCHEMATIC, 220 MHz POWER AMPLIFIER

C$_1$—15 pF (Hammarlund MAPC-15)
C$_2$—50 pF (Hammarlund MAPC-50)
C$_3$—1.75" diameter disc approximately 0.4" from plate line
C$_4$—1.50" diameter disc approximately 0.375" from plate line
L$_1$—0.8 μH. Four turns #14, 5/16" diameter, 1/2" long

L$_2$—0.03 μH. 1 turn #14, 1/2" diameter, 1/2" long
L$_3$, L$_4$—10 turns #16, closewound, 5/16" diameter
L$_5$—see figures 19, 20
RFC$_1$—Z-235 MHz r-f choke
RFC$_2$—10 turns #14, 3/8" diameter, 7/8" long
B—Blower, Dayton 2C782 or equivalent. 2 1/2" diameter, 3160 rpm
T$_1$—6 volt, 4-amperes. Stancor P-86376

therefore, a shorter line was narrowed at the high current point to increase the inductance and make the line electrically longer (figure 15).

The plate line is resonated by capacitor C$_3$ while the loading is adjusted by C$_4$. These two adjustments interact to a certain extent and the proper operating point must be found by adjusting both controls several times.

A type-N coaxial fitting is connected to the movable disc of the loading capacitor. The fitting is centered in a special tubular assembly which permits the whole connector to slide in and out of the chassis mounting fixture. This allows the variable disc of the

loading capacitor to move with respect to the fixed plate mounted on the tube anode clamp. When the final loading adjustment has been set the sliding fitting is clamped by a fixture similar to the slider on an adjustable wirewound resistor.

The disc is mounted on a threaded shaft which moves in and out through the threaded bushing on the front panel. To avoid jumpy tuning a fine thread is used.

The plate contact assembly (figure 16) is made from two copper rings and a special collet clamped together with 4-40 brass machine screws. One of the rings in the clamp has a flange to provide a mounting bracket for the plate-blocking capacitor.

Figure 14

ANODE ENCLOSURE OF AMPLIFIER

The 8874 is at the left with the plate line extending to right. The disc near the end of the cover plate is loading capacitor C_4. A Teflon chimney for the 8874 sits on top of the tube anode and protrudes out of the waveguide-beyond-cutoff air pipe in the cover plate. Note the r-f filter in the exhaust port of the blower.

Note that an additional bypass capacitor is placed in parallel with the plate circuit feedthrough capacitor to remove all the r-f energy from the plate voltage lead. The feedthrough capacitor by itself did not do the job.

Amplifier Construction The amplifier is built within two standard aluminum chassis. The plate compartment is made from a $3'' \times 4'' \times 17''$ chassis; the grid enclosure is made from a $2'' \times 4'' \times 6''$ chassis assembly that is attached to a standard $5\frac{1}{4}''$ relay rack panel. The front panel also supports the grid and plate current meters (figure 17). Behind the panel

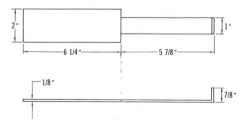

Figure 15

PLATE LINE FOR THE 220 MHz AMPLIFIER

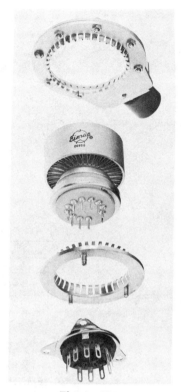

Figure 16

PLATE ANODE CONNECTOR

An exploded view of the tube socket, grid collet, tube, and tube anode assembly. The anode assembly is made of two copper rings which encircle the tube. The rings are clamped together with the collet in between. The bottom ring has a flange which is attached to the plate-blocking capacitor.

and next to the grid compartment is a flat aluminum plate that supports various resistors and the zener diode and related components (figure 18).

The pictorial drawing of figure 19 shows how the whole chassis assembly is put together. The back panel of the plate compartments is used to make an r-f shield, an air tight enclosure, and to serve as the mounting deck for the blower, filament transformer, the r-f output connector and the exhaust port for the cooling system. The blower forces air into the plate compartment and the air escapes through the anode cooling fins of the 8874, the teflon air chim-

Figure 17

OBLIQUE VIEW OF AMPLIFIER ASSEMBLY

The coaxial input receptacle and two adjustments for the cathode circuit are on the wall of the smaller compartment at the lower right. The type-N coaxial output connector is to the side of the blower. Filament transformer and air duct are on the end of the chassis, just adjacent to the high-voltage terminal. Power connections and line fuse are on a small plate in the foreground.

Figure 18

AMPLIFIER WITH PANEL REMOVED

The cathode input circuit is in the box at left with tuning capacitors mounted on the bottom. At center of the chassis are the various resistors, zener diode, zener protection fuse, and meter diodes. The shaft at the right is the plate-circuit tuning control.

ney and, finally, out through a waveguide-beyond-cutoff air pipe.

The cooling air inlet hole is shielded by a piece of copper honeycomb material similar to a radiator core which is soldered into the center of a ring and the assembly then mounted between the blower outlet and the backplate of the chassis (figure 20).

Amplifier Tuning and Adjustment Amplifier operation is completely stable with no parasitics. The circuits tune smoothly and plate-current dip occurs at the same time the power output peaks. As with all grounded-grip amplifiers, excitation should never be applied when plate voltage is removed.

The first step is to resonate the input (figure 21) and output circuits with the aid of a vhf dip meter. The 8874 should be in the socket. An SWR meter is placed in series with the input line and the input network adjusted for lowest SWR on the line from the exciter.

Tuning and loading follow the same sequence as any grounded-grid amplifier. Connect an SWR meter and load to the output of the amplifier and apply a small amount of drive power. Quickly tune the plate circuit to resonance.

The cathode circuit should now be resonated. Final adjustment on the cathode circuit for minimum SWR is done at full-power level because the input impedance of a cathode-driven amplifier is a function of the plate current of the tube.

Increase the drive in small increments along with output loading until the desired power level is reached. By simultaneously adjusting the drive and loading it will be possible to attain the operating conditions outlined in Table 2. Always tune for maximum plate efficiency, that is, maximum power output for minimum input power. It is quite easy to load heavily and underdrive to get the desired power input, but power output will be down if this is done.

Table 2. 8874 Typical Operating Conditions, 220 MHz

Plate Voltage	2000 Vdc
Plate Current (zero signal)	20 mA
Plate Current (single tone)	500 mA
Grid Voltage	—12 Vdc
Grid Current (single tone)	75 mA
Power Input	1000 W
Power Output	580 W
Drive Power (single tone)	29 W
Power Gain	13 dB

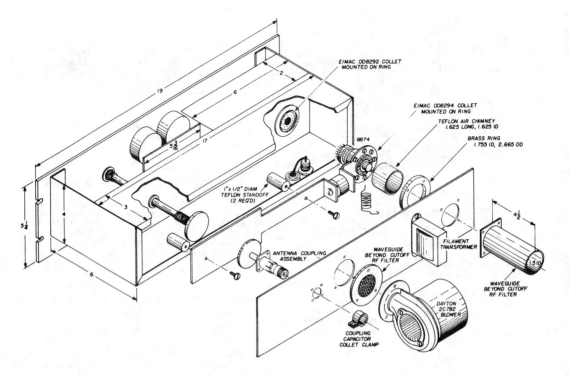

EIMAC 008292 COLLET
MOUNTED ON RING

EIMAC 008294 COLLET
MOUNTED ON RING

TEFLON AIR CHIMNEY
1.625 LONG, 1.625 ID

BRASS RING
1.755 ID, 2.665 OD

8874

1" x 1/2" DIAM
TEFLON STANDOFF
(2 REQ'D)

ANTENNA COUPLING
ASSEMBLY

WAVEGUIDE
BEYOND CUTOFF
RF FILTER

FILAMENT
TRANSFORMER

WAVEGUIDE
BEYOND CUTOFF
RF FILTER

1.510

DAYTON
2C782
BLOWER

COUPLING
CAPACITOR
COLLET CLAMP

Figure 19

PICTORIAL DRAWING OF AMPLIFIER ASSEMBLY

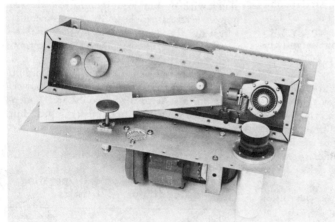

Figure 20

**ANODE COMPARTMENT
WITH PLATE LINE
REMOVED**

Line has been displaced from
its two Teflon support pillars
to show the movable disc of
the plate-tuning capacitor.
Cooling air inlet has an r-f filter at center. Air is exhausted
through tube anode and out
via Teflon chimney and air
port.

22-15 A 2-kW PEP Power Amplifier for 220 MHz

This high-power amplifier is designed for
continuous duty operation at the 1-kW dc
input level and can develop 2-kW PEP
input for SSB operation with ample reserve.
At a plate potential of 2500 or 3000 volts,
the amplifier will deliver 1230 watts output. With the higher plate voltage, up to
14 dB gain can be obtained with an ampli-

Figure 21

THE 8874 CATHODE INPUT CIRCUIT

The cathode terminals of the socket are interconnected by a small metal plate to which input coil L_2 is soldered. The filament chokes connect directly to the feedthrough capacitors mounted in the wall of the inclosure. Cathode input matching circuit is on inner wall of box.

fier efficiency of 61 percent (see Table 3). Designed and built by W6PO, the amplifier is currently being used for moonbounce communication.

The amplifier uses an 8877 high-μ power triode in a cathode-driven circuit which combines good intermodulation distortion characteristics and good gain. The plate circuit is a transmission-line resonator and the cathode input circuit is a lumped-constant T network. The amplifier is fully shielded and is built on a $10\frac{1}{2}$-inch-high rack panel (figure 22). No neutralization is required and the amplifier is completely stable.

The Amplifier Circuit A schematic of the amplifier is shown in figure 23. The grid of the 8877 is operated at dc and rf ground, the grid ring at the base of the tube providing a low inductance path between the grid element and the chassis. Plate and grid currents are measured in the cathode return lead, a 12-volt 50-watt zener diode in series with the negative return sets the desired value of idling current. Two diodes are shunted across the meter circuit to protect the instruments in case of a high-voltage arc to ground.

The standby plate current of the 8877 is reduced to a very low value by a 10K cathode resistor which is shorted out in the

transmit mode by the station control circuit. The resistor must be in the circuit when receiving to eliminate the noise generated in the station receiver if electron flow is permitted in the amplifier tube.

Table 3.
Performance of the 220-MHz Grounded Grid 8877 R-F Power Amplifier

	3000 V	2500 V	2500 V
Plate voltage	3000 V	2500 V	2500 V
Plate current (single tone)	667 mA	800 mA	400 mA
Plate current (idling)	54 mA	44 mA	44 mA
Grid voltage	−12 V	−12 V	−12 V
Grid current (single tone)	48 mA	50 mA	29 mA
Power input	2000 W	2000 W	1000 W
Power output	1230 W	1225 W	621 W
Efficiency (apparent)	61%	61%	62%
Drive power	48 W	69 W	20 W
Power gain	14 dB	12.4 dB	15 dB

Figure 22

2 kW PEP AMPLIFIER FOR 220 MHz

This compact 220 MHz amplifier uses a single 8877 running at 2 kW PEP input for heavy-duty performance. The amplifier is built upon a $10\frac{1}{2}''$ relay rack panel. The counter dial for the plate tuning capacitor is centered on the panel with the grid and plate meters above and to the side of it. The input circuit tuning control is below the counter dial. The amplifier enclosure is suspended behind the main panel by means of 3" long metal posts. Filament transformer and control wiring is in the space between panel and enclosure. Main terminal strip is just visible at left. Air vent for the amplifier is at top of the box and blower is to the rear.

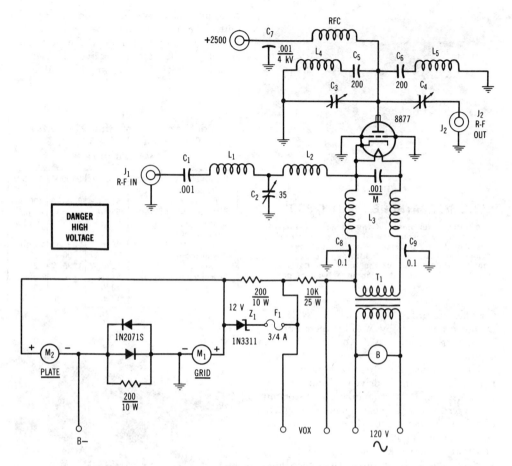

Figure 23

**SCHEMATIC, 220 MHz AMPLIFIER
WITH 8877**

C_1—1000 pF ceramic, 4 kV. Centralab 858S-1000 or equiv.
C_2—35 pF variable. Millen 22035 or equiv.
C_3—Plate tuning capacitor, see text.
C_4—Output loading capacitor, see text.
C_5, C_6—Each consists of two parallel-connected 100 pF, 5 kV ceramic capacitors, Centralab 850S-100, or equiv.
C_7—1000 pF, 4 kV feedthrough capacitor, Erie 2498 or equiv.
C_8, C_9—0.1 μF, 600 V feedthrough capacitor. Sprague 80P3 or equiv.

L_1—3 turns No. 14 wire, 1/4-inch diam., 5/8-inch long.
L_2—Copper strap 1/4-inch wide, 2 1/2 inches long, bent into a U shape, 5/8-inch wide.
L_3—7 bifilar-wound turns No. 12 enamel wire, 1/2-inch inside diameter.
L_4, L_5—Plate resonators, see fig. 30
RFC—6 turns No. 14, 1/2-inch diam., 1-inch long.
T_1—Filament transformer. 5 V, 10A Stancor P-6433 or equiv.
B—Blower. 20 cu. ft/min. Dayton 4C-446 or Ripley LR-81 or equiv.

A 200-ohm safety resistor ensures that the negative side of the power supply does not go below ground potential by an amount equal to the plate voltage if the positive side is accidentally grounded. A second safety resistor across the zener diode pre-vents the cathode potential from rising if the zener should accidentally burn open.

The Input Circuit—The cathode matching circuit is a T-network which transforms the input impedance of the tube (about 54 ohms in parallel with 40 pF)

Figure 24

CLOSEUP OF CATHODE INPUT CIRCUIT

The air-wound bifilar filament choke is at the left with the input tuning capacitor mounted directly to a socket retaining screw. Each socket screw has a clip which grounds the grid of the tube directly to the chassis. Johnson 122-247-202 ceramic socket is used, or EIMAC SK-2210 may be substituted. Grid terminals of socket are shorted together with 1/2" copper strap and inductor L_2 is soldered to crossover point of straps. Inductor L_1 is mounted between capacitor and ceramic post which supports blocking capacitor C_1. Note the chassis cutout around the socket for passage of air to tube anode.

Figure 25

INTERIOR TOP VIEW OF AMPLIFIER

The top plate is removed showing the 8877 tube and anode assembly. Parallel-connected blocking capacitors are at the side of the anode collet with inductors L_4 and L_5 bolted to the chassis. The plate tuning capacitor is at the bottom and the loading capacitor at the top, connected to the type-N coaxial receptacle. The plate r-f choke and feedthrough capacitor are in the front side of the enclosure. Note placement of the filament transformer on the outside of the box.

to 50 ohms at the coaxial input connector. The network consists of two series connected inductors and a shunt variable capacitor (figure 24). The inductors are fixed and have a very low value of inductance; in fact, the r-f return path through the chassis has about the same inductance value.

To design the input circuit, many values of circuit Q were tried in the calculations. When the design equations yielded physically realizable inductance values, then several combinations were tried in the actual amplifier. Since the stray inductances in the chassis and connecting leads were not included in the calculations, the final inductors were smaller in value than the calculated size. The actual inductors which provided a good input match are specified in the schematic drawing. Some minor variations in these coils may be expected to attain an adequate input match if the amplifier is duplicated.

The Plate Circuit—The plate circuit of the amplifier is a transmission-line type resonator. The line (L_5 plus L_6) is one half-wavelength long with the tube placed at

the center (figure 25). This type of circuit is actually two quarter-wavelength lines in parallel. One of the advantages of this design is that each of the quarter-wavelength lines is physically longer than if only one is used. This is because only half of the tube output capacitance loads each quarter-wavelength section. Another advantage to this layout is a better distribution of r-f currents around the tube anode seals.

The dc blocking capacitors are ceramic units. Two are used on each line to handle the r-f currents. The variable capacitor (C_5) tunes the plate circuit to resonance. Note that this type of capacitor structure has no wiping contacts which might increase r-f loss. All the r-f current flows through a fixed path which provides very smooth tuning with no jumping meter readings. The load capacitor (C_6) is constructed in a similar manner.

The plate r-f choke (L_7) is visible in the photograph of the plate compartment. It is connected to the plate collet assembly with the high-voltage feedthrough capacitor C_7.

Amplifier Construction　The amplifier is built in an enclosure measuring 8″ × 12″ × 7¼″ (figure 26). The 8877 socket is centered on an aluminum deck five inches from the top of the en-

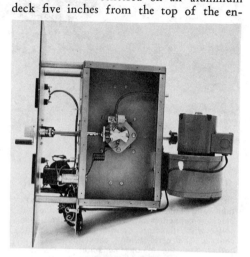

Figure 26

UNDERCHASSIS VIEW OF THE AMPLIFIER

The bottom plate has been removed for this view. Blower location and placement of control wiring can be seen. Tube socket is centered on the chassis.

Figure 27

BACK VIEW OF THE AMPLIFIER

The type-N connector is the r-f power output and the BNC connector beneath it is the connection for drive power. The knob centered on the box is the loading adjustment. Power terminal strip is at the right. A Millen high-voltage connector is used for the plate voltage terminal.

closure. A centrifugal blower forces cooling air into the underchassis area and the air escapes through the air system socket, the teflon anode chimney (SK-2216), and then the tube. The warm air is exhausted through a waveguide-beyond-cutoff air outlet (figure 27). This is an assembly which has expanded metal about ½-inch thick mounted in a frame. A perforated aluminum cover plate may suffice in most cases, although it restricts air flow slightly more and is not a very good r-f shield at 220 MHz.

The plate tuning mechanism is shown in figure 28. This simple apparatus will operate with any variable plate capacitor, providing a back-and-forth movement of about one inch. It is driven by a counter dial and provides a quick and inexpensive means of driving a vhf capacitor. It is used in other amplifier designs in this Handbook. The ground return is through a wide beryllium-copper or brass shim stock which provides spring tension for the drive mechanism.

The variable output coupling capacitor is located at the side of the 8877 anode. The type-N coaxial output connector is connected to the moveable capacitor plate by a wide beryllium-copper strap. The capacitor plate is driven in the same manner as the tuning capacitor as shown in figure 29.

The plate line is made up of two inductors, L_5 and L_6 (figure 30), and the anode collet and capacitor assembly shown in figure 31. With the inductor sizes given, the amplifier can be tuned from 220 to 222.5 MHz; no tests were run above 222.5 MHz.

The plate r-f choke is mounted between the junction of the anode collet and a pair of dual blocking capacitors. The high-voltage feethrough capacitor is mounted on the front wall of the plate compartment. The blocking capacitors are rated for r-f service and inexpensive television-type capacitors are not recommended for this amplifier. Complete structural details are shown in figure 32.

Amplifier Tuning and Adjustment　Amplifier operation is completely stable with no parasitics. The unit tunes up exactly as if it were on the hf

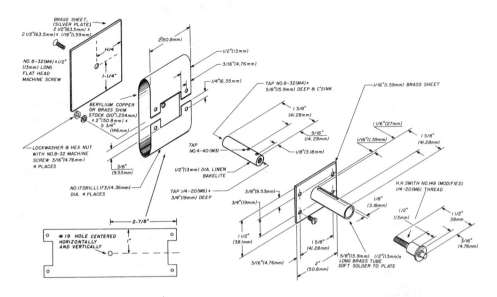

Figure 28

PLATE CIRCUIT TUNING CAPACITOR

The variable plate portion of the plate tuning capacitor. Since there are no moving or sliding contacts which carry heavy r-f current, this design permits the capacitor to be adjusted under full power without erratic tuning.

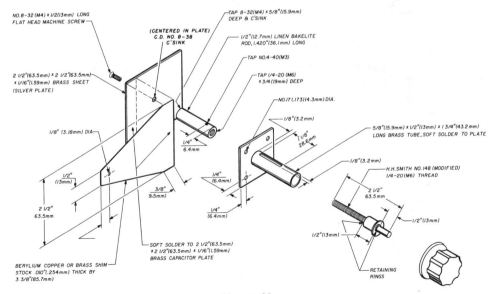

Figure 29

PLATE CIRCUIT LOADING CAPACITOR

The variable plate portion of the loading capacitor. The beryllium-copper portion carries the r-f current to the type-N connector as well as providing spring tension on the tuning mechanism. Because of the constant r-f conducting path, the loading is very smooth with no jumpiness.

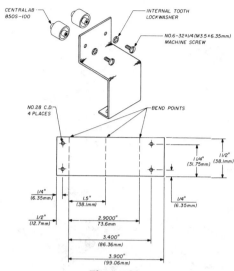

Figure 30

THE PLATE-CIRCUIT INDUCTORS

The plate-line inductor pattern and bending layout for L₄ and L₅. Two assemblies are required for the plate circuit. Copper strap 1/16″ thick is used.

bands. As with all grounded-grid amplifiers, excitation should never be applied unless the plate voltage is on the amplifier.

The first step is to resonate the input circuit with the aid of a dip meter and with the 8877 in the socket. The plate circuit should be brought into near-resonance next. An SWR meter is placed in series with the

input line and a dummy load connected to the output port of the amplifier. Filament voltage is checked and plate voltage applied.

Tuning and loading follow the same sequence as with any standard hf amplifier. Connect a second SWR indicator before the load and apply a small amount of drive. Quickly, tune the plate circuit to resonance. Load and increase drive until the parameters outlined in Table 3 are approximated. Final adjustment of the cathode input circuit for minimum SWR should be done at full power level because the input impedance of a cathode-driven amplifier is a function of the plate current of the tube.

Increase the r-f drive in small increments along with the output coupling until the desired power level is reached. Always tune for maximum efficiency (maximum output power combined with minimum input power). It is easy to load heavily and underdrive to get the desired input but efficiency will suffer if this is done.

22-16 A Tripler/Amplifier for 432 MHz

An efficient tripler or amplifier for 432-MHz operation may be designed around the

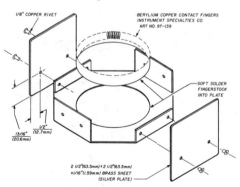

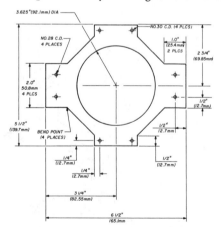

Figure 31

ANODE COLLET AND CAPACITOR PLATE SUPPORT ASSEMBLY

The two fixed capacitor plates for C₃ and C₄ are mounted to the assembly using copper pop rivets and then soldered. The two remaining bent-up edges are for mounting the blocking capacitors. The finger stock is soft-soldered into the large hole in the center. A tight fitting aluminum disc helps hold the finger stock in place while soft soldering with a hot plate. Pattern of the capacitor plate before bending is shown.

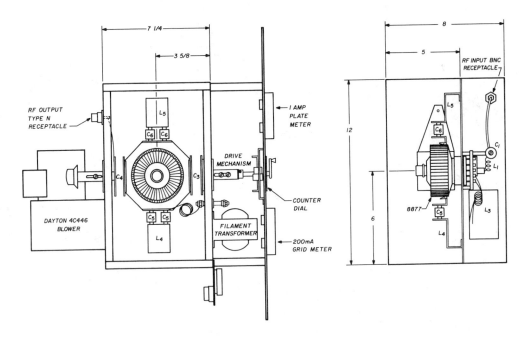

Figure 32
STRUCTURAL DETAILS OF 8877 AMPLIFIER

Relative size and position of the main components are shown. The chamber is made up of aluminum panels.

4X150A or 4CX250B external-anode tetrode. Rated at 250 watts anode dissipation
(the late production 4X150As also have
the higher rating) this high-perveance
tetrode is one of the few tubes that performs well as a tripler from 144 MHz
or as a straight amplifier at 432 MHz. A
power output of better than 60 watts may
be obtained as a tripler, and over 200 watts
output may be achieved in amplifier service.

Two units such as described may be built;
one acting as a tripler to drive the second
one as an amplifier at a power input up to
500 watts (figure 33).

The Tripler/
Amplifier Circuit
The general schematic of
the amplifier is shown in
figure 34. An easily built
coaxial plate-tank circuit provides high
efficiency at 432 MHz and the unit operates
in the same manner as if it were on the
lower-frequency bands. The circuit consists
of a short, loaded resonant cylindrical line
which uses the amplifier case as the outer
conductor. Plate voltage is fed through the
line to the anode of the tube, which is insulated from the cylindrical line by means
of a thin *teflon* sheet wrapped about the
anode.

For tripler service, the grid circuit is
tuned to 144 MHz, with the input capacitance of the tube and tuning capacitor C_2
forming a balanced tank circuit. The isolation choke (RFC_1) is at the center, or
"cold" point of the grid inductor. A series-
tuned link circuit couples the unit to the
external exciter.

In amplifier service, the grid circuit is
tuned to 432 MHz and takes the form of a
half-wavelength line, tuned to resonance by
a small capacitor placed at the end of the
line opposite the tube.

A special air-system socket designed for
the external-anode tetrode must be used. For
tripler service, the builder has the choice of
either the *EIMAC SK-600, SK-610, SK-620,*
or *SK-630* socket, together with the appropriate air chimney. The SK-606 chimney is

Figure 33

432-MHz TRIPLER/AMPLIFIER
USING 4X150A OR 4CX250B

This compact unit functions either as a tripler to 432 MHz, or as an amplifier on that band. It uses an external-anode tetrode in a modified cavity plate circuit. Enclosure is made up of side pieces held together with sheet-metal screws or "pop" rivets. In this oblique view, the B-plus connector is at the left side of the unit, with the coaxial antenna receptacle immediately adjacent to it. The antenna tuning capacitor is mounted to the end piece of the box, which may be removed by loosening the holding screws and the capacitor nut. At the center of the box is the spring-loaded tuning capacitor and at the right end is the coaxial input receptacle and the input tuning capacitors.

to be used with the *SK-606* or *SK-610* socket, and the *SK-626* chimney is to be used with the *SK-620* or *SK-630* socket.

For amplifier service at 432 MHz only, the *EIMAC SK-620* or *SK-630* sockets are recommended, as the other versions have screen terminals exposed to the plate-circuit field and exhibit more r-f feedthrough than do the suggested sockets, which have shielded screen terminals. Using the proper sockets, intrastage feedthrough at 432 MHz is sufficiently low so that stage neutralization is not required under normal, loaded operating conditions.

Not shown in the schematic is the fact that an external centrifugal blower is required to adequately cool the filament and plate seals of the external anode tube. Approximately 6.4 cubic feet per minute of cooling air at a pressure drop of 0.82 inch of water is required for full, 250-watt anode dissipation. For operation at reduced voltages and a limitation of 150 watts dissipation,

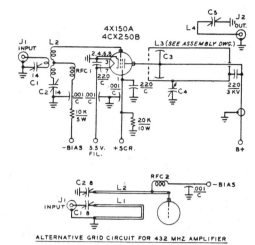

ALTERNATIVE GRID CIRCUIT FOR 432 MHZ AMPLIFIER

Figure 34

SCHEMATIC—432-MHz
TRIPLER/AMPLIFIER

C_1, C_2, C_5—14 pF, Johnson 160-107
C_3, C_4—See text
Note: Use 8 pF, Johnson 160-104 for 432-MHz alternative grid circuit
L_1—1 turn hookup wire, 3/4" diam., inside L_2 (432 MHz)
L_2—3½ turns #14, 1" diameter, 3/4" long (432 MHz)
L_3—See text
L_4—3/8"-wide copper strap form inductor 3/8" long × 1/8" high
RFC_1—1.7 µH. J. W. Miller RFC-144, or Ohmite Z-144
RFC_2—0.2 µH. J. W. Miller RFC-420 or Ohmite Z-420
J_1—Coaxial receptacle. UG-290A/U
J_2—Coaxial receptacle. UG-58A/U
Blower—6 cfm at 0.4" back pressure. Use #2½ impeller at 6000 rpm

the cooling-air requirement is 3.4 c.f.m. at a pressure drop of 0.15 inch of water.

At a frequency of 432 MHz, cathode backheating is observed in tubes of this type, and to maintain proper cathode temperature, the filament voltage should be reduced to 5.5 volts and held within plus or minus five percent of this value.

Finally, it should be noted that under certain operating conditions, the screen current of a tetrode may become negative. In order to protect the tube from excessively high screen voltage under certain negative-current conditions, it is mandatory to connect a bleeder resistor at the tube that will draw a value of current greater than drawn by the tube under negative-current operating conditions.

Figure 35

INSIDE VIEW OF
TRIPLER/AMPLIFIER

Tetrode tube socket mounts on small partition placed across interior of box. Plate-tuning capacitor and antenna capacitor are at right of assembly. The anode line of the 4X150A (4CX250B) is slotted and slips over the tube, insulated from it by a teflon wraparound insulator. B-plus passes down through the tube to a spring that makes connection to the anode. Below the partition are the grid circuit and various bypass capacitors. Power leads pass through feedthrough capacitors mounted in the rear wall of the enclosure (left). Aluminum fitting at the bottom of the box matches air-hose connection to external centrifugal blower. Blower should be turned on when filament voltage is applied to the tube.

Figure 36

CLOSEUP OF ANTENNA CIRCUITRY

Small series-tuned loop is made of copper strap connected between coaxial output receptacle and stator rods of antenna tuning capacitor. Coupling is adjusted by setting of capacitor, and link is fixed about ⅛-inch away from plate line. Plate line is soldered to brass end plate.

Tripler/Amplifier Construction   The tripler/amplifier is constructed within a metal box measuring 7″ × 2¾″ × 2⅝″. The top and bottom of the box are flat pieces of aluminum or brass measuring 7″ × 2¾″. The two side pieces are identical in size with matching holes for sheet-metal screws. Each side has small flanges along the edge which match the sides to the top and bottom pieces. The end section of the box which makes up the plate-circuit assembly is made of brass so that the brass quarter-wavelength plate line may be soldered to it. The opposite end of the box has a hole drilled off center in it to accept a fitting for an air hose or blower orifice (figure 35).

The plate line is made of a 3¾″ length of brass tube having a 1¾″ outside diameter. The line is soldered to the brass end to accept the anode of the 4X150A or 4CX250B.

An internal partition separates the grid and plate circuits and supports the socket for the tetrode. The socket is bolted atop the partition, as shown in figure 35. Connection is made to the anode for the supply voltage by means of an extension shaft run from the high-voltage connected mounted on the top plate of the box. The shaft has a section of spring steel bolted at the end to make a press fit to the top of the anode of the tube.

The plate-blocking capacitor is made of a length of 3-mil *teflon* tape, wrapped twice around the tube anode. The tape is cut to a width of one inch to allow overlap on both sides of the anode. The tape is carefully wrapped around the metal anode before the tube is pressed into the open end of the plate line, as shown in figure 37.

The top-plate of the box, in addition to the plate-line and high-voltage connector, supports the antenna receptacle (J_2) and the series antenna-tuning capacitor. The antenna pickup loop (L_4) is soldered between the receptacle and the stator of the capacitor, and is spaced away from the plate line about ⅛ inch.

Figure 37

EXPLODED VIEW OF PLATE-LINE ASSEMBLY

The high-voltage receptacle, plate bypass capacitor, and anode connector spring are at left. The brass end plate of the box and plate-circuit assembly are at the center, with the 4X150A tetrode at the right. The tube anode is wrapped with teflon tape to form a bypass capacitor, removing the dc voltage from the coaxial line. Copper line makes press fit over the anode of the tube.

Plate-tuning capacitor is a 1¾″ disc made of brass material soldered to the smooth end of a shaft that is threaded to match a panel bushing. The outer portion of the shaft is ¼-inch diameter to fit the dial drive. Tension is maintained on the shaft and bearing by placing a spring between the shaft extension and the panel bushing, as shown in the side view photograph.

Tripler/Amplifier Operation After the unit has been assembled, it should be tested for operation at reduced voltages. The first step is to grid-dip the input and output circuits to resonance to make sure they tune properly. An r-f output meter or SWR bridge should be used in conjunction with a dummy load for the initial tests. A good dummy load for 432 MHz is 500 feet of RG-58/U coaxial cable.

The far end should be shorted and water-proofed and the cable may be coiled up in a tub of water.

As with any tetrode, plate current is a function of screen voltage, and screen current is a function of plate loading. Screen voltage, therefore, should never be applied before plate voltage, and screen current should be monitored for proper plate loading. The amplifier should never be tested or operated without a proper dummy load.

To operate as a tripler, the following electrode voltages are suggested: plate voltage, 1000; screen voltage, 250; grid bias, −90 volts. The bias may be obtained from a small voltage-regulator tube or zener diode. Cooling air is applied with filament voltage which should be 5.5 volts. When these voltages are applied to the tube, plate current will be near-zero with no drive, and the screen current will be about 10 milli-amperes, or less. The screen current noted will be the sum of the positive current flowing through the bleeder resistor and the negative screen current of the tube.

A small amount of excitation at 144 MHz is applied and the grid circuit resonated, as noted by a small rise in plate current. The plate circuit should be brought into resonance. Excitation is boosted, and the tripler tuned for maximum power into the dummy load. Loading and grid drive may be increased until a plate current of 250 mA is achieved. At this level, total screen current will be about 15 to 20 mA, and grid current will be about 12 mA. Power input is about 250 watts and power output, as measured at the antenna receptacle with a vhf wattmeter, is about 70 watts. Overall tripler efficiency is about 28 percent and plate dissipation is nearly 180 watts.

Screen current is a sensitive indicator of circuit loading. If the screen current falls below 10 to 12 mA (including bleeder current), it is an indication that plate loading is too heavy or grid drive too light. Screen current readings of over 30 mA indicate drive is too heavy or plate loading is too light. A plate voltage as low as 800 volts may be used on the tripler stage, with an output of about 55 watts at a plate current of 250 mA. Plate voltages below this value are not recommended as screen current starts to climb rapidly at low plate potentials. For amplifier service, the alternate grid circuit is employed. The amplifier may be operated either class C or class AB$_1$. Operating data for both classes of service is given in the 4X150A data sheet.

22-17   A 500-Watt Amplifier for 420-450 MHz

This compact and reliable amplifier is designed for c-w, SSB or f-m service in the 420-MHz amateur band. Power input is 500 watts PEP or continuous service, with a peak drive power of less than 15 watts. Power output is better than 250 watts at a plate potential of·2000 volts. The unit (figure 38) shown was designed and built by W6PO.

The amplifier uses a single 8874 high-μ ceramic power triode in a cathode-driven circuit. A rectangular output cavity circuit is used, together with a stripline half-wavelength tuned input circuit. The amplifier is fully shielded and fits on a standard 19-inch relay rack panel as a companion unit to the 2-Meter Power Amplifier described in the previous section of this Handbook. The amplifier is completely stable and requires no neutralization. Tuning is easy and uncomplicated.

The Amplifier Circuit   The schematic of the amplifier is shown in figure 39. The 8874 is operated in a cathode driven circuit with the grid at dc and r-f ground potential. The grid ring at the base of the tube provides a low inductance path to ground between the grid element and the chassis. Plate and grid currents are measured in the cathode return lead and an 8.2-volt, 50-watt zener diode in series with the negative return sets the bias for the proper value of zero-signal plate current. The diode is fused for protection and shorted with a 200-ohm resistor to make sure the cathode remains at, or near, ground potential. Two small diodes are reverse-connected across the metering circuit to protect the meters in case of an inadvertent short circuit or heavy flow of current.

Standby plate current of the 8874 is reduced to a very low value by the 10K cathode resistor which is shorted out when the VOX relay is activated, permitting the tube to operate in a normal fashion.

The Input Circuit—The cathode input matching circuit is a modified half-wavelength line which matches the 50-ohm nominal input impedance of the amplifier to the input impedance of the 8874 which is about 160 ohms in parallel with 20 pF. A simplified drawing of the network is shown in figure 40. Illustration A shows a lumped, split-stator input circuit with the drive tapped on at a 50-ohm point in the circuit. Illustration B shows the same configuration redrawn to adapt it to the stripline circuit of illustration C. The latter assembly is used in this amplifier. The vertical reference line indicates the electrical center of the stripline, which is physically very close to the socket pins of the tube. The r-f

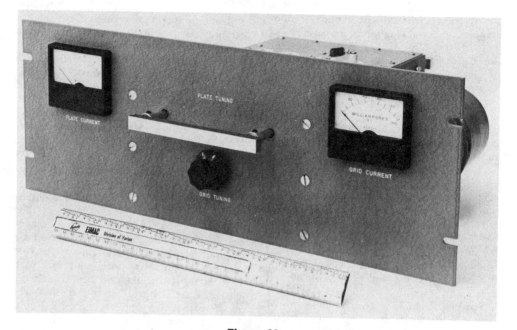

Figure 38

500-WATT AMPLIFIER FOR THE 420-MHz BAND

This compact, high-gain amplifier operates at 1-kW input, PEP or continuous rating, for c-w, SSB or f-m service. A single 8874 power triode is used in a cathode-driven circuit. The amplifier is built on a 7" relay rack panel. The "push-in, pull-out" handle for plate cavity tuning is at center panel, with the input tuning control beneath it. Plate and grid current meters are on the outer corners of the panel. The high-voltage terminal on the plate cavity is just visible over the top of the panel.

choke for the cathode return is connected at this point. The end of the line opposite the tube is tuned with a variable capacitor, and the capacitor indicated at the tube end of the line represents the input capacitance of the 8874. The tube places a low impedance load on the input circuit and tuning is extremely broad.

The cathode of the 8874 is electrically insulated from the filament; however, filament chokes are required as the filament/cathode structure of the tube is an appreciable fraction of a wavelength at 450 MHz and an r-f potential appears in the filament circuit to some degree.

The Plate Circuit—The amplifier plate circuit is a rectangular cavity which has two movable sides ("drawers") for plate tuning and antenna loading adjustments. The natural resonant frequency of such a cavity for the dominant mode is considerably less than a half-wavelength on a side because of

the output capacitance of the tube, which is at, or near, the center of the cavity (figure 41). It is difficult to equate the capacitance of the tube, which is distributed over an area large in comparison to a fraction of a wavelength, so that conventional loaded cavity equations cannot be used to mathematically determine cavity dimension. In this case, a cavity with movable drawers was constructed and "cold" resonance tests were conducted to determine the approximate volume of the cavity at resonance. An attempt was made to use a standard size aluminum chassis for the cavity to save money and construction time, and this goal was achieved with the design shown.

Power is extracted from the resonant cavity through the magnetic field, a coupling loop being introduced into the cavity so that it encloses some of the magnetic lines of force. The degree of coupling is determined by the cavity area enclosed by the loop and

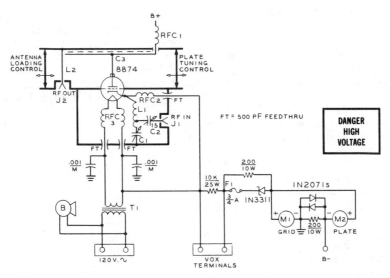

Figure 39

SCHEMATIC OF 420-MHz AMPLIFIER

C₁—Disc capacitor. See text and photographs for assembly

C₂—15 pF.

C₃—Planar capacitor, approximately 1800 pF. See text and photographs for assembly

L₁—Input stripline. See figure 36 for dimensions

L₂—Plate circuit cavity. See text and figures 37 and 41

RFC₁—10 turns #18, ¼ inch diameter, close-wound

RFC₂—Ohmite Z-450

RFC₃—Dual winding, 6 turns #18, ¼ inch diameter

M₁—0-100 mA dc Simpson 1227

M₂—0-1 Adc. Simpson 1227

Blower—Dayton 2C782. 2½ diameter, 3160 rpm

Socket—11 pin, Johnson 124-311-100 or Erie 9802-000.

Grid Collet—Eimac 882-931

Note: Filament voltage is set at 5.7 volts

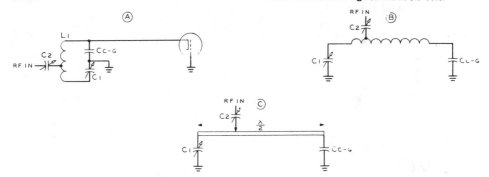

Figure 40

THE AMPLIFIER INPUT CIRCUIT

A—Split stator circuit for hf use with input placed at 50 ohm tap point on inductor. Capacitance C_{c-G} is cathode-grid capacitance of tube.

B—The same circuit adapted to stripline configuration.

C—Stripline circuit using half-wave line in place of inductor.

this area is controlled by moving one of the cavity walls, rather than by moving the loop itself.

Cavity resonance is established by changing the volume of the cavity through the use of a second sliding drawer, as seen in the photographs. The two sliding walls are adjusted in unison, much like the tuning and loading controls of a conventional high-frequency amplifier, until the adjustment presents the proper impedance match to a 50-ohm external load.

Amplifier Construction   The amplifier consists of two cavities made out of readily available aluminum chassis boxes. The cavities are supported from a 7-inch-high relay rack panel by means of side braces. The output circuit cavity measures 11″ × 7″ × 2″ and the input circuit cavity measures 7″ × 5″ × 2″. The flat surfaces of the two chassis are placed adjacent to each other so that the removable bottom plates form the outer surfaces of the cavities. Both cavities are seated firmly against the front panel and the tube is centered in the plate cavity. This places the center of the tube socket 5½″ from the front of the grid cavity.

The chassis are held together by the 8874 socket which is mounted inside the grid compartment. The 8874 requires forced-air cooling for its anode which is obtained from a blower mounted in the wall of the pressurized anode compartment, with air passing through the anode cooler and then exhausting vertically through the lid of the cavity. A small quantity of air is bled past the socket to provide base cooling, as the socket hole is about ¼ inch larger in diameter than the center portion of the socket.

For this class of service, a maximum anode dissipation of 300 watts is recommended

which requires an air flow of 6 c.f.m. at a pressure drop of 0.22 inch of water. The specified blower, or equivalent, will handle this requirement with a good safety factor.

Dimensions for the input inductor (L_1) are shown in figure 42. This device is supported by pins 1 through 3 and 8 through

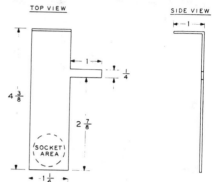

TOP VIEW   SIDE VIEW

Figure 42

INPUT LINE DIMENSIONS

10 of the tube socket and by a small teflon post placed at the far end of the line. The top surface of the line is spaced ⁵⁄₁₆″ away from the chassis deck. Tuning is accomplished by disc capacitor (C_1) which is mounted in a threaded block of copper fastened to the chassis near the front partition. A threaded panel bushing provides an addi-

Figure 41

TOP VIEW OF 420-MHz POWER AMPLIFIER

The cover has been removed from the plate cavity to show placement of the tube socket, the antenna coupling rod and the movable drawers. The drawer at the left controls antenna loading and the one on the panel determines cavity resonance. The socket for the 8874 is submounted to allow a free flow of air around the base of the tube. The grid flange is grounded to the chassis by means of a special collet (Eimac 135-305). The filament transformer and zener diode are mounted on a small chassis flange on one side of the amplifier. The cooling blower is mounted to the opposite wall. A screen across the blower opening prevents r-f loss through the orifice. The fingerstock visible on the movable drawers contacts the cover and walls of the cavity.

tional bearing so that the capacitor disc moves to and fro with an easy, rotational movement. A closeup of the input cavity is shown in figure 43.

Filament and cathode return leads are brought out of the grid enclosure via small, ceramic feedthrough capacitors as shown in the illustrations and the small filament chokes are air-wound coils mounted between the socket pins and the feedthrough capaci-

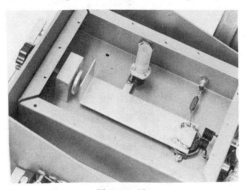

Figure 43

INTERIOR VIEW OF INPUT COMPARTMENT

The input line is soldered to the six cathode pins of the tube socket. Matching holes are drilled in the line and the pins pass through the line. Matching capacitor C_2 is soldered to the "ear" on the line. A short section of teflon rod is drilled and mounted on the capacitor shaft and adjustments are made through a hole in the perforated bottom plate, which has been removed for the photograph. At the far end of the line is disc capacitor C_1 mounted in a copper slug bolted to the chassis. The shaft of the capacitor is threaded, as is the mounting slug. The filament chokes of the 8874 are at the right end of the enclosure.

tor terminals. The input inductor has a "foot" on it that is positioned in such a way as to provide a support for the series input tuning capacitor (C_1). The capacitor is adjusted through a hole drilled in the cavity bottom plate. A short length of hollow teflon rod is slipped over the capacitor shaft to serve as a guide for an insulated screwdriver used for adjustment.

An underview of the amplifier is shown in figure 44. The cover for the input cavity is made of perforated aluminum stock held in place with screws and captive nuts. Directly behind the rear wall of the box is r-f output connector (J_2). Two sets of mount-

Figure 44

UNDERSIDE OF 420-MHz AMPLIFIER

The input cavity is covered with perforated aluminum stock held in place with 4-40 hardware and captive nuts. At one side is the teflon extension shaft for capacitor C_2. The antenna receptacle (J_2) is immediately in the foreground. Next to it is the plate covering the auxiliary mounting hole. The center of the first hole is 2½" from the rear edge of the plate cavity. At the side of the cavity is a small shelf that supports the zener diode, safety resistors, terminal strip, and filament transformer. The handle for the loading drawer is in the foreground.

ing holes are drilled for the connector, one behind the other. The unused mounting hole is covered with a small plate. The holes represent the limits of adjustment of the pickup loop (L_2) mounted in the plate cavity. The loop, which consists of a length of ¼-inch copper tubing, is soldered at one end to the center conductor of the receptacle. The opposite end is grounded to the top plate of the cavity. When the coaxial fitting is placed in the rear hole, coupling is at minimum and when it is in the front hole, coupling is maximum. For intermediate values of coupling, a rod with a slight offset is used in place of the straight rod.

The auxiliary components for the amplifier are mounted on brackets at the side of the cavity assembly. On one side is the centrifugal blower which exhausts through a screened port into the plate cavity. On the opposite side are mounted the filament transformer, zener diode, and the various components of the metering circuit as well as a

Figure 45

THE CAVITY PLATE AND COUPLING CAPACITOR

The anode of the 8874 is encircled by a fitting made of a copper ring lined with finger stock (see text). The adaptor is soldered to a copper plate insulated from the cover plate by a thickness of isomica insulation. The resulting capacitor is held together by screws and teflon insulating washers. The isomica sheet is a product of 3M Company, catalog number 0-11-S having a thickness of .005". Isomica is available on order through the branch offices of Minnesota Mining and Manufacturing Co., Schenectady, N.Y. It is available in standard sheets 18" x 36". Breakdown is about 700 volts per mil of thickness. Other equivalent materials may be used for the dielectric material.

Figure 46

REAR VIEW OF AMPLIFIER

The 8874 can be seen through the ventilation hole in the plate of the upper cavity. The teflon insulators and plate r-f choke are visible atop the plate. The loading drawer handles project out the rear of the enclosure. Note the metal rings in the push-in, pull-out rods to limit the distance of travel so that the shorting plate does not contact the output coupling loop. The aluminum chassis are "alodized" (not anodized) for protection against dirt and corrosion.

large terminal strip. Placement of components outside the cavities is not critical.

The plate coupling capacitor (C_3) is a planar device made of a sheet of $\frac{1}{16}$-inch copper $3\frac{1}{4}" \times 5\frac{1}{4}"$ in area insulated from the cover plate by a sheet of 5 mil (.005") *Isomica*. The capacitance is about 1800 pF. The connection to the anode of the tube is made by a circular spring collet (*Eimac* part number 008-294) which is soldered into a heavy copper ring. The outer diameter of the ring is 2" and the height is $\frac{3}{8}$ inch. The ring, in turn, is soldered to the capacitor plate, as shown in figure 69. The plate is held firmly to the cavity cover by means of teflon bushings and 6-32 bolts. The feedline choke (RFC_1) is mounted on the outside of the cover plate between one of the capacitor mounting bolts and an extra teflon standoff insulator (figure 46).

A hole $1\frac{7}{8}"$ diameter is cut in the coverplate and *Isomica* sheet directly above the anode of the tube to allow the cooling air to escape.

The Plate Tuning Mechanism The tuning and loading mechanisms are shown in figures 41 and 47. The tuning drawer is driven from the front panel and the loading drawer from the rear of the amplifier. Each sliding drawer consists of an aluminum plate measuring about $6\frac{1}{2}" \times 1\frac{3}{4}"$ and $\frac{1}{8}$-inch thick. The plate is lined on four sides with *Eimac* finger stock (type CF-800) which provides a low inductance, sliding contact with the walls of the cavity. In order to make sure the contact is firm and does not vary with pressure, a second aluminum plate is cut slightly smaller than the first, and the finger stock is "sandwiched" between the two plates. This assembly, in turn, is driven from the exterior of the cavity by means of two $\frac{5}{16}$-inch diameter "push-and-pull" rods and a connecting handle as seen in figure 41. A bearing plate is made of $\frac{1}{2}$-inch-thick *micarta* sheet which is bolted in place at the end of the cavity. Closely matching holes permit the driving rods to pass smoothly through the plate. A second set of guide holes in the end of the cavity permit proper alignment of the rods. An assembly drawing of this mechanism is shown in figure 46. Note that small, metal rings are slipped on each rod. Each ring has a set screw in it and the rings can be locked in position to prevent the tuning drawers from being moved too close to the tube and to the antenna pickup loop.

Amplifier Tuning and Adjustment As with all grounded-grid amplifiers, excitation should never be applied when plate voltage is removed from the amplifier.

Filament voltage is applied and the voltage at the socket pins is checked, as well as blower action. An SWR meter should be placed in the coaxial line to the exciter and a 50-ohm 500-watt dummy load connected to the amplifier. Reduced plate voltage (about 1800 volts) is applied, along with a very low drive level. The cathode circuit is adjusted for maximum drive and the plate cavity tuning drawers adjusted for maximum power output. The plate voltage is now increased to 2 kV and additional drive ap-

plied until the amplifier is delivering a few hundred watts. By adjusting drive level and loading, it should be possible to duplicate the operating conditions listed in Table 4.

Plate loading adjustment is limited by the placement of pickup rod L_2 in the plate cavity. Two adjustments are possible when a straight rod is used, depending on the placement of coaxial antenna receptacle J_2. If an intermediate loading position is required, a second rod is made up with a slight offset in it to provide an intermediate value of coupling. Always tune for maximum plate efficiency; that is, maximum output power for minimum input power. Do not underload, as grid dissipation may become excessive.

The last step is to adjust the input circuit for minimum SWR on the coaxial line to the exciter.

Table 4. 8874, Typical Operating Conditions, 420 mHz

DC Plate Voltage	2000 Vdc
Grid Bias	−8.2 Vdc
Filament Voltage	5.7 Vac
Plate Current	250 mA
Grid Current	20-40 mA
Power Output	250 Watts
Drive Power	13 Watts

Figure 47

INTERIOR OF PLATE CAVITY

The 8874 and grid grounding ring are at the center, with the output coupling "loop" immediately adjacent to the socket. The top end of the "loop" is bolted to the cover of the cavity. To vary the degree of loading beyond that permitted by the sliding drawer, extra loops can be made up that have a U-bend in them to increase the inductance. Optimum coupling distance from center of the tube to center of the pickup rod was determined to be about 2½ inches. In this particular amplifier, this worked out to be halfway between the two coaxial receptacles. A new "loop" was made out of soft copper tubing that had an L-shaped foot on each end so that the vertical section was spaced the proper distance from the tube.

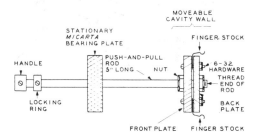

Figure 48

ASSEMBLY OF SLIDING DRAWER FOR PLATE CAVITY

22-18 A Practical 2-kW PEP Amplifier for 432 MHz

This high power vhf amplifier was built by W3HMU and used in the successful "moonbounce" expedition to South America

Figure 49

VHF LINEAR AMPLIFIER WITH 8938

This heavy duty amplifier runs 1 kW continuous duty or 2 kW PEP in the 420-450 MHz band. Using a single 8938 high-μ ceramic-metal triode, the amplifier provides maximum output with a peak grid drive requirement of about 60 watts. The amplifier compartment is mounted to a standard relay rack panel. At left are the plate- and grid-current meters with the amplifier plate-tuning dial centered on the panel. To the right and above the dial is the loading control. The r-f input receptacle and cathode tuning controls are to the right and left of the the plate-tuning dial. The r-f output receptacle is reached through a hole in the panel directly above the tuning dial.

in 1976. The amplifier combines reliability, high efficiency and ease of assembly in a small package (figure 49).

An 8938 high-μ ceramic power triode is used in a cathode-driven, grounded-grid, stripline circuit. No neutralization is required. The 8938 is a coaxial-base version of the 8877 and is rated at 1500 watts plate dissipation. The amplifier is designed for continuous duty operation at the 1 kW input level as well as the 2 kW PEP level for SSB operation. A plate potential of 2500 volts is required.

The Amplifier Circuit and Assembly   A schematic of the amplifier is shown in figure 50. The plate circuit is a half-wave stripline tuned by a single-plate variable capacitor. A second similar capacitor is used for antenna loading. The power triode is cathode driven by a half-wave stripline circuit tuned by a movable disc capacitor. A good impedance match between the driver and the tube is achieved by a second variable disc capacitor tapped on the cathode line.

Bias and metering circuits are much like the design shown for other amplifiers in this chapter. A 10K resistor in the dc cathode return circuit provides cutoff bias during standby periods. In the operating mode this resistor is shorted out by the VOX relay. Operating bias is established by a 27-volt zener diode in the cathode circuit.

Grid current is monitored in the cathode-to-ground circuit and plate current is measured in the B-minus power lead. A 200-ohm safety resistor provides reference to ground should the grid meter open up and a second resistor prevents the cathode from rising to a high dc potential in the event the zener diode burns open.

The Input Circuit—The input (cathode) circuit is half-wave stripline counterpoised against one wall of the cathode compartment. It is tuned to resonance by disc capacitor (C_1) to ground at the open end of the line. Near this same end of the line a second disc capacitor (C_2) couples drive power into the cathode circuit (figure 51).

The cathode line (L_1) is made of brass sheet 4½″ long and 1¼″ wide. The mate-

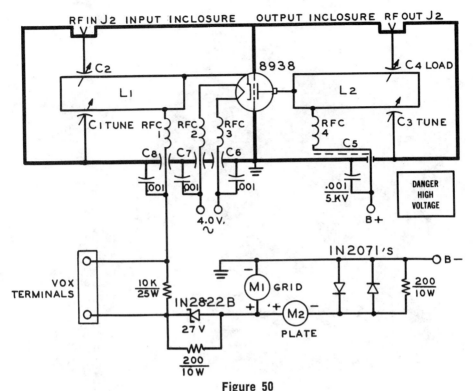

Figure 50

SCHEMATIC, 432 MHz POWER AMPLIFIER

C_1—One inch disc on piston tuner (see text)
C_2—One inch disc mounted on center conductor of transmission line.
C_3—Brass shim stock "flapper" (see text)
C_4—Brass or beryllium copper "flapper" (see text)
C_5—Plate bypass capacitor (see text)
C_6-C_8—1000 pF feedthrough capacitor (Erie)
L_1, L_2—Stripline assemblies (see text)
M_1—0-100 mA dc meter. Simpson
M_2—0-1A dc meter. Simpson
RFC_1—5½ turns #18e., ⅜" diameter, closewound. Ferrite beads slipped over lead from choke
     to cathode.
RFC_2, RFC_3—4½ turns #10e., ⅞" diameter, spaced wire diameter.

rial is $\frac{1}{16}$" thick. This plate is soldered at one end of the tube socket cathode terminal. A teflon post 1¼" high supports the midpoint of the plate.

Tuning capacitor C_1 is a 1-inch diameter brass disc soldered to the end of a surplus piston tuner. The tuner has finger stock contacts that ensure good r-f grounding of the rotating shaft. The stator of the capacitor is the end of the cathode line. The coupling capacitor (C_2) is a 1-inch diameter disc of brass soldered to the center conductor of a section of 1½-inch diameter foam-

filled, semirigid coaxial line. The inner conductor of the line projects ⅜ inch beyond the outer conductor which passes into the cathode chamber through a flange mount. The mount is slotted so that the coaxial line can be moved back and forth a short distance for preliminary adjustment. The line is then clamped firmly in place.

The socket for the 8938 is made from a surplus 2C39 socket. Looking down into the plate compartment (figure 52), the outer ring of the socket is the grounded-grid ring. It is made of finger stock (*Instru-*

Figure 51

CATHODE CIRCUIT COMPARTMENT

At top of the compartment is the tube socket and filament assembly. The filament and cathode chokes are wound of heavy wire, while the lighter cathode choke is below and closer to the deck. The input stripline is supported at the top end by the tube socket and at the center by a teflon insulator. Tuning capacitor C_1 is at left of the line with loading capacitor C_2 to the right. Cover has been removed from the compartment for the photograph.

Figure 52

PLATE COMPARTMENT WITH TUBE REMOVED

Tube and plate line have been removed to show tube socket assembly. Plate by-pass capacitor is affixed to rear wall of enclosure. Tuning capacitor plate C_3 is at right with string drive control. Air inlet hole can be seen at rear of enclosure.

ment Specialties 97-135). The finger stock is soldered to a brass sheet around the interior of a hole $2\frac{3}{16}''$ in diameter. The sheet is $3\frac{5}{8}''$ in diameter and $\frac{1}{8}''$ thick.

The grid ring assembly is bolted to the floor of the plate compartment. The grid ring mates over a $1\frac{3}{4}''$ diameter hole cut in the compartment material. The remaining coaxial rings in the socket (moving inward in order) are the cathode ring, outer and inner filament rings and center cathode pin collet. The central rings are supported by the tube socket which is mounted between two layers of plexiglas, $\frac{1}{4}$-inch thick. These plastic layers insulate cathode and filament terminals from ground.

The cathode ring is a 2C39 plate ring (*Instrument Specialties 90-70*) which is soldered inside a $1\frac{3}{8}$-inch (outside diameter) length of copper tubing. The tubing is shimmed to fit by copper flashing material. The filament ring is made from a short length of $\frac{5}{8}$-inch (outside diameter) copper tubing, 0.049″ wall, which is slotted with a hacksaw. The filament socket fits over the filament ring of the tube.

The Plate Circuit—The plate circuit is a half-wave stripline with "flapper" capacitors for tuning and loading. The line is made from a 5″ × 8″ piece of double-clad, glass epoxy PC board, $\frac{1}{8}$-inch thick. The corners are rounded to minimize voltage discontinuities.

The line is located approximately midway between the compartment base and cover, producing a short transmission line with a characteristic impedance of about 56 ohms. A finger stock cup is soldered into a hole in the plate line which passes the anode of the tube. Capacitor (C_3), which is about 0.5 pF at resonance, tunes the line. The capacitor plate is made of $\frac{1}{32}$-inch-thick brass shim stock and measures $3\frac{1}{4}$″ long by 3″ wide. The capacitor plate overlaps the outer $\frac{1}{2}$ inch of the plate line (figure 53).

Loading capacitor C_4 is made of thin beryllium copper material 2″ long by $\frac{1}{2}$-inch wide. It is soldered at one end to the center contact of a type-N receptacle. A plexiglass block bolted to the side wall of the plate compartment prevents rotation of the output coupling capacitor and the center connector pin of the receptacle. The capacitor overlaps the line by about one inch.

Both tuning and loading capacitors are adjusted by dial cord strings running to the panel controls. The cord for the tuning capacitor passes down through a clearance hole in the base plate to a shaft which is rotated by the plate tuning control, which is a planetary-drive reduction device. The string from the loading capacitor plate passes directly upward, then over a stand-off-mounted fairlead and exits the compartment at the end. It then wraps around a shaft which is also driven by a reduction dial. The capacitor plate is retracted by the tension on the string and returns to maximum position when string tension is released. Action is smooth and positive. A string with good dielectric must be used, fly fishing line was tried but melted in the strong r-f field!

The Plate Bypass Capacitor—The plate-circuit bypass capacitor (C_5) is a critical item in the upper reaches of the vhf region. In this case, a homemade bypass capacitor is used. It consists of a teflon sheet (0.015″ thick) sandwiched between a single-sided sheet of PC board and the compartment wall. The PC board measures $3\frac{1}{2}$″ × $4\frac{1}{2}$″ and has a ground lug soldered to the copper foil at one end. The lug is the high-voltage connection. The foil side of the board is in contact with the teflon dielectric which overlaps the PC board by $\frac{1}{4}$-inch on all sides. The plate r-f choke is connected between the lug on the PC board and the

Figure 53

ASSEMBLED PLATE COMPARTMENT

Tube and plate line are in place in this view. Air chimney fits over tube anode and is fitted into finger stock cup. Hole in top plate is aligned with the chimney.

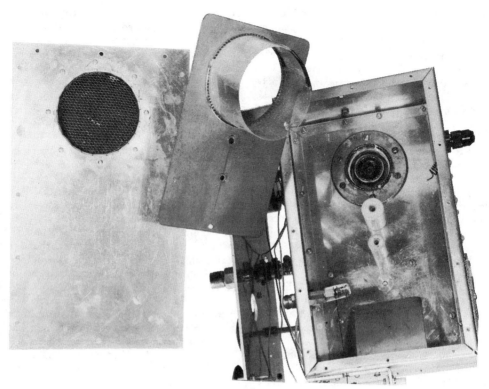

Figure 54

"EXPLODED" VIEW OF PLATE COMPARTMENT

The tube is installed in the socket and plate line is pushed down over anode of tube and bolted to the insulators. R-f choke is attached to plate line. Nylon chimney is placed over tube anode and fitted into finger stock cup (center). Top plate with air vent is aligned with chimney and bolted in place making airtight compartment.

plate line at the low r-f voltage point, which is located approximately at the inboard edge of the tube anode.

Amplifier Cooling—The 8938 requires 28 cfm of air at a pressure drop of 0.24" for 1000 watts anode dissipation at sea level. For full 1500 watts dissipation, 50 cfm of air is required at a pressure drop of 0.70". To achieve the proper cooling, a blower is mounted at the plate-tuning end of the plate circuit compartment. The air-inlet opening is covered with an aluminum insect screen. Air is forced into the plate circuit compartment by the blower and is restricted to exiting through the anode fins by the plate line and a mylar chimney, which is placed between the outer ring of the anode and the top plate air exit. A *Dayton model IC-180* blower will satisfy the 1000-watt

requirement of the 8938 under most conditions. For extended operation in a high temperature environment a *Dayton 2C-782* should be used.

The mylar chimney si 3½" in diameter and 2½" high. It is formed from 5-mil material. The plate circuit finger stock is mounted upside down on the plate line creating a convenient cup for the bottom of the mylar chimney.

Plate Line Assembly—The interior of the plate compartment is shown in figure 54. The tube is installed in the socket and the plate line is gently pushed down over the tube anode and bolted to the ceramic insulators. The mylar chimney is next placed over the tube anode and fitted into the finger stock cup. The top plate is aligned with the chimney and fastened into position, using sheet-metal screws. Remember to fasten the plate r-f choke to the line before securing the cover.

The top cover is fitted with an air outlet assembly shown in figure 55. The air outlet is made of metal honeycomb material fitted into a collar which was bent around the tube anode for sizing, and then trimmed and cut. A hose clamp holds the honeycomb disc in position in the collar. The mylar chimney fits inside the air outlet cut in the cover, forming an airtght and r-f-tight enclosure. This assembly serves as a waveguide-beyond-cutoff opening which effectively prevents energy from escaping from the plate compartment.

Amplifier Assembly—The general arrangement of the amplifier can be seen in the photographs. The plate circuit enclosure is an aluminum box measuring 12″ × 7″ × 4″ deep. The input circuit is in the lower box which measures 7″ × 5″ × 3″. An aluminum center plate divides the two chambers and provides a support for the tube socket and stripline supports.

The assembly is supported behind a 19″ relay rack panel on which the tuning controls and meters are placed. At one side of the assembly is a small chassis which contains the various metering resistors, the zener diode, and the terminal strip for control wires. The blower is mounted to the rear wall of the plate circuit box.

Figure 55

TOP VIEW OF AMPLIFIER

Air outlet from anode compartment is at top, left of assembly. Blower is mounted to rear wall. At right is subchassis with zener diode and power terminal strip. Amplifier assembly is spaced behind panel to allow room for meters and drive cables.

Amplifier Adjustment and Tuning Because of cathode backheating in the vhf region, the recommended filament voltage for the 8938 at 432 MHz is 4.0 volts. This should be measured at the socket with a 1-percent voltmeter. Cathode warmup time is 4 minutes under these conditions.

As with other high-μ triodes in cathode driven service, the 8938 should not be operated with r-f drive unless plate voltage is applied.

Once the blower has been checked for operation, reduced plate voltage is applied to the amplifier (1000 Vdc) and the idling plate current is checked. It should be about 35 mA. At 2 kV the idling plate current should rise to about 70 mA.

Drive power is applied slowly and plate loading and tuning capacitors are adjusted for maximum power output into a dummy load. The tuning capacitor must be reresonated each time the loading is changed. Once maximum output is achieved, the r-f drive coupling capacitor C_2 is adjusted until minimum SWR occurs on the coaxial line to the exciter. Once these adjustments have been completed, the plate voltage may be increased to the operating level and the controls again touched up for maximum power output.

When closing down, the plate and filament voltages are removed but the blower is left on for a few minutes to remove the heat from the tube. Representative operating conditions are given in the data sheet for the 8938.

22-19　　A 600-Watt Amplifier for the 920-MHz band

A recent decision of the International Telecommunications Union has provided a new amateur band in the 902 to 928-MHz range on a shared basis. As of late 1981, the band had not yet been authorized for amateur use in the United States.

This band is in the midportion of the uhf region and specially designed negative-grid tubes will work efficiently at these frequencies when used in a cavity-type amplifier.

Figure 56

HIGH POWER AMPLIFIER FOR 920-MHz BAND

This compact amplifier provides 200 watts output in the 915–970 MHz range with about 12 watts drive power. Using a single 3CX400U7 high mu, uhf triode, the amplifier is cathode-driven and requires no neutralization. The square output cavity is tuned by movable drawers, the controls of which are at the top and bottom of the cavity in this view. At the left and right are the multiple air vents The input and output coaxial receptacles are in the foreground. The tube plugs into the opposite side of the cavity. This is a commercial, heavy-duty cavity type CV-2810 manufactured by Varian EIMAC, San Carlos, CA 94070. The 3CX400U7 triode is shown in front of the cavity.

Shown in this section is a 600-watt input, 200-watt output amplifier having a power gain of about 12 dB and capable of operating over the frequency range of 915 to 970 MHz (figure 56). The amplifier employs a 3CX400U7/8961 high mu, uhf transmitting triode rated for 400 watts anode dissipation. The tube is employed in a TEM-mode cavity, the plate circuit having adjustable sliding drawers for tuning. Amplifier plate potential is 1500 volts and the unit may be used in the cw, fm or SSB mode.

The 3CX400U7 Transmitting Triode The 3CX400U7/8961 is designed for operation above 200 MHz, particularly in the 806- to 950-MHz portion of the spectrum allocated to land mobile services and to the amateur service. General characteristics of the tube are outlined in Table 5. The combination of high amplification factor and minimum grid interception provides good power gain in cathode-driven (grounded-grid) service. Coaxial terminals and continuous cone-shaped internal conductors for the grid and cathode elements permit the lowest possible inductance between tube elements and external circuitry.

Amplifier Design At frequencies in the 900-MHz region the amplifier circuit may be a quarter-wave resonator for the anode, and a three-quarter-wave coaxial line section between ground and cathode (note that heater terminals are separate from the cathode). An electrical diagram of a representative cavity and auxiliary circuitry is shown in figure 57 and an exploded view of a commercial cavity is shown in figure 58.

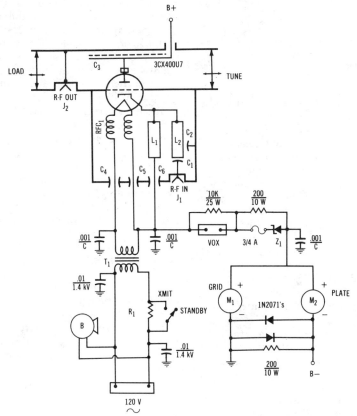

Figure 57

SCHEMATIC, 920-MHz POWER AMPLIFIER

C_1—Piston tuning capacitor mounted on receptacle J_1.
C_2—Piston tuning capacitor mounted on L_2 (see text)
C_3—Plate bypass capacitor made of .001″ thick Kapton sheet sandwiched between anode connector and cavity lid
C_4-C_6—Feedthru capacitors and Mylar disc capacitors
B—Blower. Dayton 2C782. 2½″ diam., 3160 rpm or better
M_1—25-0-25 mA, zero center
M_2—0-500 mAdc
T_1—6.3 v, 3A
J_1, J_2—Type N receptacles
Recommended collets: Anode, 154418. Grid, 882931. Cathode, 008292. Outer heater, 008291. Inner heater, 008290. All made by Varian EIMAC.
Note: RFC_1, L_1, and L_2 are part of input assembly of CV-2810 cavity.

TABLE 5

Operating Characterstics of 3XC400U7/8961 900 MHz

DC Plate Voltage	1500 V dc
Grid Bias*	−2 V dc
Filament Voltage, warmup	6.3 Vac
operating	4.5 Vac
Plate Current	400 mAdc
Grid Current**	−5 mAdc
Drive Power**	12.3 W
Useful Power Output**	190 W
Power Gain**	12.1 dB
Modulation	f-m

*Varies with class of service
**Approximate

The Plate Circuit—A quarter-wave adjustable rectangular anode cavity is used. Output coupling is magnetic. A loop is formed between the cavity walls and a post which terminates in the coaxial output connector. Coupling between loop and cavity is varied by moving a wall, or drawer, of the cavity, much in the manner of the 420-MHz amplifier described earlier in this chapter. The degree of coupling is determined by the cavity area enclosed by the loop and walls.

Resonance is established by changing the volume of the cavity by virtue of a second sliding drawer. The two drawers are adjusted in unison, much like the conventional tuning and loading controls of a high-frequecy amplifier.

The Input Circuit—Referring to figures 57 and 58, the inner conductors of the coaxial section are filament chokes (RFC_1), items A and B. The third concentric conductor is the cathode line (L_1, item C) which is a broadly tuned three-quarter-wave line. The next diameter of tubing is a sleeve (item D) tuned by an adjustable capacitor (C_2). Current flows on the inside as well as the outside of this sleeve, thereby coupling energy to the cathode line. This dvice is three-quarters wavelength long as there is approximately a loaded quarter wavelength within the 3CX400U7 itself.

Cooling the Cavity—Suitable cowling (not shown) must be provided to introduce cooling air through three short tubes on each side of the output cavity for anode cooling. The air then exhausts through the finned anode. The short tubes are dimensioned to serve as waveguide-above-cutoff frequency filters in the air openings. Approximately 11.5 cfm of air is required when the tube is operating at full rated dissipation (400 watts). The pressure drop across the anode cooler only at this flow rate is about 0.2 inch of water. These figures are based on an incoming air temperature of 50°C, and a maximum tube anode temperature of 225°C, at sea level, with the air flowing in a base-to-anode direction. Additional base cooling may be required, and this may be verified by means of temperature sensitive paints, or other equivalent means.

Heater-Cathode Operation—The nominal heater voltage for the tube is 6.3 volts. For c-w or f-m operation above 300 MHz the heater voltage should be reduced as the cathode receives additional heat from r-f charging currents and the transit-time effect. In the case of this design, operating heater voltage is 4.5 Vac. During warmup and standby heater voltage is returned to 6.3 Vac. Nominal heater voltage is applied for a minimum of 60 seconds before operation commences. For best life expectancy and most stable performance, it is desirable to regulate the heater voltage and to hold it to the final desired value within plus or minus 2 percent.

The Metering Circuit—Conventional plate and grid metering is used in this amplifier, with protection provided for the meters by means of back-to-back diodes. For the grid current, a zero-center meter is suggested because the normal grid current indication can be negative. This negative current is the result of tube characteristics and transit-time effects at the frequency of operation.

Amplifier Adjustment Before any operation is attempted, unless the operator is assured the cavity has been pretuned to the desired frequency, the unit should be checked for coarse tuning. Cavity frequency rises as the drawer is moved inward toward the tube. During tuning an r-f input directional coupler should be connected in the drive line. A thruline wattmeter, or equivalent monitor, is placed in the output line. The bias voltage and cooling air are applied to the cavity.

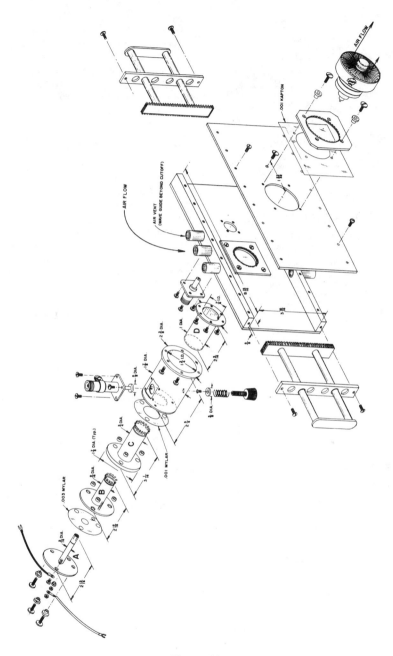

Figure 58

EXPLODED VIEW OF 920-MHz CAVITY

The input assembly is at left showing the linear filament chokes (A and B), the cathode line
(C) and the input tuning arrangement. The plate cavity is at the right with the two sliding
drawers for tuning (left) and loading (right). Anode collet assembly is at the right. Note that
output receptacle is on the bottom of the cavity, adjacent to the grid collet.

Figure 59

TOP VIEW OF CAVITY AMPLIFIER

The 3CX400U7 plugs into collets in the amplifier cavity. The anode of the tube projects through the top of the plate cavity and cooling air passes through the louvers. The plate bypass capacitor is held to the chassis by the circular insulating disc. The tuning and loading controls are at the end of the assembly. Air inlets are on the sides of the cavity.

Filament voltage of 6.3 volts is applied for 60 seconds, followed by anode voltage of 1500 volts, maximum. Plate current will read between 50 and 150 mA, depending on the bias. Grid current indication should be near zero. About 10 watts of r-f drive is applied and the plate current should rise to between 300 and 400 mA. There should be an indication of output power on the watt-meter.

The tuning and loading controls are now adjusted for maximum output, and both then adjusted until maximum output is achieved. The filament voltage is now dropped to 4.5 volts.

The next step is to adjust the input tuning and matching under full power conditions. The output coaxial probe capacitor (C_1) and the tuning control (C_2) are adjusted for minimum reflected power. These two adjustments are interlocking so they must be made alternately, adjusting for minimum power reflection. When this is achieved, the output tuning drawer should be reset for best power output.

CHAPTER TWENTY-THREE

Power Supplies

The *basic power supply* is an energy source which provides power to an electronic unit. The most common type of power supply changes ac to dc and maintains a constant voltage output within current limits. A basic supply is shown in figure 1. The ac voltage is applied to the primary winding of a transformer. The secondary winding provides an ac voltage of appropriate value, depending upon the dc voltage needed for operation of the equipment. The secondary ac voltage is rectified by diodes which pass current in

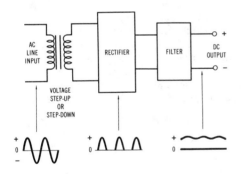

Figure 1

BASIC AC-TO-DC POWER SUPPLY

The basis for all electronic ac-to-dc power supplies. Other elements may be added to achieve special characteristics such as voltage or current regulation.

only one direction. Thus, only a portion o. the ac is passed during each cycle. This produces a pulsating dc voltage which must be filtered to smooth out the pulsations if a steady dc voltage is required. A slight amount of residual pulsation ("ripple") is usually present in the output.

23-1 Types of Power Supplies

The regulated dc supply maintains a steady dc output voltage regardless of changes in line voltage or load current. It does this by sensing the dc output voltage and automatically reacting to cancel out any detected change. The actual change in voltage can thus be reduced to extremely small values. There are two main classes of regulators; the series regulator and the switching regulator.

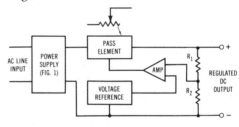

Figure 2

SERIES-REGULATOR DC POWER SUPPLY

An electronic variable resistance element is connected in series with the output line, and a voltage detector and stable voltage reference control the variable element to establish a stable condition. Voltage drop across the pass element is changed to maintain constant voltage at the output terminals.

The *series regulator* (figure 2) uses a dc supply as described and an electronic variable resistance (a tube or transistor) connected in series with the output line, a voltage detector-amplifier, and a stable voltage reference. The detector senses a voltage change at the junction of R_1 and R_2, compares it with the reference voltage, and adjusts the series resistance element accordingly to establish a stable condition. Some power is wasted as heat in the series element equal to the voltage drop across it times the current through it.

The *switching regulator* (figure 3) avoids the heat-dissipating series element by using an electronic switching element in its place.

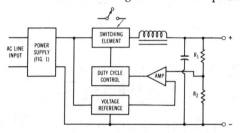

Figure 3

**SWITCHING-REGULATOR
DC POWER SUPPLY**

An electronic switching element in series with the output lead is opened and closed about 20,000 times per second. The switch is controlled by a detector amplifier to maintain constant dc output voltage in spite of line or load variations. A filter at the regulator output absorbs on-off pulsations.

The output of the sensing detector-amplifier controls an electronic switch that switches as high as 20 kHz. Increasing the amount of time the switch is closed increases the output voltage, decreasing the closed time decreases the voltage. The supply filter ab-

sorbs the on-off pulsations and provides a smooth dc output. Virtually no power is lost in the switch since its offers very little resistance to the flow of current when it is closed.

The *inverter power supply* (figure 4) has a dc source that drives switching transistors in a low-frequency oscillator circuit. An auxiliary winding of the power transformer is used to alternately switch the transistors on and off. The current flow in the transformer primary winding reverses its direction with each alternation, producing a squarewave ac voltage in the transformer secondary winding.

The operating frequency of the inverter is determined by the circuit constants. Trimming resistors and capacitors ensure that the circuit starts promptly and prevents generation of voltage spikes. (For additional information see Chapter 19).

23-2 The Primary Circuit

Electronic equipment, regardless of purpose, requires a primary power source of energy. Aside from portable equipment and small devices, the primary source of consumer power is the home electrical system which, in the United States, is nominally 120/240 volts, 60 Hz, in a 3-wire, grounded-neutral circuit. For mobile or portable equip-

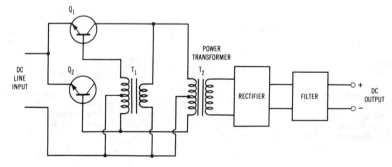

Figure 4

INVERTER POWER SUPPLY

The dc line drives a low-frequency oscillator composed of Q_1, Q_2, and T_1. Squarewave output voltage is impressed on power transformer T_2. The operating frequency of the squarewave is determined by circuit constants.

ment, the primary power source is usually a 6-, 12-, or 24-volt battery system.

The various dc voltage levels required for communication equipment are commonly supplied from the primary source via a transformer, rectifier, and filter network used in conjunction with a control and overload protection device.

In the case of vacuum tubes, the filament power can be either ac or dc and in some cases the primary power is rectified and applied directly to the high voltage circuits of the equipment, without the necessity of a voltage changing transformer.

Power-Line "Standards" A confusion of power-line voltages and frequencies, as well as a multiplicity of plugs and connectors exists throughout the world. In the United States and Canada the nominal design center for consumer equipment is 117 volts, 60 Hz. Voltages between 110 and 125 are commonly encountered. In many overseas countries, 220 or 240 volts at 50 Hz may be found. In addition, unique combinations, such as 137 volts at 42 Hz, or 110 volts at $16\frac{2}{3}$ Hz may exist as a result of special circumstances. Operation of equipment on one phase of a three-phase 240-volt power system calls for a design center of 208 volts.

Aside from the primary power complexity, an endless number of plug and receptacle designs harass the experimenter. Recently, the *National Electrical Manufacturers Association* in the United States has announced standards covering general-purpose receptacles designed for the consumer wiring system, based on a design center of 117 volts for the "120-volt" system and 234 volts for the "240-volt" system used in the majority of new homes.

A clear distinction is made in all specifications between *system ground* and *equipment ground*. The former, referred to as a grounded conductor, normally carries line current at ground potential. Terminals for system grounds are marked *W* and are color-coded *white*. Terminals for equipment grounds are marked *G* and are color-coded *green*. In this standard, the equipment ground carries current only during short circuit conditions.

A summary of some of the more common NEMA receptacle configurations, and other configurations still in popular use, are shown in figure 5. A complete chart covering all standard NEMA plugs and receptacles may be obtained for twenty-five cents from: *The Secretary, NEMA Wiring Device Section, 155 East 44th Street, New York, N.Y., 10017.*

Checking an Outlet with a Heavy Load To make sure that an outlet will stand the full load of the entire transmitter, plug in an electric heater rated at about 50 percent greater wattage than the power you expect to draw from the line. If the line voltage does not drop more than 5 volts (assuming a 120-volt line) under load and the wiring does not overheat, the wiring is adequate to supply the transmitter. About 800 watts total drain is the maximum that

SAFETY PRECAUTIONS

Voltages developed in some of the power supplies shown in this chapter are lethal. The supplies should be constructed in a cabinet or closed framework in such fashion that the components cannot be touched. All steps must be taken to prevent accidental contact with power leads of any voltage. Power leads from the supplies to the transmitting equipment should be run in high voltage cable.

Before any work is done on a power supply it is imperative that the supply be disconnected from the primary source and the filter capacitors discharged. A shorting stick may be used for this purpose. The stick is made by mounting a metal hook on one end of a long, dry stick of wood. A length of high voltage cable is run from the hook on the stick to a common ground point on the power supply. After the power is turned off, the shorting stick is used to short the filter capacitors and high voltage leads to insure that there is no voltage at these points.

All equipment should contain interlocks to open primary circuits of the power supply.

Always remember that high voltage can kill.

should be drawn from a 120-volt *lighting* outlet or circuit even though the standard baseboard outlet is rated at 15 amperes (1800 watts). For greater power, a separate pair of heavy conductors should be run right from the meter box. For a 2-kW PEP transmitter, the total drain is so great that a 240-volt "split" system ordinarily will be required. Most of the newer homes are wired with this system, as are homes utilizing electricity for cooking and heating.

With a three-wire system, be sure there is no fuse in the neutral wire at the fuse box. A neutral fuse is not required if both "hot" legs are fused, and, should a neutral fuse blow, there is a chance that damage to the radio transmitter will result.

Relay Control   Primary and secondary power circuits may be controlled by manually operated switches or remotely operated relays. A *relay* is an electrical switch operated by an independent electrical circuit. It permits a low voltage circuit to control a high voltage or current circuit by opening or closing appropriate contacts (figure 6). Because of construction requirements, most relays are double-throw with single or multiple poles. The simple control relay has one normal open and one normal closed position. When the relay is energized, the pole opens from the normally closed circuit before a contact is established with the normally open contact. Typically, a gen-

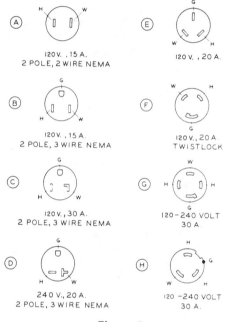

Figure 5
COMMON RECEPTACLE STANDARDS IN THE UNITED STATES

The front view of various common 120-volt and 240-volt standard receptacles is shown. 120-volt circuits have one wire (neutral) at about ground potential and the other wire (hot) above ground. The neutral wire (white W, with nickel screw terminal) is unfused while the hot wire (H, black, red, or blue with brass screw terminal) is fused. The switch should be in the hot line. The neutral is grounded at the distribution transformer and should not be grounded at any other point. Neutral is often referred to as system ground and is coded white. Equipment ground (G) is separately grounded at the electrical device and is coded green (circuits A, B, and C). 240-volt single phase receptacles are polarized so that 120-volt plugs cannot be used by error. Duplex (E) and Twistlock (F) are common industrial plugs, while the plugs of figures G and H are used with electric stoves, motors, air conditioners, etc.

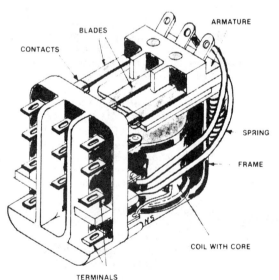

Figure 6
CONTROL RELAY

A relay is an electric switch operated by an independent electrical circuit. The relay blades are attached to a movable armature which is actuated by a magnetic coil. Normally open and normally closed contacts are mounted on a terminal frame insulated from the relay structure. The relay coil may be either ac or dc operated.

eral purpose power relay will close in 10 to 15 milliseconds and drop out in 5 to 10 milliseconds. Special, fast action relays are available for keying circuits and other rapid changeover systems.

The *contact rating* of a relay refers to the electrical limit permitted at the contacts. These are frequently stated as 2, 5, or 10 amperes at 120 volts, 0.8 power factor, or 28 volts dc, resistive load. If the relay is designed to handle a motor or other inductive load, the contact rating may be expressed in terms of horsepower; for example, ½ hp at 120 volts ac.

Many different mounting options are offered for a relay having the same electrical and mechanical characteristics. For example, the basic structure may have plug-in

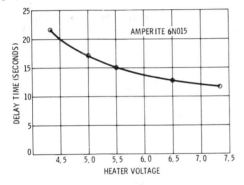

Figure 7

TYPICAL TIME DELAY OF THERMOSTATIC RELAY

Thermostatic relay is actuated by a heater and can be run on either ac or dc. Delay time is a function of heater voltage, as shown for this 6-volt model. Inexpensive delay relays are sealed in a glass bulb, making them impervious to moisture.

termination, plus a matching socket, or soldering lugs.

A thermostatic *time-delay relay* is commonly used to allow warmup time, or time for circuit stabilization after a primary circuit is energized. Compact, inexpensive delay relays provide a delay of 2 to 180 seconds and operate at various values of heater voltage (figure 7). Motor operated time-delay relays are used for high power equipment, or to achieve longer delay periods. Thermostatic relays have a recycle time of 3 to 7

seconds, and after the heater is disconnected, the contacts may remain closed for as long as 10 seconds, depending upon relay design.

Primary Circuit Transients The primary power source often contains transient voltages that could pose a damage to certain electronic equipment. High level switching of industrial loads or lightning strikes on a nearby power system can create primary transients as high as 5 kV on the ordinary 120 volt line. The average residential circuit receives more than one transient a day in excess of 200 volts and can expect at least one a year in excess of 1000 volts. Some ordinary home motor loads, such as sump pumps and oil burners, regularly introduce transients of over 1500 volts into residential circuits.

Though the power system's protection system limits the transient voltage at a suppressor built into the power network, reflections and other interactions may permit high crest voltages at other points in the system. Transients can couple secondary transients through a power transformer, not by the turns ratio, but by the transformer's often high value of primary to secondary capacitance, thus permitting a high voltage transient to be present in a low voltage circuit, regardless of the step-down effect of the transformer. In addition to primary circuit transients, large voltage peaks are often built up in a power supply when it is turned on or off. These transients are created by the release of energy stored in an inductor or capacitor passing through other inductors or capacitors. These peak voltages may be far in excess of the voltage rating of the components or the rectifier units, leading to arc over and eventual breakdown of insulation or components in the circuit.

An expensive solution to the transient problem is to ensure that the peak voltage rating of all components is higher than expected voltage transients. A more economical solution is to employ a *transient suppressor (varistor)* in the circuit to protect the components from voltage peaks. Various such units have been developed to provide transient protection, and most of the more modern communication equipment incorporates transient protection.

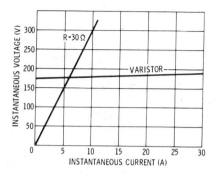

Figure 8

VOLT-AMPERE CHARACTERISTIC OF VARISTOR FOR 120-VOLT CIRCUIT

The voltage-current plot of a representative varistor shows that the device provides an almost constant voltage across the terminals over a wide range of currents. Standby power dissipation of 120-volt unit is about 0.5 watt. Curve of 30-ohm resistor is shown for comparison.

A transient suppressor is a nonlinear device that is voltage sensitive (figure 8). The higher the voltage across the suppressor the lower will be its resistance. The device is usually rated in terms of energy absorption for a single transient pulse and the voltage clamping ratio at which transient suppression becomes effective. For a 120 volt rms suppressor, a common clamping voltage is 170.

Nonlinear resistors, semiconductor devices, and spark gaps are commonly used as transient suppressors. One inexpensive suppressor consists of two zener diodes connected in series-opposition (figure 9). A simple RC

Figure 9

ZENER DIODE TRANSIENT SUPPRESSOR

network placed across the power line is also an effective transient suppressor (figure 10).

Step-Start Circuitry   When a large power supply is energized, the incoming current (*inrush* current) can be many times the steady state current until the supply reaches a state of equilibrium. The inrush current is composed of the current re-

quired to charge the filter capacitors in the power supply and, for large equipment, the

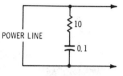

Figure 10

RESISTANCE-CAPACITANCE TRANSIENT SUPPRESSOR

heavy starting current required by power tubes during the short period when the filament temperature reaches operating level.

Inrush current may be limited by inclusion of a current-limiting resistor in the primary circuit which is shorted out after the time period required for the supply to reach a steady state condition (figure 11).

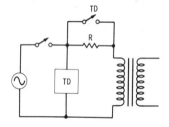

Figure 11

TIME-DELAY RELAY AND SURGE RESISTOR PROVIDE PROTECTION FROM INRUSH CURRENT

Filament Inrush Current   A cold thoriated-tungsten filament has about one-tenth the resistance of one that has reached operating temperature. The cold inrush current (over a period of a few cycles) can thus reach as high as ten times the normal value of filament current. It is good engineering practice to limit the inrush filament current to no more than twice the normal current value. Limiting may be done by the circuit of figure 11 or it may be accomplished by the use of a filament transformer that saturates at the high current level, thus causing a momentary reduction in filament voltage.

A high value of inrush current can crack the glass seals about the filament leads of a large power tube, or can cause the filament structure to warp as a result of the high magnetic field about it.

An alternative means of reducing filament inrush current is to bring the filament voltage up slowly by means of a variable-ratio autotransformer placed in the primary circuit of the filament transformer.

Interlock Protection, In order to protect the
Fuses, and operator from the high
Circuit Breakers voltages normally present in transmitting equipment, it is common practice to *interlock* the primary circuit in such a manner that turn-on is impossible until the interlock is activated (figure 12). When the interlock is broken, or incomplete, the primary circuit of the equipment cannot be completed. Door or cabinet interlocks are common devices that remove the high voltage when access is desired to the equipment. The interlock can also short the high voltage supply to ground to make sure that the filter capacitors in the supply are discharged.

Communications equipment must be protected against overload or improper tuning

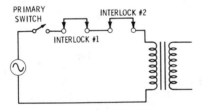

Figure 12

PRIMARY INTERLOCK CIRCUIT

Series-connected interlock switches prevent circuit from being activated until equipment doors are closed.

and the simplest form of protection is the *fuse*, a thermally operated link which blows when the current through it reaches a specified value. Most fuses are either fast action, medium-lag, or slow-blow (figure 13). The fast fuses are used to protect instruments and measuring devices, the medium action fuses are used for primary and secondary circuit protection and the slow-blow, or delayed action fuse, is for use in circuits having a high inrush current.

A fuse is normally capable of carrying a 10% overload indefinitely but will fail after a few thousand hours when operated at 100 percent of its rated load because of cyclic fatigue caused by mechanical stresses set up in the fuse element by current changes. Fuses loaded to about 50 percent of their rating will give a safety margin against cyclic failure and yet provide good protection for the equipment.

The *circuit breaker* is a mechanical switch that depends on the generation of heat to operate a bimetallic strip which trips the breaker mechanism. The thermal breaker, therefore, is a relatively slow acting device, opening the circuit after an overload period of 0.1 to 10 seconds, depending on design. A fast action, magnetic breaker can open an overload in as little as 10 ms.

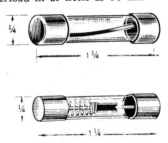

Figure 13

INSTRUMENT AND EQUIPMENT FUSES

Fast action and medium-lag fuses (top) can carry a 10-percent overload and will open at various time intervals under specific overload conditions. Typically, a 1-ampere fuse will open in 2 to 4 seconds at 200-percent overload. Slow-blow fuse (bottom) will open after one hour at 135-percent overload. Special instrument fuses will blow in milliseconds after overload.

Variable Ratio There are several types of
Autotransformers variable-ratio autotransformers available on the market. Of these, the most common are the *Variac* manufactured by the *General Radio Company,* and the *Powerstat* manufactured by the *Superior Electric Company* (figure 14). Both these types of variable-ratio transformers are excellently constructed and are available in a wide range of power capabilities. Each is capable of controlling the line voltage from zero to about 15 percent above

the nominal line voltage. The maximum power-output capability of these units is available only at approximately the nominal line voltage, and must be reduced to a maximum current limitation when the output voltage is somewhat above or below the input line voltage. This, however, is not an important limitation for this type of application since the output voltage seldom will be raised above the line voltage, and when the output voltage is reduced below the line voltage the input to the transmitter is reduced accordingly.

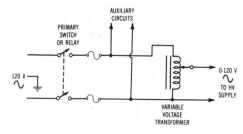

Figure 14

**VARIABLE RATIO
AUTOTRANSFORMER**

The variable ratio transformer permits the line voltage to the power supply to be reduced or otherwise adjusted. To reduce inrush current in the load, the circuit is turned on with the variable transformer set at low voltage. The voltage is then gradually increased to the proper operating potential.

23-3 Transmitter Control Methods

The majority of modern amateur transmitters are composed of an exciter unit and a separate power amplifier. Control circuits for the exciter, including VOX and time-delay circuitry, are incorporated in the exciter. The amplifier has its own power control circuitry which may be activated by the controls in the exciter, if convenient (figure 15). In this example, the high voltage supply is activated by a transmit-receive relay (K_2) controlled by the exciter circuitry.

Primary power for the amplifier is taken from a 240-volt line whose system ground (neutral) is separate from the ground circuitry of the amplifier. Power plug (P_1) can be a four pin type, such as shown in figure 1G. The supply is fused in all lines except the ground and neutral, and the "hot" legs pass through interlocks which remove the voltage if the power supply cabinet is opened.

A panel switch (*Power On*) energizes the primary relay (K_1) and also illuminates a green pilot lamp on the panel of the supply. The filament circuitry is now on, with filament transformer (T_2) activated through a time-delay circuit which reduces filament inrush current. Smaller tubes are connected to transformer (T_3), as they do not require this protection.

The supply is activated by relay K_2 which completes the primary circuit of transformer T_2. For best regulation this transformer has a 240-volt primary and is connected across the complete line. The relay and a red *Transmit* warning lamp are activated by the transmit-receive circuitry of the exciter, thus making changeover from receive to transmit automatic. The operating sequence is broken if a protective fuse blows or an interlock is opened.

Transmitter Power Control   It is convenient to be able to rapidly switch between a 2-kW PEP power input condition and 1-kW dc input for c-w operation. A linear amplifier adjusted for 2-kW PEP will show a very low level of efficiency when the drive level and antenna loading are adjusted for the 1-kW power level condition. The transition can be accomplished at high efficiency, however, by reducing the dc plate voltage of the amplifier when switching from the SSB to the c-w mode.

For example, a 2-kW PEP linear amplifier may be operating at a plate potential of 3 kV and a peak dc plate current of 666 mA. Power input is 2 kW PEP and power output is, typically, 1.2 kW, PEP. Efficiency is about 60 percent. Switching to c-w, the operator drops excitation and readjusts antenna loading to provide a dc input of 1 kW which corresponds to 3 kV at 333 mA. Unfortunately, amplifier efficiency will drop to about 30 percent under these conditions, providing a power output of 300 watts. Unless the plate tank circuit has sufficient range to provide the proper plate impedance

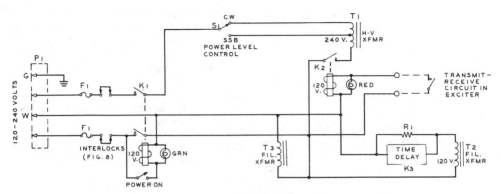

Figure 15

TRANSMITTER CONTROL CIRCUIT

Linear amplifier is controlled by transmit-receive circuit in exciter. The high-voltage supply is actuated by relay K_2. Primary power is taken from 240-Vac line whose system ground (neutral) is separate from the ground circuit of the amplifier. A step-start circuit composed of resistor R_1 and time delay relay K_3 reduces filament inrush current for the amplifier tube. Changeover from 2 kW to 1 kW is accomplished by a voltage tap on the primary of the power transformer.

for the 1 kW mode—and most simple pi- or pi-L networks do not—plate efficiency will drop badly.

If however, the plate potential under c-w operating conditions is dropped to about 65 percent of that employed in the SSB mode, plate efficiency will remain high in both conditions. For the above example, dropping the plate potential to about 2 kV and boosting the plate current to 500 mA will provide approximately the same degree of efficiency at the 1-kW dc power level as will the 3-kV potential and 666 mA peak plate current at the 2-kW PEP power level. Many manufactured linear amplifiers accomplish the SSB to c-w switchover by dropping the plate potential on the amplifier tubes in the manner described. This is easily accomplished by the use of a tapped primary or secondary winding on the plate power transformer.

Relay Sequence It is important that the antenna changeover relay be activated before r-f power flows through the relay contacts. Certain VOX or key-operated sequences do not provide this protection. As a result, the contacts of the antenna relay may be damaged from making and breaking the r-f current, or eventual damage may occur to the transmitting equipment because of repeated operation without r-f load during the periods of time

necessary for the antenna relay to close. The proper relay sequence can be achieved by actuating the antenna relay by the control system, then, in turn, actuating the transmitter by a separate set of control contacts

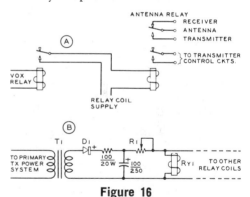

Figure 16

ANTENNA-RELAY CONTROL SYSTEM

A—Antenna relay should be actuated before r-f power flows through contacts. Extra set of contacts are used to control transmitter circuits after antenna relay closes. B—Ac relays may be operated from simple dc power supply to reduce hum and chatter. Transformer T_1 may be a 1:1 isolation transformer of 50 watts capacity, with D_1 a 1 ampere, 600 volt p.i.v. diode. Series resistor R_1 is adjusted to provide proper relay action and may be of the order of 500 to 5000 ohms, 50 watts. Additional relay coils may be placed in parallel across coil Ry_1. Relay may be energized by applying primary power (with due regard to time-lag in filter system) or by completing secondary circuit between resistor R_1 and relay coil.

on the antenna relay, as shown in figure 16A. In this manner, the antenna relay must be closed before r-f is applied to the contacts.

DC Relay Operation   Relays designed to operate from an ac source are often troublesome sources of audible hum and chatter. Cleaning the relay striker and pole pieces will alleviate this annoyance somewhat, but operation of the relay from a dc source will eliminate this difficulty. Ac relays may be operated without damage from a dc source capable of supplying a dc voltage equal to about 70 percent of the ac design voltage. Thus an 85-volt dc supply will be proper to operate 120-volt ac relays. A suitable supply for such service is shown in figure 16B.

23-4   Power-Supply Requirements

A power supply for a transmitter or for a unit of station equipment should be designed in such a manner that it is capable of delivering the required current at a specified voltage, that it has a degree of regulation consistent with the requirements of the application, that its ripple level at full current is sufficiently low for the load which will be fed, that its internal impedance is sufficiently low for the job, and that none of the components shall be overloaded with the type of operation contemplated.

The meeting of all the requirements of the previous paragraph is not always a straightforward and simple problem. In many cases compromises will be involved, particularly when the power supply is for an amateur station and a number of components already on hand must be fitted into the plan.

The power-supply requirements needed to establish the design of a satisfactory unit include the full-load output voltage; minimum, normal and peak current drain; the required voltage regulation; ripple voltage limit, and type of rectifier circuit to be used.

Once these requirements have been ascertained, the actual components for the supply may be selected. It is prudent, however, to design a supply in such a manner that it will have the greatest degree of flexibility; this will allow the supply to be used without change as a portion of new station equipment or as a bench supply to run experimental equipment.

Current-Rating Considerations   The *minimum current drain* which will be taken from a power supply will be, in most cases, merely the bleeder current. There are many cases where a particular power supply will always be used with a moderate or heavy load on it, but when the supply is a portion of a transmitter it is best to consider the minimum drain as that of the bleeder. The minimum current drain from a power supply is of importance since it, in conjunction with the nominal voltage of the supply, determines the minimum value of inductance which the input choke must have to keep the voltage from soaring when the external load is removed.

The *normal current rating* of a power supply usually is a round-number value chosen on the basis of the transformers and chokes on hand or available from the catalog of a reliable manufacturer. The current rating of a supply to feed a steady load such as a receiver, a speech amplifier, or a continuously operating r-f stage should be at least equal to the steady drain of the load. However, other considerations come into play in choosing the current rating for a keyed amplifier, an amplifier of SSB signals, or a class-B modulator. In the case of a supply which will feed an intermittent load such as these, the current ratings of the transformers and chokes may be *less* than the maximum current which will be taken; but the current ratings of the rectifier system to be used should be at least equal to the maximum current which will be taken. That is to say that 300-mA transformers may be used in the supply for an amplifier whose resting current is 20 mA but whose maximum current at peak signal will rise to 500 mA. However, the rectifier system should be capable of handling the full 500 mA.

The iron-core components of a power supply which feeds an intermittent load (such as demanded by an SSB transmitter) may be chosen on the basis of the current averaged over a period of several minutes, since it is the heating effect of the current which is of greatest importance in establishing the rating of such components. Since iron-core components have a relatively large amount of thermal inertia, the effect of an intermittent heavy current is offset to an extent by a resting period between words and syllables, or by key-up periods in the case of c-w transmission. However, the current rating of a rectifier tube is established by the magnitude of emission available from the filament of the tube, and the rating of a semiconductor rectifier is established by the maximum temperature limit of the rectifier element, both of which cannot be exceeded even for a short period of time or the rectifier will be damaged.

The above considerations are predicated, however, on the assumption that none of the iron-core components will become saturated due to the high level of intermittent current drain.

Voltage Regulation   Since the current drain of a power supply can vary over a large magnitude, it is important to determine what happens to the output voltage of the supply with regard to change in current. Power-supply regulation may be expressed in terms of *static* and *dynamic* regulation. Static regulation relates to the regulation under long-term conditions of change in load whereas dynamic regulation relates to short-term changes in load conditions. Regulation is expressed as a change in output voltage with respect to load:

$$\text{Percent Regulation} = \frac{(E_1 - E_2) \times 100}{E_2}$$

where,
    E_1 is no-load voltage,
    E_2 is full-load voltage.

Thus static regulation concerns itself with the "on" and "off" voltages of the power supply and dynamic regulation concerns itself with syllabic or keyed fluctuations in load. Static regulation is expressed in terms of average voltages and currents, whereas dynamic regulation takes into account in-

stantaneous voltage variations caused by peak currents, or currents caused by undesired transient oscillations in the filter section of the power supply. In particular, c-w and SSB transmissions having a high peak-to-quiescent ratio of current drain are affected by poor dynamic regulation in the power system.

Examples of static and dynamic regulation are shown in figure 17. In example A, the no-load power-supply voltage is 1000 and the full-load voltage is 875. Static regulation is therefore 14.3 percent. If an oscilloscope is used to examine the supply voltage during the first fractions of a second when the full load is applied, the instantaneous voltage follows the erratic plot shown in curve A of figure 17. The complex pattern of voltage fluctuations, or transients, are related to resonant frequencies present in the power-supply filter network and are of sufficient magnitude to distort the waveform of c-w signals, or to appreciably increase intermodulation distortion and alter the first syllable of speech in an SSB system. Proper design of the filter system can reduce dynamic voltage fluctuations to a minimum and, at the same time, greatly improve the static regulation of the power supply.

Static and dynamic regulation values of about 10 percent or so are considered to be limits of good design practice in amateur transmitting equipment, as illustrated by voltage curve B in figure 17.

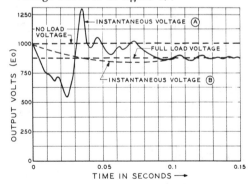

Figure 17

STATIC AND DYNAMIC REGULATION

A—Dynamic regulation illustrates voltage peaks caused by transient oscillations in filter network.
B—Static regulation is expressed in terms of no-load and full-load currents and voltages.

Ripple Voltage The alternating component of the output voltage of a dc power supply is termed the *ripple voltage*. It is superimposed on the dc voltage, and the effectiveness of the filter system can be expressed in terms of the ratio of the rms value of the ripple voltage to the dc output voltage of the supply. Good design practice calls for a ripple voltage of less than 5 percent of the supply voltage for SSB and c-w amplifier service, and less than 0.01 percent of the supply voltage for oscillators and low-level speech amplifier stages.

Ripple frequency is related to the number of pulsations per second in the output of the filter system. A full-wave rectifier, having two pulses of 60 Hz, for example, produces a 120-Hz ripple wave. A simple capacitive filter will reduce 120-Hz ripple as shown in figure 18. Ripple is an inverse ratio with capacitance, so doubling the capacitance will halve the ripple.

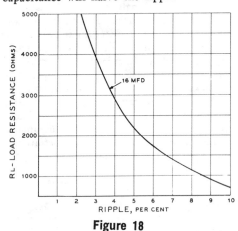

Figure 18

120-Hz RIPPLE ON 16-μF CAPACITOR AS FUNCTION OF LOAD RESISTANCE

Ripple Filter Circuits The percentage of ripple found in representative LC filter circuits is shown in figure 19. The approximate ripple percentage for filter components may be calculated with the aid of the following formulas, assuming the power line frequency to be 60 Hz and the use of a full-wave or full-wave-bridge rectifier circuit. The ripple at the output of

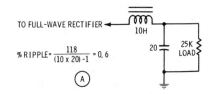

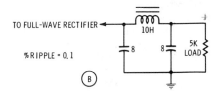

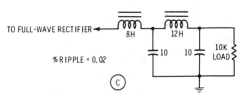

Figure 19

RIPPLE VOLTAGE FOR VARIOUS FILTERS

A—Single section filter with choke input.
B—Capacitance input filter.
C—Two section filter with choke input.

the first section of a two-section choke input filter is:

$$\text{Percent Ripple} = \frac{118}{(L \times C) - 1}$$

where,

L is the input choke inductance in henrys (at the operating current to be used),

C is the capacitance which follows the choke, expressed in microfarads.

This percentage is multiplied by the filter reduction factor of the following section of filter. This reduction factor is determined through the use of the following formula:

$$\text{Filter reduction factor} = \frac{1.76}{LC - 1}$$

where LC again is the product of the inductance and capacitance of the filter section. The reduction factor will turn out to be a decimal value, which is then multiplied by the percentage ripple obtained from the use of the preceding formula.

Resistance-Capacitance Filters   In many applications where current drain is relatively small, so that the voltage drop across the series resistor would not be excessive, a filter system made up of resistors and capacitors only may be used to advantage. In the normal case, where the reactance of the shunting capacitor is very much smaller than the resistance of the load fed by the filter system, the ripple reduction per section is equal to $1/(2\pi RC)$. In terms of the 120-Hz ripple from a full-wave rectifier the ripple-reduction factor becomes: $1.33/RC$ where R is expressed in thousands of ohms and C in microfarads. For 60-Hz ripple the expression is: $2.66/RC$ with R and C in the same quantities as above.

Filter System Resonance   The inductance of the filter choke in an LC filter network is dependent to an extent on the current drawn through it. At some values of inductance, it is possible for a 60-Hz or 120-Hz resonant circuit to be set up if the filter capacitance value is low. Filter resonance imposes a heavy peak load on the rectifier system and diodes or mercury-vapor rectifier can be damaged by such undesired currents.

A 120-Hz resonance is achieved when the product of inductance and capacitance is 1.77. Thus, a 1-μF capacitor and a 1.77-henry choke will resonate at 120 Hz. The LC product for resonance at 60 Hz is about 7.1. This latter value may occur when a 2-μF capacitor is used with a 3.55-henry choke, for example. The LC products of 1.77 and 7.1 should be avoided to prevent resonance effects, which can result in destructive transient voltages in the power-supply system.

Back-EMF   It is possible to place the filter choke in the B-minus lead of the power supply, reducing the voltage potential appearing from choke winding to ground. However, the *back-emf* of a good choke is quite high and can develop a dangerous potential from center tap to ground

on the secondary winding of the plate transformer. If the transformer is not designed to withstand this potential, it is possible to break down the insulation at this point.

23-5   Power-Supply Components

The usual components which make up a power supply, in addition to rectifiers which have already been discussed, are filter capacitors, bleeder resistors, transformers, and chokes. These components normally will be purchased especially for the intended application, taking into consideration the factors discussed earlier in this chapter.

Filter Capacitors   There are two principal types of filter capacitors: (1) paper-dielectric type, (2) electrolytic type. *Paper capacitors* consist of two strips of metal foil separated by several layers of special paper. Some types of paper capacitors are wax-impregnated, but the better ones, especially the high-voltage types, are oil-impregnated and oil-filled. Some capacitors are rated both for *flash* test and normal operating voltages; the latter is the important rating and is the maximum voltage which the capacitor should be required to withstand in service.

The capacitor across the rectifier circuit in a capacitor-input filter should have a working-voltage rating equal at *least* to 1.41 times the rms voltage output of the rectifier. The remaining capacitors may be rated more nearly in accordance with the dc voltage.

The *electrolytic capacitor* consists of two aluminum electrodes in contact with a conducting film which acts as an *electrolyte*. A very thin film of oxide is formed on the surface of one electrode, called the *anode*. This film of oxide acts as the dielectric. The electrolytic capacitor must be correctly connected in the circuit so that the anode is always at a positive potential with respect to the electrolyte, the latter actually serving as the other electrode (plate) of the capacitor. A reversal of the polarity for any length of time will ruin the capacitor.

The high capacitance of electrolytic capacitors results from the thinness of the film which is formed on the plates. The maximum voltage that can be safely impressed across the average electrolytic filter capacitor is between 450 and 600 volts; the working voltage is usually rated at 450. When electrolytic capacitors are used in filter circuits of high-voltage supplies, the capacitors should be connected in series. The positive terminal of one capacitor must connect to the negative terminal of the other, in the same manner as dry batteries are connected in series.

Electrolytic capacitors can be greatly reduced in size by the use of etched aluminum foil for the anode. This greatly increases the surface area, and the dielectric film covering it, but raises the power factor slightly. For this reason, ultramidget electrolytic capacitors ordinarily should not be used at full rated dc voltage when a high ac component is present as would be the case for the input capacitor in capacitor-input filter.

Bleeder Resistors A heavy-duty resistor should be connected across the output of a filter in order to draw some load current at all time. This resistor avoids

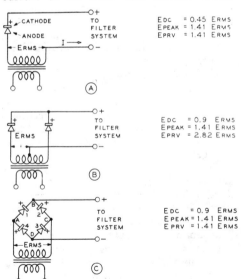

Figure 20

COMMON RECTIFIER CIRCUITS

A—Half-wave rectifier. Ripple is 121%.
B—Full-wave rectifier. Ripple is 48%.
C—Bridge rectifier. Ripple is 48%.

soaring of the voltage at no load when swinging-choke input is used, and also provides a means for discharging the filter capacitors when no external circuit load is connected to the filter. This *bleeder* resistor should normally draw approximately 10 percent of the full load current.

The power dissipated in the bleeder resistor can be calculated by dividing the square of the dc voltage by the resistance. This power is dissipated in the form of heat, and, if the resistor is not in a well-ventilated position, the wattage rating should be higher than the actual wattage being dissipated. High-voltage, high-capacitance filter capacitors can hold a dangerous charge if not bled off, and wirewound resistors occasionally open up without warning. Hence it is wise to place carbon resistors in series across the regular wirewound bleeder.

Transformers Power transformers and filament transformers normally will give no trouble over a period of many years if purchased from a reputable manufacturer, and if given a reasonable amount of care. Transformers must be kept dry; even a small amount of moisture in a high-voltage unit will cause quick failure. A transformer which is operated continuously, within its ratings, seldom will give trouble from moisture, since an economically designed transformer operates at a moderate temperature rise above the temperature of the surrounding air. But an unsealed transfomer which is inactive for an appreciable period of time in a highly humid location can absorb enough moisture to cause early failure.

Filter Choke Coils Filter inductors consist of a coil of wire wound on a laminated iron core. The size of wire is determined by the amount of direct current which is to flow through the choke coil. This direct current magnetizes the core and reduces the inductance of the choke coil; therefore, filter choke coils of the *smoothing* type are built with an air gap of a small fraction of an inch in the iron core, for the purpose of preventing saturation when maximum current flows through the coil winding. The "air gap" is usually in the form of a piece of fiber inserted between the ends of the laminations. The air gap reduces the initial inductance of the choke coil, but keeps it at a higher value under maximum load conditions. The coil must have a great many more turns for the same initial inductance when an air gap is used.

The dc resistance of any filter choke should be as low as practical for a specified value of inductance. Smaller filter chokes, such as those used in radio receivers, usually have an inductance of from 6 to 15 henrys, and a dc resistance of from 200 to 400 ohms. A high dc resistance will reduce the output voltage, due to the voltage drop across each choke coil. Large filter choke coils for radio transmitters and class-B amplifiers usually have less than 100 ohms dc resistance.

23-6 Rectification Circuits

There are a large variety of rectifier circuits suitable for use in power supplies. Figure 20 shows the three most common circuits used in supplies for amateur equipment.

Half-Wave Rectifier   A *half-wave rectifier* (figure 20A) passes current in one direction but not in the other. During one-half of an applied ac cycle when the anode of the rectifier is positive with respect to the cathode the rectifier is in a state of conduction and current flows through the rectifier. During the other half of the cycle, when the anode is negative with respect to the cathode, the rectifier does not conduct and no current flows in the circuit. The output current, therefore, is of a pulsating nature which can be smoothed into direct current by means of an appropriate filter circuit. The output of a half-wave rectifier is zero during one-half of each ac cycle; this makes it difficult to filter the output properly and also to secure good voltage regulation for varying loads. The *peak inverse voltage* with a resistive or inductive load is equal to the peak ac voltage of the transformer ($1.41 \times E_{\text{rms}}$) and is equal to twice the peak ac voltage with a capacitive load.

Full-Wave Rectifier   A *full-wave rectifier* (figure 20B) consists of a pair of half-wave rectifiers working on opposite halves of the ac cycle, connected in such a manner that each portion of the rectified wave is combined in the output circuit, as shown in figure 21. A transformer with a center-tapped secondary is required. The transformer delivers ac to each anode of each rectifier element; one anode being positive at any instant during which the other anode is negative. The center point of

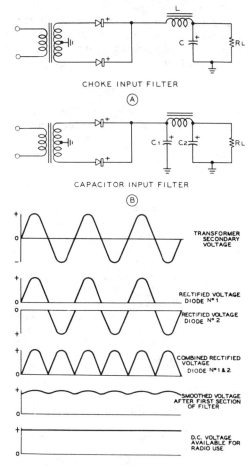

CHOKE INPUT FILTER

(A)

CAPACITOR INPUT FILTER

(B)

TRANSFORMER SECONDARY VOLTAGE

RECTIFIED VOLTAGE DIODE N° 1

RECTIFIED VOLTAGE DIODE N° 2

COMBINED RECTIFIED VOLTAGE DIODE N° 1 & 2

SMOOTHED VOLTAGE AFTER FIRST SECTION OF FILTER

D.C. VOLTAGE AVAILABLE FOR RADIO USE

Figure 21

RECTIFICATION AND FILTER ACTION

Showing transformer secondary voltage, the rectified output of each diode, the combined output of the rectifiers, the smoothed voltage after the choke-input filter, and the dc output voltage of the capacitor input filter.

the high-voltage winding of the transformer is taken as the negative (B-minus) connection.

The cathodes of the rectifier units are always positive in polarity with respect to the anode of this type of circuit, and the output current pulsates 120 times per second for a 60-Hz supply. The peak output voltage is 1.4 times the rms transformer voltage and the inverse voltage across each rectifier unit is 2.8 times the rms voltage of the transformer (as measured across one half of the secondary winding). For a given value of ripple, the amount of filter required for a full-wave rectifier is half that required for a half-wave rectifier, since the ripple frequency of the former is twice that of the latter.

Bridge Rectifier A *bridge rectifier* (figure 20C) has four rectifier elements operated from a single ac-source. During one half-cycle of the applied ac voltage, *point A* becomes positive with respect to *point C* and conduction takes place through rectifiers 3 and 1 when *point C* is positive with respect to *point A*. On one half of the cycle, therefore, rectifiers 4 and 2 are in series with the output circuit and on the other half-cycle, rectifiers 3 and 1 are in series with the circuit. The bridge circuit is a full-wave system since current flows during both halves of a cycle of the alternating current.

One advantage of a bridge-rectifier connection over a full-wave, two-rectifier system is that with a given transformer voltage the bridge circuit produces a voltage output nearly twice that of the conventional full-wave circuit. In addition, the peak inverse voltage across any rectifier unit is only 1.4 times the rms transformer voltage. Maximum output voltage into an inductive or resistive load is about 0.9 times the rms transformer voltage.

The center point of the high-voltage winding of the bridge transformer is not at ground potential. Many transformers having a center-tapped high voltage winding are not designed for bridge service and insulation between this point and the transformer core is inadequate. Lack of insulation at this point does no harm in a full-wave circuit when the center tap is grounded, but may cause breakdown when the transformer is used in bridge configuration.

Rectifier Circuits *Choke input* is used in many filter systems because it gives good utilization of both rectifier and power-transformer capability (figure 21A). In addition, it provides much better voltage regulation than does a *capacitor input* system. A minimum value of choke inductance exists, and this critical value is equal to $R_L/1000$, where R_L is the load resistance. Inductance above the critical value will limit the no-load output voltage to above the average value (E_{dc}) in contrast to the capacitor-input filter circuit (figure 21B) wherein the no-load output voltage may rise as high as the peak value of the transformer voltage. The capacitor-input filter, at full load, provides a dc output voltage that is usually slightly above the rms voltage of the transformer.

When capacitor input is used, consideration must be given to the peak value of the ac voltage impressed on the filter capacitor, which usually runs equal to the peak transformer voltage $(1.41\,E_{rms})$. The input capacitor, therefore, must have a voltage rating high enough to withstand the peak voltage if breakdown is to be avoided.

Special Single-Phase Rectification Circuits Figure 22 shows three circuits which may prove valuable when it is desired to obtain more than one output voltage from one plate transformer or where some special combination of voltages is required. Figure 22A shows a common method for obtaining full voltage and half voltage from a bridge rectification circuit. With this type of circuit, separate input chokes and filter systems are used on both output voltages. If a transformer designed for use with a full-wave rectifier is used in this circuit, the current drain from the full-voltage tap is doubled and added to the drain from the half-voltage tap to determine whether the rating of the transformer is being exceeded.

The circuit of Figure 22B is similar to that of 22A except that the negative return point is shifted to the centertap of the trans-

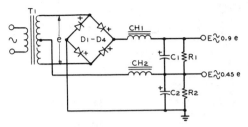

(A) HALF- AND FULL-VOLTAGE BRIDGE SUPPLY

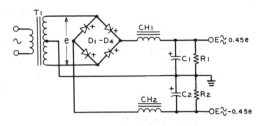

(B) DUAL POSITIVE AND NEGATIVE SUPPLY

Figure 22

SINGLE-PHASE RECTIFICATION CIRCUITS

A description of each of these circuits is given in the accompanying text.

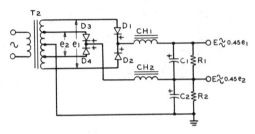

(C) TWO-VOLTAGE FULL-WAVE SUPPLY

former. The same voltages are derived from circuit B as from circuit A with the exception that the voltage derived from one full-wave rectifier is negative with respect to the ground point. The choke for this circuit is placed in the positive lead.

Illustration C shows conventional full-wave rectifiers as used with a dual-voltage transformer. Each set of transformer taps has its own rectifier set. The output voltages are proportional to the taps on the transformer secondary winding.

Polyphase Rectification Circuits   It is usual practice in commercial equipment installations when the power drain from a plate supply is to be greater than about one kilowatt to use a *polyphase rectification system.* Such power supplies offer better transformer utilization, less ripple output and better power factor in the load placed on the ac line. However, such systems require a source of three-phase (or two-phase with Scott connection) energy. Several of the more common polyphase rectification circuits with their significant characteristics are shown in figure 23. The increase in ripple frequency and decrease in percentage of ripple is apparent from the

figures given in figure 23. The circuit of figure 23C gives the best transformer utilization as does the bridge circuit in the single-phase connection. The circuit has the further advantage that there is no average dc flow in the transformers, so that three single-phase transformers may be used. A tap at half voltage may be taken at the junction of the star transformers, but there will be dc flow in the transformer secondaries with the power-supply center tap in use. The circuit of figure 23A has the disadvantage that there is an average dc flow in each of the windings.

Peak Inverse Voltage and Peak Current   In an ac circuit, the maximum peak voltage or current is $\sqrt{2}$, or 1.41 times that indicated by the ac meters in the circuit. The meters read the *root mean square* (rms) values, which are the peak values divided by 1.41 for a sine wave.

If a potential of 1000 rms volts is obtained from a high voltage secondary winding of a transformer, there will be 1410 volts peak potential from the rectifier anode to ground. In a single-phase supply the rectifier has this voltage impressed on it, either positively when the current flows or

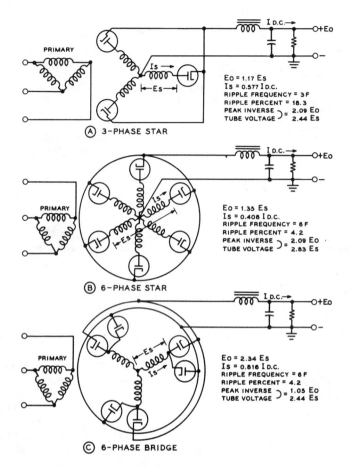

(A) 3-PHASE STAR

$Eo = 1.17\ Es$
$Is = 0.577\ I\ D.C.$
RIPPLE FREQUENCY = 3F
RIPPLE PERCENT = 18.3
PEAK INVERSE $\}= \begin{array}{l} 2.09\ Eo \\ 2.44\ Es \end{array}$
TUBE VOLTAGE

(B) 6-PHASE STAR

$Eo = 1.35\ Es$
$Is = 0.408\ I\ D.C.$
RIPPLE FREQUENCY = 6F
RIPPLE PERCENT = 4.2
PEAK INVERSE $\}= \begin{array}{l} 2.09\ Eo \\ 2.83\ Es \end{array}$
TUBE VOLTAGE

(C) 6-PHASE BRIDGE

$Eo = 2.34\ Es$
$Is = 0.816\ I\ D.C.$
RIPPLE FREQUENCY = 6F
RIPPLE PERCENT = 4.2
PEAK INVERSE $\}= \begin{array}{l} 1.05\ Eo \\ 2.44\ Es \end{array}$
TUBE VOLTAGE

Figure 23

COMMON POLYPHASE-RECTIFICATION CIRCUITS

These circuits are used when polyphase power is available for the plate supply of a high-power transmitter. The circuit at B is also called a three-phase full-wave rectification system. The circuits are described in the accompanying text.

"inverse" when the current is blocked on the other half-cycle. The *peak inverse voltage* which the device will stand safely is used as a rating for rectifiers. At higher voltages the rectifier is liable to arc back, thereby destroying or damaging it. The relations between peak inverse voltage, total transformer voltage, and filter output voltage depend on the characteristics of the filter and rectifier circuits (whether full- or half-wave, bridge, single-phase or polyphase, etc.)

Rectifiers are also rated in terms of *peak current*. The actual direct load current which can be drawn from a given rectifier depends on the type of filter circuit. A full-wave rectifier with capacitor input passes a peak current several times the direct load current.

In a filter with choke input, the peak current is not much greater than the load current if the inductance of the choke is fairly high (assuming full-wave rectification).

Mercury-Vapor Rectifier Tubes   The inexpensive *mercury-vapor* type of rectifier tube is sometimes used in the high-voltage plate supplies of amateur and commercial transmitters. When new or long-unused tubes are first placed in service, the filaments should be operated at normal temperature for approximately twenty minutes before plate voltage is applied, in order to remove all traces of mercury from the cathode and to clear any mercury deposits from the top of the envelope. After this preliminary warmup with a new tube, plate voltage may be applied within 20 to 30 seconds after the time the filaments are turned on, each time the power supply is used. If plate voltage should be applied before the filament is brought to full temperature, active material may be knocked from the oxide-coated filament and the life of the tube will be greatly shortened.

Small r-f chokes must sometimes be connected in series with the plate leads of mercury-vapor rectifier tubes in order to prevent the generation of radio-frequency hash.

Voltage Multiplying Circuits   Practical *voltage multiplying circuits* can be built up using silicon rectifiers and filter capacitors. The rectifier delivers alternating half-cycles of energy to the filter capacitor and successive rectifier/capacitor stages may be connected to provide very high values of voltage from a low voltage source.

A common voltage multiplier is the half-wave *series amplifier* circuit (figure 24). On one half the ac cycle capacitor C_1 is charged to nearly the peak source voltage through rectifier D_1. On the opposite half of the cycle, rectifier D_2 conducts and capacitor C_2 is charged to nearly twice the source peak voltage. At the same time, the next rectifier conducts and with the charge in C_2 as the source, C_3 is charged to the peak input voltage, and so on. Ripple in the output circuit is governed by:

$$E_r = \pm \frac{I_{load}}{16\,fC}\,(N^2 + N/2)$$

Ripple thus increases with the square power of the number of stages.

Regulation is governed by:

$$E_R = \frac{I_{load}}{12fC}\,(N^3 + 9/4\,N^2 + N/2)$$

The N^3 term indicates a practical limitation as to the number of stages in a practical circuit in that the internal impedance of the multiplier rises very fast.

The half-wave *parallel multiplier* circuit is shown in figure 25. The operation of the parallel multiplier follows that of the series design with the exception that each capacitor in the string is charged up to higher voltages instead of each capacitor having the same potential across it as in the series configuration. Ripple in the output circuit is independent of the number of stages and

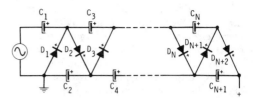

Figure 24

HALF-WAVE SERIES MULTIPLIER CIRCUIT

A single stage consists of one capacitor and one rectifier unit and provides a dc voltage at no load nearly equal to the peak ac voltage. The internal impedance of the multiplier is quite high and rises as the third power of the number of stages. Variations of this circuit are common in power supplies for electronic equipment.

is a function of capacitance, load current and frequency:

$$E_r \cong \frac{I_{load}}{fC}$$

Regulation is proportional to:

$$E_r \cong \frac{I_{load}}{fC}\,(N)$$

which indicates better regulation than provided by the series circuit, as N increases linearly instead of by the third power as in the series mode.

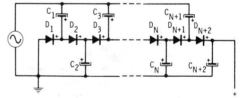

Figure 25

HALF-WAVE PARALLEL MULTIPLIER CIRCUIT

The operation of the parallel multiplier follows that of the series design with the exception that each capacitor in the string is charged up to higher voltages instead of each capacitor having the same potential across it as in the series configuration. Ripple is independent of the number of stages and is a function of capacitance, load current, and frequency. Regulation is better than that of the series design and proportional directly to the number of stages.

Series and parallel multipliers provide practical voltage multiplier circuits up to about twelve times the input voltage.

Two half-wave rectifiers may be connected in reverse sequence to provide a full-wave circuit, as shown in figure 26. The ripple frequency is 120 Hz instead of 60 Hz, as with the half-wave configuration. Various simple half- and full-wave multipliers used in communication equipment are shown in figure 27.

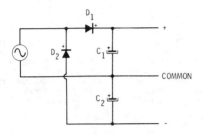

Figure 26

FULL-WAVE VOLTAGE DOUBLER

The half-wave rectifiers may be connected in reverse sequence to provide full-wave rectification. Diode D_1 charges capacitor C_1 to the peak transformer voltage on one-half the cycle and diode D_2 charges capacitor C_2 on the other half-cycle. The output is taken across the two capacitors connected in series. Ripple frequency is 120 Hz. This "building block" may be made into a voltage quadrupler or sextupler.

Diode Noise　The silicon diode which is widely used in these circuits does not conduct until the applied forward potential exceeds the threshold voltage, which is about 0.5 volt. At this voltage the diode conducts abruptly, creating a steep wavefront, capable of generating radio-frequency interference. The interference is often eliminated if a transient suppression capacitor is placed across the diode (figure 28). In some cases, especially with the use of controlled-avalanche diodes, the capacitor is omitted and the *white-noise* interference generated by the diode may be found as an annoying "rush" on the sidebands of the transmitted signal, or as an annoying noise in the receiver. Suppression capacitors and additional lead filtering in the power supply may be required to eliminate the interference created by the abrupt conduction characteristic of the diode rectifier.

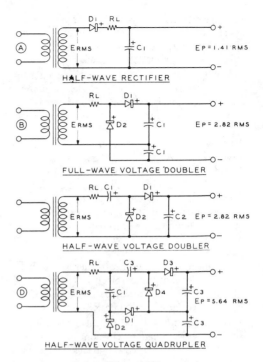

Figure 27

VOLTAGE-MULTIPLYING CIRCUITS

Voltage-multiplying circuits can be built up using silicon diode rectifiers or vacuum diodes. The basic "building block" is the half-wave rectifier (A). Capacitor C_1 is rated for twice the rms voltage of the transformer, and for a receiver supply, should be about 150 μF. Capacitor C_2 in the voltage doubler circuit of (C) is rated for four times the rms voltage of the transformer. Capacitor C_3 in the quadrupler circuit of (D) is rated for three times the rms voltage of the transformer.

23-7　Series Diode Operation

Series diode operation is commonly used when the peak-inverse voltage of the source is greater than the maximum PIV rating of a single diode. For proper series operation, it is important that the PIV be equally divided among the individual diodes. If it is not, one or more of the diodes in the stack will be subjected to a PIV greater than its maximum rating and, as a result, may be destroyed. As most failures of this type result in a shorted junction, the PIV on the remaining diodes in the stack is raised, mak-

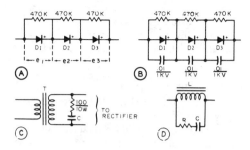

Figure 28

PROTECTION CIRCUITS FOR SEMICONDUCTOR POWER SUPPLIES

A—Peak inverse voltage should be distributed equally between series-connected diodes. If diodes do not have matched reverse characteristics, shunt resistors should be placed across the diodes.

B—Series-connected diodes are protected against high-voltage switching transients by shunt capacitors which equalize and absorb the transients uniformly along the stack.

C—Transient suppressor placed across the secondary of the high-voltage transformer protects diode stack from transients often found on the ac power line or created by abrupt change in the magnetizing current of the power transformer.

D—Suppressor network across series filter choke absorbs portion of energy released when magnetic field of choke collapses, thus preventing the surge current from destroying the diode stack.

ing each diode subject to a greater value of PIV. Failure of a single diode in a stack can lead to a "domino effect" which will destroy the remaining diodes if care is not taken to prevent this disaster. Forced voltage distribution in a stack is necessary when the individual diodes vary appreciably in reverse characteristics. To equalize the steady-state voltage division, shunt resistors may be placed across the diodes in a stack (figure 28A). The maximum value of the shunt resistor to achieve a 10-percent voltage balance, or better is:

$$Shunt\ resistance = \frac{PIV}{2 \times Max.\ Reverse\ Current}$$

Six-hundred-volt PIV diodes, for example, having a reverse current of 0.3 mA at the maximum PIV require a shunt resistance of 1 megohm, or less.

Transient Protection Diodes must be protected from voltage transients which often are many times greater than the permissible peak-inverse voltage. Transients can be caused by dc switching at the load, by transformer switching, or by shock excitation of LC circuits in the power supply or load. Shunt capacitors placed across the diodes will equalize and absorb the transients uniformly along the stack (figure 28B). The shunt capacitor should have at least 100 times the capacitance of the diode junction, and capacitance values of 0.01 μF or greater are commonly found in diode stacks used in equipment designed for amateur service.

Controlled avalanche diodes having matched zener characteristics at the avalanche point usually do not require RC shunt suppressors, reducing power-supply cost and increasing overall reliability of the rectifier circuit.

It should be noted, however, that leaving out the RC suppressors brings back the problems of "white noise," mentioned previously.

In high-voltage stacks, it is prudent to provide transient protection in the form of an RC suppressor placed across the secondary of the power transformer (figure 28C). The suppressor provides a low-impedance path for high-voltage transients often found on ac power lines, or generated by an abrupt change in the magnetizing current of the power transformer as a result of switching primary voltage or the load. The approximate value of the surge capacitor in such a network is:

$$Capacitance\ (\mu F) = \frac{15 \times E \times I}{e^2}$$

where,

E is the dc supply voltage,

I is the maximum output current of the supply in amperes,

e is the rms voltage of the transformer secondary winding.

High-voltage transients can also be caused by series filter chokes subject to abrupt load changes. An RC suppressor network placed across the winding of the choke can absorb a portion of the energy released when the magnetic field of the choke collapses, thus preventing the current surge from destroying the diode stack (figure 28D). The approximate value of the transient capacitor is:

$$Capacitance\ (\mu F) = \frac{L \times I^2}{10 \times E^2}$$

where,

L is the maximum choke inductance (henrys),

I is the maximum current passing through the choke (amperes),

E is the maximum dc supply voltage (kV)

The resistance in series with the capacitor should equal the load impedance placed across the supply.

23-8 Solid-State Supplies for SSB

Shown in figure 29 are three semiconductor power supplies. *Circuit A* provides 530 volts (balanced to ground) at 0.15 ampere. If the supply is isolated from ground by a 1:1 transformer of 250 watts capacity *point A* may be grounded and *point B* will provide half-voltage. *Circuit B* is a half-wave tripler that delivers 325 volts at 0.45 ampere. In this circuit, one side of the power line is common to the negative side of the output. *Circuit C* is a 900-watt, 0.5 ampere supply composed of two voltage doublers supplied from a "distribution" transformer having dual 120/240-volt windings.

Power Supply Rating for SSB Service The *duty cycle* (ratio of duration of maximum power output to total "on" time) of a power supply in SSB and c-w service is much smaller than that of a supply used for a-m equipment. While the power supply must be capable of supplying peak power equal to the PEP input of the SSB equipment for a short duration, the average power demanded by SSB voice gear over a period of time usually runs about one-half or less of the PEP requirement. Then, too, the intervals between words in SSB operation provide

periods of low duty, just as the spaces in c-w transmission allow the power supply to "rest" during a transmission. Generally speaking, the average power capability of a power supply designed for *intermittent voice service* (IVS) can be as low as 25 percent of the PEP level. C-w requirements run somewhat higher than this, the average c-w power level running close to 50 percent of the peak level for short transmissions. Relatively small power transformers of modest capability may be used for intermittent voice and c-w service at a worthwhile saving in weight and cost. The power capability of a transformer may be judged by its weight, as shown in the graph of figure 30. It must be remembered that the use of alc or voice compression in SSB service raises the duty, thus reducing the advantage of the IVS power rating. The IVS rating is difficult to apply to very small power transformers, since the dc resistance of the transformer windings tends to degrade the voltage regulation to a point where the IVS rating is meaningless. Intelligent use of the IVS rating in choosing a power transformer, stacked silicon rectifiers, and "computer" type electrolytic capacitors can permit the design and construction of inexpensive, lightweight high-voltage power supplies suitable for SSB and c-w service.

The Design of IVS Power Supplies The low duty of SSB and c-w modes can be used to advantage in the design of high-voltage power supplies for these services.

The Power Transformer—Relatively low-voltage transformers may be used in voltage-doubler service to provide a kilowatt or two of peak power at potentials ranging from one to three thousand volts. Most suitable power transformers are rated for commercial service and the IVS rating must be determined by experiment. Figure 30 shows a relationship between various services as determined by extensive tests performed on typical transformers. The data illustrates the relationship between transformer weight and power capability. Transformer weight excludes weight of the case and mounting fixtures. Thus, a plate transformer weighing about 17 pounds that is rated for 400

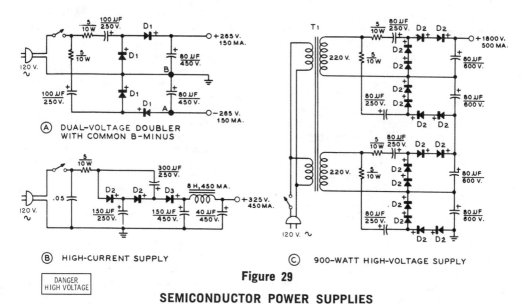

Figure 29

SEMICONDUCTOR POWER SUPPLIES

A—Voltage-quadrupler circuit. If point "A" is taken as ground instead of point "B," supply will deliver 530 volts at 150 mA from 120-volt ac line. Supply is "hot" to line.

B—Voltage tripler delivers 325 volts at 450 mA. Supply is "hot" to line.

C—900-watt supply for sideband service may be made from two voltage quadruplers working in series from inexpensive "distribution-type" transformer. Supply features good dynamic voltage regulation.

D_1, D_2, D_3—1N4005. Use .01 μF capacitor and 100K resistor across each diode.

T_1—Power distribution transformer, used backwards. 240/460 primary, 120/240 secondary, 0.75 KVA. Chicago PCB-24750.

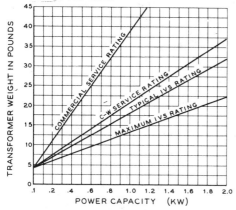

Figure 30

INTERMITTENT VOICE SERVICE IN SSB PERMITS LARGE PEAK POWER TO BE DRAWN FROM POWER TRANSFORMER

Peak-to-average ratio of nearly four to one may be achieved with maximum IVS rating. Power capacity of transformer may be determined from weight.

watts commercial or industrial service should have an 800-watt peak capacity for c-w service and a 950-watt peak capacity for intermittent SSB service. A transformer having a so-called "two-kilowatt PEP" rating for sideband may weigh as little as 22 pounds, according to this graph.

Not shown in the graph is the effect of amplifier idling (standby) current taken from the supply, or the effect of bleeder current. Both currents impose an extra, continuous drain on the power transformer and quickly degrade the IVS rating of the transformer. Accordingly, the IVS curves of figure 30 are limited to the bleeder current required by the equalizing resistors for a series capacitor filter and assume that the idling plate current of the amplifier is cut to only a few milliamperes by the use of a VOX-controlled cathode bias system. If the idling plate current of the amplifier assumes an appreciable fraction of the peak plate current, the power capability of the supply decreases to that given for c-w service.

Most small power transformers work reliably with the center tap of the secondary winding above ground potential. Some of the larger transformers, however, are designed to have the center tap grounded and lack sufficient insulation at this point to permit their use in either a bridge or voltage doubling configuration. The only way of determining if the center-tap insulation is sufficient is to use the transformer and see if the insulation breaks down at this point! It is wise to ground the frame of the transformer so that if breakdown occurs, the frame of the transformer does not assume the potential of the secondary winding and thus present a shock hazard to the operator.

The Silicon Rectifier—A bewildering variety of "TV-type" silicon rectifiers exists and new types are being added daily. Generally speaking, 600-volt PIV rectifiers, having an average rectified current rating of 1 ampere at an ambient temperature of 75°C with a maximum single-cycle surge-current rating of 15 amperes or better are suitable for use in the power supplies described in this section. Typical rectifiers are packaged in the *top-hat* configuration as well as the epoxy-encapsulated assembly and either type costs less than a dollar per unit. In addition, *potted* stacks utilizing controlled-avalanche rectifiers are available at a cost less than that of building a complete RC stack of diodes. The silicon rectifier, if properly used, is rarely the limiting factor in the design of steady-state IVS power supplies, provided proper

transient protection is incorporated in the supply.

The Filter Capacitor—Compact, "computer"-type aluminum-foil electrolytic capacitors combine high capacitance per unit of volume with moderate working voltage at a low price. Capacitors of this type can withstand short-interval voltage surges of 15 percent over their dc working voltage. In a stack, the capacitors should be protected by voltage-equalizing resistors, as shown in the power supplies in this section. The capacitors are sheathed in a *Mylar* jacket and may be mounted on the chassis or adjacent to each other without additional insulation between the units. The stack may be taped and mounted to a metal chassis with a metal clamp, as is done in some of the units described here.

Inrush Current Protection — When the power supply is first turned on, the filter capacitors are discharged and present a near short circuit to the power transfomer and rectifier stack. The charging current of a high-capacitance stack may exceed the maximum peak-recurrent current rating of the rectifiers for several cycles, thus damaging the diodes. Charging current is limited only by the series impedance of the power-supply circuit which consists mainly of the dc circuit resistance (primarily the resistance of the secondary winding of the power transformer) plus the

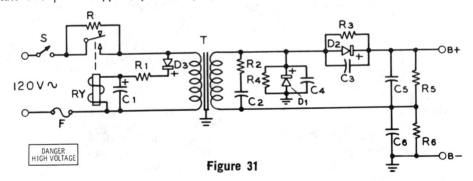

Figure 31

INRUSH CURRENT PROTECTION
FOR POWER SUPPLY

Charging current of capacitor filter may be limited by series impedance of the power supply. in voltage-doubler circuit shown here, primary resistor R limits inrush current to within the capability of the diodes. Limiting resistor is shorted out after sufficient time has elapsed to partially charge the filter capacitors. Delay time of 0.5 second is usually sufficient. R_1-C_1 combination determines time delay. Secondary surge suppression (R_2-C_2) is used, and shunt RC equalizing networks are employed across each diode stack. Filter capacitors (C_5, C_6) are "computer-grade" electrolytic capacitors in series with 10K, 10-watt wirewound resistor placed across each capacitor.

leakage reactance of the transformer. Transformers having high secondary resistance and sufficient leakage reactance usually limit the inrush current so that additional inrush protection is unnecessary. This is not the case with larger transformers having low secondary resistance and low leakage reactance. To be on the safe side, in any case, it is good practice to limit inrush current to well within the capability of the diode stack. A current-limiting circuit is shown in figure 31 which can be added at little expense to any power supply. The current-limiting resistor (R) is initially in the circuit when the power supply is turned on, but is shorted out by the relay RY after a sufficient time has elapsed to partially charge the filter capacitors of the power supply. The relay coil is in a simple time-delay circuit composed of R_1-C_1. The delay may be adjusted by varying the capacitance value, and need only be about one-half second or so. Surplus 24-volt dc relays used in dynamotor starting circuits work well in this device, as they have large low-resistance contacts and reasonable coil resistance (250 ohms or so).

Practical An IVS voltage-doubler power
IVS Supplies supply may be designed with the aid of figures 30 and 32. A typical doubler circuit, such as shown in figure 31, is to be used. The full-wave voltage doubler is preferred over the half-wave type, as the former charges the filter capacitors in parallel and discharges them in series to obtain a higher dc voltage than the peak voltage of the secondary winding of the power transformer. This saves transformer weight and expense.

Referring to figure 31, filter capacitors C_5 and C_6 are charged on alternate half cycles, but since the capacitors are in series across the load, the ripple frequency has twice the line frequency.

A second advantage of the full-wave doubler over the half-wave type is that the former tends to be self-protecting against switching transients. One diode stack is always in a conducting mode, regardless of the polarity of a transient, and the transient is therefore discharged into the filter-capacitor stack.

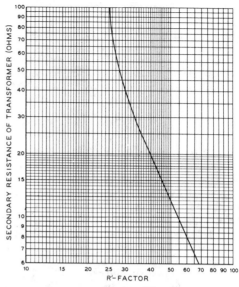

Figure 32

R' FACTOR GRAPH FOR IVS POWER SUPPLIES

The full load dc voltage of an IVS-rated voltage-doubler power supply may be determined with the aid of this graph. The secondary resistance of the transformer is measured and the R' Factor is found. For example, a transformer having a secondary resistance of 20 ohms has an R' Factor of about 40. The factor is used in the formula to calculate the full load dc voltage of the power supply. For use with bridge circuits, the R' Factor derived here should be divided by 2.5 before being used in the formula.

The filter-capacitor stack is rated for the peak no-load voltage (plus a safety factor), while the diode rectifiers must be able to withstand twice the peak no-load voltage (plus a safety factor). Good engineering practice calls for the dc *working voltage of each portion* of the capacitor stack to be equal to the peak ac voltage of the power transformer (1.41 $\times$ rms secondary voltage) plus 15 percent safety factor.

The R' Factor—The ac secondary voltage, secondary resistance, circuit reactance, and IVS capability of a transformer will determine its excellence in voltage-doubler service. The end effect of these parameters may be expressed by an empirical *R' factor* as shown in figure 32. As an example, assume a power transformer is at hand weighing 25 pounds, with a secondary winding of

840 volts (rms) and a dc secondary resistance of 8 ohms. The IVS rating of this transformer (from figure 27) is about 1.5 kW, PEP, or more. The appropriate dc no-load voltage of an IVS supply making use of this unit in voltage-doubler service, such as the circuit of figure 31, is:

$$E_{\text{NO LOAD}} = 2.81 \times e$$

where,
    e is the rms secondary voltage.

For this transformer, then, the no-load dc supply voltage is about 2360 volts. The full load voltage will be somewhat less than this value. For a maximum power capability of 1.5 kW, a full-load current of about 0.75 ampere is required if the full load dc voltage is in the vicinity of 2000. This is a realistic figure, so a "target" full-load voltage of 2000 is hopefully chosen.

The projected full-load voltage for a doubler-type supply may be determined with the aid of the R' factor and is calculated from:

$$E_{\text{LOAD}} = E_{\text{NO LOAD}} - R' \, (I \times R)$$

where,
    R' is determined from figure 32,
    I is the full current in amperes.
    R is the secondary resistance of the transformer.

For this example, R' is about 60 for the secondary resistance of 8 ohms, and the full-load dc voltage of the supply is found to be just about 2000.

The peak rectified voltage across the complete filter-capacitor stack is equal to the no-load dc voltage and is 2360 volts. Six 450-volt "computer"-type 240-μF electrolytic capacitors in series provide a 40-μF effective capacitor, with a working voltage of 2700 (peak voltage rating of 3000), a sufficient margin for safety. Each capacitor is shunted with two 100K, 2-watt resistors in parallel.

The total PIV for the diode stack is twice the peak rectified voltage and is 4720 volts. A 100-percent safety factor is recommended for the complete stack, whose PIV should thus be about 9440 volts. The number of individual diodes in a suitable stack is:

$$\text{Number of diodes} = \frac{11.2 \times \textit{rms voltage}}{\textit{Diode PIV}}$$

For this example, 600-volt PIV rectifiers are chosen and 16 are required, eight in each half of the stack.

The charging current of the capacitor stack may be safely ignored if the power supply is energized through a series primary resistor (R) such as shown in figure 31. One-ampere diodes having a single-cycle surge-current rating of 15 to 30 amperes are recommended for general use. The diffused silicon rectifiers (1N3195 and 1N-4005, for example) have a single-cycle surge-current rating of 30 amperes.

Capacitor Filters   Power supplies for SSB service whose current requirements have a large peak-to-average ratio often make use of capacitor filters (figure 30). This simple circuit eliminates the resonant transients that are often found in LC filter systems and, if the capacitance is sufficiently large, provides adequate voltage regulation. In the case of a 2-kW PEP supply (2500 volts at 0.8 ampere) the load resistance is 3100 ohms and the required capacitance for 5-percent regulation is 55 μF. Dynamic regulation of this degree is satisfactory for SSB and c-w service, as well as for amplitude modulation. As dis-

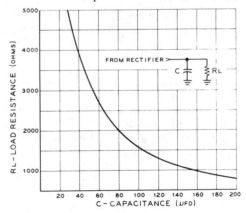

Figure 33

CAPACITOR FILTER

Capacitor filter is often used for SSB linear-amplifier power supplies. For 5-percent regulation, chart shows capacitance required for various values of dc plate-load resistance.

cussed earlier, the rectifier and power transformer must be protected from the inrush charging current of the filter capacitor.

23-9   A 1-Kilowatt IVS Power Supply

Shown in figures 34 and 35 is a typical 1-kilowatt IVS power supply designed from the above data. This supply is based on a 40-percent duty cycle and may be used for c-w service at 1-kilowatt level, or up to 1200 watts PEP or so for SSB service. The regulation of the supply is shown in the graph (figure 35), and the unit is capable of delivering 2300 volts at 0.5 ampere in IVS operation. The no-load voltage rises to 2750. The power supply is suitable for running a single 3-500Z at maximum rating, or it may be used for a pair of 8873 or 4CX250B tubes at the kilowatt level. A transformer having less secondary resistance and slightly less secondary voltage would provide improved voltage regulation. The 840-volt transformer having an 8-ohm secondary winding discussed earlier would be ideal in this application.

The power supply is constructed on a steel amplifier foundation chassis and dust cover. The diode stack is mounted on a per-forated phenolic board under the chassis. The electrolytic capacitors are taped together and held in position atop the chassis by a clamp cut from an aluminum sheet. The interior of the clamp is lined with a piece of plastic material salvaged from a package of frozen vegetables. The voltage-equalizing resistors are wired across the terminals of the capacitors. Normally, it takes 10 seconds or so to fully discharge the filter capacitors when no external load is connected to the supply. It is recommended that the supply be discharged with a 1000-ohm, 100-watt resistor before any work is done on the unit. Power-supply components and all terminals should be well protected against accidental contact. The voltage delivered by this supply is lethal and the filter capacitors hold a considerable charge for a surprising length of time. This is the price one pays for an intermittent-duty design, and care should be exercised in the use of this equipment.

To reduce the standby current and power consumption, it is recommended that cathode bias be applied to the linear amplifier stage shown in various designs in this Handbook. During transmission, the cathode resistor may be shorted out by contacts of the VOX relay, restoring the stage to proper operation.

Figure 34

COMPACT ONE-KILOWATT IVS SUPPLY FOR SSB AND C-W SERVICE

This power supply delivers 2250 volts at 500 mA for SSB operation and 2400 volts at 400 mA for c-w operation. The supply is constructed on a covered foundation unit measuring 12″ × 7″ × 9″ high (Bud CA-1751). The electrolytic capacitors are held in positon by a bracket cut from aluminum sheet. Primary power receptacle, power switch, and neon pilot light are on the front apron of the chassis, with primary fuse and Millen high-voltage connector on the rear apron. High-voltage diode stack is mounted beneath the chassis on a phenolic board.

Using the alternative 1100-volt transformer, the supply delivers 2600 volts at a c-w rating of 380 mA. Peak IVS voice rating is 500 mA. (1.25 kW, PEP). No-load voltage is about 3100, and eight electrolytic capacitors are required in the stack instead of six.

23-10 A 2-Kilowatt PEP Supply for SSB

The power supply described in this section is designed for the maximum power rating for amateur service. It is capable of 1.2 kilowatts power for c-w (50 percent duty cycle) and 2 kilowatts IVS for SSB service.

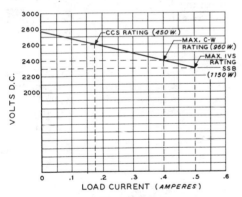

Figure 35

REGULATION CURVE OF ONE KILOWATT IVS SUPPLY

The power supply uses the circuit of figure 28. Primary surge resistor (R) is 5 ohms, 50 watts. Secondary surge-voltage resistor (R₂) is 200 ohms, 10 watts. Surge capacitor (C₂) is .02 μF, 3 KV (Aerovox P89-M). Sixteen type 1N2071 (600-volt PIV) diodes are used in an assembly such as shown in figures 36 and 37. The diode shunt capacitors are .01-μF, 600-volt ceramic discs, and the shunt resistors are 470K, ½-watt units. Six 450-volt (working), 240-μF filter capacitors are used in series, each capacitor shunted with two 100K, 2-watt resistors in parallel. The time delay relay (RY) has a 24-volt dc coil with a resistance of about 280 ohms (Potter-Brumfield PR5-DY). Contacts are rated at 25 amperes. Delay time is about 0.5 second and is determined primarily by the time constant of R₁-C₁. Suggested values are 800 μF (50 working volts) for C₁ and 600 ohms, 10 watts for R₁. Diode D₃ may be a 1N2070. The power transformer shown is a surplus unit having a 120/240-volt primary and a 960-volt secondary. The transformer weight is 18 pounds and it has an IVS rating of 1.2 kW.

The supply is ideally suited for a grounded-grid amplifier using a single 3-1000Z, 4-1000A, or a pair of 3-500Z's. Regulation of the supply is shown in figure 36. A voltmeter is incorporated in the supply to monitor the plate voltage at all times. The supply makes use of the circuit of figure 31. Twenty 600-volt PIV diodes are used in

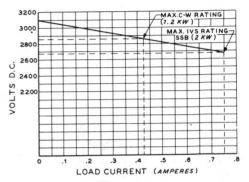

Figure 36

REGULATION OF THE 2-kW SUPPLY

The power supply uses the circuit of figure 28. Surge components are as given, except that the surge capacitor (C₂) has a rating of 5 kV. Twenty type 1N2071 (600-volt PIV) diodes are used in an assembly similar to that shown in figures 36 and 37. Eight 240 μF, 450-working-volt (500-volt peak) capacitors are used to provide 30 μF effective capacitance. Two 100K, 2-watt resistors are shunted across each capacitor. Time-delay circuit components are as suggested in figure 28. The transformer used has a 120/240 volt primary and an 1100-volt secondary, with an ICAS rating of 1.2 kW.

the rectifier stack to provide a total PIV of 12 kV, which allows an ample safety factor. Eight 240-μF, 450-volt capacitors are used in the filter stack to provide 30-μF effective capacitance at 3600 volts working voltage. The voltage across the "bottom" capacitor in the stack is monitored by a 0-to-1 dc milliammeter recalibrated 0 to 4 kV and which is used with a series multiplier to provide a 0 to 5000-volt full-scale indication. A 0-to-1 dc ammeter is placed in series with the negative lead to the high-voltage terminal strip.

The supply is built on a steel amplifier foundation chassis in the same style as the 1-kW supply described previously. All safety precautions outlined earlier should be observed with this supply.

23-11   IVS Bridge-Rectifier Supplies

The bridge-rectifier circuit is somewhat more efficient than the full-wave circuit in that the former provides more direct current per unit of rms transformer current for a given load than does the full-wave circuit. Since there are two rectifiers in opposite arms of the bridge in the conducting mode when the ac voltage is at its peak value, the remaining two rectifiers are back-biased to the peak value of the ac voltage. Thus the bridge-rectifier circuit requires only half the PIV rating for the rectifiers as compared to a center-tap full-wave rectifier. The latter circuit applies the sum of the peak ac voltage plus the stored capacitor voltage to one rectifier arm in the maximum inverse-voltage condition.

A 500-Watt IVS Bridge Power Supply   Shown in figure 37 is a 500-watt bridge power supply designed around an inexpensive "TV-replacement" type power transformer. The secondary winding is 1200 volts center-tapped at a current

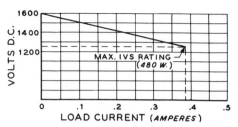

Figure 38

VOLTAGE-REGULATION CURVE OF 500-WATT BRIDGE POWER SUPPLY

rating of 200 mA. The weight of the transformer is 8 pounds, and the maximum IVS rating is about 500 watts or so. Secondary resistance is 100 ohms. Used in bridge service, the transformer makes practical an inexpensive power supply providing about 1250 volts at an IVS peak current rating of 380 mA. The no-load voltage is about 1600. For c-w use, the current rating is 225 mA at 1400 volts (about 300 watts). Maximum PIV is nearly 1700 volts so each arm of the bridge must withstand this value. Allowing a 100-percent safety factor requires 3400 volts PIV per arm, which may be made up of six 600-volt PIV diodes in series with an appropriate RC network across each diode. The diode assembly is constructed on two

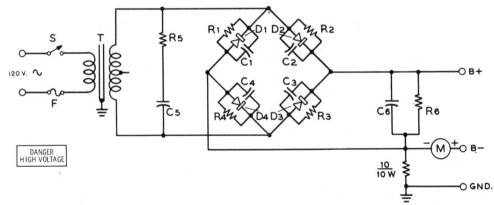

Figure 37

SCHEMATIC OF 500-WATT IVS BRIDGE POWER SUPPLY

Diode package (C_1-D_1-R_1 etc.) is composed of six each: 1N2071 diode in parallel with .01 μF, 600-volt ceramic capacitor and a 470K, ½-watt resistor. Each bridge arm requires six packages, made as shown in figures 36 and 37. The secondary voltage-surge network (C_5-R_5) is a 100-ohm, 10-watt resistor in series with a .02 μF, 3 kV capacitor (Aerovox P89-M). The power transformer has a 1200-volt center tapped 200-ma rating. The filter stack uses four 120-μF, 450-volt electrolytic capacitors in series, with 10K, 10-watt resistors across each capacitor. Meter (M) is a 0-500 dc milliammeter. A 10-ampere fuse (F) is used. Transformer core is grounded as a safety measure.

phenolic boards, one of which is shown in figures 36 and 37. A total of 24 rectifiers are required. Four 120-μF, 450-volt electrolytic capacitors in series provide 30 μF at a working voltage of 1800. The negative of the supply is above ground by virtue of the 10-ohm, 10-watt resistor which permits plate-current metering in the negative power lead while the supply and amplifier remain at the same ground potential.

This supply is designed for use with two 811A's in grounded-grid service. The tubes are biased to plate-current cutoff in standby mode by a cathode resistor which is shorted out by contacts on the push-to-talk or VOX circuitry. The power supply is built in an enclosed amplifier cabinet, similar to the one shown in figure 34. The B-plus lead is made of a length of RG-8/U coaxial cable, used in conjunction with a high-voltage coaxial connector.

23-12 A Heavy-Duty Primary Supply

This husky power supply provides a nominal 12 volts dc at a maximum continuous current of 10 amperes. It is useful as a shop supply to test mobile gear, as a battery charger, and as a general-purpose low-voltage power pack. The supply is unregulated and depends solely on the single-section filter for ripple reduction. Regulation is quite good at a current drain over one ampere, as seen in figure 41. The output voltage is controlled by the primary *powerstat*. To alert the user to the unloaded supply voltage (which may rise as high as 30 volts when the primary voltage is high) a meter protection and "alert" circuit is added. The red lamp is lit when more than 20 volts is present at the output terminals of the supply. Below 20 volts, the zener diode is nonconducting. Above 20 volts, the 10-volt zener conducts and the current through it turns the NPN transistor on and lights the warning indicator.

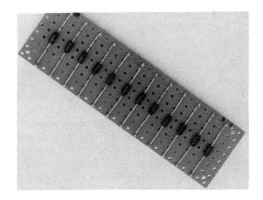

Figure 39

ASSEMBLY OF HIGH-VOLTAGE DIODE STACK

Inexpensive "TV-type" diodes may be connected in series to provide a high value of peak-inverse voltage. Shown here are twelve type-1N2071 diodes mounted on a Vectorbord (64AA32 cut to size). The diodes are soldered to Vector terminals (T9.6) mounted in the prepunched holes in the phenolic board. A pair of long-nose pliers should be used as a heat sink when soldering the diode leads. Grasp the diode lead between the diode body and the joint, permitting the pliers to absorb the soldering heat.

Figure 40

REAR VIEW OF HIGH-VOLTAGE DIODE STACK

The shunt capacitors and resistors are mounted on the rear of the phenolic board. Each diode-resistor-capacitor package has an individual pair of mounting terminals, which are jumpered together to connect the diodes in series. This arrangement provides greatest available heat sink for the components. The assembly is mounted an inch or so away from the chassis by means of 4-40 machine screws and ceramic insulators placed in corners of the board.

23-13 Regulated Power Supplies

Zener diodes or *voltage-regulator tubes* are commonly used to regulate power supplies to discrete voltages. Electronic voltage regulators have been developed that will handle higher voltage and current variations than the tube and diode devices are capable of handling. The electronic circuits, moreover, may be varied over a wide range of output voltage.

Electronic voltage regulators, in the main, are based on feedback circuits, such as discussed in Chapter 8, Section 7 whereby an error signal is passed through the feedback loop in such a manner as to cause an adjustment to reduce the value of the error signal.

Special integrated circuits have been developed for voltage-regulator service such as the LM300 and the μA-723. The IC regulator provides the gain required for the feedback loop and an auxiliary power transistor passes the major portion of the regular current. The μA-723 and the improved LM305 are shown as series positive regulators with built-in current limiting in figure 43A-B. A negative regulator using an LM304 is shown in figure 43C.

A positive regulator circuit capable of handling several hundred milliamperes (if properly heat-sinked) is shown in figure 44. No external pass transistor is required. This IC regulator is designed for floating regulation and can be powered by a small secondary 25-volt supply that "floats," such as shown in figure 45. In this configuration, the IC never has the main supply voltage across it and the only semiconductor that must stand-off the main supply voltage is the series pass transistor (usually a Darlington Pair). In this manner, the MC1466 may be used to regulate any voltage, high or low, and it also allows the output voltage to be varied from zero to maximum.

Three Terminal IC Regulators A number of three-terminal IC regulators having fixed output voltages for the more commonly used circuit supply busses are available. The National *LM-309* was the first of these, providing +5 volts at up to 1A dc for DTL and TTL logic IC supplies. More recently, both *National Semiconductor* and *Fairchild Semiconductor* have expanded the range of output voltages available in their LM-340 and μA-7800 families. In addition, *negative output* three-

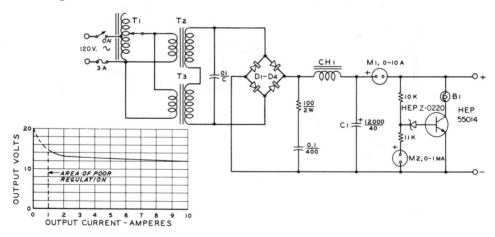

Figure 41

12-VOLT, 10-AMPERE GENERAL PURPOSE PRIMARY SUPPLY

B_1—1 amp, 28 volts. Chicago #327
C—12,000 μF, 40 volts. Sprague 123G040BC
CH_1—.03 Henry, 10-ampere. Triad C-49U
D_1-D_4—Two 1N3209 and two 1N3209R. Use two

Thermalloy heatsinks, 6500B-2
T_1—Powerstat, 200 watts, Superior 10B
T_2, T_3—11 volts, 10 amperes. Stancor P-3020
Meters: Weston model 301

terminal regulators are available in the *LM-320* and *μA-7900* series. The total ensemble of three–terminal regulators is shown in

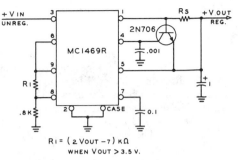

$$R_1 = (2\,V_{OUT} - 7)\,k\Omega$$
WHEN $V_{OUT} > 3.5\,V.$

Figure 44

MEDIUM CURRENT IC REGULATOR FOR POSITIVE VOLTAGE

Heat sinked MC1469R provides regulated current for voltages above 3.5 volts. No external pass transistor is trequired.

Figure 42

PRIMARY POWER SUPPLY

Handy to test mobile equipment, charge batteries or run surplus equipment, this supply provides 12 volts at 10 amperes with good regulation. Over-voltage lamp for meter protection is included.

Tables 1 through 6. The circuitry for these regulators is shown in figure 46. Note that the case of all the *positive* regulators is normally grounded for both heat sink and common electrical connection, but is never electrically grounded with the negative regulators.

Another essential requirement for stability when using these devices is that an input capacitor must be used on the positive regulators and both an input and output capacitor are required on the negative regulators. These capacitors serve much the same function as does the compensating capacitor on some operational amplifiers. The input capacitor requirement can be waived if the filter capacitor of the rectifier that provides unregulated dc to the three terminal regulator is closer than two inches (wire length) from the regulator input pin.

For greater flexibility, variable output three-and four-terminal regulators are available. The Fairchild *μA-78MG* and *μA-79MG* are positive and negative regulators capable of carrying or regulating current up to about 500 mA. The larger versions, capable

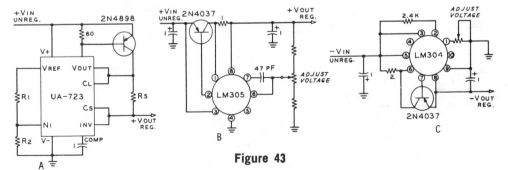

Figure 43

IC REGULATED POWER SUPPLIES

A—μA-723 integrated circuit provides gain for feedback loop to 2N4898 pass transistor for series positive regulator. B—LM305 and 2N4037 provides simple adjustable positive voltage regulator. C—LM304 and 2N4037 serve as adjustable negative voltage regulator.

TABLE 1.
Positive Three-terminal Regulators

Manufacturer	2.6 v	5 v	6 v	8 v	10 v	12 v	15 v	18 v	20 v	24 v	28 v
FAIRCHILD	—	μA 7805	μA 7806	μA 7808	—	μA 7812	μA 7815	μA 7818	—	μA 7824	—
LAMBDA	—	LAS 1505	LAS 1506	LAS 1508	LAS 1510	LAS 1512	LAS 1515	LAS 1518	LAS 1520	LAS 1524	LAS 1528
MOTOROLA	—	MC 7805	MC 7806	MC 7808	—	MC 7812	MC 7815	MC 7818	MC 7820	MC 7824	—
MOTOROLA-HEP	—	C 6110P	C 6111P	C 6112P	—	C 6113P	C 6114P	C 6115P	—	C 6116P	—
NATIONAL	—	LM 340-5	LM 340-6	LM 340-8	—	LM 340-12	LM 340-15	LM 340-18	—	LM 340-24	—
NATIONAL	—	LM 7805	LM 7806	LM 7808	—	LM 7812	LM 7815	LM 7818	—	LM 7824	—
RAYTHEON	—	RC 7805	RC 7806	RC 7808	—	RC 7812	RC 7815	RC 7818	—	RC 7824	—
SIGNETICS	—	μA 7805	μA 7806	μA 7808	—	μA 7812	μA 7815	μA 7818	—	μA 7824	—
SILICON GENERAL	—	SG 7805	SG 7806	SG 7808	—	SG 7812	SG 7815	SG 7818	—	SG 7824	—

RADIO HANDBOOK

TABLE 2.
Medium Current, Positive, Three-terminal Regulators

Manufacturer	2.6 v	5 v	6 v	8 v	10 v	12 v	15 v	18 v	20 v	24 v	28 v
FAIRCHILD	—	μA 78M05	μA 78M06	μA 78M08	—	μA 78M12	μA 78M15	—	μA 78M20	μA 78M24	—
MOTOROLA	—	MC 78M05	MC 78M06	MC 78M08	—	MC 78M12	MC 78M15	MC 78M18	MC 78M20	MC 78M24	—
NATIONAL	—	LM 341-5	LM 341-6	LM 341-8	—	LM 341-12	LM 341-15	LM 341-18	—	LM 341-24	—
SIGNETICS	—	μA 78M05	μA 78M06	μA 78M08	—	μA 78M12	μA 78M15	μA 78M18	μA 78M20	μA 78M24	—
MOTOROLA	—	MC 7705	MC 7706	MC 7708	—	MC 7712	MC 7715	MC 7718	MC 7720	MC 7724	—
TELEDYNE	—	78M05	78M06	78M08	—	78M12	78M15	—	78M20	78M24	—

TABLE 3.
Low Current, Positive, Three-terminal Regulator

Manufacturer	2.6 v	5 v	6 v	8 v	10 v	12 v	15 v	18 v	20 v	24 v	28 v
TELEDYNE	μA 78L026	—	—	—	—	829	830	—	—	—	—
FAIRCHILD	—	μA 78L05	μA 78L062	μA 78L82	—	μA 78L12	μA 78L15	—	—	—	—
MOTOROLA	—	MC 78L05	—	MC 78L08	—	MC 78L12	MC 78L15	MC 78L18	—	MC 78L24	—
NATIONAL	—	LM 342-5	LM 342-6	LM 342-8	LM 342-10	LM 342-12	LM 342-15	LM 342-18	—	LM 342-24	—
NATIONAL	—	LM 3910-5	LM 3910-6	LM 3910-8	LM 3910-10	LM 3910-12	LM 3910-15	LM 3910-18	—	LM 3910-24	—
NATIONAL	—	LM 78L05	—	LM 78L08	—	LM 78L12	LM 78L15	LM 78L18	—	LM 78L24	—
SIGNETICS	μA 78L02	μA 78L05	μA 78L06	—	—	μA 78L12	μA 78L15	—	—	—	—
PLESSEY	—	SL 78L05	SL 78L06	SL 78L08	—	SL 78L12	SL 78L15	SL 78L18	SL 78L20	SL 78L24	—

TABLE 4.
Negative Three-terminal Regulators

Manufacturer	−2.0 v	−5.0 v	−5.2 v	−6 v	−8 v	−12 v	−15 v	−18 v	−24 v
FAIRCHILD	—	μA 7905	—	μA 7906	μA 7908	μA 7912	μA 7915	μA 7918	μA 7924
MOTOROLA	MC 7902	MC 7905	MC 7905.2	MC 7906	MC 7908	MC 7912	MC 7915	MC 7918	MC 7924
MOTOROLA (HEP)	C 6117P	C 6118P	C 6119P	C 6120P	C 6121P	C 6122P	C 6123P	C 6124P	C 6125P
NATIONAL	—	LM 320-5	LM 320-5.2	LM 320-6	—	LM 320-12	LM 320-15	LM 320-18	LM 320-24
SILICON GENERAL	—	SG 320-5	SG 320-5.2	—	—	SG 320-12	SG 320-15	—	—

TABLE 5.
Negative Three-terminal Regulators, Medium Current

Manufacturer	−5 v	−5.2 v	−6 v	−8 v	−12 v	−15 v	−18 v	−20 v	−24 v
FAIRCHILD	μA 79M05	—	μA 79M06	μA 79M08	μA 79M12	μA 79M15	—	μA 79M20	μA 79M24

TABLE 6.
Negative Three-terminal Regulators, Low Current

Manufacturer	−3 v	−5 v	−5.2 v	−6 v	−8 v	−12 v	−15 v	−18 v	−20 v	−24 v
MOTOROLA	MC 79L03	MC 79L05	—	—	—	MC 79L12	MC 79L15	MC 79L18	—	MC 79L24

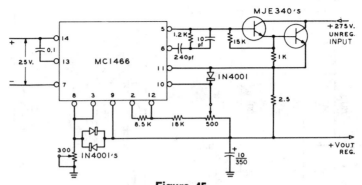

Figure 45

"FLOATING" IC REGULATOR

High-voltage IC regulator uses "floating" 25-volt supply. Series-pass transistors stand-off the main supply voltage. This circuit also allows the output voltage to be varied from zero to maximum value.

of about 1A current are the μA-78G and the μA-79G respectively.

In addition, the National adjustable regulators LM-317 (positive) and LM-337 (negative) are capable of regulating current up to about 1.5A. A larger regulator, the LM-350, is rated up to 3A. The application of these devices is shown in figure 47.

Voltage-Regulator Tubes A voltage-regulator tube (VR tube) is a gaseous device which maintains a constant voltage across its electrodes under conditions of varying supply current. A number of tube types are available which stabilize the voltage across their terminals at 75, 90, 105, or 150 volts. The regulator tube is connected in series with a current-limiting resistor of such value that will permit the regulator tube to draw from 8 to 40 mA under normal operating conditions. The tube must be supplied from a potential source that is higher than the starting, or ignition voltage of the tube (figure 48). Regulator-tube currents greater than 40 mA will shorten the life of the tube and currents lower than 5 mA or so will result in unstable regulation. A voltage excess of about 15 percent is required to ignite the tube and this is usually taken care of by the no-load voltage rise of the source supply.

The value of the limiting resistor must permit minimum tube current to flow, and at the same time allow maximum regulator-tube current to flow under conditions of no load current, as shown in the illustration.

The Vacuum-Tube Regulator Voltage regulation may be accomplished by the use of a *series control tube* and a voltage sensing and comparison circuit, as shown in figure 49. The series tube must be capable of dissipating power represented by the difference between the input voltage from the supply and the output voltage from the regulator at the maximum current flow to the load. In many cases, tubes are operated in parallel to obtain the required plate dissipation. The output voltage of the electronically regulated supply may be changed over a wide range by varying the grid voltage of the dc amplifier tube. The reference voltage may be supplied from a battery or voltage-regulator tube.

The dc amplifier compares the output voltage to that of the reference source. When the output voltage drops, the dc amplifier is unbalanced and the tube draws less plate current, thus raising the grid voltage on the series-connected control tube. The voltage drop through the control tube becomes less and the output voltage from the supply is raised, compensating for the original voltage reduction.

Practical electronic regulated supplies usually employ pentode tubes in the dc amplifier for higher amplifier gain and low-μ triode series control tubes for better control of

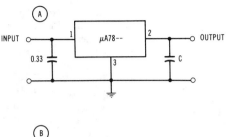

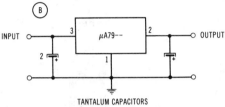

TANTALUM CAPACITORS

Figure 46

REPRESENTATIVE THREE-TERMINAL REGULATORS

A—Positive Regulator. Although no output capacitor is (C) required for stability, it improves transient response.

B—Negative Regulator. Bypass capacitors should be ceramic or solid tantalum which have good high-frequency response characteristics. If electrolytics are used, their values should be 10 μF, or larger. Bypass capacitors should be mounted with short leads, directly across regulator terminals.

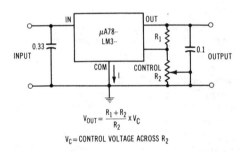

$$V_{OUT} = \frac{R_1 + R_2}{R_2} \times V_C$$

V_C = CONTROL VOLTAGE ACROSS R_2

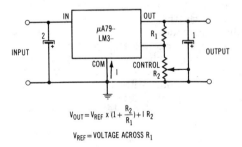

$$V_{OUT} = V_{REF} \times (1 + \frac{R_2}{R_1}) + I\,R_2$$

V_{REF} = VOLTAGE ACROSS R_1

Figure 47

ADJUSTABLE THREE-
TERMINAL REGULATORS

regulation, providing regulation of the order of plus or minus 1 percent or so.

Three Regulated Supplies Shown in this section are three small, inexpensive regulated power supplies designed by W6GXN that are useful for work with solid-state equipment. The first low-voltage supply (figure 50) provides regulated 9 volts and may be used to power the whole gamut of little transistorized consumer electronic devices normally powered by batteries as well as some specialized f-m and vhf receivers operating in this power range. The supply provides a nominal 9 volts, regulated to 0.2 volt up to approximately 300 mA current drain.

A compact 5-volt, 1-ampere regulated supply suitable for operating digital IC circuits is shown in figure 51. Since DTL (diode-transistor-logic) and TTL (transistor-transistor-logic) both operate from +5 volts and represents the most popular two of the various IC logic families, this supply should take care of powering most digital systems. The supply includes current limiting at 1 ampere. The Fairchild μA-7805 regulator is the heart of the supply and yields more "regulation per dollar" than al-

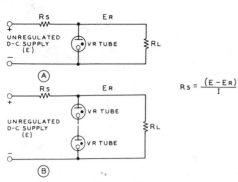

$$R_S = \frac{(E - E_R)}{I}$$

Figure 48

VOLTAGE-REGULATOR TUBE CIRCUITS

A—Single regulator tube stabilizes voltage at
discrete intervals between 90 and 150 volts.
B—Series-connected tubes offer stabilization up
to 300 volts. Series resistor (R_s) is a function of supply voltage (E) and regulated voltage (E_R).

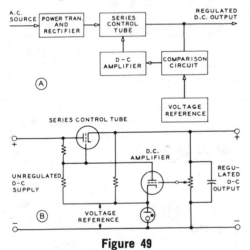

Figure 49

SERIES-REGULATED
DC POWER SUPPLY

Dc amplifier compares the output voltage of
power supply to a voltage reference source.
Voltage drop through series control tube is
adjusted to balance circuit, providing voltage
regulation of 1% or better.

most any discrete circuit that can be built.
The current-limit point is about 1 ampere.

For powering a wide variety of linear ICs,
especially operational amplifiers, the supply
of figure 52 provides plus and minus 15
volts at 300 mA. A dual regulator IC is
used. Current limiting is provided for each

of the two outputs. The two 2-ohm series
resistors in the circuit are the controlling
elements for current limiting, which is set
at 300 mA because of the current capability
of the particular transformer used. Note the
use of the IC silicon bridge rectifier as a
plus-and-minus full-wave rectifier. The center tap of the transformer is used, unlike the
ordinary bridge connection.

In both the 5-volt and the plus-and-
minus 15-volt regulated supplies the voltage output is constant until the current-
limit point is reached, then the voltage value
decreases abruptly.

A Variable-Voltage
Supply With
Current Limiting

Although the simpler
supplies described in the
previous section are
very useful for the spe-
cific voltage requirements most often
encountered, it is helpful to have a continuously variable power supply for experimental purposes. Shown in figure 53 is a
"bench supply" which provides 0 to 20 volts
with current limiting up to 200 mA. The
small size of the supply makes it convenient
to use even if the builder has only a tiny
corner of his operating desk on which to
make experimental gear.

The supply is designed around the MC-
1466L regulator IC which operates from a
"floating" 25-volt source to control another
supply of arbitrary voltage. This concept is
especially useful where the supply covers the
range down to zero volts. A small dual-
winding transformer that mounts on a
printed-circuit board is used (figure 54).

Switch S_{2A} places a 39-ohm resistor in
series with the pass transistor, Q_1, which
limits the collector dissipation of the device
when operating at low voltage and high
current. The other section of the switch
selects the correct multiplier for the voltmeter to provide either 10 or 20 volts full
scale. The switch should be set to the lower
voltage when the supply is used below a
10-volt output level.

The supply is placed within a 4" × 4" ×
4" aluminum utility box chassis. The Darlington Pair pass transistor (Q_1) is heat-
sinked to the front panel of the box with a
mica washer and a nylon 4-40 screw, while
the fuse holder and ac power switch are on
the rear of the box to keep their field away

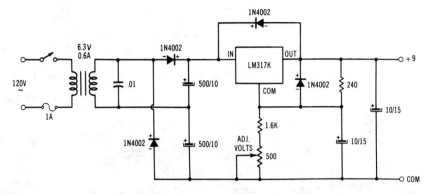

Figure 50
REGULATED 9-VOLT POWER SUPPLY

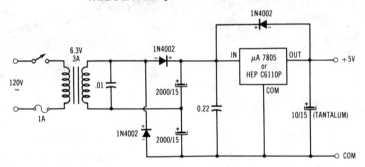

Figure 51

REGULATED 5-VOLT POWER SUPPLY

Output capacitor should be tantalum.

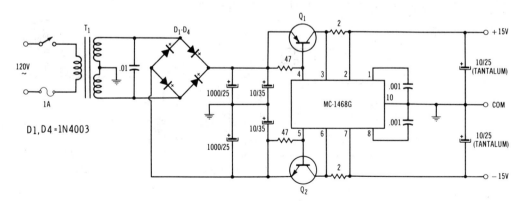

Figure 52

DUAL REGULATED FIFTEEN-VOLT SUPPLY

Transistors Q₁ and Q₂ are heat-sinked with insulating washer and silicone grease. Output capacitors should be tantalum.

from the high-gain circuitry at the front of the assembly (figure 55).

A "Mobile" Power Supply   This compact, regulated power supply provides 13.8 volts at 2 amperes and is designed to be used with 10-watt, 2- and 6-meter f-m transceivers, auto radios, and other dc powered devices in the 20-watt primary power range (figure 56).

The Fairchild μA-78CB fixed regulator is used in exactly the service for which it was intended: a regulator to provide auto elec-

Figure 53

COMPACT 20-VOLT REGULATED SUPPLY FOR LABORATORY WORK

The supply provides 0 to 20 volts at 200 milliamperes with current limiting. Meter range may be switched between 10 and 20 volts full scale. A "floating" regulator circuit is used to allow the range to be extended down to zero volts.

Figure 55

INTERIOR OF VARIABLE-VOLTAGE SUPPLY

Small components are mounted on printed-circuit board. Darlington Pair transistor Q_1 is heat-sinked to front panel with a mica washer and nylon screw.

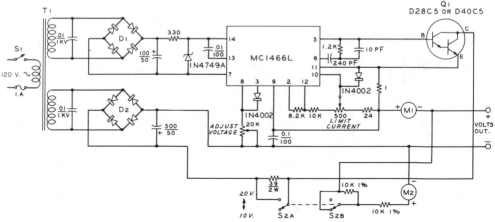

Figure 54

VARIABLE-VOLTAGE SUPPLY WITH CURRENT LIMITING

D_1, D_2—HEP 176. T_1—29, 20 volts, 250 milliamperes. Signal PC-40-250. Box chassis—LMB 444N.

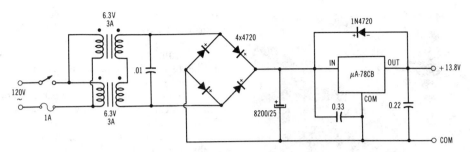

Figure 56
REGULATED POWER SUPPLY FOR MOBILE EQUIPMENT

trical system voltages for home operation of mobile equipment. The supply is constructed upon a 6″ × 6″ × 2″ chassis.

A Medium-Voltage Regulated Supply (150 to 250 volts) A stable, voltage-regulated power supply is a useful adjunct to the experimenters workshop for use with receivers, test equipment, and other devices requiring controlled voltage. Shown in figure 57 is a small power supply that is well suited to this task. The unit delivers 250 volts at 60 mA and may be controlled down to 150 volts, at which point the maximum current is limited to 40 mA. A single 6JZ8 *Compactron* tube serves as a series regulator and dc amplifier. A small NE-2 neon lamp connected in the cathode circuit of the triode section of the 6JZ8 provides reference voltage and may be used as a pilot light.

23-14 Transceiver Power Supplies

Single-sideband transceivers require power supplies that provide several values of high voltage, bias voltage, filament voltage, and dc control-circuit voltage. The supply may provide up to 600 watts of dc power in intermittent voice service. The use of high-storage "computer"-type electrolytic capacitors permits maximum power to be maintained during voice peaks, while still permitting the power transformer to be operated within an average power rating of about 50-percent peak power capability, even for extended periods of time.

Two transceiver power supplies are shown in this section. The first is designed around a power transformer specially built for SSB service. The second supply is designed around a heavy-duty "TV replacement" type power transformer. The former supply is capable of a PEP power level of better than 600 watts, while the latter design is limited to about 300 watts PEP.

A schematic of the 600-watt PEP power supply is shown in figure 59. A multiple-winding transformer is used which has sufficient capacity to run the largest transceivers on a continuous voice-operated basis. The transformer weighs 16 pounds and has great reserve capacity. The power supply provides 800 volts at an intermittent cur-

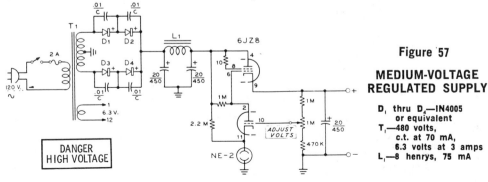

Figure 57

MEDIUM-VOLTAGE REGULATED SUPPLY

D₁ thru D₄—IN4005 or equivalent
T₁—480 volts, c.t. at 70 mA, 6.3 volts at 3 amps
L₁—8 henrys, 75 mA

Figure 58

600-WATT IVS POWER SUPPLY
FOR SSB TRANSCEIVERS

Special transceiver power supply provides heavy-duty capacity to run largest of SSB transceivers. Power transformer and filter choke are to the left, with bias-adjustment potentiometer in foreground. Multiwire cable connects supply to transceiver.

rent of 800 milliamperes, 250 volts at an intermittent current of 200 milliamperes, an adjustable bias voltage at a continuous current of 100 milliamperes, and either 6.3 volts or 12.6 volts filament supply at 12 or 6 amperes, respectively. An additional circuit provides 12 volts dc for operation of auxiliary VOX or switching relays. Controlled-avalanche diodes are used in the bridge-rectifier circuit, in conjunction with RC shunt networks and transient suppression across the power-supply secondary winding.

Additional transient protection is afforded by large bypass capacitors placed on the primary winding of the power transformer. The supply is actuated by a remote-power-line switch, usually located in the transceiver.

The construction of the supply is shown in figure 58. The aluminum chassis is small enough to fit within the speaker cabinet of the transceiver, and parts layout is not critical. The rectifier bridge is assembled on a phenolic board, and mounted below the

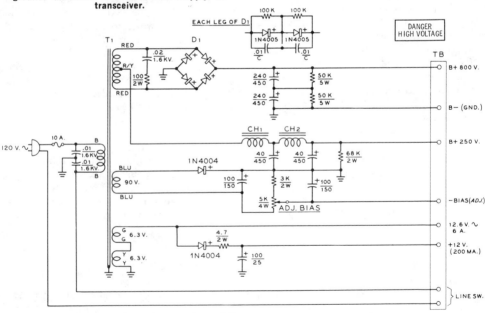

Figure 59

SCHEMATIC, 600-WATT TRANSCEIVER SUPPLY

T₁—600 volts, 400 mA; 250 volts, 100 mA; 6.3 volts, 6 amps; 6.3 volts, 6 amps, 120-volt primary. Triad P-31A
CH₁—1 henry, 300 mA
CH₂—3 henrys, 300 mA

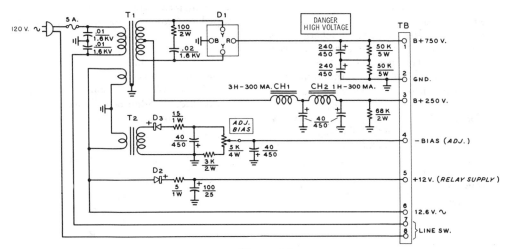

Figure 60

SCHEMATIC, 300-WATT IVS TRANSCEIVER POWER SUPPLY

Various replacement power transformers may be used with this power supply. Suggested units are: (1) 650-volt c.t. at 225 mA; 12.6-volt at 5.25 amp. (Stancor P-8339), for 650-volt dc output. (2) 750-volt c.t. at 325 mA.; 12.6 volt at 6.0 amp. (Stancor P-8365), for 750-volt dc output. (3) 540-volt c.t. at 260 mA.; 6.3-volt at 8.8 amp. (Stancor P-8356), for 600-volt dc output and 6.3 volt filament supply.

Transformer T$_2$ 6.3 volts at 1 amp. (Stancor P-8389). CH$_1$: 3 henrys at 300 mA (Stancor C-2334). CH$_2$: 1 henry at 300 mA (Stancor C-2343). D$_1$: Diode bridge, 1400-volt rms, 1.5 amp (2000-volt PIV). Diodes Inc. #BR-820Á. D$_2$, D$_3$: 1N2070.

chassis in a clear area. The filter capacitors are mounted to a phenolic board, their terminals protruding into the under-chassis area.

All voltage connections are terminated on a connector strip, and a single power cable may be run from the power supply to the

transceiver. The leads carrying the filament voltage should be doubled up, using two wires for each lead to reduce voltaged drop within the cable to a minimum. The 6.3-volt filament windings of the transformer may be arranged in either series or parallel con-

Figure 61

300-WATT IVS POWER SUPPLY FOR SSB TRANSCEIVERS

This compact IVS-rated power supply provides all operating voltages necessary to operate most popular SSB transceivers. The supply uses a "TV-replacement" power transformer in conjunction with a bridge-rectifier circuit. The unit is designed to be placed in the speaker cabinet of the transceiver, and the chassis should be shaped to custom-fit the particular speaker cabinet in use. If desired, the supply may be built on a chassis with a dust cover and placed beneath the station console.

The power transformer is to the left, with the 240-μF, 450-volt filter capacitors in the foreground. The capacitors are mounted to a phenolic plate which is bolted to the chassis. The two filter chokes are to the rear, along with the low-voltage filter capacitors and the "adjust-bias" potentiometer. The reverse-connected filament transformer is at the rear of the chassis. Semiconductor rectifiers are placed beneath the chassis.

figuration, according to the requirements of the transceiver.

Complete filter-capacitor discharge takes about 10 seconds once the supply is turned off, and it is recommended that the capacitor stack be shorted with a 1000-ohm 100-watt resistor before any work is done on the supply.

An inexpensive utility power supply may be constructed about a "TV replacement" transformer, using auxiliary transformers, as needed, for filament and bias supplies, as shown in figure 60. The filament voltage is stepped up to 117 volts by a reverse-con-

nected filament transformer (T_2) and is rectified to provide adjustable bias voltage. The power supply delivers 600 to 750 volts at 400 milliamperes peak current, and about 250 volts at 200 milliamperes peak current. Depending on choice of power transformer, either 6.3- or 12.6-volt filament supply may be provided, in addition to low-voltage dc for operation of VOX or control relays. Layout of the supply is shown in figure 61. The unit is constructed on a home-made aluminum chassis contoured to fit within a speaker cabinet.

CHAPTER TWENTY-FOUR

Radiation and Propagation

PART I
RADIATION AND THE ANTENNA

Electromagnetic waves (radiant energy) encompass a number of familiar types of radiation, such as light, radio waves, X-rays, heat and Cosmic waves. Despite this variety, all these forms of radiation are similar in that they obey the same physical laws, differing only in wavelength and frequency. The *electromagnetic spectrum* may be defined in terms of wavelength, or frequency, ranging from extremely long waves of low frequency inherent in the magnetic field of the earth; through long, short and microscopic radio waves; infrared waves; light waves; ultraviolet and X-rays; and into the infinite region of gamma and Cosmic waves of unknown origin. Of this vast range of electromagnetic radiation bathing the earth, those waves of immediate interest in the field of communication are of a dimension comparable with the size of man himself—*radio waves*.

Every electric system that carries alternating or pulsating current radiates a certain amount of this energy into space in the form of electromagnetic waves. The amount of radiated energy is small when the waves are large compared to the radiating system, as in the case of 60-Hz industrial current. As the frequency of alternation is raised and the corresponding wavelength is shortened, a portion of the electromagnetic spectrum is reached wherein the radiated energy becomes of some practical use for long-distance radio communication. This region is termed the *communication region* of the electromagnetic spectrum and is composed largely of radio waves.

Of greatest interest to the communication specialist is that portion of the spectrum falling between 10-kHz and 10,000-MHz limits, for within these arbitrary boundaries worldwide, coordinated communication takes place, and the use of this portion of the spectrum is governed by international treaty, to which the United States is a signatory power. That slice of the communications spectrum between 1800 kHz and 30 MHz is of immediate interest, since ionospheric-reflected radiocommunication over long distances takes place in this region. The spectrum between 30 MHz and 450 MHz, in addition, is widely used for shortrange communication and experimental long-distance communication making use of anamolies of propagation. These frequency ranges are discussed at length in the following chapters.

24-1 The Antenna System

The earliest use of a radio antenna was recorded by Hertz in his communication experiments of 1884. As the knowledge of radio grew, so the knowledge and use of radio antennas grew likewise. Today's antennas represent modern design applied to ideas nearly 100 years old. And while the antennas of today bear little resemblance to those of yesteryear, so the antennas of tomorrow will seem strange and unusual to the communicator of today. In a sense, while the basic theory of antennas remains the same, the application of the theory is constantly expanded and modified, in step with the growth of modern technology.

The *antenna* (aerial) is made up of a system of conductors designed to radiate or intercept electromagnetic waves. Antennas come in many shapes and sizes, but they all have one factor in common—they are made up of conducting material and require a *feed system* to extract or accept radio energy. Many antennas are complicated, but most of them are not. Some of the better hf and vhf antennas are described in this Handbook.

The antennas shown are practical and may be duplicated from the dimensions given. However, it is necessary for proper understanding of antenna operation and use to briefly examine the outer limits of antenna theory. Whenever possible, this will be done by the use of formulas, charts, and illustrations which minimize the mathematical processes involved.

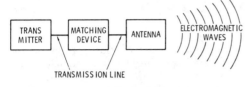

Figure 1
REPRESENTATIVE ANTENNA SYSTEM

The antenna is a device for converting guided electric waves into electromagnetic waves in free space. A matching device is often used to ease this abrupt transition, and a transmission line guides the electric waves from the transmitter to the antenna.

The Complete The *antenna* is a device for
Antenna converting guided electric
waves into electromagnetic
waves in free space. A *matching device* of some sort is generally employed to ease this abrupt transition, and a *transmission line* is often used to efficiently guide the electric waves from the transmitter to the antenna (figure 1). It is understood, moreover, that the antenna system follows the general laws of reciprocity and can extract electromagnetic waves from free space and convert them to electric waves capable of being detected by a radio receiver.

The range of frequencies (*bandwidth*) over which a reasonable match or transformation between guided waves and free waves can be achieved depends to a degree on the amplitude and nature of the mismatch in the antenna system. If the transformation is gradual so that wave parameters do not undergo a sudden change, but vary gradually between the guided and the free condition, the transition is smooth and the frequency span of efficient operation may be quite large. Accordingly, the disturbance or unwanted reflection of the guided wave may be quite small.

If, on the other hand, the transition between the guided and the free-space waves is abrupt, a region of reflection exists in the system such that a portion of the wave is sent back down the transmission line. The *reflected wave* may be compensated for, to a degree, by adjustments made to a matching device which creates equal and opposite reflections to annul the original reflection generated by the abrupt transition in the antenna system. In any case, the frequency span, or bandwidth, of the antenna system is considerably reduced over that achieved by a perfect transformation between guided and free waves.

The bandwidth of an antenna system is relative, and one way of specifying it is to define the limit of wave reflection allowed on the transmission line feeding the antenna. For example, if it is specified that the *reflected wave* shall be limited in amplitude to one quarter the value of the *incident* (direct) *wave* on the line, the overall system bandwidth may be defined by this limit, as measured under actual operating conditions.

It is common practice to specify antenna system bandwidth in terms of the amplitude of the reflected wave with respect to the incident wave. This specification may be expressed as a *voltage standing-wave ratio* (abbreviated VSWR, or simply SWR) which is measurable by an inexpensive instrument placed in series with the transmission line. The SWR figure bears a definite relationship to the amplitude of the reflected wave, and it is simpler to measure and plot the SWR of an antenna and then to define the operating limits by SWR readings than it is to interpret the SWR in terms of the amount of reflection. Generally speaking, SWR values up to 3 are acceptable in simple antenna systems, while a somewhat lower SWR value of 2 is often specified as a maximum limit for various forms of beam antennas. On the other hand, some antennas employ so-called *tuned feeders* which operate with SWR values as high as 100. Strictly speaking, the maximum value of SWR acceptable in a system is often limited by the economics of the problem and is subjective rather than objective, being a relative concept rather than an absolute limitation arbitrarily imposed.

In practice, the maximum acceptable SWR limit of an antenna system may be decreed by the greatest allowable line loss, the desired operating bandwidth, or perhaps be expanded beyond credibility by an aggressive advertising department of a particular antenna manufacturer, or it may merely be decided by whim. In any event, the SWR values mentioned earlier are acceptable for the various antenna designs commonly used by radio amateurs and are specified as arbitrary system bandwidth parameters in this Handbook.

24-2 The Electromagnetic Wave

A time-varying electromagnetic field, or wave, may be propagated through empty

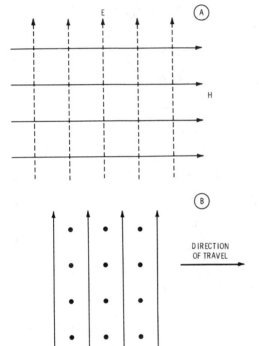

Figure 2
THE PLANE ELECTROMAGNETIC WAVE

When a wave has travelled far enough from the source the wavefront appears flat and it is called a plane wave. The plane contains the perpendicular electric (E) and magnetic (H) lines representing the wave front which is always perpendicular to the direction of wave travel. In (A) the wave is travelling out of the page toward the reader. A cross-section of a travelling wave is shown in (B). Arrows which go into the plane of the page are shown by small "X"s for the tail, and those which come out of the page are shown by dots for the points of the arrows. The particular configuration of an electromagnetic field is termed a "mode."

space at the velocity of light. The wave is considered to be made up of interrelated *electric (E)* and *magnetic (H)* fields at right angles to each other and lying in a plane, as pictured in figure 2. The wave energy is divided equally between the two fields. If the wave is pictured as originating at a point source in space, the wave spreads out in an ever-growing sphere with the source as the center. The path of an energy ray from the source to any spot on the sphere is a straight line and, at a large distance from the source, the wavefront does not appear to be spherical, but is assumed to be a flat surface, as shown in the illustration.

The plane electromagnetic wave may be represented in terms of its fields, with the vertical arrows representing the direction and strength of the electric field and the horizontal arrows the direction and strength of the magnetic field. The wave shown is said to be *vertically polarized* because the electric field is vertical. If the electric field were horizontal, the wave would be *horizontally polarized*. Other waves may be *circularly polarized,* corresponding to left-handed and right-handed helices.

The abstract concept of an electromagnetic wave travelling through space is difficult to comprehend without the assistance of mathematical proof. Viewed from the theory of electron flow in a conductor, there is no suggestion of energy radiation into space. A set of relationships termed *Maxwell's equations* form the basic tools for the analysis of most electromagnetic wave problems.

Maxwell's Equations James C. Maxwell (1831-1879), a brilliant student of the natural sciences, derived a breathtaking concept of nature and revealed a set of striking equations that encompassed the various laws of electricity derived by Faraday, Ampere, Ohm, and others. Maxwell's unified field equations of electric and magnetic behavior form today's basis of electromagnetic theory. Not only did Maxwell's equations describe all known electromagnetic phenomena, but in the broader sense predicted electromagnetic radiation, simultaneously introducing into physics the general concept of fields to describe interactions between one body and another.

The abstract concept of a radio wave travelling through space is difficult to comprehend without the assistance of Maxwell's equations. Viewed from the simple concept of electron flow in a conductor there is no suggestion of radiation of energy into space in the form of electromagnetic waves. Maxwell's assumptions that an electric field changing in time is a form of current which sets up a magnetic field about itself, and the latter, also changing in time, sets up the electric field, is the basis for the further assumption that the two interact and propagate energy from one place to another. These assumptions provide the necessary bridge between simple electron flow and an electromagnetic field about the conductor.

$$(1) \quad \mathit{div}\ E = 0 \qquad\qquad\qquad (3) \quad \mathit{curl}\ E = \frac{1}{c} \cdot \frac{\partial H}{\partial t}$$

$$(2) \quad \mathit{div}\ H = 0 \qquad\qquad\qquad (4) \quad \mathit{curl}\ H = \frac{1}{c} \cdot \frac{\partial E}{\partial t}$$

Maxwell's equations (above) form the basis of modern electromagnetic theory. The first equation states that, in the absence of electric charges, electric lines of force can neither be created nor destroyed. The second equation states the same principle for magnetic lines of force and, in addition, states that magnetic charges do not exist. The third equation is a generalized statement of Faraday's Law that a changing magnetic field produces an electric field and that the ratio of the electro-static units to the electromagnetic units is a constant (c) related to the speed of light. The fourth equation is derived from Ampere's Law and states that a changing electric field produces a magnetic field by virtue of the sum of the conduction and displacement currents and that the time rate of change of the electric field has properties related to the displacement current.

E and H represent the electric and magnetic field strengths. *Div* (divergence) and *curl* (an abbreviation for rotation) represent mathematical operations expressing rate of change and vorticity. The symbol ∂ indicates a partial differentiation with respect to time, t.

Maxwell showed that an electric charge which is accelerated or decelerated is accompanied by a magnetic field which pulsates and, with the passage of time, is propagated outward through the surrounding medium. The increase of energy, of course, has been supplied by the force responsible for the acceleration of the charge. During acceleration and deceleration, the magnetic field energy does not simply flow outward and again inward. Rather, this energy is radiated and perma- nently lost to the charge and its field. The electromagnetic field thus created is in the form of an energy wave travelling radially outward from the source, with electric and magnetic components identical in form and mutually perpendicular. The electric and magnetic components become weaker as the wave travels outward because both are inversely proportional to the radius of the wave from the point of origin.

Figure 3
MAXWELL'S FAMOUS EQUATIONS

Maxwell's equations (figure 3) picture the interplay of energy between electric and magnetic fields which is self-maintained, with the energy radiating outward from the point of origin. The equations express the continuous nature of the fields and define how changes in one field bring about changes in the other. The compound disturbance described by Maxwell's equations was proven in fact by Hertz, who generated, radiated and intercepted electromagnetic waves in 1888, fifteen years after Maxwell had predicted their existence. A complete discussion of Maxwell's equations and electromagnetic waves may be found in *Electromagnetics*, by John D. Kraus, McGraw-Hill Book Co., New York.

Radiation From An Antenna Radiation and interception of electromagnetic energy is explained by Maxwell's equations. The equations provide the link between electron motion in a conductor and electromagnetic waves in space. In addition, the equations show that the electromagnetic field, in ebb and flow, provides a quantity of energy which is propagated outward and is detached from the field of the moving electron, or charge, in the antenna.

The somewhat obscure concept of radiation from a current-carrying conductor may be pictured with the aid of an imaginary bit of antenna termed an *oscillating doublet* (figure 4). Two equal electric charges of opposite polarity spaced a fixed distance apart in space comprise this configuration. This concept allows for the regular, periodic linear displacement of charges along the axis of the doublet when excited by an alternating current. If the charges move up and down along the axis with equal and opposite velocities so that the system is in a continuous state of acceleration or deceleration, a current is said to flow in the doublet and the system must radiate energy.

The principles of radiation of electromagnetic energy are based on Maxwell's laws that a moving electric charge creates an electric field. The created field at any instant is in step with the parent field, but is perpendicular to it in space. These laws hold true whether a conductor is present or not.

At the start of oscillation (figure 4A) the doublet is neutral and the charges are just beginning to move apart. Flux lines are drawn between the charges. An electromagnetic field is created with the direction of the magnetic field in a loop around the doublet, perpendicular to the page. The electric field is in the plane of the page. As the doublet

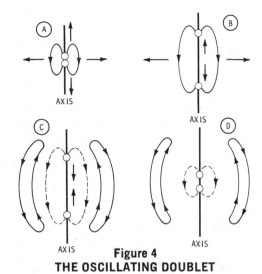

Figure 4
THE OSCILLATING DOUBLET

The creation of a closed electric field about an oscillating doublet is illustrated here. The radiation of electromagnetic energy takes place from an oscillating doublet composed of charges moving sinusoidally with respect to each other along a common axis. Current flow (movement of charges) causes a magnetic field to be created, which is perpendicular to the page and not shown. Separation of charges causes an electric field to be set up, which is shown here by electric lines of force in the plane of the page. Since the currents and charges producing these fields are out of phase, the fields are also out of phase and constitute an induction field, the energy of which cannot be detached from the doublet. The electric field, however, in a radiated wave, does not terminate on a charge, and when the charges move together (C), the field closes upon itself in the polar regions. The independent electric field, in turn, generates a magnetic field and both fields constitute a radiated electromagnetic wave flowing outward from the doublet.

moves toward its full displacement (figure 4B) energy in both magnetic and electric fields is propagated outward. The intensity of the electromagnetic field is approximately $E \times H$, showing that as the charges separate, stored energy is increasing in the space around the doublet. Maxwell's first equation, moreover, states that the electric lines of force in a radiated wave do not terminate on a charge but are closed curves ($div\ E = 0$) in the polar regions of the doublet, as shown in figure 4C.

An instant after the independent field has been formed, the doublet charges start to move together, producing lines of force opposite to the recently formed independent electric field (figure 4D). At first thought it would appear as though the periodic reversal of charge would result in a periodic reversal of the energy flow and no net energy would flow outward. This would be so if the field at a point away from the doublet at a given instant depended on the charge distribution of the doublet at that instant. However, here is a time lag between the creation of a particular current in the doublet, the charge distribution, and the consequent electromagnetic field at a given point. It is this time lag that allows some of the energy in the region around the doublet to continue to travel outward in a closed electric field even when conditions of charge at the doublet indicate a flow of energy directed inward toward the doublet. The closed, moving electric field generates a magnetic field in accord with Maxwell's third law and the detached electromagnetic field moves away from the doublet at the speed of light. The cycle starts to repeat itself with the collapse of the field when the charges move together and then separate once again.

With sinusoidal doublet motion there must, therefore, be a continuous radiation of energy over and above the amount required to establish a steady-state field. Maxwell's equations describe a beautifully simple electromagnetic wave travelling radially outward from the doublet, becoming weaker with distance since the two component fields are inversely proportional in strength to the distance travelled from the doublet. There is no loss of energy, it is merely dissipated in area as the wave spreads. Once having been produced, the expanding wave travels and propagates itself for an unlimited time, as do the light waves reaching the earth from an extragalactic nova, millions of years after the star that created them has ceased to exist.

24-3   The Standing Wave

A previous paragraph touched on the voltage standing-wave ratio (SWR) and its relation to antenna system discontinuity, and to the coefficient of reflection. This is an important concept and deserves additional elaboration.

When an electromagnetic wave travels through space, there is a balance between the electric and magnetic fields, with half the energy in each field (figure 5). If the wave

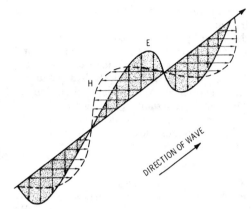

Figure 5
OBLIQUE VIEW OF TRAVELING WAVE

The traveling electromagnetic wave is represented in terms of its electric and magnetic components, identical in form, and perpendicular in direction to each other and to the direction of travel of the wave. The fields vary sinusoidally along the axis of travel and at any fixed point, the fields vary sinusoidally with time. As the wave travels, the whole pattern moves to the right with the velocity of light.

enters a new medium, or encounters a discontinuity in the medium, there must be a new redistribution of energy. Whether the new medium is a conducting, semiconducting, or nonconducting material, there will have to be a readjustment of energy relations as the wave reaches the surface of the discontinuity.

Since no new energy can be added to the wave as it passes through the boundary surface, the only way that a new balance may be achieved is for some of the energy to be rejected. The rejected energy constitutes a *reflected wave*. In this manner, the observer sees reflection of light from a conducting metal surface or from a nonconducting glass surface.

The electromagnetic wave, if unimpeded, will travel indefinitely in free space. In the hypothetical case of an infinitely long conducting medium, the travelling wave could voyage onward forever. But if the medium is broken at a point, and a *load,* or absorptive device (a discontinuity) of the correct magnitude replaces the rest of the medium, the energy is completely absorbed and converted to heat in the load. If the medium is terminated by a discontinuity having reflective properties, the discontinuity will reflect energy back through the medium toward the source. The reflected energy will combine with the forward energy in such a way as to produce a pattern in the medium known as a *standing wave*.

Wave Reflection An example of a simple discontinuity is a perfectly conducting plane surface (figure 6). A wave falling on the surface is totally reflected. Both the electric and magnetic components of the travelling wave are reflected, but while the electric component is reflected with reversal of sign (A) thus leaving the electric field at the reflecting surface zero, the magnetic component is reflected with unchanging sign (B) and is so doubled at the reflecting surface. The sum of the forward and reflected travelling waves is a standing wave which is continually changing in magnitude but is fixed in space, resembling the vibration of a string on a musical instrument. The total electric intensity at the reflecting surface is always zero, and also zero at distances that are multiples of half-wavelengths from the surface. These points of zero electric field are termed *nodes*. There are also nodes in the intensity of the magnetic field, at one-fourth wavelength and odd multiples thereof from the reflector. If there were no loss of energy, for example, in

the form of friction in the case of the vibrating string, or energy lost in the travelling wave, the standing wave would persist indefinitely.

Derivations of Maxwell's equations show that where there are nodes of magnetic fields, maximum electric fields (*loops*) occur. In addition, the standing waves of magnetic and electric fields pulse out of phase in time, so that when the magnetic field is zero, the electric field is maximum, and vice versa. Thus, the standing wave has a very different appearance from a traveling wave, although it is nothing more than the sum of two traveling waves.

The Reflection Coefficient When an electromagnetic wave falls on the surface of a dielectric or insulating material, or meets a discontinuity, there is a partial reflection and partial transmission of the incident energy. That fraction of the incident wave that is reflected, when expressed as a ratio of the original wave, is termed the *reflection coefficient*. If the reflection coefficient is low (the discontinuity of the medium possessing poor reflective qualities), there is very little reflected energy, and the total field about the reflecting surface is only slightly modified from that of a traveling wave. If, on the other

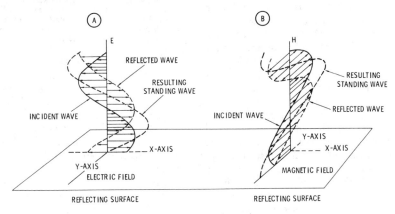

Figure 6

REFLECTION OF THE ELECTROMAGNETIC WAVE FROM A CONDUCTING SURFACE

When an electromagnetic wave is reflected from a conducting surface the electric field is reflected with reversed sign (A) so that the electric field at the reflecting surface is zero. The magnetic field is reflected with unchanging sign and is so doubled at the reflecting surface (B). The resulting wave in each case is the sum of the two travelling waves and oscillates in magnitude, but is fixed in space. It is termed a "standing wave."

hand, the reflection coefficient is near unity (the discontinuity possessing good reflective qualities), the maximum field strength will vary as a function of the distance from the surface, with well-defined nodes and loops. The resulting wave bears a definite relationship to the amplitude of the reflected wave and to the reflection coefficient, as expressed by:

$$\text{Coefficient of reflection} = \frac{\sigma - 1}{\sigma + 1}$$

where,

σ = the voltage standing-wave ratio

Finally, it should be noted that if the medium is terminated by a load of the proper magnitude, no discontinuity or reflection will exist in the medium, and the medium is considered to be *matched*. The degree of *mismatch* between the medium and the load can be defined in terms of the amplitude of the reflected wave, or in terms of the standing-wave ratio (SWR), which may be readily measured by inexpensive instruments.

24-4   General Antenna Properties

All antennas have certain general properties which apply both to receiving and transmitting modes. Thus, the more efficient the antenna is for transmitting, the more effective it is for receiving. Directive properties will be the same for transmission as for reception and, in the case of directive antennas, the gain will be the same on both transmitted and received signals. In long distance, high-frequency communication, it should be noted, the often observed odd behavior and seeming perversity of antennas which often occurs, is due to the fact that the waves may not take exactly the same paths through the ionosphere when going in opposite directions, the two waves utilizing different portions of the directive pattern of the antenna. Even so, the concept of reciprocity between transmission and reception still stands correct.

Antenna Resonance   The strength of the radio wave radiated by an antenna depends on antenna size and the amount of current flowing in it. It is rea-

sonable to expect the largest amount of current that can be achieved from the power available will provide the best radiation from a given antenna. The greatest amount of current flows when the reactance of the antenna is cancelled and the antenna made *resonant* at the operating frequency. The shortest conductor that will be self-resonant at a given frequency is one that is about half as long as the size of the radio wave. The half-wavelength antenna is used as a basis for all antenna theory and is a fundamental building block in antenna design (figure 7).

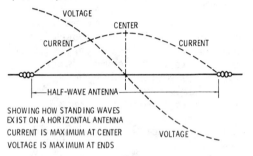

SHOWING HOW STANDING WAVES EXIST ON A HORIZONTAL ANTENNA
CURRENT IS MAXIMUM AT CENTER
VOLTAGE IS MAXIMUM AT ENDS

Figure 7

THE RESONANT ANTENNA

The greatest amount of current flows in the antenna when it is resonant. The shortest conductor that is self-resonant at a given frequency is one that is about a half-wavelength long. The reflection pattern on the antenna creates a standing wave of both voltage and current. The half-wave, center-fed antenna is often called a "doublet."

Two practical methods exist to make a conductor self-resonant. First; the frequency of the radio wave may be changed to suit the conductor length; second the electrical length of the conductor may be altered to suit the given frequency of the wave.

The electrical length of a half-wave of electromagnetic energy is related to the speed of travel of the wave (the same velocity as the speed of light) and also to the frequency of the wave by an equation that is similar to equations dealing with other waves (such as waves in the ocean, or the vibrations of a piano string). In the case of a radio wave in free space, the metric formula is:

Half wavelength (meters) =

$$\frac{150,000,000}{\text{Frequency in Hz}} = \frac{150}{\text{Frequency in MHz}}$$

The formula in the English system is:

Half wavelength (ft) =

$$\frac{492}{\text{Frequency in MHz}}$$

The physical length of an antenna element varies slightly from this fundamental electrical length because the element has thickness and is affected by nearby objects. Information will be presented in a later Section defining this relationship in practical terms.

Radiation Resistance and Reactance   When r-f power is applied to an antenna, it is radiated into space, the antenna acting as a load, or sink, for the transmitter. In order to establish a frame of reference, the power dissipated in a dummy load (such as a resistor) may be compared in terms of voltage and current with the power radiated by a real antenna. This reference frame is defined in terms of the *radiation resistance* of the antenna. Simply stated, the radiation resistance of an antenna is that imaginary resistance exhibited which seems to dissipate the power the antenna actually radiates into space. Radiation resistance is expressed in ohms and is normally measured at a point in the antenna which has the maximum value of current flowing in it. A more general term used in this connection is *antenna impedance* which, in addition to implying radiation resistance, also implies the presence of reactance in the antenna circuit.

In addition to radiation resistance, practical antennas also exhibit *loss resistance* which is energy dissipated in heat loss in the antenna element and nearby dielectrics. The total resistance of the antenna, which is the sum of these two figures, is often referred to as *feedpoint resistance*, although in popular usage the term "radiation resistance" usually encompasses the two separate entities.

The radiation resistance and resonant frequency of an antenna depend on the antenna

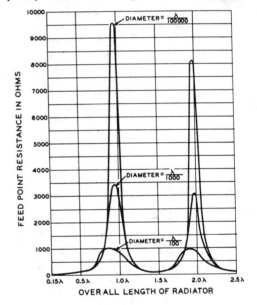

Figure 9

IMPEDANCE OF ANTENNA VARIES ALONG THE LENGTH AND EXPRESSES THE RATIO BETWEEN VOLTAGE AND CURRENT AT ANY POINT ON THE ANTENNA

The feedpoint resistance of a center-fed antenna is a function of the physical length. For example, a half-wave antenna has a center feedpoint resistance of about 73 ohms, while an antenna one wavelength long has a center feedpoint resistance of 1000 ohms to 9500 ohms (depending upon the diameter of the element). As the length of the radiator increases, the impedance excursions become less drastic, especially for "fat" radiators.

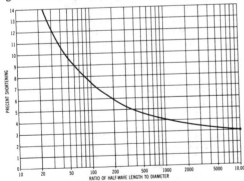

Figure 8

LENGTH-TO-DIAMETER RATIO OF ANTENNA AFFECTS RADIATION RESISTANCE

As the antenna becomes thicker with respect to length, the radiation resistance decreases and the antenna must be shortened to reestablish resonance. This chart illustrates the amount of shortening required with a resonant half-wavelength antenna in the frequency range of 2 MHz to 30 MHz.

size with respect to the radio wave and the proximity of the antenna to nearby objects which either absorb or reradiate power, such as the ground, or other antennas or conductors. The length-to-diameter ratio of the antenna also affects the radiation resistance; as the antenna becomes thicker with respect to the length, the radiation resistance decreases (figure 8).

The feedpoint resistance of a resonant antenna is the load for the transmitter and its value is important in determining the method used to couple the two together.

Antenna Impedance Because the power at any point in an antenna is the same at any other point, the impedance at any point along the antenna expresses the ratio between voltage and current at that point (figure 7). Thus, the lowest impedance occurs where the current is highest

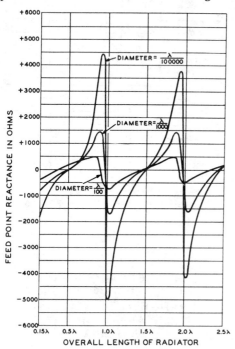

Figure 10

REACTIVE COMPONENT AT FEEDPOINT OF CENTER-FED ANTENNA

Feedpoint reactance rises rapidly when antenna is in nonresonant condition and also increases as the length-to-diameter ratio of the antenna decreases. "Fat" antennas exhibit less reactance than "thin" ones. Reactance varies rapidly for center-fed antenna one wavelength long.

and the impedance rises uniformly toward the ends of the antenna, where it can reach a value as high as 10,000 ohms for a thin dipole remote from ground (figure 9).

Like a tank circuit, an antenna may exhibit reactance at the feedpoint. Since the antenna, by definition, is nonreactive at resonance, antenna reactance implies a state of nonresonance. Antenna reactance rises rapidly off-resonance and the manner in which the reactive component varies is illustrated in figure 10. The rate-of-change of the reactance increases as the antenna length departs from resonance and also increases as the length-to-diameter ratio decreases. The reactive component of an antenna is zero when the overall antenna length is slightly less than a multiple of quarter-wavelengths long. Near resonance, the resistance and reactance terms of an antenna vary much in the manner shown in figure 11.

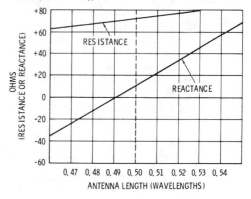

Figure 11

FEEDPOINT RESISTANCE AND REACTANCE AS FUNCTION OF ANTENNA LENGTH

Near resonance, the resistance and reactance of a dipole antenna vary in this typical manner. Reactance is zero when the antenna is slightly less than one-half wavelength long. The reactance changes more rapidly for "thin" antennas than for "fat" ones.

Both feedpoint resistance and reactance change more slowly with frequency for a fixed radiator length with "fat" elements than with "thin" elements, indicating that the effective antenna Q is lowered as element diameter increases. Lower Q is desirable, because it permits the use of a radiator over a wide frequency range without resorting to means for eliminating the reactive component. If the antenna Q is low enough, the radiator is termed a *broadband* antenna.

The curves of figure 12 indicate the theoretical feedpoint resistance of a dipole antenna for various heights above a perfect ground plane. In free space, the feedpoint resistance of a thin dipole is approximately 73 ohms. The modifying effects of the

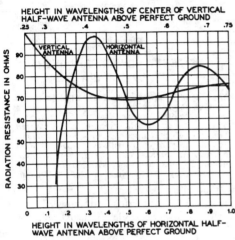

Figure 12

FEEDPOINT RESISTANCE OF DIPOLE SUSPENDED ABOVE A PERFECT GROUND

In free space the feedpoint resistance of a half-wave dipole is about 73 ohms. The modifying effects of the ground change this, as shown above, with the value approaching 73 ohms as the dipole is far removed from the ground. The ground has less effect on the feedpoint impedance of a vertical antenna.

ground change this nominal value as shown, with the value approaching 73 ohms as the dipole is removed from the ground by more than a wavelength.

Antenna Directivity Because of the manner in which current flows in an antenna, radiation from practical antennas is not uniform, but is directive to a certain degree. The amount of directivity can be altered or enhanced through the use of extra radiating elements, reflecting planes or curved surfaces; or, in the microwave portion of the radio spectrum, by the use of electromagnetic horns, lenses, and slotted devices.

The directive pattern of an antenna may also be modified by wave reflection from the

ground or from nearby objects. Structures which lie within a few wavelengths of the antenna have the greatest influence on the directivity of the antenna. The change in directivity is caused by the ability of the nearby conducting structure to reradiate energy emitted by the antenna. This reradiation may either reinforce or cancel the direct radiation of energy from the antenna, thus producing a distortion of the free-space pattern of the antenna (figure 13). By using properly adjusted conducting objects (called *driven elements, reflectors,* or *directors*) the

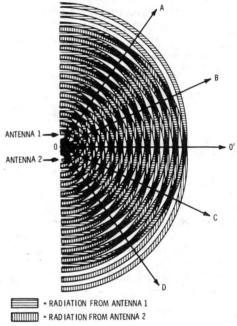

    ▭ = RADIATION FROM ANTENNA 1
    ▭ = RADIATION FROM ANTENNA 2

Figure 13

RADIATION PATTERN FROM TWO ANTENNAS

Wave interference patterns created by two adjacent antennas. Radio waves from two adjacent sources of the same frequency reinforce or cancel each other to provide wave pattern in space adjoining the antennas. In this representation, the waves reinforce each other along radial lines OA, OB, OO', OC, and OD. Midway between these lines the waves cancel each other. This pattern represents an antenna array having five lobes.

antenna radiation pattern may be deliberately distorted to produce an enhanced signal in a desired direction (figure 14). The signal gain varies with the adjustment and spacing of the various elements and the radiation resistance of the parent antenna, as well as its tuning, is affected as well.

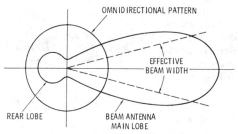

Figure 14

ANTENNA PATTERN OF DIRECTIONAL ARRAY

Polar plot shows antenna radiation as compared to an omnidirectional antenna. Signal gain varies with the number and adjustment of antenna elements in the array. The directive pattern is termed the "main lobe" of the antenna, with the unwanted lobe termed the "rear lobe." The ratio between the two lobes is called the "front-to-back ratio" of the array.

The Isotropic Radiator Directivity of an antenna is the ability of the antenna to concentrate radiation in a particular direction. All practical antennas exhibit some degree of directivity. A completely nondirectional antenna (one which radiates equally well in all directions) is known as an *isotropic radiator*, and only exists as a mathematical concept. Such an antenna, if placed at the center of a sphere, would "illuminate" the inner surface of the sphere uniformly.

Antenna Signal Gain The effective signal gain, or power gain, of an antenna is the ratio between the power required in the antenna and the power required in an isotropic radiator to achieve the same field strength in the favored direction of the antenna under measurement (figure 15). Directive gain may be expressed as the power ratio, in units called *decibels* (dB). Referring to the illustration, the power gain of the antenna under test, placed at the center of the sphere, illuminates only a portion of the sphere and the power gain is the ratio of the surface area illuminated by the isotropic antenna to that area illuminated by the test antenna. Since the field pattern of radiation of any antenna is not clear, but blends into nothingness at the extremities, the practical pattern is defined

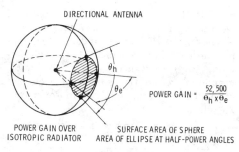

$$\text{POWER GAIN} = \frac{52,500}{\theta_h \times \theta_e}$$

Figure 15

ANTENNA POWER GAIN OVER ISOTROPIC RADIATOR

The effective power gain of an antenna is the ratio of power required in the antenna and the power required in an isotropic radiator to achieve the same field strength in the favored direction of the antenna under measurement. The power gain of a half-wave dipole over an isotropic radiator is 1.64. The gain of a directional antenna over an isotropic radiator is expressed by the formula in the illustration.

as that illuminated portion of the sphere which lies between the "half-power" angles of the radiator field. On the usual polar plot of an antenna pattern, these points are the " − 3 dB" power points.

The power gain over an isotropic radiator, or over a simple dipole, is the measuring stick for antenna performance. The power gain over a dipole may be computed from the formula shown in the illustration, which provides a quick method of determining the power gain of an antenna by measuring the radiation pattern at the − 3 dB power points.

Closely allied to the concept of power gain is the problem of suppressing unwanted radiation from the sides and rear of a directive antenna system. Unwanted energy radiated to the rear of the directional antenna may be compared to the energy radiated from the front of the array and is expressed as a power ratio in decibels termed the *front-to-back ratio*.

Simple antennas often have a symmetrical radiation pattern and may even possess modest gain without having appreciable front-to-back ratio. More complex antenna arrays exhibit higher gain and front-to-back ratio, but seldom will maximum power gain and maximum front-to-back ratio occur at the same condition of antenna adjustment.

Power gain implies *horizontal* or *vertical directivity* in the antenna pattern which can be best expressed as a directive pattern which is a graph showing the relative radiated field intensity expressed in terms of the *azimuth angle* for horizontal directivity and in terms of the *elevation angle* for vertical directivity (figure 16).

Antenna Bandwidth The *bandwidth* of an antenna is a measure of its ability to operate over a specified range of frequencies. Unlike other antenna properties, bandwidth does not have a unique definition, as it depends on the operational requirement of the antenna. Bandwidth may be limited

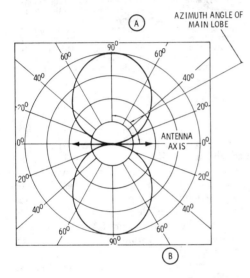

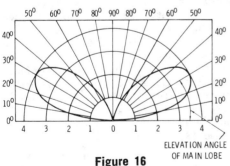

Figure 16

DIRECTIVITY PATTERNS FOR DIPOLE ANTENNA ONE-HALF WAVELENGTH ABOVE PERFECT GROUND PLANE

Plotted field intensity for dipole antenna. Azimuth angle for horizontal directivity is shown at (A). Vertical angle (elevation angle) is shown at (B).

by loss in gain, change of antenna pattern, excessive SWR on the feed system, or change in input impedance. One of these factors, such as gain or impedance, usually limits the low-frequency limit of operation, whereas change of pattern shape might determine the high-frequency limit. In amateur practice, bandwidth is usually specified in terms of a maximum SWR limit on the transmission line feeding the antenna system.

Mutual Impedance A conductor placed in the field of an antenna will have a current induced in it by virtue of the voltage applied to the antenna. In the case of two adjacent antennas, if a voltage is applied to the terminals of the first antenna and the induced current measured at the terminals of the second antenna, then an equal current will be found at the terminals of the first antenna if the original voltage is applied to the terminals of the second antenna.

This classic theory can be expanded into the concept of *mutual impedance* between two coupled antennas and accounts for the fact that the feed impedance of an individual element in an array of antennas may differ considerably from its free-space impedance because of the effect of mutual coupling with the other elements of the array. In an antenna array where the current distribution in the elements is critical because of pattern requirements, it is necessary to adjust the coupling system between the elements to provide correct current distribution and to match the *input impedance* of the array, rather than the self-impedance of the input element.

The input impedance is the sum of the self-impedance of the fed element and the mutual impedance with all other elements in the array. The magnitude and phase of the mutual impedance depends on the amplitude of the current induced in the fed antenna by the other elements and this, in turn, is a function of the spacing and tuning of the additional elements. Induced currents in the fed element are greatest when the elements of the array are close together, resonant, and parallel.

The induced current may be in phase, or out of phase, with the fed-element current and the impedance of the array may be higher, or lower, than that of the fed element. In addition, the elements may introduce reactance into the fed element, detuning it from a resonant condition. All of these effects are interlocking, and changes in spacing or tuning can create vast differences in the performance of an antenna array.

The mutual impedance between antennas of an array is important as this factor determines the current that flows in the system for a given amount of power. The current determines the power in a given array and if the mutual impedance between the elements of an array is such that the resulting currents are greater (for the same amount of power) than if the antenna elements were not coupled, then the power gain of the system is greater.

24-5   The Antenna Above A Ground Plane

The properties of an antenna placed near a large conducting ground plane will be modified by the effect of *ground reflection.* In the hf region, the ground is a basic part of the antenna system and affects both the radiation pattern of the antenna as well as its radiation resistance. To estimate the effects of the ground plane, an *image antenna* is introduced below the ground plane as shown in figure 17. The electric charges of the master antenna above the ground are reversed in the imaginary ground image antenna. In addition, the vertical components of the image are in the same direction as those in the master antenna, while the horizontal components are reversed in direction. The radiated field of the master antenna above the ground plane can be determined by replacing the ground plane with the image antenna and computing the resulting field of the two antennas. In a similar manner, the effect of the ground on the radiation resistance of the antenna can be determined by image theory.

(Of interest is the case where one end of the master antenna terminates on the ground. For the case of the *Marconi antenna* (figure 18), the input impedance of the antenna is one-half of the value of the antenna plus its image when driven in free space. The impedance of a quarter-wave Marconi, then, is one-half that of a half-wave dipole in space, or about 36.5 ohms.)

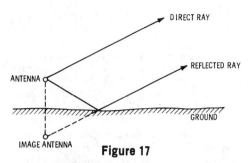

Figure 17

GROUND PLANE PROVIDES MIRROR-IMAGE ANTENNA

The effects of a nearby conducting ground may be estimated by laws of optical reflection from a mirror. An image antenna is introduced below the ground plane at the same distance from it that the master antenna is above the plane. At a distant point the field strength of the antenna is the resultant of two rays, one direct from the antenna and the other reflected from the ground.

A reflected ray is assumed to radiate from the image antenna and is combined with the direct ray, the resultant ray depending upon the orientation of the antenna with respect to the earth. The reflected, or image, ray travels a longer distance to a given point than does the direct ray and this difference in path length results in a distant field pattern that is dependent on the height of the antenna above the ground and the characteristic of the ground. At some vertical angles above the horizon the direct and reflected rays may be in phase, additive, and at other angles the rays may be out of phase with the resultant field being the difference between the two.

In summary, then, the effect of the reflecting ground plane is different for horizontal and vertical antennas because of the reversal of electric charges in the image antenna. Vertically polarized waves are reflected with no change in phase and horizontally polarized waves have their phase shifted 180 degrees on reflection. These effects produce profound differences in the field pattern of the antenna, as will be discussed in a subsequent chapter.

The "Perfect" Antenna   A simple antenna capable of covering an immense frequency span and having a smooth electrical transition between guided and free waves is shown in figure 19. A

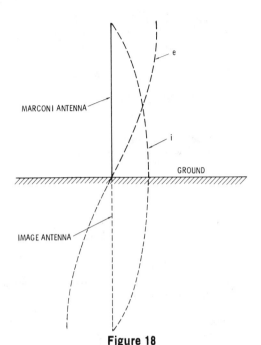

Figure 18

MARCONI ANTENNA AND GROUND IMAGE

The missing half of the dipole antenna is supplied by the ground image for the case of the Marconi antenna. Antenna feedpoint impedance is one-half that of dipole, or about 36.5 ohms.

coaxial transmission line gradually diverges in such a way as to hold constant the natural line dimension ratios, expressed as an impedance (illustration A). If the divergence is smooth, gradual, and small in terms of wavelength, relatively little reflection will exist at any point along the diverging system. A guided wave traveling along the expanding line will expand smoothly over a larger and larger area, and when reaching the end of the line, will simply proceed into free space with little, if any, reflection. This simple antenna is relatively insensitive to the frequency of the emitted wave, provided the antenna is large in relation to wavelength.

A more practical and less bulky broadband antenna which holds true to the concept of gradual, smooth dimensional change per wavelength, is shown in illustration B. If the structure modification is more severe introducing a sudden change in system cross-section, additional sources of reflection are

introduced and the bandwidth of the antenna is reduced accordingly (illustrations C and D).

For very practical reasons it is economical to hold the volume occupied by any antenna

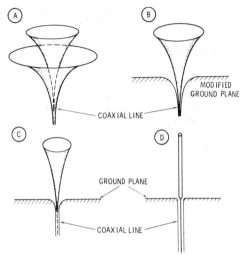

Figure 19

EVOLUTION OF A BROADBAND ANTENNA

A coaxial transmission line gradually diverges in such a way as to hold constant the natural impedance of the line (A). The wave travelling along the line will expand smoothly over a larger and larger area and, when reaching the open end of the line, will pass into free space with little reflection. This infinitely broad structure can be modified (B) while still holding to the concept of gradual dimensional change per unit of wavelength, now resembling a broadband conical antenna working against a modified ground plane. More severe modification (C) produces a true conical antenna of moderate bandwidth and more severe change in system cross-section. The ultimate modification is reached when the center structure is reduced to a monopole (D) having a very restricted bandwidth and minimum reflection only over a restricted frequency range.

to the very minimum. Wideband antennas such as those discussed are uneconomical, except in the uhf region, since they occupy more space than other designs that have acceptable bandwidth. Smaller antenna structures can be built by permitting a greater degree of reflection to occur in the transformation of radio energy from the guided to the free state, and then compensating for the undesired reflection by introducing a compensating reflection somewhere in the feed system, or transmission line.

In the hf and vhf spectrums, in particular, very thin wire or tubing elements are commonly used to assemble relatively narrow-bandwidth antenna systems having high gain, suitable only for operation over a quite restricted frequency region.

PART II

ELECTROMAGNETIC WAVE PROPAGATION

Radio waves may be propagated from a transmitting antenna to a receiving antenna along the surface of the earth, through the atmosphere, or by reflection or scattering from natural or artificial reflectors. At the lower end of the communications spectrum, the ground wave may be propagated for several hundred miles. At high frequencies, however, the ground losses are so great that the ground wave can be propagated for less than one hundred miles. Propagation in the medium and high portion of the hf band is therefore primarily by *ionospheric reflection.*

The refractive index of the atmosphere is an important factor in radio propagation, especially above 100 MHz. Scattering of the radio waves by inhomogeneities in the atmosphere is used to provide satisfactory communication up to several times the line-of-sight distance. At higher frequencies, atmospheric absorption limits propagation to an extent, but the use of high-gain beam antennas makes the use of such frequencies practical.

Propagation up to 30 kHz Propagation of *very low frequency* (vlf) radio waves over short distances is by a ground or surface wave. Attenuation of the wave is quite low. At great distances the field intensity falls rapidly because of losses in the ground and because of the curvature of the earth. These losses increase with frequency.

At sufficiently great distances propagation is chiefly due to propagation in the earth-ionosphere "waveguide" composed of earth and ionospheric multiple reflections.

Propagation From 30 kHz to 2000 kHz Propagation at these frequencies is a combination of surface and sky waves reflected from the ionosphere. The attenuation of surface wave propagation over land is shown in figure 20. Skywave reflection causes fading at medium

distances, particularly at night and on the lower frequencies during the day. Skywave field strength is subject to various irregular fluctuations due to changing properties of the ionosphere.

24-6 Propagation— 2 to 30 MHz

At frequencies between about 2 and 30 MHz and for distances greater than 100 miles, transmission depends chiefly on sky waves reflected from the *ionosphere.* This is a region high above the earth's surface where the rarefied air is sufficiently ionized by ultraviolet light from the sun to reflect or absorb radio waves. The ionosphere is considered to be that region lying between 30 to 250 miles (50-400 km) above the sur-

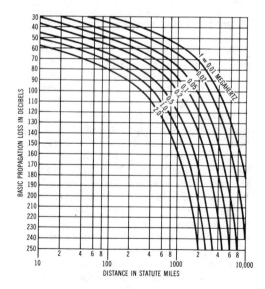

Figure 20

PROPAGATION LOSS

Basic propagation loss expected for surface waves propagated over a smooth spherical earth.

face of the earth and consists of a number of layers:

The F₂ Layer   The higher of the two major reflection regions of the ionosphere is called the F_2 *layer*. This layer has a virtual height ranging from 130 to 250 miles (200-400 km) and is the principal reflecting region for long distance high-

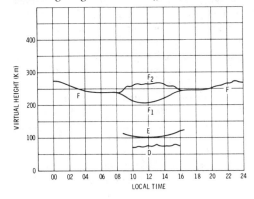

Figure 21

REPRESENTATIVE HOUR-TO-HOUR CHANGES IN THE IONOSPHERE

Ionized regions are referred to as layers, but they are not completely separated from one another. Each region overlaps the adjoining one, to some extent, forming a continuous but non-uniform area with at least four levels of peak density designated D, E, F₁ and F₂ layers. Summertime F₂ critical frequencies are lower than winter values but F₂ nighttime critical frequencies during the summer months are higher than in winter. Thus the difference between day and night critical frequencies is much smaller in the summer than during the winter.

frequency communication. Height and ionization density vary diurnally, seasonally, and with the sunspot cycle. At night, the F_2 layer merges with the F_1 layer and reduction in absorption of the E layer causes nighttime field intensities and noise to be generally higher than during daylight hours.

The F_2 layer appears about sunrise, local time, the critical frequency rising sharply, reaching a maximum a few hours after the sun is at its highest elevation, then decreasing exponentially from this value, reaching minimum during nighttime hours (figure 21).

The F₁ Layer   The F_1 *layer* has a virtual height of about 100 to 150 miles (160-240 km) and exists only during

the daylight hours. This layer occasionally is the reflecting region for hf transmission, but usually waves that penetrate the E layer also penetrate the F_1 layer, to be reflected by the F_2 layer. The F_1 layer introduces additional absorption of such waves. At night the F_1 layer is nonexistent, merging with the F_2 layer to form the single nighttime F layer.

The E Layer   Below the F layer at a height of about 60 miles (100 km) is an absorptive layer termed the E *layer*, which exists during daylight hours, reaching

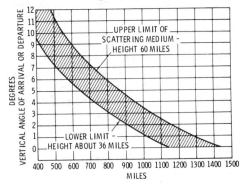

Figure 22

E LAYER SCATTER RANGE

E layer scatter range may be as great as 1400 miles for low angle, single-hop transmission. A high antenna (several thousand feet high, such as on a mountain top), combined with a sea level horizon, is ideal. The scatter occurs at layer height of about 36 to 60 miles.

a diurnal maximum at noon. For all practical purposes, the E layer disappears at night, although weak traces of it are often observed. This layer is important for daytime hf propagation at distances less than 1000 miles (1600 km), and for occasional medium-frequency nighttime propagation at distances in excess of 100 miles (160 km). Irregular cloud-like areas of unusually high ionization, called *sporadic E*, may occur up to more than half of the time on certain days or nights. A large percentage of *sporadic E* propagation is attributed to visible bombardment of the atmosphere by the sun.

Layer height and electron density of the atmosphere determine the skip-distance of *sporadic E* propagation for a given signal angle (figure 22), and distances of 400 to 1200 miles (650-1930 km) are common on

50 MHz. Multiple-hop propagation is often possible up to about 2500 miles (4000 km) on occasion. *Sporadic E* propagation has been observed in the 144 MHz band, but is not as common as on the lower frequency bands.

E layer propagation on the vhf bands is most common during the summer months, with a shorter season during the winter, with the periods reversed in the southern hemisphere.

The D Layer   Below the *E* layer, the *D* layer exists at heights of 30 to 50 miles (50-80 km). It is absorptive and exists in the middle of the day during the warmer months. Not much is presently known about the characteristics of this layer, as it is so weakly ionized that the usual pulse-probing techniques do not produce meaningful echos. It is known that the *D* layer remains ionized as long as the atmosphere receives solar radiation and disappears quickly at sundown. It is thought this layer causes high absorption of signals in the medium- and high-frequency range during the middle of the day.

The Critical   The *critical frequency* (f_c) of
Frequency   an ionospheric layer is the highest frequency which will be reflected when the wave strikes the layer at

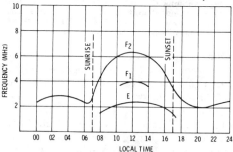

Figure 23

VIRTUAL HEIGHT OF IONOSPHERE IS PRESENTED IN AN IONOGRAM

Point of reflection of radar echo in ionosphere is measured and presented in graphic form, showing height as a function of frequency for specific times. Frequencies higher than a critical frequency will pass through the ionosphere and not be reflected, when a vertical pulse is used as a measuring device. At oblique angles, frequencies higher than the critical frequency will be reflected back to earth, creating a skip distance zone for a given circuit.

vertical incidence. Frequencies higher than f_c pass through the layer. The critical frequency of the most highly ionized layer of the ionosphere may be as low as 2 MHz at night and as high as 10 to 15 MHz in the middle of the day.

The critical frequency and height of the layers are measured by a pulse technique, the pulse and its return echo being observed on a cathode-ray tube, as in a radar set. The *virtual height*, or point of reflection in the ionosphere determined by this technique is presented in an *ionogram*, showing height as a function of frequency for specific periods of time (figure 23).

The critical frequency is of interest in that a *skip distance* zone will exist on all frequencies greater than the highest critical frequency at a given time for a given circuit. The higher the critical frequency, the greater the density of ionization and the higher the maximum usable frequency.

The Maximum   High-frequency radio
Usable Frequency   waves travel from the
(MUF)   transmitter to a distant point by reflection from the ionosphere and earth in one or more

Figure 24

THE MAXIMUM USABLE FREQUENCY

In order for a radio signal to be reflected from T to R, the electron density at B must be high enough to support reflection. As the frequency of the signal is raised, at some point the electron density will not be great enough to bend the wave back to earth and it will continue through the ionosphere into space. The upper frequency limit, or maximum usable frequency, can be calculated from ionospheric measurements by determining the critical frequency at point E. The vertical critical frequency determined is multiplied by a factor to provide the value of the oblique incident MUF for a particular distance (D) and layer height (h).

hops, as indicated in figure 24. For a radio signal to travel from T to R via the ionosphere, its frequency must be less than a maximum value. Above this frequency, the electron density at B will not be great enough to bend the signal back to earth and it will continue on through the ionosphere into space. There is, therefore, an upper limit to the range of frequencies that will be reflected by the ionosphere between any two fixed points. This upper limiting frequency is called the *maximum usable frequency (MUF) for a given circuit*. The MUF is highest near noon or in the early afternoon and is highest during periods of greatest sunspot activity, often going to frequencies higher than 30 MHz (figure 25). The MUF often drops below 5 MHz in the early morning hours. Ionospheric losses are at a minimum near the MUF and increase rapidly for lower frequencies during daylight. MUF data is published periodically in radio amateur magazines.

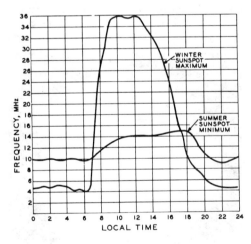

Figure 25

MUF IS HIGHEST DURING PERIODS OF MAXIMUM SUNSPOT ACTIVITY

MUF extremes are greatest during periods of high sunspot activity. Ionospheric losses are at a minimum near the MUF and increase rapidly for lower frequencies, especially during daylight. The recommended upper limit of frequency for maximum circuit reliability is called the Optimum Traffic Frequency and is selected somewhat below the MUF to provide margin for ionospheric irregularities.

The Optimum Traffic Frequency (FOT) The recommended upper limit of frequency for maximum reliability is called the *optimum traffic frequency (FOT)* and is selected somewhat below the MUF to provide some margin for ionospheric irregularities and turbulence, as well as for day-to-day deviations from the predicted monthly median values of MUF. The FOT is usually about 15-percent less than the MUF for a particular communication circuit. As far as practicable, the FOT is chosen in close proximity to the MUF in order to reduce absorption loss.

The Lowest Usable High Frequency (LUF) The *lowest usable high frequency (LUF)* is the lowest frequency that can be used for a satisfactory communication circuit over a particular path at a particular time. The LUF depends primarily on atmospheric noise and static at the receiving site for a determined signal-to-noise ratio. At frequencies below the LUF, reception will not be possible since the received signal is lost in the prevailing noise level. As the operating frequency is raised above the LUF, the signal-to-noise ratio improves.

Unlike the MUF, which is dependent entirely upon ionospheric characteristics, the LUF can be controlled to an extent by adjustments in effective radiated power and circuit bandwidth. Generally speaking, the LUF can be lowered approximately 2 MHz for each 10-decibel increase in effective radiated power.

Long Path Propagation The great circle path is the shortest distance between two points on the earth's surface. Most hf communication is via the great circle route. However, *long path propagation* along the reciprocal path is common in the hf region. Generally speaking, the long path travels through a zone of darkness while the short path travels through sunlight.

The illuminated short path implies high ionospheric absorption in addition to path attenuation. The long path passes through the dark hemisphere so that considerably less signal absorption is encountered. Moreover, the maximum usable frequency (MUF), which also depends on the position of the sun, is lower for the long path than for the short path.

Representative open hours of the long path from Europe to Australia are shown in figure 26. The spring months of the year are the best for this circuit, with May showing a long path opening of nearly four hours. The shortest openings occur in the winter months.

Many times both the short and long paths between two points are open for communication. When this happens, a distinct echo can be heard on a signal which results from the delay in arrival time of the long path signal as compared to the arrival time of the short path signal.

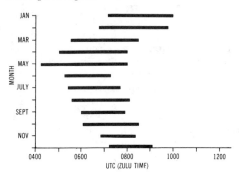

Figure 26

LONG PATH OPENINGS

Opening hours of the long Great Circle path between Western Europe and Australia at 7 MHz. (From the Journal of the ITU)

Multihop Propagation Long distance hf propagation occurs by means of a number of reflections from ionospheric layers known as *hops*. Radio signals may traverse a path by means of several simultaneous modes involving a different number of hops. The nearer the signal is to the operational MUF, the smaller is the number of modes involved.

Since there are several ionospheric layers that will support propagation, notably E, F_2, and to some extent F_1, it is possible to have modes which are combinations of hops reflected from the various layers and also modes which bounce internally between layers.

Because propagation time for each mode is different, a single short radio pulse transmitted at a given time may be manifest as a series of broken, short pulses at different arrival times. *Multipath distortion* may cause serious errors in frequency-shift keying or frequency modulation, as well as in forms of amplitude-modulated transmission.

24-7 Cycles in Ionospheric Activity

The first recorded observations of *sunspot* activity were made by Chinese observers more than 2000 years ago (figure 27). Centuries later, in 1901, Marconi was unaware that his successful spanning of the Atlantic Ocean by radio for the first time was possible only because of the existence of sunspots which, the astronomers of that time thought, might be holes cut in the sun's surface by solar hurricanes, exposing the cooler layers below.

Experiments conducted by Heaviside (1902), Appleton (1924), and Naismith (1927) proved the existence of an electrified reflecting region in the atmosphere, measured the characteristics of it and reached the con-

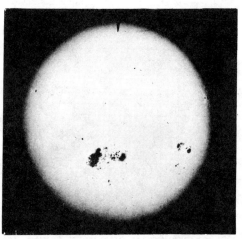

Figure 27

SUNSPOTS IN ACTION

Sunspots have been observed and recorded for more than 2000 years. In this U.S. Navy photograph, a large group of sunspots is seen, moving from east to west, as the sun rotates. Sunspot activity has direct bearing on radio transmission.

clusion that the principal solar factor in the production of ionization in the atmosphere was ultraviolet radiation from the sun. Later

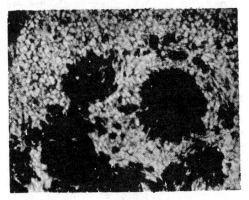

Figure 28

SPOTS ON THE PHOTOSPHERE OF THE SUN

Large spots embedded in the solar surface are seen in this NASA photograph made from an unmanned research balloon at an altitude of 80,000 feet. The granular composition of the sun's surface can be seen clearly.

investigators discovered a direct relationship between the ultraviolet radiation, the degree of ionization in the atmosphere, and its relationship with long distance radio communication.

Sunspots in Action With the aid of suitable instruments, sunspots can be seen to develop from small dark areas on the brilliant surface of the sun. Studies indicate that the inner portion of the sunspot is a depression in the sun's surface having an average depth of several thousand miles (figure 28). The temperature of the sunspot is several thousand degrees cooler than that of the general surface of the sun and gives off about one-half as much light as the same area of the *photosphere*, or surface of the sun.

Sunspots almost always appear in groups, some spots as large as 80,000 miles (128,000 km) in diameter. The groups move parallel to the equator of the sun in an east to west direction in accord with the sun's rotation. Many terrestrial phenomena which are influenced by localized sunspot activity on the sun tend to occur at intervals of about

27 days, which is the period of rotation of the sun.

The Sunspot Cycle The number of sunspot groups, and individual sunspots, visible on the sun's surface vary between wide limits over a period of time. Sunspot activity follows an approximate 11-year cycle, steadily rising from very few to a maximum amount, then slowly receding to a minimum amount again.

The sunspot count is recorded in *Zurich Sunspot Numbers* on a daily and monthly basis, and 12-month, smoothed running numbers are published in *CQ magazine* and various astronomical publications. The recordings began in 1750 and 19 complete cycles have been recorded to date. No two cycles have been exactly alike, although a definite repetitive behavior is established. Basic characteristics of the cycle, such as duration, height of maximum, depth of minimum and ascent and descent time are observed, and vary from cycle to cycle. No explanation of the sunspot cycle has yet proven to be completely satisfactory and current estimates of future performance are open to speculation. The present search for empirical laws governing solar activity has

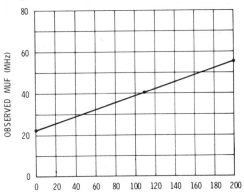

Figure 29

RELATION BETWEEN OBSERVED MUF AND SMOOTHED SUNSPOT NUMBER

When the sunspot count is high, ionization of the earth's atmosphere is heavy and the MUF is correspondingly high, opening up additional frequencies for long distance communication. Predictions for cycle 21 indicate a maximum sunspot count of about 55, thus limiting the MUF to approximately 32 MHz for the next 15 years.

proceeded according to two different schools of thought, one holding that solar activity is a periodic phenomenon, the other considering each solar cycle as an independent event.

Since hf radio transmission is dependent on the ionosphere, which varies with the sunspot cycle, the action of the cycle is of extreme interest to communicators (figure 29). When the sunspot count is high, ionization of the earth's atmosphere is heavy and the *MUF* is correspondingly high, opening up additional frequencies for long-distance communication. During cycle 19, which peaked at a count of over 200, the *MUF* regularly exceeded 50 MHz. Cycle 20, which ended in 1975, was considerably lower, limiting the *MUF* to something over 30 MHz at the peak of sunspot activity.

Cycle 21 reached its peak of about 150 during the late summer of 1979, making it among the most intense recorded since sunspot observations began. In the twenty cycles observed since 1755, only three have exceeded a smoothed sunspot number of 150; cycle 3 with a peak of 159 in 1778, cycle 18 with a peak of 152 in June, 1947 and cycle 19 with a record-breaking level of 201 in November, 1957. An extended prediction indicates that sunspot numbers in the vicinity of 100 to 150 may be observed

during the following sunspot cycle. Thus, the next 40 years may be characterized by medium to high values of sunspot activity comparable to the activity of the last 40 years.

The implication of low sunspot activity is that the *MUF* will be considerably lower, long distance propagation will be more infrequent and will occur for shorter periods of time, and with reduced signal levels. Frequencies below 8 MHz, however, may show improvement even though the higher frequencies may show marginal performance.

Thus communication using ionospheric reflection in the hf bands will continue to react to the influence of the sun and undoubtedly more vital communication circuits will be switched to satellites to overcome the vagaries of ionospheric reflection.

Geographical Variations in the MUF   At any specific time of day the sun's zenith angle varies with geographical latitude, and the intensity of ionizing radiation sweeping across the earth's upper atmosphere varies accordingly. The critical frequency and MUF, therefore, vary with geographical location, being highest in equatorial regions, where the sun is more directly overhead, and decreasing proportionately north and south of these latitudes (figure 30).

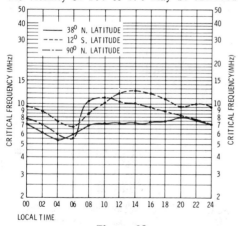

Figure 30

LATITUDE VARIATION IN F₂ CRITICAL FREQUENCIES

Values of critical frequency are generally highest in equatorial regions and lowest in high-latitude regions. Frequency also varies with time of day.

24-8   Ionospheric Disturbances

The diurnal, seasonal and solar cycle variations of the ionosphere discussed previously are dependent on the regular, more-or-less predictable behavior of the ionizing solar radiation. From time to time, however, the normal behavior of the ionosphere is upset by disturbances of a transistory or short-duration character. It is believed that these are the result of abnormal radiations from the sun. These disturbances give rise to abnormal radio propagation conditions, sometimes leading to a temporary "radio blackout," or complete failure of hf radio communications.

Ionospheric disturbances fall into two main categories: the *sudden ionospheric disturbance* (SID) and the *ionospheric storm.* The SID commences suddenly and lasts from a few minutes to an hour or so. The ionospheric storm develops over a period of a day or two and generally continues for several days. In either case, the normal behavior of the ionosphere is upset, with critical frequencies dropping, and ionospheric absorption increasing as the intensity of the disturbance increases.

The SID has a spectacular effect on hf propagation. A near-simultaneous radio fadeout occurs over a large portion of the hf spectrum, from approximately 2 MHz to 30 MHz, with even background noise sometimes disappearing. The only signals that can be heard during an SID are those from stations within the ground-wave range. The fadeout lasts for a short period, then conditions slowly return to normal.

It is thought that the SID is a result of a solar flare, a sudden, short-lived, bright eruption on the face of the sun. The incidence of solar flares varies with the solar cycle and is most prominent during years of very high solar activity.

The SID takes place about 11 minutes after a solar flare commences, and occurs only in those areas of the world in complete daylight. Not all flares produce SIDs, indicating that the SID is only one manifestation of the release of solar energy.

The typical change in a communication circuit during an SID is shown in figure 31. Signal drop-off is approximately 40 decibels in a matter of a few minutes, with the signal returning to normal in about 40 minutes.

A second type of disturbance is the ionospheric storm. While not as spectacular as the SID, the storm actually constitutes a more serious communications problem because of its much greater duration. During a storm, hf signals (from approximately 3 MHz to 30 MHz) drop to a very low level and may even disappear entirely for periods of several days. Measurements indicate that the F layer is usually at an abnormally great height during the disturbance and is subject to considerable turbulence. Unlike the SID, the higher frequencies are most affected, and the storm occurs in both daylight and darkness regions of the world. Ionospheric absorption

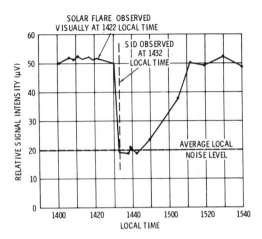

Figure 31

SID SIGNAL DROP-OUT IN A COMMUNICATION CIRCUIT

Solar flare causes a sudden ionospheric disturbance about 11 minutes later in areas of the world in complete daylight. Signal returns to normal in 30 to 40 minutes after a drop-off of about 40 decibels in strength.

increases and signals are subject to considerable fading, often of an unusual type known as *flutter fading.*

It is thought that the ionospheric storm is caused by corpuscular radiation of ionized calcium emitted from solar flares at the same time the flare emits ultraviolet and X-ray radiation which produce the SID. Corpuscular radiation travels at a velocity much lower than the speed of light because of its greater energy content and arrives at the earth at a later period of time. The radiation is so confined that unless the emission is pointing directly at the earth, it may miss the earth entirely (figure 32).

Besides radiant energy, solar flares also emit bursts of electromagnetic energy in the form of radio "noise." These bursts, occurring over a wide range of frequencies above about 10 MHz, are strongest in the vhf region of the radio spectrum. They can be received as a hissing sound on a sensitive receiver. The flares also violently disrupt the earth's magnetic field for short periods of time as they disrupt the ionosphere. These *magnetic storms* are most intense in high latitudes and often last for several days.

As satellites and space vehicles probe further into space, many of the secrets of the

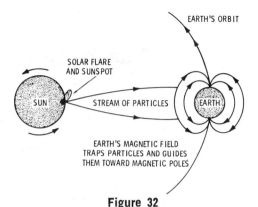

Figure 32

SOLAR PARTICLES CAUSE IONOSPHERIC STORMS ON EARTH

An ionospheric storm is caused by corpuscular radiation emitted from solar flares at the same time flare emits ultraviolet and X-ray radiation which produce the SID. Corpuscular radiation travels at lower velocity than light and arrives at the earth at a later time period. Particles cause long-lasting ionospheric storm which disrupts long distance radio communication.

solar flare, the SID and the magnetic storm will be revealed, and in the future the prediction of these phenomena may be made with greater accuracy than is possible at the present time.

Atmospheric Noise   The usefulness of a radio signal is limited by the total noise in the receiver which may be either unwanted, external noise or the internal noise of the receiver.

Atmospheric *static* is usually the limiting factor in receiver sensitivity at frequencies below 30 MHz, while receiver noise is the primary limitation at higher frequencies, especially those above 200 to 500 MHz. In the hf band, the controlling factor depends upon the location of the receiver, time of day, man-made noise and atmospheric static.

Static is caused by lightning and other natural electrical disturbances and is propagated worldwide by ionospheric reflection. Static levels are generally stronger at night than in the daytime and the levels are higher in the warm tropical areas than in the cooler northern regions, which are far removed from most lightning storms.

The average static level in the tropics may be as much as 15 decibels higher than for the temperate zones, while in the Arctic

regions the static level may be 15 to 25 decibels lower. In all areas, typical summer averages are a few decibels higher than the winter values.

External noise is an important factor in receiver design, and this subject is discussed further in the receiving section of this Handbook.

24-9   Propagation in the VHF Region

As a result of the tremendous increase in vhf activity since World War II, much has been learned about the different modes of radio propagation at these frequencies. The boundary between the hf and the vhf region is variable, falling between 30 MHz and 50 MHz and is generally taken to be the *MUF*, above which normal ionospheric reflection ceases. Deviations from this simple definition are numerous. Interestingly, certain types of vhf propagation provide the only *reliable* means of long distance radio communication known today. These types will be discussed in detail later in this chapter.

Ionospheric Scatter Propagation   *Ionospheric scatter propagation* permits communication in the frequency range of about 30 MHz to 300 MHz over

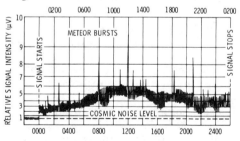

Figure 33

IONOSPHERIC SCATTER SIGNAL LEVEL IS LOW, PUNCTUATED BY METEOR BURSTS

Because only a small proportion of the radiated energy is scattered and returned to earth, scatter signals are very weak. Lower limit of ionospheric scatter is determined by masking action of normal ionospheric skip distances. Regular sky wave propagation can create selective interference on a scatter link circuit.

distances ranging from 600 miles (1000 km) to nearly 1200 miles (2000 km). It is believed that this type of propagation is due to scattering of the signal from the lower D layer, or possibly the E layer. Because only a small portion of the radiated energy is scattered and returned to earth, such scatter signals are very weak (figure 33). The lower limit of ionospheric scatter is determined by the masking action of normal ionospheric skip distance. Regular sky-wave propagation will create undesirable interference to a scatter signal and produce selective fading on a scatter link circuit.

Ionospheric scatter seems limited to a single-hop distance. Theoretically, it would be possible to communicate via double-hop scatter, which could extend the range to 2000 miles (3200 km) or so, but circuit attenuation would be extreme.

Meteor-Burst Propagation Meteors have been observed for centuries, but until recently they were assumed to be relatively few in number. Recent studies, however, have shown that the earth is constantly colliding with innumerable particles as it sweeps on its annual journey around the sun. Over ten billion particles are esti-

Name of Shower	Date of Peak Intensity	Duration (Days)	Meteors Per Hour
Quandranids	January 3	1	35-40
Lyrids	April 21	2	12-15
Eta Aquarids	May 5	9	12-20
Delta Aquarids	July 29	10	20-30
Perseids	August 12	5	50
Orionids	October 21	4	20-25
Taurids	Nov. 5; Nov. 12	20	12-15
Leonids	November 17	4	20-25
Geminids	December 13	5	40-50
Ursids	December 22	2	15

Figure 34

MAJOR METEOR SHOWERS

List of major meteor showers. The spring showers peak between midnight and 0600, the Ursids peak during the early afternoon hours. Others generally peak during hours of darkness. Seasonally, more meteors occur during May and July than at any other time.

mated to reach the earth each 24 hour period, with the largest number of these less than 0.016 cm in diameter. Only a very few are large enough to be noticed, and only an extremely small percentage of the latter are large enough to reach the ground before they are burned up by friction with the earth's atmosphere (figure 34).

When a meteor strikes the earth's atmosphere, a cylindrical region of free electrons is formed at about the height of the E layer. This slender, ionized column is quite long, and when first formed is sufficiently dense to reflect radio waves back to the earth. Frequencies in the range of 50 MHz to 80 MHz have been found best for meteor-burst transmission.

The effect of a single meteor of medium size (1 cm) shows up as a sudden "burst" of signal of short duration at a point not normally reached by the transmitter. The aggregate effect of many meteors impinging on the earth's atmosphere, while perhaps too weak to provide long-term ionization, is thought to contribute to the existence of the nighttime E layer.

Aurora Propagation At the earth's poles, where the atmosphere is more rarefied than elsewhere, radiation from the sun not only causes ionization, but often causes the air molecules to ignite. This phenomenon is called an *aurora* (or "northern" or "southern" lights). The action is similar to that which takes place in a neon tube. The aurora is a spectacular observance, with lights arcing across the night sky as yellowish-green dancing ribbons, or curtains, or great draperies which appear to fold and unfold. They occur at E layer height in the ionosphere and can be seen on the horizon as far as 600 miles (960 km) from the zenith point.

In the northern hemisphere, the zone of maximum occurrence (*auroral zone*) swings across northern Norway, Greenland and central Canada and back across Alaska, Siberia and northern European USSR (figure 35). Both north and south of this belt the occurrence of auroras decreases.

Auroras play havoc with high frequency radio communication and cause severe absorption of any hf wave that passes near or through the auroral zone. Besides absorption, the aurora superimposes an *auroral flutter* on hf signals.

Auroral propagation of vhf signals is common at frequencies between 100 MHz and 450 MHz. The propagation involves reflection of the wave from the auroral display. The reflection properties of the aurora vary quite rapidly, with the result that the reflected vhf signal is badly distorted by multipath effects. Voice modulation becomes very rough and c-w telegraphy is usually employed for auroral communication in the vhf amateur bands.

Since aurora is caused by emission of charged particles from the sun, it is natural to find that aurora propagation follows the sunspot cycle and reaches a peak at the same time as the cycle. In addition, auroras follow a seasonal pattern, peaking around March and September, although they may occur at any time.

Because of the shallow nature of the aurora belt, east-west transmission paths are usually favored. At times it is possible to communicate up to 2000 miles (3200 km) or more, via aurora propagation, but ranges of a few hundred miles are more common. Aurora propagation seems to reach a peak around sundown or early evening, and again around 0200, local time. The farther north a station is situated, the more frequently it will encounter aurora propagation, but dur-

ing rare occasions it may be possible to employ this mode of transmission in the southernmost portions of the United States.

Vhf aurora propagation may be predicted by monitoring signals in the 2-MHz to 5-MHz range for the characteristic aurora distortion. This is evidence that vhf propagation may soon be possible.

Tropospheric Scatter Propagation *Tropospheric scatter (troposcatter)* is thought to be caused by random irregularities in the atmosphere in which the refractive index differs from the mean value of surrounding areas. The scattering effect seems to take place by partial reflection where there is a rapid change of reflective index over a small range in height associated with temperature and humidity changes. The result of scatter refraction is a faint signal illumination of the ground well beyond the horizon (figure 36).

The forward-scattering mechanism involves a large transmission loss and it becomes necessary to use high gain, narrow beam antennas for both transmission and reception. The effect of the scatter angle between the receiving and transmitting beam antennas is significant and is kept as small as possible by choosing transmitting and receiving sites so as to have an unobstructed view of the horizon.

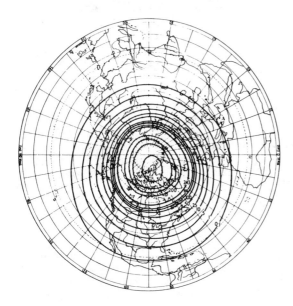

Figure 35

AURORA DISPLAY IS MOST PREVALENT AT NORTHERN LATITUDES

Aurora can be seen on occasion as far south as Mexico City. The average number of nights per year having aurora displays are shown in this polar chart. Auroral propagation of vhf signals is common at frequencies between 100 and 450 MHz, but aurora disrupts hf radio communication at the same time.

The received scatter signal fluctuates continuously due to the large number of randomly varying components; hourly, daily and monthly variations may reach 10 to 20 decibels or more. However, *consistently usable signals* are obtainable at ranges exceeding 400 miles (700 km).

The scattering mechanism may be compared to the scattering of a light beam in a heavy fog, or mist, which results in a heavy glare of light caused by miniature water droplets, leaving the background weakly illuminated. No critical frequency is involved in the scattering mechanism, though the intensity of the scattered reflections decreases with increasing frequency.

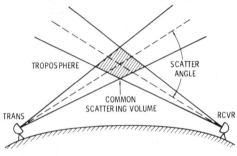

Figure 36

GEOMETRY OF TROPOSPHERIC SCATTER SYSTEM

Forward scatter mechanism involves a large transmission loss and requires high gain, narrow beam antennas at both ends of the circuit. The scatter angle is kept as small as possible by proper choice of transmitting and receiving sites.

Trans-Equatorial Scatter Propagation — *Trans-equatorial scatter (T-E scatter)* has been observed on the 50-MHz amateur band during periods of moderate and high solar activity, over long north-south paths spanning the magnetic equator at times when the expected *MUF* is considerably lower for the paths involved.

T-E scatter is believed to be due to a highly ionized distortion known to exist in the ionosphere over the magnetic equator. Waves entering this area at a favorable angle are reflected considerable distances between the sides of the bulge, resulting in a long, single-hop opening, without intermediate ground reflection, of up to 5000 miles (8000 km).

T-E scatter is a nighttime propagation phenomenon, with most openings occurring between 2000 and 2300 hours, local time at the path midpoint. The signals must cross the magnetic equator in a north-south direction or propagation will not take place. The T-E maximum usable frequency is approximately 1.5 times greater than the daylight *MUF* observed on the same path. To date, no T-E scatter propagation has been observed over 100 MHz.

Sporadic E Propagation — Sporadic *E* propagation has been mentioned earlier in this chapter. It is a popular form of communication for radio amateurs on the hf and vhf frequencies as it calls for no special station equipment. Sporadic *E* openings on the higher frequency bands may often be predicted by observing the characteristics of the 28-MHz band. The geometry of propagation is such that as the skip distance decreases on the 28-MHz band, the highest frequency that will be reflected by a sporadic *E* cloud is increasing. Experience has shown that when skip signals are heard less than 500 miles (800 km) away on 10 meters, the chances are very good that sporadic *E* propagation will be noted on the 50-MHz band over the same general direction.

Tropospheric Ducting — *Tropospheric ducting* of vhf signals is quite common and is the result of change in the refractive index of the atmosphere at the boundary between air masses of differing temperatures and humidities. Using a simplified analogy, it can be said that the denser air at ground level slows the wave front a little more than does the rarer upper air, imparting a downward curve to the wave travel.

Ducting can occur on a very large scale when a large mass of cold air is overrun by warm air. This is termed a *temperature inversion*, and the boundary between the two air masses may extend for 1000 miles (1800 km) or more along a stationary weather front.

Temperature inversions occur most frequently along coastal areas bordering large bodies of water. This is the result of natural onshore movement of cool, humid air

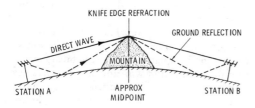

Figure 37

KNIFE-EDGE DIFFRACTION

A ridge of hills or mountains may exhibit diffraction of a vhf wave traveling over the crest. An obstacle gain as high as 20 decibels may be realized when transmitting and receiving sites are optimized for maximum diffraction.

shortly after sunset when the ground air cools more quickly than the upper air layers. The same action may take place in the morning when the rising sun warms the upper air layers.

Tropospheric communication as a result of ducting is rare below 144 MHz, but occurs commonly in the 144-MHz to 450-MHz range. Less spectacular communications are possible as a result of simple temperature inversion, where ducting is not believed possible. Ducting over water, particularly between California and Hawaii, and Brazil and Africa, has produced vhf communication in excess of 3000 miles (4500 km).

Knife-Edge Under certain conditions, it is
Diffraction possible for a ridge of hills or mountains to exhibit noticeable diffraction of a vhf wave traveling over the crest. This phenomena of wave propagation is known as *knife-edge bending*, and has been demonstrated for years with light rays. The transmission path over a practical knife-edge diffraction path depends critically on the shape of the edge, the distance separating the stations and the angle from the stations to the obstacle. Ground reflection patterns may hinder the knife-edge path, but when all factors are optimized, an *obstacle gain* as high as 20 decibels may be realized (figure 37).

Moon Reflection Since 1953, radio amateurs
Propagation have been experimenting with lunar communication (*moonbounce*). Moonbounce allows communication on earth between any two points that can observe the moon at a common time and has recently attracted the attention of growing numbers of experimentally minded vhf amateur experimenters.

The *earth-moon-earth* (EME) path varies from 442,000 miles (680,000 km) to 504,000 miles (750,000 km) for a round-

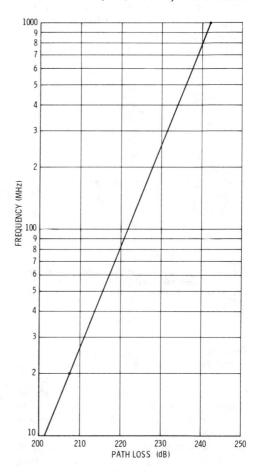

Figure 38

EME PATH LOSS

The average path loss, assuming 500 watts of radiated power, and a moon reflectivity of 7 percent is given. Path loss varies about 1 decibel plus or minus this figure as the monthly range to the moon changes. For 2-meter work, the total path loss is about 225 decibels.

trip signal, which takes approximately 2.5 seconds to make the journey. The moon subtends an angle of only one-half degree, as viewed from the earth and has a coefficient of reflection of only 7 percent for vhf energy that strikes its surface. In spite of these tremendous obstacles, EME radio amateur circuits are in almost daily operation on 144 and 432 MHz. The attenuation of the EME path is shown in figure 38. Attenuation may vary as much as plus or minus one decibel during each month as range to the moon changes. For 144-MHz moonbounce work, the total path loss is about 225 decibels.

The requirements for the amateur station interested in moonbounce experiments are well known. For 144 MHz, as an example, with a transmitter running maximum legal power, an antenna gain of 20 decibels, or more, is required, along with a receiver having a high degree of selectivity and a noise figure of 2 decibels, or better. (The cosmic noise level is about 1.9 decibels, so a system

noise figure much better than this only allows the listener to hear more noise.)

Because the moon may be moving toward or away from the EME stations at speeds up to 980 miles per hour, *Doppler shift* will change the received frequency, according to the formula:

Doppler shift (Hz) $= 2.966 \times f_{(MHz)}$

when the shift is measured at the equator of the earth.

When the moon is rising, the Doppler effect increases the received frequency; at moonset the frequency is decreased.

In addition to the normal path attenuation, additional problems are caused by *Faraday rotation* of the polarization of the received signal. Because of the reflection of the signal, the polarization sense is reversed on the received signal, along with a "twist" in polarization along the path, out and back. A plane-polarized vhf signal passing through the ionosphere is gradually rotated in phase, and may go through several rotations before passing through the ionosphere into space. After reflection and phase reversal, the signal re-enters the ionosphere and rotates once again on the return path to the receiving antenna. The overall rotation may produce a 20 to 30 decibel signal loss when received on an antenna having incorrect polarization.

Line-of-Sight Propagation Under normal propagation conditions, the refractive index of the atmosphere decreases with height so that waves travel more slowly near the ground than at higher altitudes. This variation in velocity with height results in bending of the wave toward the earth's surface. Under unusual atmospheric conditions, the refractive index may increase with height, causing the wave to bend upwards, resulting in a decrease in the line-of-sight path.

Over most of the time, uniform, downward bending is present in the vhf and uhf region and may be represented by straight-line propagation, but with the radius of the earth modified so that the relative curvature remains unchanged. The new radius is known as the *effective earth radius* (K). The average value of K in temperate climates is about 1.33.

The distance to the *radio horizon* over smooth earth, when the height h is very

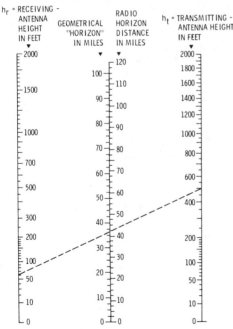

Figure 39

RADIO HORIZON NOMOGRAPH

Example shown: height of receiving antenna, 60 feet; height of transmitting antenna, 500 feet; maximum radio path length, 41.5 miles. Effective earth radius is taken as 1.33.

small compared with the earth's radius, is given with a good approximation by:

$$d = \sqrt{\frac{3\,K\,h}{2}}$$

where,

    h = height in feet above the earth,
    d = distance to radio horizon in miles,
    K = effective earth radius in miles.

The nomograph of figure 39 gives the radio horizon distance between a transmitter at a height h_t and a receiver at height h_r.

24-10   Forecast of High-Frequency Propagation

From theory and experimentation, constantly advancing hand in hand since the first ionospheric experiments of 1925, techniques have been evolved for applying cer-

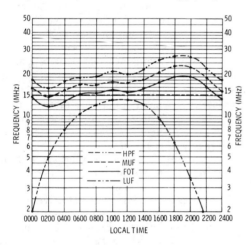

Figure 40

PROPAGATION ANALYSIS CHART FOR NEW YORK TO LONDON PATH

This analysis chart shows the propagation path for a frequency of 14 MHz and an estimated radiated power of 1000 watts. The highest probable frequency (HPF) is that value of MUF that will occur on less than 10 percent of the days of the month. The lowest usable frequency (LUF) is dependent upon the local noise level at the receiving site. The path will be closed when the LUF is greater than the HPF.

tain measurable ionospheric data to the solution of propagation and other engineering problems encountered in establishing hf radio circuits. It is possible, therefore, to estimate the *MUF* and *FOT* for a particular smoothed sunspot number for a given communication circuit. A representative propagation analysis chart for the New York to London circuit for a sunspot number of 150 is shown in figure 40.

World maps with overlay frequency contours are available for making frequency estimates manually and MUF estimations for months in advance may be made, if a predicted value of smoothed sunspot number is known. The maps are available in a set of four volumes: *Ionospheric Predictions*, OT-TER 13, obtainable from the Superintendent of Documents, U.S. Government Printing Office, Washington, DC 20402. The Institute of Telecommunication Sciences of the Environmental Sciences Services Administration (ESSA) issues forecasts which may be used to determine the *MUF* and *FOT* for high-frequency communication paths. A handy source of propagation information is broadcast by the National Bureau of Standards station WWV during part of every 15th minute period on the standard frequencies in the hf range. Finally, the headquarters station of the American Radio Relay League, W1AW, rebroadcasts Propagation Forecast Bulletins on a regularly, weekly scheduled basis to all radio amateurs.

The best estimates indicate that the usable hf spectrum is expected to dwindle to half that space available during 1980 and that between the years 1985 to 2005 the amount of usable hf spectrum may never exceed 70 percent of that available during 1980. On the other hand, the steady use of the hf spectrum is expected to continue, even in spite of the transfer of large volumes of traffic to space satellites. Spectrum conservation and improved propagation knowledge are two actions that must be taken to prevent the high-frequency spectrum from becoming less useful for communications as a result of decreasing solar activity.

Propagation Bulletins   Propagation bulletins are issued by the National Bureau of Standards radio stations WWV and WWVH and are updated four

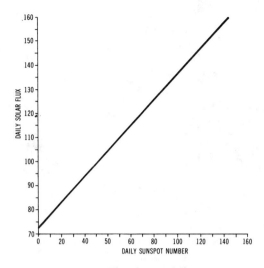

Figure 41

**THE APPROXIMATE RELATIONSHIP
BETWEEN 2800 MHz SOLAR FLUX
AND SUNSPOT NUMBER**

times daily, usually at 18 minutes after each hour. A propagation quality forecast is given along with conditions of the geomagnetic field, the K-index and the solar flux index.

The K-index is a statement of geomagnetic activity and provides an insight of propagation quality on high latitude communication paths. The solar flux index is a measure of solar radiation and is related to the sunspot number and hence the maximum usable frequency (figure 41).

Both indices tend to follow a 27-day pattern as a result of the period of rotation of the sun. They are used to make short-term propagation forecasts for circuits of interest. Generally speaking, the higher the value of solar flux and the lower the level of the K-index, the better will be the propagation conditions.

A complete overview of high frequency propagation is available in *The Shortwave Propagation Handbook*, by Jacobs and Cohen, published by Cowan Publishing Corp., NY.

CHAPTER TWENTY-FIVE

The Transmission Line

A *transmission line* is a conducting system used to guide electrical energy from one point to another. Transmission lines are used to couple antennas to transmitters and receivers, or to establish proper phase relationships between the various elements of an antenna array. Of interest to the user is the distribution of voltage and current along the transmission line for a radio wave of a single frequency, as this knowledge is necessary to achieve maximum energy transfer along the line.

Any transmission line has an *input* (generator) *end* and a *load end*. The electrical properties of the line are specified by its distributed parameters which are the *series im-*

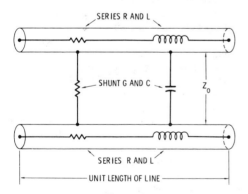

Figure 1

REPRESENTATIVE CIRCUIT OF A SHORT SECTION OF TWO-WIRE TRANSMISSION LINE

The distributed properties of resistance, inductance and capacitance may be lumped in a unit section of line, short compared to the length of the line in wavelengths. This approximation is accurate enough for practical purposes and the properties may be specified in terms of the characteristic impedance of the line.

pedance per unit of length (composed of the series resistance and the series reactance) and the *shunt capacitance* and *conductance* per unit of length. These parameters are functions of the position and diameter of the conductors, the spacing between them, and the structure of the conductors and of the surrounding medium. The two-wire transmission line serves as a generalized example for discussion, and a lumped equivalent of a line section is shown in figure 1.

Distributed Line Properties As predicted by Maxwell's equations (Chapter 24, part I), a magnetic field is set up about the conductors by the flow of current along the line, and energy is stored in, or released from, the field about the line, providing the line with the property of *inductance*. In addition, as the conductors are placed near each other, with air or other medium between them, they exhibit *capacitance* (C) to each other and, if the medium or dielectric is imperfect, a leakage path exists between the conductors, which is expressed in terms of *conductance* (G).

The illustration shows these properties as lumped constants in a sample section of two-wire transmission line. All of the properties may be expressed in terms of the series impedance and shunt admittance per unit length of line and are summed up in terms of the *characteristic impedance* (surge impedance) (Z_0) which, in turn, may be specified in terms of the physical characteristics making up the line.

25-1 Characteristic Impedance

A transmission line is described in terms of its characteristic impedance which has little, if anything, to do with line length, resistance of the conductors, or the frequency of operation of the line . In short, the characteristic impedance is equal to that value of impedance measured at the input end of the line , when the other end is terminated in an impedance of like value. This definition may seem confusing, but the validity is emphasized when it is found that raising the load impedance at the end of a certain length of transmission line may actually reduce the impedance measured at the input end. It can be seen, therefore, that it is possible for a transmission line to exhibit impedance transformations that, if understood and properly applied, can be extremely useful, but if ignored, can be catastrophic in their results.

The Equivalent Load   To demonstrate the rather intangible concept of characteristic impedance, assume a given transmission line is terminated by a resistance with a small capacitance connected in parallel and a small inductance in series with the resistance, somewhat analogous to the lumped constant situation pictured for a short length of cable (figure 2).

By mathematical conversion, the parallel RC load may be replaced by an equivalent

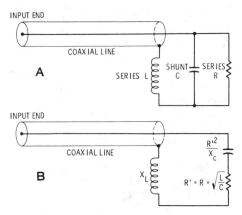

Figure 2
CHARACTERISTIC IMPEDANCE OF A TRANSMISSION LINE

A—The coaxial transmission line is terminated by a network analogous to the lumped-constant equivalent

B—Mathematical equation of circuit A with components in series connection. The equivalent load circuit is electrically equal to the characteristic impedance of the line.

series RC circuit, as shown in figure 2B. If the two reactances are equal, they cancel each other (a condition of resonance) and the following definition of the terminating resistance (R') is achieved:

Let R' equal the series terminating resistance. At resonance,

$$X_L = \frac{R'^2}{X_c} \text{ and } \begin{array}{l} X_L = 2\pi f L \\ X_c = \dfrac{1}{2\pi f C} \end{array}$$

Substituting and simplifying:

$$L = R'^2 C, \text{ or } R'^2 = \frac{L}{C}$$

$$\text{and } R' = \sqrt{\frac{L}{C}} = R$$

Thus the equivalent load circuit of figure 2B appears to a measuring instrument to be identical to the circuit of figure 2A, regardless of frequency and may therefore serve as a substitute for the terminating load of figure 2A. The input impedance of the equivalent circuit is still equal to the original impedance. There is no reason why this substitute process cannot be repeated indefinitely to build up an electrical equivalent of any transmission line, and it can be said that the input impedance of such an artificial line will always be the same, regardless of its length and the frequency of operation, provided that the far end of the artificial transmission line is always terminated in a load resistance equal to $\sqrt{L/C}$. Further, the input measurement of the line will always equal this exact amount and is apparently a resistance, termed the characteristic impedance of the line. The only difference between a real line and the artificial line is that the real one is bound to have some loss resistance as well as inductance and capacitance. Good transmission lines, however, have very little loss resistance in the hf region.

25-2 Transit Time and Wave Reflection

While electromagnetic waves travel approximately 186,240 miles per second in space, it takes more time for a wave to progress along a transmission line, from one end to the other, as the energy must charge the distributed capacitance of the line and induce an electric field along the distributed inductance of the cable (figure 3). For many solid-dielectric coaxial cables, the wave travels at about 66 percent as fast as in air, and the cable is said to have a *velocity of propagation* (V_p) of 0.66.

As the energy passes down the transmission line from generator to load, it is interesting to note that the generator has no means of determining the load conditions at the end of the line, nor does it "know" if the proper terminating condition $R = \sqrt{L/C}$ is fulfilled or not. Thus, during the short interval the wave initially travels along the line, the current supplied by the generator is determined only by the characteristic impedance of the line. The power supplied by the generator is used exclusively to create a pattern of electric and magnetic fields speeding along the line. Since the characteristic impedance of the line is a resistance (neglecting cable losses), the current and voltage along the line are in phase. Until the energy reaches the end of the transmission line, it would seem that Ohm's law has been placed in suspended animation.

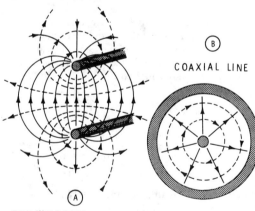

TWO-WIRE LINE   **Figure 3**

ELECTRIC AND MAGNETIC FIELDS ABOUT TRANSMISSION LINES
Lines of electric field (solid) terminate on conductors and lines of magnetic field (dashed) curve about conductors.

The "Suspension" of Ohm's Law   The transit time required for the wave to pass the length of the transmission line may be compared to quarterly income tax payments made before the annual

amount finally due has been determined. In such cases, it is necessary to make an estimated payment subject to later adjustment if the total sum is found to be in error. In a similar fashion, the generator has to "pay" current into the transmission line before it "knows" how much current the terminating load resistance will take. Ohm's law is, in effect, held in suspense until the current reaches the load at the end of the transmission line. During this finite period of transit, the only load the generator "sees" is that load caused by the creation of the electromagnetic field about the line.

If, when the energy reaches the end of the transmission line the load is a resistance, and the ratio of load voltage to line current is equal to the characteristic impedance of the line, then Ohm's law is fulfilled and the power arriving at the load is absorbed at exactly the same rate as it is being fed into the generator end of the line. The only effect of the transmission line, assuming it is lossless, is the transit time-lag of the electromagnetic wave along the line.

On the other hand, if the line energy arrives at the load and "finds" a load resistance unequal to the characteristic line impedance, Ohm's law is not fulfilled and a portion of the energy is sent back down the line toward the generator in opposition to the normal line current and voltage, the remainder of the energy being absorbed by the load in accordance with Ohm's law.

Phase Shift   The finite period of time the radio wave takes to flash along the transmission line at near the velocity of light may be expressed in terms of *phase shift* along the line. The amount of phase shift introduced by the line is a function of the velocity of propagation of the wave and the distance of the point of reference from the end of the line.

Phase shift is commonly expressed in electrical degrees and to determine the phase of the current at any point along the line, it is only necessary to determine the number of electrical wavelengths and fractions thereof between the point of investigation and one end of the line and divide the result into 360 degrees; this gives the phase shift in degrees per unit length (figure 4).

The current and voltage in a transmission line exhibit a phase shift of 360 degrees

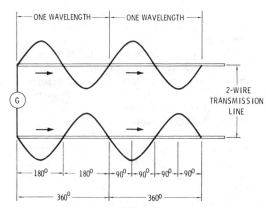

Figure 4

TRANSMISSION LINE LENGTH IN ELECTRICAL DEGREES

A conductor may be divided into electrical degrees, expressing length referenced to the input or output end of the line. One wavelength, or electrical cycle, is expressed as 360 electrical degrees. If the wave is, for example, 30 meters long, it will take 0.1 microsecond to pass one wavelength along the line. During this time, the phase has shifted through 360 degrees.

(one cycle) with respect to the source over a one-wavelength line segment. A second 360-degree phase shift will take place over a second electrical wavelength of line. The total phase shift over a transmission line two wavelengths long, then, is 720 degrees, or two complete cycles of source current.

Since the radio energy travels at constant velocity along the transmission line, the line may be divided into electrical degrees, as shown in the illustration. A quarter-wavelength of line is referred to as a 90-degree line, a half-wavelength line as a 180-degree line, and so on. In effect, the phase shift along the line may be explained in the terms used for phase shift in lumped constants, as discussed earlier in this Handbook, and as will be further discussed in the following chapters.

Wave Reflection on a Transmission Line   Before wave reflection is viewed in terms of fields and waves, it is interesting to observe it in terms of Ohm's law and simple r-f circuits. Figure 5A shows a 200-volt generator coupled to a 50-ohm load through a section of transmission line having a characteristic impedance of 50 ohms ($Z_o = 50$). The current flowing in the circuit is 4 amperes and the power dis-

sipated at the load is 800 watts. Accordingly, the generator delivers 4 amperes at 800 watts and the circuit satisfies Ohm's law in all details.

Assume the load resistance is changed to 300 ohms, designated as R'. If reflection does occur, let:

I equal generator current sent down the line,

E equal generator voltage at input end of the line,

i equal the current reflected back down the line toward load,

e equal the voltage reflected back down the line toward load.

The characteristic impedance is common to all voltages and currents, so:

$$\frac{E}{I} = Z_o = \frac{e}{i}$$

It follows from Maxwell's equations and the previous discussion that the net current in the load is $(I - i)$ and the total voltage across the load is $(E + e)$, as shown in figure 5B. To fulfill Ohm's law, then:

$$\frac{E + e}{I - i} = R'$$

when R' is any value of load resistance.

Solving these simultaneous equations, the inclusive expression for the general load condition, R, when the value of the load resistance is not equal to Z_o:

$$\frac{i}{I} = \frac{e}{E} = \frac{R - Z_o}{R + Z_o}$$

Now, if $R = 300$ ohms, then:

$$\frac{i}{4} = \frac{e}{200} = \frac{300 - 50}{300 + 50} = \frac{250}{350} = 0.715$$

and $i = 2.86$ amperes and $e = 143$ volts.

In summation, then:

Power leaving the generator: $4 \times 200 = 800$ watts.

Power arriving at load: 800 watts.

Power absorbed in load: $(E + e) \times (I - i) = 343 \times 1.14 = 391$ watts.

Power reflected by load: $e \times i = 143 \times 2.86 = 409$ watts.

If the generator has an internal impedance (and all of them do), and the impedance happens to be the same as the characteristic

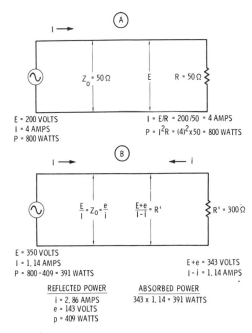

E = 200 VOLTS
I = 4 AMPS
P = 800 WATTS

$I = E/R = 200/50 = 4$ AMPS
$P = I^2R = (4)^2 \times 50 = 800$ WATTS

E = 350 VOLTS
I = 1.14 AMPS
P = 800 - 409 = 391 WATTS

E+e = 343 VOLTS
I - i = 1.14 AMPS

REFLECTED POWER	ABSORBED POWER
i = 2.86 AMPS	$343 \times 1.14 = 391$ WATTS
e = 143 VOLTS	
p = 409 WATTS	

Figure 5

**WAVE REFLECTION ON A
TRANSMISSION LINE**

**A—A matched line (Z_0 = R = 50 ohms) delivers
800 watts to a load and satisfies Ohm's law.
B—Unmatched line (Z_0 = 50 ohms, R' = 300
ohms) delivers 391 watts to load and returns
409 watts to the generator, thus satisfying
Ohm's law.**

impedance of the line, the generator will completely accept the returned power, which in this case is 409 watts. As a result, the net outgoing power from the generator is reduced to: 800 − 409 = 391 watts.

Thus, the mismatch at the load has dropped the system power from 800 to 391 watts. If, however, the internal impedance of the generator is other than equal to the characteristic impedance of the transmission line (the usual case), it will rereflect a portion of the reflected power reached it. In turn, a portion of the rereflected power will once again travel down the line, to be reflected in the load, the total power traveling in each direction along the line being a summation of all incident and reflected powers. The net outgoing power at the generator, of course, is a function of the mismatch of the generator to the line, looking backward toward the generator.

In this fashion, a system mismatch at the terminating load can seriously affect the loading of the generator, and the power in the transmission system. A considerable portion of system power can be reflected and rereflected along the line causing undesirable characteristics to appear on the line.

Of immediate interest to the operator of a transmitter which is working into a mismatched transmission line is that the mismatch at the input end of the line may be so great that the tuning system of the equipment is unable to accomodate the load. Damage to the equipment may be the end result of trying to load into a badly matched antenna system.

25-3 Waves and Fields Along a Transmission Line

Maxwell's equations define the action of a transmission line as expressed in terms of field theory. A simplified discussion of fields and waves on a line may help clarify the previous discussion.

The current along, and voltage between, the conductors of a line produce magnetic and electric fields about the line containing the energy which has left the generator but which has not yet arrived at the load. In a sense, the transmission line guides and confines an electromagnetic field, as well as conducting the energy in a form of alternating current. The former concept is of great use in explaining the action of uhf *waveguides* (hollow pipes that conduct radio energy by propagating it as a traveling electromagnetic field within the pipe). At the same time, this field concept is equally correct in the investigation of hf transmission lines. Figure 3 showed end-on views of a two-conductor line and a coaxial line. The currents flowing in the conductors produced a magnetic field and the voltage difference between the conductors produced an electrostatic field. It is impossible to have current and voltage at a point on a transmission line without the existence of a corresponding electromagnetic field, and vice versa. The two concepts are so interrelated that it is immaterial whether at a point along the line the r-f current within and the voltage between the conductors are due to the electromagnetic field, or that the field is a product of the voltage and current, or that they are simply two manifestations of the same phenomenon. The expanding series of energy transfers from an electric field to a magnetic field, and so on, to propagate the energy along the line in the same manner electromagnetic energy is propagated through space.

As mentioned earlier, the electrical characteristics of a line are expressed as a characteristic impedance, based on the assumption that the capacitance and inductance of a short unit length of line may be considered independently of the rest of the line. As a result, the properties of the unit line are considered as lumped constants, and Ohm's law applies to these constants.

In the case of a transmission line whose length is comparable to the wavelength of energy flowing along the line, this assumption is not valid, as the time-flow (transit time) of electromagnetic energy is finite and a phase difference exists between separate points along the line. This difference is significant, since at a given instant the current at one point in the line may be passing through its maximum value, while at another point it may be near zero (figure 6). In such a case, the line must be considered as a complete system of distributed impedances, and it is more convenient and correct to view the system from the field-theory concept rather than from the more conventional, lumped-constant interpretation, utilizing Maxwell's series of equations. The simpler, lumped-constant approach will be used, in a modified form, in this Handbook, since it is sufficiently correct for the problems concerned with the hf and vhf antenna systems discussed herein.

Wave Motion on a Finite Transmission Line   If a line has infinite length, or if the line is terminated in a characteristic load, incident energy traveling down the line will continue indefinitely in the first case, or will be completely absorbed by the load in the second case. In either example, only one value of impedance is measured at the input terminals (or at any other point along the line) and this value is the characteristic line impedance.

When the far end of a finite transmission line is terminated with a load other than the characteristic load, a discontinuity exists at

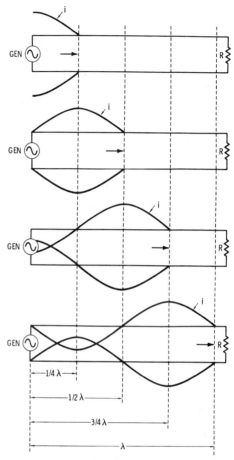

Figure 6

INSTANTANEOUS CURRENT ALONG A TRANSMISSION LINE

Transit time of electromagnetic energy is finite and a phase difference exists between separate points along the line. This example shows that the current wave passes from maximum to minimum values at successive quarter-wavelength points along the line as wave travels from left to right. The AVERAGE value of the current along the line, however, is constant.

this point and wave reflection occurs, as predicted by Maxwell's equations. Picture a finite transmission line connected through a switch to an r-f generator. Assume the switch is closed for a time equal to the period of wave energy, and then opened. As a result, one cycle of energy is sent down the line to the far end. If, for example, the line is open at the load end, the pulse of energy can go no further and the current at the end of the line collapses to zero. In doing so, a collapse also occurs in the magnetic field, creating an electric field which acts in the manner of a reverse generator, inducing a new current equal to that of the incident wave, traveling back along the line toward the input end. The reinforced electric field at the end of the line is in phase with the incident voltage, while the current component of the reflected wave at this point is equal in amplitude and opposite in phase to that of the incident wave, giving a resultant current of zero (figure 7A).

If the generator switch is again closed during the reflected wave cycle, a condition then exists in which energy is traveling both ways on the line. If the switch is held closed, both incident and reflected waves are present on the transmission line, in the manner suggested in figures 5 and 6, chapter 24.

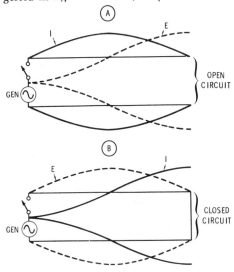

Figure 7

WAVE REFLECTION FOR OPEN- AND CLOSED-END LINES

A—Reflection on a half-wavelength open line. Current and voltage waves are 90 degrees offset after reflection. The current is zero at the open end of the line and voltage is maximum at this point. Both waves exist as standing waves, each being the resultant of conflicting incident and reflected waves. A line having a standing wave on it may be regarded as a "storehouse" of energy similar to a lumped circuit.

B—Reflection on a half-wavelength closed line. Current and voltage waves are 180 degrees out of phase from condition (A). Open- and closed-end lines are used as tank circuits in vhf and uhf equipment as well as in matching devices.

Wave reflection also occurs along a transmission line shorted at the load end for reasons comparable to the open-end situation. The voltage at the short circuit collapses because a potential difference cannot exist across zero resistance, the current at this point is doubled, the current and voltage roles being reversed from the case of the open-end termination (figure 7B). The voltage undergoes a phase reversal upon reflection and a reflected wave flows back along the line toward the generator. The line does not have to be of any particular length to allow reflections to be created on it; the only requirement is that the line be finite in length and not terminated in its characteristic impedance.

Reflection and Standing Waves Hertz's early experiments show that when a radiated wave strikes an abrupt change in medium, or a sharp boundary, some of the wave is reflected, and all of it is reflected in the case of meeting a conducting sheet or plane of perfect conductivity. Hertz also observed that the reflected wave tended at some points along the path to interfere destructively with the incident wave, while at other points it tended to interfere constructively. The net effect was the apparent creation of a third wave, termed a *standing wave*, which remained fixed in position, while the incident and reflected waves traveled along the antenna, or transmission line, at near the speed of light. Hertz concluded that an interference pattern of waves had been set up along the path.

An analogy may be drawn between a standing wave of electromagnetic energy and the vibrations of a violin string when it is plucked at some point. The string vibrates along its length, but the amplitude of vibration is a function of position along the string. The standing wave on the string, in fact, is a trapped wave that cannot escape because of the barriers created by the ends of the string. As far as the transmission line is concerned, voltmeters and ammeters placed along the line will provide visual evidence of the standing wave condition (figure 8). The r-f *power* at any point along the line remains constant, regardless of excursions of voltage and current caused by the wave interference pattern. The consecutive points of maximum current of the standing wave,

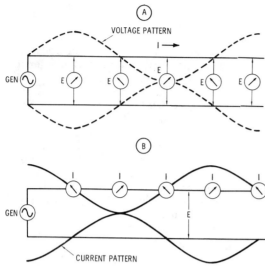

(A)

VOLTAGE PATTERN

(B)

CURRENT PATTERN

Figure 8

STANDING WAVE PATTERNS OF VOLTAGE AND CURRENT ON A TRANSMISSION LINE

Mismatched two-wire line has reflected wave which interferes with incident wave, creating a third wave which remains fixed in position, while incident and reflected waves travel along the line.
A—Representation of voltage standing wave.
B—Representation of current standing wave.

moreover, are separated by a quarter wavelength along the line from the maximum voltage points, and the pattern is repeated at half-wavelength intervals.

25-4 The Standing-Wave Ratio

Wave interference creates standing waves of voltage and current on a transmission line and measurement of these waves provides useful information concerning the electrical condition of the line. The condition may be defined in terms of the *reflection coefficient* (*k*) and the *standing wave ratio* (SWR).

The Reflection Coefficient   The reflection coefficient expresses the ratio of the reflected wave voltage (E_r) to the incident, or forward wave voltage (E_t):

$$k = \frac{E_r}{E_t}$$

If the terminating load on the line is resistive, the reflection coefficient is:

$$k = \frac{R - Z_o}{R + Z_o}$$

where,
R is the terminating load,
Z_o is the characteristic line impedance.

For example, assume a 50-ohm line is terminated in a 25-ohm load. Then,

$$k = \frac{25 - 50}{25 + 50} = \frac{-25}{75} = -0.33$$

Thus, the reflected wave is of opposite phase to the incident wave and has one-third the voltage amplitude.

The Standing-Wave Ratio   The ratio of maximum rms voltage or current to minimum rms voltage or current along a transmission line defines the standing wave ratio:

$$SWR = \frac{I_{max}}{I_{min}} = \frac{E_{max}}{E_{min}}$$

The SWR may have a range of values from unity to infinity, and is an indicator of the line properties. The voltage standing-wave ratio (VSWR) can be measured with an inexpensive instrument (SWR meter) and is a convenient quantity in making calculations of line performance. The general case for a line terminated in a resistive load of any value is:

$$SWR = \frac{R}{Z_o}$$

when R is greater than Z_o, and

$$SWR = \frac{Z_o}{R}$$

when R is less than Z_o.

Input Impedance   The value of impedance seen at the input end of a transmission line is important as this is the value that the transmitting equipment must work with. The *input impedance* must be within the limits imposed by the output matching network of the equipment in order to achieve proper loading.

The input impedance of the line depends not only on the load impedance at the far end of the line, but also on the electrical length of the transmission line. Thus, the input impedance is a function of frequency, as the electrical length of the transmission line changes in relation to the physical length with a change in frequency.

When the load impedance is not matched to the line, the input impedance of the line may be inductive, capacitive, resistive, or a combination of all three of these qualities. The magnitude and phase angle of these qualities depends on the line length, the SWR, and the characteristic impedance of the line.

An antenna system of the type used by most amateurs, is resistive at the resonant frequency and is reactive at frequencies off- resonance, exhibiting various combinations of resistance and reactance to the transmission line (figure 9). Some combination of these qualities is the rule, rather than the exception, although the resistive term of the combination is predominant in most cases.

25-5 Impedance Matching With Resonant Lines

A transmission line exhibiting wave reflection is termed a *resonant line* since it assumes many of the characteristics of a resonant circuit. Variations of formulas that apply to LC circuits also apply to resonant lines. Sections of such lines can be economically substituted for lumped tuned circuits in wave filters, impedance-matching devices, phase shifters, line-balance converters, and frequency control circuits.

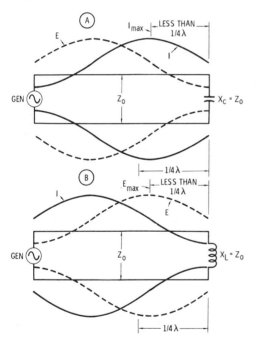

Figure 9

STANDING WAVE PATTERNS OF VOLTAGE AND CURRENT FOR REACTIVE LINE TERMINATIONS

A—With capacitive reactance termination, the maximum point of current is closer than a quarter-wavelength to the termination.
B—With inductive reactance termination, the maximum point of voltage is closer than a quarter-wavelength to the termination.

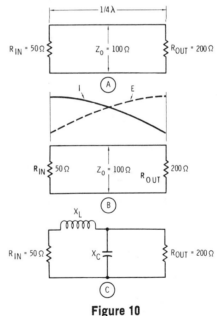

Figure 10

THE QUARTER-WAVE TRANSFORMER

The impedance inverting property of a quarter-wave line provides a good match between a high impedance and a low one. The transformer impedance is equal to the geometric mean between the two impedances. A—The matching transformer, sometimes called a "Q-section". B—The reversal of voltage and current on the transformer. C—The lumped equivalent circuit of the quarter-wave transformer. For a balanced transformer, X_L is divided into two inductors, one placed in series with each line and each having the value $X_L/2$.

Open-end and closed-end resonant lines are useful as matching devices between different impedance levels in antenna systems. Short, resonant lines (*stubs*, or *matching stubs*) can approximate capacitance or inductance and may be used to compensate for, or match out, unwanted reactive components in an antenna system.

The Quarter-Wave Transformer   The input impedance (Z_i) of a quarter-wave line terminated in a load impedance of Z_1 is:

$$Z_i = \frac{(Z_o)^2}{Z_1}$$

where, Z_o is the characteristic line impedance.

The equivalent, lumped circuit is shown in figure 10C.

The impedance inverting property of the line provides a good match between a high impedance circuit and a low impedance one. By inverting the formula, the impedance of the *matching transformer* (Z_o) required to match two different impedances is:

$$Z_o = \sqrt{Z_1 \times Z_1}$$

showing that the transformer impedance is equal to the geometric mean of the two impedances to be matched.

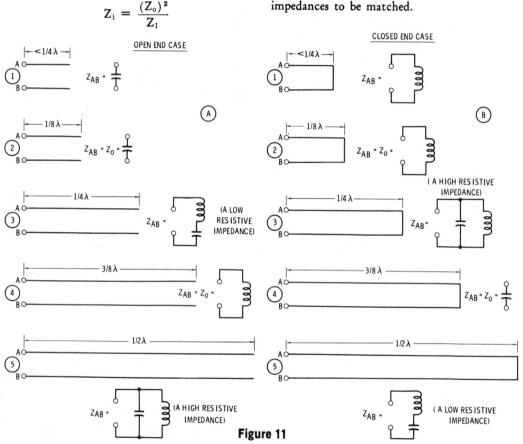

Figure 11

LUMPED CIRCUIT EQUIVALENTS
FOR RESONANT LINES

A—Open-end lines. The one-eighth wave line transforms the line impedance into an equal value of capacitive reactance. The quarter-wave line functions as an impedance inverting device and the three-eighths-wave line transforms the line impedance into an equal value of inductive reactance.

B—Closed-end lines. Conditions are exactly reversed from the open-end lines although the basic transformation remains the same. The one-eighth wave line, for example, transforms the line impedance into an equal value of inductive reactance. The two cases are 90 degrees out of phase with each other in all respects.

The Resonant Stub A shorted, quarter-wave line is equivalent to a parallel resonant circuit, making it possible to substitute the line for a lumped LC circuit (figure 7B). For the general case, the *open-end impedance* of a shorted or open line varies with line length and may be capacitive, inductive, or present a low or high resonant impedance. The open-end inductive reactance of a loss-free, shorted line, less than a quarter-wavelength long is:

$$X_L = Z_0 \tan l$$

where,

l is the electrical length in degrees,
Z_0 is the characteristic impedance of the line.

The open-end capacitive reactance of a loss-free, open line, less than a quarter-wavelength long is:

$$X_c = Z_0 \cot l$$

Figure 11 illustrates the manner in which the input reactance of a transmission line

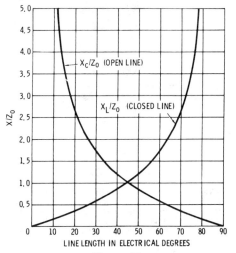

Figure 12

REACTANCE CURVES FOR OPEN- AND CLOSED-END LINES

The quantity X/Z₀ multiplied by the characteristic impedance of the line is equal to the input reactance. The behavior of the two types of line is complementary, the quarter-wave open-end line acting as a series-resonant circuit and the closed-end line acting as a parallel-resonant circuit. Such sections of line are often used as chokes or tuned circuits in the vhf/uhf region.

varies with length for the open- and closed-end cases. Figure 12 represents the reactance curves for the two types of line.

Stub Matching A line segment less than a quarter-wavelength long presents a value of reactance at the measuring end that can be used to match out unwanted reactance in an antenna system. Either open- or closed-end stubs may be used, depending on the circuit requirements.

If a transmission line is connected directly to an antenna at, or near, a current loop or node, the chances are that the antenna will present other than a matched load, and standing waves will exist on the line. At a point on the line, less than one-half wavelength from the antenna, the resistive component of the antenna load will equal the characteristic line impedance, and a reactance whose value is equal and opposite placed at this point will cancel the unwanted reactance on the transmission line (figure 13). Stub dimension and placement is a function of the SWR on the line, as measured at the load. In some cases, lumped constants are substituted for the matching stub, and the resulting device is called an *impedance matching* network.

Balancing Networks Most hf antennas are balanced systems in that equal and out-of-phase voltages to ground exist at each input terminal. The Marconi antenna, discussed in the previous chapter, is an example of an exception to this statement.

When a balanced antenna is used, the two-conductor transmission line feeding the antenna should carry equal and opposite currents throughout its length to maintain the electrical symmetry of the antenna system. The popular coaxial transmission line (discussed later in this chapter) is an unbalanced device, with one conductor normally operated at ground potential. An electrical unbalance exists when such a line is connected to a balanced antenna.

In addition, a transmission line in the near-field of an antenna is coupled to the antenna by virtue of its proximity, and induced currents can flow in the outer conductor of the coaxial line. This current is called an *antenna current* and it tends to upset the

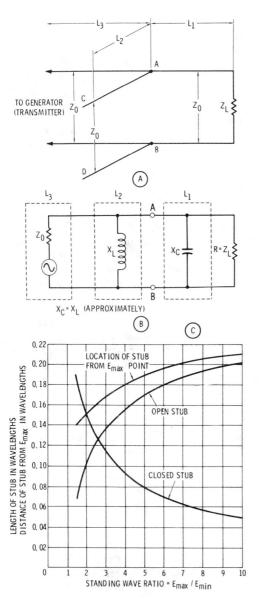

true SWR condition on the line, rendering meaningful SWR readings impossible.

To maintain proper current balance in the coaxial transmission line and also to reduce

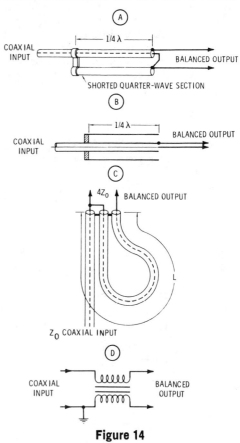

Figure 13

STUB MATCHING

An open-end or shorted-end stub connected across the main transmission line is used to reduce the SWR on the line and provide a good match between line and load impedance. The location of the stub in relation to the load (dimension L_1) and the length of the stub (dimension L_2) may be set from the measured value of the SWR. It is not necessary for the antenna to be resonant at the operating frequency as the stub matching system can compensate for the physical antenna length.

Figure 14

BALANCING DEVICES

A—Quarter wave section made of parallel transmission line serves as balancing device. Coaxial line is run through one leg of the section. Equal and opposite currents flowing in legs of quarter wave section cancel each other and resultant current flowing on outside of transmission line is zero.

B——Coaxial version of quarter-wave section. The outer sleeve acts much like an rf choke to suppress antenna currents from flowing on the coaxial transmission line.

C—Half-wavelength coaxial balun provides impedance step-up of four to one. Balancing line is 0.66 of half-wavelength if solid dielectric coaxial line is used, taking the velocity factor of the line into account.

D—Balun coil consists of bifilar winding which may be considered as a portion of the transmission line. Currents in windings are in phase and act to choke off antenna current tending to flow along coaxial transmission line.

antenna currents on the line to a minimum, the line should be brought away from the antenna at right angles to it, reducing inductive coupling to a minimum, and a balancing device should be placed between the transmission line and a balanced antenna. A suitable device is termed a *balancing network*, or *balun*. Linear and lumped-constant baluns are illustrated in figure 14.

Wideband Baluns Wideband baluns may be made of a section of transmission line wound into an inductance. A frequency span of 10 or 20 to 1 is achieved with the proper design and the device may be balanced or unbalanced with various transformation ratios. Shown in figure 15 are representative designs for phase-reversal, balance-to-unbalance and impedance-transforming baluns.

The bandwidth of a particular balun is determined at the low-frequency end of the operating range by the inductance of the windings and at the high-frequency end by the distributed capacitance of the design. If a ferrite core is used in the device, care must be taken to limit the signal level so that saturation does not occur.

A two-winding balun may be used for phase reversal, or balance to unbalance. A 1-to-4 balun requires either 3 or 4 windings, depending on the state of balance, as

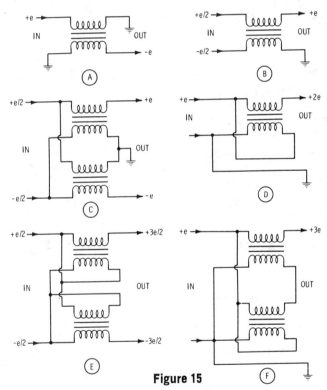

Figure 15

WIDEBAND BALUN TRANSFORMERS

A—Phase reversing, 1-to-1 transformation ratio.
B—Balance to unbalance, 1-to-1 transformation ratio.
C—Balance to balance, 1-to-4 transformation ratio.
D—Unbalance to unbalance, 1-to-4 transformation ratio.
E—Balance to balance, 1-to-9 transformation ratio.
F—Unbalance to unbalance, 1-to-9 transformation ratio.

shown in the illustration. A 1-to-9 impedance transformation may be accomplished
with the same basic design, with windings
series- instead of parallel-connected.

Two or more baluns may be interconnected to provide unusual transformation
ratios. A balance-to-unbalance device providing transformation and unbalance in two
steps is shown in figure 16. This configuration is often used to match power transistors
to a 50-ohm load.

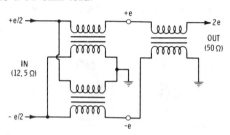

Figure 16

**SERIES CONNECTED BALUNS TO
PROVIDE LOW IMPEDANCE,
BALANCED INPUT TO 50-OHM LOAD
TERMINATION**

Left balun provides balance to balance termination with 1 to 4 transformation ratio. Right
balun provides balance to unbalance termination with 1-to-1 transformation ratio. This configuration is often used to match push-pull
VHF power transistors to a 50-ohm load, with
a strip-line balun design.

25-6 Transmission Lines

Practical transmission lines for the hf and
vhf region are composed of two conductors
separated by a dielectric. Two classes of
lines, parallel-conductor and coaxial, are in
wide use, although there are many styles of
each class of line. Transmission lines may be
either air-insulated or may be embedded in
a solid dielectric.

**Two-Wire
Open Line** A typical open transmission line
is shown in figure 17. The conductors are held in position by
means of insulating rods, the spacing varying from less than one inch for vhf service, to over one foot for high-power, hf
service.

Expressed in physical terms, the characteristic impedance of a two-wire open line
is:

$$Z_0 = 276 \log_{10} \frac{2S}{d}$$

where,

S = spacing between conductor centers,
d = conductor diameter.

Since the formula is expressed as a ratio,
the units of measurement may be in any
convenient units, so long as the same units
are used for each dimension. The equation is
accurate so long as the conductor spacing is
relatively large compared to the conductor
diameter.

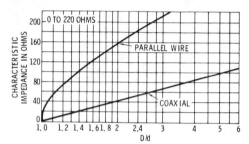

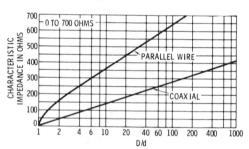

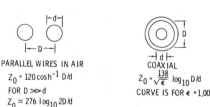

PARALLEL WIRES IN AIR
$Z_0 = 120 \cosh^{-1} D/d$
FOR D >> d
$Z_0 \approx 276 \log_{10} 2D/d$

COAXIAL
$Z_0 = \frac{138}{\sqrt{\epsilon}} \log_{10} D/d$
CURVE IS FOR ε = 1.00

Figure 17

**TWO CONDUCTOR
TRANSMISSION LINES**

**"Ribbon" and
Tubular Line** Flexible, prefabricated parallel-conductor line is widely
available for television lead-in
cable. The majority of this line has a nominal characteristic impedance of 300 ohms.

Receiving types and transmitting types having power levels of up to one kilowatt in the hf range are listed, with their pertinent characteristics in Table 1.

Table 1. Ribbon and Tubular Line

Impedance (Ohms)	Amphenol Type Number	Velocity of Propagation	Power Rating (30 MHz) In Watts
75	214-023	0.71	1000
300 (Flat)	214-056	0.82	–
300 (Oval)	214-022	0.82	–
300	214-271	0.82	500
(Tubular)	214-076	0.82	1000
(Foamed)	214-103	–	–

Coaxial Line The coaxial line has advantages that make it very practical for efficient operation in the hf and vhf regions. It is a perfectly shielded line and has a minimum of radiation loss. It may be made with braided conductors to gain flexibility and is impervious to weather. Since the line has little radiation loss, nearby metallic objects have minimum effect on the line as the outer conductor serves as a shield for the inner conductor (figure 17).

As in the case of a two-wire line, power lost in a properly terminated coaxial line is the sum of the effective resistance loss along the length of the cable and the dielectric loss between the two conductors. Of the two losses, the resistance loss is the greater; since it is largely due to the skin effect and the loss (all other conditions remaining the same) will increase directly as the square root of the frequency.

The coaxial cable used in the majority of amateur installations is a flexible type, the outer conductor consisting of a braid of copper wire, with the inner conductor supported within the outer by means of a semi-solid dielectric of exceedingly low-loss characteristics called *polyethylene*. The characteristic impedance of the cable is about 50 ohms, but other cables are available in an impedance of 75 ohms (Table 2).

In order to preserve the waterproof characteristic of the flexible, coaxial line, special coaxial fittings are available as well as less-expensive nonwaterproof fittings (Table 3).

New Coaxial Cables The new military procurement numbers for two of the popular coaxial cables in use are: M17/028-RG058 for RG-58C/U and M17/028-RG213 for RG-213/U. These cables pass rigid tests that ensure they can be used reliably as high as 1 GHz.

Theoretically, the attenuation of a coaxial cable follows a predictable curve and the SWR of the cable itself should be negligible. Manufacturing variances, however, occur

Table 2. Coaxial Cables. Six Digit Type Numbers Are Amphenol Foamed Dielectric Cables.

Impedance (Ohms)	Type Number	Velocity of Propagation	Diameter (Inches)	Power Rating (Watts) At 30 MHz
52.0	RG-8A/U	0.66	0.405	1720
50.0	RG-213/U	0.66	0.405	1720
50.0	621-111	0.80	0.405	–
53.5	RG-58/U	0.66	0.195	580
50.0	RG-58A/U	0.66	0.195	550
50.0	RG-58C/U	0.66	0.195	580
75.0	RG-11A/U	0.66	0.405	1400
75.0	621-100	0.80	0.405	–
73.0	RG-59/U	0.66	0.242	720
73.0	RG-59B/U	0.66	0.242	720
73.0	621-186	0.80	0.242	–
93.0	RG-62A/U	0.84	0.242	850
125.0	RG-63/U	–	0.405	–

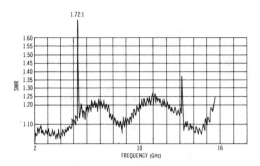

Figure 18

REPRESENTATIVE SWR MEASUREMENTS

Measurements made on a length of RG-214U uhf coaxial cable showing an SWR spike at 5.5 GHz. An attenuation "suck out" would be noticed if this cable were used at that frequency.

and may result in large reflections over narrow frequency bands which may upset equipment operation (figure 18). The new cables limit such excursions over the entire useful frequency band of the cable.

Military Standard MIL-C-17D lists various classes of coaxial line in use today. Class 1 polyvinyl chloride- (PVC) jackets were used on older cables such as RG-8/U (no longer a military-approved type), RG-58/U, and RG-59/U. These cables have a plasticizer added to the jacket in manufacture to improve cable flexibility. Eventually the plasticizer migrates through the shield and attacks the inner insulation, greatly increasing the r-f loss of the dielectric. Lifespan of this class of cable is two years from date of manufacture.

Class 2A noncontaminating jackets are used on newer cable, such as RG-213/U (RG-8/U replacement), RG-58C/U, and RG-59B/U. This class of cable has a lifespan of 15 years from date of manufacture. Class-2 jackets are similar to 2A but they are gray, not black.

RG-8/U cable is no longer manufactured to military specifications and manufacturers often produce cheapened versions of RG-8/U. The braid on these cables is usually reduced to the point where any bend in the cable causes an undesired gap in the sparse shielding. This results in r-f leakage even at relatively low frequencies. Sometimes the

Table 3. Coaxial Cable Connectors
FOR RG-8/U, RG-11/U AND RG-213/U COAXIAL LINES (0.405" DIAM.)

"UHF"-TYPE CONNECTORS

Description	Type Number	Amphenol Number
Plug	PL-259 PL-259A UG-111/U	83-1SP 83-756 83-750
Solderless Plug	—	83-151
Splice	PL-258	83-1J
Reduction Adapter: RG-58/U RG-59/U	UG-175/U UG-176/U	83-185 83-168
Receptacle	SO-239	83-1R

TYPE-N CONNECTORS (50-OHM CABLES)

Description	Type Number	Amphenol Number
Plug	UG-21B/U UG-21C/U	82-61 82-96
Splice	UG-29A/U	83-65
Receptacle	UG-58A/U	82-97
UHF to Type-N	UG-146/U	—
Type-N to BNC	UG-201A/U	31-216

TYPE-BNC CONNECTORS

Description	Type Number	Amphenol Number
Plug	UG-88/U UG-88B/U	31-002 31-018
Splice	UG-914/U	31-219
Receptacle	UG-290/U UG-625B/U	31-003 31-236
BNC to UHF	UG-273/U	31-028

Table 3. Coaxial Connectors

This partial list covers the most widely used coaxial connectors of the UHF, type-N and type-BNC families. The UHF type is considered obsolete, although by far the most widely used hardware on amateur equipment. The type-N family has superceded the UHF connectors and provides a constant impedance at cable joints and is weatherproof. The BNC-family of fittings is designed for small diameter cables, such as RG-58/U and feature a quick-disconnect bayonet lock arrangement. Most BNC fittings are weatherproof. Many other connecting devices, such as right-angle and T-adapters are available in all types, as well as special fittings to match one style of connector to another. In addition to these families, type-HN, type-C and type-MHV families of connectors exist, as well as special connectors for twinax cables.

inner conductor is reduced in area, thus increasing the resistive loss of the cable.

Available on the market is *foamed coaxial line* that has an inner dielectric containing bubbles. This increases the velocity factor of the line and presumably reduces the line loss at the same time. However none of the present common foamed line is manufactured to military specifications and none presently has Class-2A jackets. Most of the foam insulations are not impregnated with an inert gas, but with air, and moisture accumulates in the air bubbles. This increases cable attenuation and also changes the velocity factor as the cable heats up with prolonged use. Finally, the foamed cable is more susceptible to physical damage at bend points.

Cable Power Rating   The factor controlling the power capability of a coaxial line is heat, most of which is generated in the center conductor. The inner dielectric ability to withstand the heat and its effectiveness in transferring heat to the outer shield and jacket are the limiting agents in the heating process. Table 2 lists the power handling capability of common cables. For vhf/uhf use, coaxial cables utilizing a *teflon* inner dielectric are used which permit a center conductor operating temperature as high as 250°C.

SWR and Line Loss   A high value of SWR on a cable indicates a mismatched load. The mismatch increases cable loss as the portion of the reflected power makes two complete trips through the cable—once toward the load, the other back toward the transmitter. Since the reflected power makes two trips through the line, each with attenuation, it receives twice as much loss as that portion of the power which makes only one trip down the line.

This slight additional loss is not serious on the hf bands where cable attenuation is low but increases with frequency. When cable attenuation is high, as is the case at vhf, or if long runs of cable are used at lower frequencies, high SWR can drastically increase power loss.

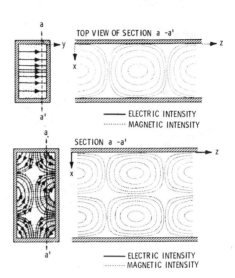

— ELECTRIC INTENSITY
······· MAGNETIC INTENSITY

— ELECTRIC INTENSITY
······· MAGNETIC INTENSITY

Figure 19

REPRESENTATIVE FIELD CONFIGURATIONS IN WAVEGUIDE

Top—Configuration for a TE$_{1,0}$ wave.
Bottom—Configuration for a TE$_{2,1}$ wave.

Waveguides   Electromagnetic energy at microwave frequencies may be propagated through a hollow metal tube under fixed conditions. Such a tube is called a *waveguide*. Any surface which separates distinctly two regions of different electrical properties can exert a guiding effect on electromagnetic waves and the surface may take the form of a hollow pipe, generally rectangular or circular in cross section, with an air dielectric.

A hollow waveguide has lower loss than a two-wire or a coaxial line since it has no dielectric or radiation loss, and the copper loss is low, because the area of current flow in the waveguide is great.

Energy may be propagated along a waveguide in several modes which are described by the relation between the electric (E) and magnetic (H) fields and the walls of the guide. The configuration of the electromagnetic fields in a waveguide can take many forms, and each is called a *mode* of operation. In all cases, either the magnetic or electric field must be perpendicular to the direction of wave travel. The modes, therefore, are classified as either *transverse elec-*

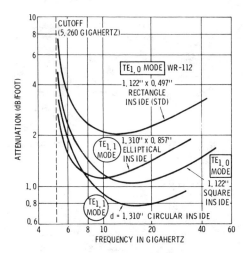

Figure 20

ATTENUATION OF WAVEGUIDE

Relationship between transmission modes and cross section of waveguides shows cutoff frequency below which guide will not propagate energy efficiently.

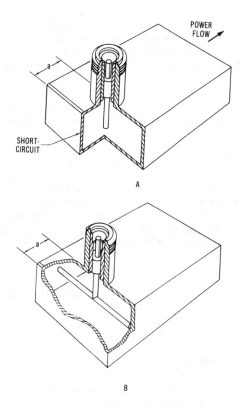

Figure 21

METHODS OF COUPLING A COAXIAL LINE TO A WAVEGUIDE FOR TE$_{1.0}$ MODE

tric, or *transverse magnetic*, abbreviated *TE* and *TM*, respectively. In addition to the letters TE or TM, subscript numbers are used to complete the description of the field pattern of the wave. The first number indicates the number of half-wave patterns of transverse lines which exist along the short dimension of the guide through the center of the cross section. The second number indicates the number of transverse half-wave patterns that exist along the long dimension of the guide through the center of the cross section (figure 19). In case there is no change in the field intensity, a *zero* is used.

Unlike coaxial and two-wire lines, the waveguide has a *cutoff frequency* below which it will not propagate energy efficiently (figure 20). The minimum frequency of operation of a particular guide is reached when, for a particular mode of transmission, the dimensions of the guide approach a half wavelength. Actually, propagation with high attenuation does take place for a small distance, and a short length of waveguide operating below cutoff is often used as a calibrated attenuator.

Energy is coupled into and removed from waveguides by the use of a coupling loop (which cuts, or couples, the lines of the magnetic field) or a probe (antenna) which is placed parallel to the electric lines. A third method is to link or contact the field of the guide by an external field through the use of a common slot or hole between guide and the external circuit.

Two representative ways of coupling from a coaxial line to a rectangular waveguide to excite the TE$_{1,0}$ mode are shown in figure 21. The technique permits an SWR less than 1.15 to 1 at the junction over a 10- to 15-percent frequency band.

Antenna Matching Systems

Some antennas, such as the half-wave dipole, can often be attached to a low-impedance transmission line for direct connection to the station equipment without the need of impedance matching devices at either end of the line.

In all antenna systems using a resonant antenna and a transmission line, however, the load presented to the transmitter is that value of impedance present at the antenna, modified by the transforming action of the

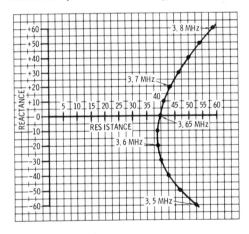

Figure 1

REACTANCE PLOT OF
80-METER DIPOLE

At the resonant frequency of 3.65 MHz this dipole has a radiation resistance figure of 40 ohms. The feedpoint impedance rises rapidly each side of the resonant frequency, reaching 57 +j60 at 3.8 MHz and 52 −j60 at 3.5 MHz. These measurements were taken at the input end of a half-wavelength feedline having an impedance of 50 ohms. At no point over the operating range does the antenna match the line impedance so a SWR value of unity is never achieved. At the resonant frequency, the minimum value of SWR is 1.25.

transmission line which is a function of line impedance, line length, load mismatch, and the operating frequency.

Most antennas, even the simple ones, exhibit a marked change in feedpoint impedance when operated off-frequency and, even at resonance, offer a feedpoint load of other than 50 ohms. Off-resonance, the feedpoint impedance shifts rapidly, producing a substantial mismatch to the transmission line and a consequent high value of SWR on the line (figure 1). The load presented to the transmitter, then, can fluctuate over an extremely large range of impedance which the equipment may be incapable of matching. Thus, for other than spot-frequency operation, most antenna systems require some type of impedance transformation or reactance compensation to provide a nominal match to the universally used 50- or 75-ohm coaxial transmission line. In some cases, additional compensation is required at the station end of the line to afford a good match to the transmitting equipment over a desired frequency span. The maximum value of mismatch permitted at the station usually defines the limits of SWR on the antenna system for a given frequency range.

26-1 SWR and Impedance Compensation

Antenna resonance is that electrical state in which the antenna presents a nonreactive load at the feedpoint. Some antennas exhibit moderate values of feedpoint reactance as the frequency of operation is moved away from the resonant frequency; others, such as short whip antennas or closely spaced parasitic arrays, are quite frequency sensitive, showing large values of reactance for small frequency changes from resonance. The *frequency sensitivity*, or Q, of the antenna determines to an extent the parameters and complexity of the matching circuit to be used.

The feedpoint reactance of an antenna varies with frequency and cannot be matched perfectly over a wide frequency band. For practical purposes, the bandwidth obtainable for a given value of SWR on the transmission line is of importance, or conversely, the minimum limit of SWR that may be achieved for a given bandwidth. In the general case, the feedpoint impedance of a resonant antenna takes the form of a series-resonant circuit whose Q, or figure of merit, is:

$$Q = \frac{f_r}{\Delta f}$$

where,

  f_1 is the resonant frequency,
  Δf is the frequency difference between the half-power points.

In this case, the half-power points are defined as the two frequencies at which the series reactance of the antenna is equal to the series resistance.

Once the Q, the feedpoint resistance, and the operating bandwidth are specified, it is possible to design a compensating network to provide the lowest value of SWR over the operating range. The network may be made up of lumped, LC circuits or may be sections of a transmission line, as the situation demands. Generally speaking, lumped constants are used at the lower frequencies to conserve size and linear circuits at the higher frequencies as pure inductance or capacitance is not simple to obtain from practical components in the vhf/uhf range.

For hf operation, the output circuit of most amateur equipment can accommodate a highly reactive load and may even include a compensation circuit to cancel large values of reactance. In many instances, a maximum value of SWR is stipulated, above which damage may occur to the components of the equipment. Commonly, a value of 2 or 3 is specified as a safe limit. Vhf equipment, on the other hand, often includes SWR protective circuitry wherein the input level to the amplifier stages is a function of the SWR—

the greater the SWR, the more the input level being limited.

In the great majority of cases, compensating circuits of some type are employed at one end or the other of the transmission line to provide a low value of SWR on the line and to provide a convenient load for the transmitter than might otherwise be provided by a high-Q antenna operated at, or near, the resonant frequency.

The characteristics of transmission lines and basic impedance matching systems have been described in a previous chapter and this, and the following chapters, describe impedance compensation devices and their practical application.

26-2   The Smith Chart

Creating an impedance match between antenna and transmission line is not difficult for spot frequency operation. In amateur operation over frequency bands, a satisfactory match is achieved by matching the impedance at the resonant frequency and allowing the SWR on the line to increase off-frequency to a predetermined value, often chosen as 3. This defines the operational bandwidth of the antenna.

The feedpoint impedance of the antenna at a given frequency may be expressed as a complex number $R \pm jX$ and may be plotted on an R-X diagram, as shown in figure 2. The antenna impedance (Z_a) determined at each frequency covered by the antenna may be plotted and the points connected with a curve. The excursions of R and X determine the SWR on the antenna feed system and a *definition circle* can establish a predetermined SWR, as shown on the graph. For example, assuming a 50-ohm line is used and the SWR limit is 3, the intercepts of the definition circle on the R axis are:

$$\frac{Z_o}{SWR} = \frac{50}{3} = 16.6 \text{ and,}$$

$$Z_o \times SWR = 50 \times 3 = 150$$

Inspection of figure 2 shows that by adding inductive reactance or capacitive reactance in series with the antenna the impedance curve can be moved up or down through the definition circle. Thus the

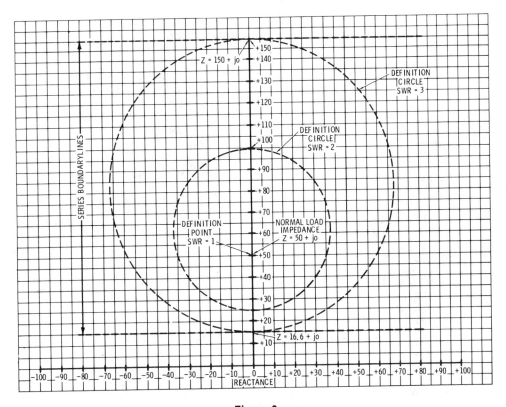

Figure 2

R-X DIAGRAM SHOWING DEFINITION CIRCLES

Complex impedances can be plotted on an R-X diagram such as this. The X-axis represents positive and negative reactance and the Y-axis represents resistance. SWR definition circles are plotted on the diagram for SWR values of 2 and 3. The series boundary lines are limits within which the antenna reactance may be modified by adding a series reactance to the circuit. While useful, this form of representation has been supplanted by the Smith Chart, in which curved, rather than straight, lines are used to form the coordinate system.

dashed *lines* represent the series boundaries, that is, if the impedance curve falls within these lines, by adding a series component, the curve can be shifted to be within the definition circle.

The Impedance The impedance circle dia-
Circle Diagram gram, or *Smith Chart* is a
specialized graph having a curved coordinate system. The system is composed of two families of circles, the *resistance circles* and the *reactance circles* (figure 3). These circles are curves of constant resistance and constant reactance. The complete coordinate system of the Smith Chart is shown in figure 4. *Wavelength* and *phase-angle scales* are plotted around the

perimeter of the chart in terms of the electrical wavelength along a transmission line, one scale running clockwise, the other counter-clockwise. The complete circle, in either case, represents a half-wavelength.

The scaled vertical line of the chart represents the ratio of the resistive component of the antenna (R) to the impedance of the transmission line (Z_o), measured at a particular frequency. *SWR circles* may be added to the Smith Chart by the user, centered at 1.0 on the vertical resistance scale. A circle centered at 1.0 and which passes through 3.0 on the same scale, for example, encloses all values of impedance which will cause a SWR of 3 or less when they terminate a transmission line having a characteristic imped-

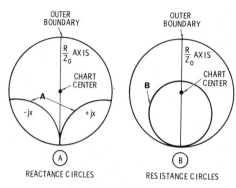

Figure 3

CONSTRUCTION OF COORDINATES OF THE SMITH CHART

The curves are constant reactance circles providing coordinates for both positive and negative reactance (A). Curves of constant resistance (B). Families of such curves are superimposed on the Smith Chart. The chart center is usually 50 ±j0 ohms, or may be normalized to 1.0 ±j0 ohms. Peripheral and radial scales are added to make a complete Smith Chart.

ance of Z_0. Charts with a center impedance of 50 (for use with 50-ohm lines) or a center impedance of 1.0 (the normalized case, for general use) are available. With this configuration, the point at which the

SWR on the transmission line is 1 is at the center of the chart and the locus of unity reflection coefficient (SWR is infinity) is the circumference of the chart.

Moving counterclockwise from the vertical resistance component line locates the negative (capacitive reactance) component, which is the ratio of the $-jX$ component to Z_0 and moving clockwise locates the positive (inductive reactance) component, which is the ratio of the $+jX$ component to Z_0, at a particular frequency.

As an example of the use of a Smith Chart, a plot of a high-frequency antenna is shown in figure 5. Various transmission line problems can be solved graphically with the use of the Smith Chart and the design of networks is considerably simplified by this technique. For additional information on the chart, the reader is referred to *Electronic Applications of the Smith Chart*, P. H. Smith, McGraw-Hill Book Co. catalog number 58930.

Use of the Smith Chart   The Smith Chart has innumerable uses and is particularly valuable in the uhf region, in conjunction with a slotted line, for translating voltage measurements along the line

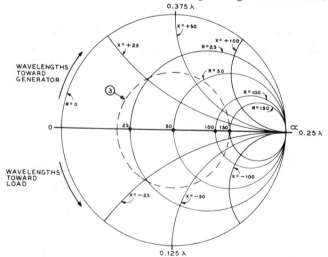

Figure 4

COORDINATE SYSTEM OF THE SMITH CHART

In this construction, the resistive component line is horizontal, running from zero ohms at the left, to infinity at the right. Reactance circles, in intervals of 25 ohms are shown, as well as resistive circles. A single SWR circle is indicated for SWR = 3. The circumference of the chart is one-half wavelength. A complete chart also has a phase angle notation around the circumference, which is not shown here.

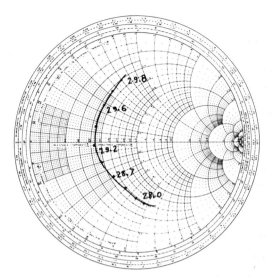

Figure 5

IMPEDANCE PLOT OF 10-METER BEAM ANTENNA

A three element, wide spaced 10-meter beam antenna plotted on the Smith Chart. Measurements were made through a half wavelength of coaxial line. Chart is normalized, with 50 ohms taken as 1.0 at the center of the chart. Frequency of resonance is slightly higher than 29.2 MHz and input impedance of antenna at that frequency is 0.45 × 50 = 22.5 ohms. SWR at resonance, without addition of impedance matching network at the antenna, is 50/22.5 = 2.23.

into SWR and impedance values. In the hf region, the characteristics of an antenna system may be determined with the aid of an *r-f impedance bridge* (see *Electronic Test Equipment chapter*) and the figures derived from these measurements transferred to a Smith Chart. Properly interpreted, the chart then can tell the user what impedance transformations must be made in the system to achieve the desired end results. Smith charts may be obtained at most college book stores or at many technical book stores. A comprehensive article on the use of the Smith Chart for amateur applications appeared in the November, 1970 and December, 1971 issues of *ham radio* magazine.

26-3 Wideband Balun Transformers

A summary of simple balun transformers was given in the previous chapter. Impedance matching transformers are often called baluns, even though they may match between systems having the same state of balance although the term *balun* implies a transformation from the balanced to the unbalanced state.

Simple and inexpensive balun transformers having wideband characteristics can be wound of wire or coaxial cable on either air or ferrite cores, as shown in this section.

The Four-to-One Balun　The balun may be thought of as a transmission line wrapped into an inductance. The low-frequency response of the device is largely determined by the reactance of the shunt winding and the current flowing through it (figure 6). If the reactance is large enough, only transmission line current (I_L) will flow and this is related to the balun load and the characteristic impedance of the winding (figure 7). A high permeability core is thus useful for extended low frequency range of operation. The high-frequency limit of the balun is determined by the transformation ratio, distributed capacitance, and the SWR existing across the ports of the balun.

The optimum value for the characteristic impedance of the balun transmission line is

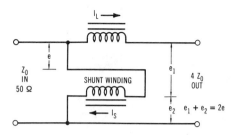

Figure 6

ANALYSIS OF 4-TO-1 TRANSMISSION LINE BALUN

The low-frequency response of the transmission line balun is determined by the reactance of the lower winding to the flow of a shunting current (I_s) due to potential e_2. If the reactance is large enough, only transmission line current flows at the low-frequency end of the operating range, thus maintaining the impedance transformation ratio. (Original analysis by J. Sevick, W2FMI).

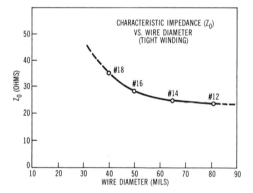

Figure 7

CHARACTERISTIC IMPEDANCE OF BALUN WINDING

Impedance of transmission line wound into a balun is a function of wire size (enamel wire used). Increasing the spacing between adjacent wires raises the impedance.

the geometric mean of the input and output load resistances. For low characteristic impedance, minimum spacing should be used between the windings and for the lowest impedance values, the line is twisted. When working into higher values of impedance, larger winding separations are necessary. This can be done by separating the turns. For example, tightly wound turns of #18 enamel wire separated by an extra layer of insulation provide a characteristic impedance close to 70 ohms.

Other Impedance Transformations   A three-winding balun is shown in figure 8. Illustration A shows a one-to-one transformation ratio and illustration B shows a nine-to-one ratio. If only the two bottom windings of the A design are used, a four-to-one impedance ratio results. The top winding can serve as an autotransformer and provide an impedance stepup of nine to one. If the top winding is tapped (illustration C) an impedance transformation of less than nine to one is achieved. (Original analysis by J. Sevick, W2FMI).

Practical Balun Transformers   A ferrite core balun for 3.5.- to 29.7-MHz service is shown in figures 9 and 10. It may be built for either a 1-to-1 or 4-to-1 transformation and a balance-to-unbalance condition. The power level is a function of core saturation and distributed capacitance, and is limited to about 250 watts average power at the high and low frequency limits, rising to about 600 watts average power over the midfrequency range. The unit may be used at impedance levels as low as 20 ohms and still provide good balance.

The 1-to-1 balun employs a trifilar winding, the three separate coils placed on the core in parallel and connected as shown in the drawing. The input terminals of the balun are nonsymmetrical, point A at the input end being taken as ground. The 4-to-1 balun has a bifilar winding and provides an unbalance-to-balance condition at impedance levels down to about 20 ohms.

When completed, the ferrite balun can be protected from moisture by placing it in a waterproof, nonmetallic container. A plastic "squeeze bottle" cut to size is suggested, with wood discs cut for the end pieces and held in place with small screws.

Air Core Baluns for HF Service—An air core balun similar to the balance-to-unbalance type described in a previous chapter is useful over a 5-to-1 frequency range in the hf spectrum. Described here are various 1-to-1 designs for 20- to 70-ohm service in the amateur bands. Power capability of the designs is about 1000 watts, average power. The balun is wound on a polyvinyl chloride (PVC) tube of the proper diameter and *Formvar* wire is used for the windings.

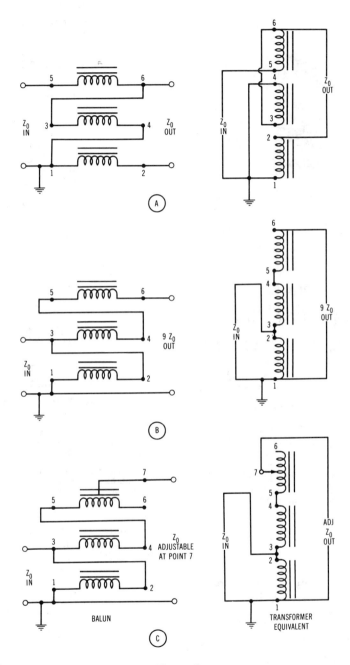

Figure 8

BALUN TYPES

Three balun types are shown at left with their transformer equivalents at right. Illustration A shows a one-to-one design with unbalanced input and balanced output. Illustration B shows a nine-to-one design with unbalanced output. Illustration C shows an adjustable unit with unbalanced output. With tap at point 5 the ratio is four to one and with tap at point 6 the ratio is nine to one (unbalance to unbalance).

Figure 9

FERRITE-CORE BALUN
FOR 3.5 TO 29.7 MHz

This balun has a trifilar winding and provides a 1-to-1 transformation. The center winding is cross-connected. The ferrite slug is Q-1 material, rated for r-f application to 10 MHz (Indiana-General Corp. core CF-503). The core is ½" (1.26 cm) diameter and 3¼" (8.25 cm) long. The ferrite material is broken to length by nicking it with a file around the circumference and breaking it with a sharp blow. When completed, the assembly is given a thin coat of Krylon, or coil dope at the ends. Do not coat the windings. The 4 to 1 balun is similar in construction but has only two windings.

The first design (figure 11) covers the 6.0- to 30.0-MHz range. The unit consists of 10 trifilar turns wound on a $1\frac{1}{16}$" (2.7 cm) diameter form. The winding is closewound. Ends of the winding are held in place by 4-40 hardware.

The second design covers the range of 2.5 to 15.0 MHz and consists of seven trifilar turns, closewound, of *Formvar* wire on a $2\frac{3}{8}$" (6.0 cm) diameter form. A third design, covering 0.54 to 2.5 MHz, consists of 18 trifilar turns of wire wound on a $3\frac{1}{2}$" (9.0 cm) diameter form.

A *Broadband Coaxial Balun*—Shown in figure 12 is a broadband balun made of coaxial cable that covers the range of 6 to 30 MHz. It is designed to be installed directly at the terminals of a triband beam antenna (7-14-21 MHz).

The balun coil is self-resonant near the center design frequency which, in this case

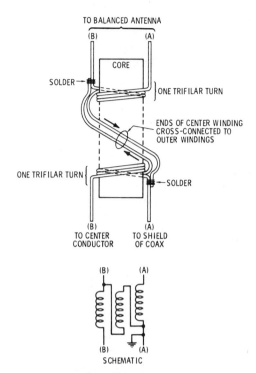

Figure 10

WINDING DATA FOR
FERRITE BALUN

The balun consists of 3 coils, each 6 turns of #14 enamel or Formvar wire. The wires are wound side by side, the ends being held in a vise. The center winding is cross-connected at the ends. Either end may be taken as input or output, but the common connection between the inner and the outer winding at one end must be taken as the ground point and attached to the shield of the coaxial line (A).

is 15 MHz. The coil is made from a 16' 8" (5.08 m.) length of 50-ohm coaxial line (RG-8A/U or RG-213/U) closewound into a coil of nine turns having an inside diameter of $6\frac{3}{4}$" (17.15 cm.). At one end of the coil the inner and outer conductors of the line are shorted together and grounded to the common ground point of the antenna assembly. The unbalanced coaxial transmission line is attached to the other end of the coil and a ground jumper is run between the outer ends of the braided connector. At the center of the coil, the outer braid is severed for a distance of about one inch, and a connection is made to the inner conductor at this point. In addition, the inner conductor is jumpered to the outer braid of the *shorted*

Figure 11

THE AIR-CORE BROADBAND BALUN

The baluns of this design have the same winding technique as described for the ferrite balun.
The units are wound on plastic tubing (poly-
vinyl-chloride) of the proper diameter. Three
windings of #14 enamel or Formvar wire are
used, the ends anchored in place with 4-40
hardware. The windings are cross-connected by
additional lengths of copper wire run between
the connecting points. The common connection
of outer and inner windings is taken as ground
at one end of the balun. The text describes
two designs which cover the 2.5- to 10-MHz
range and the 6- to 30-MHz range.

coil section. The connections are wrapped
with vinyl tape and coated with an aerosol
plastic spray to protect the joints against
the weather. A coaxial plug may be attached
to the input terminals of the balun. Connection to the balanced antenna is made at the
center connections of the coil, using short,
heavy straps.

An Adjustable Balun Transformer—The
lumped constant baluns described in the previous section have a fixed impedance transformation of 1 to 1 or 4 to 1. In many
cases some other ratio is desirable. The balun
described in this section will match a 50-ohm
coaxial line to a balanced load over the range
of 10 ohms to 50 ohms.

The balun is composed of a shorted section of transmission line having a coaxial
line running down one leg of the assembly.
Points A and B (figure 13) are balanced to
ground and the inner conductor of the co-

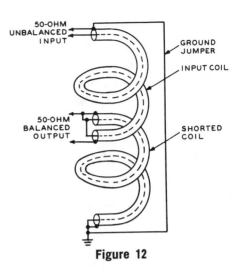

Figure 12

AN EFFECTIVE BROADBAND BALUN FOR MULTIBAND BEAMS

axial line is cross-connected to the opposite
balun leg to provide the proper phase reversal. The impedance transformation is
achieved by varying the length of the balun
and also the length of the center-fed antenna
connected to it. The antenna termination
must appear capacitive at the balun terminals for proper operation and this is achieved
by shortening the antenna element slightly
past the point of resonance.

The balun transformer is made of two
lengths of ³⁄₈-inch diameter, hard drawn
copper rod about four feet long. Spacing
between the rods is 3 inches. The balun is
designed for 20-, 15-, and 10-meter operation.

A modification of this balun design is
shown in figure 14. This balun is designed
for 14 MHz for use with a 3-element parasitic beam having a driving impedance of
approximately 20 ohms. A variable capacitor is placed across the antenna end of the
balun to permit the user to adjust the input
reactance. The length of the balun and the
value of capacitance are the variables that
determine the impedance match to the resonant driven element. The variable capacitor
should be mounted in a waterproof box to
protect it from moisture.

The Inducto-Match—A dipole element
may form a portion of a network whose
input impedance is close to 50 ohms over a

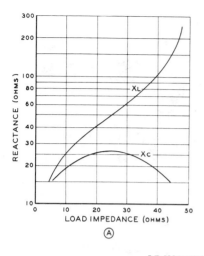

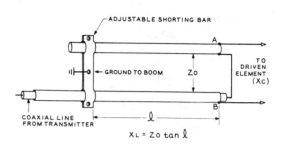

Figure 13 Ⓐ Ⓑ

ADJUSTABLE BALUN TRANSFORMER

A practical balun transformer to match a 50-ohm coaxial line to the low-impedance balanced load presented by a beam antenna is shown here. Coaxial line passes through one leg of balun. Outer conductor of the line is trimmed short to the point where the line enters balun tube, and is soldered to tube at this point. Inner conductor of the line passes along the balun tube and emerges at the antenna end, where it is cross-connected to the opposite tube as shown in the illustration. If the load impedance is known, the balun transformer may be set to length by the use of chart (A) and formula (B).

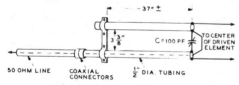

Figure 14

COAXIAL STUB BALUN FOR 14-MHz BEAM

Matching stub and balun are combined to provide balanced feed point for a 50-ohm transmission line to match low-impedance driven element. Balun is designed to be mounted on beam, at the center of driven element using short, heavy interconnecting leads.

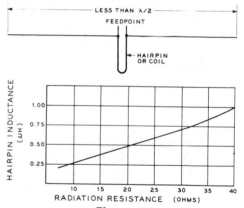

Figure 15
THE INDUCTO-MATCH

Dipole element acts as matching transformer by placing inductor at the center and shortening element to provide capacitive reactance across feedpoint. Typical three-element Yagi antenna has feedpoint impedance of about 20 ohms and calls for 0.5 µH inductor. Impedance match is made by varying inductor and length of dipole. Above chart is for 20 meters.

small frequency range (figure 15). It is necessary that the radiation resistance of the element be less than the impedance of the transmission line, and this condition is met under most circumstances.

The radiation resistance of the antenna element is made to appear as a capacitive reactance at the driving point by shortening the element past the normal resonant length. The inductive portion of the network takes the form of a hairpin or coil placed across the terminals of the driven element. The L/C ratio of the combination determines the transformation ratio of the network when

the LC product is resonant at the center frequency of antenna operation. Inductance of the hairpin or coil is best determined by experiment. Measurements made at 14 MHz, point to a shortening effect of about six inches in the overall length of the driven element, and an inductance of about 0.5 μH in the hairpin. Complete information on this compact and efficient matching system is given in the *Beam Antenna Handbook*, published by Radio Publications, Inc., Wilton, Conn.

A Broadband LC Balun—The derivation of a broadband lumped constant balun is given in figure 16. Illustration A shows two pi-network circuits with the inputs connected in parallel and the outputs series-connected. For this example, the balun is assumed to match a 50-ohm unbalanced line to a 20-ohm balanced load, a common condition for a Yagi beam antenna. One network is the conjugate of the other. The circuit can be redrawn, as in illustration B, omitting the components C_1 and L_2, as they form a resonant circuit at the design frequency. The final revision is redrawing the circuit as a bridge, as shown in illustration C. There is no coupling between the coils and they should be mounted at right angles to each other. The bandwidth of the balun is inversely proportional to the transformation ratio, and a balun having a transformation ratio of unity has a theoretically infinite bandwidth.

26-4 Antenna Matching Devices

The feedpoint impedance of a hf or vhf antenna can vary from a few ohms to as high as several hundred ohms, and can exhibit either capacitive or inductive reactance during off-resonance operation. Modern amateur stations use either low-impedance coaxial lines in the feed system or balanced lines of medium impedance.

These transmission systems require some sort of matching device at the antenna to make an efficient transition from the impedance of the line to that of the antenna, otherwise severe standing waves can occur on the transmission line. It must be remembered that no adjustment made at the transmitter end of the line will change the magnitude of the standing waves on the transmission line. Matching devices such as the baluns shown in the previous section and the matching systems shown in this section can provide a good transition between the differing imped-

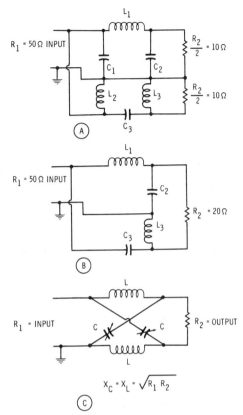

Figure 16

BROADBAND UNIVERSAL BALUN TRANSFORMER

This lumped-constant impedance transformer can be used in either balanced, or unbalanced-to-balanced condition. It may be used either for stepdown or stepup transformation. The circuit is derived from two pi networks (A), each of which has a 90-degree phase shift through it. The combination provides a 180-degree phase shift across the series-connected load terminals. The circuit is redrawn at (B), eliminating C_1 and L_2, which are parallel resonant at the design frequency. The circuit is redrawn as a bridge at (C). To match a 50-ohm load to a 20-ohm load at 14 MHz, for example, the value of X_L or X_C is 31.6 ohms. C is thus 360 pF and L is 0.36 μH. The capacitors are made adjustable so as to provide some variance in transformation and balance.

ance levels existing between feedline and antenna.

The Delta-Matched Dipole   The *delta-type matched-impedance dipole* antenna is shown in figure 17. The impedance of the transmission line is trans-

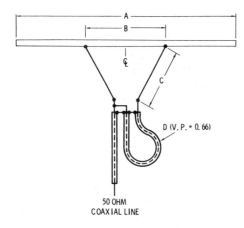

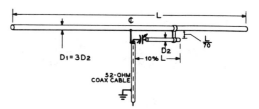

BAND	A		B		C		D	
MHz	IN	CM	IN	CM	IN	CM	IN	CM
50	111.0	281.9	22.0	55.9	15.0	38.1	79.0	200.7
144	38.0	96.5	6.5	16.5	4.0	10.2	27.0	68.6
220	25.4	64.6	4.2	10.7	2.7	6.9	18.0	45.7
432	12.9	32.8	3.0	7.6	1.5	3.8	9.1	23.1

Figure 17

DELTA MATCH DIMENSIONS
FOR VHF SERVICE

The delta dimension B is adjusted to provide a 200-ohm termination point for use with a 50-ohm coaxial line and a four-to-one coaxial balun transformer. Antenna length may have to be readjusted slightly to achieve lowest value of SWR because of reactive effect of delta wires.

formed gradually into a higher value by the fanned-out Y portion of the feeders, and the Y portion is tapped on the antenna at points where the Y portion is a compromise between the impedance at the antenna and the impedance of the line.

The delta match has become quite popular in the vhf region as a simple feed system for a high gain Yagi antenna. The device does not increase the diameter of the driven element as do other systems and its use is preferred when bandwidth is a consideration. The delta works well with a half-wave coaxial balun and provides a good match to a coaxial line. The delta wires are attached to the driven element with clips for quick and easy adjustment. Representative dimensions for the vhf bands are given in the illustration.

Figure 18

THE GAMMA MATCHING SYSTEM

See text for details of resonating capacitor

The Gamma Match   The *gamma match* is an unbalanced, single-ended version of the T-match (figure 18). One resonating capacitor is used, placed in series with the gamma rod. The capacitor should have a maximum capacitance of 8 pF per meter of wavelength. The length of the gamma rod determines the impedance transformation between the feedline and the fed element. By adjustment of the length of the rod and the value of capacitance, the SWR on the feedline may be reduced to a very low value at the resonant frequency of the driven element. Approximate dimensions for the hf bands are given in the illustration.

The Omega Match   The *omega match* is a modification of the gamma match, incorporating a shunt capacitor which permits the use of a shorter rod in the matching section. The impedance transformation is adjusted by variation of the omega capacitor, and resonance is established by adjustment of the series capacitor. Representative dimensions for the hf bands are shown in figure 19.

The Folded Dipole   The *folded dipole* is an antenna element which incorporates its own impedance transformation system (figure 20). It has the same directional properties as the simple dipole but provides a convenient method of varying the basic feedpoint impedance of the dipole.

When a dipole consists of more than one conductor, the current in the device divides between the conductors which are connected in parallel. The feedpoint resistance of such a radiator is increased by a factor of N^2, where N is the number of conductors placed

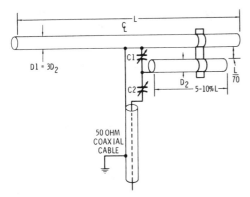

Figure 19

THE OMEGA MATCHING SYSTEM

The omega match incorporates a shunt capaci-tor (C₁) which permits electrical adjustment of the length of the gamma rod. The greater the value of C₁, the shorter will be the length of the rod. System resonance is established with the aid of series capacitor C₂. Dimensions are based on the length of the dipole element (L).

in parallel, all of the same diameter. Thus, if two equal-diameter conductors are used in a folded dipole, the feedpoint resistance will be multiplied by 2^2, or 4, and if three con-ductors are used, the feedpoint resistance will be multiplied by 3^2, or 9. As more con-ductors are added in parallel, the current

continues to divide between them and the feedpoint resistance is raised still more.

Even greater impedance transformation ratios may be achieved by varying the rela-tive size and spacing of the conductors in the folded dipole, since the impedance trans-formation ratio is dependent both on the ratio of conductor diameters and on their spacing.

The following equation may be used for the determination of the impedance trans-formation when using different diameters in the two sections of a folded element:

Transformation ratio =

$$\left(1 + \frac{Z_1}{Z_2} \right)^2$$

In this equation Z_1 is the characteristic im-pedance of a line made up of the smaller of the two conductor diameters spaced the center-to-center distance of the two con-ductors in the antenna, and Z_2 is the char-acteristic impedance of a line made up of two conductors the size of the larger of the two. This assumes that the feedline will be connected in series with the *smaller* of the two conductors so that an impedance step-up of greater than four will be obtained. If

Figure 20

FOLDED-ELEMENT MATCHING SYSTEMS

Drawing A above shows a half-wave made up of two parallel wires. If one of the wires is broken as in B and the feeder connected, the feed-point impedance is multiplied by four; such an antenna is commonly called a "folded dipole." The feed-point impedance for a simple half-wave dipole fed in this manner is approximately 300 ohms, depending on antenna height. Drawing C shows how the feedpoint impedance can be mul-tiplied by a factor greater than four by making the half of the element that is broken smaller in diameter than the unbroken half. An exten-sion of the principles of B and C is the arrange-ment shown at D where the section into which the feeders are connected is considerably shorter than the driven element.

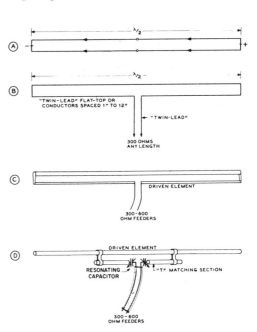

an impedance step-up of less than four is desired, the feedline is connected in series with the *larger* of the two conductors and Z_1 in the above equation becomes the impedance of a hypothetical line made up of the larger of the two conductors and Z_2 is made up of the smaller. The folded vhf unipole is an example where the transmission line is connected in series with the larger of the two conductors. Representative transformation ratios are shown in figure 21.

The Q-Section One of the earliest forms of impedance transformer used by amateurs is the *Q-section*, or quarter-wave transformer (figure 22). An impedance match between a dipole element and a balanced transmission line is obtained by utilizing a matching section, the characteristic impedance of which is the geometric mean between the input and output terminal impedances. An equivalent device for an unbalanced system can be made up of an electrical quarter-wavelength section of coaxial transmission line.

Decoupling Devices In the *Transmission Line* chapter it was stated that when a balanced antenna is used, the two conductor transmission line feeding the antenna should carry equal and opposite currents throughout its length to maintain the electrical symmetry of the system. To maintain this balance when a coaxial transmission line is used, various balance-to-unbalance devices have been described earlier in this chapter. The purpose of these devices (in addition to providing an impedance

match) is to prevent unwanted antenna current from flowing on the outside of the coaxial line.

The unwanted current may be choked off by forming the transmission line into an r-f

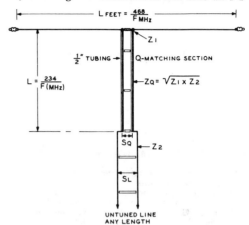

Center to Center Spacing in Inches	Impedance in Ohms for 1/2" Diameters	Impedance in Ohms for 1/4" Diameters
1.0	170	250
1.25	188	277
1.5	207	298
1.75	225	318
2.0	248	335

Figure 22

HALF-WAVE RADIATOR FED BY A Q-SECTION

The **Q** matching section is simply a quarter-wave transformer whose impedance is equal to the geometric mean between the impedance at the center of the antenna and the impedance of the transmission line to be used to feed the bottom of the transformer.

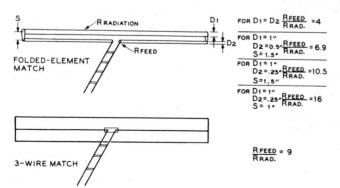

Figure 21

DATA FOR FOLDED-ELEMENT MATCHING SYSTEMS

In all normal applications of the data given the main element as shown is the driven element of a multi-element parasitic array. Directors and reflectors have not been shown for the sake of clarity.

choke which will present a high impedance to currents flowing on the outer surface of the line, while allowing the current within the line to proceed unimpeded. This can be done by coiling the line into a "doughnut" about a foot in diameter. Six turns of line, coiled in this fashion and held in place with electrical tape will provide a satisfactory r-f choke for the high frequency bands. An even more effective choke can be made by passing three turns of the line through a large ferrite core, as shown in figure 23.

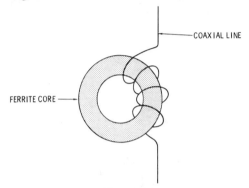

Figure 23

HF DECOUPLING CHOKE FOR COAXIAL LINE

Three turns of the transmission line passed through a large ferrite core will effectively decouple the transmission line from antenna currents. The choke is placed near the antenna terminals. Coil diameter is about 10" (25.4 cm) for RG-8A/U line. The core is Indiana General CF-124. For catalog and list of distributors, write Indiana General Corp., Crow Mills Road, Keasby, NJ 08832. Core is Q-1 material with a permeability of 125 at 1 MHz.

In the vhf region, an effective decoupling device can be constructed around a linear r-f choke, as shown in figure 24. A quarter-wave line, surrounding, or adjacent to, the outer shield of the transmission line as shown in the illustration effectively provides a balancing action to nullify antenna currents flowing on the outer shield of the transmission line.

The Matching Stub The subject of stub matching was briefly covered in the *Transmission Line* chapter. Short, open- or closed-end stubs are used for matching purposes, providing a react-

ance value that will cancel the unwanted reactance on the transmission line.

A variety of the matching stub is very useful, particularly at vhf, when it is desired to match a transmission line to an

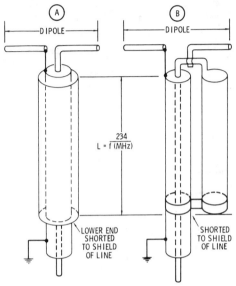

$$L = \frac{234}{f \,(\text{MHz})}$$

Figure 24

QUARTER-WAVE DECOUPLING DEVICE

A linear decoupling device is commonly used to choke off unwanted current from flowing on the outside of the coaxial line and also to provide a balanced feedpoint from an unbalanced transmission line. (A) Section of coaxial line acts as a quarter-wave resonant circuit providing a high impedance at the top end. Any Antenna current flowing on the line creates equal and opposite current flowing on the inner surface of coaxial shield. (B) Two conductor transmission line serves as balancing section.

antenna system. By connecting a resonant stub to either a current or voltage loop and attaching a transmission line to the stub at a suitable impedance point, the stub may be made to serve as an impedance matching transformer (figure 25). Illustration A shows a half-wavelength, closed stub attached to the center of a dipole element. The far end of the stub is shorted, and the low impedance at this point is reflected across the antenna terminals at the other end of the stub. A closed, resonant circuit consisting of antenna and stub is thus formed. The stub and antenna can be resonated by sliding the shorting bar up and down the stub, noting resonance with a grid-dip oscillator coupled to the stub. The feedline is moved about in the same manner until the SWR on the line is at the lowest possible value.

Illustration B shows a quarter-wavelength closed-end stub used in conjunction with a two-half-wavelengths antenna, which provides a high impedance at the feedpoint.

In any one of these examples, the balanced, open-wire line may be removed and the matching stub fed with a coaxial line and half-wave balun transformer.

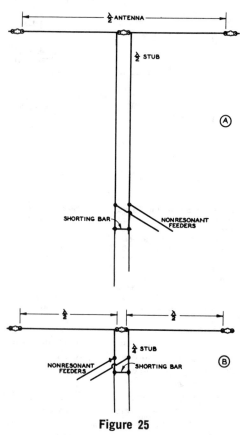

Figure 25

MATCHING STUB APPLICATIONS

Illustration A shows the use of a half-wave shorted stub to feed a relatively low impedance point such as the center of the driven element of a parasitic array, or the center of a half-wave dipole. B shows the conventional use of a shorted quarter-wave stub to voltage-feed two half-wave antennas with a 180° phase difference.

26-5   Coupling to the Antenna System

When coupling an antenna feed system to a transmitter the most important considerations are as follows: (1) means should be provided for varying the load on the amplifier; (2) the load presented to the final amplifier should be resistive (nonreactive) in character; and (3) means should be provided to reduce harmonic coupling between the final amplifier plate tank circuit and the antenna or antenna transmission line to an *extremely low* value.

Transmitter Loading and TVI   The problem of coupling the power output of a high-frequency or vhf transmitter to the radiating portion of the antenna system has been complicated by the virtual necessity for eliminating interference to TV reception. However, the TVI elimination portion of the problem may *always* be accomplished by adequate shielding of the transmitter, by filtering of the control and power leads which enter the transmitter enclosure, and by the inclusion of a harmonic attenuating filter between the output of the transmitter and the antenna system.

Although TVI may be eliminated through inclusion of a filter between the output of a shielded transmitter and the antenna system, the fact that such a filter should be included in the link between transmitter and antenna makes it necessary that the transmitter-loading problem be re-evaluated in terms of the necessity for inclusion of such a filter.

Harmonic attenuating filters must be operated at an impedance level which is close to their design value; therefore they must operate *into* a resistive termination substantially equal to the characteristic impedance of the filter. If such filters are operated into an impedance which is not resistive and approximately equal to their characteristic impedance: (1) the capacitors used in the filter sections will be subjected to high peak voltages and may be damaged, (2) the harmonic attenuating properties of the filter will be decreased, and (3) the impedance at the input end of the filter will be different from that seen by the filter at the load end (except in the case of the half-wave type filter). It is therefore important that the filter be included in the transmitter-to-antenna circuit at a point where the impedance is close to the nominal value of the filter, and at a point where this impedance is likely to remain fairly constant with variations in frequency.

Block Diagrams of Transmitter-to-Antenna Coupling Systems There are two basic arrangements which include all the provisions required in the transmitter-to-antenna coupling system, and which permit the harmonic attenuating filter to be placed at a position in the coupling system where it can be operated at an impedance level close to its nominal value. These arrangements are illustrated in block diagram form in figures 26 and 27.

The arrangement of figure 26 is recommended for use with a single-band antenna system, such as a dipole or a rotatable array, wherein an impedance matching system is included within or adjacent to the antenna. The feedline coming down from the antenna system should have a characteristic impedance equal to the nominal impedance of the harmonic filter, and the impedance matching at the antenna should be such that the standing-wave ratio on the antenna feedline is less than 3 over the range of frequency to be fed to the antenna.

The arrangement of figure 26 is more or less standard for commercially manufactured equipment for amateur and commercial use in the hf and vhf range.

The arrangement of figure 27 merely adds an antenna coupler between the output of the harmonic attenuating filter and the antenna transmission line. The antenna coupler will have some harmonic attenuating action, but its main function is to transform the impedance at the station end of the antenna transmission line to the nominal value of the harmonic filter.

26-6 Practical Antenna Couplers

The *antenna coupler*, or *antenna tuner*, is a matching device that translates the electrical characteristics of the antenna and feedline into values more compatible with the communication equipment attached to the antenna. Some form of antenna coupler may be necessary with modern solid-state transmitters. The coupler matches the antenna system impedance and SWR to a value such that the transmitter does not suffer reduced power output caused by operation into a mismatched load.

The transmitter employing vacuum tubes in the final amplifier stage and a pi-network output circuit is considerably more tolerant of a high SWR antenna load than is equivalent solid-state equipment and may not require an antenna tuner in the case where a tuner is necessary to make the solid-state equipment operable.

The Line Flattener The *line flattener* is a network inserted in a 50-ohm feed system to reduce the SWR on the line to near unity. This efficient and low cost tuner is recommended to amateurs who have solid-state transmitters and who wish to achieve a good antenna match with a minimum of adjustment. The schematic is shown in figure 28. The device is an adjustable T-section network which, in addition

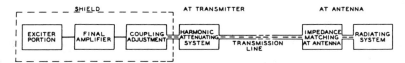

Figure 26

ANTENNA COUPLING SYSTEM

The harmonic suppressing antenna coupling system illustrated above is for use when the antenna transmission line has a low standing-wave ratio, and when the characteristic impedance of the antenna transmission line is the same as the nominal impedance of the low-pass harmonic-attenuating filter.

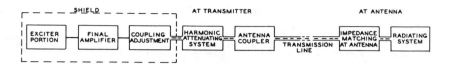

Figure 27

ANTENNA COUPLING SYSTEM

The antenna coupling system illustrated above is for use when the antenna transmission line does not have the same characteristic impedance as the TVI filter, or when the standing-wave ratio on the antenna transmission line is high.

to achieving the required match, provides up to 20 dB of attenuation to transmitter harmonics falling in the TV channels. The tuner is rated for 1 kW output power level and is built in an aluminum box. The approximate setting of the tap switch for each amateur band is given in the caption. The fixed mica capacitor is required only for operation on the 80 meter band.

For transmitter powers up to 200 watts output, capacitor C_1 may be a receiving type and for low-power transmitters, a compression-type mica capacitor may be used.

An SWR meter between the line flattener and transmitter is required for proper adjustment. Using reduced power, the tap switch is set for the band of operation and the capacitor adjusted for minimum SWR consistent with proper transmitter loading.

The Transmatch   The *transmatch* (popularized by W1ICP) is an adjustable network that can function as a line flattener, or as a matching device for an end-fed wire antenna. When combined with a balun such as described earlier in this chapter, the transmatch can also be used with a two-wire transmission line system.

The split-stator capacitor (figure 29) provides good harmonic rejection as the capacitive reactance to ground in the TV channels is very low.

The transmatch is built in an aluminum box to achieve maximum harmonic rejection and should be used in conjunction with an SWR meter placed between the tuner and the transmitter.

For preliminary adjustment, both capacitors are set at maximum value. The variable inductor is then adjusted for minimum SWR

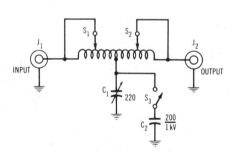

Figure 28

COAXIAL LINE FLATTENER

Coil L is 32 turns, 2½ inches in diameter, 8 turns per inch. For 80 and 40 meters, the whole coil is used. For 20 meters, 12 turns are shorted out at the outer ends of the winding. For fifteen meters 25 turns are shorted out and for 10 meters 28 turns are shorted out. Total inductance of the coil is 31 microhenrys. Switches S_1, S_2, and S_3 are separate ceramic-deck switches; each is panel controlled. Capacitor C_2 is a transmitting-type mica unit.

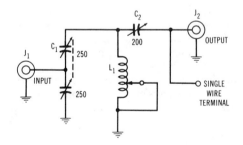

Figure 29

THE TRANSMATCH ANTENNA COUPLER

This device works over the range of 3.5 to 29.7 MHz. Capacitor C_1 is a 250 pF per section split-stator unit with 0.171 inch spacing for high power. Capacitor C_2 is insulated from the assembly and is panel-driven through an insulated coupling. Spacing of this capacitor is 0.171 inch. The inductor is a roller coil having a maximum inductance of 18 μH. For 160-meter operation the inductance should be increased to 28 μH.

on the line to the transmitter. Once this point is reached, the capacitors are tuned to decrease the SWR, with possibly a slight readjustment of the inductor. The correct setting of the controls is the one which provides a good match with maximum capacitance setting for both C_1 and C_2.

A Single-Wire Tuner A simple tuner for an end-fed wire antenna is shown in figure 30. This adjustable network will match a wide range of impedance values from 1.8 to 29.7 MHz. The components of the tuner are placed in an aluminum box and the unit is rated at 1 kW transmitter output power. Connection to the wire antenna is made by means of a large ceramic feedthrough insulator mounted at a convenient point on the box.

For proper adjustment, an SWR bridge between the tuner and the transmitter is required. Capacitor C_1 is set at minimum value and the tap switch adjusted for a drop in SWR reading. Once this has been found, the adjustable inductor and capacitor are tuned for a further reduction in SWR. Adjustment of the three controls will drop the SWR to near-unity. The transmitter controls are then readjusted for proper loading.

An Inexpensive SWR Bridge for the Tuners Some transmitters incorporate an SWR-reading circuit in their metering arrangement. For those transmitters that do not have such a convenience, the SWR meter shown in figure 31 is useful. This bridge indicates tuner resonance at the operating frequency. The resistive arm of the bridge is made up of ten 10-ohm composition resistors soldered to two 1-inch diameter copper rings made of heavy wire (figure 32). The bridge capacitors are attached to this assembly with very short leads. The diode mounts at right angles to the resistor bank to ensure minimum capacitive coupling between the resistors and the detector.

The bridge must be calibrated for 50-ohm service. This can be done by connecting a 2-watt 52-ohm (nominal value) composition resistor or other dummy load at the

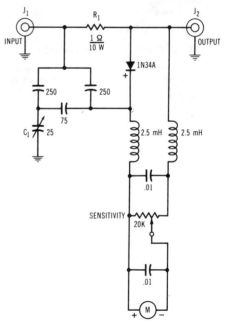

Figure 31

SIMPLE SWR METER FOR TUNERS

This resistive bridge makes use of a special series resistor made up of ten 10-ohm, 1-watt composition resistors connected in parallel. Silver-mica capacitors are used for the other bridge legs. The meter has a 0–1 mA dc movement. See figure 32 for bridge assembly.

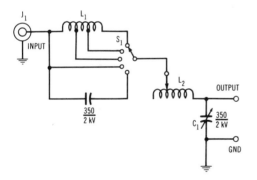

Figure 30

SINGLE-WIRE TUNER

Coil L_1 is 35 turns, 2 inches in diameter, 3½ inches long-tapped at 15 and 27 turns from the input (left) end. Coil L_2 is a rotary inductor, 10 µH. The fixed capacitor is a transmitting-type mica unit. This tuner should be used in conjunction with a good ground connection.

Figure 32

CLOSE-UP OF SWR BRIDGE

Simple SWR bridge is mounted below the chassis of the tuner. Carbon resistors are mounted to two copper rings to form low-inductance one ohm resistor. Bridge capacitors form triangular configuration for lowest lead inductance. Balancing capacitor C_1 is at lower right.

output terminals of the bridge. A small amount of r-f energy is fed to the input of the bridge until a reading is obtained on the r-f voltmeter. The 25 pF bridge balancing capacitor is then adjusted with a fiber blade screwdriver until a zero reading is obtained on the meter. The sensitivity control is advanced, as the meter null grows, in order to obtain the exact point of bridge balance.

The SWR bridge is placed in the coaxial line between the tuner and the transmitter. The transmitter is turned on and the sensitivity control of the bridge adjusted for near full scale reading. As tuner resonance is approached, the meter reading will decrease and the sensitivity control is advanced. When the tuner is in adjustment, the meter reading will be near zero. The meter need not be calibrated in terms of SWR as all tuning adjustments are conducted to provide a zero reading on the instrument.

CHAPTER TWENTY-SEVEN

HF General Purpose Antennas

An antenna is a system of conductors that radiates and intercepts electromagnetic waves. The general characteristics of hf and vhf antennas were outlined in an earlier chapter. This chapter, and the following ones, deal with the practical aspects of designing, building, and adjusting antennas for optimum performance.

Under normal circumstances, long distance hf transmission is propagated along a *Great Circle path* to the target area. Ionospheric reflection for this path is most effective when the wave is propagated at a certain definite *angle of radiation* (A) above the horizon, as shown in figure 1. Energy radiated in other directions and at other elevation angles performs no useful function. Hf directional antennas are commonly used by the various communication services.

Long distance vhf propagation is generally over a straight-line route to the target area, but the mode of propagation may be one or more of many types. Directive vhf antennas are effective for all of the common propagation modes and also help to reduce fading and interference arriving from unwanted directions. Thus, the directional character- istics and angle of radiation above the hori- zon of the antenna are of great importance to the hf or vhf operator. Other antenna attributes, such as bandwidth, power gain and front-to-back ratio are equally important.

27-1 The Angle of Radiation

The angle of radiation of an antenna is the angle above the horizon of the axis of the main lobe of radiation. With practical hf antennas of moderate size, the radiation

pattern of the main lobe is quite broad and occupies a large area in front of the antenna. The antenna "sprays" a great section of the ionosphere with energy, ensuring that even with a large change in layer height and vari- ations in propagation along the path, a cer-

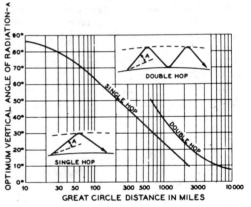

Figure 1

OPTIMUM ANGLE OF RADIATION WITH RESPECT TO DISTANCES

Shown above is a plot of the optimum angle of radiation for one-hop and two-hop com- munication. An operating frequency close to of radiation for one-hop and two-hop com- munication distance is assumed.

tain amount of the radiated signal has a good chance of reaching the target area. Multiele- ment vhf antennas that are large compared to the wavelength of the radiated wave, on the other hand, are capable of providing a sharply defined pattern at a specific angle of radiation, and their aiming may prove to be quite critical.

The angle of radiation above the horizon for a typical antenna close to the earth is dependent on the antenna height above the surface of the earth, the polarization of the antenna, and the frequency of operation. In calculating the vertical angle of radiation for a particular antenna, the image concept (Chapter 24-5) is used to establish the effects of wave reflection. The surface of the earth in the vicinity of the antenna is assumed to be flat and perfectly conductive. The angle of radiation of the vertical field pattern maximum is created by addition and cancellation of the fields from the antenna and the hypothetical image antenna. Similarly, the image antenna concept is also used to calculate the impedance and current distribution characteristics of the actual antenna. The effect of reflection from a conducting surface can be expressed as a factor which, when multiplied by the free space radiation pattern of the antenna, gives the resultant pattern for various angles above the surface. The limiting conditions are those when the direct and reflected waves are in phase or out of phase, and the resulting field strength at a distant point will be either twice the field strength from the antenna alone, or zero.

By changing the height of the antenna above the reflecting ground, the vertical

angle of the reflection and cancellation patterns may be readily changed. Ground reflection patterns have been developed by which the free space pattern of a dipole antenna can be modified to show the true vertical pattern of the antenna at any height above the ground, as shown in figure 2. These plots are multiplying factors that represent the effect of ground reflection on a horizontal antenna.

Because the current relationships between the actual antenna and the image antenna are reversed in the case of vertical polarization, the ground reflection patterns for a vertical dipole are different from those of a horizontal dipole (figure 3).

Ground Characteristics   The ground reflection charts are based on the assumption that the earth is a perfect conductor, which it is not. Under actual conditions, ground conductivity varies widely with locale. In areas of poor surface conductivity, the actual reflection surface may seem to be several feet below the actual surface and the layer of earth near the surface acts as a lossy dielectric to the radio wave. If the amplitude of the reflected wave is reduced through ground losses, the vertical pattern of reflection will be affected, as

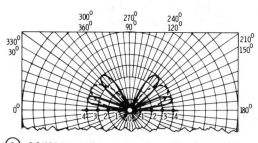

(A)   0.1 (SOLID) AND 0.25 (BROKEN) WAVELENGTH ABOVE GROUND.

(B)   0.5 (BROKEN) AND 1.0 (SOLID) WAVELENGTH ABOVE GROUND.

Figure 2

GROUND REFLECTION PATTERNS FOR A DIPOLE ANTENNA

The vertical directivity patterns of a horizontal half-wave dipole are shown here. Illustration A indicates the relative intensity of radiation at 0.1 and 0.25 wavelength above ground, and illustration B shows the increase in low-angle radiation at 0.5 and 1.0 wavelength above ground. As antenna height is increased, more lobes appear in the pattern with the lower lobes approaching the horizontal plane. A perfectly conducting ground plane is assumed for these patterns.

Figure 3

GROUND REFLECTION PATTERNS FOR A VERTICAL HALF-WAVE ANTENNA

The vertical directivity patterns of a vertical antenna are shown here. Illustration A indicates the relative intensity of radiation at 0.25 and 0.375 wavelength above ground and illustration B shows the radiation patterns for 0.5 and 0.75 wavelength above ground. These plots represent multiplying factors representing the effect of ground reflection. Note that the nulls and maxima are interchanged with those of the horizontal antenna.

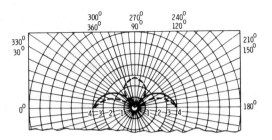

(A) 0.25 (SOLID) AND 0.375 (BROKEN) WAVELENGTH ABOVE GROUND.

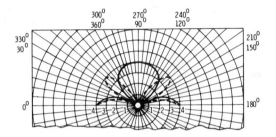

(B) 0.5 (SOLID) AND 0.75 (BROKEN) WAVELENGTH ABOVE GROUND.

will the feedpoint impedance of the antenna. The chief effect of the lossy dielectric is to absorb a large portion of the energy radiated at low angles to the earth. In addition, the magnitude of the main lobes is decreased by the amount of energy lost, or dispersed, and the nulls of the pattern tend to become obscured (figure 4).

In the vhf region, the antenna is usually several wavelengths above the surface of the earth and the direct wave from the antenna travels to the target area without benefit of the portion of the wave that travels along the ground. The loss of energy at low angles due to a lossy ground is quite low and wave attenuation is limited to that normal amount caused by path attenuation and spreading.

A perfectly conducting ground can be simulated by a *ground screen* placed under the antenna. The screen should have a small mesh compared to the size of the radio wave and should extend for at least a half wavelength in every direction from the antenna. Unless the screen is extremely large (several wavelengths in every direction) the screen will affect only the high angle radiation from a horizontal antenna and will not materially aid the effect of the earth on low angle radiation which is useful for long distance hf communication.

Figure 4

GROUND LOSS ALTERS VERTICAL PATTERN OF ANTENNA

If the amplitude of the ground reflected wave is reduced through ground losses, the vertical pattern of reflection will be affected. Chief effect of lossy ground is to absorb a large portion of the energy radiated at low angles and to fill in the nulls of the pattern.

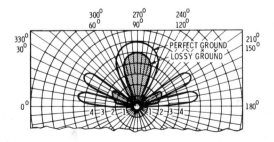

Optimum Angle of Radiation   The *optimum angle of radiation* for hf propagation between two points is dependent upon a number of variables, such as height of the ionospheric layer providing the reflection, the distance between the two stations and the number of hops necessary for propagation between the stations. It is often possible for different modes of

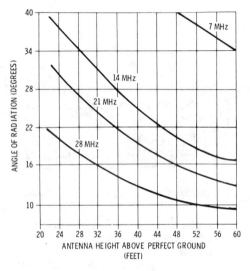

Figure 5

OPTIMUM VERTICAL ANGLE OF RADIATION FOR HF TRANSMISSION

The optimum vertical angle for hf transmission lies between 5° and 40°, depending on frequency used and path length. The optimum angle of radiation for the 7-MHz band occurs at an antenna height of 45 feet or greater above ground, for the 14-MHz band at a height of 40 feet or above, for the 21-MHz band at a height of 35 feet or above, and for the 28-MHz band at a height of 30 feet or above. Experience has shown that heights of 40 to 70 feet are a good compromise for long-distance communication on the various hf amateur bands.

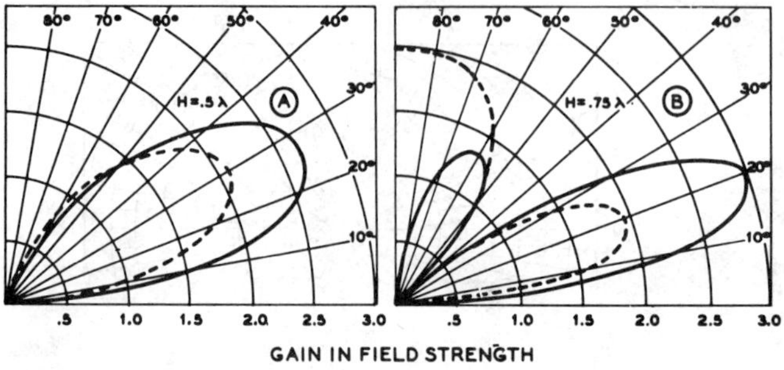

propagation to simultaneously provide signals between two points. This means, of course, that more than one angle of radiation is effective. If no elevation directivity is used under this condition of propagation, selective fading will take place because of interference between waves arriving over the different paths.

Measurements have shown that the optimum angles useful for long distance hf communication lie between 5° and 40°, the lower angles being more effective for the higher frequencies (figure 5). These figures assume normal propagation by virtue of F_2 layer reflection.

The radiation available at useful, low angles from any antenna is of interest. The reflection plots of figures 2 and 3 apply to a dipole antenna. Other antennas which concentrate radiation in certain directions and suppress it in others provide modified vertical radiation patterns because some lobes that show up in the dipole pattern do not show up to as great a degree in the pattern of a different antenna type. In the case of a beam antenna, the resultant pattern may not be symmetrical since the beam tends to suppress radiation in certain directions. An example of this is shown in figure 6, wherein the high angle radiation of a dipole placed 0.75 wavelength above the ground is greatly attenuated in the case of a beam antenna located at the operating height. Placement of the two antennas at 0.5 wavelength height, on the other hand, produces nearly identical patterns. The angle of radiation of representative beam antennas will be discussed in the next chapter.

It should be noted that the beam antenna does not lower the angle of radiation of the main lobe, as compared to a dipole. The angle of radiation is a function of antenna height

Figure 6

VERTICAL RADIATION PATTERNS

Showing vertical radiation patterns of a horizontal two element beam (solid curves) and a horizontal dipole (dashed curves) when both are 0.5 wavelength (A) and 0.75 wavelength (B) above ground. Note the suppression of the high angle radiation in the latter case.

GAIN IN FIELD STRENGTH

above ground and the operating frequency, and has little to do with antenna configuration, at least in the case of the simpler antenna arrays.

Horizontal Directivity *Horizontal directivity* is desirable for hf or vhf operation, but it is not easily obtainable with reasonable antenna dimensions at the lower frequencies. Arrays having extremely high horizontal directivity are cumbersome, but the smaller designs can be rotated for point-to-point work. As in the case of ground reflection, the effect of a nearby conducting surface can alter the horizontal directivity of an antenna. The result is that the radiation pattern loses symmetry. In some cases, pattern distortion is deliberate, as in establishing the front-to-back ratio of a beam antenna; in other cases it is unintentional.

Dipole Antenna Types The most popular and least expensive antenna for general usage is the dipole. Antennas for the lower-frequency portion of the hf range and temporary or limited use antennas for the upper portion, usually are of a relatively simple type in which directivity is not a prime consideration. Also, it is often desirable that a single antenna system be capable of operation on various bands, or on frequencies outside the amateur band (MARS, etc.). Variations of the dipole and Marconi antenna designs are well qualified for this usage and the first portion of this chapter is devoted to a discussion of such antenna systems. The latter portion of the chapter is devoted to matching systems and antenna installation.

27-2 The Center-Fed Antenna

A center-fed half-wave antenna system is usually to be desired over an end-fed system since the center-fed system is inherently balanced to ground and is therefore less likely to be troubled by feeder radiation.

A number of center-fed systems are illustrated in figure 7.

The Dipole Antenna The center-fed dipole with an open-wire transmission line is an inherently balanced antenna system if properly built. The antenna is matched to the transmitter by means of an antenna tuner and a coaxial line, as discussed in Chapter Twenty-Six. If the dipole is cut for the lowest operating frequency, it may be used on any higher amateur band by proper adjustment of the tuner. Figure 7A shows a representative antenna.

In figure 7B a half-wave shorted transmission line is used to resonate the antenna system and an open wire line (or coaxial line with balun) is tapped on the line at a point which provides a low value of SWR. This feed system is often used in vhf beam antenna designs.

The average feedpoint impedance of a center-fed dipole is about 75 ohms. The actual value varies with antenna height and construction. In figure 7C a quarter-wave matching transformer is used to accomplish an impedance transformation to a high impedance, open wire transmission line. This system is popular in the vhf region as the use of the open wire line reduces transmission line losses as compared to a conventional coaxial cable.

An alternative method for increasing the feedpoint impedance of a dipole so that a medium impedance, low loss transmission line may be used is shown in figure 7D. This dipole uses more than one wire for the radiating element. The two wires are parallel connected but only one wire is broken for the feedpoint. Since the total antenna current is divided between the wires, the impedance at the center of the broken element is four times higher than that of a single wire.

The antenna shown in the illustration is made of 300 ohm tv "ribbon line" for ease of assembly. The dipole is made slightly shorter than the conventional length $(462/F_{MHz})$ instead of $(468/F_{MHz})$ and the two wires of the twin lead are joined together at each end. The center of one of the conductors is broken and the ribbon feedline is spliced into the dipole leads.

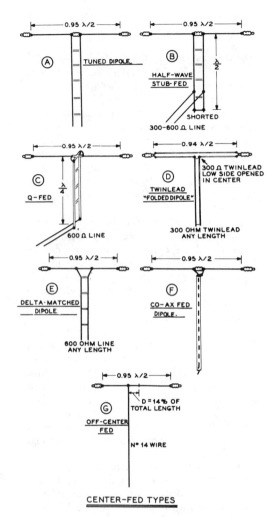

CENTER-FED TYPES

Figure 7

FEED SYSTEMS FOR A HALF-WAVE DIPOLE ANTENNA

The half-wave dipole antenna may be either center- or end-fed, as discussed in the text. For the hf region (below 30 MHz), the length of a simple dipole is computed by: length (feet) = 468/f, with f in MHz. For the folded dipole, length is computed by: length (feet) = 462/f, with f in MHz. Above 30 MHz, the length of the dipole is affected to an important degree by the diameter of the element and the method of supporting the dipole.

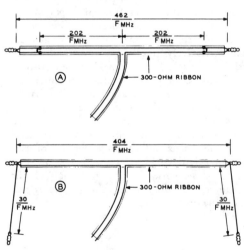

Figure 8

FOLDED DIPOLE WITH SHORTING STRAPS

The impedance match and bandwidth characteristics of a folded dipole may be improved by shorting the two wires of the ribbon a distance out from the center equal to the velocity factor of the ribbon times the half-length of the dipole as shown at A. An alternative arrangement with bent down ends for space conservation is illustrated at B.

Better bandwidth can be obtained with a folded dipole made of ribbon line if the conductors are shorted a distance of 0.82 (the velocity factor of the line) of a free space quarter wavelength from the center of the antenna (figure 8).

The delta-matched dipole is shown in figure 7E and described in Chapter Twenty-Six. It is used principally in vhf beam arrays where it is desired to have a small diameter element, unbroken at the center.

The popular coaxial-fed dipole is shown in figure 7F. For vhf operation, or for use as the driven element of a beam antenna, the feedline is run through a balun to provide proper current distribution in the antenna. The use of a balun on the lower frequencies is not generally necessary.

The single-wire fed antenna is shown in figure 7G. The feeder wire is tapped on the dipole at a point which provides an approximate impedance match. This system requires a good ground for the return current.

Dimensions for hf dipole antennas are tabulated in Table 1.

**Table 1.
Length of Wire Dipole Antenna**

FREQUENCY OR BAND (MHz)	DIPOLE LENGTH TIP-TO-TIP	
	Feet	Meters
1800 - 1900 kHz	253.0	77.16
1900 - 2000 kHz	240.0	73.20
3.5 - 3.8 MHz	125.25	38.20
3.7 - 4.0 MHz	121.0	36.90
7.0 - 7.3 MHz	65.5	19.97
10.1 MHz	46.3	14.12
14.0 - 14.35 MHz	33.0	10.06
21.0 - 21.45 MHz	22.1	6.74
28.0 - 29.7 MHz	16.3	4.97
50.0 - 52.0 MHz	9.6	2.93
52.0 - 54.0 MHz	9.2	2.81

27-3   The Vertical Antenna

The vertical antenna is of interest because its ground reflection patterns are reversed as compared to those of a horizontal antenna and because it may be supported in a minimum amount of ground space. In addition, the vertical is well suited to low-frequency service, wherein the groundwave range is used for communication. The vertical antenna is also popular in the vhf field, as much vehicular communication is vertically polarized.

The electrical equivalent of the dipole is the half-wave vertical antenna (figure 9). Placed with the bottom end from 0.01 to 0.2 wavelength above ground, it is an effective transmitting antenna for low-angle radiation in areas of high ground conduc-

tivity. The vertical antenna, in one form or another, is widely used for general broadcast service and for point-to-point work up to about 4.0 MHz. Generally speaking, the vertical antenna is susceptible to man-made interference when used for receiving, as a great majority of noise seems to be vertically polarized.

The vertical antenna produces high current density in the ground beneath and around it and ground conduction currents return to the base of the antenna. Ground system losses can dissipate a major portion of the antenna power and reduce the radiated field accordingly unless precautions are taken to ensure a low resistance ground return path for the induced currents.

The best ground surface, or ground plane, is an infinite copper sheet placed beneath the antenna. This may be approximated in the medium- and high-frequency region by a system of radial wires. Broadcast specifications call for 120 radials, each approximately 0.25 wavelength long. The radials may be buried a few inches beneath the surface of the earth for protection from damage, or laid atop the surface.

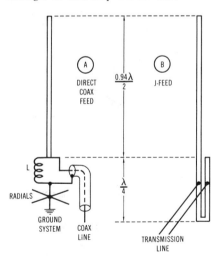

Figure 9

**HALF-WAVE VERTICAL ANTENNA
SHOWING ALTERNATIVE
METHODS OF FEED**

In the amateur service, few enthusiasts can go to the trouble and expense of installing an elaborate ground system and must be content with fewer radials in their installation. The absolute minimum number of radial wires is one, which will provide a ground point at the base of the antenna. Common usage is four and many amateurs have settled on 12 radials as a good compromise between performance and expense. Tests have indicated that reducing the number of radials drops the radiated field of the antenna, and dropping from 120 radials to 4 can result in a decrease in the radiated field as much as 8 dB, if the ground conductivity beneath the antenna is poor and the antenna is short.

Vhf vertical antennas, mounted many wavelengths above ground, are less susceptible to ground losses and experience has shown that 4 radial wires usually do a good job on antennas of this category.

The Ground-   An effective form of Mar-
Plane Antenna coni antenna is the quarter-
wave *ground-plane antenna,*
so named because of the radial ground wires. The ground plane may be mounted with the radial wires a few inches above the ground, or elevated with the radials well above the

surface of the ground. Since the radials are resonant, the ends are at a high voltage potential and they should be insulated to prevent accidental contact. In a like manner, the radials should not be grounded or buried, as this would destroy their resonance. A typical ground-plane antenna for the hf bands is shown in figure 10 along with suggested dimensions.

The base impedance of the ground plane is of the order of 30 to 35 ohms, and it may be fed with a 50-ohm coaxial line with only a slight impedance mismatch. For a more exact match a version of the hairpin match shown in Chapter Twenty-Six may be used. The match consists of a small coil placed directly across the feedpoint to the radials. A coil of 10 turns, 1½ inches in diameter, turns spaced the wire diameter can be used. The number of turns in the coil is adjusted to provide the lowest SWR on the feedline.

The feedpoint impedance of a ground plane may be raised to about 50 ohms by

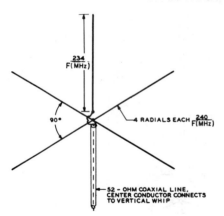

Figure 10

THE HIGH-FREQUENCY GROUND-PLANE ANTENNA

This antenna is used on the hf amateur bands, usually in the form of a vertical whip, with the radials acting as guy wires for the assembly. The whip may be mounted on a post or tower, or on the roof of a building. The wire radials often slope downward. If the antenna is mounted near the ground, a ground connection may be added at the junction of the radials. A self-supporting version of the ground plane is popular for use in the vhf spectrum.

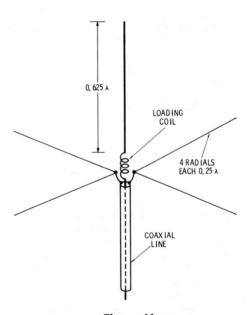

Figure 11

THE ⅝-WAVELENGTH VERTICAL ANTENNA

The extended vertical antenna provides about 3 dB gain over a quarter-wave groundplane. To establish resonance, a base loading coil is used to tune the antenna to ¾-wave resonance. Base impedance is very close to 50 ohms. Standard, quarter-wave radials are used.

drooping the radials down at a 45° angle. Some horizontally polarized radiation from the radials will take place, raising the radiation resistance of the antenna. The radials can serve as guy wires when they are brought down in this fashion.

The ⅝-Wave Vertical   The field strength of a short vertical antenna reaches a maximum figure when the antenna is ⅝ wavelength high, as opposed to ¼ wavelength. A power agin of about 3 dB over a quarter-wave vertical is achieved with the extended design. The feedpoint of a ⅝-wave vertical is reactive and a series inductance is required to establish a non-reactive termination (figure 11). Quarter-wave radials are used with this antenna configuration.

Short Vertical Antennas   An antenna that is electrically small (the length small with respect to the wavelength of operation) can perform as an efficient radiator *provided* power can be efficiently applied to the antenna. Generally speaking, very short antennas have low values of radiation resistance and very high Q. Typically, an 8-foot base-loaded whip antenna at 3.8 MHz exhibits a load resistance value as shown in figure 12A. At all frequencies below self-resonance, the equivalent circuit of the short antenna is composed of a low value of resistance in series with a large value of capacitive reactance. In order to establish a state of resonance, and to match the whip antenna to a 50-ohm source, the reactance must be cancelled out and an impedance transformation effected. Both requirements demand high-Q networks, such as the type shown in figure 12B. Even with care, a substantial portion of the available power may be lost in such networks. Generally, the higher the radiation resistance value of the whip antenna is, the easier it is to match and the higher will be the efficiency of the network.

An 80-Meter Compact Ground Plane—A 66-foot high vertical antenna for 80-meter operation presents a problem on a small lot, as the supporting guy wires tend to take up a large portion of the property. It is possible to reduce the height of the antenna by the inclusion of a loading coil near the center of

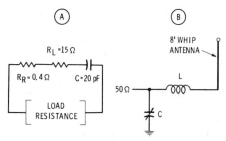

Figure 12

THE SHORT VERTICAL ANTENNA

An electrically short antenna exhibits very low radiation resistance and high Q (selectivity). Typically, an 8-foot whip operating at 3.8 MHz exhibits a load resistance that is capacitive and about 15 ohms, of which only 0.4 ohm is radiation resistance. The other 14.6 ohms represents loss resistance (A). A suitable matching network for this antenna is shown in (B). The loss figure includes network loss, assuming a coil having a Q of over 250 is used.

the vertical section (figure 13). Overall antenna height is cut to about 25 feet and the radiation resistance of the antenna is reduced to approximately 15 ohms.

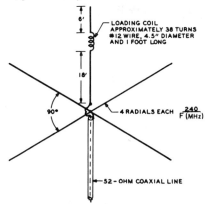

Figure 13

80-METER LOADED GROUND-PLANE ANTENNA

Number of turns in loading coil to be adjusted until antenna system resonates at desired frequency in 80-meter band.

The antenna has high Q and a rather narrow bandwidth; an operating range of about 100 kHz with an SWR of less than 2 is possible at the design frequency. The radial wires may be bent back upon themselves to conserve space, if necessary and the use of a suitable L-network to match the antenna to the transmission line is suggested.

An "All-band" Vertical Antenna—A short vertical antenna can be used on several amateur bands by employing an adjustable base-loading inductor. Sets of radial wires are used for the bands of interest. Shown in figure 14 is a 22-foot vertical antenna designed for operation on the amateur bands from 10 through 80 meters. The height is chosen to present a ¾-wavelength vertical for low-angle radiation at the highest frequency of operation. Multiple radial wires are used for the 10, 15, and 20 meter bands, and a single radial wire is used for either 40- or 80-meter operation. A ground connection may be used at the junction of the radial wires for lightning protection. If the antenna is roof mounted, it may be possible to use the metal gutter system as a ground.

Four-wire TV rotator cable is used to construct the hf radial system, each cable including a radial wire for one of the three bands. The fourth radial wire may be extended for 40- or 80-meter operation. At least three such radial assemblies should be used. These can be laid out on the roof, hidden in the attic, or passed about the yard (if the antenna is ground-mounted).

The vertical radiator is made of two ten-foot sections of aluminum TV mast, plus one five-foot section cut to the proper length. The sections are assembled with self-tapping sheet metal screws. The antenna and base coil are attached to ceramic insulators mounted on the upright support post.

The antenna is resonated to the operating frequency in each band with the aid of an SWR meter in the coaxial feedline. The two taps are adjusted for lowest value of SWR reading. The approximate tap positions are indicated in the illustration.

Phased Vertical Antennas—Two or more vertical antennas can be operated in an array to obtain additional power gain and directivity. The antennas may be in broadside, end-fire, or collinear configuration (figure 15). In illustration 15A, the broadside antennas are fed in-phase by two coaxial lines to produce a figure-8 pattern broadside to the plane of the antennas. The length of the lines from the line junction to the antennas is unimportant as long as both lines are of equal length. Illustration 15B shows the same antennas in end-fire connection, with the antennas fed out-of-phase. The pattern is in-line with the plane of the antennas. The interconnecting coaxial line must be an elec-

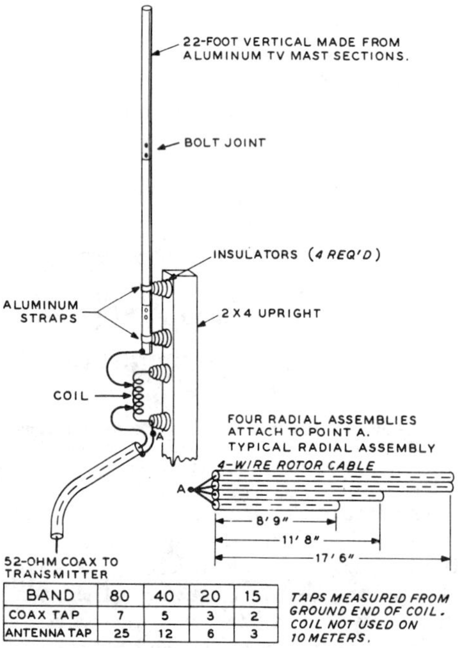

Figure 14

"ALL-BAND" VERTICAL ANTENNA

Base-loaded whip and multiple radial system may be used on all bands from 80 through 10 meters. Loading-coil taps are adjusted for lowest SWR on each band. The SWR on 10 meters may be improved by placing a 250-pF capacitor in series with the feedline connection to the base of the antenna and adjusting the capacitor for minimum SWR. Coil is 40 turns, 2″ in diameter, 4″ long (Air-Dux 1610).

BAND	80	40	20	15
COAX TAP	7	5	3	2
ANTENNA TAP	25	12	6	3

TAPS MEASURED FROM GROUND END OF COIL. COIL NOT USED ON 10 METERS.

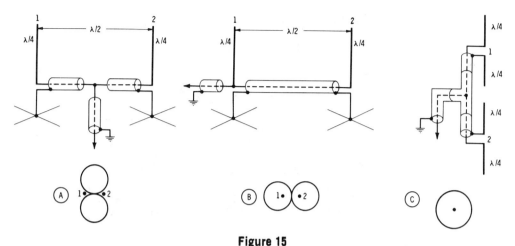

Figure 15

PHASED VERTICAL ANTENNAS

Antennas may be arranged in broadside (A), end-fire (B), or collinear (C) configuration depending on phase difference between the two antennas. Antennas are spaced one-half wavelength apart. The collinear vertical stack antenna produces an omnidirectional pattern.

trical half-wavelength long (or multiple thereof) to provide the figure-8 pattern. A collinear, vertically stacked array is shown in illustration 15C. The pattern is omnidirectional and a configuration of this type is popular on the vhf amateur bands.

The end-fire array can be modified to produce a unidirectional pattern (figure 16). The antennas are spaced a quarter wavelength with a 90° phase reversal between the antennas. The pattern is in-line with the plane of the antennas and in the direction of the vertical receiving the lagging excitation. The interconnecting line is an electrical quarter wavelength (or odd multiples thereof) long.

A good ground system is required for proper operation of a phased array and experimenters have reported satisfactory results with radial systems composed of 60 radials, each 0.25 wavelength long.

Typical radiation patterns for two vertical antennas employing different spacing and phasing are summarized in figure 17.

27-4 The Marconi Antenna

On the lower-frequency amateur bands there is often insufficient space to erect a

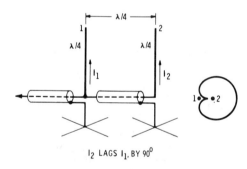

I_2 LAGS I_1, BY 90°

Figure 16

PHASED VERTICALS PRODUCE UNIDIRECTIONAL PATTERN

Two vertical antennas, spaced one-quarter wavelength apart and fed with a 90° phase reversal between them produce a unidirectional, cardioid pattern, as shown. The pattern is in line with the antennas and in the direction of the vertical receiving the lagging current.

half wavelength antenna and some form of Marconi antenna is used. This is essentially a vertical, or inverted-L antenna working against a ground or radial system.

The fundamental Marconi antenna is a quarter-wavelength radiator having an impedance transforming device to match a coaxial transmission line. Since most amateur antennas for the 160- and 80-meter

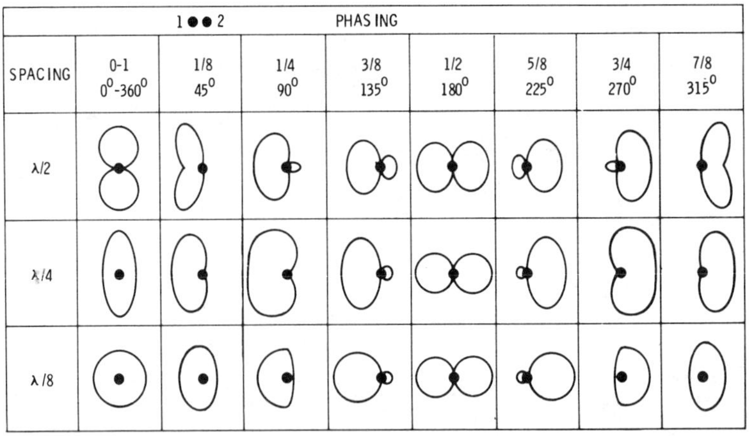

				PHASING				
1 ●● 2								
SPACING	0-1 0°-360°	1/8 45°	1/4 90°	3/8 135°	1/2 180°	5/8 225°	3/4 270°	7/8 315°
λ/2								
λ/4								
λ/8								

Figure 17

RADIATION PATTERNS FOR 2-ELEMENT PHASED ARRAY

A variety of patterns can be obtained by selection of spacing and phasing between two vertical antennas. The deep null of the phased array is of great help in the broadcast service, where protection must be given to a distant station working on the same channel.

bands are less than one quarter wavelength in height above ground (in the case of a inverted-L arrangement or a short vertical antenna) the feedpoint impedance is quite low, typically 5 to 10 ohms for a Marconi antenna 50 feet high operating at 1.8 MHz. The theoretical feedpoint resistance for an inverted-L or top-loaded vertical antenna is shown in figure 18. A sine wave current distribution in the antenna is assumed.

Variations on the basic Marconi antenna are shown in figure 19. The vertical antenna is shown in illustration 19A and the inverted-L in illustration 19B. Top loading techniques are shown in illustrations 19C through 19F. The object of all loading techniques is to produce an increase in the effective length of the radiator, and thus to raise the point of maximum current in the radiator as far as possible above the ground. The arrangement in illustration 19F provides the maximum amount of loading for a given antenna height.

Amateurs primarily interested in the higher-frequency bands, but liking to work 80 or 160 meters occasionally, can usually manage to resonate one of their hf antennas as a Marconi by working the whole system (feeders and all) against a ground system, resorting to a loading coil, if necessary.

Water-Pipe Grounds Copper water pipe, because of its comparatively large surface and cross section, has a relatively low r-f resistance. If it is possible to attach to a junction of several water pipes a satisfactory ground connection will be obtained. If one of the pipes attaches to a lawn or garden sprinkler system in the immediate vicinity of the antenna, the effectiveness of the system will approach that of buried copper radials.

The main objection to iron water-pipe grounds is the possibility of high-resistance joints in the pipe, due to the "dope" put on the coupling threads. By attaching the ground wire to a junction with three or more legs, the possibility of requiring the main portion of the r-f current to flow through a high resistance connection is greatly reduced.

Marconi Dimensions A Marconi antenna is an odd number of electrical quarter waves long (usually only one quarter wave in length), and is always resonated to the operating frequency. The correct loading of the final amplifier is accomplished by varying the coupling, rather than by detuning the antenna from resonance.

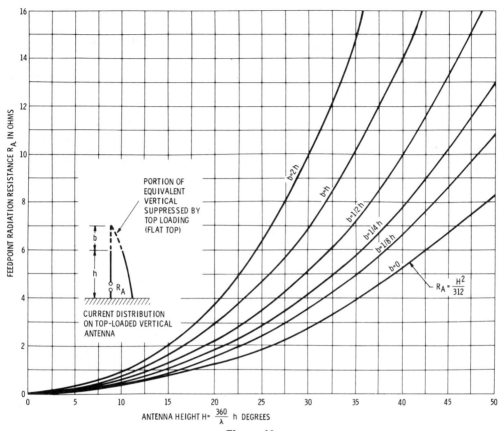

Figure 18

FEEDPOINT RADIATION RESISTANCE OF LOADED VERTICAL ANTENNA

The theoretical radiation resistance for a top-loaded vertical antenna is quite low, if any degree of loading is assumed. For an eighth-wave vertical antenna with full top loading, the radiation resistance is about 20 ohms. Practical loading conditions provide a lower value of radiation resistance than indicated here. (Graph adapted from "Performance of Short Antennas," Smith & Johnson, Proceedings of the IRE, October, 1947).

Physically, a quarter-wave Marconi may be made anywhere from one-eighth to three-eighths wavelength overall, including the total length of the antenna wire and ground lead from the end of the antenna to the point where the ground lead attaches to the junction of the radials or counterpoise wires, or where the water pipe enters the ground. The longer the antenna is made physically, the lower will be the current flowing in the ground connection, and the greater will be the overall radiation efficiency. However, when the antenna length exceeds three-eighths wavelength, the antenna becomes difficult to resonate by means

of a series capacitor, and it begins to take shape as an end-fed Hertz, requiring a method of feed such as a pi-network.

The Radial Ground Wire   The ground termination for a Marconi or other unbalanced antenna system can be improved by the addition of a radial ground wire which is connected in parallel with the regular ground connection. The radial wire consists simply of a quarter wavelength of insulated wire connected to the ground terminal of the transmitter. The opposite end of the radial wire is left disconnected,

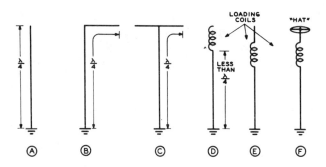

Figure 19

VARIATIONS OF THE MARCONI ANTENNA

The Marconi uses the ground image as the missing half of the half-wavelength antenna. (A) Simple quarter-wave vertical. (B) Inverted-L Marconi. (C) Top-loaded Marconi. (D) Top-loaded Marconi, using loading inductance at top of structure. (E) Loaded Marconi with inductor placed near midpoint of structure. (F) Optimum loading configuration combining loading inductor with capacitive "hat" at top of antenna. This arrangement provides maximum degree of loading for a given antenna height.

FIGURE 8 RADIATION PATTERN

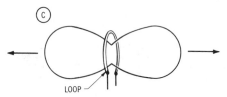

DIRECTIONAL RADIATION PATTERN

Figure 20

RADIATION PATTERNS OF LOOP ANTENNAS

A loop antenna fed from a balanced feed system provides various field patterns, depending on the size of the loop. (A) Very small loop provides nulls above and below plane of loop, with maximum response in the plane of the loop. The half-wavelength loop (B) has no nulls in the pattern, and exhibits a directional response perpendicular to the loop plane and away from the feedpoint. The full-wave loop (C) exhibits nulls in its plane, with a bidirectional response perpendicular to the loop plane. This configuration is widely used in the popular Quad antenna, and provides a power gain of about 2 dB over a dipole antenna.

or "floating." The radial wire may be run about the baseboard of the operating room or out the window and a foot or two above the ground. A high-impedance point is established at the end of the wire and a corresponding low-impedance (ground) point at the transmitter end which simulates a ground connection. While it may be used by itself as a ground termination, the radial ground wire works best when used in combination with a regular ground connection. Its use is highly recommended with all the antennas shown in this Handbook which require an external ground connection. Since the radial wire is a tuned device, separate radial wires cut to length are required for each amateur band. Several such radials can be connected in parallel at the transmitter ground point for multiband operation.

27-5 The Loop Antenna

The *loop antenna* is a radiating coil of one or more turns. A loop whose dimensions are small compared to the wavelength of operation has a figure-8 radiation pattern identical with that of a dipole oriented normal to the plane of the loop, with the electric and magnetic fields interchanged (figure 20A). For a small closed, circular loop structure, the approximate value of radiation resistance is:

$$R_r = 197 \, L^4 \text{ (for } L \text{ less than } 0.1 \text{ wavelength)}$$

where,

L equals the perimeter of the loop in wavelengths.

The radiation resistance of a small square loop is practically the same as for the circular loop if they have equal area.

When the perimeter of the loop is one-half wavelength, a resonance point is reached and the feedpoint impedance is very high (of the order of 10,000 ohms). The radiation resistance of the loop, however (referred to the current loop opposite the terminals) is very low—approximately 5 ohms. The radiation pattern of the half-wavelength loop is shown in figure 20B.

The full-wave loop (Quad loop) has a pattern similar to that shown in figure 20C and provides a power gain of approximately 2 dB over a dipole. This configuration is widely used in the popular Quad beam antenna. The feedpoint impedance of the Quad loop is of the order of 120 ohms. Practical Quad beam antennas will be discussed in a later chapter.

The Demi-Quad Loop Antenna Shown in figure 21 is a simple one-band vertical loop antenna which provides almost 2 dB gain over a dipole. The radiation pattern of the loop is a figure-8 at right angles to the plane of the wires. The demi-Quad may be square or rectangular in shape, with the feedpoint either at the center of the bottom wire or at a corner. A diamond configuration is shown in the drawing. Antenna polarization is horizontal.

The feedpoint impedance of the loop is about 120 ohms and a short section of 75-ohm coaxial line is used as a transformer to match the loop to a 50-ohm transmission line. All joints in the line, plus the connections to the loop wire should be made waterproof by coating the connections with bathtub caulk or other moisture-resistant sealant, such as *General Electric RTV-102*. If water enters the connections, or the end of the line, it can cause damage to the conductors.

The Mini-Loop Antenna A half-wave dipole can be bent into a square to form a compact loop antenna (figure 22). The loop is placed in the horizon-

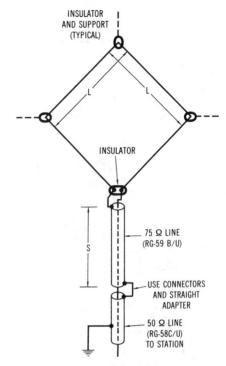

Figure 21

THE DEMI-QUAD LOOP ANTENNA

The loop is made of No. 16 enamel wire supported with small glass insulators and nylon rope. The bottom insulator is cut from a small length of lucite or plexiglass rod. Top end of 75-ohm line and straight adapter should be waterproofed.

BAND MHz	DIMENSIONS			
	L		S	
	FEET	METERS	FEET	METERS
3.5-4.0	70'8"	21.60	45'0"	13.68
7.0-7.3	35'4"	10.80	22'6"	6.84
10.1	24'7"	7.49	15'9"	4.8
14.0-14.35	17'8"	5.45	11'3"	3.42
21.0-21.45	11'10"	3.51	7'6"	2.28
28.0-29.7	8'9"	2.66	5'7"	1.71

tal plane and exhibits a slight degree of directivity in the direction of the feedpoint as shown in the illustration. Antenna feed impedance is about 20 ohms and a small matching coil shunted across the feedpoint serves to raise the impedance to about 50 ohms.

Care must be taken to reduce the capacitance across the insulator opposite the feed-

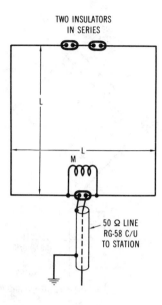

Figure 22

THE MINI-LOOP ANTENNA

This compact loop is shown from above. The loop is in the horizontal plane. Dimension L is about 0.935 of a free-space half-wavelength. Coil M consists of 10 turns, 1 inch in diameter, spaced twice the wire diameter. It is adjusted for lowest SWR on the transmission line.

BAND (MHz)	DIMENSION L	
	FEET	METERS
10.1	11'6"	3.50
14.0-14.35	8'2"	2.50
21.0-21.45	5'6"	1.68
28.0-29.7	4'0"	1.22

point, otherwise the resonant frequency of the loop will be altered. Two small insulators in series will do the job.

80-Meter Loop Antenna   A loop antenna may be used to advantage on 80 meters. The passband is quite broad and the loop may be mounted close to the ground and still provide good results. Shown in figure 23 is a loop designed by G3AQC, cut for 3.8-MHz operation. The loop is mounted in the vertical plane and employs a 4-to-1 air core balun to match a 50-ohm coaxial line. Operational bandwidth is 250 kHz between the 2-to-1 SWR points on the feedline.

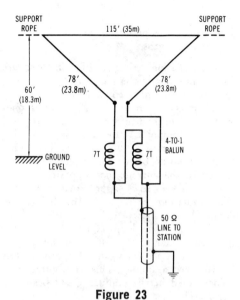

Figure 23

A VERSION OF THE G3AQC 80-METER LOOP ANTENNA

This antenna is supported in the vertical plane on 60-foot poles. The bottom of the loop is about six feet above ground. The 4-to-1 balun consists of 7 turns of No. 14 enamel (or Formvar) bifilar wound on a 2⅜ inch diameter form.

Two 80-meter loop antennas designed by GI3ZXM are shown in figure 24. The larger loop provides the broadest frequency response. Both loops are fed with open-wire ladder line and a balanced antenna tuner of the type shown later in this chapter. The passband of the smaller loop is quite sharp, requiring tuner readjustment when the frequency of operation is moved over 50 kHz.

The W9LZX *Lazy-Quad* is shown in figure 25. This is a standard quad loop laid on its side and fed at one corner by a 50-ohm coaxial line. Height of the loop is about 30 feet above ground. The radiation angle of this loop is high so that a strong signal is put out within a radius of 500 miles. At long distances, the loop performs much in the manner of a dipole at an equivalent height.

The W6TC loop for 80 and 40 meters is shown in figure 26. This loop is fed with an open-wire line which acts as a short matching transformer for 40 meter operation. On 80 meters, the effect of the line

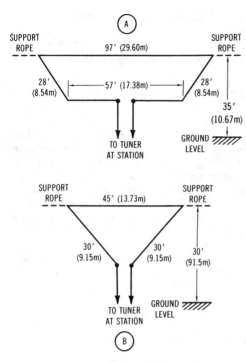

Figure 24

THE 80-METER LOOP ANTENNAS OF GI3ZXM

A—The large loop provides good frequency coverage. A random length open-wire line is fed from a tuner, such as the one shown later in this chapter. B—A smaller loop design. The frequency response is quite sharp, requiring readjustment of the tuner when the frequency of operation is moved over 50 kHz.

is negligible, except to establish loop resonance. As shown, the loop is resonant at 3.7 MHz and 7.15 MHz. A second matching transformer made of 75-ohm coax provides a good match to a 50-ohm coaxial line. The matching transformer is coiled up into a simple balun to reduce line currents that flow on the outside of the coaxial line.

To move the 80-meter resonance point lower in the band, six feet of line is added to the open wire section. The line should be removed for 40-meter operation.

The shape of the W6TC two-band loop is not as important as the total length of wire in the loop plus the open-wire line. The total length of wire is 283 feet. Better bandwidth can be achieved on 80 meters by forming the wire into a rectangle instead of a triangle, the rectangle being about 30 feet

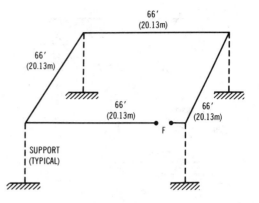

Figure 25

THE W9LZX LAZY-QUAD FOR 80 METERS

Perimeter of the loop is 264 feet. Height above ground is about 30 feet (9.15m). Length of the individual sides is not critical as long as the total length of wire is held constant. Antenna is fed at F with a 50-ohm coaxial line.

high and 71'6" long. This configuration will cover most of the 80-meter band without the addition of an open stub.

Loop Configuration Studies have been made on different loop configurations to determine loop gain and bandwidth. A summary of this information is given in figure 27. Generally speaking, loop gain is greatest when the area within the loop is greatest. That is, a circular loop provides slightly higher power gain than a square shape, and a square provides higher gain than a triangle. With regard to bandwidth, a squat, wide loop provides better bandwidth than a square, and a square better bandwidth than a tall, thin configuration. Likewise, the input impedance of the squat, wide loop is the highest of the three models. The gain of the tall, thin loop is slightly higher than the gain of the square (quad) configuration, and the gain of the squat, wide loop is lowest of the three models.

The antenna designer, then, is faced with various sets of tradeoffs in loop performance. Luckily, none of the tradeoffs is of great importance except bandwidth. The variation in loop gain is not great and variation in input impedance can be accommodated by

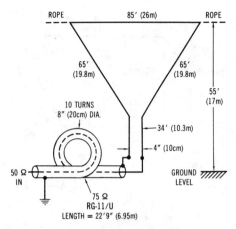

Figure 26

THE W6TC QUAD-LOOP ANTENNA FOR 80-40 METERS

This two-band loop delivers maximum performance on 80 and 40 meters. For lengths shown,
resonance is at 7.15 MHz and 3.7 MHz. Adding
6.5 feet (2 m) to open-wire line will decrease
80-meter resonance to 3.5 MHz. Extra line is removed for 7-MHz operation. The 75-ohm coaxial
matching line is coiled up to form a simple
isolation transformer.

any number of matching systems. Bandwidth, however, is important, especially on
the 80- and 10-meter bands which are quite
wide in terms of the center frequency.

27-6 The Sloper Antenna

A simple and effective radiator for the
low-frequency bands is the *sloper antenna*
(figure 28). The most common version is a
quarter-wavelength radiator, fed at the top,
using the existing metal tower structure as
a ground plane. The sloper is fed with a
coaxial line, with the shield of the line
grounded to the tower at the point where
the sloper is fed.

The sloper wire is approximately half the
length of the equivalent dipole. The exact
length is determined by the angle the wire
makes to the tower and the height of the
wire end above ground. Exact resonance
may be adjusted by varying the end height
or by trimming the wire.

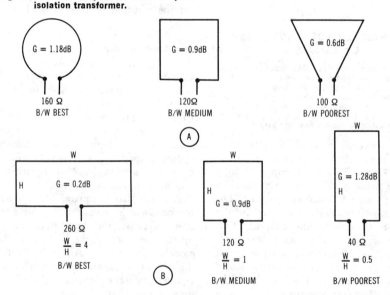

Figure 27

LOOP CONFIGURATION DETERMINES PERFORMANCE

Gain, bandwidth, and input impedance are traded when shape of quad loop is changed. All loops
shown are one wavelength in circumference. A—Bandwidth and gain are best for circular loop
as compared to square and triangular designs. B—Gain is best for vertically oriented design but
bandwidth is poor and input impedance is low compared to traditional square loop (center).
Horizontally oriented loop (left) provides good bandwidth at the expense of gain. Input impedance is high. Gain is expressed with reference to a dipole.

The sloper tends to exhibit a small amount of directivity through the tower in the direction the wire points. Some amateurs have experimented with a number of sloper wires mounted around a single support, switching from wire to wire for optimum directivity. In some cases, certain of the unused sloper wires have been used as reflectors, by lengthening them slightly using a remote relay located on the tower.

Experimenters have also tried a full dipole as a sloper, with one end tied to a high tower and the other end near ground level. Again, some directivity has been established in the general direction of the low end of the antenna, but the radiation pattern still resembles the figure-8 pattern of the dipole.

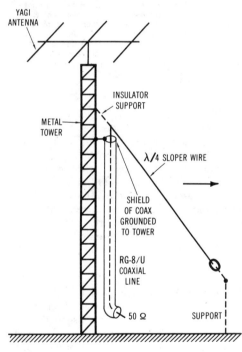

Figure 28

THE SLOPER ANTENNA FOR 160, 80, OR 40 METERS

The single wire sloper is hung from an existing tower for low-band operation. The wire is approximately a quarter wavelength long. Exact length depends on the angle the wire makes with the tower and the height of the end of the sloper above the ground. Good results have been obtained with tower heights as low as 40 feet and as high as 100 feet. The shield of the coaxial line is grounded to the tower at the top.

27-7 Space-Conserving Antennas

In many cases it is desired to undertake a considerable amount of operation on the 80- or 40-meter band, but sufficient space is simply not available for the installation of a half-wave radiator for the desired frequency of operation. This is a common experience of apartment dwellers.

One technique of producing an antenna for lower-frequency operation in restricted space is to erect a short radiator which is balanced with respect to ground and which is therefore independent of ground for its operation. Several antenna types meeting this set of conditions are shown in figure 29. Figure 29A shows a conventional center-fed dipole with bent-down ends. This type of antenna can be fed with coaxial line in the center, or it may be fed with a resonant line for operation on several bands. The overall length of the radiating wire will be a few percent greater than the normal length for such an antenna since the wire is bent at a position intermediate between a current loop and a voltage loop.

Figure 29B shows a method of using a half-length dipole. It is recommended that spaced open conductor be used for the radiating portion of the folded dipole. The reason for this lies in the fact that the two wires of the flat top are *not* at the same potential throughout their length when the antenna is operated on one-half frequency.

The antenna system shown in figure 29C may be used when not quite enough length is available for a full half-wave radiator. The dimensions in terms of frequency are given on the drawing. An antenna of this type is 93 feet long for operation on 3600 kHz and 86 feet long for operation on 3900 kHz. This type of antenna has the additional advantage that it may be operated on the 7- and 14-MHz bands, when the flat top has been cut for the 3.5-MHz band, simply by changing the position of the shorting bar and the feeder line on the stub.

A sacrifice which must be made when using a shortened radiating system (as for example the types shown in figure 29) is in the bandwidth of the radiating system.

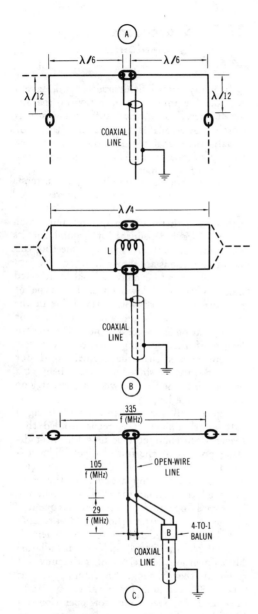

Figure 29

THREE EFFECTIVE SPACE CONSERVING ANTENNAS

A—The ends of a dipole are folded down to conserve space. B—Dipole is folded back upon itself to make an antenna only one quarter wavelength long. Frequency response of this antenna is much sharper than that of a full dipole. Inductor (L) across feedpoint is 10 turns #12 wire, 1½″ diameter, spaced wire diameter. Adjust number of turns for lowest SWR at resonant frequency of antenna. C—Center portion of dipole is folded back into transmission line. A 4-to-1 balun is tapped on the open-wire line at a point which provides lowest SWR at resonant frequency of antenna. Dimension given for antenna C is in feet and is approximate.

is raised, the amount of power lost in the ground resistance is proportionately less. If a Marconi antenna is made out of 300-ohm TV-type ribbon line, as shown in figure 30, the radiation resistance of the antenna is raised from a low value of 10 or 15 ohms to a more reasonable value of 40 to 60 ohms. The ground losses are now reduced by a factor of 4. In addition, the antenna may be directly fed from a 50-ohm coaxial line, or directly from the unbalanced output of a pi-network transmitter.

The Inverted-V   A close relative of the sloper
Antenna   is the inverted-V antenna.
   This design consists of a dipole supported at the center from a tower with the ends sloping down to near-ground level. The inverted-V is a popular low band antenna as it may be made up from two guy wires for the tower. The included angle between the wires of the V should be as large as possible; not less than 90 degrees or bandwidth will suffer.

Normally, the wires of the V lie in one plane, but some amateurs cramped for space have moved the wires together at the ends, forming a miniature V-beam antenna. A dipole compressed in this shape, of course,

The frequency range which may be covered by a shortened antenna system is approximately in proportion to the amount of shortening which has been employed.

The Twin-Lead   Much of the power loss in
Marconi Antenna   the Marconi antenna is a
   result of low radiation resistance and high ground resistance. If the radiation resistance of the Marconi antenna

shows no gain and little directivity. But it *will* work, and sometimes it is the only configuration that will fit in a restricted space.

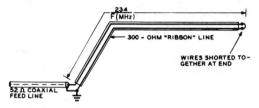

Figure 30

TWIN-LEAD MARCONI ANTENNA FOR THE 80- AND 160-METER BANDS

The length of the inverted-V is somewhat greater than that of a linear dipole and may be computed from the following formula:

$$\text{Overall length (feet)} = \frac{485}{f_{(MHz)}}$$

The Fan Dipole for 80 Meters   Two dipoles may be connected in parallel and trimmed for operation at the ends of the 80-meter band (figure 31). The ends of the antennas must be well separated for proper operation. The SWR curve of the parallel-connected dipoles resembles a *W* with the points of minimum SWR falling near the band edges.

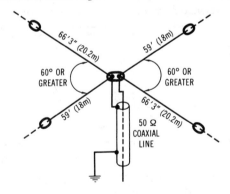

Figure 31

A BROADBAND FAN DIPOLE FOR 80 METERS

This dipole will cover the range of 3.5 to 4.0 MHz with a low SWR on the transmission line. The angle between the separate dipoles should be 60 degrees, or greater. Dipoles are fed in parallel at the center point. The wires of the dipole lie in the horizontal plane.

The Loaded Antenna   A shortened dipole or vertical antenna is often the only answer to a "tough" antenna location. Amateurs living in apartments, town houses or condominiums often find that covenants or restrictions in the lease or deed prohibit the erection of outdoor antennas of any type. It is possible to erect an *"invisible" antenna* of #26 enameled copper wire, strung to a nearby tree or lamp post, and used in conjunction with a radial ground wire inside the dwelling. A second alternative is an indoor antenna, artificially loaded to fit into the available space.

The indoor antenna will work well in a wood frame building, provided it is not electrically coupled to the electrical wiring of the building. Placement of the antenna is a "cut-and-try" process, moving the antenna about until the least interaction with the wiring of the building is noticed.

A simple loaded antenna design is shown in figure 32. The illustration shows a simple dipole installation, making use of similar loading coils in each half of the antenna. The ends of the dipole may be dropped down to conserve more space. Suggested values for coils are given in the drawing. The antenna can be resonated to the operating frequency by adjusting the loading coils for the minimum value of SWR on the transmission line at the design frequency. The coils are adjusted ½ turn at a time or by trimming the antenna tips until resonance is established. At any given coil setting, a low value of SWR will be maintained only over a narrow frequency range, depending on the amount of loading required in the installation.

An antenna that will operate on more than one band is a great convenience to the amateur operator. Various types of multiband antenna designs are available, and the choice depends on factors such as the amount of space at hand and the bands desired for the majority of operation. A number of recommended multiband antennas are shown in this section.

Long Wire Multiband Antennas   One of the simplest multiband antennas is the long wire, either end-fed, or fed at the center.

Two practical designs are shown here, along with compact models suitable

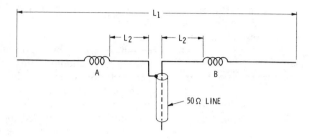

DESIGN FREQ. (MHz)	L_1		L_2		COILS A AND B
	FT	MTRS	FT	MTRS	
3.6	90' 0"	27.45	51' 3"	15.63	17µH = 48 TURNS #16, 3" LONG, 1" DIA. SPACED 16 TURNS PER INCH
3.8	85' 6"	26.0	48' 9"	14.87	
7.15	45' 9"	14.0	26' 3"	8.00	9µH=SAME AS ABOVE EXCEPT 25 TURNS, 1-1/2" LONG.

Figure 32

SHORT DIPOLE FOR 80- OR 40-METER OPERATION

This center loaded dipole design is suitable for operation over 100 kHz of the 80-meter band or over 200 kHz of the 40-meter band. The antenna is resonated to the operating frequency by varying the inductance of the loading coils or by trimming the antenna tips. A 1-to-1 balun may be used at the feedpoint, if desired. A 10-turn coil, 1" diameter and 2" long, placed across the dipole feedpoint can be used to reduce the SWR on the coaxial line. Adjust the number of turns for lowest value of SWR.

for operation on all hf bands from a small lot.

The End-Fed Long Wire—A random length, long wire makes an inexpensive multiband antenna. It may be matched to the transmitter with a simple network and tuned to resonance with the aid of an SWR meter (figure 33). For operation on all bands from 160 through 6 meters, the recommended wire length is about 136 feet. In practice, the length of the antenna can be compensated for by the tuning unit, and any length that is at least 0.2 wavelength long at the lowest operating frequency will be found to be satisfactory. A good ground system is recommended and tuned radials for each band, plus a connection to ground, are suggested. On the lower frequencies, the antenna is essentially omnidirectional, but on the higher frequencies it tends to have a cloverleaf pattern, exhibiting directivity off the ends.

The Center-Fed Long Wire—The center-fed antenna requires no ground return for proper operation and has good rejection to harmonics. For ease of tuning, certain antenna and feeder lengths operate better than

others, and suggested combinations are listed in figure 34. Other lengths will work as well, as the total wire length in flattop plus feeder is resonated by means of the compact tuning

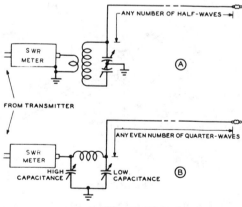

Figure 33

THE END-FED HERTZ ANTENNA

Showing the manner in which an end-fed Hertz may be fed through a low-impedance line and SWR meter by using a resonant tank circuit as at A, or through the use of a reverse-connected pi-network as at B.

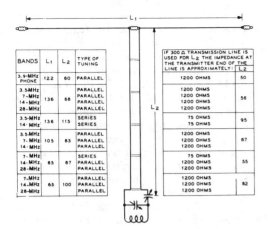

BANDS	L_1	L_2	TYPE OF TUNING	IF 300 Ω TRANSMISSION LINE IS USED FOR L_2 THE IMPEDANCE AT THE TRANSMITTER END OF THE LINE IS APPROXIMATELY:	L_2
3.9-MHz PHONE	122	60	PARALLEL	1200 OHMS	50
3.5-MHz	136	68	PARALLEL	1200 OHMS	56
7-MHz			PARALLEL	1200 OHMS	
14-MHz			PARALLEL	1200 OHMS	
28-MHz			PARALLEL	1200 OHMS	
3.5-MHz	136	115	SERIES	75 OHMS	95
14-MHz			SERIES	75 OHMS	
3.5-MHz	105	83	PARALLEL	1200 OHMS	67
7-MHz			PARALLEL	1200 OHMS	
14-MHz			PARALLEL	1200 OHMS	
7-MHz	65	67	SERIES	75 OHMS	55
14-MHz			PARALLEL	1200 OHMS	
28-MHz			PARALLEL	1200 OHMS	
7-MHz	65	100	PARALLEL	1200 OHMS	82
14-MHz			PARALLEL	1200 OHMS	
28-MHz			PARALLEL	1200 OHMS	

Figure 34

DIMENSIONS FOR CENTER-FED MULTIBAND ANTENNA

unit located at the operating position. Since the flattop does all the radiating, it would be prudent to place as much wire in the flattop as possible and leave the remainder to make up the two-wire, balanced feed system.

A flexible antenna tuner is shown in figure 35. A 50-ohm coaxial line and SWR meter connect the tuner to the transmitter. Proper antenna adjustment is achieved by observing the SWR reading and adjusting the variable capacitors for the lowest SWR reading consistent with proper transmitter loading. The switch connects the primary coils in either series or parallel. In general, the coils are series-connected for the 80-meter band and parallel connected for the higher bands.

The Windom Antenna—The single-wire-fed, or *Windom*, antenna is widely used for portable installations and locations where an unobtrusive antenna is required (figure 36). A single-wire feeder is used, having a characteristic impedance of about 300 ohms. The feeder is tapped at a point on the antenna that approximates this value on more than one band. An external ground system is required for proper operation of the antenna. Since the feeder wire radiates, it is necessary to bring it away from the antenna at right angles to the wire for at least one-half the length of the antenna. The antenna is fed with a simple L-network, such as described earlier in this section, and an SWR meter. The network is adjusted for minimum SWR on the coaxial line from network to transmitter.

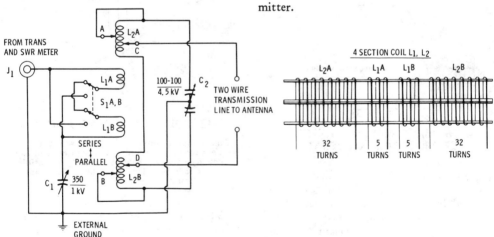

Figure 35

ANTENNA TUNER FOR CENTER-FED ANTENNA

The four section coil is made from a single length of coil stock (I-core Air Dux 2008, or equivalent). The coil is 2½" diameter, 8 turns per inch of #14 wire. Leave a 6" lead on one end and count 32 turns. Break the 33nd turn at the center to make the leads for L_2A and L_1A. Five more turns are counted and the coil broken at the 6th turn to make the opposite lead for coil L_1A and the lead for coil L_2B. Adjacent leads from the center coils are connected to the arms of the ceramic-insulated switch. Coil clips are Mueller #88. Capacitor C_1 is Johnson 154-2, or equivalent. Capacitor C_2 is Johnson 154-510, or equivalent. (Circuit and diagram courtesy of "Wire Antennas for Radio Amateurs," Orr, Radio Publications, Inc.).

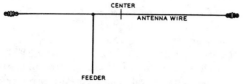

Figure 36

SINGLE-WIRE-FED ANTENNA FOR ALL-BAND OPERATION

An antenna of this type for 40-, 20- and 10-meter operation would have a radiator 67 feet long, with the feeder tapped 11 feet off center. The feeder can be 33, 66 or 99 feet long. The same type of antenna for 80-, 40-, 20- and 10-meter operation would have a radiator 134 feet long, with the feeder tapped 22 feet off center. The feeder can be either 66 or 132 feet long. This system should be used only with those coupling methods which provide good harmonic attenuation.

The 160- 80-Meter Marconi Antenna—A three-eighths wave Marconi can be operated on its harmonic frequency, providing two-band operation from a simple wire. Such an arrangement for operation on 160-80 meters, and 80-40 meters is shown in figure 37. On the harmonic frequency, the antenna acts as a three-quarter wavelength radiator, operat-

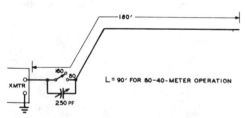

Figure 37

A TWO-BAND MARCONI ANTENNA FOR 160-80 METER OPERATION

ing against ground. Tuned radial wires, as discussed earlier in this chapter, are recommended for use with this antenna.

Overall antenna length may be varied slightly to place the self-resonant frequency at the second harmonic at the chosen spot in the band.

The Multee Antenna—A two-band antenna for 160/80 or 80/40 meters is an important adjunct to a beam antenna for the higher-frequency bands. The *multee antenna* (figure 38) is sufficiently compact to fit on a small lot and will cover two adjacent low-frequency bands and performs this task in an

efficient manner. The antenna evolves from a vertical multiwire radiator, fed on one leg only. On the low-frequency band, the top portion does little radiating so it may be folded horizontally to form a radiator for the high-frequency band. On the lower band, the antenna acts as a top-loaded vertical antenna, while on the higher band, the flattop does the radiating, rather than the vertical portion. The vertical portion, instead, acts as a quarter-wave linear transformer, matching the 6000-ohm nominal antenna impedance of the 50-ohm impedance of the coaxial transmission line.

A radial ground system should be installed beneath the antenna, two or three quarter-wave radials for each band being recommended.

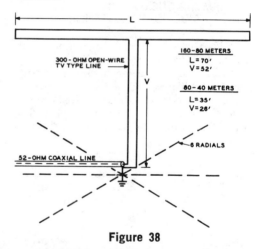

Figure 38

THE MULTEE TWO-BAND ANTENNA

This compact antenna can be used with excellent results on 160/80 and 80/40 meters. The feedline should be held as vertical as possible, since it radiates when the antenna is operated on its fundamental frequency.

When operating on either band, the transmitter should be checked for second harmonic emission, since this antenna will effectively radiate this harmonic.

The Low-Frequency Discone Antenna—The *discone* antenna is widely used on the vhf bands, but until recently it has not been put to any great use on the lower-frequency bands. Since the discone is a broadband device, it may be used on several harmonically related amateur bands. Size is the limiting

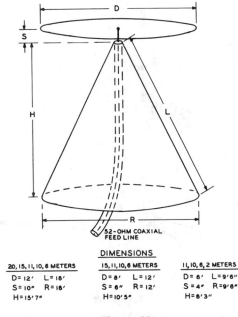

Figure 39

DIMENSIONS OF DISCONE ANTENNA FOR LOW-FREQUENCY CUTOFF AT 13.2 MHz, 20.1 MHz, AND 26 MHz

The Discone is a vertically polarized radiator, producing an omnidirectional pattern similar to a ground plane. Operation on several amateur bands with low SWR on the coaxial feed line is possible.

DIMENSIONS

20, 15, 11, 10, 6 METERS	15, 11, 10, 6 METERS	11, 10, 6, 2 METERS
D= 12' L= 18'	D= 8' L= 12'	D= 6' L=9'6"
S= 10" R= 18'	S= 6" R= 12'	S = 4" R=9'6"
H= 15'7"	H= 10'5"	H= 8'3"

For minimum wind resistance, the top "hat" of the discone is constructed from three-quarter-inch aluminum angle stock, the rods being bolted to an aluminum plate at the center of the structure. The tips of the rods are all connected together by lengths of No. 12 enameled copper wire. The cone elements are made of No. 12 copper wire and act as guy wires for the discone structure. A very rigid arrangement may be made from this design, one that will give no trouble in high winds. A 4" × 4" post can be used to support the discone structure.

The discone antenna may be fed by a length of 50-ohm coaxial cable directly

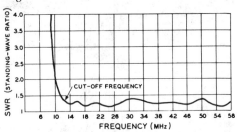

Figure 40

SWR CURVE FOR A 13.2-MHz DISCONE ANTENNA. SWR IS BELOW 1.5 TO 1 FROM 13.0 MHz TO 58 MHz

from the transmitter, with a very low SWR on all bands.

The Trap Vertical Antenna—The trap technique described in a later chapter can be employed for a three-band vertical antenna as shown in figure 42. This antenna is designed for operation on 10, 15, and 20 meters and uses a separate radial system for each band. No adjustments need be made to the antenna when changing frequency from one band to another. Substitution of a ground connection for the radials is not recommended because of the high ground loss normally encountered at these frequencies. Typical trap construction is discussed in the reference chapter, and the vertical radiator is built of sections of aluminum tubing, as described earlier.

Each trap is built and grid-dipped to the proper frequency before it is placed in the radiator assembly. The 10-meter trap is self-resonant at about 27.9 MHz and the 15-meter trap is self-resonant at about 20.8 MHz. Once resonated, the traps need no

factor in the use of a discone, and the 20-meter band is about the lowest practical frequency for a discone of reasonable dimensions. A discone designed for 20-meter operation may be used on 20, 15, 11, 10, and 6 meters with excellent results. It affords a good match to a 50-ohm coaxial feed system on all of these bands. A practical discone antenna is shown in figure 39, with an SWR curve for its operation over the frequency range of 13 to 55 MHz shown in figure 40. The discone antenna radiates a vertically polarized wave and has a very low angle of radiation. For vhf work the discone is constructed of sheet metal, but for low-frequency work it may be made of copper wire and aluminum angle stock. A suitable mechanical layout for a low-frequency discone is shown in figure 41. Smaller versions of this antenna may be constructed for 15, 11, 10, and 6 meters, or for 11, 10, 6, and 2 meters as shown in figure 39.

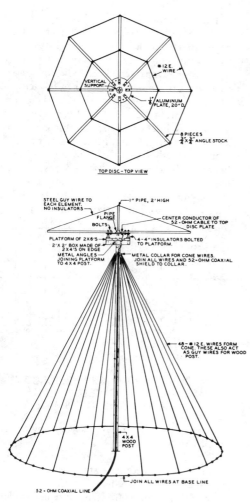

Figure 41

MECHANICAL CONSTRUCTION OF 20-METER DISCONE

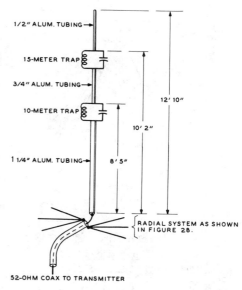

Figure 42

TRIBAND TRAP VERTICAL ANTENNA

Parallel-tuned trap assemblies are used in this vertical antenna designed for 20-, 15- and 10-meter operation. A radial ground wire set, such as described earlier in the chapter is used. Automatic trap action electrically switches antenna for proper operation on each band.

further adjustment and do not enter into later adjustments made to the antenna. The complete antenna is resonated to each amateur band by placing a single-turn coil between the base of the vertical radiator and the radial connection and coupling the grid-dip oscillator to the coil. The coaxial line is removed for this test. The lower section of the vertical antenna is adjusted in length for 10-meter resonance at about 28.7 MHz, followed by adjustment of the center section for resonance at 21.2 MHz. The last ad-

justment is to the top section for resonance at about 14.2 MHz.

It must be remembered that trap, or other multifrequency antennas, are capable of radiating harmonics of the transmitter that may be coupled to them via the transmission line. It is well to check for harmonic radiation with a nearby radio amateur. If such harmonics are noted, an antenna tuner similar to the one described later in this chapter should be added to the installation to reduce unwanted harmonics to a minimum.

The Trap Dipole Antenna—The trap principle may be applied to a dipole as well as to a vertical antenna. Shown in figure 43 are designs for various hf amateur bands. For portable, or Field Day use, the antennas may be fed directly with 50-ohm coaxial line. For fixed station use, insertion of a 1-to-1 balun between the trap antenna and the coaxial transmission line is recommended.

A 20- and 15-meter trap is shown in figure 44. It is designed to be left unprotected and is water-resistant. If desired, it may be

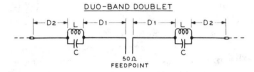

DUO-BAND DOUBLET

BANDS	D1	D2	L(μH)	C(PF)	F R
80-40	32' 0"	22' 0"	8.2	60	6.95
40-20	16' 8"	10' 6"	4.7	25	13.8
20-15	10' 5"	3' 7 1/2"	2.9	20	20.7
15-10	8' 0"	1' 11"	1.65	20	27.8

TRI-BAND DOUBLET

BAND	D1	D2	D3	L1(μH)	C1(PF)	L2(μH)	C2(PF)
20-15-10	8" 0"	1' 10"	2' 9"	2.9	20	1.65	20

Figure 43

MULTIBAND TRAP DIPOLES

Trap dipoles for duoband operation and a tri-band dipole are shown above. Traps are assembled as shown in the photograph. Antenna dimensions are based on an overall trap length of two inches. Highest band resonant frequency may be shifted by changing dimension D_1. Lower band is also affected and dimension D_2 must be adjusted to compensate for change in D_1. Sequence of adjustment is D_1, D_2, and then D_3. Dimensions listed are for center-of-band resonance. Parallel-tuned traps are adjusted to trap frequency outside the low frequency end of each band. Dipoles may be fed with a 1-to-1 balun, if desired.

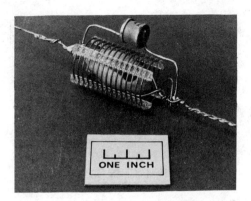

Figure 44

TRAP CONSTRUCTION

Fifteen-meter trap is shown here. Trap is designed for power level of 500 watts, PEP. Trap is built around strain insulator which removes pull of antenna from coil and capacitor. Capacitor is Centralab 853A-20Z (20 pf) and coil is 14½ turns #16, 1" diameter and 2" long (8 turns per inch), Air-Dux 808T. Trap is about 2" long with 1½" leads. Before placement in the antenna, it is grid-dipped to 20.7 MHz on the bench and adjusted to frequency by removal or addition of a fraction of a turn. Traps for other bands are constructed in similar manner. For 2 kW PEP level, coil should be #12 wire, about 2" diameter, and capacitor should be Centralab type 850S.

MHz of the 10-meter band with an SWR figure of less than 1.5/1.

Data is also given in figure 43 for a tri-band doublet covering the 20-, and 15-, and 10-meter amateur bands. Operational bandwidth is sufficient to cover all the included bands with a maximum SWR figure at the band edges of less than 2/1 on the transmission line. As with any antenna configuration, bandwidth and minimum SWR indication are a function of the height of the antenna above the ground.

The Linear Trap The parallel-tuned trap circuit used in multifrequency antennas operates as an electrical switch, connecting and disconnecting portions of the antenna as the frequency of operation is changed. The lumped trap may be replaced by a quarter-wavelength section of transmission line, shorted at the far end with equal results. Because of the problem of constructing a waterproof inductor and procuring a high-voltage capacitor, the

covered with a plastic "overcoat" made from a section of a flexible squeeze bottle, such as bleach or laundry soap containers.

Operational bandwidth on the lower-frequency band is somewhat less than that of a comparable dipole, since a portion of the antenna is wound up in the trap element and does not radiate. Typical bandwidth for an 80- and 40-meter dipole, as measured between the 2/1 SWR points on the transmission line, is: 80 meters, 180 kHz; 40 meters, 250 kHz.

Operational bandwidth of the 40- and 20-meter antenna is typically: 40 meters, 300 kHz; 20 meters, 350 kHz. In addition, the antenna may be operated over the lower 1

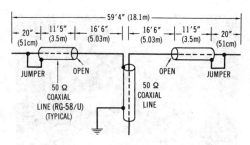

Figure 45

THE G3TKN TRIBAND LINEAR TRAP DIPOLE

This inexpensive three-band antenna for 40, 20, and 15 meters is made of wire and coaxial cable. Because of the trap design, maximum power input to the antenna should be limited to 250 watts. (Antenna design by G3TKN, courtesy of "Radio Communications" magazine.)

transmission-line form of trap is occasionally used in commercial multiband antennas. In this particular design, the trap extends parallel along the element to conserve space and is termed a *decoupling stub.*

A resonant coaxial section may take the place of the decoupling stub and this configuration is often used in multifrequency vhf beam antennas.

It is important that the end sections of the coaxial line are properly sealed to pre-

vent moisture from entering the cable; caulking material will do the job. Because of the added weight of the additional coaxial sections, it is suggested that the antenna be mounted as an inverted-V with center support and with the ends tied off to prevent sagging. The tip sections may be adjusted to provide antenna resonance at any point in the 40-meter band. (Antenna design by G3TKN, England).

Linear Trap Dipole for 40, 20, and 15 Meters   The inexpensive linear trap dipole shown in figure 45 is designed for operation on three amateur bands. It uses coaxial cable sections as linear traps which are cut to an electrical quarter wavelength at 20 meters. If RG-58/U line is used, the velocity factor is 0.66. The coaxial sections act in the same manner as a conventional antenna trap. The center section of the antenna functions as a dipole on 20 meters and on 40 meters the antenna acts as a loaded dipole, the coaxial sections acting as inductors. On 15 meters, the 40-meter dipole is resonant on the third harmonic.

CHAPTER TWENTY-EIGHT

High-Frequency Fixed Directive Antennas

It is important in most types of radio communication to be capable of concentrating the radiated signal from the transmitter in a certain desired direction and to be able to discriminate at the receiver against reception from directions other than the desired one. Such capabilities involve the use of directive antenna arrays.

Few simple antennas, except the single vertical element, radiate energy equally well in all azimuth (horizontal or compass) directions.

The directive properties of an antenna depend on length, height above ground, and the proximity of other radiators or parasitic elements. By combining antenna elements so that the radiated fields are additive, a directive, or *beam antenna*, can be designed. The power gain of a beam antenna provides worthwhile gain on received signals as well as on transmission and also provides rejection of signals coming from directions not favored, thus reducing the interference level on a given path to a great degree.

Power gain expresses the power increase in the radiated field of one antenna over a standard comparison antenna of the same orientation. The comparison antenna is commonly a half-wave dipole or an *isotropic* antenna. The isotropic antenna is one that radiates equally well in all directions and only exists as a mathematical concept. Such an antenna if placed at the center of an imaginary sphere would illuminate the inner surface of the sphere uniformly.

Power gain is the product of directivity and radiation efficiency and is measured in decibels in the direction of maximum field energy. It is common practice to use the decibel as an absolute unit of measure by fixing an arbitrary level of reference. For an isotropic reference, the symbol is *dBi*,

and for a dipole the symbol is *dBd*. Antenna gain expressed in decibels without mention of the reference level is meaningless.

Since a dipole antenna has a power gain over isotropic of 1.64 (2.1 dBi), gain expressions in one form can be converted to the other. (For a discussion of the decibel, refer to Chapter 34.)

The power that escapes from the sides and back of a directive antenna detracts from the gain future. The ratio of power radiated in the desired direction as compared to that radiated in the opposite (back) direction is called the *front-to-back ratio* and is expressed in decibels. Power radiated to the sides of an array can be expressed in terms of *front-to-side ratio*.

28-1 Directive Antennas

When a multiplicity of radiating elements are located and phased so as to reinforce the radiation in certain desired directions and to neutralize radiation in other directions, a *directive antenna array* is formed.

The function of a directive antenna when used for transmitting is to give an increase in signal strength in some direction at the expense of radiation in other directions. For reception, one might find useful an antenna giving little or no gain in the direction from which it is desired to receive signals if the antenna is able to discriminate against interfering signals and static arriving from other directions. A good directive transmitting antenna, however, can also be used to good advantage for reception.

If radiation can be confined to a narrow beam, the signal intensity can be increased a great many times in the desired direction of transmission. This is equivalent to increasing the power output of the transmitter. On the higher frequencies, it is more economical to use a directive antenna than to increase transmitter power, if more than a few watts of power is being used.

Directive antennas for the high-frequency range have been designed and used commercially with gains as high as 23 dB over a simple dipole radiator. Gains as high as 35 dBd are common in direct-ray microwave communication and radar systems. A gain of 23 dB represents a power gain of 200 times and a gain of 35 dB represents a power gain of almost 3500 times. However, an antenna with a gain of only 15 to 20 dBd is so sharp in its radiation pattern that it is usable to full advantage only for point-to-point work.

The increase in radiated power in the desired direction is obtained at the expense of radiation in the undesired directions. Power gains of 3 to 12 dBd seem to be most practical for amateur communication, since the width of a beam with this order of power gain is wide enough to sweep a fairly large area. Gains of 3 to 12 dBd represent effective transmitter power increases from 2 to 16 times.

Note: In this chapter antenna gain is *referenced to a dipole* unless otherwise stated.

Types of
Directive Arrays
     There is an enormous variety of directive antenna arrays that can give a substantial power gain in the desired direction of transmission or reception. However, some are more effective than others which require the same space. In general it may be stated that long-wire antennas of various types, such as the single long wire, the V beam, and the rhombic, are less effective for a given space than arrays composed of resonant elements, but the long-wire arrays have the significant advantage that they may be used over a relatively large frequency range while resonant arrays are usable only over a quite narrow frequency band.

While fixed wire beams have been eclipsed by the more glamorous rotatable Yagi and Quad antennas, dollar-for-dollar the wire beam is hard to beat. The wide-spaced collinear arrays, moreover, have low-Q and wide bandwidth and may be used over a

larger frequency range than their higher-Q counterparts. Finally, in some instances, the wire beam is more unobtrusive than one made of aluminum and may be erected in areas where more conventional antennas are frowned upon.

This chapter covers long-wire and multi-element fixed arrays, while the next chapter covers rotary arrays of the Yagi and Quad configuration.

28-2   The Long-Wire Antenna

A harmonically operated antenna radiates more energy in certain directions than others and can be considered to have power gain and directivity when it is several wavelengths long, or longer. The long wire has a natural current distribution in the form of either a travelling or a standing wave with the radiation adding in certain directions and cancelling in others. Long wires having a standing-wave pattern will be discussed first.

The radiation patterns of long-wire antennas are well known and available informa-

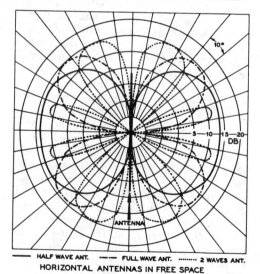

―― HALF WAVE ANT.   ―‒‒ FULL WAVE ANT.   ········ 2 WAVES ANT.
HORIZONTAL ANTENNAS IN FREE SPACE

Figure 1

**FREE-SPACE FIELD PATTERNS OF
LONG-WIRE ANTENNAS**

The presence of the earth distorts the field pattern in such a manner that the azimuth pattern becomes a function of the elevation angle.

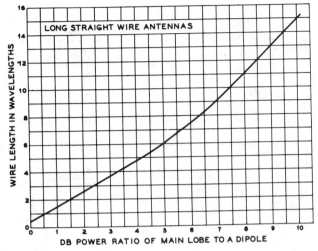

Figure 2

DIRECTIVE GAIN OF LONG WIRE ANTENNA

tion is based on theory and field measurements. The dipole antenna in free space has a figure-8 pattern in the plane of the wire. The full-wave antenna has four lobes at angles to the wire, an antenna two wavelengths long has eight lobes, etc. (Figure 1). When the antenna is more than two wavelengths long, the end lobes begin to exhibit noticeable power gain while the broadside lobes become smaller, even though more numerous.

The actual directivity of the main lobes of radiation is not particularly well defined and the minor lobes tend to fill in the nulls and a real-life long-wire antenna provides nearly omnidirectional characteristics, especially when the ground beneath it has poor conductivity. This is due to the fact that the antenna is affected by the r-f resistance of the wire and the ground and also because the current amplitude drops off at successive current loops as a result of attenuation along the wire. The attenuation is due to radiation of power from the wire and also the resistance of the wire. As the length of the long wire increases, these factors become predominant and the tuning of the antenna becomes quite broad.

The physical length of a long wire antenna is related to its electrical length by the following formula:

$$\text{Length}_{(feet)} = \frac{492 \ (N - 0.025)}{F_{(MHz)}}$$

or,

$$\text{Length}_{(meters)} = \frac{150 \ (N - 0.05)}{F_{(MHz)}}$$

where,

N is the number of *half wavelengths* on the wire.

The directive gain of a long wire antenna in terms of wire length is shown in figure 2. Suggested antenna lengths are listed in Table 1.

Practical Long Wire Antennas   The simplest long wire antenna is one that is end-fed, with the wire brought directly to the equipment site, at which point it is connected to the transmitter through an antenna tuner. Care must be taken to suppress transmitter harmonics before they reach the antenna, since an end-fed antenna offers no discrimination against harmonic energy. The antenna can operate on its harmonic frequencies with good efficiency and can also be operated against ground at half frequency as a quarter-wave Marconi antenna.

When a long wire antenna is center-fed, the antenna radiation lobes are symmetrical on each side of the feedpoint (figure 3), but if the antenna is fed near one end, the lobes on the longer leg become stronger, as compard to the shorter leg lobes.

End-Fed Wire Antenna for 20, 15, or 10 Meters—Shown in figure 4 is a single wire antenna for operation on any one of these bands. The feedline is placed one-quarter wavelength from one end of the antenna at a low-impedance point. The impedance of the antenna is of such a value that a 4:1

Table 1. Long Wire Antenna Design Table

Length in feet. To convert to meters, multiply by 0.305

Frequency In MHz	1λ	1½λ	2λ	2½λ	3λ	3½λ	4λ	4½λ
29	33	50	67	84	101	118	135	152
28	34	52	69	87	104	122	140	157
21.4	45	68	91 ½	114 ½	136 ½	160 ½	185 ½	209 ½
21.2	45 ¼	68 ¼	91 ¾	114 ¾	136 ¾	160 ¾	185 ¾	209 ¾
21.0	45 ½	68 ½	92	115	137	161	186	210
14.2	67 ½	102	137	171	206	240	275	310
14.0	68 ½	103 ½	139	174	209	244	279	314
7.3	136	206	276	346	416	486	555	625
7.15	136 ½	207	277	347	417	487	557	627
7.0	137	207 ½	277 ½	348	418	488	558	628
4.0	240	362	485	618	730	853	977	1100
3.8	252	381	511	640	770	900	1030	1160
3.6	266	403	540	676	812	950	1090	1220
3.5	274	414	555	696	835	977	1120	
2.0	480	725	972	1230	1475			
1.9	504	763	1020	1280				
1.8	532	805	1080					

balun is used to provide a good termination for a 50-ohm transmission line.

Center-Fed Wire Antenna for 15 and 10 Meters—The single-wire antenna can provide good two-band performance when fed at the center, as shown in figure 5. A 4:1 balun is used and, as in the case of the end-fed design, the antenna has a multiplicity of lobes and may be considered to be generally omnidirectional, except for appreciable nulls off the ends of the wire.

The Inverted-V Antenna—The *Inverted-V* or *Sloper* antenna is a popular design and a long wire may be modified by dropping the ends, or raising the center of the antenna, as shown in figure 6. The total included angle of the V should be not less than 90°, with angles as high as 120° providing good per-

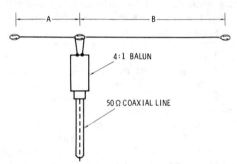

ANTENNA DIMENSIONS (A + B = 5/2 λ)

BAND	A		B	
	FEET	METERS	FEET	METERS
20	16' 6 "	5.03	154' 0"	46. 97
15	11' 0"	3.35	103' 6"	31. 57
10	8' 3"	2.51	77' 6"	23. 64

Figure 4

LONG WIRE ANTENNA FOR 20, 15 OR 10 METERS

This 5/2-wavelengths antenna provides maximum radiation at angles of about 30 degrees to the wire, with minor lobes filling in the areas at right angles to the wire. A 4-to-1 balun, such as described in Chapter 26, figures 6 and 7 is suggested for use with this antenna.

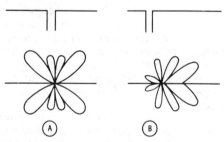

Figure 3

FEED SYSTEM AFFECTS RADIATION PATTERN OF LONG WIRE ANTENNA

When a long wire antenna is center-fed, the radiation pattern is symmetrical on each side of the feed point (A). When the antenna is fed at a low impedance point near one end (B), the pattern symmetry is destroyed and the lobes on the longer leg become stronger, as compared to the shorter leg lobes.

formance. The inverted-V provides less directivity off the ends of the antenna than a comparable horizontal installation.

It is practical to construct a multiband inverted-V antenna by end-tuning the legs. A pair of copper alligator clips and two segments of wire permit resonance on each of the three bands, as shown in figure 7. Since

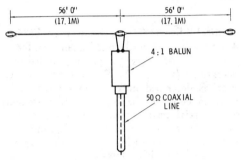

Figure 5

CENTER-FED ANTENNA FOR 15- AND 10-METER BANDS

This simple antenna provides good 2-band performance, operating as five half-wavelengths on 15 meters and seven half-wavelengths on 10 meters. The 4-to-1 balun is described in Chapter 26.

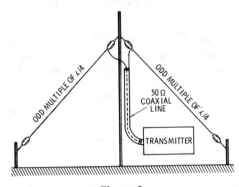

Figure 6

THE INVERTED-V ANTENNA

The inverted-V antenna may be a dipole, or each leg may be an odd multiple of half-wavelengths long. Use of a balun is optional.

the ends of the antenna are at a high r-f potential, care should be taken to keep them high enough so that they cannot be touched accidentally.

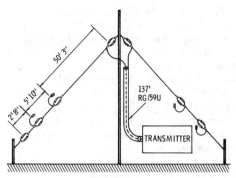

Figure 7

LONG WIRE INVERTED-V ANTENNA FOR 20, 15, AND 10 METERS

Each leg of the antenna is 50′ 3″ (15.33 meters) for 20 meter operation. A jumper 5′ 10″ (1.77 meters) is added for 15 meters, and a second jumper 2′ 8″ (0.82 meters) is added for 10-meter operation. The antenna is 3/2-wavelengths on 20, 5/2-wavelengths on 15, and 7/2-wavelengths on 10. Use of a balun is optional, but a 1-to-1 device may be added if desired. The transmission line is cut to provide a minimum value of reactance at the transmitter. Either RG-8A/U or RG-59C/U may be used.

The V Antenna If two long wires are formed into a V, it is possible to make two of the maximum radiation lobes additive in a line bisecting the V (figure 8). The resulting pattern is bidirectional with minor lobes to the sides of the major lobes.

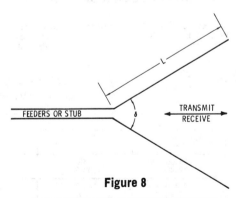

Figure 8

A HORIZONTAL V-BEAM ANTENNA

Two long wires in a V pattern, starting from a common apex, both parallel to the ground, form a V-beam. The pattern is bidirectional in a line bisecting the V.

Each leg of the V-beam can be any number of quarter wavelengths long, the feed system depending on whether the feedpoint is a high or low impedance. Representative gain figures are shown in figure 9, with suggested dimensions tabulated in Table 2. Best directivity and gain for beams having legs shorter than 3 wavelengths are obtained with a somewhat smaller angle than that determined by the lobes. Optimum directivity for a one wave-length V-beam, for example, is obtained at an included angle of 90° (figure 10).

The V-beam may be made unidirectional by placing another V-beam behind it and feeding it with a phase difference of 90°. The system will be directive through the V antenna having the lagging current. While such an antenna system is large for the hf bands, it is often used for long distance TV or f-m reception.

Practical V-Beam Antennas   The V-beam may be fed with tuned, open-wire feeders to permit multiband operation (figure 8). An antenna tuner, such as described in a previous chapter, is used to convert the balanced feed system to a coaxial termination, common to most transmitting equipment. For single-band operation, the V-beam is often fed directly at a current loop with a low-impedance coaxial line and a 4:1 balun (figure 11). Short V-beam antennas have an impedance close to 50 ohms and a direct coaxial feed may be employed, as shown in some of the following designs.

V-Beam for 20-15-10 Meters—Shown in figure 12 is a beam antenna designed for operation on the three popular hf DX bands. The antenna is 5 wavelengths long at 14 MHz and provides a power gain over a dipole of 7.5 dB at 14 MHz, 9 dB at 21 MHz and 10 dB at 28 MHz. A compromise apex angle is used to enhance multiband operation. The beam may also be used on the 80- and 40-meter bands, but provides little, if any, power gain on these lower frequencies. The beam is fed at the apex with a

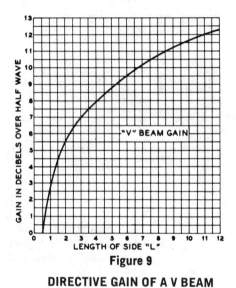

Figure 9

DIRECTIVE GAIN OF A V BEAM

This curve shows the approximate directive gain of a V beam with respect to a half-wave antenna located the same distance above ground, in terms of the side length L.

Table 2. V-Beam Design Table

Length in feet. To convert to meters, multiply by 0.305

V-ANTENNA DESIGN TABLE				
Frequency in kHz	$L = \lambda$ $\delta = 90°$	$L = 2\lambda$ $\delta = 70°$	$L = 4\lambda$ $\delta = 52°$	$L = 8\lambda$ $\delta = 39°$
28000 29000	34'8'' 33'6''	69'8'' 67'3''	140' 135'	280' 271'
21100 21300	45'9'' 45'4''	91'9'' 91'4''	183' 182'6''	366' 365'
14050 14150 14250	69' 68'6'' 68'2''	139' 138' 137'	279' 277' 275'	558' 555' 552'
7020 7100 7200	138'2'' 136'8'' 134'10''	278' 275' 271'	558' 552' 545'	1120' 1106' 1090'

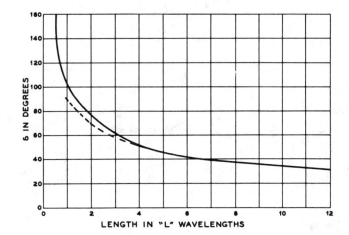

Figure 10

INCLUDED ANGLE FOR A V BEAM

Showing the included angle between the legs of a V beam for various leg lengths. For optimum alignment of the radiation lobe at the correct vertical angle with leg lengths less than three wavelengths, the optimum included angle is shown by the dashed curve.

balanced, two-wire line and an antenna tuner.

Mini-V-Beam for 20-15-10 Meters—The compact, space-saving V-beam shown in figure 13 provides about 3.0 dB gain for three-band operation. This is a compromise design which allows rather large minor lobes falling in line with the legs of the beam. Dimensions for the twin wires are provided in the illustration.

V-Beam for 80-40-20 Meters—Shown in figure 14 is a V-beam designed for operation on the lower frequency DX bands. The apex angle should be chosen for one particular band to provide best directivity. Operation on other bands will be possible, but large secondary lobes will impair the antenna directivity.

as high above level ground as possible. A minimum height of 50 feet is recommended. The wires of the beam may be allowed to slope down toward the ground to a final height of about 20 feet to improve the low-angle radiation in the forward direction.

28-3   The Rhombic Antenna

Two V-beams placed end-to-end form a *rhombic antenna* (figure 15). The simple, resonant rhombic antenna is bidirectional and provides approximately the same power gain and radiation pattern as a V-beam of equivalent size.

A variation of this design is the *nonresonant rhombic* antenna (figure 16) which is

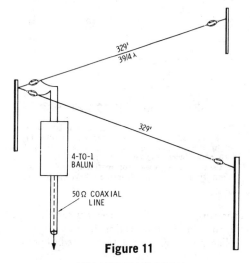

Figure 11

TEN-METER V-BEAM

This antenna provides about 11.5 decibels gain over a dipole on the 10-meter band. The included angle between the wires is 32°. The beam is fed at the apex with a 4-to-1 balun and a coaxial transmission line. See Chapter 26, figures 6 and 7 for balun design.

terminated at the end opposite the feedpoint. Resonance is not a necessary condition for antenna operation although some antennas are made resonant in order to provide a convenient, nonreactive feedpoint to eliminate complex matching networks. The nonresonant, or traveling-wave rhombic antenna provides wideband operation over an octave or more of frequency spectrum and provides a medium value of feedpoint impedance which remains relatively constant over the

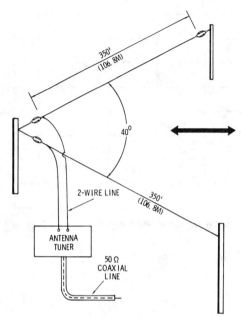

Figure 12

V-BEAM FOR 20-15-10 METERS

This V-beam provides good gain for DX work on 3 bands. Height above ground should be 50' to 70' for best results. The antenna is fed with a two wire line (300-ohm TV "ribbon" may be used for low power applications) and an antenna tuner, such as described in Chapter 26.

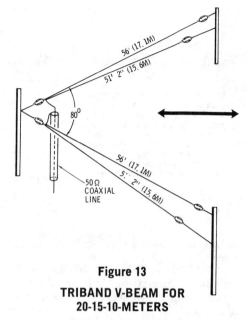

Figure 13

TRIBAND V-BEAM FOR 20-15-10-METERS

A compact design provides good gain for 3-band operation. Two V-beams are connected in parallel at the apex. The 56' legs are cut for operation on 10 and 15 meters, while the 51-foot legs provide 20-meter operation. The included angle is 80°. The shorter legs run under the longer ones, and are separated by about 10' (3 meters) distance at the tips. A 1-to-1 balun may be used with the antenna, if desired.

operating range. When properly terminated, the rhombic is unidirectional.

The power gain of a terminated rhombic antenna over a dipole is shown in figure 17, which includes an allowance of 3 dB for the power lost in the terminating resistor. This power can be considered to be that power which normally would have been radiated in the reverse direction had the resistor not been there.

When the far end of a rhombic antenna is terminated in a noninductive resistance of about 800 ohms, the input resistance of the rhombic is about 700 ohms. The terminating resistance should be capable of dissipating about one-half the average power output of the transmitter. Small composition resistors may be used, in combination with a "lossy" transmission line. Typically, a two-wire line made up of #25 nichrome wire and terminated with a number of 2-watt resistors will usually serve except for very high power.

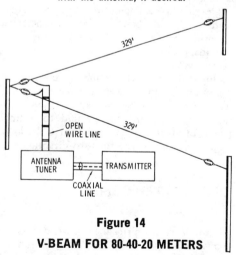

Figure 14

V-BEAM FOR 80-40-20 METERS

This V-beam performs well on the lower DX bands. The included angle should be 80° for 80 meters, 60° for 40 meters and 45° for 20 meters. Gain varies from 3 dB at 80 meters to about 7.5 dB at 20 meters. The beam is fed with a two-wire transmission line and an antenna tuner, such as described in Chapter 26.

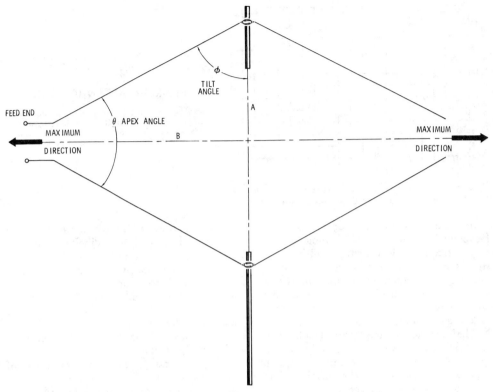

Figure 15

THE RESONANT RHOMBIC ANTENNA

Two V-beams, placed end-to-end in the horizontal plane form a rhombic antenna. The radiation pattern is bidirectional, bisecting the apex angle of the array. Because of wave interference between the two V's, the lobe amplitudes and apex angle are not the same as for equivalent, separate V antennas of the same size.

Figure 16

TYPICAL RHOMBIC ANTENNA DESIGN

The antenna system illustrated may be used over the frequency range from 7 to 29 MHz without change.

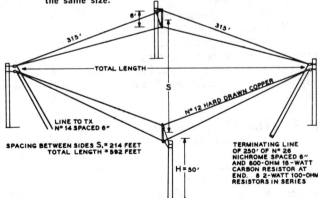

The attenuating line may be folded back upon itself to conserve space.

The rhombic antenna can be fed with a two-wire line having a characteristic impedance of about 600 ohms. An antenna tuner is commonly used at the station end of the line to match the antenna system to a 50-ohm coaxial termination.

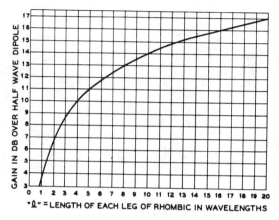

Figure 17

RHOMBIC ANTENNA GAIN

Showing the theoretical gain of a rhombic antenna, in terms of the side length, over a half-wave antenna mounted at the same height above the same type of soil.

In order to minimize fluctuations of the feedpoint impedance of the rhombic antenna as the frequency is varied, a multiwire design is commonly used, as shown in figure 16. Parallel conductors are used in the assembly, joined together at the ends, but with increasing separation as the midpoint of the legs is approached.

The rhombic antenna transmits a horizontally polarized wave at a relatively low angle above the horizon. The vertical angle of radiation decreases as the height above ground is increased, in the same manner as with a dipole antenna. For best results on long distance circuits, the rhombic antenna should be at least one wavelength above ground. The antenna should be erected over level ground, with the plane of the antenna parallel to the ground.

Figure 18 provides design information for a rhombic antenna using either the "maximum output" technique or the "alignment" technique. The alignment method is about 1.5 dB below the other method, but requires only about 75% as much leg length.

28-4 The Multielement Fixed Array

Power gain and directivity may be achieved by combining antenna elements into an array of elements. The characteristics of the half-wave dipole antenna have been described, and this element is commonly used in building antenna arrays.

When a second dipole is placed in the vicinity of a fed-dipole, and excited directly or parasitically, the resultant radiation pattern will depend upon dipole spacing and phase differential, as well as the relative magnitude of the current in the two dipoles. The dipole elements may be placed parallel to each other or end-to-end (*collinear*), or a combination of both arrangements may be used. The elements may be either horizontal or vertically polarized, depending on their relationship to the earth.

The array can have the maximum field of radiation *broadside* to the elements, (perpendicular to the axis of the array and to the plane containing the elements) or the maximum field may be *end-fire* (in line with the direction of the array axis, and through the elements). The radiation pattern for simple driven arrays is commonly controlled by holding element spacing and current constant and adjusting the phase relationship between the elements. With spacing less than 0.65 wavelength, the radiation is mainly broadside to the dipoles (bidirectional) when the phase difference is zero, and through the wires (end-fire) when the phase difference is 180° (figure 19). With a phase difference between 0° and 180° (45°, 90°, and 135°, for instance), the pattern is unsymmetrical, the radiation being greater in one direction than in the opposite direction.

Multielement Collinear Arrays   Two or more dipole elements may be arranged in *collinear* fashion and fed in phase to provide power gain and restricted beam width (figure 20). This is a form of long-wire antenna having a modi-

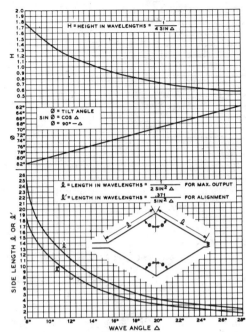

Figure 18

RHOMBIC ANTENNA DESIGN TABLE

Design data is given in terms of the wave angle (vertical angle of transmission and reception) of the antenna. The lengths l are for the "maximum output" design; the shorter lengths (l') are for the "alignment" method which gives approximately 1.5 dB less gain with a considerable reduction in the space required for the antenna. The values of side length, tilt angle, and height for a given wave angle are obtained by drawing a vertical line upward from the desired wave angle,

fied current distribution accomplished by inserting a phase-reversing network every half-wavelength. The network may be a lumped LC circuit, but usually takes the form of a quarter-wavelength transmission line. The antenna is generally known as a *Franklin antenna*, after its inventor. Additional dipole elements may be added in symmetrical fashion to form three- and four-element collinear arrays (Table 3). If the dipole tips are adjacent, the power gain over a dipole of a two-element collinear array is 1.9 decibels, that of a three-element array is 3.2 dB and that of a four-element array is 4.3 dB.

Additional gain may be achieved at the expense of simplicity by increasing the spacing between the collinear elements. A two-

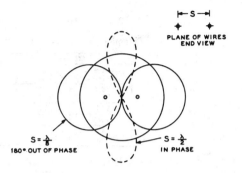

Figure 19

RADIATION PATTERNS OF A PAIR OF DIPOLES OPERATING WITH IN-PHASE EXCITATION, AND WITH EXCITATION 180° OUT OF PHASE

If the dipoles are oriented horizontally most of the directivity will be in the vertical plane; if they are oriented vertically most of the directivity will be in the horizontal plane.

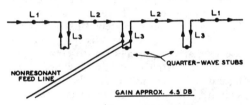

GAIN APPROX. 4.5 DB

Figure 20

THE FRANKLIN OR COLLINEAR ANTENNA ARRAY

An antenna of this type, regardless of the number of elements, attains all of its directivity through sharpening of the horizontal or azimuth radiation pattern; no vertical directivity is provided. Hence a long antenna of this type has an extremely sharp azimuth pattern, but no vertical directivity.

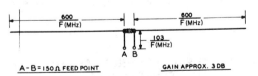

A – B = 150 Ω FEED POINT   GAIN APPROX. 3 DB

Figure 21

DOUBLE EXTENDED ZEPP ANTENNA

For best results, antenna should be tuned to operating frequency by means of grid-dip oscillator.

Table 3. Collinear Antenna Design Chart

Length in feet. To convert to meters, multiply by 0.305

COLLINEAR ANTENNA DESIGN CHART			
Frequency in MHz	L_1	L_2	L_3
28.5	16'8''	17'	8'6''
21.2	22'8''	23'3''	11'6''
14.2	33'8''	34'7''	17'3''
7.15	67'	68'8''	34'4''
4.0	120'	123'	61'6''
3.6	133'	136'5''	68'2''

element, spaced array is termed *a double extended array* and provides about 3 dB power gain (figure 21).

The elevation radiation pattern for a collinear array is essentially the same as for a dipole and this consideration applies whether the elements are of normal length or extended.

A three-element precut array for 40-meter operation is shown in figure 22. It may be fed with a 50-ohm coaxial line and a 4:1 balun transformer. The dimensions of the antenna may be doubled for 80-meter operation, or reduced by half for 20-meter work. Power gain is approximately 3 dB over a dipole. A simple spaced array is shown in figure 23 that also provides a 3 dB power gain. The pattern is changed to a cloverleaf configuration by means of the phase-reversing switch. The dipoles are fed with 72-ohm, balanced twinlead.

Multielement Broadside Arrays An antenna array may be constructed of parallel elements having in-phase currents. The power gain of such a *broadside*

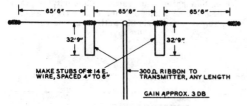

Figure 22

PRECUT COLLINEAR ARRAY FOR 40 METERS

The collinear array may be fed with a 4-to-1 balun placed at the feedpoint (see Chapter 26) and a 50-ohm coaxial line in place of the 300-ohm line. If the 300-ohm line is used, an antenna tuner is needed at the station end to convert the system to 50 ohms termination.

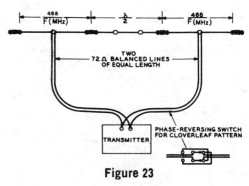

Figure 23

TWO COLLINEAR HALF-WAVE ANTENNAS IN PHASE PRODUCE A 3 DB GAIN WHEN SEPARATED ONE-HALF WAVELENGTH

array is a function of the number of elements and the spacing between the elements. Typically, gain with 0.5 wavelength spacing is 4 dB for 2 elements, 5 dB for 3 elements, and 6 dB for 4 elements. Practical limitations limit the number of in-phase elements in the hf range, since phase control becomes difficult as the number of elements are increased.

Additional gain is achieved when the element spacing is increased to 0.7 wavelength, but proper phasing becomes more of a problem at this spacing.

Collinear elements may be stacked above or below another string of collinear elements to form a *curtain array*. One of the most popular arrays of this type is the *Lazy-H* array of figure 24. Horizontal collinear elements stacked two above two make up this beam, which provides a power gain of nearly 6 dB. The pattern is bidirectional and the antenna has a high value of radiation resistance at a current loop. The high radiation resistance results in low peak voltages in the assembly and provides a broad bandwidth,

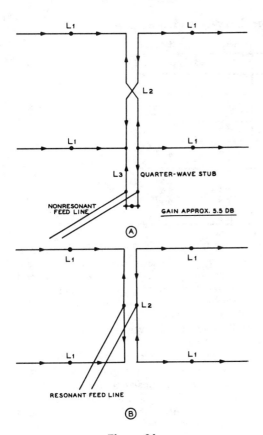

Figure 24

THE LAZY-H ANTENNA SYSTEM

Stacked, collinear pairs provide both horizontal
and vertical directivity. The Lazy-H provides
about 5.5 dB gain over a dipole. The nonresonant, balanced feedline may be replaced with
a 1-to-1 balun and 50-ohm coaxial line, if
desired.

permitting the array to be used over a wide
frequency range. Antenna dimensions are
tabulated in Table 4. The antenna may be
fed with a balanced line, as shown, or with
a coaxial line and balun arrangement. The
line, or balun, is tapped on the matching
stub at the point which provides the lowest
SWR on the transmission line.

Practical	An array of collinear and stacked
Broadside	dipole elements is termed a *Sterba*
Arrays	curtain, after the inventor of this
	antenna type. Figure 25 shows

two simple Sterba curtains. Illustrations A

and B show two methods of feeding a small
curtain antenna, and an alternative method
is shown in illustration C. A coaxial line and
1:1 balun transformer may be used with the
design shown at B in place of the balanced
feedline.

In the case of either the Lazy-H or the
Sterba curtain, the array can be made uni-

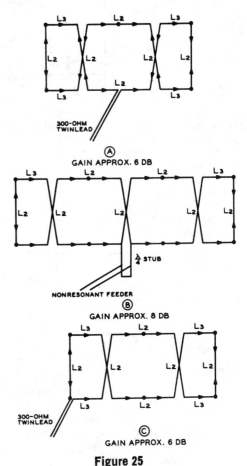

Figure 25

THE STERBA-CURTAIN ARRAY

Approximate directive gains along with alternative feed methods are shown.

directional and the gain increased by 3 dB
if a similar array is placed 0.25 wavelength
behind the driven array to act as a parasitic
reflector. The wires in the reflector should
be parallel to the radiating elements. For vhf
work, a screen or mesh of wires, slightly
greater in area than the antenna may be

Table 4. Design Chart for Lazy-H and Sterba Arrays

Length in feet. To convert to meters, multiply by 0.305

Frequency in MHz	L_1	L_2	L_3
7.0	68'2''	70'	35'
7.3	65'10''	67'6''	33'9''
14.0	34'1''	35'	17'6''
14.2	33'8''	34'7''	17'3''
14.4	33'4''	34'2''	17'
21.0	22'9''	23'3''	11'8''
21.5	22'3''	22'9''	11'5''
27.3	17'7''	17'10''	8'11''
28.0	17'	17'7''	8'9''
29.0	16'6''	17'	8'6''
50.0	9'7''	9'10''	4'11''
52.0	9'3''	9'5''	4'8''
54.0	8'10''	9'1''	4'6''
144.0	39.8''	40.5''	20.3''
146.0	39''	40''	20''
148.0	38.4''	39.5''	19.8''

substituted for the wire configuration. In some instances, parasitic elements are used as reflectors or directors with a broadside array, but these have the disadvantage that their operation is selective with respect to frequency and a relatively small change in operating frequency will seriously affect the parasitic elements.

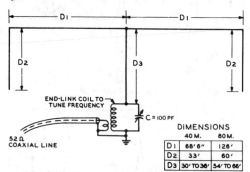

Figure 26

BOBTAIL BIDIRECTIONAL BROADSIDE CURTAIN FOR THE 7-MHz OR THE 4.0-MHz AMATEUR BANDS

This simple vertically polarized array provides low angle radiation and response with comparatively low pole heights, and is very effective for DX work on the 7-MHz band or the 4.0-MHz phone band. Because of the phase relationships, radiation from the horizontal portion of the antenna is effectively suppressed. Very little current flows in the ground lead to the coupling tank; so an elaborate ground system is not required, and the length of the ground lead is not critical so long as it uses heavy wire and is reasonably short.

The Bobtail Beam for 80 and 40 Meters—
A truncated version of a Sterba curtain is shown in figure 26. This Bobtail beam provides vertical polarization in order to obtain low-angle radiation at the lower end of the hf spectrum without resorting to unreasonable pole heights. When precut to the dimensions shown, this bidirectional array will perform well over the 7-MHz band, or the 4-MHz phone band. For the latter, the required pole height is about 70 feet, and the array will provide a signal gain of about 5 dB over a comparison vertical antenna when the path length exceeds 2500 miles.

The horizontal directivity is only moderate, the beam width at the half-power points

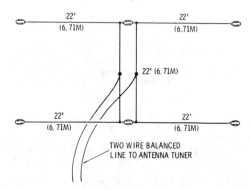

Figure 27

LAZY-H ARRAY FOR 20-15-10 METERS

This 3-band beam provides good results when mounted so that the lower wires are at least 20 feet off the ground. Gain is 4 dB on 20 meters, 5.5 dB on 15 meters, and over 6 dB on 10 meters. Feedline may be any length and can be 300-ohm TV "ribbon" for low-power applications. An antenna tuner, such as described in Chapter 26 is used with this array.

being slightly greater than that obtained from three co-phased vertical radiators fed with equal currents.

Lazy-H Array for 20, 15, and 10 Meters— The basic Lazy-H design can be modified for three band operation, as shown in figure 27. The array is fed at the center to preserve the proper phase relationship between the upper and lower radiators. Dimensions are not critical as long as symmetry is maintained. The array is bidirectional, at right angles to the plane of the array. A parallel wire feedline is used, in conjunction with an antenna tuner at the station, which converts the balanced feed system to an unbalanced, 50-ohm line, suitable for connection to the majority of modern transmitters.

The Bi-Square Broadside Array—Illustrated in figure 28 is a simple, one element broadside array that may be suspended from a single pole. The power gain is approximately 4 dB over a dipole placed at the same average elevation.

Two Bi-Square arrays may be suspended from a single pole, at right angles to each other for general coverage. Alternatively, arrays for separate bands may be suspended, with the smaller one inside the larger.

The Bi-Square is fed with a quarter-wave matching transformer made of parallel wire line. The balun and 50-ohm coaxial line are attached to a movable insulator with clips on it, which is adjusted up and down the stub for minimum SWR on the line. A balun is used between the line and the stub in this simple design to achieve good antenna current balance.

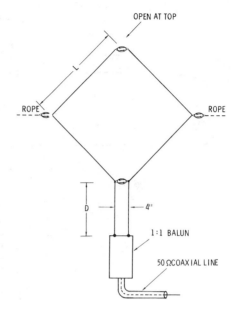

DIMENSIONS OF BI-SQUARE ANT.

BAND	10		15		20	
	FEET	METERS	FEET	METERS	FEET	METERS
L	17'	5.19	25'	7.63	34'	10.37
D	8' 3"	2.52	12' 3"	3.74	16' 6"	5.03

Figure 28

THE BI-SQUARE BIDIRECTIONAL BEAM ANTENNA

The Bi-Square antenna provides about 4.5 dB gain over a dipole antenna. Shown here is a design for 20, 15, or 10 meters. The loop is open at the top and fed with a quarter-wavelength, two-wire line at the bottom. The 50-ohm line is attached to the two-wire line through a 1-to-1 balun (Chapter 26). The attachment point of the balun on the two-wire line is varied until the lowest SWR is achieved on the coaxial line.

Practical End-Fire Arrays   The *flat-top beam*, designed by W8JK of Ohio State University is a simple and effective end-fire array (figure 29). This antenna consists of two closed-spaced dipoles or collinear arrays in the horizontal plane. Because of the close spacing, it is possible to obtain the proper phase relationships by crossing the wires at the current or voltage loops rather than by resorting to phasing stubs. This greatly simplifies the array.

The flat-top beam may be operated on a harmonic frequency, the radiation pattern being the same as a single-wire antenna of the same length. If the beam is to be used on more than one band, tuned feeders, with an antenna tuner must be used. For single-band operation, a matching stub is preferred, fed with a 50-ohm line and 1:1 balun.

The exact dimensions for the radiating elements are not critical, as slight deviations can be compensated in the match stub or tuned feeders. The antenna is adjusted to frequency by adjusting stub length (A) with a dip meter.

TOP VIEW

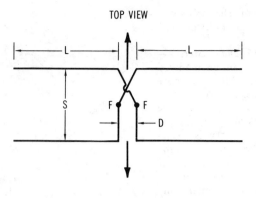

MATCHING SECTION

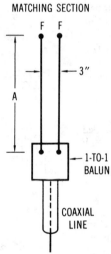

BAND	L		S		A		D	
(MHz)	FEET	METERS	FEET	METERS	FEET	METERS	FEET	METERS
7.0-7.3	59'0"	18	17'0"	5.19	26'0"	7.93	2'0"	0.61
10.1	41'6"	12.7	12'0"	3.66	18'4"	5.59	1'6"	0.45
14.0-14.35	30'0"	9.15	8'8"	2.65	13'0"	3.96	1'0"	0.30
21.0-21.45	22'6"	6.86	7'10"	2.4	10'6"	3.20	8"	0.21
28.0-29.7	15'0"	4.58	5'2"	1.55	7'0"	2.14	6"	0.15

Figure 29

FLAT-TOP BEAM DESIGN DATA

Top view of flat-top beam and plan view of matching section. Made of wire, the array is spaced by means of fiberglass or bamboo poles at ends and center. Rope bridles are attached at the end points. Matching section is attached to antenna at F-F. Resonance is established by short-ing matching section at bottom with a single turn loop and coupling antenna to a dip meter. Short is then removed and replaced with balun. Length of matching section is approximate. Typical antenna gain is 6 dBd.

28-5   Log Periodic Wire Antennas

The *log periodic* antenna is widely used in the vhf spectrum and also as an hf rotary antenna in commercial service. Theory of this interesting antenna is covered in Chapter 30.

A rotatable hf log periodic antenna is large, complex, costly, and impractical for the average amateur. However, equal performance can be obtained from a lightweight, inexpensive wire log periodic fixed in one direction.

The main feature of the log periodic antenna is the unique combination of high gain and broad frequency coverage. The antenna can be designed to span a frequency range of 10-to-1, but such an array is very large in terms of the lowest operating frequency. The designs shown in this chapter

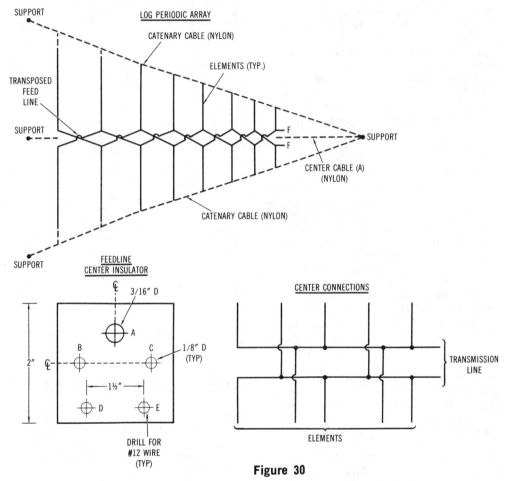

Figure 30

ASSEMBLY DRAWING FOR LOG PERIODIC WIRE ANTENNA

Plan view of antenna shows array is supported by nylon catenary cables between four supports. An open-wire feedline runs the length of the array at the center. The elements are cross-connected to the feedline. Directivity is through the array in the direction of the shortest element (arrow). Hole A in the center insulator passes the nylon center cable; the elements are attached to holes B and C. The open-wire feedline passes through holes D and E. One insulator is required for each antenna element. The feedline is not crossed as shown in the top illustration, rather the wires run parallel to each other and the elements are cross-connected. This makes a simpler and rugged installation.

are much more compact and only cover a small segment of the hf spectrum.

Construction of Log Periodics   Construction of these arrays is shown in figure 30. The antennas are made of 15 gauge aluminum electric fence wire. Connections to the wire are mechanical, with joints made with plated hardware. Bolts, nuts, and lockwashers are used and the joint is then covered with waterproof mastic, such as *RTV*. The center insulators are made of lucite or plexiglass. The insulators support the elements and also the open wire feeder which runs down the center of the array.

The log periodic antenna may be supported by nylon catenary cables run between poles or other supports. Nylon rope or similar insulating material is used to tie the elements to the catenary line. Height of the poles should be 40 to 60 feet above ground level.

Note that every other set of dipole elements is transposed on the feedline. Short jumpers at the center insulators cross-connect the dipoles as shown in the illustration. The insulators are used both as spacers and stringers for the open-wire feeder which runs down the center of the array.

The log periodic antenna is usually suspended from the center with a $\frac{3}{16}$-inch diameter nylon line, plus two side nylon catenary cables. The center line carries most of the weight of the antenna, including the feedline and the insulators. Use 800-pound-test line. The line is strung through the center hole (A) in the insulator and the array hung at shoulder height for ease in assembly. The antenna elements are passed through holes B and C and the wire twisted back upon itself. The feedline passes through holes D and E. Starting at the longest element, cross connections are made between the wires in holes B and D and between holes C and E. The next insulator is cross-connected between holes B and E, and C and D. Each insulator thereafter is connected in the same sequence so that the halves of the antenna are alternately connected first to one feed wire and then to the other.

The elements are attached to the center insulators, starting at one end of the antenna

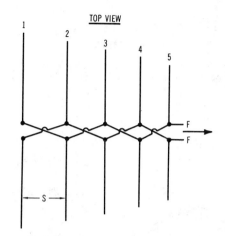

TOP VIEW

TIP-TO-TIP LENGTH OF ELEMENTS			SPACING BETWEEN ELEMENTS		
	FEET	METERS	S	FEET	METERS
1	70	21.3	1-2	14	4.27
2	64	19.5	2-3	13	3.96
3	56	17.0	3-4	12	3.66
4	49	14.9	4-5	9	2.75
5	40	12.2			

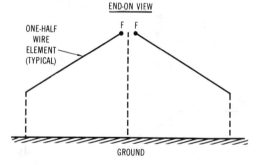

END-ON VIEW

Figure 31

INVERTED-V LOG PERIODIC BEAM FOR 7-10 MHZ

This wire array covers both the 7-MHz band and the new 10-MHz band. Element tips are tied off near ground to simplify erection of antenna. Antenna is fed at the apex (F–F) with a balun and coaxial line. The lower illustration shows an end-on view of the assembly.

and working toward the other. After all the elements are in place the catenary cables are positioned and the ends of the elements looped around the cables. The center cable should bisect the center line of the array. When everything is in position, each element is affixed to the catenary cable with

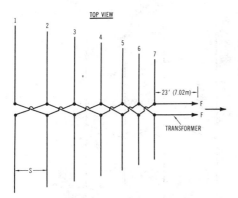

TIP-TO-TIP LENGTH OF ELEMENTS			SPACING BETWEEN ELEMENTS		
EL. NO.	FEET	METERS	S	FEET	METERS
1	36	11.0	1-2	8.0	2.44
2	32	9.76	2-3	7.25	2.21
3	28	8.54	3-4	6.25	1.91
4	24	7.32	4-5	6.0	1.83
5	21.5	6.56	5-6	5.5	1.68
6	18	5.5	6-7	4.25	1.30
7	16.25	4.96			

Figure 32

LOG PERIODIC BEAM FOR 14-22 MHZ

This array covers the span of 14-22 MHZ with an average gain of 8 dB over a dipole. Antenna is fed with an open-wire transformer at the apex and a balun (at F-F) which provides a good match to a coaxial line. Antenna height should be 40 to 60 feet for best results.

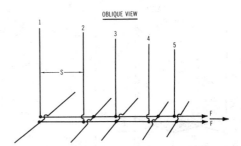

VERTICAL ELEMENT LENGTH			SPACING BETWEEN ELEMENTS		
EL. NO.	FEET	METERS	S	FEET	METERS
1	65	19.8	1-2	26	7.86
2	62	18.9	2-3	24	7.25
3	55	16.8	3-4	23	7.0
4	45	13.7	4-5	18	5.5
5	40	12.2			

Figure 33

VERTICAL LOG PERIODIC BEAM FOR 80-METER BAND

The antenna is made up of five vertical log periodic elements working against a ground screen composed of five full size elements. Radiation is in the vertical pattern. The horizontal radial elements should be at least five feet clear of the ground. Antenna is fed at F-F with a balun and coaxial line. Note that the vertical elements are transposed along the line. The center of each horizontal element is then attached to the opposite feeder wire.

nylon twine. Use temporary knots, as it may be necessary to adjust the tension after all elements are installed. (Note: The designs shown in the following section are adopted from W4AEO).

An Inverted-V Log Periodic Beam for 7–10 MHz—The antenna design shown in figure 31 covers the frequency range of 7 to 10.3 MHz. Power gain at 7 MHz is about 8 dB and is about 5 dB at 10.3 MHz.

The array is formed in the shape of a V; that is, the tips of the elements are at a lower level than the center, and is fed with open-wire feedline. This permits the structure to be supported from two poles, one at each end of the center line. Element tips should clear the ground by about 10 feet and the center of the array should be from 40 to 60 feet high. Overall length of the array is 48 feet. The open-wire line is fed at the apex of the antenna with a 4-to-1

balun and a 50-ohm coaxial line. The line is made of No. 12 wire.

A Log Periodic Beam for 14–22 MHz— Shown in figure 32 is a wire beam that provides about 8 dB gain over the range of 14 to 22 MHz. Array length is about 86 feet. To provide a good match to a 50-ohm coaxial line, the array is fed at the apex with an open-wire matching transformer 23 feet long, and a 4-to-1 balun. The spacing between the wires of the transformer is 1½ inches. The transmission line and transformer are made of No. 12 wire.

A Vertical Log Periodic Beam for 80 Meters—A relatively compact 80-meter array is shown in figure 33. This five-element, vertically polarized antenna is hung from two poles: one 70 feet high and the other 45 feet high. Antenna gain is about 8 dB over a vertical ground plane. Two horizontal radial wires are used at each vertical element. The radials should be about 5 feet off ground for best results. Note that radials

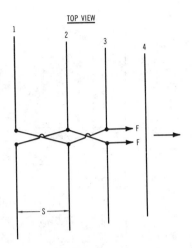

TIP-TO-TIP LENGTH OF ELEMENTS			SPACING BETWEEN ELEMENTS		
EL. NO.	FEET	METERS	S	FEET	METERS
1	136.7	41.7	1-2	48.0	14.6
2	128.5	39.0	2-3	45.0	13.7
3	120.7	36.8	3-4	25.0	7.6
4	120.0	36.6			

Figure 34

COMPACT LOG PERIODIC YAGI FOR 80 METERS

A three-element log periodic "cell" is combined with a parasitic director to form a compact 80-meter beam giving constant gain of about 5 dB across the band. The SWR on the coaxial transmission line is less than 1.6 to 1 across the band when the beam is mounted fifty feet above the ground.

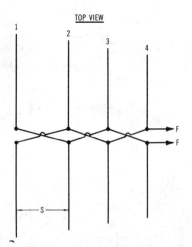

TIP-TO-TIP LENGTH OF ELEMENTS			SPACING BETWEEN ELEMENTS		
EL. NO.	FEET	METERS	S	FEET	METERS
1	69.5	21.0	1-2	25.2	7.7
2	66.0	20.0	2-3	23.9	7.3
3	62.7	19.0	3-4	22.7	7.0
4	59.6	18.0			

Figure 35

LOG PERIODIC BEAM FOR 40 METERS

The antenna elements can be tilted toward the ground forming an inverted-V configuration to conserve space. Center height of the array should be 40 to 70 feet for best results. Beam is fed with a 4-to-1 balun and coaxial line at F–F.

and verticals are cross-connected at each element with both segments of the radial connected to the same wire of the transmission line. The transmission line is made of No. 12 wire and the antenna is fed with a 4-to-1 balun and coaxial line.

A Compact Log Periodic Yagi Beam for 80 Meters—The log periodic principle can be combined with yagi parasitic elements to form a compact single-band beam, such as shown in figure 34. In this design, a three-element log periodic "cell" is combined with a single parasitic director to provide high gain and excellent bandwidth. The SWR

on the transmission line is less than 1.6 to 1 across the whole 80-meter band when the antenna is mounted at a height of fifty feet above ground, or greater. The antenna is fed at F–F with a 4-to-1 balun and a coaxial line.

A High-Gain Log Periodic Antenna for 40 Meters—Shown in figure 35 is a relatively compact wire beam for 40-meter operation. It should be mounted in a horizontal position about 40 to 70 feet in the air for best results. The tips of the elements can be allowed to tilt toward the ground, forming an inverted-V arrangement such as the antenna shown in figure 31.

CHAPTER TWENTY-NINE

HF Rotary Beam Antennas

The rotary beam antenna has become standard equipment for the vhf and upper-hf amateur bands. The rotary array offers many advantages, such as power gain, reduction in interference from undesired directions, compactness and the ability to quickly and easily change the azimuth direction.

The majority of hf rotary antennas are horizontally polarized, unidirectional parasitic type designs while the vhf rotary antenna may be either horizontally or vertically polarized, depending on local usage and the mode of communication desired. In most cases the arrays are self-supporting, being constructed of aluminum or wire elements with a metal or wood framework. The electrical design is mainly end-fire, with parasitic elements lying in a single plane. This design is chosen because it provides a maximum gain figure for a given antenna volume, without the need of interconnecting feedlines between array elements. The *parasitic beam antenna* makes use of elements whose currents are derived by radiation from a nearby driven element.

29-1 The Parasitic Beam

A beam antenna may be composed of a radiator, or driven element, plus an additional number of parasitic elements, unconnected to the driven element. The magnitude of current in the parasitic elements falls off rapidly with increasing distance from the driven element and thus there is a tendency to use relatively close spacing between the elements of a parasitic array.

The parasitic element intercepts and re-radiates energy from the driven element. The distance between the parasitic and the driven elements and the length of the parasitic element determine how the field about the elements is modified by the presence of the parasitic. Both spacing and parasitic element length determine the phase difference between the intercepted and reradiated energy and proper adjustment of these variables can produce an array which exhibits power gain in a favored direction at the expense of radiation in unwanted directions. An infinite number of combinations of element spacing and parasitic length exists, which makes the problem of designing a multi-element parasitic array a complex one. As a result, many of the existing array designs are based on experimental data collected from the study of model antennas on an antenna range.

A *parasitic director element* is one that provides power gain in a direction through the element. It is generally shorter in length than the driven element and thus capacitive in reactance and leading in phase. A *parasitic reflector element* is one that provides power gain in a direction away from the element. It is generally longer than the driven element and thus inductive in reactance and lagging in phase.

The presence of a parasitic element tends to reduce the feedpoint resistance of the driven element for close spacings and to increase it for spacings greater than one-half wavelength. Optimum dimensioning of spacing and element lengths, moreover, can only be obtained over a very narrow frequency range, and the parasitic beam will work only over a relatively restricted band

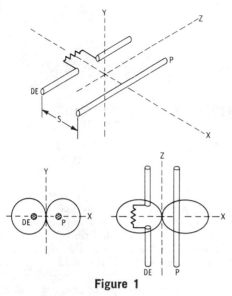

Figure 1

DIRECTIONAL RESPONSE OF 2-ELEMENT PARASITIC BEAM

Close spaced, two-element parasitic beam having a resonant parasitic element provides a bidirectional pattern with 3 dB gain. A pronounced null exists along the Y-axis. Spacing between elements is approximately 0.04 wavelength. The radiation resistance of such a beam is about 2 ohms.

since in the X-Y plane only one-half as much energy is radiated as compared to a dipole. The *front-to-back (F/B) ratio* is unity.

If the length of the parasitic element is increased a few percent, the parasite now acts as a reflector, reducing radiation to the rear and providing a greater forward power gain (figure 2). A front-to-back ratio is

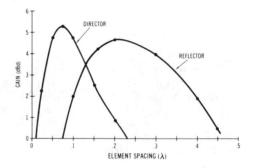

Figure 2

POWER GAIN OF TWO-ELEMENT YAGI BEAM

Power gain over a dipole of a two-element Yagi is about 5.5 dB when the parasitic element serves as a director. Maximum gain occurs with less than 0.1 wavelength spacing. When the parasitic element is a reflector, maximum gain of about 4.7 dB occurs at 0.2 wavelength spacing.

of frequencies. In most cases, the bandwidth of such an array is compatible with the width of the hf amateur bands.

The compactness of a parasitic beam antenna more than outweighs the disadvantage of the critical performance and no other antenna exists that can compare, size for size, with the power gain and directional characteristics of the parasitic array.

29-2 The Two-Element Parasitic Beam

The parasitic beam, or *Yagi-Uda* array (named after Drs. Yagi and Uda of Tokyo University), was invented in 1926 and first placed in service by radio amateurs about 1934. The simplest form of Yagi is a two element configuration with a very close spaced, resonant parasitic element (figure 1). This array provides bidirectional directivity with a power gain of about 3 dBd,

now evident. By decreasing the length of the parasitic element from resonance by a few percent, the parasite serves as a director, providing essentially the same directive pattern as before. Finally, both a reflector and a number of directors may be combined to form a multielement Yagi beam providing impressive gain over a comparison dipole. (Note: In this chapter antenna gain is *referenced to a dipole* unless otherwise noted.)

Element Spacing An infinite combination of element spacing and length exists for the two-element Yagi beam and no one specific combination provides highest gain, best front-to-back ratio, and highest driving impedance. Measurements made on antenna ranges with model Yagi antennas and comprehensive computer runs have shown that a tradeoff must be made for the two-element beam between efficiency and bandwidth, in one case, and gain and front-to-back ratio in the other.

The gain characteristics of a two-element Yagi are shown in figure 2. The case when the parasite is a reflector provides maximum gain with an element spacing of about 0.2 wavelength at the design frequency. When the parasite is properly adjusted, the gain figure is about 4.6 dB. When the parasite is a director, a maximum gain figure of about 5.4 dB occurs at an element spacing of 0.075 wavelength at the design frequency. The practical difference in gain between the two examples is minimal.

The front-to-back ratio of the two element Yagi is a complex picture. While the gain-versus-frequency response of a properly adjusted beam is quite good, the front-to-back ratio is poor for designs that are practical (figure 3). In the case of a two-ele-

25 dB at the design frequency. The feed-point resistance of such an array, however, is of the order of 4 ohms and the F/B ratio drops rapidly as the operating frequency is removed from the design frequency.

The feedpoint resistance of the two-element Yagi is shown in figure 4. Experience has shown that efficiency and bandwidth suffer when the resistance is less than about 20 ohms. Antennas exhibiting low values of feedpoint resistance require an extensive matching system which introduces loss and has a greatly restricted operating bandwidth when the SWR on the feed system is examined. Thus, from a practical point of view, the excellent front-to-back ratio shown for an element spacing of .05 wavelength shown in figure 3 is impractical.

Since a feedpoint resistance of about 20 ohms is not unduly hard to match, it can be argued that the best all-around performance may be obtained from a two-element parasitic beam employing .15 wavelength element spacing, with the parasite tuned to act as a reflector. This antenna will provide a power gain of about 4.6 dB power gain over a dipole with a F/B ratio of about 10 dB. If it is desired to use a director, opti-

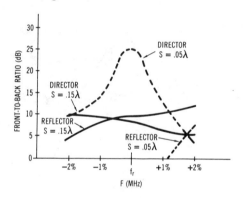

Figure 3

FRONT-TO-BACK RATIO OF TWO-ELEMENT YAGI BEAM

Front-to-back ratio is poor for two-element beam at practical element spacings. Compromise spacing of 0.15 wavelength provides F/B ratios of 5 to 12 dB. Note that F/B ratio varies with respect to the design frequency, f$_r$. The high F/B ratio at a spacing of 0.05 wavelength is obtained at the expense of bandwidth and very low feedpoint resistance.

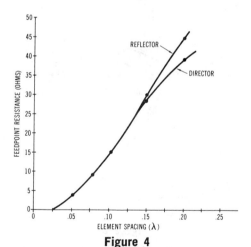

Figure 4

FEEDPOINT RESISTANCE OF TWO-ELEMENT YAGI BEAM

For practical spacing of 0.1 to 0.15 wavelength, the feedpoint resistance varies from 15 to 30 ohms at the design frequency. The excellent F/B ratio obtained at a spacing of 0.05 wavelength results in a feedpoint impedance of less than 5 ohms.

ment beam having spacing of 0.15 wavelength the F/B ratio runs less than 10 dB when the parasite is adjusted either as a reflector or director. When spacing is reduced to 0.05 wavelength, the F/B ratio of a reflector parasite is maximum outside the passband of the beam. On the other hand the F/B ratio of a director parasite is nearly

mum spacing is about .1 wavelength. This will provide equivalent power gain and F/B ratio. The two designs thus provide approximately equal performance. The reflector-type array will have a feedpoint resistance of about 30 ohms and the director-type array will have a resistance of about 16 ohms.

29-3 The Three-Element Parasitic Beam

The three-element Yagi is made up of a director, driven element, and reflector. As in the case of the two-element beam, infinnite combinations of element spacings and lengths exist and no one specific combination of these parameters provides optimum gain, front-to-back ratio, and driving impedance. The simplest case (and, as it turns out, the best choice) is where all elements are uniformly spaced on the boom. Array gain is a function of boom length and the F/B ratio peaks at a particular boom length, all else being equal.

The three-element Yagi can be tuned for maximum gain or for maximum F/B ratio (figures 5 and 6). Maximum gain occurs

at a boom length of about 0.45 wavelength. Two gain curves are shown in the illustration: one when the parasites are tuned for maximum gain and the other when they are tuned for maximum F/B ratio. The difference in gain between the two conditions is about 0.7 dB. When tuned for maximum F/B ratio, a peak ratio of about 28 dB can be obtained somewhat lower in frequency than the design frequency. As the operating frequency is raised, the F/B ratio deteriorates.

A representative three-element Yagi on a 0.3 wavelength boom will provide about 7 to 8 dB gain over a dipole and display a F/B ratio from 15 to 28 dB, depending upon element tuning. In all cases design parameters for maximum F/B ratio are more critical than those for maximum forward gain. The feedpoint resistance is about 18 to 25 ohms.

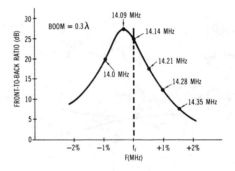

Figure 6

FRONT-TO-BACK RATIO OF THREE-ELEMENT YAGI BEAM

The F/B ratio reaches a maximum figure of about 27 dB slightly below design frequency f_r. The F/B ratio for other boom lengths resembles this curve. Front-to-back ratio increases as element spacing decreases and drops off with increased element spacing. Example shown is for 14 MHz with f_r — 14.14 MHz.

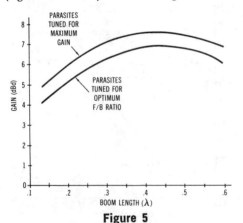

Figure 5

POWER GAIN OF THREE-ELEMENT YAGI BEAM

Power gain of three-element beam over a dipole varies from 6.5 to 7.5 dB when parasitic elements are adjusted for maximum gain. When tuned for optimum F/B ratio, power gain is about 0.7 dB less. Ground reflection can provide appreciably higher gain than indicated by curves.

Element Lengths   The length of any antenna element is a function of the wavelength, the speed of light and the element diameter. The basic relationship is:
Length (feet) for a half-wavelength =

$$\frac{492}{f_{(MHz)}}$$

where,
f is the antenna design frequency.

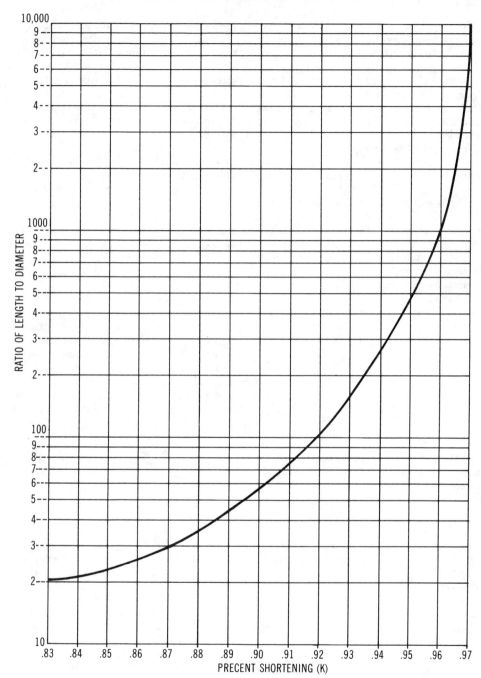

Figure 7

GRAPH OF ELEMENT DIAMETER VERSUS LENGTH

The relative wave velocity on any element is a function of the length-to-diameter ratio (after Schelkunoff and Friis). Greater shortening is required for thick elements; however, even very thin elements require appreciable shortening. The k-factor (x-axis) indicated the degree of shortening.

The length of a metallic element is less than this theoretical relationship because a practical element has thickness and because radio energy travels more slowly along the element than in free space. As the thickness (diameter) of the element increases, the element must be made shorter to establish resonance. Even for very thin wires the shortening effect is appreciable (figure 7).

At 7 MHz, for example, the length-to-diameter ratio of No. 12 wire is about 10,000 and the element length must be reduced 3 percent from the theoretical value. At the same frequency, the length of a dipole made of 3-inch diameter tubing must be reduced 6 percent. Shortening is described in terms of the *k-factor* and expressed as a percentage of original length, where k is the ratio of wavelength to element diameter.

Thus, in the first instance, $k = .97$ and in the second, $k = .94$.

In the case of a wire antenna suspended by end insulators, k is approximately .95 and the length of a half-wave antenna is:

$$\text{Length (feet)} = \frac{492 \times .95}{f_{(\text{MHz})}}$$

or,

$$\text{Length (feet)} = \frac{468}{f_{\text{MHz}}}$$

In the case of a tubing element supported at the center with no end insulators, the half-wavelength is approximately:

$$\text{Length (feet)} = \frac{463}{f_{\text{MHz}}}$$

In the case of a Yagi antenna where the length of the driven element is affected by the parasites, element length depends largely on antenna design.

Element Taper   Practical hf antennas made of aluminum tubing have tapered elements. That is, each element is made up of sections of telescoping tubing. The element diameter thus varies from center to tip. The element taper can introduce a significant change in required length. If the average diameter of a tapered element is taken as standard, then the larger diameter portion of the element has less induct-

ance per unit length than average and must be made longer. The smaller diameter portion, on the other hand, has smaller capacitance per unit length than the standard section, and must also be made longer. Taper correction, therefore, must be applied to the element as a whole and is quite significant. A representative taper correction factor is shown in figure 8. This is only approximate as the rate of taper can vary depending upon the number of lengths and diameters of concentric tubing used. This chart assumes the taper is linear from the center of the element to the tips.

Operating Bandwidth of the Yagi Antenna   The operating bandwidth of any directional antenna may be specified in terms of SWR on the feedline, pattern deterioration or loss of gain. In the case of amateur arrays, the effective bandwidth is commonly specified as a maximum value of SWR and is usually limited to a figure of 2 to 1. In most instances, bandwidth is limited by the matching device between antenna and feedline, rather than by the antenna characteristics. When adjusted for maximum gain, the bandwidth of a typical three element Yagi is about 2.5 percent of the design frequency, as defined by the SWR limitation. This means that an array cut to 14.15 MHz would have a bandwidth between the 2-to-1 SWR points on the transmission line of about 350 kHz, centered on the design frequency. In like fashion, a beam cut for 10-meter operation at 28.5 MHz would have an effective bandwidth of 700 kHz. Since the band is 1700 kHz wide, the array should be either cut for low- or high-frequency operation in the band. Operation of the Yagi outside the effective bandwidth will result in a high SWR on the transmission line and a degradation of forward gain and F/B ratio.

The bandwidth on the high-frequency side of the design frequency is limited by director resonance. That is, when the operating frequency approaches the resonant frequency of the director, the directive pattern of the array reverses itself and the parasite director tends to act as a reflector.

The Yagi arrays described in this Handbook are a compromise between bandwidth, F/B ratio and power gain. In all cases, the power gain at the design frequency is within one decibel or less of maximum theoretical gain.

Yagi Dimensions　For the general case, with no element taper, the lengths of the *three-element Yagi* hf antenna made of nontapered aluminum tubing may be calculated from:

$$\text{Length of director (feet)} = \frac{458}{f_{(\text{MHz})}}$$

$$\text{Length of driven element (feet)} = \frac{472}{f_{(\text{MHz})}}$$

$$\text{Length of reflector (feet)} = \frac{504}{f_{(\text{MHz})}}$$

$$\text{Element spacing (feet)} = \frac{148}{f_{(\text{MHz})}}$$

If the element is made of more than one section of aluminum tubing and has a taper from the center toward the tips, the correction factor shown in figure 8 must be applied to the above formulas.

For example, assume a three-element Yagi is to be built at a design frequency of 7.02 MHz for 40-meter DX work at the low-frequency end of the band. Each element is made of aluminum tubing, starting with $1\frac{3}{4}$" tubing at the center, tapering to $\frac{1}{2}$" tubing at the tips. This provides a minimum of element sag. The tubing diameters used are: $1\frac{3}{4}$", $1\frac{5}{8}$" (a reducer to $\frac{3}{4}$"), $\frac{3}{4}$", $\frac{5}{8}$" and $\frac{1}{2}$". The taper correction (figure 8) is the ratio of the element diameters ($D_1/D_2 = 1.75/0.5 = 3.5$) and the taper correction factor is 1.053. Element lengths, therefore, are:

$$\text{Length of director (feet)} = \frac{458}{7.02} \times 1.053$$
$$= 68.7'$$

$$\text{Length of driven element (feet)} = \frac{472}{7.02} \times$$
$$1.053 = 70.8'$$

$$\text{Length of reflector (feet)} = \frac{504}{7.02} \times 1.053$$
$$= 75.6'$$

Element spacing is not affected by element taper and remains at 21 feet. A 42-foot

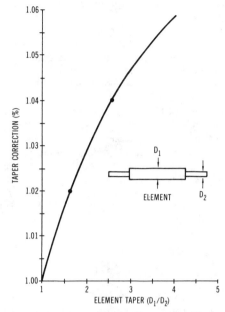

Figure 8

TAPER AFFECTS OVERALL ELEMENT LENGTH

Once the element lengths have been determined by formula, they must be increased to correct for element taper as shown by this graph. For example, if the taper (D_1/D_2) is 3, the taper correction factor is about 1.046.

boom is thus required. A 40-meter beam built to these dimensions checked out within 15 kHz of the design frequency when completed.

29-4 The Miniature Beam

A parasitic array may be built of short, electrically loaded elements in place of the more common half-wavelength elements. In addition, element spacing may be reduced severely to make the overall beam dimensions small in terms of the operating wavelength. In order to obtain the benefits of small physical size, the miniature beam must sacrifice both power gain and bandwidth to some degree. The overall loss of performance is dependent to a large extent on the r-f losses incurred in the loading system.

The usual technique is to employ high-Q loading coils or stubs, as shown in figure 9.

Some designs have used helical wound elements to achieve reduced size.

It is difficult to predict the result of reducing the size of a beam antenna as much depends on the mechanical construction and electrical efficiency of the loading system, and this must be determined on a case by case basis. Linear loading techniques have met with some success, as attested to by the popularity of this design as employed with compact 40-meter beam antennas.

Design information for a compact, 3-element Yagi antenna for 40-, 20-, 15-, or 10-meter operation is given in figure 10. This design uses full size elements and minimum boom length to achieve high gain with the smallest possible overall size.

29-5 Three-Band Beams

A popular form of beam antenna is the so-called *three-band beam*. An array of this type is designed to operate on three adjacent

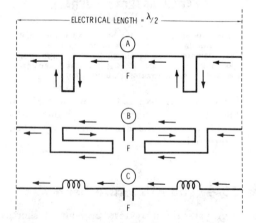

Figure 9

THE LOADED HALF-WAVE ELEMENT

Short, electrically loaded elements may be substituted for a full size element in a parasitic array. (A)—A portion of the dipole on each side of the center feedpoint is folded down to conserve space. The folded portion does not radiate as the wires carry opposite and nearly equal currents. (B)—The folded portions of the elements are laid back against the center section of the element. (C)—An inductor is substituted for the folded portion of the element. The farther from the center the loading devices are, the larger they must be for a given resonant frequency of the element.

amateur bands, such as the 10-, 15-, and 20-meter group. The principle of operation of this form of antenna is to employ parallel-tuned circuits placed at critical positions in the elements of the beam which serve to electrically connect and disconnect the outer sections of the elements as the frequency of excitation of the antenna is changed. A typical three-band element is shown in figure 11. At the lowest operating frequency, the tuned *traps* exert a minimum influence on the element which resonates at a frequency determined by the electrical length of the configuration, plus a slight degree of loading contributed by the traps. At some higher frequency (generally about 1.5 times the lowest operating frequency) the outer set of traps is in a parallel resonant condition, placing a high impedance between the element and the tips beyond the traps. Thus, the element resonates at a frequency 1.5 times higher than that determined by the overall length of the element. As the frequency of operation is raised to approximately 2.0 times the lowest operating frequency, the inner set of traps becomes resonant, effectively disconnecting a larger portion of the element from the driven section. The length of the center section is resonant at the highest frequency of operation. The center section, plus the two adjacent inner sections, are resonant at the intermediate frequency of operation, and the complete element is resonant at the lowest frequency of operation.

The efficiency of such a system is determined by the accuracy of tuning of both the element sections and the isolating traps. In addition the combined dielectric losses of the traps affect the overall antenna efficiency. As with all multipurpose devices, some compromise between operating convenience and efficiency must be made with antennas designed to operate over more than one narrow band of frequencies. Taking into account the theoretical difficulties that must be overcome it is a tribute to the designers of the better multiband beams that they perform as well as they do.

The Isolating Trap The parallel-tuned circuit which serves as an isolating trap for a multiband antenna should combine high circuit Q with good

Figure 10

COMPACT THREE-ELEMENT BEAM

This precut three-element beam performs well and provides a good front-to-back ratio. Dimensions are given for all-band coverage, except for 10 meters, where the beam is cut for the 28.0- to 29.0-MHz portion. Dimensions are given for elements having a minimum taper. For a 2-to-1 diameter taper, all lengths should be increased by about three percent. Length of the driven element may vary slightly depending on the matching system used. Power gain is better than 1 dB and front-to-back ratio is greater than 25 dB.

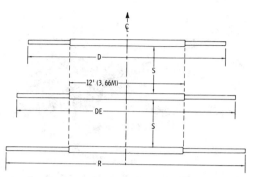

DIMENSIONS

BAND	$D = \dfrac{465}{f(MHz)}$		$DE = \dfrac{473}{f(MHz)}$		$R = \dfrac{501}{f(MHz)}$		$S = \dfrac{130}{f(MHz)}$	
	FEET	METERS	FEET	METERS	FEET	METERS	FEET	METERS
40	65' 4''	19.91	66' 10''	20.32	70' 8''	21.60	18' 0''	5.50
20	32' 8''	9.96	33' 5''	10.16	35' 4''	10.82	9' 0''	2.75
15	21' 10''	6.68	22' 4''	6.81	23' 7''	7.25	6' 2''	1.87
10	16' 4''	4.98	16' 8''	5.08	17' 8''	5.41	4' 6''	1.37

environmental protection. A highly satisfactory trap configuration based on the original design of W3DZZ is shown in figure 12. The trap capacitor, which has a value of about 25 pF is made of two sections of aluminum tubing which form a portion of the antenna element. The capacitor dielectric is molded lucite, or similar plastic material, given a coat of epoxy to help resist crazing and cracking caused by exposure to sunlight. The coil is wound of No. 8 aluminum wire and, with the capacitor placed within it, has a Q of nearly 300. The leads of the coil are bent around the tubing and a small aluminum block is used to form an inexpensive clamp. If desired, an aluminum cable clamp may be substituted for the homemade device.

The isolating trap is usually tuned to the lower edge of an amateur band, rather than to the center, to compensate for the length of the unit. In general, the 15-meter trap is tuned to approximately 20.8 MHz and the 10-meter trap is tuned near 27.8 MHz. The trap frequency is not critical within a few hundred kilohertz. Resonance is established by squeezing or expanding the turns of the coil while the trap is resonated on the bench with a grid-dip oscillator and a calibrated receiver.

A substitute for the molded capacitor may be made up of two 40 pF, 5-kV ceramic capacitors connected in series (*Centralab* 850S-50Z) and mounted in a length of phenolic tubing of the proper diameter to slip within the aluminum antenna sections. The trap coil is then wound about the capacitor assembly in the manner shown in the photograph.

A complete discussion of Yagi antennas of all types, including the tribander, is contained in the *Beam Antenna Handbook*, available from Radio Publications, Inc., Wilton, CT 06897.

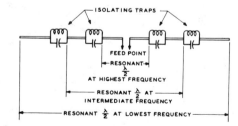

Figure 11

TRAP-TYPE "THREE-BAND" ELEMENT

Isolating traps permit dipole to be self-resonant at three widely different frequencies.

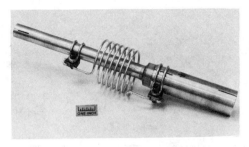

Figure 12

HIGH-Q ISOLATING TRAP

This trap has a Q of nearly 300 and is well suited for multiband antennas. The coil is wound of No. 8 aluminum clothesline wire and is 3″ in diameter and 3″ long. The 15-meter trap has seven turns (illustrated) and the 10-meter trap has five turns. The capacitor is made from two lengths of aluminum tubing, coaxially aligned in a lucite dielectric. Capacitor length is about five inches and tubing sizes are ¾ inch and 1¼ inch. Capacitance is about 25 pF. Lucite projects from end of capacitor to form ½-inch collar which is coated with epoxy to prevent deterioration of the dielectric under exposure to sunlight. Similar traps have been made using teflon as a dielectric material. Ends of aluminum tubes are slotted to facilitate assembly to antenna elements.

29-6   The Multielement Yagi Beam

Additional gain may be obtained from a Yagi array through the use of more than two parasitic elements. Gain is proportional to element spacing and tuning and hence proportional to boom length. In fact, gain is nearly independent of the number of elements along the length of a given boom as long as there are enough. And there is a practical limit of the upper number of elements for a given boom length. Figure 13 illustrates representative gain figures for various boom lengths and number of elements. The dashed gain-vs-boom length line is a compromise derived from various conflicting measurements made over a period of years by knowledgeable experimenters. The actual curve is not smooth and exhibits bumps of up to 1 dB along the boom length figure. These variations are probably within the expected accuracy of such measurements and do not detract from the generalized data.

F/B ratio of the multielement beam design is not appreciably better than that of the three-element configuration, running between 25 dB and 30 dB, depending on adjustment. In all cases the F/B ratio peaks sharply slightly lower than the design frequency. Table I provides dimen-

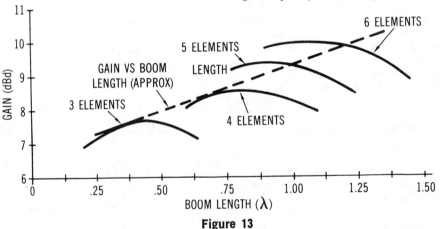

Figure 13

GAIN VERSUS BOOM LENGTH FOR MULTIELEMENT YAGI BEAMS

Gain is proportional to boom length and nearly independent of the number of elements on the boom as long as there are enough. The dashed line shows average gain figure in terms of boom length. The results for an individual beam depend on tuning, height above ground, and other mitigating factors.

sional data for several hf, multielement Yagi arrays.

Band (MHz)	Director	Driven Element	Reflector	Spacing
7.0-7.3	65'0'' (19.8)	67'0'' (20.42)	71'6'' (21.80)	21'0'' (6.4)
10.1	45'4'' (13.83)	46'9'' (14.25)	49'11'' (15.22)	14'8'' (4.47)
14.0-14.35	32'4'' (9.87)	33'4'' (10.17)	35'6'' (10.86)	10'5'' (3.19)
18.1	25'4'' (7.7)	26'1'' (7.95)	27'10'' (8.48)	8'2'' (2.49)
21.0-21.45	21'6'' (6.57)	22'3'' (6.77)	23'9'' (7.23)	7'0'' (2.12)
24.9	18'4'' (5.60)	19'0'' (5.78)	20'3'' (6.17)	5'11'' (1.81)
28.0-29.0	16'1'' (4.88)	16'6'' (5.03)	17'8'' (5.37)	5'2'' (1.58)
29.0-29.7	15'8'' (4.78)	16'2'' (4.93)	17'3'' (5.26)	5'0'' (1.55)

f_R = 7.05, 14.15, 21.25, 28.6 and 29.2 MHz

Dimensions in parentheses are in meters

Feedpoint resistance at f_R approx. 23 ohms

Table 1. Dimensional Data for HF Yagi Beam Antennas

Element length and spacing are given in feet and meters for the high frequency bands. For a four or five element beam the spacing and length of the additional directors are as shown for the three element case. These dimensions are for nontapered elements. For taper, apply the correction factor given in figure 8.

29-7   Feed Systems for Parasitic Arrays

Table 1 gives, in addition to other information, the approximate feedpoint resistance referred to the center of the driven element of multielement parasitic Yagi arrays. It is obvious, from this low value of radiation resistance, that special care must be taken in materials used and in the construction of the elements of the array to ensure that ohmic losses in the conductors will not be an appreciable percentage of the radiation resistance. It is also obvious that some method of impedance transformation must be used in many cases to match the low

feedpoint resistance of these antenna arrays to the normal range of characteristic impedance used for antenna transmission lines.

The various matching systems described in Chapter 26 apply to Yagi beams in general. Many homemade beams employ either the gamma or the omega match for ease of adjustment, whereas commercial arrays generally employ a balun matching system for economic reasons. In most cases, it is not mechanically desirable to break the center of the driven element for feeding the system, especially in the hf beam assemblies. Breaking the driven element rules out the practicability of building an all-metal array, and imposes mechanical limitations with any type of construction.

29-8   Building the Yagi Beam

The majority of hf Yagi beam antennas make use of elements made up of lengths of telescoping aluminum tubing. This assembly is easy to construct and avoids the problem of getting sufficiently good insulation at the ends of the elements, as the elements may be supported at the center with a minimum amount of sag.

Available tubing comes in 12-foot sections and 6061-T6 alloy is recommended as a good compromise between strength and ability to resist corrosion.

The element diameter depends on the size of the element. Generally speaking, a 20-meter Yagi element may be made of a center section of tubing about 1½" (3.81 cm) diameter, with end-sections made of 1⅛" (2.86 cm) diameter tubing. Alternatively, the element may be made of a 1½" diameter center section, intermediate sections of 1⅛" diameter tubing, and end sections of 1" (2.54 cm) tubing. Overall element length is determined by the distance the smaller sections are extended beyond the end of the center section (figure 15). For ease in telescoping, the difference in diameters (clearance) between the sections should be about 0.01" (0.025 cm).

For 15- and 10-meter beams, the center section of a typical element may be made of 1" diameter tubing, with end-sections of ⅞" (2.23 cm) diameter tubing.

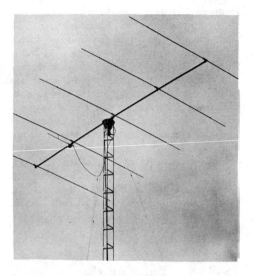

Figure 14

FIVE ELEMENT 28-MHz BEAM ANTENNA AT W6SAI

Antenna boom is made of twenty foot length of three-inch aluminum irrigation pipe. Spacing between elements is five feet. Elements are made of twelve foot lengths of ⅞-inch aluminum tubing, with extension tips made of ¾-inch tubing. Beam dimensions are taken from Table 1.

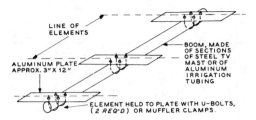

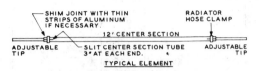

Figure 15

3-ELEMENT ALL-METAL ANTENNA ARRAY

All-metal configuration permits rugged, light assembly. Joints are made with U-bolts and metal plates for maximum rigidity.

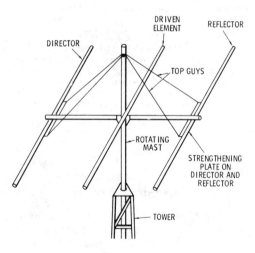

Figure 16

GUY WIRES FORM TRUSS TO STRENGTHEN ROTARY BEAM

A top truss will reduce boom sag and strengthen the beam assembly. Mast is extended above the beam and two guy wires are run to reflector and director elements. The elements can be strengthened by adding short plates at the center, the guy wires attaching to holes in the ends of the plates. The guy wires should be broken up with strain insulators.

In all cases, the greater the taper of the element from center to tip, the greater will be the element length.

Material for the Boom   The stability and ruggedness of the Yagi antenna is primarily determined by the supporting boom. Small vhf arrays and small arrays for 6 and 10 meters can be built using a boom diameter of 1″ (2.54 cm), or possibly less. For larger arrays, or for 15- and 20-meter 3-element Yagi arrays, a boom diameter of at least 2″ (5.1 cm) is recommended, with a diameter of 3″ (7.62 cm) suggested for areas having a harsh climate, or heavy wind conditions.

Regardless of boom diameter, the overall strength and stability of the assembly can be improved by the addition of a top truss, such as shown in figure 16.

For large hf arrays, including 40-meter beams, a 3″ diameter boom having a wall thickness of 0.065″ (0.17 cm) is suggested.

29-9 Stacking Yagi Antennas

Any gain antenna may be stacked to provide additional gain in the same manner that dipoles may be stacked. Thus, if an array of two dipoles would provide a power gain over a single dipole of 3 dB, the substitution of Yagi arrays for each of the dipoles would add the gain of one Yagi to the gain obtained with the dipoles. In other words, doubling the number of arrays provides 3 dB gain under optimum conditions.

In order to obtain this theoretical improvement in gain, the spacing between the arrays is critical. For small arrays stacked in the vertical plane (figure 17) the optimum array spacing is about 0.6 wavelength. Smaller spacing provides less gain and larger spacing is of no advantage. As array gain increases, however, the spacing must increase to achieve maximum stacking gain. A good rule of thumb is that stacking spacing must be at least as great as the boom length of one of the arrays. The gain curve for two, three-element stacked Yagi beams is shown in figure 18. The solid curve is for vertical stacking (one beam above another) and the dashed curve is for horizontal stacking (side by side). Vertical stacking provides a sharp lobe in the vertical plane and horizontal stacking narrows the lobe in the horizontal plane. The stacking mode usually depends on whether the operator desires enhanced azimuthal or elevation directivity.

29-10 The Cubical Quad Beam

The *Cubical Quad* beam is a parasitic array whose elements consist of closed loops having a circumference of one wavelength at the design frequency. The loops may take a diamond, square, or triangular shape (figure 19). The Quad beam has proven to be a very effective antenna and provides somewhat enhanced gain over a Yagi having an equal number of elements.

One advantage of the Quad configuration is that a smaller array for a higher frequency band can be readily placed within a larger, lower frequency array, facilitating the construction of a compact, high gain beam for 20, 15, and 10 meters on a small frame.

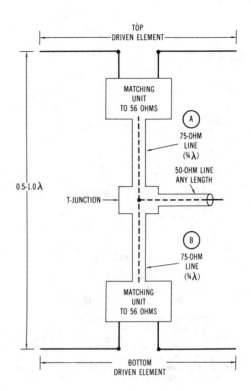

Figure 17

DRIVEN ELEMENTS OF STACKED YAGI BEAM ANTENNAS

Elements of stacked beams must be driven in phase. This is accomplished by driving at a common junction point through equal length lines (A and B). The feedpoint resistance of each driven element is adjusted to 56 ohms at the design frequency by means of a matching unit and each feedline has an impedance of 75 ohms. The lines are cut to ¾ wavelength and act as an impedance matching transformer (¼ wavelength) in series with a 1-to-1 transformer (½ wavelength). The parallel impedance at the junction point is 50 ohms. Array spacing may be 0.5 to 1.0 wavelength, depending on the size of the Yagi antennas.

The wave polarization of a Quad array is a function of the placement of the feedpoint on the driven loop. When fed at the center of the horizontal side, the Quad is horizontally polarized and is vertically polarized when fed at the center of a vertical side. The parasitic elements, being closed loops, function equally well regardless of the polarization of the driven element.

The power gain of a driven Quad loop is about 1.2 dB over a dipole and the addition

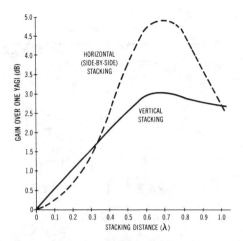

Figure 18

STACKING GAIN VERSUS STACKING DISTANCE FOR YAGI ANTENNAS

The stacking gain for three-element Yagi antennas is about 3 dB for vertical stacking and occurs at an array spacing of 0.6 wavelength. Maximum gain of about 4.8 dB is achieved for horizontal stacking when the element tips are about 0.7 wavelength apart. Optimum stacking distance increases as gain of array increases.

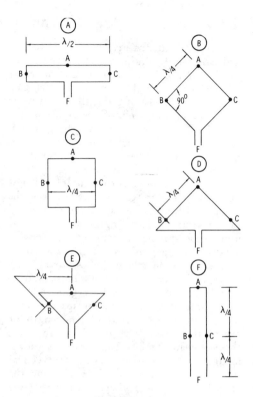

Figure 19

EVOLUTION OF THE QUAD LOOP

The Quad loop evolves from the folded dipole (A) and may take a diamond, square, or triangular shape (B, C, D, E). The Quad is fed at a high current point with a balanced feed system. If the feedpoint F is closed, and the Quad loop opened at either point B or C, vertical polarization will result. The limiting case is a two wire, shorted transmission line (F), which represents the folded dipole pulled open to the maximum amount. The configurations of (D) and (E) provide vertical as well as horizontal polarization.

of parasitic loop elements brings the gain up in the same ratio as adding the equivalent elements to a Yagi array. Thus, element for element, the Quad exhibits about 1.2 dB more overall gain than an equivalent Yagi on the same length boom.

The Quad Loop   The Quad loop may be compared to a "pulled open" folded dipole as shown in figure 19. If the loop is stretched past the Quad configuration, it ultimately becomes a two-wire transmission line, one-half wavelength long, shorted at the far end. The input impedance of the loop is about half that of the folded dipole, or approximately 140 ohms. The loop exhibits a figure-8 radiation pattern similar to the dipole.

The Two-Element Quad   The conventional two-element Quad for hf operation is horizontally polarized and the parasitic element is tuned as a reflector (figure 20). At a spacing of about 0.13 wavelength, the Quad provides

a power gain of nearly 6.5 dB over a dipole mounted at the median height of the Quad. The gain curve for a change in element spacing is quite flat for spacings between 0.1 and 0.2 wavelength. The radiation resistance, at a spacing of 0.13 wavelength is very close to 60 ohms.

The angle of radiation above the horizontal of a two-element Quad resembles that of a dipole or Yagi at the higher elevation angles. However, because of the effect of stacking, the angle of radiation for a Quad antenna at the lower heights is appreciably below that

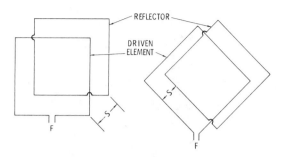

| BAND | LENGTH OF SIDE | | | | SPACING(S) | |
| | DRIVEN ELE. | | REFLECTOR | | | |
	FEET	METERS	FEET	METERS	FEET	METERS
40	35' 2"	10.76	36' 4"	11.08	17' 0"	5.2
20	17' 8"	5.38	18' 2"	5.36	8' 6"	2.6
15	11' 8"	3.56	12' 3"	3.74	5' 7"	1.7
10	8' 8"	2.65	9' 1"	2.77	4' 2"	1.3

Figure 20

THE TWO-ELEMENT QUAD BEAM

This simple, 2-element Quad provides a power gain of nearly 6.5 dB over a dipole. The antenna may be fed with either a 50- or 75-ohm coaxial line and 1-to-1 balun. Spacing (S) is not critical. The framework shown in figure 21 may be used with this array.

of a dipole or Yagi array. At a height of ½ wavelength, for example, the angle of radiation of the main lobe of a Quad antenna is about 4° below that of a dipole. At an elevation of ⅜ wavelength, the angle of radiation of a Quad is nearly 10° below that of a dipole and, at a height of ¼ wavelength the dipole is almost useless as most of the radiation is directed upwards. The Quad antenna, however, at the same height holds the main lobe at an angle of 40° above the horizon. Thus for low heights, the Quad antenna provides an appreciably lower angle of radiation than does either the dipole or the Yagi array.

Element Dimensions   Element lengths for the Quad antenna may be expressed in terms of the circumference of the loop, regardless of whether the shape of the element is square, diamond, triangular or circular. The following formulas apply to hf Quads made of wire:

Circumference of driven element:

$$\text{Feet} = \frac{1005}{f_{(\text{MHz})}}$$

$$\text{Meters} = \frac{306.5}{f_{(\text{MHz})}}$$

Circumference of director element:

$$\text{Feet} = \frac{975}{f_{(\text{MHz})}}$$

$$\text{Meters} = \frac{297.4}{f_{(\text{MHz})}}$$

Circumference of reflector element:

$$\text{Feet} = \frac{1030}{f_{(\text{MHz})}}$$

$$\text{Meters} = \frac{314.2}{f_{(\text{MHz})}}$$

A Simple Quad Framework   Shown in figure 21 is an all-metal support structure for a 2-element Quad. Built of thin wall conduit pipe and angle iron, this "spider" will accommodate bamboo or *Fiberglas* arms of sufficient length for a 20-, 15-, or 10-meter Quad, or an interlaced triband version. The "spider" is made in two parts so the elements may be assembled on the ground and carried to the top of the tower for final assembly. Boom length is only two feet, so the entire Quad can be easily supported by a single person.

When the structure is completed, it should be given a good coat of antirust paint, followed by a coat of heavy duty, outdoor paint to retard rust and corrosion. All hardware should be either stainless steel, or heavily plated.

The Multielement Quad   The three-element Quad provides improved gain and front-to-back ratio over a two-element design but few antennas of this type are used since the center element tends to interfere with the rotational and support system of the antenna. The four-element Quad, on the other hand, is quite popular as it is symmetrical with respect to the supporting structure and does not interfere with the rotating system.

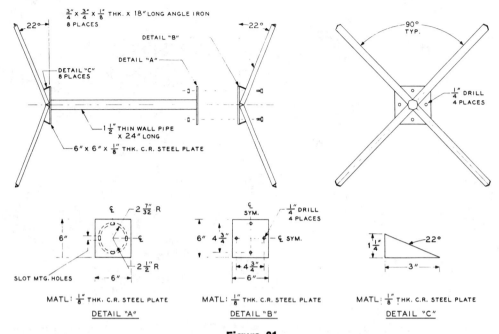

Figure 21

SPIDER CENTER STRUCTURE FOR QUAD ANTENNA

The four-element Quad provides a power gain of about 10.5 dB over a dipole, and about 1.2 dB over a four-element Yagi beam mounted on the same length boom. Dimensions for a typical four-element Quad are shown in figure 22. The boom length is 30 feet, made up of two sections of 2½ " (6.4 cm) aluminum tubing having an 0.065" (0.17 cm) wall. Material is 6061-ST6. The sections are joined by a short section of tubing machined to slip-fit within the boom sections.

The elements are supported on *Fiberglas* arms mounted to the boom with cast aluminum fittings. A simple gamma match is used to provide adjustment and the antenna is fed with a coaxial transmission line.

A complete discussion of Quad antennas of all types is contained in the book *All About Cubical Quad Antennas*, available from Radio Publications, Inc., Wilton, Conn., 06897.

29-11 The Driven Array

Multielement beams may be composed of driven elements, rather than parasitically excited elements. This arrangement provides somewhat greater frequency coverage than does the parasitically excited array. Shown in figure 23 is the so-called ZL-*Special*, two-element driven array. Half-wave elements are used, fed at the center with a transposed feedline. The antenna provides a cardioid pattern with a power gain of about 4 dB over a dipole.

Various other types of unidirectional driven arrays are illustrated in figure 24. The array shown in figure 24A is an end-fire system which may be used in place of a parasitic array of similar dimensions when greater frequency coverage than is available with the Yagi type is desired. Figure 24B is a combination end-fire and collinear system which will give approximately the same gain as the system of figure 24A, but which requires less boom length and greater total element length. Figure 24C illustrates the familiar lazy-H with driven reflectors (or directors, depending on the point of view) in a combination which will show wide bandwidth with a considerable amount of forward gain and good front-to-back ratio over the entire frequency coverage.

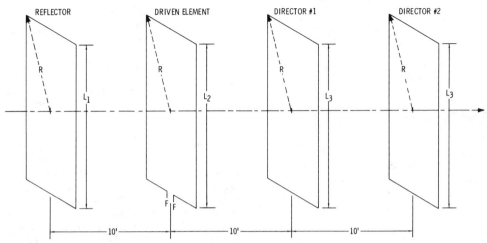

Figure 22

FOUR-ELEMENT QUAD BEAM PROVIDES 10.5 dB GAIN

BAND	REFLECTOR (L₁)		DRIVEN ELEMENT (L₂)		DIRECTOR (L₃)	
	FEET	METERS	FEET	METERS	FEET	METERS
20	18' 2"	5.58	17' 7"	5.37	17' 2"	5.23
15	12' 1"	3.69	11' 9"	3.58	11' 5"	3.47
10	9' 0"	2.75	8' 8"	2.65	8' 6"	2.59

DIMENSIONS

DISTANCE FROM CENTER OF BOOM TO LOOP SUPPORT (R)

BAND	REFLECTOR		DRIVEN ELEMENT		DIRECTOR	
	FEET	METERS	FEET	METERS	FEET	METERS
20	12' 9 3/4"	3.90	12' 5 1/2"	3.80	12' 1 1/2"	3.70
15	8' 6 1/2"	2.60	8' 3 1/2"	2.53	8' 1"	2.46
10	6' 4"	1.93	6' 2"	1.88	6' 0"	1.83

Mounted on a 30-foot boom for 20 meters, this antenna provides a power gain of over ten times. A multiband Quad for 20, 15, and 10 meters may be built on the boom, using these dimensions. Alternatively, the boom may be shortened to 22 feet for a 15-meter Quad, or to 15 feet for a 10-meter version. For 3-band operation, the driven loops are connected in parallel at the feedpoint (F-F) and fed with a 1-to-1 balun and 50-ohm coaxial line. Additional feed information is given in the Quad handbook, discussed in the text.

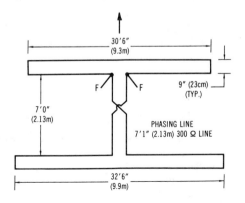

Figure 23

TOP VIEW OF TWO ELEMENT "ZL-SPECIAL" PHASED ARRAY

The two-element phased array provides about 4 dB power gain over a dipole with a F/B ratio of nearly 30 dB. The cross-connected 300-ohm feedline provides a 135-degree phase difference between the elements. Since the line is transposed, the actual electrical length is 45 degrees. Dimensions shown are for 20-meter operation with element diameters of ½ inch. Feedpoint (F-F) resistance is about 100 ohms. A balun may be used to match the antenna to a 50-ohm coaxial line.

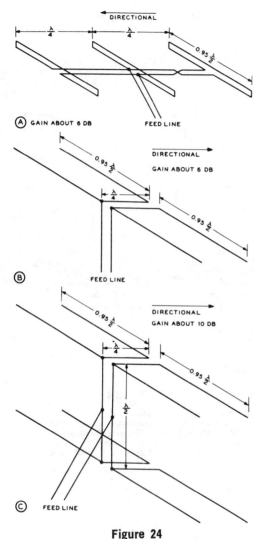

Figure 24

UNIDIRECTIONAL ALL-DRIVEN ARRAYS

A unidirectional all-driven end-fire array is shown at A. B shows an array with two half waves in phase with driven reflectors. A lazy-H array with driven reflectors is shown at C. Note that the directivity is through the elements with the greatest total feedline length in arrays such as shown at B and C.

Unidirectional Stacked Broadside Arrays Three practical types of unidirectional stacked broadside arrays are shown in figure 25. The first type, shown at figure 25A, is the simple lazy-H type of antenna with parasitic reflectors for each element. In figure 25B is shown a more complex array with six half waves and six reflectors which will give a very worthwhile amount of gain.

In both of the antenna arrays shown the spacing between the driven elements and the reflectors has been shown as one-quarter wavelength. This has been done to eliminate the requirement for tuning of the reflector, as a result of the fact that a half-wave element spaced exactly one-quarter wave from a driven element will make a unidirectional array when both elements are the same length. Using this procedure will give a gain of 3 dB with the reflectors over the gain without the reflectors, with only a moderate decrease in the radiation resistance of the driven element. Actually, the radiation resistance of a half-wave dipole goes down from 73 ohms to 60 ohms when an identical half-wave element is placed one-quarter wave behind it.

A very slight increase in gain for the entire array (about 1 dB) may be obtained at the expense of lowered radiation resistance, the necessity for tuning the reflectors, and decreased bandwidth by placing the reflectors 0.15 wavelength behind the driven elements and making them somewhat longer than the driven elements. The radiation resistance of each element will drop approximately to one-half the value obtained with untuned half-wave reflectors spaced one-quarter wave behind the driven elements.

Antenna arrays of the type shown in figure 25 require the use of some sort of lattice work for the supporting structure since the arrays occupy appreciable distance in space in all three planes.

Feed Methods The requirements for the feed systems for antenna arrays of the type shown in figure 25 are less critical than those for the close-spaced parasitic arrays shown in the previous section. This is a natural result of the fact that a larger number of the radiating elements are directly fed with energy, and of the fact that the effective radiation resistance of each of the driven elements of the array is much higher than the feedpoint resistance of a parasitic array. As a consequence of this fact, arrays of the type shown in figure 25 can be expected to cover a somewhat greater fre-

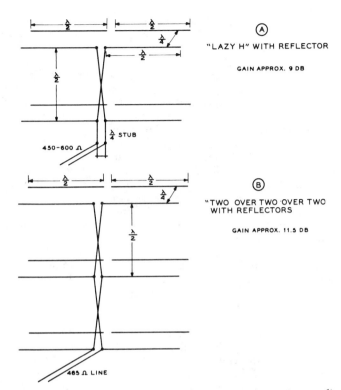

Figure 25

BROADSIDE ARRAYS WITH PARASITIC REFLECTORS

The apparent gain of the arrays illustrated will be greater than the values given due to concentration of the radiated signal at the lower elevation angles.

quency band for a specified value of standing-wave ratio than the parasitic type of array.

In most cases a simple open-wire line may be coupled to the feedpoint of the array without any matching system. The standing-wave ratio with such a system of feed will often be less than 2 to 1. However, if a more accurate match between the antenna transmission line and the array is desired a conventional quarter-wave stub, or a quarter-wave matching transformer of appropriate impedance, may be used to obtain a low standing-wave ratio.

29-12 Tuning the Parasitic Array

Although satisfactory results may be obtained in most cases by precutting the antenna elements to the dimensions given earlier in this chapter, the occasion might arise when it is desired to retune the parasitic beam, or to check on the operation of the

antenna. The following information applies to the Yagi antenna, but the same general process applies to any parasitic array, such as the Quad.

The process of tuning an array may satisfactorily be divided into two more or less distinct steps: the actual tuning of the array for best front-to-back ratio or for maximum forward gain, and the adjustment to obtain the best possible impedance match between the antenna transmission line and the feedpoint of the array.

Tuning the Array   The actual tuning of the array for best front-to-back ratio or maximum forward gain may best be accomplished with the aid of a low-power transmitter feeding a dipole antenna (polarized the same as the array being tuned) at least four or five wavelengths away from the antenna being tuned and located at the same elevation as that of the antenna under test. A calibrated field-strength meter of the remote-indicating type is then coupled to the feedpoint of the antenna array being tuned. The transmissions from the portable transmitter should be made as short as possible and the call sign of the station making the test should be transmitted at least every ten minutes.

One satisfactory method of tuning the array proper, assuming that it is a system with several parasitic elements, is to set the directors to the dimensions given in Table 1 and then to adjust the reflector for maximum forward signal. Then the first director should be varied in length until maximum forward signal is obtained, and so on if additional directors are used. Then the array may be reversed in direction and the reflector adjusted for best front-to-back ratio. Subsequent small adjustments may then be made in both the directors and the reflector for best forward signal with a reasonable ratio of front-to-back signal. The adjustments in the directors and the reflector will be found to be interdependent to a certain degree, but if small adjustments are made after the preliminary tuning process a satisfactory set of adjustments for maximum performance will be obtained. It is usually best to make the end section of the elements smaller in diameter so that they will slip inside the larger tubing sections. The smaller sliding sections may be clamped inside the larger main sections.

Matching to the Antenna Transmission Line The problem of matching the impedance of the antenna transmission line to the array is much simplified if the process of tuning the array is made a substantially separate process as just described. *After* the tuning operation is complete, the resonant frequency of the driven element of the antenna should be checked, directly as the center of the driven element if practical, with a grid-dip meter. It is important that the resonant frequency of the antenna be at the *center* of the frequency band to be covered. If the resonant frequency is found to be much different from the desired frequency, the length of the driven element of the array should be altered until this condition exists. A relatively small change in the length of the driven element will have only a second-order effect on the tuning of the parasitic elements of the array. Hence, a moderate change in the length of the driven element may be

made without repeating the tuning process for the parasitic elements.

When the resonant frequency of the antenna system is correct, the antenna transmission line, with impedance-matching device or network between the line and antenna feedpoint, is then attached to the array and coupled to a low-power exciter unit or transmitter. Then, preferably, a standing-wave meter is connected in series with the antenna transmission line at a point relatively much closer to the transmitter than to the antenna.

If the standing-wave ratio is below 1.5 to 1 it is satisfactory to leave the installation as it is. If the ratio is greater than this range it will be best when twin line or coaxial line is being used, and advisable with open-wire line, to attempt to decrease the SWR.

It must be remembered that no adjustments made at the *transmitter* end of the transmission line will alter the SWR on the line. All adjustments to better the SWR must be made at the *antenna* end of the line and to the device which performs the impedance transformation necessary to match the characteristic impedance of the antenna to that of the transmission line.

Before any adjustments to the matching system are made, the resonant frequency of the driven element must be ascertained, as explained previously. If all adjustments to correct impedance mismatch are made at this frequency, the problem of reactance termination of the transmission line is eliminated, greatly simplifying the problem. The following steps should be taken to adjust the impedance transformation:

1. The output impedance of the matching device should be measured. An Antennascope and a grid-dip oscillator are required for this step. The Antennascope is connected to the output terminals of the matching device. If the driven element is a folded dipole, the Antennascope connects directly to the split-section of the dipole. If a gamma match or T-match is used, the Antennascope connects to the transmission-line end of the device. If a Q-section is used, the Antennascope connects to the bottom end of the section. The grid-dip oscillator is cou-

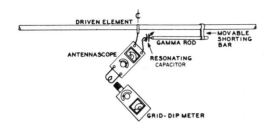

Figure 26

ADJUSTMENT OF GAMMA MATCH BY USE OF ANTENNASCOPE AND GRID-DIP METER

pled to the input terminals of the Antennascope as shown in figure 26.

2. The grid-dip oscillator is tuned to the resonant frequency of the antenna, which has been determined previously, and the Antennascope control is turned for a null reading on the meter of the Antennascope. The impedance presented to the Antennascope by the matching device may be read directly on the calibrated dial of the Antennascope.

3. Adjustments should be made to the matching device to present the desired impedance transformation to the Antennascope. If a folded dipole is used as the driven element, the transformation ratio of the dipole must be varied as explained previously in this Handbook to provide a more exact match. If a T-match or gamma match system is used, the length of the matching rod may be changed to effect a proper match. If the Antennascope ohmic reading is *lower* than the desired reading, the length of the matching rod should be *increased*. If the Antennascope reading is *higher* than the desired reading, the length of the matching rod should be *decreased*. After each change in length of the matching rod, the series capacitor in the matching system should be re-resonated for best null on the meter of the Antennascope.

(See Chapter 31 for details of the instruments.)

Raising and Lowering the Array A practical problem always present when tuning up and matching an array is the physical location of the structure. If the array is atop the mast it is inaccessible for adjustment, and if it is located on stepladders where it can be adjusted easily it cannot be rotated. One encouraging factor in this situation is the fact that experience has shown that if the array is placed 8 or 10 feet above ground on some stepladders for the preliminary tuning process, the raising of the system to its full height will not produce a serious change in the adjustments. So it is usually possible to make preliminary adjustments with the system located slightly greater than head height above ground, and then to raise the antenna to a position where it may be rotated for final adjustments. If the position of the matching device as determined near the ground is marked so that the adjustments will not be lost, the array may be raised to rotatable height and the fastening clamps left loose enough so that the elements may be slid in by means of a long bamboo pole. After a series of trials a satisfactory set of adjustments can be obtained.

The matching process does not require rotation, but it does require that the antenna proper be located at as nearly its normal operating position as possible. However, on a particular installation the standing-wave ratio on the transmission line near the transmitter may be checked with the array in the air, and then the array may be lowered to ascertain whether or not the SWR has changed. If it has not, and in most cases if the feeder line is strung out back and forth well above the ground as the antenna is lowered, they will not change, the last adjustment may be determined, the standing-wave ratio again checked, and the antenna re-installed in its final location.

29-13   Indication of Direction

The most satisfactory method for indicating the direction of transmission of a rotatable array is that which uses *Selsyns* or *Synchros* for the transmission of the data from the rotating structure to the indicating

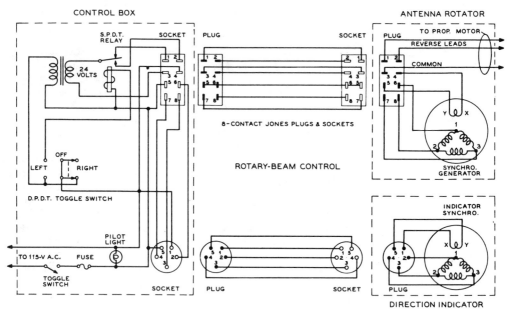

Figure 27

SCHEMATIC OF A COMPLETE ANTENNA CONTROL SYSTEM

pointer at the operating position. A number of Synchros and Selsyns of various types are available. A heavy-duty commercial Selsyn indicating system is shown in figure 27.

The majority of TV rotators and heavy-duty rotators designed for amateur service have built-in direction-indicating systems. In most instances, this is a modified form of indicating potentiometer and a milliammeter connected so that the rotation of the antenna is represented by the current flowing through the meter.

CHAPTER THIRTY

VHF and UHF Antennas

The *very-high-frequency* or *vhf* domain is defined as that range falling between 30 and 300 MHz. The *ultrahigh-frequency* or *uhf range* is defined as falling between 300 and 3000 MHz. This chapter will be devoted to the design and construction of antenna systems for operation on the amateur 50-, 144-, 220-, and 420-MHz bands. Although the basic principles of antenna operation are the same for all frequencies, the shorter physical length of a wave in this frequency range and the differing modes of signal propagation make it possible and expedient to use antenna systems different in design from those used in the range from 3 to 30 MHz.

30-1 Antenna Requirements

Any type of antenna design usable on the lower frequencies *may* be used in the vhf and uhf bands. In fact, simple nondirective half-wave or quarter-wave vertical antennas are very popular for general transmission and reception from all directions, especially for short-range repeater work. But for serious vhf or uhf work the use of some sort of directional antenna array is a necessity. In the first place, when the transmitter power is concentrated into a narrow beam the apparent transmitter power at the receiving station is increased many times. A "billboard" array or a Yagi beam having a gain of 16 dB will make a 25-watt transmitter sound like a kilowatt at the other station. Even a much simpler and smaller three- or four-element parasitic array having a gain of 7 to 10 dB will produce a marked improvement in the received signal at the other station.

However, as all vhf and uhf workers know, the most important contribution of a high-gain antenna array is in reception. If a remote station cannot be heard it obviously is impossible to make contact. The limiting

factor in vhf and uhf reception is in almost every case the noise generated within the receiver itself. Atmospheric and solar noise are quite low, and ignition interference can almost invariably be reduced to a satisfactory level through the use of an effective noise limiter. Even with a low noise front-end in the receiver, the noise contribution of the first tuned circuit will be relatively large. Hence it is desirable to use an antenna system which will deliver the greatest signal voltage to the first tuned circuit for a given field strength at the receiving location.

Since the field intensity being produced at the receiving location by a remote transmitting station may be assumed to be constant, the receiving antenna which intercepts the greatest amount of wave front (assuming that the polarization and directivity of the receiving antenna is proper) will be the antenna which gives the best received signal-to-noise ratio. An antenna which has two square wavelengths of effective area will pick up twice as much signal power as one which has one square wavelength area, assuming the same general type of antenna and that both are directed at the station being received. Many instances have been reported where a frequency band sounded completely dead with a simple ground-plane receiving antenna but when the receiver was switched to a three-element or larger array a considerable amount of activity from 80 to 160 miles distant was heard.

VHF Antenna Types The vhf directional antennas most used by serious experimenters fall into four characteristic groups: collinear, broadside, end-fire, and frequency-independent (figure 1). All of these, except the last, have been discussed in earlier chapters of this handbook. It is common vhf practice to combine antennas of one type into a large directional array of an-

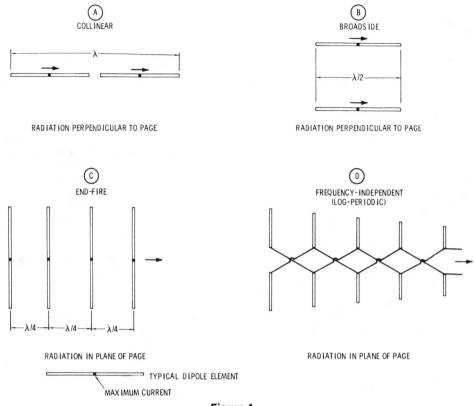

Figure 1

GENERAL TYPES OF VHF ANTENNA ARRAYS

A—Collinear array. Elements lie in same plane, end to end and are fed with equal, in-phase energy. Maximum radiation is at right angles to line of array.

B—Broadside array. Elements lie in same plane, parallel to one another and are fed with equal in-phase energy. Maximum radiation is perpendicular to axis of array and to plane containing the elements.

C—Endfire array. Elements lie in same plane, parallel to one another. Radiation coincides with direction of axis of array. Elements may be fed with progressive phase shift, or may be parasitically excited from one driven element.

D—Frequency-independent array. Elements lie in same plane, parallel to one another. Radiation coincides with direction of axis of array. Elements fed out-of-phase with progressive phase shift. Element lengths are function of the angle they subtend from the apex point of array and whose distance from the apex is such to provide wideband behavior.

tennas, exhibiting a high gain figure and very narrow beam width. Such assemblies are usually impractical on the lower frequencies because of the excessively large size of the antenna combination. However, at 50 MHz and higher, a high-gain antenna is quite small, and "back yard" arrays can be built for moonbounce communication, meteor reflection, or tropospheric scatter work that provide upward of 20 dB gain. Arrays for the higher bands of comparable power gain are, of course, much smaller.

This chapter covers some of the more popular antenna designs that have been proven in service and can be easily built, assembled and adjusted without the use of complicated test equipment.

The choice of antenna to use depends on the type of communication the experimenter is primarily concerned with and usually involves a trade-off between operational bandwidth and power gain. No "universal" vhf antenna exists that will satisfy the requirements of every vhf operator.

Because vhf beams are relatively small compared to hf beam antennas of equivalent gain, it is possible for the experimenter to easily build and evaluate various vhf antenna designs. The broadside, curtain style beams are generally simpler to get working than the Yagi, although the Yagi can be made with fewer elements for a given amount of power gain. Unless the Yagi is accurately adjusted, the broadside array may end up with more signal gain than the Yagi. Stacking Yagi antennas, in addition, can raise additional problems not always encountered in the broadside array. The expected 3-dB power gain expected for the addition of a second Yagi may not be realized unless the antennas are spaced far enough apart so that the apertures do not overlap. This may cause the minor lobes of the pattern to enlarge, which may lead to undesired signal pickup and degraded front-to-side ratio.

In spite of the design problems associated with the Yagi, many successful designs have been worked out and some of the better ones are described in this chapter. The log-periodic Yagi (LPY) beam overcomes some of the difficulties associated with the Yagi and a wideband, LPY array for 50 MHz is shown. Other practical beams for 220 and 420 MHz are shown in later sections.

Generally speaking, an omnidirectional antenna pattern with vertical polarization and moderate gain is desired by the f-m enthusiast who wishes to work into numerous repeaters at various distances and directions from his station. A rotary beam antenna in this instance would be a nuisance. The moonbounce enthusiast requires a high gain antenna, movable in both azimuth and elevation so that he can track the moon. The OSCAR experimenter perhaps requires a medium-gain antenna having a broad lobe that will allow the satellite to move a distance across the sky before it becomes necessary to realign the antenna. Selecting the proper vhf antenna is the first important step, then, in the order of priorities that faces the vhf operator.

VHF Antenna Placement   For hf DX work, the higher the antenna, the lower the angle of radiation will be and, presumably, the better the DX results. In the vhf region, height is a virtue, especially for extended, line-of-sight contacts. However, the antenna height must be balanced against the increase in transmission line loss, which can be quite high, especially in the upper vhf and lower uhf range. Large vhf antennas, too, are often damaged by winter weather, especially when mounted high and in the clear. For specialized communication, such as moonbounce, antenna height is of little importance as long as the antenna can "see" the moon. Satellite work with OSCAR does not require great antenna height either, as long as the satellite path is in the clear, with regard to the transmitting and receiving antennas.

VHF Antenna Polarization   Vhf mobile operation generally implies vertical polarization and base stations in general contact with mobiles use vertical polarization exclusively. Long range vhf operation, however, seems to show no preference for one form of polarization over another. Manmade noise seems to be vertically polarized and many amateurs avoid vertical polarization if they live in an area having a high level of "r-f smog." Generally speaking, horizontal polarization seems to hold a slight edge over vertical polarization for long distance vhf communication and construction problems seem to be less with horizontal elements when a large antenna array is assembled. Cross polarization (horizontal to vertical, and vice versa) entails a circuit loss of about 20 dB, so it is wise to check what type of polarization is in use in your area before the construction of a large antenna array is undertaken.

VHF Transmission Lines   Both parallel-conductor air line and coaxial line having a solid dielectric are commonly used in the vhf region. In cases where line loss is a limiting factor, air-insulated coaxial line may be used. It is wise to use the very minimum amount of transmission line possible since line loss mounts rapidly at frequencies above 50 MHz (figure 2). Generally speaking, the popular 50-ohm coaxial line (RG-8A/U and RG-58A/U) are commonly used for short cable runs on the vhf bands up to 450 MHz. Longer runs require the larger, expensive RG-17/U cable, or open-wire line. Foam-dielectric coaxial line may be used for less

CABLE	Z_o	V_p	ATTENUATION IN dB/100 FT.				POWER RATING (WATTS)			
			50 MHz	144 MHz	220 MHz	432 MHz	50 MHz	144 MHz	220 MHz	432 MHz
RG-58C/U	52.5	.66	3.0	6.0	8.0	15.0	350	175	125	90
RG-58(F)	50	.79	2.2	4.1	5.0	7.1	450	230	160	120
RG-59B/U	73	.66	2.3	4.2	5.5	8.0	500	250	180	125
RG-59(F)	75	.79	2.0	3.4	4.6	6.1	650	320	230	160
RG-8A/U RG-213/U	52	.66	1.5	2.5	3.5	5.0	1500	800	650	400
RG-8(F)	50	.80	1.2	2.2	2.7	3.9	1950	1100	850	520
RG-11A/U	75	.66	1.55	2.8	3.7	5.5	1500	800	650	400
RG-17A/U RG-218/U	52	.66	0.5	1.0	1.3	2.3	4500	2300	1800	1200

Figure 2

COAXIAL CABLES FOR VHF USE

The popular RG-8A/U and RG-58C/U are recommended for general purpose vhf use. Foam-dielectric cable, having somewhat lower loss than the solid-dielectric cable, is indicated by (F). The impedance (Z_o) of these cables is about 50 ohms. The velocity of propagation (V_p) of the wave along the cable is a function of the dielectric material. Where line loss is an important factor, air-insulated rigid coaxial line may be used.

loss, as compared to solid-dielectric line, and many amateurs prefer this newer type of cable.

Since most vhf f-m gear is designed for use with coaxial cable, the use of low-loss, open line is impractical. Some amateurs, however, use an antenna tuner or balun and convert their station equipment to use either TV-style 300-ohm twin lead, or open-wire line in order to reduce line losses (figure 3). While the initial cost of the TV line is low and the overall efficiency of the line is high, the line loss increases rapidly when the line is wet or covered with ice and snow. In addition, the line must be installed well clear of metallic objects and run in straight lengths with gradual turns. Heavy-duty transmitting line is better than the smaller receiving line, but either type must be carefully installed to hold signal loss to a minimum.

Air-insulated parallel-conductor line must also be carefully installed or its low loss characteristic may be lost. The line should be taut, with no sharp turns. The line must be symmetrical with respect to ground and nearby metallic objects which might unbalance it. Even a slight electrical unbalance can cause the line to radiate energy and to decrease the power delivered to the load.

As a result of the installation problems, most vhf amateurs settle for flexible coaxial line in spite of its higher loss. Properly pre-pared, a coaxial line is waterproof and may be run anywhere, since the r-f energy is mainly contained within the cable. In addition, vhf SWR meters are available, or easily made, to be used with standard 50- or 75-ohm coaxial line.

While the coaxial line is waterproof, the ends of the line are not, and water can easily get inside the exposed end of the line if precautions are not taken. To protect the line, it is necessary to coat the coaxial fittings with a waterproofing sealant, such as *General Electric RTV-102*. As a substitute, white bathtub calking compound may be used. The coaxial plugs should be coated on the interior with *Dow Corning DC-4* paste, or equivalent, to prevent moisture from entering the plug.

VHF Coaxial Hardware Most amateur equipment is fitted with the so-called *uhf-style* coaxial fittings, which are a relic of the "forties." The plug is known as the PL-259 and the receptacle is the SO-239. These items are not waterproof and are not suited for use above 150 MHz, since they introduce an appreciable SWR "bump" in the transmission line. The newer and more efficient *type-N*, or type *BNC* coaxial connectors are now used in up-to-date vhf equipment (figure 4). These families of vhf hardware are considered to be

DESCRIPTION AND MAKE	V_p	ATTENUATION dB/100 FEET		
		50 MHz	100 MHz	400 MHz
GENERAL PURPOSE 7×28 WIRE AM-214-056 C-4506 CL-01004	0.85	0.72	1.3	2.6
HEAVY DUTY 7×26 WIRE TUBULAR AM-214-076 C-4523	0.82	0.7	1.1	2.3
GENERAL PURPOSE HOLLOW OR FOAM CORE 7 × 28 WIRE AM-214-022 C-4527	0.84	0.55	0.8	1.8
SHIELDED FOAM DIELECTRIC C-4535 CL-05720	0.76	2.0	2.8	5.9
UNSHIELDED FOAM DIELECTRIC C-4532 CL-05790	0.80	—	1.4	—
C=Consolidated Wire CL=Columbia Wire		AM=Amphenol		

Figure 3

"RIBBON" LINE FOR VHF USE

Attenuation varies somewhat between different cable manufacturers. Types having foam dielectric have lower loss than equivalent types having solid dielectric. Amphenol also makes a heavy-duty 75-ohm "ribbon" line for transmitting service (214-023).

"constant impedance" and do not appreciably affect the SWR on the transmission line at least up to 500 MHz.

Generally speaking, RG-8A/U line and type-N fittings are recommended for high power operation and or long cable runs from equipment to antenna. The smaller, light duty RG-58/U cable and associated BNC hardware are suggested for low power and short cable runs.

Antenna Changeover It is recommended that the same antenna be used for transmitting and receiving in the vhf and uhf range. An ever-present problem in this connection, however, is the antenna changeover relay. Reflections at the antenna

changeover relay become of increasing importance as the frequency of transmission is increased. When coaxial cable is used as the antenna transmission line, satisfactory coaxial antenna changeover relays with low reflection can be used.

On the 220- and 420-MHz amateur bands, the size of the antenna array becomes quite small, and it is practical to mount two identical antennas side by side. One of these antennas is used for the transmitter, and the other antenna for the receiver. Separate transmission lines are used, and the antenna relay may be eliminated.

TYPE-N CONNECTORS		
DESCRIPTION	MILITARY TYPE	AMPHENOL TYPE
PLUG	UG-21/U	3900
SPLICE	UG-29/U	15000
RECEPTACLE	UG-58/U	82-97
TYPE BNC CONNECTORS		
PLUG	UG-88/U	31-002
SPLICE	UG-914/U	31-219
RECEPTACLE	UG-625/U	5575
ADAPTER TO UHF	UG-273/U	31-028

Figure 4

VHF COAXIAL HARDWARE

Type-N and type BNC coaxial hardware are used on up-to-date vhf and uhf equipment. These units are constant-impedance design and do not appreciably affect the SWR on the transmission line at least up to 500 MHz. Many type numbers exist, and these listed are representative. Adapters are available to convert from one system of hardware to another.

Effect of Feed System on Radiation Angle It is important that line radiation be held to a minimum or the radiation pattern of a high gain vhf antenna may be adversely affected. Military-style cables having the "RG" nomenclature exhibit radiation loss through the outer braided shield of about −35 dB below the power in the line. Less expensive cables having a looser outer braid, or having less wires in the braid, often show a radiation loss in the neighborhood of −20 dB. Line radiation not only robs the antenna of some power, but can distort the radiation pattern of the antenna and dilute the front-to-back ratio of an otherwise good antenna pattern. In addition, the radiation angle of the main lobe of the antenna may be bent upward by the effect of line radiation, and if the transmission line passes through a noisy area, line pickup may mask the weak-signal ability of the receiving system.

Thus, the best grade of coaxial line should be used to minimize line radiation through the braid, and the line itself should be led away from the antenna at right angles to the radiating elements of the antenna. Lastly, a balun or other balancing device should be placed between a balanced antenna and a coaxial feedline to keep antenna current from flowing on the outside of the shield of the line.

Element Diameter   In the vhf region, aluminum tubing is commonly used for antenna construction since element length is short and the material is inexpensive and readily available. The diameter of

the various elements in a vhf array must be sufficient so that they are self-supporting even in severe weather and so that their surface conductivity is not low enough to degrade the performance appreciably. If, on the other hand, the diameter of the elements is too large, the circuit Q of the element is lowered and its effectiveness, particularly as a parasitic in a Yagi array, is decreased. Most vhf antenna designs are based on ⅜" (0.9 cm) or 1" (2.54 cm) diameter tubing for 50-MHz work, ¼" (0.6 cm) or ³⁄₁₆" (0.5 cm) diameter tubing for 144-MHz work, and ⅛" (0.3 cm) diameter rod for 220- or 420-MHz work. If smaller diameter elements are used, the length of the elements must be increased accordingly to maintain resonance. The relationship between element diameter and length is difficult to ascertain, beyond actual measurements made on an antenna range, and variations in element diameter from a given design should be approached

Table 1. Wavelength and Antenna Dimensions (Rounded)

F (MHz)	λ/2 (SPACE)		λ/4 (SPACE)		λ/2 DIPOLE		0.2λ (SPACE)	
	Inches	cm	Inches	cm	Inches	cm	Inches	cm
50.5	116.5	295.9	58.3	148.0	110.5	280.7	46.8	118.8
51.5	114.5	290.8	57.3	145.5	108.5	275.6	45.8	116.3
52.5	112.5	285.7	56.3	143.0	106.5	270.5	44.5	113.0
53.5	110.5	280.6	55.3	140.5	104.5	265.4	44.0	111.7
144	41.0	104.1	20.5	52.0	38.9	98.8	16.4	41.7
145	40.8	103.6	20.4	51.7	38.6	98.0	16.3	41.4
146	40.5	102.8	20.2	51.3	38.3	97.2	16.2	41.1
147	40.2	102.1	20.1	51.0	38.0	96.6	16.1	40.8
148	40.0	101.6	20.0	50.7	37.8	96.0	16.0	40.6
220	26.8	68.0	13.4	34.0	25.5	64.7	10.7	27.2
221	26.7	67.8	13.4	34.0	25.3	64.3	10.7	27.2
222	26.6	67.5	13.3	33.7	25.2	64.0	10.6	26.9
223	26.5	67.3	13.3	33.7	25.1	63.7	10.6	26.9
224	26.4	67.0	13.2	33.5	25.0	63.5	10.5	26.6
420	14.0	35.6	7.0	17.8	13.3	33.8	5.6	14.2
430	13.7	34.8	6.8	17.4	13.0	33.0	5.5	13.9
440	13.4	34.0	6.7	17.0	12.7	32.2	5.4	13.7
450	13.1	33.2	6.5	16.6	12.4	31.4	5.3	13.5

CHANGE Per MHz	FORMULAS—
50 MHz = 2.0"	$\lambda/2 \text{ (Space)} = \dfrac{5905}{f \text{ (MHz)}}$
144 MHz = 0.3"	$\lambda/2 \text{ (Dipole)} = \dfrac{5600}{f \text{ (MHz)}}$

—For parasitic director, multiply dipole length by 0.95.
—For parasitic reflector, multiply dipole length by 1.05.
—For additional directors, multiply dipole length by 0.94.
—1" tubing for 50 MHz, ¼" tubing for 144 MHz, ⅛" tubing for 220 and 420 MHz arrays.

with caution. As an example, reducing element diameter by a factor of four at 50 MHz requires an increase in element length by about 8 percent to maintain resonance. Representative lengths for a dipole element for the vhf bands is given in Table 1.

Since the length-to-diameter ratio of antennas above 100 MHz or so is somewhat smaller than that of high-frequency arrays and because the arrays are physically smaller dimensions are generally given in inches, based on the following formula:

$$\text{Dipole length (inches)} = \frac{5600}{f_{\mathrm{MHz}}}$$

The metric equivalent is:

$$\text{Dipole length (cm)} = \frac{14,224}{f_{\mathrm{MHz}}}$$

The dimensions for small (3, 4, or 5 element) Yagis may be derived from Table 1, based on elements of the listed diameters and using a nominal spacing of 0.2 wavelength. If other element spacings are to be used, the reflector and director elements will have to be readjusted accordingly. Closer reflector driven-element spacing will call for a slightly shorter reflector for optimum gain. Closer director driven-element spacing will call for a slightly longer director for optimum gain. Generally speaking, anything closer to 0.2-wavelength spacing in Yagi arrays tends to reduce the bandwidth, reduce the driven-element impedance, and increase the front-to-back ratio.

The parasitic element should not be painted as this tends to detune the element. A light coat of *Krylon* plastic spray may be used to protect the element against weather.

30-2   Base Station Antennas

Vhf mobile communication makes use of vertical polarization and most vertical antennas are omnidirectional in the azimuth plane unless the pattern is modified by the addition of parasitic elements. In the great majority of cases, the desired base station coverage is omnidirectional and simple vertical dipoles form the basic antenna element.

Various vertical antennas are shown in figure 5. Antenna A is known as the *sleeve antenna*, the lower half of the radiator being a large piece of pipe up through which the concentric feedline is run. At B is shown the *ground-plane* vertical, and at C a modification of this latter antenna. In many cases, the antennas of illustrations A and C have a set of quarter-wavelength radials placed beneath the array to decouple it from the transmission line.

The radiation resistance of the ground-plane vertical is approximately 30 ohms, which is not a standard impedance for coaxial line. To obtain a good match, the first quarter wavelength of feeder may be of 50 ohms impedance, and the remainder of the line of 75 ohms impedance. Thus, the first quarter-wavelength section of line is used as a matching transformer, and a good match is obtained.

In actual practice the antenna would consist of a quarter-wavelength rod, mounted by means of insulators atop a pole or pipe mast. Elaborate insulation is not required, as the voltage at the lower end of the quarter-wavelength radiator is very low. Self-supporting rods 0.25 wavelength long are extended out, as shown in the illustration, and

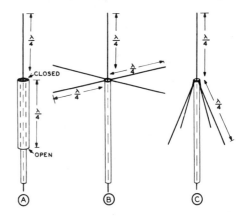

Figure 5

THREE VERTICALLY POLARIZED LOW-ANGLE RADIATORS

Shown at A is the "sleeve" or "hypodermic" type of radiator. At B is shown the ground-plane vertical, and C shows a modification of this antenna system which increases the feed-point impedance to a value such that the system may be fed directly from a coaxial line with no standing waves on the feedline.

connected together. Since the point of connection is effectively at ground potential, no insulation is required; the horizontal rods may be bolted directly to the supporting pole or mast, even if of metal. The coaxial line should be of the low-loss type especially designed for vhf use. The shield connects to the junction of the radials, and the inner conductor to the bottom end of the vertical radiator. An antenna of this type is moderately simple to construct and will give a good account of itself when fed at the lower end of the radiator directly by the 50-ohm RG-8/U coaxial cable. Theoretically the standing-wave ratio will be approximately 1.5-to-1 but in practice this moderate SWR produces no deleterious effects.

The modification shown in figure 5C permits matching to a standard 50- or 75-ohm flexible coaxial cable without a linear transformer. If the lower rods hug the line and supporting mast rather closely, the feedpoint impedance is about 75 ohms. If they are bent out to form an angle of about 30° with the support pipe the impedance is about 50 ohms.

The number of radial legs used in a vhf ground-plane antenna of either type has an important effect on the feed-point impedance and on the radiation characteristics of the antenna system. Experiment has shown that three radials is the minimum number that should be used, and that increasing the number of radials above six adds substantially nothing to the effectiveness of the an-

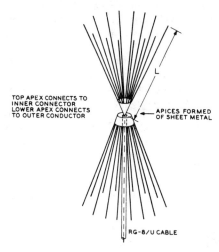

Figure 6

THE DOUBLE SKELETON CONE ANTENNA

A skeleton cone has been substituted for the single element radiator of figure 5C. This greatly increases the bandwidth. If at least 10 elements are used for each skeleton cone and the angle of revolution and element length are optimized, a low SWR can be obtained over a frequency range of at least two octaves. To obtain this order of bandwidth, element length L should be approximately 0.2 wavelength at the lower frequency end of the band and the angle of revolution optimized for the lowest maximum SWR within the frequency range to be covered. A greater improvement in the impedance-frequency characteristic can be achieved by adding elements than by increasing the diameter of the elements. With only 3 elements per "cone" and a much smaller angle of revolution a low SWR can be obtained over a frequency range of approximately 1.3 to 1.0 when the element lengths are optimized.

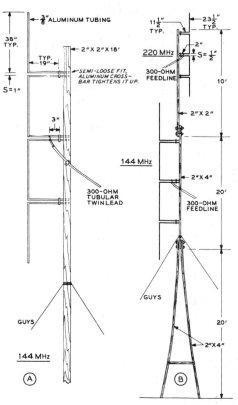

Figure 7

NONDIRECTIONAL ARRAYS FOR 144 AND 220 MHz

On right is shown a two-band installation. For portable use, the whole system may easily be dissembled and carried on a luggage rack atop a car.

tenna and has no effect on the feedpoint impedance.

Double Skeleton Cone Antenna The bandwidth of the antenna of figure 5C can be increased considerably by substituting several space-tapered rods for the single radiating element, so that the "radiator" and skirt are similar. If a sufficient number of rods are used in the skeleton cones and the angle of revolution is optimized for the particular type of feedline used, this antenna exhibits a very low SWR over a 2-to-1 frequency range. Such an arrangement is illustrated schematically in figure 6.

A Nondirectional Vertical Array Half-wave elements may be stacked in the vertical plane to provide a nondirectional pattern with good horizontal gain. An array made up of four half-wave vertical elements is shown in figure 7A. This antenna provides a circular pattern with a gain of about 4.5 dB over a vertical dipole. It may be fed with 300-ohm TV-type line. The feed line should be conducted in such a way that the vertical portion of the line is at least one-half wavelength away from the vertical antenna elements. A suitable mechanical assembly is shown in figure 7B for the 144- and 220-MHz amateur bands.

A Stacked Sleeve Antenna for 144 MHz The sleeve antenna makes a good omnidirectional array for 144 MHz in areas where vertical polarization is used. A double stack, such as illustrated in figure 8, will provide low-angle radiation and a power gain of about 3 decibels. The array is designed to be fed with a 50-ohm coaxial transmission line.

The antenna is built on an eight-foot length of aluminum TV mast section, 1⅛″ diameter. A quarter-wavelength whip extends from the top of the assembly, and two sleeves are mounted to the mast section below the whip. Both sleeves are electrically connected to the mast at their tops, and the bottom sleeve is shock-excited by the top antenna array, which functions as a simple dipole. Directly below the sleeves

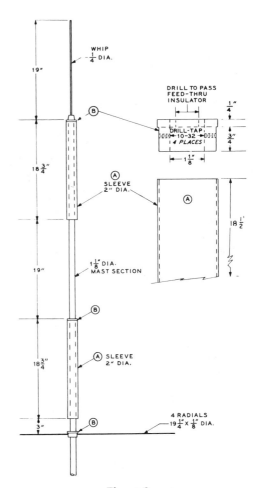

Figure 8

SLEEVE ANTENNA FOR 144 MHz

Stacked dipoles provide nondirectional coverage with low-angle radiation. The top whip is fed by a coaxial line passed up through the mast section and is insulated from remainder of the antenna structure. Lower dipole is composed of mast section and matching skirt which is grounded to the mast at the top. Bottoms of both skirts are free. Radials beneath bottom section impede flow of antenna current on outside of coaxial line.

are mounted four quarter-wave horizontal radials which decouple the stacked antenna from the outer shield of the coaxial transmission line.

Antenna construction is straightforward and simple. The top of the mast is closed with an aluminum plug (B) having a ceramic feedthrough insulator mounted in it. The vertical whip attaches to the insulator, as does the center conductor of the coaxial feedline. The outer shield of the line is grounded to the mast section at the insulator. The outer sleeve (A) is attached to the mast section by means of machine screws tapped into the aluminum plug.

The lower sleeve is attached to the mast in a similar manner, as shown in the drawing. The radials, made of aluminum clothesline wire are threaded and screwed to an aluminum mounting cylinder (similar to B) which encircles the mast.

Three aluminum fittings (B) are required: one for the top sleeve, one for the lower sleeve, and one for the radials. The top fitting is shown in figure 8. The center one is similar, except that it is drilled to pass the mast section. The fitting for the radials is similar to the center one, except that the 1/4-inch lip at the top is omitted.

The length of the fitting is such so that the inner resonant portion of the sleeve is slightly shorter than the outer section. The outer section acts as a portion of the antenna and the inner section acts as a decoupling transformer. The resonant lengths are different for each case, and the length of the fitting makes up the electrical difference.

The sleeves are free at the lower ends, with no connection or support at this point. Care must be taken to make the assembly waterproof, as an accumulation of moisture in the sleeve may detune it. Plugs at the bottom of the sleeves, therefore, are not advised.

The 50-ohm coaxial transmission line runs up the inside of the mast to the top fitting where the outer shield is grounded to the structure by means of a washer placed beneath the feedthrough insulator. The shield is soldered to a lug of the washer, which may be cut from thin brass or copper shim stock.

When fed with a 50-ohm transmission line, the measured SWR across the 144-MHz band is less than 2/1, and better than 1.5/1 at the center frequency of 146 MHz.

The J-Pole Antenna   A half-wavelength vertical makes a good general purpose base station antenna as it requires no radials for proper operation and provides a slight power gain over a ground-plane antenna.

Shown in figure 9A is a J-Pole antenna for 50 MHz. It comprises a vertical dipole fed at the base with a quarter-wave matching transformer and a coaxial line. The assembly is quite rugged and can be mounted atop an existing tower, or can be formed from an existing whip antenna.

The 144MHz J-Pole antenna is shown in figure 9B. The antenna is basically the same as the 6-meter version, except that a gamma match system is used to match the coaxial line to the quarter-wave transformer. The tap point of the gamma and the setting of the series capacitor are adjusted for lowest SWR on the coaxial transmission line.

The Discone Antenna   The *Discone* antenna is a vertically polarized omnidirectional radiator which has very broadband characteristics and permits a simple, rugged structure. This antenna presents a substantially uniform feedpoint impedance, suitable for direct connection of a coaxial line, over a range of several octaves. Also, the vertical pattern is suitable for groundwave work over several octaves, the gain varying only slightly over a very wide frequency range.

A Discone antenna suitable for multiband amateur work in the uhf/vhf range is shown schematically in figure 10. The distance (D) should be made approximately equal to a free-space quarter wavelength at the lowest operating frequency. The antenna then will perform well over a frequency range of at least 8 to 1. At certain frequencies within this range the vertical pattern will tend to rise slightly, causing a slight reduction in gain at zero angular elevation, but the reduction is very slight.

Below the frequency at which the slant height of the conical skirt is equal to a free-space quarter wavelength the standing-wave ratio starts to climb, and below a frequency approximately 20 percent lower than this the standing-wave ratio climbs very rapidly. This is termed the *cutoff frequency* of the antenna. By making the slant height approximately equal to a free-space quarter wavelength at the lowest frequency employed (refer to figure 11), an SWR of less than 1.5 will be obtained throughout the operating range of the antenna.

The Discone antenna may be considered as a cross between an electromagnetic horn and an inverted ground-plane unipole an-

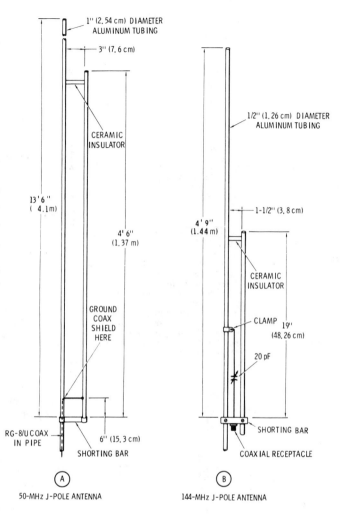

Figure 9

THE J-POLE ANTENNA FOR 50 AND 144 MHZ

The half wave vertical antenna is a popular omindirectional installation for the vhf bands. A—50 MHz J-pole antenna. The coaxial cable is brought up inside the aluminum pipe, which serves as antenna and mast. The outer shield of the line is grounded to the pipe 6 inches above the mounting clamp for the matching section. The inner conductor is tapped on the matching section as shown. B—144 MHz J-pole antenna. A coaxial receptacle is mounted on the shorting bar. The inner terminal is tapped on the vertical radiator through a 20-pF variable capacitor. Adjustment of the capacitor and the tap point (about 2″ above the bar) permit a very low value of SWR to be achieved on the transmission line. This is a simple version of the gamma match.

(A) 50-MHz J-POLE ANTENNA

(B) 144-MHz J-POLE ANTENNA

tenna. It looks to the feed line like a properly terminated high-pass filter. The top disc and the conical skirt may be fabricated either from sheet metal, screen (such as "hardware cloth"), or 12 or more "spine" radials. If screen is used, a supporting framework of rod or tubing will be necessary for mechanical strength except at the higher frequencies. If spines are used, they should be terminated on a stiff ring for mechanical strength, except at the higher frequencies.

The top disc is supported by means of three insulating pillars fastened to the skirt. Either polystyrene or low-loss ceramic is suitable for the purpose. The apex of the conical skirt is grounded to the supporting mast and to the outer conductor of the coaxial line. The line is run down through the supporting mast. An alternative arrangement, one suitable for certain mobile applications, is to fasten the base of the skirt directly to an effective ground plane such as the top of an automobile.

Horizontally Polarized Antennas   On occasion, horizontal polarization is desired in a base station. Shown in figure 12 are two simple, omnidirectional horizontally polarized antennas. A set of crossed dipoles, fed 90° out of phase is shown in illustration A. This *turnstile* antenna is the basic antenna element used in many f-m broadcast arrays. The antennas

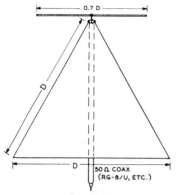

Figure 10

**THE DISCONE BROADBAND
RADIATOR**

This antenna system radiates a vertically polarized wave over a very wide frequency range.
The "disc" may be made of solid metal sheet,
a group of radials, or wire screen; the "cone"
may best be constructed by forming a sheet
of thin aluminum. A single antenna may be used
for operation on the 50-, 144-, and 220-MHz amateur bands. The dimension D is determined by
the lowest frequency to be employed, and is
given in figure 11.

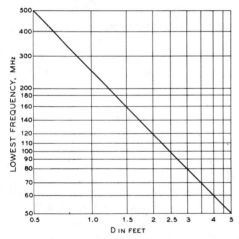

Figure 11

**DESIGN CHART FOR THE DISCONE
ANTENNA**

are displaced 90° and the vector sum of the
patterns is essentially omnidirectional.

A second antenna producing a uniform,
horizontally polarized pattern is shown in
illustration B. Three dipoles are curved to
form a circle and are excited in phase. A
"bazooka" balun is included in the system

to prevent antenna currents from flowing on
the outer surface of the coaxial conductor.

The *halo antenna* (figure 13) is a third
popular form of horizontally polarized radiator. Basically, the halo is a dipole element
formed into a circle and end-loaded by a
capacitor to establish resonance. Any conventional feed system may be used with this
antenna.

**The Vhf Rhombic
Antenna**   For vhf transmission and
reception in a fixed direction, a horizontal *rhombic*
permits 10 to 16 dB gain with a simpler construction than does a phased dipole array, and
has the further advantage of being useful
over a wide frequency range.

Except at the upper end of the vhf range
a rhombic array having a worthwhile gain is
too large to be rotated. However, in locations 75 to 150 miles from a large metropolitan area a rhombic array is ideally suited
for working into the city on extended (horizontally polarized) ground wave while at
the same time making an ideal antenna for
TV reception.

The useful frequency range of a vhf
rhombic array is about 2 to 1, or about plus
40% and minus 30% from the *design frequency*. This coverage is somewhat less than
that of a high-frequency rhombic used for
sky-wave communication. For ground-wave
transmission or reception the only effective
vertical angle is that of the horizon, and a
frequency range greater than 2 to 1 cannot
be covered with a rhombic array without an
excessive change in the vertical angle of
maximum radiation or response.

The dimensions of a vhf rhombic array are
determined from the design frequency
and figure 14, which shows the proper *tilt
angle* (see figure 15) for a given leg length.
The gain of a rhombic array increases with
leg length. There is not much point in constructing a vhf rhombic array with legs
shorter than about 4 wavelengths, and the
beam width begins to become excessively
sharp for leg lengths greater than about 8
wavelengths. A leg length of 6 wavelengths
is a good compromise between beam width
and gain.

The tilt angle (68°) given in figure 15 is
based on a wave angle of zero degrees. For leg
lengths of 4 wavelengths or longer it will

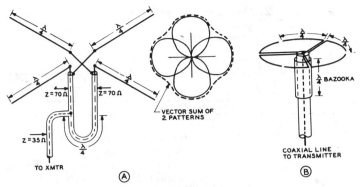

Figure 12

HORIZONTAL POLARIZED, OMNIDIRECTIONAL VHF ANTENNAS

A—Turnstile antenna is widely used in f-m broadcast service.
B—Modified turnstile using circular elements. A series of antennas of this type may be mounted in a stack on a single tower to provide power gain without sacrificing the omnidirectional pattern.

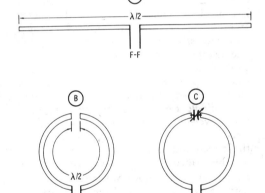

Figure 13

EVOLUTION OF THE HALO ANTENNA

A—Half wave dipole antenna fed at F-F
B—Dipole bent into circle
C—Short dipole bent into circle and end-loaded to establish resonance. Halo antenna is placed parallel to the earth to establish horizontal polarization and essentially omnidirectional pattern. Conventional feed system, such as a gamma match, may be used. Circuit-Q of a Halo is quite high and operational bandwidth is less than that of equivalent dipole.

be necessary to elongate the array a few percent (pulling in the sides slightly) if the horizon elevation exceeds about 3 degrees.

Table 2 gives dimensions for two dual purpose rhombic arrays. One covers the 6-meter amateur band and the "low" tele-

vision band. The other covers the 2-meter amateur band, the "high" television band, and the $1\frac{1}{4}$-meter amateur band. The gain is approximately 12 dB over a matched half-wave dipole and the beam width is about 6 degrees.

The recommended feedline is an open-wire line having a surge impedance between 450 and 600 ohms. With such a line the SWR will be less than 2 to 1. A line with two-inch spacing is suitable for frequencies below 100 MHz, but one-inch spacing is recommended for higher frequencies.

If the array is to be used only for reception, a suitable termination consists of two

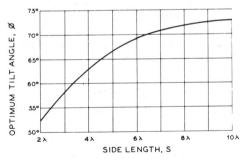

Figure 14

VHF RHOMBIC ANTENNA DESIGN CHART

The optimum tilt angle (see figure 15) for "zero-angle" radiation depends on the length of the sides.

Table 2. Rhombic Antenna Dimensions

DIMENSION	50 MHz AND LOW-BAND TV		144-200 MHz AND HIGH-BAND TV	
	Feet	Meters	Feet	Meters
SIDE (S)	90'0"	27.45	32'0"	9.76
LENGTH (L)	166'10"	50.88	59'4"	18.09
WIDTH (W)	67'4"	20.53	23'11"	7.29
TILT ANGLE = 68°				

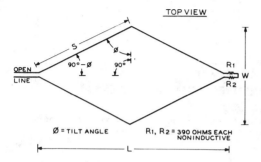

Figure 15

VHF RHOMBIC ANTENNA CONSTRUCTION

390-ohm carbon resistors in series. If 2-watt resistors are employed, this termination also is suitable for transmitter outputs of 10 watts or less. For higher powers, however, resistors having greater dissipation with negligible reactance in the upper vhf range are not readily available.

For powers up to several hundred watts a suitable termination consists of a "lossy" line consisting of stainless-steel wire (corresponding to No. 24 or 26 gauge) spaced 2 inches, which in turn is terminated by two 390-ohm 2-watt carbon resistors. The dissipative line should be at least 6 wavelengths long.

30-3 The Log-Periodic Antenna

Frequency-independent antennas, of which the *Log-periodic* array is an example, are structures that have the same performance at different frequencies by virtue of the fact that the array is self-scaling and has no dimensions that are frequency sensitive. A basic self-scaling structure (shown in figure 16) is described by angles alone, with no characteristic length. Practical structures of

this type are finite in size, thus limiting the frequency-independent behavior. Variations of this basic design may take the form of toothed structures, such as illustrated.

An outgrowth of this form of wideband antenna is the *log-periodic dipole array* (figure 17) which is well suited to vhf and uhf work. This interesting antenna is made up of dipole elements whose lengths are determined by the angle they subtend from the apex point, and whose distance from the apex is such as to provide the log-periodic behavior. The dipoles are fed at the center from a parallel-wire line in such fashion that successive dipoles come out from the line in opposite directions, equivalent to a 180° phase shift between elements. A broadband log-periodic structure is thus formed, with most of the radiation coming from those dipole elements in the vicinity of a half-wavelength long. The bandwidth of the structure is thus limited by the length of the longest and shortest elements, which must be approximately a half-wavelength long at the extreme frequency limits of the antenna array. Gain and bandwidth of the log-periodic antenna thus bear a definite relationship to the included angle of the structure and the length.

An easily constructed log-periodic antenna is the *log-periodic dipole array*, a two-dimensional structure made up of a series of dipoles, fed at the center in such a way that adjacent dipoles are out of phase. The array is fed at the apex and the elements are excited from a parallel-wire transmission line which, if properly designed, may serve as the support structure for the dipoles. The dipole array, in effect, is a balanced transmission line with elements fed from each line, each set of elements reversed in feed polarity. The limiting structure, is a two-element array, and amateur versions of this device are often termed the "ZL-Special" antenna.

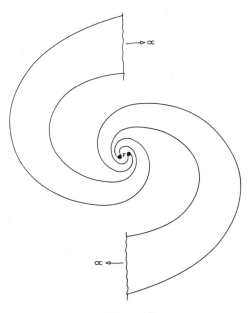

Figure 16

SPIRAL ANTENNA STRUCTURE

This equiangular spiral antenna structure serves as a frequency independent antenna as its shape is entirely specified by angles. The shape of the antenna, when expressed in terms of operating wavelength, is the same for any frequency. The structure is fed at the center (point F) and the arm length is infinite.

The balanced log-periodic dipole structure may be fed with an unbalanced coaxial line by using the support structure as a balun, feeding the coaxial line back from the feed-point through the structure toward the rear.

A L-P Dipole Array for 140—450 MHz A practical L-P dipole array for the vhf spectrum is shown in figure 17. The antenna has a power gain over a dipole of about 7 decibels and may be fed with a 50-ohm coaxial transmission line. The maximum SWR on the transmission line, after adjustment of the boom spacing is better than 2.5/1 over the entire range. The L-P array is built on a twin boom made of ½-inch diameter, heavy-wall aluminum tubing. Two lengths of material are clamped together to form a low-impedance transmission line 84″ (213 cm) long. The clamps may be made of hard wood, or other good insulating material. An impedance match

between the array and the transmission line is effected by varying the spacing of the boom, which changes the impedance of the transmission line created by the proximity of the booms to each other.

Alternate halves of successive dipole elements are fastened to a boom section by threading the element, and affixing it to a clamp, as shown in the illustration. Element spacings are measured from the rear of the array and are rounded off to the nearest quarter inch.

When the array is completed, all elements lie in the same plane, with successive elements off center from the supporting structure by virtue of the alternate feed system employed. Boom spacing should be set as shown in the drawing, and later adjusted for minimum SWR on the coaxial transmission line at the various frequencies of interest.

The coaxial line is passed through one boom from the rear and connection to both booms is made at the nose of the array. The outer braid of the line is connected to the boom through which the line passes, and the center conductor connects to the opposite boom. Type-N coaxial connectors are recommended for use in this frequency region.

A L-P Yagi for 50 MHz A yagi antenna consists of a driven element plus parasitic elements to increase the gain and directivity of the radiation pattern over that of a dipole. The number of parasitic elements, their length and spacing with respect to the driven element determine the characteristics of the parasitic yagi antenna. As gain and directivity increase, bandwidth decreases, limiting the ultimate usefulness of this antenna over a complete amateur band, especially at 10 meters and above. To increase the bandwidth of the array, the log-periodic principle used for broadband antennas may be applied to the parasitic beam. The log-periodic yagi array consists of log-periodic elements, interspersed with parasitic reflectors and directors to form individual cells, differing in size by a geometric constant. The driven element in each cell is fed by a common balanced transmission line.

A variation of the log-periodic principle is used in the parasitic antenna described in this section. This L-P yagi antenna is com-

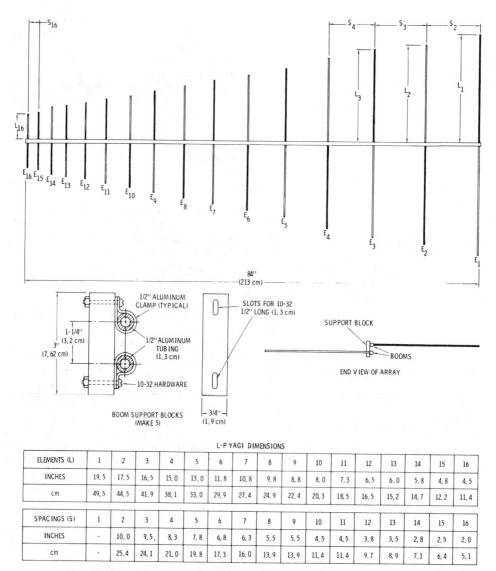

L-P YAGI DIMENSIONS

ELEMENTS (L)	1	2	3	4	5	6	7	8	9	10	11	12	13	14	15	16
INCHES	19.5	17.5	16.5	15.0	13.0	11.8	10.8	9.8	8.8	8.0	7.3	6.5	6.0	5.8	4.8	4.5
cm	49.5	44.5	41.9	38.1	33.0	29.9	27.4	24.9	22.4	20.3	18.5	16.5	15.2	14.7	12.2	11.4

SPACINGS (S)	1	2	3	4	5	6	7	8	9	10	11	12	13	14	15	16
INCHES	-	10.0	9.5	8.3	7.8	6.8	6.3	5.5	5.5	4.5	4.5	3.8	3.5	2.8	2.5	2.0
cm	-	25.4	24.1	21.0	19.8	17.3	16.0	13.9	13.9	11.4	11.4	9.7	8.9	7.1	6.4	5.1

Figure 17

LOG-PERIODIC ANTENNA FOR 140 TO 450 MHz

Vhf log-periodic dipole array is built on double-boom structure made of two lengths of aluminum tubing spaced by insulated support blocks. Elements coded black are attached to the top boom and elements coded white are attached to the lower boom. The coaxial transmission line is inserted in the rear of one boom and passed through the boom, which acts as a balancing device. Center conductor is attached to opposite boom, and shield is attached to balancing boom.

posed of a five element log-periodic section designed to cover the 50- to 52-MHz range and is used in conjunction with three parasitic director elements mounted in front of the log-periodic section. A top view of

the antenna is shown in figure 18. The antenna exhibits about 12 decibels forward gain and compares nearly identically with an 8-element yagi mounted on a 30-foot boom. The overall length of the L-P yagi is only

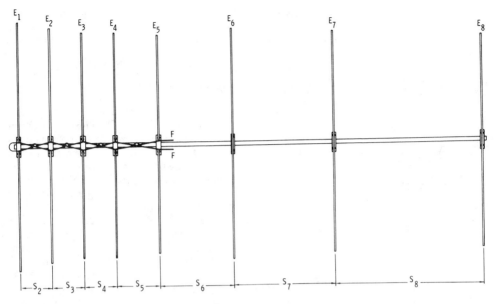

L-P YAGI DIMENSIONS (TIP-TO-TIP)

ELEMENTS	1	2	3	4	5	6	7	8
INCHES	116.5	110	106	104	102	108	103	100
cm	296	279	269	264	259	274	261	254

SPACING	1	2	3	4	5	6	7	8
INCHES	-	15.5	15.7	15.7	20	35	49	71.5
cm	-	39.4	40	40	51.1	88.3	124.5	181.6

Figure 18

L-P YAGI ANTENNA FOR SIX METERS

This design combines bandwidth of log-periodic structure with gain of yagi antenna. L-P yagi may be built on 1½-inch (4.0 cm) diameter boom, about 19 feet (5.8 m) long. L-P elements are insulated from boom by mounting on insulating blocks. Yagi elements are grounded to boom at their center point. The antenna is fed with a balanced 70-ohm ribbon line at the feedpoint and the L-P transmission line is made up of No. 8 aluminum clothesline wire, criss-cross connected between the elements. Rear element is shorted with six-inch loop of aluminum wire. The spacing between the inner tips of the L-P elements is 3½ inches (8.9 cm).

about 18½ feet (5.64 meters) and it provides improved bandwidth performance and smaller size than the comparable yagi array.

30-4   The Helical Beam Antenna

Most vhf and uhf antennas are either vertically polarized or horizontally polarized (plane polarization). However, *circu-*larly polarized antennas having interesting characteristics which may be useful for certain applications. The installation of such an antenna can effectively solve the problem of horizontal versus vertical polarization.

A circularly polarized wave has its energy divided equally between a vertically polarized component and a horizontally polarized component, the two being 90 degrees out of phase. The circularly polarized wave may be either "left handed" or "right handed," depending on whether the vertically polarized component leads or lags the horizontal component.

A circularly polarized antenna will respond to any plane polarized wave whether horizontally polarized, vertically polarized,

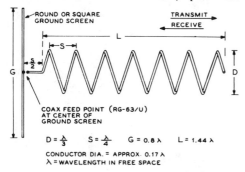

$$D = \frac{\lambda}{3} \qquad S = \frac{\lambda}{4} \qquad G = 0.8\,\lambda \qquad L = 1.44\,\lambda$$

CONDUCTOR DIA. = APPROX. 0.17 λ
λ = WAVELENGTH IN FREE SPACE

Figure 19

THE HELICAL BEAM ANTENNA

This type of directional antenna system gives excellent performance over a frequency range of 1.7 to 1.8 to 1. Its dimensions are such that it is ordinarily not practical, however, for use as a rotatable array on frequencies below about 100 MHz. The center conductor of the feedline should pass through the ground screen for connection to the feedpoint. The outer conductor of the coaxial line should be grounded to the ground screen.

or diagonally polarized. Also, a circular polarized wave can be received on a plane polarized antenna, regardless of the polarization of the latter.

When using circularly polarized antennas at *both* ends of the circuit, however, both must be left handed or both must be right handed. This offers some interesting possibilities with regard to reduction of interference. At the time of writing, there has been no standardization of the "twist" for general amateur work.

Perhaps the simplest antenna configuration for a directional beam antenna having circular polarization is the *helical beam* which consists simply of a helix working against a ground plane and fed with coaxial line. In the uhf and the upper vhf range the physical dimensions are sufficiently small to permit construction of a rotatable structure without much difficulty.

When the dimensions are optimized, the characteristics of the helical beam antenna are such as to qualify it as a broadband antenna. An optimized helical beam shows

little variation in the pattern of the main lobe and a fairly uniform feed-point impedance averaging approximately 125 ohms over a frequency range of as much as 1.7 to 1. The direction of "electrical twist" (right or left handed) depends on the direction in which the helix is wound.

A six-turn helical beam is shown schematically in figure 19. The dimensions shown will give good performance over a frequency range of plus or minus 20 percent of the design frequency. This means that the dimensions are not especially critical when the array is to be used at a single frequency or over a narrow band of frequencies, such as an amateur band. At the design frequency the beam width is about 50 degrees and the power gain about 12 dB, referred to a nondirectional circularly polarized antenna.

For the frequency range 100 to 500 MHz a suitable ground screen can be made from "chicken wire" poultry netting of 1-inch mesh, fastened to a round or square frame of either metal or wood. The netting should be of the type that is galvanized *after* weaving. A small, sheet-metal ground plate of diameter equal to approximately D/2 should be centered on the screen and soldered to it. Tin, galvanized iron, or sheet copper is suitable. The outer conductor of the RG-63/U (125-ohm) coax is connected to this plate, and the inner conductor contacts the helix through a hole in the center of the plate. The end of the coax should be taped with *Scotch* electrical tape to keep water out.

It should be noted that the beam proper consists of six full turns. The start of the helix is spaced a distance of S/2 from the ground screen, and the conductor goes directly from the center of the ground screen to the start of the helix.

Aluminum tubing in the 2014 alloy grade is suitable for the helix. Alternatively, lengths of the relatively soft aluminum electrical conduit may be used. In the vhf range it will be necessary to support the helix on either two or four wooden longerons in order to achieve sufficient strength. The longerons should be of the smallest cross section which provides sufficient rigidity, and should be given several coats of varnish. The ground plane butts against the longerons and the whole assembly is supported from the balance point if it is to be rotated.

Aluminum tubing in the larger diameters ordinarily is not readily available in lengths greater than 12 feet. In this case several lengths can be spliced by means of short telescoping sections and sheet-metal screws.

The tubing is closewound on a drum and then spaced to give the specified pitch. Note that the length of one complete turn when spaced is somewhat greater than the circumference of a circle having the diameter D.

Broad-Band 144- to 225-MHz Helical Beam   A highly useful vhf helical beam which will receive signals with good gain over the complete frequency range from 144 through 225 MHz may be constructed by using the following dimensions (180 MHz design center):

D	22 in. (55.8 cm)
S	16½ in. (41.9 cm)
G	53 in. (134.6 cm)
Tubing o.d.	1 in. (2.5 cm)

The D and S dimensions are to the center of the tubing. These dimensions must be held rather closely, since the range from 144 through 225 MHz represents just about the practical limit of coverage of this type of antenna system.

Note that an array constructed with the above dimensions will give unusually good highband TV reception in addition to covering the 144- and 220-MHz amateur bands and the taxi and police services.

On the 144-MHz band the beam width is approximately 60 degrees to the half-power

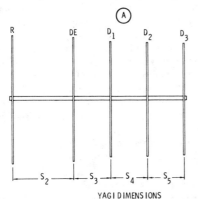

YAGI DIMENSIONS

ELEMENT	REFLECTOR (R)	DRIVEN ELEMENT	DIRECTOR #1	DIRECTOR #2	DIRECTOR #3
INCHES	40.0	38.5	36.0	35.5	35.0
cm	101.6	97.8	91.4	90.1	89.0

SPACING	1	2	3	4	5
INCHES	-	20	12	12	12
cm	-	50	30	30	30

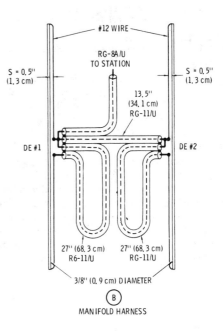

Figure 20

BEAM ANTENNA FOR OSCAR SATELLITE

Two 144-MHz Yagi beams, mounted at right angles to each other on the same boom and fed 90° out of phase will provide circular polarization. (A)—Dimensions of one Yagi array are shown. The element lengths are cut for a wood boom. The second set of elements are mounted on the same boom, displaced by 90° and moved along the boom about one inch so that the elements do not touch when they pass through the boom. The parasitic elements are cut from ⅛" (0.3 cm) diameter aluminum rod or wire. (B)— The manifold harness uses a single length of RG-11/U (75-ohm) coaxial line as a phasing section and lengths of RG-11/U as half-wave length baluns. The coaxial transmission line is RG-8/U (50 ohms).

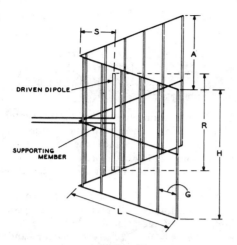

Figure 21

CONSTRUCTION OF THE CORNER REFLECTOR ANTENNA

Such an antenna is capable of giving high gain with a minimum of complexity in the radiating system. It may be used either with horizontal or vertical polarization. Design data for the antenna is given in the Corner-Reflector Design Table.

points, while the power gain is approximately 11 dB over a nondirectional circularly polarized antenna. For high-band TV coverage the gain will be 12 to 14 dB, with a beam width of about 50 degrees, and on the 220-MHz amateur band the beam width will be about 40 degrees with a power gain of approximately 15 dB.

The antenna system will receive vertically polarized or horizontally polarized signals with equal gain over its entire frequency range. Conversely, it will transmit signals over the same range, which then can be received with equal strength on either horizontally polarized or vertically polarized receiving antennas. The standing-wave ratio will be very low over the complete frequency range if RG-63 U coaxial feed line is used.

A Circularly Polarized 144-MHz Yagi Beam for OSCAR The advantages of circular polarization are obvious when communication with (or through) an OSCAR satellite is attempted. The random, tumbling motion of the satellite provides an ever-changing signal of random polarization at the ground station. Even though a circularly polarized antenna exhibits a loss of 3 dB over a comparable linearly polarized antenna when a linearly polarized signal of the correct polarization is received, the normal reflection and diffraction of most vhf signals tend to mask out this difference.

The polarization shift of a space satellite will cause a slow regular fading of the received signal, the maximum signal being received when the signal is in phase with the polarization of the receiving antenna, and fades up to 20 dB can be noticed when the signal is 90° out of phase. Circular polarization provides a much more uniform coverage under these circumstances.

Finally, circular polarization may be used advantageously for communication between a base station and a mobile station, the "flutter" caused by the polarization shift due to the motion of the mobile station being greatly reduced when circular polarization is used at the base station.

Two 144-MHz Yagi beam antennas mounted at right angles to each other on the same boom and fed 90° out of phase will provide circular polarization (figure 20). The phase shift is obtained by using two feedlines, one a quarter-wavelength longer than the other. The two lines are parallel-connected to a common transmission line which goes to the station.

Each Yagi has a folded dipole driven element designed to match a 300-ohm load. A four-to-one balun at each antenna transforms this impedance down to approximately 75 ohms. The antennas are interconnected by a short phasing line to obtain the proper 90° phase shift. The line is an electrical quarter-wavelength long. The direction of polarization (clockwise or counterclockwise) depends on which dipole is directly energized by the transmission line, and which is energized by the phasing line. It is possible to switch polarization rotation by means of a coaxial relay placed at the antenna.

The impedance presented at the feedpoint of the two antennas is half the feedpoint impedance of each antenna, or about 37 ohms. A quarter-wave transformer made of 50-ohm line will match this to a 75-ohm transmission line, or a 50-ohm line may be used, with a resulting SWR of about 1.4 at the antenna resonant frequency.

The antenna may be tested by aiming it at a linearly polarized signal (such as from a repeater). Rotating the array on its axis should produce no more than 1 dB signal variation if the phasing is correct. Power gain of the array is approximately 8 dB.

30-5 The Corner-Reflector and Horn-Type Antennas

The corner-reflector antenna is a good directional radiator for the vhf and uhf region. The antenna may be used with the radiating element vertical, in which case the directivity is in the horizontal or azimuth plane, or the system may be used with the driven element horizontal, in which case the radiation is horizontally polarized, and most of the directivity is in the vertical plane. With the antenna used as a horizontally polarized radiating system the array is a very good low-angle beam array although the nose of the horizontal pattern is still quite sharp. When the radiator is oriented vertically the corner reflector operates very satisfactorily as a direction-finding antenna.

Design data for the corner-reflector antenna is given in figure 21 and in Table 3, *Corner-Reflector Design Data*. The planes which make up the reflecting corner may be made of solid sheets of copper or aluminum for the uhf bands, although spaced wires with the ends soldered together at top and bottom may be used as the reflector on the lower frequencies. Copper screen may also be used for the reflecting planes.

The values of spacing given in the corner-reflector chart have been chosen such that the center impedance of the driven element would be approximately 75 ohms. This means that the element may be fed directly with 75-ohm coaxial line, or a quarter-wave matching transformer such as a Q-section may be used to provide an impedance match between the center impedance of the element and a 460-ohm line constructed of No. 12 wire spaced 2 inches (5 cm).

In many uhf antenna systems, waveguide transmission lines are terminated by *pyramidal horn* antennas. These horn antennas (figure 22A) will transmit and receive either horizontally or vertically polarized waves. The use of waveguides at 144 and 220 MHz, however, is out of the question because of the relatively large dimensions needed for a waveguide operating at these low frequencies.

A modified type of horn antenna may still be used on these frequencies, since only one particular plane of polarization is of interest to the amateur. In this case, the horn antenna can be simplified to two triangular sides of the pyramidal horn. When these two sides are insulated from each other, direct excitation at the apex of the horn by a two-wire transmission line is possible.

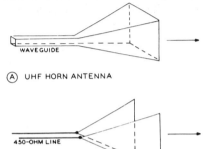

(A) UHF HORN ANTENNA

(B) VHF HORIZONTALLY POLARIZED HORN

Figure 22

TWO TYPES OF HORN ANTENNAS

The "two-sided horn" of illustration B may be fed by means of an open-wire transmission line.

In a normal pyramidal horn, all four triangular sides are covered with conducting material, but when horizontal polarization alone is of interest (as in amateur work) only the *vertical* areas of the horn need be used. If vertical polarization is required, only the *horizontal* areas of the horn are employed. In either case, the system is unidirectional, away from the apex of the horn. A typical horn of this type is shown in figure 22B. The two metallic sides of the horn are insulated from each other, and the sides of the horn are made of small mesh "chicken wire" or copper window screening.

A pyramidal horn is essentially a high-pass device whose low-frequency cutoff is reached when a side of the horn is ½ wavelength. It will work up to infinitely high frequencies, the gain of the horn increasing by 6 dB every time the operating frequency is doubled. The power gain of such a horn compared to a half-wave dipole at frequencies higher than cutoff is:

$$\text{Power gain (dB)} = \frac{8.4\ A^2}{\lambda^2}$$

where A is the frontal area of the mouth of the horn. For the 60-degree horn shown in figure 23 the formula simplifies to:

$$\text{Power gain (dB)} = 8.4\ D^2, \text{ when } D \text{ is}$$
expressed in terms of wavelength.

When D is equal to one wavelength, the power gain of the horn is approximately 9 dB.

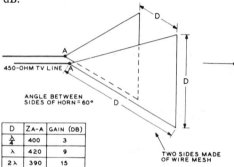

D	Z$_{A-A}$	GAIN (DB)
$\frac{\lambda}{4}$	400	3
λ	420	9
2λ	390	15

Figure 23

THE 60° HORN ANTENNA FOR USE ON FREQUENCIES ABOVE 144 MHz

30-6   VHF Yagi Beam Antennas

The multielement rotary beam is undoubtedly the most popular type vhf antenna in use. In general, the design, assembly and tuning of these antennas follows a pattern similar to that used for the larger rotary arrays used on the lower-frequency amateur bands. The characteristics of the latter antennas are discussed in a previous chapter of this Handbook, and the information contained in that chapter applies in general to the vhf beam antennas discussed herewith.

Element Lengths   Optimum length for parasitic elements in vhf arrays is a function of element spacing and the diameter of the element. To hold a satisfactory length/diameter ratio, the diameter of the element must decrease as the frequency of operation is raised. At very-high frequencies, element length is so short that the diameter of a self-supporting element becomes a large fraction of the length. Short, large-diameter elements have low Q and are not practical in parasitic arrays. Thus the yagi array becomes critical in adjustment and marginal in operation in the upper reaches of the vhf spectrum. Yagi antennas can be made to work at 432 MHz and higher, but their adjustment is tedious, and preference is given to broadside arrays having relatively large spacings between elements and high impedance. The yagi antenna, however, remains "the antenna to beat" for the 50-, 144-, and 220-MHz amateur bands.

The yagi antennas shown in this section are of all-metal construction with the elements directly grounded to the boom. Either a gamma-match system, T match, or folded-dipole element may be used on the arrays. For short lengths of transmission line, 50-ohm low-loss coaxial cable is recommended for use with a gamma match, or with folded dipole or T match and a coaxial balun. Longer line lengths should be made up of 300-ohm TV-type "ribbon" line or open-wire TV-type transmission line. Care should be taken to keep the ribbon or open-wire lines clear of nearby metallic objects.

Table 3. Corner Reflector Dimensions

BAND	CORNER	R		S		H		A		L	
(MHz)	ANGLE	in	cm	in	cm	in	cm	in	cm	in	cm
50	60°	110	279.4	115	292	140	355.6	230	584.2	230	584
144	60°	38	96.5	40	101.6	48	121.9	100	254	100	254
220	60°	24.5	62.2	25	63.5	30	76.2	72	183	72	183
420	60°	13	33.0	14	35.6	18	45.7	36	91.5	36	91.5

(1)—DIMENSION G IS 18" (45.7 cm) FOR 50 MHz, 3" (7.6 cm) FOR 144-220 MHz AND MESH SCREEN FOR 420 MHz
(2)—ANTENNA GAIN ON ALL BANDS IS 12 dB
(3)—FEEDPOINT IMPEDANCE IS ABOUT 75 OHMS

The Yagi Assembly   Mechanical assembly of a vhf Yagi is critical since the boom and mounting hardware approach a fraction of the operating wavelength. Multielement Yagi beams built on wood booms provide confusing results in recent tests. It was found that moisture absorption and shrinkage of the wood made repeatability of measurements almost impossible, despite various coatings applied to the wood. Metal beam Yagis, however, were entirely repeatable if the elements were lengthened to compensate for the boom structure. The amount of change was a function of how the element was mounted to the boom (figure 24). In general, small diameter booms have less effect on element length than larger booms. Mounting hardware also affected element length to a small degree. Element taper usually does not enter the picture as most vhf beam elements are constructed of single sections of tubing that have no taper.

When the element is run through the middle of the boom, the element length should be increased by about 0.7 times the diameter of the boom to compensate for the shunting effect caused by the metallic boom structure.

When the element is mounted directly above the boom, but in contact with it, the element length should be increased by about .06 of the boom diameter, but when the element is mounted only a very short distance above the boom, no correction factor is required.

Yagi Beams for 6 and 2   All-aluminum beam antennas are easy to construct for the 6- and 2-meter amateur bands. The three-element array is very popular for general 6-meter operation, and up to ten elements are often used for DX work on this band. The four-element array is often used on 2 meters, either horizontally or vertically polarized, and arrays having as many as twelve to fifteen elements are used for meteor-scatter and over-horizon work on 144 MHz.

Shown in figures 25 and 26 is a simple three-element array for the 6-meter band. The design frequency is 50.5 MHz, and the beam is capable of operation over the 50- to 51-MHz frequency span. The antenna

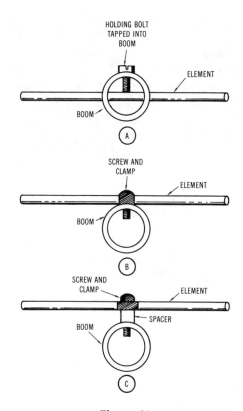

Figure 24

MOUNTING VHF ELEMENT TO THE BOOM

A—Element passed through hole in boom. Element length should be increased by 0.7 boom diameter to compensate for shunting effect of the boom. B—When element is mounted directly above the boom, element length should be increased by 0.06 boom diameter. C—When element is mounted clear of the boom no correction factor is required.

may be fed from a 50-ohm coaxial line with a half-wave balun and T match as shown in the illustration. The supporting boom is made of a length of 1⅛-inch diameter aluminum TV mast section, and the elements are made of ½-inch diameter aluminum tubing. The elements are mounted in position by drilling the boom to pass the element and then clamping the point as shown in the drawing.

The T-match system must be properly resonated at the center frequency of antenna operation. To do this, the antenna is temporarily mounted atop a step ladder, in the clear, and fed with a few watts of power

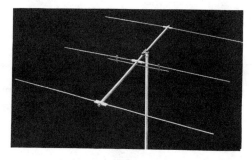

Figure 25

THREE-ELEMENT YAGI BEAM
FOR SIX METERS

This all-aluminum array is a popular six-meter antenna. Available in kit form, it also may easily be constructed from available aluminum tubing. Elements are clamped to the boom and either a T match, Gamma match or split-driven-element feed system used. T match with half-wave coaxial balun is recommended system for ease in adjustment. Brass or aluminum hardware should be employed to prevent corrosion of elements due to weather.

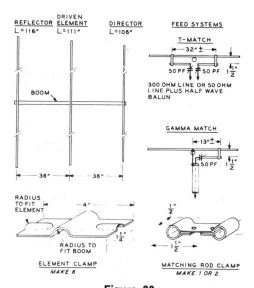

Figure 26

SIX-METER BEAM ASSEMBLY

Element clamps are fabricated from soft aluminum strip. All joints should be cleaned and covered with Penetrox paste to prevent corrosion. Elements may be made of sections of telescoping tubing. Diameters between one inch and one-half inch are recommended.

from the station transmitter. An SWR meter or reflectometer is placed in the line near

the antenna and the length of the T sections and the series capacitors are adjusted to provide the lowest value of SWR on the transmission line. The capacitors are varied in unison to preserve the symmetry of balance. The capacitors should be enclosed in a weatherproof box and mounted at the center of the T section.

A four-element array for the 2-meter band is shown in figures 27 and 28. Dimensions are given for a center frequency of 146 MHz. The antenna provides a power gain of about 9 decibels over a dipole and is capable of good operation over the complete 2-meter band. For optimum operation at the low end of the band, all element lengths should be increased by one-half inch.

Antenna construction is similar to the 6-meter array in that an aluminum section of tubing is used for the boom and the elements are passed through holes drilled in the boom. One-quarter inch aluminum tubing is used for the elements. The T match and coaxial balun are used to match the antenna to a 50-ohm coaxial transmission line.

Long Yagi Antennas   For a given power gain, the *Yagi antenna* can be built lighter, more compact, and with less wind resistance than any other type. On the

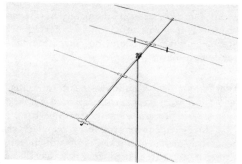

Figure 27

FOUR-ELEMENT YAGI BEAM
FOR TWO METERS

Light aluminum is employed for easy-to-build two meter beam. Reynolds "Do It Yourself" aluminum, available at many hardware and building supply stores, may be used. Construction is similar to six-meter array. If boom diameter is about one inch, the boom may be drilled for the elements, which are then held in place by a sheet-metal screw through boom and element.

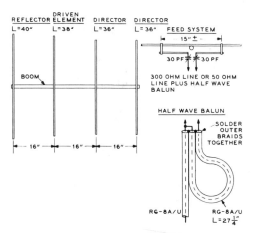

Figure 28

TWO-METER BEAM ASSEMBLY

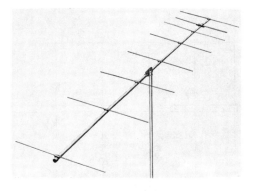

Figure 29

TWO-METER LONG YAGI ARRAY

Elements are mounted atop boom by means of small clamps made of soft aluminum strap. Either folded dipole or T-matching device may be used with antenna. Eight-element beam similar in construction is manufactured and sold in kit form.

other hand, if a Yagi array of the same approximate size and weight as another antenna type is built, it will provide a higher order of power gain and directivity than that of the other antenna (figure 29).

The power gain of a Yagi antenna increases directly with the physical length of the array. The maximum practical length is entirely a mechanical problem of physically supporting the long series of director elements, although when the array exceeds a few wavelengths in length the element lengths, spacings, and Q's become more and more critical. The effectiveness of the array depends on a proper combination of the mutual coupling loops between adjacent directors and between the first director and the driven element.

Shown in this section are several Yagi beam antenna designs based on a design technique developed by the *National Bureau of Standards* and popularized by W1JR.

A Six-Element Yagi for 6 Meters—This antenna design provides a power gain of 10.2 dBd and is built on a 24-foot boom (figure 30). All elements are cut from ½-inch diameter aluminum tubing and are mounted on insulating blocks attached to a 1½-inch diameter boom. The antenna is designed for a center frequency of 50.1 MHz. Measured F/B ratio at the design frequency is about 18 dB.

The driven element is attached directly to the boom with a U-bolt while the parasites are insulated by means of phenolic blocks attached to the boom by U-shaped clamps. The elements are bolted to the blocks.

The driven element is fed with a gamma match, the gamma capacitor being about 12 inches of RG-8A/U coaxial cable with the outer jacket and shield removed. The cable is inserted in the ⅜-inch diameter gamma tube. The shorting bar at the end of the gamma rod is adjusted for lowest SWR on the feedline at the design frequency of the antenna.

A 13-Element Yagi for 2 Meters—Shown in figure 31 is a long Yagi design centered at 144.2 MHz. The array provides over 14 dB gain compared to a dipole and has a F/B ratio of about 22 dB. The antenna is designed to be fed with either a 300-ohm ribbon line or a 50-ohm coaxial line and a half-wave coaxial balun. The elements are passed through holes drilled through a 1¼-inch diameter boom and are compensated for the boom diameter. If the elements are mounted atop the boom on insulated blocks, they should all be shortened about ¾ inch. SWR on the transmission line is adjusted by varying the spacing between the driven element and the wire yoke beneath it.

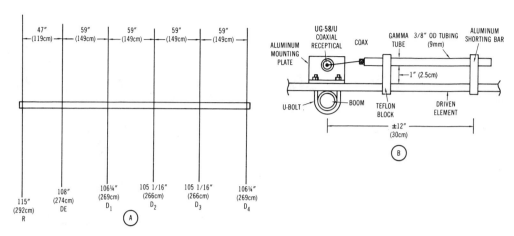

Figure 30

SIX-ELEMENT YAGI BEAM FOR 6 METERS

This antenna, designed by W1JR, provides a power gain of 10.2 dB over a dipole. It is built on a 24-foot long boom. The parasitic elements are insulated from the boom by small phenolic blocks held to the boom with U-bolts. The driven element is attached to the boom directly. The length of the driven element may be adjusted to achieve lowest SWR after the initial adjustments are made to the length of the gamma matching rod. Gain of this design drops off quite quickly on the high side of the design frequency but much more slowly on the low-frequency side.

Plan view of the antenna is shown at A and details of the gamma matching section are shown at B.

A 15-Element Beam for 432 MHz—This high gain beam was designed by W1JR and four of them were used for the record-breaking contact between Hawaii and California. Power gain is about 13.6 dB over a dipole and the F/B ratio is about 22 dB. The boom is 116 inches long and is made of ¾-inch diameter tubing. The elements are ³⁄₃₂-inch diameter brass rods which are knurled and tapped into undersize holes in the boom. Element spacing of all directors is 8⁷⁄₁₆ inches, center to center.

The ends of the directors are filed flat to close tolerance and have been compensated for boom diameter. A delta match is used on the driven element and the array is fed with a 4-to-1 coaxial balun and 50-ohm transmission line. Assembly information is given in figure 32.

A Loop Yagi for 1300 MHz—The simple Yagi design is impractical at 1300 MHz because it is difficult to design a practical antenna that is not extremely fragile because of the thinness of the elements. Some experimenters have tried loop elements with some success and shown in this section is a 25-element loop Yagi designed by G3JVL. The antenna shows a measured power gain of about 17 dB over a dipole at the design frequency (figure 33). The antenna consists of 23 loop directors, a loop driven element and reflector, plus a second backup reflector made of an aluminum sheet. The whole array is built on an aluminum tube 6'6" long and ¾ inch in diameter.

The loops are made of metal strips 0.32" thick and 0.25" wide. Aluminum is used for all elements except for the reflector and the driven element. Length of the strips is shown in the illustration. The parasitic element strips are drilled at the ends for 4-40 hardware (0.125" diameter), folded into a loop and bolted to the boom.

The driven element is soldered to the end of a short length of semirigid coaxial line (teflon or air insulated line is to be preferred). The cable passes through a hole drilled in the boom. Cable loss is very high at this frequency and the cable run to the equipment should be as short as possible. The loop is screwed to the boom with a ⅛-inch spacer.

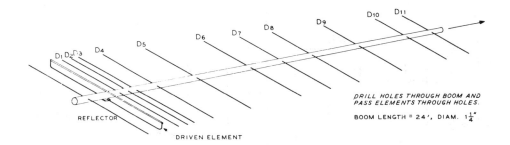

DRILL HOLES THROUGH BOOM AND
PASS ELEMENTS THROUGH HOLES.

BOOM LENGTH = 24', DIAM. $1\frac{1}{4}$"

REFLECTOR

DRIVEN ELEMENT

ELEMENT DIMENSIONS, 2-METER BAND

ELEMENT	LENGTH				SPACING FROM DIPOLE
(DIAM. 1/8")	144 MHz	145 MHz	146 MHz	147 MHz	
REFLECTOR	41"	$40\frac{3}{4}$"	$40\frac{7}{16}$"	$40\frac{3}{16}$"	19"
DIRECTORS	$36\frac{3}{4}$"	$36\frac{1}{2}$"	$36\frac{3}{8}$"	$36\frac{3}{16}$"	D1 = 7"
					D2 = 14.5"
					D3 = 22"
					D4 = 38"
					D5 = 70"
					D6 = 102"
					D7 = 134"
					D8 = 166"
					D9 = 198"
					D10 = 230"
					D11 = 242"

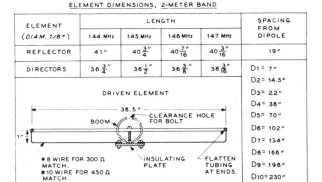

DRIVEN ELEMENT

38.5"

CLEARANCE HOLE FOR BOLT

BOOM

1"

#8 WIRE FOR 300 Ω MATCH.
#10 WIRE FOR 450 Ω MATCH.

INSULATING PLATE

FLATTEN TUBING AT ENDS.

Figure 31

LONG YAGI BEAM ANTENNA FOR 2 METERS

This design provides 16 dB gain over a dipole and covers about 1 MHz at 144 MHz. Dimensions are provided for four frequencies in the 2-meter band. Multiply dimensions by 2.54 to obtain element lengths and spacings in centimeters. Antenna may be fed by 50-ohm coaxial line and half-wave balun.

30-7   Stacking vhf Antennas

By stacking, it is meant that two or more single antennas of any type form a broadside array, so that antennas can be stacked horizontally as well as vertically. Any number of antennas, within reason, may be stacked and coupled together to provide enhanced gain and directivity.

The optimum stacking distance for two dipoles is 0.67 wavelength for maximum gain, but this is not generally true for high-gain beam antennas. By spacing the beams so that their apertures just "touch," power gain will increase directly as the number of antennas used.

The beamwidth of the stacked array will change according to the direction of the stacking. If the array is made four antennas wide, the beamwidth in the horizontal plane will be one-fourth of the beamwidth of one antenna. If the array is made two antennas high, the vertical beamwidth will be one-half that of one antenna alone.

As a simple rule of thumb in stacking extended Yagi antennas, or other arrays having high gain, it is suggested that stacking distance be equal to $\frac{3}{4}$ of the length of the antenna. This figure will be quite close to the aperture size of a single antenna. Examples of good stacking technique are shown in figures 34 and 35.

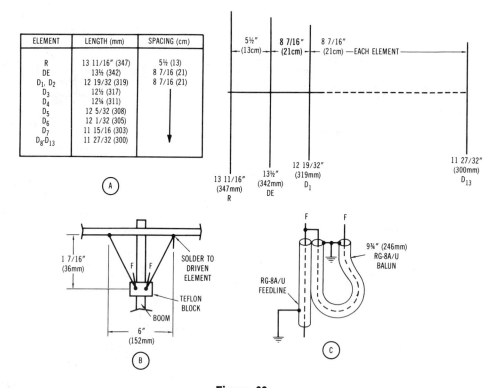

ELEMENT	LENGTH (mm)	SPACING (cm)
R	13 11/16" (347)	5½ (13)
DE	13½ (342)	8 7/16 (21)
D_1, D_2	12 19/32 (319)	8 7/16 (21)
D_3	12½ (317)	
D_4	12¼ (311)	
D_5	12 5/32 (308)	
D_6	12 1/32 (305)	
D_7	11 15/16 (303)	
D_8-D_{13}	11 27/32 (300)	

Figure 32

FIFTEEN-ELEMENT YAGI BEAM FOR 432 MHz

This design provides a power gain of about 13.6 dB over a dipole. The antenna is cut for 432.0 MHz. Array is built on a boom 116 inches long made of ¾-inch diameter aluminum tubing. Elements are ³⁄₃₂-inch diameter brass rods pressed into undersized holes in the boom. A delta match is made of no. 14 wire tapped each side of the driven element. A 4-to-1 coaxial balun matches the system to a 50-ohm coaxial line. Antenna layout is shown at A, the delta match system at B, and the coaxial balun at C.

The Manifold Feed System   Most high gain vhf antennas used for moonbounce or meteor scatter work are made up of many antennas arranged into a large array. The power applied to the array must be divided equally among the antennas and be in the proper phase to permit the individual fields to add vectorially. A *manifold feed system,* such as shown in figure 36 may be used to feed a large number of antennas. The manifold harness may use open wire line, or coaxial cable. The antennas must be identical and well matched to the impedance of the interconnecting phasing line. In this example, each driven element is adjusted to provide a 200-ohm feedpoint. Typically, this may be done with a gamma match or a folded dipole having the proper transformation ratio. Each element is equipped with a half-wave balun to provide a 50-ohm termination for the coaxial phasing line. The length of the lines is unimportant, as long as they are equal. At the junction point (A), the lines are connected in parallel to provide a nominal impedance of 25 ohms. Two 50-ohm quarter-wave transformers change this impedance level to 100 ohms in each case, and the 100-ohm points are connected in parallel at (B) to provide a 50-ohm termination for the coaxial transmission line to the station.

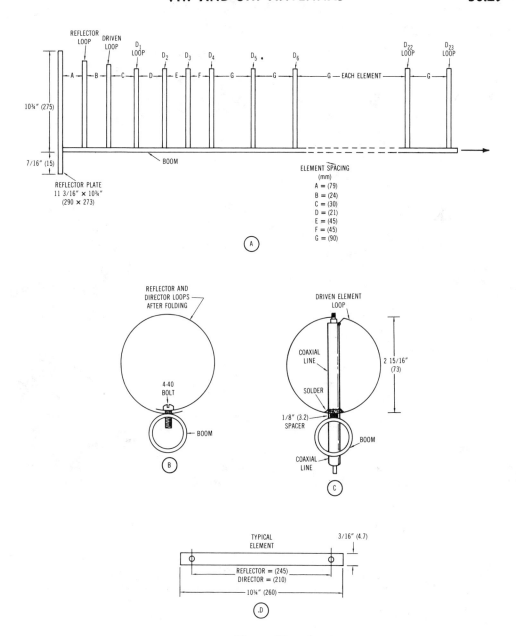

Figure 33

LONG LOOP YAGI BEAM FOR 1296 MHz

This antenna design by G3JVL solves some of the problems inherent in the Yagi antenna at very high frequencies. Quad-type loops are used as the elements, backed up by a reflector plate. A—Side view of Yagi showing placement of loops. Loop spacing is critical and given in millimeters for greatest accuracy. B—Parasite loops are made of copper straps folded into a circle. The circle diameter is determined by the dimension between the mounting holes. Loops are bolted directly to the boom. C—The driven-element loop is soldered to the coaxial line at the bottom and split at the top for connection to the line. The loop is bolted to the boom and spaced about ⅛ inch above it. D—Strap element is bent into loop. Note: The critical dimensions are given in millimeters.

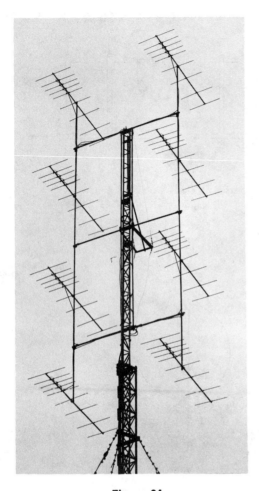

Figure 34

**72 ELEMENT, 144 MHz
ARRAY OF W3OLV**

Eight nine-element, log-periodic Yagi antennas are stacked four high and two wide in this impressive array. Antenna structure can be tilted from vertical for tracking the moon or for meteor scatter work. Antenna provides over 21 dB power gain as compared to a dipole. Beams are fed with a coaxial manifold harness so that length of feedline to each beam is of equal length from common feedpoint at center of array.

Figure 35

**160-ELEMENT, 144-MHz
ARRAY OF K8III**

Thirty-two, five-element, extended-expanded arrays are stacked eight high and four wide in this rugged installation. Each individual beam is a "lazy-H," expanded configuration, backed up by two reflectors and a single director. Array may be tilted for moonbounce or meteor scatter work. Power gain is better than 23 dB over a dipole.

30-8 Extended, Expanded vhf Arrays

Two collinear elements 0.64 wavelength long make a simple array that provides about 3 dB power gain over the dipole. This is known as an *extended double Zepp* antenna among old-timers. The directional pattern of this simple antenna resembles a dipole, except that it is somewhat sharper and has minor lobes at an acute angle to the line of the antennas.

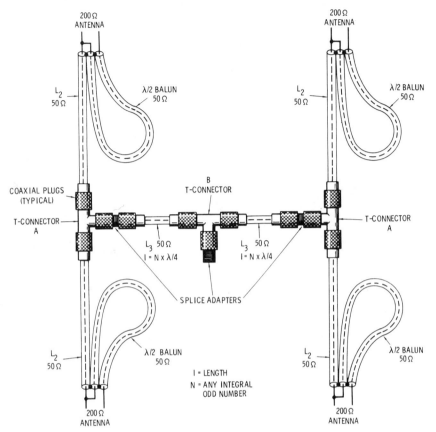

Figure 36

COAXIAL MANIFOLD FEED SYSTEM

Four antennas are fed from a central feedpoint (B). Each branch feeds two other antennas at secondary feedpoint (A). Each antenna has a folded-dipole driven element and a half-wave coaxial balun to provide an unbalanced 50-ohm feedpoint. A gamma match may be substituted for the folded dipole and balun. For the larger, more complicated arrays, many vhf experimenters use balanced, open-wire line in preference to coaxial line.

Figure 37

THE W6GD EXTENDED-EXPANDED FEED SYSTEM FOR STACKED ARRAYS

Four Yagi antennas may be fed with this simple system to provide improved power gain. The driven elements only are shown in this drawing. The Yagis are stacked two above two, with the center line of the Yagi boom marked as shown. The array is fed at F-F with a half-wavelength, shorted stub. The assembly is grid-dipped to frequency and the movable short soldered in position. A coaxial line and half-wavelength balun feed the balanced stub a few inches above the shorting bar. Dimensions are for 144 MHz. Normal length reflectors and directors are used.

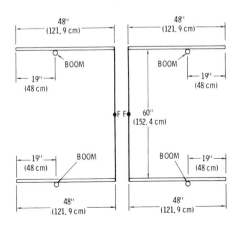

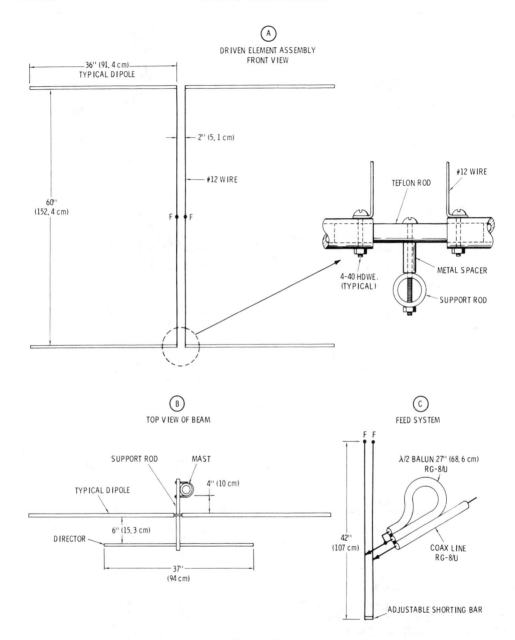

Figure 38

SIX-ELEMENT BEAM FOR 144 MHz

A broadside array with two directors provides about 10 dBd power gain. Each set of collinear elements has a single director, the combination providing more gain than if a double set of directors were used. A—The four dipole elements are stacked two above two, with a two-wire phasing line connecting the stack. The assembly is fed at point F-F. Elements are shown for horizontal polarization. B—The top view of the array shows placement of the director elements in front of the collinear dipoles. The support rod is plugged with a wood dowel at the points where the dipole and director elements are attached. C—A half-wavelength stub is attached to the phasing line at point F-F and a coaxial balun and 50-ohm line are tapped on the stub to provide the lowest value of SWR.

A simple manifold feed system designed by ex-W6GD is shown in figure 37. Four Yagi antennas, stacked 2 above 2 may be driven by this easily built harness. Each driven element is extended in length to about 5/8 wavelength, which places the horizontal Yagi beams about 3/4 wavelength apart when the driven elements are placed tip to tip. The spacing is a little less than optimum, but gain and antenna pattern are not seriously affected. Vertical stacking is about 3/4 wavelength. Dimensions for the 144 MHz band are given in the drawing.

Four of these collinear arrays may be fed from a single transmission line, as shown in figure 37 to provide a simple driving element for more complex arrays. Because of the gain of the collinear elements, a stack of four provides a power gain of about 6 dB over a dipole.

The extended, expanded antenna stack may be used in Yagi arrays or in broadside arrays. Shown in this section are representative antennas for the 144, 220, and 420 MHz bands that make use of this principle.

A Six-Element Broadside Beam for 144 MHz This compact array provides about 10 dB power gain. Only six elements are used, four in a collinear broadside configuration, plus two added directors. A single director for each set of collinear elements provides more gain than separate directors placed in front of each element (figure 38). Director spacing is quite close and the director element is longer than usual. The array is fed with a half-wavelength, shorted stub and coaxial balun, and may be mounted in either a horizontal or vertical position. Horizontal polarization is shown in the illustration. The collinear dipoles are physically connected by a short length of *teflon* rod slipped inside the ends of the tubing. The dipole pair is then supported from a horizontal support rod by means of 4-40 hardware and metal spacers affixed at the center of the rod. The elements are made of 1/2" (1.2 cm) diameter tubing, as is the support rod. The interconnecting phasing line is made of No. 12 wire and is fed at the midpoint by a half-wavelength matching stub, coaxial balun and 50-ohm line (illustration C).

Only two directors are required for the array, centered between the collinear elements (illustration B). The directors are 1/2" tubing mounted to the support rod with a small clamp.

The antenna is adjusted to the design frequency with the coaxial balun and transmission line removed. The wire stub is coupled to a dip oscillator and the short adjusted to provide resonance. The balun and line are then attached near the shorting bar and power applied to the array. The tap position of the balun and the shorting bar are then adjusted to provide the lowest value of SWR on the transmission line. The coaxial line and balun are brought back to the mast and the line is run down the mast to make sure it does not enter the active field of the antenna.

A 24-Element Expanded Yagi Array for 220 MHz The extended, expanded concept developed by ex-W6GD works well with Yagi beams, as illustrated by this "four-over-four" array for 220 MHz. The antenna provides about 14 dB gain over a dipole and performs well plus or minus 1 MHz of the design frequency.

Four, six-element beams are arranged in a square (figure 39). The driven elements of each Yagi extend inwards to about 5/8 wavelength and are fed with a two-wire phasing line. A half-wavelength, shorted stub and coaxial balun (as shown in figure 38) are connected to the phasing line at point F-F.

The elements are made of 1/8-inch diameter rod or tubing passed through a wood boom. The elements are held in position by means of a woodscrew passing into the boom and pressing against the element.

The shorting stub is about 26" (66 cm) long and the half-wavelength coaxial balun is 17 1/2" (44.5 cm) long. The balun and coaxial transmission line use RG-8A/U coaxial line. The balun taps on the stub about 3" (7.6 cm) from the shorting bar.

Adjustment is similar to that of the 144-MHz array. The stub length and tap point are adjusted for lowest SWR on the coaxial transmission line.

Element lengths are chosen for a wood structure and will have to be lengthened if a metal structure is substituted.

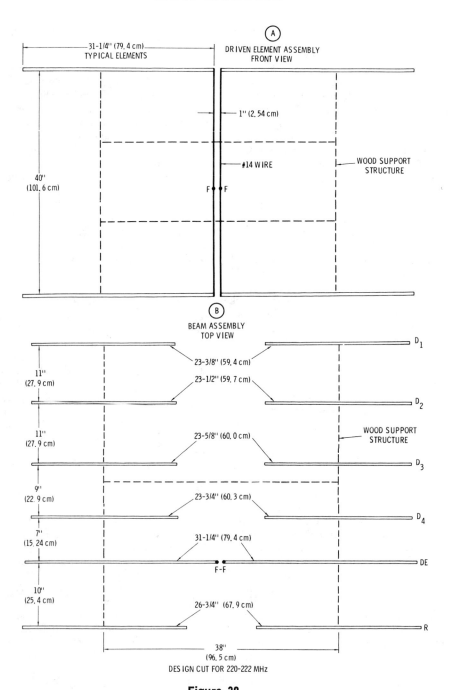

Figure 39

"FOUR-BY-FOUR" ARRAY FOR 220 MHz

This compact and inexpensive beam provides about 14 dB gain over a 2-MHz segment of the
220-MHz band. The array uses the extended, expanded concept for maximum gain and simplicity of feed. The array is built on a wood framework made of 1″ × 1″ lumber, well painted
to protect it from the weather.

The W6GD Broadside Array for 432 MHz The extended-expanded broadside array was designed by the late W6GD of Stanford University and has consistently out-performed larger and more sophisticated antennas at 432 MHz. The W6GD beam is a 16-element beam and has been measured to have 12 decibels power gain over a dipole. Extended elements are used with ¾-wavelength spacing. The array has a sharp front lobe, with nulls at 19° and 42° each side of center and must be aimed carefully for best results.

All elements are made of 0.175-inch diameter brass rod. The active elements are made of square "U"s bent from four lengths of rod, each 51½ inches long. The half-wavelength reflectors are cut of the same material and are 13⅛ inches long. The W6GD array is built on a wooden frame-work, so designed as to keep the supporting structure in back of the array. The driven elements are self-supporting except for four insulating blocks placed at low-voltage points. The blocks and spacers are drilled and slipped on the brass rods before the assembly is bent into shape (figure 40).

After assembly, the matching stubs are silver-soldered to the driven elements and the balun and the interconnecting transmission line temporarily connected in place. The line is tapped up each stub to attain a low value of SWR on the coaxial or open-wire transmission line. Placement of the taps is determined by experiment.

A complete discussion of vhf antennas is contained in the *VHF Handbook*, available from Radio Publications, Inc., Wilton, CT 06897.

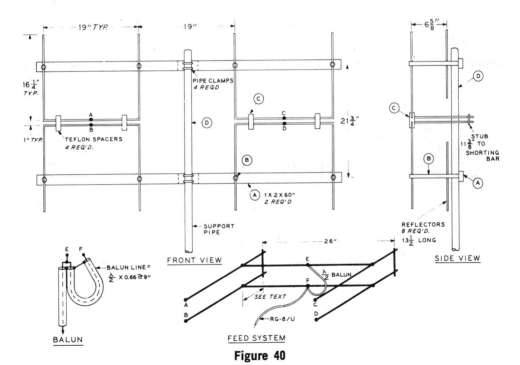

Figure 40

W6GD EXPANDED BROADSIDE ARRAY FOR 432-MHz

The 16-element beam is made of brass rod suspended from a wooden frame at low-voltage points on the antenna. Small ceramic insulators are used to mount the rods. Antenna elements and lines are aligned by means of small teflon or polystyrene spacer blocks passed over the rods before they are bent into shape. Half-wave lines are employed in feed system together with a full-wavelength transformer and balun to provide a close match to a 50-ohm transmission line. Lines and transformer are made up of brass rod and adjustable shorting bars are used.

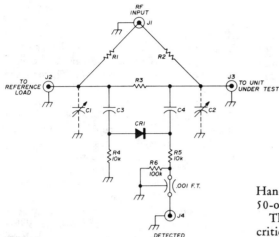

Figure 41

VHF SWR BRIDGE

C_1, C_2—Small capacitive tab for balance
  (see text)
C_3, C_4—.001 μF disc ceramic or chip capacitor
CR_1—1N82 or equivalent germanium diode
J_1, J_4—UG-290A/U type BNC connector
J_2, J_3—UG-58/U type N connector
R_1, R_2—47 to 55 ohms, ¼-watt carbon-composition
R_2—51 ohms, ¼ watt as above

30-9   A VHF SWR Meter

Shown in figures 41 and 42 are construction details for an inexpensive SWR meter that functions well through 450 MHz. It can be used for adjusting the antennas shown in this chapter. The device is based upon the designs given in Chapter 31 of this Handbook and is intended for use with a 50-ohm coaxial line.

The values of resistors R_1 and R_2 are not critical but both should be the same type and matched for best accuracy. This can be done by comparing a dozen similar resistors on an ohmmeter and choosing the two which are closest in value. Capacitors C_1 and C_2 are small copper tabs that can be added close to J_2 and J_3 if the ultimate in balance is desired.

When building such a bridge short leads and symmetry are prime considerations as long leads and stray capacitance can obscure bridge balance. A recommended layout is shown in the illustration.

To check the bridge identical loads are placed at J_1 and J_2. The dc output at J_4 should be zero when an r-f signal is applied at J_1. If the identical loads are swapped in position, the output at J_4 should remain zero. A simple homemade 50-ohm load is shown in figure 43. Two of these can be used to check the bridge.

For operation, one 50-ohm load is plugged into J_2 and the antenna or other device under test is attached to J_3. A 0-100 μA may be used for relattive readings at J_4.

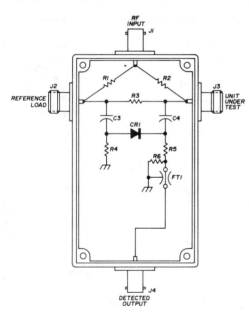

Figure 42

BRIDGE LAYOUT

Recommended parts layout for vhf SWR bridge. Enclosure is a Pomona 2417 shielded box.

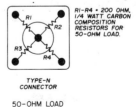

RI-R4 · 200 OHM,
1/4 WATT CARBON
COMPOSITION
RESISTORS FOR
50-OHM LOAD.

TYPE-N
CONNECTOR

50-OHM LOAD

Figure 43

**VHF 50-OHM
LOAD**

CHAPTER THIRTY-ONE

Electronic
Test Equipment

All amateur stations are required by law to have certain items of test equipment available within the station. A c-w station is required to have a frequency standard or other means, in addition to the transmitter frequency control, for ensuring that the transmitted signal is on a frequency within one of the frequency bands assigned for such use. An SSB station is required in addition to have a means of determining that the transmitter is not being modulated in excess of its modulation capability, and in the case of an a-m transmitter, not more than 100 percent. Further, any station operating with a dc power input greater than 900 watts is required to have a means of determining the exact input to the final stage of the transmitter, so as to ensure that the dc power input to the plate circuit of the output stage does not exceed 1000 watts.

The additional test and measurement equipment required by a station will be determined by the type of operation contemplated. It is desirable that all stations have an accurately calibrated voltohmmeter for routine transmitter and receiver checking and as an assistance in getting new pieces of equipment into operation. An oscilloscope and an audio oscillator make a very desirable adjunct to a station using f-m transmission, and are recommended items of test equipment if single-sideband operation is contemplated. A calibrated signal generator is almost a necessity if much receiver work is contemplated, although a noise generator will serve in place of the signal generator. Extensive antenna work invariably requires

the use of some type of standing-wave meter. Lastly, if much construction work is to be done, a simple, solid-state dip meter will be found to be one of the most used items of test equipment in the station.

Other modern pieces of test equipment such as digital voltmeters, counters and frequency synthesizers are becoming common items of station equipment as the amateur operator advances rapidly into today's world of solid-state equipment.

31-1 Voltage, Current, and Resistance Measurements

The measurement of *voltage, current,* and *resistance* in electronic circuits is very important in the design, operation, and maintenance of equipment. Solid-state devices and vacuum tubes of the types used in communications work must be operated within rather narrow limits in regard to electrode voltages and they must be operated within certain maximum and minimum limits with regard to the voltage and current flowing in the circuit elements.

Analog Instruments Both direct current and voltage may be measured with the aid of an *analog instrument* which indicates a measurement smoothly and continuously as the voltage or current passes through an infinite number of different values. The most common instrument of this type consists of a coil that is

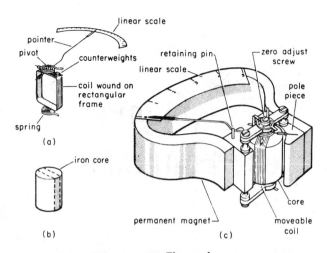

Figure 1

THE D'ARSONVAL METER MOVEMENT

Pointer deflection is proportional to the current flowing through the instrument.

free to move in a varying magnetic field (*d'Arsonval*-type instrument). This consists of a coil of fine wire suspended in the field of a permanent magnet (figure 1). A pointer is attached to the coil and the coil is held at rest by spiral springs. When the voltage or current being measured causes a current to flow in the coil, an electromagnetic field is produced about the coil. A rotational force in the coil in opposition to the perma-

nent magnet repels the field of the coil creating spring force, causing the pointer to move along a graduated scale mounted on the meter frame. For a given design, the pointer movement is proportional to the coil current and therefore proportional to the voltage or current being measured.

Several types of coil suspension mechanisms are in use. The coil may be suspended between pivots or may be held in

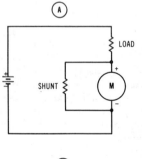

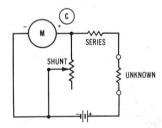

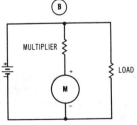

Figure 2

D'ARSONVAL METER MEASURES VOLTS, AMPERES, OR OHMS

A—Meter measures proportion of current flowing in the circuit
B—Meter is calibrated in volts but measures current flowing in multiplier resistor
C—Meter measures current flowing through unknown resistance

position by taut metal bands that flex and twist as the movement turns.

The d'Arsonval movement is directly used for dc measurements (figure 2A). The current to be measured passes through the coil and causes a proportional deflection of the pointer. The basic meter movement is very sensitive and when large currents are to be measured a parallel path, or *shunt*, is provided to bypass most of the current in the circuit around the meter coil. For high accuracy, the shunt must be a precise fraction of the coil resistance.

The d'Arsonval movement is also used for measuring voltage (figure 2B). A resistance (termed a *multiplier*) is connected in series with the meter movement so that a certain voltage is required to cause full-scale deflection.

Resistance measurements are accomplished as shown in figure 2C. This simple *ohmmeter* is representative of more sophisticated circuits. The meter movement is connected across a variable shunt resistor and the combination is connected in series with a fixed resistor, a power source and the unknown resistor. In this application, meter displacement is reversed; that is, when the unknown resistance is zero, the meter will indicate maximum (full scale) deflection and the meter is adjusted to read "zero" by means of the adjustable shunt resistor. The meter scale is calibrated backwards, with the left end of the scale being infinity and the right end being zero. The scale is nonlinear, with the graduations crowded together near the high end.

Measurement of Alternating Current and Voltage   The d'Arsonval movement can also be used for measuring ac voltages or current by the use of a diode rectifier to convert the ac to a dc value. The pointer deflection of the meter is proportional to the average value of the unidirectional, pulsating dc wave produced by the rectifier. Commonly, ac meter deflection is proportional to the rms value of the measured sine wave (see Chapter 3) and the scale is calibrated for 1.11 times the actual current or voltage that would be read by a correctly scaled dc meter. A meter calibrated in this fashion will show

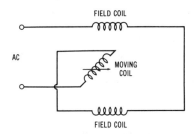

Figure 3

DYNAMOMETER-TYPE METER

Magnetic field is provided by series-connected field coils.

an *rms reading* regardless of the waveform of the ac wave.

For higher accuracy ac measurements at low frequencies (below 2000 Hz) an *electrodynamometer*-type meter (figure 3) is commonly used. It is similar to the d'Arsonval meter except that the magnetic field in which the coil moves is provided by a pair of field coils rather than by a permanent magnet. All coils are connected in series and the measured current flows through all of them, providing a true rms measurement. Power is consumed by this type of meter and it does not have the high sensitivity of the true d'Arsonval movement.

The dynamometer movement can be used for ac power measurements as shown in figure 4. The field coils are series connected with the load so that field current is proportional to load current. A resistor is connected in series with the moving coil across the line so that coil voltage is proportional to line voltage. The meter now reads the

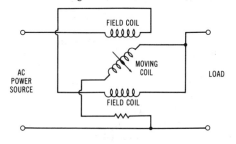

Figure 4

AC WATTMETER

Fixed coil field is proportional to load current and moving coil field is proportional to line voltage.

product of volts and amperes, or watts, assuming the power factor is unity.

A *watt-hour* meter is used to record the use of electrical power in a system. This instrument, instead of a moving coil, uses a pivot-mounted metal disc and field coils energized by the voltage and current in the circuit. The interacting magnetic fields produced by the coils cause the disc to rotate at a speed proportional to the voltage and current. The shaft of the disc is geared to a counting mechanism that records power consumption in terms of watt-hours.

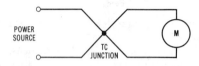

Figure 5

THERMOCOUPLE METER

Thermocouple junction develops voltage when heated by current passing through it. Thermoelectric voltage is impressed on meter movement.

High frequency alternating current can be measured by means of a *thermocouple* meter (figure 5). This provides a true rms reading. The thermocouple consists of a dissimilar-metal junction which produces a small voltage when heated. The thermoelectric voltage is impressed on the meter movement.

The Multimeter The d'Arsonval meter can provide a number of different measurements and ranges by the use of selectable shunts and multipliers. Typical multimeters provide measurement of ac and dc volts, milliamperes, amperes and ohms. In all cases, the instrument must extract current from the circuit in order to deflect the meter movement. In extracting the current, the multimeter (or *voltohmmeter*, VOM) can often load the circuit and cause inaccurate measurements. The loading effect depends upon the internal resistance of the multimeter. The resistance of a multimeter depends on the full scale current drawn, and is usually specified in *ohms per volt*. This means that the resistance of a multimeter on the 100-volt range could be 100,000 ohms, or 1000 ohms per volt. A more

sensitive instrument might exhibit 2,000,000 ohms on the 100-volt range, or 20,000 ohms per volt. For a high resistance circuit, the second multimeter would be acceptable, whereas the first would not.

Generally speaking, the resistance of the multimeter should be at least ten times the source impedance of the voltage measured in order to avoid excessive error in the reading. Circuits with low voltage and high impedance place critical demands on the multimeter. For highest accuracy, the multimeter resistance should exceed twenty-five times the source impedance, reducing the loading error to less than two percent, which is within the accuracy range of the meter movement.

Since the ohmmeter section of the multimeter contains a battery, care must be taken when measuring or otherwise checking circuits and components that are voltage sensitive. Some semiconductors and small electrolytic capacitors can be destroyed if the multimeter supplies excessive voltage or current during measurement.

31-2 The Digital Voltmeter (DVM)

The common d'Arsonval (analog) instrument is available at medium cost with an accuracy of ±2 percent of full-scale reading. Laboratory instruments accurate to ±0.5 percent are obtainable at a much higher price. The new *digital meter* provides an accuracy of ±0.5 percent in the less costly models and as high as ±0.001 percent in the laboratory models. Since the digital meter displays the measurement numerically, rather than as a pointer movement over a graduated scale, it is easier to read and reduces operator error.

The digital meter is basically an analog-to-digital converter with a numeric display. The digital meter is commonly available as a *digital voltmeter* (DVM), *digital panel meter* (DPM), or *digital multimeter* (DMM).

The heart of an electronic digital meter is the device which converts analog voltage to a digital form. This is known as analog-to-digital conversion, or ADC. Most digital meters make the conversion by one of four systems:

Successive Approximation—This instrument converts the input voltage into digital form by a series of approximations and decisions. The device consists of a digital storage register, a digital-to-analog converter, an error detector, a precision voltage reference, and control circuitry. The input voltage is compared first with the most significant reference bit. If the input voltage is less than the most significant bit of the reference, the most significant bit of the register is cleared and the next lower bit is switched in for comparison. The process of switching in the next lower significant bit is continued until a decision is made on all digits. At this point, the voltmeter has completed its measurement.

Continuous Balance—This type of meter performs a digital measurement by comparing the unknown voltage against a voltage derived from a reference source. At the beginning of a measurement, the unknown is compared to the "full-scale" reference. If a null is not reached, a voltage derived from the reference is reached by an incremental

value representing a unit of the least significant digit by automatically switching precision resistors. This process continues until a null is reached.

Ramp (Voltage to Time Conversion)— The ramp meter measures the length of time it takes for a linear ramp of voltage to become equal to the unknown input voltage after starting from a known level. This time period is measured with an electronic time-interval counter and is displayed on an in-line indicating device. A block diagram of a ramp-type DVM is shown in figure 6. A voltage ramp is generated and compared with the unknown voltage and with zero voltage. Coincidence with either voltage starts an oscillator and the electronic counter registers the cycles. Coincidence with the second comparator stops the oscillator. The elapsed time is proportional to the time the ramp takes to go between the unknown voltage and zero volts, or vice versa. The order in which the pulses come from the two comparators indicates the polarity of the unknown voltage.

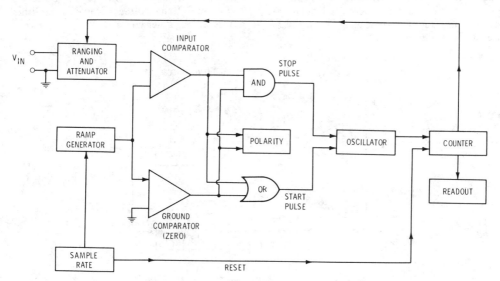

Figure 6

BLOCK DIAGRAM OF RAMP-TYPE DVM

This represents a typical ramp-type DVM. A voltage ramp is generated by a ramp generator and is compared with the unknown input signal in the input comparator and also with zero voltage. Coincidence with either voltage starts an oscillator and the counter registers the cycles. Coincidence with the other comparator stops the oscillator. The elapsed time is proportional to the time the ramp voltage takes to go between the unknown voltage and zero volts, or vice versa. The order in which the pulses come from the two comparators indicates the polarity of the unknown voltage. The accumulated reading in the counter can be used to control ranging circuits.

Integrating (Voltage to Frequency Conversion)—This device measures the true average of the input voltage over a fixed encoding time instead of measuring the voltage at the end of the encoding time as do ramp-type instruments. Conversion of a voltage to a frequency is accomplished in the manner shown in figure 7. The circuit functions as a feedback control system which governs the rate of pulse generation, making the average voltage of the rectangular pulse train equal to the dc input voltage.

A positive unknown voltage results in a negative-going ramp voltage at the output of the integrator which continues until it reaches a voltage level that triggers the level detector, which, in turn, triggers the pulse generator. The pulse from that device tends to discharge capacitor C to bring the input of the integrator back to the starting level. The entire cycle then repeats. Since the ramp slope is proportional to the input voltage, a steeper slope causes the ramp to have a shorter time duration and the pulse repetition rate is consequently higher. As the repetition rate is proportional to the input voltage, the pulses can be counted during a given time interval to obtain a digital measure of the input voltage.

A variation of the voltage-to-frequency conversion technique is the *dual-slope instrument* that makes a two-step measurement that combines integration in the first step with automatic comparison of its internal standard in the second. This technique rejects noise because of integration and achieves good stability from comparison with the standard. Direct numerical readout is accomplished with numerical display tubes or solid-state light-emitting devices.

A form of the dual-slope digital voltmeter is the *Heathkit Digital Multimeter IM-102*. This instrument measures ac and dc volts, ac and dc current, and resistance. All of the inputs are scaled to, or converted to, the basic measuring ranges of 200 millivolts or 2 volts, depending on the setting of the range switch. The measuring circuit is a high-impedance bipolar analog-to-digital converter. Resistance is measured by passing a scaled constant current through the unknown resistor and measuring the voltage drop across it. Alternating voltages are converted to dc by an average-sensing, rms-calibrated, converter assembly. Current is measured by the voltage drop it establishes across a shunt network.

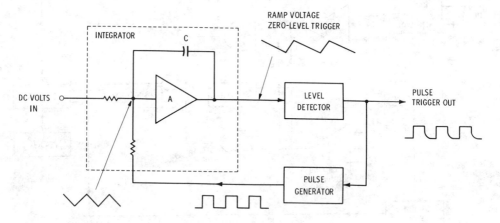

Figure 7

BLOCK DIAGRAM OF VOLTAGE-TO-FREQUENCY CONVERSION

This integrating device measures the true average of the input voltage over a fixed encoding time as do ramp-type instruments. Conversion of a voltage to a frequency takes place by a feedback control system which governs the rate of pulse generation, making the average voltage of the rectangular pulse train equal to the dc input voltage. The integrator output voltage triggers the level detector which drives the pulse generator. The pulse tends to discharge capacitor C to bring the input of the integrator back to the starting level. The repetition rate is thus proportional to the input voltage and the pulses are counted during a fixed time interval to obtain a digital measure of the input voltage.

In addition to the electronic DM, an elec-tromechanical type exists which employs stepping switches, stroboscopic devices, or analog servo systems.

The electronic DM, in addition to offering a high order of accuracy, also can provide *autopolarity,* whereby the correct polarity (either negative or positive) is automatically indicated on the display, for a measured quantity. Some instruments also feature *autoranging,* which provides switching from range to range automatically and *autozero,* whereby all zeros are displayed when no measurement is being made.

Other features include *overranging,* a feature wherein some indication (usually a blinking light or flashing display) that the quantity being measured is too high in value for the range selected and *lead-compensated resistance,* wherein the resistance of the mea-surement leads is nulled out with a front panel control.

31-3   Electronic Voltmeters

An *electronic voltmeter* is essentially a detector in which a change in the input signal will produce a change in the indicat-ing instrument (usually a d'Arsonval meter) placed in the output circuit. A *vacuum-tube voltmeter* (vtvm) may use a diode rectifier and several amplifying tubes, whereas a *solid-state voltmeter* makes use of transistors or ICs for the measurement of alternating or direct current.

When an electronic voltmeter is used in dc measurement it is used primarily because of the very great input resistance of the device. Thus, the electronic voltmeter may be used for the measurement of agc, afc, and discriminator output voltages where no loading of the circuits can be tolerated.

The electronic voltmeter requires a closed dc path for proper operation and—like the simple meter—can be overloaded and, thus,

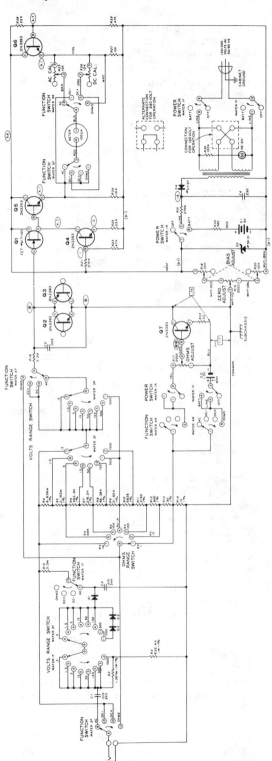

Figure 8

HEALTH SOLID-STATE VOLTMETER IM-16

is limited in the amplitude of the voltage the input circuit can handle. Modern electronic voltmeters have an input resistance of 10 megohms, or more, and usually incorporate a series resistance of 1 megohm, or more, to isolate the electronic voltmeter circuit from the circuit under test.

The Solid-State Voltmeter The circuit of a solid-state voltmeter is shown in figure 8. The three input circuits (AC Volts, DC Volts, and Ohms) are shown on the left-hand side of the schematic. These circuits perform the switching attenuation and rectification required to supply the correct voltage to the detecting and indicating circuits at the right-hand side of the schematic. Approximately 0.5 volt is required at the gate of FET input transistor Q_1 for full-scale deflection of the meter. Voltages greater than 0.5 are attenuated in the input circuits.

Input transistor Q_1 has a very high impedance gate circuit which keeps it from loading the input switching and attenuating circuits. A constant current source (Q_4), is used in place of a resistor in the source circuit of the FET. *Bias adjust* and *zero adjust* controls are provided to set the meter pointer to zero when no signal voltage is passed through the input circuits.

Transistors Q_2 and Q_3, together with a 3.3-megohm series input resistor, are used to protect the input FET from accidental overload. The reverse-connected transistors perform like a 9-volt zener diode, short circuiting higher input voltages by virtue of the drop across the series input resistor.

The meter movement is driven by the voltage applied to the output circuit by Q_1. The source of Q_1 is directly coupled to the base of Q_5. Transistors Q_5 and Q_6 are used as emitter followers to provide the power to drive the meter. When the circuit is properly adjusted, no current flows through the meter without a signal being applied to Q_1.

Since the source of current Q_1 is constant and Q_5 is a direct-coupled emitter follower, voltage variations at the input of Q_1 are transferred to the meter circuit; a negative going input signal causing the meter pointer to move backwards. Meter polarity may be reversed so that negative going input voltages cause forward meter readings. The *zero adjust* control, moreover, varies the gate bias on Q_1 by introducing a positive voltage in series with the source which is returned to a "floating" negative return bus.

31-4   Power Measurements

Audio-frequency or radio-frequency power in a resistive circuit is most commonly and most easily determined by the indirect method, i.e., through the use of one of the following formulas:

$$P = EI, \ P = E^2/R, \ P = I^2R$$

These three formulas mean that if any two of the three factors determining power are known (resistance, current, voltage) the power being dissipated may be determined. In an ordinary 120-volt ac line circuit the above formulas are not strictly true since the power factor of the load must be multiplied into the result—or a direct method of determining power such as a wattmeter may be used. But in a resistive a-f circuit and in a resonant r-f circuit the power factor of the load is taken as being unity.

For accurate measurement of a-f and r-f power, a *thermogalvanometer* or *thermocouple* ammeter in series with a noninductive resistor of known resistance can be used. The meter should have good accuracy, and the exact value of resistance should be known with accuracy. Suitable dummy-load resistors are available in various resistances in ratings up to thousands of kilowatts. These are virtually noninductive, and may be considered as a pure resistance up to 150 MHz depending on the design.

Sine-wave power measurements (r-f or single-frequency audio) may also be made through the use of a high impedance voltmeter and a resistor of known value. In fact a solid-state voltmeter of the type shown in figure 8 is particularly suited to this work. The formula $P = E^2/R$ is used in this case. However, it must be noted that some devices indicate the *peak* value of the ac wave. This reading must be converted to the rms or *heating* value of the wave by multiplying it by 0.707 before substituting the voltage value in the formula. (*Note:* Some solid-state multimeters are *peak reading* but are calibrated rms on the meter scale).

Power may also be measured through the use of a *calorimeter*, by actually measuring the amount of heat being dissipated. Through the use of a water-cooled dummy-load resistor this method of power output determination is being used by some of the most modern broadcast stations.

Power may also be determined *photometrically* through the use of a voltmeter, ammeter, incandescent lamp used as a load resistor, and a photographic exposure meter. With this method the exposure meter is used to determine the relative visual output of the lamp running as a dummy-load resistor and of the lamp running from the 120-volt ac line. A rheostat in series with the lead from the ac line to the lamp is used to vary its light intensity to the same value (as indicated by the exposure meter) as achieved as a dummy load. The ac voltmeter in parallel with the lamp and ammeter in series with it is then used to determine lamp power input by: $P = EI$. This method of power determination is satisfactory for audio and low-frequency r-f but is not satisfactory for vhf work because of variations in lamp efficiency due to uneven heating of the filament.

Finally, r-f power may be measured by means of a *directional coupler*, as discussed later in this chapter.

The Dummy Load   A suitable r-f load for power up to a few watts may be made by paralleling 2-watt composition resistors of suitable value to make a 50-ohm resistor of adequate dissipation.

A 2-kW dummy load having an SWR of less than 1.05 to 1 at 30 MHz is shown in figures 9, 10, and 11. The load consists of twelve 600-ohm, 120-watt *Globar* type CX noninductive resistors connected in parallel. A frequency-compensation circuit is used to balance out the slight capacitive reactance of the resistors. The compensation circuit is mounted in an aluminum tube 1″ in diameter and 2⅝″ long. The tube is plugged at the ends by metal discs, and is mounted to the front panel of the box.

The resistors are mounted on aluminum T-bar stock and are grounded to the case at

Figure 9

2-KILOWATT DUMMY LOAD FOR 3-30 MHz

Load is built in case measuring 22″ deep, 11″ wide and 5″ high. Meter is calibrated in watts against microampere scale as follows: (1) 22.3 µA. (5), 50 µA. (10), 70.5 µA. (15), 86.5 µA. (20), 100 µA. Scale may be marked off as shown in photograph. Calibration technique is discussed in text. Alternatively, a standing-wave bridge (calibrated in watts) may be used to determine power input to load.

Vents in top of case, and ¼-inch holes in chassis permit circulation of air about resistors. Unit should be fan-cooled for continuous dissipation.

the rear of the assembly. Connection to the coaxial receptacle is made via copper strap.

The power meter is calibrated using a solid-state voltmeter and r-f probe. Power is applied to the load at 3.5 MHz and the level is adjusted to provide 17.6 volts at "Calibration point." With the *Watts Switch* in the 200-watt position, the potentiometer is adjusted to provide a reading of 100 watts on

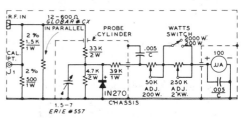

NOTE: FIXED RESISTORS ARE *OHMITE "LITTLE DEVIL"* COMPOSITION UNITS.

Figure 10

SCHEMATIC, KILOWATT DUMMY LOAD

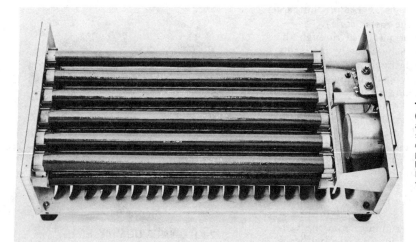

Figure 11

DUMMY-LOAD ASSEMBLY

Twelve Globar resistors (surplus) are mounted to aluminum "Tee" stock six to a side, in fuse clips. Right end is supported by ceramic pillars from front panel. Probe, meter, and potentiometers are at right.

the meter. In the 2000-watt position, the other potentiometer is adjusted for a meter reading of 200 watts. The excitation frequency is now changed to 29.7 MHz and the 17.6-volt level re-established. Adjust the frequency compensating capacitor until the meter again reads 100 watts. Recheck at 3.5 MHz and repeat until the meter reads 100 watts at each frequency when 17.6-volt level is maintained.

31-5   Measurement of Circuit Constants

The measurement of the resistance, capacitance, inductance, and Q (figure of merit) of the components used in communications work can be divided into three general methods: the impedance method, the substitution or resonance method, and the bridge method.

The Impedance Method   The *impedance method* of measuring inductance and capacitance can be likened to the ohmmeter method for measuring resistance. An ac voltmeter, or milliammeter in series with a resistor, is connected in series with the inductance or capacitance to be measured and the ac line. The reading of the meter will be inversely proportional to the impedance of the component being measured. After the meter has been cali-

brated it will be possible to obtain the approximate value of the impedance directly from the scale of the meter. If the component is a capacitor, the value of impedance may be taken as its reactance at the measurement frequency and the capacitance determined accordingly. But the dc resistance of an inductor must also be taken into consideration in determining its inductance. After the dc resistance and the impedance have been determined, the reactance may be determined from the formula: $X_L = \sqrt{Z^2 - R^2}$. Then the inductance may be determined from: L equals $X_I/2\pi f$.

The Substitution Method   The *substitution method* is a satisfactory system for obtaining the inductance or capacitance of high-frequency components. A large variable capacitor with a good dial having an accurate calibration curve is a necessity for making determinations by this method. If an unknown inductor is to be measured, it is connected in parallel with the standard capacitor and the combination tuned accurately to some known frequency. This tuning may be accomplished either by using the tuned circuit as a wavemeter and coupling it to the tuned circuit of a reference oscillator, or by using the tuned circuit in the controlling position of a two terminal oscillator such as a dynatron or transitron. The capacitance required to tune this first frequency is then noted as C_1. The circuit or the oscillator is then tuned to the *second harmonic* of this first frequency and the amount of capacitance again noted, this time as C_2. Then the distributed capacitance across the coil (including all stray capacitances) is equal to: $C_0 = (C_I - 4C_2)/3$.

This value of distributed capacitance is then substituted in the following formula along with the value of the standard capacitance for either of the two frequencies of measurement:

$$L = \frac{1}{4\pi^2 f_1^2 \, (C_1 + C_0)}$$

The determination of an unknown capacitance is somewhat less complicated than the above. A tuned circuit including a coil, the unknown capacitor and the standard capacitor, all in parallel, is resonated to some convenient frequency. The capacitance of the standard capacitor is noted. Then the unknown capacitor is removed and the circuit re-resonated by means of the standard capacitor. The difference between the two readings of the standard capacitor is then equal to the capacitance of the unknown capacitor.

31-6 Measurements With a Bridge

The Wheatstone Bridge   Experience has shown that one of the most satisfactory methods for measuring circuit constants (resistance, capacitance, and inductance) at audio frequencies is by means of the ac bridge. The *Wheatstone (dc) bridge* is also one of the most accurate methods for the measurement of dc resistance. With a simple bridge of the type shown in figure 12A it is entirely practical to obtain dc resistance determinations accurate to four significant figures. With an ac bridge operating within its normal rating as to frequency and range of measurement it is possible to obtain results accurate to three significant figures.

Both the ac and the dc bridges consist of a source of energy, a standard or reference of measurement, a means of balancing this standard against the unknown, and a means

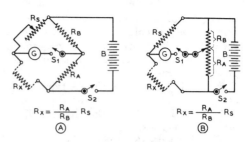

Figure 12

TWO WHEATSTONE BRIDGE CIRCUITS

These circuits are used for the measurement of dc resistance. In A the "ratio arms" R_B and R_A are fixed and balancing of the bridge is accomplished by variation of the standard R_S. The standard in this case usually consists of a decade box giving resistance in 1-ohm steps from 0 to 1110 or to 11,110 ohms. In B a fixed standard is used for each range and the ratio arm is varied to obtain balance. A calibrated slide wire or potentiometer calibrated by resistance in terms of degrees is usually employed as R_A and R_B. It will be noticed that the formula for determining the unknown resistance from the known is the same in either case.

of indicating when this balance has been reached. The source of energy in the dc bridge is a battery; the indicator is a sensitive galvanometer. In the ac bridge the source of energy is an audio oscillator (usually in the vicinity of 1000 Hz), and the indicator is usually a pair of headphones or a sensitive meter. The standard for the dc bridge is a resistance, usually in the form of a decade box. Standards for the ac bridge can be resistance, capacitance, and inductance in varying forms.

Figure 12 shows two general types of the Wheatstone or dc bridge. In A the so-called "ratio arms" (R_A and R_B) are fixed (usually in a ratio of 1 to 1, 1 to 10, 1 to 100, or 1 to 1000) and the standard resistor (R_S) is varied until the bridge is in balance. In commercially manufactured bridges there are usually two or more buttons on the galvanometer for progressively increasing its sensitivity as balance is approached. Figure 12B is the *slide-wire type* of bridge in which fixed standards are used and the ratio arm is continuously variable. The slide wire may actually consist of a moving contact along a length of wire of uniform cross section in which case the ratio of R_A to R_B may be read off directly in centimeters or inches, or in degrees of rotation if the slide wire is bent around a circular former. Alternatively, the slide wire may consist of a linear-wound potentiometer with its dial calibrated in degrees or in resistance for each end.

Figure 13A shows a simple type of ac bridge for the measurement of capacitance and inductance. It can also, if desired, be used for the measurement of resistance. It is necessary with this type of bridge to use a standard which presents the same type of impedance as the unknown being measured: resistance standard for a resistance measurement, capacitance standard for capacitance, and inductance standard for inductance determination.

The Wagner Ground For measurement of capacitances from a few picofarads to about 0.001 μF, a *Wagner-grounded substitution capacitance bridge* of the type shown in figure 13B will be found satisfactory. The ratio arms R_A and R_B should be of the same value within 1 percent; any value between 2500 and 10,000 ohms for both will be satisfactory. The two resistors R_C and R_D should be 1000-ohm wirewound potentiometers. C_S should be a straight-line capacitance capacitor with an accurate vernier dial; 500 to 1000 pF will be satisfactory. C_C can be a two- or three-gang broadcast capacitor from 700 to 1000 pF maximum capacitance.

The procedure for making a measurement is as follows: The unknown capacitor C_X is placed in parallel with the standard capacitor C_S. The *Wagner ground* (R_D) is varied back and forth a small amount from the center of its range until no signal is heard in the phones with the switch (S) in the center position. Then the switch (S) is placed in either of the two outside positions, C_C is adjusted to a capacitance somewhat greater than the assumed value of the unknown C_X, and the bridge is brought into balance by variation of the standard capacitor (C_S). It may be necessary to cut some resistance in at R_C and to switch to the other outside position of S before an exact balance can be obtained. The setting of C_S is then noted, C_X is removed from the circuit (but the leads which went to it are not changed in any way which would alter their mutual capacitance), and C_S is read-

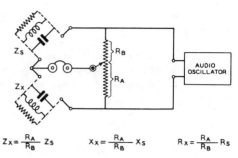

$$Z_x = \frac{R_A}{R_B} Z_S \qquad X_x = \frac{R_A}{R_B} X_S \qquad R_x = \frac{R_A}{R_B} R_S$$

Z_X = IMPEDANCE BEING MEASURED, R_S = RESISTANCE COMPONENT OF Z_S
Z_S = IMPEDANCE OF STANDARD, X_X = REACTANCE COMPONENT OF Z_X
R_X = RESISTANCE COMPONENT OF Z_X, X_S = REACTANCE COMPONENT OF Z_S

(A)

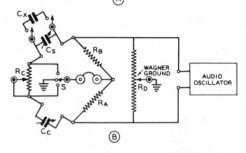

(B)

Figure 13

TWO AC BRIDGE CIRCUITS

The operation of these bridges is essentially the same as those of figure 12 except that ac is fed into the bridge instead of dc and a pair of phones is used as the indicator instead of the galvanometer. The bridge shown at A can be used for the measurement of resistance, but it is usually used for the measurement of the impedance and reactance of coils and capacitors at frequencies from 200 to 1000 Hz. The bridge shown at B is used for the measurement of small values of capacitance by the substitution method. Full description of the operation of both bridges is given in the accompanying text.

justed until balance is again obtained. The difference in the two settings of C_S is equal to the capacitance of the unknown capacitor C_X.

31-7   The R-F Bridge

The basic bridge circuits are applicable to measurements at frequencies well up into the uhf band. While most of the null circuits used from dc to about 100 MHz are adaptations of the fundamental Wheatstone

Bridge circuit, many other types of networks that can be adjusted to give zero transmission are employed at higher frequencies.

At very-high frequencies, where impedances can no longer be treated as lumped elements, null circuits based on coaxial line techniques are used. The upper frequency limit of conventional bridge circuits using lumped parameters is determined by the magnitude of the residual impedance of the elements and the leads. The corrections for these usually become unmanageable at frequencies higher than 100 MHz or so.

The "General Radio," Bridge An *r-f bridge* suitable for use up to about 60 MHz is shown in figure 14. The bridge can measure resistances up to 1000 ohms and reactances over the range of plus or minus 5000 ohms at 1 MHz. The reactance range varies inversely as the frequency in MHz. Measurements are made by a series-substitution method in which the bridge is first balanced by means of capacitors C_P and C_A with a short-circuit across the unknown terminals. The short is then removed, the unknown impedance connected in its place, and the bridge rebalanced. The unknown

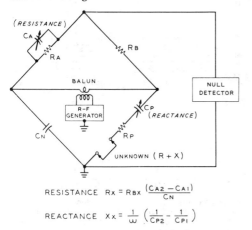

$$\text{RESISTANCE} \quad R_X = R_{BX} \frac{(C_{A2} - C_{A1})}{C_N}$$

$$\text{REACTANCE} \quad X_X = \frac{1}{\omega} \left(\frac{1}{C_{P2}} - \frac{1}{C_{P1}} \right)$$

Figure 14

THE "GENERAL RADIO" R-F BRIDGE

This bridge is suitable for r-f measurements up to 60 MHz or so. Calibrated reactance (C_P) and resistance (C_A) dials allow direct measurements at 1 MHz. At other frequencies reactance reading must be divided by the frequency in MHz. Wide band balun input transformer allows bridge to be driven from signal generator via a coaxial line.

resistance and reactance values are then read from the difference between the initial and final balances.

The bridge measures the equivalent series resistance and reactance $(r + jx)$ of the unknown impedance, whereas some other types of r-f instruments provide an answer in terms of an equivalent parallel combination of resistance and reactance or conductance and susceptance. The numerical results between equivalent series and parallel measurements will not be the same although the equivalent series impedance can be mathematically converted into the equivalent parallel impedance (or admittance) and vice versa.

A vhf variation of the r-f bridge provides direct measurements up to 500 MHz by sampling the electric and magnetic fields in a transmission line. Two attenuators are controlled simultaneously; one receives energy proportional to the electric field in the line, and the other receives energy proportional to the magnetic field. The magnitude of the unknown impedance is determined by adjusting this combination for equal output from each attenuator. The two equal signals may also be applied to opposite ends of another transmission line, and phase angle can be determined from their point of cancellation.

Above 500 MHz, impedance measurements are normally determined by inserting a detector probe in a slotted section of transmission line, as discussed in the next section of this chapter.

The R-X Meter A version of the r-f bridge is the *R-X meter*. This device is a package combination of an r-f generator and detector, plus a calibrated r-f bridge. The R-X meter reads the parallel combination of resistance and reactance over a frequency range of 500 kHz to 250 MHz. The resistance range is 15 to 100,000 ohms and the reactive range is zero to 20 pF, capacitive. Inductive reactance is measured in terms of negative capacitance, the value being equal to that required to resonate the negative capacitance reading at the test frequency. The maximum value of negative capacitance readable is 100 pF.

The latest development in impedance measuring devices is the *vector impedance meter* which provides magnitude and phase angle of an unknown at a given frequency on two panel meters. Operational range of a typical device is 500 kHz to 108 MHz and the reading is in the form of a series equivalent impedance.

31-8   Antenna and Transmission-Line Instrumentation

The degree of adjustment of any amateur antenna can be judged by a study of the standing-wave ratio on the transmission line feeding the antenna. Various types of instruments have been designed to measure the ratio of forward to reflected power by sampling the r-f incident and reflected waves on the transmission line, or to measure the actual radiation resistance and reactance of the antenna in question. The most important of these instruments are the *slotted line,* the *directional coupler,* and the *r-f imped-ance bridge.*

The Slotted Line   The relationship between the incident and the reflected power and standing wave present on a transmission line is expressed by:

$$K = \frac{1 + R}{1 - R}$$

where,

K = Standing-wave ratio,

R = Reflection coefficient, or ratio of relative amplitude of reflected signal to incident signal.

When measurements of a high degree of accuracy are required, it is necessary to insert an instrument into a section of line in order to ascertain the conditions existing within the shielded line. For most vhf measurements, wherein a wavelength is of manageable proportions, a *slotted line* is the

instrument frequency used. Such an instrument, shown in figure 15, is an item of test equipment which could be constructed in a home workshop which includes a lathe and other metal-working tools. Commercially built slotted lines are very expensive since they are constructed with a high degree of accuracy for precise laboratory work. The slotted line consists essentially of a section of air-dielectric line having the same characteristic impedance as the transmission line into which it is inserted. Tapered fittings for the transmission line connectors at each end of the slotted line usually are required due to differences in the diameters of the slotted line and the line into which it is inserted. A narrow slot from ⅛ inch to ¼ inch in width is cut into the outer conductor of the line. A probe then is inserted into the slot so that it is coupled to the field inside the line. Some sort of accurately machined track or lead screw must be provided to ensure that the probe maintains a constant spacing from the inner conductor as it is moved from one end of the slotted line to the other. The probe usually includes some type of rectifying element whose output is fed to an indicating instrument alongside the slotted line.

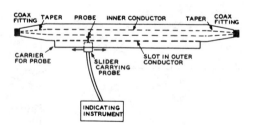

Figure 15

THE UHF SLOTTED LINE

The conductor ratios in the slotted line, including the tapered end sections, should be such that the characteristic impedance of the equipment is the same as that of the transmission line with which the equipment is to be used. The indicating instrument may be operated by the dc output of the rectifier coupled to the probe, or it may be operated by the ac components of the rectified signal if the signal generator or transmitter is amplitude-modulated at a constant percentage.

The unfortunate part of the slotted-line system of measurement is that the line must be somewhat over one-half wavelength long at the test frequency, and for best results should be a full wavelength long. This requirement is easily met at frequencies of 420 MHz and above where a full wavelength of 28 inches or less. But for the lower frequencies such an instrument is mechanically impractical.

The Directional Coupler The r-f voltage on a transmission line may be considered to have two components. The *forward component* (incident component) and the *reverse component* (reflected component). The reverse component is brought about by operation of the line when terminated on a load that is unequal to the characteristic impedance of the line.

A *directional coupler* is an instrument that can sense either the forward or reflected components in a transmission line by taking advantage of the fact that the reflected components of voltage and current are 180 degrees out of phase while the forward components of voltage and current are in phase.

The directional coupler is inserted in the transmission line at an appropriate location. For a coaxial line, the instrument consists of a short section of line containing a small loop coplanar with the inner conductor (figure 16). The loop is connected through a resistor to the outer conductor, and this resistor is capacitively coupled to the inner conductor of the line. The voltage appearing across the series arrangement of loop and resistor is measured when the voltage across the resistor and the voltage induced in the loop are aiding and again when they are in opposition to each other. By rotating the loop through 180 degrees, the readings may be used to determine the amount of mismatch and the power carried by the line. Operation is substantially independent of load impedance and meter impedance at any frequency within the useful range of the instrument.

When the directional coupler is used to measure the SWR or the reflection coefficient on the line, the value obtained for these quantities depends only on the *ratio* of the two measured voltages. Power measurements are more stringent, since the absolute value of transmission line voltage must be determined and construction of a simple, compact r-f voltmeter that presents a linear reading over a wide frequency range and at various power levels is not simple.

In order to sample forward and reverse power, it is necessary to reverse the orientation of the directional coupler in the line, or to employ two couplers built in one unit

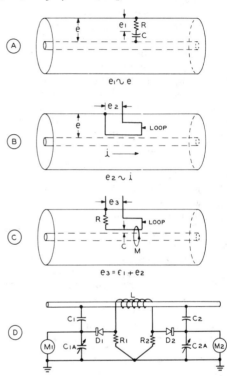

Figure 16

THE DIRECTIONAL COUPLER

The directional coupler (reflectometer) is a coaxial-line section containing an r-f voltmeter which reads the incident or reflected component of voltage, depending on the position of the pickup device in the line.

A—Voltage relationships for a series resistance-capacitance combination placed between the conductors of a coaxial line; e_1 is proportional to e.

B—Loop coupled to inner conductor will give voltage (e_2) proportional to current flowing in line (i).

C—Representation of reflectometer. Capacitance is provided by proximity of loop to inner conductor.

D—Double reflectometer provides simultaneous measurement of incident and reflected voltages. Ferrite core is placed around center conductor, with secondary winding acting as a coupling loop.

but oriented oppositely. It is necessary, moreover, to have both couplers identical in *coupling factor* and *directivity*.

Directivity   The fraction of forward power that is sampled by the coupler is termed the *coupler factor*, and the *directivity* is the ability of the coupler to discriminate between opposite directions of current flow. If, for example, one percent of the power is coupled out, the coupling factor is 20 decibels. If the coupler is now reversed to sample the power in the reverse direction, it may couple out, say 0.001 percent, of the forward power even though there may be actually no reflected power. It is thus coupling out an amount of power 50 decibels below the power in the line. The discrimination between forward and reverse power is the difference between the coupled values, or 30 decibels. A *directivity* of 30 dB is common for better types of reflectometers and SWR measurements derived from the measured reflection coefficient are sufficiently accurate for adjustment of simple beam antennas. It should be noted, however, that it is difficult to make measurements with any degree of accuracy at low SWR values with inexpensive directional couplers, because the directivity power ratio at SWR values below about 1.5/1 or so falls within the error limits of directivity capability of all but the best and most expensive reflectometers.

The SWR Bridge   The SWR bridge is a useful device for determining the standing-wave ratio on, and the power transmitter along, a transmission line. When the SWR on a given line is unity, the line is terminated in a pure resistance equal to the characteristic impedance of the line. If the line and terminating load are made part of an r-f bridge circuit, the bridge will be in a balanced condition when the SWR is unity (figure 17). A sensitive r-f voltmeter connected across the bridge will indicate balance and the magnitude of bridge unbalance, and may be calibrated in terms of SWR, power, or both. It may be seen in figure 17A that the meter reading is proportional to bridge unbalance, and is thus proportional to the reflected power and is not influenced by the forward power in the

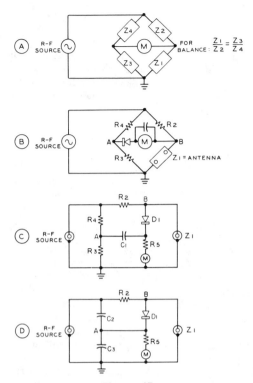

Figure 17

THE BRIDGE DIRECTIONAL COUPLER

A—When r-f bridge is balanced any change in load (Z_1) will result in bridge unbalance and cause a reading on meter M. Reading is due to reflected voltage. SWR may be derived from:

$$SWR = \frac{E_o - E_r}{E_o + E_r}$$

where,
  E_o equals incident voltage,
  E_r equals reflected voltage.
B—Equivalent bridge circuit. Bridge must be individually calibrated since performance differs from formula due to nonlinearity of voltmeter circuit loading, and line discontinuity introduced by presence of bridge.
C, D—Practical bridge circuits having one side of meter grounded to line.

circuit. The meter will read zero if, and only if, the transmission line is properly terminated in Z_1 so that $Z_1 = Z_0$ of the line, so as to have unity standing-wave ratio.

Various forms of the SWR resistance bridge exist as shown in the illustration, but all of them are based on the principle of measurement of bridge balance by means of a null-indicating meter. Circuit B consists of two resistive voltage dividers across the r-f source, with an r-f voltmeter reading the difference of potential across the points A and B. Circuit C is identical, but redrawn so as to show a practical layout for measurement in a coaxial system with one side of the generator and the r-f voltmeter at ground potential. Circuit D is similar, except that one of the voltage dividers of the bridge is capacitive instead of resistive.

SWR Bridge Designs   Various forms of the SWR bridge are shown in figure 18. Circuit A is the *Micromatch* capacitance bridge. In order to pass appreciable power through the bridge, the series resistor is reduced to one ohm, thus requiring the capacitance divided to maintain about the same ratio as set in the resistive arm. For a 50-ohm transmission line, the transformation ratio is 50/1, and the 25-pF variable capacitor must be set at a value corresponding to about fifty times the reactance of the 820-pF capacitor. The power-handling capability of the bridge is limited by the dissipation capability of the 1-ohm resistor

Circuit B incorporates a differential capacitor to obtain an adjustable bridge ratio. The capacitor may be calibrated in terms of the unknown load and may be used to indicate resistive loads in the range of 10 to 500 ohms. The bridge has an advantage over the circuits of illustrations A and C in that it may be used in the manner of a simple impedance bridge to determine the radiation resistance of a *resonant* antenna. The bridge is placed at the antenna terminals, and the frequency of the driving source and the setting of the differential capacitor are varied to produce a null indication on the meter. The null occurs at the resonant frequency of the antenna, and the radiation resistance at that frequency may be read from the instrument.

A less-expensive variation of the variable r-f bridge is shown in illustrations C and D and is called the *Antennascope*. The Antennascope is a variable bridge making use of a (relatively) noninductive potentiometer in one leg. These simple instruments are useful in antenna adjustment as they indicate the resonant frequency of the antenna

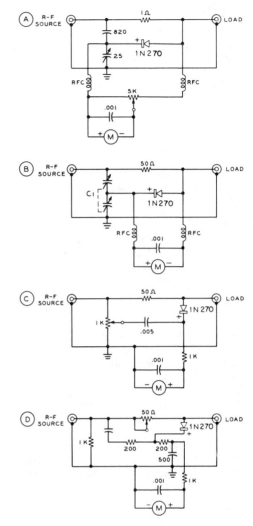

Figure 18

SWR BRIDGES

A—Micromatch bridge.
B—Capacitance ratio bridge.
C—Antennascope.
D—Antennascope with calibrating resistor in active leg of bridge.
Note: Meter M may be 0-500 dc microammeter.

and the approximate radiation resistance of the driven element at this frequency. At other than the resonant frequency, the antenna exhibits a reactive component and the null of the instrument will not be complete. Even so, at the low values of impedance encountered in most amateur beam antennas, the readings obtained at frequencies off resonance approximate the resistive component of the radiation resistance of the antenna.

Construction information for a practical Antennascope and other SWR instruments will be described in the following section of this Handbook.

31-9   Practical SWR Instruments

Simple forms of the directional coupler and the SWR bridge are suited to home construction and will work well over the range of 1.8 to 150 MHz. No special tools are needed for construction and calibration may be accomplished with the aid of a handful of 1-watt composition resistors of known d-c value resistance.

The Antennascope   The *Antennascope* is a modified SWR bridge in which one leg of the bridge is composed of a noninductive variable resistor (figure 18D). This resistor is calibrated in ohms, and when its setting is equal to the radiation resistance of a resonant antenna under test, the bridge is in a balanced state. If a sensitive voltmeter is connected across the bridge, it will indicate a voltage null at bridge balance. The radiation resistance of the antenna may then be read directly from the calibrated dial of the instrument.

When the test antenna is nonresonant, the null indication on the Antennascope will be incomplete. The frequency of the exciting signal must then be altered to the resonant frequency of the antenna to obtain accurate readings of radiation resistance. The resonant frequency of the antenna, of course, is also determined by this exercise.

The circuit of the Antennascope is shown in figure 20. A 100-ohm noninductive potentiometer (R_1) serves as the variable leg of the bridge. The other legs are composed of the 200-ohm composition resistors and the radiation resistance of the antenna. If the radiation resistance of the antenna or external load under test is 50 ohms, and the potentiometer is set at midscale, the bridge is balanced and the diode voltmeter will read

zero. If the radiation resistance of the antenna is any other value between about 10 and 100 ohms, the bridge may be balanced to this new value by varying the setting on the potentiometer, which is calibrated in ohms.

Building the Antennascope—The Antennascope is constructed within an aluminum box chassis measuring about $4'' \times 2'' \times 1\frac{1}{2}''$, and placement of the major components may be seen in the photographs. A $1\frac{1}{4}$-inch diameter hole is drilled in the lower portion of the panel and the variable potentiometer is mounted in this hole on a thin piece of insulating material such as micarta or bakelite. The terminals of the potentiometer and the case are at r-f potential, so it is essential for proper bridge operation to have a minimum of capacitance between the potentiometer and ground.

The two 200-ohm, $\frac{1}{2}$-watt resistors should be matched on an ohmmeter, and a number of the 500-pF capacitors should be checked on a bridge to find two units of equal capacitance. The exact value of resistance

Figure 19

THE ANTENNASCOPE

The Antennascope may be used to measure the resonant radiation resistance of antennas at frequencies up to 150 MHz. Grid-dip oscillator is coupled to input loop of Antennascope and antenna under test is connected to output terminals with short, heavy leads.

and capacitance in either case is not critical, it is only necessary that the companion units be equal in value. Care should be taken when soldering the small resistors in the circuit to see that they do not become overheated, causing the resistance value to permanently change. In like manner, the germanium diode should be soldered in the circuit using a pair of long-nose pliers as a heat sink to remove the soldering heat from the unit as rapidly as possible.

As shown in the photographs, copper strap cut from flashing stock is used for wiring the important r-f leads. The output leads terminate in an insulated terminal strip on one side of the box and the input coupling loop is made of a section of brass rod, which is tapped at each end for 6-32 machine nuts. The loop is bent and posi-

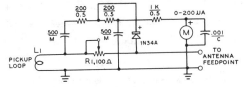

Figure 20

SCHEMATIC, ANTENNASCOPE

R₁—100-ohm composition potentiometer. Ohmite AB or Allen-Bradley type J linear taper.
L₂—2 turns brass wire to fit dip-oscillator coil. See photos
M—0 200 μa dc meter

tioned so as to slip over the coil of a dip oscillator used as the driving source.

Testing the Antennascope—When the instrument is completed, a dip oscillator may be coupled to the input link. The oscillator should be set somewhere in the 10-MHz to 20-MHz range and coupling is adjusted to obtain a half-scale reading on the meter of the Antennascope. Various values of pre-calibrated 1-watt composition resistors ranging from 10 to 90 ohms should be placed across the output terminals of the Antennascope and the potentiometer adjusted for nulls on the indicating meter. The settings of the potentiometer may then be marked on a temporary paper dial and, by interpolation, 5-ohm points can be marked on the scale for the complete rotation of the control. The dial may then be removed and inked.

This calibration will hold to frequencies well above the 2-meter band, but as the internal lead inductance of the Antennascope starts to become a factor, it will no longer be possible to obtain a complete null on the indicating meter. Wired as shown, the meter null begins to rise off zero in the region of 150 MHz.

Using the Antennascope—The Antennascope is coupled to a dip oscillator by means of the input link. Additional turns may need to be added to the link to obtain sufficient pickup below 7 MHz or so. Enough coupling should be obtained to allow at least ¾-scale reading on the meter with no load connected to the measuring terminals. For general use, the measuring terminals of the instrument are connected across the antenna terminals at the feed-

Figure 21

INTERIOR OF ANTENNASCOPE

Strap connection is made between common input and output terminals. Dip-oscillator coupling loop is at right.

point. Either a balanced or unbalanced antenna system may be measured, the "hot" lead of the unbalanced antenna connection to the ungrounded terminal of the Antennascope. Excitation is supplied from the dip oscillator and the frequency of excitation and the Antennascope control dial are varied until a complete meter null is obtained. The frequency of the source of excitation now indicates the resonant frequency of the antenna under test, and the approximate radiation resistance of the antenna may be read upon the dial of the Antennascope.

On measurements made on 40- and 80-meter antennas it may be found impossible to obtain a complete null on the Antennascope. This is usually caused by pickup of a nearby broadcast station, in which case the rectified signal of the station will obscure the null action of the Antennascope. This action is only noticed when antennas of large size are being checked.

The Antennascope is designed to be used directly at the antenna terminals without an intervening feedline. It is convenient to mount the instrument and the dip oscillator as a single package on a strip of wood. This unit may then be carried up the tower and attached to the terminal of the beam antenna. It is also possible to make remote measurements on an antenna with the use of an electrical half-wavelength of transmission line placed between the Antennascope and the antenna terminals.

The Monimatch　　The *Monimatch* is a dual reflectometer constructed from a length of flexible coaxial transmission line (figure 22). The heart of the Monimatch is a pickup line made from a 14-inch length of RG-8A/U coaxial cable. The coupling loop of this special section is a piece of No. 22 enamel or *formvar* covered wire slid under the flexible outer shield of the coaxial line for a distance of about eight inches. The coaxial pickup line is then conveniently wound around the inside walls of the mounting box so that the protruding ends of the coupling loop fall adjacent to the simple switching circuit. The coupling loop and center conductor of the coaxial line form a simple reflectometer terminated at either end by a noninductive potentiometer. Choice of termination is determined by the panel switch. When the potentiometer is adjusted to the balance point, the bridge is calibrated and ready for use. The selector switch permits reading forward or reverse power in the coaxial line and an SWR of unity is indicated by a null reading on the meter of the instrument.

The special coaxial pickup loop is easily made. A 14-inch length of RG-8A/U cable

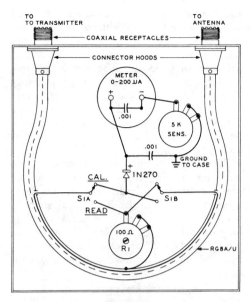

Figure 22

MONIMATCH

R₁—100-ohm composition potentiometer. Ohmite AB, or Allen Bradley type J, linear taper
S₁—Dpdt rotary switch. Centralab 1462
Case—5″ × 7″ × 2″ chassis with back plate.

is trimmed square at the ends and the outer vinyl jacket is carefully removed. Two holes to pass the pickup wire are carefully made in the outer braid of the section with the aid of an awl or needle. Be careful not to break the fine wires of the braid. The holes are made 8 inches apart, and centered on the section. The outer shield is next bunched up a bit to loosen it and a length of No. 22 insulated wire is threaded under the braid, in and out of the holes. A stiff copper wire may be threaded through the holes and used as a needle to pass the flexible copper wire under the braid. Finally, the braid is smoothed out to its original length and the pickup wire checked with an ohmmeter to make sure that no shorts exist between the braid and the wire. The braid is then wrapped with vinyl tape at the two holes. The last step is to solder connector hoods and coaxial receptacles on each end of the line, making the assembly "r-f-tight."

The special line may now be mounted in the instrument case, along with the various other components, as shown in the illustration. The calibrating potentiometer is mounted on an insulating plate in the center of a one-inch hole to reduce the capacity of the unit to ground. The coaxial line should be grounded only at the coaxial receptacles, and should otherwise be wrapped with vinyl tape to prevent it from shorting to the case or other components.

A noninductive 50-ohm dummy load is attached to the output of the Monimatch and it is driven from an r-f source. Place the panel switch in the *Calibrate* position and adjust the *sensitivity* control for a half-scale reading of the meter. Now switch to the *Read* position and adjust the sensitivity control for full-scale reading. Adjust the *Calibrate* potentiometer in the back of the Monimatch for a null in the meter reading —it should be very close to zero on the scale. Switch back to *Calibrate* again and once again adjust the sensitivity control for full-scale meter reading. Finally, switch once again to *Read* and re-null the meter with the *Calibrate* potentiometer. The Monimatch is now ready for use.

Using the Monimatch—The Monimatch is inserted in the coaxial line to the antenna, power is applied and the switch set to *Calibrate* position. The sensitivity control is adjusted for full-scale reading and the switch is thrown to the *Read* position. Adjustments to the antenna may now be made to reach an SWR of unity, at which point the meter reading will be at maximum null, or close to zero. If desired, the Monimatch may be calibrated in terms of SWR by observing the reading when various values of noninductive composition resistors of known value are measured with the device.

A Practical Reflectometer The *reflectometer* is an accurate, inexpensive and easily constructed instrument for the experimenter. Shown in this section is a practical reflectometer made from a short section of coaxial transmission line. It is designed for use with output power of up to 2000 watts and at frequencies up to 150 MHz. An easily wound toroid transformer is used for a pickup element, in conjunction with two reverse-connected diode voltmeters, affording quick indication of forward and reverse conditions within the transmission line. One voltmeter reads the incident com-

ponent of voltage and the other reads the reflected component. The magnitude of standing-wave ratio on the transmission line is the ratio of these two components.

The upper frequency limit of the reflectometer is determined by the dimensions of the pickup loop which should be a small fraction of a wavelength in size. When used to measure SWR, the resultant figure depends on the ratio of two measured voltages which are usually valid figures regardless of variations in load impedance and frequency. When used as a wattmeter, the absolute transmission-line voltage must be measured and the detection devices must have a flat frequency response with diodes operating in the square-law region for widest frequency coverage.

When used for SWR measurements, calibration of the reflectometer is not required since relative readings indicate the degree

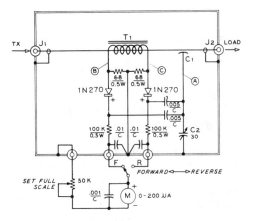

Figure 23

REFLECTOMETER

C_1—Sleeve formed of #28 tinned wire wrapped around inner dielectric of line for $3/8$-inch length. See text
T_1—40 turns #28 insulated wire equally spaced around toroid core, Q-1 material. Indiana-General CF-114, 1.25″ diameter × 0.38″ thick. See figure 25 for assembly

of mismatch and all system adjustments are conducted so as to make this ratio as high as possible, regardless of the absolute values. Power measurements may be made if the instrument is calibrated against a known dummy load in both the forward and reverse directions. The reflectometer may be left in the transmission line to indicate SWR and relative power output of the transmitter.

Building the Reflectometer—Assembly of the reflectometer is shown in figure 26. A short length of coaxial line of the chosen impedance is trimmed to length. The outer insulation and outer braid are cut with a sharp knife for a distance of about ¾ of an inch at the center of the line, exposing a section of the inner dielectric. Around the dielectric a length of No. 28 tinned wire is wound to form a sleeve about ⅜-inch long for 50-ohm cable. If 70-ohm cable is used, the sleeve should be about ⅝-inch long. The sleeve is tinned and forms capacitor C_1 to the inner conductor. A short length of insulated wire is soldered to the sleeve (*lead A*). The capacitor is now wrapped with vinyl tape. Next, a short section of thin copper shim stock is wrapped over the tape to form a simple Faraday shield which ensures that the coupling between the primary of T_1 (the inner conductor of the coaxial line) and the secondary (the winding on the ferrite core) is inductive and not capacitive. *One end* of the shield is carefully soldered to the outer braid of the coaxial line and the other end is left free.

The ferrite core is now wrapped with vinyl tape and 40 turns of No. 28 insulated wire are evenly wrapped around the core. The core is then slipped over the cable section and positioned directly above capacitor C_1. The reflectometer section is then completed by forming a copper shield around the toroid assembly. In this case, the shield is made up of two copper discs soldered to the cable braid, over which is slipped a cop-

per cylinder made of thin shim stock. The cylinder and end rings are soldered into an inclusive shield, as shown in the photograph, with the three pickup leads passing through small holes placed in the cylindrical end sections.

The reflectometer and associated components are placed in an aluminum box (figure 24) having a terminal strip attached for connection to an external reversal switch and meter. Final adjustment is accomplished by feeding power through the reflectometer into a dummy load having a low value of SWR and adjusting capacitor C_2 for minimum meter indication when the instrument is set for a reflected-power reading.

31-10 Frequency and Time Measurements

All frequency and time measurements within the United States are based on data transmitted from the *National Bureau of Standards*. Several time scales are used for time measurement: (1)—*Universal Time (UT)*. Universal time, or *Greenwich Mean Time* (GMT), is a system of mean solar time based on the rotation of the earth about its axis relative to the position of the sun. Several UT scales are used: uncorrected astronomical observations are denoted U∅; the UT time scale corrected for the earth's polar variation is denoted UT1; the UT1 scale corrected for annual variation in the rotation of the earth is denoted UT2. Time signals transmitted by standard stations are generally based on the UT2 time scale. Although UT is in common use, it is non-uniform because of changes in the earth's

Figure 24

INTERIOR VIEW OF REFLECTOMETER

Complete assembly including accessory components is placed in cast aluminum box, 4 × 2½″ × 1½″ (Pomona Electronic #2904). Calibrating capacitor is adjustable through small hole drilled in box.

speed of rotation. (2)—*Ephemeris Time* (ET). Scientific measurements of precise time intervals require a uniform time scale. The fundamental standard of constant time is defined by the orbital motion of the earth about the sun and is called Ephemeris time, and is determined from lunar observations. (3)—*Atomic Time* (AT). Molecular and atomic resonance characteristics can be used to provide time scales which are apparently constant and nearly equivalent to ET. The designation A.1 has been given to the time scale derived from the zero-field resonance of *cesium*. The U.S. Frequency Standard at Boulder, Colorado, is maintained by reference to the A.1 time scale.

Standard Radio Frequency and Time Signals High- and low-frequency time signals are broadcast on standard frequencies in the United States by the *National Bureau of Standards* over radio stations WWV and WWVB (located near Fort Collins, Colorado) and WWVH (located near Kekaha, Kauai, Hawaii). The broadcasts of WWV may also be heard by telephone by dialing (303) 499-7111, Boulder, Colorado. Station WWVH may be heard by dialing (808) 335-4363.

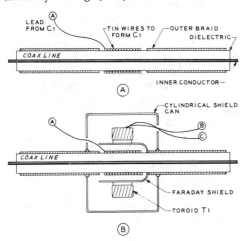

Figure 25

ASSEMBLY DETAILS OF THE REFLECTOMETER

A—Assembly of coaxial capacitor C_1.
B—Assembly of capacitor, Faraday shield and toroid transformer T_1. Leads A, B, and C connect as shown in figure 23.

Stations WWV, WWVH, and WWVB broadcast nominal frequencies and time consistent with the internationally agreed upon time scale, *Coordinated Universal Time* (UTC). WWV broadcasts on 2.5, 5, 10, 15, and 20 MHz; while WWVH broadcasts on all these frequencies except 20 MHz. Transmissions are continuous. WWVB broadcasts *Stepped Atomic Time* (SAT) on the standard frequency of 60 kHz. This station broadcasts continuously except for scheduled maintenance periods.

Frequency accuracy, offset, and effects of the propagation medium are covered in the technical bulletin *NBS Time and Frequency Dissemination Services*, NBS Special Publication 432, available from the superintendent of Documents, U.S. Government Printing Office, Washington D.C. 20402.

Time Announcements—Voice announcements are made from WWV and WWVH once every minute. To avoid confusion, a man's voice is used on WWV and a woman's voice on WWVH. The WWVH announcement occurs first, at 15 seconds before the minute, while the WWV announcement occurs at 7.5 seconds before the minute. Tone markers are transmitted simultaneously from both stations. The time referred to in the announcements is "Coordinated Universal Time," generally equivalent to "Greenwich Mean Time" (GMT). Local time may be derived from Chart 1 of Chapter 35.

Standard Time Intervals—Pulses mark the seconds of each minute, except for the 29th and 59th second pulses which are omitted. All pulses commence at the beginning of each second. In alternate minutes during most of each hour 500 Hz or 600 Hz audio tones are broadcast. A 440 Hz tone is broadcast once each hour.

Official Announcements Forty-five second announcements are available to other Federal Agencies to disseminate official and public service information. Other announcements include:

Marine Storm Warnings—Weather information about major storms in the Atlantic and eastern North Pacific are broadcast in voice from WWV at 8, 9, and 10 minutes

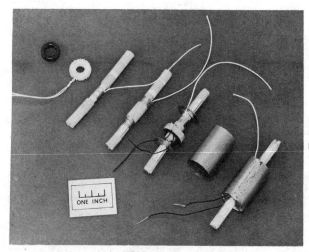

Figure 26

ASSEMBLY SEQUENCE OF REFLECTOMETER UNIT

Left-to-Right—Toroid-core transformer T₁, coaxial capacitor assembly, Faraday shield, completed unit, outer shield, transformer with attached leads.

after each hour. Similar storm warnings are given for the east and central North Pacific from WWVH at 48, 49, and 50 minutes after each hour. Information regarding these broadcasts may be obtained by writing to: National Weather Service, Silver Spring, MD 20910.

BCD Time Code—A binary coded decimal time code is transmitted continuously by WWV and WWVH on a 100 Hz subcarrier. The code provides a standard timing base for scientific observations made simultaneously at different locations. Time code information is outlined in the *NBS Special Publication 432.*

Geophysical Alerts—Current *geoalerts* are broadcast from WWV at 18 minutes after each hour. The messages are changed approximately every six hours at 1800, 0000, 0600, and 1300 UTC. Part A of the message gives the solar-terrestrial indices for the day, Part B gives the solar-terrestrial conditions for the previous 24 hours, and Part C gives the forecast for the next 24 hours. If *stratwarn* conditions exist, a brief advice is given at the end of the message. The alert covers solar activity, the geomagnetic field, the geomagnetic storms and solar flares. Inquiries regarding these messages should be addressed to NOAA, Space Environment Services Center R43, Boulder, CO 80303. Tel: (303) 499-8129.

In addition to the NBS broadcasts, the Dominion Observatory of Canada transmits time ticks and voice announcements in English and French on 3.330, 7.335, and 14.670 MHz. Many other countries of the world also transmit standard frequency and time signals, particularly on 5, 10, and 15 MHz.

The standard-frequency transmissions may be used for accurately determining the limits of the various amateur bands with the aid of the station receiver and a *secondary frequency standard* which utilizes an accurate low-frequency crystal oscillator. The crystal is zero-beat with WWV by means of its harmonics and then left with only an occasional check to see that the frequency has not drifted off with time. Accurate signals at smaller frequency intervals may be derived from the secondary frequency standard by the use of multivibrator or divider circuits to produce markers at intervals of 25, 10, 5, or 1 kHz. In addition, a variable-frequency *interpolation oscillator* may be used in conjunction with the secondary standard to measure frequencies at any point in the radio spectrum.

Shown in figure 28 is a simple 1 MHz calibration oscillator which provides marker signals up to 150 MHz or so.

31-11 A Precision Crystal Calibrator

Modern direct-reading h-f receivers require a high order of calibrator accuracy. Shown in this section is a versatile crystal-

WWV BROADCAST FORMAT

VIA TELEPHONE: (303) 499–7111
(NOT A TOLL-FREE NUMBER)

U.S. DEPARTMENT OF COMMERCE
National Bureau of Standards

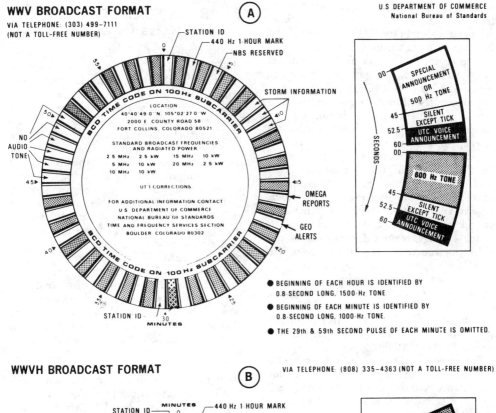

- BEGINNING OF EACH HOUR IS IDENTIFIED BY 0.8-SECOND LONG, 1500-Hz TONE.
- BEGINNING OF EACH MINUTE IS IDENTIFIED BY 0.8-SECOND LONG, 1000-Hz TONE.
- THE 29th & 59th SECOND PULSE OF EACH MINUTE IS OMITTED.

WWVH BROADCAST FORMAT

VIA TELEPHONE: (808) 335–4363 (NOT A TOLL-FREE NUMBER)

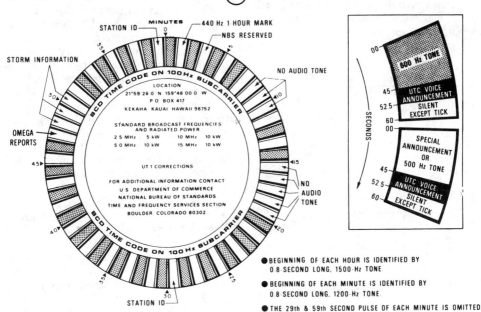

- BEGINNING OF EACH HOUR IS IDENTIFIED BY 0.8-SECOND LONG, 1500-Hz TONE.
- BEGINNING OF EACH MINUTE IS IDENTIFIED BY 0.8-SECOND LONG, 1200-Hz TONE.
- THE 29th & 59th SECOND PULSE OF EACH MINUTE IS OMITTED.

Figure 27
HOURLY BROADCAST SCHEDULE OF WWV AND WWVH

controlled secondary frequency standard utilizing a 1 MHz AT-cut crystal of excellent temperature stability. The circuit of this instrument is shown in figure 29.

The crystal is used in an FET oscillator (Q_1) having a high input impedance coupled to an amplifier (Q_2), followed by an impedance transformer (Q_3) to the logic circuit level. Integrated circuit U_1 is a quadruple TTL-type gate used as a Schmitt trigger to provide fast rise and fall time for the decade divider (U_2) and the dual flip-flop (U_3). The available outputs are: 1 MHz, 500 kHz, 100 kHz, 50 kHz, and 25 kHz. The IC (U_2) is configured as a divide-by-two and a divide-by-five combination to provide the 500-kHz and 100-kHz markers. A dual-voltage, regulated power supply provides plus fourteen and plus five volts with very low ripple and good regulation.

Frequency of the 1-MHz crystal is set by adjusting capacitor C_1 while zero-beating one of the 1-MHz harmonics with a transmission of WWV, or the frequency may be set with the aid of a frequency counter connected to the 1-MHz output.

For receiver calibration, a 5-pF capacitor at the receiver end of a short length of low capacitance coaxial cable (93 ohm) will permit maximum harmonic signal to be delivered at the antenna terminals.

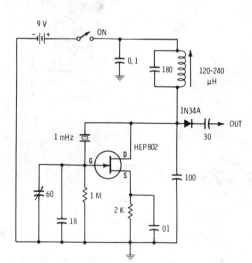

Figure 28

INEXPENSIVE CRYSTAL CALIBRATOR

31-12   A Silicon Diode Noise Generator

The limiting factor in signal reception above 25 MHz is usually the thermal noise generated in the receiver. At any frequency, however, the tuned circuits of the receiver must be accurately aligned for best signal-to-noise ratio. Circuit changes (and even alignment changes) in the r-f stages of a receiver may do much to either enhance or degrade the noise figure of the receiver. It is exceedingly hard to determine whether changes of either alignment or circuitry are really providing a boost in signal-to-noise ratio of the receiver, or are merely increasing the gain (and noise) of the unit.

A simple means of determining the degree of actual sensitivity of a receiver is to inject a minute signal in the input circuit and then measure the amount of this signal that is needed to overcome the inherent receiver noise. The less injected signal needed to override the receiver noise by a certain, fixed amount, the more sensitive the receiver is.

A simple source of minute signal may be obtained from a silicon crystal diode. If a small dc current is passed through a silicon crystal in the direction of higher resistance, a small but constant r-f noise (or hiss) is generated. The voltage necessary to generate this noise may be obtained from a few flashlight cells. The *noise generator* is a broadband device and requires no tuning. If built with short leads, it may be employed for receiver measurements well above 150 MHz. The noise generator should be used for comparative measurements only, since calibration against a high-quality commercial noise generator is necessary for absolute measurements.

A Practical Noise Generator   Described in this section is a simple silicon crystal noise generator. The schematic of this unit is illustrated in figure 30. The 1N21 crystal and .001-μF ceramic capacitor are connected in series directly across the output terminals of the instrument. Three small flashlight batteries are wired in series and mounted inside the case, along with the 0-2 dc milliammeter and the noise-level potentiometer.

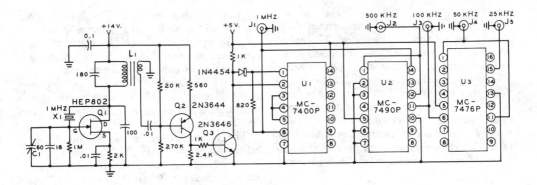

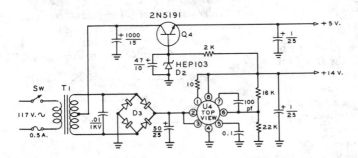

Figure 29

SCHEMATIC, PRECISION CRYSTAL CALIBRATOR

D₃—HEP 176
U₄—LM 300, SG 305T or CA 3055
L₁—120-240 μH. CTC2060-8. Secondary is 10 turns #24 insulated wire
T₁—16-volt, center-tapped. Triad F-90X

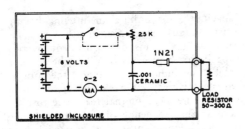

Figure 30

A SILICON DIODE NOISE GENERATOR

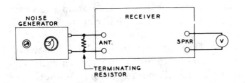

Figure 31

TEST SETUP FOR NOISE GENERATOR

To prevent heat damage to the 1N21 crystal during the soldering process, the crystal should be held with a damp rag, and the connections soldered to it quickly with a very hot iron. Across the terminals (and in parallel with the equipment to be attached to the generator) is a 1-watt carbon resistor whose resistance is equal to the impedance level at which measurements are to be made. This will usually be either 50 or 300 ohms. If the noise generator is to be used at one impedance level only, this resistor may be mounted permanently inside of the case.

Using the Noise Generator The test setup for use of the noise generator is shown in figure 31. The noise generator is connected to the antenna terminals of the receiver under test. The receiver is turned on, the avc turned off, and the r-f gain control is placed full on. The audio volume control is adjusted until the output meter advances to one-quarter scale. This reading is the basic receiver noise. The noise generator is turned on, and the noise-level potentiometer adjusted until the noise output voltage of the receiver is doubled. The more resistance in the diode circuit, the better is the signal-to-noise ratio of the receiver under test. The r-f circuit of the receiver may be aligned for maximum signal-to-noise ratio with the noise generator by aligning for a 2-1 noise ratio at minimum diode current.

31-13   The R-F Noise Bridge

Conventional impedance measurements on an antenna system usually call for an r-f impedance bridge, signal generator and a bridge detector. Such heavy and expensive devices are not suited for work on an amateur antenna where the operator and equipment may be balanced atop a ladder or hanging from the tower by one arm. A simple alternative that is light, inexpensive and accurate enough for most amateur hf antenna measurements is the *r-f noise bridge*. By combining a simple r-f bridge with a wideband noise generator in a small shielded box, an adequate impedance measuring device can be built. Since the r-f source is a wideband noise generator, the system selectivity is derived from the station receiver. This is a very important point; it is the receiver alone which establishes at what frequency the impedance measurement takes place, as the signal source may be considered to be "white noise."

R-f noise bridges are commercially available and there are three different types of

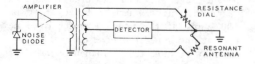

Figure 32

THE BASIC NOISE BRIDGE

The noise bridge is composed of a wideband noise source, an r-f bridge and a selective detector. The antenna forms one leg of the bridge. The bridge is balanced by nulling the noise signal in the detector. No reactive compensation is provided in the bridge.

instruments that are used. The first design has the bridge configuration shown in figure 32. The bridge is balanced by equating the resistance of potentiometer R_1 to the resistive portion of the antenna impedance. Since no provision is made for a reactive leg in the bridge, this unit is only useful for measurement of an antenna at the resonant frequency.

The second design is the modified noise bridge shown in figure 33. This unit makes it possible to measure both resistive and reactive impedance. When the 140-pF variable capacitor, C_N, is set at half-value, or 70 pF, the bridge is balanced for reactance. This allows the user to increase the capacitance for nulling an antenna that has a net parallel capacitive reactance, and to decrease the

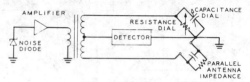

Figure 33

NOISE BRIDGE WITH PARALLEL CAPACITANCE RANGE

Addition of a capacitance dial permits noise bridge to measure both resistive and reactive impedance in antenna.

capacitance for nulling an antenna having a net inductive reactance. In either case, the capacitor dial reads out equivalent parallel capacitance in picofarads. When the antenna exhibits an inductive reactance, the equivalent negative capacitance is that value which would cause a coil of that reactance to resonate at the frequency of measurement. To obtain the equivalent series $R + jX$ values of the antenna, the operator must go through a mathematical parallel-to-series conversion.

The r-f noise bridge described in this section uses the bridge configuration shown in figure 34. Note the similarity to figure 33, except that the capacitors in this design are in series, rather than in parallel, with the resistance potentiometer and the antenna terminals. This allows (as before) inductive or capacitive reactance to be observed. The difference is that the reading of the two dials at null is $R + jX$ directly.

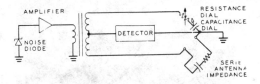

Figure 34

NOISE BRIDGE WITH SERIES CAPACITANCE RANGE

Inductive and capacitive reactance of antenna under test are read as a series impedance in the form of R + jX.

The Noise Bridge Circuit The complete noise bridge circuit is shown in figure 35. A zener diode, Z_1, is used as a "white noise" source, followed by a broadband noise amplifier (Q_1-Q_3). It is important to use devices that have good high-frequency response to provide adequate "white noise," especially at the upper frequency limits of the bridge. The transistors listed provide good results above 30 MHz.

The choice of zener diode is not critical so long as breakdown voltage is between 3.6 and 6.8 volts. The bypass and coupling capacitors in the noise amplifier are .01-μF ceramic disc types. Coupling transformer T_1 is quadrifilar-wound on a ferrite toroid. The 68-pF bridge capacitor should be a silver mica unit and potentiometer R_x is a low-inductance, composition type.

Noise Bridge Construction Construction of the noise bridge is shown in figures 36 and 37. The unit is built within a cast aluminum box measuring $4\frac{1}{2}'' \times 3\frac{1}{2}'' \times 2\frac{1}{4}''$. The *receiver* and *unknown* (antenna) coaxial receptacles are mounted between the two panel controls. All components are mounted on a piece of copper-plated circuit board which is bolted inside the lid of the box. The stator of the variable capacitor is grounded to the board surface with a short, wide length of copper strap. The various components are mounted by their leads to insulated tie points soldered to the copper plane as shown in the rear view photograph. Placement of parts is not critical, provided attention is paid to lead length. The transformer is wound by winding four wires in parallel on the core. The indicated dots on the scheamtic of T_1 in figure 35 represent the same end of all windings.

Aligning the Bridge Calibration of the bridge is simple if a capacity meter and ohmmeter are available. The resistance dial for potentiometer R_x can be calibrated with the ohmmeter. To calibrate the reactance dial, set capacitor C_x to roughly half-capacitance and temporarily disconnect the lead to R_x. Drill a $\frac{3}{8}$-inch diameter hole in the box in a spot next to the ungrounded stator terminal of C_x and close the box. Pass

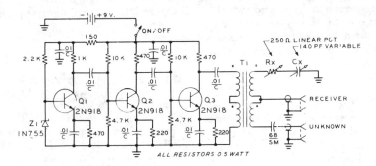

Figure 35

SCHEMATIC OF WIDEBAND NOISE BRIDGE

This simple bridge provides usable measurements up to 30 MHz. The noise generator is a zener diode (Z_1), followed by three stages of resistance coupled amplification (Q_1-Q_3). The station receiver is used as the bridge detector. Transformer T_1 is quadrifilar wound of 4 turns No. 28 formvar on an Indiana General CF-102-Q2 ferrite core. (Indiana General Corp., Keasbey, NJ 08832. Also Amidon Assoc., 12033 Otsego St. No. Hollywood, CA 91607.) Potentiometer is Allen-Bradley type JA1N-056S, or Ohmite type AB-CU2511.

Figure 36

THE R-F NOISE BRIDGE

Components are mounted on the lid of a cast aluminum box. The Series-C (capacitance) dial is at the left and the Series-R (resistance) dial is at the right. The coaxial receptacles for the detector (receiver) and antenna (unknown) are at the center of the panel. Dials are hand calibrated, as discussed in the text.

Figure 37

INTERIOR OF THE R-F NOISE BRIDGE

The series capacitor is at upper left and the series variable resistor is at upper right, with the coaxial connectors between them. The small ferrite transformer is between the potentiometer and the lower connector. The components of the noise generator are mounted on solder terminals in the foreground.

a small probe from a capacitance meter (such as the *Tektronix 130*, or equivalent) into the hole to touch the stator terminal. Using the meter as a reference, C_x is adjusted to read 70 pF to ground. The dial is marked, then C_x is varied to provide plus and minus dial markings from 70 pF to 10 pF in small steps. The values below 70 pF are plotted on the dial as being measured from zero, which is 70 pF. Thus, the dial in a typical instrument will be calibrated 70-0-70 pF, with the maximum values of capacitance marked as "inductance" on the dial.

Using the Noise Bridge   The station receiver is connected to the *receiver* port of the bridge with a length of coaxial line and the bridge is connected to the antenna directly, or through a coaxial line that is an electrical half-wavelength at the frequency of measurement. When the noise bridge is activated, the "white noise" will be heard as a strong hiss in the receiver. The *series R* and *series C* dials are adjusted to null, or balance out the hiss, and the radiation resistance and reactance of the antenna under test is read directly on the dials

of the bridge. The reactance reading is in picofarads, which may be converted directly to reactance with the aid of a slide rule or pocket computer.

In some cases, when measurements are made in the vicinity of a strong, local broadcasting station, the bridge null may be obscured, as the bridge element is reacting to the pickup of the signal by the antenna under test.

31-14   A Universal Crystal Test Unit

This simple test unit will test crystals ranging in frequency from a few hundred kHz to over 90 MHz (figure 38). Transistor Q_1 forms a variation of the Colpitts oscillator with feedback adjusted by capacitor C_1. The r-f voltage of the oscillator is rectified by two diodes and the resulting dc voltage provides forward bias for Q_2 whose emitter current lights an indicating lamp. If the crystal fails to oscillate, the lamp remains dark. Various crystal sockets can be incorporated in the tester for different styles of

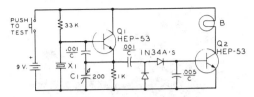

Figure 38

UNIVERSAL CRYSTAL TEST UNIT

**Crystal feedback is controlled by capacitor C₁.
An adjustable mica capacitor may be used. Bulb
is 10-volt, 14 mA (CM7-7344, Chicago Miniature
Lamp, or equivalent).**

holders. The unit is built within a small
aluminum utility box with a self-contained
battery.

31-15   An Inexpensive Transistor Tester

This compact and inexpensive transistor
checker will measure the dc parameters of
most common transistors. Either NPN or
PNP transistors may be checked. A six-position test switch permits the following parameters to be measured: (1) $I_{CO} - D_C$ collector current when collector junction is
reverse-biased and emitter is open circuited;
(2) I_{CO-20}—collector current when base current is 20 microamperes; (3) I_{CO-100}—collector current when base current is 100 microamperes; (4) I_{CEO}—collector current when
collector junction is reverse-biased and base
is open circuited; (5) I_{CES}—collector current
when collector junction is reverse-biased and
base is shorted to emitter; (6) I_{EO}—emitter
current when emitter junction is reverse-
biased and collector is open circuited.

Using the data derived from these tests,
the *static* and *ac forward-current transfer
ratios* (h_{FE} and h_{fe}, respectively) may be
computed as shown in figure 40. This data
may be compared with the information listed
in the transistor data sheet to determine the
condition of the transistor under test.

The transistor parameters are read on a
0-100 dc microammeter placed in a diode
network which provides a nearly linear scale
to 20 microamperes, a highly compressed
scale from 20 microamperes to one milliampere, and a nearly linear scale to full scale

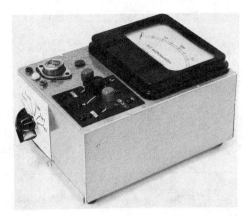

Figure 39

TRANSISTOR CHECKER

**An expanded-scale meter provides accurate meas-
urement of transistor parameters in this easily
built instrument. Six dc parameters may be
measured and with the data derived from these
tests, the ac forward-current transfer ratios
may be computed. Two transistor sockets are
mounted at the left of the tester, with the three
selector switches to the right. Six-position test
switch is mounted to bottom side of box. Tip
jacks are placed in parallel with transistor socket
terminals to permit test of transistors having
unorthodox bases.**

at 10 milliamperes. Transistor parameters
may be read to within 10 percent on all
transistor types from mesas to power alloys
without switching meter ranges and without damage to the meter movement or
transistor.

By making the sum of the internal resistance of the meter plus series resistor R_1
equal to about 6K, the meter scale is compressed only one microampere at 20 microamperes. *Meter adjust* potentiometer R_2 is
set to give 10 milliamperes full-scale meter
deflection. The scale may then be calibrated
by comparison with a conventional meter.

If the *NPN-PNP switch* (S_2) is in the
wrong position, the collector and emitter
junctions will be forward biased during the
I_{CO} and I_{EO} tests (switch positions 1 and 6).
The high resulting current may be used as
a check for open or intermittent connections
within the transistor.

The transistor checker also measures h_{FE}
with 20 microamperes and 100 microamperes base current. Depressing the h_{fe} switch
(S_1) decreases the base drive about 20 per-

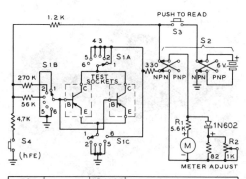

TO TEST	WHEN	ADJUST S₁ TO	RESULT
Ico	VCB = 6 V.	1	READ METER DIRECT
Ic	IB = 20 µA	2	"
Ic	IB = 100 µA	3	"
Iceo	VCE = 6 V.	4	"
Ices	VCE = 6 V.	5	"
Ieo	VEO = 6 V.	6	"
hFE	IB = 20 µA	2	CALCULATE: $hFE = \dfrac{IC}{IB} = \dfrac{METER\ READING}{20\ µA}$
hFE	IB = 100 µA	3	CALCULATE: $hFE = \dfrac{IC}{IB} = \dfrac{METER\ READING}{100\ µA}$
hfe	IB = 20 µA	2	CALCULATE: $hfe = \dfrac{IC1 - IC2}{4 \times 10^{-6}}$ WHERE: IC1 = METER READING
hfe	IB = 100 µA	3	CALCULATE: $hfe = \dfrac{IC1 - IC2}{20 \times 10^{-6}}$ IC2 = METER READING WITH S4 CLOSED
6 V. BATTERY	—	4	WITH 150 Ω RESISTOR CONNECTED TO C-E OF TEST SOCKET, FULL-SCALE METER DEFLECTION WILL RESULT WHEN S3 IS PRESSED.

Figure 40

SCHEMATIC OF TRANSISTOR CHECKER

S₁A, B, C—Three-pole 6-position. Centralab 1021
S₂, S₃, S₄—Centralab type 1400 nonshorting lever switch
M—0-200 dc microammeter. General Electric or Simpson (4½")

cent, permitting h_fe to be estimated from the corresponding change in collector current (formulas 1 and 2). All tests are conducted with a 330-ohm resistor limiting the collector current to about 12 milliamperes and the maximum transistor dissipation to about 20 milliwatts. The checker therefore cannot harm a transistor regardless of how it is plugged in or how the test switches are set.

The *battery test* provides full-scale meter deflection of 10 milliamperes when the battery potential is 6 volts. This is achieved by connecting a 150-ohm resistor from collector to emitter of a test socket.

Figure 41

INTERIOR VIEW OF TRANSISTOR CHECKER

Components of meter diode circuit are mounted to phenolic board attached to meter terminals. Other small resistors may be wired directly to switch lugs. The four 1½-volt batteries are held in a small clamp at the rear of the case. Chassis is cut out for lever-action switches and opening is covered with three-position switch plate.

Test Set Construction   The transistor checker is built in an aluminum box measuring 3" × 5" × 7", as shown in the photographs. Test switch S₁ is mounted on the end of the box; and the transistor sockets, microammeter, and the various other switches are placed on the top of the box. Three insulated tip jacks are wired to the leads of one transistor test socket so that transistors having unorthodox bases or leads may be clipped to the tester by means of short test leads. Four 1½-volt flashlight cells are mounted to the rear of the case by an aluminum clamp. Potentiometer R₂, the meter diode, and associated components are fastened to a phenolic board attached to the meter terminals. Switch S₁ has an indicator scale made of heavy white cardboard, lettered with India ink and a lettering pen.

31-16 An Inexpensive IC Capacitance Meter

Described in this section is a simple, inexpensive, and accurate capacitance meter built around a single IC (figure 42). The

Figure 42

INEXPENSIVE CAPACITANCE METER

This accurate capacitance meter makes use of a single RC 556DP (Raytheon) IC. Built in a small instrument cabinet, the device makes use of a surplus, large-scale microammeter salvaged from a war-surplus voltohmmeter. Only the 0-to-100 scale is used. The range switch is at the left, with the test capacitor terminals directly below it. The zero-set control is to the right.

device measures capacitance values ranging from 1 pF to 1 μF in five decade ranges.

The readout is a large, surplus dc microammeter. Linear scale relationship with respect to capacitance is achieved by measuring the dc value of a pulse waveform applied to the test capacitor, the value having a linear relationship with respect to the duty cycle of the waveform. If the width of the pulse is directly proportional to the value of the capacitor, the meter reading will be linear. A one-shot multivibrator provides a pulse width proportional to its timing capacitor, as shown in figure 43.

The constant-frequency trigger source is one-half an RC 556 operating as an astable oscillator. The output of the oscillator at pin 5 is about 500 Hz with a pulse amplitude of about 12 volts. The other half of the RC 556 is a one-shot multivibrator. The pulse width is proportional to RC, where R is the timing resistor selected by a panel switch, and C the capacitor under test. The meter reading is proportional to the pulse width with a given timing resistor, thus providing the required linear relationship.

Capacitance Meter Circuitry The resistors associated with the range switch should have a tolerance of five percent, or less. A 200K variable resistance in series with the meter provides a one-time calibration. Once this adjustment is made, no further calibration is required.

A *zero-adjustment* control is placed on the front panel for the lower capacitance ranges as the stray circuit and input capacitance total about 20 pF. This produces a false reading on the 100-pF scale without the bucking voltage which cancels out this stray pulse effect.

Calibration and Operation A simple IC-regulated power supply is included in the design (figure 44). The capacitance meter is turned on and the range switch set to 100 pF. With no capacitor attached to the test terminals, the zero-set control is adjusted for a meter reading of zero. A 100-pF, one-percent mica capacitor is attached to the test terminals and the 200K *calibration control* adjusted for a full-scale meter reading.

Each time the range switch is set to a new position, the zero-set control requires readjustment before the measurements are made, as is common with other capacitance meters of this general design.

Since ac is not applied to the capacitor under test, polarized capacitors may be checked, with the negative lead of the capacitor connected to the grounded test terminal. Any capacitor may be tested with this device provided it has a voltage rating of at least 8 volts.

31-17   A 2-Tone Generator for SSB Testing

To examine linearity of an amplifier by observation of the output signal some means must be provided to vary the output signal level from zero to maximum with a regular pattern that is easily interpreted. A simple means is to use two audio tones of equal amplitude to modulate the SSB transmitter. This is termed a *two-tone test*. This procedure causes the transmitter to emit two

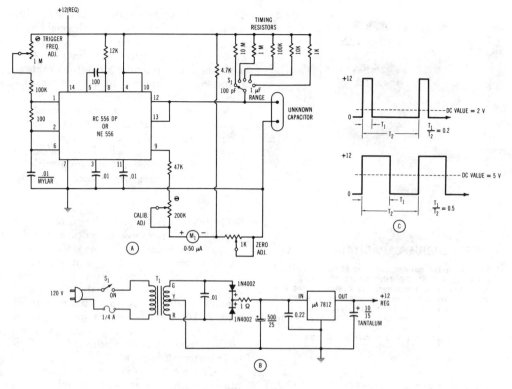

Figure 43

SCHEMATIC, IC CAPACITANCE METER

A—Schematic of meter.
B—Power supply.
   T_1—30 V center tap. Triad F-90X or equivalent.
   C—The average dc value of the pulse waveform varies linearly with the duty cycle.
Note: Two NE 555s could be used instead of one RC 556Dp, or even a single Exar XR-2556 is
      an acceptable substitute. Appropriate pin numbers must be changed, however.
      Meter ranges for the timing resistor are: 100 pF, 10 meg; 1000 pF, 1 meg; .01 μF, 100K;
      0.1 μF, 10K; 1.0 μF, 1K.

steady signals separated by the frequency difference of the two audio tones. The resultant, or beat, between the two r-f signals produces a pattern which, when observed on an oscilloscope has the appearance of a carrier 100-percent modulated by a series of half sine waves.

With a two-equal-tone test signal, the following equations approximate the relationships between two-tone meter readings, peak envelope power, and average power for class-AB or class-B operation:

Dc plate current:

$$I_b = \frac{2 \times i_{pm}}{\pi^2}$$

Plate Power Input (watts):

$$P_{in} = \frac{2 \times i_{pm} \times E_b}{\pi^2}$$

Average Power Output (watts):

$$P_o = \frac{i_{pm} \times e_p}{8}$$

Plate efficiency:

$$N_p = \left(\frac{\pi}{4}\right)^2 \times \frac{e_p}{E_b}$$

where,

i_{pm} equals peak of the plate current pulse,
e_p equals peak value of plate voltage swing,
E_b equals dc plate voltage, π equals 3.14

Figure 44

INTERIOR OF THE CAPACITANCE METER

Components of the meter are mounted on a small section of Perf board attached to the panel. The regulated dc power supply is placed on a small aluminum bracket behind the instrument.

Finally, peak-envelope-power *output* under these conditions is twice the average-power output. Thus, using a two-tone test signal, a linear amplifier may be tuned up at a power-output level of half that normally achieved at the so-called "two kilowatt PEP" input level. Power-*input* level, on the other hand, of the two-tone test condition is about two-thirds that of the single-tone condition.

The Two-Tone Generator Shown in figure 45 is the schematic of a simple two-tone audio generator which provides a pair of linearly added sine waves. The second harmonic and intermodulation products are reduced at least 35 decibels below one tone. It is designed for either a single-ended audio input circuit (common to most SSB exciters) or a balanced line input. The generator operates from an internal 9-volt battery and contains no transformers so no power-line frequency associated components are produced in the two-tone signal.

Two bridge-type audio generators and associated buffer stages are contained on a single IC. One generator is adjusted for 1000

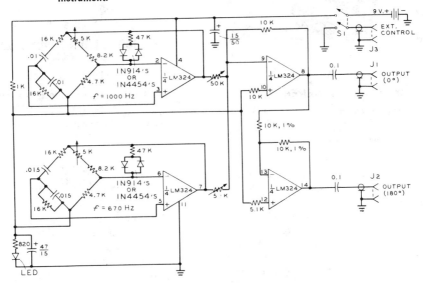

Figure 45

SCHEMATIC OF TWO-TONE GENERATOR

Single IC chip provides two oscillators and associated buffer stages. Six-volt light-emitting diode (LED) provides "on" indication. (Dialco 507-4748-3331-500). Potentiometers are Bourns Trim-pot model 3305, or equivalent.

Figure 46

TWO-TONE AUDIO GENERATOR

Figure 47

INTERIOR OF GENERATOR

Components are mounted on small section of glass-epoxy board. The IC is at center, mounted in a socket. Terminal connections are at right, rear.

transmitter is adjusted for maximum power output without waveform flattopping. Under these conditions, the power input is:

$$\text{PEP Input (watts)} = I_b \times E_b$$
$$\left(1.57 - 0.57 \, \frac{I_o}{I_b} \right)$$

where,

E_b equals dc plate voltage,
I_b equals two-tone dc plate current,
I_o equals idling plate current with no test signal.

Figure 48

VARIABLE-FREQUENCY AUDIO GENERATOR

This compact, solid-state audio generator covers the range of 20 Hz to 20 kHz with a distortion level of 0.05 percent or less. The frequency-control potentiometer is near center, with the frequency-range switch at the right. Unit is built in a small aluminum utility cabinet.

Hz and the other one for 670 Hz, although other audio tone combinations may be used.

The device is constructed within an aluminum utility box measuring 3½″ × 2″ × 1½″ (figure 46). All components are mounted on a perforated circuit board, as shown in the interior photograph (figure 47). The 9-volt battery is mounted beneath the board in a small clip.

The Two-Tone Test—The test oscillator is connected to the audio system of the SSB transmitter which is tuned up into a dummy load with an oscilloscope coupled to the load to show a typical test pattern. The

31-18   A Variable-Frequency Audio Generator

Described in this section is a high-quality, variable-frequency audio generator that covers the range of 20 Hz to 20 kHz, with a distortion level of 0.05% or less (figure 48).

Unlike the expensive laboratory oscillators which require dual (tracking) variable resistors or capacitors, this compact oscillator uses a single variable resistor for tuning. The circuit is shown in figure 49.

Three operational IC amplifiers are used. Op-amp U_2 functions as an active bandpass filter, U_1 serves as a broadband amplifier, and U_3 is used as a dual zener diode. The feedback loop that sustains oscillation involves 180 degrees of phase shift around U_1 and 180 degrees of phase shift around U_2. To permit oscillation, sufficient circuit gain occurs only at the maximum response frequency of the active bandpass filter that is designed around U_2. The frequency of oscillation is thus controlled by varying the center frequency of the bandpass filter. Level stabilization is obtained by clipping the sine wave by means of U_3, the Q of the active filter circuit removing the harmonics created by the clipping. Only the base-emitter diodes of the two input transistors of U_3 are used (figure 50), the other leads are left floating. The LM 709C was used because of its very low price in comparison to the cost of a good seven-volt zener diode.

A test point is provided for the builder to monitor the percentage of sine-wave clipping in use, the level being set by potentiometer R_1. This is normally set so that about 20 percent of the sine wave total amplitude is clipped when the frequency control potentiometer (R_2) is at the low-frequency (maximum resistance) position.

To power the audio oscillator, a simple dual-voltage regulated supply providing plus and minus 15 volts is included.

31-19   A Function Generator

One of the most recent and useful pieces of test equipment available for the amateur is the *function generator* (figure 51). The generator described in this section has three symmetrical outputs: sine, triangular and square waves. While the function generator does not replace either the sine wave oscillator or the pulse generator, it is more versatile than either because of its variety of output signals.

The schematic of the generator is shown in figure 52. It is designed around the *Intersil 8038* IC that was specifically developed

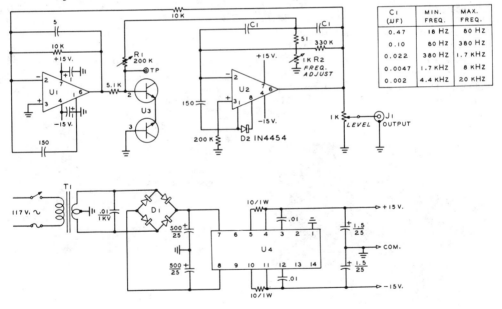

C_1 (μF)	MIN. FREQ.	MAX. FREQ.
0.47	18 HZ	80 HZ
0.10	80 HZ	380 HZ
0.022	380 HZ	1.7 KHZ
0.0047	1.7 KHZ	8 KHZ
0.002	4.4 KHZ	20 KHZ

Figure 49

SCHEMATIC, AUDIO GENERATOR

U_1, U_2—LM 301H (National Semiconductor)
U_3—LM709C used as dual zener (pins 2 and 3) (National Semiconductor)
U_4—SG 3501D (Silicon General)
D_1—HEP 176
T_1—32-volt, center-tapped, Triad F-90X

for this application. This device has three separate output ports for the waveforms in question, and no external integrators or shaping circuits are required. The only auxiliary equipment to make the IC a complete item of test equipment is a power supply and an output amplifier.

The output amplifier is comprised of an operational amplifier output driver. The driver has only a 6-ohm impedance so 43 ohms is placed in series with it to provide a 50-ohm port when it is required to drive a matched coaxial line.

Dc offset is provided by a 10K potentiometer placed across the ±15 volt supply through a series connected 100K resistor to the inverting input of the op amp. The output waveform from the 8083 IC is fed to the noninverting input of the op amp via a 10K potentiometer which serves as the *output level* control of the function generator. The voltage gain of the IC combination is about four which is enough to boost the various output signals of the 8038 to a 10 volt peak-to-peak level. Individual trim-pots on the three outputs (R_1, R_3, R_4) of the 8038 are used to assure that all three signals have the same peak-to-peak level.

Coarse frequency control is accomplished by means of switched capacitors. The given values of 1500 pF to 15 μF in 5 ranges provide an output frequency range of 1 Hz to 100 kHz. Resistors are simultaneously switched with the capacitors to relieve the necessity of using precision capacitors. The resistor values fall in the vicinity of 100 ohms and are chosen during the calibration process. The *fine frequency* control is a 10K potentiometer in series with switch S_1A.

Aside from the *fine frequency, output level, offset* and the three trim-pots for the relative levels of the three waveforms, there are three other adjustments: potentiometers R_2, R_5, and R_6. These are respectively: *square wave offset, sine wave distortion adjust,* and *symmetry adjust.* These will be discussed in detail in the section on calibration.

The power supply delivers ±15 volts, 100 mA regulated. A packaged unit may be used, or the supply shown in figure 53 can be built. The power regulator is meant to be operated on a heat sink with a mica insulated washer and the appropriate heat sink grease for thermal conductivity.

Generator Construction—The generator is built within a small aluminum cabinet measuring 7″ × 5¼″ × 4½″ (figure 54). The generator components are mounted on a small piece of perforated glass-epoxy circuit board measuring 3½″ × 2½″. The board is supported above the chassis on

Figure 51

THE FUNCTION GENERATOR

This function generator delivers sine, triangular, or square waves over the frequency range of 1 Hz to 100 kHz. Panel controls (left to right) are: DC Offset, Fine Frequency, Coarse Frequency, Waveform Selector Switch, and Output Level. Dual output connecters are provided for coaxial lead or for test leads. Output impedance is 50 ohms.

Figure 50

COMPONENTS OF AUDIO GENERATOR ARE MOUNTED ON P.C. BOARD

U_3, the inexpensive IC used as a dual zener diode, is in the foreground at left. The two op-amps are placed in sockets supported on small terminals soldered to the board.

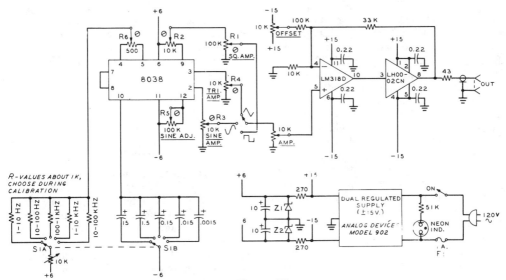

Figure 52

SCHEMATIC OF FUNCTION GENERATOR

Capacitors above 1 μF are tantalytic types
Capacitors below 1 μF are mylar types
Z_1, Z_2—6-volt zener. Motorola HEP-Z0408
Potentiometers R_2-R_6 are multiturn Trim-pots (Allen-Bradley, Bourns, or equivalent)
Power Supply is Analog Devices 902 or equivalent (see figure 53)

small metal spacers. The zener diodes, filter capacitors and 270-ohm series resistors are supported on a terminal strip in front of the board. The resistors and capacitors associated with the frequency control switch are mounted between the two decks of the ceramic switch. The remainder of the space is taken up with the compact dc power supply.

Calibrating the Generator To calibrate the function generator, first press all trimmer potentiometers (R_1-R_6) to midposition and temporarily connect a

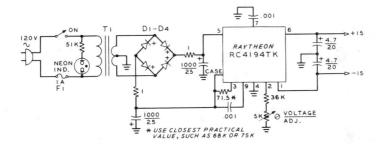

Figure 53

ALTERNATIVE POWER SUPPLY FOR FUNCTION GENERATOR

D_1-D_4—Motorola MDA 920-3 or HEP-176 bridge rectifier
T_1—34 volts, center-tapped at 100 mA. Triad F-93X
Capacitors above 1 μF are tantalytic types
Regulator is Raytheon RC4194TK

1K potentiometer in series with the #3 contact of switch S_1A and the center-tap of potentiometer R_6. Set this temporary control for maximum resistance and place the coarse frequency switch to position 3 (100 to 1000 Hz). Turn the *fine-frequency* control potentiometer to the high end and the *offset* control (R_6) to midposition (zero offset). Triangular waveform will be checked first since it will most clearly indicate proper symmetry. Connect an oscilloscope to the output terminals and apply power to the generator. With the amplitude control (*output level*) at maximum, adjust potentiometer R_6 for proper symmetry and then adjust the temporary potentiometer for a frequency of 1000 Hz. The oscillator frequency with the *fine-frequency* control set at minimum should reach 100 Hz, or less.

The sine wave output is checked next. With the *amplitude* control at maximum, adjust potentiometer R_3 for a 10-volt peak-to-peak output signal. Adjust trimmer potentiometer R_5 for minimum observed distortion of the sine wave.

Next, switch to square wave output. Both potentiometers R_2 and R_1 must be adjusted for the desired 10-volt peak-to-peak signal. Trimmer R_2 will cancel out the offset voltage present for the square wave output and trimmer R_1 will control the amplitude. There is interaction between these two adjustments and they should be reset a few times to accomplish the desired results.

Measure the resistance value of the temporary potentiometer and make a note of the required value for this range. Switch to each of the remaining ranges in turn and determine a value of resistance to calibrate each one. Remove the test potentiometer and install fixed calibrating resistors at the proper positions on the bandwswitch.

31-20 An Electronic Multimeter

Probably the most important piece of test equipment for the radio enthusiast is the multimeter. The electronic multimeter is capable of measuring a wide range of ac and dc voltages resistances and currents. The multimeter shown in this section has been designed specially for building and maintaining radio transmitting and receiving equipment, both solid state and with vacuum tubes. It is well suited for this type of service (figure 55).

An accurate d'Arsonval meter with a conventional pointer is used rather than the popular digital readout. Often it is necessary to tune or adjust circuitry for a maximum, minimum or zero reading, while the actual voltage is of secondary importance. Using a digital meter for this task is difficult at best. In addition, the voltage sensitivity of this circuit extends to 10 millivolts, full scale, on both ac and dc ranges. This is important when the instrument is used as an r-f probe at low signal levels. This sensitivity can be achieved very simply in this analog design.

Figure 54

INTERIOR VIEW OF FUNCTION GENERATOR

Trim-pots and ICs are mounted on perforated circuit board in the foreground. The associated capacitors and resistors are mounted to the terminals of switch S_1A-B (Coarse Frequency). The unitized power supply is at the rear of the chassis, with line cord and fuse directly below it. Zener diode regulators and tantalytic filter capacitors are mounted to a small terminal strip below the Output Level (amplitude) control.

Figure 55

LABORATORY-TYPE
ELECTRONIC MULTIMETER

This easily built multimeter is well suited for the home workshop. Meter ranges are: 0-10, 0-50, 0-100, 0-500 millivolts; 0-1, 0-5, 0-10, 0-50, 0-100 and 0-500 volts (ac and dc) plus 0-10, 0-100, 0-1000, 0-100K, 0-1M and 0-10 megohms. Ac response is good up to 20 kHz. Meter is fully protected from overload.

The ac response of this instrument is good up to 20 kHz and falls off smoothly above this figure. Sensitivity, however, is good enough at the same time to measure the output of a low level microphone.

The meter movement is fully protected from overload. An overload of 1000 times full scale, or 1000 volts, whichever is less, will not damage the voltage function. On the ohms function, an applied voltage of 10 volts will not damage the meter.

Even though precision parts are held to a minimum, the dc accuracy of the instrument is about ±2% and the ohms and ac accuracy is about ±3%. The input impedance is 10 megohms in parallel with 100 pF for both ac and dc ranges.

Multimeter Circuitry The circuit of the multimeter is shown in figure 56. The device consists of a power supply and four integrated circuit FET input operational amplifiers. The four op-amps provide the three basic circuit functions. Amplifier U_1 is a wideband, dc-coupled "times 70" amplifier used for all three meter modes. The output of the amplifier is ±3.5 volts on the dc functions and 10 volts peak-to-peak on the ac functions. This amplifier has an eight volt regulator (U_5) to limit the current supplied to the meter during the dc and ohms functions to about fifty percent above full scale. Integrated circuit U_2 is a precision half-wave rectifier for conversion of the signal to dc for the meter movement. The gain of this stage is set at 2.22 so that the meter (which reads the average of any pulsating voltage) is calibrated directly for the rms value of a sine wave. Circuits U_3 and U_4 constitute a precision current source for resistance measurements.

Many multimeters apply a fixed voltage to the unknown resistance and measure the resulting current. This results in a nonlinear calibration for the resistance scale of the meter, requiring a nonstandard scale difficult for the builder to calibrate. The constant current technique allows a linear resistance scale on the meter than can be read directly in ohms.

The range switch is a two-part attenuator. The input decade attenuator (S_2A) has five steps and the output 5-to-1 attenuator (S_2B) has two steps, making a total of 10 steps of attentuation requiring only five precision resistors.

The meter is protected two ways. First, a damping resistor is placed across the movement. If an overload is applied, the over-damped pointer will move slowly off the scale, with no damage. In addition, the heavy damping prevents the pointer from moving during transit of the instrument.

The second meter protection limits overload current through the meter to a safe amount, controlled by the saturation voltage of the operational amplifier. Even though the input may be overloaded by a factor of 1000 times the full-scale value, the meter will not be damaged.

Overall accuracy of the multimeter depends primarily on the accuracy of the meter movement. A new meter is expensive, but a surplus meter can often be turned up at a bargain price. The attenuator resistors are one-percent devices, but the rest of the resistors may have a five percent tolerance. Film resistors are recommended for stability.

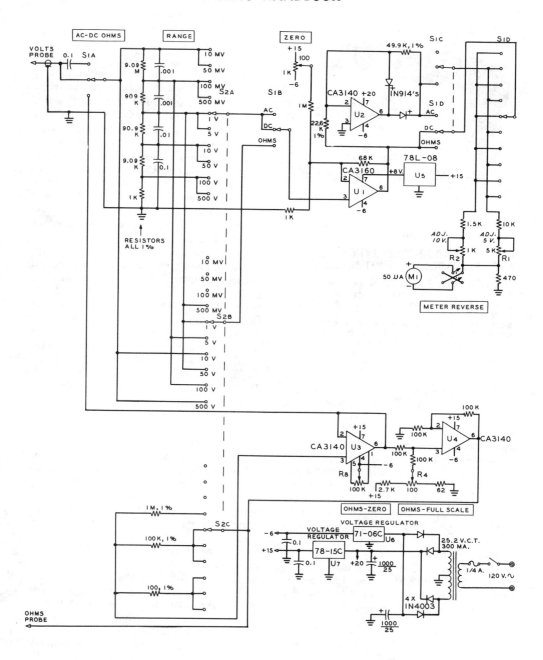

Figure 56

SCHEMATIC OF MULTIMETER

U₁-U₄—RCA integrated circuits
Voltage regulator—Fairchild or Motorola
Precision resistors—Metal oxide film units by
   IRC or Dale

Capacitors—Mylar
S₁—2 pole, 3 position
S₂—4 pole, 10 position
Meter—50 μA. Simpson or Triplett

Multimeter Assembly—The interior layout of the instrument is shown in the photographs. The entire electronics package is mounted on a perforated board except for the power transformer, rectifiers, and filters, which are mounted in an aluminum box (figure 57). Because of the very low signal levels involved, all precautions must be taken to prevent ac hum pickup. The power cord goes directly into the box and the power switch on the panel is interconnected with shielded wires. There is no power line ground connection, nor are there any line bypass capacitors.

Meter Calibration—The meter scale must be modified unless the builder is fortunate enough to have a meter with the correct markings. The scales in the multimeter progress in a 1-to-5 sequence; that is, 1, 5, 10, 50, etc., full scale, whether the reading is volts or milliamperes—ac, or dc. The ohms scale is decade; that is, 1, 10, 100, etc., full scale. Additional markings are therefore required in order to accommodate all of the multimeter ranges.

To accomplish this task, remove the glass from the meter and, using a screwdriver free of magnetic particles, remove the meter face. Using dry transfer letters, mark a new scale above the graduations of the old one,

Figure 57
INTERIOR VIEW OF MULTIMETER

Range switch is at top of the panel with instrument circuit board directly below it. The power supply is included in the aluminum box at the rear of the chassis.

placing a 1 over the 5 mark, a 2 over the 10 mark, etc., ending with a 10 over the 50 mark. Spray the meter face with a clear protective lacquer and reassemble the meter.

Multimeter Testing—After the device has been built and the wiring checked, place the function switch in the *dc volts* position and the range switch on *100 volts*. Adjust the front panel *zero* control to zero the meter once the power has been turned on. An accurate voltage source is required to calibrate the meter. Fresh flashlight cells, used in conjunction with a one-percent meter, may be used, as long as open circuit voltage is 5 volts, or slightly less. Three series-connected 1.5 volt cells will do the job. Connect the calibrating source to the *volts* probe and reduce the range switch to 5 volts. Adjust potentiometer R_1 so that the multimeter reads the same as the calibrating voltmeter. Now switch to the *10-volt* range and adjust potentiometer R_2 to provide the same voltage reading as before. Next, set the function switch to *ohms*, the range switch to *10K ohms* and short the *ohms* probe to ground (chassis). Adjust potentiometer R_3 for a zero meter reading. Now, connect a 10K, one percent resistor between the *ohms* probe and ground and adjust potentiometer R_4 for a full-scale meter deflection. This completes the alignment procedure.

The RF Probe and Volts Probe—The *volts* probe is a specially constructed shielded probe (figure 58). A commercially available VOM probe set was purchased and the black probe used unmodified for the *ohms* function. The red probe was modified by removing the wire and replacing it with a shielded lead. The shield was carried to within ¼ inch of the probe tip. A ground lead is then brought to the top of the probe body and secured with heat-shrink tubing or electrical tape. An alligator clip is attached to the end of this ground wire.

A shielded microphone connector is placed on the free end of the i-f probe cable. Shielded wire is also used inside the instrument to connect the input receptacle to the attenuator. A ground lug on the input microphone connector serves as a common ground point. Good shielding and the common ground point are necessary to prevent ac hum pickup and to prevent ground loops

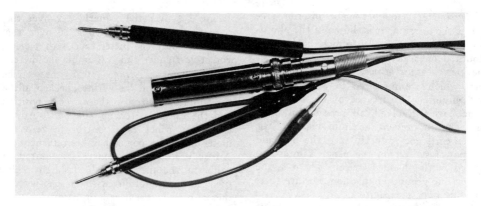

Figure 58

MULTIMETER PROBES

Ground (black) probe is at top, the ac voltage probe is at center. The dc (red) probe is at the bottom of the photograph.

from destroying the accuracy of the low voltage ranges.

The excellent sensitivity of the multimeter allows very low rf voltages to be measured with a simple r-f probe. It is possible to measure levels as low as 50 microwatts across a 50-ohm load with this probe. The circuit of the r-f probe is a half wave rectifier and filter (figure 59). For best frequency response a point-contact diode, the 1N82A, is used. Good results have been obtained up to 1000 MHz.

Very short leads should be used within the probe. The dc-blocking capacitor should be a mylar or mica unit. The probe housing should be of metallic material in order to reduce hand capacity. The calibration of the r-f probe will not be exact. Response will vary with frequency and the impedance of the circuit being measured. Furthermore, at low signal levels, the response will become "square law," that is, the input signal is

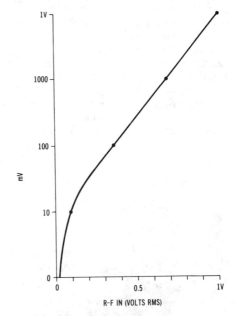

Figure 60

RESPONSE OF R-F PROBE

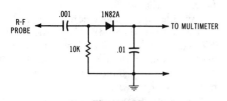

Figure 59

R-F PROBE FOR MULTIMETER

proportional to the square of the output signal. Figure 60 shows a typical response of the probe. From this response, it can be seen why a 10 mV full-scale meter reading is valuable for the probe as a 50 mV r-f signal produces only 3 mV of dc output.

CHAPTER THIRTY-TWO

The Oscilloscope

The *cathode-ray oscilloscope* is an instrument which permits visual examination of various electrical phenomena of interest to the electronic engineer. Instantaneous changes in voltage, current and phase are observable if they take place slowly enough for the eye to follow, or if they are periodic for a long enough time so that the eye can obtain an impression from the screen of the cathode-ray tube. In addition, the cathode-ray oscilloscope may be used to study any variable (within the limits of its frequency-response characteristic) which can be converted into electrical potentials. This conversion is made possible by the use of some type of *transducer*, such as a vibration pickup unit, pressure pickup unit, photoelectric

cell, microphone, or a variable impedance. The use of such a transducer makes the cscilloscope a valuable tool in fields other than electronics.

Oscilloscopes have become more versatile and complex in the last decade. A new order of measurement capability has made the 'scope into a device now used in computers, calculators, and the very heart of many complex electronic products, as well as serving as a measuring and indicating device.

Oscilloscope bandwidth, or the highest frequency signal the 'scope can display, is the most important factor indicating the degree of performance. The price of a 'scope is usually directly related to bandwidth. *Risetime* is a measure of how quickly the 'scope can

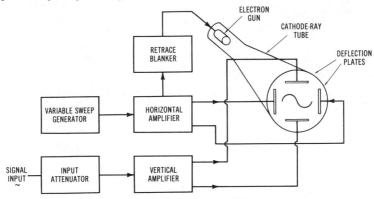

SIMPLIFIED OSCILLOSCOPE CIRCUIT

Beam deflection in a cathode-ray tube is accomplished by controlling the voltage on two sets of deflection plates in the tube. Sweep generator establishes the horizontal time base and input signal is applied to the vertical plates.

respond to an instantaneous change in voltage level at the input. It, along with bandwidth, is an important, but not sufficient judge of performance. Bandwidth and risetime are usually interrelated by a simple formula:

$$\text{Risetime} = 0.35 \text{ bandwidth}$$

Pulse risetime is an important specification, although not always given by the manufacturer. It relates to the ability of the 'scope to reproduce an ideal pulse. A pulse would be perfectly reproduced if the instrument had infinite bandwidth (or infinitely fast risetime). This is because the vertical leading edge of the pulse contains high-frequency components, which must pass through the oscilloscope amplifier system undistorted, in order to appear on the display exactly as generated.

Gain compression is a measure of the faithfulness of reproduction of a waveform on the screen. The ratio of change in signal amplitude of a waveform at different positions on the screen with respect to the waveform displayed at midscreen indicates the degree of gain compression.

Time-base accuracy represents how accurately, in terms of time period, the hori-

zontal deflection is maintained and *time-base linearity* indicates how constant the rate of travel is for the 'scope trace when moving from extreme left to extreme right.

Other parameters relating to oscilloscope performance include *screen persistence, writing speed,* and *spot size,* all of which should be described for a modern, multipurpose oscilloscope.

32-1   A Modern Oscilloscope

For the purpose of analysis, the operation of a modern oscilloscope will be described. The 'scope is completely solid state except for the cathode-ray tube. The simplified block diagram of the instrument is shown in figure 1. This oscilloscope (the *Heathkit* model IO-102) is capable of reproducing sine waves up to 5 MHz and has a rise time of 80 nanoseconds. The sweep speed is continuously variable from 10 Hz to 500 kHz in five ranges, and the electron beam of the cathode-ray tube can be moved vertically or horizontally, or the movements may be combined to produce composite patterns on the screen. As shown in the diagram, the cathode-ray tube receives signals from two

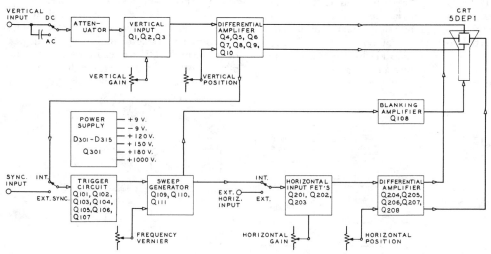

Figure 1

BLOCK DIAGRAM OF A MODERN OSCILLOSCOPE

This simplified diagram of the Heath Kit IO-102 solid-state oscilloscope features triggered sweep and a blanking circuit that permits observation of extremely short pulses. The cathode-ray tube is the only vacuum tube in the instrument.

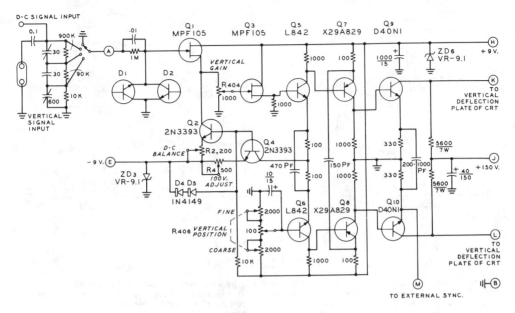

Figure 2

VERTICAL AMPLIFIER

The vertical amplifier is capable of passing sine waves up to 5 MHz. The compensated input attenuator and peaking circuits provide gain that is essentially independent of frequency. Emitter-follower Q_5 is coupled to amplifier Q_6 to provide push-pull signal necessary for the deflection plates of the cathode-ray tube. The input signal is limited in amplitude by diodes D_1 and D_2 (the junction of inexpensive bipolar transistors).

sources: the *vertical* (Y-axis) and the *horizontal* (X-axis) *amplifiers*, and also receives *blanking pulses* that remove unwanted return trace signals from the screen. The operation of the cathode-ray tube has been covered in an earlier chapter and the auxiliary circuits pertaining to signal presentation will be discussed here.

The Vertical Amplifier The incoming signal to be displayed is coupled through a frequency-compensated attenuator network (figure 2). The gain may thus be controlled in calibrated steps. A capacitor blocks the dc component of the signal when ac signals are applied to the circuit. A portion of the input signal is applied through a voltage-limiting resistor and two limiting diodes (D_1, D_2) to a FET connected as a source follower amplifier (Q_1). This device provides the high input impedance necessary to prevent circuit loading. Transistor Q_2 is a constant-current source

for the FET and diodes D_4 and D_5 hold the base of Q_2 at a constant voltage. Since Q_2 is a form of emitter follower, the emitter voltage is a function of the base voltage, and the emitter voltage also remains constant. This voltage appears across the dc *balance* control which is adjusted so that the source voltage of the FET is zero when an input signal is not present. Thus, a signal applied to the gate of Q_1 causes only voltage changes at the source because the current through Q_1 is constant. The voltage variations are applied across the *vertical gain* control and a portion of this signal is applied to the gate of source follower Q_3. Transistor Q_4 forms a constant-current source for transistors Q_5 and Q_6. Since the emitter of each device is connected to this source, the source serves as a common-emitter resistance and sets the operating point for the following stages.

Transistors Q_5 and Q_6 have a common-emitter resistance and any signal present at the Q_5 emitter is coupled to the emitter of Q_6, which functions as a common-base am-

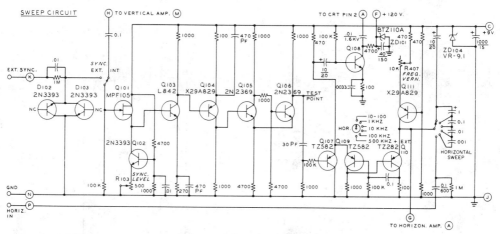

Figure 3

THE SWEEP CIRCUIT

The sweep may be triggered either by the input signal or by an external source. Schmitt trigger circuit (Q_{105} and Q_{106}) produces a regular pulse each time it is triggered driving the astable multivibrator (Q_{109} and Q_{110}). Timing capacitors and the frequency vernier potentiometer determine sweep speed. During the wait period between trigger pulses, the CRT is cut off so that the blanking waveform is not seen. Negative pulse from blanking amplifier Q_{108} is applied to pin #2 of the cathode-ray tube to perform this function.

plifier whose base is held constant by the *vertical position* potentiometer. The signal at the collector of transistor Q_6 is 180° out of phase with the signal at the collector of Q_5, thus forming a push-pull configuration required to drive the deflection plates of the cathode-ray tube.

Drive transistors Q_7 and Q_8 are common-emitter amplifiers which drive output amplifier transistors Q_9 and Q_{10} which have their collector potential derived from the +150 volt supply.

The Sweep Circuit   Investigation of electrical waveforms by the use of a cathode-ray tube requires that some means be readily available to determine the variation in the waveforms with respect to time. An X-axis *time base* on the screen of the cathode-ray tube shows the variation in amplitude of the input signal with respect to time (figure 3). This display is made possible by a *time-base generator (sweep generator)* which moves the spot across the screen at a constant rate from left to right between selected points, returns the spot almost instantaeously to its original position,

and repeats this procedure at a specified rate (referred to as the *sweep frequency*).

The Sweep-Trigger Circuit—An external *synchronizing impulse* which may be either a portion of the amplified signal or a signal applied to the *external sync* terminals is coupled to the gate of source follower Q_{101}. Two limiting diodes protect the transistor from high voltage surges. Constant-current source Q_{102} is adjusted by the *sync level* control to provide proper bias for the synchronizing circuits. This ensures that even a small signal can synchronize the sweep generator.

Transistors Q_{103} and Q_{104} amplify the signal and apply it to the *Schmitt trigger circuit* consisting of Q_{105} and Q_{106}. This trigger circuit is a regenerative bistable circuit which produces a regular pulse output each time it is triggered and reset. Devices Q_{109} and Q_{110} form an astable multivibrator whose frequency is controlled by the switchable timing capacitors. The capacitors are charged through Q_{110} and discharged through the constant-current source circuit of Q_{111}. The *frequency vernier* potentiometer determines the current flowing

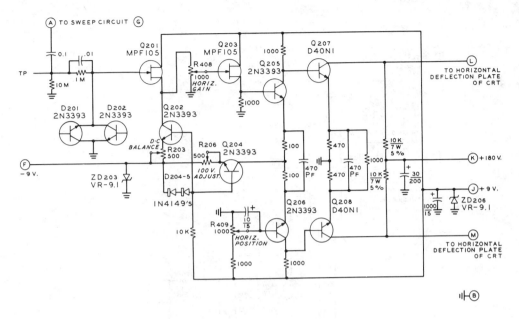

Figure 4

HORIZONTAL AMPLIFIER

The horizontal amplifier is similar to the vertical amplifier except it does not have PNP stage Q_7-Q_8 shown in figure 3. Amplified sweep waveform is applied to the horizontal-deflection plates of the CRT causing the electron beam to sweep across the face of the tube producing a visible trace. Transistor Q_{205} serves as an emitter follower to produce push-pull driving signal for Q_{207} and Q_{208}. Horizontal positioning of signal on screen of CRT is determined by the base bias of Q_{204}.

through Q_{111} which, in turn, determines the discharge current and discharge time of the timing capacitor. As the capacitor discharges, a positive-going sawtooth voltage is generated and coupled to the horizontal amplifier. The frequency of the horizontal sweep is determined by the particular timing capacitor and the discharge current.

The Blanking Circuit—During the wait period between trigger pulses, the cathode-ray tube is completely cut off so that the blanking waveform is not seen. Since transistors Q_{107} and Q_{109} have a common emitter resistor, a signal applied to the base of Q_{107} is emitter-coupled to transistor Q_{109}. The pulse output of the Schmitt trigger (Q_{106}) is coupled to Q_{109}. This causes this transistor to turn on and Q_{110} to cut off and start the sweep just prior to the time it would normally begin. When the signal at the emitter of Q_{109} goes positive, a positive pulse is applied to the base of blanking amplifier Q_{108}. A negative-going output

pulse is coupled to the grid of the cathode-ray tube which turns off the electron beam during retrace.

The Horizontal Amplifier — Since the amplitude of the sweep waveform at the output of the sweep generator is not large enough to drive the horizontal deflection plates of the cathode-ray tube, further amplification is needed. The signal from the sweep generator is applied to the horizontal amplifier, whose circuitry is similar to that of the vertical amplifier (figure 4). The major difference is that the horizontal amplifier does not have a PNP amplifier stage corresponding to Q_7 and Q_8 in the vertical amplifier. The positive-going sawtooth wave from the sweep generator is amplified and applied to the horizontal plates of the cathode-ray tube. This increasing voltage causes the electron beam to sweep across the face of the tube producing a visible trace. The sweep rate of the electron beam is determined by the sawtooth frequency.

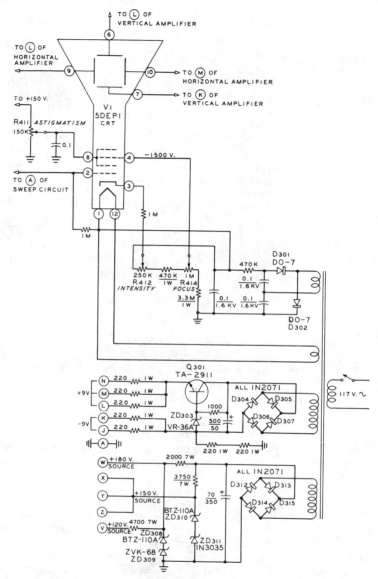

Figure 5

POWER SUPPLY

Power supply provides −1500 volts for CRT and various low voltages for solid-state circuitry of the oscilloscope. Intensity and focus voltages are supplied from a voltage-divider network. Optimum focus is obtained when the deflection plates of CRT and the astigmatism grid are at the same potential.

The Power Supply　The power supply provides positive and negative voltages for the various stages of the oscilloscope, as shown in figure 5. A high-voltage winding of the power transformer is connected to a voltage-doubler circuit to pro-

vide − 1500 volts to the cathode-ray tube. *Intensity* and *focus* voltages are also supplied from a voltage-divider network. A separate 6.3-volt winding supplies the filament voltage for the cathode-ray tube. Optimum focus is obtained when the deflection

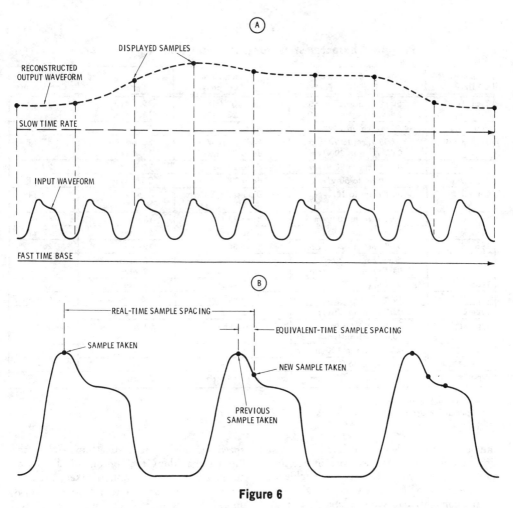

Figure 6

SYNTHESIS OF A WAVEFORM BY SAMPLING TECHNIQUE

A—The sampling technique displays a synthesized reproduction of the original signal and is similar to the stationary image of a rapidly spinning wheel produced by an optical strobe light. The display appears as a series of image-retaining dots rather than the usual continuous presentation of a conventional oscilloscope.

B—The relationship between real time and equivalent time. In practice, a large number of dots form the display so that the trace appears continuous. The new time base of the synthesized display is adjusted to provide a picture equivalent to the original wave, the trace being independent of the repetition rate of the observed signal.

plates of the cathode-ray tube and the *astigmatism* grid are at the same potential. Since the vertical-deflection plate voltages (collectors of Q_9 and Q_{10}) are adjusted to 100 volts dc by the constant-current source Q_1, the astigmatism potential is also adjusted to 100 volts. A low-voltage regulated supply provides $+9$ and -9 volts and a third supply provides the various other voltages required by the oscilloscope circuits.

32-2   The Sampling Oscilloscope

In a conventional 'scope, the visual examination and analysis of waveforms in the uhf spectrum are restricted by the gain-band-width limitations of the deflection circuits

TABLE 1

Plug-In Characteristics for the Type 545 Oscilloscope

Plug-In Unit	Calibrated Deflection Factor	Minimum Bandpass	Risetime	Input Capacitance
Type 1A1*	50 mv/cm to 20 v/cm 5 mv/cm	dc to 33 MHz dc to 23 MHz	10.6 nsec 15.2 nsec	15 pF
Type 1A2*	50 mv/cm to 20v/cm	dc to 33 MHz	10.6 nsec	15 pF
Type B	0.005 v/cm to 20 v/cm 0.05 v/cm to 20 v/cm	2 Hz to 12 MHz dc to 20 MHz	30 nsec 18 nsec	47 pF
Type CA*	0.05 v/cm to 20 v/cm	dc to 24 MHz	15 nsec	20 pF
Type D	1 mv/cm to 50 v/cm	dc to 300 kHz-2 MHz	0.18 μsec	47 pF
Type E	50 μv/cm to 10mv/cm	0.06 Hz to 20 kHz -60 kHz	6 μsec	50 pF
Type G	0.05 v/cm to 20 v/cm	dc to 20 MHz	18 nsec	47 pF
Type H	5 mv/cm to 20 v/cm	dc to 15 MHz	23 nsec	47 pF
Type K	0.05 v/cm to 20 v/cm	dc to 30 MHz	12 nsec	20 pF
Type L	5 mv/cm to 2 v/cm 0.05 v/cm to 20 v/cm	3 Hz to 24 MHz dc to 30 MHz	15 nsec 12 nsec	20 pF
Type M*	0.02 v/cm to 10 v/cm	dc to 20 MHz	17 nsec	47 pF
Type N**	10 mv/cm	dc to 600 MHz	0.6 nsec	50 Ω input Z
Type O**	0.05 v/cm to 20 v/cm	dc to 25 MHz	14 nsec	47 pF
Type Q**	10 μstrain/cm to 10,000 μstrain/cm	dc to 6 kHz	60 μsec	Adjustable
Type R**	0.5 ma/cm to 100 ma/cm	—	—	—
Type S**	0.05 v/cm to 0.5 v/cm	—	—	—
Type Z**	0.05 v/cm to 25 v/cm	dc to 13 MHz	27 nsec	24 pF

*Multichannel plug-in units.
**Special feature plug-in units.

and associated video amplifiers. Fast rise time is obtained at the expense of reduced amplifier gain and high sensitivity is achieved at the expense of fast rise time. In addition, when a conventional 'scope is used to display extremely fast changing signals of low amplitude, the trace becomes dim and the presentation may no longer be visible.

To offset these problems, *pulse sampling* techniques are used whereby fast, repetitive waveforms in the thousands of MHz range are converted into slow-speed signals of much lower frequency and identical waveform.

Sampling is the electronic equivalent of the optical stroboscope principle used for the visual examination of rapid mechanical motion. The synthesis of a recurring waveform is shown in figure 6 wherein the display appears as a series of image-retaining dots rather than the continuous presentation of a conventional oscilloscope. The dots, uniformly spaced in time, are produced by high speed sampling pulses superimposed on the input signal along the contour of the waveform. Each time a sample is taken, the spot is moved along the X-axis and is repositioned on the Y-axis to the corresponding voltage amplitude of the signal. This process is continued until a replica of the original information is presented on the screen of the instrument.

The sampling gate of the 'scope is controlled by a *strobe generator* activated by the trigger signal, which may be derived internally or externally to the instrument. The amplitude of the input signal is measured so as to control the vertical output signal of the 'scope amplifier to an amplitude equal to the sampling signal level.

At the start of each sampling pulse, the cathode-ray display tube is unblanked, the pulse height samples are mixed with the vertical input signal and the resultant signal-modulated sample is amplified, lengthened in time and applied to the Y-axis of the 'scope.

Plug-in Modules Many modern oscilloscopes use plug-in modules which offer great operational flexibility. Probably the most common, and one of the earliest of these 'scopes, is the *Tektronix 535/545* series. These units are often available to the amateur at a reasonable price in surplus electronics stores. There are 17 plug-in pre-amplifier modules for this series of instruments (Table 1).

The type CA plug-in head for the 534/535 series is the one most often seen with this 'scope and is a dual-trace, dc to 24-MHz head. With this plug-in, two waveforms can be observed on the screen, each with essentially the full bandwidth of the 'scope. Note that "dual trace" in most modern oscilloscopes is not synonymous with "dual beam;" the basic 'scope has only one electron gun and deflection system, and the dual presentation is accomplished by chopping, or alternately displaying the two inputs (every other sweep). Other plug-in heads giving greater sensitivity, differential input and other features are available.

The *Tektronix 561* uses the next logical level of modularization; the front end (pre-amplifier) is a plug-in unit and so is the sweep control system. In this way it is possible to not only use different types of amplifiers for the Y-axis input, and different sweep control modules, but it is also possible to use two identical input amplifiers in the X- and Y-axes for applications such as lissajous figure phase and frequency comparisons.

The plug-in design is carried to four units per oscilloscope in the *Tektronix 7904*. This 'scope has slots for four plug-in modules. In some advanced 'scopes, the modules have smaller slots for submodules. These complicated modular techniques are used to achieve the greater and greater range of functions demanded by today's technology.

32-3 Display of Waveforms

Together with a working knowledge of the controls of the oscilloscope, an understanding of how the patterns are traced on the screen must be obtained for a thorough knowledge of oscilloscope operation. With

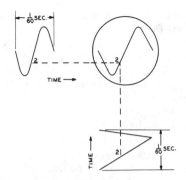

Figure 7

PROJECTION DRAWING OF A SINE WAVE APPLIED TO THE VERTICAL AXIS AND A SAWTOOTH WAVE OF THE SAME FREQUENCY APPLIED SIMULTANEOUSLY ON THE HORIZONTAL AXIS

this in mind a careful analysis of two fundamental waveform patterns is discussed under the following headings:

1. Patterns plotted against time (using the sweep generator for horizontal deflection).
2. *Lissajous figures* (using a sine wave for horizontal deflection).

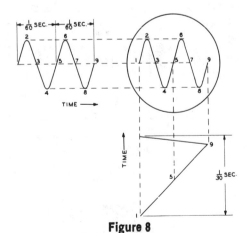

Figure 8

PROJECTION DRAWING SHOWING THE RESULTANT PATTERN WHEN THE FREQUENCY OF THE SAWTOOTH IS ONE-HALF OF THAT EMPLOYED IN FIGURE 7

Patterns Plotted Against Time   A sine wave is typical of such a pattern and is convenient for this study. This wave is amplified by the vertical amplifier and impressed on the vertical (Y-axis) deflection plates of the cathode-ray tube. Simultaneously the sawtooth wave from the time-base generator is amplified and impressed on the horizontal (X-axis) deflection plates.

The electron beam moves in accordance with the resultant of the sine and sawtooth signals. The effect is shown in figure 7 where the sine and sawtooth waves are graphically represented on time and voltage axes. Points on the two waves that occur simultaneously are numbered similarly. For example, point 2 on the sine wave and point 2 on the sawtooth wave occur at the same instant. Therefore the position of the beam at instant 2 is the resultant of the voltages on the horizontal and vertical deflection plates at instant 2. Referring to figure 7, by projecting lines from the two point-2 positions, the position of the electron beam at instant 2 can be located. If projections were drawn from every other instantaneous position of each wave to intersect on the circle representing the tube screen, the intersections of similarly timed projects would trace out a sine wave.

In summation, figure 7 illustrates the principles involved in producing a sine-

wave trace on the screen of a cathode-ray tube. Each intersection of similarly timed projections represents the position of the electron beam acting under the influence of the varying voltage waveforms on each pair of deflection plates. Figure 8 shows the effect on the pattern of decreasing the frequency of the sawtooth wave. Any recurrent waveform plotted against time can be displayed and analyzed by the same procedure as used in these examples.

The sine-wave problem just illustrated is typical of the method by which any waveform can be displayed on the screen of the cathode-ray tube. Such waveforms as square wave, sawtooth wave, and many more irregular recurrent waveforms can be observed by the same method explained in the preceding paragraphs.

32-4   Lissajous Figures

Another fundamental pattern is the *Lissajous figures*, named after the 19th-century French scientist. This type of pattern is of particular use in determining the frequency ratio between two sine-wave signals. If one of these signals is known, the other can be easily calculated from the pattern made by the two signals on the screen of the cathode-ray tube. Common practice is to connect the known signal to the horizontal channel and the unknown signal to the vertical channel.

The presentation of Lissajous figures can be analyzed by the same method as previ-

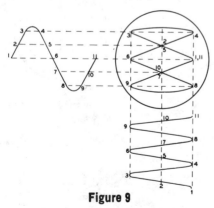

Figure 9

PROJECTION DRAWING SHOWING THE RESULTANT LISSAJOUS PATTERN WHEN A SINE WAVE APPLIED TO THE HORIZONTAL AXIS IS THREE TIMES THAT APPLIED TO THE VERTICAL AXIS

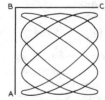

Figure 10

METHOD OF CALCULATING FREQUENCY RATIO OF LISSAJOUS FIGURES

ously used for sine-wave presentation. A simple example is shown in figure 9. The frequency ratio of the signal on the horizontal axis to the signal on the vertical axis is 3 to 1. If the known signal on the horizontal axis is 180 Hertz, the signal on the vertical axis is 60 Hertz.

Obtaining a Lissajous Pattern on the Screen; Oscilloscope Settings 1. The horizontal amplifier should be disconnected from the sweep oscillator. The

2. An audio oscillator signal should be connected to the vertical amplifier of the oscilloscope.

3. By adjusting the frequency of the audio oscillator a stationary pattern should be obtained on the screen of the oscilloscope. It is not necessary to stop the pattern, but merely to slow it up enough to count the loops at the side of the pattern.

4. Count the number of loops which intersect an imaginary vertical line *AB* and the number of loops which intersect the imaginary horizontal line *BC* as shown in figure 10. The ratio of the number of loops which intersect *AB* is to the number of loops which intersect *BC* as the frequency of the horizontal signal is to the frequency of the vertical signal.

Figure 11 shows other examples of Lissajous figures. In each case the frequency ratio shown is the frequency ratio of the signal on the horizontal axis to that on the vertical axis.

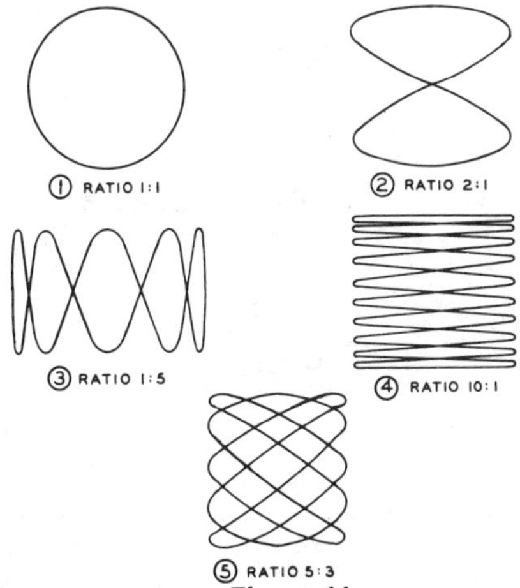

Phase Difference Patterns Coming under the heading of Lissajous figures is the method used to determine the phase difference between signals of the same frequency. The patterns involved take on the form of ellipses with different degrees of eccentricity.

The following steps should be taken to obtain a phase-difference pattern:

1. With no signal input to the oscilloscope, the spot should be centered on the screen of the tube.

2. Connect one signal to the vertical amplifier of the oscilloscope, and the other signal to the horizontal amplifier.

3. Connect a common ground between the two frequencies under investigation and the oscilloscope.

① RATIO 1:1   ② RATIO 2:1

③ RATIO 1:5   ④ RATIO 10:1

⑤ RATIO 5:3

Figure 11

OTHER LISSAJOUS PATTERNS

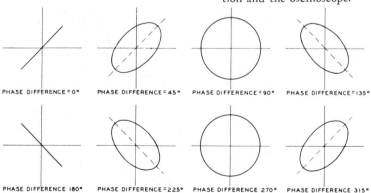

PHASE DIFFERENCE = 0°   PHASE DIFFERENCE = 45°   PHASE DIFFERENCE = 90°   PHASE DIFFERENCE = 135°

PHASE DIFFERENCE 180°   PHASE DIFFERENCE = 225°   PHASE DIFFERENCE 270°   PHASE DIFFERENCE 315°

Figure 12

LISSAJOUS PATTERNS OBTAINED FROM THE MAJOR PHASE DIFFERENCE ANGLES

4. Adjust the vertical amplifier gain so as to give about 3 inches of deflection on a 5-inch tube, and adjust the calibrated scale of the oscilloscope so that the vertical axis of the scale coincides precisely with the vertical deflection of the spot.
5. Remove the signal from the vertical amplifier, being careful not to change the setting of the vertical gain control.
6. Increase the gain of the horizontal amplifier to give a deflection exactly the same as that to which the vertical amplifier control is adjusted (3 inches). Reconnect the signal to the vertical amplifier.

The resulting pattern will give an accurate picture of the exact phase difference between the two waves. If these two patterns are exactly the same frequency but different in phase and maintain that difference, the pattern on the screen will remain stationary. If, however, one of these frequencies is drifting slightly, the pattern will drift slowly through 360°. The phase angles of 0°, 45°, 90°, 135°, 180°, 225°, 270°, and 315° are shown in figure 12.

Each of the eight patterns in figure 12 can be analyzed separately by the previously used projection method. Figure 13 shows two sine waves which differ in phase being projected on to the screen of the cathode-ray tube. These signals represent a phase difference of 45°.

Determinination of   The relation commonly
the Phase Angle   used in determining the phase angle between signals is:

$$\text{Sine } \theta = \frac{Y \; intercept}{Y \; maximum}$$

where,

θ equals phase angle between signals,
$Y \; intercept$ equals point where ellipse crosses vertical axis measured in tenths of inches (calibrations on the calibrated screen),
$Y \; maximum$ equals highest vertical point on ellipse in tenths of inches.

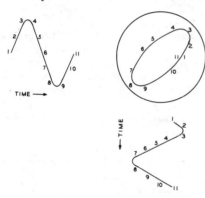

TIME →

Figure 13

PROJECTION DRAWING SHOWING THE RESULTANT PHASE-DIFFERENCE PATTERN OF TWO SINE WAVES 45° OUT OF PHASE

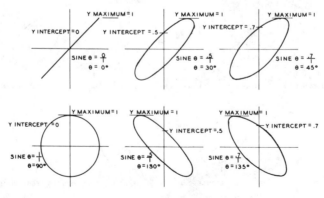

Figure 14

EXAMPLES SHOWING THE USE OF THE INTERCEPT FORMULA FOR DETERMINATION OF PHASE DIFFERENCE

Several examples of the use of the formula are given in figure 14. In each case the Y *intercept* and Y *maximum* are indicated together with the sine of the angle and the angle itself. For the operator to observe these various patterns with a single signal source such as the test signal, there are many types of phase shifters which can be used. Circuits can be obtained from a number of radio textbooks. The procedure is to connect the original signal to the horizontal channel of the oscilloscope and the signal which has passed through the phase shifter to the vertical channel of the oscilloscope, and follow the procedure set forth in this discussion to observe the various phase-shift patterns.

32-5 Receiver I-F Alignment with an Oscilloscope

The alignment of the i-f amplifiers of a receiver consists of adjusting all the tuned circuits to resonance at the intermediate frequency and at the same time permitting passage of a predetermined number of sidebands. The best indication of this adjustment is a resonance curve representing the response of the i-f circuit to its particular range of frequencies.

A representative response of a receiver i-f system is shown in figure 15. A response curve of this type can be displayed on a 'scope with the aid of a sweep generator.

The Resonance Curve on the Screen To present a resonance curve on the 'scope, a frequency-modulated signal source must be available. Some signal generators have a built in sweep circuit in the

form of a voltage-variable capacitor (VVC) which sweeps the signal frequency 5 to 10 kHz each side of the fundamental frequency. In addition, a blanking circuit in the generator is applied to the 'scope to blank out the return trace so that a double-hump resonance curve is not obtained.

32-6 Single-Sideband Applications

Measurement of power output and distortion are of particular importance in SSB transmitter adjustment. These measurements are related to the extent that distortion rises rapidly when the power amplifier is overloaded. The usable power output of an SSB transmitter is often defined as the maximum peak envelope power obtainable with a specified *signal-to-distortion* ratio. The oscilloscope is a useful instrument for measuring and studying distortion of all types that may be generated in single-sideband equipment.

Single-Tone Observations When an SSB transmitter is modulated with a single audio tone, the r-f output should be a single radio frequency. If the vertical plates of the oscilloscope are coupled to the output of the transmitter, and the horizontal amplifier sweep is set to a slow rate, the scope presentation will be as shown in figure 16. If unwanted distortion products or carrier are present, the top and bottom of the pattern will develop a "ripple" proportional to the degree of spurious products.

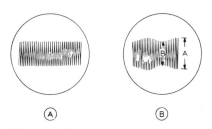

(A)   (B)

Figure 16

SINGLE-TONE PRESENTATION

Oscilloscope trace of SSB signal modulated by single tone (A). Incomplete carrier suppression or spurious products will show modulated envelope of (B). The ratio of supression is:

$$S = 20 \log \frac{A + B}{A - B}$$

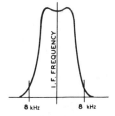

Figure 15

FREQUENCY RESPONSE OF HIGH-FIDELITY I-F SYSTEM

The Linearity Tracer The *linearity tracer* is an auxiliary detector to be used with an oscilloscope for quick observation of amplifier adjustments and parameter variations. This instrument consists of two SSB *envelope detectors* the outputs of which connect to the horizontal and vertical inputs of an oscilloscope. Figure 17 shows a block diagram of a typical linearity test set-up. A two-tone test signal is normally employed to supply an SSB modulation envelope, but any modulating signal that provides an envelope that varies from zero to full amplitude may be used. Speech modulation gives a satisfactory trace, so that this instrument may be used as a visual monitor of transmitter linearity. It is particularly useful for monitoring the signal level and clearly shows when the amplifier under observation is overloaded. The linearity trace will be a straight line regardless of the envelope shape if the

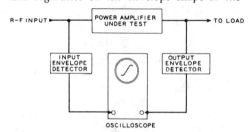

Figure 17

BLOCK DIAGRAM OF LINEARITY TRACER

amplifier has no distortion. Overloading causes a sharp break in the linearity curve. Distortion due to too much bias is also easily observed and the adjustment for low distortion can easily be made.

Another feature of the linearity detector is that the distortion of each individual stage can be observed. This is helpful in troubleshooting. By connecting the input envelope detector to the output of the SSB generator, the over-all distortion of the entire r-f circuit beyond this point is observed. The unit can also serve as a voltage indicator which is useful in making tuning adjustments.

The circuit of a typical envelope detector is shown in figure 18. Two matched germanium diodes are used as detectors. The detectors are not linear at low signal levels,

but if the nonlinearity of the two detectors is matched, the effect of their nonlinearity on

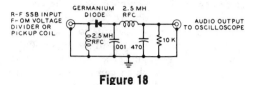

Figure 18

SCHEMATIC OF ENVELOPE DETECTOR

the oscilloscope trace is cancelled. The effect of diode differences is minimized by using a diode load of 5000 to 10,000 ohms, as shown. It is important that both detectors operate at approximately the same signal level so that their differences will cancel more exactly. The operating level should be 1 volt or higher.

It is convenient to build the detector in a small shielded enclosure such as an i-f transformer can fitted with coaxial input and output connectors. Voltage dividers can be similarly constructed so that it is easy to insert the desired amount of voltage attenuation from the various sources. In some cases it is convenient to use a pickup loop on the end of a short length of coaxial cable.

The phase shift of the amplifiers in the oscilloscope should be the same and their frequency response should be flat out to at least twenty times the frequency difference of the two test tones. Excellent high-frequency characteristics are necessary because the rectified SSB envelope contains harmonics extending to the limit of the envelope detector's response. Inadequate frequency response of the vertical amplifier may cause a little "foot" to appear on the lower end of the trace, as shown in figure 19. If it is small, it may be safely neglected.

Another spurious effect often encountered is a double trace, as shown in figure 20. This can usually be corrected with an RC network placed between one detector and the oscilloscope. The best method of testing the detectors and the amplifiers is to connect the input of the envelope detectors in parallel. A perfectly straight line trace will result when everything is working properly. One detector is then connected to the other r-f source through a voltage divider adjusted so that no appreciable change in the setting of

Figure 19

**EFFECT OF INADEQUATE
RESPONSE OF VERTICAL
AMPLIFIER**

Figure 20

**DOUBLE TRACE CAUSED
BY PHASE SHIFT**

the oscilloscope amplifier controls is required. Figure 21 illustrates some typical linearity traces. *Trace A* is caused by inadequate static plate current in class-A or class-B amplifiers or a mixer stage. To regain linearity, the grid bias of the stage should be reduced, the screen voltage should be raised, or the signal level should be decreased. *Trace B* is a result of poor grid-circuit regulation

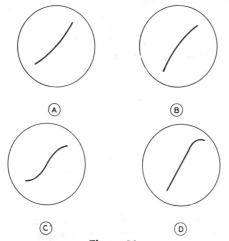

Figure 21

TYPICAL LINEARITY TRACES

when grid current is drawn, or a result of nonlinear plate characteristics of the amplifier tube at large plate swings. More grid swamping should be used, or the exciting signal should be reduced. A combination of the effects of A and B are shown in *Trace C. Trace D* illustrates amplifier overloading. The exciting signal should be reduced.

A means of estimating the distortion level observed is quite useful. The first- and third-order distortion components may be derived by an equation that will give the approximate signal-to-distortion level ratio of a *two-tone* test signal, operating on a given linearity curve. Figure 22 shows a linearity curve with two ordinates erected at half and full peak input signal level. The length of

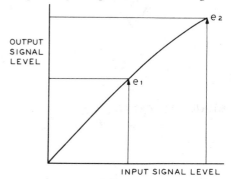

Figure 22

**ORDINATES ON LINEARITY CURVE FOR
3RD-ORDER DISTORTION EQUATION**

the ordinates e_1 and e_2 may be scaled and used in the following equation:

Signal-to-distortion ratio in dB =

$$20 \log \frac{8 e_1 - e_2}{2 e_1 - e_2}$$

32-7   A-M Applications

The oscilloscope may be used as an aid for the proper operation of an a-m transmitter, and may be used as an indicator of the overall performance of the transmitter output signal, and as a modulation monitor.

Waveforms   There are two types of patterns that can serve as indicators, the *trapezoidal pattern* (figure 23) and the *modulated-wave pattern* (figure 24). The

Figure 23

TRAPEZOIDAL MODULATION PATTERN

trapezoidal pattern is presented on the screen by impressing a modulated carrier-wave signal on the vetical deflection plates and the signal that modulates the carrier-wave signal (the modulating signal) on the horizontal deflection plates. The trapezoidal pattern can be analyzed by the method used previously in analyzing waveforms. Figure

Figure 24

MODULATED CARRIER-WAVE PATTERN

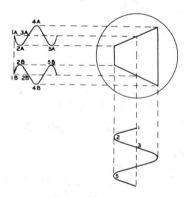

Figure 25

PROJECTION DRAWING SHOWING TRAPEZOIDAL PATTERN

25 shows how the signals cause the electron beam to trace out the pattern.

The modulated-wave pattern is accomplished by presenting a modulated carrier wave on the vertical deflection plates and by using the time-base generator for horizontal deflection. The modulated-wave pattern also can be used for analyzing waveforms. Figure 26 shows a representative modulation pattern.

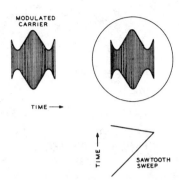

Figure 26

PROJECTION DRAWING SHOWING MODULATED-CARRIER WAVE PATTERN

The trapezoidal pattern is obtained by applying a portion of the audio signal to the horizontal input of the 'scope. This may be taken from the modulator through a small coupling capacitor and a high resistance voltage divider. Only a fraction of a volt of signal is required for the 'scope. A small amount of modulated r-f signal is coupled directly to the vertical deflection plates of the oscilloscope. This may be taken from a loop coupled to the final tank circuit or via a resonant circuit coupled to the transmission line of the transmitter.

On modulation of the transmitter, the trapezoidal pattern will appear. By changing the degree of modulation of the carrier wave the shape of the pattern will change. Figures 27 and 28 show the trapezoidal pattern for various degrees of modulation. The percentage of modulation may be determined by the following formula:

$$\text{Modulation percentage} = \frac{E_{max} - E_{min}}{E_{max} + E_{min}} \times 100$$

where,
E_{max} and E_{min} are defined as in figure 27.

An overmodulated signal is shown in figure 29.

The Modulated-Wave Pattern The modulated-wave pattern is obtained by applying a portion of the modulated r-f signal to the horizontal input circuit of the 'scope. The vertical amplifier is

TRAPEZOIDAL PATTERNS

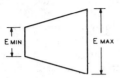

Figure 27 **Figure 28** **Figure 29**

(LESS THAN 100% MODULATION) (100% MODULATION) **(OVERMODULATION)**

CARRIER-WAVE PATTERN

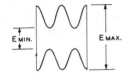

Figure 30 **Figure 31** **Figure 32**

(LESS THAN 100% MODULATION) (100% MODULATION) **(OVERMODULATION)**

connected to the internal sweep circuit of the instrument, which is synchronized with the modulating signal by applying a small portion of the audio signal to the *external sync* input terminal of the oscilloscope. The percentage of modulation may be determined in the same fashion as with a trapezoid pattern. Figures 30, 31, and 32 show the modulated wave pattern for various levels of modulation.

32-8   The Spectrum Analyzer

The *spectrum analyzer* is a receiver-oscilloscope combination that provides a convenient means of measuring the amplitude and frequency of radio signals because it can discriminate the energy of individual

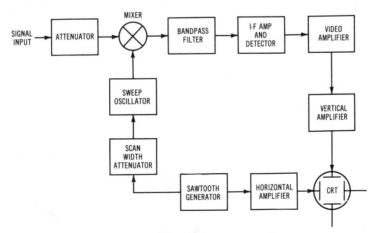

Figure 33

BLOCK DIAGRAM OF A SPECTRUM ANALYZER

Input spectrum is displayed in terms of frequency versus amplitude.

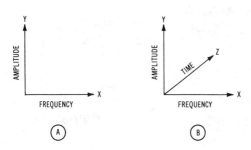

Figure 34

SPECTRUM DISPLAY

A—Two axis display.
B—Three axis display.

emissions, even at very low signal levels, on a swept-frequency basis (figure 33).

The spectrum analyzer commonly consists of an electronically swept oscillator sweeping through a range of frequencies in the radio spectrum, and an oscilloscope device whose horizontal sweep is synchronized with the sweep applied to the oscillator. The oscillator may be coupled to a mixer and receiver system so that either local signals or received signals are applied to the analyzer. The analyzer may plot either frequency versus amplitude in a two dimensional display or frequency versus amplitude versus time in a three dimensional display (figure 34).

CHAPTER THIRTY-THREE

Construction Practices

With a few possible exceptions, such as cabinets, brackets, neutralizing capacitors and transmitting coils, it hardly pays one to attempt to build the components required for the construction of an amateur transmitter. This is especially true when the parts are of the type used in construction and replacement work on receivers and TV, as mass production has made these parts very inexpensive.

Those who have and wish to spend the necessary time can effect considerable monetary saving in their equipment by building them from the component parts. The necessary data is given in the construction chapter of this handbook.

To many builders, the construction is as fascinating as the operation of the finished transmitter; in fact, many amateurs get so much satisfaction out of building a well-performing piece of equipment that they spend more time constructing and rebuilding equipment than they do operating the equipment on the air.

33-1 Tools

Beautiful work can be done with metal chassis and panels with the help of only a few inexpensive tools. The time required for construction, however, will be greatly reduced if a fairly complete assortment of metal-working tools is available. Thus, while an array of tools will speed up the work, excellent results may be accomplished with few tools, if one has the time and patience.

The investment one is justified in making in tools is dependent upon several factors. If you like to tinker, there are many tools useful in radio construction that you would probably buy anyway, or perhaps already have, such as screwdrivers, hammer, saws, square, vise, files, etc. This means that the money taken for tools from your radio budget can be used to buy the more specialized tools, such as socket punches or hole saws, taps and dies, etc.

The amount of construction work one does determines whether buying a large assortment of tools is an economical move. It also determines if one should buy the less expensive type offered at surprisingly low prices by the familiar mail order houses, "five and ten" stores, and chain auto-supply stores, or whether one should spend more money and get first-grade tools. The latter cost considerably more and work but little better when new, but will outlast several sets of the cheaper tools. Therefore they are a wise investment for the experimenter who does lots of construction work. The amateur who constructs only an occasional piece of apparatus need not be so concerned with tool life, as even the cheaper grade tools will last him several years, if they are given proper care.

The hand tools and materials in the accompanying lists will be found very useful around the home workshop. Materials not listed but ordinarily used, such as paint, can best be purchased as required for each individual job.

ESSENTIAL HAND TOOLS AND MATERIALS

1 Dual heat soldering gun, 100/140 watts
1 Spool resin core solder, 60/40 alloy
1 Set screwdrivers, $\frac{1}{8}$" and $\frac{1}{4}$" blade, 8" shaft
1 Set Phillips screwdrivers, #1, #2 and #4
1 Set nutdrivers, $\frac{1}{4}$", $\frac{5}{16}$" and $\frac{11}{32}$"
1 Hand "nibbling" tool
1 Long-nose pliers, 4"
1 Combination pliers, 6"
1 Diagonal "oblique" cutting pliers, 5"
1 Hand drill (egg-beater type)
1 Electrician's pocket knife
1 Combination steel rule and square, 1 foot
1 Yardstick, or steel tape
1 Multiple connection outlet box and extension cord
1 Set twist drills, $\frac{1}{4}$" shank, $\frac{1}{16}$" to $\frac{1}{4}$" (12 pcs.)
1 Set Allen and spline-head wrenches
1 Hacksaw and blades
1 Set medium files and handle
1 Roll vinyl electrical tape
1 Can paint thinner, or cleaner

HIGHLY DESIRABLE HAND TOOLS AND MATERIALS

1 Soldering iron, pencil type, 40 watt with interchangeable tips
1 Controlled temperature soldering stand
1 Electric drill, $\frac{1}{4}$", variable speed
1 DYMO label embosser
1 Cutting pliers, end-cut, 4"
1 Tap and die set for 4-40, 6-32, 8-32, 10-32 and 10-24
1 "Pop" rivet gun
1 Bench vise, 3" jaws
1 Metal snips
1 Center punch, spring-loaded
1 Set round punches, $\frac{5}{8}$", $\frac{3}{4}$", $\frac{7}{8}$", $1\frac{1}{8}$"
1 Fluorescent light and magnifier, 5" lens.
1 Crescent wrench, 6"
1 Set taper reamers
1 Set jeweler's screwdrivers

4 Small C-clamps
1 Wire stripper
1 Set alignment tools
1 Dusting brush
1 Small welding torch (gas)
1 Ratchet and socket set, $\frac{3}{16}$" to $\frac{1}{2}$"
1 12 drawer portable storage cabinet
1 Desoldering tool

Not listed are several special-purpose radio tools which are somewhat of a luxury, but are nevertheless quite handy, such as various around-the-corner screwdrivers and wrenches, special soldering iron tips, etc. These can be found in the larger radio parts stores and are usually listed in their mail order catalogs.

33-2   The Material

Electronic equipment may be built on a foundation of circuit board, steel, or aluminum. The choice of foundation material is governed by the requirements of the electrical circuit, the weight of the components of the assembly, and the financial cost of the project when balanced against the pocketbook contents of the constructor.

Breadboard and Brassboard   Experimental circuits may be built up in a temporary fashion termed *breadboarding*, a term reflecting the old practice of the "twenties" when circuits were built on wooden boards. Modern breadboards may be built upon circuit board material or upon prepunched phenolic boards. The prepunched boards contain a grid of small holes into which the component leads may be anchored for soldering.

A *brassboard* is an advanced form of assembly in which the experimental circuit is built up in semipermanent form on a metal chassis or copper-plated circuit board. Manufacture and use of printed-circuit boards is covered later in this chapter.

Special Frameworks   For high-powered r-f stages, many amateur constructors prefer to discard the more conventional types of construction and employ instead special metal frameworks and brackets which they design specially for the parts which they intend to use. These are usually

Figure 1

SOFT ALUMINUM SHEET MAY BE CUT WITH HEAVY KITCHEN SHEARS

arranged to give the shortest possible r-f leads and to fasten directly behind a panel by means of a few bolts, with the control shafts projecting through corresponding holes in the panel.

Working with Aluminum The necessity of employing "electrically tight inclosures" for the containment of TVI-producing harmonics has led to the general use of aluminum for chassis, panel, and inclosure construction. If the proper type of aluminum material is used, it may be cut and worked with the usual woodworking tools found in the home shop. Hard, brittle aluminum alloys such as 2024 and 6061 should be avoided, and the softer materials such as 1100 or 3003 should be employed.

Reynold's *Do-it-Yourself* aluminum, which is being distributed on a nationwide basis through hardware stores, lumber yards, and building material outlets, is an alloy which is temper selected for easy working with ordinary tools. Aluminum sheet, bar, and angle stock may be obtained, as well as perforated sheets for ventilated inclosures.

Figures 1 through 4 illustrate how this soft material may be cut and worked with ordinary shop tools, and figure 5 shows a simple operating desk that may be made from aluminum angle stock, plywood, and a flush-type six-foot door.

Figure 2

CONVENTIONAL WOOD EXPANSION BIT IS EFFECTIVE IN DRILLING SOCKET HOLES IN SOFT ALUMINUM

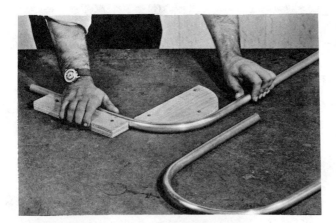

Figure 3

**SOFT ALUMINUM
TUBING MAY BE
BENT AROUND
WOODEN FORM
BLOCKS. TO PREVENT
THE TUBE FROM
COLLAPSING ON
SHARP BENDS, IT IS
PACKED WITH
WET SAND**

simple construction methods, and short cuts in producing inclosures.

The simplest type of aluminum inclosure is that formed from a single sheet of perforated material as shown in figure 6. The top, sides, and back of the inclosure are of one piece, complete with folds that permit the formed inclosure to be bolted together along the edges. The top area of the inclosure should match the area of the chassis to en-

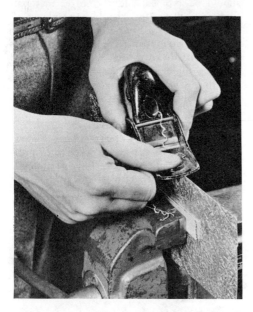

Figure 4

**A WOODWORKING PLANE MAY BE USED
TO SMOOTH OR TRIM THE EDGES OF
ALUMINUM STOCK.**

33-3   TVI-Proof Inclosures

Armed with a right-angle square, tinsnips and a straight edge, the home constructor will find the assembly of aluminum inclosures an easy task. This section will show

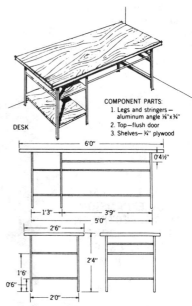

Figure 5

**INEXPENSIVE OPERATING DESK MADE
FROM ALUMINUM ANGLE STOCK, PLY-
WOOD AND A FLUSH-TYPE DOOR**

sure a close fit. The front edge of the inclosure is attached to aluminum angle strips that are bolted to the front panel of the unit; the sides and back can either be bolted to matching angle strips affixed to the chassis, or may simply be attached to the edge of the chassis with self-tapping sheet-metal screws.

A more sophisticated inclosure is shown in figure 7. In this assembly aluminum angle stock is cut to length to form a framework on which the individual sides, back, and top of the inclosure are bolted. For greatest strength, small aluminum gusset plates should be affixed in each corner of the in-

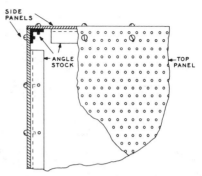

Figure 7

HOME MADE SHIELDED INCLOSURE

Perforated aluminum sheet is screwed or riveted to angle stock to form r-f tight inclosure. Small perforations in sheet provide adequate ventilation for low power equipment but do not impair quality of shielding.

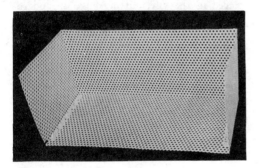

Figure 6

TVI INCLOSURE MADE FROM SINGLE SHEET OF PERFORATED ALUMINUM

Reynolds Metal Co. "Do-it-yourself" aluminum sheet may be cut and folded to form TVI-proof inclosure. One-half inch lip on edges is bolted to center section with 6-32 machine screws.

closure. The complete assembly may be held together by No. 6 sheet-metal screws or "pop" rivets.

Regardless of the type of inclosure to be made, care should be taken to ensure that all joints are square. Do not assume that all prefabricated chassis and panels are absolutely true and square. Check them before you start to form your shield because any dimensional errors in the foundation will cause endless patching and cutting *after* your inclosure is bolted together. Finally, be sure that paint is removed from the panel and chassis at the point the inclosure attaches to the foundation. A clean, metallic contact along the seam is required for maximum harmonic suppression.

33-4   Inclosure Openings

Openings into shielded inclosures may be made simply by covering them by a piece of shielding held in place by sheet-metal screws.

Openings through vertical panels, however, usually require a bit more attention to prevent leakage of harmonic energy through the crack of the door which is supposed to seal the opening. Hinged door openings, however, do not seal tightly enough to be called TVI-proof. In areas of high TV signal strength where a minimum of operation above 21 MHz is contemplated, the door probably is satisfactory as-is.

To accomplish more complete harmonic suppression, the edges of the opening should be lined with preformed, spring-alloy *finger stock* (figure 8) to act as electronic "weatherstripping." Harmonic leakage through such a sealed opening is reduced to a minimum level. The mating surface to the finger stock should be paint-free and should provide a good electrical connection to the stock.

33-5   Sheet Metal Construction Practice

Chassis Layout   The chassis first should be covered with a layer of wrapping paper, which is drawn tightly down on all sides and fastened with scotch tape. This

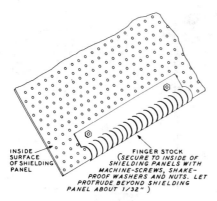

Figure 8

FINGER STOCK PROVIDES R-F TIGHT OPENING

Finger stock secured to edges of door panel will provide good electrical contact with inclosure walls.

allows any number of measurement lines and hole centers to be spotted in the correct positions without making any marks on the chassis itself. Place on it the parts to be mounted and play a game of chess with them, trying different arrangements until all the leads are made as short as possible, tubes and transistors are clear of coil fields, r-f chokes are in safe positions, etc. Remember, especially if you are going to use a panel, that a good mechanical layout often can accompany sound electrical design, but that the electrical design should be given first consideration.

All too often parts are grouped to give a symmetrical panel, irrespective of the arrangement behind. When a satisfactory arrangement has been reached, the mounting holes may be marked. The same procedure now must be followed for the underside, always being careful to see that there are no clashes between the two (that no top mounting screws come down into the middle of a paper capacitor on the underside, that the variable capacitor rotors do not hit anything when turned, etc.).

When all the holes have been spotted, they should be center-punched *through* the paper into the chassis. Don't forget to spot holes for leads which must also come through the chassis.

For transformers which have lugs on the bottoms, the clearance holes may be spotted

by pressing the transformer on a piece of paper to obtain impressions, which may then be transferred to the chassis.

Punching In cutting socket holes one should use socket punches. These punches are easy to operate and only a few precautions are necessary. The guide pin should fit snugly in the guide hole. This increases the accuracy of location of the socket. If this is not of great importance, one may well use a drill of $\frac{1}{32}$ inch larger diameter than the guide pin.

The male part of the punch should be placed in the vise, cutting edge up and the female portion forced against the metal with a wrench. These punches can be obtained in sizes to accommodate all tube sockets and even large enough to be used for meter holes. In the large socket sizes they require the use of a $\frac{3}{8}$-inch center hole to accommodate the bolt.

Transformer Cutouts Cutouts for transformers and chokes are not so simply handled. After marking off the part to be cut, drill about a $\frac{1}{4}$-inch hole on each of the inside corners and tangential to the edges. After burring the holes, clamp the piece and a block of cast iron or steel in the vise. Then, take your burring chisel and insert it in one of the corner holes. Cut out the metal by hitting the chisel with a hammer. The blows should be light and numerous. The chisel acts against the block in the same way that the two blades of a pair of scissors work against each other. This same process is repeated for the other sides. A file is used to trim up the completed cutout.

Another method is to drill the four corner holes large enough to take a hack saw blade, then saw instead of chisel. The four holes permit nice looking corners.

Removing Burrs In both drilling and punching, a burr is usually left on the work. There are three simple ways of removing these. Perhaps the best is to take a chisel (be sure it is one for use on metal) and set it so that its bottom face is parallel to the piece. Then gently tap it with a hammer. This usually will make a clean job with a little practice. If one has access to a counterbore, this will also do a nice job. A countersink will work, although it bevels the edges. A drill of several sizes larger is a much used arrangement. The third method is by filing off the burr, which does a good job but scratches the adjacent metal surfaces badly.

Mounting Components   There are two methods in general use for the fastening of transformers, chokes, and similar pieces of apparatus to chassis or breadboards. The first, using nuts and machine screws, is slow, and the manufacturing practice of using self-tapping screws or rivets is gaining favor. For the mounting of small parts such as resistors and capacitors, "tie points" are very useful to gain rigidity. They also contribute materially to the appearance of finished apparatus.

Rubber grommets of the proper size placed in all chassis holes through which wires are to be passed, will give a neater appearing job and also will reduce the possibility of short circuits.

Soldering   Making a strong, low-resistance solder joint does not mean just dropping a blob of solder on the two parts to be joined and then hoping that they'll stick. There are several definite rules that *must* be observed.

All parts to be soldered must be absolutely clean. To clean a wire, lug, or whatever it may be, take your pocket knife and scrape it thoroughly, until fresh metal is laid bare. It is not enough to make a few streaks; scrape until the part to be soldered is bright.

Make a good mechanical joint before applying any solder. Solder is intended primarily to make a good *electrical* connection; mechanical rigidity should be obtained by bending the wire into a small hook at the end and nipping it firmly around the other part, so that it will hold well even before the solder is applied.

Keep your iron properly tinned. It is impossible to get the work hot enough to take the solder properly if the iron is dirty. To tin your iron, file it, while hot, on one side until a full surface of clean metal is exposed. Im-

mediately apply rosin core solder until a thin layer flows completely over the exposed, surface. Repeat for the other faces. Then take a clean rag and wipe off all excess solder and rosin. The iron should also be wiped frequently while the actual construction is going on; it helps prevent pitting the tip.

Apply the solder to the work, not to the iron. The iron should be held against the parts to be joined until they are thoroughly heated. The solder should then be applied against the parts and the iron should be held in place until the solder flows smoothly and envelops the work. If it acts like water on a greasy plate, and forms a ball, the work is not sufficiently clean.

The completed joint must be held perfectly still until the solder has had time to solidify. If the work is moved before the solder has become *completely* solid, a "cold" joint will result. This can be identified immediately, because the solder will have a dull "white" appearance rather than one of shiny "silver." Such joints tend to be of high resistance and will very likely have a bad effect on a circuit. The cure is simple, merely reheat the joint and do the job correctly.

For general construction work, 60-40 solder (60% tin, 40% lead) is generally used. It melts at 370°F.

Finishes   If the apparatus is constructed on a painted chassis (commonly available in flat black and gray and "hammertone"), there is no need for application of a protective coating when the equipment is finished, assuming that you are careful not to scratch or mar the finish while drilling holes and mounting parts. However, many amateurs prefer to use unpainted (zinc or cadmium plated) steel chassis, because it is much simpler to make a chassis ground connection with this type of chassis. In localities near the sea coast it is a good idea to paint the edges of the various chassis cutouts even on a painted chassis, as rust will get a good start at these points unless the metal is protected where the drill or saw has exposed it.

An attractive dull gloss finish, almost velvety can be put on aluminum by sand-blasting it with a very weak blast and fine particles and then lacquering it. Soaking the aluminum in a solution of lye produces somewhat the same effect as a fine-grain sand blast.

Metal panels and inclosures may be painted an attractive color with the aid of aerosol spray paint, available in many colors. After the panel is spray-painted, press-on *decals* may be used to letter the panel. Once the decals have dried, the panel may then be given a spray coat of clear plastic or lacquer to hold the decals in position and to protect the surface.

33-6   Printed Circuits

Etched or *printed circuits* were developed to apply mass-production techniques to electronic assemblies, utilizing the processes of the graphic arts industry. On a large-volume basis, the etched-circuit technique provides uniformity of layout and freedom from wiring errors at a substantial reduction in assembly time and cost. In this assembly scheme, the methods of the photoengraving process are used to print photographic patterns representing electronic circuitry on copper-foil clad insulating board. By using an *etch-resistant* material (impervious to acid) for the pattern of conductors, the unmasked areas of the foil may be etched away, leaving the desired conducting pattern, conforming to the wiring harness of the electronic assembly.

The etched board is drilled at appropriate places to accept lead wires, thus permitting small components such as resistors and capacitors to be affixed to the board by inserting the leads in the matching holes. Larger components, such as sockets, inductors, and small transformers, are fitted with tabs which pass through matching holes in the board. The various components are interconnected by the foil conductors on one or both sides of the board. All joints are soldered at one time by immersing one side of the board in molten solder.

The foil-clad circuit board is usually made of laminated material such as phenolic, silicon, *teflon*, or *fiberglas*, impregnated with resin and having a copper foil of 0.0007- to 0.009-inch thickness affixed to the board under heat and pressure. Boards are available in thicknesses of $\frac{1}{64}$ to $\frac{1}{4}$ inch.

While large production runs of etched-circuit boards are made by a photographic process utilizing a master negative and

photosensitive board, a simpler process may be used by experimenters to produce circuit boards in the home workshop through the use of *tape* or *ink resist*, plus a chemical solution which etches away all unmasked copper, without affecting the circuit board.

Homemade Circuit Boards   Circuit boards may be easily constructed for electronic assemblies without the need of photographic equipment. The method is simple and fast and requires few special materials. The circuit board is made from a full-scale template of the circuit. Precut board is available from large radio supply houses as are the etchant and resist used in this process. This is how the board is prepared:

Step 1—A full-scale template of the desired circuit is drawn. Lead placement must be arranged so that the conductors do not cross each other except at interconnection points. Holes for component leads and terminals are surrounded by a foil area for the soldered connection. It is suggested that a trial layout be drawn on a piece of graph paper, making the conductors about $\frac{1}{16}$-inch wide and the terminal circles about $\frac{1}{8}$-inch in diameter. When conductors must cross, a point is selected where a component may be used to bridge one conductor; or a wire jumper may be added to the circuit.

Special layout paper marked with the same pattern as on perforated boards may also be used.

Step 2—The template is transferred to the foil-clad board. The board should be unsensitized and cut somewhat oversize. Either single-clad or dual-clad board may be used. For simple circuits, the complete layout can be traced on the board by eye, using a ruler and a pencil. For more complicated circuits, the template should be applied directly to the copper foil by the use of rubber cement. The circuit is traced and the board lightly centerpunched at all drill points for reference. The template and cement are now removed.

Step 3—Once the board has been punched, the board is cleaned to remove copper oxide. A bright, uniform finish is required to ensure proper adhesion of the resist and complete etching. Kitchen cleaning powder may be used for this operation, followed by a thorough washing of the board in water. Care should be taken to avoid touching the copper foil from this point on. Now, to etch out the circuit on the copper foil, the resist material is applied to areas where the copper will remain, and the areas that are not covered with resist will be etched away.

Step 4—The conductors and interconnecting points are laid down on the copper laminate using resist material (figure 10). One form of resist is liquid and is applied from a resist marking pen. A second form of resist is thin vinyl tape having adhesive backing. In an emergency, India Ink or nail polish may be used for resist. Using the original templates as a visual guide, the resist is applied to the clean foil and allowed to dry.

Suitable etchants are *ferric chloride* or *ammonium persulfate.* The etchant may be liquid or a powder which is mixed with hot water according to directions. Ready-made etchant kits using these chemicals are available from several manufacturers.

The board is now ready to be immersed in an *etchant bath,* or tank. A quick and effective etching technique makes use of a *froth etching* bath (figure 11), described as follows.

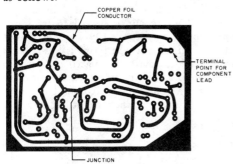

Figure 10

LIQUID RESIST MAKES PRACTICAL PRINTED CIRCUIT

Liquid resist is applied to copper foil of circuit board to protect conductor areas from etchant. Each lead hole is circled, the circle being about four times the diameter of the hole. After the holes have been circled, lines are drawn between them in accordance with the circuit sketch. Junctions are marked with a solid circle. Connecting path should be about 1/16-inch wide, ample to carry a current of about 10 amperes, if required. Tape or a "transfer" resist material provides professional appearance to board. Placement of components may be marked on reverse of board in India Ink.

DRILL NUMBER	Diameter (in.)	Clears Screw	Correct for Tapping Steel or Brass†
1	.228	—	—
2	.221	12-24	—
3	.213	—	14-24
4	.209	12-20	—
5	.205	—	—
6	.204	—	—
7	.201	—	—
8	.199	—	—
9	.196	—	—
10*	.193	10-32	—
11	.191	10-24	—
12*	.189	—	—
13	.185	—	—
14	.182	—	—
15	.180	—	—
16	.177	—	12-24
17	.173	—	—
18*	.169	8-32	—
19	.166	—	12-20
20	.161	—	—
21*	.159	—	10-32
22	.157	—	—
23	.154	—	—
24	.152	—	—
25*	.149	—	10-24
26	.147	—	—
27	.144	—	—
28*	.140	6-32	—
29*	.136	—	8-32
30	.128	—	—
31	.120	—	—
32	.116	—	—
33*	.113	4-36 4-40	—
34	.111	—	—
35*	.110	—	6-32
36	.106	—	—
37	.104	—	—
38	.102	—	—
39*	.100	3-48	—
40	.098	—	—
41	.096	—	—
42*	.093	—	4-36 4-40
43	.089	2-56	—
44	.086	—	—
45*	.082	—	3-48

*Sizes most commonly used in radio construction.

†Use next size larger for tapping bakelite and similar composition materials (plastics, etc.).

Figure 9
NUMBERED DRILL SIZES

The Froth Etching Technique The froth etcher is designed for fast etching of both single and double faced boards on which fine resolution is also important. It produces uniformly etched boards in about four minutes with very little undercutting of the foil. As a bonus, the process automatically aerates the etchant, greatly extending its life.

Constructing the froth etcher tank is quite simple. A heat-resistant glass dish (*Pyrex,* or equivalent) with cover serves as the tank. Also required are a tungsten-car-

Figure 11

**FROTH ETCHING IS QUICK
AND EASY**

The continuous air flow through the aerators creates a surface froth that "scrubs" the circuit board with constantly agitated etchant. The sliding clamp holder which is attached to the dish cover permits rapid insertion or reversal of the printed circuit board. Sample board is clamped to cover holder in foreground.

bide hacksaw blade to notch the dish cover, some two-part epoxy adhesive, some rubber air tubes and a thermometer. To provide the continuous air flow, three inexpensive ceramic aquarium aerators and an aquarium pump are used. Finally, a plexiglass holder for the boards is required.

The small ceramic aerators are cemented to the bottom of the glass dish, as shown in figure 12. The quick-change printed circuit board holder is cemented to the glass cover as shown in figure 13. The thermometer and short lengths of plastic tubing which serve as holders for the air hoses are cemented to the side of the dish and the cover is notched to provide egress for them. The complete froth bath assembly is shown in figure 13.

The continuous air flow through the aerators creates a surface froth that "scrubs" the circuit board with constantly agitated etchant. The board is held in position in the bath by the plexiglass holder shown in figure 14. The etchant used consists of ferric chloride in the proportion of 1¾ pounds of *FeCl* to every quart of water, mixed at a temperature of between 100°F and 110°F.

The froth bath is placed on an electric hot plate and filled with etchant to a level that just reaches the bottom of the copper-clad board when it is mounted in the lid holder. The etchant is heated to its lower operating temperature (100°F) and the hot plate is turned off. The board is now placed in the holder, the cover placed on the dish and the air supply is turned on, adjusting it to create a continuous, vigorous froth over the total surface of the etchant. After a few

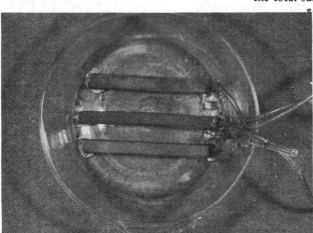

Figure 12

**INTERIOR OF ETCHING
BATH**

Aquarium aerators are cemented into the bottom of the heat-resistant glass dish, along with sections of plastic tubing to support the rubber air tubes and the thermometer. The tubes are connected by T-fittings to a single tube running to the main air supply, which is an aquarium air pump.

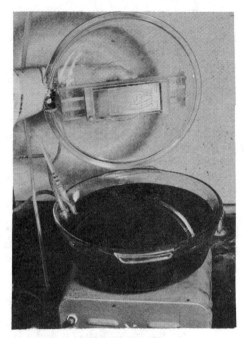

Figure 13

**BATH LID AND CIRCUIT
BOARD HOLDER**

Plexiglass holder grips the edge of the printed circuit board, assuring uniform etch of the entire surface. One clamp is threaded and fitted with a nylon screw to accomodate boards of various sizes. A rubber band around the clamp provides tension. Observing the etching process is easily done by lifting the heat-resistant glass etcher cover to which the printed circuit board is attached. Before the cover is removed, air supply must be turned off to prevent any splattering of the etchant.

minutes—anything from three to eight minutes, depending on the freshness of the solution—inspect the board by raising the cover. The air supply must be turned off first to prevent splattering of the etchant. When the process is observed to be complete, the board is removed and washed in clean water.

The resist material can be left on the board to protect the conductors until the board is cut to final size, clamped between wood blocks in a vise and trimmed with a fine hacksaw blade. The resist is then removed with soft steel wool or a solvent. The complete board is then given a final cleaning with soft steel wool and the center-punched points drilled with a #54 pilot

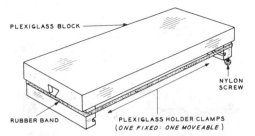

Figure 14

**PLEXIGLASS HOLDER FOR
ETCHANT TANK**

drill. The holes are then drilled out to a larger size as required for component assembly.

The components are mounted to the board on the side opposite the conductors. The leads are passed through the appropriate holes, bent slightly to hold the component in place, and then clipped close to the conductor surface. After checking placement and observing polarity where necessary, the leads may be individually soldered to the conductor with a small pencil-tip iron. Use small diameter (0.032-inch diameter or smaller) solder and take care not to overheat the board or components during this operation. The last step is to wash the circuit side of the board with solvent to remove any soldering flux and then to give the board a coating of clear acrylic (*Krylon*) plastic spray from an aerosol can.

(The Froth etching technique is reprinted from *Electronics*, July 3, 1972; copyright McGraw-Hill, Inc. 1972).

33-7   Coaxial Cable Terminations

Commercial electronics equipment usually employs *series N* and *series BNC* coaxial connectors, whereas the majority of amateur equipment employs the older *UHF series* coaxial connectors. Shown in figure 15 is a simplified and quick method of placing the UHF plug (PL-259) on RG-8A/U or RG-11/U coaxial line. The only special tools needed are a *Stanley 99A* (or equivalent) shop knife and a *General Hardware* 123 (or equivalent) midget tubing cutter.

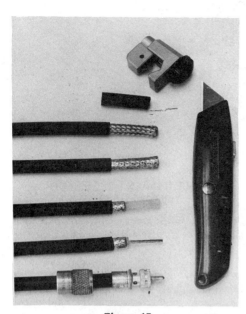

Figure 15

CABLE PREPARATION FOR PL-259 COAXIAL PLUG

Midget tubing cutter and utility knife are used to prepare RG-8/U cable for uhf-type plug. Cable jacket is removed and outer braid tinned with hot iron. Braid is then cut with tubing cutter and inner insulation trimmed with knife. PL-259 shell is twisted on cable and soldered in position through holes in shank.

The first step is to slide the coupling ring of the PL-259 plug over the coaxial line. Next, the utility knife is used to circumscribe a cut in the outer, black vinyl jacket of the cable 1¼ inches back from the end. The cut should be square, and the free jacket piece is slit and removed from the cable.

Next, using a hot iron or soldering gun, quickly tin the exposed braid of the cable. Do this quickly so the inner polyethylene insulation does not soften. Clean the flux from the braid with paint thinner after the solder cools.

The next step is to cut the solid, tinned braid with the tubing cutter so that 7/16 inch remains. Mark the cutting line with a pencil and place the cutting wheel over the mark. Tighten the wheel and revolve the cutter about the cable. The unwanted braid end may be removed, using wire cutters as snips.

Next, trim the inner polyethylene insulation with the utility knife so that 1/16 inch remains exposed beyond the braid. Using a circular cut, slice the insulation and pull the slug free with a twisting motion. Tin the inner conductor. The last step is to push the shell of the PL-259 plug on the prepared cable end. Screw it on with your fingers until the tinned braid is fully visible through

Figure 16

GOOD SHOP LAYOUT AIDS CAREFUL WORKMANSHIP

Built in a corner of a garage, this shop has all features necessary for electronic work. Test instruments are arranged on shelves above bench. Numerous outlets reduce "haywire" produced by tangled line cords. Not shown in picture are drill press and sander at end of left bench.

the solder holes of the plug. Using an iron with a small point, solder the plug to the braid through the four holes, using care that the solder does not run over the outer threads of the plug. Lastly, run the coupling ring down over the plug and solder the inner conductor to the plug tip.

33-8 Workshop Layout

The *size* of your workshop is relatively unimportant since the shop *layout* will determine its efficiency and the ease with which you may complete your work.

Shown in figure 16 is a workshop built into a 10′ × 10′ area in the corner of a garage. The workbench is 32″ wide, made up of four strips of 2″ × 8″ lumber supported on a solid framework made of 2″ × 4″ lumber. The top of the workbench is covered with hard-surface *Masonite*. The edge of the surface is protected with aluminum "counter edging" strip, obtainable at large hardware stores. Two wooden shelves 12″ wide are placed above the bench to hold the various items of test equipment. The shelves are bolted to the wall studs with large angle brackets and have wooden end pieces. Along the edge of the lower shelf a metal "outlet strip" is placed that has a 117-volt outlet every six inches along its length. A similar strip is run along the *back* of the lower shelf. The front strip is used for equipment that is being bench-tested, and the rear strip powers the various items of test equipment placed on the shelves.

At the left of the bench is a storage bin for small components. A file cabinet can be placed at the right of the bench. This neccessary item holds schematics, transformer data sheets, and other papers that normally are lost in the usual clutter and confusion.

The area below the workbench has two storage shelves which are concealed by sliding doors made of ¼-inch *Masonite*. Heavier tools, and large components are stored in this area. On the floor and not shown in the photograph is a very necessary item of shop equipment: a large trash receptacle.

A heavy duty workbench that may be bolted to a cement block or stud wall is shown in figure 17.

33-9 Components and Hardware

Procurement of components and hardware for a construction project can often be a time consuming and vexing task as smaller radio parts stores often have limited or incomplete stocks of only the most fast-moving items. Larger distributors carrying industrial stocks, however, maintain warehouse inventories of components or have facilities for obtaining them at short notice. It is recommended, therefore, that the ex-

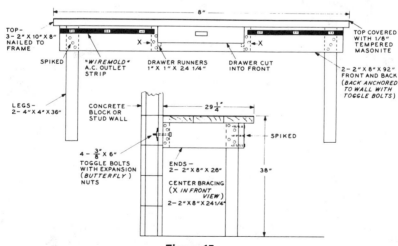

Figure 17

HEAVY-DUTY WORKBENCH TO BOLT TO CEMENT BLOCK OR STUD WALL

perimeter have at hand catalogs from some of the larger supply houses which distribute to the electronics industry. The following industrial catalogs of large mail-order distributors are suggested as part of your technical library:

Allied Electronics Co., 401 East 8th St., Fort Worth, Texas 76102; *Newark Electronics,* 500 No. Pulaski Rd., Chicago, Ill. 60624.

A complete 1700-page catalog of electronic parts and components (*The Radio Electronic Master Catalog*) may be obtained from United Technical Publications, 645 Stewart Ave., Garden City, N.Y. 11530. Copies of this master catalog are often available at large radio supply houses.

Other companies that supply components are: *Amidon Associates,* 12033 Otsego St., North Hollywood, CA 91607 (ferrite cores); *Caywood Electronics Co.,* 67 Mapleton St., Malden, MA 02148 (components and hardware); *Peter W. Dahl,* 4007 Fort Blvd., El Paso, TX 79930 (transformers); *Hammond Mfg. Co., Ltd.,* 394 Edinburgh Rd. No., Guelph, Ontario N1H 1E5, Canada (transformers); *Herbach & Rademan, Inc.,* 401 East Erie Ave., Philadelphia, PA 19134 (general components); *Jameco Electronics Co.,* 1355 Shoreway Rd., San Carlos, CA 94002 (solid state components); *J. W. Miller division of Bell Industries,* 19070 Reyes Ave., Compton, CA 90224 (inductors, ferrite cores); *Polypaks,* Box 942, Lynnfield, MA 01940 (surplus components).

CHAPTER THIRTY-FOUR

Electronic Mathematics and Calculations

Amateur radio, as well as the larger field of electronics, has advanced well beyond the point of trial-and-error design and operation. So much so that a general knowledge of mathematics is required in order to perform the calculations encountered in circuit design and in order to interpret the results obtained in operation.

One of the natural developments in the progress of radio engineering has been the quick acceptance of higher mathematical methods as an aid in solving particular circuit needs. Just as the study of mathematics in early times developed the logarithm as an aid in processing numbers (multiplying, raising to powers, etc.), so advances in electronics, circuit theory and transmission line theory have given rise to the development and use of tools such as vector analysis, Boolean algebra, and the Smith Chart. Some of these techniques lend themselves to use by amateurs and will be discussed in this chapter.

While the mathematical development of these tools is rather complex, their use is not and provided basic rules are followed, results can be achieved for involved problems with less effort than if a purely arithmetical approach were used. These tools are nothing more than processes that enable solutions to problems to be found either more quickly or more simply than by lesser means.

The subject of mathematics falls into two areas, the formation of the problem and the actual calculation of the required result, once the input parameters have been established. The calculation of a result to a formulated problem, given all the necessary input parameters, is known as *arithmetic*.

34-1   Arithmetic

The fundamental manipulation of numbers is well known and there is no need to repeat it. However, the experimenter has at his disposal one of the many developments of our time, the desktop or pocket *electronic calculator*. These are available, at low cost, in a variety of complexities from a basic four function device (addition, subtraction, multiplication, and division) to units involving a number of trigonometric and exponential functions, as well as some constants, intermediate result storage, and even the ability to perform programmed calculations of vast proportions.

Having a readout to 8 or more digits, the average calculator is accurate enough for most purposes and takes the drudgery out of calculation. As a result, the pocket calculator has become an electronic scratch pad, saving a good deal of operator time, while giving results of high accuracy. Calculators that are more complex (and thus more expensive) than the basic four function design, only simplify the overall operation in most cases.

Calculator Use   It is unnecessary to detail the operation of a general pocket calculator as they are accompanied by instructions that detail the pecularities of the respective unit. With any device, however, there are a few general rules that must be observed in the evaluation of expressions:

Consider a general expression of this kind:

Sin $\{[\frac{3}{4} \times (77.1 + 16.97)]\} \div$
$\{[(7.3^{1.9} - 3.04) + 3.3]\}$

Expressions (terms) within parenthesis must be evaluated first, then those within the brackets, and finally those in braces. Operator expressions and complete exponents have to be evaluated before these bracketed expressions can be completed.

The expression reduces to:
$$\sin (\frac{3}{4} \times 94.07) \div$$
$$[(43.943 - 3.04) + 3.3]$$
Then to:
$$\sin 70.5525 \div (40.643 + 3.3)$$
Then to:
$$0.9429 \div 43.943$$
And finally the answer
$$0.02146$$

With operator experience, and a suitable calculator, this calculation could have been performed with 7 number entries and only 9 operator entries, with a total of 33 buttons to press. Longhand methods would involve the use of tables of logarithms and sines, and certainly a lot more time.

Although not covering all the rules of computation, the above example indicates the level of care that must be taken. For example, if in the given expression, the portion $sin \ ^3_4 \times 77.1$ had been performed first, the resulting incorrect answer would have been 0.4091.

The Square   One operator that is not pro-
Root Operator   vided on the simpler pocket
               calculators is the *square root*
operator. However, there is a simple process by which a square root can be obtained using a four function calculator only. The process uses the *iteration* formula:
$$\sqrt{a} \cong \frac{(a/x + x)}{2}$$
where,
    a is the given square,
    x is an approximation.

For example, assume that the square root of 153 is required. The square of 12 is 144 and the square of 13 is 169. Thus, the square root of 153 will be about 12.5, as an approximation. Applying the formula and substituting 12.5 for *x*:

$$\sqrt{a} \cong \frac{\left(\dfrac{153}{12.5} + 12.5\right)}{2} = 12.37$$

This new value is substituted for *x* and re-iterated:

$$\sqrt{a} = \frac{\left(\dfrac{153}{12.37} + 12.37\right)}{2} = 12.369316$$

which is correct to 8 significant figures.

Two methods may be used to establish how many applications of the formula to use. Either the formula may be reapplied until there is no further change in the result, or the result, after a couple of applications of the formula can be squared and compared with the original number whose square root was required, i.e. 12.369316 squared is 152.99997 which is very close to 153, the original number.

Logarithms   It can be demonstrated that the following equations hold:

$$a^{(p+q)} = a^p \times a^q$$
$$a^{(p-q)} = a^p \div a^q$$

In these equations *a* is defined as a *fixed base* and p and q as *indices*. As an example of the first equation, let $a = 2$, $p = 3$, and $q = 4$. Then,

$$a^{(p+q)} = 2^{(3+4)} = 2^7 = 128$$
$$a^p \times a^q = 2^3 \times 2^4 = 8 \times 16 = 128$$

This represents a process that enables complex multiplication and division to be accomplished more simply by addition and subtraction. This technique has been put to convenient use in the form of *logarithms*. The logarithm is defined by the following relationship:

$$a^x = y$$

where $\log_a (y)$ equals x.

Thus, the logarithm is simply the *exponent* to which the base (a) is raised to obtain the number (y). The number (y), moreover, can never be negative.

The foregoing equations involving the exponents (p) and (q) may then be rewritten as:

$$\log_a p + \log_a q = \log_a (p \times q)$$
$$\log_a p - \log_a q = \log_a (p \div q)$$

and by expansion:

$$\log_a p^x = x \log_a p$$

In electronic calculations, base (a) usually takes one of three values: 10, e, or 2

Base (a) = 10   The *base 10* is the *common logarithm*, chosen because it is the most convenient one to use in the decimal system. It is normally written as *log n*; the value of (a) is not included in this expression as it is assumed to be 10. The convenience of the base 10 can be seen by referring to a table of logarithms (Table 1). If the logarithms of numbers from $n = 1$ to $n = 10$ (or 10 to 100) are determined, the logarithms of all other positive numbers can be found, since $\log 10 = 1$, $\log 100 = 2$, etc., and $\log 0.1 = -1$, $\log 0.01 = -2$, etc. Thus:

$$\log 346 = \log 100 + \log 3.46$$

Then, referring to figure 1, for (n) between 1 and 10:

$$\log 3.46 = 0.5391$$

and by the reasoning above,

$$\log 100 = 2$$

Therefore, $\log 346 = 2.5391$

In this example the logarithm of 3.46 from Table 1 was found by considering the first two figures in the left-hand column (10 to 99) and the third figure along the top row (0 to 9). The intersection of this row and column gives the logarithm of the number: 0.5391.

In many tables there are nine additional columns on the right-hand side. These are called *difference columns* or *proportional columns*. If desired, these can be used to provide further accuracy when the logarithm of a four-figure number is required. The method is to determine the logarithm of the first three figures, as above, and add the difference from the appropriate difference column as determined by the fourth digit.

The inverse, or *antilogarithm* is determined by the reverse of the above process, or alternatively by the use of tables of antilogarithms. This is how these methods apply to a typical numerical example:

$$64.72 \times 1.342 \div 647$$

$$
\begin{array}{lll}
\log 64.72 & = & 1.8110 \\
+ \log\ 1.342 & = & 0.1277 \\
\hline
& & 1.9387 \\
- \log 647 & = & -2.8109 \\
\hline
& & 1.1278
\end{array}
$$

A new symbol $(\bar{1})$ is introduced. The logarithm of a number between 1 and 10 is always positive and is called the *mantissa*. The whole-number part of the logarithm is called the *characteristic* and can be positive or negative depending on whether the number itself is greater or less than 1. The notation above, then, really means $(+0.1278 -1)$. Conventionally, $\bar{1}$ and $\bar{2}$ (etc.) are used in place of -1 and -2 to indicate that the mantissa is positive.

In these terms, the result of the above calculation becomes:

antilog 1.1278
$$= \text{antilog } 0.1278 \times \text{antilog } (-1)$$
$$= 1.342 \times 0.1$$
$$= 0.1342$$

The Decibel   There is a convenient convention for the comparison of electrical power levels that makes use of the common logarithm (Table 2). This is written as the *decibel* (dB). A circuit having either amplification or attenuation is said to have a power gain of *q* decibels, where,

$$q = 10 \log \left(\frac{\text{power out}}{\text{power in}} \right)$$

Thus, if an amplifier has an output of 100 watts resulting from an input of 1 watt, the amplifier is said to have a power gain of 20 dB, since:

$$10 \log \left(\frac{\text{power out}}{\text{power in}} \right) = 10 \log 100\ 1$$

$$= 10 \times 2 = 20 \text{ dB}$$

This notation is also used to express *signal level* in a circuit, but this is meaningless unless a reference is considered, since the decibel refers only to relative quantities. The usual reference of zero dB is one milliwatt (mW). This has been chosen because it is approximately the power level associated with a telephone circuit at the microphone. When this level is applied to an impedance of 600 ohms (also the nominal impedance of a telephone line) the voltage across the line is 0.775 volt, rms. Unfortunately, this terminology is often misused and terms such as "a voltage gain of x decibels" are widely used. The reference is only true if the impedances of the input and output levels are identical.

Table 1. Four-Place Logarithms

	0	1	2	3	4	5	6	7	8	9	\ 1 2 3	4 5 6	7 8 9
10	0000	0043	0086	0128	0170	0212	0253	0294	0334	0374	4 8 12	17 21 25	29 33 37
11	0414	0453	0492	0531	0569	0607	0645	0682	0719	0755	4 8 11	15 19 23	26 30 34
12	0792	0828	0864	0899	0934	0969	1004	1038	1072	1106	3 7 10	14 17 21	24 28 31
13	1139	1173	1206	1239	1271	1303	1335	1367	1399	1430	3 6 10	13 16 19	23 26 29
14	1461	1492	1523	1553	1584	1614	1644	1673	1703	1732	3 6 9	12 15 18	21 24 27
15	1761	1790	1818	1847	1875	1903	1931	1959	1987	2014	3 6 8	11 14 17	20 22 25
16	2041	2068	2095	2122	2148	2175	2201	2227	2253	2279	3 5 8	11 13 16	18 21 24
17	2304	2330	2355	2380	2405	2430	2455	2480	2504	2529	2 5 7	10 12 15	17 20 22
18	2553	2577	2601	2625	2648	2672	2695	2718	2742	2765	2 5 7	9 12 14	16 19 21
19	2788	2810	2833	2856	2878	2900	2923	2945	2967	2989	2 4 7	9 11 13	16 18 20
20	3010	3032	3054	3075	3096	3118	3139	3160	3181	3201	2 4 6	8 11 13	15 17 19
21	3222	3243	3263	3284	3304	3324	3345	3365	3385	3404	2 4 6	8 10 12	14 16 18
22	3424	3444	3464	3483	3502	3522	3541	3560	3579	3598	2 4 6	8 10 12	14 15 17
23	3617	3636	3655	3674	3692	3711	3729	3747	3766	3784	2 4 6	7 9 11	13 15 17
24	3802	3820	3838	3856	3874	3892	3909	3927	3945	3962	2 4 5	7 9 11	12 14 16
25	3979	3997	4014	4031	4048	4065	4082	4099	4116	4133	2 3 5	7 9 10	12 14 15
26	4150	4166	4183	4200	4216	4232	4249	4265	4281	4298	2 3 5	7 8 10	11 13 15
27	4314	4330	4346	4362	4378	4393	4409	4425	4440	4456	2 3 5	6 8 9	11 13 14
28	4472	4487	4502	4518	4533	4548	4564	4579	4594	4609	2 3 5	6 8 9	11 12 14
29	4624	4639	4654	4669	4683	4698	4713	4728	4742	4757	1 3 4	6 7 9	10 12 13
30	4771	4786	4800	4814	4829	4843	4857	4871	4886	4900	1 3 4	6 7 9	10 11 13
31	4914	4928	4942	4955	4969	4983	4997	5011	5024	5038	1 3 4	6 7 8	10 11 12
32	5051	5065	5079	5092	5105	5119	5132	5145	5159	5172	1 3 4	5 7 8	9 11 12
33	5185	5198	5211	5224	5237	5250	5263	5276	5289	5302	1 3 4	5 6 8	9 10 12
34	5315	5328	5340	5353	5366	5378	5391	5403	5416	5428	1 3 4	5 6 8	9 10 11
35	5441	5453	5465	5478	5490	5502	5514	5527	5539	5551	1 2 4	5 6 7	9 10 11
36	5563	5575	5587	5599	5611	5623	5635	5647	5658	5670	1 2 4	5 6 7	8 10 11
37	5682	5694	5705	5717	5729	5740	5752	5763	5775	5786	1 2 3	5 6 7	8 9 10
38	5798	5809	5821	5832	5843	5855	5866	5877	5888	5899	1 2 3	5 6 7	8 9 10
39	5911	5922	5933	5944	5955	5966	5977	5988	5999	6010	1 2 3	4 5 7	8 9 10
40	6021	6031	6042	6053	6064	6075	6085	6096	6107	6117	1 2 3	4 5 6	8 9 10
41	6128	6138	6149	6160	6170	6180	6191	6201	6212	6222	1 2 3	4 5 6	7 8 9
42	6232	6243	6253	6263	6274	6284	6294	6304	6314	6325	1 2 3	4 5 6	7 8 9
43	6335	6345	6355	6365	6375	6385	6395	6405	6415	6425	1 2 3	4 5 6	7 8 9
44	6435	6444	6454	6464	6474	6484	6493	6503	6513	6522	1 2 3	4 5 6	7 8 9
45	6532	6542	6551	6561	6571	6580	6590	6599	6609	6618	1 2 3	4 5 6	7 8 9
46	6628	6637	6646	6656	6665	6675	6684	6693	6702	6712	1 2 3	4 5 6	7 7 8
47	6721	6730	6739	6749	6758	6767	6776	6785	6794	6803	1 2 3	4 5 5	6 7 8
48	6812	6821	6830	6839	6848	6857	6866	6875	6884	6893	1 2 3	4 4 5	6 7 8
49	6902	6911	6920	6928	6937	6946	6955	6964	6972	6981	1 2 3	4 4. 5	6 7 8
50	6990	6998	7007	7016	7024	7033	7042	7050	7059	7067	1 2 3	3 4 5	6 7 8
51	7076	7084	7093	7101	7110	7118	7126	7135	7143	7152	1 2 3	3 4 5	6 7 8
52	7160	7168	7177	7185	7193	7202	7210	7218	7226	7235	1 2 2	3 4 5	6 7 7
53	7243	7251	7259	7267	7275	7284	7292	7300	7308	7316	1 2 2	3 4 5	6 6 7
54	7324	7332	7340	7348	7356	7364	7372	7380	7388	7396	1 2 2	3 4 5	6 6 7

proportional parts

Negative gain (loss) may be calculated by the same process. For example, an attenuator pad has equal input and output impedances, and the output voltage is 1 volt when the input voltage is 100 volts.

$$\text{gain} = 20 \log \left(\frac{\text{volts out}}{\text{volts in}} \right)$$

$$= 20 \log 1/100 = 20\,(-2) = -40\ \text{dB}$$

Alternatively, the pad may be said to have a loss of $+ 40$ dB. Also,

Table 1. Four-Place Logarithms

	0	1	2	3	4	5	6	7	8	9	proportional parts 1 2 3	4 5 6	7 8 9
55	7404	7412	7419	7427	7435	7443	7451	7459	7466	7474	1 2 2	3 4 5	5 6 7
56	7482	7490	7497	7505	7513	7520	7528	7536	7543	7551	1 2 2	3 4 5	5 6 7
57	7559	7566	7574	7582	7589	7597	7604	7612	7619	7627	1 2 2	3 4 5	5 6 7
58	7634	7642	7649	7657	7664	7672	7679	7686	7694	7701	1 1 2	3 4 4	5 6 7
59	7709	7716	7723	7731	7738	7745	7752	7760	7767	7774	1 1 2	3 4 4	5 6 7
60	7782	7789	7796	7803	7810	7818	7825	7832	7839	7846	1 1 2	3 4 4	5 6 6
61	7853	7860	7868	7875	7882	7889	7896	7903	7910	7917	1 1 2	3 4 4	5 6 6
62	7924	7931	7938	7945	7952	7959	7966	7973	7980	7987	1 1 2	3 3 4	5 6 6
63	7993	8000	8007	8014	8021	8028	8035	8041	8048	8055	1 1 2	3 3 4	5 5 6
64	8062	8069	8075	8082	8089	8096	8102	8109	8116	8122	1 1 2	3 3 4	5 5 6
65	8129	8136	8142	8149	8156	8162	8169	8176	8182	8189	1 1 2	3 3 4	5 5 6
66	8195	8202	8209	8215	8222	8228	8235	8241	8248	8254	1 1 2	3 3 4	5 5 6
67	8261	8267	8274	8280	8287	8293	8299	8306	8312	8319	1 1 2	3 3 4	5 5 6
68	8325	8331	8338	8344	8351	8357	8363	8370	8376	8382	1 1 2	3 3 4	4 5 6
69	8388	8395	8401	8407	8414	8420	8426	8432	8439	8445	1 1 2	2 3 4	4 5 6
70	8451	8457	8463	8470	8476	8482	8488	8494	8500	8506	1 1 2	2 3 4	4 5 6
71	8513	8519	8525	8531	8537	8543	8549	8555	8561	8567	1 1 2	2 3 4	4 5 5
72	8573	8579	8585	8591	8597	8603	8609	8615	8621	8627	1 1 2	2 3 4	4 5 5
73	8633	8639	8645	8651	8657	8663	8669	8675	8681	8686	1 1 2	2 3 4	4 5 5
74	8692	8698	8704	8710	8716	8722	8727	8733	8739	8745	1 1 2	2 3 4	4 5 5
75	8751	8756	8762	8768	8774	8779	8785	8791	8797	8802	1 1 2	2 3 3	4 5 5
76	8808	8814	8820	8825	8831	8837	8842	8848	8854	8859	1 1 2	2 3 3	4 5 5
77	8865	8871	8876	8882	8887	8893	8899	8904	8910	8915	1 1 2	2 3 3	4 4 5
78	8921	8927	8932	8938	8943	8949	8954	8960	8965	8971	1 1 2	2 3 3	4 4 5
79	8976	8982	8987	8993	8998	9004	9009	9015	9020	9025	1 1 2	2 3 3	4 4 5
80	9031	9036	9042	9047	9053	9058	9063	9069	9074	9079	1 1 2	2 3 3	4 4 5
81	9085	9090	9096	9101	9106	9112	9117	9122	9128	9133	1 1 2	2 3 3	4 4 5
82	9138	9143	9149	9154	9159	9165	9170	9175	9180	9186	1 1 2	2 3 3	4 4 5
83	9191	9196	9201	9206	9212	9217	9222	9227	9232	9238	1 1 2	2 3 3	4 4 5
84	9243	9248	9253	9258	9263	9269	9274	9279	9284	9289	1 1 2	2 3 3	4 4 5
85	9294	9299	9304	9309	9315	9320	9325	9330	9335	9340	1 1 2	2 3 3	4 4 5
86	9345	9350	9355	9360	9365	9370	9375	9380	9385	9390	1 1 2	2 3 3	4 4 5
87	9395	9400	9405	9410	9415	9420	9425	9430	9435	9440	0 1 1	2 2 3	3 4 4
88	9445	9450	9455	9460	9465	9469	9474	9479	9484	9489	0 1 1	2 2 3	3 4 4
89	9494	9499	9504	9509	9513	9518	9523	9528	9533	9538	0 1 1	2 2 3	3 4 4
90	9542	9547	9552	9557	9562	9566	9571	9576	9581	9586	0 1 1	2 2 3	3 4 4
91	9590	9595	9600	9605	9609	9614	9619	9624	9628	9633	0 1 1	2 2 3	3 4 4
92	9638	9643	9647	9652	9657	9661	9666	9671	9675	9680	0 1 1	2 2 3	3 4 4
93	9685	9689	9694	9699	9703	9708	9713	9717	9722	9727	0 1 1	2 2 3	3 4 4
94	9731	9736	9741	9745	9750	9754	9759	9763	9768	9773	0 1 1	2 2 3	3 4 4
95	9777	9782	9786	9791	9795	9800	9805	9809	9814	9818	0 1 1	2 2 3	3 4 4
96	9823	9827	9832	9836	9841	9845	9850	9854	9859	9863	0 1 1	2 2 3	3 4 4
97	9868	9872	9877	9881	9886	9890	9894	9899	9903	9908	0 1 1	2 2 3	3 4 4
98	9912	9917	9921	9926	9930	9934	9939	9943	9948	9952	0 1 1	2 2 3	3 4 4
99	9956	9961	9965	9969	9974	9978	9983	9987	9991	9996	0 1 1	2 2 3	3 3 4

Table 2. Decibel Gains Versus Power Ratios

The decibel, abbreviated dB, is a unit used to express the ratio between two amounts of power, P_1 and P_2, existing at two points. By definition number of $dB = 10 \log_{10}(P_1/P_2)$. It is also used to express voltage and current ratios: number of $dB = 20 \log_{10}(V_1/V_2) = 20 \log_{10}(I_1/I_2)$.

Strictly, it can be used to express voltage and current ratios only when the voltages or currents in question are measured at places having identical impedances.

Power Ratio	Voltage and Current Ratio	Decibels	Nepers	Power Ratio	Voltage and Current Ratio	Decibels	Nepers
1.0233	1.0116	0.1	0.01	19.953	4.4668	13.0	1.50
1.0471	1.0233	0.2	0.02	25.119	5.0119	14.0	1.61
1.0715	1.0351	0.3	0.03	31.623	5.6234	15.0	1.73
1.0965	1.0471	0.4	0.05	39.811	6.3096	16.0	1.84
1.1220	1.0593	0.5	0.06	50.119	7.0795	17.0	1.96
1.1482	1.0715	0.6	0.07	63.096	7.9433	18.0	2.07
1.1749	1.0839	0.7	0.08	79.433	8.9125	19.0	2.19
1.2023	1.0965	0.8	0.09	100.00	10.0000	20.0	2.30
1.2303	1.1092	0.9	0.10	158.49	12.589	22.0	2.53
1.2589	1.1220	1.0	0.12	251.19	15.849	24.0	2.76
1.3183	1.1482	1.2	0.14	398.11	19.953	26.0	2.99
1.3804	1.1749	1.4	0.16	630.96	25.119	28.0	3.22
1.4454	1.2023	1.6	0.18	1000.0	31.623	30.0	3.45
1.5136	1.2303	1.8	0.21	1584.9	39.811	32.0	3.68
1.5849	1.2589	2.0	0.23	2511.9	50.119	34.0	3.91
1.6595	1.2882	2.2	0.25	3981.1	63.096	36.0	4.14
1.7378	1.3183	2.4	0.28	6309.6	79.433	38.0	4.37
1.8197	1.3490	2.6	0.30	10^4	100.000	40.0	4.60
1.9055	1.3804	2.8	0.32	$10^4 \times 1.5849$	125.89	42.0	4.83
1.9953	1.4125	3.0	0.35	$10^4 \times 2.5119$	158.49	44.0	5.06
2.2387	1.4962	3.5	0.40	$10^4 \times 3.9811$	199.53	46.0	5.29
2.5119	1.5849	4.0	0.46	$10^4 \times 6.3096$	251.19	48.0	5.52
2.8184	1.6788	4.5	0.52	10^5	316.23	50.0	5.76
3.1623	1.7783	5.0	0.58	$10^5 \times 1.5849$	398.11	52.0	5.99
3.5481	1.8836	5.5	0.63	$10^5 \times 2.5119$	501.19	54.0	6.22
3.9811	1.9953	6.0	0.69	$10^5 \times 3.9811$	630.96	56.0	6.45
5.0119	2.2387	7.0	0.81	$10^5 \times 6.3096$	794.33	58.0	6.68
6.3096	2.5119	8.0	0.92	10^6	1 000.00	60.0	6.91
7.9433	2.8184	9.0	1.04	10^7	3 162.3	70.0	8.06
10.0000	3.1623	10.0	1.15	10^8	10 000.0	80.0	9.21
12.589	3.5481	11.0	1.27	10^9	31 623	90.0	10.36
15.849	3.9811	12.0	1.38	10^{10}	100 000	100.0	11.51

To convert:

    Decibels to nepers, multiply by 0.1151

    Decibels per statute mile to nepers per kilometer, multiply by 7.154×10^{-2}

    Decibels per nautical mile to nepers per kilometer, multiply by 6.215×10^{-2}

    Nepers to decibels, multiply by 8.686

    Nepers per kilometer to decibels per statute mile, multiply by 13.978

    Nepers per kilometer to decibels per nautical mile, multiply by 16.074.

Where the power ratio is less than unity, it is usual to invert the fraction and express the answer as a decibel loss.

$$\text{loss} = 10 \log \left(\frac{\text{power in}}{\text{power out}} \right) \text{ decibels}$$

Base (a) = **(e)** = **2.71828** The expression $\log_e (n)$, is known as the *natural logarithm*; it is usually written $\ln (n)$. It is called the natural logarithm because it occurs so regularly in nature, from the hanging shape of a chain to the voltage decay of an RC circuit. It is the latter case which is of most interest to the amateur (figure 1).

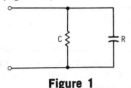

Figure 1

SIMPLE RC PARALLEL CIRCUIT HAS EXPONENTIAL VOLTAGE DECAY

In this simple RC decay circuit the voltage across the network falls to $1/e$ of its initial value in time (t) where, $t = RC$ (seconds = ohms × farads). Thus,

$$V_o/V_t = e^{\frac{t}{RC}}$$

and,

$$\ln (V_o/V_tQ) = \ln e^{\frac{t}{RC}} = \frac{t}{RC} \quad \ln e = \frac{t}{RC}$$

and thus,

$$\ln (V_o/V_t) = t/RC$$

where,

V_o is the initial voltage and V_t is the final voltage at time (t), as shown in figure 2.

For example, let $R = 1$ megohm, $C = 1$ μF; i.e. $RC = 1$ and,
$$V_o = 1000 \text{ and } V_t = 0.1$$

then,

$$t = RC \ln (V_o/V_t) = 1 \times \ln (1000/0.1)$$
$$t = \ln 10{,}000$$

Tables of natural logarithms are inconvenient and are generally not available. On the other hand, logarithms to the base (10) may be used with a suitable multiplier:

$$\ln (n) = 2.3026 \log (n)$$

and,

$$\log (n) = 0.4343 \ln (n)$$

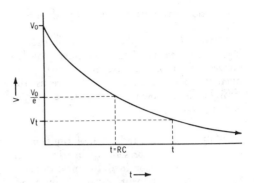

Figure 2

DECAY CURVE OF FIGURE 1

Accordingly, in the above example,

$$t = \ln (10{,}000)$$
$$= 2.3026 \log (10{,}000)$$
$$= 2.3026 \times 4$$
$$t = 9.2104 \text{ seconds}$$

The inverse of this procedure may be used to determine an unknown R or C, but commonly only for very large values (those outside the range of available test instruments).

An European usage of the natural logarithm is the *neper*. This is a unit of power comparison similar to the decibel, and is defined as:

$$\text{gain} = \ln \left(\frac{\text{power out}}{\text{power in}} \right) \qquad \text{nepers}$$

Base (a) = 2 Logarithms to the base (2) are encountered in electronics when the concept of channel or system utilization, or efficiency is considered. Similarly, they are also encountered in some noise calculations.

Other Uses of Logarithms There are various other uses of logarithms, among which are: *the slide rule*. The arithmetic part of a slide rule is based on logarithms to the base (10), and effectively multiplies and divides by adding and subtracting logarithms. The use of a slide rule is strongly advocated for all persons in electronics, but no instructions are included herewith, because there are many texts and instruction manuals supplied by slide rule manufacturers.

Graphical Uses—Many graphical uses of logarithms exist. For example, log/log curves of amplitude response versus frequency, since the plotted results are usually straight lines which are easy to construct and analyze.

Significant In most radio calculations, num-
Figures bers represent quantities which were obtained by measurement. Since no measurement gives absolute accuracy, such quantities are only approximate and their value is given only to a few significant figures. In calculations, these limitations must be kept in mind and one should not finish, for instance, with a result expressed in more significant figures than the given quantities at the beginning. This would imply a greater accuracy than actually was obtained and is therefore misleading, if not ridiculous.

An example may make this clear. Many ammeters and voltmeters do not give results to closer than $\frac{1}{4}$ ampere or $\frac{1}{4}$ volt. Thus if we have $2\frac{1}{4}$ amperes flowing in a dc circuit at $6\frac{3}{4}$ volts, we can obtain a theoretical answer by multiplying 2.25 by 6.75 to get 15.1875 watts. But it is misleading to express the answer down to a ten-thousandth of a watt when the original measurements were only good to $\frac{1}{4}$ ampere or volt. The answer should be expressed as 15 watts, not even 15.0 watts. If we assume a possible error of $\frac{1}{8}$ volt or ampere (that is, that our original data are only correct to the *nearest* $\frac{1}{4}$ volt or ampere) the true power lies between 14.078 (product of $2\frac{1}{8}$ and $6\frac{5}{8}$) and 16.328 (product of $2\frac{3}{8}$ and $6\frac{7}{8}$). Therefore, any third significant figure would be misleading as implying an accuracy which we do not have.

Conversely, there is also no point to calculating the value of a part down to 5 or 6 significant figures when the actual part to be used cannot be measured to better than 1 part in one hundred. For instance, if we are going to use 1% resistors in some circuit, such as an ohmmeter, there is no need to calculate the value of such a resistor to 5 places, such as 1262.5 ohms. Obviously, 1% of this quantity is over 12 ohms and the value should simply be written as 1260 ohms.

There is a definite technique in handling these approximate figures. When giving values obtained by measurement, no more figures are given than the accuracy of the measurement permits. Thus, if the measurement is good to two places, we would write, for instance, 6.9 which would mean that the true value is somewhere between 6.85 and 6.95. If the measurement is known to three significant figures, we might write 6.90 which means that the true value is somewhere between 6.895 and 6.905. In dealing with approximate quantities, the added cipher at the right of the decimal point has a meaning.

There is unfortunately no standardized system of writing approximate figures with many ciphers to the left of the decimal point. 69000 does not necessarily mean that the quantity is known to 5 significant figures. Some indicate the accuracy by writing 69 $\times$ 10^3 or 690 $\times$ 10^2, etc., but this system is not universally employed. The reader can use his own system, but whatever notation is used, the number of significant figures should be kept in mind.

Working with approximate figures, one may obtain an idea of the influence of the doubtful figures by marking all of them, and products or sums derived from them. In the following example, the doubtful figures have been underlined.

$$
\begin{array}{r}
60\underline{3} \\
34.\underline{6} \\
0.12\underline{0} \\
\hline
637.\underline{720} \quad \text{answer:} \quad 638
\end{array}
$$

Multiplication:

$$
\begin{array}{r}
65\underline{4} \\
0.3\underline{42} \\
\hline
1308 \\
2616 \\
196\underline{2} \\
\hline
223.668 \quad \text{answer:} \quad 224
\end{array}
\qquad
\begin{array}{r}
65\underline{4} \\
0.34\underline{2} \\
\hline
196|2 \\
26|16 \\
1|308 \\
\hline
224
\end{array}
$$

It is recommended that the system at the right be used and that the figures to the right of the vertical line be omitted or guessed so as to save labor. Here the partial products are written in the reverse order, the most important ones first.

In division, labor can be saved when after each digit of the quotient is obtained, one figure of the divisor be dropped. Example:

$$
\begin{array}{r}
1.28 \\
527\ )\ \overline{673} \\
527 \\
\hline
53\ )\ 146 \\
106 \\
\hline
5\ )\ 40 \\
40 \\
\hline
\end{array}
$$

34-2   Algebra

Algebra is not a separate branch of mathematics but is merely a form of *generalized arithmetic* in which letters of the alphabet and occasional other symbols are substituted for numbers, from which it is often referred to as *literal notation*. It is simply a short-hand method of writing operations which could be spelled out.

The laws of most common electrical phenomena and circuits (including of course radio phenomena and circuits) lend themselves particularly well to representation by literal notation and solution by algebraic equations or formulas.

While we may write a particular problem in Ohm's law as an ordinary division or multiplication, the general statement of all such problems calls for the replacement of the numbers by symbols. We might be explicit and write out the names of the units and use these names as symbols:

$$volts = amperes \times ohms$$

Such a procedure becomes too clumsy when the expression is more involved and would be unusually cumbersome if any operations like multiplication were required. Therefore as a short way of writing these generalized relations the numbers are represented by letters. Ohm's law then becomes

$$E = I \times R$$

In the statement of any particular problem the significance of the letters is usually indicated directly below the equation or formula using them unless there can be no ambiguity. Thus the above form of Ohm's law would be more completely written as:

$$E = I \times R$$

where,
  E equals e.m.f. in volts,
  I equals current in amperes,
  R equals resistance in ohms.

Letters therefore represent numbers, and for any letter we can read "any number." When the same letter occurs again in the same expression we would mentally read "the same number," and for another letter "another number of any value."

These letters are connected by the usual operational symbols of arithmetic, $+$, $-$, $\times$, $\div$, etc. In algebra, the sign for division is seldom used, a division being usually written as a fraction. The multiplication sign, $\times$, is usually omitted or one may write a dot only. Examples:

$$2 \times a \times b = 2ab$$
$$2 \cdot 3 \cdot 4 \cdot 5a = 2 \times 3 \times 4 \times 5 \times a$$

In practical applications of algebra, an expression usually states some physical law and each letter represents a variable quantity which is therefore called a *variable*. A fixed number in front of such a quantity (by which it is to be multiplied) is known as the *coefficient*. Sometimes the coefficient may be unknown, yet to be determined; it is then also written as a letter; k is most commonly used for this purpose.

The Negative Sign   In ordinary arithmetic we seldom work with negative numbers, although we may be "short" in a subtraction. In algebra, however, a number may be either negative or positive. Such a thing may seem *academic* but a negative quantity can have a real existence. We need only refer to a *debt* being considered a negative possession. In electrical work, however, a result of a problem might be a negative number of amperes or volts, indicating that the direction of the current is opposite to the direction chosen as positive. This will be illustrated later.

Having established the existence of negative quantities, we must now learn how to work with these negative quantities in addition, subtraction, multiplication, etc.

In addition, a negative number added to a positive number is the same as subtracting a positive number from it.

$$\frac{7}{-3}\ (\text{add})$$ is the same as $$\frac{7}{3}\ (\text{subtract})$$

or it might be written

$$7 + (-3) = 7 - 3 = 4$$

Similarly, we have:

$$a + (-b) = a - b$$

When a minus sign is in front of an expression in brackets, this minus sign has the effect of reversing the signs of every term within the brackets:

$$- (a - b) = -a + b$$
$$- (2a + 3b - 5c) = -2a - 3b + 5c$$

Multiplication—When both the multiplicand and the multiplier are negative, the product is positive. When only one (either one) is negative the product is negative. The four possible cases are illustrated below:

$$+ \times + = +$$
$$- \times + = -$$

$$+ \times - = -$$
$$- \times - = +$$

Division—Since division is but the reverse of multiplication, similar rules apply for the sign of the quotient. When both the dividend and the divisor have the same sign (both negative or both positive) the quotient is positive. If they have unlike signs (one positive and one negative) the quotient is negative.

$$\frac{+}{+} = +$$

$$\frac{+}{-} = -$$

$$\frac{-}{+} = -$$

$$\frac{-}{-} = +$$

Powers—Even powers of negative numbers are positive and *odd* powers are negative. Powers of positive numbers are always positive. Examples:

$$- 2^2 = -2 \times -2 = +4$$
$$- 2^3 = -2 \times -2 \times -2$$
$$= +4 \times -2 = -8$$

Roots—Since the square of a negative number is positive and the square of a positive number is also positive, it follows that a positive number has two square roots. The square root of 4 can be either $+2$ or -2

for $(+2) \times (+2) = +4$ and $(-2) \times (-2) = +4.$

Addition and Subtraction　*Polynomials* are quantities like $3ab^2 + 4ab^3 - 7\ a^2b^4$ which have several terms of different *names*. When adding polynomials, only terms of the same name can be taken together.

$$7a^3 + 8ab^2 + 3a^2b \qquad\qquad + 3$$
$$a^3 - 5ab^2 \qquad\qquad\quad - b^3$$
$$\overline{8a^3 + 3ab^2 + 3a^2b - b^3 + 3}$$

Collecting terms. When an expression contains more than one term of the same name, these can be added together and the expression made simpler:

$$5\ x^2 + 2\ xy + 3\ xy^2 - 3\ x^2 + 7\ xy =$$
$$5\ x^2 - 3\ x^2 + 2\ xy + 7\ xy + 3\ xy^2 =$$
$$2\ x^2 + 9\ xy + 3\ xy^2$$

Multiplication　Multiplication of single terms is indicated simply by writing them together.

$$a \times b \text{ is written as } ab$$

$$a \times b^2 \text{ is written as } ab^2$$

Bracketed quantities are multiplied by a single term by multiplying each term:

$$a\ (b + c + d) = ab + ac + ad$$

When two bracketed quantities are multiplied, each term of the first bracketed quantity is to be multiplied by each term of the second bracketed quantity, thereby making every possible combination.

$$(a + b)\ (c + d) = ac + ad + bc + bd$$

In this work particular care must be taken to get the signs correct. Examples:

$$(a + b)\ (a - b) = a^2 + ab - ab - b^2 =$$
$$a^2 - b^2$$

$$(a + b)\ (a + b) = a^2 + ab + ab + b^2 =$$
$$a^2 + 2ab + b^2$$

$$(a - b)\ (a - b) = a^2 - ab - ab + b^2 =$$
$$a^2 - 2\ ab + b^2$$

Division It is possible to do longhand division in algebra, although it is somewhat more complicated than in arithmetic. However, the division will seldom come out even, and is not often done in this form. The method is as follows: Write the terms of the dividend in the order of descending powers of one variable and do likewise with the divisor. Example:

Divide $5a^2b + 21b^3 + 2a^3 - 26ab^2$ by
$2a - 3b$

Write the dividend in the order of descending powers of a and divide in the same way as in arithmetic.

$$
\begin{array}{r}
a^2 + 4\,ab - 7b^2 \\
2a - 3b\,)\,\overline{2a^3 + 5a^2b - 26\,ab^2 + 21b^3} \\
2a^3 - 3a^2b \\
\hline
+\,8a^2b - 26\,ab^2 \\
+\,8a^2b - 12\,ab^2 \\
\hline
-\,14\,ab^2 + 21b^3 \\
-\,14\,ab^2 + 21b^3 \\
\hline
\end{array}
$$

Another example: Divide $x^3 - y^3$ by $x - y$:

$$
\begin{array}{r}
x - y\,)\,\overline{x^3 + 0 + 0 - y^3}\,(\,x^2 + xy + y^2 \\
x^3 - x^2y \\
\hline
+\,x^2y \\
x^2y - xy^2 \\
\hline
+\,xy^2 - y^3 \\
xy^2 - y^3 \\
\hline
\end{array}
$$

Factoring Very often it is necessary to simplify expressions by finding a factor. This is done by collecting two or more terms having the same factor and bringing the factor outside the brackets:

$$6ab + 3ac = 3a\,(2b + c)$$

In a four term expression one can take together two terms at a time; the intention is to try getting the terms within the brackets the same after the factor has been removed:

$$
30ac - 18bc + 10ad - 6bd = \\
6c\,(5a - 3b) + 2d\,(5a - 3b) = \\
(5a - 3b)\,(6c + 2d)
$$

Of course, this is not always possible and the expression may not have any factors. A similar process can of course be followed when the expression has six or eight or any even number of terms.

A special case is a three-term polynomial, which can sometimes be factored by writing the middle term as the sum of two terms:

$$
x^2 - 7xy + 12y^2 \text{ may be rewritten as} \\
x^2 - 3xy - 4xy + 12y^2 = \\
x\,(x - 3y) - 4y\,(x - 3y) = \\
(x - 4y)\,(x - 3y)
$$

The middle term should be split into two in such a way that the sum of the two new terms equals the original middle term and that their product equals the product of the two outer terms. In the above example these conditions are fulfilled for $-\,3xy - 4\,xy = -\,7\,xy$ and $(-\,3xy)\,(-\,4xy) = 12\,x^2y^2$. It is not always possible to do this and there are then no simple factors.

Working with Powers and Roots When two powers of the same number are to be multiplied, the exponents are added.

$$
a^2 \times a^3 = aa \times aaa = aaaaa = a^5 \text{ or} \\
a^2 \times a^3 = a^{(2+3)} = a^5 \\
b^3 \times b = b^4 \\
c^5 \times c^7 = c^{12}
$$

Similarly, dividing of powers is done by subtracting the exponents.

$$
\frac{a^3}{a^2} = \frac{aaa}{aa} = a \text{ or } \frac{a^3}{a^2} = a^{(3-2)} = a^1 = a
$$

$$
\frac{b^5}{b^3} = \frac{bbbbb}{bbb} = b^2 \text{ or } \frac{b^5}{b^3} = b^{(5-3)} = b^2
$$

Now we are logically led into some important new ways of notation. We have seen that when dividing, the exponents are subtracted. This can be continued into negative exponents. In the following series, we successively divide by a and since this can now be done in two ways, the two ways of notation must have the same meaning and be identical.

a^5 a^2 $a^{-1} = \dfrac{1}{a}$

a^4 $a^1 = a$ $a^{-2} = \dfrac{1}{a^2}$

a^3 $a^0 = 1$ $a^{-3} = \dfrac{1}{a^3}$

These examples illustrate two rules: (1) any number raised to "zero" power equals one or unity; (2) any quantity raised to a negative power is the inverse or reciprocal of the same quantity raised to the same positive power.

$$n^0 = 1 \qquad a^{-n} = \frac{1}{a^n}$$

Roots—The product of the square root of two quantities equals the square root of their product.

$$\sqrt{a} \times \sqrt{b} = \sqrt{ab}$$

Also, the quotient of two roots is equal to the root of the quotient.

$$\frac{\sqrt{a}}{\sqrt{b}} = \sqrt{\frac{a}{b}}$$

Note, however, that in addition or subtraction the square root of the sum or difference is *not* the same as the sum or difference of the square roots.

Thus, $\sqrt{9} - \sqrt{4} = 3 - 2 = 1$
but $\sqrt{9 - 4} = \sqrt{5} = 2.2361$
Likewise $\sqrt{a} + \sqrt{b}$ is *not* the same as $\sqrt{a + b}$

Roots may be written as fractional powers. Thus $\sqrt{a}$ may be written as $a^{1/2}$ because

$$\sqrt{a} \times \sqrt{a} = a$$
and, $a^{1/2} \times a^{1/2} = a^{1/2 + 1/2} = a^1 = a$

Any root may be written in this form

$$\sqrt{b} = b^{1/2} \quad \sqrt[3]{b} = b^{1/3} \quad \sqrt[4]{b^3} = b^{3/4}$$

The same notation is also extended in the negative direction:

$$b^{-1/2} = \frac{1}{b^{1/2}} = \frac{1}{\sqrt{b}} \qquad c^{-1/3} = \frac{1}{c^{1/3}} = \frac{1}{\sqrt[3]{c}}$$

Following the previous rules that exponents add when powers are multiplied,

$$\sqrt[3]{a} \times \sqrt[3]{a} = \sqrt[3]{a^2}$$
but also $a^{1/3} \times a^{1/3} = a^{2/3}$
therefore $a^{2/3} = \sqrt[3]{a^2}$

Powers of powers—When a power is again raised to a power, the exponents are multiplied;

$(a^2)^3 = a^6$ $(b^{-1})^3 = b^{-3}$
$(a^3)^4 = a^{12}$ $(b^{-2})^{-4} = b^8$

This same rule also applies to roots of roots and also powers of roots and roots of powers because a root can always be written as a fractional power.

$$\sqrt[3]{\sqrt{a}} = \sqrt[6]{a} \text{ for } (a^{1/2})^{1/3} = a^{1/6}$$

Removing radicals—A root or radical in the denominator of a fraction makes the expression difficult to handle. If there must be a radical it should be located in the numerator rather than in the denominator. The removal of the radical from the denominator is done by multiplying both numerator and denominator by a quantity which will remove the radical from the denominator, thus *rationalizing* it:

$$\frac{1}{\sqrt{a}} = \frac{\sqrt{a}}{\sqrt{a} \times \sqrt{a}} = \frac{1}{a}\sqrt{a}$$

Suppose we have to rationalize

$\dfrac{3a}{\sqrt{a} + \sqrt{b}}$ In this case we must multiply numerator and denominator by $\sqrt{a} - \sqrt{b}$, the same terms but with the second having the opposite sign, so that their product will not contain a root.

$$\frac{3a}{\sqrt{a} + \sqrt{b}} = \frac{3a\,(\sqrt{a} - \sqrt{b})}{(\sqrt{a} + \sqrt{b})\,(\sqrt{a} - \sqrt{b})} = \frac{3a(\sqrt{a} - \sqrt{b})}{a - b}$$

Imaginary Numbers   Since the square of a negative number is positive and the square of a positive number is also positive, the square root of a negative number can be neither positive nor negative. Such a number is said to be *imaginary*; the most common such number ($\sqrt{-1}$) is often represented by the letter i in mathematical work or j in electrical work.

$$\sqrt{-1} = i \text{ or } j \text{ and } i^2 \text{ or } j^2 = -1$$

Imaginary numbers do not exactly correspond to anything in our experience and it is best not to try to visualize them. Despite this fact, their interest is much more than academic, for they are extremely useful in many calculations involving alternating currents.

The square root of any other negative number may be reduced to a product of two roots, one positive and one negative. For instance:

$$\sqrt{-57} = \sqrt{-1} \ \sqrt{57} = i\sqrt{57}$$

or, in general

$$\sqrt{-a} = i \sqrt{a}$$

Since $i = \sqrt{-1}$, the powers of i have the following values:

$$i^2 = -1$$

$$i^3 = -1 \times i = -i$$

$$i^4 = +1$$

$$i^5 = +1 \times i = i$$

Imaginary numbers are different from either positive or negative numbers; so in addition or subtraction they must always be accounted for separately. Numbers which consist of both real and imaginary parts are called *complex* numbers. Examples of complex numbers:

$$3 + 4i = 3 + 4 \sqrt{-1}$$
$$a + bi = a + b \sqrt{-1}$$

Since an imaginary number can never be equal to a real number, it follows that in an equality like

$$a + bi = c + di$$

a must equal *c* and *bi* must equal *di*

Complex numbers are handled in algebra just like any other expression, considering i as a known quantity. Whenever powers of i occur, they can be replaced by the equivalents given above. This idea of having in one equation two separate sets of quantities which must be accounted for separately, has

found a symbolic application in vector notation. These are covered later in this chapter.

Equations of the First Degree Algebraic expressions usually come in the form of equations, that is, one set of terms equals another set of terms. The simplest example of this is Ohm's law:

$$E = IR$$

One of the three quantities may be unknown but if the other two are known, the third can be found readily by substituting the known values in the equation. This is very easy if it is E in the above example that is to be found; but suppose we wish to find I while E and R are given. We must then rearrange the equation so that I comes to stand alone to the left of the equality sign. This is known as *solving the equation for I.*

Solution of the equation in this case is done simply by transposing. If two things are equal then they must still be equal if both are multiplied or divided by the same number. Dividing both sides of the equation by R:

$$\frac{E}{R} = \frac{IR}{R} = I \text{ or } I = \frac{E}{R}$$

If it were required to solve the equation for R, we should divide both sides of the equation by I.

$$\frac{E}{I} = R \text{ or } R = \frac{E}{I}$$

A little more complicated example is the equation for the reactance of a capacitor:

$$X = \frac{1}{2 \pi fC}$$

To solve this equation for C, we may multiply both sides of the equation by C and divide both sides by X

$$X \times \frac{C}{X} = \frac{1}{2 \pi fC} \times \frac{C}{X}, \text{ or}$$

$$C = \frac{1}{2 \pi fX}$$

This equation is one of those which requires a good knowledge of the placing of the decimal point when solving. Therefore we give a few examples: What is the reactance of a 25 pF capacitor at 1000 kHz? In filling in the given values in the equation we must remember that the units used are farads, hertz, and ohms. Hence, we must write 25 pF as 25 millionths of a millionth of a farad or 25×10^{-12} farad; similarly, 1000 kHz must be converted to 1,000,000 Hz. Substituting these values in the original equation, we have

$$X = \frac{1}{2 \times 3.14 \times 1,000,000 \times 25 \times 10^{-12}}$$

$$X = \frac{1}{6.28 \times 10^6 \times 25 \times 10^{-12}} = \frac{10^6}{6.28 \times 25}$$

$$= 6360 \text{ ohms}$$

A bias resistor of 1000 ohms should be bypassed, so that at the lowest frequency the reactance of the capacitor is 1/10th of that of the resistor. Assume the lowest frequency to be 50 hertz, then the required capacity should have a reactance of 100 ohms, at 50 Hz.

$$C = \frac{1}{2 \times 3.14 \times 50 \times 100} \text{ farads}$$

$$C = \frac{10^6}{6.28 \times 5000} \text{ microfarads}$$

$$C = 32 \text{ } \mu F$$

In the third possible case, it may be that the frequency is the unknown. This happens for instance in some tone-control problems. Suppose it is required to find the frequency which makes the reactance of a 0.03 μF capacitor equal to 100,000 ohms.

First we must solve the equation for f. This is done by transposition.

$$X = \frac{1}{2 \pi f C} \qquad f = \frac{1}{2 \pi C X}$$

Substituting known values

$$f = \frac{1}{2 \times 3.14 \times 0.03 \times 10^{-6} \times 100,000} \text{Hz}$$

$$f = \frac{1}{0.01884} \text{ hertz} = 53 \text{Hz}$$

These equations are known as first degree equations with one unknown. First degree, because the unknown occurs only as a first

power. Such an equation always has one possible solution or *root* if all the other values are known.

If there are two unknowns, a single equation will not suffice, for there are then an infinite number of possible solutions. In the case of two unknowns we need *two independent* simultaneous equations. An example of this is:

$$3x + 5y = 7 \qquad 4x - 10y = 3$$

Required, to find x and y.

This type of work is done either by the *substitution method* or by the *elimination method*. In the substitution method we might write for the first equation:

$$3x = 7 - 5y \qquad \therefore x = \frac{7 - 5y}{3}$$

(The symbol $\therefore$ means, *therefore* or *hence*).

This value of x can then be substituted for x in the second equation making it a single equation with but one unknown, y.

It is, however, simpler in this case to use the elimination method. Multiply both sides of the first equation by two and add it to the second equation:

$$
\begin{array}{r}
6x + 10y = 14 \\
4x - 10y = 3 \\
\hline
10x = 17 \qquad x = 1.7
\end{array}
$$
add

Substituting this value of x in the first equation, we have

$$5.1 + 5y = 7 \therefore 5y = 7 - 5.1 = 1.9 \therefore$$
$$y = 0.38$$

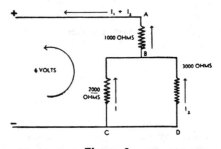

Figure 3

In this simple network the current divides through the 2000-ohm and 3000-ohm resistors. The current through each may be found by using two simultaneous linear equations. Note that the arrows indicate the direction of electron flow.

An application of two simultaneous linear equations will now be given. In figure 3 a simple network is shown consisting of three resistances; let it be required to find the currents I_1 and I_2 in the two branches.

The general way in which all such problems can be solved is to assign directions to the currents through the various resistances. When these are chosen wrong it will do no harm for the result of the equations will then be negative, showing up the error. In this simple illustration there is, of course, no such difficulty.

Next we write the equations for the meshes, in accordance with Kirchhoff's second law. All voltage drops in the direction of the curved arrow are considered positive, the reverse ones negative. Since there are two unknowns we write two equations.

$$1000\ (I_1 + I_2) + 2000\ I_1 = 6$$
$$-\ 2000\ I_1 + 3000\ I_2 = 0$$

Expand the first equation

$$3000\ I_1 + 1000\ I_2 = 6$$

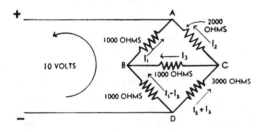

Figure 4

A MORE COMPLICATED PROBLEM REQUIRING THE SOLUTION OF CURRENTS IN A NETWORK

This problem is similar to that in figure 3 but requires the use of three simultaneous linear equations.

Multiply this equation by 3

$$9000\ I_1 + 3000\ I_2 = 18$$

Subtracting the second equation from the first

$$11000\ I_1 = 18$$
$$I_1 = 18/11000 = 0.00164\ \text{amp}.$$

Filling in this value in the second equation

$$3000\ I_2 = 3.28 \qquad I_2 = 0.00109\ \text{amp}.$$

A similar problem but requiring three equations is shown in figure 4. This consists of an unbalanced bridge and the problem is to find the current in the bridge-branch, I_3. We again assign directions to the different currents, guessing at the one marked I_3. The voltages around closed loops ABC [eq. (1)] and BDC [eq. (2)] equal zero and are assumed to be positive in a counterclockwise direction; that from D to A equals 10 volts [eq. (3)].

$$\text{(1)}$$
$$-\ 1000\ I_1 + 2000\ I_2 - 1000\ I_3 = 0$$

$$\text{(2)}$$
$$-\ 1000\ (I_1 - I_3)$$
$$+\ 1000\ I_3 + 3000\ (I_2 + I_3) = 0$$

$$\text{(3)}$$
$$1000\ I_1 + 1000\ (I_1 - I_3) - 10 = 0$$

Expand equations (2) and (3)

$$\text{(2)}$$
$$-\ 1000\ I_1 + 3000\ I_2 + 5000\ I_3 = 0$$

$$\text{(3)}$$
$$2000\ I_1 - 1000\ I_3 - 10 = 0$$

Subtract equation (2) from equation (1)

$$\text{(a)}$$
$$-\ 1000\ I_2 - 6000\ I_3 = 0$$

Multiply the second equation by 2 and add it to the third equation

$$\text{(b)}$$
$$6000\ I_2 + 9000\ I_3 - 10 = 0$$

Now we have but two equations with two unknowns.

Multiplying equation (a) by 6 and adding to equation (b) we have

$$-27000\ I_3 - 10 = 0$$
$$I_3 = -\ 10/27000 = -\ 0.00037\ \text{amp}.$$

Note that now the solution is negative which means that we have drawn the arrow for I_3 in figure 4 in the wrong direction. The current is 0.37 mA in the other direction.

Second Degree or Quadratic Equations   A somewhat similar problem in radio would be, if power in watts and resistance in ohms of a circuit are given, to find the voltage and the current. Example: When lighted to normal brilliancy, a 100-watt lamp has a resistance of 49 ohms; for what line voltage was the lamp designed and what current would it take?

Here we have to use the simultaneous equations:

$$P = EI \text{ and } E = IR$$

Filling in the known values:

$$P = EI = 100 \text{ and } E = IR = I \times 49$$

Substitute the second equation into the first equation

$$P = EI = (I) \; I \times 49 \; = 49 \; I^2 = 100$$

$$\therefore I = \sqrt{\frac{100}{49}} = \frac{10}{7} = 1.43 \text{ amp.}$$

Substituting the found value of 1.43 amp. for I in the first equation, we obtain the value of the line voltage, 70 volts.

Note that this is a *second degree* equation for we finally had the second power of I. Also, since the current in this problem could only be positive, the negative square root of $100/49$ or $-10/7$ was not used. Strictly speaking, however, there are two more values that satisfy both equations, these are -1.43 and -70.

In general, a second degree equation in one unknown has two roots, a third degree equation three roots, etc.

The Quadratic Equation   Quadratic or second degree equations with but one unknown can be reduced to the general form

$$ax^2 + bx + c = 0$$

where,

x is the unknown,

a, b, and c are constants.

This type of equation can sometimes be solved by the method of factoring a three-term expression as follows:

$$2x^2 + 7x + 6 = 0$$

$$2x^2 + 4x + 3x + 6 = 0$$

factoring:

$$2x \; (x + 2) + 3 \; (x + 2) = 0$$

$$(2x + 3) \; (x + 2) = 0$$

There are two possibilities when a product is zero. Either the one or the other factor equals zero. Therefore there are two solutions.

$$2x_1 + 3 = 0 \qquad\qquad x_2 + 2 = 0$$

$$2x_1 = -3 \qquad\qquad x_2 = -2$$

$$x_1 = -1\tfrac{1}{2}$$

Since factoring is not always easy, the following general solution can usually be employed; in this equation a, b, and c are the coefficients referred to above.

$$x = \frac{-b \pm \sqrt{b^2 - 4ac}}{2a}$$

Applying this method of solution to the previous example:

$$x = \frac{-7 \pm \sqrt{49 - 8 \times 6}}{4}$$

$$= \frac{-7 \pm \sqrt{1}}{4} = \frac{-7 \pm 1}{4}$$

$$x_1 = \frac{-7 + 1}{4} = -1\tfrac{1}{2}$$

$$x_2 = \frac{-7 - 1}{4} = -2$$

A practical example involving quadratics is the law of impedance in ac circuits. However, this is a simple kind of quadratic equation which can be solved readily without the use of the special formula given above.

$$Z = \sqrt{R^2 + (X_L - X_C)^2}$$

This equation can always be solved for R, by squaring both sides of the equation. It should now be understood that squaring both sides of an equation as well as multiplying both sides with a term containing the unknown *may add a new root*. Since we know here that Z and R are positive, when we square the expression there is no ambiguity.

$$Z^2 = R^2 + (X_L - X_C)^2$$

$$\text{and } R^2 = Z^2 - (X_L - X_C)^2$$

$$\text{or } R = \sqrt{Z^2 - (X_L - X_C)^2}$$

$$\text{also: } (X_L - X_C)^2 = Z^2 - R^2$$

$$\text{and } \pm (X_L - X_C) = \sqrt{Z^2 - R^2}$$

But here we do not know the sign of the solution unless there are other facts which indicate it. To find either X_L or X_C alone it would have to be known whether the one or the other is the larger.

34-3   Trigonometry

Trigonometry is the science of mensuration of *triangles*. At first glance triangles may seem to have little to do with electrical phenomena; however, in ac work most currents and voltages follow laws equivalent to those of the various trigonometric relations which we are about to examine briefly. Examples of their application to ac work will be given in the section on *vectors*.

Angles are measured in *degrees* or in *radians*. The circle has been divided into 360 degrees, each degree into 60 minutes, and each minute into 60 seconds. A decimal division of the degree is also in use because it makes calculation easier. Degrees, minutes and seconds are indicated by the following signs: °, ′ and ″. Example: 6° 5′ 23″ means six degrees, five minutes, twenty-three seconds. In the decimal notation we simply write 8.47°, eight and forty-seven hundredths of a degree.

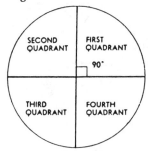

Figure 5

THE CIRCLE IS DIVIDED INTO FOUR QUADRANTS BY TWO PERPENDICULAR LINES AT RIGHT ANGLES TO EACH OTHER

The "northeast" quadrant thus formed is known as the first quadrant; the others are numbered consecutively in a counterclockwise direction.

When a circle is divided into four quadrants by two perpendicular lines passing through the center (figure 5) the angle made by the two lines is 90 degrees, known

as a *right angle*. Two right angles, or 180° equals a *straight angle*.

The radian—If we take the radius of a circle and bend it so it can cover a part of the circumference, the arc it covers subtends an angle called a *radian* (figure 6). Since the diameter of a circle equals 2 times the radius, there are 2π radians in 360°. So we have the following relations:

1 radian = 57° 17′ 45″ = 57.2958°
π = 3.14159
1 degree = 0.01745 radians
π radians = 180°
$\pi/2$ radians = 90°
$\pi/3$ radians = 60°

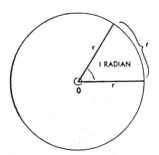

Figure 6

THE RADIAN

A radian is an angle whose arc is exactly equal to the length of either side. Note that the angle is constant regardless of the length of the side and the arc so long as they are equal. A radian equals 57.2958°.

In trigonometry we consider an angle *generated* by two lines, one stationary and the other rotating as if it were hinged at 0 (figure 7). Angles can be greater than 180 degrees and even greater than 360 degrees as illustrated in this figure.

Two angles are complements of each other when their sum is 90°, or a right angle. *A* is the complement of *B* and *B* is the complement of *A* when

$$A = (90° - B)$$

and when

$$B = (90° - A)$$

Two angles are supplements of each other when their sum is equal to a straight angle, or 180°. *A* is the supplement of *B* and *B* is the supplement of *A* when:

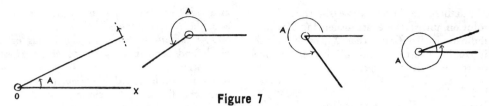

Figure 7

AN ANGLE IS GENERATED BY TWO LINES, ONE STATIONARY
AND THE OTHER ROTATING

The line OX is stationary; the line with the small arrow at the far end rotates in a counterclock-wise direction. At the position illustrated in the lefthandmost section of the drawing it makes an angle, A, which is less than 90° and is therefore in the first quadrant. In the position shown in the second portion of the drawing the Angle A has increased to such a value that it now lies in the third quadrant; note that an angle can be greater than 180°. In the third illustration the angle A is in the fourth quadrant. In the fourth position the rotating vector has made more than one complete revolution and is hence in the fifth quadrant; since the fifth quadrant is an exact repetition of the first quadrant, its values will be the same as in the lefthandmost portion of the illustration.

$$A = (180° - B)$$

and

$$B = (180° - A)$$

In the angle A (figure 8A), a line is drawn from P, perpendicular to b. Regardless of the point selected for P, the *ratio* a/c will always be the same for any given angle, A. So will all the other proportions between a, b, and c remain constant regardless of the position of point P on c. The six possible ratios each are named and defined as follows:

$$\text{sine } A = \frac{a}{c} \qquad \text{cosine } A = \frac{b}{c}$$

$$\text{tangent } A = \frac{a}{b} \qquad \text{cotangent } A = \frac{b}{a}$$

$$\text{secant } A = \frac{c}{b} \qquad \text{cosecant } A = \frac{c}{a}$$

Let us take a special angle as an example. For instance, let the angle A be 60 degrees as in figure 8B. Then the relations between the sides are as in the figure and the six functions become:

$$\sin. 60° = \frac{a}{c} = \frac{\frac{1}{2}\sqrt{3}}{1} = \frac{1}{2}\sqrt{3}$$

$$\cos 60° = \frac{b}{c} = \frac{\frac{1}{2}}{1} = \frac{1}{2}$$

$$\tan 60° = \frac{a}{b} = \frac{\frac{1}{2}\sqrt{3}}{\frac{1}{2}} = \sqrt{3}$$

$$\cot 60° = \frac{\frac{1}{2}}{\frac{1}{2}\sqrt{3}} = \frac{1}{\sqrt{3}} = \frac{1}{3}\sqrt{3}$$

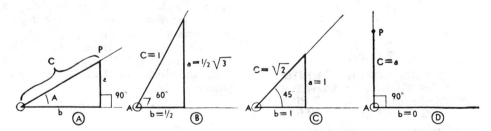

Figure 8

THE TRIGONOMETRIC FUNCTIONS

In the right triangle shown in (A) the side opposite the angle A is a, while the adjoining sides are b and c; the trigonometric functions of the angle A are completely defined by the ratios of the sides a, b and c. In (B) are shown the lengths of the sides a and b when angle A is 60° and side c is 1. In (C) angle A is 45°; a and b equal 1, while c equals $\sqrt{2}$. In (D) note that c equals a for a right angle while b equals 0.

$$\sec 60° = \frac{c}{b} = \frac{1}{1/2} = 2$$

$$\csc 60° \quad \frac{c}{a} = \frac{1}{1/2\sqrt{3}} = 2/3\sqrt{3}$$

Another example: Let the angle be 45°, then the relations between the lengths of *a*, *b*, and *c* are as shown in figure 8C and the six functions are:

$$\sin 45° = \frac{1}{\sqrt{2}} = 1/2\sqrt{2}$$

$$\cos 45° = \frac{1}{\sqrt{2}} = 1/2\sqrt{2}$$

$$\tan 45° = \frac{1}{1} = 1 \qquad \cot 45° = \frac{1}{1} = 1$$

$$\sec 45° = \frac{\sqrt{2}}{1} = \sqrt{2}$$

$$\text{cosec } 45° = \frac{\sqrt{2}}{1} = \sqrt{2}$$

There are some special difficulties when the angle is zero or 90 degrees. In figure 8D an angle of 90 degrees is shown; drawing a line perpendicular to *b* from point *P* makes it fall on top of *c*. Therefore in this case $a = c$ and $b = 0$. The six ratios are now:

$$\sin 90° = \frac{a}{c} = 1 \qquad \cos 90° = \frac{b}{c} = \frac{0}{c} = 0$$

$$\tan 90° = \frac{a}{b} = \frac{a}{0} = \infty \quad \cot 90° = \frac{0}{a} = 0$$

$$\sec 90° = \frac{c}{b} = \frac{c}{0} = \infty \quad \text{cosec } 90° = \frac{c}{a} = 1$$

When the angle is zero, $a=0$ and $b=c$. The values are then:

$$\sin 0° = \frac{a}{c} = \frac{0}{c} = 0 \quad \cos 0° = \frac{b}{c} = 1$$

$$\tan 0° = \frac{a}{b} = \frac{0}{b} = 0 \quad \cot 0° = \frac{b}{a} = \frac{b}{0} = \infty$$

$$\sec 0° = \frac{c}{b} = 1 \qquad \text{cosec } 0° = \frac{c}{a} = \frac{c}{0} = \infty$$

In general, for every angle, there will be definite values of the six functions. Conversely, when any of the six functions is known, the angle is defined. Tables have been calculated giving the value of the functions for angles.

From the foregoing we can make up a small table of our own (figure 9), giving values of the functions for some common angles.

Relations Between Functions It follows from the definitions that

$$\sin A = \frac{1}{\text{cosec } A} \qquad \cos A = \frac{1}{\sec A}$$

$$\text{and } \tan A = \frac{1}{\cot A}$$

Angle	Sin	Cos.	Tan	Cot	Sec.	Cosec.
0	0	1	0	∞	1	∞
30°	1/2	$1/2\sqrt{3}$	$1/3\sqrt{3}$	$\sqrt{3}$	$2/3\sqrt{3}$	2
45°	$1/2\sqrt{2}$	$1/2\sqrt{2}$	1	1	$\sqrt{2}$	$\sqrt{2}$
60°	$1/2\sqrt{3}$	1/2	$\sqrt{3}$	$1/3\sqrt{3}$	2	$2/3\sqrt{3}$
90°	1	0	∞	0	∞	1

Figure 9

Values of trigonometric functions for common angles in the first quadrant.

From the definitions also follows the relation cos A=sin (complement of A)=sin $(90°-A)$ because in the right triangle of figure 10 cos $A=b/c$=sin B and $B=90°-A$ or the complement of A. For the same reason:

$$\cot A = \tan (90° - A)$$

$$\csc A = \sec (90° - A)$$

Relations in Right Triangles In the right triangle of figure 10, sin $A=a/c$ and by transposition

$$a = c \sin A$$

For the same reason we have the following identities:

$$\tan A = a/b \qquad a = b \tan A$$

$$\cot A = b/a \qquad b = a \cot A$$

In the same triangle we can do the same for functions of the angle B

$$\sin B = b/c \qquad b = c \sin B$$

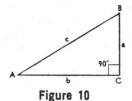

Figure 10

In this figure, the sides a, b, and c are used to define the trigonometric functions of angle B as well as angle A.

$$\cos B = a/c \qquad a = c \cos B$$
$$\tan B = b/a \qquad b = a \tan B$$
$$\cot B = a/b \qquad a = b \cot B$$

Functions of Angles Greater Than 90 Degrees In angles greater than 90 degrees, the values of a and b become negative on occasion in accordance with the rules of Cartesian coordinates. When b is measured from 0 towards the left it is considered negative and similarly, when a is measured from 0 downwards, it is negative. Referring to figure 11, an angle in the *second quadrant* (between 90° and 180°) has some of its functions negative:

$$\sin A = \frac{a}{c} = \text{pos.} \qquad \cos A = \frac{-b}{c} = \text{neg.}$$

$$\tan A = \frac{a}{-b} = \text{neg.} \qquad \cot A = \frac{-b}{a} = \text{neg.}$$

$$\sec A = \frac{c}{-b} = \text{pos.} \qquad \cosec A = \frac{c}{a} = \text{pos.}$$

For an angle in the *third quadrant* (180° to 270°), the functions are

$$\sin A = \frac{-a}{c} = \text{neg.} \qquad \cos A = \frac{-b}{c} = \text{neg.}$$

$$\tan A = \frac{-a}{-b} = \text{pos.} \qquad \cot A = \frac{-b}{-a} = \text{pos.}$$

$$\sec A = \frac{c}{-b} = \text{neg.} \qquad \cosec A = \frac{c}{-a} = \text{neg.}$$

And in the *fourth quadrant* (270° to 360°):

$$\sin A = \frac{-a}{c} = \text{neg.} \qquad \cos A = \frac{b}{c} = \text{pos.}$$

$$\tan A = \frac{-a}{b} = \text{neg.} \qquad \cot A = \frac{b}{-a} = \text{neg.}$$

$$\sec A = \frac{c}{b} = \text{pos.} \qquad \cosec A = \frac{c}{-a} = \text{neg.}$$

Summarizing, the sign of the functions in each quadrant can be seen at a glance from figure 12, where in each quadrant are written the names of functions which are positive; those not mentioned are negative.

Graphs of Trigonometric Functions *The sine wave*—When we have the relation $y = \sin x$, where x is an angle measured in radians or degrees, we can draw a curve of y versus x for all values of the independent variable, and thus get a good conception how the sine varies with the

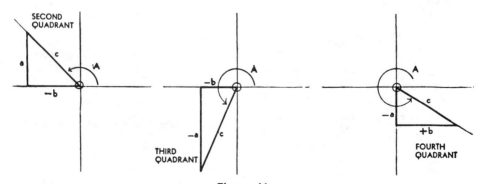

Figure 11

TRIGONOMETRIC FUNCTIONS IN THE SECOND, THIRD, AND FOURTH QUADRANTS

The trigonometric functions in these quadrants are similar to first quadrant values, but the signs of the functions vary as listed in the text.

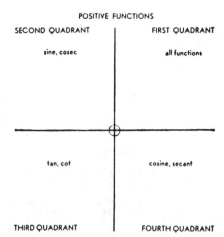

POSITIVE FUNCTIONS

SECOND QUADRANT | FIRST QUADRANT

sine, cosec | all functions

tan, cot | cosine, secant

THIRD QUADRANT | FOURTH QUADRANT

Figure 12

SIGNS OF THE TRIGONOMETRIC FUNCTIONS

The functions listed in this diagram are positive; all other functions are negative.

magnitude of the angle. This has been done in figure 13A. We can learn from this curve the following facts.

1. The sine varies between $+1$ and -1
2. It is a periodic curve, repeating itself after every multiple of 2π or $360°$
3. $\sin x = \sin (180° -x)$ or $\sin (\pi -x)$
4. $\sin x = -\sin (180° + x)$, or $-\sin (\pi + x)$

The cosine wave—Making a curve for the function $y = \cos x$, we obtain a curve similar to that for $y = \sin x$ except that it is displaced by $90°$ or $\pi/2$ radians with respect to the Y-axis. This curve (figure 13B) is also periodic but it does not start with zero. We read from the curve:

1. The value of the cosine never goes beyond $+1$ or -1
2. The curve repeats, after every multiple of 2π radians or $360°$
3. $\cos x = -\cos (180° -x)$ or $-\cos (\pi -x)$
4. $\cos x = \cos (360° -x)$ or $\cos (2\pi -x)$

The graph of the tangent is illustrated in figure 14. This is a discontinuous curve and illustrates well how the tangent increases from zero to infinity when the angle increases from zero to 90 degrees. Then when the angle is further increased, the tangent starts from minus infinity going to zero in the second quadrant, and to infinity again in the third quadrant.

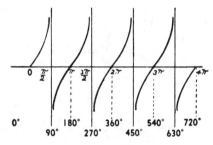

Figure 14

TANGENT CURVES

The tangent curve increases from 0 to ∞ with an angular increase of 90°. In the next 180° it increases from −∞ to +∞.

1. The tangent can have any value between $+\infty$ and $-\infty$
2. The curve repeats and the period is π radians or $180°$, not 2π radians
3. $\tan x = \tan (180° + x)$ or $\tan (\pi + x)$

Figure 13

SINE AND COSINE CURVES

In (A) we have a sine curve drawn in Cartesian coordinates. This is the usual representation of an alternating current wave with out substantial harmonics. In (B) we have a cosine wave; note that it is exactly similar to a sine wave displaced by 90° or $\pi/2$ radians.

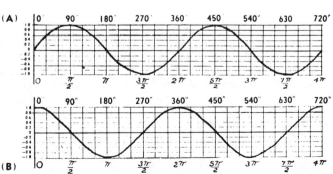

4. Tan $x = -\tan (180° - x)$ or $\tan (\pi - x)$

The graph of the cotangent is the inverse of that of the tangent, see figure 15. It leads us to the following conclusions:

1. The cotangent can have any value between $+\infty$ and $-\infty$
2. It is a periodic curve, the period being π radians or $180°$
3. Cot $x = \cot (180° + x)$ or $\cot (\pi + x)$
4. Cot $x = -\cot (180° - x)$ or $-\cot (\pi - x)$

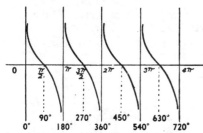

Figure 15

COTANGENT CURVES

Cotangent curves are the inverse of the tangent curves. They vary from $+\infty$ to $-\infty$ in each pair of quadrants.

The graphs of the secant and cosecant are of lesser importance and will not be shown here. They are the inverse, respectively, of the cosine and the sine, and therefore they vary from $+1$ to infinity and from -1 to $-$infinity.

Perhaps another useful way of visualizing the values of the functions is by considering figure 16. If the radius of the circle is the unit of measurement then the lengths of the lines are equal to the functions marked on them.

Trigonometric Tables   There are two kinds of trigonometric tables. The first type gives the functions of the angles, the second the logarithms of the functions. The first kind is also known as the table of *natural* trigonometric functions.

These tables give the functions of all angles between 0 and $45°$. This is all that is necessary for the function of an angle between $45°$ and $90°$ can always be written as the cofunction of an angle below $45°$

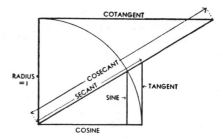

Figure 16

ANOTHER REPRESENTATION OF TRIGONOMETRIC FUNCTIONS

If the radius of a circle is considered as the unit of measurement, then the lengths of the various lines shown in this diagram are numerically equal to the functions marked adjacent to them.

Example: If we had to find the sine of $48°$, we might write

$$\sin 48° = \cos (90° - 48°) = \cos 42°$$

Tables of the logarithms of trigonometric functions give the common logarithms ($\log_{10}$) of these functions. Since many of these logarithms have negative characteristics, one should add -10 to all logarithms in the table which have a characteristic of 6 or higher. For instance, the log sin $24° = 9.60931 - 10$. Log tan $1° = 8.24192 - 10$ but log cot $1° = 1.75808$. When the characteristic shown is less than 6, it is supposed to be positive and one should not add -10.

Vectors   A *scalar* quantity has *magnitude* only; a *vector* quantity has both *magnitude* and *direction*. When we speak of a speed of 50 miles per hour, we are using a scalar quantity, but when we say the wind is northeast and has a velocity of 50 miles per hour, we speak of a vector quantity.

Vectors, representing forces, speeds, displacements, etc., are represented by arrows. They can be added graphically by well known methods illustrated in figure 17. We can make the parallelogram of forces or we can simply draw a triangle. The addition of many vectors can be accomplished graphically as in the same figure.

In order that we may define vectors algebraically and add, subtract, multiply, or divide them, we must have a logical notation system that lends itself to these operations. For this purpose vectors can be defined by

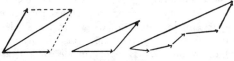

Figure 17

Vectors may be added as shown in these sketches. In each case the long vector represents the vector sum of the smaller vectors. For many engineering applications sufficient accuracy can be obtained by this method which avoids long and laborious calculations.

coordinate systems. Both the Cartesian and the polar coordinates are in use.

Vectors Defined by Cartesian Coordinates Since we have seen how the sum of two vectors is obtained, it follows from Figure 18, that the vector $\dot{Z}$ equals the sum of the two vectors x and y. In fact, any vector can be resolved into vectors along the X- and Y-axis. For convenience in working with these quantities we need to distinguish between the x- and

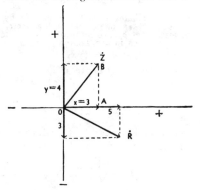

Figure 18

RESOLUTION OF VECTORS

Any vector such as z may be resolved into two vectors, x and y, along the X- and Y- axes. If vectors are to be added, their respective x and y components may be added to find the x and y components of the resultant vector.

y-component, and so it has been agreed that the x-component alone shall be marked with the letter j. Example (figure 18):

$$\dot{Z} = 3 + 4j$$

Note again that the sign of components along the X-axis is positive when measured from 0 to the right and negative when measured from 0 towards the left. Also, the component along the Y-axis is positive when

measured from 0 upwards, and negative when measured from 0 downwards. So the vector, $\dot{R}$, is described as

$$\dot{R} = 5 - 3j$$

Vector quantities are usually indicated by some special typography, especially by using a point over the letter indicating the vector, as $\dot{R}$.

Absolute Value of a Vector The absolute or scalar value of vectors such as $\dot{Z}$ or $\dot{R}$ in figure 18 is easily found by the theorem of Pythagoras, which states that in any right-angled triangle the square of the side opposite the right angle is equal to the sum of the squares of the sides adjoining the right angle. In figure 18, OAB is a right-angled triangle; therefore, the square of OB (or Z) is equal to the square of OA (or x) plus the square of AB (or y). Thus the absolute values of Z and R may be determined as follows:

$$|Z| = \sqrt{x^2 + y^2}$$
$$|Z| = \sqrt{3^2 + 4^2} = 5$$
$$|R| = \sqrt{5^2 + 3^2} = \sqrt{34} = 5.83$$

The vertical lines indicate that the absolute or scalar value is meant without regard to sign or direction.

Addition of Vectors An examination of Figure 19 will show that the two vectors

$$\dot{R} = x_1 + j\,y_1$$
$$\dot{Z} = x_2 + j\,y_2$$

can be added, if we add the X-components and the Y-components separately.

$$\dot{R} + \dot{Z} = x_1 + x_2 + j\,(y_1 + y_2)$$

For the same reason we can carry out subtraction by subtracting the horizontal components and subtracting the vertical components

$$\dot{R} - \dot{Z} = x_1 - x_2 + j\,(y_1 - y_2)$$

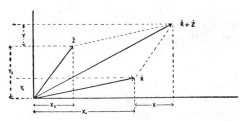

Figure 19

ADDITION OR SUBTRACTION OF VECTORS

Vectors may be added or subtracted by adding or subtracting their x or y components separately.

Let us consider the operator j. If we have a vector a along the X-axis and add a j in front of it (multiplying by j) the result is that the direction of the vector is rotated forward 90 degrees. If we do this twice (multiplying by j^2) the vector is rotated forward by 180 degrees and now has the value $-a$. Therefore multiplying by j^2 is equivalent to multiplying by -1. Then

$$j^2 = -1 \text{ and } j = \sqrt{-1}$$

This is the imaginary number discussed before under algebra. In electrical engineering the letter j is used rather than i, because i is already known as the symbol for current.

Multiplying Vectors When two vectors are to be multiplied we can perform the operation just as in algebra, remembering that $j^2 = -1$.

$$\overset{..}{R}\overset{..}{Z} = (x_1 + jy_1)(x_2 + jy_2)$$
$$= x_1 x_2 + jx_1 y_2 + j x_2 y_1 + j^2 y_1 y_2$$
$$= x_1 x_2 - y_1 y_2 + j(x_1 y_2 + x_2 y_1)$$

Division has to be carried out so as to remove the j-term from the denominator. This can be done by multiplying both denominator and numerator by a quantity which will eliminate j from the denominator. Example:

$$\frac{\overset{..}{R}}{\overset{..}{Z}} = \frac{x_1 + jy_1}{x_2 + jy_2} = \frac{(x_1 + jy_1)(x_2 - jy_2)}{(x_2 + jy_2)(x_2 - jy_2)}$$

$$= \frac{x_1 x_2 + y_1 y_2 + j(x_2 y_1 - x_1 y_2)}{x_2{}^2 + y_2{}^2}$$

Polar Coordinates A vector can also be defined in polar coordinates by its magnitude and its vectorial angle with an arbitrary reference axis. In figure 20 the vector Z has a magnitude 50 and a

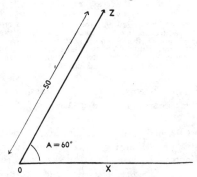

Figure 20

IN THIS FIGURE A VECTOR HAS BEEN REPRESENTED IN POLAR INSTEAD OF CARTESIAN COORDINATES

In polar coordinates a vector is defined by a magnitude and an angle, called the vectorial angle, instead of by two magnitudes as in Cartesian coordinates.

vectorial angle of 60 degrees. This will then be written

$$\overset{.}{Z} = 50 \angle 60°$$

A vector $a + jb$ can be transformed into polar notation very simply (see figure 21)

$$\overset{.}{Z} = a + jb = \sqrt{a^2 + b^2} \angle \tan^{-1}\frac{b}{a}$$

In this connection tan^{-1} means the *angle of which the tangent is*. Sometimes the notation *arc tan b/a* is used. Both have the same meaning.

A polar notation of a vector can be transformed into a Cartesian coordinate notation in the following manner (figure 22)

$$\overset{.}{Z} = p \angle A = p \cos A + jp \sin A$$

A sinusoidally alternating voltage or current is symbolically represented by a rotating vector, having a magnitude equal to the peak voltage or current and rotating with an angular velocity of $2\pi f$ radians per second or as many revolutions per second as there are cycles per second.

The instantaneous voltage (e), is always equal to the sine of the vectorial angle of this rotating vector, multiplied by its magnitude.

$$e = E \sin 2\pi f t$$

The alternating voltage therefore varies with time as the sine varies with the angle. If we plot time horizontally and instantaneous voltage vertically we will get a curve like those in figure 13A&B.

In alternating-current circuits, the cur-

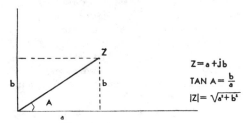

Figure 21

Vectors can be transformed from Cartesian into polar notation as shown in this figure.

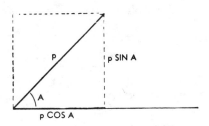

Figure 22

Vectors can be transformed from polar into Cartesian notation as shown in this figure.

rent which flows due to the alternating voltage is not necessarily in step with it. The rotating current vector may be ahead or behind the voltage vector, having a *phase difference* with it. For convenience we draw these vectors as if they were standing still, so that we can indicate the difference in phase or the *phase angle*. In figure 23 the current lags behind the voltage by the angle θ, or we might say that the voltage leads the current by the angle θ.

Vector diagrams show the phase relations between two or more vectors (voltages and currents) in a circuit. They may be added and subtracted as described; one may add a voltage vector to another voltage vector or a current vector to a current vector but not

a current vector to a voltage vector (for the same reason that one cannot add a force to a speed). Figure 23 illustrates the relations in the simple series circuit of a coil and resistor. We know that the current passing through coil and resistor must be the same and in the same phase, so we draw this current I along the X-axis. We know also that the voltage drop IR across the resistor is in phase with the current, so the vector IR representing the voltage drop is also along the X-axis.

The voltage across the coil is 90 degrees ahead of the current through it; IX must therefore be drawn along the Y-axis. E the applied voltage must be equal to the vectorial sum of the two voltage drops, IR and IX, and we have so constructed it in the drawing. Now expressing the same in algebraic notation, we have

$$\dot{E} = IR + jIX$$

$$\dot{IZ} = IR + jIX$$

Dividing by I

$$\dot{Z} = R + jX$$

Due to the fact that a *reactance* rotates the voltage vector ahead or behind the current vector by 90 degrees, we must mark it with a j in vector notation. Inductive reactance will have a plus sign because it shifts the voltage vector forward; a capacitive reactance is negative because the voltage will lag behind the current. Therefore:

$$X_L = + j\, 2\pi\, fL$$

$$X_C = - j\, \frac{1}{2\pi f C}$$

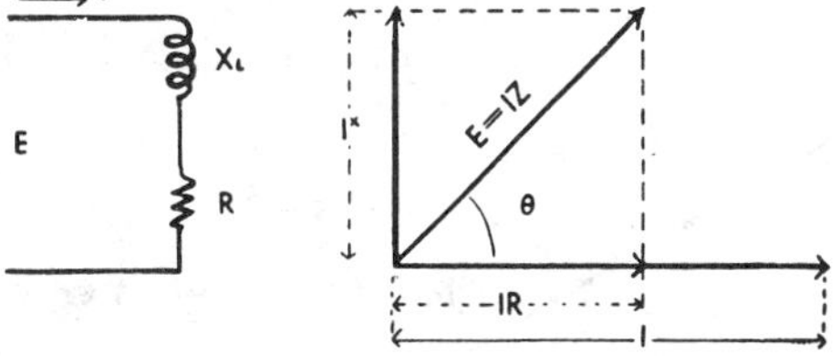

Figure 23

VECTOR REPRESENTATION OF A SIMPLE SERIES CIRCUIT

The right-hand portion of the illustration shows the vectors representing the voltage drops in the coil and resistance illustrated at the left. Note that the voltage drop across the coil (X_L) leads that across the resistance by 90°.

In figure 23 the angle θ is known as the phase angle between E and I. When calculating power, only the real components count. The power in the circuit is then

$$P = I \ (IR)$$

but $IR = E \cos \theta$

$$\therefore P = EI \cos \theta$$

The $\cos \theta$ is known as the power factor of the circuit. In many circuits we strive to keep the angle θ as small as possible, making $\cos \theta$ as near to unity as possible. In tuned circuits, we use reactances which should have as low a power factor as possible. The merit of a coil or capacitor, its Q, is defined by the tangent of this phase angle:

$$Q = \tan \theta = \frac{X}{R}$$

For an efficient coil or capacitor, Q should be as large as possible; the phase-angle should then be as close to 90 degrees as possible, making the power factor nearly zero. Q is almost but not quite the inverse of $\cos \theta$. Note that in figure 24

$$Q = \frac{X}{R} \quad \text{and} \cos \theta = \frac{R}{Z}$$

When Q is more than 5, the power factor is less than 20%; we can then safely say $Q = 1/\cos \theta$ with a maximum error of about $2\frac{1}{2}$ percent, for the worst case, when $\cos \theta = 0.2$, Q will equal $\tan \theta = 4.89$. For higher values of Q, the error becomes less.

Note that from figure 24 can be seen the simple relation:

$$\dot{Z} = R + jX_L$$

$$|Z| = \sqrt{R^2 + X_L^2}$$

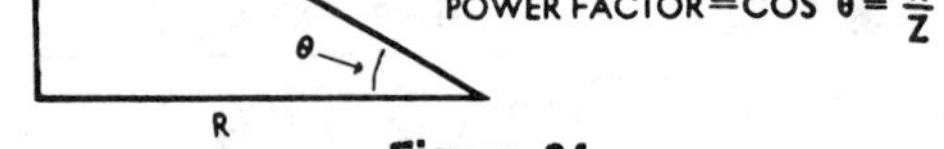

Figure 24

The figure of merit of a coil and its resistance is represented by the ratio of the inductive reactance to the resistance, which as shown in this diagram is equal to $\frac{X}{R}$, which equals $\tan \theta$.

For large values of θ (the phase angle) this is approximately equal to the reciprocal of the $\cos \theta$.

34-4   Boolean Algebra

Boolean Algebra, a language of logic, is the simplest form of mathematics possible. Each variable has only one of two discrete values. These values can be called *off* and *on*, or 0 and 1, depending on usage. The results derived from this branch of mathematics are not the familiar sums, differences, products, etc., but even more basic answers of *yes, no,* or *which*.

Classic mathematics has, in the past, ignored this type of calculation because the results were usually very easily found intuitively and problems such as "box (a), is colored green, and box (b) is the same as box (a); then box (b) must be green" need no complex mathematics to obtain a solution.

As the problems increase in complexity (as seen in modern logic usage) the need for a formal mathematics of logic also increases. Boolean algebra fulfills this requirement appropriately.

Three symbols of operation are commonly used:

$$\text{and} = a \cdot b$$
$$\text{or} = a + b$$
$$\text{not (or inverse)} = \bar{a}$$

In applying Boolean algebra, the following seven important identities apply:

$$a + 1 = 1 \quad a + \bar{a} = 1$$
$$a \cdot 1 = a \quad a \cdot \bar{a} = 0$$
$$a + a = a \quad a \cdot (b + c) = a \cdot b + a \cdot c$$
$$a \cdot a = a$$

and a theorem called *De Morgan's Theorem* states:

$$\bar{a} + \bar{b} = \overline{a \cdot b}$$
$$\text{or,} \ \bar{a} \cdot \bar{b} = \overline{a + b}$$

A major application of this theory in electronics is the minimization of relay or other logic functions, where the aim is to reduce the Boolean expression to the minimum number of components in the circuit.

Calculation using Boolean algebra consists of converting from written language to this mathematical form, a process similar to the preparation of a computer program. This form is then simplified using the above identities and rules, finally giving the desired result. This result must then be translated back to "plain language" in terms of relay contacts or other circuit blocks.

The arithmetic processing of a Boolean expression follows the same rules as normal algebraic manipulation. The usual distribution rules apply, i.e.:

$$(a + b) \cdot (c + d)$$
$$= a \cdot c + a \cdot d + b \cdot c + b \cdot d$$

Identities in these equations may be applied at any time, remembering that whenever an expression such as $(a \cdot a \cdot b)$ is found it can be simplified to $(a \cdot b)$ since, by definition $(a \cdot a = a)$. Again, the expression $(2a)$ must be written as $(a + a)$, which is equal to (a).

A Boolean Relay Circuit   Consider, for example, a relay system for a station, possibly including overload alarms, indication of drive failure, or overvoltage alarm. There are three inputs (x), (y), and (z) and an alarm is to be operated:

1. if x operates or y does not operate,
2. if x does not operate, or z does not operate,
3. and if x, y, or z operate.

This could be achieved by the circuit of figure 25. Translating this into Boolean algebra, we obtain:

Alarm $= (x$ or not $y)$ and (not x or $z)$ and
$$(x \text{ or } y \text{ or } z)$$
$$= (x + \bar{y}) \cdot (\bar{x} + z) \cdot (x + y + z)$$
expanding this expression,
Alarm $= (x + \bar{y}) \cdot (\bar{x} \cdot x + \bar{x} \cdot y$
$$+ \bar{x} \cdot z + x \cdot z + z \cdot y + z \cdot z)$$
This simplifies to:
Alarm $= (x + \bar{y}) \cdot (\bar{x} \cdot y + \bar{x} \cdot z$
$$+ x \cdot z + z \cdot y + z)$$
Again, multiplying and simplifying,
Alarm $= \bar{x} \cdot \bar{y} \cdot z + \bar{x} \cdot \bar{y} \cdot z$
$$+ x \cdot \bar{y} \cdot z + x \cdot z + \bar{y} \cdot z$$

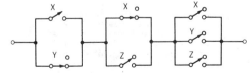

Figure 25

UNSIMPLIFIED RELAY NETWORK

Factoring gives,
$$\text{Alarm} = z \cdot (x + \bar{y})$$
Translating this into plain language:
Alarm $= z$ operates and x operates or
y does not operate

This expression is much simpler and could be constructed as in figure 26. It can be seen that this has effected a worthwhile savings in components compared to figure 25.

The theory may be expanded to systems where an output may become a new input in order to achieve memory or provide a logic sequence. It is possible to use even more complex Boolean methods to solve more complicated problems, such as a switching network with 16 inputs.

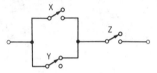

Figure 26

SIMPLIFIED OR MINIMAL CIRCUIT FOR RELAY NETWORK

34-5   The Smith Chart

There are several forms of chart-type calculators which may be used for calculations involving antennas, transmission lines, and impedance matching devices. In general, the most convenient of these, and certainly the most generally used, is the *Smith Chart* (figure 27).

The Smith Chart consists of two sets of orthogonal circles, where in one set, a given circle is the locus of points of equal real resistance, and equal reactance for the other set. Since there are both positive and negative values of reactance, the reactance loci are in reality two symmetrical sets. A breakdown of the Smith Chart into its components is shown in chapter 26, section 2 of this handbook.

In addition to resistance and reactance, additional sets of circles representing the loci of points of equal VSWR, transmission loss and reflection coefficients are present, or may be applied to the chart. With all of these parameters present in a single graphical form, it is no wonder that the Smith Chart is used

IMPEDANCE OR ADMITTANCE COORDINATES

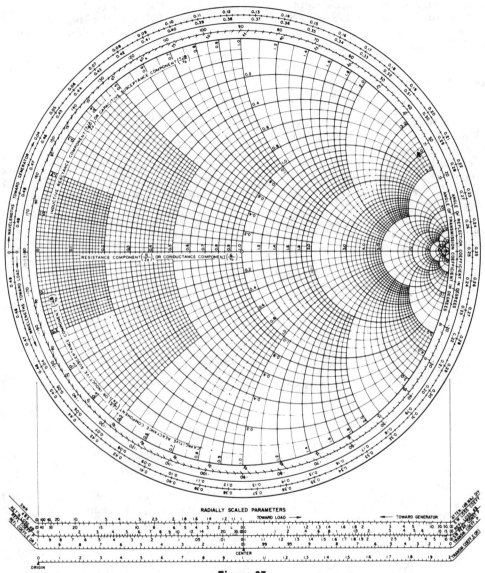

RADIALLY SCALED PARAMETERS

Figure 27

THE SMITH CHART

The Smith Chart is an impedance circle diagram having a curved coordinate system. The chart is composed of two families of circles, the resistance circles and the reactance circles. Wavelength scales are plotted around the perimeter of the chart, as well as a phase-angle scale. The perimeter of the chart represents a half-wavelength. The scaled horizontal line at the center represents the resistance scale, while the expanding arcs represent lines of reactance (positive and negative). The center point of the chart is normalized in this case to 1 + j0.

to such an extent. However, the value of the Smith Chart does not end here. Since within the zero resistance circle all values of impe-

dance are represented, for positive values of resistance, the Smith Chart is also used as a graphical representation of port impedance

IMPEDANCE OR ADMITTANCE COORDINATES

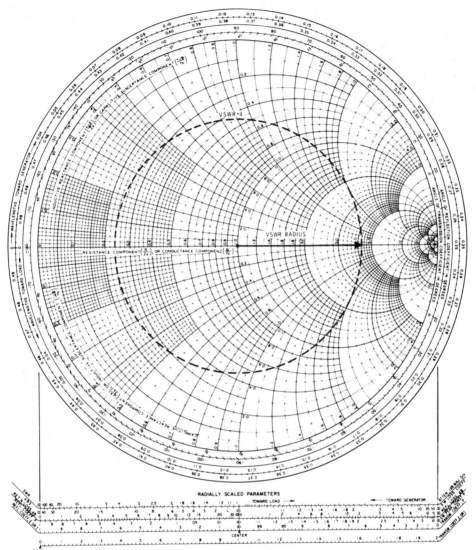

Figure 28

SWR CIRCLE ON THE SMITH CHART

SWR circles may be added to the Smith Chart, centered at 1.0 on the resistance scale. A circle centered at 1.0 and which passes through 4.0 on the resistance scale encloses all impedances which will cause a VSWR of 4 to 1, or less, on the system under examination. Charts having a center impedance value of 50 ohms are also available.

with respect to frequency, and as a method of performing vector analysis.

When understood, the Smith Chart is no more difficult to use than an engineering-model slide rule or calculator. The results obtained from the Smith Chart are, for normal purposes, very reliable for the range of accuracies commonly needed, e.g., two to three significant figures.

Some Smith Charts are *normalized*, that is, the center has the coordinate $1 + j0$. To convert an impedance to a normalized impedance, it must be divided by the characteristic impedance of the system. Thus, 90 + j 160 ohms normalized to a 50-ohm system would be determined by dividing each of the coordinates by 50, to obtain the normalized impedance of 1.8 + j 3.2.

There are variations of the normalized Smith Chart with a center reference of 75 ohms or an admittance of 20 millimho, depending upon application. Charts also exist in an expanded form for operation near the system characteristic impedance, and which allow accuracy up to four significant figures in that area.

Applications of the Smith Chart In the case of a lossless system, the VSWR is a constant over a particular part of the transmission line, assuming no stubs or other

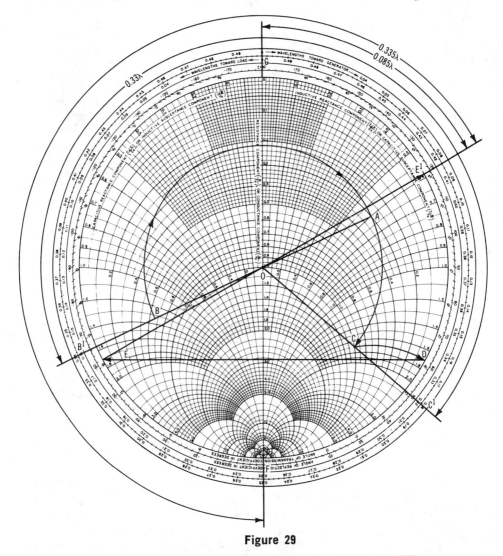

Figure 29

IMPEDANCE MATCHING WITH THE SMITH CHART

A 50-ohm transmission line is matched to a load impedance of 15 + j30 ohms (a normalized impedance of 0.3 − j0.6) by means of the chart.

discontinuities. All parameters in this part of the line are then equidistant from the center of the Smith Chart, that is, they lie on a circle whose radius is that of the normalized resistance component corresponding to the VSWR on the zero-reactance radius (figure 28).

One rotation of the Chart corresponds to a half-wavelength (180°) along the transmission line so that the impedance at all points along the line may be found directly once the impedance at a given point is determined. This may be determined from a maximum voltage measurement. At this point the real impedance is at a maximum and the point lies on the zero reactance line. Thus, to locate a transmission line point on the Smith Chart, two parameters are needed: the first to identify the radius of the operating locus (the VSWR) and the second to determine one position on that locus. At any other point on the line, its location on the Smith Chart, and thus its impedance, may be found by direct measurement on the chart, remembering that 360° around the Smith Chart corresponds to 180 electrical degrees, or a half-wavelength. It is important to note that the distance along the line must be in terms of electrical wavelengths, rather than inches or centimeters and must be increased by the square root of the dielectric constant of the insulation of the line to compensate for the velocity factor of the line.

In a practical case, it may be necessary to match a partly reactive antenna load to a transmission line. This may be achieved by means of a parallel-connected reactance at a point on the transmission line. Discrete components may be used, or open or shorted sections of transmission line may be employed to provide these reactances. The Smith Chart provides a convenient method of establishing all the parameters of such a match once the impedance at a particular point is known.

Consider the problem of providing a match on a transmission line where it is known that at point A the impedance is $15 + j30$ ohms (figure 29). The procedure is to establish another point on the line where a pure reactance may be placed in parallel, resulting in a nonreactive load having the required resistance, usually the characteristic impedance of the transmission line. For this example, a 50-ohm match for a 50-ohm transmission line is considered.

Step 1: Normalize the impedance, i.e. $0.3 + j0.6$. Plot this on a Smith Chart (point A in figure 29).

Step 2: Find the admittance at this point by constructing the diametrically opposite point (point B). This step is necessary since the solution will be two impedances in parallel. The real and imaginary parts of the admittance simply add up (see Chapter 3-2).

Step 3: Rotating toward the generator, point C is determined at which the real part of the admittance is 1.0. At this point, constructing to D, there is a positive susceptance of $j1.7$. The angle B' through G to C' locates the position of the matching reactance. The distance from the known source A to this point is 0.33 wavelength.

Step 4: If a pure susceptance opposite in sign is placed at this point, the total admittance will be unity, or one, the desired result. The value to be added is at point E. The reactance equivalent to this is diametrically opposite at point E' and is $+j0.58$, or reverting to ohmic expression, is an inductance having a value of $+j29$ ohms at the operating frequency.

Thus, a match to 50 ohms is achieved by placing an inductance of $+j29$ ohms at the point 0.33 wavelength back along the line towards the generator from point A.

This solution is not unique. There exists a second point on the chart where the real part of the admittance is unity and the whole solution repeats every half-wavelength.

The Smith Chart may be used to take this problem a little further. Open- and short-circuit lengths of transmission line are nearly purely reactive, the sign and value of the reactance being determined by line length. It is common practice to use such lengths, and the general name of *stubs* is given to them. In terms of impedance, the point G is a short circuit and the point F is an open circuit. (The conditions are reversed if admittances are considered).

Returning to the example:

Step 5: The point E' is the required stub impedance, and the distance E' toward load to G is the stub length for a short circuit stub, and is 0.085 wavelength. Similarly, an open circuit stub would have length E' toward load to F, or 0.335 wavelength. Any number of half-wavelengths may be added to these figures.

Thus, a short circuit stub 0.085 wavelength long placed at a point 0.33 wavelength back towards the generator from point A will be one of many configurations that will match the system to 50 ohms.

There are two important points to note: Matching is calculated at the frequency of interest and all lengths are in electrical terms and velocity factors must be considered.

The Coaxial Matching Stub   In the case of a coaxial line where it may be difficult to obtain specific stub positions, two stubs arranged ⅜, ¼, or ⅝ wavelengths apart may be used, their length being determined by a suitable manipulation of the Smith Chart. The explanation for this is that it is normally possible to cut the stub nearest the load so as to present an admittance at the second stub whose real part is equal to the characteristic admittance of the system. A match can now be obtained by the use of a susceptance of equal magnitude but opposite sign at the second stub. This is termed *double-stub tuning*. A requirement for the double-stub match is that the distance between the stubs not be a multiple of half-wavelengths.

For additional references on the Smith Chart and its uses, the following works are recommended: "Transmission Line Calculator," P. H. Smith, *Electronics*, January, 1944. "The Smith Chart," Hickson, *Wireless World*, January, 1960. "How to Use the Smith Chart," Fisk, *Ham Radio*, November, 1970.

34-6   Graphical Representation

Formulas and physical laws are often presented in graphical form; this gives us a "bird's eye view" of various possible conditions due to the variations of the quantities involved. In some cases graphs permit us to solve equations with greater ease than ordinary algebra.

Coordinate Systems   All of us have used coordinate systems without realizing it. For instance, in modern cities we have numbered streets and numbered avenues. By this means we can define the location of any spot in the city if the nearest street crossings are named. This is nothing but an application of *Cartesian* coordinates.

In the Cartesian coordinate system (named after Descartes), we define the location of any point in a plane by giving its distance from each of two perpendicular lines or *axes*. Figure 30 illustrates this idea. The vertical axis is called the *Y-axis*, the horizontal axis is the *X-axis*. The intersection of these two axes is called the *origin* (O). The location of a point (P) (figure 30) is defined by measuring the respective distances, x and y along the *X-axis* and the *Y-axis*. In this example the distance along the X-axis is 2 units and along the Y-axis is 3 units. Thus we define the point as P (2, 3) or we might say $x = 2$ and $y = 3$. The measurement x is called the *abscissa* of the point and the

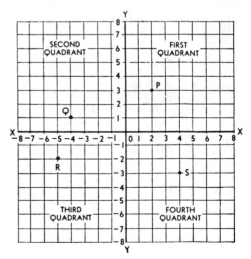

Figure 30

CARTESIAN COORDINATES

The location of any point can be defined by its distance from the X and Y axes.

distance y is called its *ordinate*. It is arbitrarily agreed that distances measured from 0 to the right along the X-axis shall be reckoned positive and to the left negative. Distances measured along the Y-axis are positive when measured upward from 0 and negative when measured downward from 0. This is illustrated in figure 30. The two axes divide the plane area into four parts called quadrants. These four quadrants are numbered as shown in the figure.

It follows from the foregoing statements, that points lying within the first quadrant have both x and y positive, as is the case with the point P. A point in the second quadrant has a negative abscissa, (x), and a positive ordinate, $(y.)$ This is illustrated by the point Q, which has the coordinates $x = -4$ and $y = +1$. Points in the third quadrant have both x and y negative. $x = -5$ and $y = -2$ illustrates such a point, (R). The point (S), in the fourth quadrant has a negative ordinate, (y) and a positive abscissa or x.

In practical applications we might draw only as much of this plane as needed to illustrate our equation and therefore, the scales along the X-axis and Y-axis might not start with zero and may show only that part of the scale which interests us.

Representation of Functions   In the equation:

$$f = \frac{300,000}{\lambda}$$

f is said to be a function of λ. For every value of λ there is a definite value of f. A variable is said to be a function of another variable when for every possible value of the latter, or *independent* variable, there is a definite value of the first or *dependent* variable. For instance, if $y = 5x^2$, y is *a* function of x and x is called the independent variable. When $a = 3b^3 + 5b^2 - 25b + 6$, then a is a function of b.

A function can be illustrated in our coordinate system as follows. Let us take the equation for frequency versus wavelength as an example. Given different values to the independent variable find the corresponding values of the dependent variable. Then plot the *points* represented by the different sets of two values.

kHz	λ_{meters}
600	500
800	375
1000	300
1200	250
1400	214
1600	187
1800	167
2000	150

Plotting these points in figure 31 and drawing a smooth curve through them gives us the *curve* or *graph* of the equation. This curve will help us find values of f for other

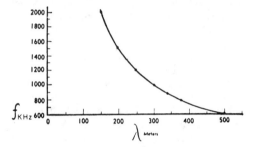

Figure 31

REPRESENTATION OF A SIMPLE FUNCTION IN CARTESIAN COORDINATES

In this chart of the function $f = \dfrac{300,000}{\lambda_{meters}}$ distances along the X axis renpresent wavelength in meters, while those along the Y axis represent frequency in kilohertz. A curve such as this helps to find values between those calculated with sufficient accuracy for most purposes.

values of λ (those in between the points calculated) and so a curve of an often-used equation may serve better than a table which always has gaps.

When using the coordinate system described so far and when measuring linearly along both axes, there are some definite rules regarding the kind of curve we get for any type of equation. In fact, an expert can draw the curve with but a very few plotted points since the equation has told him what kind of curve to expect.

First, when the equation can be reduced to form $y = mx + b$, where x and y are the variables, it is known as a *linear* or *first degree* function and the curve becomes a straight line. (Mathematicians still speak of a "curve" when it has become a straight line.)

When the equation is of the second degree, that is, when it contains terms like x^2 or y^2 or xy, the graph belongs to a group of curves, called *conic sections*. These include the circle, the ellipse, the parabola and the hyperbola. In the example given above, our equation is of the form

$$xy = c, \quad c \text{ being equal to } 300,000$$

which is a second degree equation and in this case, the graph is a hyperbola.

This type of curve does not lend itself readily for the purpose of calculation except near the middle, because at the ends a very large change in λ represents a small change in f and vice versa. Before discussing what can be done about this let us look at some other types of curves.

Suppose we have a resistance of 2 ohms and we plot the function represented by Ohm's law: $E = 2I$. Measuring E along the X-axis and amperes along the Y-axis, we plot the necessary points. Since this is a first degree equation, of the form $y = mx + b$ (for $E = y$, $m = 2$ and $I = x$ and $b = 0$) it will be a straight line so we need only two points to plot it.

(line passes through origin)

I	E
0	0
5	10

The line is shown in figure 32. It is seen to be a straight line passing through the origin. If the resistance were 4 ohms, we should get the equation $E = 4I$, and this also represents a line which we can plot in the same figure. As we see, this line also passes through the origin but has a different slope. In this

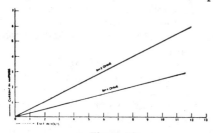

Figure 32

Only two points are needed to define functions which result in a straight line as shown in this diagram representing Ohm's law.

sents a line which we can plot in the same figure. As we see, this line also passes through the origin but has a different slope. In this

illustration the slope defines the resistance and we could make a protractor which would convert the angle into ohms. This fact may seem inconsequential now, but use of this is made in the drawing of loadlines on tube curves.

Figure 33 shows a typical, grid-voltage, plate-current static characteristic of a triode. The equation represented by this curve is rather complicated so that we prefer to deal with the curve. Note that this curve extends through the first and second quadrant.

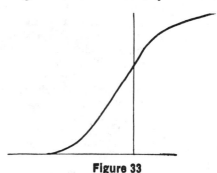

Figure 33

A TYPICAL GRID-VOLTAGE PLATE-CURRENT CHARACTERISTIC CURVE

The equation represented by such a curve is so complicated that we do not use it. Data for such a curve is obtained experimentally, and intermediate values can be found with sufficient accuracy from the curve.

Families of Curves It has been explained that curves in a plane can be made to illustrate the relation between *two* variables when one of them varies independently. However, what are we going to do when there are *three* variables and *two* of them vary independently. It is possible to use three dimensions and three axes but this is not conveniently done. Instead of this we may use a *family of curves*. We have already illustrated this partly with Ohm's law. If we wish to make a chart which will show the current through *any* resistance with *any* voltage applied across it, we must take the equation $E = IR$, having three variables.

We can now draw one line representing a resistance of 1 ohm, another line representing 2 ohms, another representing 3 ohms, etc., or as many as we wish and the size of our paper will allow. The whole set of lines is then applicable to any case of Ohm's law falling within the range of the chart. If any two of the three quantities are given, the third can be found.

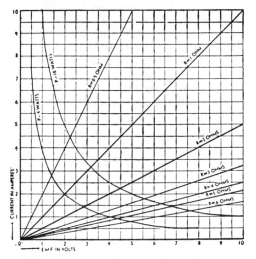

Figure 34

A FAMILY OF CURVES

An equation such as Ohm's law has three variables, but can be represented in Cartesian coordinates by a family of curves such as shown here. If any two quantities are given, the third can be found. Any point in the chart represents a definite value of E, I, and R, which will satisfy the equation of Ohm's law. Values of R not situated on an R line can be found by interpolation.

Figure 34 shows such a family of curves to solve Ohm's law. Any point in the chart resents a definite value each of *E, I,* and *R* which will satisfy the equation. The value of *R* represented by a point that is not situated on an *R* line can be found by interpolation.

It is even possible to draw on the same chart a second family of curves, representing a fourth variable. But this is not always possible, for among the four variables there should be no more than *two independent variables.* In our example such a set of lines could represent power in watts; we have drawn only two of these but there could of course be as many as desired. A single point in the plane now indicates the four values of *E, I, R,* and *P* which belong together and the knowledge of any two of them will give us the other two by reference to the chart.

Another example of a family of curves is the dynamic transfer characteristic or *plate family* of a tube. Such a chart consists of several curves showing the relation between

plate voltage, plate current, and grid bias of a tube. Since we have again three variables,

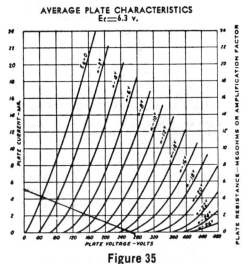

AVERAGE PLATE CHARACTERISTICS
$E_f = 6.3$ v.

Figure 35

"PLATE" CURVES FOR A TYPICAL VACUUM TUBE

In such curves we have three variables, plate voltage, plate current, and grid bias. Each point on a grid bias line corresponds to the plate voltage and plate current represented by its position with respect to the X and Y axes. Those for other values of grid bias may be found by interpolation. The loadline shown in the lower left portion of the chart is explained in the text.

we must show several curves, each curve for a fixed value of one of the variables. It is customary to plot plate voltage along the X-axis, plate current along the Y-axis, and to make different curves for various values of grid bias. Such a set of curves is illustrated in figure 35. Each point in the plane is defined by three values, which belong together, plate voltage, plate current, and grid voltage.

Now consider the diagram of a resistance-coupled amplifier in figure 36. Starting with the B-supply voltage, we know that whatever plate current flows must pass through the resistor and will conform to Ohm's law. The voltage drop across the resistor is subtracted from the plate supply voltage and the remainder is the actual voltage at the plate, the kind that is plotted along the X-axis in figure 35. We can now plot on the plate family of the tube the *loadline,* that is the line showing which part of the plate supply voltage is across the resistor and

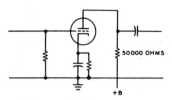

Figure 36

PARTIAL DIAGRAM OF A RESISTANCE-COUPLED AMPLIFIER

The portion of the supply voltage wasted across the 50,000-ohm resistor is represented in figure 35 as the loadline.

which part across the tube for any value of plate current. In our example, let us suppose the plate resistor is 50,000 ohms. Then, if the plate current were zero, the voltage drop across the resistor would be zero and the full plate supply voltage is across the tube. Our first point of the loadline is $E = 250$, $I = 0$. Next, suppose, the plate current were 1 mA, then the voltage drop across the resistor would be 50 volts, which would leave for the tube 200 volts. The second point of the loadline is then $E = 200$, $I = 1$. We can continue like this but it is unnecessary for we shall find that it is a straight line and two points are sufficient to determine it.

This loadline shows at a glance what happens when the grid-bias is changed. Although there are many possible combinations of plate voltage, plate current, and grid bias, we are now restricted to points along this line as long as the 50,000-ohm plate resistor is in use. This line therefore shows the voltage drop across the tube as well as the voltage drop across the load for every value of grid bias. Therefore, if we know how much the grid bias varies, we can calculate the amount of variation in the plate voltage and plate current, the amplification, the power output, and the distortion.

Logarithmic Scales   Sometimes it is convenient to measure along the axes the *logarithms* of our variable quantities. Instead of actually calculating the logarithm, special paper is available with logarithmic scales, that is, the distances measured along the axes are proportional to the logarithms of the numbers marked on

them rather than to the numbers themselves.

There is semilogarithmic paper, having logarithmic scales along one axis only, the other scale being linear. We also have full logarithmic paper where both axes carry logarithmic scales. Many curves are greatly simplified and some become straight lines when plotted on this paper.

As an example let us take the wavelength-frequency relation, charted before on straight cross-section paper.

$$f = \frac{300,000}{\lambda}$$

Taking logarithms:

$$\log f = \log 300,000 - \log \lambda$$

If we plot $\log f$ along the Y-axis and $\log \lambda$ along the X-axis, the curve becomes a straight line. Figure 37 illustrates this graph on full logarithmic paper. The graph may be read with the same accuracy at any point in contrast to the graph made with linear coordinates.

This last fact is a great advantage of logarithmic scales in general. It should be clear that if we have a linear scale with 100 small divisions numbered from 1 to 100, and if we are able to read to one tenth of a division, the possible error we can make near 100, way up the scale, is only 1/10th of a percent. But near the beginning of the scale, near 1, one tenth of a division amounts to 10 percent of 1 and we are making a 10 percent error.

In any logarithmic scale, our possible error in measurement or reading might be, say $\frac{1}{32}$ of an inch which represents a fixed amount of the log depending on the scale used. The net result of adding to the logarithm a fixed quantity, as 0.01, is that the antilogarithm is multiplied by 1.025, or the error is $2\frac{1}{2}\%$. No matter at what part of the scale the 0.01 is added, the error is always $2\frac{1}{2}\%$.

An example of the advantage due to the use of semilogarithmic paper is shown in figures 38 and 39. A resonance curve, when plotted on linear coordinate paper will look like the curve in figure 38. Here we have plotted the output of a receiver against frequency while the applied voltage is kept constant. The curve does not give enough information in this form for one might think that a signal 10 kHz off resonance would not cause any current at all and is tuned out. However, we frequently have off resonance signals which are 1000 times as strong as the desired signal and one cannot read on the graph of figure 38 how much any signal is attenuated if it is reduced more than about 20 times.

In comparison look at the curve of figure 39. Here the response (the current) is plotted in logarithmic proportion, which allows us to plot clearly how far off resonance a signal has to be to be reduced 100, 1000, or even 10,000 times.

Note that this curve is now "upside down"; it is therefore called a *selectivity* curve. The reason that it appears upside down is that

the method of measurement is different. In a selectivity curve we plot the increase in signal voltage necessary to cause a standard output off resonance. It is also possible to plot this increase along the Y-axis in decibels; the curve then looks the same although linear paper can be used because now our unit is logarithmic.

An example of full logarithmic paper being used for families of curves is shown in the reactance charts of Charts 1 and 2.

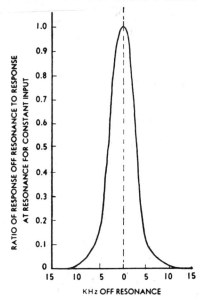

Figure 38

A RECEIVER RESONANCE CURVE

This curve represents the output of a receiver versus frequency when plotted to linear co-ordinates.

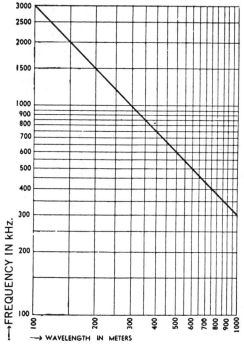

Figure 37

A LOGARITHMIC CURVE

Many functions become greatly simplified and some become straight lines when plotted to logarithmic scales such as shown in this diagram. Here the frequency versus wavelength curve of figure 33 has been replotted to conform with logarithmic axes. Note that it is only necessary to calculate two points in order to determine the "curve" since this type of function results in a straight line.

Nomograms or
Alignment Charts

An alignment chart consists of three or more sets of scales which have been so laid out that to solve the formula for which the chart was made, we have but to lay a straight edge along the two given values on any two of the scales, to find the third and unknown value on the third scale. In its simplest form, it is somewhat like the lines in figure 40. If the lines *a, b,* and *c* are parallel and equidistant, we know from ordinary geometry, that $b = \frac{1}{2}(a + c)$. Therefore, if we draw a scale of the same units on all three lines, starting

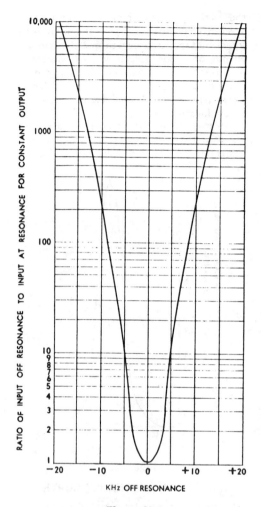

Figure 39

A RECEIVER SELECTIVITY CURVE

This curve represents the selectivity of a receiver plotted to logarithmic coordinates for the output, but linear coordinates for frequency. The reason that this curve appears inverted from that of figure 38 is explained in the text.

with zero at the bottom, we know that by laying a straightedge across the chart at any place, it will connect values of *a*, *b*, and *c*, which satisfy the above equation. When any two quantities are known, the third can be found.

If, in the same configuration we used logarithmic scales instead of linear scales, the relation of the quantities would become

$$\log b = \tfrac{1}{2}\,(\log a + \log c) \text{ or } b = \sqrt{ac}$$

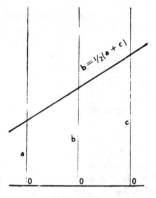

Figure 40

THE SIMPLEST FORM OF NOMOGRAM

By using different kinds of scales, different units, and different spacings between the scales, charts can be made to solve many kinds of equations.

If there are more than three variables it is generally necessary to make a double chart, that is, to make the result from the first chart serve as the given quantity of the second one. Such an example is the chart for the design of coils illustrated in Chart 3. This nomogram is used to convert the inductance in microhenrys to physical dimensions of the coil and vice versa. A pin and a straightedge are required. The method is shown under "R-F Tank Circuit Calculations" later in this chapter.

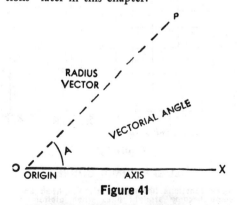

Figure 41

THE LOCATION OF A POINT BY POLAR COORDINATES

In the polar coordinate system any point is determined by its distance from the origin and the angle formed by a line drawn from it to the origin and the 0-X axis.

Chart 1. REACTANCE-FREQUENCY CHART FOR AUDIO FREQUENCIES

See text for applications and instructions for use

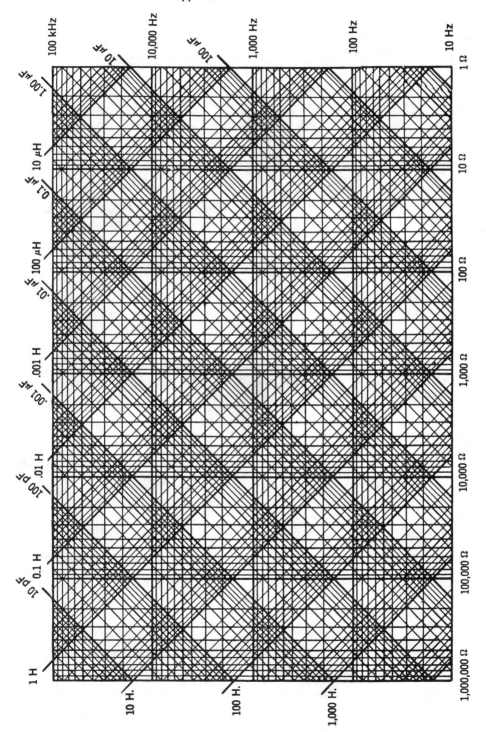

Polar Coordinates Instead of the Cartesian coordinate system there is also another system for defining algebraically the location of a point or line in a plane. In this, the polar coordinate system, a point is determined by its distance from the origin, O, and by the angle it makes with the axis O-X. In figure 41 the point P is defined by the length of OP, known as the radius vector and by the angle A the vectorial angle. We give these data in the following form

$$P = 3 \angle 60°$$

Polar coordinates are used in radio chiefly for the plotting of directional properties of microphones and antennas. A typical example of such a directional characteristic is shown in figure 42. The radiation of the antenna represented here is proportional to the distance of the characteristic from the origin for every possible direction.

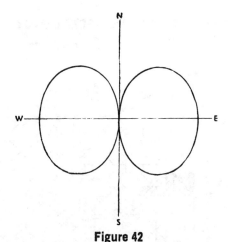

Figure 42

THE RADIATION CURVE OF AN ANTENNA

Polar coordinates are used principally in radio work for plotting the directional characteristics of an antenna where the radiation is represented by the distance of the curve from the origin for every possible direction.

Reactance In audio frequency calcula-
Calculations tions, an accuracy to better than a few percent is seldom required, and when dealing with calculations involving inductance, capacitance, resonant frequency, etc., it is much simpler to make use of reactance-frequency charts such as those in Charts 1 and 2 rather than to wrestle with a combination of unwieldy formulas. From these charts it is possible to determine the reactance of a capacitor or coil if the capacitance or inductance is known, and vice versa. It follows from this that resonance calculations can be made directly from the chart, because resonance simply means that the inductive and capacitive reactances are equal. The capacitance required to resonate with a given inductance, or the inductance required to resonate with a given capacitance, can be taken directly from the chart.

While the chart may look somewhat formidable to one not familiar with charts of this type, its application is really quite simple, and can be learned in a short while. The following example should clarify its interpretation.

For instance, following the lines to their intersection, we see that 0.1 H. and 0.1 μF intersect at approximately 1500 Hz and 1000

ohms. Thus, the reactance of either the coil or capacitor taken alone is about 1000 ohms, and the resonant frequency about 1500 Hz.

To find the reactance of 0.1 H. at, say, 10,000 Hz, simply follow the inductance line diagonally up toward the upper left till it intersects the horizontal 10,000-Hz line. Following vertically downward from the point of intersection, we see that the reactance at this frequency is about 6000 ohms.

To facilitate use of the chart and to avoid errors, simply keep the following in mind: The vertical lines indicate reactance in ohms, the horizontal lines always indicate the frequency, the diagonal lines sloping to the lower right represent inductance, and the diagonal lines sloping toward the lower left indicate capacitance. Also remember that the scale is *logarithmic*. For instance, the next horizontal line above 1000 Hz is *2000 Hz*. Note that there are 9, not 10, divisions between the heavy lines. This also should be kept in mind when interpolating between lines when best possible accuracy is desired; halfway between the line representing 200 Hz and the line representing 300 Hz is *not* 250 Hz, but approximately 230 Hz. The 250 Hz point is approximately 0.7 of the way between the 200-Hz line and the 300-Hz line rather than halfway between.

Chart 2. REACTANCE-FREQUENCY CHART FOR 100 kHz TO 100 MHz
This chart is used in conjunction with the nomograph (Chart 3) for radio frequency computations

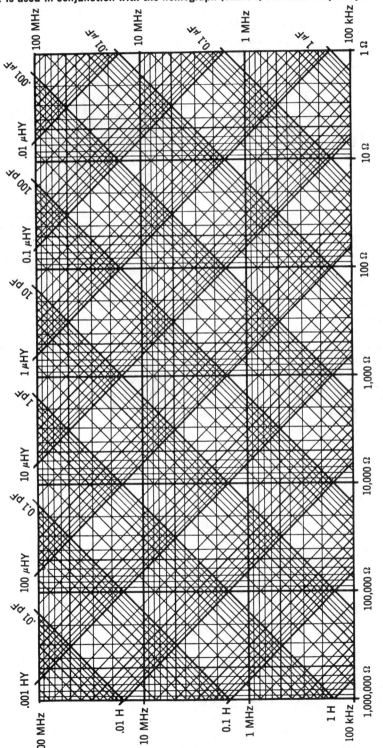

Use of the chart need not be limited by the physical boundaries of the chart. For instance, the 10-pF line can be extended to find where it intersects the 100-H. line, the resonant frequency being determined by projecting the intersection horizontally back on to the chart. To determine the reactance, the logarithmic ohms scale must be extended.

R-F Tank Circuit Calculations When winding coils for use in radio receivers and transmitters, it is desirable to be able to determine in advance the full coil specifications for a given frequency. Likewise, it often is desired to determine how much capacity is required to resonate a given coil so that a suitable capacitor can be used.

Fortunately, extreme accuracy is not required, except where fixed capacitors are used across the tank coil with no provision for trimming the tank to resonance. Thus, even though it may be necessary to estimate the stray circuit capacity present in shunt with the tank capacity, and to take for granted the likelihood of a small error when using a chart instead of the formula upon which the chart was based, the results will be sufficiently accurate in most cases, and in any case give a reasonably close point from which to start "pruning."

The inductance required to resonate with a certain capacitance is given in Chart 2. By means of the r-f chart, the inductance of the coil can be determined, or the capacitance determined if the inductance is known. When making calculations, be sure to allow for stray circuit capacitance, such as tube interelectrode capacitance, wiring, sockets, etc. This will normally run from 5 to 25 picofarads depending on the components and circuit.

To convert the inductance in microhenrys to physical dimensions of the coil, or vice versa, the nomograph in Chart 3 is used. A pin and a straightedge are required. The inductance of a coil is found as follows:

The straightedge is placed from the correct point on the turns column to the correct point on the diameter-to-length ratio column, the latter simply being the diameter divided by the length. Place the pin at the point on the plot axis column where the straightedge crosses it. From this point lay the straightedge to the correct point on the diameter column. The point where the straightedge intersects the inductance column will give the inductance of the coil.

From the chart, we see that a 30-turn coil having a diameter-to-length ratio of 0.7 and a diameter of 1 inch has an inductance of approximately 12 microhenrys. Likewise any one of the four factors may be determined if the other three are known. For instance, to determine the number of turns when the desired inductance, the D/L ratio, and the diameter are known, simply work backward from the example given. In all cases, remember that the straightedge reads either turns and D/L ratio, *or* it reads inductance and diameter. It can read no other combination.

The actual wire size has negligible effect on the calculations for commonly used wire sizes (no. 10 to no. 30). The number of turns of insulated wire that can be wound per inch (solid) will be found in a copper wire table.

34-7   Calculus

The branch of mathematics dealing with the instantaneous *rate of change* of a variable is called *calculus*. This differs from other branches of mathematics which deal with finding fixed or constant quantities when a given value changes.

As an example, using the formula,

$$i = \frac{E}{R} \cdot \varepsilon^{-t/RC}$$

the current at any given instant (i) can be found by the use of algebra. Calculus allows the solution of the problem so that the rate at which the current changes at any given instant may be found. The rate, in this instance, is a variable quantity. A variable is a quantity to which an unlimited number of values can be assigned, such as i, which varies with time. The variable may be restricted to values falling between *limits*, or it may be unrestricted. It is *continuous* if it has no breaks or interruptions over the limits of investigation. A variable whose value is determined by the first variable is called a *function* of the first variable. Thus,

Chart 3. COIL CALCULATOR NOMOGRAPH

For single layer solenoid coils, any wire size. See text for instructions.

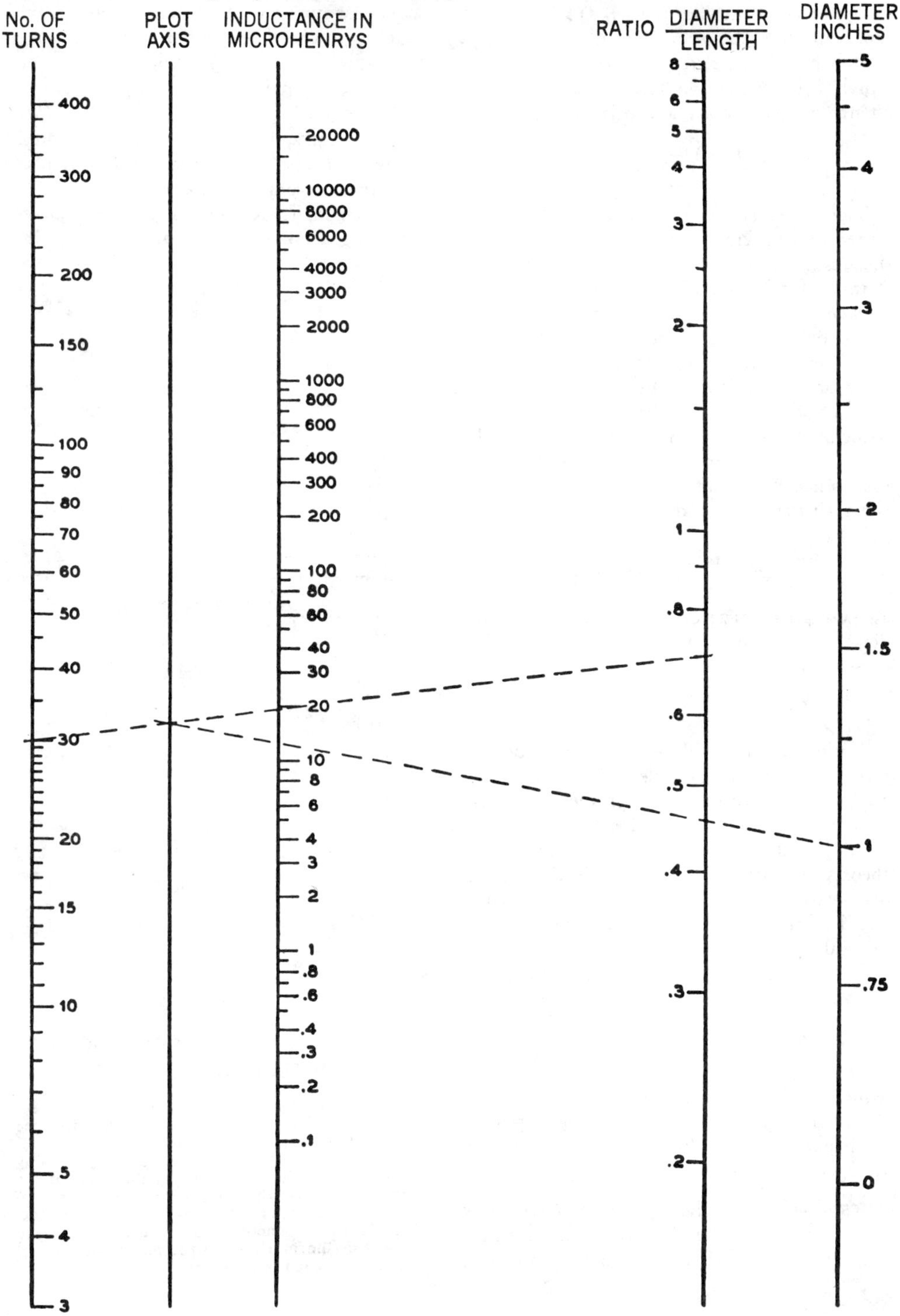

the symbol $f(x)$ indicates a function of x, so that $y = f(x)$.

If, for example, $f(a)$ and $f(b)$ are two values of the function, f, then $f(b) - f(a)$ represents the change in f brought about by the change from a to b in the number at which f is evaluated. The average rate of change of f between a and b is:

$$\frac{f(b) - f(a)}{b - a}$$

Such an equation may be graphed as discussed in a previous section and the slope of the resulting curve indicates the rate of change of the variable. The change, or *increment*, of the variable is the difference found by subtracting one value of the variable from the next, as shown above. The increment of x is denoted by Δx. In the equation $y = f(x)$, as x changes, so does the value of y. And as the increment Δx is made smaller, Δy also diminishes and the limiting case, when Δx is sufficiently small, is termed the *derivative* of x and is symbolized by:

$$\frac{dy}{dx} = \frac{\text{limit}}{\Delta x \to 0} \frac{\Delta y}{\Delta x} = f(x)$$

as Δx approaches zero. The symbol dy/dx indicates the limiting value of a fraction expressed by

$$\frac{\Delta y}{\Delta x}.$$

Various rules of differentiation may be derived from the general rule and most of these apply directly to electrical problems dealing with the rate of change of a variable, such as a capacitor discharge, transmission-line theory, etc. Maximum and minimum values of a variable can be determined by setting the derivative equal to zero and solving the general equation.

It is convenient to use differential expressions such as:

$$dy = f(x)\, dx = \frac{dy}{dx} \cdot dx$$

which states that the differential (dy) of a function (x) equals its derivative (dy/dx) multiplied by the differential (dx) of the independent variable.

Integral Calculus The derivative process may be inversed to find a function when the derivative is known, and this

is termed *integration*, or integral calculus. Integral calculus is helpful in electronic problems, especially those dealing with sine waves, or portions of waves of voltage, current or power. The symbol of integration is the capital script s: $\int$ and,

$$\int f'(x)\, dx = f(x)$$

Graphically, integration may be thought of as a process of summation and is often used in electronics in this fashion.

Shown in figure 43 is an area $ABCD$ representing a portion of a current wave. It is

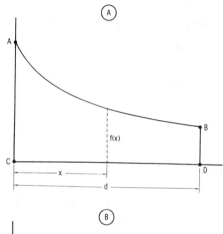

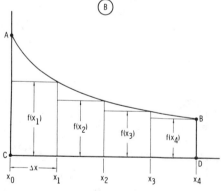

Figure 43

GRAPHICAL REPRESENTATION OF INTEGRATION PROCESS

A—Area ABCD represents a portion of a current wave. The area is a function of x, the distance along the x-axis from the point of origin.
B—The area ABCD is approximated by drawing rectangles in it and adding up the areas of the rectangles. As the number of rectangles increases, the approximation approaches the actual area. The actual area is the limit of the sums of the individual areas as the number of areas approach infinity.

required to find the value of current, which is represented by the area under the wave AB divided by the period CD. The area $ABCD$ may be approximated by drawing rectangles in it and adding up the areas of the rectangles. The height of each rectangle is defined by the function $f(x)$, which is the height of the curve x in units from the baseline. If, for example, four rectangles are drawn with equal bases, the sum of their areas is given by:

$$S_4 = f(x_1)\Delta x + f(x_2)\Delta x + f(x_3)\Delta x + f(x_4)\Delta x$$

In this equation, Δx equals $d/4$, the length of each base.

Now, if the baseline CD is divided into n equal parts by the points $x_0, x_1 \ldots . x_n$, and if n rectangles are drawn, the sum S_n of the areas of the n rectangles is given by:

$$S_n = f(x_1)\Delta x + f(x_2)\Delta x + \ldots . + f(x_n)\Delta x$$

And $\Delta x = \dfrac{d}{n}$, or the length of each base.

S_n thus is an approximation of the area $ABCD$ and as n gets larger, S_n becomes closer and closer to the actual area. The actual area (A) of region $ABCD$ is the limit of S_n as n approaches infinity:

$$A = \underset{n \to \infty}{\text{limit}}\ S_n$$

The limit of S_n as n approaches infinity is the *definite integral* of the function f from zero to d, written in the integral form as:

$$\int_0^d f(x)\, dx$$

Integration commonly takes place between limits to restrict the scope of the problem. A full study of calculus is beyond the scope of this chapter and for more information on this subject, the reader is referred to *Electronics Mathematics* (Volumes 1 and 2), Nunz and Shaw, McGraw-Hill Book Co., New York, NY.

34-8 Electronic Computers

Mechanical computing machines were first produced in the seventeenth century in Europe although the simple Chinese *abacus* (a digital computer) had been in use for centuries. Until the last decade only simple mechanical computers (such as adding and bookkeeping machines) were in general use.

The transformation and transmission of the volume of information required by modern technology requires that machines assume many of the information processing systems formerly done by the human mind. Computing machines can perform routine operations more quickly and more accurately than a human being, processing mathematical and logistical data on a production line basis. The computer, however, cannot create, but can only follow instructions. If the instructions are in error, the computer will produce a wrong answer.

Computers may be divided into two classes: the *digital* and the *analog*. The digital computer *counts*, and its accuracy is limited only by the number of significant figures provided for in the instrument. The analog computer *measures*, and its accuracy is limited by the percentage errors of the devices used, multiplied by the range of the variables they represent.

Digital Computers The *digital computer* operates in discrete steps. In general, the mathematical operations are performed by combinations of additions. Thus multiplication is performed by repeated additions, and integration is performed by summation. The digital computer may be thought of as an "on-off" device operating from signals that either exist or do not exist. The common adding machine is a simple computer of this type. The "on-off" or "yes-no" type of situation is well suited to switches, electrical relays, or to solid-state circuitry.

A simple electrical digital computer may be used to solve the old "farmer and river" problem. The farmer must transport a hen, a bushel of corn, and a fox across a river in a small boat capable of carrying the farmer plus one other article. If the farmer takes the fox in the boat with him, the hen will eat the corn. On the other hand, if he takes the corn, the fox will eat the hen. The circuit for a simple computer to solve this problem is shown in figure 44. When the switches are moved from "south shore" to "north shore" in the proper sequence the warning

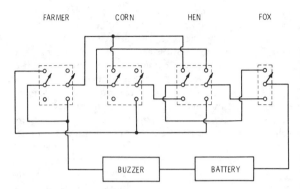

Figure 44

SIMPLE PUZZLES IN LOGIC MAY BE SOLVED BY ELECTRIC COMPUTER. THE "FARMER AND RIVER" COMPUTER IS SHOWN HERE.

buzzer will not sound. An error of choice will sound the buzzer.

A second simple "digital computer" is shown in figure 45. The problem is to find the three proper push buttons that will sound the buzzer. The nine buttons are mounted on a board so that the wiring cannot be seen.

Each switch of these simple computers executes an "on-off" action. When applied to a logical problem "yes-no" may be substituted for this term. The computer thus can act out a logical concept concerned with a simple choice. An electronic switch may be substituted for the mechanical switch to increase the speed of the computer. The early computers, such as the ENIAC (*Electronic Numerical Integrator and Calculator*) employed over 18,000 tubes for memory and registering circuits capable of "remembering" a 10-digit number.

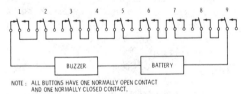

NOTE: ALL BUTTONS HAVE ONE NORMALLY OPEN CONTACT
AND ONE NORMALLY CLOSED CONTACT.

Figure 45

A SEQUENCE COMPUTER

Three correct buttons will sound the buzzer.

Binary Notation   To simplify and reduce the cost of the digital computer it was necessary to modify the system of operation so that fewer devices were used per bit of information. The *ENIAC-type* computer requires 50 tubes to register a 5-digit number. The readout devices can be arranged in five columns of 10 each. From right to left the columns represent units, tens, hundreds, thousands, etc. The bottom device in each column represents "zero," the second represents "one," the third "two," and so on. Only one device in each column is excited at any given instant. If the number 73092 is to be displayed, number seven in the fifth column is excited, number three in the fourth column, number zero in the third column, etc. as shown in figure 46.

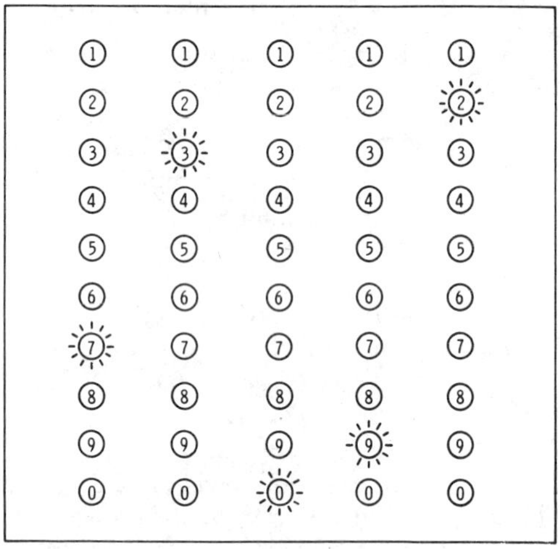

A simpler system employs the *binary decimal* notation, wherein any number from one to fifteen can be represented by four devices. Each of the four positions has a numerical

Figure 46

BINARY NOTATION MAY BE USED FOR DIGITAL DISPLAY. BINARY BOARD ABOVE INDICATES "73092."

value that is associated with its place in the group. More than one of the group may be excited at once, as illustrated in figure 47. The values assigned to the positions in this particular group are 1, 2, 4, and 8. Additional devices may be added to the group, doubling the notation thus: 1, 2, 4, 8, 16, 32, 64, 128, 256, etc. Any numerical value lower than the highest group number can be displayed by the correct device combination.

DECIMAL NOTATION	BINARY NOTATION
0	0
1	1
2	1, 0
3	1, 1
4	1, 0, 0
5	1, 0, 1
6	1, 1, 0
7	1, 1, 1
8	1, 0, 0, 0
9	1, 0, 0, 1
10	1, 0, 1, 0

Figure 48

BINARY NOTATION SYSTEM REQUIRES ONLY TWO NUMBERS, "0" AND "1."

DIGIT	TUBE(S)
1	1
2	2
3	2 + 1
4	4
5	4 + 1
6	4 + 2
7	4 + 2 + 1
8	8
9	8 + 1
10	8 + 2
11	8 + 2 + 1
12	8 + 4
13	8 + 4 + 1
14	8 + 4 + 2
15	8 + 4 + 2 + 1

Figure 47

BINARY DECIMAL NOTATION. ONLY FOUR TUBES ARE REQUIRED TO REPRESENT DIGITS FROM 1 TO 15. THE DIGIT "12" IS INDICATED ABOVE.

A third system employs the *binary notation* which makes use of a *bit* (binary digit) representing a single morsel of information. The binary system has been known for over forty centuries, and was considered a mystical revelation for ages since it employed only two symbols for all numbers. Computer service usually employs "zero" and "one" as these symbols. Decimal notation and binary notation for common numbers are shown in figure 48. The binary notation represents 4-digit numbers (thousands) with ten bits, and 7-digit numbers (millions) with 20 bits. Only one device is required to display an information bit. The savings in components and primary power drain of a binary-type computer over the older ENIAC-type computer is obvious. Figure 49 illustrates a com-

puter board showing the binary indications from one to ten.

More recent technology has permitted more convenient readouts, the well-known seven segment and Nixie devices, for example. However, the theory is the same and the binary coded decimal format is extensively used.

Digital Computer Use Early digital computers were operated by keyboards, punched tapes or punched cards by means of direct instruction in terms of additions, subtractions, core locations and printing or display instructions. In all, a limited number of instructions were available to the user, and all other operators such as squares, cubes, trigonometrical ratios, etc., had to be programmed out in full, in terms of the available operators. The operation of these early computers was a difficult and tedious process, made even more difficult since the user had to know exactly how the computer operated. Early machines did not have indicators to tell the user when one part of the machine was full, and errors due to overflow were quite common.

As technology improved and computer users became more demanding, machine codes and then computer languages were developed. Effort was directed towards enabling the computer to be operated directly in terms of the English language. The early computer languages, such as ALGOL and FORTRAN, were developed as direct coding to comput-

DECIMAL NOTATION	COMPUTER NOTATION
0	● ● ● ●
1	● ● ● ○
2	● ● ○ ●
3	● ● ○ ○
4	● ○ ● ●
5	● ○ ● ○
6	● ○ ○ ●
7	● ○ ○ ○
8	○ ● ● ●
9	○ ● ● ○
10	○ ● ○ ●

● = OFF   ○ = ON

Figure 49

BINARY NOTATION AS REPRESENTED ON COMPUTER BOARD FOR NUMBERS FROM 1 TO 10.

ers in general. Standardization between manufacturers meant that a program developed for one computer could be performed on another computer, provided that the second computer had a sufficiently large vocabulary. As the languages developed, often for specific purposes, the programming became easier and the computer more user-oriented.

The *interface* (readin, readout) between the computer and the user became more convenient as the older paper tapes and punched cards gave way to the many types of computer terminals that we have today. Each is developed for its particular application so that the result is in the form most convenient to the user.

These terminals are not always paper or other visual displays of numbers and letters as they were in virtually all the early computers. Manufacturing machines, telephone exchanges and a whole host of control applications are operated directly by computers at very high speeds. The situation has now developed that a user can order a machine to perform almost any operation requiring logic and it will be available. Computer technology is evident in many areas of everyday life. The small hand-held calculators mentioned earlier are true computers, and some of the programmable, handheld and desktop models available now have the computing power of the multimillion dollar "monsters" of fifteen years ago.

The reasons for the use of computers are as diverse as their application. From the calculation point of view, the time that may be saved by the use of a computer is very great. On the other hand, computers are now used to perform operations in production line applications, for example, that would not otherwise be possible in the time period available.

Computer Inputs As time progresses, large computers are becoming readily accessible and many amateurs may obtain access to at least a *time share* terminal if they so desire. The use of multioperator time-share systems on low priority means that even the largest and most powerful computers are available at very low cost.

It is not possible to detail a single set of operating instructions that is suitable for use with all computers, as most manufacturers adopt only a limited range of computer languages such as BASIC, FORTRAN, and ALGOL, and usually only one operator access language such as CALL 360 or CANDE. Thus, it will be necessary for the prospective computer operator to consult the literature provided by the appropriate computer manufacturer and learn the language required. Some very powerful computer packages are available, providing curve plotting from limited results and actual circuit design and system optimization, as examples.

Figures 50 and 51 provide a simple example of a computation performed on a large computer via a time-share terminal. Figure 50 lists a program in FORTRAN IV which enables a printout of transmit- and receive-crystal frequencies for use in a communication system. The program operates interactively with the operator to enter the information required for computation so that variations may be made to the intermediate frequency of the system or carrier range without having to change the main program. A brief description of the operation of the program is shown on the right of the printout of figure 50.

The Microprocessor Many of the recent developments that have been put to use in radio communication equipment involve the use of digital control, as for example in frequency synthesi-

```
LIST XTAL

10 C*"PROGRAM   TO FIND CRYSTAL FREQUENCIES"
20 PRINT 10
30 10 FORMAT (2X,"ENTER I.F. FREQUENCY")
40 READ/,F
50 PRINT 20
60 20 FORMAT (2X,"ENTER TX/RX MULTIPLIERS")
70 READ/,A,B
80 PRINT 30
90 30 FORMAT (2X,"ENTER START/STEP/FINISH FREQUENCIES")
100 READ/,C,D,E
110 PRINT 40
120 40 FORMAT (8X,"FREQUENCY",6X,"TX XTAL",6X,"RX XTAL")
130 DO 50 P=C,E,D
140 TX=P/A
150 RX=(P-F)/B
160 50 WRITE(6,60)P,TX,RX
170 60 FORMAT (2X,3F14.6)
180 CONTINUE
190 END
*
```

Interactive entry of information on equipment and frequencies.

Sets up heading for output.

Performs calculations.

Lists results.

Figure 50

FORTRAN IV PROGRAM FOR CRYSTAL FREQUENCIES

This program provides printout of transmit and receive crystal frequencies for use in a communication system. A description of the program is shown at the right.

zation (Chapter 12). Any control information which can be put in logical form may be used to control a "logical" item of electronic equipment, such as a transceiver, even though that input control information is in an unsuitable form. A logic transformation system is required.

It would be possible to build a logic translation unit to record the input information from, say, a keyboard to display the output on a digital readout and to provide the required control signals to a transceiver. Units of this type have been developed and a general class of small digital computers, called *microprocessors* can perform the task. Microprocessors are very flexible and can be programmed to perform calculations within themselves or to be interfaced with calculator modules similar to those used in hand-held calculators. The imagination of the user would appear to be the limit.

Analog Computers In the period before the miniaturization of the digital computer, there was a need for a computer system of limited precision and low cost. Before the extensive use of electronics, mechanical computers were in use which made use of differential gears to add and subtract, and discs, spirals, and cams to perform multiplication, integration, and

function generation. These machines ranged from the simple slide rule through the complex World War II bomb-sight computers to even more complex laboratory machines. They are not to be confused with the office-type mechanical calculator, which is a true digital machine.

These "computing engines" or *analog computers* relied for their accuracy on their own internal mechanical precision, and on the ability of the operator to read the output results from a scale. In general, the larger and more precise the scale, the more accurate the results.

Mechanical analog computers have largely given away to electronic machines where operational amplifiers and associated components enable mathematical operations to be performed and diode matrices serve as function generators. The output of the device is read on a chart or meter.

A real system that has been translated into mathematical functions can thus be observed under laboratory conditions without the need of the real system to actually exist. Since the system is in a mathematical form, it is possible to change time scales in order to observe the behavior of a system in slow motion. However, as digital computers improve and become more readily available at low cost, the analog computer finds less application and may eventually be of historical interest only.

```
#RUNNING 5208
 ENTER I.F. FREQUENCY

#?
6
 ENTER TX/RX MULTIPLIERS
12,1
 ENTER START/STEP/FINISH FREQUENCIES
52,.1,53
         FREQUENCY        TX XTAL        RX XTAL
         52.000000        4.333333       46.000000
         52.100000        4.341667       46.100000
         52.200000        4.350000       46.200000
         52.300000        4.358333       46.300000
         52.400000        4.366667       46.400000
         52.500000        4.375000       46.500000
         52.600000        4.383333       46.600000
         52.700000        4.391667       46.700000
         52.800000        4.400000       46.800000
         52.900000        4.408333       46.900000
         53.000000        4.416667       47.000000

#ET=1:07.8 PT=0.3 ID=0.3


#RUNNING 0548
 ENTER I.F. FREQUENCY

#?
10.7
 ENTER TX/RX MULTIPLIERS
18,3
 ENTER START/STEP/FINISH FREQUENCIES
146,.05,147
         FREQUENCY        TX XTAL        RX XTAL
         146.000000       8.111111       45.100000
         146.050000       8.113889       45.116667
         146.100000       8.116667       45.133333
         146.150000       8.119444       45.150000
         146.200000       8.122222       45.166667
         146.250000       8.125000       45.183333
         146.300000       8.127778       45.200000
         146.350000       8.130556       45.216667
         146.400000       8.133333       45.233333
         146.450000       8.136111       45.250000
         146.500000       8.138889       45.266667
         146.550000       8.141667       45.283333
         146.600000       8.144444       45.300000
         146.650000       8.147222       45.316667
         146.700000       8.150000       45.333333
         146.750000       8.152778       45.350000
         146.800000       8.155556       45.366667
         146.850000       8.158333       45.383333
         146.900000       8.161111       45.400000
         146.950000       8.163889       45.416667
         147.000000       8.166667       45.433333

#ET=1:13.8 PT=0.2 ID=0.2
```

Figure 51

PROGRAM PRINTOUT

A—Printout of crystal frequencies for a communication system having a receive first intermediate frequency of 6 MHz, a transmitter multiplier of 12, and a fundamental frequency receiver crystal for input frequencies from 52 to 53 MHz in steps of 0.1 MHz.
B—Printout of crystal frequencies for a communication system having a receive first intermediate frequency of 10.7 MHz, a transmitter multiplier of 18 and a receiver multiplier of 3, for input frequencies from 146 to 147 MHz in steps of 0.05 MHz.

CHAPTER THIRTY-FIVE

Nomenclature of Components and Miscellaneous Data

35-1 Component Standardization

Standardization of electronic components or parts is handled by several cooperating agencies, among whom are the *Electronic Industries Association* (EIA), the *USA Standards Institute*, the *Joint Electron Device Engineering Council* (JEDEC) and the *National Electrical Manufacturers Associa-*tion (NEMA). International standardization is carried out through the various technical committees of the *International Electrotechnical Commission*. Additional standardization is covered by the *International Standards Organization*. Military standards (MIL) are issued by the *US Department of Defense* or one of its agencies. Standard outlines, systems of nomenclature and coding and technical characteristics of components are a few of the items standardized in electronic equipment.

Table 1. Standard Color Code of Electronics Industry

Color	Significant Figure	Decimal Multiplier	Tolerance in Percent*	Voltage Rating	Characteristic
Black	0	1	±20 (M)	—	A
Brown	1	10	±1 (F)	100	B
Red	2	100	±2 (G)	200	C
Orange	3	1 000	±3	300	D
Yellow	4	10 000	GMV‡	400	E
Green	5	100 000	±5(J)†, (0.5)§	500	F
Blue	6	1 000 000	±6, (0.25)§	600	G
Violet	7	10 000 000	±12.5, (0.10)§	700	—
Gray	8	0.01†	±30, (0.05)§	800	I
White	9	0.1†	±10†	900	J
Gold	—	0.1	±5 (J), (0.5)‖	1 000	—
Silver	—	0.01	±10 (K)	2 000	—
No color	—	—	±20	500	—

* Tolerance letter symbol as used in type designations has tolerance meaning as shown. ±3, ±6, ±12.5, and ±30 percent are tolerances for USA Std 40-, 20-, 10-, and 5-step series, respectively.
† Optional coding where metallic pigments are undesirable.
‡ GMV is −0 to +100-percent tolerance or Guaranteed Minimum Value.
§ For some film and other resistors only.
‖ For some capacitors only.

Table 2. Preferred Values*

	USA Standard Z17.1†		USA Standard C83.2‡		
Name of Series	"5"	"10"	±20%(E6)	±10%(E12)	±5%(E24)
Percent step size	60	25	≈40	20	10
Step multiplier	$(10)^{1/5}=1.58$	$(10)^{1/10}=1.26$	$(10)^{1/6}=1.46$	$(10)^{1/12}=1.21$	$(10)^{1/24}=1.10$
Values in the series (Use decimal multipliers for smaller or larger values)	10	10	10	10	10
	—	12.5 ⎫	—	—	11
	—	(12) ⎭	—	12	12
	—	—	—	—	13
	—	—	15	15	15
	16	16	—	—	16
	—	—	—	18	18
	—	20	—	—	20
	—	—	22	22	22
	—	—	—	—	24
	25	25	—	—	—
	—	—	—	27	27
	—	31.5 ⎫	—	—	30
	—	(32) ⎭	—	—	—
	—	—	33	33	33
	—	—	—	—	36
	—	—	—	39	39
	40	40	—	—	—
	—	—	—	—	43
	—	—	47	47	47
	—	50	—	—	—
	—	—	—	—	51
	—	—	—	56	56
	—	—	—	—	62
	63	63	—	—	—
	—	—	68	68	68
	—	—	—	—	75
	—	80	—	—	—
	—	—	—	82	82
	—	—	—	—	91
	100	100	100	100	100

* USA Standard C83.2 applies to most electronics components; it was formerly EIA GEN 102 and is similar to IEC Publication 63. USA Standard Z17.1 covers preferred numbers and is similar to ISO R3, R17.

† "20" series with 12-percent steps ($(10)^{1/20}=1.22$ multiplier) and a "40" series with 6-percent steps ($(10)^{1/40}=1.059$ multiplier) are also standard.

‡ Associate the tolerance ±20%, ±10%, or ±5% only with the values listed in the corresponding column. Thus, 1200 ohms may be either ±10 or ±5, but not ±20 percent; 750 ohms may be ±5, but neither ±20 nor ±10 percent.

The Color Code In general, the color code of Table 1 is used for marking equipment. The *tolerance* specification is the maximum deviation allowed from the specified nominal value of the component, though for very small values of capacitance the tolerance may be specified in picofarads (pF). Where no tolerance is specified, components are likely to vary ± 20 percen from the nominal value.

Preferred Values To maintain an orderly progression of sizes, preferred numbers are frequently used for the nominal values. Each preferred value differs from its predecessor by a constant multiplier, and the final result is rounded to two significant figures. The USA standard of preferred numbers, widely used for fixed resistors, capacitors, and time-delay relays is listed in Table 2.

Distinction must be made between the *breakdown voltage rating* (test volts) and the *working voltage rating*. The maximum continuous voltage determines the working voltage rating. Application of the test voltage for more than a few seconds may result in permanent damage or failure of the component.

The *characteristic* term of the specification is frequently used to include various qualities of a component, such as temperature coefficient, Q value, maximum operating temperature, etc. One or two letters are assigned in the EIA or MIL type designations and the characteristic may be indicated by color coding on the part.

bers, band J is black. Band N is used to designate the suffix letter as shown in Table 3. This band may be omitted in 2- or 3-digit coding if not required. A single band indicates the cathode end of a diode.

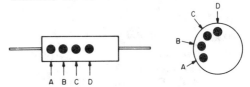

Figure 2

ALTERNATIVE METHODS OF COMPONENT VALUE CODING
Table 3.

Color	Suffix Letter	Number
Black	—	0
Brown	A	1
Red	B	2
Orange	C	3
Yellow	D	4
Green	E	5
Blue	F	6
Violet	G	7
Gray	H	8
White	J	9

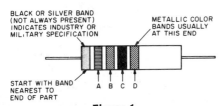

Figure 1

COMPONENT VALUE CODING

The code of Table 1 determines values. Band A color = First significant figure of value in ohms, picofarads, or microhenries. Band B color = Second significant figure of value. Band C color = Decimal multiplier for significant figures. Band D color = Tolerance in % (if omitted, the broadest tolerance series of the part applies).

Component Value Coding Axial lead and some other components are often color coded by circumferential bands to indicate value and tolerance. Usually the value may be decoded as indicated in Table 1 and figure 1. Sometimes instead of circumferential bands, colored dots are used as shown in figure 2.

Semiconductor diodes have a color code system as shown in figure 3. The sequential number portion (following the "1N" of the assigned industry type number) may be indicated by the color bands. Colors have the numerical significance given in Table 1. Bands J, K, L, and M represent the digits in the sequential number. For 2-digit num-

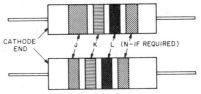

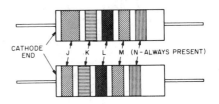

Figure 3

SEMICONDUCTOR-DIODE VALUE CODING

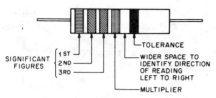

Figure 4

RESISTOR VALUE COLOR CODE FOR 3 SIGNIFICANT FIGURES. COLORS OF TABLE 1 DETERMINE VALUES

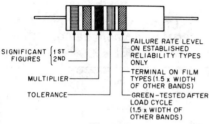

Figure 5

RESISTOR COLOR CODE PER MIL-STD-221D.

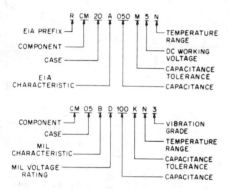

Figure 6

TYPE DESIGNATION FOR MICA-DIELECTRIC CAPACITORS. EIA AT TOP; MIL AT BOTTOM

EIA standard and *MIL specification* requirements for color coding of composition resistors are identical (see figure 1). Colors have the significance shown in Table 1 and figure 4 shows the *EIA standard resistor markings.* The *MIL-standard resistor* markings are shown in figure 5. Small *wirewound resistors* in $\frac{1}{2}$-, 1- or 2-watt ratings may be color coded as described, but band A will be twice the width of the other bands.

A comprehensive numbering system, the type designation, is used to identify *mica capacitors.* Type designations are of the form shown in figure 6. Fixed mica dielectric capacitors are identified by the symbol *CM.* For EIA, a prefix letter *R* is always included. The case designation is a two-symbol digit that identifies a particular size and shape of case. The MIL or EIA characteristic is indicated by a single letter in accordance with Table 4.

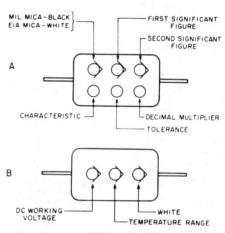

Figure 7

STANDARD CODE FOR FIXED MICA CAPACITORS

See color code in Table 1. A is the basic 6-dot form. The 9-dot form with B on the other side of the capacitor is used if the additional data are required.

Type	Top Row			Bottom Row			Description
	Left	Center	Right	Left	Tolerance Center	Multiplier Right	
RCM20A221M	white	red	red	black	black	brown	220 pF ±20% EIA class A.
CM30C681J	black	blue	gray	red	gold	brown	680 pF ±5%, MIL characteristic C.

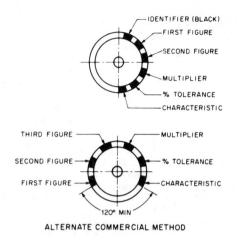

NOMINAL CAPACITY-μF				RATED DC VOLTAGE
COLOR	FIRST DIGIT	SECOND DIGIT	DOT	
BLACK	-	0	×1	10
BROWN	1	1	×10	-
RED	2	2	×100	-
ORANGE	3	3	-	-
YELLOW	4	4	-	6.3
GREEN	5	5	-	16
BLUE	6	6	-	20
VIOLET	7	7	-	-
GRAY	8	8	×0.01	25
WHITE	9	9	×0.1	3
PINK	-	-	-	35

1ST DIGIT — — MULTIPLIER DOT
2ND DIGIT
VOLTAGE

Figure 8
COLOR CODING OF BUTTON MICA CAPACITORS

See Table 1 for color code. Commercial color code for characteristic not standardized; varies with manufacturer.

The significance of the various colored dots for EIA and MIL specification mica capacitors is shown in figure 7, with the colors having the meaning as explained in Table 1. Examples of EIA and MIL type designations are shown below.

Button mica capacitors are color coded in various ways, of which the two most widely used methods are shown in figure 8.

Tantalum Capacitors   The small "tear-drop" tantalum capacitors are compact units widely used in modern solid-state equipment. They are available in a

Figure 9
Standard Code for Tantalum Capacitors

range of dc voltage from 3 to 35 volts and in a capacitance range of 0.1 μF to 100 μF. The voltage-capacity product of representative units is about 300. That is, the larger capacitances are only available in lower voltage ratings. The capacitors are polarized and the sum of the dc and peak ac voltages should not exceed the rated voltage, nor should reverse voltage in excess of 0.3 volts be applied to a capacitor. Nominal capacitance tolerance is ±20 percent and maximum operating temperature is 85°C (185°F). The color code used for these units is shown in figure 9.

Film and Mica Capacitors   Film and mica capacitors are manufactured with a hard, dipped coating and are available in various voltage and capacitance ranges. Some types have short, crimped leads

First digit of capacitor's value:
Second digit of capacitor's value:
Multiplier: Multiply the first & second digits by the proper value from the Multiplier Chart.
To find the tolerance of the capacitor, look up this letter in the Tolerance columns.

1 5 1 K

MULTIPLIER		TOLERANCE OF CAPACITOR		
FOR THE NUMBER:	MULTIPLY BY:	10pF OR LESS	LETTER	OVER 10pF
0	1	±0.1pF	B	
1	10	±0.25pF	C	
2	100	±0.5pF	D	
3	1000	±1.0pF	F	±1%
4	10,000	±2.0pF	G	±2%
5	100,000		H	±3%
			J	±5%
8	0.01		K	±10%
9	0.1		M	±20%

Figure 10
CODE FOR FILM AND MICA CAPACITORS

RADIO HANDBOOK

Table 4. Fixed-Mica-Capacitor Requirements by MIL Characteristic and EIA Class. *

MIL Characteristic or EIA Class	MIL-Specification Requirements†			EIA-Standard Requirements		
	Maximum Capacitance Drift	Maximum Range of Temperature Coefficient (ppm/°C)		Maximum Capacitance Drift	Maximum Range of Temperature Coefficient (ppm/°C)	Minimum Insulation Resistance (megohms)
A	—	—		$\pm(5\%+1\text{ pF})$	±1000	3000
B	—	—		$\pm(3\%+1\text{ pF})$	±500	6000
C	$\pm(0.5\%+0.1\text{ pF})$	±200		$\pm(0.5\%+0.5\text{ pF})$	±200	6000
I	—	—		$\pm(0.3\%+0.2\text{ pF})$	-50 to $+150$	6000
D	$\pm(0.3\%+0.1\text{ pF})$	±100		$\pm(0.3\%+0.1\text{ pF})$	±100	6000
J	—	—		$\pm(0.2\%+0.2\text{ pF})$	-50 to $+100$	6000
E	$\pm(0.1\%+0.1\text{ pF})$	-20 to $+100$		$\pm(0.1\%+0.1\text{ pF})$	-20 to $+100$	6000
F	$\pm(0.05\%+0.1\text{ pF})$	0 to $+70$		—	—	—

* Maximun dissipation factors are given in the section on Dissipation Factor. Where no data are given in this table, such characteristics are not included in that particular standard.

† Insulation resistance of all MIL capacitors *must* exceed 7500 megohms.

Figure 11

AIRWOUND INDUCTORS									
COIL DIA. *INCHES*	TURNS PER INCH	B & W	I-CORE	INDUCTANCE *μH*	COIL DIA. *INCHES*	TURNS PER INCH	B & W	I-CORE	INDUCTANCE *μH*
$\frac{1}{2}$	4	3001	404T	0.18	$1\frac{1}{4}$	4	—	1004	2.75
	6	—	406T	0.40		6	—	1006	6.30
	8	3002	408T	0.72		8	—	1008	11.2
	10	—	410T	1.12		10	—	1010	17.5
	16	3003	416T	2.90		16	—	1016	42.5
	32	3004	432T	12.0	$1\frac{1}{2}$	4	—	1204	3.9
$\frac{5}{8}$	4	3005	504T	0.28		6	—	1206	8.8
	6	—	506T	0.62		8	—	1208	15.6
	8	3006	508T	1.1		10	—	1210	24.5
	10	—	510T	1.7		16	—	1216	63.0
	16	3007	516T	4.4	$1\frac{3}{4}$	4	—	1404	5.2
	32	3008	532T	18.0		6	—	1406	11.8
$\frac{3}{4}$	4	3009	604T	0.39		8	—	1408	21.0
	6	—	606T	0.87		10	—	1410	33.0
	8	3010	608T	1.57		16	—	1416	85.0
	10	—	610T	2.45	2	4	—	1604	6.6
	16	3011	616T	6.40		6	—	1606	15.0
	32	3012	632T	26.0		8	3900	1608	26.5
1	4	3013	804T	1.0		10	3907-1	1610	42.0
	6	—	806T	2.3		16	—	1616	108.0
	8	3014	808T	4.2	$2\frac{1}{2}$	4	—	2004	10.1
	10	—	810T	6.6		6	3905-1	2006	23.0
	16	3015	816T	16.8		8	3906-1	2008	41.0
	32	3016	832T	68.0		10	—	2010	108.0
NOTE: COIL INDUCTANCE APPROXIMATELY PROPORTIONAL TO LENGTH. I.E., FOR 1/2 INDUCTANCE VALUE, TRIM COIL TO 1/2 LENGTH.					3	4	—	2404	14.0
						6	—	2406	31.5
						8	—	2408	56.0
						10	—	2410	89.0

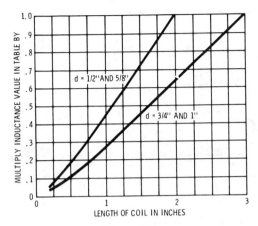

Figure 12

**FACTOR TO BE APPLIED TO
AIR-WOUND INDUCTORS OF
½", ⅝", ¾", AND 1" DIAMETER**

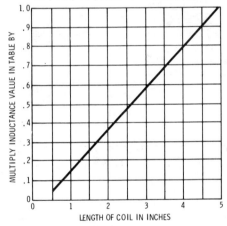

Figure 13

**FACTOR TO BE APPLIED TO
AIR-WOUND INDUCTORS OF
1¼" DIAMETER, OR GREATER**

Table 5. Requirements for Button Mica Capacitors.

Characteristic		Max Range of Temp Coeff (ppm/°C)	Maximum Capacitance Drift
MIL	Commercial		
—	C	±200	±0.5%
D	—	±100	±0.3% or 0.3 pF, whichever is greater
—	E	(−20 to +100)+0.05 pF	±(0.1%+0.10 pF)
—	F	(0 to +70)+0.05 pF	±(0.05%+0.10 pF)

* EIA Standard REC-109-C.

Table 6. Properties of Typical Printed-Circuit Dielectric Base Materials.

Material	Comparable MIL Type	Punchability	Mechanical Strength	Moisture Resistance	Insulation	Arc Resistance	Abrasive Action on Tools	Max Temperature (°C)*
NEMA type XXXP paper-base phenolic	—	Good	Good	Good	Good	Poor	No	105
NEMA type XXXPC paper-base phenolic	—	Very good	Good	Very good	Good	Poor	No	105
NEMA type FR-2 paper-base phenolic, flame resistant	—	Very good	Good	Very good	Good	Poor	No	105
NEMA type FR-3 paper-base epoxy, flame resistant	PX	Very good	Very good	Very good	Very good	Good	No	105
NEMA type FR-4 glass-fabric-base epoxy, general purpose, flame resistant	GF	Fair	Excellent	Excellent	Excellent	Very good	Yes	130 (125)
NEMA type FR-5 glass-fabric-base epoxy, temperature and flame resistant	GH	Fair	Excellent	Excellent	Excellent	Very good	Yes	155 (150)
NEMA type G-10 glass-fabric-base epoxy, general purpose	GE	Fair	Excellent	Excellent	Excellent	Very good	Yes	130 (125)
NEMA type G-11 glass-fabric-base epoxy, temperature resistant	GB	Poor	Excellent	Excellent	Excellent	Very good	Yes	155 (150)
Glass-fabric-base polytetrafluoroethylene	GT	—	Good	Excellent	Excellent	Excellent	—	(150)
Glass-fabric-base fluorinated ethylene propylene	FEP	—	Good	Excellent	Excellent	Excellent	—	(150)

* MIL-STD-275B rating shown in parentheses if different from industry rating.

35-2 Useful Reference Data
Table 7. Conversion Table—Units of Measurement

CONVERSION TABLE — UNITS OF MEASUREMENT		
MICRO = (μ) ONE-MILLIONTH	KILO = (K) ONE THOUSAND	
MILLI = (m) ONE-THOUSANDTH	MEGA = (M) ONE MILLION	
TO CHANGE FROM	TO	OPERATOR
UNITS	MICRO-UNITS	$\times$ 1,000,000 or $\times$ 10^6
	MILLI-UNITS	$\times$ 1,000 or $\times$ 10^3
	KILO-UNITS	$\div$ 1,000 or $\times$ 10^{-3}
	MEGA-UNITS	$\div$ 1,000,000 or $\times$ 10^{-6}
MICRO-UNITS	MILLI-UNITS	$\div$ 1,000 or $\times$ 10^{-3}
	UNITS	$\div$ 1,000,000 or $\times$ 10^{-6}
MILLI-UNITS	MICRO-UNITS	$\times$ 1,000 or $\times$ 10^3
	UNITS	$\div$ 1,000 or $\times$ 10^{-3}
KILO-UNITS	MEGA-UNITS	$\div$ 1,000 or $\times$ 10^{-3}
	UNITS	$\times$ 1,000 or $\times$ 10^3
MEGA-UNITS	KILO-UNITS	$\times$ 1,000 or $\times$ 10^3
	UNITS	$\times$ 1,000,000 or $\times$ 10^6

for automatic insertion in printed-circuit assemblies. Often the capacitors have a code printed on them that indicates the capacitance, the tolerance and the multiplier (figure 10). Note that the letter R may be used at times to signify a decimal point; as in $2R2 = 2.2$ pF or 2.2 μF.

A typical capacitor may be coded 151K which indicates $15 \times 10 = 150$ pF with a tolerance of $\pm$ 10 percent.

Printed Circuit Boards Rigid printed-circuit base materials are available in thicknesses varying from $\frac{1}{64}''$ to $\frac{1}{2}''$. The important properties of the usual materials are given in Table 6. For special applications, other materials are available such as glass-cloth teflon, Kel-F, or ceramic. The most widely used material is NEMA-XXXP paper base phenolic.

Air Wound Inductors Commercial, air-wound inductors suitable for r-f circuitry are available. Two of the more available types are summarized in figure 11. In order to determine the inductance of a short length of coil stock, the factor to be applied to the inductance of the $\frac{1}{2}''$ diameter and $\frac{3}{4}''$ diameter coils is shown in figure 12. The factor for the larger diameter coils (for coil lengths up to $5''$) is shown in figure 13.

PRINCIPAL LOW-VOLTAGE POWER SUPPLIES IN THE WORLD

Territory (Frequency) Voltage

North America:

Alaska (60) 120/240
Bermuda (60) 115/230; some 120/208
Belize (60) 110/220
Canada (60) 120/240; some 115/230

Territory (Frequency) Voltage

Costa Rica (60) 110/220
El Salvador (60) 110/220
Guatemala (60) 110/240; some 220, 120/208
Honduras (60) 110/220
Mexico (50, 60) 127/220 and other voltages
    Mexico City (50) 125/216

Territory (Frequency) Voltage

Nicaragua (60) 120
Panama (60) 110/220; some 120/240, 115/230
United States (60) 120/240 and 120/208

West Indies:

Antigua (60) 230/400
Bahamas (60) 115/200; some 115/220
Barbados (50) 120/208; some 110/200
Cuba (60) 115/230; some 120/208
Dominican Republic (60) 115/230
Guadeloupe (50) 127/220
Jamaica (50, some 60) 110/220
Martinique (50) 127/220
Puerto Rico (60) 120/240
Trinidad (60) 115/230
Virgin Islands (60) 120/240

South America:

Argentina (50) 220/380; also 220/440 dc
Bolivia (50, also 60) 220 and other voltages
Brazil (50, 60) 110, 220; also other voltages and dc
    Rio de Janeiro (50) 125/216
Chile (50) 220/380; some 220 dc
Colombia (60) 110/220; also 120/240 and others
Ecuador (60) 120/208; also 110/220 and others
French Guiana (50) 127/220
Guyana (50, 60) 110/220
Paraguay (50) 220/440; some 220/440 dc
Peru (60) 220; some 110
Surinam (50, 60) 127/220; some 115/230
Uruguay (50) 220
Venezuela (60, some 50) 120/208, 120/240

Europe:

Austria (50) 220/380; Vienna also has 220/440 dc
Azores (50) 220/380
Belgium (50) 220/380 and many others; some dc
Canary Islands (50) 127/220
Denmark (50) 220/380; also 220/440 dc
Finland (50) 220/380
France (50) 120/240, 220/380, and many others
Germany (Federal Republic) (50) 220/380; also
    others, some dc
Gibraltar (50) 240/415
Greece (50) 220/380; also others, some dc
Iceland (50) 220; some 220/380
Ireland (50) 220/380; some 220/440 dc
Italy (50) 127/220, 220/380 and others
Luxembourg (50) 110/190, 220/380
Madeira (50) 220/380; also 220/440 dc
Malta (50) 240/415
Monaco (50) 120/240, 220/380
Netherlands (50) 220/380; also 127/220
Norway (50) 230
Portugal (50) 220/380; some 110/190
Spain (50) 127/220; also 220/380, some dc
Sweden (50) 127/220, 220/380; some dc
Switzerland (50) 220/380
Turkey (50) 220/380; some 110/190

Territory (Frequency) Voltage

United Kingdom (50) 240/415 and others, some dc
Yugoslavia (50) 220/380

Asia:

Afghanistan (50) 220/380
Burma (50) 230
Cambodia (50) 120/208; some 220/380
Sri Lanka (50) 230/400
Cyprus (50) 240
Hong Kong (50) 200/346
India (50) 230/400 and others, some dc
Indonesia (50) 127/220
Iran (50) 220/380
Iraq (50) 220/380
Israel (50) 230/400
Japan (50, 60) 100/200
Jordan (50) 220/380
Korea (60) 100/200
Kuwait (50) 240/415
Laos (50) 127/220; some 220/380
Lebanon (50) 110/190; some 220/380
Malaysia (50) 230/400; some 240/415
Nepal (50) 110/220
Okinawa (60) 120/240
Pakistan (50) 230/400 and others, some dc
Philippines (60) 110, 220, and others
Saudi Arabia (50, 60) 120/208; also 220/380,
    230/400
Singapore (50) 230/400
Syria (50) 115/200; some 220/380
Taiwan (60) 100/200
Thailand (50) 220/380; also 110/190
Vietnam (50) 220/380 future standard
Yemen Arab Republic (50) 220
Yemen, Peoples Democratic Republic (50) 230/400

Africa:

Algeria (50) 127/220, 220/380
Angola (50) 220/380
Dahomey (50) 220/380
Egypt (50) 110, 220 and others; some dc
Ethiopia (50) 220/380; some 127/220
Guinea (50) 220/380; some 127/220
Kenya (50) 240/415
Liberia (60) 120/240
Libya (50) 125/220; some 230/400
Malagasy Republic (50) 220/380; some 127/220
Mauritius (50) 230/400
Morocco (50) 115/200; also 230/400 and others
Mozambique (50) 220/380
Niger (50) 220/380
Nigeria (50) 230/400
Rhodesia (50) 220/380; also 230/400
Senegal (50) 127/220
Sierra Leone (50) 230/400
Somalia (50) 220/440; also 110, 230
South Africa (50) 220/380; also others, some dc

Sudan (50) 240/415
Tanganyika (50) 230/400
Tunisia (50) 220/380; also others
Uganda (50) 240/415
Upper Volta (50) 220/380
Zaire (50) 220/380

Oceania: .

Australia (50) 240/415; also others and dc
Fiji Islands (50) 240/415
Hawaii (60) 120/240
New Caledonia (50) 220/440
New Zealand (50) 230/400

Notes:

1. Abstracted from "Electric Power Abroad," issued 1963 by the Bureau of International Commerce of the US Department of Commerce. This pamphlet is obtainable from the Superintendent of Documents, US Government Printing Office, Washington, D.C. 20402.

2. The listings show electric (residential) power supplied in each country; as indicated, in very many cases other types of supply also exist to a greater or lesser extent. Therefore, for specific characteristics of the power supply of particular cities, reference should be made to "Electric Power Abroad." This pamphlet also gives additional details such as number of phases, number of wires to the residence, frequency stability, grounding regulations, and some data on types of commercial service.

3. In the United States in urban areas, the usual supply is 60-hertz 3-phase 120/208 volts; in less densely populated areas it is usually 120/240 volts, single phase, to each customer. Any other supplies, including dc, are rare and are becoming more so. Additional information for the US is given in the current edition of "Directory of Electric Utilities," published by McGraw-Hill Book Company, New York, N.Y.

4. All voltages in the table are ac except where specifically stated as dc. The latter are infrequent and in most cases are being replaced by ac. The lower voltages shown for ac, wye or delta ac, or for dc distribution lines, are used mostly for lighting and small appliances; the higher voltages are used for larger appliances.

COMPONENT COLOR CODING

POWER TRANSFORMERS

PRIMARY LEADS —————— BLACK
    IF TAPPED:
    COMMON ————— BLACK
    TAP ————————— BLACK/YELLOW
    END ——————— BLACK/RED

HIGH VOLTAGE WINDING ——— RED
    CENTER-TAP ————— RED/YELLOW

RECTIFIER FILAMENT WINDING—YELLOW
    CENTER-TAP ————— YELLOW/BLUE

FILAMENT WINDING N° 1 ——— GREEN
    CENTER-TAP ————— GREEN/YELLOW

FILAMENT WINDING N° 2 ——— BROWN
    CENTER-TAP ————— BROWN/YELLOW

FILAMENT WINDING N° 3 ——— SLATE
    CENTER-TAP   SLATE/YELLOW

I-F TRANSFORMERS

PLATE LEAD ———————— BLUE
B+ LEAD ——————————— RED
GRID (OR DIODE) LEAD——— GREEN
A-V-C (OR GROUND) LEAD——— BLACK

AUDIO TRANSFORMERS

PLATE LEAD (PRI.) ————— BLUE OR BROWN
B+ LEAD (PRI.) —————— RED
GRID LEAD (SEC.) ————— GREEN OR YELLOW
GRID RETURN (SEC.) ——— BLACK

Chart 1. World Time

Aleutian Islands, Tutuila, Samoa	1:00	2:00	3:00	4:00	5:00	6:00	7:00	8:00	9:00	10:00	11:00	Midnite	1:00	2:00	3:00	4:00	5:00	6:00	7:00	8:00	9:00	10:00	11:00	Noon	1:00
Anchorage, Fairbanks, Hawaiian Islands, Tahiti	2:00	3:00	4:00	5:00	6:00	7:00	8:00	9:00	10:00	11:00	Midnite	1:00	2:00	3:00	4:00	5:00	6:00	7:00	8:00	9:00	10:00	11:00	Noon	1:00	2:00
Los Angeles, San Francisco, Seattle, Juneau	4:00	5:00	6:00	7:00	8:00	9:00	10:00	11:00	Midnite	1:00	2:00	3:00	4:00	5:00	6:00	7:00	8:00	9:00	10:00	11:00	Noon	1:00	2:00	3:00	4:00
Chicago, Central America (except Panama), Mexico, Winnipeg	6:00	7:00	8:00	9:00	10:00	11:00	Midnite	1:00	2:00	3:00	4:00	5:00	6:00	7:00	8:00	9:00	10:00	11:00	Noon	1:00	2:00	3:00	4:00	5:00	6:00
New York, Montreal, Miami, Havana, Panama, Bogota, Lima, Quito	7:00	8:00	9:00	10:00	11:00	Midnite	1:00	2:00	3:00	4:00	5:00	6:00	7:00	8:00	9:00	10:00	11:00	Noon	1:00	2:00	3:00	4:00	5:00	6:00	7:00
Bermuda, Puerto Rico, Caracas, La Paz, Asuncion	8:00	9:00	10:00	11:00	Midnite	1:00	2:00	3:00	4:00	5:00	6:00	7:00	8:00	9:00	10:00	11:00	Noon	1:00	2:00	3:00	4:00	5:00	6:00	7:00	8:00
Buenos Aires,* Rio de Janeiro, Santos, Sao Paulo, Montevideo	9:00	10:00	11:00	Midnite	1:00	2:00	3:00	4:00	5:00	6:00	7:00	8:00	9:00	10:00	11:00	Noon	1:00	2:00	3:00	4:00	5:00	6:00	7:00	8:00	9:00
Iceland	11:00	Midnite	1:00	2:00	3:00	4:00	5:00	6:00	7:00	8:00	9:00	10:00	11:00	Noon	1:00	2:00	3:00	4:00	5:00	6:00	7:00	8:00	9:00	10:00	11:00
Lisbon, Dublin, Algiers, Dakar, Ascension Island	Midnite	1:00	2:00	3:00	4:00	5:00	6:00	7:00	8:00	9:00	10:00	11:00	Noon	1:00	2:00	3:00	4:00	5:00	6:00	7:00	8:00	9:00	10:00	11:00	Midnite
Greenwich Civil Time (GCT) or Universal Time (UT)	0000	0100	0200	0300	0400	0500	0600	0700	0800	0900	1000	1100	1200	1300	1400	1500	1600	1700	1800	1900	2000	2100	2200	2300	2400
London,* Paris,* Madrid,* Brussels, Rome, Berlin, Vienna, Oslo, Stockholm, Copenhagen, Amsterdam, Tunis, Warsaw	1:00	2:00	3:00	4:00	5:00	6:00	7:00	8:00	9:00	10:00	11:00	Noon	1:00	2:00	3:00	4:00	5:00	6:00	7:00	8:00	9:00	10:00	11:00	Midnite	1:00
Athens, Israel, Ankara, Cairo, Capetown	2:00	3:00	4:00	5:00	6:00	7:00	8:00	9:00	10:00	11:00	Noon	1:00	2:00	3:00	4:00	5:00	6:00	7:00	8:00	9:00	10:00	11:00	Midnite	1:00	2:00
Moscow,* Ethiopia, Iraq, Madagascar	3:00	4:00	5:00	6:00	7:00	8:00	9:00	10:00	11:00	Noon	1:00	2:00	3:00	4:00	5:00	6:00	7:00	8:00	9:00	10:00	11:00	Midnite	1:00	2:00	3:00
Bombay, Ceylon, New Delhi	5:30	6:30	7:30	8:30	9:30	10:30	11:30	12:30	1:30	2:30	3:30	4:30	5:30	6:30	7:30	8:30	9:30	10:30	11:30	12:30	1:30	2:30	3:30	4:30	5:30
Bangkok, Chungking, Chengtu, Kunming	7:00	8:00	9:00	10:00	11:00	Noon	1:00	2:00	3:00	4:00	5:00	6:00	7:00	8:00	9:00	10:00	11:00	Midnite	1:00	2:00	3:00	4:00	5:00	6:00	7:00
Hong Kong, Manila, Shanghai, Saigon, Taipeh, Celebes	8:00	9:00	10:00	11:00	Noon	1:00	2:00	3:00	4:00	5:00	6:00	7:00	8:00	9:00	10:00	11:00	Midnite	1:00	2:00	3:00	4:00	5:00	6:00	7:00	8:00
Japan, Adelaide, Korea, Manchuria	9:00	10:00	11:00	Noon	1:00	2:00	3:00	4:00	5:00	6:00	7:00	8:00	9:00	10:00	11:00	Midnite	1:00	2:00	3:00	4:00	5:00	6:00	7:00	8:00	9:00
Sydney, Melbourne, Brisbane, Guam, New Guinea, Khabarovsk	10:00	11:00	Noon	1:00	2:00	3:00	4:00	5:00	6:00	7:00	8:00	9:00	10:00	11:00	Midnite	1:00	2:00	3:00	4:00	5:00	6:00	7:00	8:00	9:00	10:00
Solomon Islands, New Caledonia	11:00	Noon	1:00	2:00	3:00	4:00	5:00	6:00	7:00	8:00	9:00	10:00	11:00	Midnite	1:00	2:00	3:00	4:00	5:00	6:00	7:00	8:00	9:00	10:00	11:00
Wellington,* Auckland*	11:30	12:30	1:30	2:30	3:30	4:30	5:30	6:30	7:30	8:30	9:30	10:30	11:30	12:30	1:30	2:30	3:30	4:30	5:30	6:30	7:30	8:30	9:30	10:30	11:30

Notes: (1) Light-face figures designate AM, bold figures PM. (2) Time is that used at places indicated. In general, this is standard time but for places marked with asterisks it is permanent daylight saving time. Temporary daylight saving time is commonplace but not indicated above. (3) When passing the heavy line going up or to the left, subtract 1 day. When passing the heavy line going down or to the right, add 1 day.

Table 8. Copper Wire Table

Gauge No. B. & S.	Diam. in Mils[1]	Circular Mil Area	Turns per Linear Inch[2] Enamel	Turns per Linear Inch[2] S.S.C.	Turns per Linear Inch[2] D.S.C. or S.C.C.	Turns per Linear Inch[2] D.C.C.	Turns per Square Inch[2] S.C.C.	Turns per Square Inch[2] Enamel	Turns per Square Inch[2] D.C.C.	Feet per Lb. Bare	Feet per Lb. D.C.C.	Ohms per 1000 ft. 25° C.	Correct Capacity at 1500 C.M. per Amp.[3]	Diam. in mm.
1	289.3	82690	—	—	—	—	—	—	—	3.947	—	.1264	55.7	7.348
2	257.6	66370	—	—	—	—	—	—	—	4.977	—	.1593	44.1	6.544
3	229.4	52640	—	—	—	—	—	—	—	6.276	—	.2009	35.0	5.827
4	204.3	41740	—	—	—	—	—	—	—	7.914	—	.2533	27.7	5.189
5	181.9	33100	—	—	—	—	—	—	—	9.980	—	.3195	22.0	4.621
6	162.0	26250	—	—	—	—	—	—	—	12.58	—	.4028	17.5	4.115
7	144.3	20820	—	—	—	—	—	—	—	15.87	—	.5080	13.8	3.665
8	128.5	16510	7.6	—	7.4	—	—	—	—	20.01	19.6	.6405	11.0	3.264
9	114.4	13090	8.6	—	8.2	7.1	87.5	84.8	80.0	25.23	24.6	.8077	8.7	2.906
10	101.9	10380	9.6	—	9.3	7.8	110	105	97.5	31.82	30.9	1.018	6.9	2.588
11	90.74	8234	10.7	—	10.3	8.9	136	131	121	40.12	38.8	1.284	5.5	2.305
12	80.81	6530	12.0	—	11.5	9.8	170	162	150	50.59	48.9	1.619	4.4	2.053
13	71.96	5178	13.5	—	12.8	12.0	211	198	183	63.80	61.5	2.042	3.5	1.828
14	64.08	4107	15.0	—	14.2	13.6	262	250	223	80.44	77.3	2.575	2.7	1.628
15	57.07	3257	16.8	—	15.8	14.7	321	306	271	101.4	97.3	3.247	2.2	1.450
16	50.82	2583	18.9	18.9	17.9	16.4	397	372	329	127.9	119	4.094	1.7	1.291
17	45.26	2048	21.2	21.6	19.9	18.1	493	454	399	161.3	150	5.163	1.3	1.150
18	40.30	1624	23.6	23.6	22.0	19.8	592	553	479	203.4	188	6.510	1.1	1.024
19	35.89	1288	26.4	26.4	24.	21.8	775	725	625	256.5	237	8.210	.86	.9116
20	31.96	1022	29.4	29.4	27.0	23.8	940	895	754	323.4	298	10.35	.68	.8118
21	28.46	810.1	33.1	32.7	29.8	26.0	1150	1070	910	407.8	370	13.05	.54	.7230
22	25.35	642.4	37.0	36.5	34.1	30.0	1400	1300	1080	514.2	461	16.46	.43	.6438
23	22.57	509.5	41.3	40.6	37.6	31.6	1700	1570	1260	648.4	584	20.76	.34	.5733
24	20.10	404.0	46.3	45.3	41.5	35.6	2060	1910	1510	817.7	745	26.17	.27	.5106
25	17.90	320.4	51.7	50.4	45.6	38.6	2500	2300	1750	1031	903	33.00	.21	.4547
26	15.94	254.1	58.0	55.6	50.2	41.8	3030	2780	2020	1300	1118	41.62	.17	.4049
27	14.20	201.5	64.9	61.5	55.0	45.0	3670	3350	2310	1639	1422	52.48	.13	.3606
28	12.64	159.8	72.7	68.6	60.2	48.5	4300	3900	2700	2067	1759	66.17	.11	.3211
29	11.26	126.7	81.6	74.8	65.4	51.8	5040	4660	3020	2607	2207	83.44	.084	.2859
30	10.03	100.5	90.5	83.3	71.5	55.5	5920	5280	—	3287	2534	105.2	.067	.2546
31	8.928	79.70	101.	92.0	77.5	59.2	7060	6250	—	4145	2768	132.7	.053	.2268
32	7.950	63.21	113.	101.	83.6	62.6	8120	7360	—	5227	3137	167.3	.042	.2019
33	7.080	50.13	127.	110.	90.0	66.3	9600	8310	—	6591	4697	211.0	.033	.1798
34	6.305	39.75	143.	120.	97.0	70.0	10900	8700	—	8310	6168	266.0	.026	.1601
35	5.615	31.52	158.	132.	104.	73.5	12200	10700	—	10480	6737	335.0	.021	.1426
36	5.000	25.00	175.	143.	111.	77.0	—	—	—	13210	7877	423.0	.017	.1270
37	4.453	19.83	198.	154.	118.	80.3	—	—	—	16660	9309	533.4	.013	.1131
38	3.965	15.72	224.	166.	126.	83.6	—	—	—	21010	10666	672.6	.010	.1007
39	3.531	12.47	248.	181.	133.	86.6	—	—	—	26500	11907	848.1	.008	.0897
40	3.145	9.88	282.	194.	140.	89.7	—	—	—	33410	14222	1069	.006	.0799

[1] A mil is 1/1000 (one thousandth) of an inch.

[2] The figures given are approximate only, since the thickness of the insulation varies with different manufacturers.

[3] The current-carrying capacity at 1000 C.M. per ampere is equal to the circular-mil area (Column 3) divided by 1000.

Table 9. Fractions of an Inch With Metric Equivalents

Fractions of an inch	Decimals of an inch	Millimeters		Fractions of an inch	Decimals of an inch	Millimeters
1/64	0.0156	0.397		33/64	0.5156	13.097
1/32	0.0313	0.794		17/32	0.5313	13.494
3/64	0.0469	1.191		35/64	0.5469	13.891
1/16	0.0625	1.588		9/16	0.5625	14.288
5/64	0.0781	1.984		37/64	0.5781	14.684
3/32	0.0938	2.381		19/32	0.5938	15.081
7/64	0.1094	2.778		39/64	0.6094	15.478
1/8	0.1250	3.175		5/8	0.6250	15.875
9/64	0.1406	3.572		41/64	0.6406	16.272
5/32	0.1563	3.969		21/32	0.6563	16.669
11/64	0.1719	4.366		43/64	0.6719	17.066
3/16	0.1875	4.763		11/16	0.6875	17.463
13/64	0.2031	5.159		45/64	0.7031	17.859
7/32	0.2188	5.556		23/32	0.7188	18.256
15/64	0.2344	5.953		47/64	0.7344	18.653
1/4	0.2500	6.350		3/4	0.7500	19.050
17/64	0.2656	6.747		49/64	0.7656	19.447
9/32	0.2813	7.144		25/32	0.7813	19.844
19/64	0.2969	7.541		51/64	0.7969	20.241
5/16	0.3125	7.938		13/16	0.8125	20.638
21/64	0.3281	8.334		53/64	0.8281	21.034
11/32	0.3438	8.731		27/32	0.8438	21.431
23/64	0.3594	9.128		55/64	0.8594	21.828
3/8	0.3750	9.525		7/8	0.8750	22.225
25/64	0.3906	9.922		57/64	0.8906	22.622
13/32	0.4063	10.319		29/32	0.9063	23.019
27/64	0.4219	10.716		59/64	0.9219	23.416
7/16	0.4375	11.113		15/16	0.9375	23.813
29/64	0.4531	11.509		61/64	0.9531	24.209
15/32	0.4688	11.906		31/32	0.9688	24.606
31/64	0.4844	12.303		63/64	0.9844	25.003
1/2	0.5000	12.700		—	1.0000	25.400

INDEX